Quaternary Vegetation Dynamics of Europe

Haupt

Quaternary Vegetation Dynamics of Europe

Gerhard Lang, Brigitta Ammann,
Karl-Ernst Behre, Willy Tinner (Editors)

Haupt Verlag

FSC
www.fsc.org
MIX
Paper | Supporting responsible forestry
FSC® C004592

1st Edition 2023

Design and typesetting: Katarina Lang Book Design
with Kezia Stingelin and Angela Reinhard
Lithos: FdB – Für das Bild
Cover picture: Christoph Schwörer 2020, Spatio-temporal dynamics
of *Fagus* (beech) in Europe during the past 20 000 years

This publication is listed in the German National Bibliography.
More information can be found at http://dnb.dnb.de.

Haupt Verlag is supported by the
Swiss Federal Office of Culture for the years 2021-2024.

ISBN 978-3-258-08214-1 (E-Book 978-3-258-48214-9)

Printed in Germany

www.haupt.ch

Table of contents

Preface

To the memory of Max Welten (1904–1984) and Werner Trautmann (1924–1989)

A summary overview of the Quaternary history of European vegetation had been lacking until 1994, if one disregards geographically or temporally narrower or much broader surveys. However, during the 20th century, Quaternary botanical research, including neighbouring fields, had brought so many new insights that in the 1990s the time seemed ripe to attempt a synthesis in the form of a basic outline. For one person, however, this was an ambitious undertaking in view of the enormous amount of material, if all aspects needed to be even half covered. Nevertheless, fully aware of the risk of such an inevitably subjective presentation for the whole of Europe, Gerhard Lang undertook this venture. After decades of his own research in a geographically limited framework, he felt the need, to broaden the horizon of vegetation history and palaeoecology and to communicate this to others in as comprehensible a form as possible. The result was the book "Quartäre Vegetationsgeschichte Europas", published in 1994 in German by the publisher Gustav Fischer in Jena. His attempt at a synthesis was intended as an introduction primarily for palaeoecologists and neoecologists, but also served as a source of information for researchers from neighbouring disciplines.

In the structure of his book Gerhard Lang followed the outline which, in his opinion, had proved didactically successful in the lectures he gave at the Universities of Karlsruhe and Bern. This began with a concise overview of the most important events in the history of vegetation in the European Quaternary and was followed by an orientation on the main methods which are important for understanding them. Then came the central part covering the major periods of the Quaternary with the emphasis on the Late-glacial and the Holocene, followed, in conclusion, by a consideration of various general problems of fundamental importance for palaeoecology and neoecology. In accordance with the title of the book, the focus of interest was on the plant world itself and its changes. Thus, vegetation and floral history did not primarily serve as an instrument for determining other, climatic or landscape-historical

Fig. I. Gerhard Lang (*21 October 1924, †19 June 2016) on the "International Moor-Excursion" to the Bodensee and Oberschwaben - 1985.

events. These were, therefore, only followed insofar as they were important for understanding plant responses. Gerhard Lang attached great importance to rich and meaningful illustrations. If possible, no section of the text should remain without graphic illustration. To this end, many illustrations taken from original publications were redrawn, usually in a somewhat modified form, or completely new drafts were produced. He designed almost all the final artwork for his book himself – at that time without computer graphics - and drew it by hand. This was certainly a method that seems old-fashioned today, but it gave him great pleasure, although it also took much more time than originally planned. A basis of Gerhard Lang's book were the excursion meetings of the European Quaternary botanists in various European countries, which Gerhard Lang attended from 1954 onwards. Of particular interest were the field meetings of participants in the IGCP 158 (International Geological Correlation Programme) led by Björn Berglund and Leszek Starkel, which started in the second half of the 1970s. Finally, the "International Moor-Excursions" to important regions of Europe, organised every year since 1976 by the Institute of Geobotany of the University of Bern, deepened his horizon of vegetation history. All these events provided opportunities for personal contacts and discussions, many of them in the field at the study sites.

Students and teachers of many disciplines welcomed his comprehensive book, which in the German speaking countries of Europe is still an important reference. In 2006 Brigitta Ammann, Karl-Ernst Behre, Pim (W.O.) van der Knaap and Willy Tinner made the ambitious plan to republish Gerhard Lang's book, in expanded form, in English. This long-term publication project proved to be extremely demanding. Very sadly Gerhard Lang (1924-2016) passed away before our project came to an end, after 6 years of common work.

The new English version relies in numerous parts on the original book of Gerhard Lang. New text, topics and figures have been added to reflect the progress achieved in palaeoecology during the past now almost 30 years. We would like to thank all our colleagues, in particular the co-authors of this book. Among others, very special thanks go to Pim (W.O.) van der Knaap, who designed and drew the great number of pollen diagrams and Christoph Schwörer, who made many new colourful figures and compiled the first version of this new book from a myriad of text fragments. Linguistic corrections and proofreading by Sheila Hicks and Cathy Jenks as well as technical corrections by Lieveke van Vugt are gratefully acknowledged. We would also like to thank our colleagues at Institute of Plant Sciences and Oeschger Centre for Climate Change Research at University of Bern, Peter von Ballmoos, César Morales del Molino, Erika Gobet, Martin Grosjean, Andy Lotter and Thomas Stocker for their great support and suggestions. Special thanks go to Haupt Verlag Bern and its managing director Matthias Haupt, for their favourable reception of the book and their almost infinite patience until the completion of the manuscript. We also thank Katarina Lang and her team for carefully designing and typesetting this book. The Oeschger Centre for Climate Change Research of University of Bern, the Fondation Johanna Dürmüller-Bol, the Burgergemeinde Bern, the Sebastiana Stiftung and the UniBern Forschungsstiftung are gratefully acknowledged for their financial support.

In his original book of 1994 Gerhard Lang wrote: "At the end of many years of work, the work still seems to me to be highly incomplete, apart from the fact that it can of course never be finished at all because of the ongoing research". We still have just this same feeling. New literature is only included up to 2020, in some parts to 2018. Some researchers may be disappointed to find that their work has been included too little or not at all, but due to the limited scope of the work and the huge body of literature, a selection had to be made which is in no way intended to be taken as a judgement on the quality. Any criticism and suggestions for improvement are welcome.

Bern and Wilhelmshaven, March 2023

Brigitta Ammann, Karl-Ernst Behre
and Willy Tinner

1.

An overview of the general framework of the development of flora and vegetation

Karl-Ernst Behre[1]

[1]Lower Saxony Institute for Historical Coastal Research, Wilhelmshaven, Germany

1.1

Geology and geomorphology

The Quaternary is the latest stage of Earth's history. As the name implies, there were three preceding periods in the old nomenclature, the Primary (now Palaeozoic), the Secondary (now Mesozoic) and the Tertiary (now Cenozoic) but, according to modern knowledge, there was still a very long phase before the Palaeozoic. Compared with the older stages, the Quaternary is extremely short and comprises a timespan of only 2.6 million years (Ma).

According to modern knowledge, Earth was formed around 4.5 billion years ago. Isolated continents soon formed on its surface and moved on the Earth's crust. Several times they collided or met and reshaped into large supercontinents where all the world's landmasses were joined together. The position and contours of these early supercontinents are still vague and open to discussion. The penultimate supercontinent, however, broke up and its parts later re-assembled to the better-known supercontinent called Pangaea by Alfred Wegener. It was he, who already as early as 1915 proposed the theory of continental drift, the details of which were largely worked out on the basis of palaeobotanical and palaeozoological investigations (Wegener, 1915).

This theory, however, was strongly disputed and eventually rejected by the scientific community. Only in the 1960s, when the theory of plate tectonics arose, did geologists once more become aware of Wegener's ideas and nowadays the moving of continents is accepted without question.

The Pangaea supercontinent existed from around 250 Ma BP and comprised all continents including Antarctica. Around 200 Ma BP Pangaea began to split up and its parts became the modern continents that moved around the Earth's surface to their present positions. The time of separation of the various continents can be determined by the distribution of fossil plants and animals and the different development of their taxa. The *Glossopteris* flora, which denotes the southern part of this supercontinent, Gondwana, is well known in this respect.

Now that the theory of plate tectonics has gained acceptance it is recognized that there are eight large and many minor plates that comprise not only the mainlands but also the marine areas where sea-floor spreading takes place. Europe is a part of the Eurasian plate, which borders to the south on the African plate. As at other plate boundaries, strong volcanic activity and mountain-building took place along this boundary. Of eminent and long lasting importance was the formation of a huge oceanic trench between the two plates. It is called Tethys and during the Cretaceous extended as far as East Asia. In the course of the Tertiary it became reduced to the Mediterranean sea in the west, while other remnants are the Black Sea and the Caspian Sea. The calcareous deposits in the Tethys and its surrounding shelf seas and bays constitute the huge areas of limestone which extend northwards to the Alps that still determine large parts of the modern landscape.

During the Tertiary, these broad-scale processes continued and formed the geology of Europe with notable consequences for the following Quaternary period. Due to the tectonic movements of the plates north of the Tethys, the long west-east running range of the alpidic mountains was formed, which became a strong barrier for the migration of plants during the subsequent climate changes of the Quaternary. Another outcome of the plate tectonics in southern Europe is the volcanic activity that led to the emergence of many volcanoes, not only in the Mediterranean but as far north as France and central Germany, where they have long been inactive but often form isolated mountains with a specific vegetation. Several crater lakes formed in the Pleistocene became excellent archives for vegetation history.

The severe climatic deterioration at the start of the Quaternary – the Ice Age – led to strong morphological changes on top of the Tertiary subsoil. Huge ice sheets were formed in the north as well as smaller ice caps on the Alps and several other mountain areas. The movement of these ice masses led to intense glacial erosion and the removal of bedrock and soil material: large areas of northern Europe and of the mountainous areas became periodically completely bare until new soils developed or even remained bare. The adjoining areas

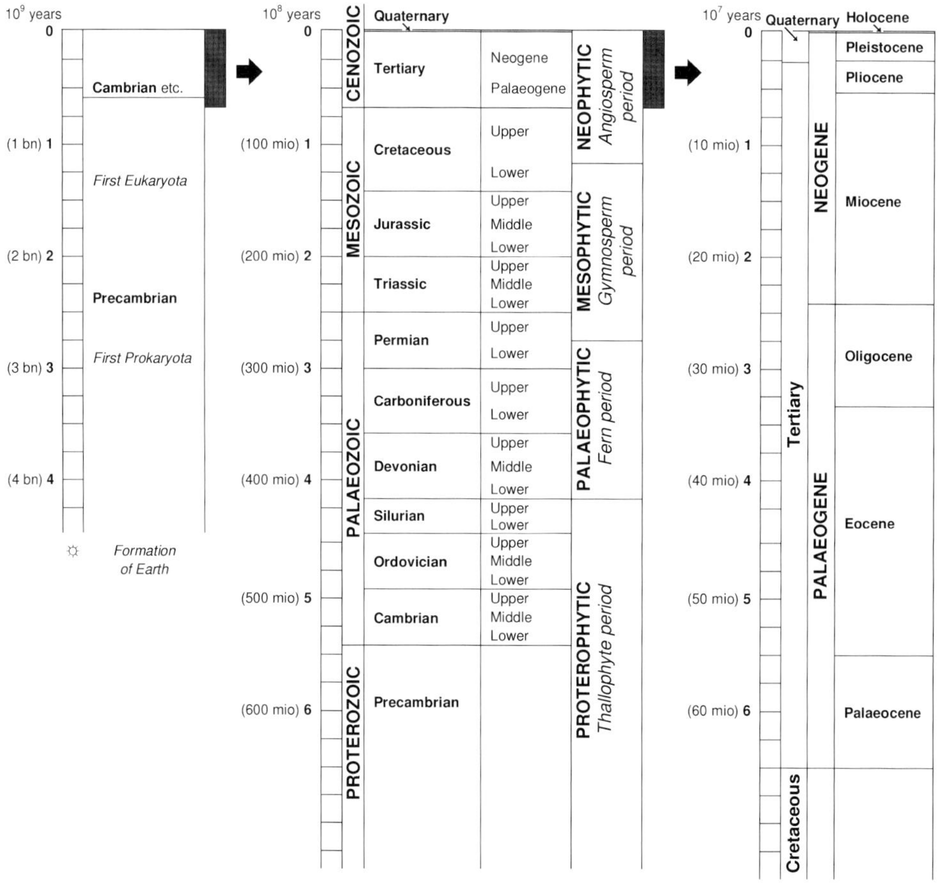

Fig. 1-1. Geochronological subdivision of the Earth's history. The scale between the three parts differs by a factor of ten (after Lang, 1994, modified and updated).

of the formerly glaciated regions were covered with thick layers of till and at their boundaries, end moraines were formed, sometimes more than a hundred metres in height. Large numbers of lakes were formed in the glaciated areas ranging from small kettle-hole depressions, which came into existence by the melting of dead ice, up to large ones formed behind moraine dams, for instance. Areas that were glaciated during the last glaciation (Weichselian or Würmian) usually display existing lake landscapes, for example in Finland, Sweden or north-east central Europe, while the lakes of earlier glaciations were filled up during the following interglacials and glacials. Nowadays these are very important as archives for Pleistocene vegetation history.

Outside of the glaciated region there was a large periglacial area, where various cold-climate processes created special landscapes. Large quantities of gravel, sand and clay were deposited and formed outwash plains or sandurs immediately adjacent to the glaciated areas. Along the rivers, Rhine or Rhone for example, fluvial accumulation with elevated river banks took place for several hundred kilometres. All these geological processes formed a great variety of different environments which resulted in vegetation units adapted specifically to those conditions. Another important feature along the rivers are the terraces that were formed due to interglacial down-cutting through the deposits aggregated during glacial periods.

The most important deposit in the periglacial area and beyond is loess – windblown dust – finer than sand. It was formed during dry and cold steppe phases and often attains many metres in thickness. The loess belt stretches from western Europe across central to eastern Europe and is particularly well developed in Austria and large regions of the Ukraine and southern Russia. Its nutrient content is high and, therefore, the loess areas are preferred for farming.

The stratigraphy of the loess often shows buried palaeosoils, reflecting interstadial or interglacial phases. This indicates that soil formation, not only in loess, takes place only in warm periods. During cold phases there is no development of soils: the surface in the permanently frozen areas, the so-called permafrost region, is constantly disturbed by frost action such as squeezing and upheaval of the uppermost layers, called cryoturbation, and the downslope movement of sediments, called solifluction.

The formation of soils is an important precondition for the development of vegetation. At the beginning of the warm phases many plant species, which could grow in the area given the climatic conditions, are unable to spread due to the lack of mature soils while others, such as several pioneer plants, are already able to occupy the raw habitats and so contribute to the formation of productive soils.

1.2

Early development of flora and vegetation

After the formation of the Earth it still took a very long time until the first traces of life, single-celled bacteria, appeared around 3.8 billon years ago (Fig. 1-1). It took, however, more than another billion years before multicellular life began. After this very long Precambrian period, the Palaeozoic started at around 542 Ma ago with the Cambrian and from this point onwards it is mainly palaeontology, the science of plants and animals, that is used for the subdivision of the geological development.

Fossil spores have been found as early as the late Ordovician and, in the late Silurian, the land plants started their development. The following Devonian was already dominated by lycopods and early ferns, which represent the earliest woody plants. They proliferated strongly in the tropical climate of the Carboniferous (which existed in most parts of Europe) and dense woodlands spread over wide areas forming the coal deposits that we see today. They consist mainly of species of the ancient lycopods such as *Sigillaria* or the scale tree *Lepidodendron* that produce tall trees. Also the tree-like *Calamites*, which belongs to the Equisetaceae, became quite common in the coal forest flora.

Eventually, towards the end of the Carboniferous, the first seed plants appeared. The first step was made by the Pteridospermae, followed by the Progymnospermae and the Cordaits, closely related to the gymnosperms.

Most of the coal-forming trees became extinct during the Permian: *Glossopteris*, a prominent genus of the seed fern, appeared while the lycopods were replaced by the conifers, which soon evolved into many systematic groups. In addition to many extinct genera, *Ginkgo* and *Cycas* appeared: living fossils in our present flora. Evolution proceeded further with the Benetittales, seed plants and early progenitors of the angiosperms.

In the following Triassic and Jurassic, conifers dominated the vegetation and became highly diversified. Already at that stage all modern families of the conifers existed. The ensuing Cretaceous is characterized by the development of the angiosperms, which must have come into being earlier but, apart from their pollen found in the Jurassic, specific fossils are still missing. As early as the Lower Cretaceous, the angiosperms split up into the Monocotyledoneae and Dicotolydoneae and from

Fig. 1-2. Contours of last glacial ice margins and Early Holocene shorelines, in cal ka BP (modified after Lang, 1994).

the Middle Cretaceous, angiosperms dominated the vegetation worldwide (Kadereit, 2008).

From the Upper Cretaceous until the Upper Miocene the European vegetation is characterized by paratropical rainforests, which are closely related to tropical rainforests but include members of some genera typically thought of as 'temperate' such as *Alnus, Acer* and others. In some regions, subtropical rain- and laurel forests, as well as temperate laurel forests, were present in accordance with the climatic conditions with high temperatures in the Eocene and later in the Miocene (see Fig. 3.1-1). During these two periods most of the huge brown coal deposits were formed.

Overall, during the Tertiary the vegetation of Europe evolved from paratropical to warm-temperate and temperate forms in response to a progressive but undulating cooling. There were mainly two separate ecological units: the evergreen, laurophyll 'palaeotropical geoflora' and the deciduous, broadleaved 'Arctotertiary geoflora'. During the Tertiary, the Arctotertiary geoflora occupied the northern parts of the Holarctic, the 'Greenland region' and has advanced into Europe in waves since the Palaeocene. This flora formed the basis for the Tertiary mixed mesophytic forests and became the roots of the Quaternary temperate mixed oak forests. Considerable differences in their composition allow us to separate several floral regions in Eurosiberia so that there are four well-defined floral regions by the end of the Pliocene. A Mediterranean region, however, cannot be recognized although Mediterranean floral elements appear from the Eocene onwards and the Mediterranean xeromorphic vegetation has its origin in the Tertiary laurophyll vegetation. The Mediterranean sclerophyll forests, in their present form, are a very young phenomenon and probably arose only after the destruction of the laurophyll forests during the Pleistocene (Mai, 1989, 1995).

1.3

Quaternary climate development and the consequences in terms of the position of coastlines

The Quaternary flora and vegetation are determined by climatic oscillations at a much higher magnitude than during the Tertiary. There were alternating cold periods (glacials) and warm periods (interglacials). Oscillations at a smaller scale, with cool to slightly temperate climate conditions, occurred during the glacial periods and are called interstadials. By definition the interglacials are classified as temperate periods with a climatic optimum at least as high as the present interglacial (Holocene) in the same region. Interstadials are assumed to have been either too short or too cold to reach the climate level of interglacial status in the same region (Litt et al., 2008). On the basis of the marine isotope record (see Fig. 3.3-1) there have been around 100 climate oscillations since the beginning of the Quaternary.

Within the Quaternary there has been an overall shift to cooler conditions. During the long Lower Pleistocene the cold phases became gradually colder and the amplitude of the oscillations increased: during the Middle and Upper Pleistocene the temperatures in the cold phases became very low, resulting in extensive ice advances in the north and smaller ones in and around high mountains in more southern areas such as the Alps, Carpathians and Pyrenees.

The duration of the cold phases in the Lower Pleistocene was also markedly shorter than in the Middle and Upper Pleistocene when the warm periods lasted around 15 000 years while the cold phases exceeded 100 000 years. Taken as a whole the Pleistocene was a cool to cold period with intercalated warm phases.

Penck and Brückner (1901–1909) were the first to present a chronological order for the Alpine glaciations and called them Günz, Mindel, Riss and Würm, but even before that Penck (1879) had denominated the large north European glaciations as Elster, Saale and Weichsel. There are also indications of one or more preceding glaciations in northern central and eastern Europe as well as in the Alps. Large areas of Europe were, nevertheless, spared from an ice cover and, therefore, the term 'cold stage' is often used instead of glacial.

Knowledge of the last glaciation in Europe, the Weichselian (Fig. 1-2), is good. It did not extend as far as the preceding large glaciations but is clearly the geological basis for large parts of central and northern Europe. Several ice advances formed clear end moraines such as the Pomeranian, Frankfurt and Brandenburg phases and the lakes and outwash plains of this glaciation created diversified landscapes across large areas. This is complemented by a huge loess belt covering large parts of Europe outside the glaciated regions.

The glaciations mentioned above were, however, not the first in Earth's history. A very early Permo-Carboniferous glaciation is known from the Gondwana continent and there were also Pliocene glaciations, but these were restricted to the polar regions.

During the glaciations huge amounts of water were shifted from the sea into the glaciers, which often reached up to 3000 m in thickness, and this resulted in global sea-level changes of great magnitudes. This kind of sea-level oscillation is called eustatic. Very little is known about the older fluctuations of sea-level but the evidence increases from the Upper Pleistocene to the Holocene, in particular with respect to the North Sea and the Baltic Sea. During the glaciations, most parts of the North Sea and the Baltic Sea basins became dry or temporally glaciated. The fall in global sea-level around the maximum of the last glaciation is estimated at about 130 m.

The melting of the glaciers led to the refilling of the shallow North Sea and Baltic Sea basins during the interglacials. The coastlines of these periods, however, deviate from the modern ones such that the area of the Holsteinian Sea extended far into northern Germany but did not reach the western Netherlands (Ehlers, 1994). The coastline of the Eemian Sea was similar to that of the modern North Sea (Streif, 1990). There was, however, in southern Jutland a connection to the Baltic Sea (Konradi et al., 2005) which did not come into existence during the Holocene. The configuration of the Eemian Sea in the Baltic was very variable as it is in the Holocene and the position of the coastlines is still uncertain. During the first 2000 years of the Eemian interglacial there was, however, a connection to the White Sea so that Scandinavia became an island for a time (Andrén et al., 2011).

In addition to the eustatic sea-level changes the coastlines were displaced by another factor: isostasy. Due to the very heavy ice load in the glaciated regions, in particular in Scandinavia, the Earth's crust sank considerably. The unloading of this weight in the interglacials, due to the ice melting, caused uplift which is still occurring at a magnitude of around 100 cm per century in the centre of the Scandinavian shield. These two factors determined the varied configuration of the Baltic Sea in the past. Thus its coastline is mainly determined by both the eustatic sea-level change and the glacio-isostatic adjustment (Harff et al., 2017).

These opposing trends result in various stages of the Baltic Sea, with or without connections to the North Sea, and that means fresh or brackish-marine. Five main stages of the Baltic are distinguished:

- Baltic Ice Lake (freshwater) from around 14 500 to 9700 BC
- Yoldia Sea (brackish) from 9700 to 8800 BC
- Ancylus Lake (freshwater) from 8800 to 6550 BC
- Littorina Sea (brackish-marine) 6550 BC to AD 1500
- Mya Sea (marine-brackish) from AD 1500 (after Björck, 2008).

As a consequence of the isostatic uplift of the Scandinavian shield in the postglacial, the area around it is sinking and, in particular, the southern coastline of the Baltic in north-east Germany was, and still is, displaced by the sinking coast together with the general eustatic rise in sea-level.

In contrast to the special development of the Baltic Sea, the North Sea is a marginal sea, open to the Atlantic Ocean. In the Early and Mid Holocene, when parts of the North Sea basin were still dry, forests occupied those areas, but these were later drowned in the course of the rising sea level.

For the Holocene North Sea there are many investigations to reconstruct former sea-levels and shorelines (Shennan and Andrews, 2000; Behre, 2007a). Both basal and intercalated peats provide sea-level points to construct a sea-level curve from around 10 000 BC to the present. After a long strong increase, sea-level rise decelerated around 5000 BC and from 3000 BC there occurred temporary drops in sea-level, during which large areas fell dry and became influenced by fresh water so that peat formed, which was later covered again with marine sediments. Altogether, seven of these regressions of various magnitudes have been identified in the southern North Sea during the course of the last 5000 years, all having severe consequences for the coastal vegetation as well as habitation.

In accordance with the rise in sea-level, shorelines moved very fast. After 8000 BC the Dogger Bank was cut off and became a large island in the middle of the North Sea: it was inundated around 2000 years later. Around 7000 BC, Britain was separated from the European mainland and as early as around 6000 BC the sea reached the foreland of the Frisian Islands (Fig. 1-2). While for a long time the coastal habitation reacted only passively to the shoreline changes, people started active resistance against the sea in the late 11th century by building dikes. They started with ring dikes which were interconnected in the 13th century to a continuous closed dike line. For the coastal vegetation this was a dramatic change. Prior to diking, the coastal zone was characterized by a great variety of vegetation units ranging from salt marshes to freshwater communities with many intermediate forms. Now the dikes sharply separated the seaward marine and brackish area from the landward freshwater area (Behre, 2002).

Similarly in the Mediterranean Sea, eustatic sea-level changes resulted in shoreline displacements, in particular in wide parts of the shallow Adriatic Sea the rising sea level disconnected the Italian Peninsula from the Balkans. Similarly, Malta was disconnected from Sicily, also Sardinia and Corsica became two separate islands. Numerous other displacements have been described as well-defined raised beaches at fixed levels in several coastal areas around the Mediterranean Sea. The Mediterranean basin, however, is a tectonically strongly disturbed region and many of the suggested sea-level fluctuations can be explained by differential tectonic movements rather than by local relative sea-level changes (Pirazzoli, 1991).

Great progress has been made during recent decades in dating the organic and inorganic layers of the Quaternary. For the younger periods, mainly the Holocene, the radiocarbon method has been refined to be used for very small samples. More precise dating started earlier with varve counting, particularly in the Baltic region, and this was later extended to the investigation of annually laminated sediments in many freshwater lakes in all parts of Europe. With the addition of dendrochronological data, very precise dating of many Holocene sites is now possible. Quite a number of physical and chemical methods have been developed in order to date Pleistocene sequences. In particular they have been applied to the key profiles for modern Quaternary records: the deep sea cores and the long ice cores from Greenland and Antarctica. These continuous cores span large parts of the Quaternary and allow global correlations of the climatic sequence. The Marine Isotopic Stages (MIS) that now cover the whole Quaternary have become the framework for the detailed subdivision of the Quaternary period.

1.4

Changes of vegetation belts and decrease of floral elements in the course of the Pleistocene

The numerous climate oscillations during the Pleistocene resulted in multiple radical changes of the European flora and vegetation. During the cold phases the forests were pushed to the south or south-east while treeless tundra or steppe spread onto the non-glaciated regions. During the extreme cold glacials of the Middle and Upper Pleistocene there were no longer forests north of the Alps. Some woody species survived in favourable or sheltered areas in southern and south-eastern Europe, while others withdrew much further. The manifold repeated expansions of cold periods of increasing strength over time, forced the thermophilous species to migrate to far and sometimes remote areas and some of them failed to come back to the European areas they had occupied before. A very severe obstacle for plant migrations was the long west-east running mountain range from the Pyrenees to the Carpathians and many species did not manage to cross this barrier at each climate change. This is one reason why the number of plant species is much higher in America or east Asia where the direction of the mountain ranges is north-south instead of west-east. As a consequence the flora of the various interglacials show considerable differences. Many species of the Pliocene flora had already become extinct in Europe at the Tertiary/Pleistocene boundary but persisted in other parts of the world. Thus in Europe, *Sciadopitys,* for example, did not survive the oldest part of the Pleistocene while *Eucommia* reached at least the lower Middle Pleistocene: both trees are nowadays native only in east Asia. Another tree, *Pterocarya*, nowadays very popular in European parks and naturally occurring in the Caucasus and east of it, was common in European forests until the end of the penultimate interglacial and became an important indicator for the Holsteinian interglacial. The same is also true for herbs and aquatics: for instance, *Brasenia schreberi (=purpurea)*, a well-known aquatic that occurs all over the world except in Europe, disappeared here only at the end of the last interglacial. Other components of the Plio-Pleistocene flora, such as *Fagus*, have a varied history in that they occurred in some interglacials but were absent at least over large regions in others.

Overall, the number of tree and other species began to decrease in the Lower Pleistocene, at first slowly but at an increasing rate towards the Upper Pleistocene (see Table 3.3-1). One has to take account, however, of the very limited amount of data from the Lower Pleistocene. Future sites may well reveal that some 'old' species survived longer than is assumed today.

1.5

The Holocene

The current interglacial – the Holocene – began about 11 600–11 700 cal yr BP. According to Table 4.1-2 the Holocene starts at 11.7 cal ka BP (e.g. Litt et al., 2008) and so has about the same duration as the last interglacial but, in contrast to that interglacial, it has strong climatic and vegetational oscillations during the immediately previous Late-glacial. The Holocene started with a strong climatic improvement, so that particularly north of the Alps many species could not immediately use the climate possibilities because of a time lag in their remigration.

In addition to the barriers formed by the mountain ranges, the development of the coastal configuration caused by sea-level changes also had strong consequences for the spreading of plant and animal species. A good example is provided by the British Isles. In the Early Holocene they were connected to the European mainland and there were no obstacles for plants and animals to spread into this region. Once Britain was separated from the continent, around 7000 BC, the possibilities for spreading became reduced and the late-comer trees from the south, such as *Acer pseudoplatanus* and *A. platanoides* as well as *Picea* and *Abies,* were no longer able to reach the British Isles. Other trees, such as *Fagus* and *Carpinus* reached Britain in spite of their late expansion, probably with the help of Bronze Age people, but they did not reach Ireland which had already been separated earlier so that *Tilia* is also lacking there. Similar effects must have occurred in earlier periods as can be seen by the Pleistocene distribution of plants and animals.

After the *Betula/Pinus*-phase of the Early Holocene, mesocratic deciduous forests soon spread northwards and beyond their present distribution areas. While *Fagus* advanced to the north slowly and late, the important trees *Abies* and *Picea* did not even reach the North Sea coast as they had done earlier in the Eemian. *Picea*, however, found its way into Scandinavia from north-western Russia. The climatic optimum was reached in the Atlantic period, when *Tilia* or the aquatic *Trapa natans* reached their northernmost distribution, far beyond their modern boundaries. In the course of the climatic deterioration in the younger Holocene, the natural vegetation adapted to the changing conditions. Another consequence of the climatic deterioration was the formation and extension of huge raised bogs around the southern North Sea as well as in the Alpine foreland which led to distinctive changes in the landscape.

The increasing activity of humans became the main factor influencing the vegetation development in the course of the Holocene and created vegetation types and landscape features that had never occurred before. At first the hunter, gatherer and fishers acted only passively as permanent settlements did not yet exist. The resulting impact on the vegetation was very small and often invisible.

A totally new epoch started with the immigration of the first farming cultures: now the human relationship to the natural environment changed abruptly – passive adaptation changed to active encroachment. Arable farming with the domestication of the first cultivated plants was developed at about 9000 BC in the Near East in the so-called Fertile Crescent and, at about the same time, the domestication of the first animals took place. This new economy had, as a precondition, the establishment of permanent settlements, so this signifies the beginning of a non-nomadic life-style. This total change in the economic base, from hunting and gathering to farming, was called very appropriately the "Neolithic Revolution" by Childe (1951) and the term Neolithic is bound to this form of economy, not to a particular time. Consequently the date for the beginning of the Neolithic is metachronous and varies between regions, depending on when animal and plant husbandry were first initiated.

The new form of economy soon spread across the Balkans to central Europe and another movement took place westwards across the Mediterranean Sea. The expansion of the Neolithic economy led to an increasing opening of the wooded landscapes by arable farming, forest grazing and general woodcutting. So many new biotopes were created and gave native species as well as introduced weeds, which arrived together with their cultivated host plants, the chance to spread. The number of plant species increased steadily and this became still stronger after the contact with America and the establishment of modern trade routes all over the world.

Agriculture, forest grazing, burning, shredding, pollarding (see Rackham, 1976) as well as selective wood cutting and the introduction of exotic, mainly American, trees changed the composition of the forests in many ways. Even more consequences ensued from several special types of economy that led to the formation of large heaths and dry grasslands. Until around AD 1900 the hugely widespread Atlantic dwarf-shrub heaths that existed from Belgium to the Netherlands, Germany and Denmark were regarded as natural, but it transpired that they were of anthropogenic origin (see Behre, 1988) and this also applies to areas in Great Britain (Pennington, 1974) as well as Brittany and the 'lands' in the north-west of France (van Zeist, 1964) and along the coasts of the Iberian Peninsula. The same is true for the Norwegian coastal heaths, where fire was an important factor for their formation (Kaland, 1986; Prøsch-Danielsen and Simonsen, 2000). The huge unwanted areas of sclerophyll woodlands with macquis and garrigue in the Mediterranean also came into existence as a result of people's activities and meadows and pastures all over Europe, except for salt marshes and above the tree line in the mountains, would not exist without human activities.

The transformation of the landscapes from natural to anthropogenic occurred steadily but with peaks, for example in the Roman period, but the final disruption of landscapes and vegetation started in the medieval period and reached its maximum in modern times.

Compared with preceding interglacials, the Holocene is a period on its own. The strong human impact in this period could account for most of the phenomena mentioned above. As can be shown by physical palaeotemperature investigations as well as by pollen analysis, there were also climatic changes. Both had considerable and manifold impact on the vegetation but the real cause of some change is often difficult to illucidate.

1.6

Fauna and humans

Dramatic changes also occurred in the animal kingdom during the Quaternary in that faunal assemblages are closely related to the climate and certain types of vegetation. This is particularly visible in the large mammals that followed the environmental changes and migrated northwards during warm periods and southwards during colder ones. Over the long period of the Quaternary there were extinctions as well as evolution of species. A good example of the latter is shown by the elephant's lineages from the *Mammuthus meridionalis* to *M. trogontherii* and finally *M. primigenius* (van Kolfschoten, 1992).

The interglacials are characterized by a forest fauna consisting of, for example, the straight-tusked elephant *Palaeoloxodon (Elephas) antiquus*, Kirchberg's rhinoceros *Stephanorhinus kirchbergensis*, wild boar *Sus scrofa*, roe deer *Capreolus capreolus*, fallow deer *Cervus dama* and also water buffalo *Bubalus murrensis* and even the Mediterranean immigrant hippopotamus *Hippopotamus amphibius*, which occurred along the Rhine and as far north as England and indicates relatively high winter temperatures there.

During the cold phases there were steppe or tundra faunas in central Europe, including the steppe mammoth *Mammuthus trogontherii*, the ancestor of the later woolly mammoth *M. primigenius*, reindeer *Rangifer tarandus*, musk oxen *Ovibus mos-*

chatus as well as lemmings, various species of *Lemmus* and *Dicrostonyx* of the small mammal fauna.

A striking and much discussed phenomenon is the extinction of many of the large mammals at the end of the Pleistocene after the Eemian and around the end of the Weichselian, including both herbivores and predators. Most authors blame people for these extinctions but others blame climate change. However, similar changes in climate had happened several times before without such consequences.

The first humans appeared in Europe relatively late. The evolution of humans took place in Africa. After various forms of development *Homo erectus* was the first to be accepted as a real human in that they were able to produce artefacts: they came into being almost two million years ago. In Europe, finds of *Homo erectus* are very sparse and are lacking in the long Lower Pleistocene. The oldest bones in Europe have been found in Spain and are dated to around 780 000 years BP while the oldest find from north of the Alps was discovered at Heidelberg in Germany with an age of 600 000 years BP. Around 300 000 years BP, Neanderthals (*Homo neanderthalensis*) evolved from *Homo erectus* and this group spread far across Europe until Neanderthals disappeared around 30 000 years BP. Several millennia before that, modern humans, *Homo sapiens,* had immigrated to Europe, also coming from Africa via the Orient. These people evolved independently from *Homo erectus* and gradually replaced the Neanderthals (Terberger, 2009). Now a cultural revolution took place, expressed in tool industries as well as in artistic cave paintings and sculptures and leading gradually to modern periods. Only much later did the sharp increase in population result in anthropogenic environmental changes.

2.

Palaeoecological materials and methods

Gerhard Lang[1], Shinya Sugita[2], Brigitta Ammann[1,3]

[1] Institute of Plant Sciences, University of Bern, Bern, Switzerland
[2] Institute of Ecology, Tallinn University, Tallinn, Estonia
[3] Oeschger Centre for Climate Change Research, University of Bern, Bern, Switzerland

2.1

Introduction

"At one extreme are its philosophical underpinnings. At the opposite, practical end of the spectrum, are the techniques used to obtain scientific results – radioactive tracing, remote sensing, controlled laboratory experiments, the deployment of instrumental arrays or pollen analysis, for example." Oldfield (2005)

The spectrum of methods used in Quaternary palaeoecology and vegetation history is very broad and can only be briefly summarized in this chapter. A number of more detailed books and book chapters exist for various specific topics. A comprehensive introduction by Birks and Birks (1980), reprinted in 2004, offers much more than methods: namely principles, research questions and results. Decades ago the need to standardize methods across Europe was felt and, within the International Geological Correlation Programme, the project on Palaeohydrology (IGCP 158a on rivers, IGCP 158b on lakes) first produced a handbook on methods edited by Berglund (1986), reprinted in 2003, and subsequently a synthesis by Berglund et al. (1996a). These, in turn, led to the establishment of the European Pollen Database (EPD). More recently two new major methodological sources have become available: (1) Tracking Environmental Change Using Lake Sediments now published in the Developments in Paleoenvironmental Research series, especially volumes 1, 3 and 5 and (2) the sections about pollen and plant macrofossils in volume 3 of the Encyclopedia of Quaternary Science (2007 and 2nd edition in 2013 edited by S.A. Elias and C.J. Mock).

Methods help to address the basic questions of 'where', 'how', 'when' and 'why' did changes in flora and vegetation occur? In other words, what and where are the natural archives, how do we extract botanical information from them, how can we develop a chronology and what processes may explain the changes (or the stability)? Answers to these questions will help to formulate the next research questions of why, where and how to continue. Concise introductions to the principles are offered for pollen analysis by Seppä (2007a, 2013) and for plant macrofossil analysis by H.H. Birks (2007a, 2013), Jackson and Booth (2013) and Birks (2014).

2.2

Where: natural archives and their sampling

2.2.1

Deposits in lakes and mires

In cultural archives we hate humidity because it destroys paper and paintings through fungi, microbes and other agents. In natural archives, however, water conserves the plant remains (pollen, macrofossils and wood, all of which are susceptible to oxidation) by reducing or even excluding oxygen from the air. Therefore, the best natural archives are the wettest spots in a landscape, for example lakes and mires.

Site selection depends primarily on the research question being asked. In addition, it is worth considering what is the origin of the lake or mire basin. Hutchinson (1957, 1975) provides an overview of 76 types of lake origin, of which we show a selection in Table 4.6-1. Concerning the importance of these possible origins there is a difference between number and relevance: most lakes are of glacial origin (as in Fennoscandia and northern North America) but lakes of special relevance to studies of environmental history may be those outside formerly glaciated areas, either because of their spatial coverage (such as lakes of karstic, deflational, fluvial or geomorphological origin) or because they hold long records (volcanic lakes or meteoritic impact craters).

Basin analysis may be useful for choosing the best coring sites. Even basins that look simple from a landscape perspective may turn out to be more complex. Depending on the research question (and often in accordance with the available coring equipment), we may want to choose littoral or profundal coring spots. For tracing lake-level changes, littoral cores are more informative but for complete sequences (that is without hiatuses) the profundal

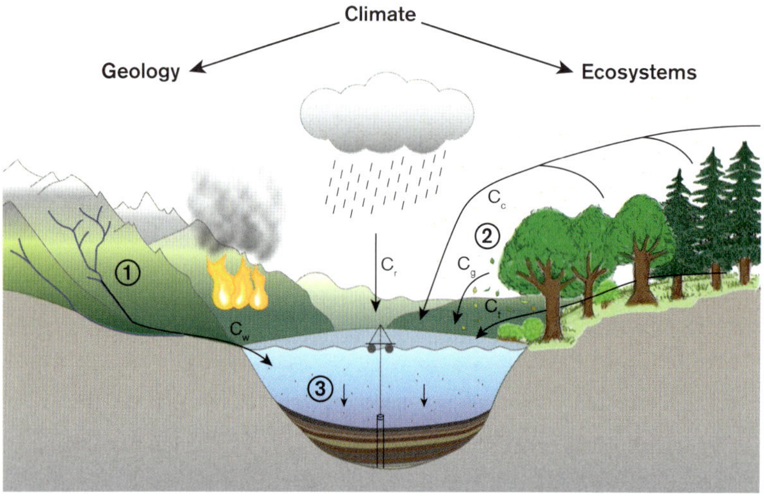

1: Abiotic input from catchment via erosion
2: Biotic terrestrial input (e.g. pollen, plant macrofossils, insects)
3: Biotic aquatic input (e.g. pollen of waterplants, diatoms, insects, cladocera)

Fig. 2.2-1. Lake basins are natural archives for abiotic and biotic particles that record geological and ecological conditions and their changes over time. Biotic remains of both terrestrial and aquatic origin are deposited in lakes. The processes of pollen (and macrofossil) production, dispersal and deposition are highly complex, and some of them are taxon-specific: for example pollen productivity of entomophilous plants is usually much smaller than that of anemophilous plants; pollen dispersal to the free atmosphere (and the fall-out curve) is higher for tall trees than for small herbs. Tauber (1965) distinguished five ways of pollen transport: Cg = gravity component, Cc= canopy component, Cr= wash-out by rain, Ct = trunk-space component, Cw= runoff component. The ultimate goal is to reconstruct flora and vegetation (and their patterns) around the lake at various distances and to understand the causative factors for change (see chapters 4 and 5). Plant macrofossils are more heterogenous than pollen in several respects: various parts of the plants can be found (fruits, seeds, bud-scales, wood and more), accordingly their dispersal and preservation are complex, but the possibility to identify macrofossils to the species level and their more local origin makes them very valuable for palaeoecology, see Box 1 by Hilary H. Birks. In addition non-pollen palynomorphs (NPPs) may be important, see Box 2 by Bas van Geel below.

is more appropriate (see vol. 1 of Tracking Environmental Change Using Lake Sediments: Scholz, 2001; Moorman, 2001; King and Peck, 2001). Differences in sediment accumulation may also be found between wind-exposed and sheltered parts of the basin as has been demonstrated by Odgaard (1993).

Some lakes may become completely filled and then overgrown by peat, usually first by fen peat and later by *Sphagnum*-dominated raised bog (see Fig. 2.2-2 and chapter 4.7). Ombrotrophic peat can be used as a special archive for purely atmospheric input (that is without input from surface run-off): this is important for the reconstruction of atmospheric pollution history (Shotyk et al., 1998, 2000; Martínez-Cortizas et al., 1999; Weiss et al., 1999; Roos-Barraclough et al., 2002).

2.2.2

Other archives

In addition to lakes and mires many other archives exist, such as marine sediments, glacier ice or cave deposits. Each type brings its own possibilities and difficulties.

Marine cores are very important for time-scales longer than the Holocene. Their lower sedimentation rates have an advantage – reaching further back in time – and a disadvantage – even refined sampling may not provide high temporal resolution. In spite of the vast pollen-source area of marine cores their comparability to terrestrial records has been shown in some cases, for example by Tzedakis et al. (1997) or Sánchez-Goñi et al. (2002).

Cores in glacier ice can provide very interesting records if the pollen concentration can be enhanced by appropriate laboratory treatments (Brugger et al., 2018a). Again the pollen-catchment area is very large but landscapes can be studied that otherwise have hardly any archives (Brugger et al., 2018b).

In archaeological contexts, samples within excavations are often compared with records from outside the settlements (on-site records versus off-site records; see chapter 4.5).

Soils of certain types such as mor (raw-humus top of podzols) or shallow organic layers formed in small forest hollows without permanent open water can provide refined spatial records, namely the stand-scale development of vegetation. Whereas the source area for pollen and macrofossils may be large for lakes and mires (see chapters 4.6 and 4.7) it is often small and spatially restricted for small forest hollows due to the very local deposition of

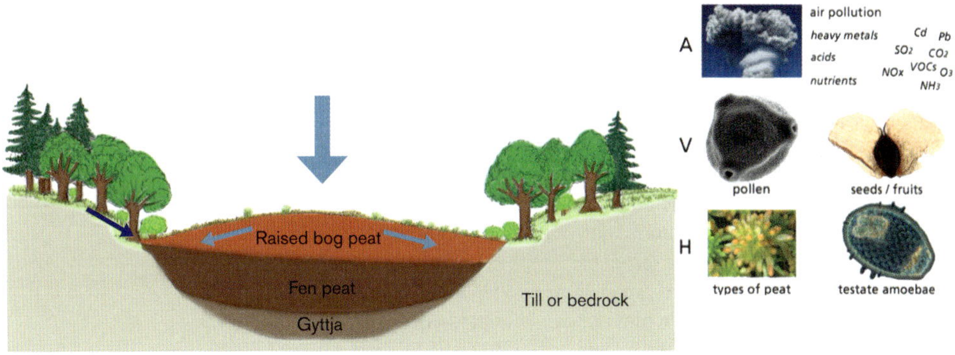

Fig. 2.2-2. Schematic system of how a site may develop from a lake to a fen and a raised bog, incorporating environmental components through time: A from the atmosphere, V from vegetation, H from hydrology and aquatic ecosystems. Mires can also have other origins than lakes, see chapter 4.7. Because raised bogs are in their centre disconnected from the hydrology of the surrounding landscape they offer the possibility of separating atmospheric input (large blue arrow) from erosional input; thus studies of air pollution are possible ("pollution history", see Box 4 by Neil Rose). Hydrological changes may be recorded in the composition and preservation of the peat and in the biostratigraphy of the testate amoebae, see Box 3 by Katarzyna Marcisz.

plant remains (Iversen, 1964; Andersen, 1970; Aaby, 1983; Bradshaw, 1988, 2013; Bradshaw et al., 2005).

Cave deposits may be difficult because of poor pollen preservation but of great interest because they often represent long and/or archaeologically important periods. Cave sediments may include pollen, plant macrofossils, charcoal and fossilized animal droppings (coprolites, see Gale and Carruthers, 2001).

In rare cases **archives controlled by animals** can be used, for example droppings from bats or other mammals, as shown by van der Knaap (1989) or Yll et al. (2006). Fossilized droppings can be important in archaeological contexts, be it for the Palaeolithic or the Neolithic (Rasmussen, 1993; González-Sampériz et al., 2003; McGarry and Caseldine, 2004; Carrión et al., 2008). In caves, coprolites and sediments may provide complementary evidence. By comparing pollen spectra in surface samples with those in dung pellets from goat, sheep, rabbits and other animals, Carrión (2002a) could show that a high diversity of insect-pollinated species (rare or absent in lake sediments) can be traced in this way. Birds and mammals as collectors of modern pollen by means of their feathers and coats were studied by Groenman-van Waateringe (1998): if such pollen is brought into lakes or bogs and becomes incorporated in the sediment it may provide a chance to record rare taxa but it may also be a source of long-distance contamination.

Table 2.2-1. Components of lake sediments and peat after Troels-Smith (1955) and Aaby and Berglund (1986, 2003).

1. Turfa (T)		Peat: Macroscopic, mainly subterranean parts of phanerogams and mosses
	T. bryophytica (Tb)	Moss-peat: T. sphagni, T. hypnacea, T. polytrichi
	T. lignosa (Tl)	Wood, roots, twigs of trees, shrubs (e.g. Ericaceae)
	T. herbacea (Th)	Rhizomes, roots of non-woody plants (e.g. ferns, *Phragmites,* Cyperaceae)
2. Detritus (D)		Above-ground parts of phanerogams
	D. lignosus (Dl)	Fragments of wood and bark, > 2 mm
	D. herbosus (Dh)	Fragments of herbaceous parts, > 2 mm
	D. granosus (Dg)	Fragments of woody and herbaceous parts, 2 – 0.1 mm
3. Limus (L)		Gyttja: Remains of micro-organisms (animals, plants), < 0.1 mm
	L. detrituosus (Ld)	Detritus-gyttja, algal gyttja: organic remains of micro-organisms (animals, plants, silica exclusively of organic origin)
	L. siliceus organogenes (Lso)	Remains of micro-organisms built of silica (diatoms, needles of spongi)
	L. calcareus (Lc)	Marl: precipitated calcium carbonates, particles < 0.1 mm
	L. ferrugineus (Lf)	Precipitated Fe-oxides or Fe-sulfides, if reduced: black, if oxydized: yellow or red
4. Substantia humosa (Sh)		Strongly decomposed organic matter, without structure, blackish, not peat, detritus or limus
5. Argilla (A)		Clay and silt: Mineral particles, < 0.06 mm
	A. steatodes (As)	Clay: Mineral particles, < 0.002 mm
	A. granosa (Ag)	Silt: Mineral particles, 0.06 – 0.002 mm
6. Grana (G)		Sand and gravel: Mineral particles, > 0.06 mm
	G. minora (Gmin)	Sand: Mineral particles, 2 – 0.06 mm
	G. majora (Gmaj)	Gravel: Mineral particles, 60 – 2 mm

2.2.3

Coring and core description

A clean and complete core is a *conditio sine qua non* for a successful palaeoecological study as mistakes during the coring operation may be very difficult to discover and impossible or time-consuming to mend. Glew et al. (2001) present more than 10 hand-operated types of corer (for lake sediments) together with their advantages and disadvantages. Wright (1991) offers good advice for frequently encountered problems. Some of the corer types (Fig. 2.2-3) may also be used on peat (piston corers on rods, chamber-type samplers such as the Russian-Belarusian peat sampler). In many cases the top of a sequence is especially difficult to recover as an undisturbed core. In lakes, freeze-corers usually provide a clean sediment-water interface, while in peat sections the top may best be collected as a monolith (for example in a long, square sided box) because the fibres of only partially decomposed *Sphagnum* or other mosses make undisturbed coring difficult. If basins in remote or hardly accessible areas are cored a trade-off between light and/or fancy equipment becomes obvious (unless helicopters or mules are available). More technical coring techniques for deep lakes are described by Leroy and Colman (2001). A general overview including coring, sampling and much more is reported by Glew et al. (2001) and Coleman (1996).

The **storage** of cores is an important concern. Ideal storage conditions can be summarized as 'wet, cool, and dark'. The material must be protected against drying out and against fungi and other microbes. Therefore, proper wrapping in plastic contact-foil and 'sleeves' and storage at cool temperatures (c. 4°C, not below zero) are essential. Dark storage is important because, when stored in the light, algae may develop that capture modern CO_2 from the air thus contaminating the sediment with respect to radiocarbon sampling.

Description of sediments and peats

The description of the cores obtained is a strong basis for any further studies. Aaby (1979), based on Troels-Smith (1955), provides a strong and inexpensive tool which is orientated towards the estimation of the mineral and biological components of sediment or peat. Aaby and Berglund (1986) offer a somewhat refined form and Birks and Birks (2004) provide the connection between the sediment or peat types and hydrological conditions, especially water depth and trophic state. Our figures 2.2-4 and 2.2-5 and Table 2.2-1 visualize the components and their origin. Colour description can be standardized by using the colour charts developed by Munsell for soil descriptions (Munsell, 1975).

2.3

How: analysis of samples

Strategies for laboratory sampling of the cores must be planned carefully to provide enough material for all the studies envisaged (e.g. pollen, plant macrofossils, loss-on-ignition, diatoms, chironomids, Cladocera and radiocarbon or other dating techniques). Non-destructive methods such as photography and geochemistry by XRF should be carried out before the core is subsampled, that is, on the intact core (just cleaned on the surface). If possible, sampling should be done on a single core or a composite master based on the correlation between parallel and overlapping cores. Some analyses, however, require large sample volumes, for example for plant macrofossils and especially for beetles, so, in those cases, the correlation of additional neighbouring cores may become necessary. In addition, a variety of mineralogical and geochemical techniques are very useful.

Samples for analyses of pollen (and plant macrofossils) should be taken as known volumes rather than weights (even if volumes are less precise). The volume needed per sample depends on the pollen concentration (which is not known beforehand). In sediments or peat of Holocene age, 1 cm^3 (ml) is usually enough for pollen, and 10–100 cm^3 for plant macrofossils. Before any chemical treatment, marker grains should be added to samples for pollen analysis as described by Stockmarr (1971) and Maher (1981). This is important because, with a well-established chronology and age-depth model, we may later want to calculate pollen (or macrofossil) influx (grains per surface and year), also known as PAR (pollen accumulation rates). Its advantage compared to percentages is that it is not closed data (not internally dependent). For optimal calculations the number

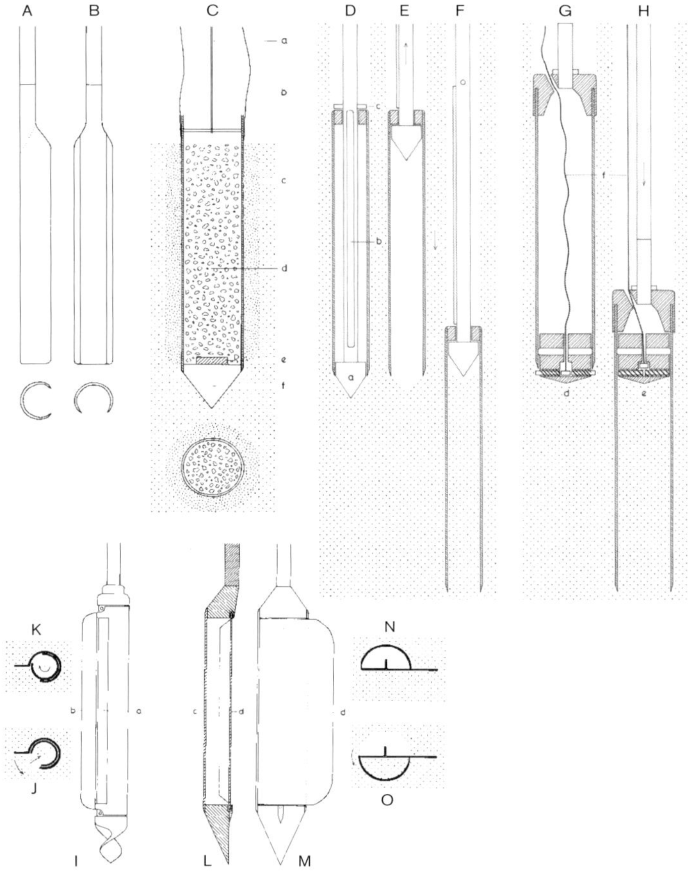

Fig. 2.2-3. Different corer types. **A** and **B:** Dutch 'marsh-spoon' in side view and front view. **C:** 'fridged finger' or 'frozen finger' in side view and cross-section; a: metal cable. b: soft plastic tube fixed to the metal tube; c: frozen sediment; d: refrigerating mixture of dry ice and a mixture of alcohols (ethanol, butanol, trichloraethylene); e: lead weight; f: point of corer that is not affected by cold temperatures. **D, E, F:** Piston-sampler Dachnowski, D: position at the beginning of coring; a: piston; b: fixing-rod for the piston; c: pin to fix the chamber. E: tube with piston fixed after withdrawal to the upper position (chamber open). F: corer in lowered position (chamber filled). **G, H:** Livingstone corer, G = corer empty, before coring; d: piston fixed by rubber stopper and springs at the lower end; f: steel cable (loose in G and tight in H); H = corer with filled chamber; e: piston unlocked, fixed by the tight cable, allows gliding of the chamber into the sediment by pushing the rods. The metal parts in the cross-section are hatched. **I, J, K:** Swedish side-sampler or Hiller-corer, I = side view; a: chamber (inner tube, fixed to a rod at the top and with an auger point at the bottom); b: external tube with slit and sharpened flange (rotating between pins). J = cross-section when open, K = cross-section when closed (as in I). **L, M, N, O:** Russian-Belarusian side-sampler, L = side view as cross section; c: chamber (half-tube fixed to a rod at the top and with a point at the bottom); d: metal-plate that can be turned. M = front view with d = metal plate, on which the sediment will remain after turning (opening). N = cross-section when chamber is empty, the position in which it is pushed into the sediment. O = cross-section with chamber full of sediment after the corer has been turned. The metal parts in the cross-section are black or hatched.

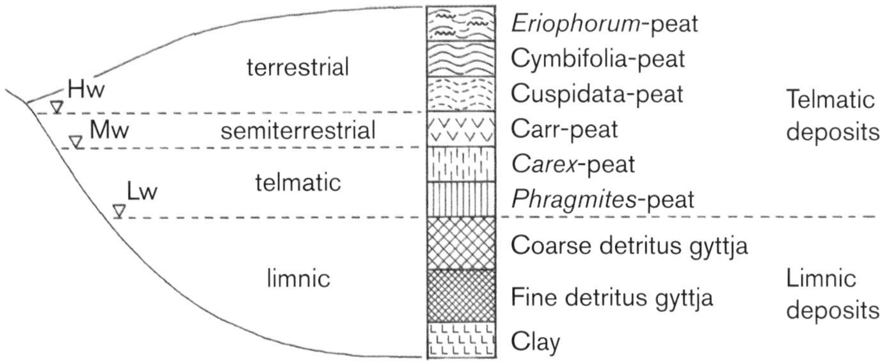

Fig. 2.2-4. Types of sediment and peat in a basin and the relationships to low, middle and high water (Lw, Mw, Hw) and their signatures in the Troels-Smith system (Troels-Smith, 1955)

Fig. 2.2-5. Signatures for sediments and peat according to Troels-Smith (1955) and Aaby and Berglund (1986).

of added exotic markers should be similar to the number of pollen and spores in the same volume of sample residue (which cannot be known at the beginning but can be initially estimated and later adapted).

The sampling resolution depends on the research question and the processes that need to be understood. Often, sampling resolution is refined in an iterative approach, but re-opened and re-sampled cores frequently provide less certainty about the exact position of the samples (if the sediment was disturbed during the initial sampling). A good way out of this dilemma is to cut the core lengthwise in two halves, one of which goes to a cool archive, the other one gets completely sub-sampled in as fine a resolution as possible for as many types of analyses as wished. A device for refined sub-sampling was introduced by Joosten and de Klerk (2007).

Laboratory preparation techniques (and safety concerns) are described for pollen by Moore et al. (1991) and Bennett and Willis (2001) and for plant macrofossils by Birks (2001). All preparation techniques aim at concentrating the biological particles to be identified (e.g. pollen, plant macrofossils, charcoal, stomata and other cells) by either dissolving unwanted parts of the sediment or wet-sieving according to size classes. In order to potentially overcome the problems associated with pollen percentages, marker grains in known concentrations must be added (see above).

A puzzling biological paradox is if the living cells within the pollen grain can survive for only hours or days, why do plants invest in substances (sporopollenin) in the outer wall of the pollen grain (ektexine) that can survive for millions of years and resist such violent chemical treatments as being immersed in hydrofluoric acid? It is pertinent to remember that there may be evolutionary reasons for this, such as the development from isospores in ferns, to heterospores in, for example Lycopods, to pollen and ovules in Angiosperms.

For the analysis of plant macrofossils H.H. Birks (2007a, 2013) provides a list of relevant literature grouped into manuals, books and papers on seeds and fruits, on mosses, on wood and on cuticles, bud scales and leaves (see also Box 1 by Hilary Birks).

For the analysis of pollen a strong understanding of morphology is needed for the best possible identification. Europe is in a privileged position in having a number of good reference books for pollen morphology, especially the series of the Northwest

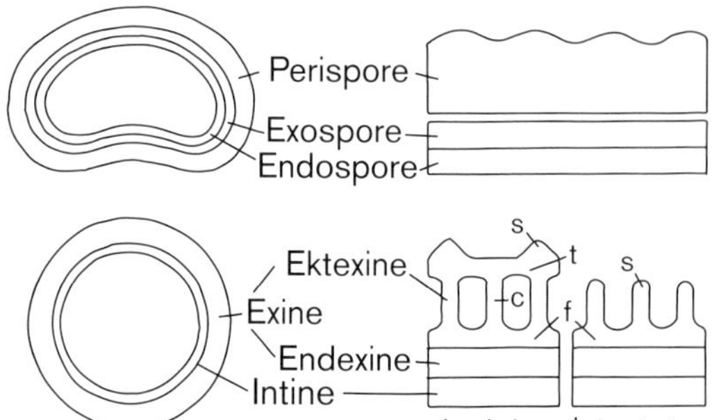

Fig. 2.3-1. Structure of the cell walls in spores (A) and pollen (B). c: columella, f: foot layer, s: sculptural element, t: tectum. Pollen-morphological nomenclature after Iversen and Troels-Smith (1950) and Fægri et al. (1989).

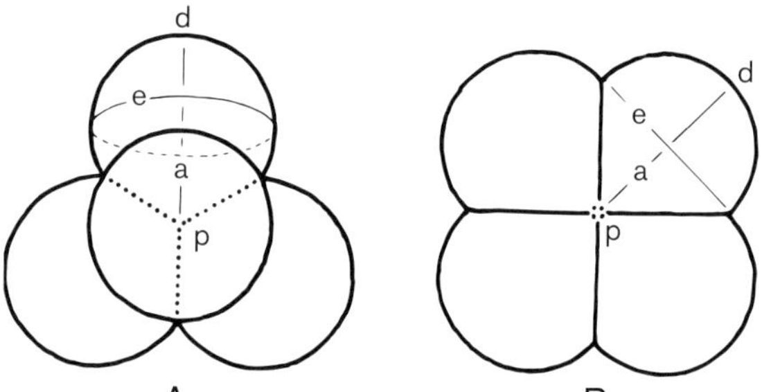

Fig. 2.3-2. Two types of tetrads after meiosis in Cormophytes. A: Tetraedric tetrad (two cell divisions at right angles to each other) in Bryophytes, Pteridophytes and most Dicotyledons. B: Tetragonal tetrad (two cell divisions in the same plane) in Pteridophytes, Gymnosperms, Monocotyledons and primitive Dicotyledons. a: polar axis, d: distal pole, e: equatorial plane, p: proximal pole. Modified from Birks and Birks (1980).

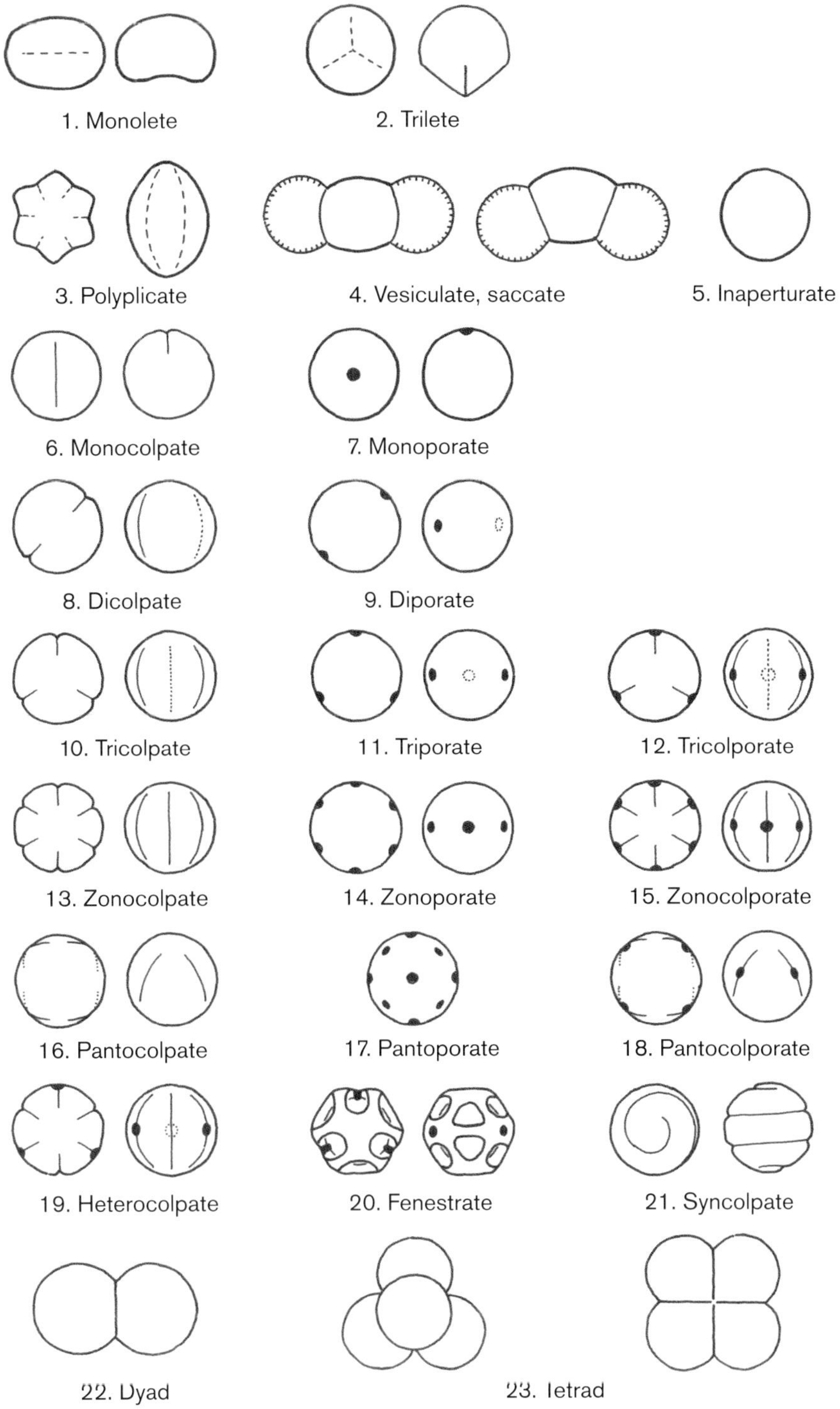

Fig. 2.3-3. Types of pollen and spores in the European flora. 1 and 2: Pteridophytes (spores). 3–5: Gymnosperms (pollen). 5–23: Angiosperms (pollen), inaperturate pollen occur both in Gymnosperms and Angiosperms. The left figures show the view from the pole; the right figures show the equatorial view, distal pole at the top, proximal pole at the bottom. Schematic, modified with additions after Iversen and Troels-Smith (1950).

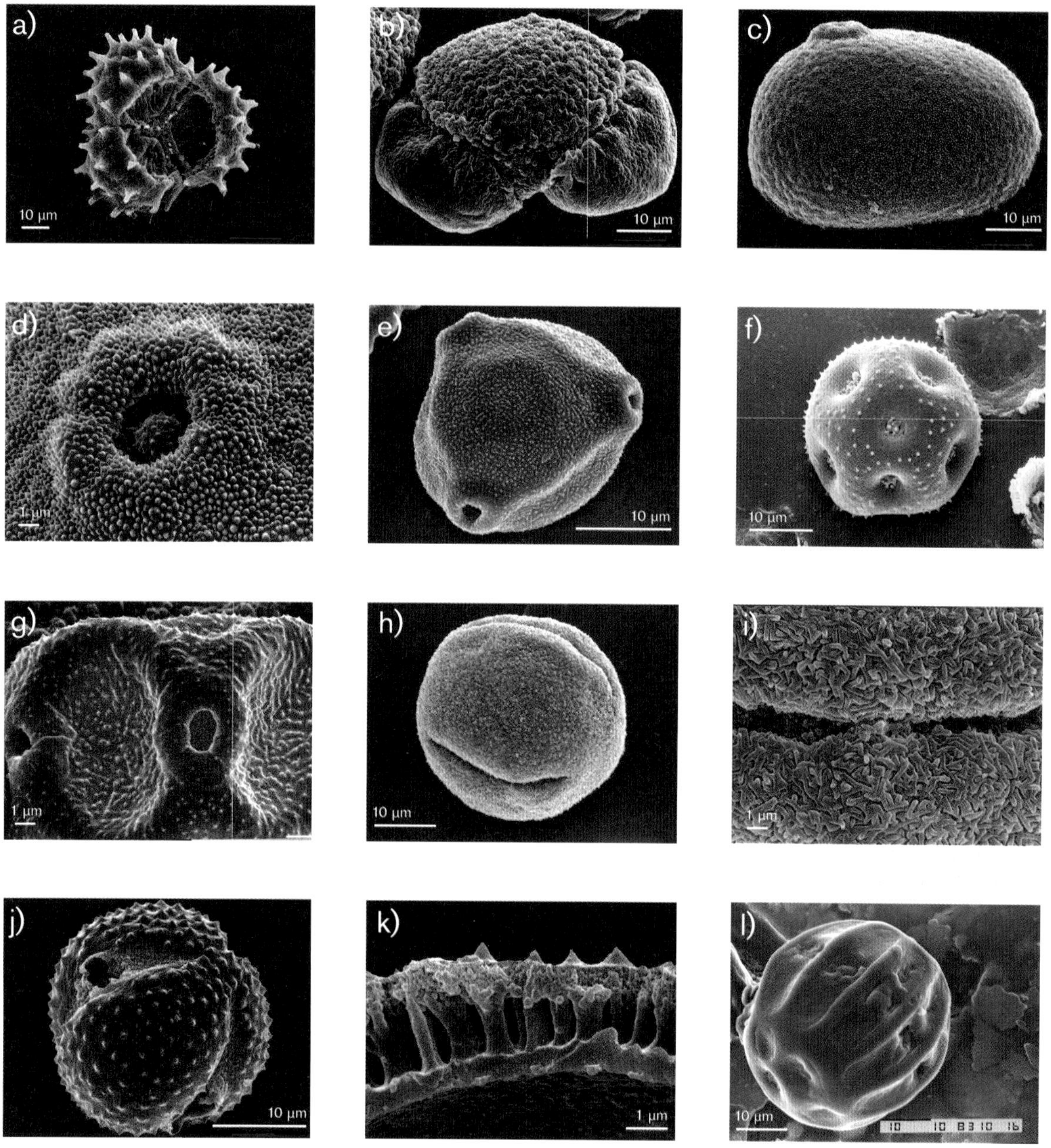

Fig. 2.3-4. Scanning electron microscope (SEM) pictures of selected pollen and spores. a) *Selaginella selaginoides:* a tetrad (the 4 grains remain together; surface is echinate); b) *Pinus mugo,* a bisaccate grain (found in many but not all Gymnosperms); c) *Secale cereale,* rye, a monoporate grain (like all Poaceae); d) *Secale cereale,* rye, close-up of the pore; e) *Betula pubescens,* a triporate, scabrate grain; f) *Scleranthus* sp., a polyporate, echinate grain; g) *Alnus viridis,* equatorial view, a stephanoporate, scabrate grain; h) *Fagus sylvatica*, a tricolporate, scabrate grain (pores hidden, easy to see in light microscope; i) *Fagus sylvatica,* close-up of the hidden pore (in light microscope a clear ring); j) *Artemisia mutellina* (synonyms *A. umbelliformis* and *A. laxa*), a tricolporate, echinate grain; k) *Artemisia mutellina,* cross-section through the wall: tectum with spines and columellae; l) *Polygala vulgaris,* a stephanocolporate, psilate grain. Photographs by Lucia Wick, IPS, University of Bern.

European Pollen flora edited by Punt et al. (1976–2009), Moore et al. (1991), Beug (2004, second edition in 2015) and the atlases by Reille (1992b, 1995, 1998).

It is of interest that even among tree pollen there are types where two species of a genus belong to different pollen types, for example *Carpinus betulus* is different from *Carpinus orientalis*, the latter belonging to the *Ostrya* type (Beug, 2004). Terminology of pollen can be found in Hesse et al. (2009). Pollen morphology is also an important tool in melissopalynology, in aeropalynology and sometimes in forensic studies.

The colour of fresh pollen is often (but not always) yellow and Flenley (2011) and many phytochemists explain it with the content of flavonoids that are often (but not always) yellow (*flavus* meaning yellow in Latin). These pigments in pollen get mixed with the colourless wax from the glands in the abdomen of bees and masticated to beeswax: this explains the colour of beeswax and our candles.

On a pollen slide at least five additional types of subfossils can be identified:

- non-pollen palynomorphs (NPPs), see Box 2 by Bas van Geel.
- testate amoebae, see Box 3 by Katarzyna Marcisz.
- spheroidal carbonaceous particles (SCPs), see Box 4 by Neil Rose.
- conifer stomata (keys by Trautmann, 1953; Sweeney, 2004), see figure 2.3-5.
- microcharcoal (Whitlock and Larsen, 2001; Tinner and Hu, 2003; Finsinger and Tinner, 2005); macrocharcoal analysis is usually carried out alongside macrofossil analysis (Finsinger et al., 2014).

For the identification of wood (and charred wood) books on wood anatomy (Schweingruber, 1990 or, for central European taxa, Schoch et al., 2004, also available online) can be used.

The **sum to which pollen should be counted** per sample depends on the research question. A minimum requirement is about 300 grains. If car ly human impact is the question then sums >500 grains are usually needed in order to trace rare indicator taxa such as weeds. Since time is always at a premium it is often more efficient to invest time in counting a higher density of samples in the core (giving a more continuous record or better temporal resolution) than to increase the counting sum for fewer samples. Green (1983) wrote:

"Studies of pollen analytical techniques have rightly emphasized the need to minimize the large variances that arise from sampling, processing and counting. Most efforts to reduce spurious variance have involved increasing the number of grains counted per sample, but extra counting yields diminishing returns beyond about 1000 grains per sample (Fægri and Iversen, 1964; Maher, 1981). It is better to count more samples than more grains per sample (Walker, 1966): in statistical, sedimentological and biological terms a change in pollen abundance that is consistent in (say) five samples extracted at 1-cm intervals and counted to a sum of 1000 grains per sample is vastly more significant than the same change seen in only one sample (with 5 cm interval) counted to a sum of 5000 grains."

The **reference pollen sum**, used as 100% for the percentage presentation, is most often the sum of all pollen counted excluding that of aquatic plants, the spores of ferns and mosses, and NPPs. The reference pollen sum can be adapted according to local conditions or to specific research questions, for example by excluding Cyperaceae or *Alnus* that may be locally overrepresented.

The **presentation** of the identified and counted pollen and plant macrofossils is usually done through diagrams in which the y-axis is a depth scale – which can ideally be transformed into a timescale. Most pollen diagrams present results as **percentages**, although the internal interdependence of closed data is a disadvantage. Diagrams expressed in pollen **concentrations** are rare because the quantity of pollen is heavily dependent on sedimentation rates and their changes. For many research questions pollen diagrams presented in terms of **pollen accumulation rates** (or **influx**, pollen cm^{-2} $year^{-1}$) are useful. Giesecke et al. (2010) summarize the history of the two terms 'influx' and 'accumulation rates', here we use them as synonyms. The quality of influx values depends heavily on a good age-depth model. The relevance of influx versus percentages becomes particularly obvious at ecotones such as timberlines (see chapter 4.9), where tree-pollen may dominate the surface samples of the tundra if presented as percentages, but be low in concentrations or influx (Aario, 1941; Hicks, 2001). Interpretation of influx diagrams can benefit from research involving annual pollen monitoring by means of traps, as standardized by the PMP-programme (Hicks et al., 2001; see Box 5 by Sheila Hicks).

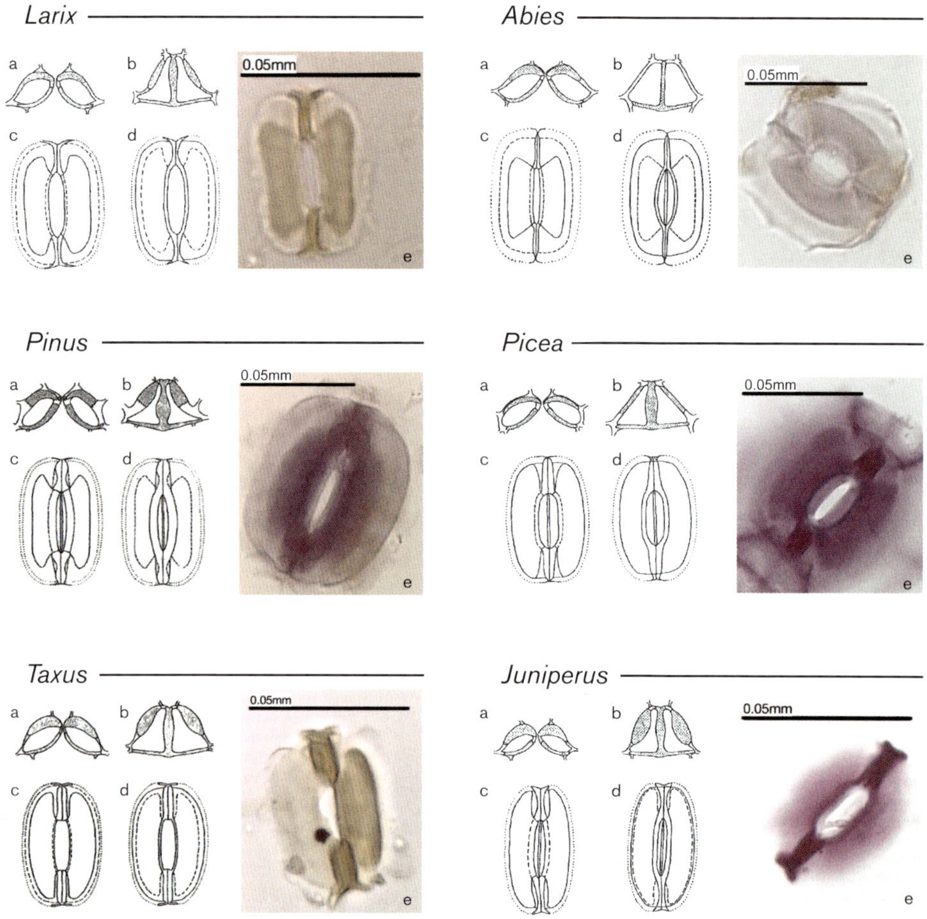

Fig. 2.3-5. Schematic views of stomata for six conifer genera: a) median cross-section, b) polar cross-section, woody parts are dotted. c) upper side: dotted line for the non-woody cell wall; continuous lines for the upper woody lamellae, dashed lines for the lower woody lamellae (the latter is thinner and sometimes not preserved); d) lower side: dotted line for the non-woody cell wall; the upper woody lamellae are on a lower level (and in *Picea* not visible). Because of the two views the continuous and the dashed lines are opposite in c) and d). Diagnostic are also the stems of the woody lamellae (for example broad in *Pinus,* narrow in *Abies*) and their hooks or horns. Drawings from Trautmann (1953), photographs from the Institute of Plant Sciences, Bern. Gerhard Lang dedicated his book (1994) to Werner Trautmann, who pioneered stomata analyses on pollen slides.

Box 1

Macrofossils and pollen

Hilary H. Birks

Department of Biological Sciences, University of Bergen, N-5020 Bergen, Norway

What is a plant macrofossil?

A plant macrofossil can be defined as a plant fossil that is visible to the naked eye and that can be manipulated by hand. Macrofossils have become an essential tool in palaeoecological studies (Birks and Birks, 2000) and have long been the foundation of palaeoethnobotany and archaeobotany. The sizes of macrofossils range from 0.5 mm to enormous tree trunks ('megafossils'). Most are 0.5–2.0 mm. They are usually studied using a stereomicroscope at 10× - 40× magnification. A high-power microscope and a scanning electron microscope (SEM) give magnifications 100× and more. Plant remains are not only fruits and seeds but also vegetative parts, such as leaves, cuticles, buds, bud scales, anthers, flower parts, rhizomes, twigs, wood and bark. Lower-plant remains include sporangia and megaspores, mosses and Characeae oospores.

Where do we find macrofossils?

Plant remains are preserved in anoxic, waterlogged, arid or frozen environments. *In situ* plant remains in mires reflect the peat-forming vegetation. Macrofossils in lake sediments reflect local vegetation and neighbouring communities, such as littoral swamps and nearby upland vegetation, especially trees (H.H. Birks, 1973a). Remains can be transported into lakes by wind, birds, inflowing streams, solifluction and slope-wash.

How do we interpret macrofossil data?

Individual taxa can be used as indicator species using modern ecological tolerances. A macrofossil assemblage is derived from vegetation at different scales and its composition depends on the taphonomic processes of production, dispersal, deposition (including sorting and redeposition) and preservation. These vary from site to site. Surface sample studies can indicate the relation of the abundance of the fossils to their abundance in the vegetation (H.H. Birks, 1973a; Zhao et al., 2006). Quantitative reconstructions of vegetation, environment and climate are difficult from macrofossil assemblages (Birks, 2014).

Macrofossils complement pollen data.

The combination of macrofossils and pollen in studies of vegetation history and successions, both aquatic and terrestrial, provides much more information than that provided by pollen analysis alone (Birks and Birks, 2000; Eide et al., 2006; Birks, 2015). Some ways that macrofossils can complement pollen are given below and illustrated in figure B1-1. Figure B1-2 illustrates some of the important macrofossils mentioned in figure B1-1.

A pollen assemblage has a wide and undefined source area. Relatively heavy macro-remains are generally not dispersed far from the parent plant (H.H. Birks, 1973a) which allows past local vegetation change and dynamics to be reconstructed in some detail (Birks and Birks, 2003; Birks et al., 2001; Birks, 2015). Macrofossils can indicate the local occurrence of taxa with abundant, widely dispersed pollen, such as *Betula* and *Pinus* (Birks and Mathewes, 1978; Birks and Birks, 2000; Birks, 2003) with corresponding climatic implications (van Dinter and Birks, 1996; Birks, 2015).

Pollen has limited taxonomic resolution, especially of ecologically important taxa such as trees, Poaceae and Cyperaceae. Macro-remains can usually be identified to a lower taxonomic resolution, often to species level (for example the separation of *Betula* spp., particularly *B. nana* and *B. pubescens*, van Dinter and Birks, 1996; conifer needles for example *Larix, Pinus, Picea,* Cupressaceae; Dunwiddie, 1987). The identification of bud scales provides records of some deciduous trees (for example *Quercus, Ulmus, Corylus, Ostrya/Carpinus*) that otherwise produce few macrofossils (Eide et al., 2006; Baker, 2013).

Some plants produce small quantities of fragile or nondescript pollen and non-vascular cryptogams do not produce pollen. Macrofossils can provide definite palaeoecological records for taxa such as

Populus, Cupressaceae, *Juncus*, *Dryas*, *Salix herbacea* and many others, and for cryptogams. Their macro-remains can be recovered, sometimes in abundance (for example mosses, Characeae, *Juncus* spp., *Najas* spp.), thus refining reconstruction of the ecosystem (Jonsgard and Birks, 1995; Birks, 2000).

In treeless situations with low pollen production such as arctic and alpine tundra, steppe and desert, long-distance pollen can bias interpretations. Macrofossils reflect the local vegetation and hence palaeoenvironmental and palaeoclimatic changes (Birks, 2001; Birks and Birks, 2000; Baker, 2013). This is particularly important in treeline reconstruction (Tinner and Theurillat, 2003; Eide et al., 2006; Birks and Bjune, 2010; Orbán et al., 2018), for geographical movements of taxa (Birks, 2015), for full- and Late-glacial vegetation and climate reconstructions (Birks and Mathewes, 1978; Goetcheus and Birks, 2001; Birks, 2003) and vegetation and climate history in arctic regions (H.H. Birks, 1991a; Heggen et al., 2010). The Holocene macrofossil record at Sägistalsee, Switzerland, was used to test hypotheses about the relative impacts of catchment vegeta-

Fig. B1-1.
Comparison of pollen or spore and macrofossil influx for taxa that produce sufficiently large numbers of both types of fossil in the Late-glacial and Early Holocene at Abernethy Forest, Scotland. Each pollen or spore taxon is followed by the equivalent macrofossil taxon (taxa). The values are plotted against radiocarbon age of the sediment. The pollen zones are drawn across the pollen curves and the macrofossil zones across the macrofossil curves. The Late-glacial/Holocene transition is placed between zones 3 and 4. Macrofossils show local presence, for example the actual arrival time of *Pinus* compared with its pollen curve. Macrofossils show occurrence of several taxa within a pollen type, for example *Betula* tree type *(B. pubescens)* and *B. nana* in *Betula* pollen, several taxa within Cyperaceae pollen. Macrofossils and pollen can be derived from different species, for example *Salix* pollen and *S. herbacea* macrofossils. Sometimes macrofossil and pollen records coincide well, for example *Nymphaea*. Sometimes, macrofossils show changes in abundance but pollen does not, for example *Calluna vulgaris*. From Birks and Birks (1980) after Birks and Mathewes (1978).

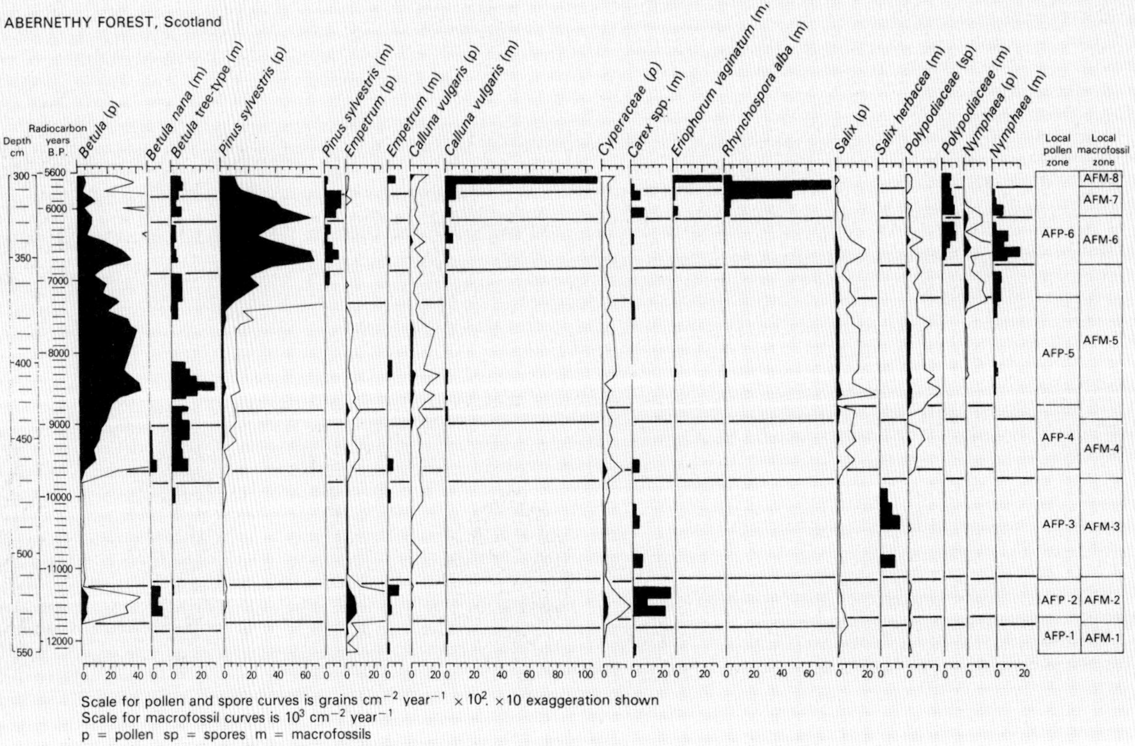

tion and climate changes on the palaeolimnology of the alpine lake (Lotter and Birks, 2003a).

Macrofossils are less abundant than pollen and larger sediment volumes are needed for analysis.

Other applications of macrofossil analysis

Macrofossils are necessary for AMS radiocarbon dating of terrestrial plant material, to avoid errors produced by old carbon reservoirs in lake sediments and by non-contemporaneous material such as deep-growing roots in peat. Ancient DNA from fossil plant material may give information about past vegetation composition which is not already apparent from the macrofossils themselves (Birks and Birks, 2016; Pedersen et al., 2016). Multi-disciplinary studies involving macrofossils reveal vegetation and climate changes (Birks et al., 2000) and human impact on lake ecosystems through acidification, eutrophication, and pollution (Birks et al., 2001).

A fuller discussion of plant macrofossil methods and interpretations can be found in Birks (2001, 2013, 2017).

Fig. B1-2.

Photographs of some of the seeds and fruits compared with pollen in figure B1-1. A: *Betula nana* fruit. B: *Betula nana* female catkin scale. C: *Betula nana* leaf. D: *Betula pubescens* fruit. E: *Betula pubescens* female catkin scale. A-E are all represented by *Betula* pollen. *Betula nana* occurred in the Allerød and the earliest Holocene. Tree birch (*B. pubescens*) arrived and expanded in the Early Holocene. F: *Carex demissa*, a typical trigonous nutlet (3 stigmas). G: *Carex nigra*, a typical lenticular nutlet (2 stigmas). H: *Rhynchospora alba*. I: *Eriophorum vaginatum*. F-I are all represented by Cyperaceae pollen. The pollen influx gives no suggestion of the number of macrofossils and importance of Cyperaceae in the hydrosere. J: *Nymphaea alba*. Pollen and macrofossils conform well.

Box 2

Non-Pollen Palynomorphs for improved understanding of past environmental change

Bas van Geel

Institute for Biodiversity and Ecosystem Dynamics, University of Amsterdam, The Netherlands.
E-mail: B.vanGeel@uva.nl

In pollen slides we analyze the pollen grains, but often many other microfossils are preserved. Over the past years a variety of 'extra' microfossils (non-pollen palynomorphs or NPPs; Fig. B2-1) were recorded in Quaternary deposits. This research has resulted in some additional ecological indicators, useful especially for the reconstruction of local conditions. Among useful NPPs are fungal spores, algal remains and cyanobacteria. Hundreds of NPPs were described and illustrated and their indicator value discussed (van Geel, 1978, 2001; for an overview of references see Miola, 2012). Among the NPPs there still are taxa which have not been properly identified, but some of them nevertheless can be used as palaeo-environmental indicators. In such cases, the ecological information is inferred from the co-occurrence (curve matching) with identified taxa (for example pollen, seeds, bryophytes). Most algae, fungi, and cyanobacteria have a much longer geological history – as recognisable microfossils – than angiosperms and gymnosperms, and, therefore, some of the NPP studies are also of interest for palynologists specialised in the analysis of pre-Quaternary deposits.

Three examples of useful NPPs:

1) The observation of ascospores of coprophilous fungi (for example *Podospora*, *Sordaria*, *Sporormiella;* van Geel et al., 2003; van Geel and Aptroot, 2006) allows the reconstruction of population density changes of herbivores in the past. Fruit-bodies of coprophilous fungi from inside the intestinal tracts of mammoths point to a behavioural aspect: coprofagy (van Geel et al., 2011)

2) Prehistoric pollution (phosphate eutrophication) of lakes by farmer communities can be reconstructed using remains of cyanobacteria (*Gloeotrichia*, *Anabaena*, *Aphanizomenon*; van Geel et al., 1994).

3) Spores of Zygnemataceae (*Mougeotia*, *Spirogyra*, *Debarya*, *Zygnema*) are often well-preserved. The presence of spores points to shallow, stagnant freshwater (van Geel, 1976). Zygnemataceae have a long, pre-Quaternary history (van Geel and Grenfell, 1996).

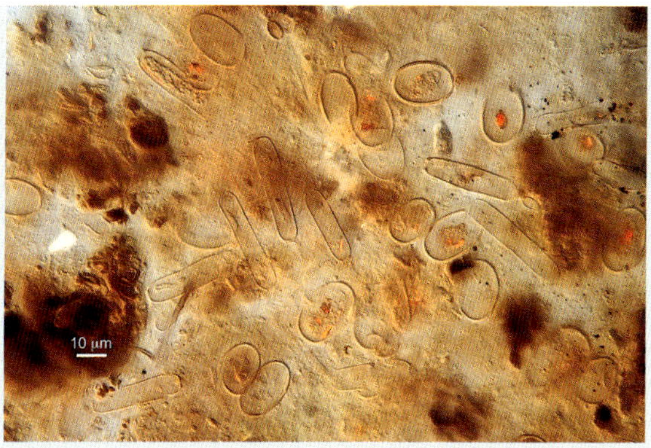

1.

2.

3.

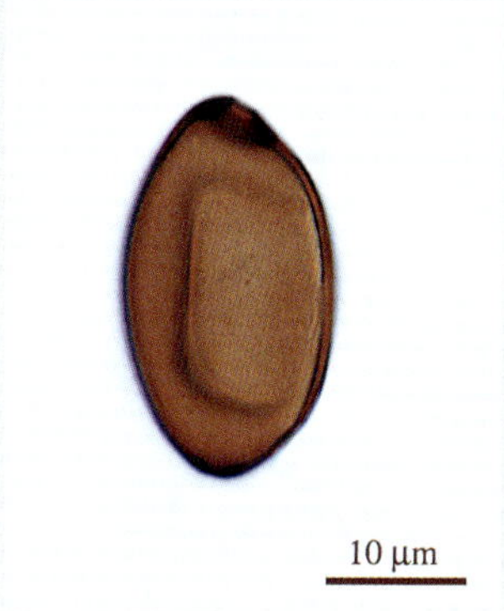

4.

5.

Fig. B2-1.

Examples of different Non-Pollen Palynomorphs.

1: Akinetes of *Anabaena* (broad) and *Aphanizomenon* (narrow)

2: *Podospora*, ascospore

3: *Mougeotia*, zygospore

4: *Sordaria*, ascospore

5: *Spirogyra*, zygospore or aplanospore

Box 3

Testate amoebae are a proxy for past hydrological dynamics and trophic status in peatlands

Katarzyna Marcisz

Adam Mickiewicz University in Poznań, Poland.
E-mail: marcisz@amu.edu.pl

Testate amoebae (TA) are a diverse group of amoebae that produce a decay-resistant shell ('a test'). TA belong to at least three unrelated groups of organisms: Arcellinida, Euglyphida and Amphitrematidae. However, due to their similar functional roles and production of shells, TA are treated as one group when used in ecological, experimental and palaeoecological studies (Mazei and Tsyganov, 2006). Different groups of TA produce different types of shells, i.e. shells made of biosilica plates (idiosomes), idiosomes with organic coating, proteinaceous shells and shells built from recycled organic or mineral particles (xenosomes) (Mitchell et al., 2008b, see Fig. B3-1). The identification of TA is based on their species-specific shells. TA occur in wetland ecosystems – lakes, soil, peatlands and rivers. Species ecology is known best for the peatland TA (Charman, 2001; Mitchell et al., 2008a).

In palaeoecological studies of peatlands TA are used as proxies for changes in past hydrology and trophy. Sample preparation for the analysis of TA involves sieving the peat under running water and analysing the filtrate under a light microscope. Using training sets and transfer function models, TA are the only proxy providing quantitative depth-to-water table reconstructions. pH can also be inferred (Lamentowicz et al., 2008; Birks et al., 2012d; Amesbury et al., 2016). The past environmental changes inferred from TA are often used in multi-proxy high-resolution studies of peatlands (for example along a bog-to-fen gradient; van der Knaap et al., 2011; Marcisz et al., 2015). Common wet indicator species are *Amphitrema wrightianum, Archerella flavum, Hyalosphenia elegans* and *Hyalosphenia papilio*. Dry indicator species include *Assulina* sp., *Corythion* sp. and *Trinema* sp. Presence of *Centropyxis aerophila* indicates minerotrophic conditions on the mire, whereas *Euglypha* sp. indicate acidic conditions. Additionally, morphological and functional traits of TA have been shown to greatly support palaeoecological reconstructions and improve interpretations of environmental studies (Marcisz et al., 2020).

(1) *Cylindrifflugia bacillariarum*

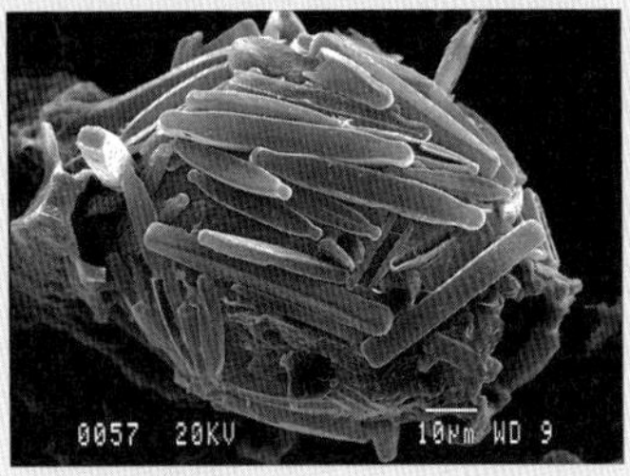

(2) *Assulina muscorum*

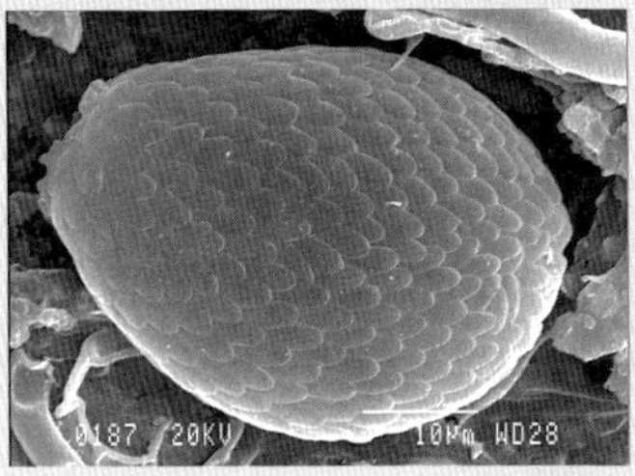

(3) *Galeripora catinus*

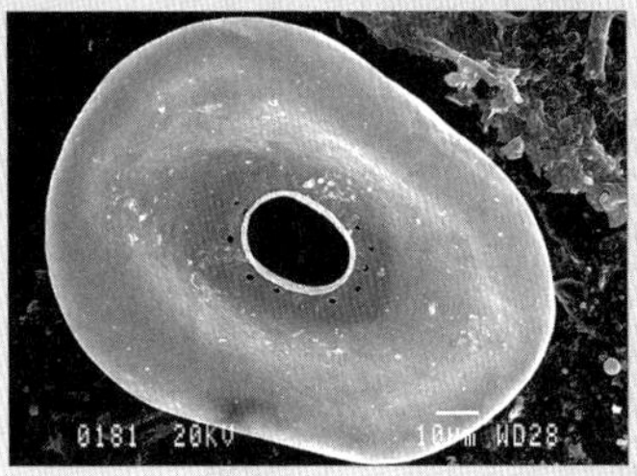

(4) *Euglypha strigosa*

Fig. B3-1.
Examples of testate amoebae with different shell types:
(1) *Cylindrifflugia bacillariarum* with xenosomic shells that includes diatom shells;
(2) *Assulina muscorum*, idiosomic shell with organic coating;
(3) *Galeripora catinus* proteinaceous shell;
(4) *Euglypha strigosa*, idiosomic shell with spines.
Photos: Mariusz Lamentowicz.

Box 4

Spheroidal carbonaceous particles

Neil Rose

Environmental Change Research Centre, Department of Geography, University College London, Gower Street, London WC1E 6BT UK. n.rose@ucl.ac.uk

Spheroidal carbonaceous particles (SCPs) are a component of fly-ash, produced solely by the industrial combustion of coal-series and oil fuels. They have no natural sources and so provide an unambiguous indicator of atmospherically deposited contamination from these emissions. Because SCPs are composed of elemental carbon, they are chemically very resistant and strong acids may be used in their extraction from natural archives such as peats and sediments (Rose, 1994). As a consequence, SCPs are resistant to palynological preparation techniques and may be found on pollen slides. Indeed, in the 1960s, their association with spores and pollen led them to be misidentified as a microfossil and given the genus *Linotolypa*, although this was later corrected (Reid and Allitt, 1981).

SCPs have a characteristic morphology (Fig. B4-1) and may be identified under the light microscope using the criteria of shape, colour, 'depth' and porosity (Rose, 2008). They have been recorded in a number of archives including peats (Yang et al., 2001), ice cores (Hicks and Isaksson, 2006) and marine sequences (Thornalley et al., 2018) but mainly lake sediments (Rose, 2001). As they have no natural sources the earliest presence of SCPs occurs in the mid-19th century. This start date is locally variable but a more widespread signal occurs in the 1950s, when a dramatic increase in SCP accumulation reflects the demand for electricity following the Second World War. This mid-20th century increase appears to be a global phenomenon and may act as a useful stratigraphic marker for the proposed Anthropocene Epoch (Rose, 2015). Furthermore, in many areas, SCP concentration profiles show robust and repeatable characteristics and so may be used to provide chronological information, once regionally calibrated (Rose and Appleby, 2005). More recently, with the introduction of particle arrestor technology and switches in fuels to natural gas, nuclear and renewable energy there has been a dramatic decline in SCP deposition. For two weeks in 2019, the UK generated all its electricity without the use of coal combustion for the first time since 1882 and SCP fluxes in annual sediment traps in UK upland lakes are now at levels not seen since the early-20th century.

Ultimately, if climate change is to be addressed, alternative energy sources will replace the industrial combustion of coal and oil, the production of SCPs will cease and their records will eventually form a discrete layer in archives across the world. As industrial artefacts they have been in existence for less than 200 years and so the full extent of their preservation is currently unknown. However, SCP-like particles (resembling non-combusted SCPs) have been identified from Bronze Age archives. If SCPs survive similarly then palaeoecologists, including palynologists, may find them for many centuries to come.

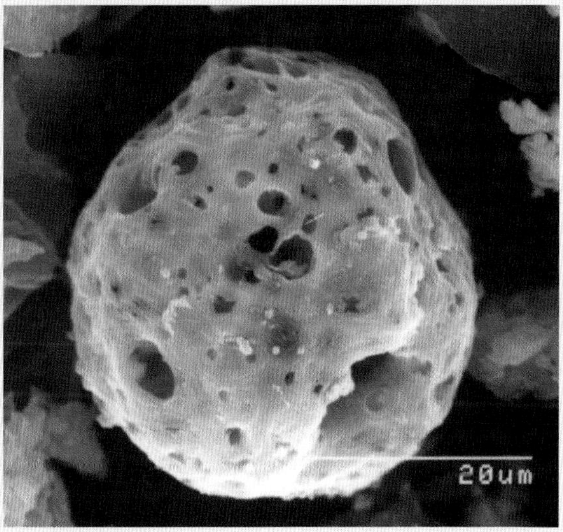

Fig. B4-1.
SEM micrograph of a spheroidal carbonaceous particle. Image: Neil Rose

Current computer programs for drawing pollen diagrams are TILIA (Grimm, 1993), psimpoll (Bennett, 2000) and PolPal (Walanus and Nalepka, 1997).

The presentation – and interpretation – of plant macrofossil data needs special care because the data are of a more heterogeneous nature than pollen data, namely binary (presence/absence), ordinal (on an abundance scale, often 1 to 4 or 1 to 10) or quantitative (numbers of taxon-specific and organ-specific findings). Birks (2014) presents useful and detailed advice.

2.4

When: dating techniques

In order to understand the dynamics of the history of flora and vegetation we need to know not only the age of events but also their duration and especially the rates of change involved. It is, therefore, advisable to transform the depth scales of the sediment series into age scales. A strong chronology is also a precondition for the transformation of pollen concentrations to influx, as correct influx is sometimes superior to percentage values which have the problem of 'closed data'. A concise overview of the various dating techniques available can be found in Bradley (2015). An in-depth treatment of dating techniques with interesting examples is provided by Walker (2005). Table 2.4-1 presents a short overview.

The most widely used dating technique in vegetation history (at least for the last c. 50 000 years) is the radiocarbon dating of organic matter. Of the three carbon isotopes, ^{14}C is by far the rarest and the only one that decays (with a half-life of 5730 years). A physical and ecological background to the method is provided by Olsson (1986), Björck and Wohlfahrt (2001), Walker (2005) and Jull (2007). The earliest technique was based on decay counting per time (Libby, 1952; Cook and van der Plicht, 2007), and needed large samples (about 10–15 g dry weight of peat or wood; more material for older ages). However, in the 1980s, direct mass spectrometry was developed, or rather isotope ratio mass spectrometry in the tandem accelerator, AMS (Walker, 2005; Jull, 2007). The greatest advantage of this method is that about 1000 times smaller samples are needed and so stratigraphic resolution can be refined (much thinner samples from the cores). Moreover, the organic particles submitted for dating can be identified. This is especially important for the distinction between aquatic and terrestrial plants because some water plants can incorporate old carbon from the sediment (the so-called hard-water error; Andrée et al., 1986; Hatté, 2007; Grimm et al., 2009; Finsinger et al., 2019).

2.4.1

Old carbon should be avoided in radiocarbon samples

There are several pathways via which old carbon can be incorporated into the organic matter that is to be sampled for radiocarbon dating:

1. In limestone areas the carbon source of some water plants can be (in addition to atmospheric CO_2) the HCO_3^- of dissolved carbonates from the catchment (Maberly and Madsen, 1998; Schippers et al., 2004; see Table 2.4-2). In oceans the sediment or mollusc shells also contain carbonatic carbon (another 'reservoir effect'; Stuiver and Braziunas, 1993; Mangerud, 1972; see also Marine Reservoir Correction Database, Reimer and Reimer 2001).
2. In bogs methanotrophic symbionts in *Sphagnum* can feed on old carbon from methane sources that gets built into tissue of living plants (Raghoebarsing et al., 2005).
3. In glacier forelands old carbon released from the glaciers may become incorporated into pioneer predators and other parts of the ecosystems (Hågvar and Ohlson, 2013; Hood et al., 2015).
4. Near volcanoes plants may take up old carbon dioxide from fumes, but this effect may be relevant for only short distances up to 200 m (Bruns et al., 1980; Weninger, 1990).
5. Reworked material can be a problem either near the bottom of a core (Late-glacial material) or when anthropogenic erosion in the catchment caused input of older organic material.

Sampling: Whenever possible, bulk sediment samples should be avoided for radiocarbon dating (Finsinger et al., 2019). From a botanical viewpoint an emphasis must be put on possible ways of contaminating the samples and how to avoid this.

(1) Too young material can be brought into samples of sediment or peat by penetrating roots or rhizomes. Identification of these organs or tissues

Table 2.4-1. Overview of various dating techniques. A brief introduction to all these techniques is provided by Bradley (2015) and a more detailed one by Walker (2005).

	Measurement	Principle	Time range (yrs)	Material needed	Major problems
Radio-isotopes	Direct (AMS) or decay counting	^{14}C	200 – 50 000	Organic matter, optimal: terrestrial macros, wood	Production of ^{14}C not constant: helio- and geomagnetic Δ. Exchange of ^{14}C in compartments of global C-cycle, esp. ocean. Δ of total CO_2, natural and anthropogenic
		Other cosmogenic nuclides ^{10}Be, ^{36}Cl			
	Equilibrium, by mass spectrometry: activity ratios in decay series	$^{230}Th/^{238}U$	1000 – 350 000	$CaCO_3$ in corals, speleothems, sediments (molluscs)	
		$^{230}Pa/^{235}U$	5000 – 50 000		
		^{210}Pb	last 150	sediment	Supported and unsupported models
		^{137}Cs	After 1950 AD	sediment	
	Integrated effects	TL	10^2 – 10^8	Fine quartz, feldspar	
		OSL			
		Fission tracks	10^5 – 10^9	Tephra, glass	
Palaeo-magnetic	Reversals				Distinction between reversals and events
	Secular dipol variation	Inclination, declination, intensity	10^2 – 10^8	Volcanic and fine grained sediments	Regional master curves needed
Chemical	Amino acid racemisation		10 – 10^6	Molluscs, other shells with proteins	Racemisation is dependent on both time and temperature
	Tephra	Geochemical markers	varies		Not continuous; rare in non-volcanic areas
Sedimen-tological	Varves (annual laminations)	Anorganic: clastic or geochemical		In proglacial or in deep lakes	Sometimes hard to prove the annual nature of laminations. At many sites only certain periods show varved sediment
		Biological: algae and carbonates as seasonal deposits		In deep anoxic lakes (without bioturbation)	
Biology	Lichenometry	Lichen diameter		Glacier forefields	Species specific calibration needed
	Dendrochronology	Early and late wood		Well preserved wood	Species specific, valid for certain regions
	Palynology	Correlation of assemblages			Relative dating, valid only within restricted areas
Archaeo-logy	Types of artefacts			Palaeolithic stone tools, ceramics, etc.	Artefact specific, valid for certain regions

Table 2.4-2. Waterplants and their potential use of old carbon (HCO_3^-). Modified from Maberly and Madsen (1998) and references therein.

Species restricted to CO_2	Species able to use CO_2 and HCO_3^-
Aldrovanda vesiculosa	*Ceratophyllum demersum*
Berula erecta	*Elodea canadensis*
Callitriche cophocarpa, C. stagnalis	*Hydrilla verticillata*
Eurhynchium rusciforme	*Ranunculus peltatus*
Fontinalis antipyretica	*Vallisneria spiralis*
Myriophyllum verticillatum	*Myriophyllum spicatum*
Nitella flexilis	*Potamogeton crispus*
Nuphar lutea	*Potamogeton lucens*
Potamogeton natans	*Potamogeton pectinatus*
Sparganium emersum (S. simplex), S. erectum	
Juncus bulbosus	
Lemanea mammillosa	

to family, genus or species may help to avoid such material.

(2) Too old material can be brought into a sample either with reworked organic particles or if old carbon (other than the atmospheric carbon dioxide) was taken up by the plants (the reservoir effect). Reworked organic particles are often encountered either at the bottom of cores (just above the till but also outside formerly glaciated areas) or during Mid- and Late-Holocene periods with intensive human impact that caused soil erosion in the catchment. In both cases a clear indication of reworked material is often the presence of pre-Quaternary pollen and spores. Wood fragments of thermophilous trees in early Late-glacial sediments are also indicative of reworking.

Carbon sources for photosynthesis other than atmospheric CO_2 include old CO_2, bicarbonates and methane.

Old CO_2 emanating from decomposing organic matter in deeper layers of sediment or peat may occur in relevant concentrations near the sediment-water interface or near the ground surface of mire vegetation, where plants may use it for photosynthesis.

HCO_3^- can be taken up by many but not all waterplants (see Table 2.4-1). Its concentration depends on the geology of the catchment and on the water chemistry (pH etc.). The difference in radiocarbon ages measured on material with and without the uptake of HCO_3^- is therefore called the 'hard-water error'. Marine sediment samples are also subject to a hard-water error (often called the reservoir effect) which is not constant through time. A botanically and ecophysiologically interesting question remains as to how much some submerged and floating plants in freshwater can use bicarbonates as a carbon source (Sand-Jensen and Krause-Jensen, 1997) and thus re-use old carbon eroded from the catchment (causing the hard-water error). For a long time it was thought that aquatic mosses did not use bicarbonates and would, therefore, be useful material for dating. However, some exceptions were found (Peñuelas, 1985) and, additionally, many grow on the sediment surface in a microzone enriched with CO_2 that comes from respiratory processes in the uppermost sediment (Maberly, 1985). This means that the two carbon sources (old CO_2 and HCO_3^-) may be combined.

Whenever possible, plant material of terrestrial origin should be selected for dating in order to avoid hard-water errors. Identification of plant macrofossils is necessary to distinguish terrestrial from aquatic remains.

CH_4 needs to be oxidized by bacteria to CO_2 before it can be taken up by plants. There are at least three ways for this uptake:

- Through uptake by fungi feeding on CH_4 and on CO_2 escaping from deeper levels in bogs (Kilian et al., 2000).
- Through mycorrhizal fungi that provide C to the higher plants but feed on CH_4 (Kilian et al., 2000).

- Through symbiotic methanotrophic bacteria living inside *Sphagnum* leaves (Raghoebarsing et al., 2005).

Such organisms may subsequently form part of the radiocarbon sample.

An additional botany-based improvement to radiocarbon-sampling is feasible if wood is available for dating. It is sometimes possible to distinguish the youngest rings just below the bark and the cambium from the oldest rings in the centre of the trunk, in which case specific sets of rings can be dated. In long-living trees this may be a noticeable difference.

Calibration: the atmospheric concentration of ^{14}C (or the ratio $^{14}C/^{13}C/^{12}C$) has not been constant over time. Past differences are expressed in $\Delta^{14}C$ in ‰ as the deviation from the standard ^{14}C activity (Stuiver et al., 1998; Scott, 2007). The atmospheric ^{14}C-concentration depends on two different processes.

1. The production of ^{14}C is controlled by cosmic rays, in other words by how much the Earth gets shielded from these by helio- and geomagnetics (hence the link between solar forcing and the calibration of radiocarbon dates; Bleicher, 2013).

2. The repartition of ^{14}C in the reservoirs of the global carbon cycle, that is ocean, atmosphere and biosphere and their processes of exchange (Burr, 2007; van der Plicht, 2007). Therefore, sidereal years and radiocarbon years are not identical and a calibration is needed (Reimer and Reimer, 2007). Original calibrations worked with tree rings of *Quercus* in Ireland (Pilcher et al., 1984; Baillie, 1994). North of the Alps, *Quercus* was also used (Becker, 1980, 1993) and before the re-immigration of oak north of the Alps *Pinus sylvestris* was used (Friedrich et al., 2001, 2004; Muscheler et al., 2008). In North America, the bristle-cone pine *(Pinus longaeva)* and some other taxa provided early success in calibration (Stuiver et al., 1986, 1998). Annually laminated (varved) lake sediments have also been used for calibration (Goslar et al., 1998, 2000; Kitagawa and van der Plicht, 1998a,b; Brauer et al., 2000; Nakagawa et al., 2012) as well as the varved marine core from Cariaco Basin (Hughen et al., 2004a,b).

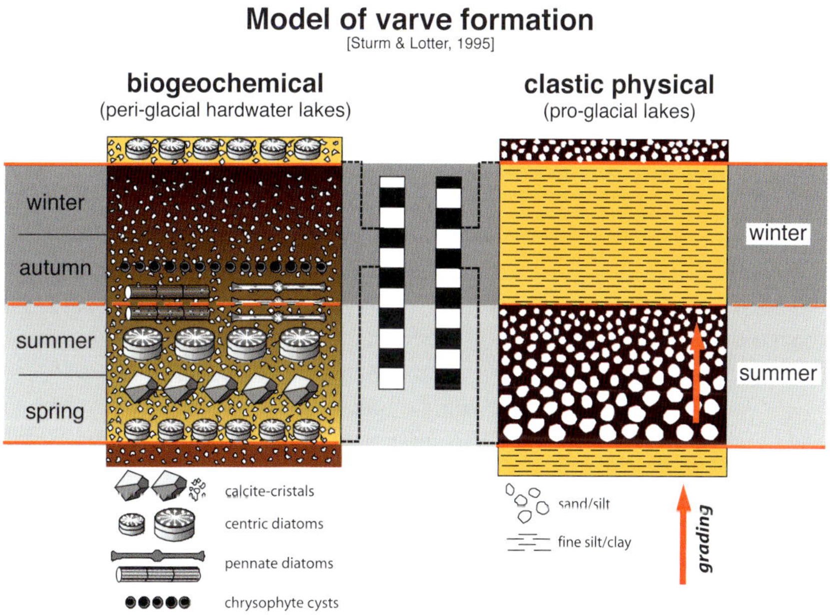

Fig. 2.4-1. Models of two contrasting varve types formed in different lake types. Note that the light and the dark layers are formed in opposite seasons (Sturm and Lotter, 1995).

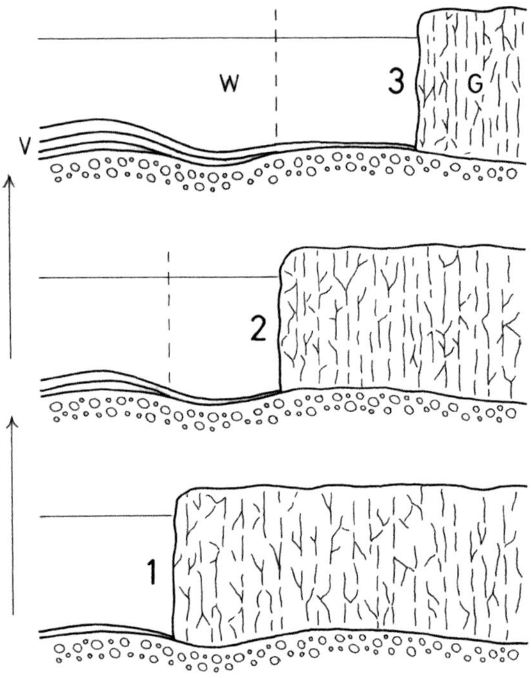

Fig. 2.4-2. Formation of glacial varves V, when a retreating glacier G deposits coarser material in summer and finer material (dark clay settling down under the surface-ice) in winter into a meltwater lake W. Modified after Woldstedt (1961).

2.4.2

Age-depth modelling

There are two schools of thought about how to select levels for radiocarbon dating: either the emphasis is put on levels of change, be it sedimentological or palynological change, or the aim is to get samples fairly evenly spaced down the whole core. Even better than samples evenly distributed over the depth scale would be samples evenly spaced over the timescale - which, of course, is unknown when samples are taken for dating, unless some dates are available prior to sampling. Combinations of both approaches are possibly best.

An important goal for many questions is to put an age on each pollen sample. For this, age-depth modelling is necessary and this must be done on calibrated ages (Bartlein et al., 1995; Telford et al., 2004). Uncertainties and errors affect not only the dated levels (the within-object variance of Heegaard et al., 2005) but also the core sections between them (the between-object variance of Heegaard et al., 2005). Bayesian techniques as demonstrated by Blaauw et al. (2003) are particularly suited when many dates are available in order to deal with age inversals and for wiggle matching (Blaauw and Heegaard, 2012; Blaauw et al., 2018). Unprecedented chronological precisions were reached by applying radiocarbon wiggle matching on yearly laminated sediments, delivering age-depth models with 2σ age uncertainties < ±20 cal yr (Rey et al., 2019b). Although wiggle-matching is less efficient if varves are lacking or difficult to recognize, it can still considerably reduce age uncertainties.

2.4.3

Other dating techniques (not based on radiocarbon)

Various techniques depend on the availability of the special records needed:

- Tephrochronology relies on volcanic ash (tephra) that needs to be identified to a specific eruption (Pilcher and Hall, 1992, 1996; van den Bogaard et al., 1994; Turney and Lowe, 2001; van den Bogaard and Schmincke, 2002; Wulf et al., 2004). Tephra can form 'marker-beds' that provide spatially widespread levels (that need to be dated via other methods).
- Varve chronology depends on the presence of laminations, the nature of which need to be shown as annual (glacigenic or biogenic varves; Lamoureux, 2001). Two different types are illustrated in figure 2.4-1; more types can be found in Heinz et al. (1993). The onset and end of varve formation can provide clues about the history of the catchment (for instance, erosion) or the history of oxygenation of the bottom waters (Dräger et al., 2017). Clastic varves formed in proglacial lakes were the basis for an early chronology of deglaciation in southern Sweden (De Geer, 1912; Litt et al., 2009).
- If laminations can be proven to be annual then many possibilities open up for high-resolution studies as was shown by Welten (1944) for Faulenseemoos (Switzerland), by Peglar (1993b) for Diss Mere (UK), by Zolitschka and Negendank (1996) for Lago Grande di Monticchio (Italy), by Ralska-Jasiewiczowa et al. (1998) for Lake Gościąż (Poland), by Dörfler et al. (2012) in

northern Germany or in Maar lakes of the Eifel in Germany (Zolitschka et al., 2000; Zolitschka, 2007, 2013).

- For recent sediment (of the last c. 150 years) ^{210}Pb (and sometimes younger artificial fallout radionuclides) can be useful (Appleby, 2001).
- For older ages a number of dating techniques exist, partly using long-lived radio-isotopes especially of the 3 Uranium-Thorium series, or Argon (as Potassium-Argon or Argon-Argon dating), partly using luminescence phenomena (TL, OSL, and ESR; Lian and Huntley, 2001; Walker, 2005; Lian, 2007).
- Dendrochronology is a very precise dating method that has its strongest application in historical and archaeological studies (Baillie, 2002; Hughes, 2002; Sass-Klaasen, 2002; Walker, 2005). However, in biostratigraphical studies, such as analyses of pollen and plant macrofossils that are based on sediment or peat it can rarely be used because two major prerequisites need to be fulfilled: firstly pieces of wood large enough for tree-ring counting need to be present and secondly the wood needs to be of a species for which a dendrochronological master curve of that area exists so that the local finding can be linked with the master curve. In some archaeological contexts the dendrochronological frame work is very strong and useful for connected biostratigraphical work.

2.5

What happened? Qualitative and quantitative changes: patterns

Data storage and data handling becomes increasingly important because of the increasing complexity of multi-disciplinary research (Juggins, 2004; Grimm et al., 2007).

Data handling includes presentation, error estimation, data analysis and aspects or methods for interpretation: concise overviews of this rapidly growing field in terms of pollen are provided by Bennett and Willis (2001) or Seppä (2007) and for plant macrofossils by H.H. Birks (2001, 2007a, 2013) and H.J.B. Birks (2014).

When trying to understand what happened in the past it is best to distinguish three phases sensu Ball (1975, 1980). Summarized in the words of Birks (1997): "These are (1) the empirical-descriptive phase when patterns are detected, described, and classified, resulting in local, descriptive accounts; (2) the narrative phase when plausible inductively-based circumstantial explanations or generalisations are proposed for the observed patterns in terms of consistent, coherent narratives or plausible but untestable scenarios; and (3) the analytical phase with a hypothetico-deductive basis in which multiple working hypotheses are presented, evaluated, tested, and rejected."

Description of patterns: geologists have a long tradition of defining zones in their stratigraphic descriptions (Hedberg, 1972, 1976). For Quaternary studies the biostratigraphies based on assemblage zones are very useful and will usually become connected with chronostratigraphies (depending on dating quality) or climato-stratigraphies (depending on interpretation of the litho- and biostratigraphies).

Limits between bio-zones, whether they are pollen-zones or macrofossil-zones, have traditionally been drawn subjectively 'by eye and hand'. However, because pollen counts are essentially a matrix (depths and pollen-types), numerical methods can be applied to make the establishment of zonations more reproducible and less subjective. A number of techniques exists that make use of the fact that datasets of pollen or macrofossil results are a stratigraphically constrained classification, as described by Birks and Gordon (1985), Birks (1986b) and Birks et al. (2012d), all with a wealth of references. Especially the "optimal partitioning using the sum of squares criterion" is very robust (H.J.B. Birks, 2007b). The number of statistically significant zones can then be diagnosed with the 'broken-stick method' according to Bennett (1996).

A study at a single site provides 'local assemblage zones'. If several sites have been studied in an area then 'regional pollen zones' may be extracted or postulated as illustrated for several European countries in the synthesis book by Berglund et al. (1996a).

Numerical treatments of plant-macrofossil data are explained in a comprehensive and concise way by Birks (2014).

The 'narrative phase' (sensu Ball, 1980; H.J.B. Birks, 1997a, 2007b) is also based on numerical

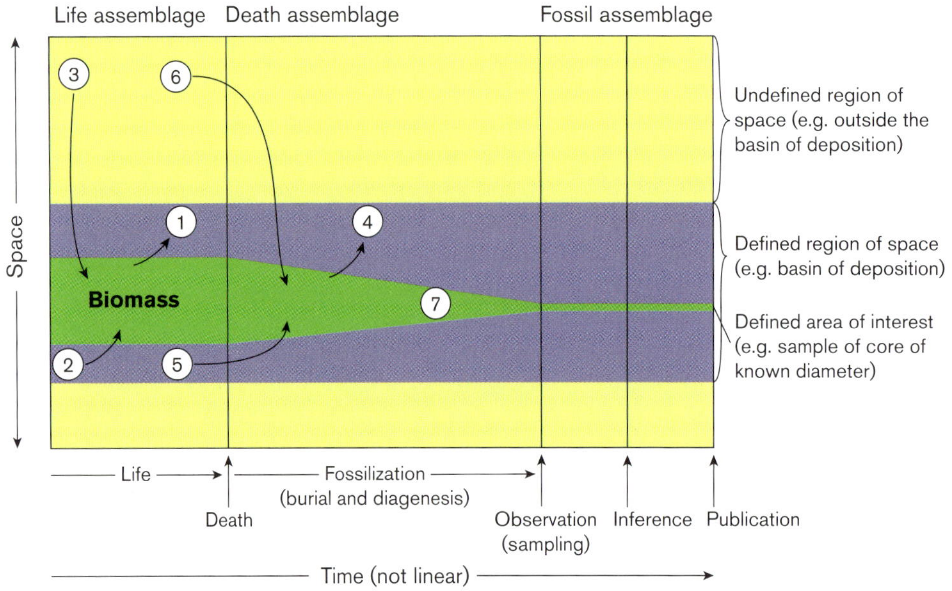

Fig. 2.5-1. Processes affecting biotic assemblages and their subfossil remains, that is transitions from living ecosystems to death assemblages (thanatocoenoses): the green field contains organisms in the basin (lake), the yellow field contributes particles such as pollen and plant macrofossils or insects from terrestrial ecosystems around the basin. Processes include (1) non-preservation (decay), (2) preservation, (3) input from near and far, (4) losses during sub-fossilization, (5) resuspended and redeposited material, (6) reworked material by erosion, (7) reduction of taxa during sub-fossilization. We then take cores at certain points, we analyse certain groups of taxa but never all – and from the patterns found we infer processes; later we may or may not publish the results. Redrawn from Birks and Birks (1980), based on an unpublished diagram by E.J. Cushing.

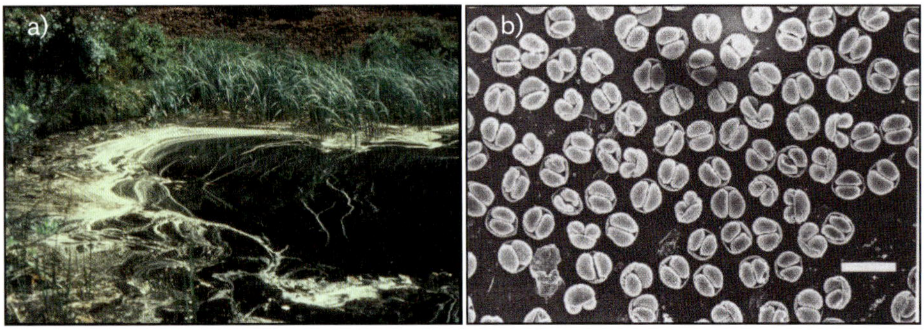

Fig. 2.5-2. a) Today's accumulation of saccate pollen near the shore of a small lake. b) Modern *Pinus mugo* pollen floating on a water drop showing the predominant orientation: most grains have their corpi in the water and their sacci in the air. White bar: 100 µm. Prepared for SEM and photographed by Micaele Schnitzler, University of Massachusetts.

techniques. Since all ecological and palaeoecological processes are multidimensional it is necessary to summarize compositional change in a consistent way in order to extract major trends that could reflect major directions of variance. Stratigraphically constrained ordination includes a number of tools, depending on the assumed linear or unimodal response models, judged from whether their compositional change (gradient length) is smaller or larger than 2 standard deviations (in a preliminary detrended correspondence analysis, DCA). As elegantly summarized by H.J.B. Birks (2007c), the techniques can be applied to single datasets, to several stratigraphic datasets from a single site, or to several datasets in a spatial context.

From a biological point of view important questions concern at least four aspects.

1. Biodiversity or palynological richness, estimated by rarefaction analysis according to Birks and Line (1992) and discussed by Birks (1996c), Odgaard (1999, 2007), van der Knaap (2009) and Giesecke et al. (2014a).

2. Population increase or decrease are best based on influx (Bennett, 1983b; Birks and Gordon, 1985). Related to population size is the reconstruction of taxon-specific biomass (Seppä et al., 2009b; Blarquez et al., 2011; Mazier et al., 2012a).

3. Rates of compositional change (H.J.B. Birks, 2007c).

4. Vegetation reconstruction is, in a way, the ultimate goal of vegetation history, but all its various techniques are based on comparisons with present vegetation (modern analogue technique (MAT); see chapter 2.7.3), and there may have been vegetation types (or patterns) with no modern analogue, for example under no-analogue climates or because species with long migrational lags had not arrived yet, or because of the absence of human impact especially before the Neolithic (H.J.B. Birks, 1993, 1995, 2007b).

Taphonomy, the post-depositional processes, are summarized in figure 2.5-1 which reflects several possible 'caveats'.

In some cases blessings may come in disguise: irregularities or unexpected records that at first seem to be a nuisance may turn out to provide interesting information. Examples include reworked (for example, Tertiary) pollen that indicates erosion in the catchment, which decreases with soil formation (Pokorný, 2002; Hošek et al., 2014), hiatuses that may be caused by low lake levels and littoral erosion (Schmidt et al., 2000; Väliranta et al., 2005; and subchapter 4.6.2), or short peaks in saccate conifer pollen that indicate more littoral situations (lower lake levels) because the strongly buoyant saccate pollen tends to accumulate near the shores (see Fig. 2.5-2; Hopkins, 1950; Ammann, 1994).

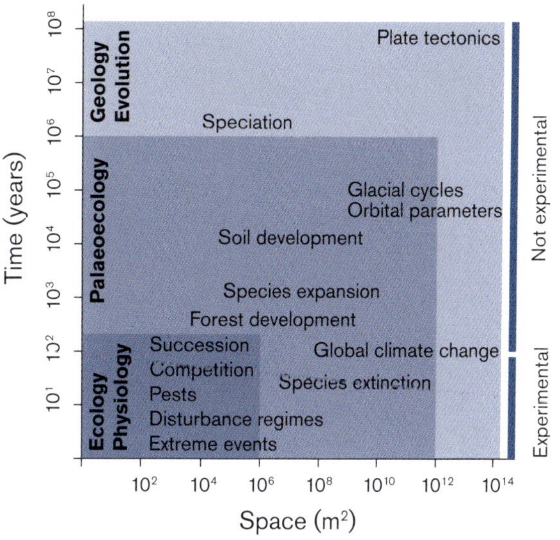

Fig. 2.6-1. Scales of natural processes on logarithmic axes of time and space. The processes in the corner of small scales can – at least partly – be observed and tested with experiments. Processes at the large scales are also studied by neighbouring disciplines, such as geography and geology. Processes at intermediate scales are the topics of ecology and palaeoecology. Redrawn from Birks (1986a) with modifications.

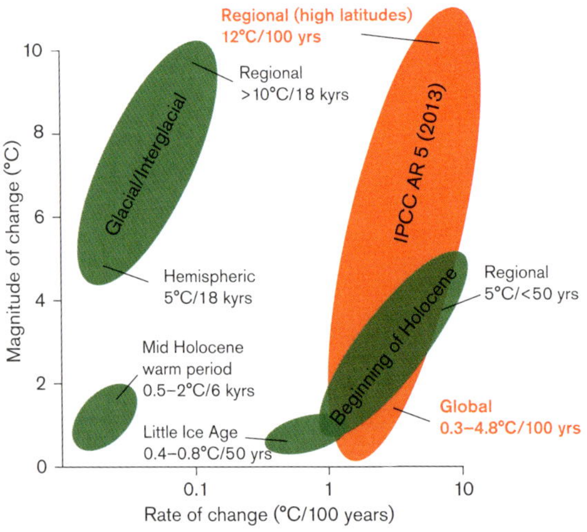

Fig. 2.6-2. Summary comparison of rates and magnitudes of possible future climate change (estimated in terms of mean annual temperature) with those associated with several well-known periods of past changes in regions that were vegetated. Rates of future regional temperature change could far exceed any widespread change in the late Quaternary. Reproduced with permission from Jackson and Overpeck (2000).

2.6.

Why did it happen? Methods of inference: processes

This is the 'analytical phase' in the theoretical concept of Ball (1980) and ways of implementing it in palaeoecology are proposed by Birks and Gordon (1985) and H.J.B. Birks (1997, 2007b) through trying to find possible causes of environmental change. Hypothesis generation and hypothesis testing are a major challenge that separates forcing factors or 'predictors' (such as climate change, soil development and deposition of volcanic ash) from 'responses' as recorded in a biostratigraphy. The goal is, in the sense of Popper, to formulate and to test falsifiable hypotheses (Lotter and Birks, 1993). Numerical techniques can thus be very important at various stages of palaeoecological research (H.J.B. Birks, 2007b) for:

- the identification of pollen types that may be difficult to separate (as for example *Betula*; Birks, 1968)
- error estimation (Bennett and Willis, 2001): as summarized by H.J.B. Birks (2007b), errors may be large for pollen accumulation rates (PAR) or influx
- zonation of biostratigraphies from single sites (Birks and Gordon, 1985) and the number of statistically significant zone-boundaries (Bennett, 1996)
- several biostratigraphies from a single site (a single core; Birks et al., 2000)
- several biostratigraphies in a region (spatial coherence or differentiation).

By linking several spatial and temporal scales (Fig. 2.6-1), vegetation history and palaeoecology in general do not only look backwards in time but by increasing our understanding of long-term ecological processes can also help to estimate ecological responses to future changes such as global warming (see Fig. 2.6-2).

An extensive treatment of data handling and numerical techniques can be found in Birks et al. (2012d).

Data storage and management is becoming an increasing concern. A solution for biostratigraphically oriented palaeoecologists is offered by Juggins (2004) and by many public data repositories such as the Neotoma Paleoecology Database (Williams et al., 2018) or the European Pollen Database (EPD).

2.7.

Environmental reconstructions from pollen spectra

2.7.1.

Introduction

To reconstruct past vegetation, its composition and structure, is one of the primary goals of the analysis of pollen and plant macrofossils. Other goals include the reconstruction of biogeographical dynamics or other environmental factors and their changes but these will not be summarized in this methodological chapter. Environmental factors may concern the reconstruction of soils (Brubaker, 1975; Pennington, 1986; Willis et al., 1997), land-use (Gaillard et al., 1992), and, especially, climate (Webb and Bryson, 1972; Birks, 1981b,1995; Birks and Seppä, 2004; Gonzales et al., 2009b). For climate reconstructions, multi-proxy approaches rather than solely pollen- and macrofossil-based ones are important, for example combining with chironomid stratigraphies as in Lotter et al. (2000, 2012), or with several biostratigraphies as in Birks et al. (2000). For the sake of brevity we will here omit two major issues, namely climate reconstruction per se, and vegetation reconstructions at the largest spatial scale, namely biome reconstruction (but see chapters 5.2 and 5.3). A comprehensive discussion of methods and the results of climate reconstructions can be found in Bartlein (2013) and Shuman (2013) or in the special volumes of Quaternary Science Reviews (vol. 31 and vol. 36, 2012; vol. 106, 2014). Major publications about biomization and biome reconstruction can be found in Prentice et al. (1996), Prentice and Webb (1998), Tarasov et al. (1998) and Williams et al. (2004). Continental to global scale vegetation reconstructions have important links to modelling changes in both vegetation and in climate because of the feedbacks between the land surface and the atmosphere through the exchanges of energy (albedo), water vapour and greenhouse gases.

For pollen studies there are at least three quite different approaches for interpreting pollen spectra and their temporal and spatial variability in terms of flora and vegetation: the modern analogue technique (MAT; 2.7.3), the indicator-taxa approach (2.7.4) and the representation factor approach (2.7.5). Moreover, the three approaches are not mutually exclusive but may be combined.

Before we briefly describe these three approaches, two aspects of data collection must be considered: what type of surface samples and what type of vegetation survey can be used?

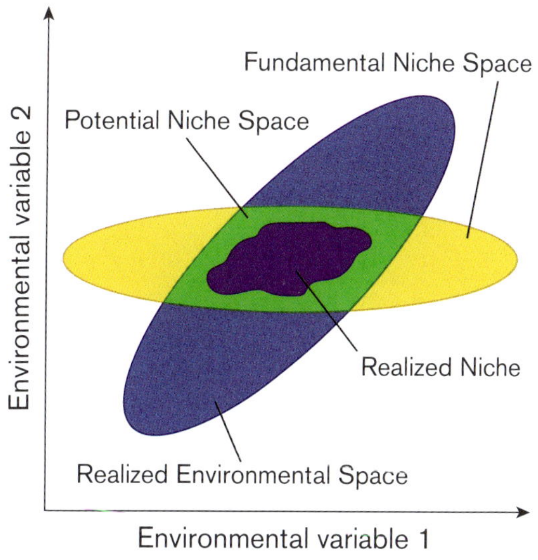

Fig. 2.7-1. The niche concept: The Realized Environmental Space comprises the combinations of two environmental variables that exist at a particular time (independent of taxa). The Fundamental Niche Space is the tolerance or survival envelope for a particular species as defined by the two variables. Parts of the Fundamental Niche Space may lie outside of the Realized Environmental Space in one period of time but inside it during another period (for example, after climatic change). The Potential Niche Space of a species is the intersection between the first two realms and depends on them. The Realized Niche is the subset of the Potential Niche Space in which a species actually occurs at a given time. It is constrained by biotic and other factors. Reproduced with permission from Jackson and Overpeck (2000).

Box 5

Monitoring modern pollen deposition

Sheila Hicks

University of Oulu, Finland

Modern pollen deposition is monitored in order to get a better understanding of the factors influencing pollen dispersal, deposition and incorporation in the sediment. It provides a set of reference data for interpreting fossil pollen diagrams in a quantitative manner. Pollen deposition monitored via pollen traps has the advantage over that from surface samples (mosses or raw humus) in that the time factor can be controlled, which enables Pollen Accumulation Rates (PARs, grains cm^{-2} $year^{-1}$) to be calculated and the record of individual taxa to be considered independently, rather than as a percentage of the pollen sum.

The collection of such data requires several years since it is the long-term average signal (over certainly more than 5 years) which reflects plant abundance (Hicks, 2001). The annual variation in pollen production can be great and is determined by weather conditions. The abundance of *Pinus* pollen in northern Fennoscandia, for example, is directly related to July temperature of the year before flowering (Huusko and Hicks, 2009). Pollen trap records, therefore, can be used to address both vegetation change and climate change histories depending upon the time resolution of the record.

In Europe the most commonly used trap is a modification of the type designed by Tauber. This is a passive trap, unlike the Burkhard trap used in aerobiology studies, which turns into the wind and actively sucks in air. Ideally it is located in the centre of a mire (open area c. 50–100 m) and is sunk into the ground so that the opening is at ground level (to simulate the surface of the mire; Fig. B5-1). It collects pollen for a full calendar year and is emptied once a year at the end of the flowering season. This sampling strategy gives data which reflect the fossil record from peat and the temporal resolution is the best that one could expect to obtain from peat cores.

It is important to record the vegetation relevant to the trap in order to calibrate it with the pollen deposited and this should be done in concentric circles around the trap out to beyond the Relevant Source Area of Pollen (RSAP; *see modelling section 2.7.5.*) so that distance-weighted plant abundance can be calculated.

In addition to the relationship between pollen quantity and plant abundance and pollen quantity and temperature/moisture the results of monitoring over the past 30–40 years (for a selection see Giesecke et al., 2010) have demonstrated that if the pollen source area is known (quantitatively determined) then PARs can be used to calculate tree volume within that area (Sugita et al., 2010). However, it should be remembered that trees can increase in volume without increasing in cover and trees can continue to occupy space (keeping the same volume and cover) without producing pollen if weather conditions are unsuitable for flowering (Hicks, 2006).

It is always important when interpreting the data to be aware of the temporal, spatial and ecological scale of the results.

The data produced should be stored in a format similar to that of the EPD/Neotoma so as to be easily compared with other long-term records. A common format also enables the abundance of specific taxa or assemblages of taxa to be compared over distance and through time (Abraham et al., 2021).

Fig. B5-1.

Features of a 'Tauber type' pollen trap. It is the size of the hole through which the pollen enters and the fact that this is surrounded by a sloping 'collar' that is significant. The size and type of container can vary but there should be an exchangeable complete lid for sealing it on collection. A handle, so that it can be easily and safely carried, is also an advantage. Two alternatives are illustrated.

2.7.2.

Basic data needed

Surface samples are the basis for most pollen-based reconstructions of past flora and vegetation. Surface samples have always been used as an informal approximation of the pollen-vegetation relationship at all spatial scales (Fægri and Iversen, 1950, 1989). They are also important for inferences beyond the vegetation, such as climate reconstructions, because the modern flora is the only one for which we can know anything about the ecological requirements of the taxa involved (methodological uniformitarianism, principle of actualism).

Early studies mostly used moss polsters as the material for surface samples, sometimes also surface litter or – when moss polsters were lacking in dry or high-elevation conditions – just surface soils (Janssen, 1966, 1967, 1973; Ritchie and Lichti-Federovich, 1967; Wright, 1967; Wright et al., 1967; Heim, 1970; Gaillard et al., 1992, 1994; Broström et al., 2004, 2005). Later surface samples from lake sediments ('core tops') were also used (Broström et al., 1998; Gaillard et al., 1998; Nielsen and Sugita, 2005; Soepboer et al., 2007). With moss polsters, either only the green parts were used, in order to have the record of just the last few years (Boyd, 1986; Cundill, 1986) or the green parts were avoided and the lower brown parts were used to avoid interannual fluctuations (the "moss humus" of Andersen, 1970). Certainly the 'pollen sample view of the landscape', that is the area represented in a sample, is very different for a moss sample than it is for a lake-sediment sample (see Relevant Source Area of Pollen (RSAP) below). When designing a study involving surface samples we have to consider not only today's vegetation but also aspects of the pollen-dispersal processes (see Fig. 2.2-1).

With respect to the water-borne component, Cw, Wilmshurst and McGlone (2005) show that palynomorph assemblages may differ between moss polsters and lake sediment-surface samples, because of both differential preservation (especially relevant for soil-surface samples) and inwash or fluviatile input (especially in some larger lakes). European studies on input from surface run-off, streams and rivers can be found in Peck (1973), Pennington (1973, 1979) and Bonny (1978). The theoretical and modelling aspects of Cw are treated by Sugita (2007a).

While both moss polsters and lake-core tops can usually only provide relative pollen values (proportions, percentages) a third way of surface sampling, namely pollen traps, makes it possible to calculate pollen influx (see Hicks, 2007; and Box 5 by Sheila Hicks). A milestone in research organized around pollen traps is the special issue of Vegetation History and Archaeobotany dedicated to Sheila Hicks (Giesecke et al., 2010), which provides a review of the development of both pollen trapping in the air (aerobiology) and pollen deposition (on the ground) in modified Tauber traps. The comparison of pollen records in moss polsters with those in near-by traps for different vegetation situations is presented by Cundill (1991), Räsänen et al. (2004), Pardoe et al. (2010) and Lisitsyna and Hicks (2012). Aspects of climatic control on pollen productivity in the same and/or in the previous year are discussed by Hicks (1996), Kuoppamaa et al. (2009), Pidek et al. (2010), Nielsen et al. (2010) and van der Knaap et al. (2010). The advantage of combining absolute concentration and influx values (from traps) with percentages led to improvements in studies at ecotones, as demonstrated by Hicks (2001), Seppä and Hicks (2006) and Birks and Bjune (2010), and to the quantification of pollen production, dispersal and deposition, all of which are needed for vegetation reconstruction (Sjögren et al., 2010).

For tropical forests and mountains, pollen traps of a different type from the modified Tauber traps were developed because heavy rainfall may cause flooding of the containers. Jantz et al. (2013) provide an overview of Oldfield traps (Flenley, 1973) and Behling traps (Behling et al., 2001; Krisai et al., 1991).

For Europe, access to data from surface samples is available through the European Pollen Database (EPD). On a continental to sub-continental scale the various origins of the samples (e.g. lakes, bogs, soils) may not matter too much. Overpeck et al. (1985) successfully used about 66% from lakes and 33% from moss polsters but for finer spatial scales it may be important to be cautious in the selection of surface data in order to avoid a meaningless mixture of various pollen-source areas.

The **vegetation survey** can be conducted using a number of methods. Forest inventories may provide information about basal areas (m^2/m^2 or m^2 per hectare) of the woody taxa involved. More difficult to obtain but very useful are estimations of crown volume or foliage (m^3/m^2) as an approximation of the potential flowering surface of a can-

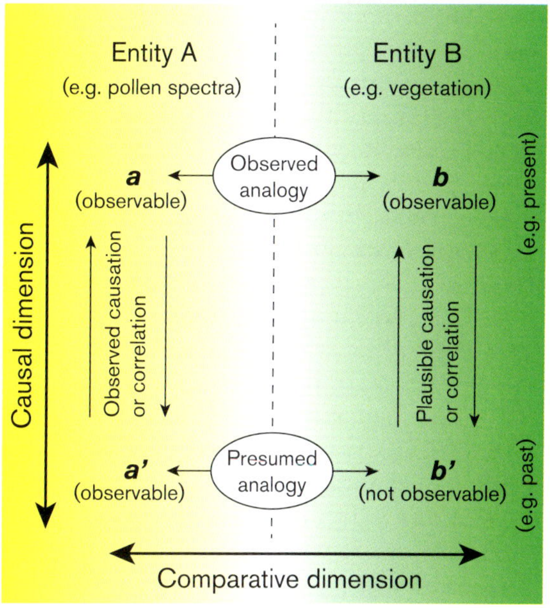

Fig. 2.7-2. Reasoning by analogy has two dimensions, the comparative axis and the causal axis. The concept of Hesse (1966) can be applied to the modern analogue technique (MAT; Jackson and Williams, 2004). Entity A represents the pollen spectra, entity B the vegetation, (a) and (b) for the present, (a') and (b') for the past; (b') cannot be observed; it represents the fossil vegetation to be reconstructed. On the causal dimension the similarity or dissimilarity of (a) and (a') are supposed to be controlled by, for example, climate or human impact. By analogy causes for the differences between (b) and (b') can be concluded. Reproduced with permission from Jackson and Overpeck (2000).

opy (Sugita, 1993, 2007b). On small spatial scales, records based on classical plant-sociology in many cases provide useful estimates of abundance of both woody and herbaceous taxa. Because distance weighting of the vegetation surrounding a sampling point will be of importance (see distance-weighted plant abundance below), survey techniques involving plots (that need to be specifically located; Gaillard et al., 1992, 1994) or concentric rings ('walking in circles') may be useful in the first few hundred metres around the site. On broader spatial scales, aerial photographs or vegetation maps may be used for areas beyond that (Sugita, 1994; Broström et al., 2004, 2005). The incremental radii of the vegetation survey will influence the estimation of the distance-weighted plant abundance or distance-weighted basal area of a taxon in the relevant source area (Sugita, 1994; Calcote, 1995; Jackson and Kearsley, 1998). For five non-arboreal taxa in a Norwegian heathland three methods of vegetation survey were compared by Bunting and Hjelle (2010): whereas the estimates of the relevant source area of pollen (RSAP, here <4 m) was not affected by the survey strategy, the estimates of relative pollen productivity (rel PPE) and the background component (usually labelled ω) differed markedly. Caution is therefore needed when studies are combined that may be based on different vegetation-survey techniques.

In the following, three main approaches to reconstruct flora and vegetation from fossil pollen assemblages and their changes through time are briefly discussed.

2.7.3. The approach of the modern analogue technique (MAT)

Informally and intuitively all early interpretations of pollen diagrams were based on a comparison between modern and fossil samples (Fægri et al., 1989). For more quantitative techniques, H.J.B. Birks (2007b) gives a concise overview of methods of vegetation reconstruction. Early approaches to formalize the comparison included ordinations and classifications (Birks and Gordon, 1985) while later, an alternative approach was the comparison of modern and fossil samples through dissimilarity coefficients.

Already in 1985 Overpeck et al. compared eight different dissimilarity coefficients from three classes on two long transects of surface samples in North America. The squared chord distance proved to be an especially robust measure. For Europe, see Huntley (1990a). Later Gavin et al. (2003) and Lytle and Wahl (2005) have refined the statistical techniques of distance metrics.

If not only vegetation but other environmental aspects are to be reconstructed (for example climatic variables) the concept of environmental gradients that form an n-dimensional space becomes very important. For an example with two environmental variables, see figure 2.7-1.

Vegetation ecologists and biogeographers have long worked on concepts of gradients and the responses of taxa (Austin, 1976, 1987; Austin and Gaywood, 1994; Jeffree and Jeffree, 1994). Examples are also given in figures 2.7-5 and 2.7-6 after Kühl et al. (2002). Jackson and Williams (2004) discuss the two logical steps in the search for analogues: first, the comparative element (modern and fossil assemblages, their goodness of fit and the decision as to whether a modern analogue exists) and second, the causal element (Fig. 2.7-2).

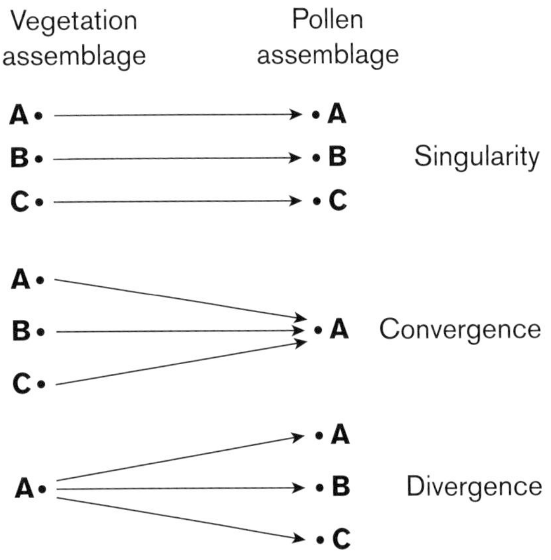

Fig. 2.7-3. Three possible relationships between vegetation (its composition and structure) and the pollen assemblage it produces. The case of **singularity** does not cause problems in interpretation, but it needs to be recognized as such – and may be rare. **Convergence**, when multiple vegetation types produce similar pollen assemblages, may be caused by (1) taxonomic imprecision (for example boreal and Mediterranean pines), (2) the dominance of a palynologically 'silent' taxon (for example *Juncus* or shrubby Rosaceae) and (3) various different vegetation patches at various distances from the pollen sample. **Divergence** occurs when from a single vegetation type various pollen assemblages are recorded, which can be caused by differential pollen production (depending on e.g. individual fitness, age of stand, masting behaviour), pollen dispersal (for example pines near timberlines), deposition (see Fig. 2.5-2) and preservation (as with 'silent *Juncus*'). Reproduced with permission from Jackson and Williams (2004).

The relationship between vegetation composition and pollen assemblage is often not a one-to-one relationship, as illustrated in figure 2.7-3.

MAT has been used for the reconstruction of past land-use and soil conditions by Gaillard et al. (1996) applying methods of quantification such as (1) weighted averaging regression and calibration and (2) estimation of dissimilarity by squared chord distance.

In palaeoecology the reconstruction of past climates became a hot topic and, therefore, climatic gradients are widely used as 'climatic envelopes' or 'response surfaces' by the three classes of methods mentioned (MAT, indicator taxa, multivariate transfer functions as summarized in H.J.B. Birks, 2007b; see also chapter 5.2). In climate reconstructions, MAT results were also compared to results from geochemistry (Martin et al., 2020).

The non-analogue situation. The assumption that inevitably some modern vegetation analogue for any fossil sample exists somewhere in today's vegetation is certainly not justified. Overpeck et al. showed already in 1985 that non-analogue vegetation samples exist for pollen samples older than 10 500 cal yr BP at Kirchner Marsh in Minnesota and for samples older than 13 000 cal yr BP at Wintergreen Lake in Michigan. In the discussion around climate reconstructions, non-analogue situations are especially important (Birks and Seppä, 2004).

On a broader scale, Whitmore et al. (2005) provide an overview for North America and Greenland with more than 4600 surface samples and their relationship to vegetation characteristics and climate variables.

The causes of non-analogue vegetation types are complex and may be climatic, non-climatic or a combination of both.

1. Non-analogue climates on a global scale due to differences in seasonal insolation as, for example, during the Late-glacial and the Early Holocene, affecting seasonality and gradients in temperature and precipitation (Jackson and Williams, 2004).
2. Non-analogue climates on a regional scale as, for example, near a residual continental ice-sheet, where anti-cyclonic or föhn-wind circulations affected both temperature and precipitation (COHMAP, 1988).
3. Biological/ecological causes such as migrational lags (e.g. due to long life spans, slow migration and

slow soil development) that may have caused various species pools in which taxa which are important today were missing.

Points (1) and (2) are based on the hypothesis that vegetation was in a dynamic equilibrium with climate, whereas point (3) is based on the disequilibrium hypothesis.

4. The surface samples may be affected too much by human impact. The latter is a major problem in Europe which has experienced strong human impact over millennia. Additionally, in many parts of Europe during the last 200 years agriculture has become industrialized, wetlands drained and some forests planted with introduced conifers. For Denmark, Odgaard and Rasmussen (1998, 2000) found a solution in using not modern analogues but historical analogues (based on historical maps). Nielsen (2004), Nielsen and Odgaard (2004) and Nielsen and Sugita (2005) developed this approach by using samples from AD 1800 (that is before industrialization) for estimating the relevant pollen source area of pollen or distance-weighted plant abundance.

Based on an improved set of surface samples, Gonzales et al. (2009a,b) introduced the method of extended response-surfaces to address the non-analogue problem.

2.7.4.

The indicator-taxa approach

Because ecological information is strongest at the species level, the taxonomic resolution of a sample plays an important role. The latter is often higher in plant macrofossils than in pollen (for example dwarf birch versus tree birches, or *Juncus* that is virtually absent as pollen). For climate reconstructions, the indicator-taxa approach is thoughtfully commented on by Birks (1981b). An early and now classic example for climate reconstruction was presented by Iversen (1944). A newer development and quantitative refinement is the method of probability density functions (PDF; Kühl et al., 2002), which, so far, has only been used for climate reconstructions. It does not need modern analogues and is somewhat close to an indicator-taxa approach since it uses presence/absence (rather than quantities) of many taxa as extracted from chorological maps (Hultén, 1964, 1971; Meusel et al., 1965, 1978, Meusel and Jäger, 1992).

Figure 2.7-4 gives examples of taxa which are well-suited and unsuited, to this approach.

One- and two-dimensional PDFs for four selected species are presented in figure 2.7-5 and an example of an application of the method to a last-interglacial assemblage rich in *Carpinus* is provided in figure 2.7-6.

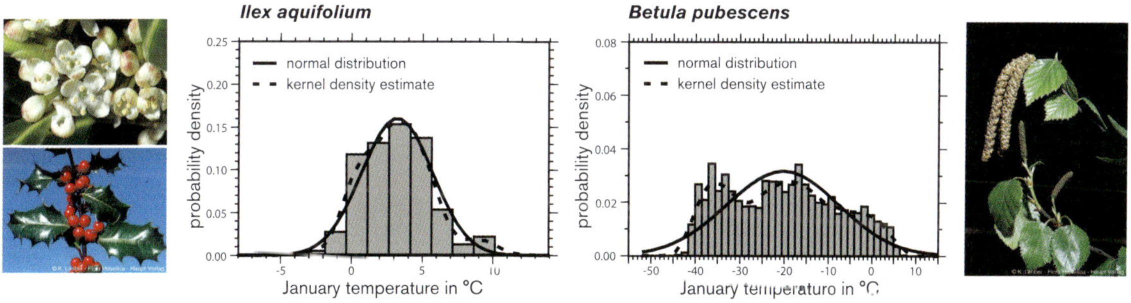

Fig. 2.7-4. Estimated probability density functions (PDFs) of the relationship to January temperatures for two contrasting species expressed in a parametric form (the solid line indicates normal distribution) and in two nonparametric forms (the dashed line shows the histogram and kernel). Left: *Ilex aquifolium*: the normal distribution (parametric) fits well with the kernel and the histograms (both non parametric). This species may be used for the reconstruction of January temperatures (as already shown by Iversen, 1944). Right: *Betula pubescens*. The normal distribution shows a poor fit with both the histograms and the kernel and should not be used as an estimator function. Modified after Kühl et al. (2002).

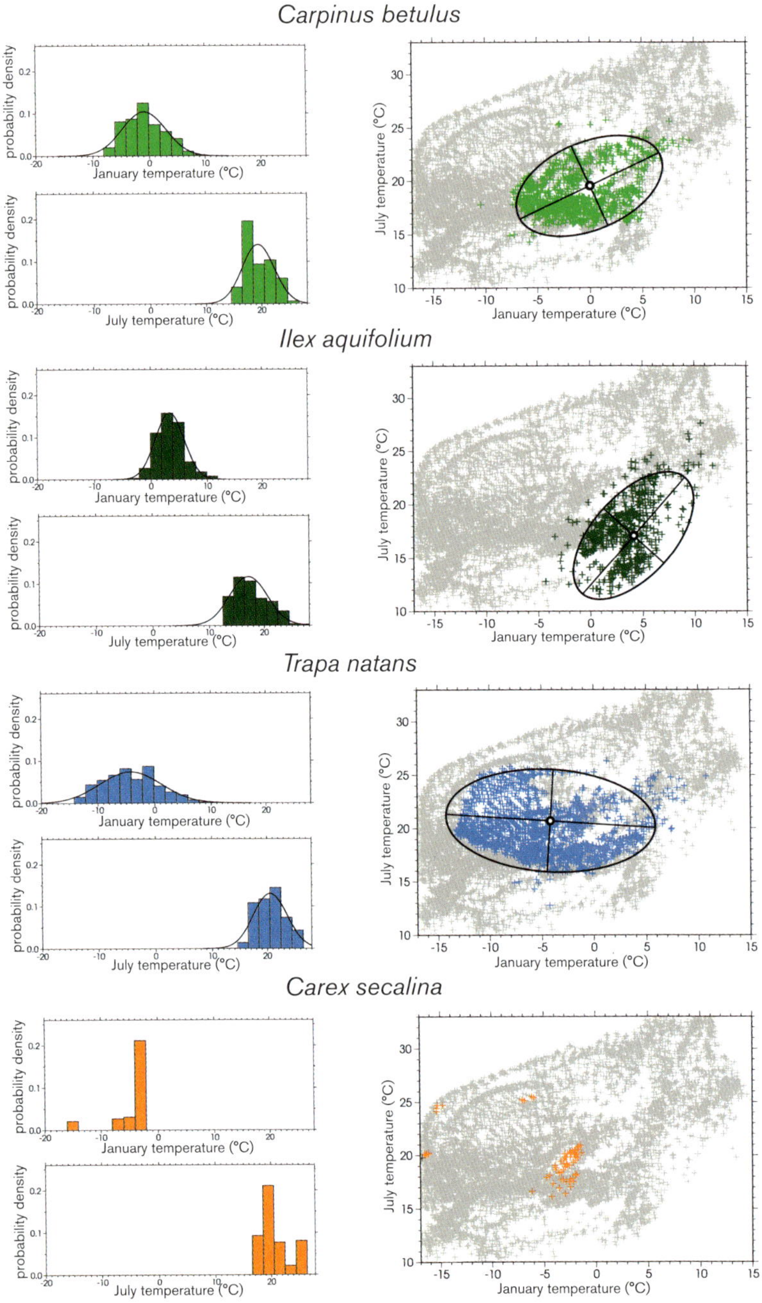

Fig. 2.7-5. Estimated one-and two-dimensional PDFs of four selected species. Left: The histograms (non-parametric PDFs) and normal distributions (parametric PDFs) represent one-dimensional PDFs. Right: Coloured crosses indicate presence and grey crosses absence of the taxon. A small circle marks the mean temperature vector of the corresponding normal distribution and the ellipses represent 90% of the integral of the normal distribution centred by this mean. Most sample points lie within this range. The interval, however, may not necessarily include 90% of the data points. *Carex secalina* as an example of an azonally distributed species is an exception. A normal distribution does not appear to be an appropriate estimating function for this species and, therefore, no normal distribution is indicated on the left. Modifed from Kühl et al. (2002).

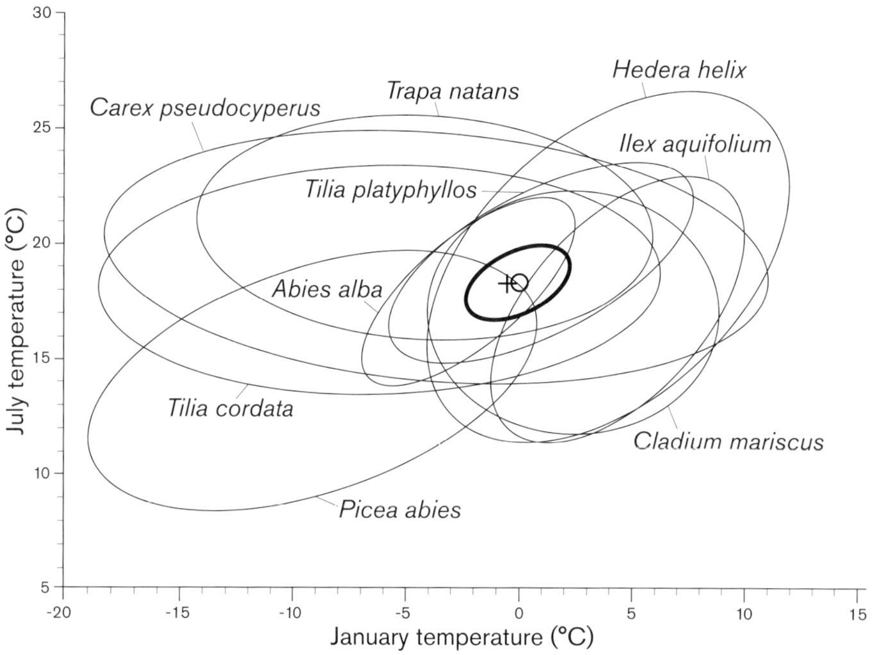

Fig. 2.7-6. Temperature reconstruction for an assemblage during the *Carpinus*-phase at the Eemian site of Gröbern in Germany (data from Menke and Tynni, 1984): the thin ellipses indicate the PDFs of the individual taxa included in the reconstruction; the thick ellipse shows the 90% uncertainty range of the reconstruction result. Redrawn from Kühl et al. (2002).

Such studies will help to estimate the ecological tolerances of taxa and therefore their potential response to ongoing and future climate change (see chapter 5).

2.7.5.

The representation factor approach: mechanistic modelling for calibration of the pollen-vegetation relationship

In addition to modern analogue techniques (section 2.7.3), indicator species approaches (section 2.7.4) and applications of multivariate statistical methods (e.g. PCA, DCA, CCA, etc.), a different approach is based on mechanism-based modelling. This has developed since the 1950s for pollen-vegetation calibration. It is often referred to as the representation-factor approach. This section describes briefly the main concepts and models important in this approach with some historical anecdotes about its development.

Since the inception of palynology (von Post, 1916a,b, 1967) palynologists have been aware of possible factors and mechanisms that affect the pollen-vegetation relationship, such as inter-taxonomic differences in pollen productivity and dispersal and the spatial structure of the surrounding vegetation (Davis, 2000). Various concepts and models have been proposed to clarify the impacts of such factors on the relationship. The methodological and conceptual developments up until the mid 1980s are summarized in Birks and Gordon (1985; 182–204 pp) and Prentice (1988b). Since then, a simple but elegant modelling approach proposed by Prentice (1985, 1988b) has stimulated discussion and accelerated the development of this approach further (Sugita, 1993, 1994; Gaillard et al., 2008). This section uses the Prentice-Sugita model of the pollen-vegetation relationship (Prentice, 1988b; Sugita, 1993, 1994) and its variants as a guide to summarize the overall development of the approach and its implications.

2.7.5.1

The pollen-vegetation relationship at the regional scale

Based on the work by Chamberlain (1975), Prentice (1985, 1988b) proposed a model that describes theoretically the relationship between pollen loading on a sedimentary basin and abundances of plants growing around the basin as a function of distance from it. This one-dimensional model, although highly simplified, captures some of the important mechanisms and factors affecting the relationship and opens ways for a new generation of reconstruction methods of the past vegetation. Sugita (1993, 1994) refined and modified Prentice's model (1985, 1988) as

$$y_{i,k} \equiv P_i \int_R^{Zmax} x_{i,k}(z)\ g_i\,(z)\,dz \quad \text{Eq. (1).}$$

Here

$y_{i,k}$: pollen loading of taxon *i* at site *k* in an absolute unit,
P_i: pollen productivity of taxon *i* in an absolute unit,
z: distance from the center of site *k*,
$x_{i,k}(z)$: mean plant abundance of taxon *i* at distance z around site k in an absolute unit,
$g_i(z)$: pollen dispersal and deposition of taxon *i* as a function of distance *z*,
R: radius of site *k*, and
Z_{max}: maximum spatial extent of the regional vegetation in a radius within which most of the pollen of taxon *i* comes from; in theory and practice Z_{max} is assumed to be 50-400 km.

Dimensions of these symbols, parameters and functions are explained in detail in Sugita (1993, 1994, 2007a,b) and Sugita et al. (1999, 2010a,b). The mean plant abundance $x_{i,k}$ is expressed as a function of distance from a given site *k*, effectively making it possible to describe the spatial heterogeneity and patchiness of the surrounding vegetation along the distance axis from the centre of the site. By selecting a proper function for pollen dispersal and deposition $g_i(z)$, the integral part of the right-hand side of Eq. (1) is linearly related to the pollen loading, and the slope represents pollen productivity of the taxon in an absolute unit. The integral part is termed the 'distance-weighted plant abundance' (DWPA; Sugita, 1994). Using a Gaussian Plume Model (GPM) from Sutton (1953) and used by Tauber (1965), Prentice (1985, 1988b) defines an appropriate form of $g_i(z)$ for pollen deposition on a point at the centre of the basin, and Sugita (1993, 1994) and Sugita et al. (1999) for pollen deposition on the entire surface of the basin. The former is useful to describe pollen deposition on bogs and mires and the latter on lakes and ponds. Different forms of the GPM (Kabailiene, 1969; Bunting and Middleton, 2005) and a Lagrangian Stochastic Model (Kuparinen et al., 2007; Theuerkauf et al., 2012) have been proposed as $g_i(z)$. Major assumptions and conditions important for Eq. (1) are discussed in some detail later in this section.

The most important implication from Eq. (1) is that pollen loading is linearly related to the DWPA, not plant abundance alone around the sites: in other words the spatial structure of the surrounding vegetation and the locations of the source plants matter. There are many possible ways to define the DWPA in theory. For example, palynologists often use the average abundance of plants within a given radius to assess the pollen-vegetation relationships in either percentage/proportion or an absolute unit. For the DWPA in Eq. (1) to be the average plant abundance within Z_{max} the pollen dispersal-deposition function is required to be

$$g_i(z) = {}^{2z}/_{(Z_{max}^2 - R^2)}.$$

This implies that the plant abundances further away from the site are more 'weighted' than those nearby; this contradicts our theoretical and empirical understanding, however. According to Eq. (1) we always assume a certain weighting function, consciously or unconsciously, to measure the surrounding plant abundance to describe the pollen-vegetation relationship. It is also important to realize that selection of $g_i(z)$ affects the empirical estimates of pollen productivity *Pi* based on field measurements of the pollen and vegetation data; that is, the estimate of P_i is model-dependent in empirical calibration studies.

Historical perspective: use of pollen percentages and introduction of the R-value model

Eq. (1) is informative to evaluate the significance of the contribution of Fagerlind (1952) and Davis (1963) to the current understanding of the representation factor approach. Palynologists have discussed the potential and limit for interpretation of pollen percentage data, assuming a linear pollen-vegetation relationship in absolute units similar to Eq. (1) as the basis for their discussion. Fagerlind (1952) describes in detail the complex nonlinear nature of the pollen-vegetation relationship in percentages and Davis (1963) introduces the concept of the R-value model as a way to circumvent the nonlinear relationship in percentages for an objective evaluation of the relationship. For the sake of simplicity these two papers use the mean plant abundance of taxon *i*, X_i , as a substitute for the DWPA part in Eq. (1) with the assumption that all pollen grains of the taxon come from within the region (i.e. within a radius Z_{max}.). Assuming that X_i is a taxon-specific constant regardless of sampling sites in the region, Eq. (1) is rewritten as

$$y_i \equiv P_i X_i \qquad \text{Eq.(1.1).}$$

Using pollen proportion p_i and plant composition in proportion v_i for taxon *i*, Davis (1963) proposes the concept of the 'R-value' and shows that it is a taxon-specific constant as follows:

$$R_i \equiv \frac{p_i/v_i}{p_{ref}/v_{ref}} = \frac{P_i}{P_{ref}} \qquad \text{Eq. (1.2),}$$

where $p_i = \frac{y_i}{y.} \approx \frac{n_i}{n.}$, $v_i = \frac{X_i}{X.}$,

and 'ref' means a reference taxon: here, n_i is pollen count of taxon *i*, n. total sum of pollen counts of all taxa, *y.* total sum of pollen loading of all taxa, and *X.* total sum of the mean plant abundances of all taxa in the region. The important point is that the R-value of taxon ***i*** as defined equals the absolute pollen productivity of taxon *i* relative to that of a reference taxon as long as all the assumptions are satisfied. Inversion of Eq. (1.2) leads to the following equation between plant composition v_i and pollen proportion p_i,

$$v_i = \frac{p_i/R_i}{\sum_{j=1}^{t} p_j/R_j} \qquad \text{Eq. (1.3),}$$

where t is the total number of taxa included in the pollen assemblages at the study sites. Thus, once reliable R-values are available, this approach is potentially useful for an objective reconstruction of the past regional vegetation from fossil pollen proportions.

Since the publication of Davis (1963) the R-value model had been criticized in the US and Europe and was mostly abandoned at the end of the 1960s and the early 1970s. However, most of the criticisms have focused on the technical sides of the applications. The number of sites used for estimating R-values is so limited that the uncertainties of the estimates are too large to be useful for a reasonable reconstruction of the past plant composition (Livingstone, 1968; Parsons and Prentice, 1981, 1983). The areas of vegetation survey used for calibration in Davis (1963), Livingstone (1968) and others are too small: for example Davis (1963) used the plant abundance data within a radius of 1 km from the study site for data analysis. For the R-value model and Eq. (1) to be valid, the radii for the vegetation survey need to be much larger, for example ≥50–400 km (Prentice, 1988b; Sugita, 1993, 1994, 2007a; Hellman et al., 2008a). The measure of plant abundance in the surrounding vegetation was not properly weighted as explained above: for example, the mean plant abundances in the study regions were used in the R-value applications in the 1960s–1980s. When the assumptions are met and site-selection strategies well considered, the concept of the R-value model is still sound and useful. For instance, the REVEALS model (Sugita, 2007b; Theuerkauf et al., 2016) is a modified and generalized version of Eq. (1.3) for quantitative reconstruction of regional vegetation using proper forms of the pollen dispersal-deposition function g_i *(z)* in Eq. (1).

2.7.5.2.

The pollen-vegetation relationship at the landscape to local scales

Appropriate forms of the pollen dispersal-deposition function, such as the Gaussian Plume Model (Sutton, 1953; Tauber, 1965) and Lagrangian Stochastic Model (Kuparinen et al., 2007; Theuerkauf et al., 2012) have a long tail (or 'fat' tail) as the distance from a source increases (Prentice, 1985). Because of this property of g_i *(z)*, pollen loading coming from beyond a certain distance in a given region – the 'background pollen' from the regional source – tends to become consistent among similarly and small-sized study sites, even when the spatial structure of the vegetation is highly patchy and heterogeneous (Sugita, 1994, 2007a,b). This distance – the relevant source area of pollen (RSAP) – is originally defined as the area, or distance, beyond which correlations do not continue to improve between pollen loading and the DWPA for all the plant taxa of interest in the region (Sugita, 1994): simulations demonstrate that pollen loading coming from beyond the RSAP becomes consistent regardless of site locations (Sugita, 2007b). Accordingly, Eq. (1) can be rewritten as

$$y_{i,k} \equiv P_i \int_R^{Z_{RSAP}} x_{i,k}(z)\ g_i\ (z)\ dz + P_i \int_{Z_{RSAP}}^{Z_{max}} x_{i,k}(z)\ g_i\ (z)\ dz = P_i\ DWPA(R|Z_{RSAP})_{i,k} + B_i \quad \text{Eq. (2)}$$

where Z_{RSAP} is the distance at the RSAP, $DWPA(R/Z_{RSAP})_{i,k}$ the distance-weighted plant abundance of taxon i from basin radius R to Z_{RSAP} defined as $\int_R^{Z_{RSAP}} x_{i,k}(z)\ g_i\ (z)\ dz$ around site k, and B_i the 'background' pollen loading of taxon i coming from beyond the RSAP, defined as

$$P_i \int_{Z_{RSAP}}^{Z_{max}} x_{i,k}(z)\ g_i\ (z)\ dz.$$

Definitions of all other symbols are the same as described before.

Eq. (2) means that, when B_i is a known species-specific constant in a given region, differences in pollen loading between sites show the between-site differences in the DWPAs. Sugita (2007b) demonstrates that B_i can be approximated from the mean abundance of taxon i within Z_{max} in the region. This information is crucial for the development of the Landscape Reconstruction Algorithm (LRA) for quantitative reconstruction of past vegetation, derived from both Eqs. (1) and (2) (Sugita, 2007a,b).

Historical perspective: Andersen's (1970) linear model based on relative pollen loading

Andersen (1970) proposed a linear model of the pollen-vegetation relationship that is isomorphic to Eq. (2). His model assumes that total pollen loading of the 'exotic' plant types is constant among sites and that the 'exotic' plant types do not grow in the area of comparison: the area is assumed to be the RSAP here. Thus, pollen loading of the 'exotic' plants coming from beyond the RSAP is expressed after Eq. (2) as:

$$y_{exotic,k} \equiv P_{exotic} \int_{Z_{RSAP}}^{Z_{max}} x_{exotic,k}(z)\ g_{exotic}\ (z)\ dz = P_{exotic}\ DWPA(Z_{RSAP}|Z_{max})_{exotic} = \text{const.}$$

Here, $DWPA(Z_{RSAP}/Z_{max})_{exotic}$ represents the distance-weighted plant abundance of the exotic plants in the region beyond the RSAP. Pollen loading of taxon i at site k relative to that of the exotic plants becomes

$$\frac{y_{i,k}}{y_{exotic,k}} \equiv \frac{P_i\ DWPA(R|Z_{RSAP})_{i,k} + B_i}{P_{exotic}\ DWPA(Z_{RSAP}|Z_{max})_{exotic}} = P_i^*\ DWPA(R|Z_{RSAP})_{i,k} + B_i^* \quad \text{Eq. (2.1)},$$

where P^*_i represents

$$P_i / P_{exotic}\ DWPA(Z_{RSAP}|Z_{max})_{exotic}$$

and B^*_i the background pollen loading of taxon i relative to that of the exotic plants. Andersen (1970) then defines $P_{rel,\,i}$ – the relative pollen productivity of taxon i against a reference taxon – as

$$P_{rel,i} \equiv P_i^* / P_{ref}^* = P_i / P_{ref} \quad \text{Eq.(2.2).}$$

Andersen (1970) obtained the total sum of the crown areas for each taxon within radii of 20 m and 30 m from each moss polster and examined the relationships between the relative pollen loading and plant abundance data. The results showed the pollen-vegetation relationships to be linear for all the tree taxa and B^*_i to be close to zero in both radii, suggesting that pollen grains of the constituent tree taxa are mostly coming from within 20–30 m radii. $P_{rel,\,i}$ is termed the "correction factor" in Andersen (1970) and is applied for reconstructing the past local forests within a 20–30 m radius around forest hollows in Denmark (Andersen, 1973). Because of the way the plant abundance data are quantified and compared with pollen data, the pollen dispersal-deposition function assumed is

$$g_i(z) = 2z / (Z_{max}^2 - R^2)$$

as in Davis (1963). Accordingly, it would be better to be cautious about the correction factors of Andersen (1973) as the substitutes for the relative pollen productivity of the tree taxa studied. Also, the assumption of the constant background pollen loading, B^*_i in Eq. (2.1), throughout the Holocene is hard to justify for vegetation reconstruction in Denmark and elsewhere. Despite these shortcomings the concept and approach by Andersen (1970) is creative and informative in both evaluation/calibration of the pollen-vegetation relationships and applications of the 'representation factor' approach for vegetation reconstruction when pollen and vegetation data are in absolute, or semi-absolute, units.

Historical perspective: use of pollen percentages and introduction of the Extended R-value (ERV) model

Taking into account the concept behind Andersen's linear model, Parsons and Prentice (1981) and Prentice and Parsons (1983) proposed an approach that uses pollen-vegetation datasets in proportions to estimate the "representation coefficients" (Parsons and Prentice, 1981) of the constituent taxa. Hereafter, let α_i represent the representation coefficient as in Parsons and Prentice (1981, 1983). When an appropriate pollen dispersal-deposition function $g_i(z)$ is applied in DWPA calculation in Eq. (2), α_i is equivalent to the relative pollen productivity of taxon i against that of a reference taxon: $\alpha_i \approx P_i / P_{ref}$ (Sugita, 1994, 2007a,b; Sugita et al., 2010a,b). This approach is considered as an extension of the R-value model of Davis (1963) in such a way as to include 'background' terms of the pollen-vegetation relationships that are affected by long-distance transport of pollen in the region. Parsons and Prentice (1981) derive two sub-models of the pollen-vegetation relationships in proportions from Eq. (2) with model-specific assumptions of the background term. Sugita (1994) proposes a third sub-model that is applicable when plant abundance data are expressed in an absolute unit and pollen data in proportions. Following Parsons and Prentice (1981) and Prentice and Parsons (1983), here after 'z_i' represents the background term of each sub-model for taxon i, not the distance from pollen sites as used in Eqs. (1) and (2).

Sub-model 1: the background term of taxon i is defined as $z_i \equiv B_i / y_{\cdot k} = const.$ Then, the relationship between pollen proportion $p_{i,k}$ and vegetation proportion $v_{i,k}$ is expressed for taxon i at site k as,

$$p_{i,k} = \alpha_i \, v_{i,k} \, f_{1,k} + z_i \text{ , where}$$

$$f_{1,k} \equiv \left(1 - \sum_{j=1}^{t} z_j\right) \Big/ \sum_{j=1}^{t} (\alpha_j \, v_{j,k}) \quad \text{Eq. (2.3).}$$

Sub-model 2: the background term is defined as

$$z_i \equiv \left(B_i / P_{ref} \right) \Big/ DWPA(R|Z_{RSAP})_{\cdot k} = const.$$

for taxon i. Pollen proportion $p_{i,k}$ and vegetation proportion $v_{i,k}$ are related as

$$p_{i,k} \, f_{2,k} = \alpha_i \, v_{i,k} + z_i \text{, where}$$

$$f_{2,k} \equiv \sum_{j=1}^{t} (\alpha_j \, v_{j,k} + z_j) \quad \text{Eq. (2.4).}$$

Sub-model 3: the background term of taxon i is defined as

$$z_i \equiv B_i / P_{ref} = const.$$

The relationship between pollen proportion $p_{i,k}$ and plant abundance $DWPA(R/Z_{RSAP})_{i,k}$ becomes

$$p_{i,k} \, f_{3,k} = \alpha_i \, DWPA(R|Z_{RSAP})_{i,k} + z_i,$$

where

$$f_{3,k} \equiv \sum_{j=1}^{t} (\alpha_j \, DWPA(R|Z_{RSAP})_{j,k} + z_j) \quad \text{E q . (2.5).}$$

Definitions of symbols in Eqs. (2.3), (2.4) and (2.5) are:

$$p_{i,k} = \frac{y_{i,k}}{y_{\cdot k}} \approx \frac{n_{i,k}}{n_{\cdot k}}, \quad v_{i,k} = \frac{DWPA(R|Z_{RSAP})_{i,k}}{DWPA(R|Z_{RSAP})_{\cdot k}};$$

t is the number of taxa used for analysis,
$n_{i,k}$ pollen count of taxon i at site k,
$n_{\cdot k}$ total sum of pollen counts of all taxa at site k,
$y_{\cdot k}$ total sum of pollen loading of all taxa at site k,
and $DWPA(R/Z_{RSAP})_{\cdot k}$ total sum of the DWPAs of all taxa at site k.

$f_{1,k}$, $f_{2,k}$ and $f_{3,k}$ are site-specific coefficients for sub-models 1, 2 and 3, respectively. All 'α's and 'z's are obtained by the maximum likelihood method after Parsons and Prentice (1981, 1983/1984) using an appropriate optimization technique with iteration processes.

The Extended R-value model is a sophisticated extension of the R-value model for an objective description of the pollen-vegetation relationships in proportions. Its application requires careful examination of the basic assumptions and conditions of Eq. (2), however. In their series of papers in the 1980s, Prentice and Parsons treated vegetation data in proportion as given: no specific instruction was provided for the way the plant abundance is measured and distance-weighted. For example, Prentice and Webb (1986) used the mean plant abundances of major tree taxa to estimate the representation factors in the Midwest of the United States. Thus, use of their estimates as a substitute of relative pollen productivity estimates needs caution. The lack of appropriate estimates of the DWPA data could also be one of the contributing factors for the RSAP estimates that tended to be larger (Prentice et al., 1987) than those obtained since the 1990s–2000s (Sugita, 2013). All things considered, however, the ERV model is an invaluable and practical tool for obtaining empirical estimates of relative pollen productivity of major plant types. With proper and standardized protocols of vegetation survey methods (Bunting et al., 2013), well-defined criteria for site selection (Hjelle and Sugita, 2012) and further improvement in pollen dispersal-deposition function (Bunting and Middleton, 2005; Theuerkauf et al., 2016), the number of taxa, for which pollen productivity estimates are obtained by this approach, has increased dramatically since the 2000s in many parts of Europe, China and elsewhere (Broström et al., 2008; Mazier et al., 2012b; Li et al., 2017; Mariani et al., 2016). These estimates are essential for quantitative reconstruction of the past vegetation (Sugita, 2007a,b).

2.7.5.3.

Assumptions and conditions for the Prentice-Sugita model (specifically Eqs. (1) & (2))

This section discusses the major assumptions necessary for Eqs. (1) and (2) and implications for empirical applications of the model.

1. *Pollen productivity of each taxon is constant in space and time.* The pollen productivity of each taxon depends, to some degree, on (a) weather conditions such as temperature of the current or the previous year and/or interannual variability in weather parameters (Hicks et al., 1994; Huusko and Hicks, 2009; Sjögren et al., 2006), (b) the vegetation structure (closed canopy v. free-standing trees) and age class, (c) plant fitness, which depends not only on climate but also on soil types and light, for example pollen productivity may be lowered near the range limits of a species or at timberlines and (d) the range of species included in a composite pollen taxon, for example Poaceae. Long-term monitoring of pollen productivity in various environmental conditions is essential for a better understanding of the extent to which this assumption holds.

2. *Pollen dispersal and deposition is approximated by a diffusion model of small particles under neutral atmospheric conditions.* Prentice (1985, 1988b) and Sugita (1993, 1994) use one version of the Gaussian Plume Model (GPM) for a ground-level source proposed by Sutton (1953) and used by Tauber (1965). A GPM with elevated source has also been applied (Kabailiene, 1969), but the application was less sensitive to basin size in terms of pollen deposition (Prentice, 1988b; Jackson and Lyford, 1999). The neutral atmospheric conditions are discussed in detail by Tauber (1965) and Jackson and Lyford (1999). Changing wind speed influences the DWPAs for all pollen types (Nielsen and Sugita, 2005): however, the RSAP does not change significantly within the range of wind speed between 1 m s^{-1} and 25 m s^{-1} (Nielsen and Sugita, 2005; Sugita, 2007a,b; Gaillard et al., 2008). In general, the GPMs are less reliable for long-distance transport of small particles >1–10 km (Gregory, 1973; Pasquill and Smith, 1983). A Lagrangian Stochastic model of Kuparinen et al. (2007) – a new and more advanced dispersal model – has been

applied to improve the description of pollen dispersal, particularly for long-distance pollen (Theuerkauf et al., 2012, 2016; Mariani et al., 2016, 2017).

3. *Pollen deposited in the sampling basin comes from all directions evenly.* Pollen dispersal and deposition is expressed along the distance axis from the centre of the basin with this assumption for the sake of simplicity. In reality wind roses vary from site to site, as do frequency, direction and strength of wind. Major directions of wind can be season-dependent in different regions. Topographically complex terrains, such as mountains and hilly areas, have convoluted wind fields. This assumption needs to be evaluated further (Bunting and Middleton, 2005).

4. *Wind above the canopy is the dominant agent of pollen transport.* Tauber (1967, 1977) distinguished five components of pollen dispersal (Fig. 2.2-1): 1) over the canopy, 2) through the trunk space in a forest, 3) the 'rain-out' of pollen in the atmosphere washed out by precipitation, 4) waterborne pollen by streams, rivers and slope run-off and 5) a gravity component. All of these mechanisms of pollen transport affect the pollen-vegetation relationships jointly or separately in different vegetation types, topographic conditions and seasonal weather patterns. Whenever necessary, modification of the pollen dispersal-deposition function would be required to take into account those different components of pollen transport for site-specific patterns of pollen deposition, such as trunk-space component on small forest hollows.

The Prentice-Sugita model does not consider the impacts of waterborne pollen on pollen deposition in the sampling basin. Pollen deposition in large lakes, in particular, would be strongly influenced by inflowing rivers (Peck, 1973; Pennington, 1973, 1979). Site selection is critical when lakes and ponds are used as the sampling basins for either the model-based calibration of the pollen-vegetation relationship or vegetation reconstruction with the Landscape Reconstruction Algorithm (Sugita, 2007a,b) that is derived from the model.

5. *The sampling basin is a circular opening in the canopy of vegetation cover.* To make the one-dimensional expression (that is, along the distance axis from the centre of the basin) of the pollen-vegetation relationship simple and tractable, the basin size is expressed as the radius of an opening in the canopy of any vegetation cover that produces pollen. Some regions are rich in round basins, such as kettle holes caused by dead-ice in formerly glaciated areas: however, shapes of lakes and bogs are generally far from round in many parts of the globe. Bunting and Middleton (2005) and Hjelle and Sugita (2012) provide preliminary assessments of the impacts of this simplification of basin size and shape on the pollen-vegetation relationships.

Box 6

Roles of REVEALS and LOVE in the Landscape Reconstruction Algorithm (LRA)

Shinya Sugita

Institute of Ecology, Tallinn University, Estonia

The framework of the LRA (Sugita, 2007a,b; Fig. B6-1) was designed in such a way to make the process of pollen-based reconstruction of past vegetation and land cover transparent, spatially explicit and quantifiable across different spatial scales. One of the key concepts behind the LRA is the relevant source area of pollen (RSAP): it is defined as the area, or distance, beyond which correlations do not continue to improve between pollen loading and plant abundance for all the plant taxa of interest in the region (Sugita, 1994). This implies that pollen loading coming from beyond RSAP, that is the background pollen, is consistent and quantifiable at similarly sized sites in the region even if the spatial structure of vegetation is heterogeneous and patchy (Sugita et al., 2010b).

The LRA consists of two steps. The first is to estimate regional vegetation composition using the REVEALS (Regional Estimates of Vegetation Abundance from Large Sites) model (Sugita, 2007a) and the second is to apply the LOVE (Local Vegetation Estimates) model for reconstruction of local vegetation composition within RSAP at smaller-sized sites using the REVEALS estimates as inputs (Sugita, 2007b). The LRA requires reliable estimates of pollen productivity and its standard error, pollen counts, estimates of fall speed of pollen for plant taxa of interest and other parameters necessary for the dispersal-deposition function selected.

The REVEALS model is the inverse of a generalized form of the R-value model of Davis (1963) and approximates the mean vegetation composition within a 50–100-km radius. Biases caused by the size difference in sedimentary basins and the inter-taxonomic differences in pollen dispersal and deposition are corrected according to the Prentice-Sugita model of the pollen-vegetation relationship (Prentice, 1985, 1988b; Sugita, 1993, 1994). Pollen records from large sites >100–500 ha reflect the regional vegetation better and have smaller site-to-site variations than those from smaller sites (Sugita, 1994, 2007a). Accordingly, REVEALS applications using pollen records from 'large sites' are more robust than using those from smaller-sized sites. A number of smaller-sized sites can be used together to average out the between-site differences in the REVEALS results: however, uncertainties in the results are expected to be larger than those using 'large sites' (Sugita, 2007a; Mazier et al., 2012b; Fyfe et al., 2013; Trondman et al., 2016). REVEALS has been evaluated in several regions in Europe (Hellman et al., 2008a,b; Soepboer et al., 2010; Mazier et al., 2012b; Fyfe et al., 2013; Abraham et al., 2014; Pirzamanbin et al., 2014; Hjelle et al., 2015; Marquer et al., 2020) and has been widely applied as a stand-alone method for reconstruction of the regional vegetation and land cover in many parts of the globe (Gaillard et al., 2010; Nielsen et al., 2012; Abraham et al., 2014; Trondman et al., 2015; Marquer et al., 2014, 2017; Mariani et al., 2017; Dawson et al., 2018; Li et al., 2018; Roberts et al., 2018; Cao et al., 2019).

Conceptually the LOVE model first approximates the background pollen based on the REVEALS results and then subtracts it from the total pollen deposition to estimate vegetation composition within the RSAP (Sugita, 2007b; Sugita et al., 2010). Considering pollen counts as input variables, the equation derived for LOVE is isomorphic to the inverse form of sub-model 2 of the Extended R-value (ERV) model (Parsons and Prentice, 1981; Prentice and Parsons, 1983). The main difference is the definition of the background pollen. The ERV model assumes the background pollen to be constant through time for each taxon; the LOVE model uses the REVEALS estimates of regional vegetation composition to estimate the background pollen as a function of time in the past (Sugita, 2007b). As in REVEALS, the Prentice-Sugita model is modified and incorporated into LOVE to reduce biases caused by the inter-taxonomic differences in pollen dispersal and deposition, and basin size. The LOVE model requires RSAP as an input parameter: but RSAP in the past is difficult to estimate. Various factors affect RSAP, such as (1) the size of sedimentary basins, (2) the type and spatial structure of vegetation, and interactions between (1) and (2) (Sugita, 1994, 2013; Bunting et al., 2004; Hellman et al., 2009). Forward and inverse modelling approaches have been proposed, evaluated and practiced to overcome this difficulty in RSAP estimate (Sugita, 2007b; Hellman et al., 2009; Sugita et al., 2010; Marquer et al., 2020). The entire LRA approach – REVEALS and LOVE together – has been tested in a number of regions in North America and northern Europe (Nielsen and Odgaard, 2010; Sugita et al., 2010; Fyfe et al., 2013; Hjelle et al., 2015) and has been applied for reconstruction of vegetation and land cover at stand to landscape scales in several areas (Overballe-Petersen et al., 2012; Cui et al., 2013, 2014; Fyfe et al., 2013, 2018a; Poska et al., 2014; Mazier et al., 2015; Alenius et al., 2017; Fredh et al., 2017, 2019; Marquer et al., 2020).

Further improvements of the LRA approach are expected, including selection of more realistic and updated models of pollen dispersal and deposition than the Gaussian Plume Model (Kuparinen and Schurr, 2007; Theuerkauf et al., 2012, 2016), compilation of reliable estimates of pollen productivity in various parts of the world (Broström et al., 2008; Mazier et al., 2012a; Bunting et al., 2013; Li et al., 2017; Wada et al., 2018) and reconstruction of vegetation and land cover in topographically complex terrains (Hjelle et al., 2015; Marquer et al., 2020). To facilitate constructive criticisms it is critical to keep the LRA approach as simple as possible to make it easier to see objectively how the reconstruction process works.

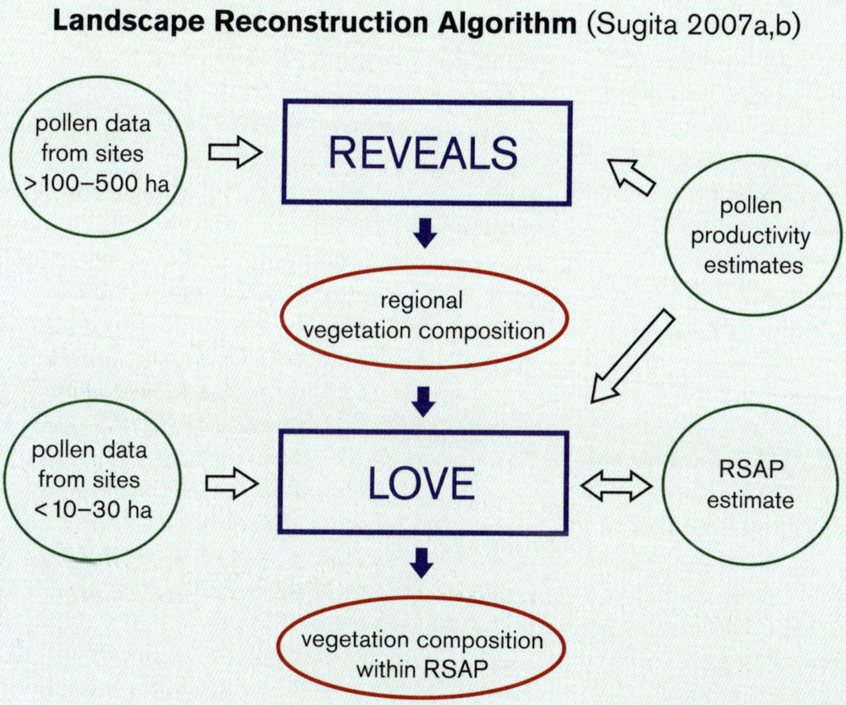

Fig. DC 1.
The roles of the REVEALS model and the LOVE model on the Landscape Reconstruction Algorithm (LRA).

Box 7

STEPPS: A Bayesian approach to quantifying trends and uncertainty in forest composition

Andria Dawson[1], Jack Williams[2], Mathias Trachsel[2], Christopher Paciorek[3], Jason McLachlan[4]

[1]Mount Royal University, Canada; [2]University of Wisconsin - Madison, USA; [3]University of California, Berkeley, USA, [4]University of Notre Dame, USA

Estimating changing vegetation from fossil pollen assemblages entails understanding uncertain processes, from pollen production and dispersal to inference about continuous vegetation from sparse networks of pollen sites. However, most pollen-vegetation models (PVMs) do not explicitly account for such uncertainties in a formal, coherent way. STEPPS (Spatio-Temporal Empirical Prediction from Pollen in Sediments) is a Bayesian hierarchical spatial PVM that models pollen counts from a network of pollen records as a function of gridded forest-composition data, taking into account local pollen dispersal, regional pollen dispersal, differential pollen production across tree taxa and uncertainty in both observations and processes. As opposed to working with individual sites, STEPPS treats spatiotemporal networks of pollen data as a cohesive unit to reconstruct the relative abundance of tree taxa in space and time.

There are two stages to the STEPPS modelling approach: calibration and prediction. In the calibration stage, process parameters that describe the relationship between pollen and vegetation are estimated. The calibration stage requires spatially continuous forest composition data and contemporaneous pollen counts from a network of depositional sites. In the prediction stage, the process parameters estimated in the calibration stage are used to infer vegetation for times where there are pollen data but no vegetation data. As a spatiotemporal model, STEPPS estimates forest composition with uncertainty even at locations without extant pollen data. A complete description of the model can be found in Dawson et al. (2016, 2019) and Paciorek and McLachlan (2009).

Encouragingly, estimates of the pollen-vegetation relationship from the STEPPS calibration stage, such as pollen productivity, are generally consistent with estimates from independent parameter estimates in the literature (Fig. B7-1). When supported by a dense network of pollen data, STEPPS is able to capture the major features of vegetation structure at regional scales (Fig. B7-2).

At the prediction stage, inference about changing vegetation in STEPPS is informed by both the central tendency and the uncertainty in estimates of forest composition. Comparisons between maps of taxonomic distribution over time, for instance, can identify the locations where composition is stable, where it changes and where the uncertainty in the data is too large to allow confident conclusions (Fig. B7-3).

Statistical estimates using STEPPS of vegetation change over the last 2000 years in the US Upper Midwest challenge the common assumption in Earth system modelling of steady-state ecosystems prior to the Industrial Revolution c. AD 1800. Key features include ongoing range infilling by hemlock in northern Wisconsin and expansion of mesic hardwood species in the Big Woods region of southern Minnesota (Dawson et al., 2019).

The ability of statistical models, such as STEPPS, to identify the strength of such patterns in palaeoecological data points to new directions in research: STEPPS can be used to optimize future field campaigns by targeting locations or times with high uncertainty. Posterior estimates of composition from STEPPS can be used to constrain and improve modelled reconstructions of vegetation change. And the modular nature of hierarchical models like STEPPS facilitates the exploration of alternative model structures, including components found in other PMVs, such as LOVE/REVEALS.

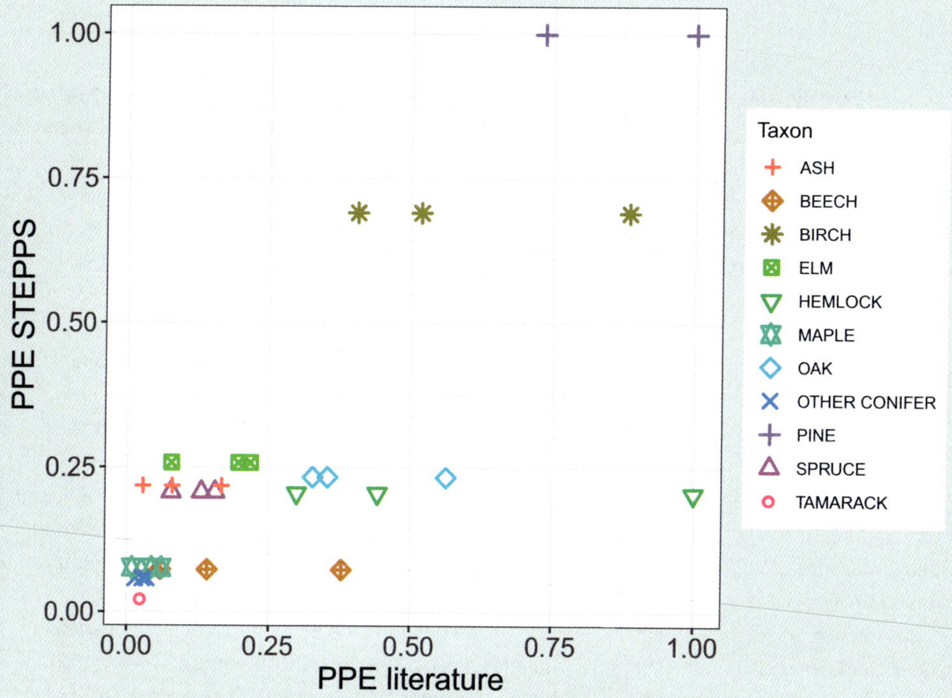

Fig. B7-1.

Pollen productivity estimates (PPEs) predicted by STEPPS compared to PPEs in the literature for the upper midwestern USA.

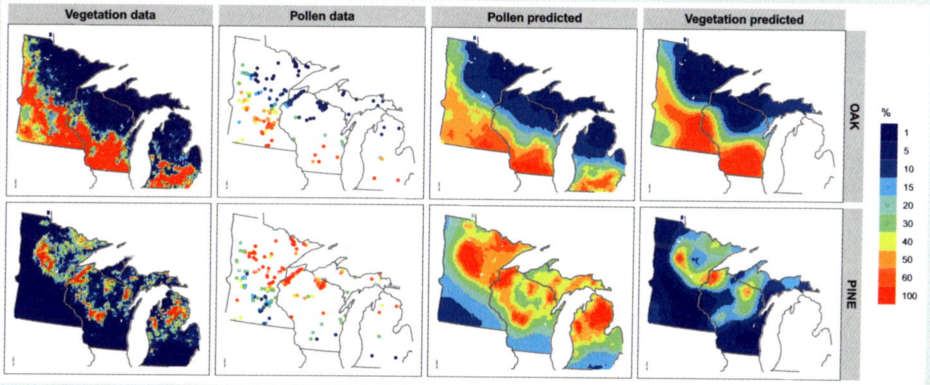

Fig. B7-2.

Heat maps of the data (vegetation and pollen) and the STEPPS predictions (pollen and vegetation) in the upper midwestern USA c. AD 1800. The percentages in the legend indicate the upper limit of the interval.

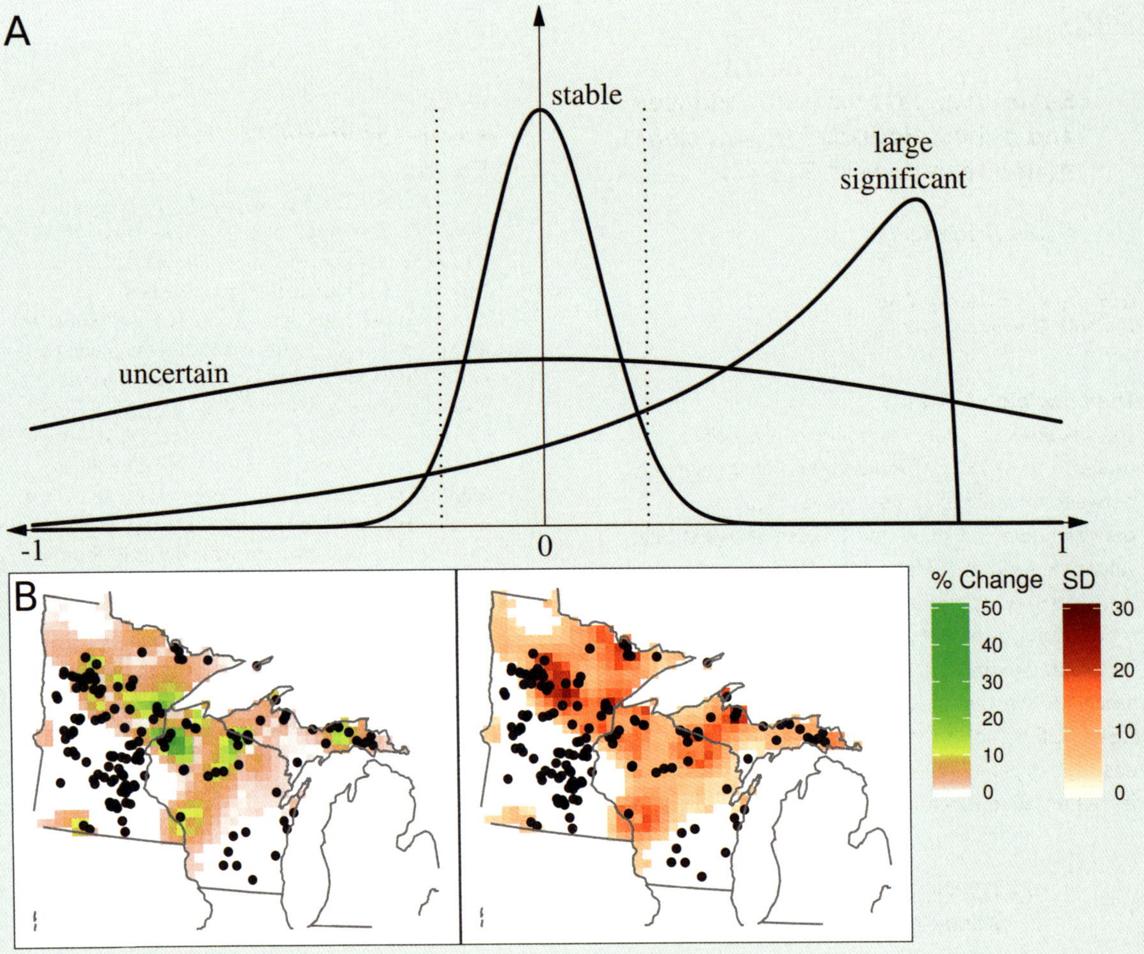

Fig. B7-3.

A) Differences in taxonomic abundance between time periods can be distributed in a way that indicates stability over time, a large and significant change or uncertain change. B) Mean magnitude of change (left) and standard deviation of the magnitude of change (right) of pine abundance between 300 yr BP and 2100 yr BP.

Box 8

Estimating past plant abundances and pollen productivity from single pollen records with ROPES

Martin Theuerkauf

Greifswald Mire Centre GMC,
D-17489 Greifswald, Germany

The principle of ROPES

The link between plant abundances and pollen values is biased by differences in pollen productivity between taxa, differences in pollen dispersal between taxa and, of course, the use of percentage values. Already von Post recognized this limitation in his seminal lecture in 1916. The bias can be reduced by correcting pollen counts with suitable factors and models, for example (relative) pollen productivity estimates (PPEs or RPPs) and the REVEALS model. Such correction factors are not easy to assess and are therefore only available for a few selected regions. Moreover, pollen productivity

Fig. B8-1.

Principles of the ROPES approach, illustrated with *Pinus* values from the Lake Tiefer See pollen record (NE Germany). Left: Past cover reconstructed with REVEALS and arbitrary PPEs (all PPEs=1). In this case, changes in the reconstructed cover (silhouettes) are very different from the changes in the pollen accumulation rate (PAR) values (red line) and the ratio of PAR values over reconstructed cover (PoR; grey line) shows distinct deviations from the mean (dashed line). This reconstruction is hence obviously not suitable. Right: Cover reconstructed with ROPES, that is PPEs have been adjusted in an iterative process to give the best possible match between changes in PAR values and reconstructed cover. Now changes in this reconstruction match changes in PAR values, indicating that this reconstruction is robust. The PoR ratio still shows variations around the mean because of noise in the PAR values, but no diverging trends.

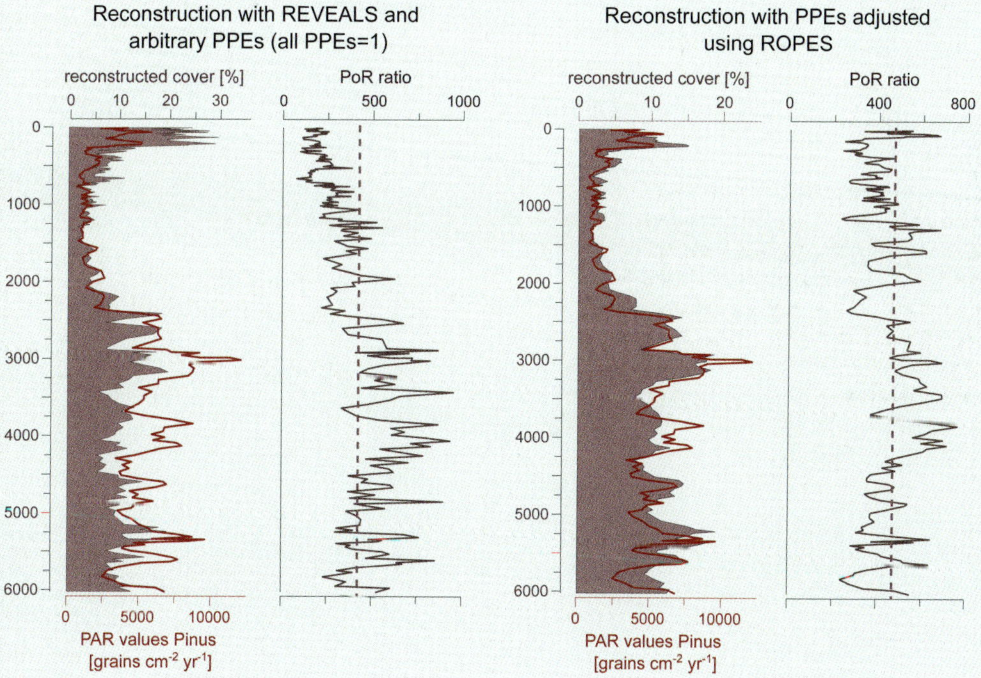

may change in response to changes in climate or land use or immigration of taxa affecting competitive niches. For example, pollen productivity of grasses was clearly higher before high-intensity grassland management was introduced (Theuerkauf et al., 2015). PPEs are commonly derived in modern settings and may thus not be suitable to reconstruct vegetation of the past. Pollen accumulation rates (PAR, or influx) represent past plant abundances for each taxon independently and in a linear fashion. As such, they could provide an alternative basis for quantitative reconstructions. Yet PAR values are very sensitive to accurate dating and, more importantly, they are affected by basin size as well as (in the case of lakes) sediment redeposition and focussing. As a result, PAR values tend to differ considerably across single sites and among nearby sites and they thus appear too noisy for calibration in terms of plant abundances.

ROPES builds upon the idea that while the actual PAR values may be too noisy, changes in PARs should be more robust: they should be largely similar across single sites and among neighbouring sites. Changes in PARs represent changes in past plant abundances in a one-to-one fashion. ROPES uses these changes in PAR values to infer actual past plant abundances with the help of the REVEALS model. The rationale is that a REVEALS reconstruction, when applied with suitable PPEs, should show the same changes for each taxon along a pollen record as the PAR values (Fig. B8-1). To find suitable PPEs, ROPES starts with a random set of PPEs, which is then permuted and optimised to find PPEs that produce the desired REVEALS reconstruction.

To find this best set of PPEs, ROPES employs the PoR ratio of PAR values divided by REVEALS-derived plant abundances. The ratio is calculated for each taxon and sample in a pollen record. Variations in the ratio along the record show how well changes in PARs match changes in the reconstructed abundances – variation declines as the match improves. Hence, ROPES searches for PPEs that minimize variation in the PoR ratio using numerical optimization.

Output

ROPES first of all estimates past plant abundances from single pollen records without presupposed correction factors (PPEs). The only input besides pollen data is a pollen dispersal model. Second, ROPES calculates PPEs. This feature can be used to explore changes in pollen productivity in space and time (Fig. B8-2). The calculated PPEs can be used for application of REVEALS in other sites that are not suited for ROPES. Finally, the PoR ratio is a diagnostic tool that indicates changes in pollen productivity of one or several taxa through time.

Requirements and limits

ROPES requires pollen data (counts + PARs) from large lakes or peatlands, with dominantly atmospheric pollen deposition. As the method builds upon changes in PARs, records should show sufficient change for all taxa. Performance increases with the number of samples; typically records with at least 100 samples are needed. To ensure that the PAR values are reliable, records should originate from stable sedimentary environments. REVEALS and hence also ROPES assume homogenous regional vegetation composition, that is similar composition near the site and further away. For sites near ecotones or in regions with prominent vegetation patterns due to, for example, mountain ranges, the reconstructed regional cover may differ substantially from the true cover.

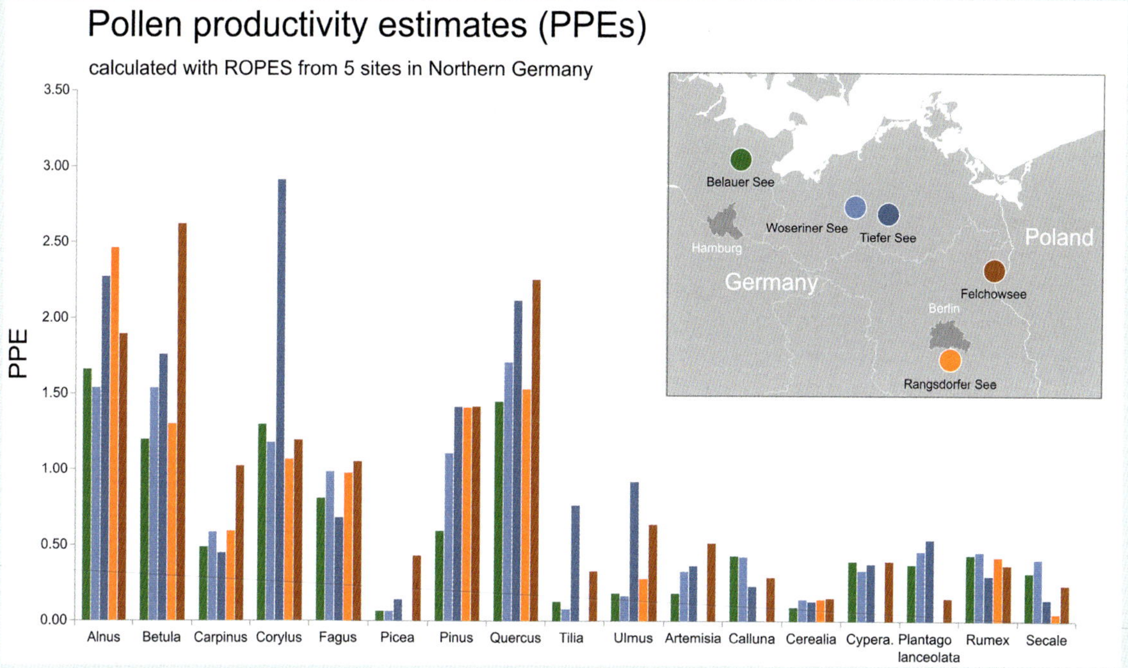

Fig. B8-2.

Pollen productivity estimates produced from six Holocene pollen records across northern Germany with ROPES.

3.

Pleistocene History of Vegetation and Flora

Karl-Ernst Behre[1], W.O. (Pim) van der Knaap[2,3]

[1] Lower Saxony Institute for Historical Coastal Research, Wilhelmshaven, Germany
[2] Institute of Plant Sciences, University of Bern, Bern, Switzerland
[3] Oeschger Centre for Climate Change Research, University of Bern, Bern, Switzerland

3.1

Before the Pleistocene: the Tertiary

Most of the stock of species that dominated European vegetation in the Quaternary developed during the Tertiary but the first steps had already been made in the preceding Cretaceous. Ferns and gymnosperms prevailed in the lower Cretaceous and from the Middle Cretaceous the Angiosperms emerged and evolved rapidly. At the transition to the Tertiary the latter were already present with many species and had often replaced the archaic plant communities. Most of the modern families and genera already existed at the onset of the Tertiary. Their differentiation into species resulted in the situation that, already in the Miocene, fossil plant remains show only slight differences from modern species.

In the course of the Tertiary, European vegetation development was for the most part determined by changes in geography, geology and climate (Fig. 3.1-1). In this process, plate tectonics played an important role. In the preceding geological formations the Eurasian Plate and the African Plate were separated by sea-floor spreading. This led to a wide, west-east, world-spanning sea-belt: the Tethys that reached as far as east Asia. It is in the Tethys and its lateral seas that the calcareous Mesozoic and Old-Tertiary sediments, which are frequently encountered in central Europe and the Mediterranean region, were deposited. In the course of crustal movements starting in the Middle Cretaceous the African Plate shifted northwards.

At this point the alpine-type orogenesis started and the large mountain chains from the Pyrenees across the Alps and Carpathians as far as the Caucasus, as well as the Appenines and the Balkan and Anatolian mountains were upfolded. The maximum of this orogenesis was attained in the late Miocene, that is about 10 million years ago. As a result, the Tethys was closed in the east but remained open in the west and became the Mediterranean Sea.

As a result of these geological changes, on the one hand, plant species from the south-east could find their way to Europe more easily while, on the other hand, their migration within Europe was hampered by the west-east running mountain ranges. This became particularly obvious later, during the Pleistocene, with the multiple changes of glacials and interglacials.

The thermal maximum of the Tertiary took place soon after its beginning in the Eocene and this is clearly reflected in the vegetation. Long ago the flora of that time was named palaeotropic. With a closer look, however, it is clear that this term is not accurate since of all the genera investigated only the palm genus *Nypa* is of genuine tropical origin. Most of the recorded genera that occurred in the tropics at that time grew only in mountainous areas, therefore, nowadays the term paratropical is often used to characterize them and, in most cases, they are in fact subtropical genera.

During the Eocene, the evergreen paratropical rainforests and likewise the evergreen laurel forests were the most important vegetation types in Europe and their structure was highly differentiated. Characteristic for them were tropical-subtropical families, such as Anacardiaceae, Annonaceae,

Bombacaceae, Moraceae with the large genus *Ficus*, or the palms of the Arecaceae and the Lauraceae with *Cinnamomum*. They can be compared best with modern subtropical lowland rainforests in south-east Asia (Mai, 1989).

Alongside them there existed extensive subtropical swamp forests and fens, which later became the brown coal deposits (Fig.3.1-2). These were formed mainly in the warm epochs, the Eocene and Miocene. *Taxodium*, *Glyptostrobus* and *Nyssa* constitute a major part of the Miocene brown coal (Mai, 1981).

During the Eocene, in the south of what was then Europe, along the Tethys Sea, mangroves grew, together with Rhizophoraceae and *Nypa* palms. A similar vegetation type can be found nowadays in south-east Asia. Until the Oligocene, the freshwater vegetation was also dominated by subtropical representatives such as the Hydrocharitaceae with *Ottelia* and *Vallisneria* or the Pontederiaceae with *Eichhornia*.

In the course of the Eocene, and particularly during the Oligocene, the tropical-subtropical rainforests were gradually replaced by subtropical-warm temperate laurel forests. In addition to the evergreen woods, composed of numerous subtropical Angiosperm families such as Lauraceae, Magnoliaceae, Hamamelidaceae, Theaceae and conifers of the Taxodiaceae family (for example *Arthrotaxis*, *Cryptomeria, Sequoia*), the forests also included several families of deciduous trees. These forests can best be compared with the modern subtropical-warm temperate laurel forests of east Asia or the Canary Islands and Madeira, some of them with modern redwoods as in Pacific North America.

In the early Tertiary, the palaeogeographic contour of Europe was quite different from that of today (Fig. 3.1-3 for Upper Eocene). Three large mainland regions existed, each of them housing its own flora. From Spain in south-west Europe across France and central Europe as far as Poland extended the central European region with a paratropical flora. This was also called the 'Gelinden region' after the famous leaf flora from the Upper Pliocene at Gelinden in Belgium. To the north-west, separated by the sea, was situated the Greenland region, which was connected with North America and housed the arcto-tertiary flora. To the south this extended through to, and included, the British Isles. The large region in the north-east was called the Volgo-Mugojar region and reached from Scandinavia as far as the Ural mountains (Mai, 1989).

In the Miocene, the Mediterranean Sea dried out for some time and the climate became distinctly arid. As a consequence the laurophyll flora, which was bound to subtropical-humid conditions, was pushed way to the south-west, to the Canary Islands and Madeira. Even today genuine laurophyll relics, such as *Hedera canariensis, Laurus canariensis, Persea indica* and other species of the laural forests, exist on these islands.

From the Upper Miocene some of these laurophyll taxa developed xeromorphs which now dominate the sclerophyllic Mediterranean flora. This Mediterranean flora was complemented by addi-

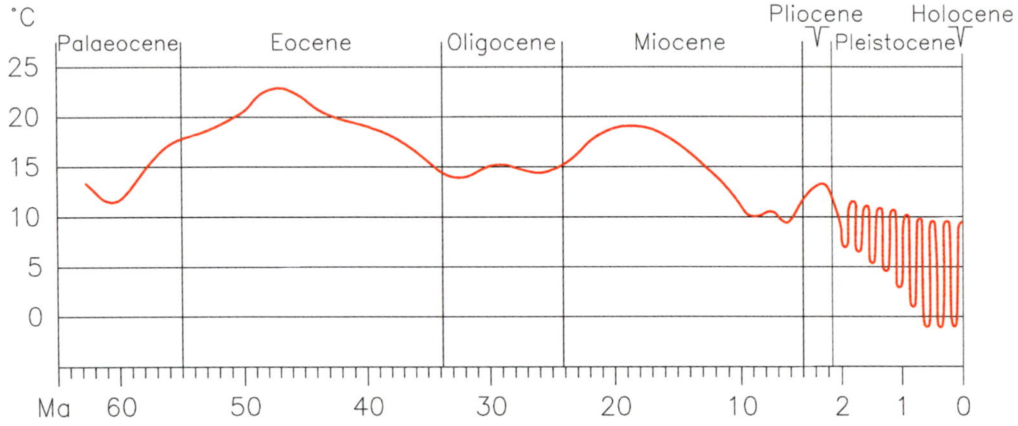

Fig. 3.1-1. Climatic oscillations in the course of the Tertiary and Quaternary. Estimated annual mean temperatures for West and Central Europe. After Woldstedt (1961) and Mai (1995), updated.

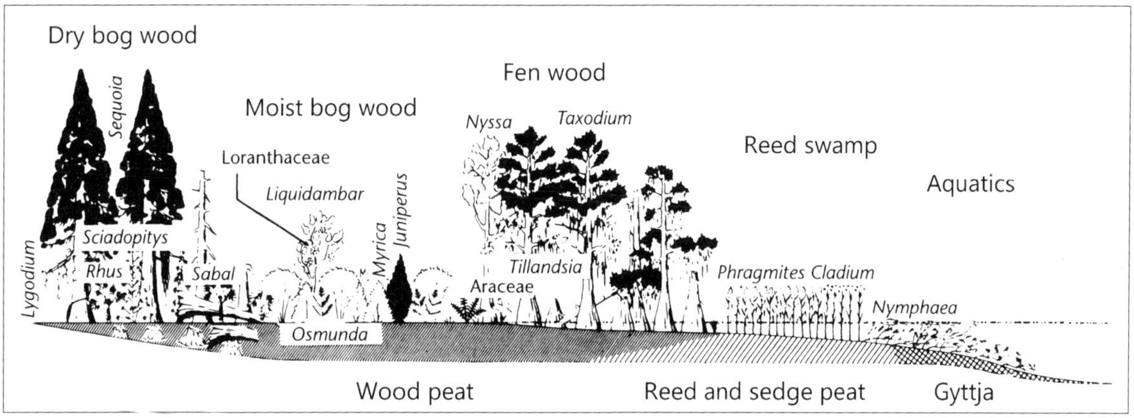

Fig. 3.1-2. Reconstruction of the vegetation zonation of a middle Tertiary browncoal mire in Central Europe. After Teichmüller, from Duvigneaud (1974), modified.

tional species that evolved from arcto-tertiary taxa. As early as the Upper Miocene in some sites there appeared for the first time numerous hardwood genera which today characterize the Mediterranean region. These fossil floras are called 'dry floras' and their main taxa are *Arbutus, Ceratonia, Nerium, Olea, Phillyrea, Pistacia*, the group of *Quercus ilex* and others.

The deciduous forests, which are now typical for Europe, have their origin in the circumpolar area of the northern hemisphere. There, in the Greenland region, the arcto-tertiary flora has developed since the Upper Cretaceous. Decisive for the evolution of the deciduous trees was the strong seasonal light change in the polar region. Already in the Palaeocene the classical sites in Greenland, Spitsbergen and Alaska hosted genera such as *Liriodendron, Platanus, Metasequoia, Acer, Tilia, Pterocarya, Corylus* and others (Mai, 1981).

Similarly, as early as the Palaeocene, the expansion of arcto-tertiary elements southwards in the direction of Europe started but was, however, slowed down for some time by the climate warming in the Eocene. From this period a very rich flora has been recovered in the famous Messel pit fossil site in southern Germany. During the Oligocene, a distinct climate deterioration took place, connected with a marine regression that concurrently led to more continental conditions over large parts of Europe. As a result, large numbers of temperate and deciduous species invaded everywhere into the European forest communities to form mixed mesophytic forests. The transformation from a palaeo-subtropical to an arcto-tertiary flora and their mutual expansion are the aspects characterising the floristic development during the Oligocene and Miocene.

Between the Upper Miocene and the start of the Pliocene, that is long before the onset of the Quaternary in Europe, most of the evergreen taxa that were bound to a subtropical-humid climate became extinct in Europe. Only a few evergreen representatives of palaeotropic origin have survived in Europe until today: among others, *Buxus, Hedera, Ilex, Laurus, Rhododendron* and *Smilax*.

From the Upper Miocene, also at the species level, trees and herbs which are common today appeared which do not differ morphologically from the modern species. In the broadleaved forests these were the maple species *Acer campestre, monspessulanum, opalus, platanoides* and *pseudoplatanus, Alnus glutinosa* and *incana, Betula pendula, Carpinus betulus* and *orientalis, Castanea sativa, Fagus orientalis, Juglans regia*, the oak species *Quercus cerris, castanifolia, pontica, pubescens* and *robur*, as well as many others.

Strong changes also occurred in the undergrowth. The herbaceous aquatic flora had already attained a great geological age and consisted of numerous taxa, but the herbaceous terrestrial angiosperms only developed much later: a feature which is regarded as a consequence of climatic deterioration in the Upper Tertiary. At that point the mosses, Sellaginellas and evergreen ferns, which until then

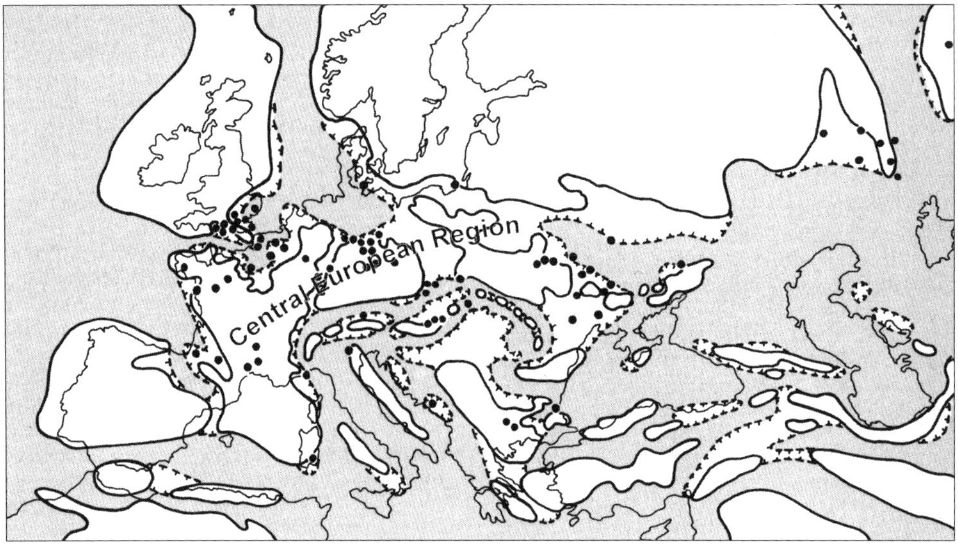

Fig. 3.1-3. Palaeogeography of western Eurasia in the Upper Eocene (42-34 my BP) with locations of fossil macrofloras. Full lines = Eocene boundaries of land and sea, broken lines = Eocene boundaries of epicontinental seas. After Mai (1995).

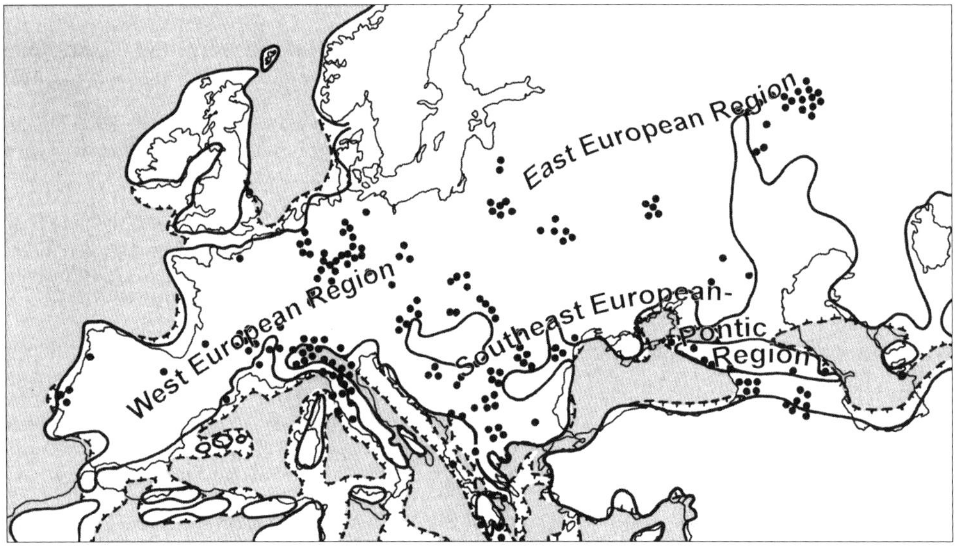

Fig. 3.1-4. Palaeogeography of western Eurasia in the Pliocene with locations of fossil macrofloras. Full lines = maximum sea expansion, broken lines = minimal sea expansion. After Mai (1995).

had covered most soils, were replaced by herbaceous angiosperm shade-tolerant plants. Almost all of the important modern herbaceous genera are already documented in the Pliocene (Mai, 1981), a period when the modern palaeogeography was virtually reached (Fig. 3.1-4).

At the end of the Tertiary, over large parts of Europe, a warm-temperate forest vegetation, extraordinarily rich in species existed, composed of numerous deciduous broadleaf genera and many conifers. The modern sclerophyllous flora in the Mediterranean had not yet formed and developed only in the following Quaternary, while the newly upfolded high mountains were occupied by an alpine flora.

3.2

The Tertiary/Quaternary (Pliocene/Pleistocene) boundary and the subdivision of the Pleistocene

Intense debates about the chronology of the younger part of Earth's history have taken place in recent decades. From 1978 the division of the Tertiary into two periods Palaeogene and Neogene was requested by geologists. While the Palaeogene was to include the epochs Palaeocene, Eocene and Oligocene, the Neogene was to comprise not only the Miocene and Pliocene epochs but also the Quaternary, which formerly followed the Tertiary. Although this terminology was passed at the Geological Congresses and was, therefore, formally valid it was heavily disputed and the old terms, in particular the Quaternary, continued to be used. Special controversies arose about the onset of the Pleistocene, which had become an epoch and replaced the whole Quaternary. The start of this was placed at the top of the palaeomagnetic Olduvai event, that is 1.8 Ma BP. With this boundary there was a conflict with the central and north-west European opinion that regarded the first clearly cooler stage (Praetiglian), which led to a distinct floral change, as the basis of the Quaternary: this occurred around 2.6 Ma BP (Litt, 2007).

After some time of considerable confusion, which is visible in the chronologies of the papers published during this period, there eventually came a reconsideration in 2009, which was also formally adopted by the International Geological Congress (IUGS, see Cohen et al., 2013, updated 2016). Now the Quaternary was again recognized as a period on its own, coequal with the Palaeogene and Neogene periods, with all three together forming the Cenozoic era. At the same time the Quaternary's lower boundary was placed at the beginning of the Gelasium, a stage that had been created in the meantime. So the Quaternary finally starts as a chronostratigraphical unit in absolute years at 2.58 Ma BP just above the palaeomagnetically defined Gauss/Matuyama boundary (Gibbard et al., 2010), which means that it is well reproducible and can be globally correlated across far distances.

At this level a strong climatic change with all its consequences can be demonstrated (Zagwijn, 1992, Pillans and Naish, 2004). Marine oxygen isotope and ice-rafting records together provide a very detailed and complete picture of the progressive deterioration of the global climate. After a period of transition over which the intensity of glaciations gradually increases, the first expansion of continental ice sheets culminates around 2.6 Ma, first of all in North America, but in the Andes and Patagonia in South America as well and on the other side of the Atlantic. In Iceland there are also indications of glaciation at this time. The shift of arctic marine faunas to the southern part of the North Sea as well as to the Atlantic is recorded in the bottom sediments. Another important feature with a strong climatic background is the start of loess formation in China at about the Gauss/Matuyama boundary, with loess covering huge areas up to 200 m in thickness on the Chinese loess plateau.

In central Europe, the uppermost Tertiary, represented by the Reuverian layers, is followed by the cold Praetiglian complex. In this stage, not only typical Tertiary trees such as *Taxodium, Sequoia, Sciadopitys* or *Nyssa* disappeared, but also numerous other changes took place that were caused by the climatic deterioration.

With respect to geochronology the Pleistocene has been divided into three parts: the Lower, Middle and Upper Pleistocene. The boundaries of these geological units were also ratified at the IUGS conference in 2009. The boundary between Lower and Middle Pleistocene is fixed by the palaeomagnetically defined Brunhes-Matuyama boundary at 0.781 Ma BP, while the boundary between Middle

Fig. 3.2-1. Geochronology and stratigraphy around the onset of the Quaternary. 1: Period, 2: Epoch, 3: Stages general (standard), 4: Age, 5: Magnetozone.

and Upper Pleistocene is represented by the beginning of the last interglacial, the Eemian, at 0.126 Ma BP. Finally, the end of the Pleistocene and the start of the Holocene were placed at about 600–11 700 cal yr BP (Litt et al., 2007). In vegetation history, as well as in zoological palaeontology, the temporal terms Early instead of Lower and Late instead of Upper Pleistocene are often preferred.

Even more than by a general temperature decline, the Quaternary as a whole, is characterized by a kind of climatic unrest that led to continuous periodical oscillations between warm and cold phases with all their consequences, in particular with respect to the vegetation. And last but not least the evolution of humans is a Quaternary phenomenon. Starting from the initial genus *Australopithecus,* between 3 and 2.5 Ma BP, the genus *Homo* evolved which, for the first time, were able to produce primitive stone artefacts – the onset of human culture. One of the oldest humans was therefore called *Homo habilis* – the skilful human – who evolved further to *Homo erectus* (Terberger, 2019).

3.3

The Pleistocene in general

3.3.1

Climatic oscillations

The main feature of the Pleistocene is the numerous climatic oscillations, which are clearly reflected in the Marine Isotopic Stages (MIS) determined by the $^{16}O/^{18}O$ ratios. Altogether there are around 50 interglacials but there is no exact number as this depends on the definition: how many MIS peaks can be regarded as full interglacials? In the course of the Pleistocene the magnitude of these oscillations increased considerably, particularly over the last one million years. While the maximum temperature of the warm phases decreased only slightly, the minimum temperature of the cold phases declined considerably (Fig. 3.3-1) and this led eventually to the Nordic glaciations. A large ice sheet was formed over Scandinavia which during the Middle and Upper Pleistocene expanded several times into central Europe. Also, from the Alps and other mountains, glaciers advanced into their forelands. At the time of the largest ice extent, during the Elsterian and Saalian glaciations, around two thirds of the European continent was covered by ice, which was up to 3000 m thick.

The increasing cooling in the younger part of the Pleistocene can also be seen in the duration of the warm and cold periods: in the Lower Pleistocene the ratio between both is around 1:1 while in the upper Middle Pleistocene and the Upper Pleistocene the cold phases (now with glaciations) are about ten times longer than the warm ones. As can be seen from the names in Fig. 3.3-1, knowledge of the Lower Pleistocene is still very fragmentary and most of the named periods consist of several warm and cold phases.

The traditional subdivision of the Pleistocene period is into warm interglacials and cold glacials. Because these terms derive from the field of glaciology, they are inappropriate in non-glaciated areas or in southern regions where glaciations never occurred. Therefore the use of the terms 'warm stage' and 'cold stage' is nowadays often preferred. In addition to the full interglacials or warm periods there are minor ones, called interstadials with lower temperature. The distinction of both was defined by Jessen and Milthers (1928) on the basis of palaeobotanical criteria as follows: an interglacial is a temperate period with a summer temperature at least as high as that during the Holocene climatic optimum of the area in question and expressed by an adequate vegetation; an interstadial is defined as a warmer period with a summer temperature essentially lower than the climatic optimum in the Holocene period.

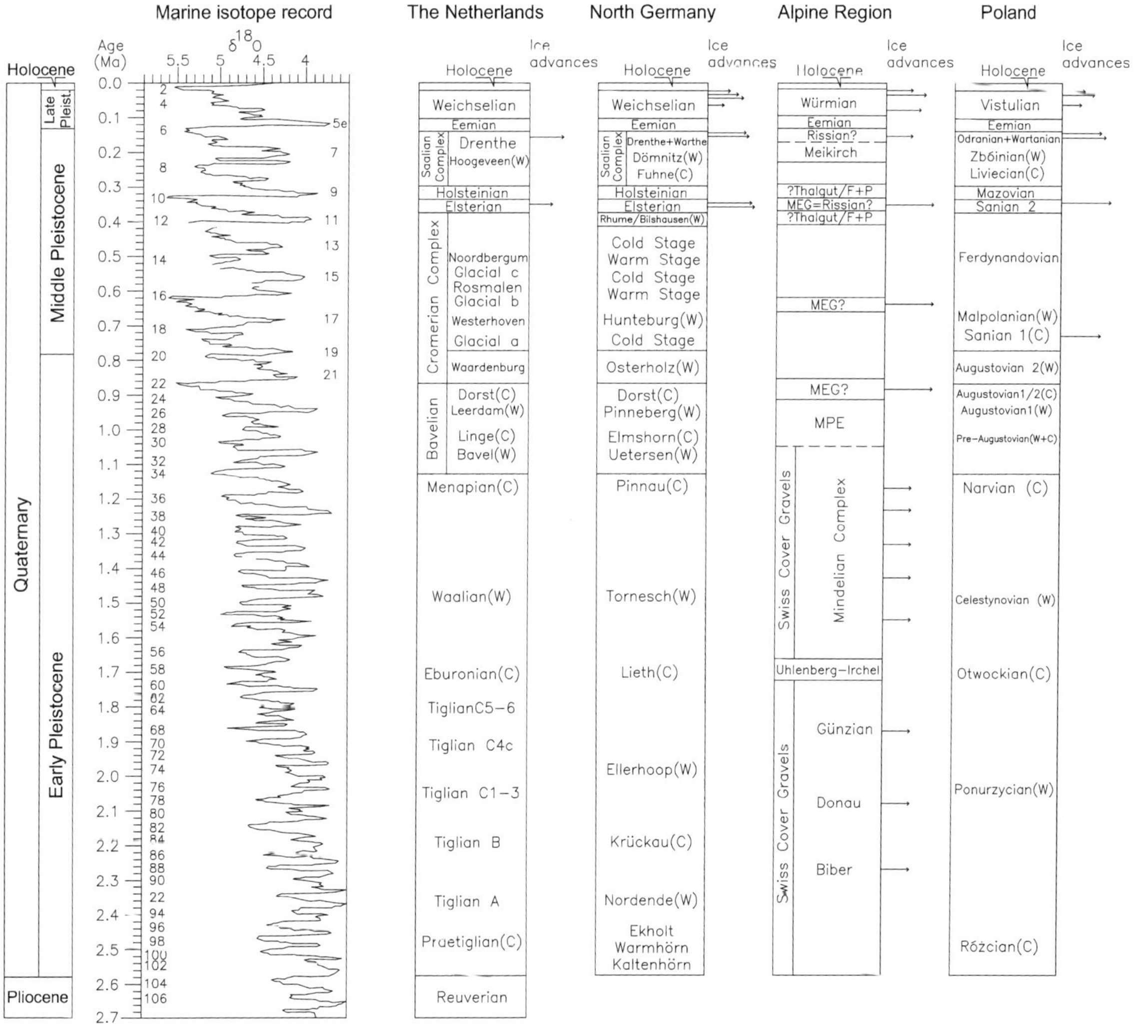

Fig. 3.3-1. Stratigraphic subdivision of the Quaternary in central Europe. After Litt et al. (2008), modified.

It is very important to always compare interglacial and interstadial vegetation in the same region, otherwise central Europe would nowadays represent an interglacial while northern Europe would be an interstadial from the central European viewpoint. This definition is independent of glacier advances and retreats, which are not always synchronous and can be applied everywhere.

In accordance with the climatic conditions there are several stages of interstadial vegetation, which also differ between latitudes. An interstadial with boreal forests in central Europe, such as the Brörup in the Weichselian, is represented in Scandinavia as a slight climatic improvement without any trees or even shrubs. Interstadials in the middle latitudes of Europe may, for instance, have tundra vegetation in the Weichselian (for example Glinde, Denekamp) as well as boreal forests (for example Brörup) in the north and even thermophilous deciduous forests (for example St. Germain) further south. To avoid problems in the use of the term *interstadial* in vegetation history, Behre (1989) proposed an interstadial as a period of climatic amelioration separated from the preceding and following phases and expressed by distinct changes in flora and vegetation, but, independent of the form of the vegetation, be it forest, shrubs, heath or grassland. This relative definition has the consequence that, using an absolute scale in a certain area, the interstadial vegetation at one time may be the same as the stadial vegetation of another period. For example, in north-west Germany the interstadial shrub tundra of the Oerel and Glinde are comparable in terms of climate with the stadial shrub tundra of the Younger Dryas.

In order to characterize cool phases and warm phases independent of their intensity or duration the terms cryomere and thermomere are also in use.

3.3.2

General floristic development in the course of the Quaternary

The transition from the Tertiary to the Quaternary marks a deep incision into the flora, in particular in Europe north of the Alps. Many Tertiary species and genera became extinct: only a few of them such as *Tsuga* or *Eucommia* survived for some time but their number quickly decreased. From this point on the vegetation reflects primarily the numerous climatic oscillations.

While the floristic composition of the vegetation during the various Pleistocene cold periods was similar there was a strong variability in the forest species and also in the herb and aquatic vegetation in the interglacials. Several reasons are responsible for this. The main cause is the orientation of the mountain ranges which in Europe run west-east from the Pyrenees via the Massif Central and the Alps to the Carpathians, thus forming a barrier for any south to north migration of plants and animals. This barrier became particularly effective when the climatic oscillations in the Pleistocene led to shifts of the vegetation belts in the direction north to south and vice versa. In the course of the large number of cold and warm stages not all species found their way back to the north of the mountain chain through the few available gaps for example along the Rhone or Danube rivers. This led to a considerable loss of floral and faunal species. The reduction of species can best be seen through the trees and shrubs that are represented in pollen analyses but is also relevant for the other groups of plants. The effectiveness of these barriers can be demonstrated very well by comparing Europe with North America or East Asia, where such transverse mountains are lacking, resulting in a much higher species diversity in these regions.

Another reason for the reduction of species towards the late Pleistocene and Holocene was the increasing duration and the decreasing temperature of the cold stages in the course of the Pleistocene. As a consequence, apart from possible hidden refugia, no forests existed north of the European mountain range and there were only a few areas south of it where woody and other species could survive. On the other hand the thermophilous species escaped further and further to the south (Fig. 4.11-2) and, during the comparatively short time of the warm periods, some of them were unable to return from their distant refugia into their old distribution areas.

The Mediterranean Sea was also a certain barrier for plant migrations, much in contrast to the situation in Asia and North America where subtropical and temperate species could reach low and thus warm latitudes (< 35° N) without any sea obstacle. The large number of Mediterranean species north of it, however, indicate that they were able to survive locally or to cross it much more easily than the mountains.

During the glaciations huge quantities of water accumulated in the glaciers and this resulted in a decline in global sea level. In the Weichselian, for example, this was of around 130 m. Large areas became dry during these periods, in particular the shallow parts of the northern Adriatic Sea and most of the North Sea basin, while the Baltic sea was covered by ice. Afterwards, when sea level rose again, at first Ireland and later Britain once again became islands and the sea formed a new barrier to the migration of plants and animals. The first species returning had no problems to occupy these areas but, after the separation, many latecomers had no chance to overcome this obstacle and reach these islands. Therefore the number of native species there is much lower than on the European continent. According to Mitchell (1986) the number of wild plants in Ireland nowadays is 815 and in England and Scotland 1172, compared with 2674 in Germany and 3500 in France (including Mediterranean species). *Acer pseudoplatanus* and *A. platanoides* as well as *Picea* and *Abies* did not come to England in the Holocene while *Fagus* and *Carpinus*, in spite of their late expansion, reached the country probably helped by humans. On the way to Ireland the number of tree species decreased further: *Tilia*, *Fagus* and *Carpinus* are lacking there. Animals often experienced this in the same way so, for instance, the common viper (*Vipera berus*) is lacking from Ireland, but according to a legend, it was in this case the holy St. Patrick who drove it out of the country.

Because of the increasing number of pollen diagrams that have been produced the general occurrence of the main plant species within the Quaternary is more or less known. This is particularly the case for the trees and the water plants.

In Table 3.3-1 the occurrence of the main plant taxa in Europe north of the west to east running mountain ranges are arranged in the order of their disappearance. To illustrate the general picture only the best known interglacials are mentioned and several of them have been grouped, as in the Cromerian, which includes at least 5 interglacials. The evidence for the younger part, starting with the Holsteinian, is considerably better than for the older periods, where it becomes increasingly worse. In the Lower Pleistocene and the lower Middle Pleistocene the records are often only fragmentary.

Looking into all available data it can be seen that the general reduction in species is often interrupted: some of them fail to return during one or several interglacials then come back later but the general tendency is clearly downwards. The individual species probably did not always have the same refugia in which to survive. When a cold stage began, they were pushed southwards, mainly by the cooling, but their way was also influenced by other climatic factors such as precipitation and drought, late frosts and so on that are not always connected with general cooling. Often underestimated is the competition with other species which varied at different times and in different areas. Some species may, by chance, have found a sheltered niche nearby, out of which they could later spread very early while others had to migrate very far and some became extinct. This resulted in their presence or absence in the following warm period and is the explanation for the apparent irregularities in the general plant record of the Pleistocene.

Beside the very important migration factor described above, the varying mean annual temperature of the warm stages is another reason for the reduction of trees and other plants. Annual temperature decreases slightly, in particular from the Reuverian and the Lower Pleistocene to the Middle Pleistocene (see the Marine Isotopic Record in Fig. 3.3-1), and delimits the distribution areas of some species that are still present in Europe south of the Alps. Others became extinct. Therefore authors such as Jäger (1969) or Meusel and Niedermaier (1985) argue that the modern distribution areas of most southern and south-eastern European woody taxa are determined by the present climatic conditions rather than by obstacles to migration.

Apart from the general explanation, however, many details of the floral development support the suggestion that migration processes were primarily responsible for the composition of the vegetation in the interglacial periods. For instance, *Pterocarya* and *Fagus* are lacking from most of the different Cromerian interglacials known to date (of varying ages) but they appear again in many Holsteinian sites in central Europe, though not at the thermal optimum but at the very end of the interglacial. In the Eemian, *Fagus* reached south-eastern Europe in its youngest phase but not central Europe, while in the Holocene it is very common there.

On the other hand, *Abies* was common during a long part of the Holsteinian in central Europe and

Table 3.3-1. Occurrence of woody plants and aquatics in West and Central Europe from the end of the Tertiary to the Holocene based on records of pollen and macro-remains. Re: Reuverian (Tertiary), Te: Tiglian, Wa: Waalian, Ba: Bavelian, Cr: Cromerian, Ho: Holsteinian, Ee: Eemian, Hl: Holocene. The first column indicates the modern distribution of taxa that became exticnt in West and Central Europe. *: totally extinct, +: rare, EA: East Asia, NA: North America, SE: South and Southeast Europe (including Colchis). After van der Hammen et al. (1971), from Lang (1994), updated.

		Pliocene	Pleistocene						
			Lower			Middle		Upper	
		Re	Te	Wa	Ba	Cr	Ho	Ee	Hl
Conifers									
Pseudolarix (Pinaceae)	EA	O	.	.	.	.	.	.	.
Sciadopitys (Sciadopityaceae)	EA	O	O	.	+	.	.	.	.
Sequoia (Cupressaceae)	NA	O	+	.	.	.	.	.	.
Taxodium (Cupressaceae)	NA	O	+	.	.	.	.	.	.
Chamaecyparis (Cupressaceae)	EA/NA	O	O	.	.	.	.	.	.
Thuja (Cupressaceae)	EA/NA	O	O	.	.	.	.	.	.
Tsuga (Pinaceae)	EA/NA	O	O	O	O	+	.	.	.
Abies (Pinaceae)		O	O	O	O	O	O	O	O
Picea (Pinaceae)		O	O	O	O	O	O	O	O
Pinus (Pinaceae)		O	O	O	O	O	O	O	O
Taxus (Taxaceae)		O	O	O	O	O	O	O	O
Deciduous broadleaved trees									
Aesculus (Sapindaceae)	SE	O	.	.	.	.	.	.	.
Halesia (Styracaceae)	EA/NA	O	.	.	.	.	.	.	.
Liquidambar (Hamamelidaceae)	EA/NA	O	+	.	.	.	.	.	.
Stewartia (Theaceae)	EA/NA	O	.	.	.	.	.	.	.
Styrax (Styracaceae)	EA	O	.	.	.	.	.	.	.
Zelkova (Ulmaceae)	SE	O	.	.	.	.	.	.	.
Nyssa (Cornaceae)	EA/NA	O	+	.	.	.	.	.	.
Actinidia (Actinidiaceae)	EA	O	O	.	.	.	.	.	.
Liriodendron (Magnoliaceae)	NA	O	O	.	.	.	.	.	.
Magnolia (Magnoliaceae)	EA/NA	O	O	.	.	.	.	.	.
Phellodendron (Rutaceae)	EA	O	O	.	.	.	.	.	.
Carya (Juglandaceae)	EA/NA	O	+	+	+	.	.	.	.
Castanea (Fagaceae)	SE	O	O	O	+	.	.	.	.
Juglans (Juglandaceae)	SE	O	O	O	O	.	.	.	.
Ostrya (Betulaceae)	SE	O	+	+	+	.	.	.	.
Celtis (Cannabaceae)	SE	O	O	O	+	O	.	.	.
Eucommia (Eucommiaceae)	EA	O	O	O	O	O	.	.	.
Parthenocissus (Vitaceae)	EA/NA	O	O	O	O	O	.	.	.
Pterocarya (Juglandaceae)	SE	O	O	O	O	O	O	.	.
Fagus (Fagaceae)		O	O	+	+	O	O	+	O
Quercus (Fagaceae)		O	O	O	O	O	O	O	O
Betula (Betulaceae)		O	O	O	O	O	O	O	O
Alnus (Betulaceae)		O	O	O	O	O	O	O	O
Carpinus (Betulaceae)		O	O	O	O	O	O	O	O
Tilia (Malvaceae)		O	O	O	O	O	O	O	O
Ulmus (Ulmaceae)		O	O	O	O	O	O	O	O
Evergreen trees									
Rhododendron (Ericaceae)	SE	O	.	.	.	.	O	.	.
Buxus (Buxaceae)		O	.	.	O	O	O	O	+
Hedera (Araliaceae)		O	.	.	.	.	O	O	O
Ilex (Aquifoliaceae)		O	.	.	O	O	O	O	O
Water- and swamp plants									
Azolla tegeliensis (Azollaceae)	*	O	O	.	.	.	.	.	.
Proserpinaca (Haloragaceae)	NA	O	O	.	.	.	.	.	.
Euryale ferox (Nymphaeaceae)	EA	O	O	.	.	.	.	.	.
Azolla filiculoides (Azollaceae)	NA	O	O	O	O	O	O	.	.
Brasenia schreberi (Nymphaeaceae)	EA/NA	O	O	O	.	O	O	O	.
Dulichium arundinaceum (Cyperaceae)	NA	O	O	O	.	O	O	O	.
Salvinia natans (Salviniaceae)		O	O	O	O	O	O	O	+
Trapa natans (Trapaceae)		O	O	O	.	O	O	O	O

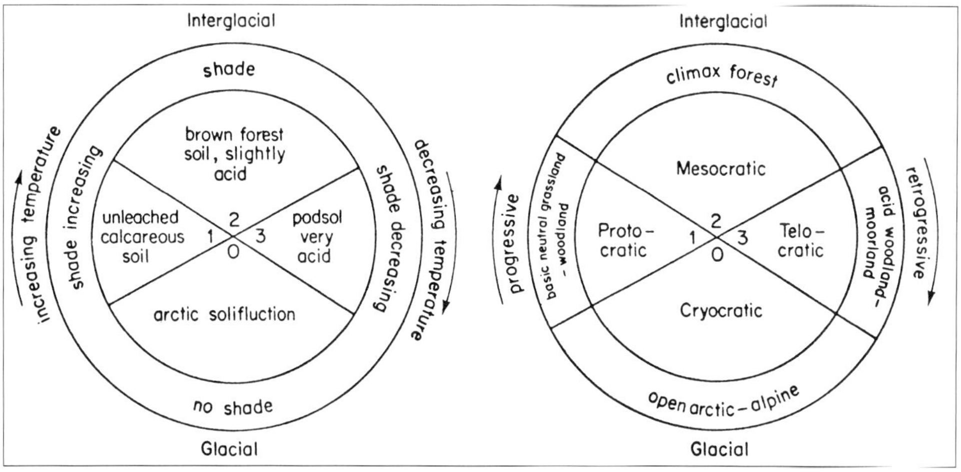

Fig. 3.3-2. The glacial-interglacial cycles. After Iversen (1958).

had its northern boundary in England and Ireland, while in the Eemian it was a latecomer and reached the North Sea only in the final phase of the Eemian. In the Holocene, *Abies* has not proceeded beyond the central German uplands, probably because it has to compete strongly with *Fagus*, which was absent in this area in the Eemian (but see chapter 4.10 for human disturbance).

While the extirpation of trees in the course of the Pleistocene was comparably fast in Europe north of the Alps (Table 3.3-1), a number of them persisted for a much longer time in southern Europe (Tzedakis et al., 2006). For example, species of Taxodiaceae disappeared after the Praetiglian stage (c. 2.5 Ma) in the north but survived in the south until about 1 Ma and are locally recorded in, for eample, Crotone (Calabria, Italy) up to MIS 11. *Pterocarya* disappeared from Europe north of the Alps after MIS 9 but persisted in Valle di Castiglione (central Italy) until MIS 7, while *Zelkova* became extinct in northern Europe in the Praetiglian but survived in central Italy until the first part of the Weichselian and is still present on the islands of Sicily and Crete (Follieri et al., 1986).

3.3.3

Glacial-interglacial cycles

The rhythmic climatic cycles that occurred during the Quaternary had consequences for the whole ecosystem. In particular they affected the development of vegetation and soils but they also influenced the trophic status of lakes and the faunal spectrum. Taking these effects as a basis, it was Iversen (1958) who first formally elaborated a template of the recurring changes in glacial-interglacial cycles (see also Grichuk, 1964; Watts, 1988). This scheme (Fig. 3.3-2) integrates three concepts: (1) a cycle of increase and decrease in temperature, (2) a cycle of progressive closing of the vegetation to high forest and regressing to open forest with heath and (3) a unidirectional leaching of the soils. It was developed for Denmark but can be applied to most of Europe north of the Alps, particularly to the glaciated boreal regions. Van der Hammen et al. (1971) used a similar form of presentation to show the development of vegetation in unglaciated regions such as the Mediterranean.

In glaciated regions there was a succession of four phases. First, A) the **cryocratic** phase, a glacial stage with severe climate, frost-disturbed soils with strong solifluction and open arctic-alpine vegetation in full sunlight. This was followed by B) the **protocratic** phasc, the onset of an interglacial stage with rising temperatures: the yet unleached and often calcareous raw soils became stable and were occupied by pioneer woodland with increasing shade. Third, C) the **mesocratic** phase which represents the climax of the interglacial with maximum temperature and dense shade-casting broadleaved forests: the brown soils were slightly leached. The final **telocratic** phase D) shows climatic deterioration with opening up of the forests and acid podsolised soils while heath and moorland expanded.

This general scheme of glacial-interglacial cycles can well be applied to the younger Pleistocene which has very strong climatic cycles and a similar climatic development within each. When looking more closely at the very long cold or glaciated phases, however, they are not as uniform as they appear at first sight but are interrupted by longer or shorter interstadials. Nevertheless, compared with the last two cycles, the older Pleistocene periods are more variable in climate and therefore also in its consequences.

In the Pleistocene pollen diagrams in this volume only the true interglacials of the Holsteinian and Eemian are subdivided into the phases described by the letters A,B,C and D. The subdivision of the older diagrams is made according to their respective authors.

3.3.4

Fauna and humans

The fauna also reflects the climatic changes in the course of the Pleistocene. While the small vertebrates and the molluscs are often very helpful for chronological problems, the most impressive species are the large mammals of the Ice Age and only these will be mentioned here.

Up until now there are no known typical cold faunas from the Lower Pleistocene, which is partly explained by the very few finds but also by the fact that the mammalian fauna had not yet adapted to a cold environment (von Koenigswald, 2007).

Increasing adaptation can best be seen by the mammoths, starting with the southern mammoth *Mammuthus meridionalis,* which was followed in the early Middle Pleistocene, up until the Saalian, by the steppe mammoth *M. trogontherii*, the ancestor of the woolly mammoth *M. primigenius*, which became the symbol of cold climates.

The younger cold phases are characterized by a number of other large mammals, mainly herbivores, such as the woolly rhinoceros *Coelodonta antiquitatis*, the musk ox *Ovibos moschatus* or the reindeer *Rangifer tarandus*.

Important species living in the woodlands of the warm periods were the straight-tusked elephant *Paleoloxodon (Elephas) antiquus,* closely related to the living Indian elephant *Elephas maximus,* the European water buffalo *Bubalus murrensis,* Kirchberg's rhinoceros *Stephanorhinus kirchbergensis*, the fallow deer *Cervus dama*, the roe deer *Capreolus capreolus* and the wild boar *Sus scrofa*. Other species such as the giant elk *Megaloceros giganteus,* wild horses *Equus* spp., the red deer *Cervus elaphus,* the aurochs *Bos primigenius* or the steppe bison *Bison priscus* as well as the cave bear *Ursus spaelaeus* are indifferent to climate. Of particular interest is the occurrence of the hippopotamus *Hippopotamus amphibius*, which was common until the last interglacial in the western part of Europe along the Rhine and has even been found at Trafalgar square in London but was not present in the more eastern Elbe catchment and beyond.

Many of the species mentioned above became extinct after the last interglacial, the Eemian, and the last glacial, the Weichselian. There is a still ongoing discussion as to whether this was caused by natural events or by humans. Most arguments blame humans for these extirpations because they strongly spread and proliferated at this time and, moreover, had developed effective weapons while there is no plausible assumption for naturally occurring changes which took place only at this interglacial-glacial transition and not at earlier ones and which only affected certain species.

From this time humans had also appeared in the environment. Earliest humans had developed in Africa around 3 million years ago and entered Europe as *Homo erectus*, the species which was able to produce the first artefacts, around 1 Million years ago. Numerous bone remains of them, as well as artefacts, were excavated in Gran Dolina, Spain, where they date back to 780 000 yr BP. The oldest central European find is a lower jaw from Heidelberg-Mauer with an age of around 600 000 years. Some time later, at about 320 000 yr BP, humans had learnt to hunt in groups as is shown by the famous finds of spears together with many horse skeletons at Schöningen in central Germany that date to the Holsteinian interglacial. At this time Neanderthals *Homo neanderthalensis* evolved from *Homo erectus* and spread across Europe. For a long period this species was the only human in Europe until another species, modern humans, *Homo sapiens* arrived. *Homo sapiens* represents a later evolution of *Homo erectus* that immigrated to Europe at about 40 000 yr BP. Until around 30 000 yr BP, when Neanderthals became extinct, both species lived together (Terberger, 2019). However, besides the reduction of animal species, particularly large mammals, humans did not affect the natural landscape until the Holocene, when they created the cultural landscape.

3.4

The Lower Pleistocene

The biostratigraphic subdivision of the Lower (Early) Pleistocene has until now been made mainly on the basis of the deposits in the Lower Rhine region. However, the vegetation history in this region is often only fragmentarily preserved and, in most cases, the decisive layers are stratigraphically not in superposition. The chronological correlation has been done by a variety of methods and has sometimes proved difficult. Nevertheless, although still incomplete in parts, a reasonable framework has been established. Up until now the only long sequence with several interglacials in superposition has been provided by the Lieth series in Schleswig-Holstein. Due to the different environment, however, its vegetation development is considerably different from that of the Lower Rhine sites, where the first subdivision was made.

Temporally, the Lower Pleistocene period spans the time from the beginning of the Gelasium (at 2.6 Ma BP) until the start of the Cromerian and lasted around 1.8 Ma, that is about 70% of the whole Pleistocene (see Fig. 3.2-1).

The smaller magnitude of the climatic oscillations in the Lower Pleistocene compared with the Upper Pleistocene has led to less sharply defined separations between the warm and cold stages and to transitional forms in the vegetation. This means that the terminology with respect to interglacials, interstadials and even stadials sometimes also raises problems.

Especially in comparison with the Upper Pleistocene, there are generally no characteristic successions of the immigration and culmination of individual species that could be used to distinguish the various interglacials of the Lower Pleistocene. In this period the trees returned in the warm phases,

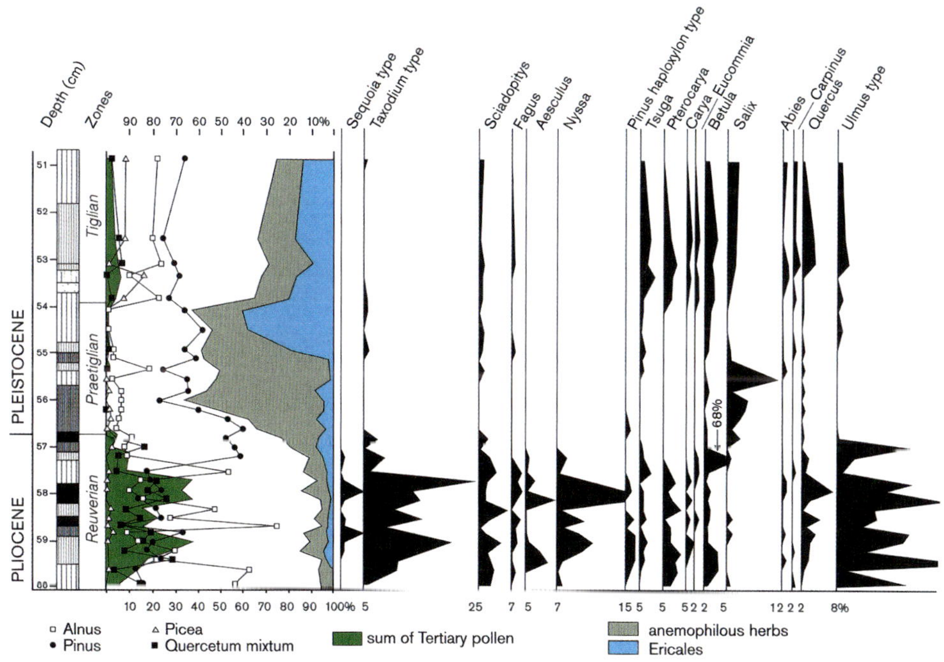

Fig. 3.4-1. Simplified original pollen diagram Meinweg/NL from the boundary Tertiary/Quaternary (type section). After Zagwijn (1960), from Janssen (1974).

starting with the less thermophilous species and followed by the more thermophilous ones. The forest composition was mainly caused by the climate rather than by the different immigration speed of the trees (de Jong, 1988). The probable reason is that the cold stages were considerably milder than in the Upper Pleistocene, which is documented in the $\partial^{18}O$ values in the MIS curves (see Fig. 3.3-1). The trees did not have to retreat to far distant areas and there were less often difficult conditions for individual species, but refugia were relatively close and many species remained in them together.

In the following, the Rhine system is described first, because its names were introduced into the international literature earliest. Afterwards the Lieth series will be described.

The type locality for the transition from the Pliocene to the Pleistocene is Meinweg (Fig. 3.4-1), situated in the Maas valley immediately at the German/Dutch border (Zagwijn, 1960). It starts with the final phase of the Tertiary, the **Reuverian**, when deciduous forests prevailed, mixed with conifers. In addition to *Picea*, mixed oak forest trees and *Alnus*, the Reuverian forests contained considerable numbers of trees which are characteristic of the central European Upper Tertiary, such as the conifers *Sequoia, Taxodium, Sciadopitys* and *Tsuga* as well as the dicotyledons *Carya, Pterocarya, Eucommia, Nyssa, Liquidambar* and *Aesculus*. Additional taxa had already been recorded earlier by means of macroremains at the Reuverian type locality (Reid and Reid, 1915), of which the most important are *Actinidia, Magnolia, Meliosma, Phellodendron* and *Styrax*. This still thermophilous vegetation was followed by a considerable cooling which resulted in the disappearance of the Tertiary plants and a strong decrease in the members of mixed oak forests, all replaced by the expansion of *Pinus*. The end of the pollen curves of most of the Tertiary trees also mark the end of the Tertiary itself.

The start of the Quaternary – the Pleistocene – is marked by the continuation of the climatic deterioration during which *Pinus* now decreased considerably and an open pine forest developed, where dwarf-shrub heaths, dominated by Ericales, played an important role. This first cold stage is called the **Praetiglian**: however, apart from single pollen grains of *Sanguisorba minor* and *Artemisia* typical glacial plants are missing, which indicates that its climate was not yet subarctic but cold.

After this cold stage the **Tiglian** follows. This is a very long period, in absolute age around 600 000 years and extends until the top of the Olduvai palaeomagnetic zone at 1.8 Ma. The Tiglian (Tegelen, which linguistically means clay) type area is situated in the Dutch province of Limburg. The most important profiles are located in Tegelen, the type section, and Eindhoven and were investigated by Zagwijn (1963), the Maalbeek pit (Fig. 3.4-2) by Westerhoff et al. (1998) and some additional ones from the youngest part of the sequence and further to the west by Kasse (1988), all situated in the lowlands, filled up by deposits of the Rhine system. Due to the very long duration of the Tiglian, the pollen profiles represent only fragments of this period.

As a whole, the Tiglian was initially regarded as a warm stage but now the available palynological data, although still scarce, show that it is a very complex unit with alternations of warm, temperate and cool phases. Zagwijn (1963, 1975) divided it into three phases: A with a temperate climate, B had a cool character but is known only very fragmentarily, while C was very long and is subdivided into six pollen zones of which C3 and C5 were warm and C4 and C6 cool. Nevertheless Zagwijn (1963) regarded the Tiglian to be "still within limits that would permit its inclusion within one interglacial stage, not uniform in climate but with fluctuations from boreal to warm temperate". This status is still kept in the Pleistocene chronology of most authors.

In the meantime, however, it turned out that at least in one of the cool phases of the Rhine system subarctic conditions prevailed. Kasse (1988) described periglacial structures including frost cracks and small ice wedges from Beerse close to Turnhout in north-east Belgium that had to be assigned to Tiglian C4. This is supported by pollen of glacial plants and, therefore, this was called the Beerse glacial. As can be seen from the strong setbacks in the Marine Isotopic Stages, with 35 oscillations alone within the Tiglian, still more subarctic phases can be expected in this period. The time resolution in the pollen diagrams is very low in particular in comparison with the very long duration of the period and the profiles consist of clay with some sand layers so that hiatuses are also common.

A main feature of the floristic development in the Tiglian is the absence over large parts of Europe of important Tertiary species which were still frequent in the Reuverian. Several of them disappeared at once in the Praetiglian, while *Sequoia, Taxodium, Liquidambar* or *Nyssa* returned in small quantities and became extinct in the course of the Tiglian. Others survived longer (see Tab. 3.3-1).

The vegetation of Tiglian A is notably different from the younger Tiglian phases. Only here is *Fagus* (which is rare in the Lower Pleistocene) encountered, being frequently present on dry ground accompanied by the species of the mixed oak forest and *Pinus*. The forests of wet areas were composed of *Alnus* together with *Pterocarya* and *Carya* but with some *Taxodium* still present. Recently, further south in the Rhine region, an interglacial has been found in a deep core in Schifferstadt at Ludwigshafen which, in addition to *Sciadopitys* and *Taxodium*, has *Fagus* values up to 18%, and was assigned to Tiglian A (Knipping, 2008). Only at Maalbeek are higher percentages of *Fagus* (up to 40%) found. Tiglian B is a cool phase, where heathland and pine forests prevailed.

The longest part of the Tiglian is Tiglian C which is also known only fragmentarily. The available pollen data point to forestation for most of the period, apart from the subarctic C4. The dry ground was occupied by mixed oak forest and pine-spruce forest but accompanied by trees that are exotic or southern European nowadays such as *Eucommia, Ostrya* and *Magnolia* as well as the climbers *Actinidia* and *Parthenocissus*. Wet land or river bank forests were also widespread elements in the environment and were first of all settled by *Pterocarya*, the pollen of which reaches up to 45% in C5 in Tegelen, and *Alnus*. In these forests *Vitis* was also common.

Deposits of the Praetiglian and the Tiglian also occur in the adjoining German territory along the lower Rhine and were investigated by Urban (1978), who called them Frechen cold and warm phases. The floristic composition is similar to the Dutch sites with *Fagus* at the beginning of the sequence as well as *Nyssa, Aesculus, Parthenocissus*, large amounts of *Pterocarya* and, in particular, *Eucommia* (up to 48.5%).

The cold stage following the Tiglian is called **Eburonian** in the Rhine region. As in the case of the Tiglian, there are only very few profiles where parts can be assigned to the Eburonian by means of stratigraphy and supported by pollen analysis. The longest sequences are from Eindhoven (Zagwijn, 1963) and Appelenberg (Kasse, 1988). The pollen diagrams show that most of the parts detected were forested with boreal forests, where *Pinus* and some *Picea* prevailed. In the upper parts *Quercus* and other temperate species such as *Corylus* and *Alnus* were also present. In the middle parts, open grass vegetation together with *Artemisia* is indicated. A zonation of the Eburonian period, has been attempted but is not really possible, because the available pollen data are too few and there are inevitably gaps in the profiles. The longest diagrams consist of 18 to 22 pollen samples and have to be compared with an estimated duration of the Eburonian between 0.2 and 0.5 Ma: this means a very low time resolution. Like the Tiglian the Eburonian has to be regarded as a very complex period, in which, according to the evidence from the Marine Isotopic Stages, there were several cold and warm or temperate stages. Neither the absolute age nor the relation to the MIS are known so far.

The Eburonian is followed in the Rhine region by the **Waalian**, which is regarded as a long warm stage. This period, too, is still little known. The best profiles in the lower Rhine region are from Leerdam (Fig. 3.4-3, from Zagwijn and de Jong, 1984), Eindhoven (Zagwijn, 1963) and Gilze (Kasse, 1988), while another one from a deep coring in the Upper Rhine region close to Heidelberg has been analysed by Hahne et al. (2008).

The sites in the Lower Rhine region were obviously situated in mires or wetland areas as is shown by the pollen data. Grasses (probably mainly *Phragmites*) and sedges are frequent and the forest vegetation is dominated by trees of this environment: *Alnus* and *Pterocarya* as well as *Pinus* that may have partially also grown there. The dry ground was mainly occupied by mixed oak forest, in which especially *Quercus*, but also *Ulmus* and *Corylus* were common. Of the Tertiary species in addition to *Pterocarya, Tsuga* and *Eucommia* also occurred regularly, as well as the southern European *Ostrya* and *Celtis*, while all the others are rare and may have been redeposited in some cases.

The pollen diagrams of Leerdam and Eindhoven I show two warm-temperate parts, at the beginning and at the end, and in between considerably less thermophilous species and more Ericaceae but also an interruption with sandy deposits. This induced Zagwijn to propose a tripartite zonation of the Waalian into A, B and C. However, considering an estimated duration of the Waalian of 200–400 ka and the extreme low time resolution in the available pollen diagrams, as well as possible hiatuses,

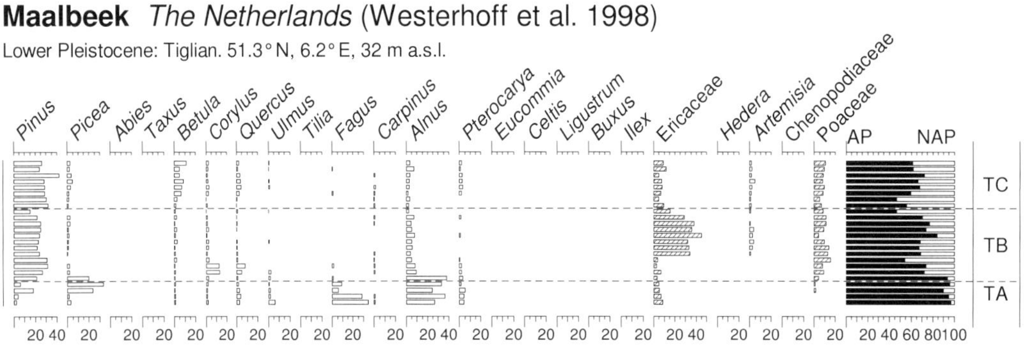

Fig. 3.4-2. Pollen diagram covering Tiglian A–C (beginning of Pleistocene) from Maalbeek, southern Netherlands.

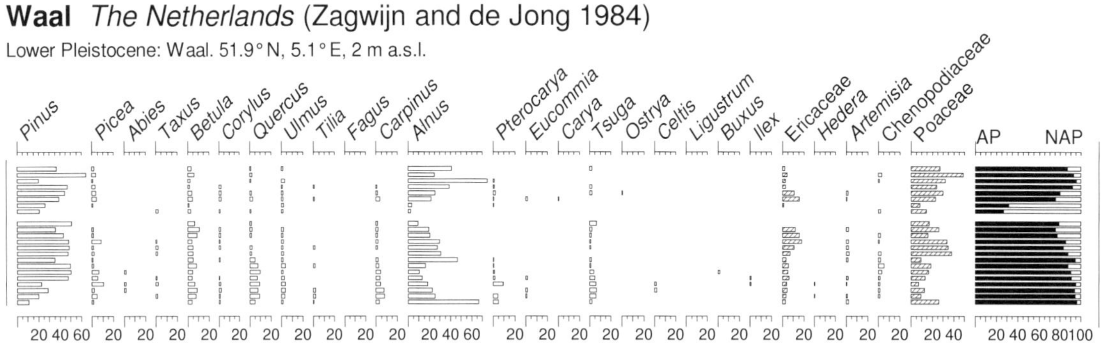

Fig. 3.4-3. Pollen diagram of the Waal interglacial from the long sequence of Leerdam, Netherlands.

this is only a rough division and seems to be a general trend which is so far based on very little evidence. Further investigations will probably reveal more oscillations as can be seen by looking at the MIS curve.

The Waalian was followed by a new distinct cold stage, the **Menapian**, which is also known only fragmentarily. The best profile so far is from Kinderlaan, situated south of Tilburg in the Netherlands (Kasse, 1988). Here the clay of the Waalian is followed by silt and sand that represent at least parts of the Menapian cold stage. The pollen spectra are totally dominated by Poaceae and Cyperaceae, supplemented by some heliophilous herbs, such as *Thalictrum* and *Armeria*. Open tundra vegetation shows cold climatic conditions. Two sites in the Netherlands and Lower Saxony with Scandinavian erratics are interpreted as evidence of the first large ice advance from the north (Ehlers et al., 2004) but this has still to be confirmed.

Up until 1975 the Menapian was regarded as one long cold stage separating the Waalian from the Cromerian. Then several new sites were investigated in the area east of Breda in the Dutch Lower Rhine district and it turned out that, in addition to the first cold stage which kept its name Menapian, two interglacials and another two cold stages were hidden in this period: Bavel, Linge, Leerdam and Dorst, altogether called the **Bavelian** stage (Zagwijn and de Jong, 1984).

As can be seen in the marine isotopic record the intensity of the Pleistocene coolings increased as the Pleistocene progressed and the duration of the cold phases also became longer. This resulted in more distant and more diverse refugia during the cold phases for the plant and animal species and, as a consequence, a differentiation in the immigration of the species during the warm phases. So, from the Bavelian onwards the warm stages show a characteristic succession in the appearance and expansion of the thermophilous trees. Most of the exotic elements characteristic of the Lower Pleistocene warm stages could, however, survive in Europe up until the Leerdam interglacial and *Eucommia* was still present in the first interglacial of the Cromerian.

This may indicate that, during the preceding cold stages these 'exotics' could maintain themselves in Europe and could immigrate once again in the next interglacial. This was no longer the case during the later cold stages of the Middle Pleistocene (de Jong, 1988).

The first warm stage in the Bavelian, the **Bavel** interglacial, is characterized by the last certain occurrence of *Tsuga* in Europe, when it reached up to 30% in the pollen diagrams. In the middle of the interglacial *Eucommia* is also well represented and in the same zone *Carpinus* gains particularly high percentages. In addition to *Alnus*, which together with *Typha*, *Azolla filiculoides*, *Salvinia* and others indicates the wet environment along the Rhine, the trees of the mixed oak forests from the uplands and all the important conifers were also present.

The Bavel interglacial is followed by the **Linge** cold phase. There are up to now only very few pollen spectra from this period, which is dated by its position between the Bavel and Leerdam interglacials. In the Linge cold phase open grass vegetation with many herb species including considerable amounts of *Artemisia* prevailed, accompanied by *Pinus*, *Betula* and *Juniperus* (Zagwijn and de Jong, 1984).

The second interglacial of the Bavelian is the **Leerdam**. The most complete pollen diagram of this period is from the type locality Leerdam itself. It shows a fully developed interglacial, where, apart from *Pinus* and *Picea*, *Quercus*, *Ulmus* and *Corylus* are the main trees of the upland areas. *Carpinus* occurs in smaller percentages, much less than in the preceding interglacial and there are only a very few records of *Tsuga* pollen grains, which may have been redeposited as the sediments consist entirely of clay and sand. *Eucommia*, however, is still present in low quantities. High percentages of Poaceae and Cyperaceae together with *Alnus* and *Pterocarya* indicate the local environment of river-bank and carr forests. *Vitis*, a typical climber in the river-bank forests, is also regularly present.

The end of the Bavelian is formed by the deposits of the **Dorst** cold stage of which only very few pollen spectra are available to date. The strong dominance of Cyperaceae and Poaceae together with some *Pinus* pollen indicates that there must have been a very open grass tundra.

If the assignment of the Bavelian to the Marine Isotopic Stages as shown in Fig. 3.3-1 is correct, no other interglacials can be expected within this time period and the duration of the Bavelian stage can be estimated to around 260 ka.

Following this presentation of the Lower Pleistocene model region along the lower Rhine, other parts of Europe need to be reviewed.

The only long sequence where the Lower Pleistocene interglacials are found in superposition came to light in the late 1960s, when most of the chronological nomenclature was already established. It is the site Lieth near Elmshorn in Schleswig-Holstein/north Germany. There is a sinkhole in the surface of a Zechstein salt dome, which subsided continuously throughout the Lower Pleistocene resulting in the consistent deposition of sand and silt in the cold stages and the formation of organic layers as peat (which later became brown coal) in the warm stages. Exposures of this sequence have been opened by a deep lime pit (Fig. 3.4-4) and several additional corings provided the connections between the different layers. Numerous pollen profiles have been analysed by Menke (1969, 1975, 1980) and the description below follows these publications. Special definitions of the various periods are compiled in Litt et al. (2007).

The Lieth sequence extends down until the Praetiglian, when 30 m of sand and silt were deposited. It is not clear whether the Pliocene was reached here, but the Praetiglian layers can be connected with those of Oldenswort nearby, which are underlain by more than 800 m of Pliocene deposits that end with a well developed Reuverian, so the transition from the Neogene to the Quaternary is confirmed.

The vegetation development from the Reuverian to the Praetiglian at Lieth is very similar to that at the Dutch type locality at Meinweg (Zagwijn, 1960). The thermophilous and Tertiary elements disappear and open tundra vegetation dominated by Ericales and Poaceae and accompanied by some *Pinus* and *Alnus* spp. developed. As in the Netherlands there are no certain indications of subarctic conditions. According to the Marine Isotopic Stages the Praetiglian must have had a long duration (see Fig. 3.3-1).

At Lieth the Praetiglian and the Tiglian could be classified in more detail than in the Rhine region. The Praetiglian can be subdivided into three phases: the first cold stage, described above, Menke called the **Kaltenhörn**, which corresponds to the lower part of the Dutch Praetiglian. High pollen values of grass vegetation with other herbs and dwarf-shrub Ericales together with little tree

Fig. 3.4-4. Exposure of the long Lower Pleistocene sequence of Lieth, northern Germany with several interglacial and glacial layers. (In the foreground the investigator, B. Menke).

pollen point to a boreal-subarctic climate. This was followed by a clear amelioration: the **Warmhörn** that can also be observed in the Meinweg profile in the Netherlands. These temperate climatic conditions are represented by forests with *Pinus, Alnus, Quercus* and so on but also some *Juglans* and this period has to be connected with the middle part of the Dutch Praetiglian. Then followed a cold stage not known so far from the Rhine region, the **Ekholt**. The detailed pollen diagram of this cold stage shows grass tundra with numerous herb species, including *Artemisia, Thalictrum, Selaginella selaginoides* and *S. helvetica* as well as some dwarf-shrub Ericales. The Ekholt cold stage represents the coldest period in the whole Lieth series with clear subarctic to arctic conditions.

The Ekholt cold stage was succeeded by the **Nordende** warm stage which represents a fully developed interglacial with warm-temperate climate during which peat was formed in the depression (Fig. 3.4-5). After a *Pinus* phase the forests were dominated by *Quercus* together with some *Carpinus* and *Ostrya* type. In the younger part *Eucommia* and *Pterocarya* were also common and, as a special Tertiary relic, *Symplocos* is also recorded. Of particular interest is the expansion of the Ericales in the course of the warm stage which together with much *Sphagnum* point to oligotrophic conditions, caused by the sandy soils around the site. The Nordende can be correlated with the first part of the Tiglian, the Tiglian A. In the Rhine region this is characterized by the occurrence of *Fagus*, which is also recorded in Lieth, however, only with single pollen grains.

The subsequent cold period, when mainly silt was deposited, is called the **Krückau**. The vegetation is characterized by more or less treeless tundra, where grasses and other herbs prevailed but the Ericales, mainly of *Empetrum* type, were considerably more frequent than in the preceding Ekholt cold stage. The climatic conditions were subarctic, but not as severe as in the Ekholt.

On top of the Krückau there follows a long period that was first regarded as a forested interstadial but more detailed analyses upgraded it to a full interglacial, the **Ellerhoop** warm stage. On the basis of the pollen evidence, as well as the lithostratigraphy, it has to be divided into two warm parts, one at the beginning and the other at the end with a cooler part in between. The pollen diagrams (Fig. 3.4-6) are dominated by high quantities of various Ericales species which, together with *Sphagnum,* document the development of a raised bog at the site. The tree vegetation around the bog shows a clear immigration sequence starting with birch, followed by a pine zone, and then by the expansion of *Ulmus, Alnus glutinosa* type, *Quercus* and *Ostrya* type. Later *Carpinus, Corylus* and *Eucommia* appeared. While most of this profile consists of peat and some gyttja, in the middle part sand and silt were deposited which contain only few pollen that point to boreal climate conditions. The Ellerhoop is presumably equivalent to the Tiglian C in the Dutch system and the subarctic phase TC4, described there, is probably hidden in the sand layers of Lieth.

Above the Ellerhoop, sand layers were again deposited and these belong to the subsequent **Lieth** cold stage. Only a very few pollen spectra are available which indicate the presence of several glacial species. An important feature is the record of cryoturbations in the sand layers of the Lieth outcrops, pointing to subarctic conditions at that time. Because of the position above the Ellerhoop and below the next warm stage the Lieth cold stage is correlated with the Eburonian in the Rhine region.

The Lieth was followed by the **Tornesch** warm stage (Fig. 3.4-7), which was very long and included several climatic oscillations. Menke subdivided it, like the Ellerhoop, into three parts, A, B and C. The pollen diagrams show that the whole period was dominated by Ericales with average percentages of more than 50%, much of it is of the *Empetrum* type but *Bruckenthalia*, *Calluna* and *Andromeda* were common as well. Together with plenty of *Sphagnum* spores they indicate that poor environmental conditions also existed on the sandy soils around the site. Intercalated gyttja layers contain *Isoëtes, Brasenia* and *Pilularia*, thus confirming their oligotrophic character. In both warm parts, Tornesch A and C, these sands were occupied by well-developed warm-temperate forests. As in the preceding interglacials of the Lieth series a clear immigration sequence of the trees can be observed. After a pine-birch zone the sequence starts with *Quercus,* that became the most important forest constituent, and was soon followed by the appearance of *Ulmus, Alnus glutinosa* type, *Carpinus* and *Corylus.* Exotics such as *Ostrya* type, *Eucommia* and *Pterocarya* were also common. The thermophilous species reached their maximum in Tornesch C when *Vitis* and *Viscum* were also present.

In the cool phase, Tornesch B, the thermophilous trees decreased while pine and birch expanded but there was no distinct deforestation. Another climatic setback took place within the warm stage, Tornesch C, when the thermophilous species disappeared and *Betula* and *Pinus* spread again: there was also some sand deposition in this part.

As a whole, the Tornesch is climatically a very variable complex. It can be correlated with the Waalian in the Rhine region, which is also quite a variable stage. As already mentioned, a comparison with the MIS curves suggests that further oscillations can be expected in this period.

Above the Tornesch stage several metres of sand were deposited, which were in most parts empty of pollen. However, at the beginning and the end the transition phases, another cold stage could be recognized by some pollen spectra containing *Pinus* and *Betula* as well as grasses and other herbs. This cold stage was called **Pinnau** but it is not certain whether clear arctic or subarctic conditions were reached in this period. According to its position it should be correlated with the cold Dutch Menapian.

The following **Uetersen** warm stage (Fig. 3.4-8) is also a period with several climatic oscillations. Two sequences of brown coal and gyttja, formed in warm periods, are separated by an intercalated sand layer, which documents a cool phase. After the first part, where boreal forests with mainly *Pinus* and some *Alnus viridis* type developed, there is a slight climatic deterioration with more open forests and numerous herb species followed by the main part which represents the climatic optimum. Here mixed oak forests developed but they were still rich in pine. Exotics or southern European trees, such as *Pterocarya, Castanea* and *Juglans*, occur in traces only, which is much less than in the preceding interglacials. It is not sure, whether subarctic or arctic conditions were reached in the cool phase because they may be hidden in the sand layer. The area around the site was still very poor in nutrients, which is indicated by the high values of *Isoëtes* and the occurrence of *Pilularia* in the gyttja layers.

The Uetersen warm stage has to be correlated with the Bavel interglacial in the Rhine system. It was followed by sand and silt layers, the pollen content of which points to an open forest-tundra. This cold stage was called the **Elmshorn** and is overlain by the last brown coal layer, which was formed during the **Pinneberg** warm stage.

In the optimum of the Pinneberg, mixed oak forest species were present but only in low quanti-

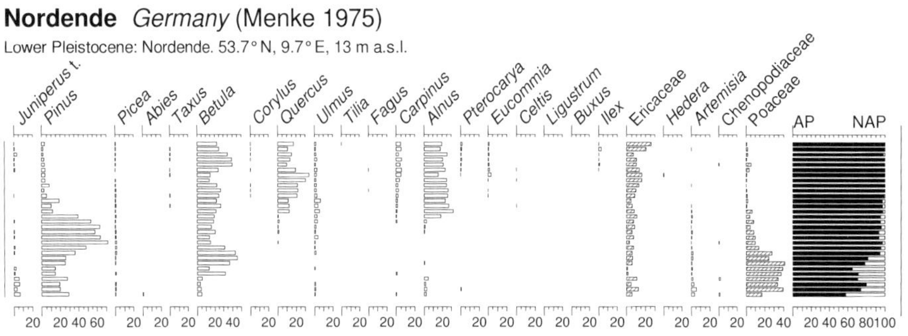

Fig. 3.4-5. Pollen diagram of the Nordende interglacial at the beginning of the Lieth series, northern Germany (type section).

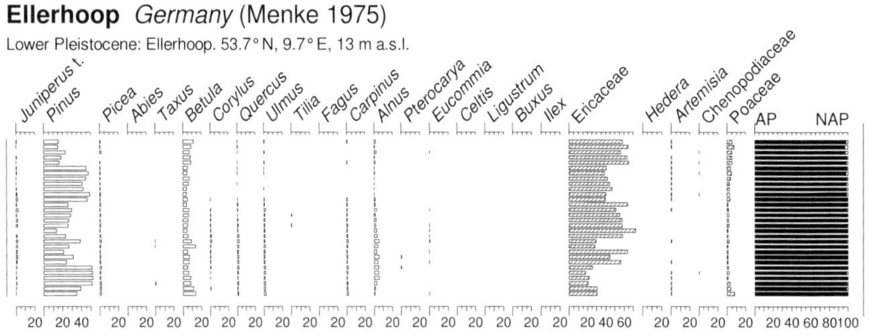

Fig. 3.4-6. Pollen diagram of the Ellerhoop interglacial in the Lieth series, northern Germany (type section).

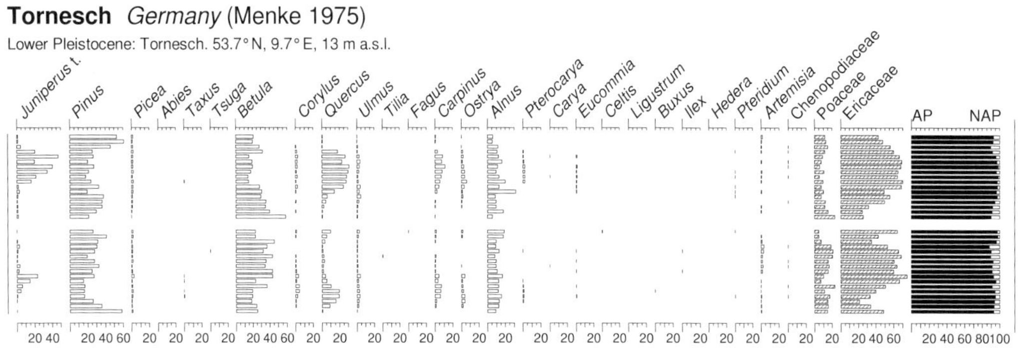

Fig. 3.4-7. Pollen diagram of the Tornesch interglacial in the Lieth series, northern Germany (type section).

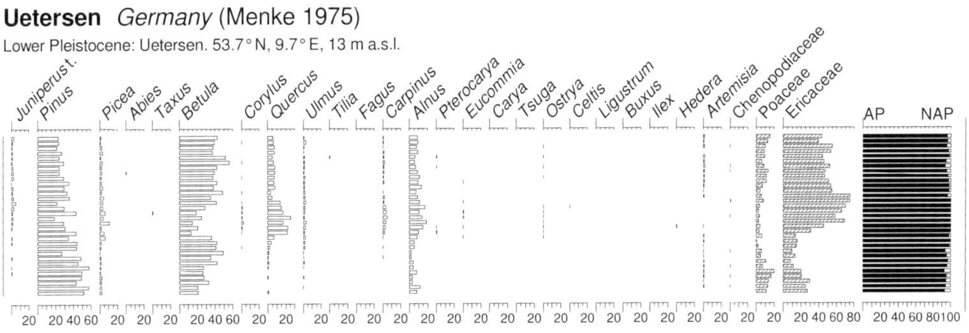

Fig. 3.4-8. Pollen diagram of the Uetersen interglacial in the Lieth series, northern Germany (type section).

ties. The main trees were pine and birch and this has to be seen in connection with the high values of Ericales pollen, consisting predominantly of *Empetrum*. In the younger part of the stage Ericales reach average percentages of around 70% and reflect the poor nutrient conditions on the bog itself as well as in the sandy surroundings. This must have hampered the spread of more demanding species. Exotic elements were rare in this stage although *Ostrya* type and *Eucommia* were still present. *Pterocarya* is missing completely.

Only the sites in the Rhine region and the sequence of Lieth, both described above, cover the whole of the Lower Pleistocene. In addition to these there are some isolated localities of limited temporal range that show only small parts of this long period. Although the distance between the lower Rhine and Lieth is only around 400 km their vegetation was very different because they show totally diverse environments and pollen catchment areas.

Because both the Rhine region and Lieth series are indispensable for the Lower Pleistocene vegetation history, their advantages and disadvantages have to be discussed and assessed.

The pollen profiles from the Rhine region in most cases document fluvial environments along the Rhine and Maas rivers. Although there are some upland forest species, trees from the flood plains and river banks, as well as from mire like areas, dominate the pollen diagrams. These are mainly represented by *Alnus, Pterocarya* and *Salix* but also some *Quercus, Ulmus* and *Fraxinus* belong to this group in that they occupy the upper parts of the river banks. Together with the herb vegetation the trees indicate that eutrophic conditions prevailed along the rivers. This is in accordance with the sediments of the sites, which mainly consist of clay and silt. Another important point is that a certain proportion of the pollen content in these deposits is contributed by the rivers themselves coming from the huge catchment areas of the Rhine and Maas and their tributaries. Therefore the pollen diagrams of these sites show a mixture from local to far distant pollen sources. This is a great advantage because, in that way, the general development of the vegetation for a large part of central Europe is reflected in the diagrams.

The main disadvantages of the sites in the Rhine region are, however, the lack of continuous sequences. There are many sites, but only occasionally do a few warm and cold stages appear in superposition so that the correlation between sites is sometimes difficult. Furthermore, the minerogenic sediments are frequently interrupted by layers of very coarse material, which means that it is often difficult to recognize gaps in the profiles.

The contrast between both investigation areas is great in many ways. The continuous and almost uninterrupted sequence at Lieth is unique and provides an indisputable record of the order of the alternating warm and cold stages. The open outcrops of the site offered excellent possibilities for stratigraphic observations as well as for sampling and the densely spaced pollen analyses provided a good time resolution. The clastic material deposited in the sinkhole during the cold phases derives from around the site and the peat layers of the warm stages were formed in the sinkhole itself. This means that far distance transport of pollen by water can be excluded and that the pollen assemblages are of local to regional origin.

The pollen catchment area around Lieth remained the same for about one million years and the vegetation within this area was characteristic of the poor sandy soils that covered the preceding Tertiary surface and had been leached further during the course of the Lower Pleistocene. These more or less oligotrophic conditions are registered by frequent aquatics such as *Isoëtes* or *Pilularia* and the continuous expansion of heathlands, as can be seen in the diagrams. The tree vegetation was adapted to the nutrient conditions, so *Pinus* and *Betula* also prevailed in the warm stages, not for climatic reasons but for edaphic ones. *Quercus* was also common and formed acid oak forests, while more demanding species occurred less frequently or were rare. The pollen diagrams show a rather local vegetation and therefore the correlation of this sequence with contemporaneous diagrams from the Rhine region causes some difficulties. This is even more complicated because predominantly eutrophic environments have to be compared with oligotrophic ones. Differences in the vegetation picture of the two regions are foremostly caused by factors of the environment and the catchment area rather than by climate. However, a typical immigration sequence of the tree species can be observed in Lieth from as early as the Nordende warm stage. In the Rhine region such a sequence is only observable from the Bavel warm stage onwards. This probably has to be ascribed to the longer migration distance to Schleswig-Holstein.

The correlation between the various stages of the lower Rhine and the Lieth series is made on the basis of the occurrence of some exotic species, some general similarities and, with reservations, also on the stratigraphical position. The question remains, however, whether the series of cold and warm phases known so far from the Lower Pleistocene is complete. Between the Pliocene/Pleistocene boundary and the Dorst cold stage, the marine isotopic record shows not less than 81 distinct cold and warm phases (see Fig. 3.3-1). Most of them, if not all, should have led to strong reactions in the vegetation of Europe. This means that quite a number of additional oscillations still have to be expected, oscillations that are hidden in probable hiatuses in both series. Because up until now there are no absolute ages, the discovery of any new stages may lead to changes in the correlation proposed above.

Apart from the key areas along the lower Rhine and at Lieth there are only a few sites with a long sequence of Early Pleistocene vegetation history. The most important one is Leffe, a well known site in Italy. Even older are a number of sites in Italy and other parts of the Mediterranean, where shorter sequences from the Gelasian (i.e. from 2.6 to 1.8 Ma) have been investigated (Bertini, 2010). Several climatic cycles, mainly of a duration of 40 ka have been observed. In southern Italy these early warm phases are marked by open vegetation (dry conditions) and the cold phases by forests (humid conditions). In northern Italy the scenario is different and the climatic cycles are characterized by the contrast between subtropical to warm temperate forests and higher-elevation coniferous forests (mainly *Picea*) without any considerable steppe-like vegetation.

The long sequence of Leffe, which is situated at the southern margin of the Lombardian Pre-Alps in Italy, starts in the middle part of the Lower Pleistocene. The **Leffe** basin contains a very long continuous archive with a duration of almost one million years from the late Pliocene to the late Lower Pleistocene including a continuous biogenic deposit that was formed over 700 ka from 1.8 to 1.1 Ma (c. MIS 60-32). Pollen as well as botanical macroremains are well preserved and there is also a rich faunal assemblage. A unit 50 m thick encompasses twelve climatic cycles of, on average, 40 ka in length. The vegetation shows that the amplitude of the climatic oscillations was moderate, as could be expected from the marine isotope record. There is a cyclic alternation of conifer-dominated phases of cold, temperate climate and broadleaved dominated phases characterising warm temperate climates. The main constituents of the forests were *Pinus, Picea, Carya, Carpinus* and *Pterocarya* but *Eucommia, Juglans, Fagus* and *Vitis* were also present all through the period (Ravazzi and Moscariello, 1998; Ravazzi, 2003; Ravazzi et al., 2009). Figure 3.4-9 shows the vegetation development across seven of these cycles.

Pollen diagrams of similar climatic cycles were presented by Fusco (2007) from marine deposits in the **Lamone** valley about 230 km south-east of Leffe. They were tentatively correlated with MIS from 64 to 46 (with gaps). From these fragmentary records a complete theoretical vegetation cycle was proposed to consist of a temporal succession of four main vegetation formations: mixed-deciduous forests (dominated by *Quercus*), followed by Juglandaceae forests (dominated by *Carya* and indicating the climatic optimum) which were replaced by submountain-

Leffe *Italy* (Ravazzi 2003)

Lower Pleistocene: seven glacial-interglacial cycles. 45.8° N, 9.9° E, 455 m a.s.l.

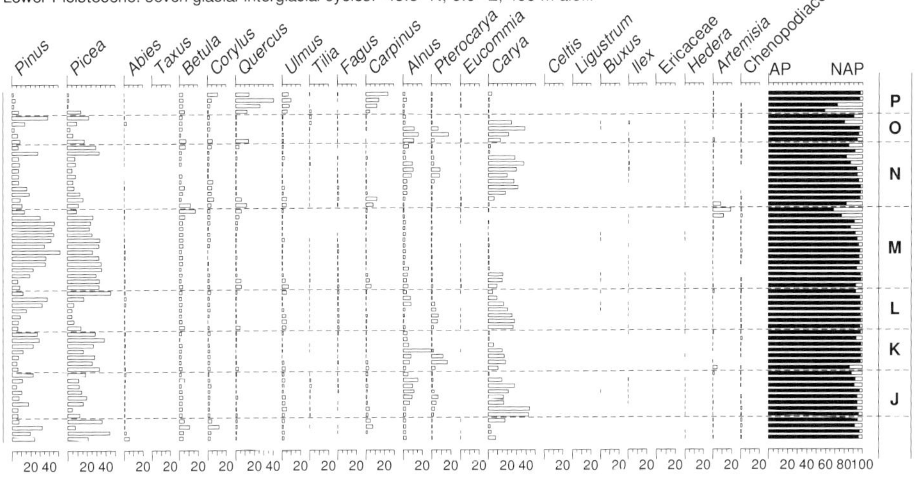

Fig. 3.4-9. Pollen diagram of seven glacial-interglacial cycles in the long Lower Pleistocene sequence of Leffe, pre-Alps, Italy.

Fornaci di Ranica *Italy* (Ravazzi 2003; Ravazzi et al. 2005)

Lower Pleistocene: Bavelian. 45.7° N, 9.7° E, 310 m a.s.l.

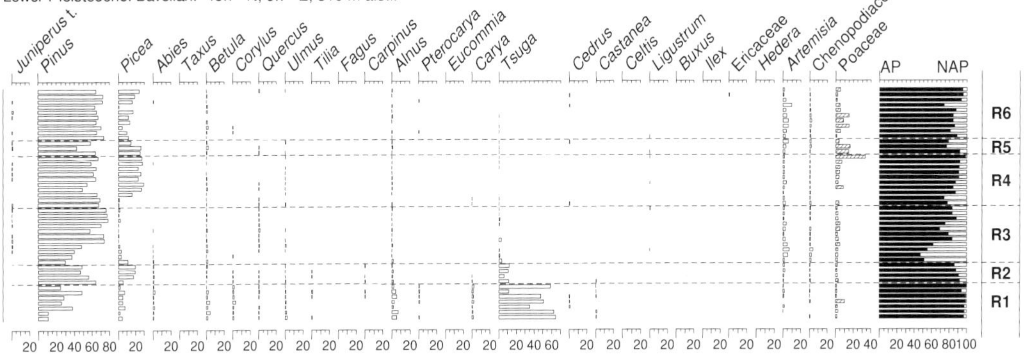

Fig. 3.4-10. Pollen diagram of the end of the Lower Pleistocene (Bavelian) showing a warm stage (R1 + R2) followed by a cold stage (from R3) at Fornaci di Ranica, pre-Alps, Italy.

Tenaghi Philippon *Greece* (van der Wiel and Wijmstra 1987)

Lower Pleistocene. 41.0° N, 24.3° E, 40 m a.s.l.

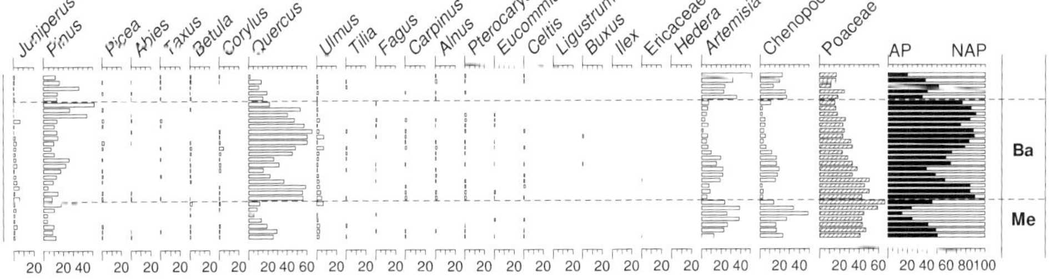

Fig. 3.4-11. Pollen diagram of the end of the Menapian (Me) cold stage followed by the Bavelian (Ba) warm stage and another cold stage from Tenaghi Philippon, northern Greece.

Table 3.4-1. Important Pre-Eemian Quaternary sites in Europe.

		Lower Pleistocene	Cromerian	Holsteinian	Long sequences	Literature
Northern Europe and Central European Lowland						
1	Gort (IR)			Ho		Jessen et al., 1959
2	Cromer Forest Bed (GB)		Cr			Reid, 1890; West, 1980
3	Hoxne (GB)			Ho		West, 1956
4	Marks Tey (GB)			Ho		Turner, 1970
5	Hoogeveen (NL)			WD		Zagwijn, 1973
6	Leerdam (NL)	LP				Zagwijn and de Jong, 1984
7	Waardenburg (NL)		Cr			Zagwijn et al., 1971
8	Bavel (NL)	LP				Zagwijn and de Jong, 1984
9	Gilze (NL)	LP				Kasse, 1988
10	Eindhoven (NL)	LP				Zagwijn, 1963
11	Tegelen (NL)	LP				Zagwijn, 1963, 1975
12	Reuver (NL)	LP				Zagwijn, 1960
13	Maalbeek (NL)	LP				Westerhoff et al., 1998
14	Meinweg (NL)	LP				Zagwijn, 1957
15	Kärlich (D)		Cr			Bittmann, 1992, 1996
16	Döttingen (D)			Ho		Diehl, 2007
17	Bilshausen (D)			Cr		Müller, 1965
18	Osterholz (D)		Cr			Grüger, 1967
19	Schöningen (D)			Ho,WD		Bittmann, 2012; Urban and Sierralta, 2012
20	Dethlingen (D)			Ho		Koutsodendris et al., 2010
21	Munster-Breloh (D)			Ho		Müller, 1974b
22	Bossel (D)			Ho		Müller and Höfle, 1994; Geyh and Müller, 2005
23	Gorleben (D)		Cr		x	Müller, 1992
24	Hamburg-Dockenhuden (D)			Ho		Linke and Hallik, 1993
25	Wacken (D)			WD		Menke, 1968
26	Lieth (D)	LP			x	Menke, 1969, 1975
27	Dömnitz/Pritzwalk (D)			Ho,WD		Erd, 1973
28	Olgod (DK)		Cr			Andersen, 1965
29	Harreskov (DK)		Cr			Andersen, 1965
30	Vejlby (DK)			Ho		Andersen, 1965
31	Nowiny Zukowskie (PL)			Ho		Hrynowiecka-Czmielewska, 2010
32	Brus (PL)			Ho		Pidek, 2003
33	Zdany (PL)		Cr			Pidek, 2003
34	Ossówka (PL)			Ho		Nitychoruk et al., 2005
35	Ferdinandów (PL)		Cr			Jancyk-Kopikowa, 1963
36	Butenai (LT)			Ho		Kondratiene, 1996
37	Strante (LV)			Ho		Kalnina, 2001
38	Ulmale (LV)			Ho		Kalnina, 2001
Southwestern and southern Central Europe including the Alps						
39	Velay craters (F)			Ho,WD	x	de Beaulieu and Reille, 1992b; Reille et al., 2000
40	Schifferstadt (D)	LP				Knipping, 2008
41	Thalgut (CH)			Ho		Drescher-Schneider, 2000a
42	Meikirch (CH)			WD		Welten, 1982a; Preusser et al., 2005
43	Samerberg (D)			Ho,WD	x	Grüger, 1979, 1983
Southern Europe						
44	Leffe (I)	LP			x	Ravazzi, 2003
45	Pianico-Sèllere (I)		Cr		x	Ravazzi, 2003; Brauer et al. 2007
46	Ranica (I)	LP				Ravazzi, 2003; Ravazzi et al., 2005
47	Lamone (I)	LP				Fusco, 2007
48	Valle di Castiglione (I)			Ho,WD	x	Follieri et al., 1988
49	Valle di Manche (I)	LP	Cr			Capraro et al., 2005
50	Ioannina (GR)		Cr	Ho,WD	x	Tzedakis, 1994
51	Tenaghi Philippon (GR)	LP	Cr	Ho,WD	x	van der Wiel and Wijmstra, 1987a,b; Tzedakis et al., 2006

ous and mountainous forests (dominated by *Picea* together with *Tsuga*) and finally by open vegetation.

Another important site of the late Lower Pleistocene in the Italian pre-Alps is represented by the lacustrine deposits of **Fornaci di Ranica** (Ravazzi et al., 2005). They span the final stage of a temperate period and the subsequent transition to a cold phase, followed by a slow climatic improvement during the palaeomagnetic Jaramillo subchron (MIS 30–28, in the Bavelian). Figure 3.4-10 shows high pollen maxima of *Tsuga*, followed by a marked forest withdrawal caused by a considerable cooling and leading to a *Larix*-taiga parkland with *Salix, Hippophaë, Juniperus* and considerable amounts of herbs, including cold-adapted taxa such as *Saxifraga oppositifolia* type and *S. stellaris* type in addition to *Larix* itself. Later a re-establishment of mixed conifer forests took place. This fossil plant record thus provides the first evidence of mean summer temperatures that were near or below the thermal tree limit for the region south of the Alps.

From the end of the Lower Pleistocene there are several palynological records from southern Italy (compiled in Bertini, 2010). At these sites the climatic fluctuations were already more distinct than in the older ones, which corresponds with the increasing climate amplitude in the MIS curves. In the warm phases, temperate forests prevailed with mainly

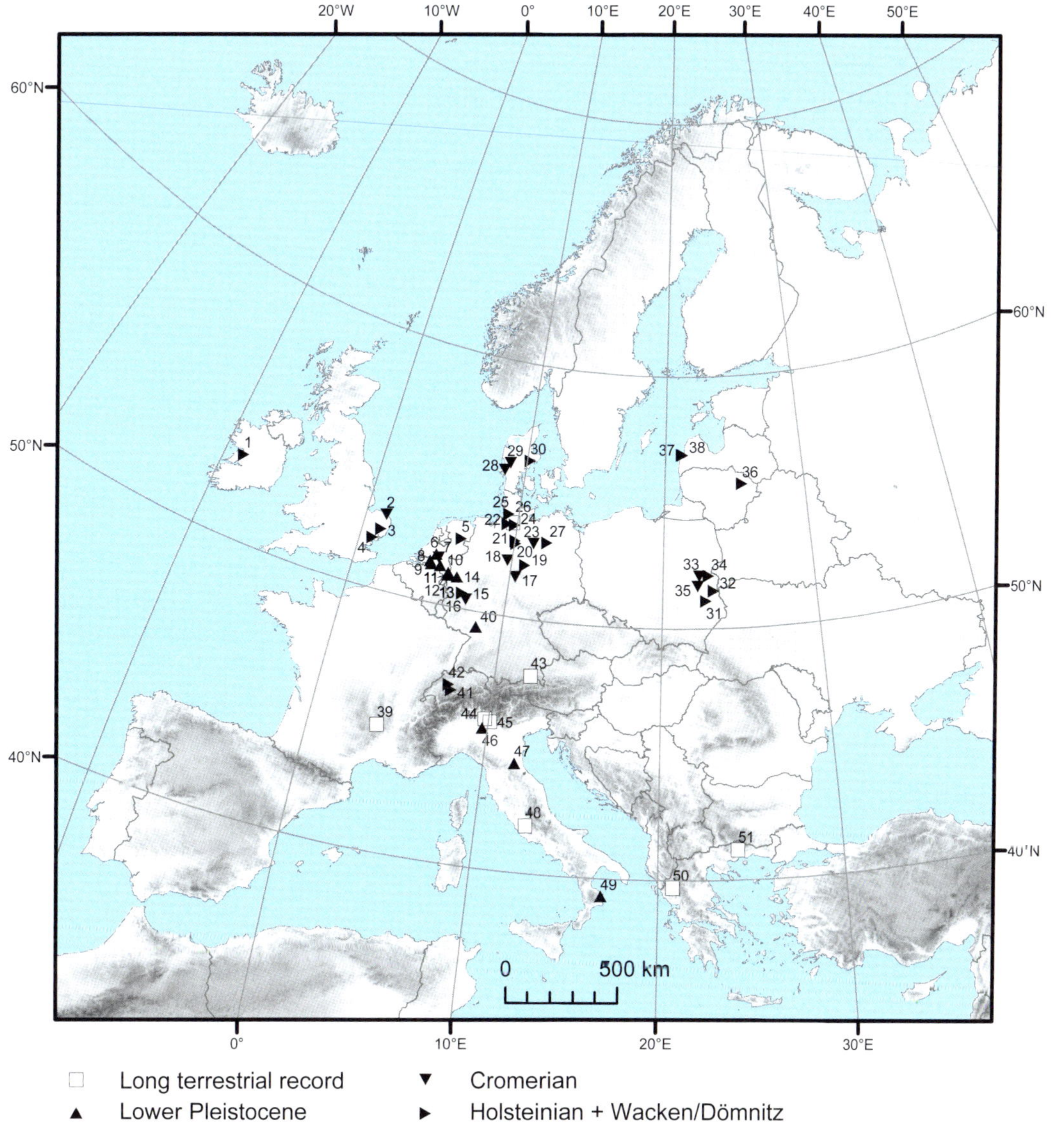

Fig. 3.4-12. Location of important pre-Eemian Quaternary sites in Europe. Numbers correspond to Table 3.4-1.

deciduous taxa such as *Quercus*, *Carya*, *Carpinus* or *Zelkova* while the most warmth-demanding taxa such as Taxodiaceae gradually disappeared. These forests alternated with open vegetation that was dry but still warm. It was dominated by many steppe taxa, in particular *Artemisia* and *Ephedra* but thermophilous taxa such as *Cistus* and *Phlomis fruticosa* were also present.

In the late Early Pleistocene localities of northern Italy (for example Leffe or the marine Lamone) cold phases are often characterised by expansions of conifer forests, mainly formed by *Picea* and containing some *Cedrus* and *Tsuga*, while in the warm phases thermophilous trees occupied the region without any considerable occurrence of steppe vegetation.

The most important terrestrial sequence in Europe has been retrieved from the basin of **Tenaghi Philippon** in Greek Macedonia. According to the revised chronostratigraphy of Tzedakis et al. (2006) it extends back to 1.35 Ma and thus includes the upper part of the Early Pleistocene. Figure 3.4-11 shows the end of the Menapian 'glacial' and the Bavel 'interglacial' (in the Duch terminology). At Tenaghi Philippon a continuous series of alternating vegetation periods could be detected, which are in close correspondence with the marine isotopic stages starting from MIS 43. The 'interglacial' phases were forested mainly by *Pinus* and *Quercus* and, in the lowermost one, notable amounts of *Tsuga*, *Castanea* and *Celtis* were also recorded, while the 'glacial' phases show AP minima with NAP values during MIS 42, 40 and 38 exceeding 90%, with *Artemisia* and Chenopodiaceae at over 80%, indicating extreme aridity (van der Wiel and Wijmstra, 1987a).

3.5

The Cromerian

As early as the 19th century several exposed sections from sites along the coast of Norfolk and Suffolk in England were described by Reid (1890) as the Cromer Forest Bed series. They are very rich in macroscopic plant remains and were first regarded as Pliocene but later it was found that they are Middle Pleistocene in age. They are situated below the Anglian (Elsterian) till. Still later the monograph of West (1980) showed that there are two warm periods at the type locality, West Runton: the older Pastonian and the younger Cromerian interglacial. In the meantime numerous interglacial deposits all over Europe, most of them only fragmentary, were included in the Cromerian and it became apparent that this period was a long stage that contains several interglacials belonging to the early Middle Pleistocene.

The geological boundary between the Lower and Middle Pleistocene has been linked to the Brunhes/Matuyama palaeomagnetic boundary at 0.781 Ma BP, which is in late MIS 19. According to biostratigraphical criteria, however, the onset of the Cromerian complex is defined slightly earlier to the start of the interglacial that corresponds with MIS 19.

A first subdivision of the Cromerian was made by Zagwijn et al. (1971) for the sites in the Netherlands, which he grouped into interglacials numbered I–IV. Since then quite a number of new and sometimes complete profiles have been discovered in Europe but their correlation is often very difficult. It became clear, however, that the interglacials I–IV were followed by another interglacial, V (called the Rhume-interglacial) with important sites at Bilshausen (Fig. 3.5-1; Müller, 1965) and Kärlich (Bittmann, 1992) in north Germany and Praclaux (Reille et al., 2000) in south-eastern France (MIS 11). Countings of annually laminated sediments at Bilshausen provided a duration for this interglacial of 25 000 years, extended by a Late-glacial period of 8400 years (Geyh and Müller, 2007).

Up until now there is only one long sequence north of the Alps where several of these interglacials occur in superposition. This is the case in the long corings at Gorleben/Lower Saxony, where four Cromerian interglacials above the Bavelian stage were encountered (Müller, 1992). Much longer is the well-known sequence of Tenaghi Philippon in Greece, with numerous interglacials: its duration was recently revised to be 1.35 million years (Tzedakis et al., 2006) but a detailed correlation with the northern regions is not yet consolidated.

Table 3.5-1 shows the current correlation of the major sites of central and north-western Europe that have been included in the Cromerian. One of the Cromerian cold phases shows evidence of far reaching glaciations, described as south Polish and Don glaciations which probably occurred in MIS 18, that is after the first Cromerian interglacial. Another ear-

Table 3.5-1. Subdivision of the Cromerian and position of important interglacial sites.

Interglacials	Glacials	Britain and Denmark	Netherlands	Germany	Poland	France	Others	Stages (MIS)
Holsteinian		Hoxnian/Holsteinian	Holsteinian	Holsteinian	Masovian	(Landos)	Lichwin	9
	Elsterian	Anglian/Elsterian	Elsterian	Elsterian	Sanian 2			10
V				Bilshausen, Kärlich		Praclaux		11
	D							12
IV		Cromer: Ostend	Noordbergum	Thermomer Gorleben				13
	C							14
III			Rosmalen	Thermomer Gorleben	Ferdinandow 2		Schklovsk 2	15
	B							16
II		Cromer: West Runton Olgod, Harreskov	Westerhoven	Hunteburg, Morsleben	Ferdinandow 1 Malopolanian		Schklovsk 1	17
	A				Sanian 1		Don Glaciation	18
I			Waardenburg	Osterholz	Augustovian 2		Pianico-Sellere?	19

ly glaciation in Denmark may perhaps also be placed here. A special period is the Ferdinandowian (Zdany, Fig.3.5-2) of which quite a number of sites have been detected, mainly in south-east Poland. They show two climatic optima, characterised by broadleaved trees (the lower one with a continuous *Celtis* curve), separated by a moderately cold phase with *Pinus, Betula* and also some *Larix* (Rzechowski, 1996; Pidek, 2003, 2015; Pidek and Malek, 2010). The stratigraphic position of this bipartite Ferdinandowian is still disputed: it probably correlates with Cromerian II and III. Quite a number of Cromerian sites from different interglacials are also found in Lithuania. Most of these, however, are only fragmentary and their correlation with other European deposits is difficult (Kondratiene, 1996b).

The main biostratigraphic difference in comparison with the Lower Pleistocene is the disappearance of specific Tertiary species. Only *Eucommia*, the gutta-percha tree, which is nowadays restricted to China, was still common in the first Cromerian interglacial, where it reached 13% in Osterholz (Fig. 3.5-3; Grüger, 1967) and 4% in Waardenburg (Zagwijn, 1996b). Small amounts of *Tsuga* were recorded even later in the Gorleben interglacials (Müller, 1992) but may have been redeposited. In south Europe this disappearance of species occurred more gradually. *Eucommia* survived longer and could be traced in Tenaghi Philippon in small quantities until the middle Cromerian (van der Wiel and Wijmstra, 1987b) and in the south Italian site Valle di Manche even *Carya* still occurred regularly in the first Cromerian interglacial (Capraro et al., 2005). Several Mediterranean species, such as *Celtis*, existed in Europe throughout the Cromerian but disappeared afterwards north of the Alps. During the cold phases with no or little tundra vegetation in the north, steppe vegetation prevailed in the Mediterranean dominated by *Artemisia*.

The plant-rich Cromerian sites in Britain also have *Eucommia* and *Tsuga* in the oldest (Pastonian) phase, as well as *Azolla filiculoides, Salvinia natans, Trapa natans* and other species which continue to be present until the late Cromerian, but which did not reach Britain in the postglacial period (West, 1980).

Trees such as *Pinus, Alnus* and *Picea* and most species of the *Quercetum mixtum* were the main constituents of the forests in all the Cromerian interglacials north of the Alps, while, at least according to our present knowledge, other taxa were often conspicuously lacking. The latter is the case with *Taxus* and *Abies*, which are lacking at most sites in Cromerian I and III. The occurrence of *Carpinus* is rather peculiar as it is quite common in most of these interglacials but completely lacking in middle Cromerian localities such as the lower warm phase of the Ferdinandowian in eastern Poland (but also elsewhere). During Cromerian II, *Abies* and *Carpinus* did not reach the Danish sites of Harreskov (Fig. 3.5-4) and Olgod (Andersen, 1965), which are correlated to it and occurred only scarcely in Dutch and northern German localities, probably caused by delayed immigration. It is remarkable that *Fagus* is

Bilshausen *Germany* (Müller 1965)

Middle Pleistocene: Cromerian. 51.7° N, 10.2° E, 150 m a.s.l.

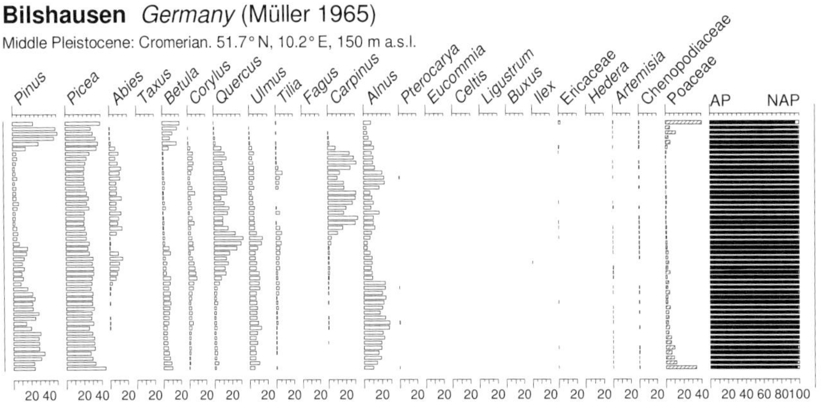

Fig. 3.5-1. Pollen diagram of the youngest Cromerian, the Rhume interglacial, from Bilshausen, northern Germany.

Zdany *Poland* (Pidek 2003)

Middle Pleistocene: Cromerian. 52.1° N, 22.4° E, 175 m a.s.l.

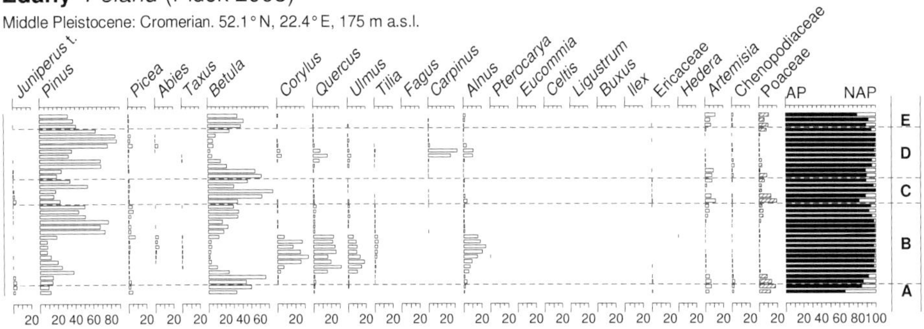

Fig. 3.5-2. Pollen diagram of the Ferdinandowian in the lower Cromerian with two warm phases from Zdany, eastern Poland.

Osterholz *Germany* (Grüger 1967)

Middle Pleistocene: Cromerian. 52.1° N, 9.8° E, 130 m a.s.l.

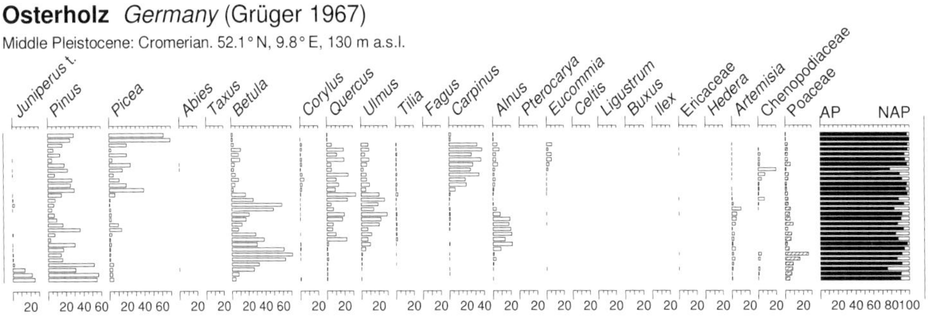

Fig. 3.5-3. Pollen diagram of the first Cromerian interglacial from the type section Osterholz, northern Germany.

Harreskov *Denmark* (Andersen 1965)

Middle Pleistocene: Cromerian. 56.1° N, 9.0° E, 51 m a.s.l.

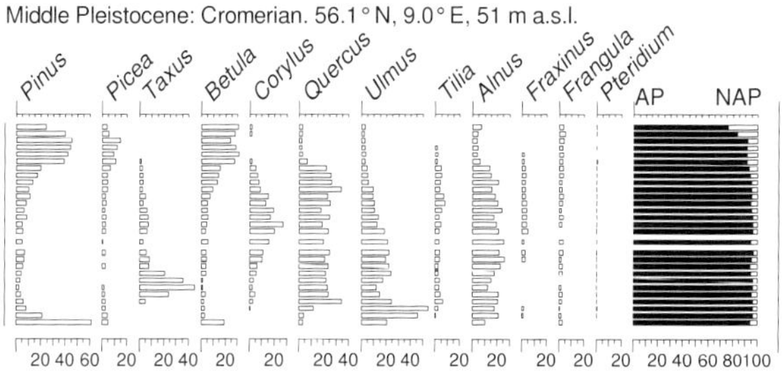

Fig. 3.5-4. Pollen diagram of the second Cromerian interglacial from Harreskov, Denmark.

Pianico Sèllere *Italy* (Rossi 2003)

Middle Pleistocene: Cromerian. 45.8° N, 10.1° E, 320 m a.s.l.

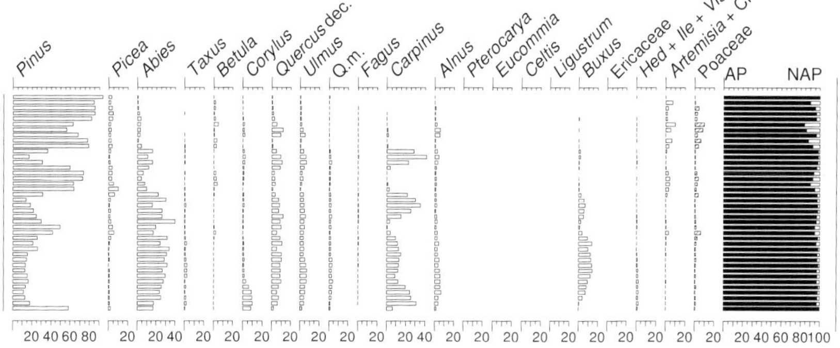

Fig. 3.5-5. Pollen diagram most probably of the beginning of the Cromerian with several strong climatic oscillations, from Pianico-Sèllere, Italian pre-Alps. *Quercus* dec. – *Quercus* deciduous; Q.m. – Quercetum mixtum (here, *Tilia* and some other deciduous trees); *Hed* – *Hedera*, *Ile* – *Ilex*, *Vis* – *Viscum*, *Vit* – *Vitis*, Chenop. – Chenopodiaceae.

Tenaghi Philippon *Greece* (van der Wiel and Wijmstra 1987)

Middle Pleistocene: Cromerian. 41.0° N, 24.3° E, 40 m a.s.l.

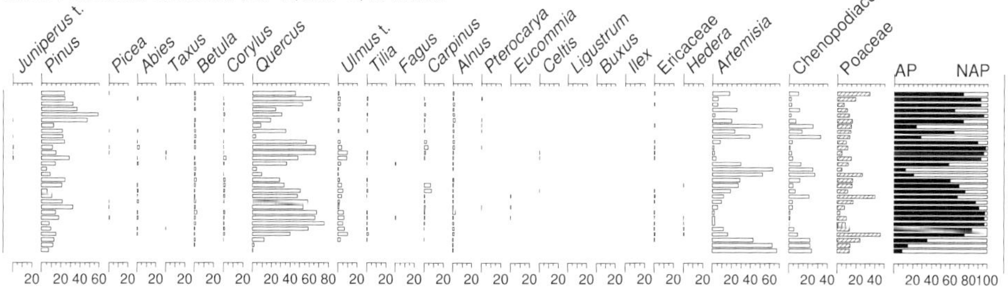

Fig. 3.5-6. Pollen diagram of a Cromerian interglacial, situated between Cromerian II and III from Tenaghi Philippon, northern Greece. *Ulmus* t. includes *Ulmus* and *Zelkova*.

missing from all Cromerian interglacials throughout the whole area north and north-west of the Alps. It is only present, but very rarely, in south-eastern Europe in several interglacials at Tenaghi Philippon (van der Wiel and Wijmstra, 1987a,b).

Most interesting are the records of *Rhododendron ponticum* from Pianico-Sèllere in the Italian southern Alps, already recorded in the 19th century. This is the most important Cromerian site in the southern Alps and southwards. It is rich in excellently preserved macroremains and famous for its 15 500 years of annual laminations, though the whole series is still longer. Its absolute age, however, is still under dispute: presumably it has to be placed in MIS 19 at the beginning of the Cromerian (Roulleau et al., 2009). This placement is supported by traces of *Tsuga*-pollen but, according to Brauer et al. (2007), MIS 11 is still also possible. The pollen diagram (3.5-5; Ravazzi, 2003) shows several strong oscillations with warm mesic forests, dominated by *Carpinus*, mixed with remarkable amounts of *Buxus,* and open pine forests with steppic elements such as *Artemisia, Ephedra* and Chenopodiaceae (Rossi, 2003).

The eminent site of Tenaghi Philippon in northern Greece has provided a continuous sequence for the Cromerian (van der Wiel and Wijmstra, 1987b). As in the Lower Pleistocene it shows an alternation of steppe phases (representing 'glacial' periods) characterised by high values of *Artemisia* and Chenopodiaceae pollen and forested 'interglacial' periods with *Quercus* as the dominating trees mixed with considerable amounts of Ulmaceae (*Ulmus* and *Zelkova*). The number of Cromerian 'interglacials' is, however, greater than those so far described north of the Alps: an additional one, situated between Cromerian II and III in north-west Europe, is presented in Fig. 3.5-6 (van der Wiel and Wijmstra, 1987b).

Altogether the knowledge of the Cromerian vegetation history and stratigraphy is still fragmentary. Quite a lot of faunal remains have been found in Cromerian contexts, for example in the Cromer forest beds in England, in Voigtstedt, Kärlich and Miesenheim in Germany and at important French and Italian sites (Turner, 1996). Characteristic species were the straight-tusked elephant, *Palaeoloxodon antiquus*, medium-sized cave bear, *Ursus deningeri*, the rhino *Dicerorhinus etruscus*, horse, *Equus* spp. and, in the late Cromerian, the shrew *Sorex savini*, Mosbach-wolf, *Canis lupus mosbachiensis* and roe deer, *Capreolus capreolus*. The steppe mammoth, *Mammuthus trogontherii*, was found only in deposits below the Matuyama/Brunhes boundary whereas the southern mammoth *M. meridionalis* has been recorded from above this boundary. Very important for stratigraphic correlations in the Middle Pleistocene is the occurrence of the two genera *Mimomys* and *Arvicola* which follow each other. The boundary between these rodents is situated between the Cromerian II and III (von Koenigswald and van Kolfschoten, 1996).

In the Cromerian the first humans entered Europe, *Homo erectus* coming from Africa. More than 1000 remains have been found in Sima de los Huesos and Gran Dolina in northern Spain, the oldest are from MIS 19, that is around 780 000 yr BP. The only other human find from the Cromerian is the well-known lower jaw of *Homo heidelbergensis* from Mauer (near Heidelberg) estimated to be about 600 000 years old.

3.6

The Holsteinian and Wacken/Dömnitz

The term Holsteinian was first used to define marine deposits of a large transgression between the Elsterian and the Saalian glaciations which advanced far into the northern German mainland. Later the Lower Elbe area, where limnic and marine sequences could be linked, became the type region for the Holsteinian. Several type sections provided pollen diagrams from Hamburg-Dockenhuden, Bossel and Pritzwalk for limnic sequences (Litt et al., 2007). Now there are quite a number of complete pollen diagrams that provide good knowledge of Holsteinian vegetation history, showing a concurrent vegetation development in central Europe (Bittmann, 2012). In the lowlands this is represented in particular by the sites Munster-Breloh (Müller, 1974a), Dethlingen (Koutsodendris et al., 2010), Pritzwalk (Erd, 1973) and Gröbern-Schmerz (Eißmann et al., 1995). After a Late-glacial phase with herbs and a *Betula* peak many diagrams are dominated by *Pinus* and *Alnus* throughout the interglacial. Together with these two genera *Corylus* and *Quercus* spread strongly. Other Quercetum mixtum taxa were *Fraxinus* and

Ulmus while *Tilia* occurs in almost all localities and throughout the whole interglacial in only small quantities. In central and eastern Europe *Picea* expanded strongly and early, followed by a distinct *Taxus* maximum. Later in the interglacial *Carpinus* spread and, with some delay, *Abies* immigrated and reached as far north as Denmark (Vejlby: Andersen, 1965). During this later period *Quercus* achieved its maximum expansion.

Special biostratigraphic markers of the Holsteinian are *Fagus* and *Pterocarya* which immigrated rather late but show a characteristic increase towards the end of this interglacial. In northern Germany *Pterocarya* attains 8% of tree pollen (Granzin: Erd, 1973) and even 12% (Wuthenow: Cepek and Erd, 1975): however, in the later Pleistocene as well as in the Holocene *Pterocarya* failed to reach Europe. This tree, together with the water fern *Azolla filiculoides*, is the most important species for correlating Holsteinian sites over large distances. *Azolla filiculoides* was still present during the following Wacken/Dömnitz interglacial; afterwards, it also became extinct in Europe but has recently been reintroduced as green manure. The interglacial ends with an increase in *Pinus* and *Betula* together with some *Larix*, after which open vegetation replaced the forests.

During the climatic optimum of the interglacial *Buxus* immigrated and soon spread strongly, far beyond its postglacial distribution. Besides *Buxus* there are low but regular records of *Celtis, Vitis* and *Ligustrum* as far north as Lithuania (for example Butenai, Fig. 3.6-3: Kondratiene, 1996). Altogether the vegetation indicates higher temperatures in the Holsteinian compared with the Eemian and the Holocene (Kühl and Litt, 2007).

A significant feature of the Holsteinian is a sudden setback in the vegetation in the middle part of the interglacial, followed by a short-lived expansion of *Betula* and *Pinus*. A tephra layer occurs at this level so this setback was probably caused by volcanic dust. It is followed for 1500 years by a characteristic gap in the occurrence of *Carpinus,* while other trees do not show an interruption during this period (Diehl, 2007). The reason for this long absence of *Carpinus* is unknown because it cannot fully be attributed to the preceding volcanic event.

In the low mountains of northern Germany (Döttingen, Fig. 3.6-1; Diehl, 2007) as well as along the northern edge of the Alps (Samerberg: Grüger, 1983; Thalgut; Drescher-Schneider, 2000a) there are deviations in the development of the vegetation. Here the mountainous species, *Abies,* spread strongly before *Picea* and widely occupied the habitat of *Pinus*. The pollen diagrams from these areas also show high values of *Buxus* and *Fagus* in some parts. Because of deviations in the vegetation development within the Holstein complex at Schöningen a second interglacial was proposed by Urban (1995) and called Reinsdorf. This, however, turned out to be a younger part of the Holsteinian itself (MIS 9), (Urban and Sierralta, 2012; Bittmann, 2012).

Further to the east there are several complete Holsteinian profiles, in particular from eastern Poland, for example Brus (Fig. 3.6-2; Pidek, 2003), Ossówka (Nitychoruk et al., 2005) and Nowiny Żukowskie (Hrynowiecka-Czmielewska, 2010) as well as from the Baltic Countries, for example Butenai from Lithuania (Fig. 3.6-3; Kondratiene, 1996) and Ulmale from Latvia (Fig. 3.6-4; Kalnina, 2001). They show a similar vegetation development to that in central Europe: *Corylus*, however, spreads later there and does not reach the typical early maxima that it achieves in the west. On the other hand *Carpinus* is much more abundant in the east but *Fagus* is only sporadically recorded even though it reached Latvia.

The vegetation development with its setback and also the duration of the interglacial are very similar all over central and eastern Europe. On the basis of the U/Th-dates from the type site Bossel (Geyh and Müller, 2005), as well as from Schöningen (Urban et al., 2011), the Holsteinian belongs to MIS 9.

In the long Quaternary sequences in France two interglacials are encountered that are very similar to each other: the Praclaux and the Landos (Fig. 3.6-5) above it (Reille et al., 2000). In their optimum phases they start with maxima of *Corylus* and *Quercus*, followed by *Taxus* and finally *Abies* and *Fagus*. *Pterocarya* appears only sporadically in the Praclaux interglacial and is completely lacking from the Landos interglacial. This is probably due to the elevation of the profiles, which is around 1100 m asl, because *Pterocarya* is a tree of riverbank forests. The strong early expansion of *Abies* in the Praclaux as opposed to its late expansion in the Landos, as well as in central Europe, and in particular the complete lack of *Carpinus* in the Praclaux led to the correlation of the Landos interglacial with the Holsteinian. Another important argument is the

Döttingen *Germany* (Diehl 2007)

Middle Pleistocene: Holsteinian. 50.3° N, 7.0° E, 460 m a.s.l.

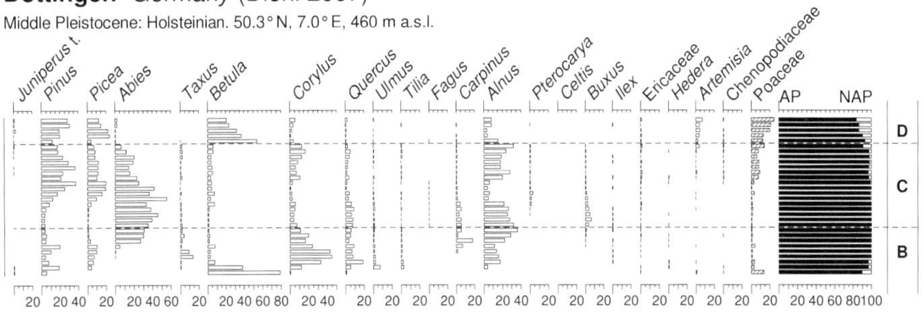

Fig. 3.6-1. Pollen diagram of the Holsteinian interglacial from Döttingen, western Germany.

Brus *Poland* (Pidek 2003)

Middle Pleistocene: Holsteinian. 51.5° N, 23.3° E, 158 m a.s.l.

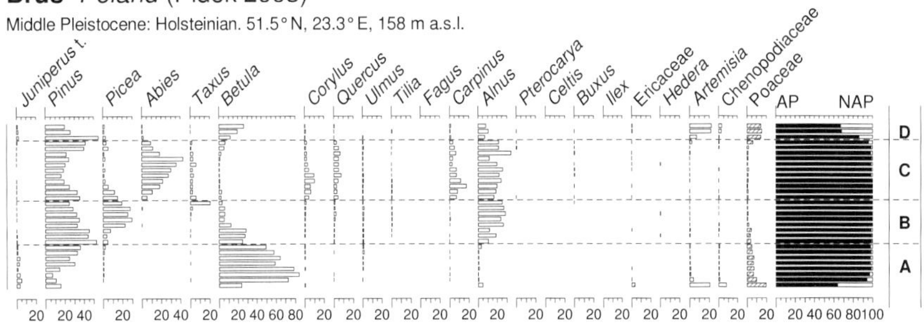

Fig. 3.6-2. Pollen diagram of the Holsteinian interglacial from Brus, south-east Poland.

Butenai *Lithuania* (Kondratiene 1996)

Middle Pleistocene: Holsteinian. 55.7° N, 23.8° E, 104 m a.s.l.

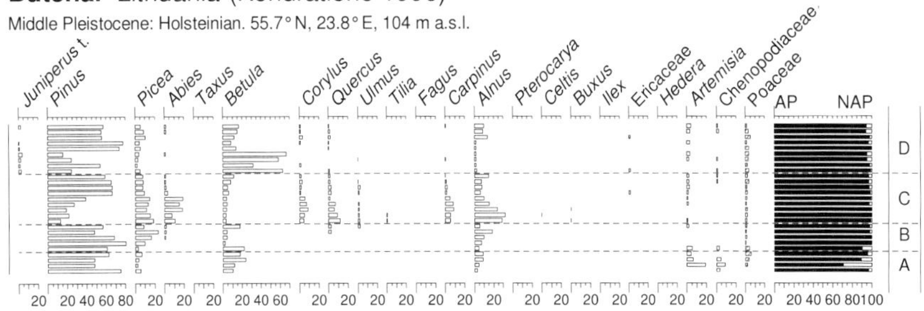

Fig. 3.6-3. Pollen diagram of the Holsteinian interglacial from Butenai, Lithuania.

Ulmale *Latvia* (Kalnina 2001)

Middle Pleistocene: Holsteinian. 56.9° N, 16.7° E, 13 m a.s.l.

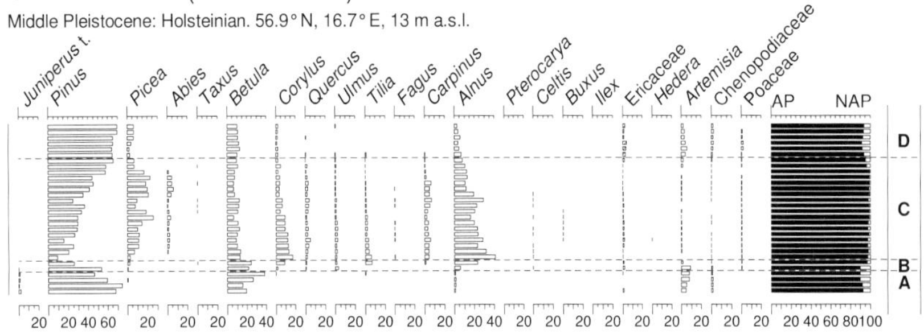

Fig. 3.6-4. Pollen diagram of the Holsteinian interglacial from Ulmale, Latvia.

Landos *France* (Reille et al. 2000)

Middle Pleistocene: Holsteinian. 44.8° N, 3.8° E, 1100 m a.s.l.

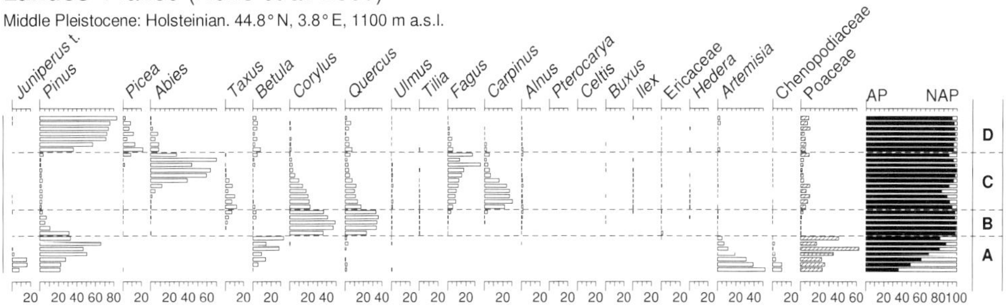

Fig. 3.6-5. Pollen diagram of the Landos (Holsteinian) interglacial from the Praclaux crater, southern France.

Marks Tey *Great Britain* (Turner 1970)

Middle Pleistocene: Holsteinian. 51.9° N, 0.8° E, 33 m a.s.l.

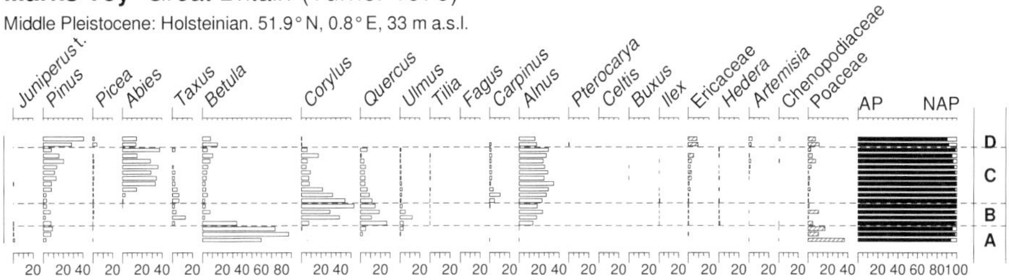

Fig. 3.6-6. Pollen diagram of the Hoxnian (Holsteinian) interglacial from Marks Tey, south-east England.

Gort *Ireland* (Jessen et al. 1959)

Middle Pleistocene: Holsteinian. 53.1° N, 8.8° W, 50 m a.s.l.

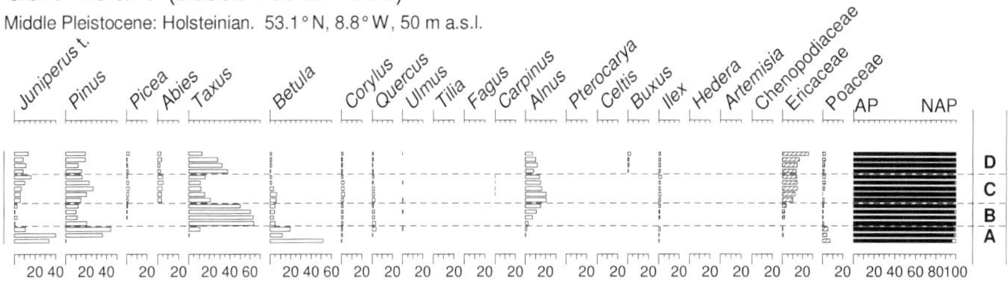

Fig. 3.6-7. Pollen diagram of the probably Holsteinian interglacial from Gort, Ireland.

Munster-Breloh & Hetendorf *Germany* (Müller 1974b (top); Meyer 1974 (bottom))

Middle Pleistocene: Holsteinian. 53.0° N, 10.1° E, 74 m a.s.l.

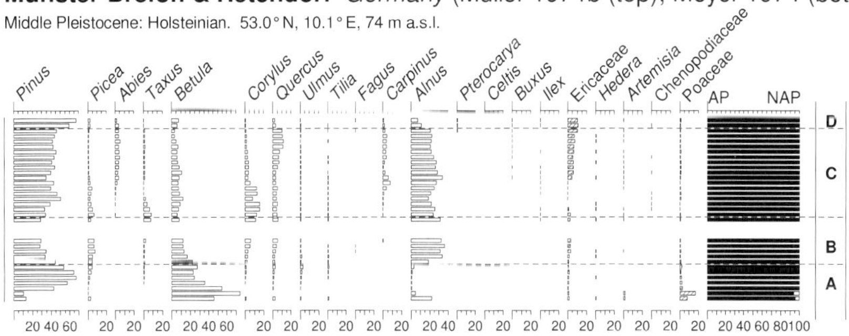

Fig. 3.6-8. Composite pollen diagram of the Holsteinian interglacial from Munster-Breloh & Hetendorf, northern Germany.

setback of the vegetation, which is a characteristic feature of the central European Holsteinian sites. It appears in the same zone in the Landos, expressed by high values of Poaceae, while it cannot be recognized in the Praclaux. This is confirmed by their absolute ages: the Landos as well as the Holsteinian have to be correlated with MIS 9, while the Praclaux cannot be younger than MIS 11 (Reille et al., 2000; Geyh and Müller, 2005).

In Britain the important interglacial sites Hoxne and Marks Tey (Fig. 3.6-6; Turner, 1970) and probably also Gort in Ireland (Fig. 3.6-7; Jessen et al., 1959) have to be linked with the Holsteinian (in Britain called Hoxnian). Their vegetation development is similar to that on the continent. During this interglacial, *Abies* reached the British Isles for the last time and even spread to Ireland. *Picea* is also present there but rare and only locally. The single finds of *Fagus* pollen, however, give no unambiguous evidence for the presence of this tree on these islands. *Pterocarya*, *Azolla filiculoides* and the very low pollen values of *Tilia* are characteristic of the Holsteinian (Hoxnian) in Britain too. Even exotic species, such as *Rhododendron ponticum*, far outside its present distribution in northern Turkey and the Caucasus, reached Ireland and testify to warm and distinctly oceanic climatic conditions at this time, as is the situation to-day (Turner, 1970). Also remarkable is the occurrence of *Cassiope tetragona* in Ireland, a species only present today in the extreme west of Europe. The same is true of the aquatic *Eriocaulon septentrionale*, the main distribution area of which today is North America.

Very important for the correlation with the continental Holsteinian is the setback of the vegetation at exactly the same level in the pollen diagrams as can be seen at Marks Tey and several other British sites where the vegetation succession is suddenly interrupted by a temporary phase with high herbaceous pollen (Turner, 1970, 2005).

There are, however, still uncertainties with the absolute ages. At present the two interglacials which are assigned to the Hoxnian are very similar in their vegetation but, according to amino acid datatings, correlate with MIS 11 and MIS 9, respectively (Penkman et al., 2013; Roe et al., 2009). As a consequence of these dates, the far reaching glacial advance of the Anglian, which corresponds to the continental Elsterian is sometimes put into MIS 12. On the continent, however, two differently dated interglacials of Holsteinian type are not known. Open questions still remain.

The duration of the Holsteinian has been determined by varve countings. On the basis of these the forest phase of this interglacial in England (Turner, 1975) and at Munster-Breloh (Fig. 3.6-8; Müller, 1974a) is 15 000–16 000 years long. On the other hand the preceding Rhume-interglacial, assigned to MIS 11, has a length of about 25 000 years. The different durations of these two interglacials support a correlation of the Holsteinian and Rhumian with MIS 9 and 11, respectively, rather than with MIS 11 and MIS 13, if compared with the time scale of the Vostok and Dome ice cores (Geyh and Müller, 2007).

Altogether the Holsteinian sites to the north and west of the Alps can be correlated well by their similar course of vegetation development and the characteristic setback. According to the U/Th-dates in the type region in north-west Germany, their stratigraphic position is in MIS 9, although some authors favour MIS 11. As a consequence, however, the strong and far-reaching glaciation of the Elsterian (Lowestoft, Mindelian) would have to be correlated with MIS 12.

A distinct interstadial was recorded at Hamburg-Dockenhuden (Linke and Hallik, 1993) in the early Saalian and indications of probably the same interstadial were presented by Kalnina (2001) from Latvia. A somewhat younger warm phase of interglacial character has been discovered within the Saalian complex at the German sites Wacken (Menke, 1968) and Dömnitz (Erd, 1973) and consequently named the Wacken/Dömnitz warm phase. It has been designated to MIS 7. At both localities their deposits are situated in superposition to the Holsteinian and in between there is the Fuhne non-glaciated cold phase which belongs to the early Saalian complex. In the Wacken/Dömnitz warm phase the vegetation development is very similar to that of the Holsteinian. It, however, differs from the Holsteinian in the late appearance of *Picea* that clearly spreads after *Carpinus* as well as in the obvious lack of *Abies*, at least in northern Germany. *Azolla filiculoides* is still present but *Pterocarya* has already disappeared. Our knowledge of the vegetation sequence of the Wacken/Dömnitz interglacial is still limited because there are hardly any sites for comparison north of the Alps. Despite certain differences, the Schöningen interglacial in Germany (Urban and

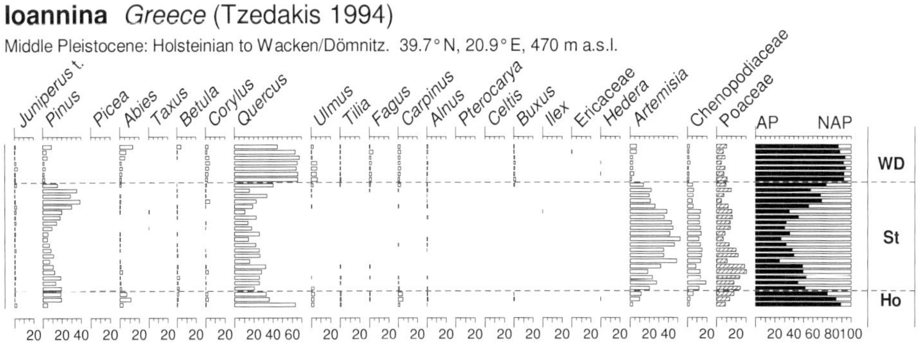

Fig. 3.6-9. Pollen diagram from the Holsteinian (Ho) to the Wacken/Dömnitz (WD) interglacial with a steppe phase (St) in between in the long record of Ioannina, north-west Greece.

Sierralta, 2012), the Hoogeveen interstadial in the Netherlands (Zagwijn, 1973) and the alpine site of Meikirch in Switzerland (Preusser et al., 2005) have been assigned, together with the Wacken/Dömnitz, to MIS 7, and this interglacial is also recorded in the long sequence of Samerberg (Grüger, 1983). Several sites of this interglacial have been described from Lithuania where it was called Snaigupėlė. Its vegetation development there is similar to the Eemian but *Azolla* is present (Kondratiene, 1996; Kondratiene and Damušytė, 2009). Further to the east the temperate Kostroma interstadial that separates the Dnieper and Moscow stages of the Saalian glaciation may be an equivalent of the Wacken/Dömnitz interstadial (Sudakova et al., 2007).

According to the Marine Isotopic stratigraphy there are three separate warm phases in MIS 7. These were found in superposition in the long terrestrial record in the French Velay region and have been described as Bouchet I-III (Reille et al., 1998). Their vegetation shows cooler conditions than in the preceding Holsteinian (Landos). The warmest is Bouchet I, which is dominated by *Quercus* and *Corylus* with only small proportions of the other Quercetum mixtum species, and *Carpinus* arrives later and finally *Picea*. Both younger temperate phases are still cooler, however, with considerable amounts of *Fagus*. So the vegetation is in accordance with the climatic course of the three peaks of MIS 7, of which the first one was the warmest.

In the Mediterranean, MIS 7 is well developed in the long sequences of Tenaghi Philippon and Ioannina in Greece where the forests are mainly dominated by *Quercus* and *Pinus* as well as *Ulmus/Zelkova*. At the most typical Mediterranean site of Ioannina, temperate *Abies* and *Corylus* show a characteristic decrease from considerable amounts in MIS 11 to low percentages in MIS 7 (Tzedakis et al., 2001). In the cold phases of this region steppe and desert-steppe prevail, at least at lower elevations. *Pterocarya,* which is important for the correlation with the sites north and west of the Alps, occurs for the last time with a marked peak in the Litochoris interglacial of Tenaghi Philippon which is assigned to MIS 9 and also correlated with the French Landos interglacial (Tzedakis et al., 2006) thus supporting the all European correlation. During the 'glacial' phases, steppe vegetation and even desert-steppe prevailed over large areas, particularly at low elevations, which is well illustrated for the period between the forested interglacials MIS 9 and MIS 7 in Ioannina (see Fig. 3.6-9; Tzedakis, 1994). The climate setback in MIS 9, typical of central European sites, is also clearly documented in Tenaghi Philippon in the opening up of the forest to a forest steppe (Fletcher et al., 2013).

Slightly outside south-east Europe, in eastern Turkey, there is a unique new long terrestrial record from Lake Van that provides a key record for the eastern Mediterranean. Here continuous sedimentation took place for over 600 000 years, spanning MIS 15 to 1. A reliable chronology has been achieved by combining several methods. The climax vegetation of the interglacial stages in the Lake Van region is characterized by an oak steppe-forest with *Pistacia* and *Juniperus*. *Pinus,* as an additional tree, only played a major role during the last and penultimate interglacials as well as in the Brörup interstadial. During the glacial stages and stadial substages cold-adapted vegetation prevailed, which was very similar in each stage and can be described as dwarf-shrub steppe and desert steppe with *Ephedra*, *Artemisia*, chenopods, grasses and forbs (Litt et al., 2014).

Human remains of the period from MIS 9-7 (Holsteinian to Wacken/Dömnitz) are extremely rare. However, the skull from Steinheim and several skull fragments from Swanscombe, on the Thames, belong to this period. By far the most numerous human remains, more than 30 altogether (among them 23 skull fragments), have been excavated at Bilzingsleben in Thuringia, central Germany (Kahlke and Mania, 1994). They belong to a late form of *Homo erectus* and occurred together with a forest fauna dominated by *Palaeoloxodon antiquus*. Plant remains from this site, such as *Celtis australis* and *Syringa josikaea*, indicate a warm environment for these early humans.

Of great general importance are the finds made in the open cast mine of Schöningen, east of Hannover, central Germany (Thieme, 2007). Although human remains are still missing, many stone artefacts have been excavated. Some unique finds comprised well-preserved wooden artefacts from waterlogged deposits. Most important are eight neatly made wooden spears, found together with numerous bones of wild horses in a hunting camp of Holsteinian age. Thousands of bones of large mammals, most of them from wild horses, do not only show that humans had been present there but that their cognitive abilities were much higher than has so far been assumed. These large mammals, which are both fast and quick to react, were collectively hunted, which requires a systematically prepared coordination.

The fauna of the Holsteinian is quite well known (von Koenigswald, 2007). Numerous sites from this period have provided large quantities of bones as well as teeth. Apart from the wild horse (*Equus* spp.) they, in particular, include other large mammals such as the forest or straight-tusked elephant (*Palaeoloxodon antiquus*), forest rhinoceros (*Stephanorhinus kirchbergensis*), aurochs (*Bos primigenius*) and wild water buffalo (*Bubalus murrensis*), in addition to wild boar (*Sus scrofa*), giant elk (*Megaloceros giganteus*), red deer (*Cervus elaphus*), roe deer (*Capreolus capreolus*) and cave bear (*Ursus spelaeus*), as well as the large rodent, the Eurasian beaver (*Castor fiber*), which has been frequently recorded.

3.7

The Eemian

The last interglacial or Eemian is the best known Pleistocene warm period with regard to vegetation and climate. Hundreds of sites provide complete pollen diagrams as well as excellent macroremains, in particular from northern central Europe and around the Alps. So a mapping of Hermsdorf and Strahl (2008) in the young moraine area of the German Land Brandenburg provided not less than 536 sites, of which 188 were confirmed by pollen analysis. Only a small number of the many Eemian pollen diagrams in Europe are presented in Figures 3.7-1 to 3.7-21. It is, however, remarkable that up until now there is no complete pollen diagram for the Eemian in Britain and also in the Netherlands, with their eponymous locality, Eemian sites are rare. Several former 'Eemian' diagrams from Britain are now dated to MIS 7. Therefore an old composite Eemian diagram from Britain had to be used in Fig. 3.7-1 (West, 1980). The interglacial is called after the small river Eem in the Netherlands but there are also regional names such as Ipswichian in Britain and Mikulino in Russia.

The Eemian is correlated with the MIS 5e. On the basis of a marine record off Portugal (Shackleton et al., 2002) its duration was calculated to around 17 000 years. In central Europe, however, for many years its duration was assumed to be only 11 000 years as shown by varve counts of lacustrine sediments at Bispingen, northern Germany (Müller, 1974b; Fig. 3.7-3). On the other side, in the Mediterranean region a clear increase in woody vegetation, in particular *Quercus robur* type, started considerably earlier and also persisted longer. Eemian thermophilous forest trees continued to be present for a further 5000 years, as can be seen in the Lago Grande di Monticchio record from southern Italy. Annually laminated sediments were also available at this site and, on the basis of this count, the duration of the interglacial was estimated to have been from 127 200 yr BP to 109 500 yr BP, resulting in a length of 17 700 years (Brauer et al., 2007; Allen and Huntley, 2009). This severe discrepancy to central Europe was solved by varve-datings of new cores in Bispingen which indicated the existence of major gaps in the old profile and obtained a length of the Eemian of 17 000 years, which is now in accordance with other estimations (Lauterbach et al., 2012).

East Anglia [10 sites] *Great Britain* (West 1980)

Upper Pleistocene: Eemian. Around 52.2° N, 0.5° E, 5-100 m a.s.l.

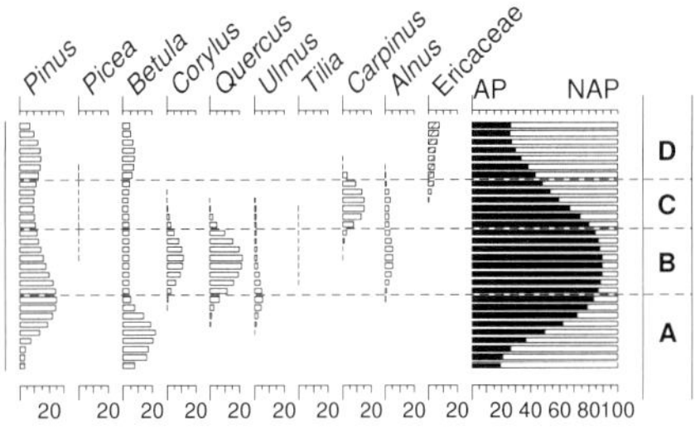

Fig. 3.7-1. Schematic pollen diagram based on several sites of the Eemian interglacial for East Anglia, south-east England.

Fugla Ness *Great Britain* (Birks and Ransom 1969)

Upper Pleistocene: Eemian. 60.5° N, 1.4° W, 20 m a.s.l.

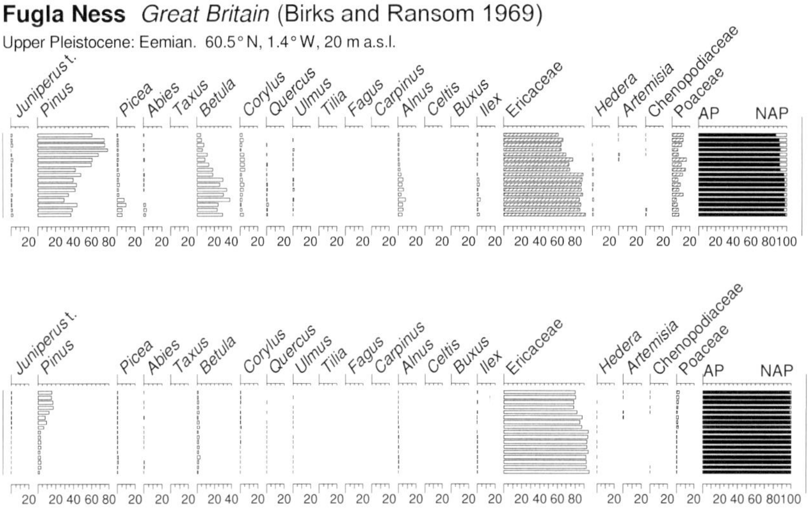

Fig. 3.7-2. Two versions of the pollen diagram of the Eemian interglacial from Fugla Ness, Shetland Islands, Scotland. Ericaceae are excluded in the calculation sum in the upper version, included in the lower.

Bispingen *Germany* (Müller 1974b)

Upper Pleistocene: Eemian. 53.1° N, 10.0° E, 75 m a.s.l.

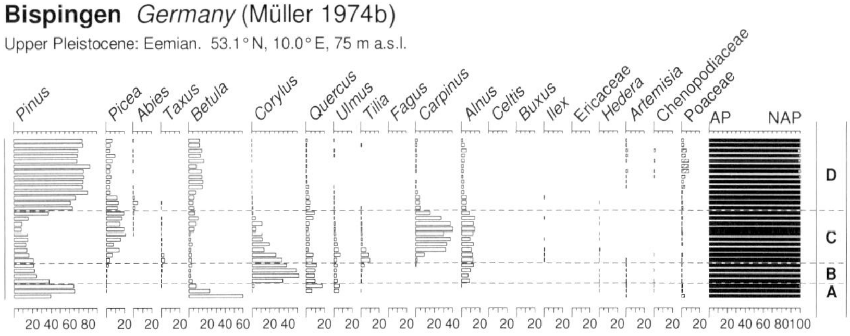

Fig. 3.7-3. Pollen diagram of the Eemian interglacial from Bispingen, northern Germany.

Oerel *Germany* (Behre and Lade 1986)

Upper Pleistocene: Eemian. 53.5° N, 9.1° E, 12 m a.s.l.

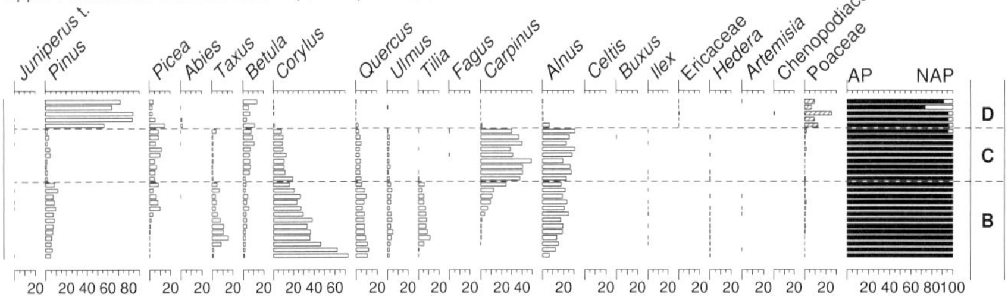

Fig. 3.7-4. Pollen diagram of the Eemian interglacial from Oerel, north-west Germany.

Hollerup *Denmark* (Andersen 1965)

Upper Pleistocene: Eemian. 56.4° N, 9.9° E, 49 m a.s.l.

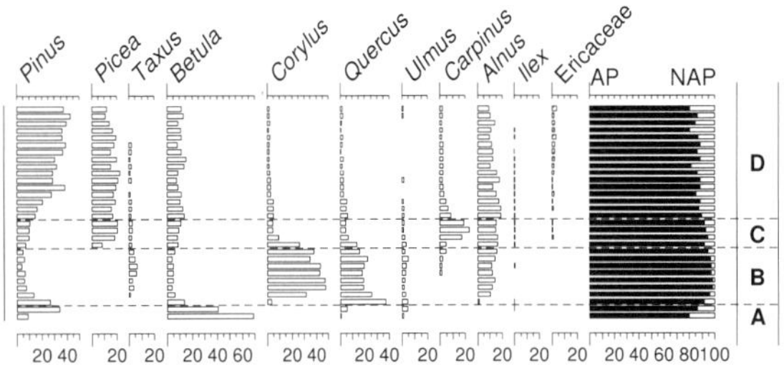

Fig. 3.7-5. Pollen diagram of the Eemian interglacial from Hollerup, Denmark.

Gröbern *Germany* (Litt 1990)

Upper Pleistocene: Eemian. 51.8° N, 12.6° E, 91 m a.s.l.

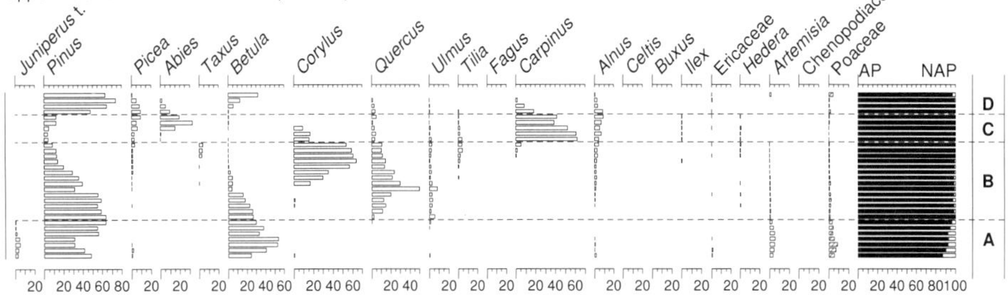

Fig. 3.7-6. Pollen diagram of the Eemian interglacial from Gröbern, central Germany.

In several profiles, in particular Brokenlande, northern Germany (Menke and Ross, 1967) and Dziewule, eastern Poland (Fig. 3.7-8; Binka and Nitychoruk, 2001), the Saalian Late-glacial, that is the transition from the glacial to the interglacial period, can be traced in great detail.

The vegetation cover started as a herb tundra with grasses and a high proportion of *Artemisia*, after which further climatic amelioration is indicated by the spread of rich shrub communities. A special feature is the rapid increase in *Hippophaë* that dominated this vegetation until it was replaced by *Juniperus* and increasing *Salix* and *Betula nana* together with herbaceous species. In eastern Poland *Pinus cembra* and *P. sylvestris*, tree birches and the continental *Picea abies* ssp. *obovata* also spread. These trees, together with the sparse occurrence of *Larix*, point to forests similar to those of the modern taiga. Most of them, in particular *Picea*, disappeared from most areas completely with the start of the Eemian.

The pollen diagrams from the Late Saalian point to a steady and gradual vegetation development from

Imbramowice *Poland* (Mamakowa 1989)
Upper Pleistocene: Eemian. 50.9° N, 16.6° E, 175 m a.s.l.

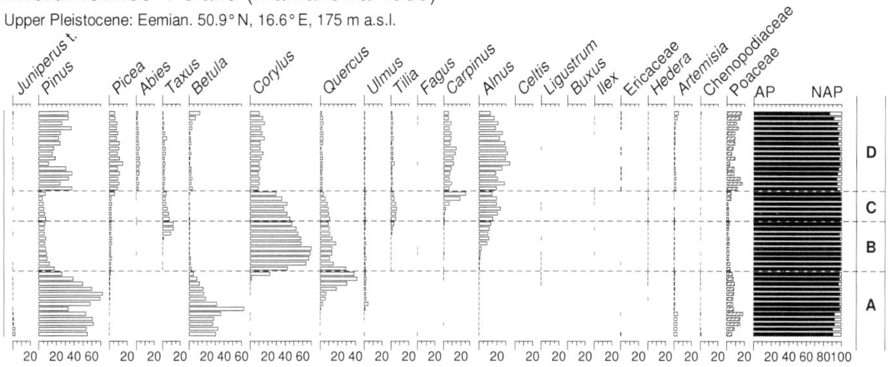

Fig. 3.7-7. Pollen diagram of the Eemian interglacial from Imbramowice, western Poland.

Dziewule *Poland* (Bińka and Nitychoruk 2001)
Upper Pleistocene: Eemian. 52.0° N, 22.5° E, 159 m a.s.l.

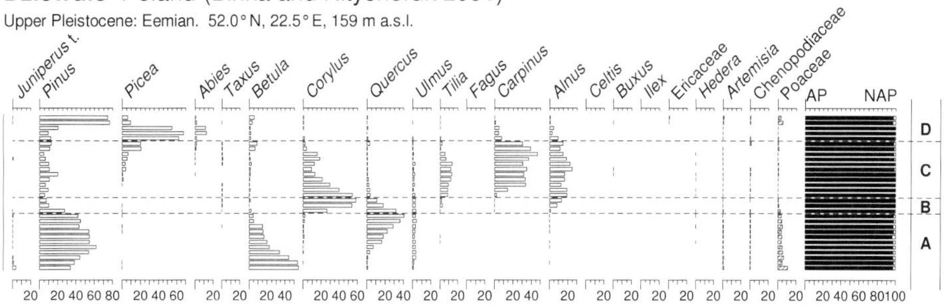

Fig. 3.7-8. Pollen diagram of the Late Saalian and the Eemian interglacial from Dziewule, eastern Poland.

Naklo *Poland* (Noryskiewicz 1978)
Upper Pleistocene: Eemian. 53.2° N, 17.6° E. 61 m a.s.l.

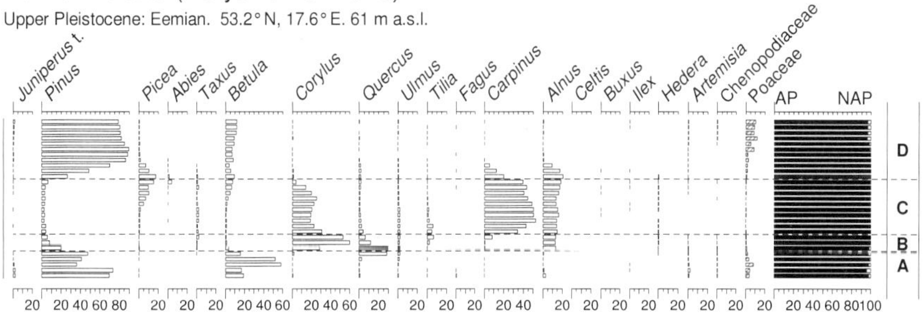

Fig. 3.7-9. Pollen diagram of the Eemian interglacial from Naklo, northern Poland.

open vegetation to forests without any climate setbacks. This is in sharp contrast to the oscillations in climate and vegetation during the Late Weichselian.

The Saalian Late-glacial phase ends with a strong increase in *Juniperus* which, at the onset of the interglacial, was followed by tree birches that were soon replaced by a dominance of *Pinus sylvestris*. The immigration of the thermophilous trees followed rather rapidly. The pollen diagrams show a characteristic succession with often high peaks of *Ulmus*, which was always the first to arrive, followed by *Quercus* and *Fraxinus* and then by very high values of *Corylus* which sometimes exceeded even the high postglacial peaks.

There is an important contrast with regard to the immigration and expansion of *Corylus* between the Eemian and the Holocene. In the Holocene, apart from Poland and eastern Europe, *Corylus* is the first of the thermophilous species to expand north of the Alps and its shrubs preceded the spread of *Ulmus* and *Quercus* but eventually became undergrowth in the mixed oak forests. *Corylus* immigrated in a remarkably different way in the early Eemian, arriving only after the expansion of *Ulmus, Quercus* and *Fraxinus* but being able to gain very high shares within these forests. This is not a local feature but a general one and occurred throughout the whole region north of the Alps: good examples being provided through the pollen diagrams from England (East Anglia, Fig. 3.7-1: West, 1980), Denmark (Hollerup, Fig. 3.7-5: Andersen, 1965), Germany (Jammertal, Fig. 3.7-13: Müller et al., 2005), Poland (Imbramowice, Fig. 3.7-7: Mamakowa, 1989) and Lithuania (Druskininai, Fig. 3.7-11: Kondratiene, 1996). An explanation for the high percentages, which are often even higher than in the Holocene, may be that in the Eemian the pollen record also includes the tree hazel, *Corylus colurna* which, in the postglacial, is restricted to south-east Europe to the Balkans. So far, however, there is no confirmation of this from macroremains.

The middle part of the Eemian started with the expansion of *Alnus* which was soon followed by *Tilia* and *Taxus*. This is another important difference from the Holocene, where *Taxus* is generally rare. In particular in Germany and eastern France *Taxus* reached a high portion in the forests, with pollen percentages up to 65% in Zeifen (southern Germany: Jung et al., 1972) and 40% on the island of Helgoland in the North Sea (Behre, 1970). Its occurrence, however, was only temporary and it declined strongly with the expansion of *Carpinus*. *Tilia* also shows different behaviour in the Eemian than in the Holocene in that it was almost lacking from France and, in the warm southern region in Germany, its occurrence was much lower than today.

A long phase in which *Carpinus* dominates is most characteristic for the Eemian interglacial. *Carpinus* is regarded as replacing *Fagus,* which failed to expand north of the Alps in the Eemian. It is, however, not just a replacement, since *Carpinus* has different ecological demands than *Fagus* and therefore influences the conditions of competition in the forests and, through that, the spread and frequency of several tree species. The distribution area of *Carpinus* in the Eemian was considerably larger than its modern one, for example in Russia (see Fig. 3.7-11; Gey et al., 2001) and Britain. The spread of *Carpinus* was accompanied by the expansion of *Picea*, its maximum following that of *Carpinus* in the lowlands. In the northern Alpine foreland, however, the expansion of *Picea* started considerably earlier (for example Samerberg, Fig. 3.7-14: Grüger, 1979; Eurach, Fig. 3.7-12: Beug, 1979; Gondiswil, Fig. 3.7-15: Wegmüller, 1992). At moist sites in particular, as well as in mountainous areas, *Picea abies* was equally abundant and contemporaneous with *Carpinus*.

The geographical range of *Picea* is an important feature of the Eemian vegetation as the genus extended far beyond its modern distribution area. Spruce was common all over the region north of the Alps as far north as Denmark and spread, in the west, to France (La Grande Pile, Fig. 3.7-17: de Beaulieu and Reille, 1992a) and also in low frequencies to Britain.

Abies has a similar history to *Picea* in the Eemian. Today it is considered to be a mountainous species and, in the Eemian, it also occurred early in the Alps (Mondsee, Fig.3.7-16: Drescher-Schneider, 2000b; Samerberg, Fig. 3.7-14; Gondiswil, Fig. 3.7-15) and the Alpine foreland (Eurach, Fig. 3.7-12). Its spread to other areas was, however, delayed and occurred mainly after the strong expansion of *Carpinus*. Like *Picea, Abies* extended its area into the lowlands, considerably beyond its modern distribution range, actually reaching the North Sea coast in Germany (Oerel, Fig. 3.7-4: Behre and Lade, 1986) and the Netherlands, although in small frequencies, and even extending its area to northern and eastern Poland (Kupryjanowicz et al., 2018a). However, it was probably lacking from Britain, where it had been common during the Holsteinean interglacial.

After the decline of *Carpinus* and *Alnus, Picea* and *Abies* conclude the warm interglacial phase and are replaced by *Pinus* and *Betula* forests which mark the end of the Eemian interglacial.

Because the Eemian had only a slightly warmer climate than the Holocene it is often taken as an example of the natural development of vegetation without any human impact. This, however, is misleading and should be avoided. Apart from the fact that in the Eemian *Carpinus* occupies the place that *Fagus* has in the Holocene the distribution history of the three important conifers *Picea, Abies* and *Taxus* is totally different and at the moment there is no overall explanation for these differences.

Domozerovo *Russia* (Gey et al. 2001)

Upper Pleistocene: Eemian. 59.3° N, 39.2° E, 100 m a.s.l.

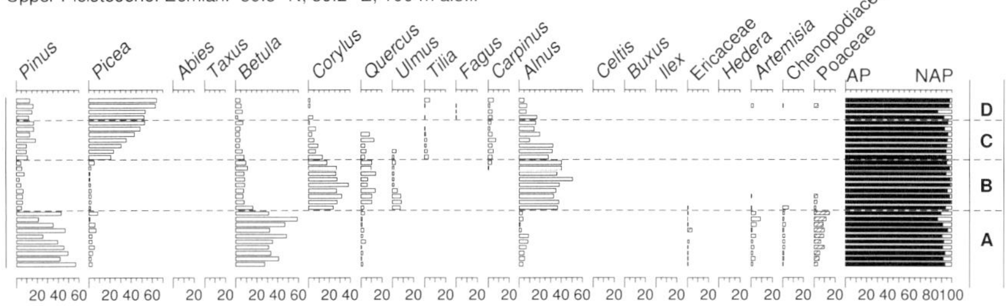

Fig. 3.7-10. Pollen diagram of the Eemian interglacial from Domozerovo, north-west Russia.

Druskininai *Lithuania* (Kondratiene 1996)

Upper Pleistocene: Eemian. 541° N, 24.0° E, 90 m a.s.l.

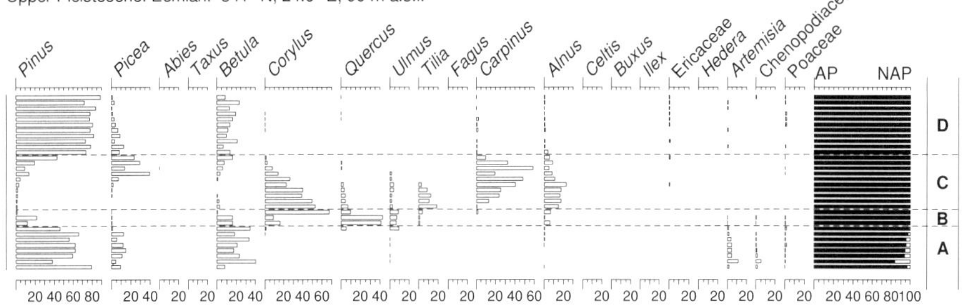

Fig. 3.7-11. Pollen diagram of the Eemian interglacial from Druskininai, southern Lithuania.

Eurach *Germany* (Beug 1979)

Upper Pleistocene: Eemian. 47.8° N, 11.3° E, 615 m a.s.l.

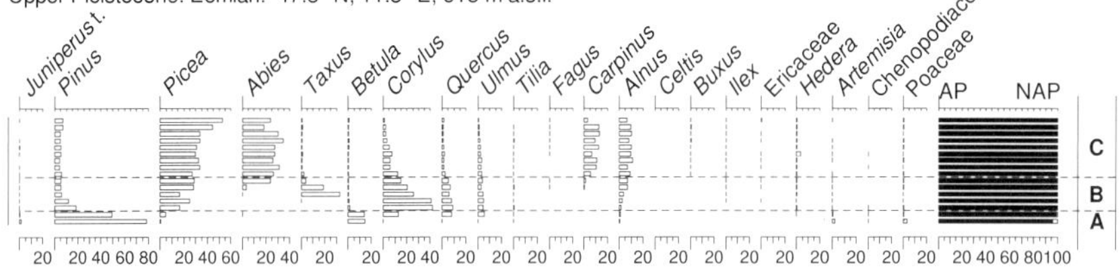

Fig. 3.7-12. Pollen diagram of the Eemian interglacial from Eurach, southern Germany.

Jammertal *Germany* (Müller et al. 2005)

Upper Pleistocene: Eemian. 48.1° N, 9.7° E, 578 m a.s.l.

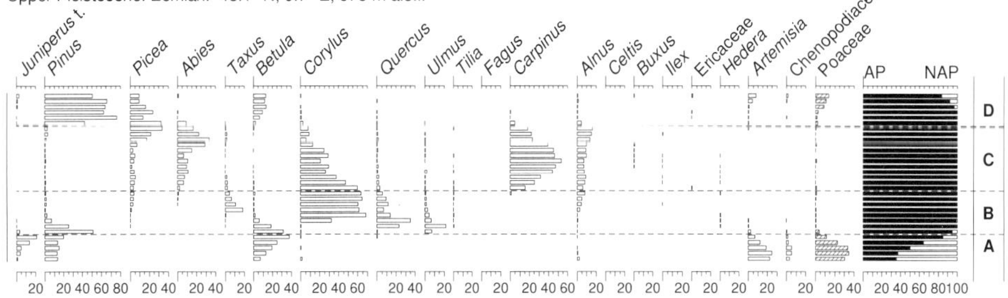

Fig. 3.7-13. Pollen diagram of the Eemian interglacial from Jammertal, south-west Germany.

Samerberg *Germany* (Grüger 1979)

Upper Pleistocene: Eemian. 47.8° N, 12.2° E, 594 m a.s.l.

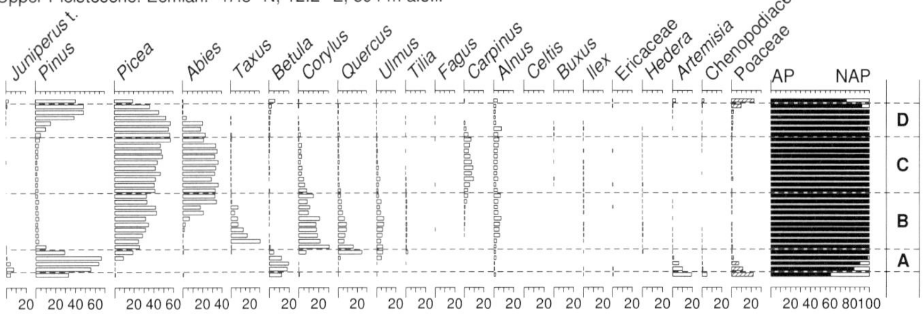

Fig. 3.7-14. Pollen diagram of the Eemian interglacial from Samerberg, Alps, Germany.

Gondiswil *Switzerland* (Wegmüller 1992)

Upper Pleistocene: Eemian. 47.3° N, 7.8° E, 625 m a.s.l.

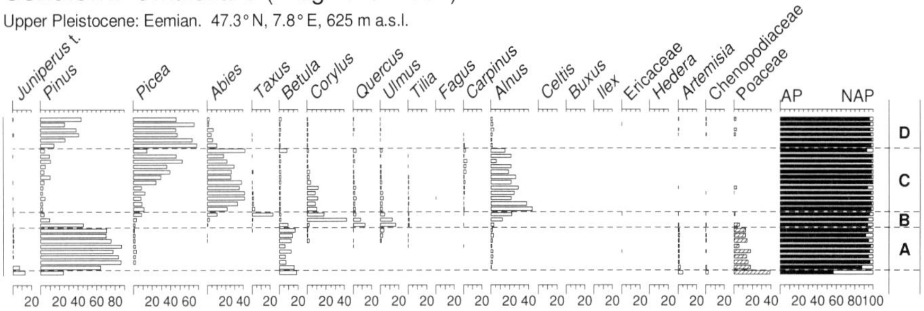

Fig. 3.7-15. Pollen diagram of the Eemian interglacial from Gondiswil, pre-Alps, Switzerland.

Mondsee *Austria* (Drescher-Schneider 2000b)

Upper Pleistocene: Eemian. 47.8° N, 13.3° E, 534 m a.s.l.

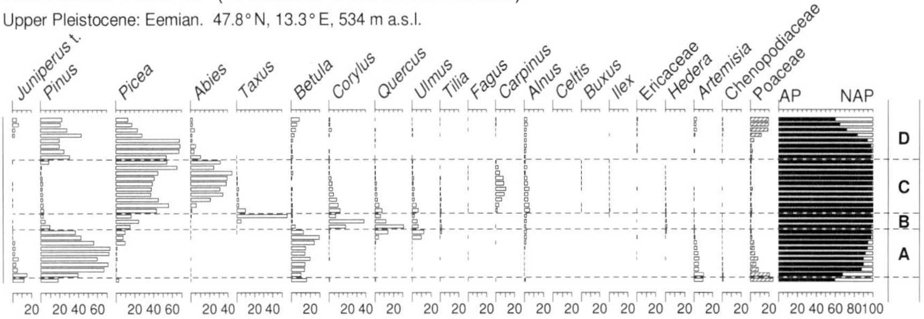

Fig. 3.7-16. Pollen diagram of the Eemian interglacial from Mondsee, Alps, Austria.

Compared with the Holsteinian interglacial the vegetation development of the Eemian was surprisingly uniform all over Europe north of the Alps and the Carpathians, apart from the fact that, in Scandinavia, boreal forests dominated rather than temperate forests (Menke and Tynni, 1984). This uniformity is explained by the situation that at least part of the climate warming of north-western and northern Europe was related to the access to warm Atlantic waters through the Straits of Dover, into the southern North Sea and even further into the Baltic with a connection also to the White Sea (Turner, 2000).

The global nature of the Eemian climate is shown by the evidence from ice cores and marine records. Palaeoclimatic reconstructions from pollen and plant macrofossils indicate an increase in temperature of, on average, two degrees compared with Holocene temperatures for Europe (Kühl, 2003). The highest values were reached at the beginning of

La Grande Pile *France* (de Beaulieu and Reille 1992a)

Upper Pleistocene: Eemian. 47.7° N, 6.5° E, 330 m a.s.l.

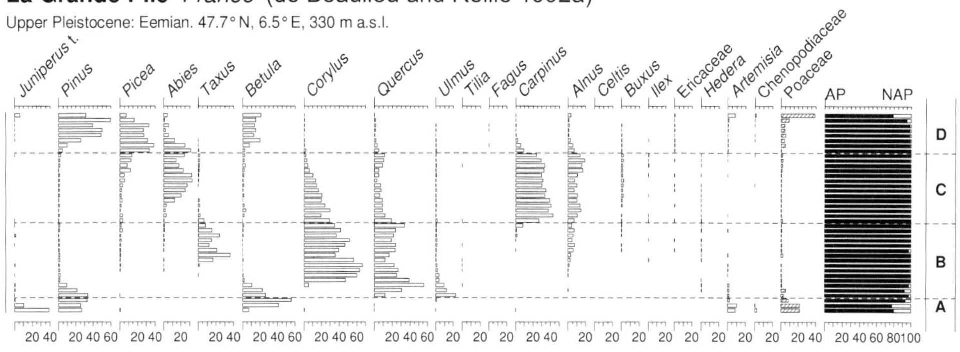

Fig. 3.7-17. Pollen diagram of the Eemian interglacial from La Grande Pile, eastern France.

Les Echets *France* (de Beaulieu and Reille 1984)

Upper Pleistocene: Eemian. 45.8° N, 4.9° E, 267 m a.s.l.

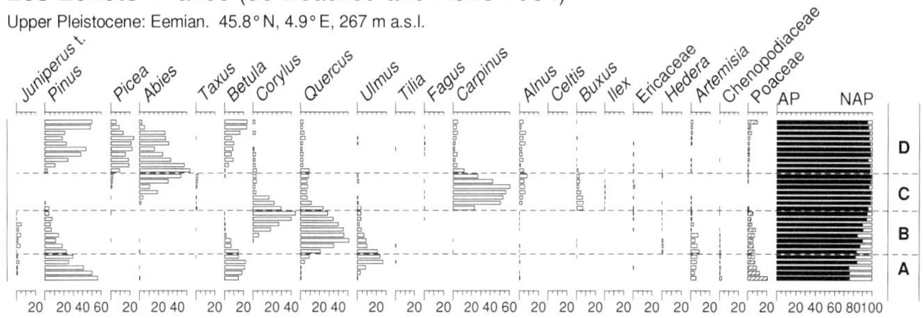

Fig. 3.7-18. Pollen diagram of the Eemian interglacial from Les Echets, eastern France.

Lake Fimon *Italy* (Pini et al. 2010)

Upper Pleistocene: Eemian. 45.3° n, 11.3° E, 23 m a.s.l.

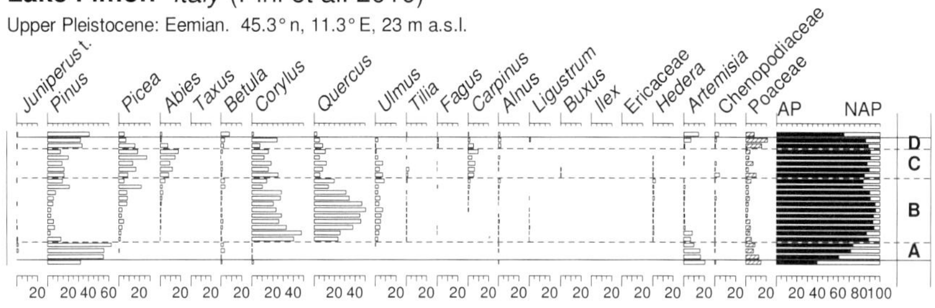

Fig. 3.7-19. Pollen diagram of the Eemian interglacial from Lake Fimon, southern Alpine foreland, Italy.

Staro Orjachovo *Bulgaria* (Bozilova and Djankova 1976)

Upper Pleistocene: Eemian. 43.0° N, 27.8° E, 50 m a.s.l.

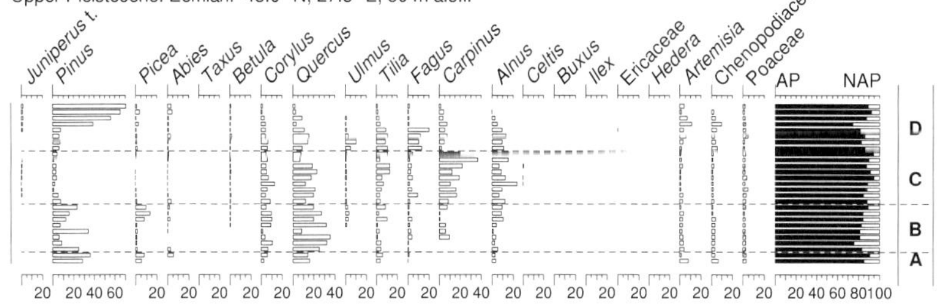

Fig. 3.7-20. Pollen diagram of the Eemian interglacial from Staro Orjachovo, eastern Bulgaria.

the interglacial, after which temperature decreased slightly but humidity increased. At one point there was a discussion about severe climate setbacks during the Eemian but later results indicated that the Eemian climate was relatively stable without any strong deterioration (Litt et al., 1996; Cheddadi et al., 1998).

This warmer climate of the Eemian interglacial is reflected in the vegetation in that some thermophilous taxa expanded northwards considerably beyond their Holocene ranges. Good examples of this are the sub-Mediterranean *Buxus*, which was common over large parts of central Europe and *Ilex*, which spread far to the east covering the whole of Poland and beyond, thus indicating that winter temperatures had also been higher than in the present postglacial. *Ligustrum* and *Syringa* also occurred considerably further north than today and *Hedera* was common as far north as Latvia (Kalnina, 2001). Of particular interest are the records of macroremains of the sub-Mediterranean *Acer monspessulanum* in several Eemian sites in Britain (Godwin, 1975) and the finding of *Acer tataricum* in several continental localities as far north as central Germany (Litt, 1994). These records suggest that the pollen of *Acer*, which can be identified only to genus level, may also have derived from these two species. *Tilia tomentosa*, which nowadays is restricted to the Balkan Peninsula, reached northern Germany and Poland more than 800 km north of its modern extent (Kupryjanowicz et al., 2018b). Oceanic species such as *Ilex aquifolium* and *Hedera helix* spread eastwards into what are now regarded as more continental areas as did *Taxus baccata* and *Buxus sempervirens* (Turner, 2000). The same is true for some thermophilous aquatics such as *Trapa natans*, *Aldrovanda vesiculosa* and *Salvinia natans*, which extended considerably further north than they do to-day, the latter as far north as Jutland and even Lithuania (Kondratiene, 1996). Two other aquatics, *Brasenia schreberi* and *Dulichium arundinaceum* (=*spathaceum*), have been recorded in several sites but became extinct in Europe after the Eemian (Drzymulska, 2018). Nowadays they are restricted to North America.

As in the Holocene, the Eemian vegetation in the Mediterranean was quite different from that north of the Alps. While the southern Alpine foreland vegetation still resembled that of central Europe (for example Lake Fimon, Fig. 3.7-19: Pini et al., 2010), sites further south provide pollen diagrams that reflect dry summers and mild, wet winters as in the Holocene. In two areas there are complete Eemian sequences from different elevational belts with different amounts of precipitation: in Greece (Tenaghi Philippon (Fig. 3.7-21), Kopais and Ioannina: Tzedakis, 2000; Milner et al., 2013) and in central/southern Italy (Valle di Castiglione and annually laminated Lago Grande di Monticchio: Follieri et al., 1988; Allen and Huntley, 2009).

At the Mediterranean sites, the end of the Saalian glaciation period shows steppe vegetation, dominated by Chenopodiaceae, *Artemisia* and grasses. While north of the Alps the following Eemian is characterized by a rather consistent vegetation development over large distances, this is much more variable in southern Europe, mainly due to the stronger differences in morphology and geology.

The general course of vegetation development in Italy, as well as in Greece, shows characteristic

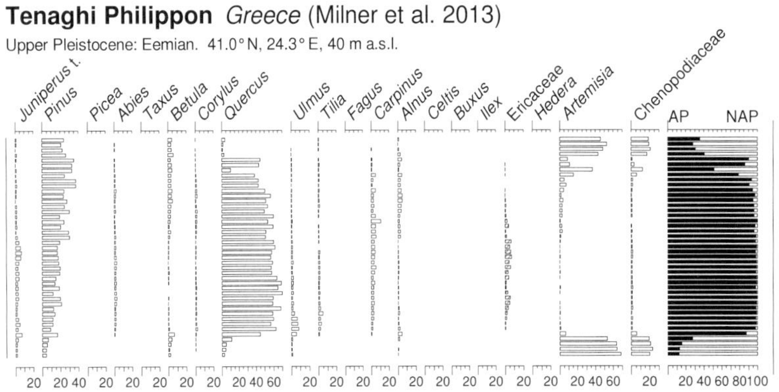

Fig. 3.7-21. Pollen diagram of the Eemian interglacial and its transitions from and to steppe formation from Tenaghi Philippon, northern Greece.

Table 3.7-1. Important Eemian and Weichselian sites in Europe. E: Eemian, B: Brörup, O: Odderade, Oe: Oerel, G: Glinde, PW: Pleni Weichselian.

		Eemian	Lower Weichselian	Pleni-Weichselian	long sequences	Literature
Northern Europe and Central European Lowland						
1	Fugla Ness (GB)	E				Birks and Ransom, 1969; Duller et al., 2002
2	Allt Odhar (GB)		B			Walker et al., 1992
3	Bobbitshole (GB)	E				West, 1957
4	Chelford (GB)		B			Simpson and West, 1958; Heijnis and v.d. Plicht, 1992
5	Wing (GB)	E				Hall, 1980
6	Alesund (N)			PW		Mangerud et al., 1981
7	Brumunddal (N)		B			Helle et al., 1981
8	Jämtland (S)		B			Lundqvist, 1967,1981
9	Peräpohjola (SF)		B			Korpela, 1969; Donner, 1983
10	Sokli (SF)	E	BO	Oe	x	Helmens et al., 2000; Bos et al., 2009
11	Scladina (B)	E	BO			Bastin et al., 1986
12	Amersfoort (NL)		B			Zagwijn, 1961
13	Eerbeek (NL)		BO	PW		van der Meer et al., 1984
14	Hengelo (NL)			PW		van der Hammen et al., 1967; van der Hammen, 1995
15	Denekamp (NL)			PW		van der Hammen et al., 1967; van der Hammen, 1995
16	Helgoland (D)	E				Behre, 1970
17	Brokenlande (D)	E				Menke and Ross, 1967
18	Odderade (D)	E	BO			Averdieck, 1967, 1998
19	Oerel (D)	E	BO	Oe+G	x	Behre and Lade, 1986; Behre, 1989
20	Osterwanna (D)	E	B			Behre, 1974
21	Rederstall (D)	E	BO			Menke and Tynni, 1984
22	Groß Todtshorn (D)	E	BO	Oe	x	Caspers and Freund, 2001
23	Quakenbrück (D)	E	BO	Oe		Hahne et al., 1994
24	Neuenhaus (D)	E	B	Oe		Caspers and Freund, 2001
25	Bispingen (D)	E				Müller, 1974b
26	Gröbern (D)	E	BO			Litt, 1994
27	Kittlitz (D)	E	BO			Erd, 1973
28	Brörup Hotel (DK)		B			Andersen, 1961
29	Rodebaek (DK)	E	B			Andersen, 1961
30	Hollerup (DK)	E				Andersen, 1965
31	Stenberget (S)	E	B			Berglund and Lagerlund, 1981
32	Wladyslawow (PL)	E	BO			Tobolski, 1991
33	Naklo (PL)	E				Noryskiewicz, 1978
34	Imbramowice (PL)	E				Mamakowa, 1989
35	Zgierz-Rudunki (PL)	E	BO			Jastrzębska-Mamełka, 1985
36	Dziewule (PL)	E				Bińka and Nitychoruk, 2001
36	Horoszki Duze (PL)	E	BO	Oe+G	x	Granoszewski, 2003
37	Solniki (PL)	E	B			Kupryjanowicz, 2008
38	Dzierniakowo (PL)	E	BO			Kupryjanowicz, 2008
39	Korochovchiina (BEL)	E	BO			Voznyachuk, 1978
40	Stari Bezradychy (UKR)	E	W	PW		Gerasimenko, 2000
41	Druskininai (LT)	E				Kondratiene, 1996
42	Mikulino (BEL)	E				Grichuk, 1984
43	Nizhnyaya Byarshchina (RS)	E				Grichuk, 1984
44	Chermenino (RS)		B			Zarrina et al., 1981
45	Domozerovo (RS)	E				Gey et al., 2001
46	Yamozero (RS)		O	PW		Henriksen et al., 2008

Southwestern and southern Central Europe including the Alps

47	Plateau du Devès (F)	E	BO	PW	x	Pons et al., 1992
48	Ruisseau de l'Amourette (F)	E	B			Gremmen et al., 1984
49	Les Echets (F)	E	BO	Oe	x	de Beaulieu and Reille, 1984a,b
50	Grande Pile (F)	E	BO	PW	x	Woillard, 1975; de Beaulieu and Reille, 1992a
51	Meikirch (CH)	E			x	Welten, 1982a
52	Gondiswil (CH)	E	BO	PW	x	Wegmüller, 1992
53	Sulzberg (CH)	E	BO	Oe	x	Welten, 1982a
54	Gossau (CH)		O	PW		Schlüchter et al., 1987
55	Dürnten (CH)	E	BO	Oe, PW	x	Welten, 1982a
56	Jammertal (D)	E				Müller et al., 2005
57	Wurzach (D)	E	BO	Oe		Grüger and Schreiner, 1993
58	Eurach (D)	E				Beug, 1979
59	Samerberg (D)	E	BO	Oe	x	Grüger, 1979
60	Zeifen (D)	E				Jung et al., 1972
61	Mondsee (A)	E	BO			Klaus, 1987; Drescher-Schneider, 2000b
62	Nieselach (A)	E				Fritz, 1992; Draxler, 2000
63	Azzano Decimo (I)	E	BO	PW		Pini et al., 2009
64	Lake Fimon (I)	E	BO	PW	x	Pini et al., 2010
65	Bulhary (CZ)			PW		Rybníčková and Rybníček, 2014
66	Safarka (SK)			PW		Jankovská et al., 2002

Southern Europe

67	Area Longa (SP)		BO			Ramil-Rego et al., 2000
68	Padul (SP)		BO	PW	x	Pons and Reille, 1988
69	Carihuela Cave (SP)			PW		Carrión, 1992a
70	Cova Beneito (SP)			PW		Carrión, 1992b; Carrión and Munuera, 1997
71	Navarres (SP)			PW		Carrión and van Geel, 1999
72	Lagaccione (I)		BO	PW		Magri, 1999
73	Lago di Vico (I)		O	PW		Magri and Sadori, 1999
74	Valle di Castiglione (I)	E	BO	PW	x	Follieri et al., 1988
75	Monticchio (I)			PW		Watts et al., 1996a
76	Canolo Nouvo (I)			PW		Grüger, 1977
77	Ioannina (GR)	E	BO	PW	x	Tzedakis, 1994
78	Xinias (GR)			PW		Bottema, 1979
79	Tenaghi Philippon (GR)	E	BO	PW	x	Wijmstra, 1969; Milner et al., 2013
80	Staro Orjachowo (BG)	E				Bozilova and Djankova, 1976

Table 3.7-1. continued

Mediterranean conditions in the first half of the Eemian with a typical shrub vegetation of *Olea, Phillyrea* and *Pistacia* and a considerable amount of evergreen oaks, probably mainly *Quercus ilex*. These elements decreased or disappeared from the middle of the interglacial period and were mainly replaced by deciduous oaks, which at higher elevations, such as Monticchio, were already common from the beginning of the interglacial. Other trees which increased in abundance were *Alnus, Pinus* and *Betula* and, perhaps at higher elevations, *Abies* and *Fagus*, the latter is, however, almost lacking from Greece. These changes in the vegetation indicate a climatic shift from the typical Mediterranean climate to cooler and more humid conditions in the second half of the Eemian and correspond well with the changes in the region north of the Alps.

Carpinus, the dominant tree in the younger part of the Eemian north of the Alps, was present throughout the interglacial in the south and this was also the case with *Corylus*, however, in much smaller quantities.

Most interesting is the occurrence of the deciduous tree *Zelkova*, a genus of the Ulmaceae family, which has been recorded in several sites in southern as well as northern Italy but became extinct after the early Weichselian (Follieri et al., 1986). Nowadays there are only a few relic stands of this genus on some Mediterranean islands (Sicily, Crete) and in the Caucasus.

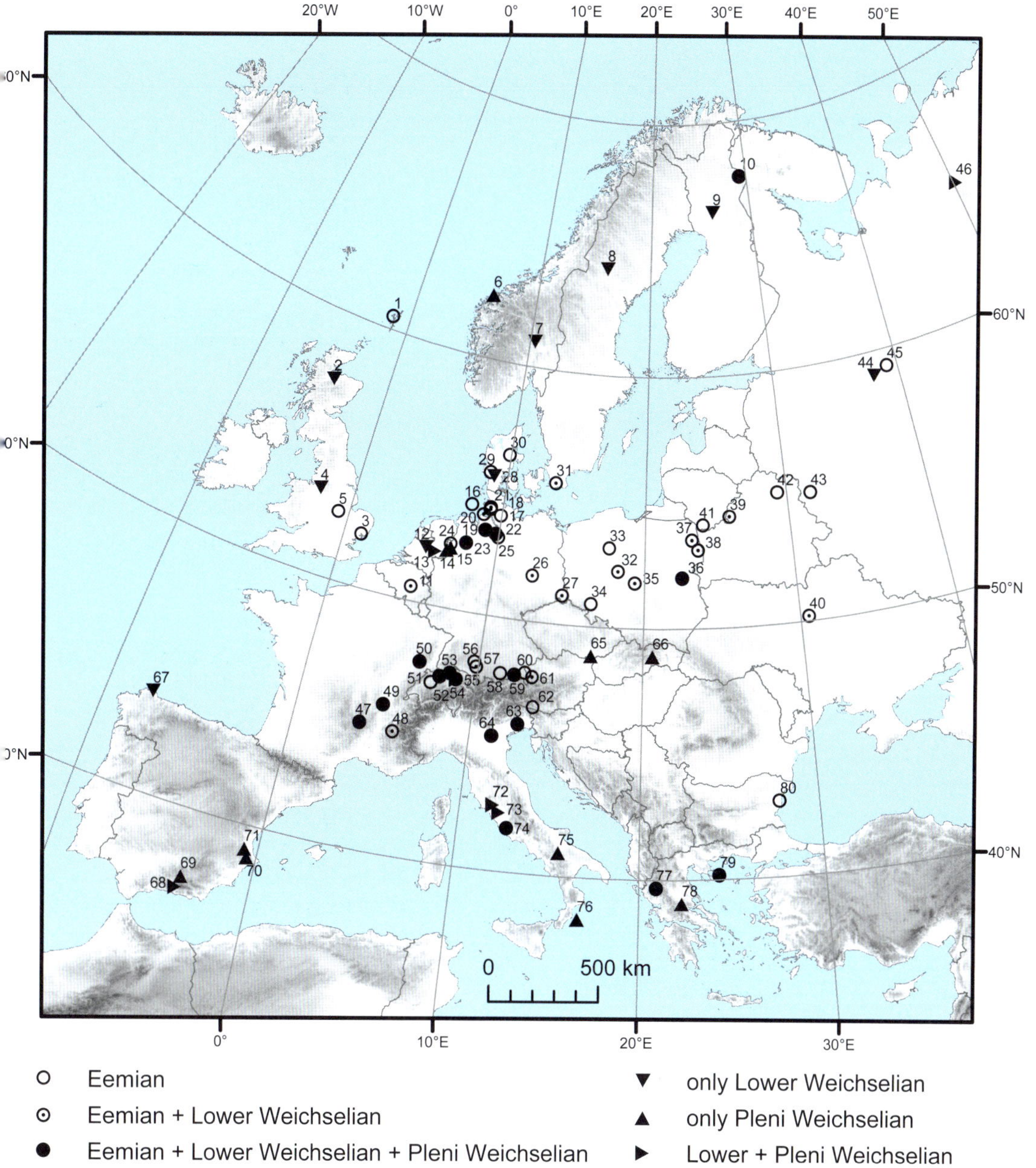

Fig. 3.7-22. Location of important Eemian and Weichselian sites in Europe. Numbers correspond to Table 3.7-1.

The Eemian is followed by the first Weichselian stadial and, in Italy, the forests became open and a wooded steppe was formed with trees scattered in a ground vegetation that was formed by grasses, *Artemisia*, Chenopodiaceae and *Ephedra* together with many heliophytic species. Due to a strong climatic gradient towards the east, the steppe in Greece was much drier and the vegetation was dominated by *Artemisia* and Chenopodiaceae.

The Eemian fauna is well-known from many sites rich in bones, including a famous one below Trafalgar Square in London (Stuart, 1976). In particular the mammalian fauna had quite a number of species which have since disappeared. At that time the forest elephant, *Palaeoloxodon antiquus*, the deer species *Megaloceros giganteus*, *Cervus elaphus* and *Cervus dama*, as well as the roe deer *Capreolus capreolus*, the forest rhinoceros *Stephanorhinus kirchbergensis*, the aurochs *Bos primigenius*, the wild boar

Sus scrofa and others lived in the European forests. The European pond turtle *Emys orbicularis*, also had a wide distribution in the Eemian that reached northwards as far as central Sweden.

In the rivers, even *Hippopotamus* entered central Europe from the Mediterranean, spread up the Rhone and down the Rhine and across the Channel into southern England. Its bones were found together with those of the water buffalo *Bubalus murrensis*, at several sites in the western parts of Europe but not in the more continental region east of the Rhine catchment (von Koenigswald, 2007).

The so-called forest elephant fauna spread from the Mediterranean to the western part of Europe and was restricted to the Atlantic climate region north of the Alps. The number of large mammals decreased to the east and there are almost no records from Russia.

These animals were hunted and eaten by *Homo neanderthalensis* who had occupied Europe in the Eemian. At Lehringen, near Bremen in Germany, an almost complete skeleton of a forest elephant has been found together with many flint artefacts and a spear made of *Taxus* wood (Adam, 1951). Also in Germany, another forest elephant had been slaughtered at Gröbern, north-east of Leipzig.

3.8

The Weichselian glacial period

The Weichselian, called Würmian in the Alpine region, represents the best known glacial period and is a good example for the interruption of a very long cold period by interstadial phases of various durations. In total the Weichselian had a length of around 100 000 years, but only parts of this period were glaciated. The Last Glacial Maximum (LGM), when the glaciers from the north and from the high mountains of the Alpine range reached their greatest extension, occurred in MIS 2 and lasted about 6000 years, culminating around 21 000 yr BP.

Fortunately, there are several long terrestrial pollen records that cover large parts of the Weichselian and, in many cases, enable the correlation of the major interstadials and stadials over great distances. The most important sites with long terrestrial sequences are La Grande Pile and Les Echets in France, Oerel and Groß Todtshorn in northern Germany, Samerberg, Gondiswil, Sulzberg and Mondsee close to and in the Alps, Sokli in northern Finland, Lago di Castiglione, Lagaccione and Monticchio in Italy, Carihuela and Villarquemado in Spain and Tenaghi Philippon and Ioannina in Greece (see Tab. 3.7-1 and Fig. 3.7-22). In the north they comprise mainly peat layers, separated by minerogenic material, while in France and in southern Europe there are crater lakes with long continuous sequences of lacustrine deposits.

3.8.1

The Lower (early) Weichselian

At the end of the Eemian there was a rapid decrease in temperature in Europe which was mainly caused by the southward displacement of the Gulf Stream, resulting in substantial cooling in northern Europe and drier conditions in the Mediterranean region. The polar timberline retreated southward from 69°N in northernmost Scandinavia to 52°N in central Europe, resulting in a southward withdrawal of the temperate plant species (Müller and Kukla, 2004). The first stadial was the Herning (Melisey I in France) in the course of which most parts of central and northern Europe became almost completely deforested, apart from some *Pinus* in the southern areas. Open vegetation with grasses, *Artemisia*, dwarf shrubs and heliophytes prevailed, complemented by shrubs, such as *Juniperus* and *Betula humilis* (Caspers and Freund, 2001).

On the other hand, in south-east France, open forests with *Carpinus*, *Abies* and *Picea* remained, as in Les Echets near Lyon and La Grande Pile, south of the Vosges, while elevations above 1000 m were occupied by grass steppe, as could be seen at Lac de Bouchet (Pons et al., 1992). Further south, in the Eu-Mediterranean zone around Rome (Val di Castiglione) and in north-east Greece (Tenaghi Philippon), dry steppe vegetation with prodigious *Artemisia* and Chenopodiaceae dominated in the lowlands, accompanied by only a little *Pinus* and some *Juniperus*, while at higher elevations in southern Italy quite a number of deciduous tree species occurred, forming a wooded steppe (Monticchio) (Follieri et al., 1988; Allen and Huntley, 2009).

After the Herning stadial, the climatic conditions improved considerably and two long, temperate forested interstadials followed that characterise the Lower Weichselian. They are called Brörup (St. Germain I in France and Italy) and Odderade (St. Germain II) and correspond to MIS 5c and 5a. They were separated by the Rederstall stadial (Melisey II in France). During both interstadials most parts of Europe as far north as Scandinavia were again occupied by forests.

There are many climate records for both the early and later parts of the Weichselian and they show that the climate of Europe was always highly determined by the influence of the North Atlantic Ocean. During the Lower Weichselian interstadials Brörup and Odderade there was a strong climate gradient from west to east but more especially from north to south, which was considerably stronger than in either the Eemian or the Holocene. This was primarily caused by two factors. The first is the southward displacement of the Gulf Stream mentioned above and the second the lowering in sea-level so that most parts of the shallow North Sea became dry, resulting in the marked increased continentality of

Allt Odhar *Great Britain* (Walker et al. 1992)

Upper Pleistocene: Brörup. 57.4° N, 4.0° W, 371 m a.s.l.

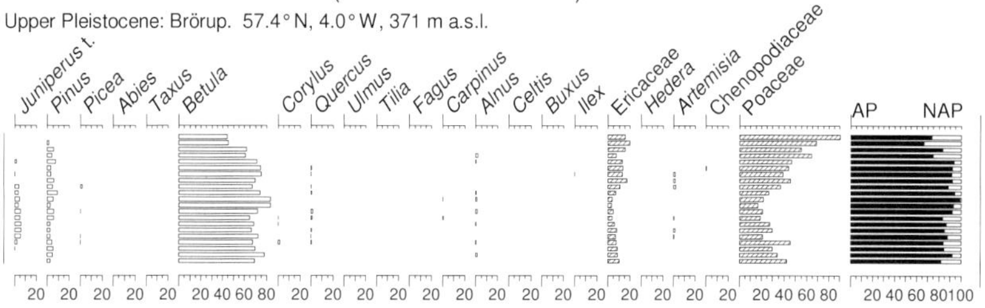

Fig. 3.8-1. Pollen diagram of the Brörup interstadial from Allt Odhar, Scotland.

Brumunddal *Norway* (Helle et al. 1981)

Upper Pleistocene: Brörup. 61.0° N, 10.4° E, 395 m a.s.l.

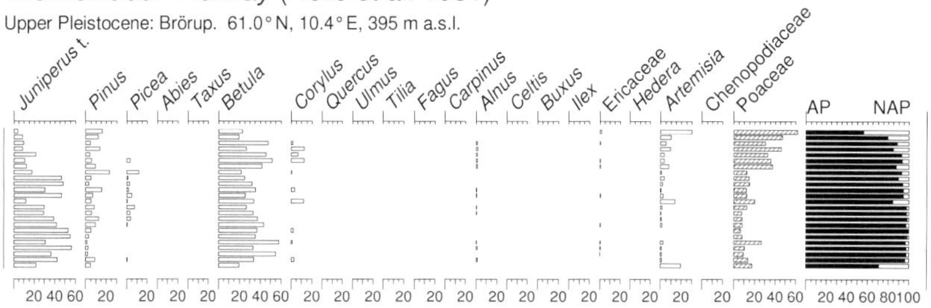

Fig. 3.8-2. Pollen diagram of the Brörup interstadial from Brumunddal, south-east Norway.

Oerel *Germany* (Behre and Lade 1986)

Upper Pleistocene: Brörup and Odderade. 53.5° N, 9.1° E, 12 m a.s.l.

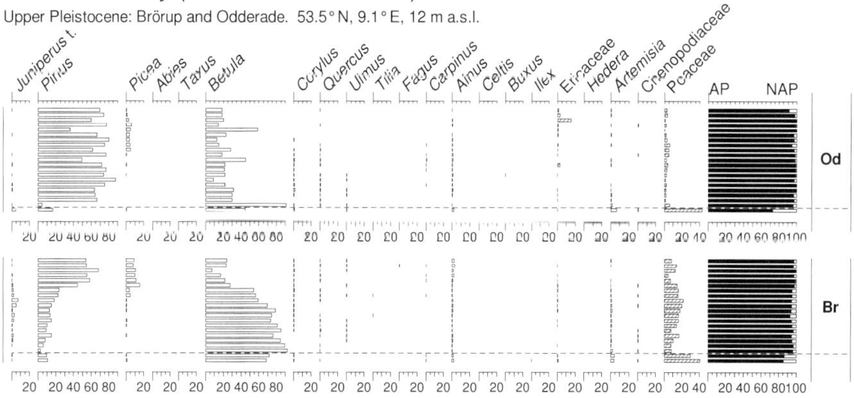

Fig. 3.8-3. Pollen diagrams of the Brörup interstadial (Br) and the Odderade interstadial (Od) from Oerel, north-west Germany.

Zgierz-Rudunki *Poland* (Jastrzebska-Mamelka 1985)

Upper Pleistocene: Brörup and Odderade. 51.8° N, 19.4° E, 201 m a.s.l.

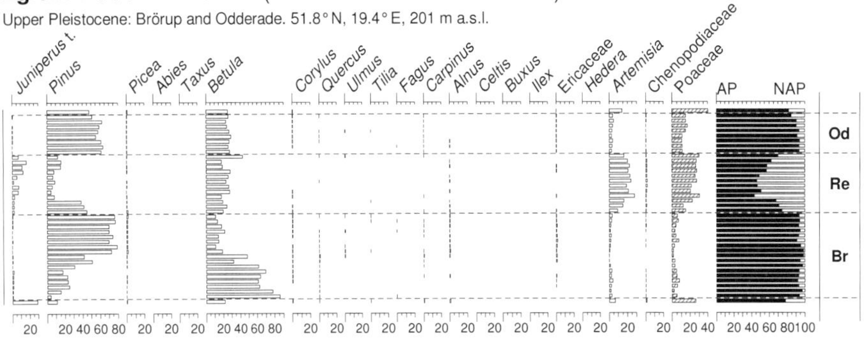

Fig. 3.8-4. Pollen diagrams of the Brörup interstadial (Br), the Rederstall stadial (Re) and the Odderade interstadial (Od) from Zgierz-Rudunki, central Poland.

Samerberg *Germany* (Grüger 1979)

Upper Pleistocene: Brörup and Odderade. 47.8° N, 12.2° E, 594 m a.s.l.

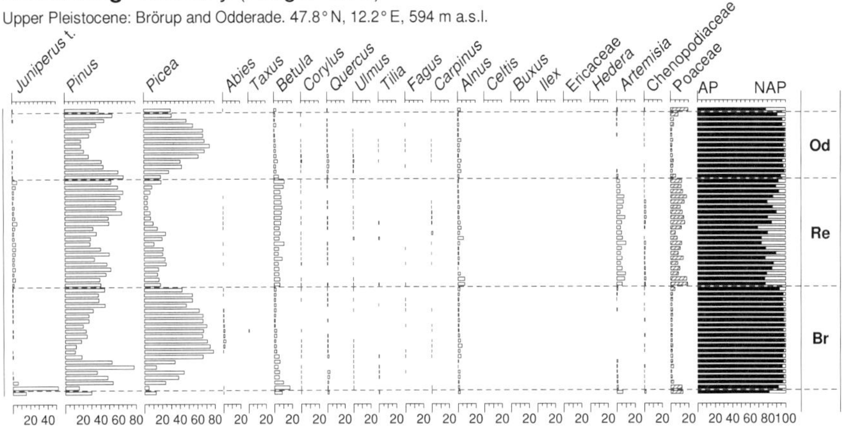

Fig. 3.8-5. Pollen diagram of the Brörup interstadial (Br), the Rederstall stadial (Re) and the Odderade interstadial (Od) from Samerberg, Alps, Germany.

Mondsee *Austria* (Drescher-Schneider 2000b)

Upper Pleistocene: Brörup and Odderade. 47.8° N, 13.3° E, 534 m a.s.l.

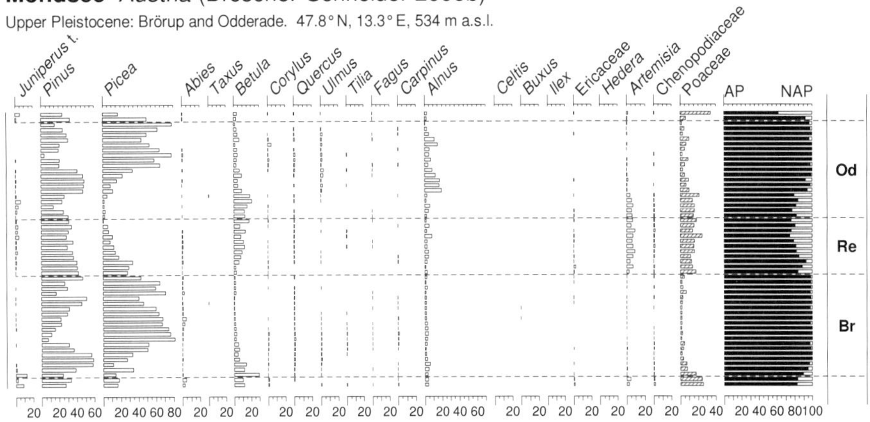

Fig. 3.8-6. Pollen diagram of the Brörup Interstadial (Br), the Rederstall stadial (Re) and the Odderade interstadial (Od) from Mondsee, Alps, Austria.

Les Echets *France* (de Beaulieu and Reille 1984)

Upper Pleistocene: Brörup and Odderade. 45.8° N, 4.9° E, 267 m a.s.l.

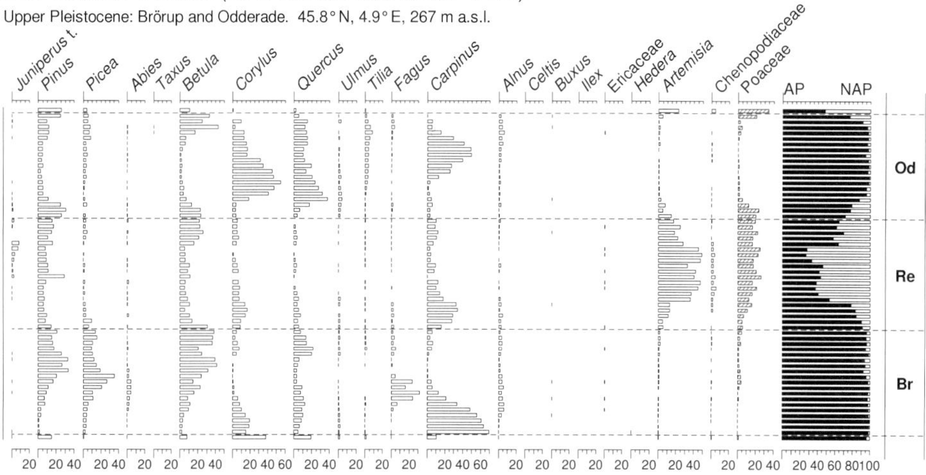

Fig. 3.8-7. Pollen diagram of the Brörup interstadial (= St. Germain I) (Br), the Rederstall stadial (= Melisey II) (Re) and the Odderade interstadial (= St.Germain II) (Od) from Les Echets, south-east France.

Lake Fimon *Italy* (Pini et al. 2010)

Upper Pleistocene: Brörup and Odderade. 45.3° N, 11.3° E, 23 m a.s.l.

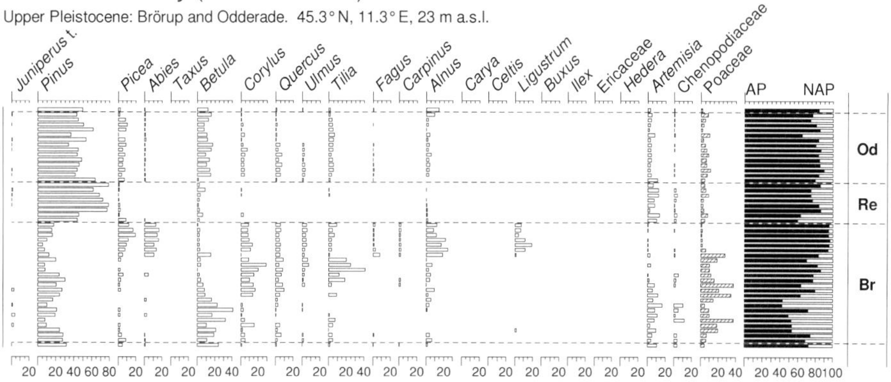

Fig. 3.8-8. Pollen diagrams of the Brörup interstadial (Br), the Rederstall stadial (Re) and the Odderade interstadial (Od) from Lake Fimon, Alpine foreland, Italy.

Lagaccione *Italy* (Magri 1999)

Upper Pleistocene: Brörup and Odderade. 42.6° N, 11.8° E, 355 m a.s.l.

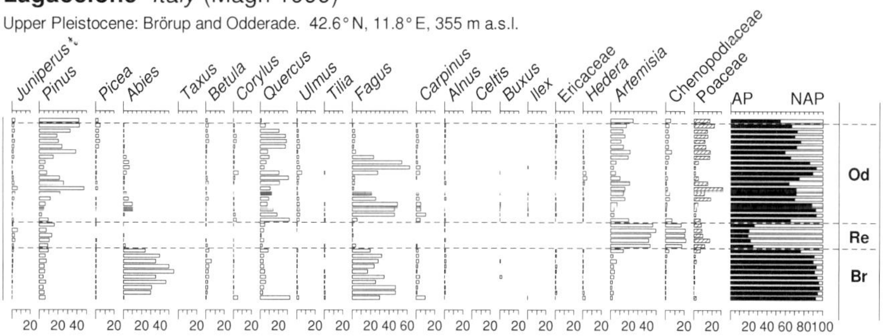

Fig. 3.8-9. Pollen diagram of the Brörup interstadial (Br), the Rederstall stadial (Re) and the Odderade Interstadial (Od) from Lagaccione, central Italy.

north-west Europe. Climate reconstructions show that winter temperatures were very low while summer temperatures were at almost the same level as today.

The absence of temperate forests in Europe north of the Alps was mainly caused by low (winter) temperatures combined with low precipitation values. Both of these Lower Weichselian interstadials lasted long enough to allow a reimmigration of temperate trees even from distant areas but, instead, the forests were boreal and their most characteristic member was *Larix decidua* which occurred in traces westwards as far as eastern France.

The type locality of the Brörup is located in Jutland. Here, as in the adjoining lowlands from the Netherlands (Amersfoort) across Germany (for example Oerel, Fig. 3.8-3: Behre and Lade, 1986; or Gröbern) to Poland (for example Zgierz-Rudunki, Fig. 3.8-4: Jastrzębska-Mamełka, 1985; or Dzierniakowo, Kupryjanowicz 2008), Belarus (Korochounchiina) and northern central Russia (Chermenino) the boreal forests were closed. The Brörup interstadial in this region is subdivided into a *Betula* zone followed by a *Pinus* zone. During the latter, temperature increased slightly and *Picea* as well as *Alnus* spread across the Baltic Sea. *Picea* also spread to north-eastern Poland and beyond. In Fennoscandia the mild summer temperatures during the Brörup allowed birch forests to grow with *Larix* and *Alnus* as far as northern Finland and the forest development was similar to the boreal forests in the Holocene in this area (Sokli, Helmens et al., 2000; Helmens, 2014).

To the more Atlantic north-west, as in Brumunddal, Norway (Fig. 3.8-2; Helle et al., 1981), *Pinus* became rare but *Larix* was still present in the open birch forests there and, even in the extreme north-west, open forests with predominantly *Betula* existed at that time, as could be demonstrated in Allt Odhar, northern Scotland (Fig. 3.8-1; Walker et al., 1992). Traces of some thermophilous trees indicate that these must have occurred in small quantities at suitable localities in the southern parts of central Europe.

In contrast to the north, south-west Europe had an Atlantic character during the Lower Weichselian. Therefore the vegetation development in eastern and south-eastern France was totally different from that of northern central Europe. In France (La Grande Pile, Les Echets, Fig. 3.8-7: de Beaulieu and Reille, 1984a,b) thermophilous trees such as *Quercus* and *Carpinus*, as well as *Corylus*, played an important role in the forests and the vegetation was similar to that of the Eemian in northern central Europe: the same can be seen in the southern Alpine foreland, where even *Tilia* played an important role (Lake Fimon, Fig. 3.8-8: Pini et al., 2010).

For many years there had been an intensive discussion as to whether these interstadials represent previously unknown interglacials until the correlation with the cooler Lower Weichselian interstadials in the north could be assured.

There is good evidence of the Brörup in the Mediterranean from the Italian crater lakes (Lagaccione, Fig. 3.8-9: Magri, 1999; Valle di Castiglione; Monticchio) and also in Greece (Tenaghi Philippon, Fig. 3.8-10: Wijmstra, 1969) which shows that temperate forests with deciduous *Quercus* and *Carpinus* among others developed and even *Buxus* occurred in this period. To the east the forests became drier and more open with *Pinus*, *Artemisia* and grasses while in Padul in southern Spain (Pons and Reille, 1988) Mediterranean evergreen *Quercus* forests were present. The most important feature of the Brörup in south and south-west Europe is, however, the strong expansion in the middle part of this interstadial of *Fagus*, which at some sites reached more than 40% of the terrestrial pollen (Magri, 1999). This is in sharp contrast to the Eemian, where *Fagus* was lacking in many areas of Europe. It was, however, common in some areas of southern Europe (e.g. Bulgaria; Fig. 3.7-20) and occurred also in north Anatolia (Shumilovskikh et al., 2013) from where it may have immigrated.

Within the Brörup interstadial a climatic setback occurred which is clearly visible in the vegetation development in France (La Grande Pile, Les Echets), where the thermophilous trees were replaced by *Pinus* and *Betula*. This setback has been called the Montaigu event (Reille et al., 1992). It is also very distinct in Italy (Val di Castiglione) and particularly in Greece (Tenaghi Philippon, Ioannina), where the vegetation cover became open with high amounts of *Artemisia* and Chenopodiaceae. In central Europe this climatic deterioration did not change the boreal forests significantly and the pollen diagrams show no or only slight evidence of a cooler phase. The earlier phase, before this climatic setback, had previously been described as a separate interstadial, called Amersfoort (Zagwijn, 1961) but was later included in the Brörup stadial l. (Behre, 1989; Caspers and Freund, 2001). Further to the east, in eastern Poland (for example Solniki, Kupryjanowicz, 2008) and Belarus (Korochounchiina, Voznyachuk et al., 1978) the Montaigu event was again very distinct but as a dry phase with more or less open forest vegetation, where *Artemisia* and some Chenopodiaceae were common.

Apart from the strong expansion of *Fagus*, which, however, did not cross the Alps, another two species that were lacking in the northern Eemian – *Picea omorika* and the dwarf shrub *Bruckenthalia spiculifolia* – spread to the north as far as to the coasts of the North Sea and the Baltic during the Brörup. Both are recorded by pollen and numerous macroremains. In the Holocene they failed to return and the modern distribution area of both is restricted to very small districts in the mountainous regions of the Balkans.

On the basis of the thickness of the deposits the duration of the Brörup must have been almost as long as the preceding Eemian. In Rederstall, northern Germany the Brörup is represented by 10.5 m diatomite layers of which some parts are annually laminated. An extrapolation on the basis of the varve counts gave a length of the Brörup of between 5800 and 10 500 years (Grüger, 1991), which is in some degree comparable to the 11 000 – 17 000 year duration of MIS 5e. As with the Eemian, the length of the Brörup in the southern European Monticchio is longer: varve countings gave a duration of 14 900 years and 8800 years for the Odderade interstadial (Allen and Huntley, 2000).

The Brörup was followed by a severe climatic deterioration, the Rederstall stadial (Melisey II in France and Italy). At the beginning of this stadial the mean January temperature in northern central Europe fell to around -18 °C while the mean July temperature still reached about +12 °C (Kühl et al., 2007). Europe, including most of the Mediterranean zone, was treeless at this time and tundra developed, dominated by *Artemisia*, Poaceae and Chenopodiaceae together with some *Juniperus*. In the north-western parts Ericaceae were also common.

The second great interstadial of the Weichselian, the Odderade, was named after a site in northern Germany (Averdieck, 1967). In many ways, including the climate gradient, it is similar to that of the Brörup and its duration is also comparable but there are some differences. North of the Alps and in Poland the initial *Betula* zone was considerably shorter than in the Brörup and likewise the spread of temperate trees in France showed no delay as in the preceding interstadial. This means that at the onset of the Odderade the climatic improvement was faster than in the Brörup. It did not, however, reach the same climatic level. In central Europe, *Alnus,* and in France, the temperate trees, especially *Fagus,* were less frequent. Generally the forests were more open than in the Brörup; in particular to the east (Belarus) the share of *Artemisia* increased considerably in the pollen diagrams. A good indicator of temperature is the thermophilous shrub *Buxus*, which regularly occurred in the Brörup as far north as France and was also present in the Austrian Mondsee (Fig. 3.8-6; Drescher-Schneider, 2000b) but did not appear in these sites in the Odderade.

In the Mediterranean region thermophilous trees prevailed in the Odderade and, also here, the vegetation was similar to the Brörup interstadial. *Fagus*, however, that had reached the remarkable value of 40% in the Brörup at Les Echets, was generally much lower in the Odderade (Fig. 3.8-7, de Beaulieu and Reille, 1984).

Table 3.8-1. Subdivision of the Weichselian period.

Holocene			
Weichselian (Würmian)	Upper (Late) Weichselian	*Younger Dryas-stadial*	
Weichselian (Würmian)	Upper (Late) Weichselian	Alleröd-interstadial	
Weichselian (Würmian)	Upper (Late) Weichselian	*Older Dryas-stadial*	
Weichselian (Würmian)	Upper (Late) Weichselian	Bölling-interstadial	
Weichselian (Würmian)	Middle (Pleni) Weichselian	*Last Glacial Maximum*	
Weichselian (Würmian)	Middle (Pleni) Weichselian	Denekamp-interval	
Weichselian (Würmian)	Middle (Pleni) Weichselian	Hengelo-interval	
Weichselian (Würmian)	Middle (Pleni) Weichselian	(Moershoofd-interval-complex)	
Weichselian (Würmian)	Middle (Pleni) Weichselian	Glinde-interstadial	
Weichselian (Würmian)	Middle (Pleni) Weichselian	*Ebersdorf-stadial*	
Weichselian (Würmian)	Middle (Pleni) Weichselian	Oerel/Dürnten-interstadial	
Weichselian (Würmian)	Middle (Pleni) Weichselian	*Schalkholz-stadial*	
Weichselian (Würmian)	Lower (Early) Weichselian	Odderade-interstadial (St. Germain II)	
Weichselian (Würmian)	Lower (Early) Weichselian	*Rederstall-stadial (Melisey II)*	
Weichselian (Würmian)	Lower (Early) Weichselian		Brörup s. str.
Weichselian (Würmian)	Lower (Early) Weichselian	Brörup-Interstadial (St. Germain I)	*Montaigu event*
Weichselian (Würmian)	Lower (Early) Weichselian		Amersfoort
Weichselian (Würmian)	Lower (Early) Weichselian	*Herning-stadial (Melisey I)*	
Eemian–interglacial			

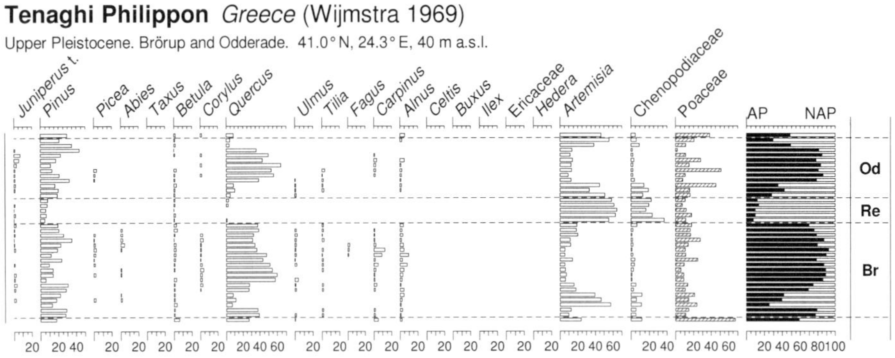

Fig. 3.8-10. Pollen diagram of the Brörup interstadial (Drama) (Br), the Rederstall stadial (Re) and the Odderade interstadial (Elevtheroupolis) (Od) from Tenaghi Philippon, northern Greece.

3.8.2

The Middle (Pleni-) Weichselian

The Pleni-Weichselian spans Marine Isotopic Stages 4-2, that is from 75 000 to 24 000 yr BP, that is before the Last Glacial Maximum (LGM) and continues until 14 500 yr BP, the start of the Late-glacial. It is characterised by a considerably colder climate than during the Lower Weichselian and culminated in the LGM around 21 000 yr BP. There were no strong climatic oscillations with significant ameliorations allowing the growth of forests in the north as in the Lower Weichselian. The climatic and vegetational differences between stadials and interstadials were often only slight and transitional. The further north, the more distinct are the separations between stadials and interstadials due to the sensitive position at the edge of the vegetation, while in the south, particularly in Italy, the pollen diagrams from Pleni-Weichselian sites often show rather uniform conditions. The chronological correlation of the various sites during this period becomes more exact than for earlier periods as we now enter the range of radiocarbon dating.

3.8.2.1

The Pleni-Weichselian north of the Alps

For a long time the general view of the glaciation history in the north was based on the opinion that there had been a gradual growth of the Fennoscandian ice sheet over several tens of thousands of years, starting in MIS 4 and culminating in the LGM (MIS 2). Recent data, however, have revealed a dynamic ice-sheet behaviour including a fast build-up and decay. For MIS 3 extensive ice-free areas over central Scandinavia and Finland can be reconstructed mainly through new vegetation evidence but also from mammoth remains (Helmens et al., 2000; Helmens, 2014). This is in good temporal agreement with the long ice-free phase in western Norway, which had already been described as the Ålesund interstadial on the basis of shell remains from several localities (Mangerud et al., 1981).

The Pleni-Weichselian started with the Schalkholz stadial (MIS 4) in the course of which a major glaciation of Fennoscandia took place. At this time arctic to subarctic conditions prevailed in most parts of central Europe.

In the subsequent MIS 3 the climatic conditions improved considerably all over Europe. A well-developed interstadial has been found in the long terrestrial record of Oerel in north-west Germany and has been called after this type section (Fig. 3.8-11; Behre and Lade, 1986; Behre, 1989). Since then the Oerel has been found above the Odderade in another three sites in that region: Groß Todtshorn, Neuenhaus and Quakenbrück (Caspers and Freund, 2001). Its organic layers are clearly delimited by sand layers: in Oerel itself this interstadial even includes well preserved *Sphagnum* peat which has been ^{14}C dated to 58–54 ka BP (Behre and van der Plicht, 1992). The net precipitation at that time must have been such that the formation of a raised bog was possible. This interstadial can be correlated in the northern pre-Alps with the third interstadial in the continuous sequences of

Sulzberg and Dürnten (Welten, 1982a) as well as with Samerberg 3 (Grüger, 1979) and Wurzach (Grüger and Schreiner, 1993). In the circum-alpine region this interstadial is usually called Dürnten.

In this third Weichselian interstadial in northern Germany open treeless shrub tundra extended over the north German lowlands, where *Betula nana* was widespread and, according to the macroremains, *Betula humilis* was also occasionally present. In the upper part of the interstadial *Juniperus* and especially Ericales spread strongly. To the south climatic conditions had been considerably more temperate and all the pre-Alpine sites mentioned above showed forest vegetation. These forests consisted mainly of *Picea* and *Pinus* in Dürnten (Fig. 3.8-12; Welten, 1982a) with a large proportion of *Larix*, while in the westernmost site, Sulzberg, notable amounts of *Corylus* and *Alnus* occurred.

Although large parts of central Europe had been treeless during MIS 3, several sites from northern Finland, the Baltic Countries and north-west Russia show that there must have been forest vegetation at these high latitudes. A key site is Sokli in northernmost Finland, where a long terrestrial record spans the period from the Eemian to MIS 3. In the Tulppio interstadial (Fig. 3.8-13; Bos et al., 2009), which is correlated with the Oerel, low-arctic tundra vegetation prevailed but considerable proportions of *Pinus* and *Betula* pollen and even some *Alnus* indicate that forests must have existed only a few hundred kilometres south or south-east of this site (Bos et al., 2009). The conclusion is that ice-free conditions with *Betula*-dominated vegetation, including tree birch, persisted over large areas of Fennoscandia during major parts of MIS 3 up until c. 35 000 BP (Helmens, 2014). Further to the north-east new investigations from Kotelny island in northernmost Siberia show at least local occurrence of *Larix* around 38 000 BP, and this is supported by macroremains (van Geel et al., 2017).

In the long sequence of the Oerel site another distinctly delimited interstadial follows. After the Ebersdorf stadial another peat layer was formed that represents the Glinde interstadial for which several ^{14}C-dates provide an age of 51 000–48 000 yr BP (Behre and van der Plicht, 1992). The Glinde was also dominated by open shrub tundra with *Betula nana*, Ericales, grasses and a great variety of heliophilous herbs thus indicating more open vegetation than in the preceding Oerel interstadial, again without any trees. Climatic conditions were very cold: the stenotherm beetles indicate average January temperatures of around -17.5° C compared with -5 – -10° C during the preceding Oerel and less than -10° C in most parts of the Brörup while the average July temperature was about 9.5° C (Behre et al., 2005). Up until now there are no certain analogues to the Glinde interstadial but on the basis of the ^{14}C-dates tentative correlations can be made with the Pile oscillation at La Grande Pile in eastern France (de Beaulieu and Reille, 1992a) and with an isolated interstadial at Reichwalde, eastern Germany (Bohncke et al., 2008).

For the younger part of the Pleni-Weichselian in central Europe there are no sequences with interstadials in superposition but a great number of isolated more or less organic layers have been found. They show a treeless interstadial character with tundra vegetation and most of these sites were put into the interstadials described in the Netherlands

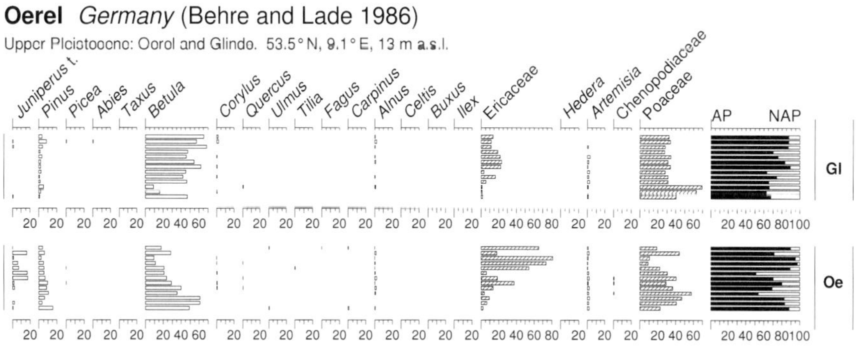

Fig. 3.8-11. Pollen diagrams of the Oerel interstadial (Oe) and the Glinde interstadial (Gl) from Oerel, north-west Germany (type section for both).

as Moershoofd, Hengelo and Denekamp (van der Hammen et al., 1967; Zagwijn, 1974). In most cases these classifications have been made on the basis of ^{14}C-determinations, since their stratigraphic position was often unclear.

The acceptance of these organic layers as interstadials has, however, to be taken with caution. Their lithological delimitation is often not clear and pure peat is rare. Moreover, the now numerous ^{14}C-dates from the key area in the Netherlands, Belgium and north-west Germany form a more or less continuous scatter which spans the period from Moershoofd to Denekamp (Ran, 1990) and this is similar for Britain and elsewhere. It has to be assumed that in central and north-west Europe the greater part of the middle Pleni-Weichselian represents a relatively mild phase during which, at edaphically favourable places, a more or less closed tundra vegetation became established and that, here and there, humus or peat formation occurred which may not necessarily reflect climatic improvements (Vandenberghe, 1985; Houmark-Nielsen et al., 2016). This long tundra period was probably interrupted by slight climatic ameliorations which should be called intervals. They led, for instance, during the Hengelo (about 39 000–36 000 yr BP) and Denekamp (about 32 000–28 000 yr BP) to shrub vegetation. Interstadial deposits from other European countries were placed into these chronostratigraphic units, originally defined in the Netherlands, only on the basis of ^{14}C-dates which are of varying reliability.

To sum up, the comparably mild phase of the Pleni-Weichselian in central and north-west Europe during MIS 3 is in good agreement with the long ice-free period in Fennoscandia at that time. However, average summer temperatures must have been below 10° C in most parts of central Europe since the climate did not allow forest vegetation during MIS3. This is in remarkable contrast with the north-east, where, as described above, boreal forests were even able to grow much further north. The climatic pattern must have been considerably different from that of today.

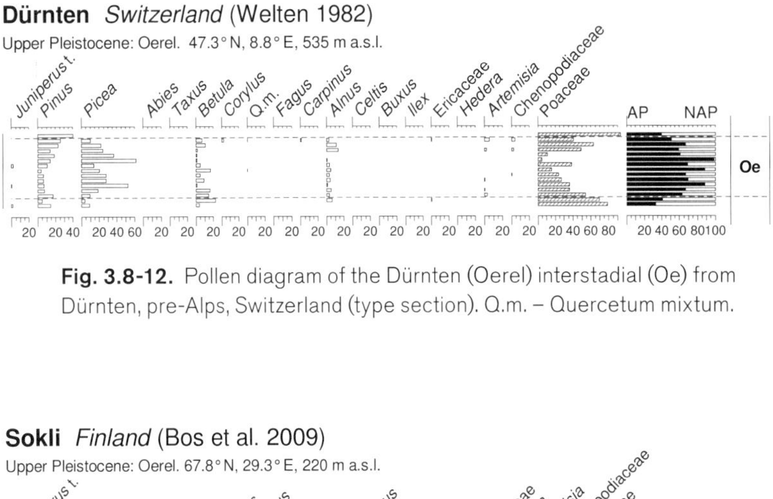

Fig. 3.8-12. Pollen diagram of the Dürnten (Oerel) interstadial (Oe) from Dürnten, pre-Alps, Switzerland (type section). Q.m. – Quercetum mixtum.

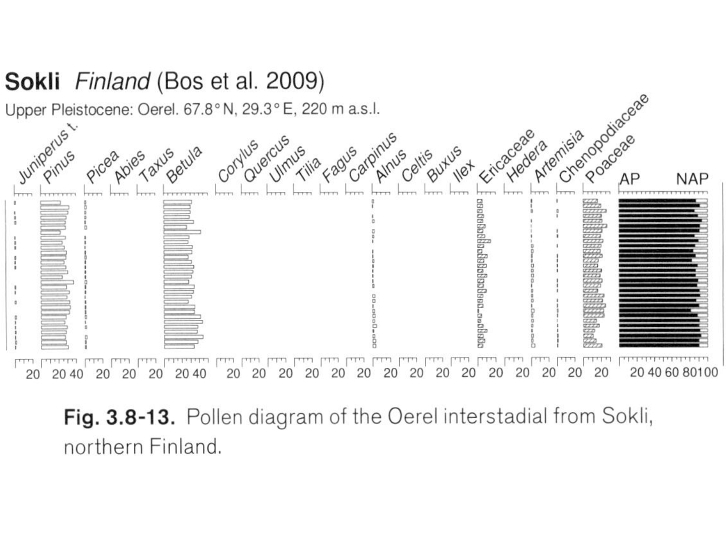

Fig. 3.8-13. Pollen diagram of the Oerel interstadial from Sokli, northern Finland.

3.8.2.2

The Pleni-Weichselian in southern Europe and France

There are quite a number of pollen profiles from southern Europe, as well as from France, mainly from crater lakes. They span the middle Weichselian and even the LGM and provide good evidence of the vegetation development. During this long period, phases of cool mixed woodland and wooded steppe alternated several times with cold or warm steppe periods (Allen and Huntley, 2000; Watts et al., 2000). There was a distinct west-east climatic gradient resulting in drier vegetation in stadial phases in Greece and culminating in desert steppe, as could be shown in the long record of the Turkish Lake Van (Litt et al., 2014).

In the Mediterranean interstadials in Italy, for which Monticchio is the best example, deciduous trees prevailed and the forests were mainly formed by *Quercus* (including even small amounts of evergreen oaks; Follieri et al., 1998), but *Fagus, Betula, Ulmus* and other deciduous trees together with the conifers *Pinus* and *Abies* are also documented in the crater lakes. The forests have been more or less open and their ground vegetation was dominated by grasses and *Artemisia* as well as by some Chenopodiaceae and other heliophytic species.

There is no problem in comparing the absolute chronology of the Weichselian, as derived from varve counts at Monticchio, with the Greenland ice core chronology. Nevertheless, the correlation of its various Pleni-Weichselian interstadials with those from La Grande Pile in France or from central/northern Europe is still difficult and tentative while the correlation of the Lower Weichselian interstadials is clear (Allen and Huntley, 2000).

On the Iberian peninsula archaeological excavations in several caves in SE Spain provided Weichselian pollen profiles from various altitudes. The longest sequence Carihuela starts already in the Eemian and covers most of the last glaciation (Carrión, 1992a,b; Carrión et al., 1999) while the peat deposit of Navarres, close to Valencia (Carrión and van Geel, 1999), spans the period across the LGM from 31 000 to the Holocene and shows a temperate period of probably more than ten thousand years in the middle Pleni-Weichselian when evergreen as well as deciduous oaks together with *Pinus pinaster* type played an important part in the vegetation.

Mediterranean species disappeared in the course of the Weichselian in the mountainous sites while in low altitudes xerothermic or mesic evergreen forests with mainly *Quercus* and *Pinus* existed but also Mediterranean shrubs like *Olea/Phillyrea* and particular in Padul *Erica arborea* were present in remarkable amounts beside of traces of Mediterranean tree species. Even taxa of the thermophilous Ibero-NW African flora like *Periploca, Withamia* and *Osiris* were present, which are extremely sensitive to frost. If these thermophilous taxa survived the LGM in sheltered pockets, the refugia of many other plants must have been much closer to central Europe than has so far been predicted.

In the Mediterranean region species such as *Erica arborea* indicate an oceanic climate in the west, while eastwards conditions became drier and steppe vegetation with *Artemisia* and Chenopodiaceae increased considerably during the middle Pleni-glacial, as can be seen at Tenaghi Philippon, Greece. Here this dry vegetation occupied the low elevations while in the mountains in northern Greece there were some forests. This forest belt likely had two treelines: besides the normal upper treeline a lower one was caused by drought.

Further to the north the important French sites La Grande Pile, Les Echets and Plateau du Deves (Reille and de Beaulieu, 1988) also cover the whole Weichselian Pleni-glacial and Late-glacial. During MIS3-4 treeless tundra or steppe prevailed here with grasses, *Artemisia*, Chenopodiaceae and a great variety of heliophytic herbs. In the diagrams several short peaks of *Pinus* or *Betula* indicate temporary local stands of small tree populations but no distinct interstadials. ^{14}C dates show that these episodes are not synchronous between the different sites and there are also only tentative analogues to the distinct interstadials of Oerel/Dürnten and Glinde in the adjoining central Europe. The vegetation of the middle Weichselian is explained in a similar way as in northern central Europe that it was adapted to a changing mosaic of strictly localized edaphic conditions.

3.8.2.3

The stadial vegetation during the Pleni-glacial up to the Last Glacial Maximum

For most of the long period of time from MIS 4–2 stadial phases existed but the climate of the individual stadials differed considerably. MIS 4, the Schalkholz stadial between the Odderade and Oerel interstadials, was very cold with ice advances to Denmark and beyond, as far as northernmost Germany and Poland (Houmark-Nielsen, 2010; Marks et al., 1995). The Ebersdorf stadial in the lower part of MIS 3 between the Oerel and Glinde interstadials was again very cold. During both stadials and the LGM in MIS 2, full arctic conditions prevailed in the periglacial parts of central Europe and its surroundings. As already described, the younger part of MIS 3 was a long period with considerably milder climatic conditions. Decisive for the absence or presence of certain plant species were the extremely low winter temperatures while the depression of summer temperatures had considerably less effect. Quite a number of plants occurred which today do not grow south of the polar or above the alpine treeline. This feature is explained by the higher altitude of the sun in Pleni-glacial central Europe compared with that in the modern arctic region.

The vegetation was adapted to the specific conditions in particular stadials and regions reflecting not only temperature but also differences in precipitation. It ranged from bare arctic areas without vegetation to tundra, shrub tundra or a kind of steppe vegetation. This great variety of habitats is also characteristic of the stadial phases in older glaciations.

Locally there was often a mosaic of various plant communities depending on the habitats with changing edaphic and local climatic conditions. The stadial vegetation of central Europe can be compared in many ways with that of Greenland today. The modern tundra vegetation in northernmost Europe is, however, not an analogue for former tundra areas. It is usually dominated by a variety of Ericales while Weichselian tundras show a strong dependence on increasing continentality resulting in Ericales being frequent in the west but decreasing distinctly to the east. The precipitation gradient from the west to the east was considerably steeper than nowadays and is clearly expressed by the absence of steppe plants in the west as in the British Isles (Bos et al., 2004) and their strong increase to the east, which in the pollen diagrams is particularly expressed by the amounts of *Artemisia* and Chenopodiaceae.

In accordance with the open landscape a general feature of the stadial vegetation is the great number of heliophytes that are nowadays encountered in arctic and subarctic areas. In addition to many grasses and sedges there were the creeping dwarf willows such as *Salix herbacea* and many characteristic herbs of which only the following are mentioned here: *Helianthemum, Armeria maritima, Sanguisorba officinalis, Gypsophila, Astragalus alpinus, Polemonium caeruleum, Dryas octopetala, Linum, Valeriana, Papaver, Plantago* and *Selaginella selaginoides.*

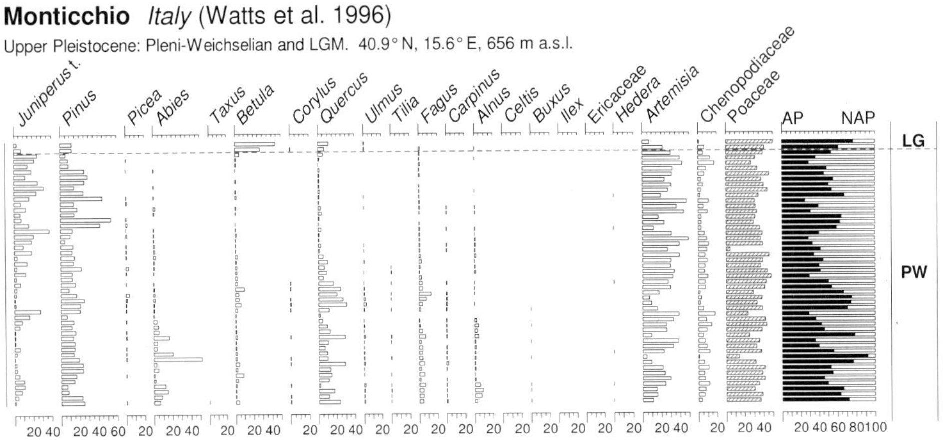

Fig. 3.8-14. Pollen diagram of the Pleni-Weichselian (PW) until the LGM from Monticchio, southern Italy. LG – Late-glacial.

During the LGM, the glaciers advanced over large areas of Britain, Germany, Poland and Russia as well as from the Alps and the high mountain ranges west and east of them into their northern and southern forelands. The periglacial area between the alpine and the northern glaciers was covered with arctic to subarctic tundra and steppe-tundra vegetation without trees. Also in the Mediterranean climatic conditions were cold and dry but trees had not disappeared completely. A unique uninterrupted sequence across the LGM has been recovered from southern Italy at Monticchio (Fig. 3.8-14; Watts et al., 1996b). Most areas of low elevation in south and south-east Europe were covered by grassland or *Artemisia* and Chenopodiaceae steppe. At higher elevations, with higher precipitation, *Pinus* was present at many places, accompanied by *Juniperus* and *Ephedra*. The pollen diagrams as well as the macroremains indicate that deciduous trees also survived in some sheltered areas from where they later spread. These refugia are dealt with in chapters 4.3 and 4.4.

In the more continental regions south-east of central Europe climatic conditions for tree growth were better and new evidence shows that open taiga forests existed there in the younger Pleni-Weichselian, probably including the LGM. In the Bulgarian Rhodopes Mountains at least *Pinus* and perhaps small patches of *Quercus* were present during the LGM (Tonkov et al., 2014). Particularly in Hungary many sites have provided charcoal from *Pinus cembra, P. sylvestris, Picea, Picea/Larix* and *Juniperus* and a bit earlier around 31 000 years PB even *Carpinus* at one site. In the neighbouring Croatia, charcoal from *Fagus, Sorbus, Ulmus* and *Populus* has been found at the end of MIS 3. Further to the north in the Czech Republic charcoal from all native conifers has been identified in the Upper Palaeolithic site of Dolni Věstonice shortly before the LGM (Willis et al., 2000; Willis and van Andel, 2004) and in southern Moravia a forest tundra with *Pinus sylvestris* and *P. cembra* type, *Picea, Juniperus, Larix* and also *Alnus* has been reconstructed by pollen analysis for around 25 000 yr BP (Bulhary: Rybničková and Rybniček, 2014). Similar evidence comes from another pollen diagram from Šafárka in the mountains of north-west Slovakia (Jankovská and Pokorný, 2008).

3.8.3

Fauna and humans

The animal world of the Weichselian is well known, in particular from many cave deposits where early humans, together with cave hyaena and cave bear, accumulated large numbers of animal bones. These show that several large mammals were present for the last time in the fauna of the periglacial zone in Europe, becoming extinct at the end of the Weichselian. They include the woolly mammoth *Mammuthus primigenius*, the woolly rhinoceros *Coelodonta antiquitatis*, the giant elk *Megaloceros giganteus*, as well as the predators of cave bear *Ursus spelaeus*, cave hyaena *Crocuta spelaea*, sabre-toothed cat *Homotherium crenatum* and cave lion *Panthera leo spelaea*. In addition to the bones, quite a number of the large mammals were illustrated in cave paintings of this period, particularly in France and Spain.

Some of the still existing animals such as the musk ox *Ovibos moschatus*, reindeer *Rangifer tarandus*, arctic hare *Lepus timidus* and arctic fox *Alopex lagotus*, point to tundra environments while others, such as the wild horse *Equus hyndruntinus*, saiga antilope *Saiga tatarica* and steppe lemming *Lagurus lagurus*, which occurred westwards as far as central Europe are native in modern steppe areas of the east (von Koenigswald, 2010).

Human speciation appears to have stopped during the Weichselian. Much earlier, *Homo erectus* had already developed to *Homo neanderthalensis* and at the beginning of the last glaciation Europe was sparsely occupied by Neanderthals. Only later, around 40 000 yr BP, did modern humans, *Homo sapiens*, immigrate from Africa and entered Europe. They had a higher stage of development and replaced the Neanderthals, the last ones of which died out around 38 000 yr BP (Terberger, 2019).

3.9

The Quaternary history of some tree species

3.9.1

Fagus

Beech is one of the most important trees in the Holocene and is widely spread across Europe. Its overall Quaternary history is, however, quite varied.

The oldest records of the genus *Fagus* are known from the east Asian Palaeocene. Since then numerous species have developed all of which occur in the northern hemisphere, while *Nothofagus*, a sideline of this development, occurs only in the southern hemisphere. Nowadays *Fagus* includes 11 species: 9 of them are restricted to east Asia and North America while *F. sylvatica* occurs in Europe and *F. orientalis* from eastern Europe to the Middle East. Neither the pollen grains nor the wood can be distinguished to species level.

The general occurrence of *Fagus* in the Tertiary continues from the Pliocene into the Lower Pleistocene, the Tiglian A, as can be seen in the Dutch Meinweg (Fig. 3.4-1; Zagwijn, 1960). Traces are also recorded from the contemporary Nordende interglacial in north Germany (Menke, 1975). After that there are no records of *Fagus* for more than 2 million years apart from the Italian pre-Alpine site, Leffe, where it was present in the middle part of the Lower Pleistocene (Ravazzi, 2003).

From around 500 ka BP the long terrestrial records from Greece show that *Fagus* was present in the upper Middle Pleistocene interglacials (MIS 11–7) in the mountainous zone (Ioannina) but not in the dry, low elevations (Tenaghi Philippon). Additionally, the long sequences in southern France register *Fagus* from the late Cromerian Praclaux (MIS 11) until the Bouchet interstadials (MIS 7a,c,e) and in Castiglione, near Rome, beech was one of the most important trees in the Holsteinian, being recorded from the beginning of this profile (MIS 9a).

Although the ecological conditions were suitable for *Fagus*, beech did not come to Europe north and north-west of the Alps until the Holsteinian (MIS 9) and its occurrence is a characteristic feature for the uppermost part of the Holsteinian in most of the pollen diagrams from this region. *Fagus* disappeared, however, at the end of the Holsteinian and was again absent from this part of Europe during the interglacial/interstadial periods MIS 7e to 7a.

During the long and extensive Saalian glaciation *Fagus* must have had to retreat very far, since its return in the following Eemian interglacial took a very long time and it did not reappear at all in most parts of Europe. Throughout the whole of the Mediterranean region the crater lake Lago di Castiglione near Rome provides the only locality where *Fagus* was found, and then only at low frequencies. Because there is no *Fagus* in the important Eemian sites in Greece, it probably returned to Europe via a more northern route across the Balkans. A good indication of this are the high beech percentages in the site Staro Orjachowo in eastern Bulgaria (Fig. 3.7-20; Bozilova and Djankova, 1976) and from a Black Sea core north of Anatolia (Shumilovskikh et al., 2013) with the proviso, however, that this pollen cannot be attributed to *Fagus orientalis* whose distribution range begins around there.

From the Balkans, *Fagus* entered south-east Austria in the second half of the Eemian, where several pollen diagrams from southern Carinthia show high quantities of *Fagus*, up to 50% in Nieselach (Fritz, 1992; Draxler, 2000). The further migration to the west took place along and through the Alps as could be shown in the southern Alpine foreland (Azzano Decimo and Lake Fimon, Fig. 3.8-8: Pini et al., 2010) and at low frequencies in the Austrian Mondsee (Fig. 3.8-6; Drescher-Schneider, 2000b) but at higher frequencies in the Swiss pre-Alpine lowlands (Meikirch, Hunzikenbrücke and others). Considerable *Fagus* stands also existed in the western Alpine foreland (R. de l'Amourette: Gremmen et al., 1984) where it obviously reached its western boundary. At Les Echets *Fagus* is almost absent and further to the south-west there are no Eemian records for the whole Iberian Peninsula for this period.

In its distribution area in and around the Alps, *Fagus* was restricted to the uplands and the sub-mountainous zone, outside of which, as in La Grande Pile in France as well as in several sites in Germany and Poland, only occasional single pollen grains from long distance transport have been traced. North of the Alps and the Carpathians *Fagus* had mostly been replaced by *Carpinus* and in the uplands by *Abies* and *Picea*. In the Eemian *Fagus* was unable to settle its postglacial homelands in these areas.

At the beginning of the Weichselian glaciation, during the first stadial (Herning), *Fagus* had to withdraw again but obviously not too far because it soon occupied most of the area that it had in the late Eemian. In central Italy, probably close to a refuge, it appears early in the Brörup and reached remarkable values, 40–50%, in Castiglione and Lagaccione (Fig. 3.8-9; Magri, 1999). Further to the north *Fagus* spread slowly but expanded dramatically in the second half of the Brörup and gained a considerable share of the forests in the eastern Italian foreland of the Alps as far as Slovenia. In the Alps themselves it was scarce and did not expand further to the north. Another important distribution area became the south-western foreland of the Alps and southern France having high percentages in R. de l'Amourette, as in the Eemian, but now also in Les Echets (Fig. 3.8-7; de Beaulieu and Reille, 1984b) and less abundantly on the Plateau du Deves.

In the following Odderade interstadial there was a similar distribution of *Fagus* as in the Brörup. *Fagus* was, however, considerably less abundant apart from several sites in central Italy, where it was the most common tree with values up to 60% in Lagaccione (Fig. 3.8-9) and now, for the first time, it also reached northernmost Spain with distinctive values in the pollen diagrams (Magri et al., 2006).

After these early interstadials, *Fagus* continued to be almost continuously present at low frequencies in several parts of Italy, for example in Monticchio, until the LGM and, in Canolo Nuovo, Calabria it even temporarily reached remarkable values of 68% (Grüger, 1977). Genetic markers of modern Italian populations have shown, however, that these are different from the populations outside of Italy and cannot have been the refuges for central and northern Europe. These regions must have been colonized from the south-east to which point several records of macroremains from southern Moravia, Slovakia and Slovenia (Magri et al., 2006).

3.9.2

Pterocarya fraxinifolia

The genus *Pterocarya* belongs to the Juglandaceae family and includes six species, five of them growing in China and Japan while the modern distribution area of the sixth, *Pterocarya fraxinifolia,* is the Caucasus and south-west Asia. This species, the Caucasian wingnut, is a large deciduous tree, up to 40 m high and is nowadays cultivated widely in European parks and gardens.

There are many finds of fruits and leaves of several species of *Pterocarya* as well as pollen from the Tertiary. In the course of the Pliocene *Pterocarya fraxinifolia* had split off from *P. limburgensis* which then became extinct (Mai, 1988). During the Lower Pleistocene and the early Middle Pleistocene *Pterocarya* had a wide distribution, in particular in southern Europe. North of the Alps it was common in the Lower Pleistocene sites along the Rhine as a member of the river bank forests, while it was absent or scarce on the dry soils represented in the Lower Pleistocene Lieth series. During the first strong glaciation, the Elsterian, it retreated far to the south-east and appeared in the following Holsteinian north of the Alps only at the end of this period when, together with *Fagus*, it forms a characteristic biostratigraphic marker for the last phase of this interglacial. After the Holsteinian, *Pterocarya* never again reached its former European distribution area: there is only one record from Valle di Castiglione in Italy from the Wacken/Dömnitz interglacial (MIS 7), if the dating is correct (Follieri et al., 1988).

3.9.3

Eucommia ulmoides

Eucommia ulmoides is the only living species of a genus of small deciduous trees. Its occurrence is nowadays confined to hills and low mountains in southern China. Called gutta-percha tree, *Eucommia* is today cultivated in east Asia as an important herbal medicine. During the Tertiary *Eucommia* was widely spread all over the northern hemisphere, belonging to the arcto-tertiary plant elements (Mai, 1995).

After the Tertiary, *Eucommia* occurred from the Tiglian onwards in many of the Lower Pleistocene sites, in particular in southern Europe but also, though less frequently, in northern Germany (Lieth-sequence). It was still clearly present in the first Cromerian interglacial (Osterholz, Fig. 3.5-3: Grüger, 1967) where it formed a community together with *Carpinus* but it disappeared in the second Cromerian interglacial from which only tiny occurrences are recorded. During the early Middle

Pleistocene this tree became extinct in Europe and the surrounding areas and is now present only in the Far East.

3.9.4

Picea omorika

The genus *Picea* came into existence during the early Palaeocene in North America, migrated to Asia and, starting in the upper Palaeocene, is recorded in Europe (Mai, 1995), where the species *P. omorika* developed in the Pliocene. Now, together with *P. abies*, it is one of the two native *Picea* species in Europe. The distinction of both species is easy if macroremains are available, while the pollen can be distinguished by its size and the diameter of the bladders but, due to a certain overlap between the two species, a quantitative distinction is difficult. As a consequence many pollen diagrams show only the genus *Picea*. Nevertheless there is enough evidence for the European history of *P. omorika*.

P. omorika is recorded from several sites from the earliest Pleistocene up to the Elsterian glaciation. After this glaciation *P. omorika* failed to expand and is absent from the Holsteinian profiles of most of Europe, being mentioned only from some sites in eastern Europe (Mai, 1988). In the Eemian *P. omorika* spread across the Alps to the north but is recorded at only low values. This changed considerably during the Lower Weichselian interstadials when *P. omorika*, together with *P. abies*, spread northwards as far as the North Sea coast. It was in Aue, north of the Saxon Erzgebirge where, for the first time, a large number of needles and some cones from this strange conifer were described as *P. omorikoides*, because it was obviously closely related to *P. omorika* (Weber, 1898). Some years later the same author found even more, excellently preserved macroremains near Lüneburg, northern Germany. Both sites are now regarded to be of Brörup age and the finds have been included as a variety of P. *omorika*. In the meantime quite a lot of sites of Brörup and Odderade age north of the Alps provided pollen and sometimes also some needles of *P. omorika*.

Later in the Weichselian, *P. omorika* became almost extinct and survived only in a very small area at the border between Serbia and Bosnia-Herzegovina but was, however, unable to expand again during the Holocene, so its modern habitat is restricted to about 80 hectares in the Tara Mountains at an elevation between 800 and 1400 m. Only in 1876 was this tree discovered and described as Serbian spruce but since then it has found its way, as a popular tree, into gardens and parks all over Europe.

4.

Late-glacial and Holocene vegetation changes and their causes

4.1

Stratigraphies and chronological subdivisions of the Late-glacial and the Holocene

Gerhard Lang[1], Willy Tinner[1,2], César Morales-Molino[1,2], Christoph Schwörer[1,2], Brigitta Ammann[1,2]

[1] Institute of Plant Sciences, University of Bern, Bern, Switzerland
[2] Oeschger Centre for Climate Change Research, University of Bern, Bern, Switzerland

4.1.1

Introduction

For comparative overviews of vegetation histories covering periods as long as the Late-glacial and the Holocene some structuring into shorter sections is necessary. Several sets of criteria have been used and have been (partly) connected. Some of these connections were, at different points in time, used in differing ways, a fact that sometimes caused confusion. A knowledge of the origin of the subdivisions and the terms used may help to understand any discrepancies. Because the subdivisions were usually derived from profiles in Late-glacial and Holocene sediments different stratigraphic methods have been applied during the 20th century:

1. A temporal division based on archaeological records (archaeostratigraphic subdivisions)
2. A temporal division based on the geomorphological records of ice-retreat (morphostratigraphic subdivisions)
3. A temporal division based on the records of pollen, plant macrofossils or other biological remains (biostratigraphic subdivisions).
4. A temporal division based on records of climate change (climatostratigraphic subdivisions)
5. Temporal subdivisions based on absolute dating (chronostratigraphic subdivisions).

Methods (1)–(3) are based on time-transgressive processes that varied across Europe; method (4) is based on climate reconstructions and, therefore, on interpretations. Methods (1)–(4) give only relative chronologies, with no fixed time units. Only method (5) can provide a basis for independent spatial correlation – but dating series alone need to be accompanied by ecological (biological), geochemical or isotopic information and this may have its own methodological uncertainties. The International Commission on Stratigraphy and its Subcommission on Quaternary Stratigraphy have helped the efforts of the INTIMATE Working Group (Integration of ice-core, marine and terrestrial records) that produced many useful papers concentrating first on the Late-glacial (for example the special volume 20 of Quaternary Science Reviews, including Lowe, 2001) and later on the Holocene (Walker et al., 2012).

In order to show the possibilities and limitations of the various approaches we briefly go through these five methods.

4.1.2

Archaeostratigraphic subdivisions

During the Late-glacial and the Holocene the cultures of *Homo sapiens* in Europe changed and, therefore, so did their influence on flora and vegetation: the hunters and gatherers of Palaeolithic and Mesolithic cultures were followed during the Neolithic by the first farmers and later by cultures that used metals so that the impact on forests and vegetation became widespread. For many reasons these cultural changes cannot be used as synchronous levels across Europe. First, the developments of these cultures were metachronous (see chapter 4.5). Second, transitions between cultures were often not sharp but included periods of overlap. Third, for some areas either the archaeological or the palaeoecological records are not sufficiently dense. For restricted areas or for certain periods the cooperation between archaeology and palaeoecology is very fruitful. Table 4.1-1 gives a simplified overview in calibrated radiocarbon years before present.

4.1.3

Morphostratigraphic subdivisions

The discontinuous ice retreat during the Late-glacial and Early Holocene not only had a strong impact on the landscapes but also left some important 'time markers': terminal moraines indicate periods of interrupted retreat or re-advance of both inland-ice (continental) and mountain glaciers (local). Within the area of the northern Inland Ice in particular, important geomorphological features provided dates via glacigenic varves which were appreciated at an early stage by De Geer (1912) and later by Litt et al. (2009) and his students. This technique can only be applied to the Late-glacial and Early Holocene in formerly glaciated regions. Comparisons with radiocarbon dating, however, indicate certain discrepancies (Wohlfarth et al., 1993).

4.1.4

Biostratigraphic subdivisions

Complete profiles of Late-glacial and Holocene deposits contain important qualitative and quantitative changes in pollen, spores, plant macrofossils and other biological remains. Such profiles can be subdivided on the basis of these remains which results in biostratigraphic zones or briefly 'biozones'. In pollen diagrams, the qualitative and quantitative records of pollen are first described for single sites as local pollen assemblage zones (LPAZ). Over

Table 4.1-1. Archaeological cultures in Europe with approximate time-boundaries (expressed in calibrated years BP). A more detailed overview for European archaeology is given in Table 4.5.4-1.

	Greece	Southern Central Europe	Northern Europe
Iron Age			
	3200	2850	2600
Bronze Age			
	5000	4300	3800
Neolithic			
	9000	7500	6000
Mesolithic			
	13 000	11 300	11 300
Upper Palaeolithic			

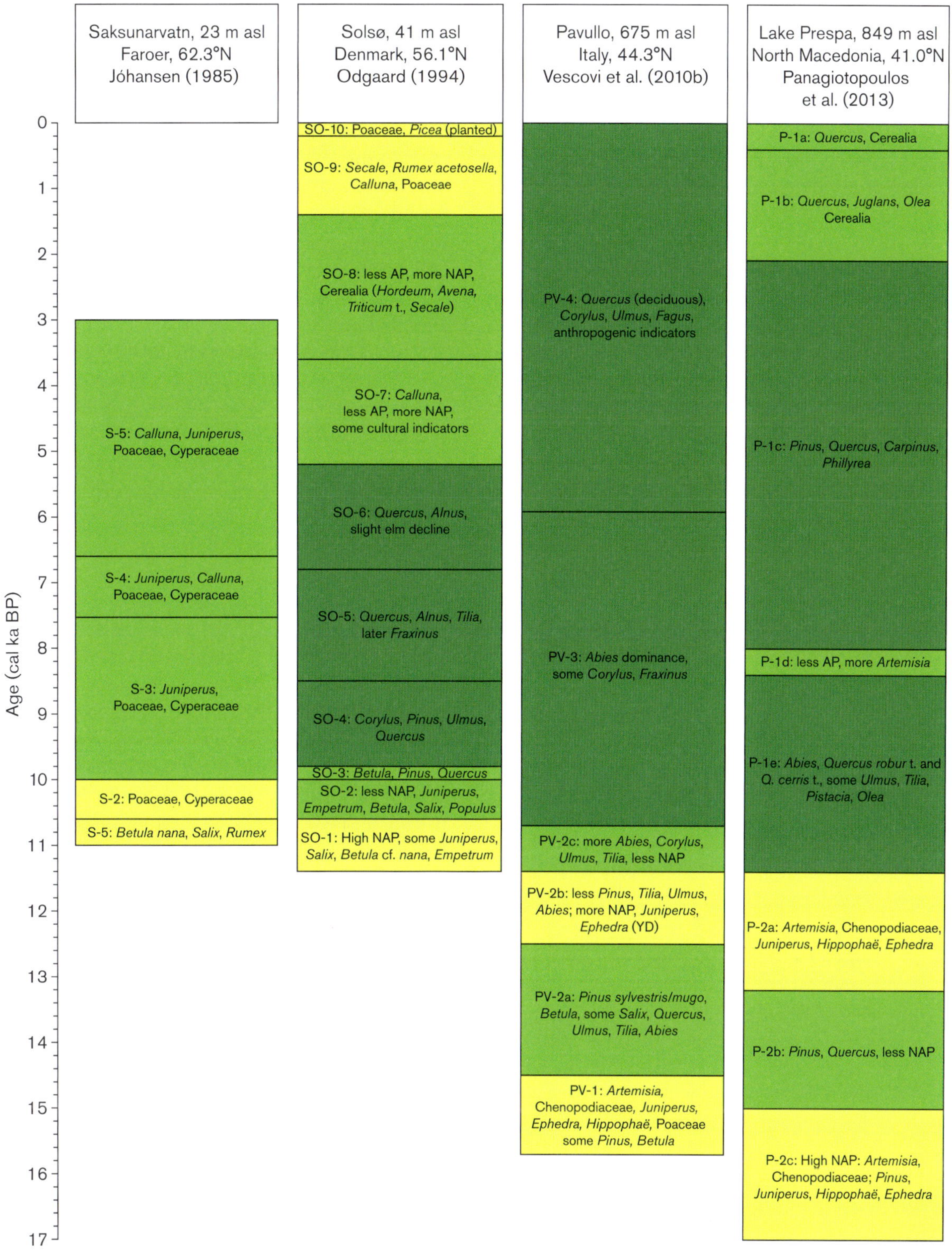

Fig. 4.1-1. Examples of biostratigraphies across Europe, from Saksunarvatn (62.3°N) to Lake Prespa (41°N). Naturally, biostratigraphies are very different across latitudes, while a chronostratigraphy is valid everywhere. For event stratigraphies see Walker et al. (2012, 2018).

larger areas where several sites have been studied, regional PAZs can be developed: both spatial levels of PAZ may be developed either visually or, if possible, by using numerical tools (Birks, 1986b, 2012a). Two early attempts to summarize the sequence of major vegetation types were (a) the Blytt-Sernander system interpreting the dominant pollen taxa in terms of climate (see below) and (b) the 'Central European basic succession' (mitteleuropäische Grundsukzession) by Rudolph (1930): NAP-zone → *Betula-Pinus* zone → *Corylus*-mixed-oak-forest zone → *Fagus* or *Fagus-Abies* zone with onset of *Picea* at various times according to regions. Refinements were then proposed for north-west Germany by Overbeck and Schneider (1938) and for all of central Europe by Firbas (1949).

Pollen zones – like all stratigraphic biozones – are time-transgressive over larger distances and are thus only valid for restricted areas. In Figure 4.1-1, this is visualized by comparing four European regions. Because of this metachronism, caused by environmental factors (e.g. climate), biotic interactions and neutral processes such as species migration (see chapter 4.11), no biostratigraphic subdivision of the Late-glacial and Holocene is possible that would be valid for the whole of Europe but, on a regional scale, biostratigraphies are still very useful.

For the Late-glacial the description of macrofossils of *Dryas octopetala, Salix reticulata* and *S. polaris* in southern Sweden by Nathorst (1870), recorded far away from their modern ranges that are restricted to high latitudes and elevations, led to the term 'Dryas clay', that is older than the Holocene sequence. Later Hartz and Milthers (1901) found an organic layer within the Dryas clay at Allerød in North Zealand, Denmark. Iversen (1942, 1954) then described at Bølling Sø an even older organic layer with *Betula* pollen below the Allerød (Birks and Seppä, 2010). Thus the Late-glacial terms Bølling and Allerød more or less conform to the requirements for the descriptions of type localities. But colleagues from neighbouring disciplines asked: What does 'Dryas' mean? Nathorst (1870) described leaves of *Dryas octopetala*. The wintergreen leaves of this arctic-alpine pioneer resemble somewhat the leaves of *Quercus* (see Fig. 4.8-1). Oak in Greek is δρῦς and the suffix -ας means 'entity' or 'belonging to', a bit like in 'Trias'. In French, *Dryas* is called Chênette (little oak), in Italian Camedrio alpino (from dwarf oak in Greek, χαμαίδρυς).

In many studies the limits between biozones (or isotope zones) are set at the points of inflection, that is at the midpoints of rapid changes of high amplitude such as the onset of the Bølling or the onset of the Younger Dryas (Broecker, 1992); but ecologically it would make more sense to separate the zone of transition because the onset of the change and its levelling out are important (see Table 4.1-4; Ammann et al., 2000). Van Raden et al. (2013) separate two transition zones in the Late-glacial $\delta^{18}O$ stratigraphy, namely at the onsets of the Bølling and of the Younger Dryas. The transition from the Younger Dryas to the Early Holocene is also separated as a distinct zone of ca. 350 years at Kråkenes (Birks et al., 2000; Felde and Birks, 2019). This second approach requires a high sampling resolution.

4.1.5

Climatostratigraphic subdivisions

Changes in climate that can be estimated for the Late-glacial and the Holocene from discontinuous ice retreat and from spatial and population expansions of taxa offer possibilities for structuring this period. If one could assume that changes in climate occur in the same direction and at the same time over large parts of Europe, a synchroneity would result. However, early on Firbas (1949) indicated sources of potential error including the spatial heterogeneity and the selection of indicator species.

The earliest attempts to subdivide the Late-glacial and the Holocene in a climatostratigraphic way were developed by Blytt (1876) and later expanded by Sernander (1909). On the basis of the species composition and the degree of oxidation of peat they developed a scheme of periods of changing temperature and degrees of humidity (see Table 4.1-2). The terms from Preboreal to Subatlantic were later often applied to biostratigraphic and, later, chronostratigraphic periods. Tables 4.1-2, 4.1-3 and 4.1-4 provide overviews that can help to understand the older literature. The Blytt-Sernander terms were developed for northern Europe but are still often applied in eastern Europe. Mangerud et al. (1974) put this terminology into context and emphasized that the Blytt-Sernander terms may still be used, provided that they are clearly understood as chronozones. Here we add the Blytt-Sernander terms only in some rare cases if the original authors have used them, but we

Tab. 4.1-2. Classical climato- and chrono-stratigraphies for north-west Europe. Dates in brackets are uncalibrated, dates without brackets calibrated radiocarbon years BP (before AD 1950); the vertical axis is not time-linear. Climato-stratigraphically the Preboreal and Boreal belong to the Early Holocene, the Atlantic to the Mid Holocene and the Subboreal and Subatlantic to the Late Holocene.

Southern Sweden Nilsson (1935, 1964) [yr BP]	**Denmark** Jessen (1935), Iversen (1954) [yr BP]	**British Isles** Jessen (1949), Godwin (1956) [yr BP]	**Central Europe** Firbas (1949, 1954) [yr BP]	**Chronozones** Mangerud (1974) [cal yr BP/ (^{14}C yr BP)]	**Event stratigraphy** [cal yr BP]	
I/II Subatlantic (SA)	IX Subatlantic	VIII Subatlantic	IX/X Subatlantic	Subatlantic (SA)	Late Holocene	Holocene
2300	2500	2700	2800/2500	2600 (2500)		
III/IV Subboreal (SB)	VIII Subboreal	VIIIb Subboreal	VIII Subboreal	Subboreal (SB)		
5300	5000	5000	4500	5700 (5000)	4200	
V/VI Atlantic (AT)	VII Atlantic	VIIa Atlantic	VI/VII Atlantic	Atlantic (AT)	Mid Holocene	
8200	8000	7500	7500	8900 (8000)	8200	
VII/VIII Boreal (BO)	V/VI Boreal	V/VI Boreal	V Boreal	Boreal (BO)	Early Holocene	
9900	9000	9500	8800/8500	10100 (9000)		
IX Preboreal (PB)	IV Preboreal	IV Preboreal	IV Preboreal	Preboreal (PB)		
10300	10300	10300	10100	11600 (10000)	11700	
X Younger Dryas (DR 3)	III Younger Dryas	III Upper Dryas	III Younger subarctic period	Younger Dryas (DR3)	Younger Dryas	Late Glacial
11100	11000	10800	11000	12900 (11000)	12800	
XI Allerød (AL)	II Allerød	II Allerød	II Middle subarctic period	Allerød (AL)	Allerød	
12000	11700	12000	12000	13800 (12000)	13900	
XII Older Dryas	Ic Older Dryas	I Lower Dryas	Ib Older subarctic period	Bølling (BØ) (incl. Older Dryas DR2)	Older Dryas	
	12000				14000	
	Ib Bølling				Bølling	
	12500			15600 (13000)	14600	
	Ia Oldest Dryas		Ia Oldest unforested period	Oldest Dryas (DR1)	Oldest Dryas	

Tab. 4.1-3. The event stratigraphy for the Late-glacial and the Holocene as developed by the INTIMATE group (from Björck et al., 1998 to Rasmussen et al., 2014b, and Walker et al., 2018), compared to the Greenland event stratigraphy (GES) and the chronostratigraphy after Mangerud et al. (1974). For the classical climato-stratigraphic position of the Holocene periods (Subboreal to Subatlantic) see Table 4.1-2. For instance, in such a climate or event-related comparison the Atlantic period falls (entirely) into the Mid Holocene.

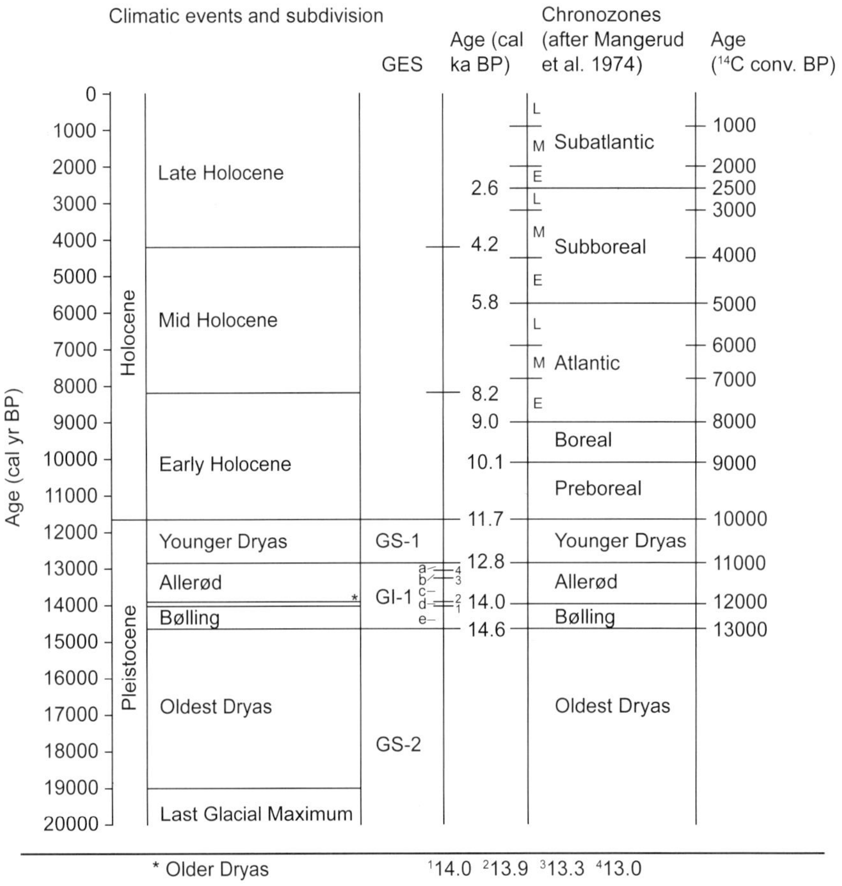

always put them in the context of the chronological Holocene subdivision used in this book (see 4.1.6). For Fennoscandia, Birks and Seppä (2010) provide a fine overview for this connection between bio- and climatostratigraphy.

4.1.6

Chronostratigraphic subdivisions

Certainly the simplest and clearest subdivison of the Late-glacial and the Holocene is a purely chronostratigraphical one but this requires an adequate number of radiocarbon (or other) dates, a condition not yet fulfilled in all areas of Europe. For dating techniques see chapter 2.4. Subdivisions in periods of 500 or 1000 years (in uncalibrated radiocarbon years) were already used by Huntley and Birks (1983). However, it will help communication and understanding if groups of such periods receive names. A proposal, initially for northern Europe, was presented by Mangerud et al. (1974) and also discussed for the rest of Europe (Mangerud, 1982). In this, the classical terms from Oldest Dryas to Subatlantic were still used but with a strictly chronostratigraphic connotation (see Tables 4.1-2 and 4.1-3). When using the Blytt-Sernander terms for the Holocene, the danger of confusing the two must be avoided by clearly indicating if the terms are used with a bio-, climato-, or chronostratigraphic meaning. In this book we use the Holocene terms only in their historical context, assigning them to the Early, Mid or Late Holocene (Table 4.1-2); the same is true for the abbreviations PB, BO, AT, SB, SB, SA, introduced by Nilsson (1964).

For the **Late-glacial** the terms Bølling, Allerød and Younger Dryas were originally defined in a biostratigraphic sense (Hartz and Milthers, 1901; Iversen, 1942, 1954; Járai-Komlódi, 1966a,b). The North Atlantic INTIMATE project inaugurated in 1995 studied possible synchronisations (versus 'leads and lags') and developed an event stratigraphy with an improved chronology (Björck et al., 1998; Lowe et al., 2001, 2008). Since the terms Bølling, Allerød and Younger Dryas now also have a chronostratigraphic meaning for northern Europe following Mangerud et al. (1974) (ages given there in conventional, uncalibrated radiocarbon dates), they could additionally be applied to Greenland. The ice cores GRIP, GISP2 and NGRIP with a GICC05-chronology (based on annual layers) and curves for $\delta^{18}O$ (proxy for temperature) and calcium ion concentrations [Ca^{2+}] (proxy for dust, plotted on a reversed logarithmic scale) are an important combination of a chronostratigraphy with a climatostratigraphy (Ravazzi, 2003; Rasmussen et al., 2014a,b). Ages are given in ka b2k according to GICC05, that is thousand years before AD 2000, not before 1950. In Rasmussen et al. (2014b) the Bølling and Allerød (B-A) are considered together as one chronozone, also labelled Dansgaard-Oeschger event 1 (DO-1). Rasmussen et al. (2014b) and Rowley-Conwy (1982) emphasize that the Bølling/Allerød, on the one hand, and the Younger Dryas, on the other hand, are 'non-archive specific' names but they are not synonyms for the GI-1 and GS-1 that are specific to the ice cores; B-A and YD can be considered 'correlative' of GI-1 and GS-1. B-A is understood as one chronozone but refined structures are found consistently within this 'Late-glacial interstadial' as indicated by the 7-fold subdivision in Rasmussen et al. (2014b) and the 9-fold subdivision in van Raden et al. (2013) (see Table 4.1-4). Even if these structures are not recognized chronozones they are of ecological interest.

The Oldest Dryas and the Older Dryas, however, are considered too weakly dated by Rasmussen et al. (2014b) to be accepted as chronostratigraphic units. For the Oldest Dryas its end is clear

Tab. 4.1-4. Late-glacial chronology after van Raden et al. (2013) providing a refined overview over cool and warm periods between Greenland, the North Atlantic, central Europe and North America. The short horizontal lines between the Bølling and Allerød indicate the various limits proposed by different authors.

Modified Greenland terminology (van Raden et al., 2013)	Switzerland, S-Germany (Lotter et al., 1992)	Eifel region, N-Germany (Merkt and Müller, 1999; Litt and Stebich, 1999)	N-America, Cariaco basin (Yu, 2007; Hughen et al., 1996)	N-America (Yu and Eicher, 1998)	Norwegian Sea (Karpuz and Jansen, 1992)	N-Atlantic (Thornalley et al., 2010)	Greenland (Lowe et al., 2008)	Age (cal ka BP)
YD	Younger Dryas	Younger Dryas	Younger Dryas	Younger Dryas	Younger Dryas	Younger Dryas	GS-1	11.7
transition								
GI-1a	Allerød	Allerød	Allerød	Allerød	Allerød	Allerød	GI-1a	12.8
GI-1b	Gerzensee Oscillation	Gerzensee Oscillation	IACP	Killarney Oscillation	Older Dryas II	IACP II	GI-1b	13.0
GI-1c1		Allerød	Allerød		Allerød	Allerød		13.3
GI-1c2	Allerød	Older Dryas	Older Dryas	Allerød	Older Dryas I	IACP I	GI-1c	
GI-1c3		Bølling	Bølling		Bølling	Allerød		
GI-1d	Aegelsee Oscillation	Oldest Dryas	IBCP	Older Dryas	BCP II	Older Dryas	GI-1d	13.9
GI-1e1					Bølling			14.0
GI-1e2	Bølling	Meiendorf	Bølling	Bølling	BCP I	Bølling	GI-1e	
GI-1e3					Bølling			
transition								
Oldest Dryas	Oldest Dryas	Pleniglacial	Oldest Dryas	Oldest Dryas	Oldest Dryas	Oldest Dryas	GS-2	14.6

because this is the onset of the Bølling but its onset has often been discussed. Inside former glaciations we may agree with Welten (1979) or his oral (smiling) summary: "That is where my hand-corer stops". This is certainly not a synchronous horizon. Less anecdotally, we propose an onset of the Oldest Dryas at 19 000 cal yr BP, following the end of the Last Glacial Maximum (LGM) (Table 4.1-3). In this book the LGM is dated at 23 000–19 000 cal yr BP. The available European ice sheet evidence shows considerable chronological variability but is generally close in time to the global LGM at 23 000–19 000 cal yr BP, as defined by sea-level and marine oxygen isotope records (Hughes et al. 2013). This chronological delimitation of the LGM is in agreement with chironomid-based summer temperature reconstructions covering the period 31 000 –11 000 cal yr BP (Samartin et al. 2012, 2016). On the basis of the available palaeoclimatic evidence (e.g. chironomids, ostracods, lake-levels, oxygen isotope records) we assume that during the LGM at ca. 23 000–19 000 cal yr BP European plant communities were affected by coldest and driest ice age conditions (for details see chapters 4.3, 4.4. and 4.11). For the position of the Older Dryas see Table 4.1-4.

Basal ages of cores are of great interest but often not easy to establish. Reworked pollen near the base can be a strong indication of reworked material including the organic matter we submit to the radiocarbon laboratory. Figure 4.1-1 gives an example and the remark 'rwp' in some of the histograms in chapter 4.3 indicate reworking, often visible as records of thermophilous taxa in glacial clay.

For the **Holocene** Walker et al. (2012) propose – as an alternative to the Mangerud-type of chronozones – a tripartition of the Holocene with two boundaries "...an Early-Middle Holocene Boundary at 8200 a BP and a Middle-Late Holocene Boundary at 4200 a BP, each linked to a Global Stratotype Section and Point (GSSP)". These two boundaries are also found outside of Europe in numerous terrestrial, marine, tree-ring, stalagmite and ice records (Rowley-Conwy, 1982). Since July 2018 these boundaries have been accepted by the International Union of Geological Sciences: as stages/ages the Early Holocene is named the 'Greenlandian', the Mid Holocene 'Northgrippian' and the Late Holocene 'Meghalayan'. In this book we use the terms Early, Mid and Late Holocene according to table 4.1-2.

4.2

Today's vegetation and woody flora

Gerhard Lang[1], Brigitta Ammann[1,2], Christoph Schwörer, Willy Tinner[1,2]

[1] Institute of Plant Sciences, University of Bern, Bern, Switzerland
[2] Oeschger Centre for Climate Change Research, University of Bern, Bern, Switzerland

4.2.1 Introduction

Considered over long time scales, today's vegetation is only the last realization of a dynamic process and not a static concept. On the one hand, knowledge of modern vegetation is essential for our understanding of vegetation history, on the other hand, considering this history allows a deeper understanding of the present vegetation. In Europe this comparison is made difficult because of the human impact on the environment over millennia: what we observe as the modern plant cover is only rarely the end product of a natural vegetation history.

A distinction was often made between 'modern natural vegetation' (plant cover without any human impact) and 'modern potential natural vegetation' (plant cover including human impact but as it would be after the cessation of this influence). Both are concepts based on soils, climate and biogeography, but they remain constructions with large uncertainties (Tüxen, 1956; Frenzel, 1968; Neuhäusl, 1991; Bohn et al., 2003; Birks, 2019).

For our purpose a map of the modern potential natural vegetation on a small scale (Fig. 4.2-1) will be sufficient showing the 'zonal vegetation' that depends on large-scale factors such as temperature and precipitation and that omits the 'azonal vegetation' depending on edaphic conditions (for example riparian forests, mires, halophytic habitats). Moreover, the elevational belts of mountain systems cannot be fully shown on this scale. A higher spatial resolution can be found in Ozenda (1979) and Bohn et al. (2003).

As an overview we can distinguish five vegetation zones (biomes) in Europe:

- The arctic and alpine zone with treeless dwarf shrub, meadow or tall herb vegetation (A)
- The boreal zone with a dominance of conifers (B)
- The temperate zone with mainly forests of deciduous trees (T)
- The Mediterranean zone with co-dominance (mesomediterranean) or dominance (thermomediterranean) of evergreen broadleaved trees and shrubs (M)
- The Pannonic-Pontic-Anatolian zone with forest steppes, steppes and semi-deserts (P)

Most limits are spatially not clearly delimited and this can be expressed in terms such as 'forest-tundra' or 'subarctic' – belts that can be very broad for example in north-west Russia.

4.2.2 Modern woody flora of Europe

The modern flora of trees and shrubs in Europe consists of over 300 species/subspecies that belong to 99 genera in 45 families. These numbers include trees and shrubs >2 m in height (282 species/subspecies), some lianas (14 species/subspecies) and epiphytes (5 species/subspecies). With regard to their role in the vegetation, these taxa are of very uneven importance because, in the zonal vegetation, often just a few genera or species are dominant (Table 4.2-1).

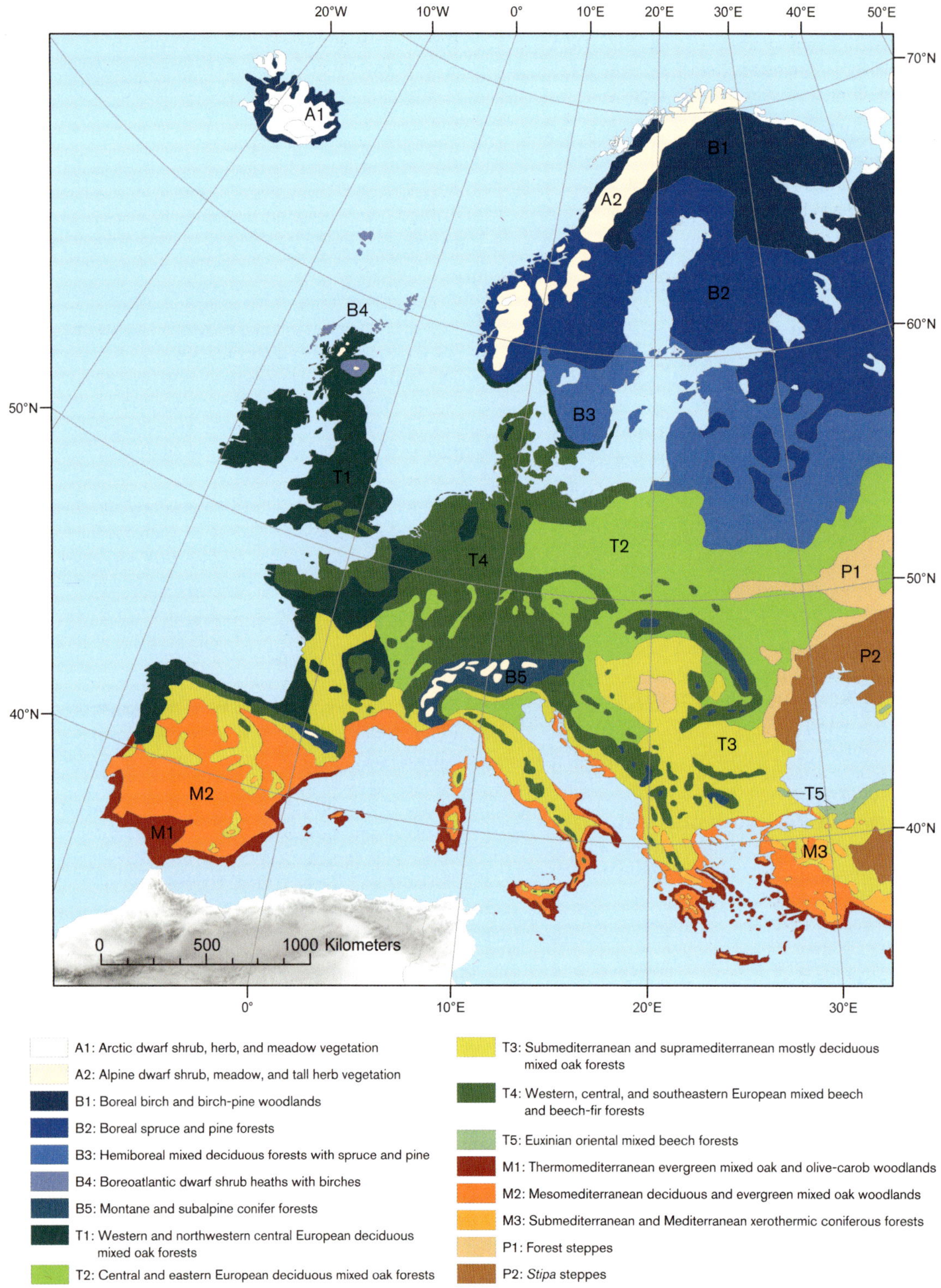

Fig. 4.2-1. Simplified map of the modern natural potential vegetation in Europe. A: Arctic and alpine vegetation zone with treeless plant communities B: Boreal and hemiboreal vegetation zone with forests of evergreen conifers and mixed deciduous-conifer forests in northern Europe, and conifer forests in montane and subalpine belts of mountain ranges. T: Temperate vegetation zone with summergreen (deciduous) broadleaved forests. M: Mediterranean vegetation zone with evergreen broadleaved forests and scrub (sclerophyllous and laurophyllous evergreen trees and shrubs as well as palms). P: Pannonic-Pontic-Anatolian vegetation zone with forest steppes, steppes and semi-deserts. Simplified and slightly modified after Bohn et al. (2003). For more information see Table 4.2-1.

Table 4.2-1. Modern vegetation zones (A to P) and vegetation regions (numbers) with their major taxa. For spatial distribution in Europe see Fig. 4.2-1.

A Arctic and alpine zone: Dwarf shrub, meadow and tall herb vegetation. Treeless associations in the arctic tundra in the far north and in the alpine belt of the European mountain ranges.

A1 Arctic dwarf shrub, herb and meadow, as well as moss and lichen vegetation. On Iceland, in northernmost Fennoscandia, on the Kola peninsula and east of it. Dominant: *Betula nana,* low *Salix* species, Ericaceae *(Arctostaphylos, Cassiope, Empetrum, Loiseleuria, Phyllodoce, Vaccinium),* Cyperaceae, *Dryas*, Poaceae.

A2 Alpine dwarf shrub, meadow and tall herb vegetation. In European mountain ranges (outside the Scandinavian mountains only in small areas, in contact with coniferous mountain forests, B5). Dominant: Ericaceae (*Arctostaphylos, Empetrum, Loiseleuria, Rhododendron, Vaccinium*), Cyperaceae, herbs, Poaceae.

B Boreal and hemiboreal zone: Evergreen coniferous forests (taiga) and conifer-deciduous mixed forests in northern Europe; coniferous forests in the montane and subalpine belts of European mountain ranges.

B1 Boreal birch and birch-pine woodlands and scrub. Boreoatlantic birch woodlands and scrub (without pine) on Iceland and in northern Norway (B1a), northern boreal birch-pine woodlands and scrub in the rest of Fennoscandia and in northern Russia (B1b). Dominant: *Betula pubescens* ssp. *tortuosa, Pinus sylvestris,* in the north-east also *Picea abies* ssp. *obovata.*

B2 Boreal spruce and pine forests. Mainly in the middle and southern boreal zone in Sweden, Finland, northern Russia. Dominant: *Picea abies, Picea abies* ssp. *obovata* (in the east), *Pinus sylvestris, Betula pubescens.*

B3 Hemiboreal mixed forests of deciduous trees with spruce and pine. A transitional zone (broadening eastwards) between the boreal coniferous zone and the temperate summergreen zone. In southern Scandinavia, the southernmost coast of Finland, the Baltic countries, western Russia and Belarus. Dominant: *Picea abies, Pinus sylvestris, Quercus robur, Carpinus betulus, Tilia cordata, Ulmus glabra, Acer platanoides.*

B4 Boreoatlantic dwarf shrub heaths with birches. In the Scottish Highlands, on the coasts of the Faroe and Shetland Islands, western Norway and southern Iceland. Dominant: Ericaceae, *Betula pubescens* (as small patches), *Pinus sylvestris* (northern Scotland).

B5 Montane and subalpine conifer forests and prostrate scrub. In the European mountains, usually in close contact to treeless vegetation (A2) of the alpine belt. Dominant in the Pyrenees: *Abies alba, Pinus sylvestris, Pinus uncinata;* in the Alps and the Carpathians: *Picea abies, Abies alba, Larix decidua, Pinus cembra, P. sylvestris, P. uncinata, P. mugo* (as scrub); locally on the Balkan Peninsula: *Pinus peuce* and *Picea omorika.*

T Temperate zone: Summergreen broadleaved forests in western, central and eastern Europe and in the mountains of southern Europe.

T1 Western and north-western central European mixed oak forests. In atlantic and subatlantic lowlands and on the British Isles, on the continent from north-western Spain to Denmark and southern Scandinavia. Dominant: *Quercus robur, Q. petraea, Fraxinus excelsior,* partially also *Carpinus betulus.*

T2 Central and eastern European mixed oak forests. In subcontinental and continental lowlands of central and eastern Europe to the Ural mountains. Dominant: *Quercus petraea* (absent in the east), *Q. robur, Carpinus betulus* (absent in the east), *Tilia cordata, Acer platanoides.*

T3 Submediterranean and supramediterranean temperate mixed oak forests. In northern parts of southern Europe (Iberian Peninsula, southern France, Apennine Peninsula, Balkan Peninsula), as well as in isolated, warm lowland areas of central Europe. Dominant: many regionally diverse summergreen oak species *(Quercus pubescens, Q. petraea, Q. cerris, Q. faginea, Q. pyrenaica, Q. dalechampii, Q. frainetto, Q. pedunculiflora, Q. virgiliana, Q. polycarpa, Q. hartwissiana), Fraxinus ornus, Ostrya carpinifolia, Carpinus orientalis* (in the south-east), *Castanea sativa, Buxus sempervirens.*

T4 Western, central and south-eastern European beech and beech-fir forests. In western and northern central Europe in lowlands, in southern central and south-western, southern and south-eastern Europe together with *Abies* in the montane belt. Dominant: *Fagus sylvatica, F. moesiaca (F. sylvatica x F. orientalis,* only on the Balkan Peninsula), *Abies alba.*

T5 Euxinian oriental beech forests. Along the southern coast of the Black Sea in the montane belt. Dominant: *Fagus orientalis, Prunus laurocerasus, Rhododendron ponticum.*

M Mediterranean zone: Evergreen broadleaved woodlands and scrub of the Mediterranean area.

M1 Thermomediterranean evergreen oak and olive-carob woodlands. In the lowest and southernmost Mediterranean belt of the European Mediter-

ranean area. Dominant: *Quercus suber, Q. coccifera, Q. ilex, Q. rotundifolia* (in Spain), *Ceratonia siliqua.*
M2 Mesomediterranean woodlands of evergreen holm oak with deciduous trees. Above the lowest belt M1 or north of it, on its northern and upper limit in contact with the summergreen mixed-oak forests of T3. Large areas especially in the southern part of the Iberian and Italian Peninsulas, in southern Greece, and in the south-western part of Turkey. Dominant: *Quercus ilex, Q. pubescens, Q. rotundifolia* (in Spain), *Q. suber, Q. coccifera, Q. brachyphylla* (in Greece), *Q. infectoria, Q. macrolepis, Phillyrea latifolia.*
M3 Submediterranean and Mediterranean xerothermic coniferous forests. In the southern parts of the Mediterranean area only in small patches (Spain, Corsica, Italy, Greece, Anatolia), partly in contact with the oro-xerophytic vegetation of the Mediterranean mountains. Dominant usually one or more of the following taxa: *Pinus sylvestris, P. nigra* aggr., *P. brutia, P. heldreichii, Abies pinsapo, A. borisii-regis, A. cephalonica, A. cilica* (in Turkey), *Cedrus libani, Cupressus sempervirens, Juniperus alpina, J. excelsa, J. sabina, J. thurifera.*

P Pannonic-Pontic-Anatolian zone:
Forest-steppes, steppes and semi-deserts. From the western coast of the Black Sea towards the east over the Ukraine as a belt that becomes larger towards the east as far as southern Siberia and Kazakhstan.
P1 Forest-steppes. A mosaic of species-rich meadow-steppes and islands of forests with summergreen oaks *(Quercus robur, Q. pubescens, Q. petraea, Q. pedunculiflora, Q. cerris, Q. virgiliana), Carpinus betulus, Tilia cordata, T. tomentosa, Acer tataricum.* In the Pannonian basin, the lower reaches of the Danube and in the south-eastern Carpathians to the southern Ural mountains.
P2 *Stipa*-steppes. From the western coast of the Black Sea (Dobrudscha) south of the belt of the forest-steppes to beyond the Ural mountains. Dominant: species of *Stipa (S. capillata, S. lessingiana, S. stenophylla, S. ucrainica, S. tirsa, S. borysthenica), Festuca valesiaca, Agropyron desertorum, A. pectinatum, Artemisia taurica,* Liliaceae (geophytes).
P3 *Artemisia*-semideserts. As large areas only outside the map, especially in Kazakhstan (Caspian lowland, Aral Sea). Dominant: *Artemisia* species *(A. austriaca, A. incana, A. pauciflora, A. taurica), Camphorosma monspeliacum, Kochia prostrata, Salsola laricina,* therophytes.

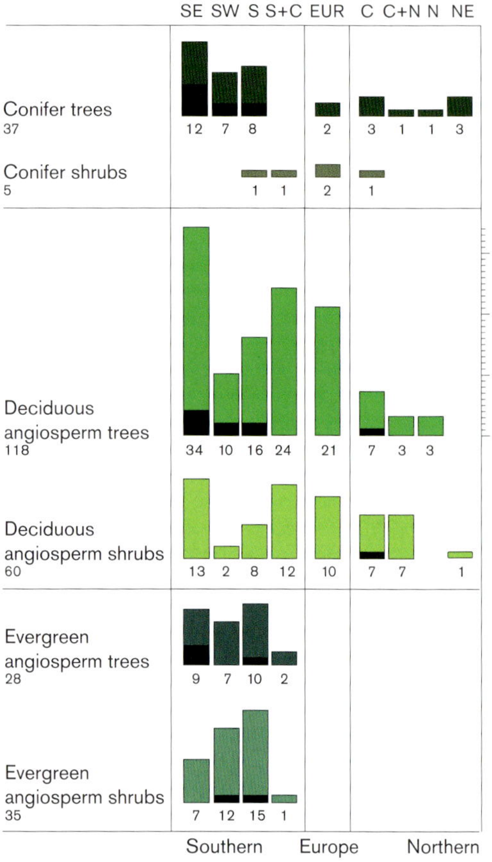

Fig. 4.2-2. Life-forms, species numbers and geographical distribution of woody taxa in Europe. C: main range in central Europe. C+N: main range in central and northern Europe. EUR: main range in all of Europe. N: main range in northern Europe (Fennoscandia). NE: main range in north-eastern Europe. S: main range in southern Europe. S+C: main range in southern and central Europe. SE: main range in south-eastern Europe. SW: main range in south-western Europe. Black: endemic species. The six shades of green refer to the life forms mentioned in the left column. For the attribution of woody species to the floristic regions of Europe see Table A-2 in Lang (1994).

The various taxa show very different distributions in Europe. Since the woody flora is not as rich as the complete flora we may subdivide Europe into a small number of florisitic regions:
1. In northern Europe, the European part of the boreal floristic zone of Meusel et al. (1965) and the northern European floristic province of Takhtajan (1986), we can separate the Fennoscandian northern Europe *sensu stricto* from the northern part of Russia (Fig. 4.2-2) as north and north-east, which

according to Meusel et al. (1965) is the Fennoscandian and the Boreorussian floristic province.

2. As central Europe we circumscribe the area from the British Isles eastwards until deep into the continent (C in Fig. 4.2-2); the central European floristic province of Meusel et al. (1965); the Atlantic, central European and eastern European floristic province of Takhtajan (1986).

3. In southern Europe, the European part of the submeridional and meridional floristic zone of Meusel et al. (1965) and Meusel and Jäger (1992), we can further divide the Mediterranean region into south-western and south-eastern Europe. South-western Europe comprises the western submediterranean and western Mediterranean floristic province of Meusel et al. (1965) and the south-western Mediterranean and Iberian floristic province of Takhtajan (1986). South-eastern Europe comprises the central submediterranean and central Mediterranean floristic province as well as the eastern submediterranean and eastern Mediterranean floristic province of Meusel et al. (1965) and the Adriatic, Balkan, eastern Mediterranean and Euxinian floristic province of Takhtajan (1986).

Most woody taxa, namely 237 species/subspecies have distributions with centres usually in just one of the regions mentioned. In addition, there are taxa with very wide distributions (EUR in Fig. 4.2-2; 36 taxa) or, in contrast, endemics (black bars in Fig. 4.2-2; 28 taxa).

Considering the geographical distribution of life-forms expressed in numbers of species (Fig. 4.2-2) the high proportion of the summer-green taxa becomes obvious: with 178 species/subspecies they reach 63% of all European woody taxa, whereas the evergreen taxa with 63 species/subspecies make up 22% and conifers, with 42 species/subspecies, make up 15%. The steep north–south gradient between the number of woody taxa is also conspicuous: 19 species/subspecies in northern Europe (less than 7%) vs 211 species/subspecies in southern Europe (c. 75%), which is more than a factor of 10. Of course such numbers do not say anything about the numbers of individuals. Within southern Europe diversity is especially high in the south-east. Endemic woody taxa are nearly restricted to southern Europe, half of them to south-eastern Europe. Endemic woody taxa are lacking in northern Europe.

On a global scale the European temperate tree species diversity is low, as Latham and Ricklefs (1993) show: the 1,166 species "are distributed among Europe (including the Caucasus), eastern Asia, North America's Pacific slope, and eastern North America respectively, approximately in the ratio 2:12:1:4". Causes are not only the barriers to migration (west–east aligned mountain ranges and the Mediterranean Sea) as already proposed by Gray (1878) and Engler (1879), but possibly also the "autochthonous production of new taxa" (Latham and Ricklefs, 1993). Also of interest are the spectra of pollination ecology and dispersal ecology: while for the whole of Europe the ratio of anemophilous to entomophilous woody taxa is 39 : 61%, in species-poor northern Europe with climate less favourable to insects it is 87 : 13%, while in southern and central Europe it is 37 : 63%. Similarly, the dispersal mode for the whole of Europe has a ratio of anemochorous to (mainly) zoochorous taxa of 35 : 65%, whereas in northern Europe it is 75 : 25%, the contrary of elswhere in Europe.

4.2.3

Modern vegetation of Europe

The modern vegetation of Europe has been described and mapped with various methods for various spatial scales, for example as formations throughout Europe (Bohn et al., 2003) or as bioclimatic zones for the Iberian Peninsula (Rivas-Martínez, 1987; Rivas-Martínez et al., 2001). On finer spatial scales phytosociology has provided a consistent scheme for many parts of Europe. Even though the methods of Braun-Blanquet (1964), Oberdorfer (1992), Weber et al. (2000) and others are less used today, we refer to associations and/or alliances in the chapter on selected taxa (4.4) to indicate where the taxa occur most frequently at present and form communities. However, we certainly do not suggest that the vegetation types of the past were similar or even identical to the ones of the present (see also chapter 5.8). For vegetation ecology in general see also van der Maarel and Franklin (2013).

In the following chapters we look into regional vegetation histories (4.3), into histories of selected taxa (4.4), into human impact (4.5), into developments of lakes, mires and soils (4.6 – 4.8), into tree- and timberline dynamics (4.9), into long-term disturbance ecology (4.10) and into the causes of vegetation dynamics (4.11).

4.3

Regional vegetation history

Gerhard Lang[1], Brigitta Ammann[1,2], W.O. (Pim) van der Knaap[1,2], César Morales-Molino[1,2], Christoph Schwörer[1,2], Willy Tinner[1,2]

[1] Institute of Plant Sciences, University of Bern, Bern, Switzerland
[2] Oeschger Centre for Climate Change Research, University of Bern, Bern, Switzerland

4.3.1

Basics and principles of vegetation history reconstruction since the Last Glacial Maximum

During the coolest periods of the Weichselian glaciation large parts of Europe were treeless and vast areas in the north and the higher mountains in the south were under glacier ice (see chapter 3). With the rapid rise of temperature during the Late-glacial and the Early Holocene – interrupted by cooling phases – these ice masses melted, except for small remnants in high latitudes and at high elevations. Various species of trees and shrubs started to spread over the continent but every species behaved differently. This means that taxa that form distinct communities today did not necessarily spread and move as communities. In addition, some of the early communities may have no modern analogue (see chapter 5.3.4). Although, at a very coarse scale, this 'migration' of woody taxa took place mainly from south to north, it was by no means a simple front-like spread (see chapter 5.2.3.3). At least four aspects caused a differential spread of taxa: (1) the location and number of glacial refugia, (2) the potential taxon-specific speed of the spread, (3) possible barriers (sea or mountain chains) and (4) species-specific ecological requirements, for example temperatures (annual means or summer means), moisture, seasonality or frequency of (late) frost, plus other non-climatic conditions such as soils. Therefore, various species became dominant or sub-dominant in different areas during the Late-glacial and the Holocene. In contrast to earlier interglacials not only climate change but also prehistoric human impact played a role in these spatially differentiated sequences of vegetation types.

As can be expected, close connections exist in each area between the regional vegetation history and the modern potential natural vegetation, that is the final result of the Holocene development. Therefore a simplified map of today's potential natural vegetation as biomes is given in Figure 4.2-1. This shows three major forest zones: boreal (B), temperate (T) and Mediterranean (M) and two regions without forest (A for arctic and P for Pontic). The biomes of this map form the basis for the map of regional vegetation history (Fig. 4.3-1). Within the three forest regions a west–east oriented differentiation from oceanic to continental (marked with small letters) is made (e.g. Ta, Tb, Tc). Naturally the map of the present vegetation is spatially more refined and all limits are certainly not sharp but include transitional zones. Apparent discrepancies between the two maps reflect differences related to the vegetation histories of the regions.

Sometimes the allocation of important palaeoecological sites to vegetation history regions is ambiguous:

1. Often sites are close to two or more borders and may therefore be important to all of these regions. Examples are Kråkenes, (Birks et al., 2000) near to Ba, Bb and Ta; Sandvikvatn (Paus, 1988) near to Ta, Bb and Bd; Ran Viken (Digerfeldt, 1973) near to Bd and Tb; Mekelermeer (Bohncke, 1991) near Ta and Tb; Svityaz (Schwörer et al., 2022) near to Tb and Tc; and Ayoó de Vidriales (Morales-Molino and García-Antón, 2014) near Ta and Ma.
2. Important sites may be situated in broad transitional vegetation zones such as the 'forest tundra' of European Russia with sites such as Liet-Ti (Väliran-

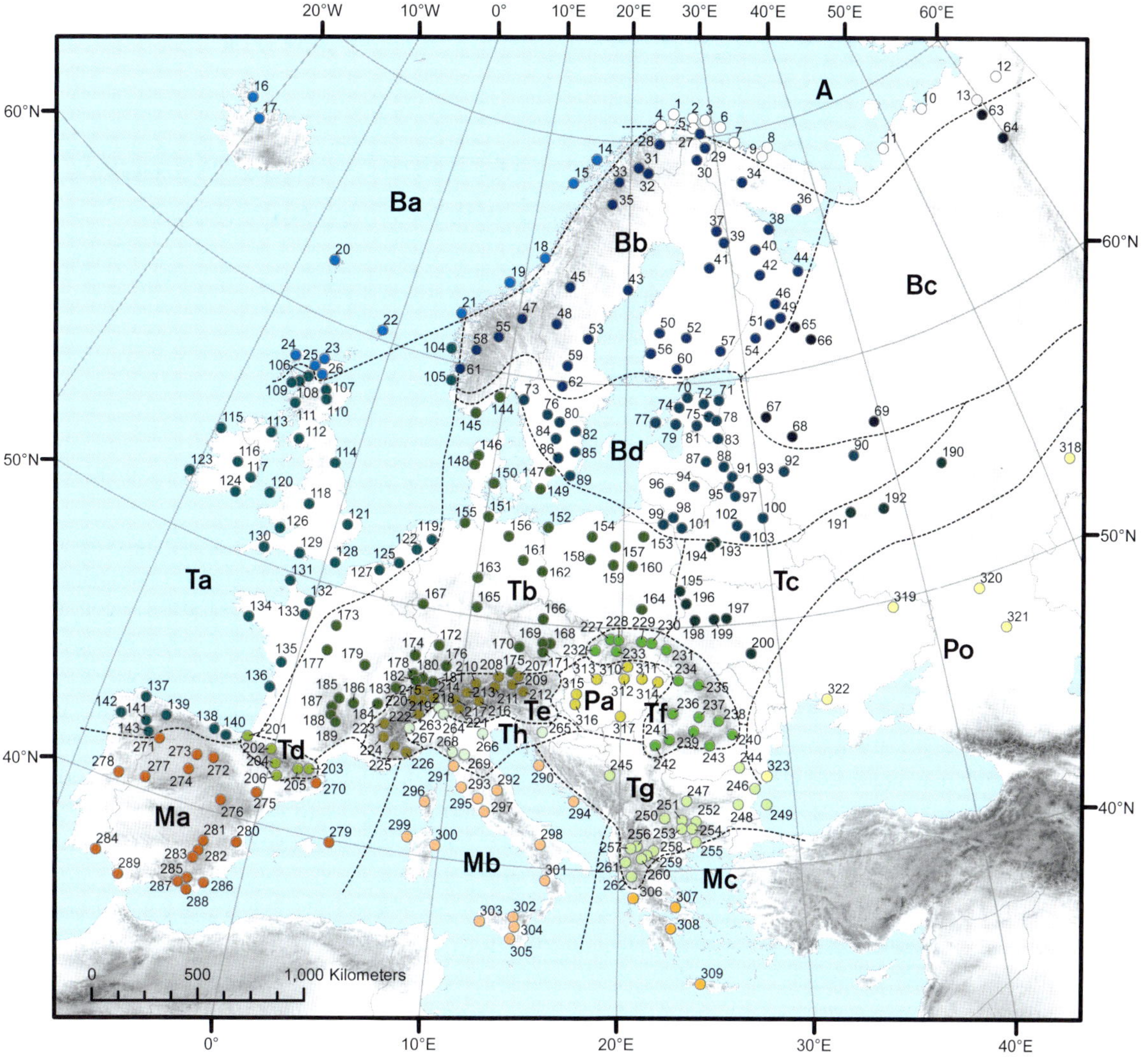

Fig. 4.3-1. European regions of vegetation history. A: arctic. Ba: boreo-atlantic. Bb: boreo-fennoscandian. Bc: boreo-russian. Bd: hemiboreal. Ta: temperate-atlantic. Tb: temperate central European. Tc: temperate eastern European. Td: Pyrenees. Te: Alps. Tf: Carpathians. Tg: temperate south-eastern European. Th: temperate Northern Italian (and adjacent areas). Ma: western Mediterranean. Mb: central Mediterranean. Mc: eastern Mediterranean. Pa: Pannonic. Po: Pontic. The coloured dots indicate palaeoecological sites used to infer the vegetation history of selected taxa (chapter 4.4), with numbers corresponding to Tab. 7-1 in the appendices.

ta et al., 2006, 2011) or Poteryanny Zub Lake (Gervais et al., 2002).

3. Coring sites with vast pollen-source areas such as marine cores may be relevant for more than one area, for example cores from the Black Sea for regions Tg and Po (Filipova-Marinova et al., 2013a).

4. Some sites may have shifted their association to regions over time, for example sites that today are north of the latitudinal or above the elevational treeline (such as in the arctic or arctic-alpine zone) were situated in the boreal forest during the Holocene Thermal Maximum (HTM, Snyder et al., 2000; Gervais et al., 2002; Väliranta et al., 2003, 2011 ; Samartin et al. 2017; Fischer et al. 2018).

A number of studies have been designed as **transects** that cross the borders between regions. Such transects may help to understand the effects of certain environmental gradients on vegetation history. Examples are provided by

- von Post (1916b, 1924) between Zealand (Denmark) in Tb and southern Sweden in Bd, accessible in Birks (2019)
- Welten (1982b) on elevational gradients between Tb and Te
- Lang (1985) on elevational and latitudinal gradients between Tb, Te and Th
- A transect emphasizing the records of human impact can be found in Berglund (1991)
- Peñalba (1994) between temperate-atlantic (Ta) and western Mediterranean (Ma)
- Seppä (1996) on a north-south transect between the Arctic (A) and Fennoscandia (Bb)
- Bjune et al. (2004) on a continental oceanic gradient (from boreo-atlantic Ba to boreo-fennoscandian Bb),
- Bjune (2005) on a north-south and elevational transect in southern Norway
- Eide et al. (2006) on an elevational gradient that crosses the regions between Tb, Bd and Bb
- Colombaroli and Tinner (2013) from Tb , Te and Th to Mb with an emphasis on diversity
- Huntley et al. (2013) on a west-east transect in the Arctic.

Graphical representation and interpretation

The basic characteristics of regional vegetation history can be traced best by comparing standardized pollen diagrams. Sites have been selected that include the entire Holocene and – where possible – the Late-glacial and have an acceptable chronology. Although, in rare cases, these criteria could not be fulfilled. Interesting and important sites which concentrate on shorter time periods, such as the Late-glacial and Early Holocene, or on archaeologically relevant periods are not included here, neither are marine records (pollen catchment too large). In some areas the spatial network of sites is incomplete, partly because basins with sediments or peat are rare (for example outside the glaciations), partly because modern studies are lacking. In order to improve the comparability we do not show complete original diagrams but concentrate on strongly simplified histograms. In some cases we use old family names because they were used in the original publications and determination keys, for example Chenopodiaceae. The selection of taxa is kept constant throughout the European regions. The calculation sum includes all pollen from dry-ground taxa, but excludes wetland taxa (e.g. Cyperaceae and Poaceae) and aquatics.

In the histograms the temporal resolution is 500 years on two time scales, calibrated years before present (cal yr BP) and BC/AD. Very rapid changes, such as the onset of the Holocene, cannot be captured accurately on this coarse resolution. Solid horizontal lines are set for the onset of the Late-glacial interstadial (Bølling-Allerød; 14 500 cal yr BP), and for the onset of the Holocene (11 500 cal yr BP). Dotted lines are set at 2500 cal yr BP, 6500 cal yr BP, 8500 cal yr BP and 16 500 cal yr BP. The bars of the histograms are kept white so colours for specific taxa can be added by the readers according to their questions.

In cases where plant macrofossils were analyzed, the results are indicated on the histograms with dots, but for many sites macrofossils were identified only for selected periods, for example the Late-glacial and Early Holocene. In these cases the macrofossil findings are not indicated because their absence in lower or higher core sections could be misunderstood ('an absence of evidence is no evidence for absence').

Of the many possible arrangements of the selected histograms the sequence chosen has a specific

Table 4.3-1. Pollen correction factors for important European plant taxa, mainly based on pollen productivity. Such factors can be applied (by multiplication) to pollen counts before calculation of pollen percentages. Very rare pollen types such as *Hedera*, *Viscum* and *Ilex* cannot be corrected because of their erratic and low occurrences. Pollen correction factors of 0.25 show overrepresented taxa, values of 4 underrepresented taxa, while values of 1 indicate that the taxa are almost correctly represented by pollen occurrences. Correction factors for Poaceae and Cyperaceae pollen are not shown, in our diagrams they were excluded from the pollen sum because potentially growing in wetlands (e.g. lakes or peatlands). In this book all diagrams show uncorrected values. Following Andersen (1970), Lang (1994), Conedera et al. (2006) and own observations.

Conifers	*Abies*	1
	Larix	4
	Picea	1
	Pinus Diploxylon (mainly *P. sylvestris*)	0.25
	Pinus Haploxylon (mainly *P. cembra*)	1
	Juniperus	1
Deciduous broadleaved trees	*Acer*	4
	Alnus	0.25
	Betula	0.25
	Carpinus	1
	Castanea	1
	Corylus	0.25
	Fagus	1
	Fraxinus	1
	Ostrya	0.25
	Quercus robur type	1
	Salix	4
	Tilia	4
	Ulmus	1
Evergreen broadleaved trees	*Buxus*	4
	Quercus ilex type	1
	Quercus suber/cerris type	1
	Olea	1
	Pistacia	1
Herbs	*Artemisia*	1
	Chenopodiaceae	1
	Plantago lanceolata type	2
	Rumex acetosa type	1
	Cerealia type	4

function: usually visualizing a gradient in latitude, longitude, or elevation.

If an in-depth consideration of an area is the goal, these strongly simplified histograms can never replace the original (often large) diagrams because of the small number of taxa and the low temporal resolution of the histograms. However, the constant sequence of taxa and synchronous time scale facilitates an overview. For the reconstruction of vegetation history pollen data need to be corrected according to taxon-specific productivity and dispersal characteristics which may vary in time and space (chapter 2.7). Given the high spatio-temporal variability in pollen productivity and dispersal this is usually done qualitatively when inferring the vegetation history. As a first approximation, correction factors for important taxa can be used (Table 4.3.1). Such rough estimates may have a general validity, but it is crucial to consider the local conditions at the study sites, e.g. by comparing surface sediment and soil samples with recent vegetation composition. Non-arboreal pollen (NAP) is often underrepresented, thus in Europe >20% of NAP is considered to indicate rather open forest conditions, 30-50% of NAP forest-steppe or parklands and > 50% of NAP steppes or grasslands (with isolated tree stands in locally favourable areas). Conversely >80% of arboreal pollen (AP) is considered to indicate forested conditions. Exceptions occur under steppic conditions, where spatially restricted azonal forest stands (e.g. around wetlands) may generate AP sums significantly < 80%. In our pollen diagrams Cyperaceae and Poaceae are excluded from the pollen sum, thus NAP >10% may already indicate rather open forests and NAP > 20% forest steppe or parkland vegetation.

4.3.2

Regional vegetation history

A, Arctic region

On the European continent the arctic region consists of a relatively narrow belt in the far north, stretching from the North Cape along the northern

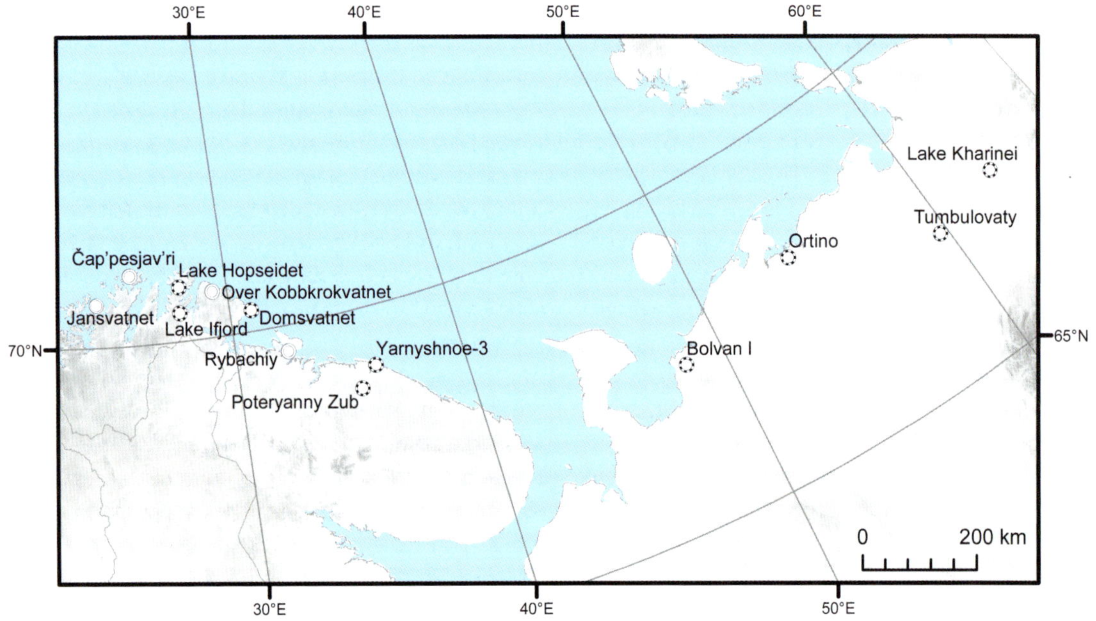

Fig. 4.3-2a. Map of selected sites in the European arctic (A) region. Dots with black borders indicate sites shown in the histograms (Fig.4.3-2b), white borders indicate sites used for the maps showing the history of selected taxa (see chapter 4.4) and dashed borders indicate sites used for both.

edge of the Kola Peninsula and becoming broader in northern Russia towards the Ural mountains. Various subdivisions of the vegetation types have been proposed. The modern vegetation is mostly dwarf shrub tundra with *Betula nana* and other dwarf shrubs (for example *Salix polaris, S. glauca* in the subarctic and southern arctic). Towards the south this vegetation grades into forest-tundra, while farther southwards are the boreal coniferous forests (Fennoscandian and boreo-russian regions). Towards the north the dwarf shrub tundra of the southern arctic extends into patchy tundra with creeping shrubs including *Salix polaris* and *Dryas octopetala* (middle arctic). Farther north are the northern arctic (or high arctic) and the 'polar desert' with hardly any vegetation. On the archipelago of Svalbard, this high arctic belt fringes the glaciers on several of the islands, whereas on the main island of Spitsbergen two vegetation types of the middle arctic can also be distinguished: the *Cassiope tetragona* zone along the somewhat warmer inner fjords, and the *Dryas octopetala* zone on the outer west coast (cooler and with more fog) and at elevations above the *Cassiope* zone (Brattbakk, 1986; Birks, 1991a).

Plant geographically the flora of the arctic region belongs to the lapponic and samojedic provinces as part of the circum-arctic flora according to Meusel et al. (1965). Takhtajan (1986) only considers the samojedan area as belonging to the arctic floristic province (together with parts of Iceland).

During the Last Glacial Maximum (LGM) the western part of the region was completely ice-covered, while the east was ice-free (Svendsen et al., 1999, 2004; Mangerud et al., 2002). Ice retreat occurred during the Late-glacial and Early Holocene but on Svalbard and Novaya Zemlya large glaciers remained.

On Svalbard, in the modern middle arctic zone, the Holocene vegetation development is difficult to determine because all tree pollen, with the exception of *Salix,* is long-distance transported from the continent (Środoń, 1960; Hyvärinen, 1970; van der Knaap, 1987, 1988). Pollen of *Salix*, most probably *S. polaris*, *S. glauca,* and other creeping willows, represents the only woody component in the local pollen assemblages. It is hard to say if these *Salix* are low pollen producers. They are so close to the ground that their pollen is not well dispersed. Values of this pollen type are low but were somewhat higher between 7000 and 3000 cal yr BP. During this period, Ericaceae are also more frequent, most probably *Cassiope*. Among the NAP, *Saxifraga*, Brassicaceae, Caryophyllaceae and Ranunculaceae are important. Because macrofossil analyses provide a more local signal than pollen and also a higher taxonomic resolution, Birks (1991a) was able to show at Skardtjørn, western Spitsbergen, that most taxa were present around 8000 cal yr BP, and that the vegetation composition has not changed substantially during the last 8000 years, although the vegetation cover has decreased. The site at Skardtjørn passed through the threshold between the modern vegetation types of the inner fjord (*Cassiope*, *Salix polaris*) to that of the outer fjord vegetation (*Dryas octopetala*) of the mid-arctic region around 2500 cal yr BP. A gradual Late Holocene cooling between about 4000 and 2500 cal yr BP can be inferred. This is compatible with van der Knaap's (1987, 1988) results from Spitsbergen and many arctic sites on Greenland (Fredskild, 1969, 1973) and in Canada (reviewed in Bradley, 1990; Birks, 1991a).

In northernmost Fennoscandia the pioneer work of Aario (1940) showed that, in surface samples, the values of pine as percentages are higher north of the modern treeline than in the pine forest. Hyvärinen used absolute pollen values (pollen accumulation rates or influx/grains cm^{-2} year^{-1}) instead of relative ones (percentages) on sediment cores on a north–south transect in Finnish Lapland, reaching from the arctic site Domsvatnet (Hyvärinen, 1976) to the boreal sites Bruvatnet, Suovalampi and Akuvaara (Hyvärinen, 1975a; see map in Fig. 4.3-2a). Seppä (1996) and Seppä et al. (2002) extended the transect and, at Lake Ifjord, even obtained a sediment core that reached as far back as the Younger Dryas (dominated by *Artemisia* pollen). At Jansvatnet, near Hammerfest, the Younger Dryas was structured in three periods (cool, very cold, cool; Birks et al., 2012c). As at Domsvatnet, the lakes Hopseidet, Ifjord and Jansvatnet, and also on three sites of a west–east transect (Čap'pesjav'ri; over Gunnardsfjorden; over Kobbkrokvatnet; Huntley et al., 2013), birch woodlands (but no pines) developed during the Early Holocene in areas where today there is tundra, illustrating the HTM. The retreat of birch started around 5500 cal yr BP (more or less simultaneously with the pine retreat further south; see Fennoscandian sites Bb in Fig. 4.3-4 b).

The arctic of north-west Russia was only under LGM-ice in the west (Kola Peninsula) and was mostly ice-free in the lowlands of the central area and in the east. According to Svendsen et al. (1999, 2004), the Scandinavian ice sheet reached its maximum position shortly after 17 000 cal yr BP, whereas the Barents and Kara ice sheets were at their maximum prior to 50 000 yr BP. Consequently, the core of Bolvan I/Timan Ridge-15a reaches beyond the Late-glacial interstadial (Bølling-Allerød) with a dwarf shrub tundra, and a colder/drier Younger Dryas with a steppe-tundra and snowbeds (Paus et al., 2003). In contrast, the cores from the northern shore of the Kola Peninsula reach either to just about the beginning (Yarnyshnoe-3) or the end (Poteryanny Zub) of the Younger Dryas (Snyder et al., 2000; Gervais et al., 2002; Paus et al., 2003). The Holocene vegetation dynamics of these two sites near the northern shore of the Kola Peninsula resembles that of arctic Fennoscandia, including a pine treeline (Gervais et al., 2002). The more continental parts of the Kola Peninsula belong to the region Bb because they resemble the boreal part of Fennoscandia, except for the fact that *Picea abies* spp. *obovata* had already reached the Kola Peninsula after 7000 cal yr BP (on two routes) and replaced the Mid Holocene pines by c. 4500 cal yr BP, thus earlier than in the west (Kremenetski et al., 1999a).

Further east the Early Holocene is characterized either by birch forests (Timan Ridge-15a; Paus et al., 2003) or birch and spruce forests as indicated by spruce stomata at Tumbulovaty and macros at Kharinei and even at Ortino (Kaakinen and Eronen, 2000; Väliranta et al., 2003, 2011; Kultti et al., 2004). This early development of woody vegetation culminates in the Mid Holocene and early Late Holocene at c. 8000–3500 cal yr BP and shows a west-east contrast (Salonen et al., 2011): in the west, near the north shore of the Kola Peninsula, it is recorded with stomata, macrofossils and wood of *Pinus sylvestris* some 100 km north of the present limit of pine; in the east a *Picea* forest was established c. 150 km north of its present limit. Around lake Kharinei, spruce started to become rare after 3500 cal yr BP, the last macro remains of trees disappeared around 2500 cal yr BP and aggradation and expansion of permafrost started (Oksanen et al., 2001; Salonen et al., 2011). At the eastern sites of Ortino, Kharinei and Tumbulovaty, *Pinus sylvestris* was either absent or only growing on the margins of mires (low percentages and PAR, absence of macrofossils and stomata). The onset of the HTM seems earlier when based on the evidence of aquatic macrofossils than it is when only tree-pollen evidence is considered. The reconstructions based on aquatic macrofossils are in accordance with those based on chironomids. This illustrates that organisms with short life cycles and fast dispersal/migration rates may not lag behind climate change as much as tree populations may do (Väliranta et al., 2015).

On Novaya Zemlya, during the warmest parts of the Holocene, a dwarf shrub tundra developed but never any woodland (Serebryanny and Malyasova, 1998; Serebryanny et al., 1998).

Overviews: For arctic-alpine sites, that is high-elevation sites in the Scandes, see Velle et al. (2005b) on Råtåsjøen at 1169 m asl and Paus (2010) on Flåfattjønna at 1110 m asl for the south, and Rosén et al. (2001), on Lake Sjuodjijaure at 826 m asl and Küttel (1984) on Vuolep Allakasjaure at 995 m asl for the north. For Holocene treelines further east (outside Europe), see Kremenetski et al. (1998) reviewing 115 sites.

Transects are presented by Hyvärinen (1975b, 1976), Seppä (1996), Kremenetski et al. (1998) and Huntley et al. (2013), including macrofossils by Väliranta et al. (2011), limnology by Duff et al. (1999) and, on a larger scale further eastwards, by Binney et al. (2009).

Data-model comparisons can be found in Fang et al. (2013).

Climate reconstructions are presented by Serebryanny et al. (1998), Andreev and Klimanov (2000), Kaakinen and Eronen (2000), Jensen et al. (2002a), Kultti et al. (2003, 2004), Paus et al. (2003), Rosén et al. (2003), Väliranta et al. (2003, 2015), Bjune et al. (2004), Velle et al. (2005a,b), Allen et al. (2007), Salonen et al. (2011), Birks et al. (2012c) and Huntley et al. (2013).

Lake Ifjord *Norway* (Seppä 1996)

A 70.4° N, 27.6° E, 317 m a.s.l.

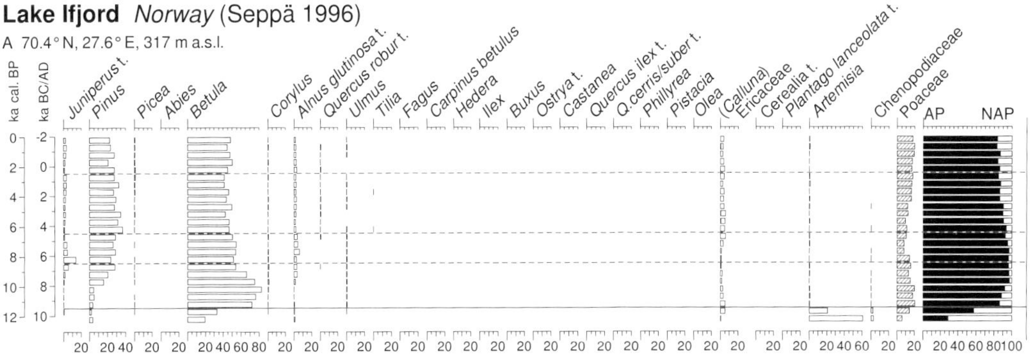

Lake Hopseidet *Norway* (Seppä 1996)

A 70.8° N, 27.7° E, 225 m a.s.l.

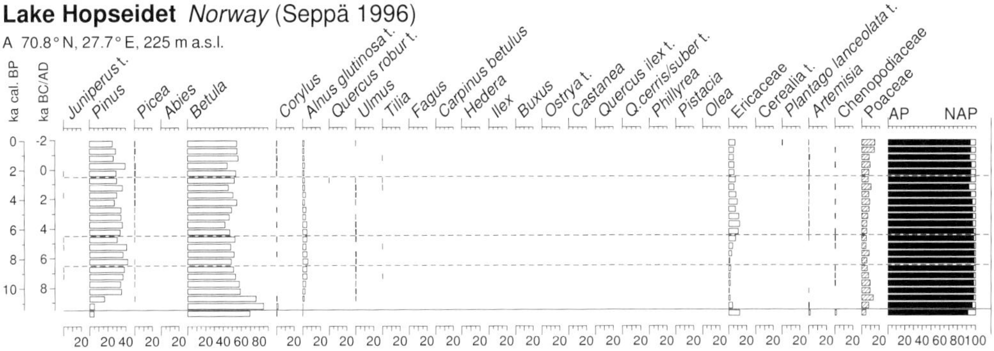

Domsvatnet *Norway* (Hyvärinen 1976)

A 70.3° N, 31.0° E, 170 m a.s.l.

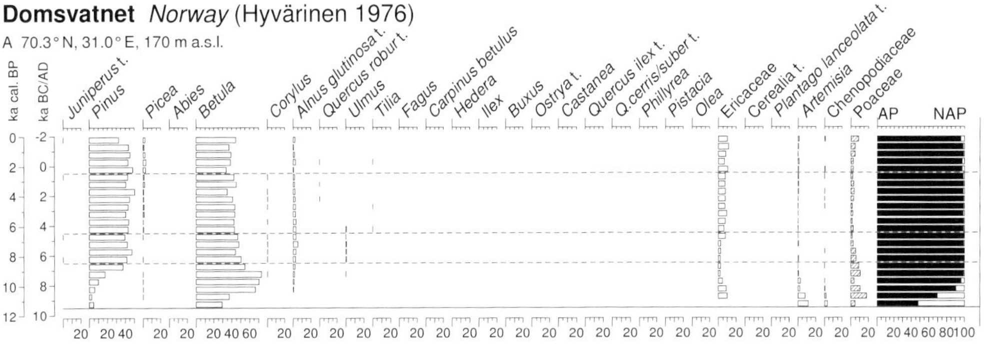

Poteryanny Zub *Russia* (Gervais et al. 2002)

A 68.8° N, 35.3° E, 131 m a.s.l.

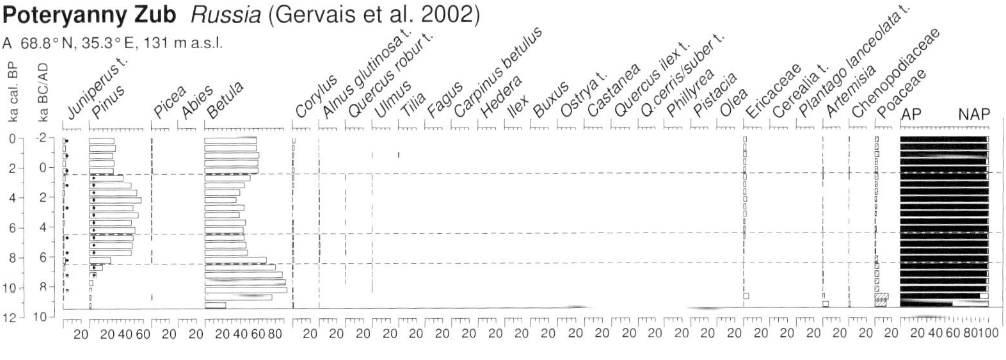

Yarnyshnoe-3 *Russia* (Snyder et al. 2000)

A 69.1° N, 36.1° E, 54 m a.s.l.

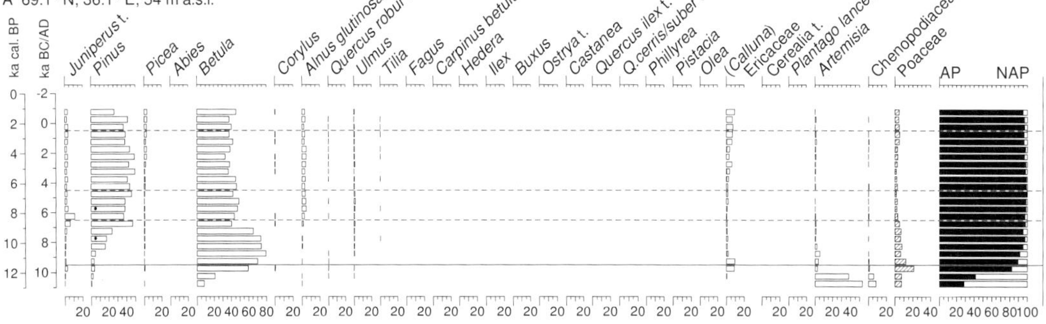

Bolvan I (Timan Ridge 15a) *Russia* (Paus et al. 2003)

A 67.4° N, 48.4° E, 168 m a.s.l.

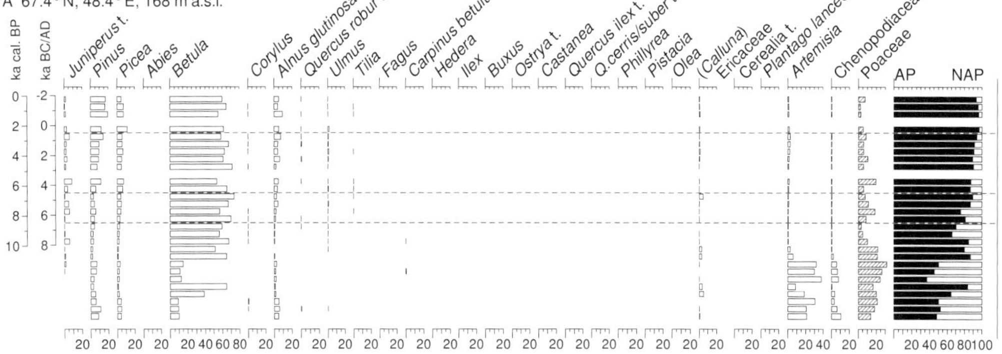

Lake Kharinei *Russia* (Salonen et al. 2011)

A 67.4° N, 62.8° E, 110 m a.s.l.

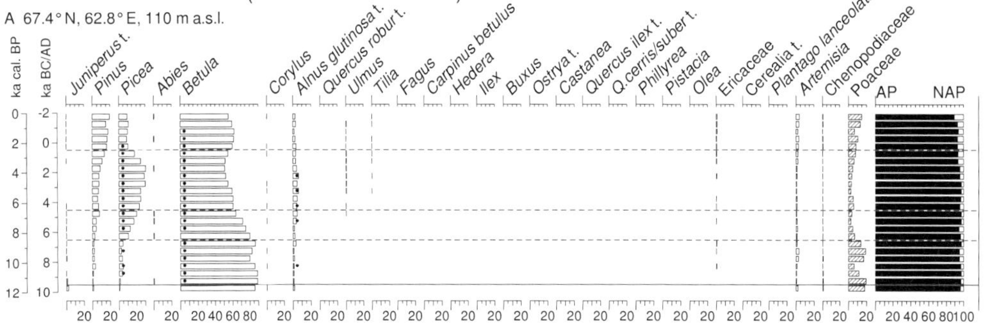

Tumbulovaty *Russia* (Kultti et al. 2004; Väliranta et al. 2011)

A 67.1° N, 59.7° E, 115 m a.s.l.

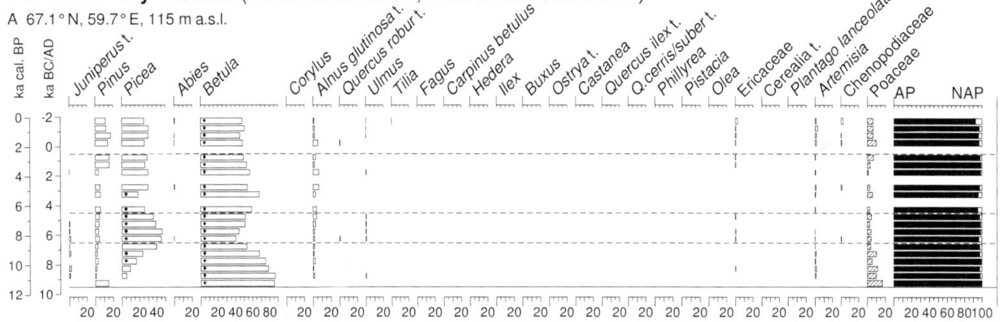

Ortino *Russia* (Kaakinen and Eronen 2000)

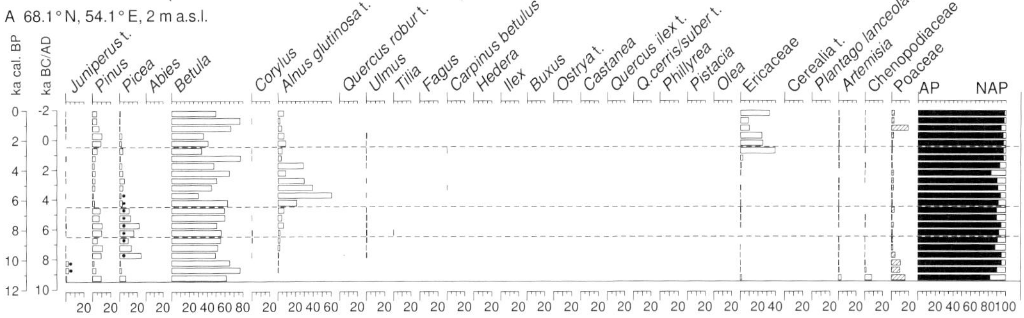

Fig. 4.3-2b. Selected sites in the European arctic (A) region. Late-glacial and Holocene pollen records as histograms at a 500-year resolution. The sites are today nearly all in arctic, that is tree-less, vegetation. Exceptions are paired sites, with the first pair on the north shore of the Kola Peninsula, where Lake Yarnyshnoe is in the tundra and Poteryanny Zub is in the forest-tundra. The second pair is further east, where Lake Kharinei and Ortino are truly arctic sites, whereas Lake Tumbulovaty is situated in the very broad belt of forest-tundra. The dominance of arboreal pollen is striking, but partly explained by the dominance of dwarf shrubs in the tundra that also produce it (e.g. *Betula, Salix, Juniperus, Vaccinium*). The overrepresentation of arboreal pollen illustrates the scenario described by Aario (1943) that low pollen productivity of the tundra vegetation and the importance of long-distance transported pollen may bias the pollen signal. Where available the occurrence of plant macrofossils is indicated by dots (e.g. Lake Kharinei, Tumbulovaty; for Ortino we combined Ortino I with pollen and stomata with Ortino II with macros). Conifer stomata (but no macrofossils) were analysed at Poteryanny Zub and Yarnyshnoe-3. The records of plant macrofossils and stomata are strong indicators of vegetation changes during the HTM (e.g. Lake Kharinei, Tumbulovaty, Poteryanny Zub, Yarnyshnoe-3). At some sites Ericaceae increase in the Late Holocene.

Ba, Boreo-atlantic region

This region comprises the western part of the boreal zone near the Atlantic Ocean: Iceland, the islands of Faroe, Shetland, Orkney, the Outer Hebrides, the northernmost part of Scotland and the west Norwegian coast north of latitude 61°N. The vegetation of this region has nearly no conifers today. Plant geographically, this area of the outermost north-west of Europe largely corresponds to the boreo-atlantic floristic province of Meusel et al. (1965) but excludes the larger part of Scotland which, in its vegetation history, more closely resembles the Atlantic region of the temperate zone.

This region became ice-free during the Late-glacial, except for parts of Iceland and small areas in western Norway, where some glaciers remain until today. In general, the Holocene vegetation history is characterized by the important role of *Betula*. The analysis of macrofossils is, therefore, especially important in order to identify the species.

Late-glacial and Holocene: On **Iceland** Vasari and Vasari (1990) showed that the vegetation on the deglaciated ground during the Late-glacial may have been barren and patchy (as in the high arctic), but then developed during the Early Holocene into a dwarf shrub tundra with *Betula nana, Juniperus* and *Salix herbacea*. Only from the Mid Holocene onwards did tree birches expand, as shown through macrofossil analyses; concurrent with this expansion juniper decreased. Woody taxa declined during the second half of the Holocene and especially after the landnam by the Vikings after AD 870.

On the **northern Norwegian coast** the Andøya-Andfjord area was deglaciated at a very early stage (before 20 000 cal yr BP; Vorren et al., 2013).

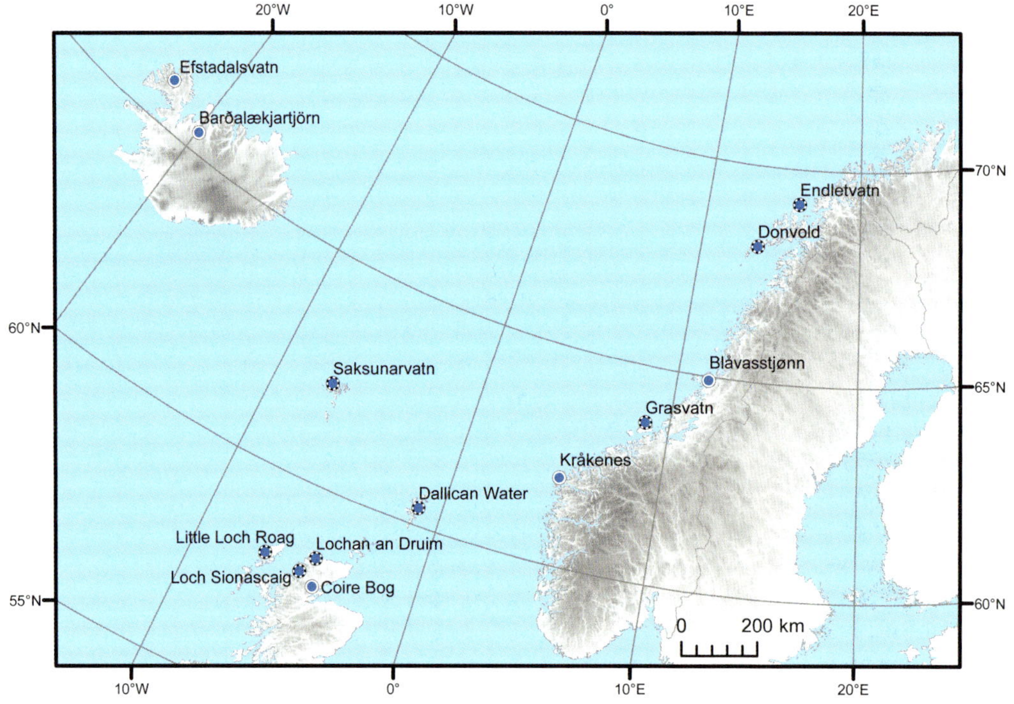

Fig. 4.3-3a. Map of selected sites in the boreo-atlantic (Ba) region. Dots with black borders indicate sites shown in the histograms (Fig. 4.3-3b), white borders indicate sites used for the maps showing the history of selected taxa (see chapter 4.4) and dashed borders indicate sites used for both.

Detailed records of the Late-glacial and its transition to the Holocene are described for several sites (Vorren, 1978; Alm and Birks, 1991; Alm, 1993; Vorren et al., 2009; Aarnes et al., 2012). The Younger Dryas is clearly indicated by an increase in high-arctic taxa (Vorren et al., 2009; Aarnes et al., 2012). During the Early Holocene tree birches (*Betula pubescens*) developed earlier in the inner fjords (likely warmed by foehn winds) than on the outer coasts, and they later dominated the vegetation (Vorren et al., 2009). In addition, *Pinus* pollen appears from the Early Holocene onwards with values up to 20%, but *Pinus sylvestris* may have been very local or absent along the coast.

On the **southern central Norwegian (outer) coast** deglaciation occurred during the Late-glacial interstadial. In the sequence of Holocene tree immigration, *Picea abies* never arrived at Grasvatn (Paus, 1982, 1996) – see also the modern distribution of spruce in Figs. 4.4.2-6 and 4.4.2-7. During the second half of the Holocene heath formation is prominent.

On the **north Atlantic archipelagos of Faroe, Shetland, Orkney and the Outer Hebrides**, but also on the northernmost part of the Scottish mainland, tree establishment during the Holocene was spatially restricted to sheltered places or was absent. On the Faroe Islands single trees of *Betula pubescens, Corylus avellana, Populus tremula* and *Sorbus aucuparia* occurred, as shown by macrofossil analyses (Jóhansen, 1985, 1996; Hannon et al., 2003). There, shrubby vegetation lasted from about 8000 cal yr BP to AD 660 and the decrease in the number of woody taxa started prior to human settlement but became intensified afterwards (Hannon and Bradshaw, 2000; Hannon et al., 2005). On Shetland (see Dallican Water in Fig. 4.3-3b), Early Holocene birch woodlands were enriched by *Corylus, Quercus* and possibly other deciduous trees. Later, however, heathlands developed; the role of herbivores and humans in this development is discussed by Bennett et al. (1992). During the Holocene other areas also had an open vegetation more or less devoid of trees but with *Betula nana, Juniperus* and *Calluna.*

In the Outer Hebrides, on the Isle of Lewis (see Little Loch Roag in Fig. 4.3-3b), forest was absent throughout the Holocene (Birks and Madsen, 1979). South Uist had a woodland history between c. 8200 and 4000 cal yr BP (Loch Lang in Bennett et al., 1990) that resembles the one of the Inner Hebrides (in temperate-atlantic Ta), for example Loch Ashik (Birks and Williams, 1983), though *Quercus, Ulmus* and *Alnus glutinosa* are absent from Loch Lang today (Bennett et al., 1990). At most sites *Calluna* plays an important role.

On the **Scottish mainland** only the northern fringe belongs to the boreo-atlantic region: for the site of Lochan an Druim, Birks (1984) included the study of macrofossils and compared the results with 9 other Scottish sites (partly in Ba, partly in Ta). The cooling and drying of the Younger Dryas was recorded as a shift from dwarf shrubs to an open grass and sedge heath (with high *Empetrum* values). The Holocene opened with an expansion of *Juniperus*, *Betula* and *Corylus*. Later, human impact is recorded by the *Calluna* and *Plantago lanceolata* curves. At Loch Sionascaig the record reaches back to the Younger Dryas and the Holocene is also complete (Pennington et al., 1972).

Human impact: Mesolithic hunters were present on Shetland (Bennett et al., 1992). Later, in connection with human settlement and the grazing of domesticated animals, the pollen diagrams show an increase in Poaceae. On the Shetland Islands this possibly started as early as around 5000 cal yr BP, on the Isle of Lewis after 4000 cal yr BP (Birks and Madsen, 1979), and on the Faroe Islands probably in three phases around 4300 cal yr BP, AD 600–650 and AD 800–850 with the Vikings (Bennett et al., 1992; Jóhansen, 1985, 1996; Hannon and Bradshaw, 2000; Hannon et al., 2005).

Overviews can be found in Berglund et al. (1996a); for the Faroe Islands by Jóhansen (1996) with 6 sites; for Scotland by Birks (1996b) covering 9 sites; and for Norway by Moe et al. (1996) covering 3 sites.

Transects: Of the three Norwegian transects two have their westernmost site in the boreo-atlantic region (Litlvatnet in KILO and Storsandvatnet in OTT, Birks et al., 2012a).

Pollen maps: Taxa responses to climate change were mapped by Birks et al. (2014).

Climate reconstructions are presented by Alm (1993), Vorren and Plassen (2002), Bjune et al. (2004), Vorren et al. (2009, 2013), Aarnes et al. (2012) and Birks et al. (2014).

Human impact has been described by Birks and Madsen (1979), Bennett et al. (1990, 1992) and Sjögren (2009).

Endlevatn *Norway* (Vorren 1978; Vorren and Alm 1985; Vorren et al. 2013)

Ba 69.2° N, 16.0° E, 36 m a.s.l.

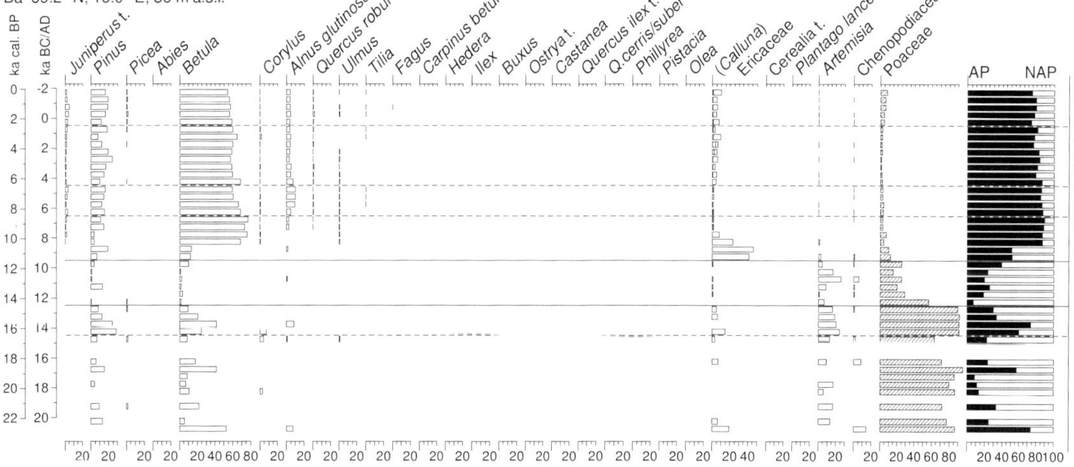

Donvold *Norway* (Nilssen 1983)

Ba 68.1° N, 13.6° E, 10 m a.s.l.

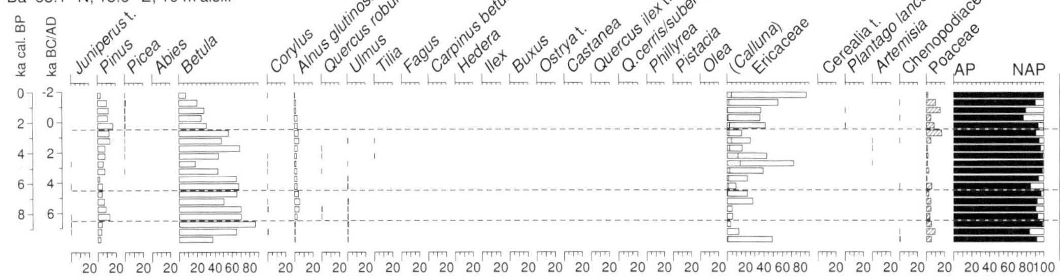

Grasvatn *Norway* (Paus 1982)

Ba 63.7° N, 8.7° E, 45 m a.s.l.

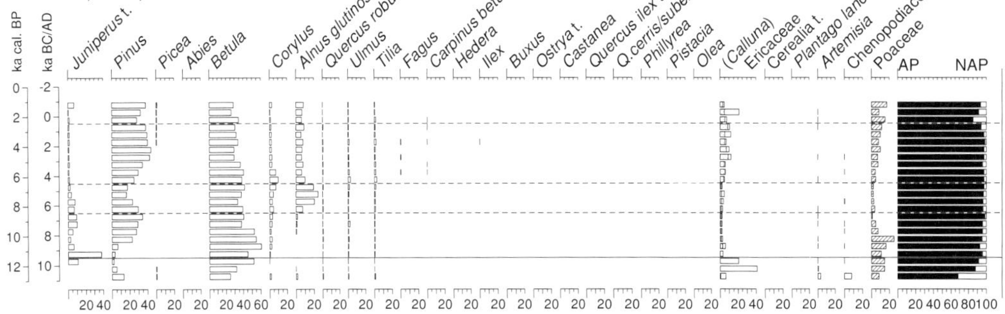

Saksunarvatn *Faroe Islands* (Jóhansen 1982)

Ba 62.3° N, 7.2° W, 23 m a.s.l.

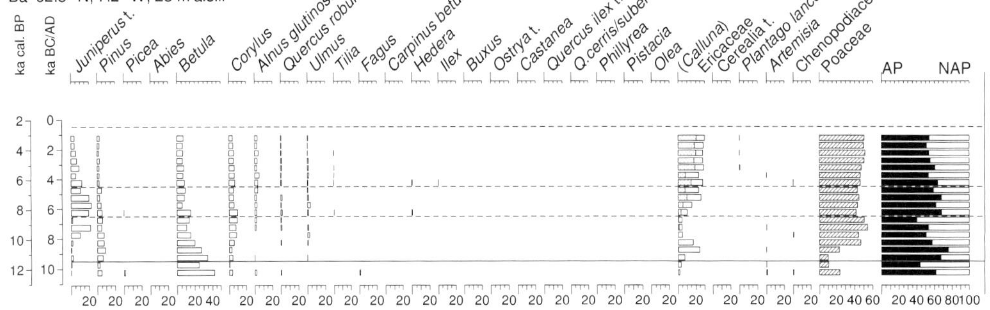

Dallican Water *Great Britain* (Bennet et al. 1992)

Ba 60.4° N, 1.1° W, 56 m a.s.l.

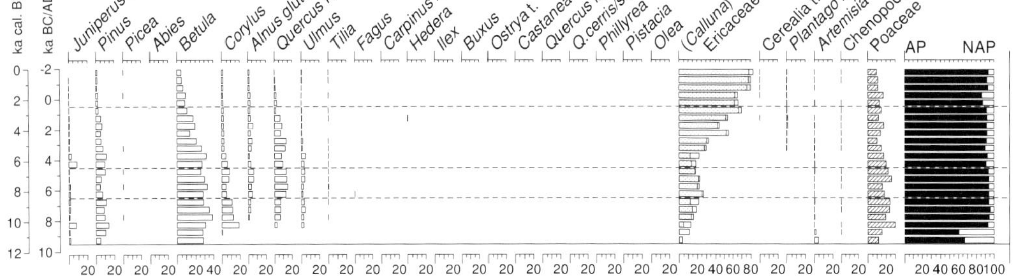

Lochan an Druim *Great Britain* (Birks HH 1980, 1984)

Ba 58.5° N, 4.7° W, 25 m a.s.l.

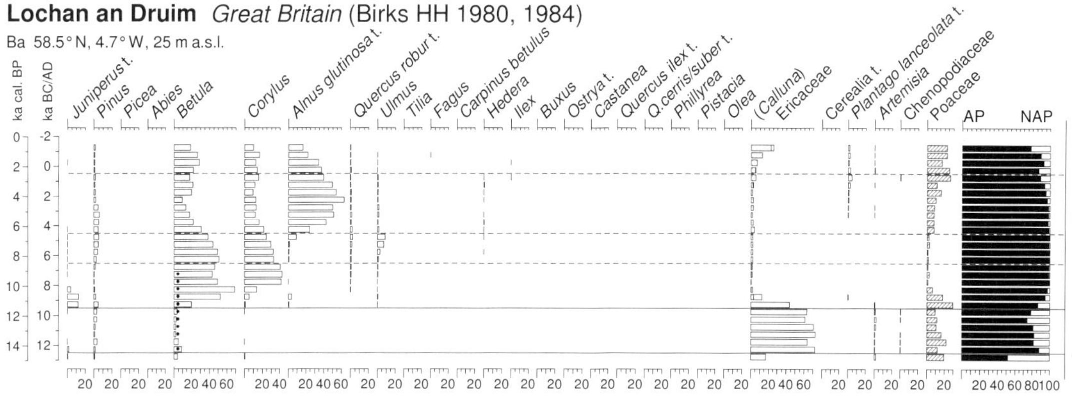

Loch Sionascaig *Great Britain* (Pennington et al. 1972)

Ba 58.3° N, 5.2° W, 74 m a.s.l.

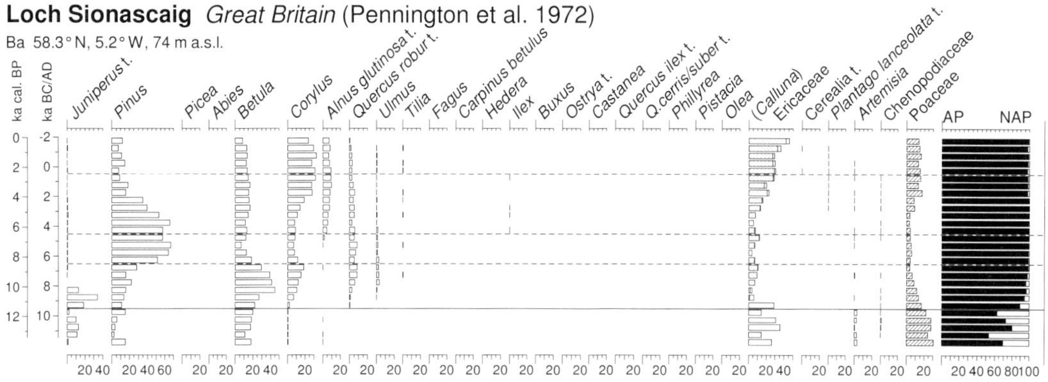

Little Loch Roag *Great Britain* (Birks HJB 1973; Birks HJB and Madsen 1979)

Ba 58.1° N, 6.9° W, 27 m a.s.l.

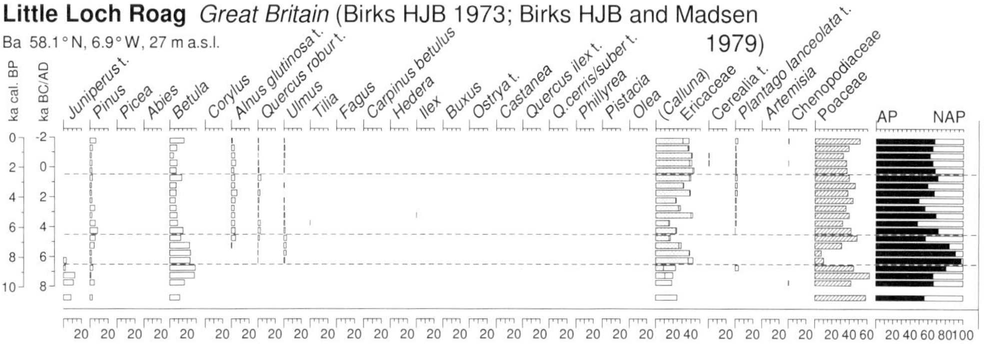

Fig. 4.3-3b. Selected sites along the boreo-atlantic (Ba) fringe of Europe arranged from south to north as histograms at a 500-year resolution. Several records go back to the early Late-glacial and may contain some reworked pollen (rwp). Among the Ericaceae the early peaks are often due to *Empetrum*; the younger peaks, in contrast, are often caused by *Calluna* (shown as dotted bars), indicating the increasing importance of heathlands.

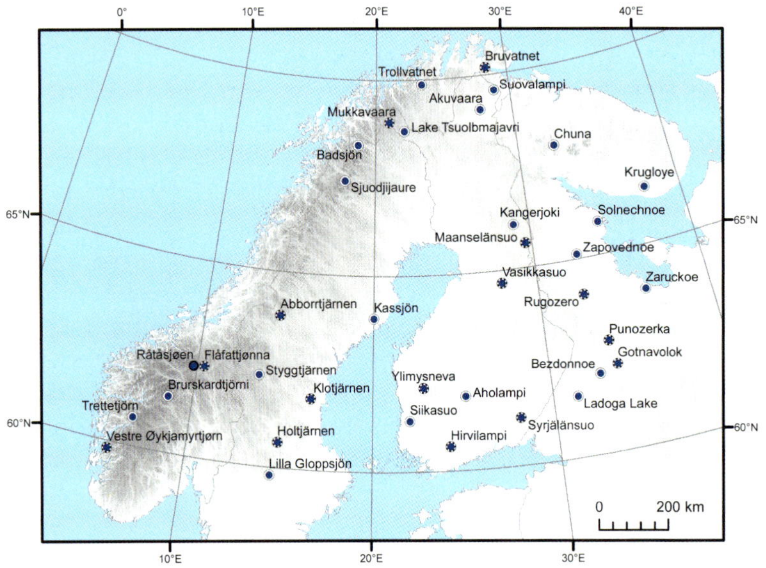

Fig. 4.3-4a. Map of selected sites in the boreo-fennoscandian (Bb) region. Dots with black borders indicate sites shown in the histograms (Fig. 4.3-4b), white borders indicate sites used for the maps showing the history of selected taxa (see chapter 4.4) and dashed borders indicate sites used for both.

Bb, Boreo-Fennoscandian region

This region includes Fennoscandia (between about 60°N and 70°N) but excludes the outermost arctic north, the Norwegian coasts, the southern third of Sweden, and the southern coast of Finland. Its eastern border runs from the White Sea through Karelia to the Gulf of Finland. Today, the vegetation of this region is dominated by coniferous forests. Plant geographically, the region largely corresponds to the Scandinavian floristic province of Meusel et al. (1965). In the west the Scandes range, with peaks up to 2500 m asl, marks a sharp climatic boundary with the boreo-atlantic region.

During the Last Glacial Maximum (LGM) almost the entire region was covered by the Fennoscandian Ice Sheet (Svendsen et al., 2004). In the north, the west and also partly in the east of the region some sites became ice-free at an early stage, see Østervatnet and Bruvatnet near Varangerfjord, Endlevatn on Andøya in region Ba, Gotnavolok near Lake Onega in region Bc (Hyvärinen, 1975a; Vorren, 1978; Prentice, 1981; Elina and Filimonova, 1996). The central parts of the region Bb only became ice-free towards the end of the Late-glacial or during the Early Holocene (Fig. 1-2). The location of the ice margin during the Younger Dryas is clearly visible in the southern landscapes as the Ra moraines in southern Norway, the middle Swedish end moraines and the Salpausselkä complex in Finland (Mangerud et al., 2007). During the melting of the ice and after its disappearance eustatic and isostatic sea-level changes led to the changing coastlines of the Baltic Sea (see chapter 3).

Late-glacial and Holocene: Late-glacial sediments have been reached at some sites; they are usually dominated by NAP (e.g. Bruvatnet, Punozerka, or Gotnavolok). The Holocene vegetation development is characterised by the dominance of *Betula* (*B. pubescens*) first and later by *Pinus* (*P. sylvestris*) (Fig. 4.3-4b). *Picea abies*, a third important tree species, immigrates in a complex fashion with a rapid Early Holocene spread from east to west and a Mid to Late Holocene front-like spread moving west and south (Giesecke and Bennett, 2004; Giesecke, 2005a, and literature therein). During the Mid Holocene some more warmth-demanding deciduous trees (*Quercus, Ulmus, Tilia*) which do not grow north of the hemiboreal zone today, possibly reached somewhat higher latitudes (Barnekow et al., 2008). *Corylus* fruits were found in peat north of its current range already in the 19th century (Andersson, 1902, 1909) and because of their weight and short dispersal distances the 'false presence problem' is of no relevance. An up-dated and methodologically refined study of the climatic control over hazel's range presented by Seppä et al. (2015) showed that during the Holocene Thermal Maximum (HTM) *Corylus avellana* occurred in Finland and Sweden 350–400 km further north than today. At sites in the north and at higher elevations the abundance of birch and/or pine was higher in the Mid Holocene and decreased after c. 5000–4000 cal yr BP (Kremenetski and Patyk-Kara, 1997; Barnekow, 1999, 2000; Barnekow and Sandgren, 2001; Gervais et al., 2002; Bjune et al., 2004, 2005; see also selected sites in Table 4.4.5-1). These results are in agreement with findings and interpretations of pine megafossils north of the present treeline (Eronen, 1979; Eronen and Hyvärinen, 1982; Kullman, 1992a, 1993, 1995; Eronen and Huttunen, 1993; Eronen and Zetterberg, 1996; see also chapter 4.9). On the Khibiny mountains (central Kola Peninsula), *Pinus sylvestris* dominated between around 7000 and 5000 radiocarbon years BP (c. 7800–5700 cal yr BP) and was then replaced by *Picea abies* spp. *obovata* (Kremenetski et al., 1999a). In Fennoscandia *Picea abies* spp. *abies* never formed the conifer treeline, in contrast to the situation in the boreo-russian region Bc.

Corresponding but not identical changes took place at the latitudinal and elevational treelines, a fact that indicates a large-scale cause such as changes in insolation (hypsithermal or HTM). In addition to these general trends (by extrinsic forcing, *sensu* Williams et al., 2011) some minor fluctuations (by intrinsic forcing, *sensu* Williams et al., 2011) are recorded at many sites, some of them caused by species-specific immigration. A simplified overview of HTM reconstructions for Fennoscandia is presented in Table 4.4.5-1.

Early Mesolithic sites are known from many places including the far north of Fennoscandia and European Russia (Möller et al., 2012). **Human impact** on the vegetation is strongly time-transgressive; it started in Fennoscandia roughly between c. 6000 cal yr BP and c. 2000 cal yr BP and increased towards the present (Josefsson et al., 2014).

Overviews can be found in Berglund et al. (1996a); for Norway by Moe et al. (1996) covering 9 sites; for Sweden by Berglund et al. (1996b) covering 15 sites; for Finland by Vasari et al. (1996) covering 52 sites; for Russian Karelia by Elina and Filimonova (1996) covering 26 sites; and for the Kola Peninsula by Kremenetski et al. (1997).

Transects are presented by Seppä (1996), Bjune et al. (2004), Bjune (2005) and Eide et al. (2006). The south–north (and elevational) transect of Eide et al. (2006) runs through the extraordinarily steep gradients of southern Norway visible on the map in figure 4.3-1 where bands of Tb, Bd and Bb are 'telescoped' together. In the Torneträsk area near Abisko, Barnekow (1999) put an elevational transect of three sites into context with five areas on a north–south transect (from northern Finland through Sweden to southern Norway). Elevational transects with six levels are presented by Barnekow (2000) and Barnekow and Sandgren (2001). An additional site nearby was analysed for pollen, diatoms and chironomids making quantitative climate reconstructions possible (Bigler et al., 2006). In central Norway a transect (four sites) is presented in Moe et al. (1996). For the transition from the Fennoscandian to the hemiboreal region, Berglund (1991) presents rich material for the last 6000 years. The very first transect based on palynology was probably the one by von Post (1916b, 1967) summarized by van der Maarel and Franklin (2013).

Pollen maps: Giesecke and Bennett (2004) and Seppä et al. (2009a) for *Picea*, Kremenetski et al. (1999b) for *Picea* and *Alnus incana*, Björse et al. (1996) for tree abundance maps for *Fagus* and *Pinus* and forest type maps for southern Sweden.

Data-model comparisons can be found in Miller et al. (2008), von Stedingk and Fyfe (2009) and Bialozyt et al. (2012).

Climate reconstructions: see Table 4.4.5-1 and Karlén (1976), Kullman (1999), Birks et al. (2000), Barnekow and Sandgren (2001), Rosén et al. (2001), Seppä and Birks (2001, 2002), Seppä et al. (2002, 2009c, 2015), Bigler et al. (2002, 2006), Heikkilä and Seppä (2003), Bjune et al. (2005), Antonsson et al. (2006), Väliranta et al. (2007, 2015), Bjune and Birks (2008), Bjune et al. (2009, 2010), Giesecke et al. (2008), Huusko and Hicks (2009), Birks and van Dinter (2010) and Aarnes et al. (2012). The history of palaeoclimatic research is presented by Birks and Seppä (2010). Various methods and results are also presented in a special issue of Climate of the Past (Renssen et al., 2010).

Bruvatnet *Norway* (Hyvärinen 1975)

Bb 70.2° N, 28.4° E, 119 m a.s.l.

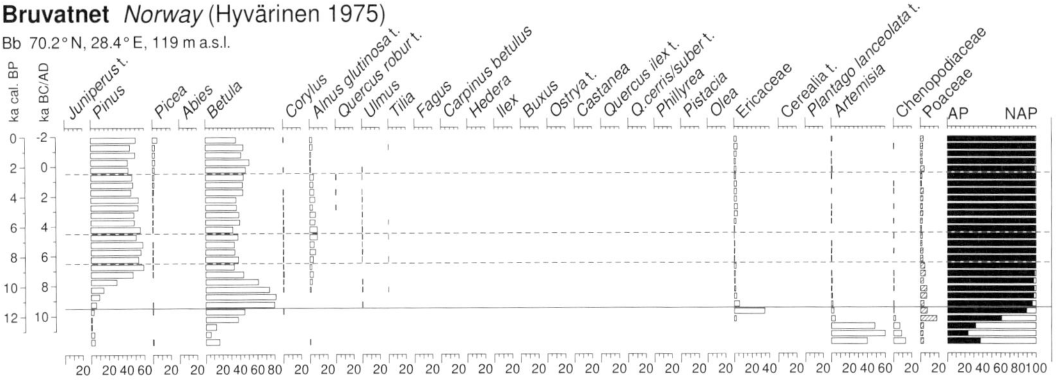

Mukkavaara *Finland* (Hyvärinen 1993)

Bb 68.9° N, 21.0° E, 535 m a.s.l.

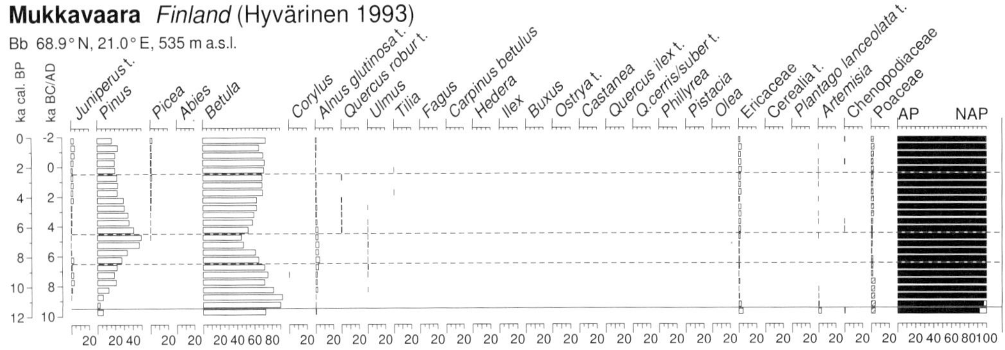

Vasikkasuo *Finland* (Vuorela and Kankainen 1991)

Bb 64.7° N, 27.9° E, 270 m a.s.l.

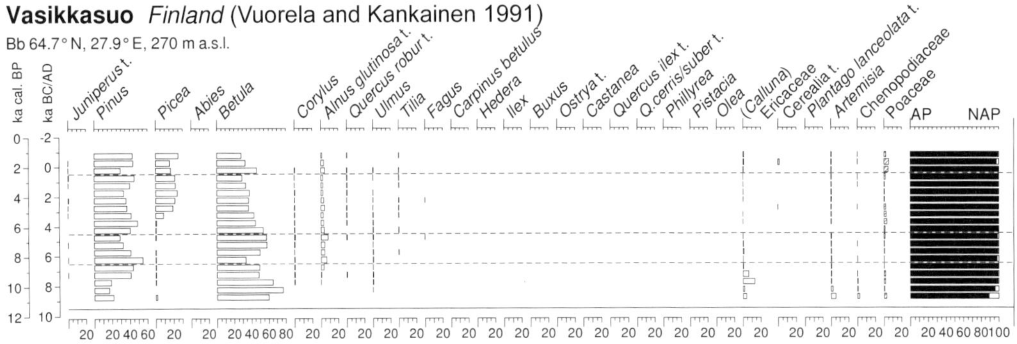

Maanselänsuo *Finland* (Vasari 1965)

Bb 65.6° N, 29.6° E, 248 m a.s.l.

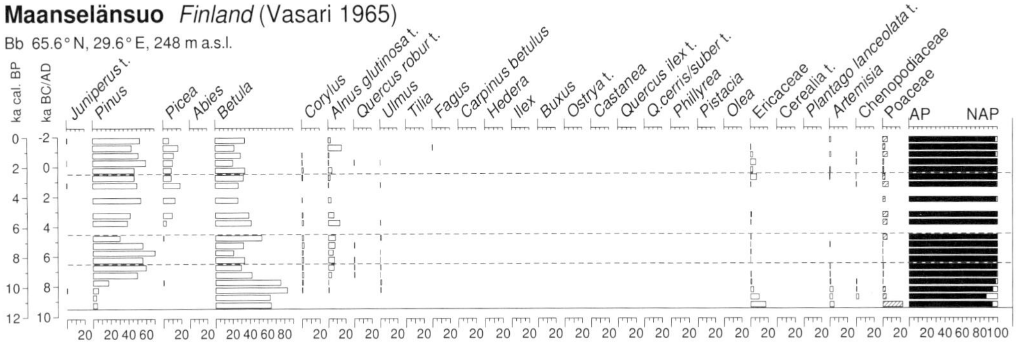

Rugozero *Russia* (Elina 1981)

Bb 64.1° N, 32.6° E, 140 m a.s.l.

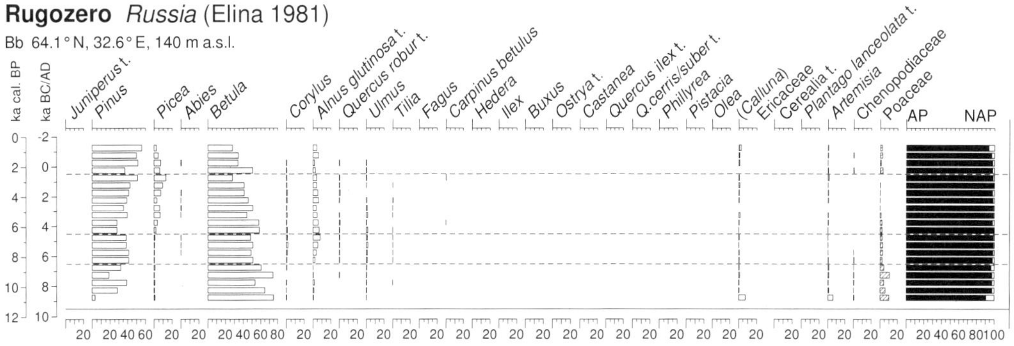

Abborrtjärnen *Sweden* (Giesecke 2005)

Bb 63.9° N, 14.5° E, 387 m a.s.l.

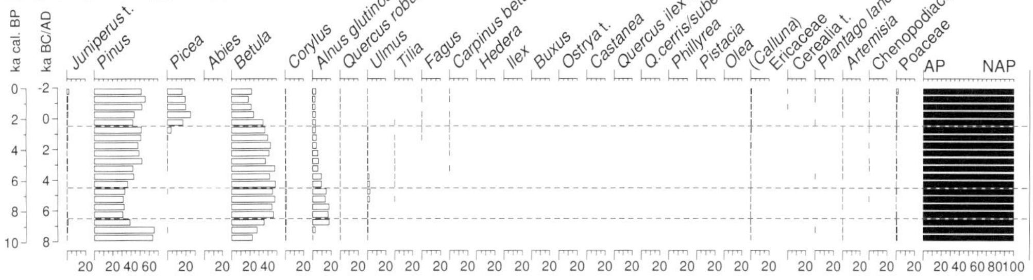

Punozerka *Russia* (Elina and Filimonova 1987)

Bb 62.2° N, 33.6° E, 157 m a.s.l.

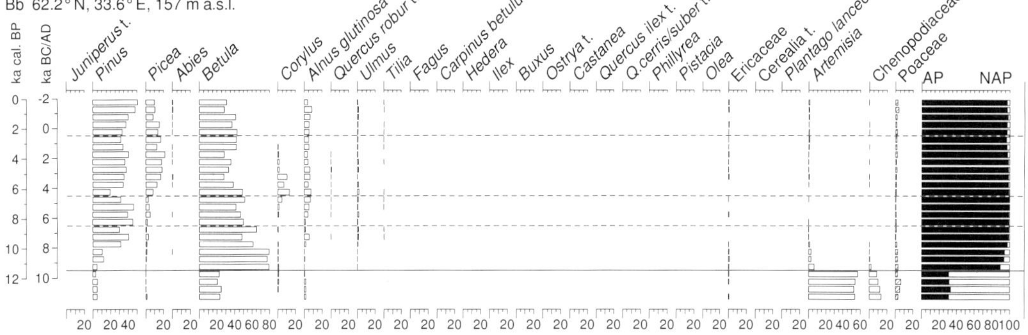

Flåfattjønna *Norway* (Paus 2010)

Bb 62.3° N, 10.4° E, 1110 m a.s.l.

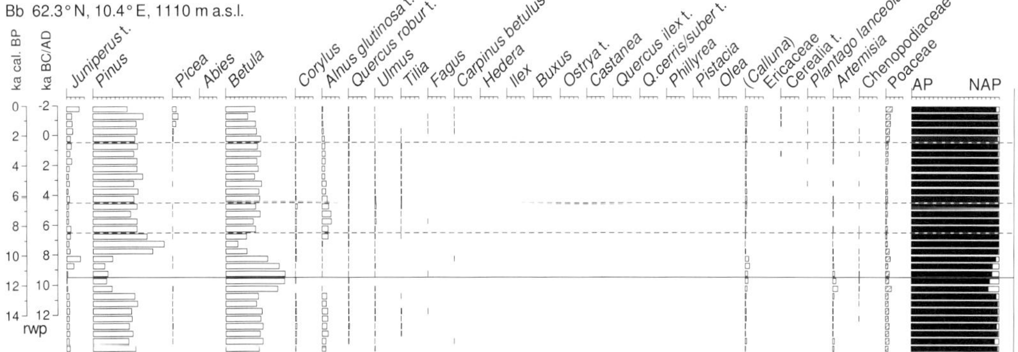

Råtåsjøen *Norway* (Velle et al. 2005)

Bb 62.3° N, 9.8° E, 1169 m a.s.l.

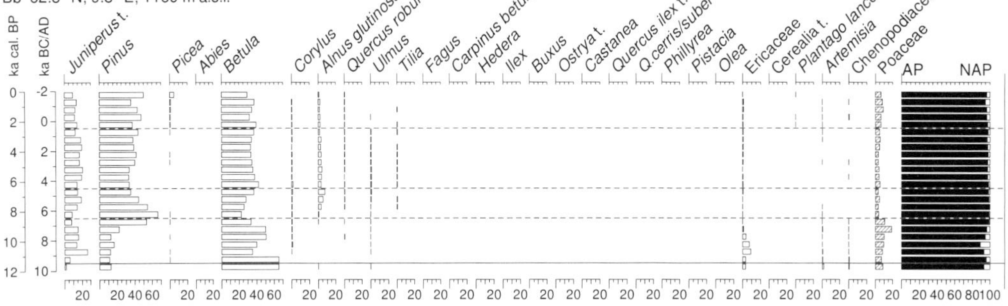

Gotnavolok *Russia* (Elina and Filimonova 1987, 1996)

Bb 62.2° N, 33.8° E, 110 m a.s.l.

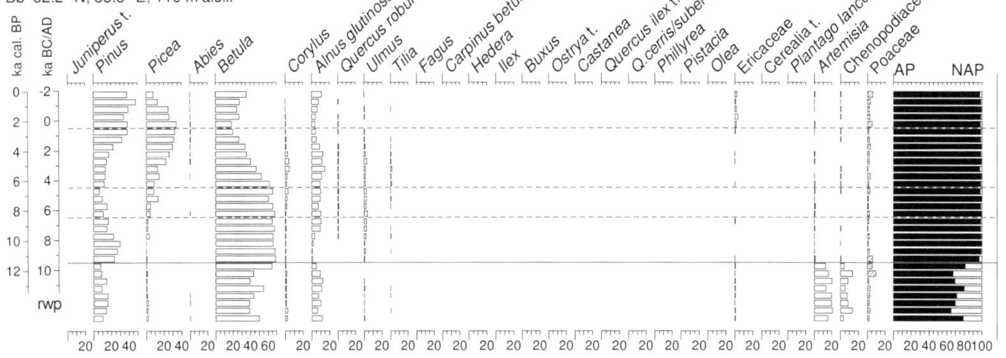

Ylimysneva *Finland* (Huttunen 1990)

Bb 62.1° N, 22.9° E, 172 m a.s.l.

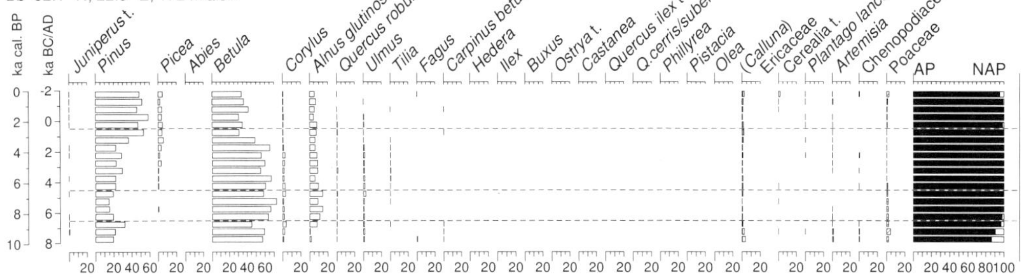

Syrjälänsuo *Finland* (Vuorela I EPD-unpubl.)

Bb 61.2° N, 28.1° E, 83 m a.s.l.

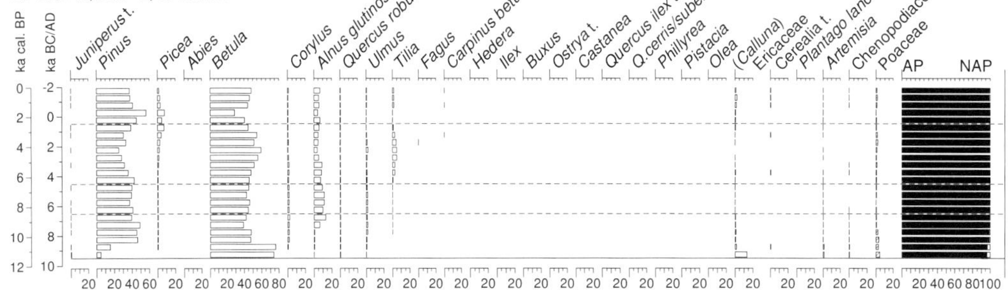

Holtjärnen *Sweden* (Giesecke 2005)

Bb 60.7° N, 14.9° E, 232 m a.s.l.

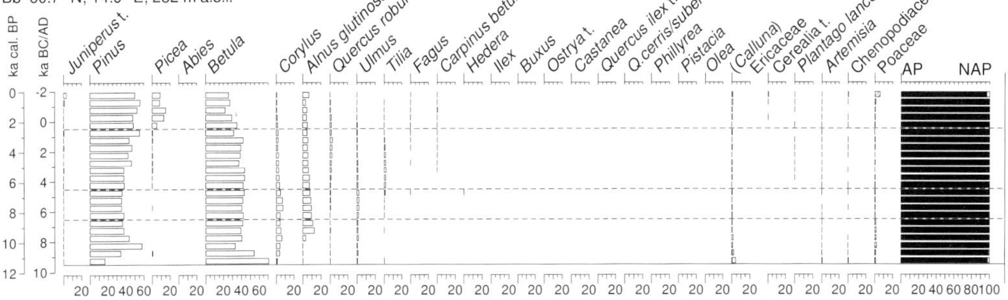

Hirvilampi *Finland* (Rankama and Vuorela 1988)

Bb 60.6° N, 24.3° E, 114 m a.s.l.

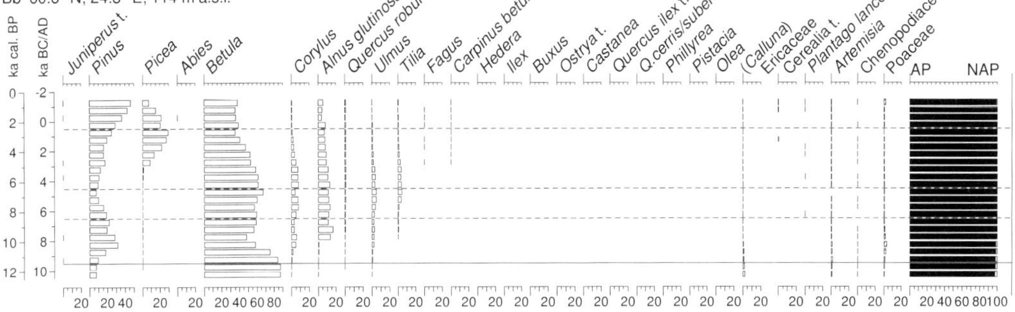

Vestre Øykjamyrtjørn *Norway* (Bjune 2005, Bjune et al. 2005)

Bb 59.8° N, 6.0° E, 570 m a.s.l.

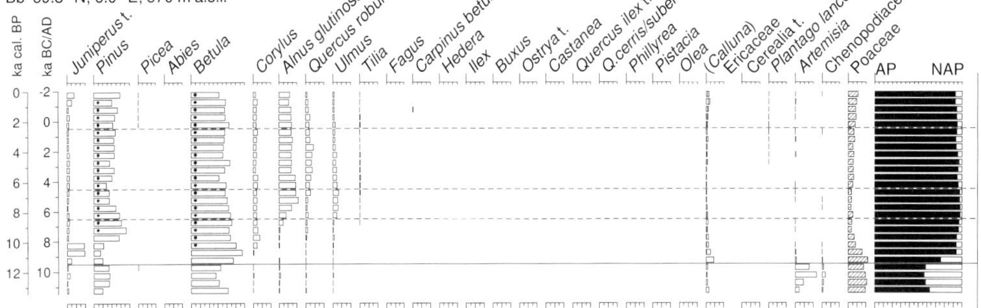

Fig. 4.3-4b. Selected sites of the boreo-fennoscandian (Bb) region as histograms at a 500-year resolution. The absence of *Juniperus* in some records may not be real but 'analyst related'. Among the Ericaceae *Calluna* never plays an important role, but small occurrences are mentioned in the headline. Reworked pollen at the base of cores are indicated by rwp. Dots in the histogram show where plant macrofossils were recorded. Here the sites are arranged from north to south. If we arrange them from east to west we observe the spatial spread of *Picea*.

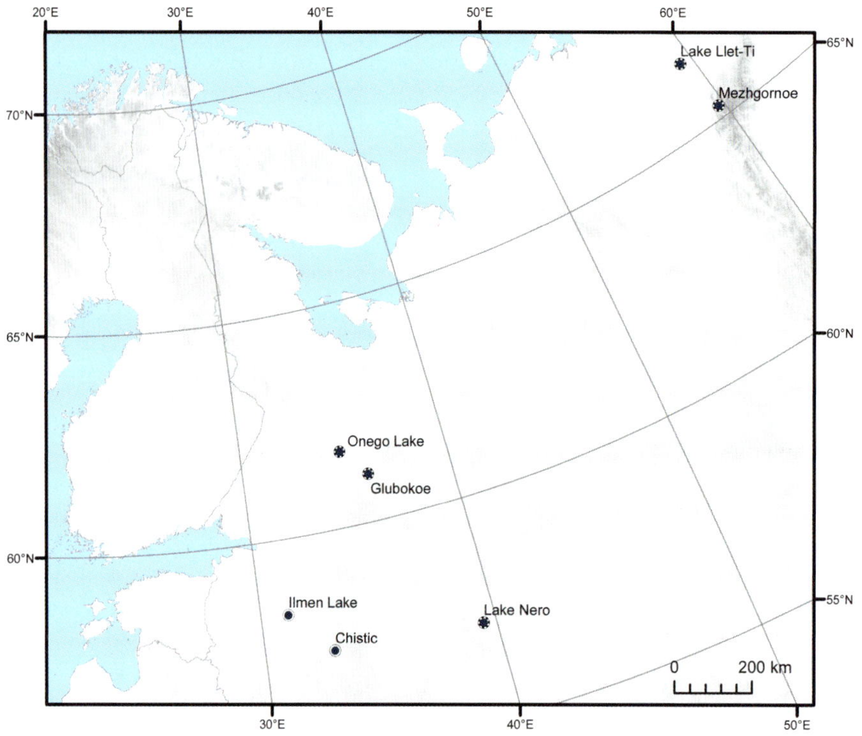

Fig. 4.3-5a. Map of selected sites in the boreo-russian (Bc) region. Dots with black borders indicate sites shown in the histograms (Fig. 4.3-5b), white borders indicate sites used for the maps showing the history of selected taxa (see chapter 4.4) and dashed borders indicate sites used for both.

Bc, Boreo-russian region

The boreo-russian region includes the north of Russia to the Ural mountains in the east, excluding the arctic region along the Barents Sea. The southern limit runs from the Gulf of Finland eastwards to south of Lake Ilmen and north of Moscow. The modern vegetation is dominated by boreal coniferous forests and numerous mires (northern and southern taiga). Plant geographically, it corresponds to the boreo-russian floristic province of Meusel et al. (1965). The region is the northern part of the vast eastern European plain, a lowland with a few hilly ranges that rarely exceed 300 m asl. The transition zone between the taiga and the tundra, the 'forest tundra' becomes very broad at the more eastern longitudes.

Only the western part of the region was covered by the Fennoscandian Ice during the LGM and even that was already ice-free during the Late-glacial. The major (eastern) part of the plain remained free of ice during the Weichselian-Valdai (Svendsen et al. 1999, 2004).

Late-glacial and Holocene: Late-glacial sediments were reached at Lake Onego in the west and at Lake Llet-Ti in the east (Khomutova, 1976, 1989; Väliranta et al., 2006, 2011). At the latter site the presence of both dwarf and tree birches, as well as spruce, is recorded by macroremains for the Younger Dryas (possibly even for the Allerød). Early Holocene records of macrofossils of *Betula* sect. *albae* (tree birches) and, somewhat later, *Picea abies* were found at five sites by Väliranta et al. (2011). The HTM or hypsithermal phase with dense mixed spruce forests lasted from about 9000 to 6400–5200 cal yr BP. After c. 5800 cal yr BP the portion of *Betula* sect. *nanae* and *Alnus* sect. *alnaster* (*Alnus viridis* ssp. *fruticosa*) increased as a consequence of the southwards shift of the treeline (Fig. 4.3-5b). At somewhat higher elevation, as at Lake Mezhgornoe (550 m asl), a downwards shift of the vegetation belts led to the modern situation where *Larix* forms the treeline above mixed forests with *Abies sibirica* at higher and *Picea abies s.l.* at lower elevations (Kultti et al. 2003). So far **human impact** by agriculture is not known in the region.

Overviews for all of northern Eurasia can be found in Kremenetski et al. (1998), MacDonald et al. (2000), Kultti et al. (2003) and Velichko et al. (1997).

Transects can be found in Väliranta et al. (2006, 2011) and Binney et al. (2009) and for limnological aspects in Duff et al. (1999).

Climate reconstructions are presented for example in Sarmaja-Korjonen et al. (2003), Kultti et al. (2003, 2004), Solovieva et al. (2005), Salonen et al. (2011) and Wohlfarth et al. (2008).

Lake Llet-Ti *Russia* (Klimaschewski A unpubl-EPD; Väliranta et al. 2006)
Bc 66.5° N, 59.3° E, 50 m a.s.l.

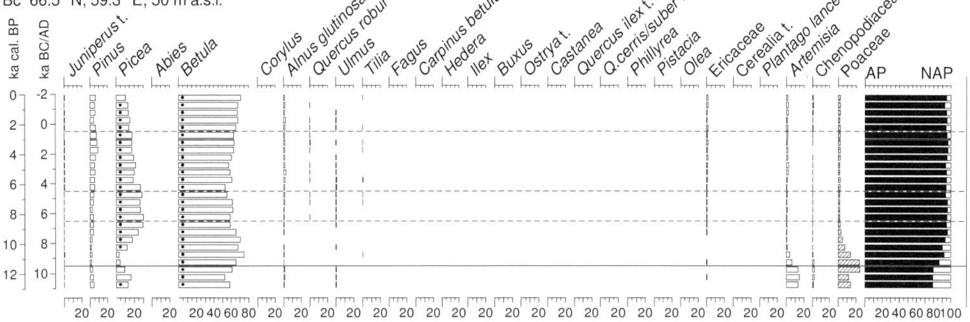

Mezhgornoe Lake *Russia* (Kultti et al. 2003)
Bc 65.3° N, 59.7° E, 550 m a.s.l.

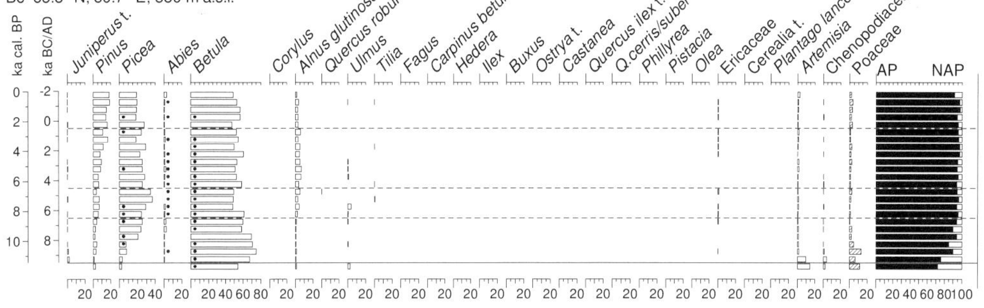

Onego Lake *Russia* (Khomutova 1976, 1989; Khomutova and Elina 1990)
Bc 61.7° N, 34.9° E, 33 m a.s.l.

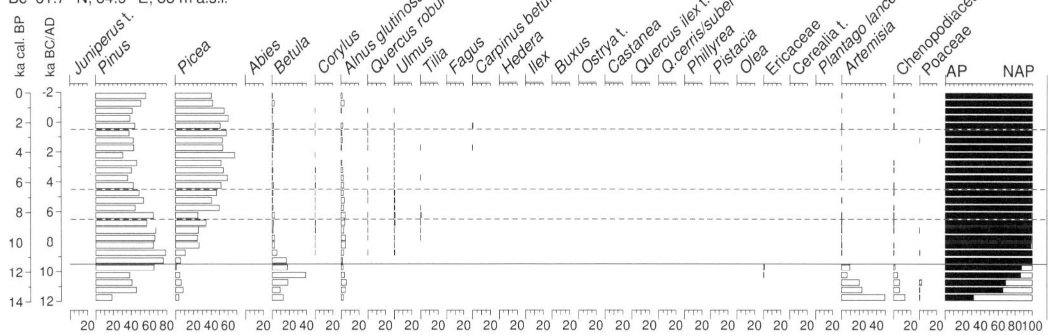

Glubokoe *Russia* (Elina 1981)

Bc 61.1° N, 36.1° E, 50 m a.s.l.

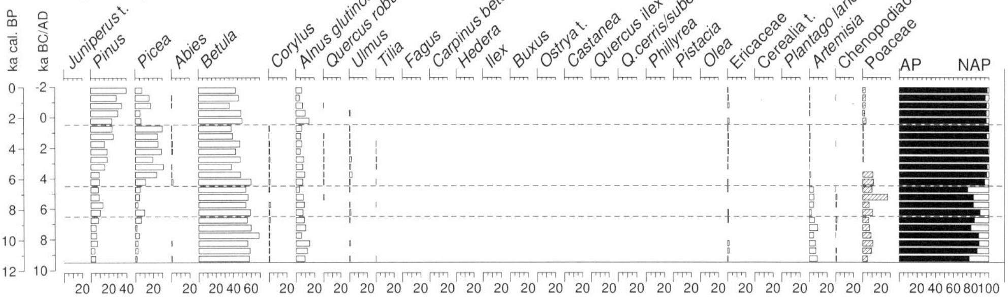

Lake Nero *Russia* (Gunova 1972, 1975)

Bc 57.2° N, 39.5° E, 95 m a.s.l.

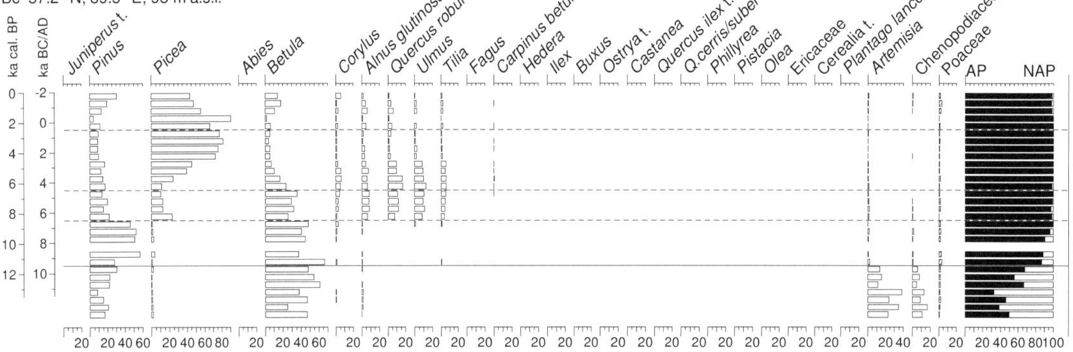

Fig. 4.3-5b. Selected sites of the boreo-russian (Bc) region as histograms at a 500-year resolution. The absence of *Juniperus* in some records may not be real but 'analyst related'. Among the Ericaceae *Calluna* was never recorded. For Onego Lake the diagrams N6 (above 9 ka BC) and N8 (below 9 ka BC) are combined. For Lake Nero the better-dated core 2P (above 6.5 ka BC) is combined with the lower part of core 274 (below 6.5 ka BC). This causes, however, a sharp transition at 6.5 ka BC, which may suggest chronological uncertainties (and note the complete absence of Ericaceae).

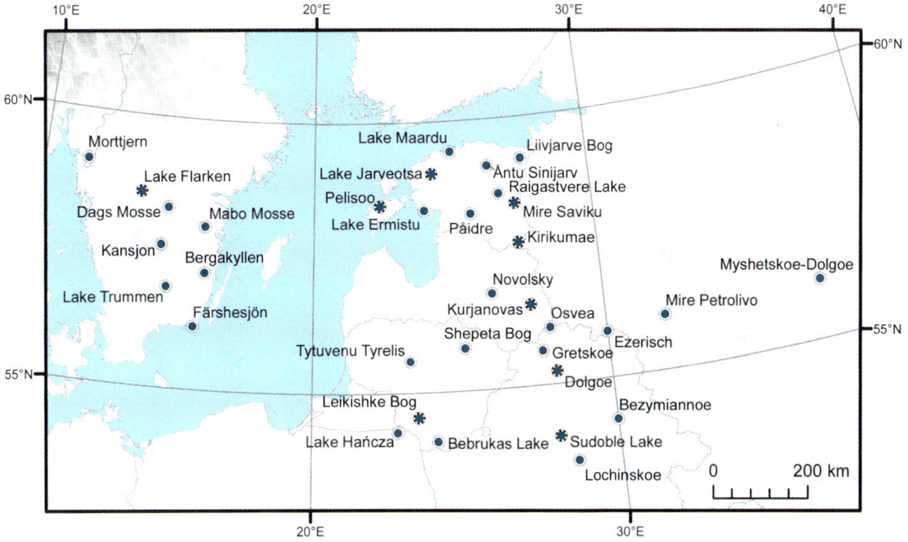

Fig. 4.3-6a. Map of selected sites in the hemiboreal (Bd) region. Dots with black borders indicate sites shown in the histograms (Fig. 4.3-6b), white borders indicate sites used for the maps showing the history of selected taxa (see chapter 4.4) and dashed borders indicate sites used for both.

Bd, Hemiboreal region

This region consists of the southern marginal area of the boreal zone. Its modern vegetation is characterized by mixed forests of conifers and deciduous broadleaved trees. It covers the southern part of Scandinavia, excluding the southernmost coasts of Norway and Sweden but including the south-west of Finland, the Baltic countries, northern Belarus, and tapering off into central Russia. With this vast extension in a west-east direction the region shows a strong gradient in continentality that would potentially make a further subdivision possible. Plant geographically, the region is not considered a unity either by Meusel et al. (1965) or by Takhtajan (1986). Meusel et al. (1965) include it as northern parts of the central European flora-province (our Tb) and of the Sarmatian province. As in the boreo-russian region (Bc) we are dealing mostly with lowlands and some hills.

During the LGM the region was largely under the Fennoscandian ice sheet that melted during the Late-glacial. Only the east (in Russia) was ice free (Svendsen et al., 2004). Maps showing the stages of deglaciation in the Baltic countries are found in Kalm (2006), Rinterknecht et al. (2006) and Veski et al. (2012). Around the Baltic Sea eustatic and isostatic sea-level changes had a strong effect on the coastlines during the Late-glacial and Holocene (see chapter 4.2).

Late-glacial and Holocene: The Late-glacial and Holocene forest development is dominated by *Betula* and *Pinus sylvestris*, as is the case further north. At the Latvian lake Lielais Svētiņu the Bølling is well recorded with macrofossils of *Betula nana*, *Dryas octopetala* and *Salix polaris*. Macroremains of pine are recorded for the Allerød (from 13 400 cal yr BP; see Veski et al., 2012).

Younger Dryas sediments were reached at several sites (Fig. 4.3-6b). At Lielais Svētiņu this period is very prominent and characterized by high values for *Artemisia* and some *Juniperus* (Veski et al., 2012). *Picea abies* spread from east to west (see Dolgoe, Sudoble, Kirikumae, where it was already present during the Younger Dryas) and from north to south. At Lake Kurjanovas (Latvia), Heikkilä et al. (2009) found stomata and bud-scales of *Pinus* as early as in the sediments of Bølling-Allerød. The Younger Dryas record showed not only more *Artemisia* but also more *Picea* (pollen and needles): this remarkable finding is discussed by Heikkilä et al. (2009) in terms of biogeography, permafrost and the amplification of the

Lake Jarveotsa *Estonia* (Poska 1994)

Bd 59.0° N, 24.3° E, 42 m a.s.l.

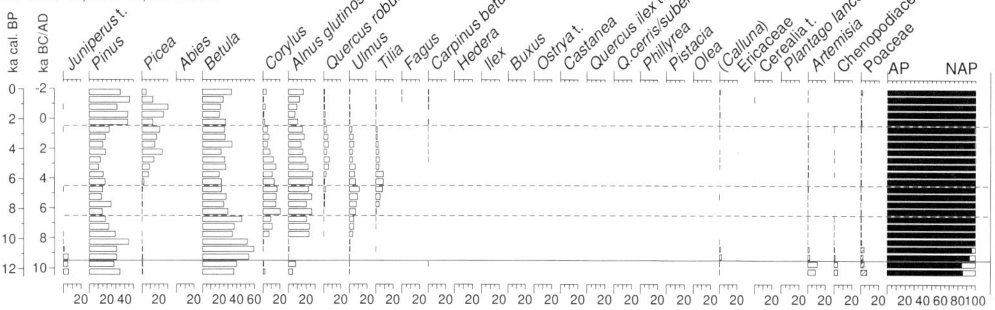

Lake Flarken *Sweden* (Digerfeldt 1977)

Bd 58.6° N, 13.7° E, 109 m a.s.l.

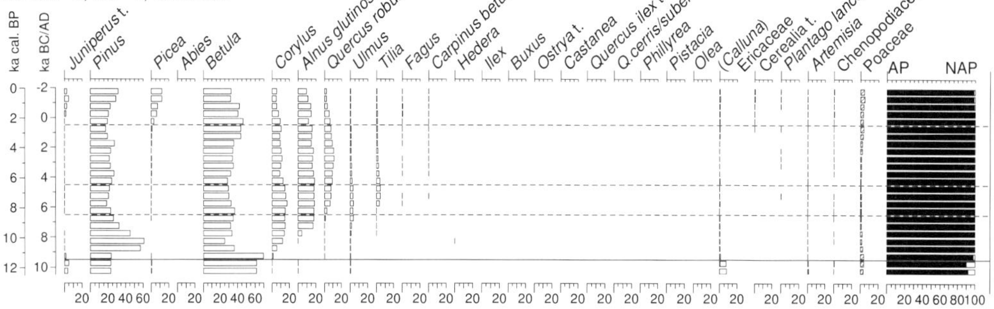

Pelisoo *Estonia* (Saarse et al. 1990)

Bd 58.5° N, 22.4° E, 33 m a.s.l.

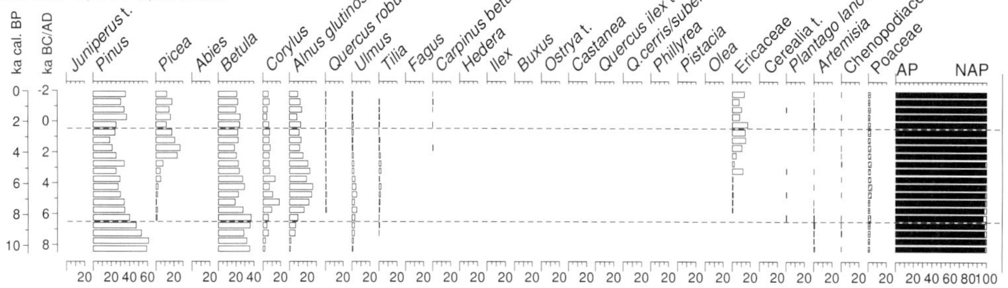

Mire Saviku *Estonia* (Sarv and Il'ves 1975)

Bd 58.4° N, 27.2° E, 30 m a.s.l.

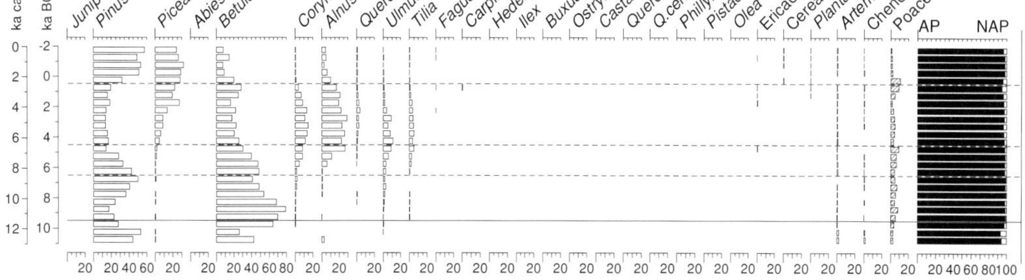

Kirikumae-63 *Estonia* (Saarse et al. 1996; Saarse and Rajamäe 1997)

Bd 57.7° N, 27.3° E, 183 m a.s.l.

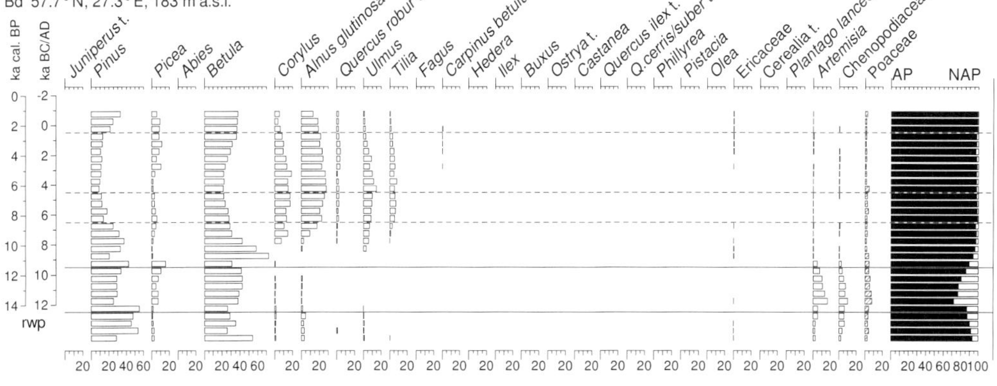

Kurjanovas *Latvia* (Heikkilä et al. 2009)

Bd 56.5° N, 27.5° E, 111 m a.s.l.

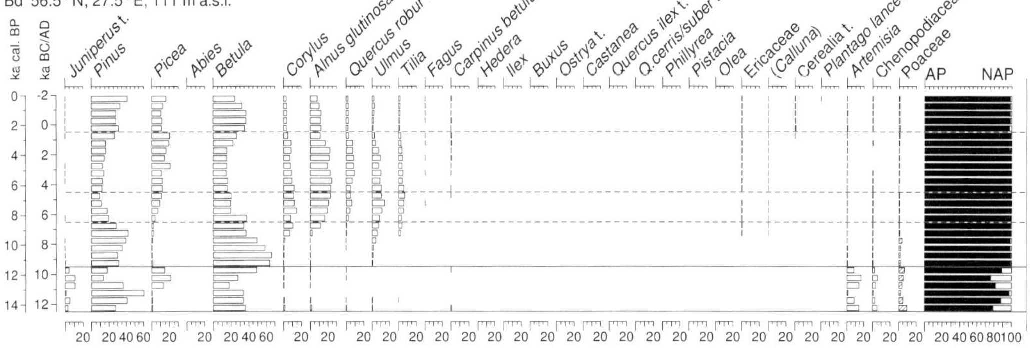

Dolgoe *Belarus* (Zernitskaya 1991)

Bd 55.2° N, 28.2° E, 173 m a.s.l.

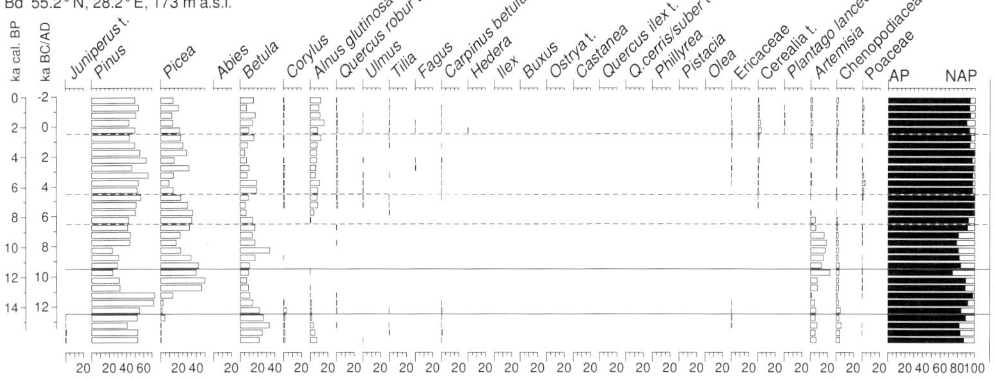

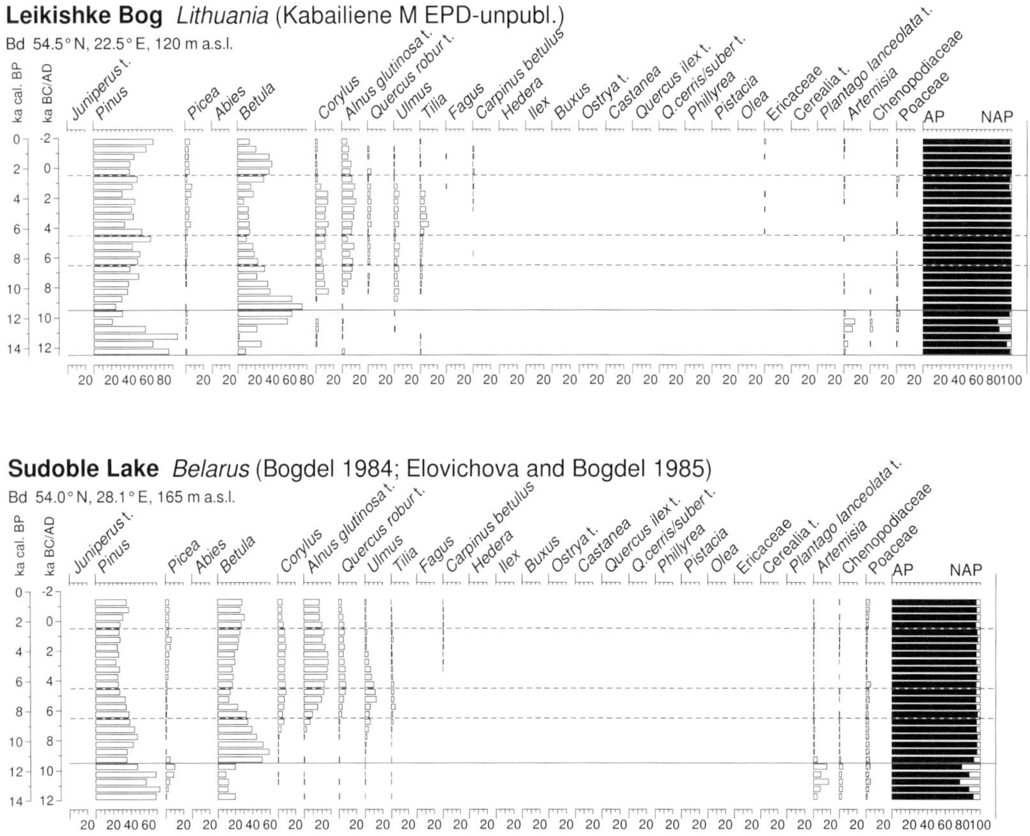

Fig. 4.3-6b. Selected sites in the hemiboreal (Bd) region arranged from south to north as histograms at a 500-year resolution. The absence of *Juniperus* in some records may not be real but 'analyst related'. At the base of some records reworked pollen of thermophilous deciduous trees is found; this is marked as 'rwp' only for Kirikumae where the timescale is left empty because it is very uncertain.

seasonality. During the Early Holocene, broadleaved deciduous trees (*Ulmus, Quercus, Tilia*) played a stronger role in Bd than in Bb and Bc. These taxa were also present during the second half of the Holocene (including the present), but they became reduced in favour of the conifers. *Carpinus* was also present at many sites during the younger part of the Holocene.

Human activities, including forest clearances, are recorded at least since the Neolithic (Kabailiene 1996; Saarse et al., 1996) and are prominent in some areas for the last two millennia (Berglund, 1966; Digerfeldt, 1972, 1977; Poska, 1994; Poska and Saarse, 2002, 2006; Niinemets and Saarse, 2006; Sillasoo et al., 2009).

Overviews are available in Berglund et al. (1996a); for southern Sweden by Berglund et al. (1996b) covering 14 sites; for southern Finland by Vasari et al. (1996) covering 12 sites; for Estonia by Saarse et al. (1996) with 14 sites; and for Lithuania by Kabailiene (1996) with 8 sites. For the Late-glacial, Heikkilä et al. (2009) provide synthetic insights; for the younger Holocene in Estonia, Reitalu et al. (2013) offer numerical analyses.

Transects can be found in Berglund (1991) and in Eide et al. (2006) and, based on macrofossils, in Amon et al. (2014).

Pollen maps are presented by Saarse and Veski (2001).

Climate reconstructions can be found in Seppä and Poska (2004), Charman et al. (2009), Heikkilä and Seppä (2010), Lauterbach et al. (2011b), Stankevica et al. (2015) and Veski et al. (2015).

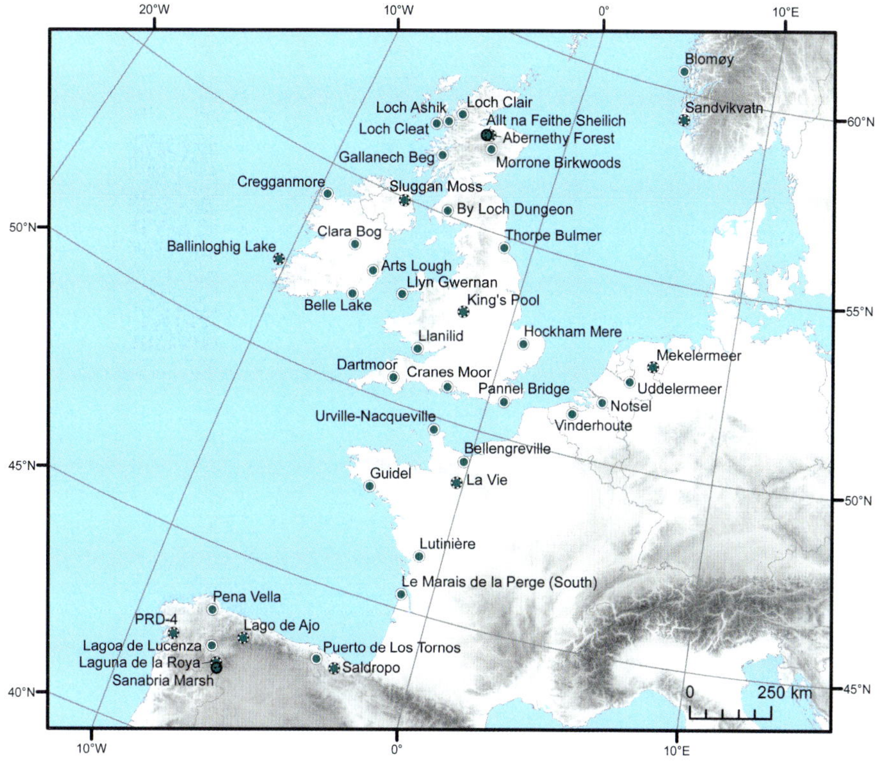

Fig. 4.3-7a. Map of selected sites in the temperate-atlantic (Ta) region. Dots with black borders indicate sites shown in the histograms (Fig. 4.3-7b), white borders indicate sites used for the maps showing the history of selected taxa (see chapter 4.4) and dashed borders indicate sites used for both.

Ta, Temperate-atlantic region

This region includes the western part of the temperate zone of deciduous broadleaved forests (primarily mixed oak forests), namely the British Isles, a narrow band along the coast of Norway south of 61°N, the west of The Netherlands, Belgium and France and the north-west of the Iberian Peninsula. Plant geographically, these limits correspond to a large degree to the atlantic floristic province of Meusel et al. (1965), but also include areas of Scotland. The flora province of Takhtajan (1986) also has similar borders, but it includes the Massif Central in France and the northernmost areas of Scotland that we assign to the boreo-atlantic region. The temperate-atlantic region has a very wide latitudinal extension which is also reflected in the diversity of trees involved, for instance *Quercus ilex* or *Castanea sativa* in the south. Mountains are found in Scotland (up to 1343 m asl), northern England (up to 978 m asl), Wales (up to 1085 m asl) and the Cantabrian region (up to 2648 m asl). The high mountains of the Pyrenees are treated as a separate region (Td).

During the Last Glacial Maximum (LGM) major parts of the British Isles and smaller parts of the Cantabrian mountains were under ice. Towards the end of the Late-glacial, that is during the Younger Dryas, the only ice left was in northern Scotland, an area that became ice-free during the Early Holocene. Because of the low sea-level, England was connected to the European continent until the onset of the Holocene.

Late-glacial and Holocene: in areas unglaciated or deglaciated at an early stage the vegetation development started during the Oldest Dryas with treeless communities into which *Juniperus, Betula* (and on the Iberian Peninsula also *Pinus*) spread and then dominated the Late-glacial interstadial. In Ireland and Scotland this establishment of woodlands occurred later, namely in the Early Holocene.

The Younger Dryas is regularly marked by high values of *Artemisia* and a depression or disappearance of *Betula* (and of *Pinus* where it had established). The *Quercus* maximum at Sanabria Marsh during the YD is still enigmatic, while the *Quercus* depression at Laguna de la Roya corresponds more with the drier and cooler conditions of the Younger Drays (and resembles some records in Th).

The Holocene forest development was largely controlled by deciduous tree taxa (Fig. 4.3-7b). During the first half of the Holocene the important roles of *Corylus*, summergreen *Quercus, Ulmus, Fraxinus* and partly of *Tilia* are conspicuous. *Pinus sylvestris* expanded over the British Isles and Ireland, as is also documented by the macrofossils, but it never played a major role and even disappeared in the second half of the Holocene, except in parts of eastern Scotland especially, where pine forest still occurs today (Birks, 1970, 1975; Birks and Mathewes, 1978). On the continent *Fagus* became conspicuous in Brittany and Normandy during the last few thousand years from where it reached southern England but only at low values (Birks, 1989; Magri et al., 2006).

The large-scale deforestations that have occurred during the last millennia are remarkable at most sites. *Calluna vulgaris* plays a major role as an indicator of the formation of heaths (see chapter on human impact and Trajectory Type 2 of Marquer et al., 2014). *Castanea* pollen has been found at some sites since Roman times, but at Sanabria Marsh and Laguna de la Roya it was also present in the Early Holocene or even the Late-glacial (possibly reworked?). This would be in agreement with the concept of a potential refuge in the Cantabrian mountains (Krebs et al., 2004).

Early overviews of vegetation history for the British Isles are found in Godwin (1956, 1975), Pennington et al. (1972), Pennington (1974) and Watts (1985). A review for the Inner Hebrides is presented by Birks and Williams (1983).

More recent overviews are provided for Ireland by Mitchell et al. (1996) with 9 sites; for Wales by Chambers (1996) with 9 sites; for England by Greig (1996) with 32 sites; and for Scotland by Birks (1996b) with 38 sites.

Transects can be found for example in Macklin et al. (2000).

Pollen maps of various types can be found in Birks (1989), Peñalba (1994) and Birks and Birks (2014).

Data-model comparisons are presented by Jeffers et al. (2012, 2015a).

Climate reconstructions can be found in Allen et al. (1996), Barber et al. (1998, 2000, 2003), Langdon et al. (2003), Langdon and Barber (2005), Charman et al. (2009), Birks and Birks (2014), Muñoz Sobrino et al. (2013) and Iriarte-Chiapusso et al. (2016).

Human impact: Macklin et al. (2000).

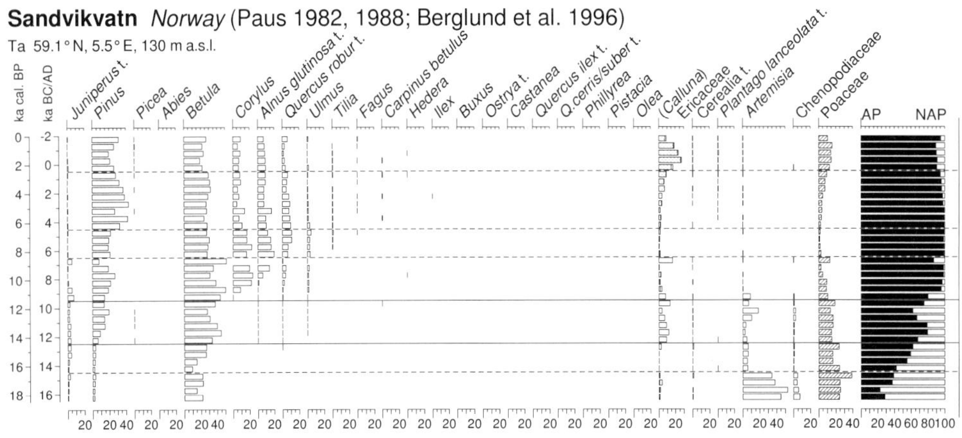

Abernethy Forest *Great Britain* (Birks HH 1970; Birks HH and Mathewes 1978)

Ta 56.2° N, 3.7° W, 221 m a.s.l.

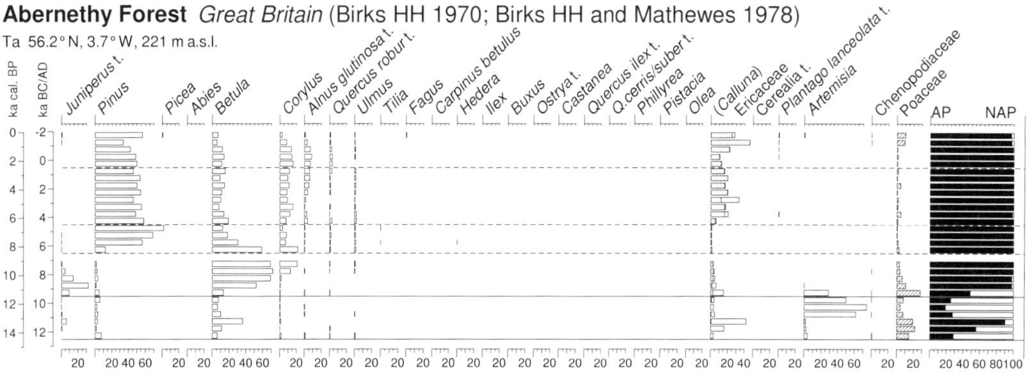

Sluggan Moss *Ireland* (Smith and Goddard 1991)

Ta 54.9° N, 6.3° W, 52 m a.s.l.

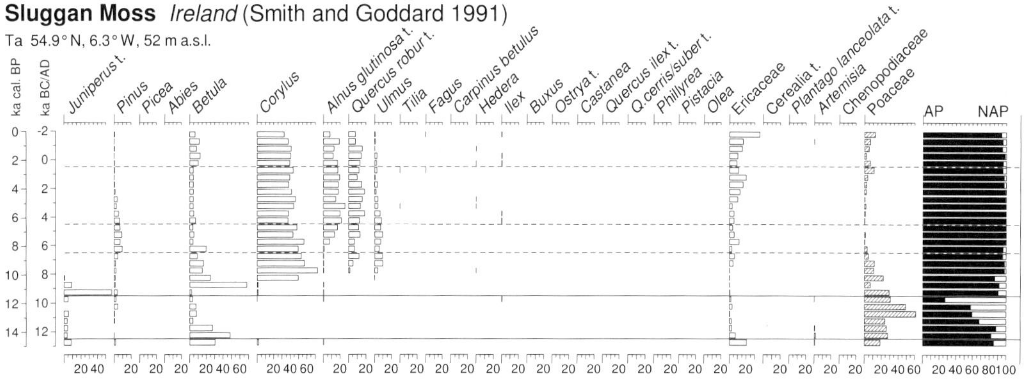

King's Pool *Great Britain* (Bartley and Morgan 1990)

Ta 52.8° N, 2.1° W, 70 m a.s.l.

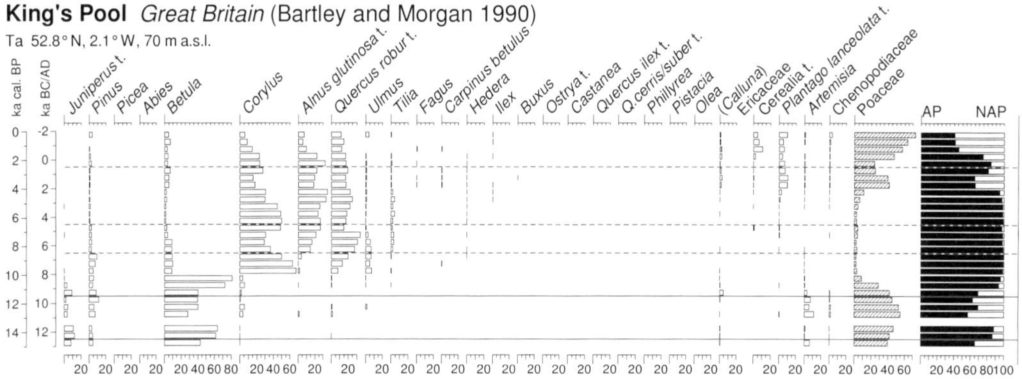

Ballinloghig Lake *Ireland* (Barnosky 1988)

Ta 52.2° N, 10.3° W, 84 m a.s.l.

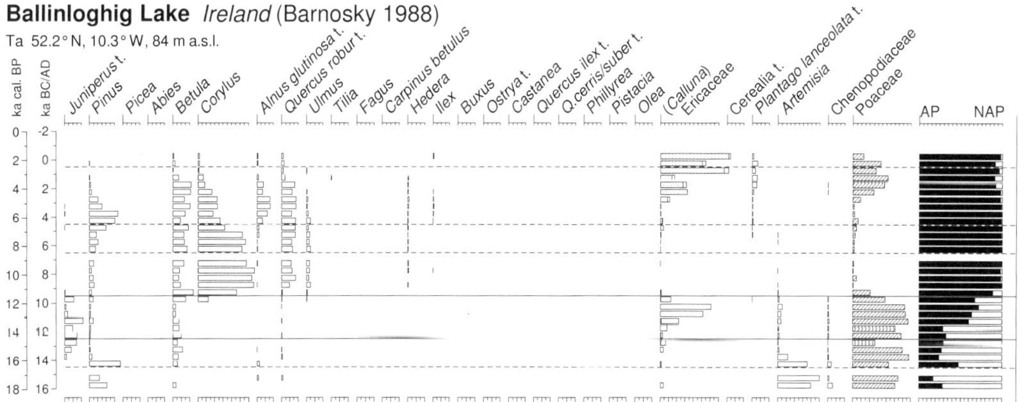

La Vie *France* (Barbier 1999)

Ta 48.5° N, 0.3° W, 238 m a.s.l.

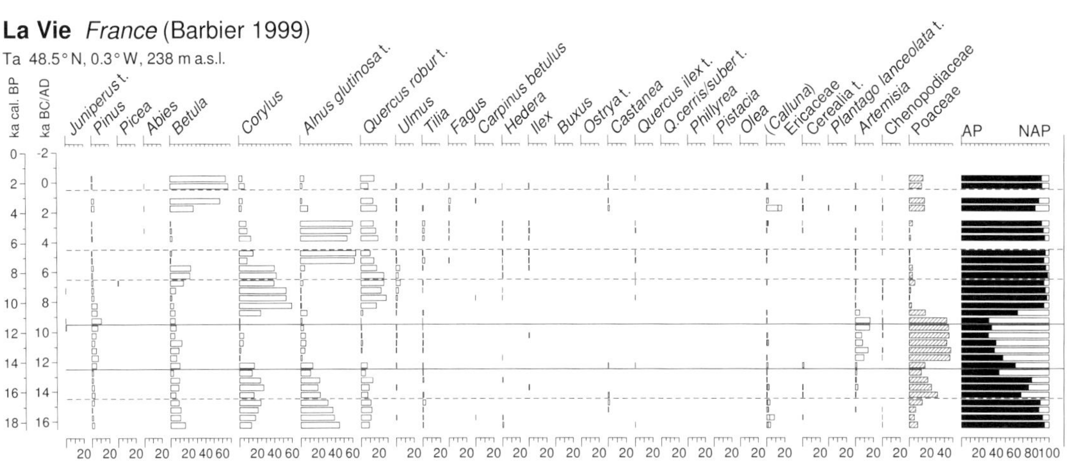

Lago de Ajo *Spain* (McKeever 1984; Allen et al. 1996)

Ta 43.1° N, 6.2° W, 1570 m a.s.l.

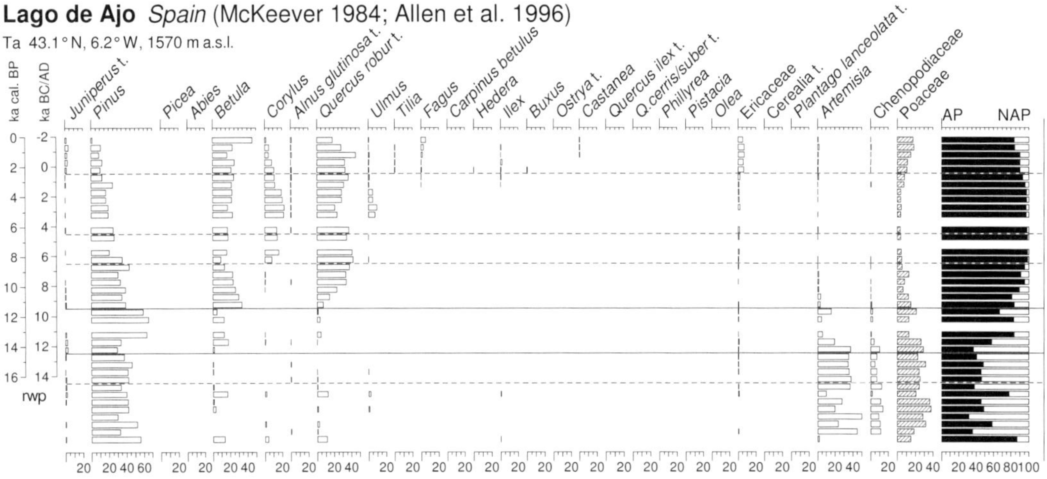

Saldropo *Spain* (Peñalba 1989, 1994)

Ta 43.1° N, 2.7° W, 625 m a.s.l.

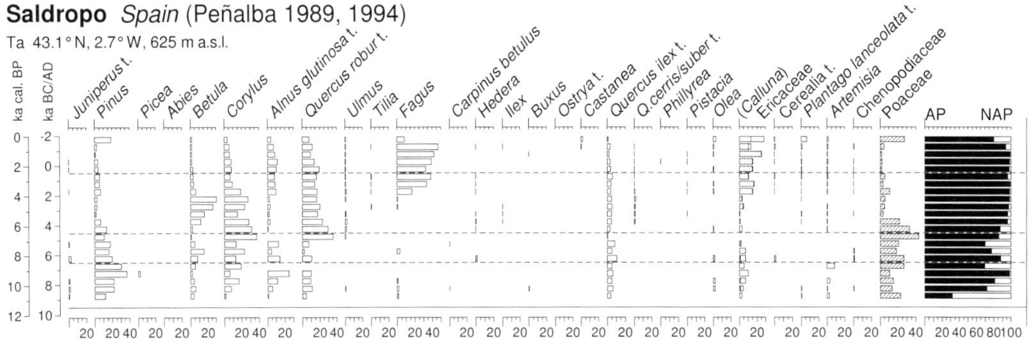

PRD-4 *Spain* (Lopez-Merino et al. 2012)

Ta 42.5° N, 8.5° W, 200 m a.s.l.

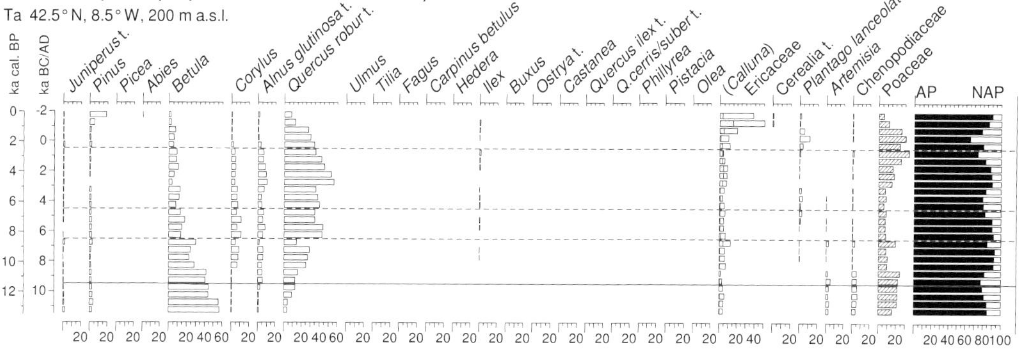

Laguna de la Roya *Spain* (Allen et al. 1996)

Ta 42.2° N, 6.8° W, 1608 m a.s.l.

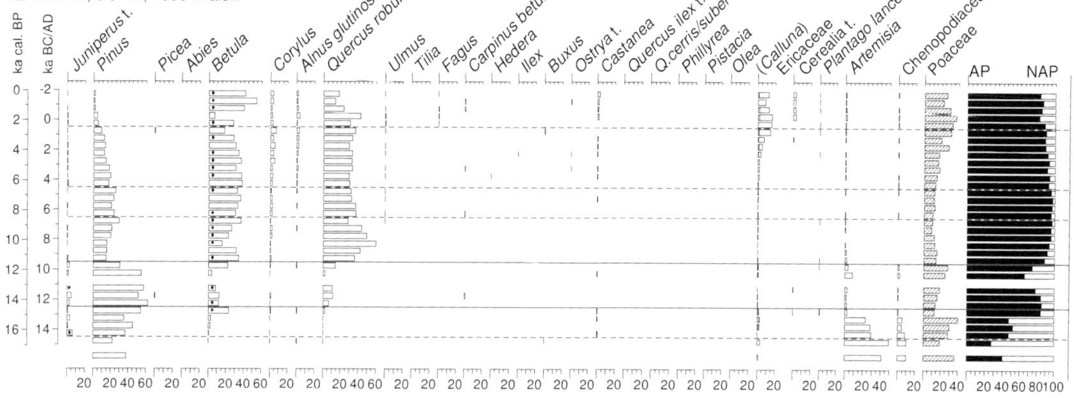

Sanabria Marsh *Spain* (Hannon 1985; Turner and Hannon 1988; Allen et al. 1996)

Ta 42.1° N, 6.7° W, 1050 m a.s.l.

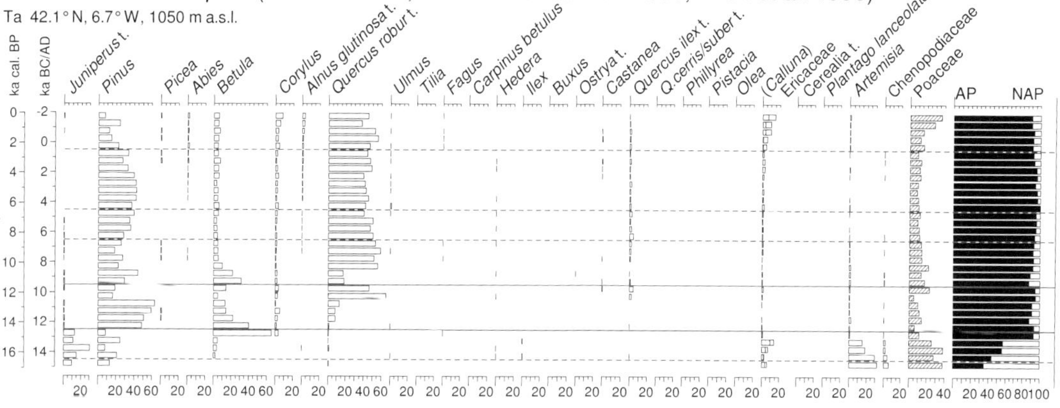

Fig. 4.3-7b. Selected sites in the temperate-atlantic (Ta) region, from north to south as histograms at a 500-year resolution. In the histograms a dot indicates a macrofossil record. In northern Spain three of the sites are in the Cantabrian mountains at high elevations (Laguna de la Roya, Lago de Ajo, Saldropo). In Ericaceae the fine-dotted part stands for records of *Calluna* (see chapter 4.5.8 on heath development). At Abernethy Forest detailed macrofossil analyses concentrate on the Late-glacial and the Early Holocene (see Fig. B1-1 in Box 1 about macrofossils in chapter 2).

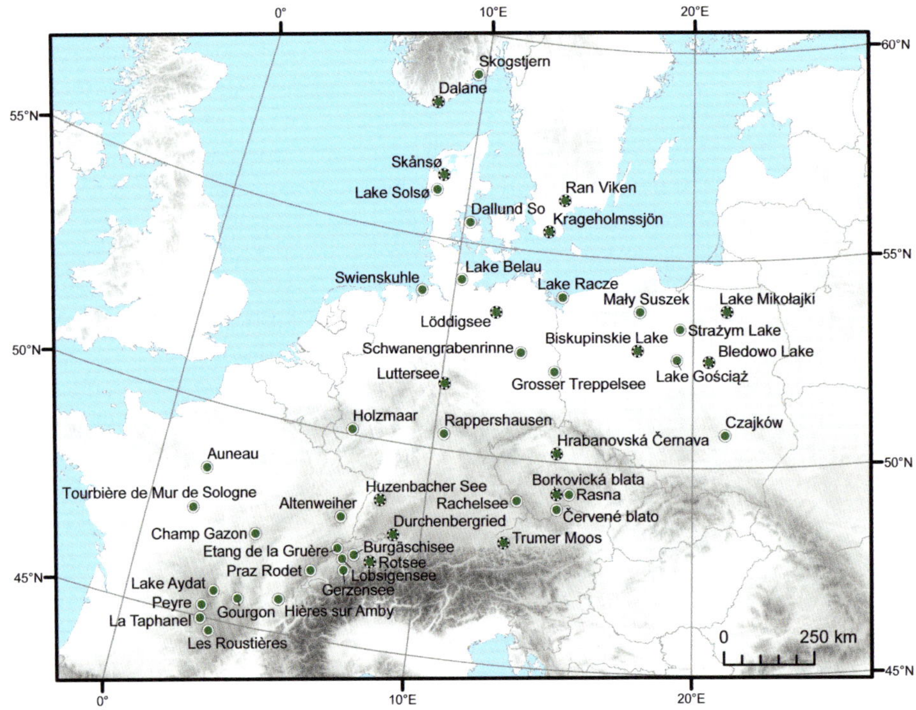

Fig. 4.3-8a. Map of selected sites in the temperate central European (Tb) region. Dots with black borders indicate sites shown in the histograms (Fig. 4.3-8b), white borders indicate sites used for the maps showing the history of selected taxa (see chapter 4.4) and dashed borders indicate sites used for both.

Tb, Temperate-central European region

This region lies east of the temperate-atlantic region (Ta) and reaches from southern Norway and Sweden (Skåne and Blekinge) to the northern limits of the mountain chains of the Pyrenees, the Alps and the Carpathians. It includes the French Massif Central in the west and the lowlands south of the Carpathians in the east. In the east and the southeast the limits are not sharp, so we propose a special region for the Pannonian basin. The region Tb comprises more-or-less the area of the potential natural vegetation of the temperate deciduous forests that are dominated or co-dominated today by *Fagus sylvatica*. *Fagus* also plays an important role in the adjacent southern European regions Tg and Th. Plant geographically, the region Tb corresponds to the southern parts of the subatlantic and the central European floristic provinces of Meusel et al. (1965). The region consist of lowlands in the north and of many low mountain ranges in the south (Harz mountains up to 1142 m asl; Giant mountains up to 1603 m asl; Ore mountains up to 1244 m asl; Bavarian-Bohemian Forest up to 1457 m asl; Black Forest up to 1493 m asl, Vosges up to 1426 m asl; French Massif Central up to 1886 m al; Jura mountains up to 1720 m asl). The Alps are considered separately as region Te.

During the LGM the northern parts of the region were covered by the Fennoscandian ice sheet. The Alpine foreland was also covered by alpine ice and the higher parts of the low mountain ranges had local ice caps (French Massif Central, Vosges, Black Forest, Bavarian-Bohemian Forest, Giant mountains, Harz mountains, Jura mountains) but the larger part of region Tb remained unglaciated. In the glaciated parts the ice melted early during the Late-glacial.

Late-glacial and Holocene: During the Oldest Dryas a 'steppe-tundra' dominated by grasses, *Artemisia* and Chenopodiaceae covered large areas. At many (southern) sites the Oldest Dryas is tripartite: characterized at the base by pioneer taxa (NAP), high pine percentages (long-distance transported?) and some reworked thermophilous taxa, in the middle by a tundra with high NAP-diversity (less pine, no reworked pollen) and, at the top, by a shrub-tundra rich in NAP with *Betula nana*, *Salix* spp. and some *Juniperus* (Vescovi et al., 2007). Forest establishment had already occurred in the lowlands during the Late-glacial: in the southern parts during the Bølling and in the northern parts during the Allerød. In the north-west and west, these forests were dominated by *Betula* and in the south-east and east by *Pinus* (*sylvestris*). The onset and end of the Younger Dryas were marked vegetational events throughout central Europe as Isarin and Bohncke (1999) showed in a synthesis incorporating 140 sites.

The Holocene development was characterized by the massive spatial spread and population expansion of *Corylus* during the Early Holocene and by the expansion of *Quercus, Ulmus, Tilia* and other deciduous woody taxa during the Early and Mid Holocene, advantaged by warm summers and relatively continental conditions during the HTM (ca. 10 000–5000 cal yr BP). The importance of *Corylus* decreased from west to east. During the second half of the Holocene, under more oceanic conditions, *Fagus* expanded mainly from the south-east to the north-west and north at the expense of the mixed deciduous forests with oak (Magri et al., 2006). In the south, *Fagus* was accompanied by and sometimes dominated by *Abies alba* (Tinner et al., 2013). *Picea abies* also gradually advanced from several refugia in low mountain ranges from east to west, but it never became as important as in the Alps and their margins (Fig. 4.3.2-8b; Tollefsrud et al., 2008). The development from a vegetation comprising herbs and dwarf shrubs through *Betula/Pinus* to *Corylus/* mixed-oak forests and eventually *Fagus* or *Fagus/ Abies* was termed by Firbas (1949) as the "Central European basic succession". In space, it shows many qualitative, quantitative and temporal differences.

Some central European sites showed annually laminated sediments which allowed the study of vegetation changes at high temporal resolution, for example at Lake Gościąż in north-eastern Poland (Ralska-Jasiewiczowa et al., 1998) and in the maar lakes of the Eifel in Germany (Stebich, 1999; Litt et al., 2001).

The density of well-studied sites is high enough in some parts of the region to look into regional aspects. In southern Germany, a synthesis of 15 sites with diagrams (and 101 sites in the literature) was published by Lang (2005) for the Black Forest while in northern Germany over 40 sites from the Harz mountains to the river Weser are synthesized by Beug (2016). Towards the north and east, differences in vegetation development controlled by soil types are shown by Mortensen et al. (2014) and Theuerkauf and Joosten (2009).

Human impact on the vegetation and other aspects of the environment are pronounced from the Neolithic onwards through records of cereals, apophytes, anthropochores, increasing NAPs and charcoal, but earlier impacts are also discussed for southern central Europe (Behre, 2007b, 2008a; Tinner et al., 2007, 2008).

Overviews in Berglund et al. (1996a) include: Denmark by Andersen et al. (1996) with 9 sites; Poland by Ralska-Jasiewiczowa and Latałowa (1996) with 39 sites; Czech and Slovak Republics by Rybníčková and Rybníček (1996) with 25 sites; Germany by Behre et al. (1996) with 20 sites; Belgium by Verbruggen et al. (1996) with 11 sites: France by Visset et al. (1996) with 29 sites; Switzerland by Ammann et al. (1996) with 61 sites (but only 16 sites in the lowlands of Tb) and Austria by Bortenschlager et al. (1996) with 14 sites.

Transects can be found at the northern limit of the region in Eide et al. (2006) and for north-east Germany in de Klerk (2008) (c. 150 sites) and Theuerkauf and Joosten (2009). An altitude-time transect for a part of the Vosges mountains was developed by de Valk (1981), see Visset et al. (1996).

Pollen maps are provided for Poland by Ralska-Jasiewiczowa et al. (2004) based on 257 sites.

Data-model comparisons are presented by Lotter and Kienast (1992) and Lischke et al. (2002, 2013).

Climate reconstructions for the Late-glacial see Isarin and Bohncke (1999), Isarin and Renssen (1999), Lotter et al. (2000, 2012), Barber et al. (2004), Peyron et al. (2005) and Feurdean et al. (2014). For sites in Denmark and southern Sweden Schenk et al. (2020) used macrofossils of 62 plant species and their minimum July temperatures at their modern northern distribution limit in Finland (Schenk and Väliranta, 2020) to estimate summer temperatures during 15–11 cal ka BP.

Dalane *Norway* (Eide et al. 2006)

Tb 58.2°N, 8.0° E, 40 m a.s.l.

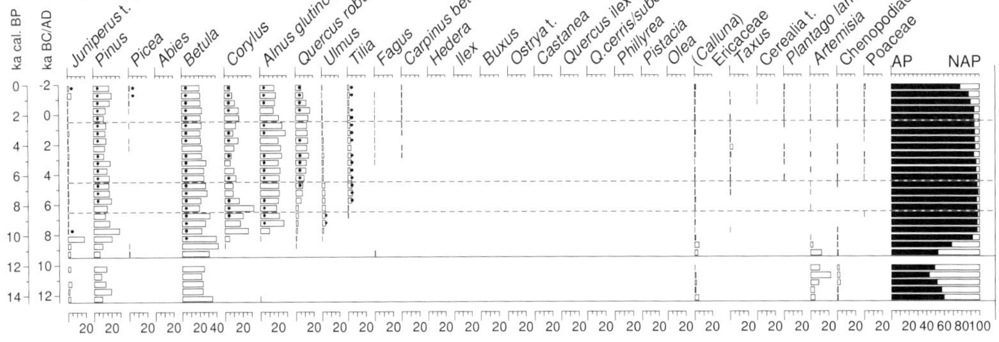

Skånsø *Denmark* (Odgaard 1994)

Tb 56.5° N, 8.8° E, 5 m a.s.l.

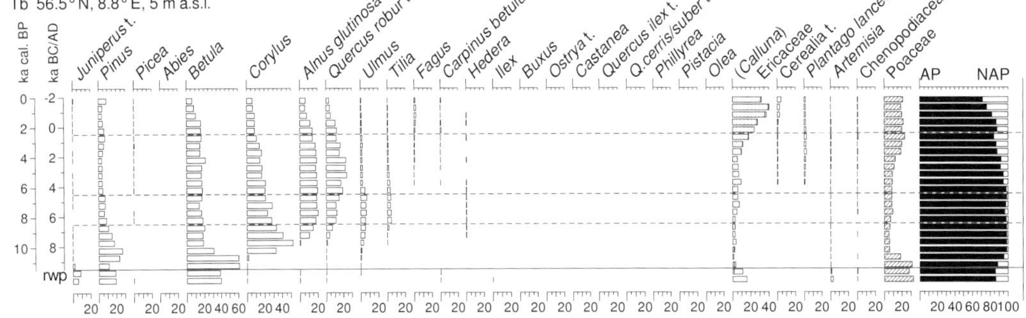

Ran Viken *Sweden* (Digerfeldt 1973)

Tb 56.3° N, 14.3° E, 213 m a.s.l.

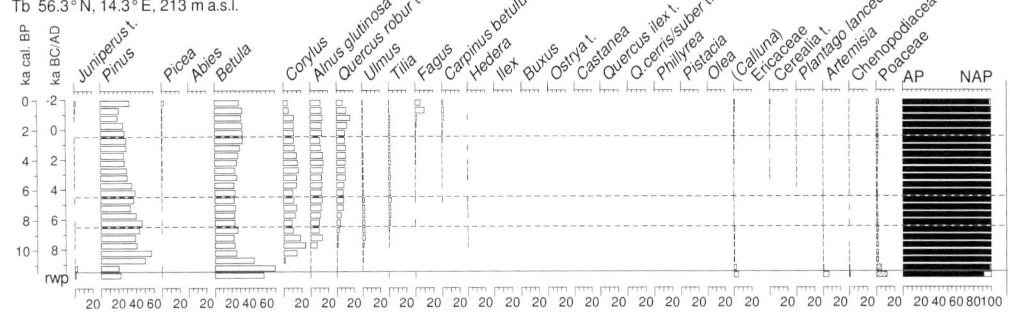

Krageholmssjön *Sweden* (Gaillard-Lemdahl 1984; Gaillard and Digerfeldt 1991)

Tb 55.5° N, 13.7° E, 43 m a.s.l.

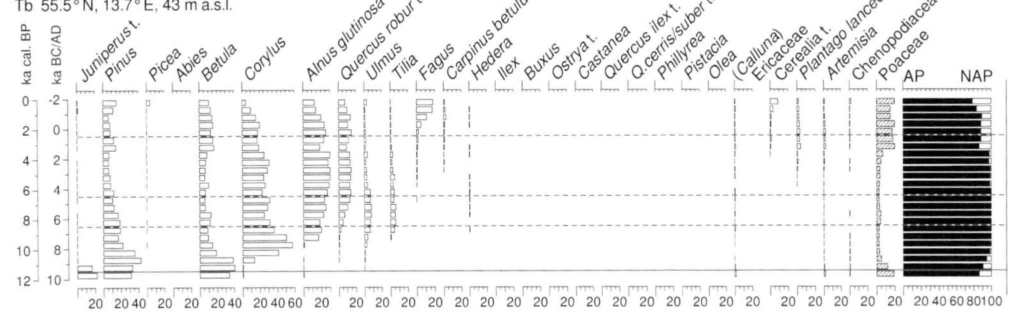

Lake Mikołajki *Poland* (Ralska-Jasiewiczowa 1966, 1989)

Tb 53.8° N, 21.4° E, 116 m a.s.l.

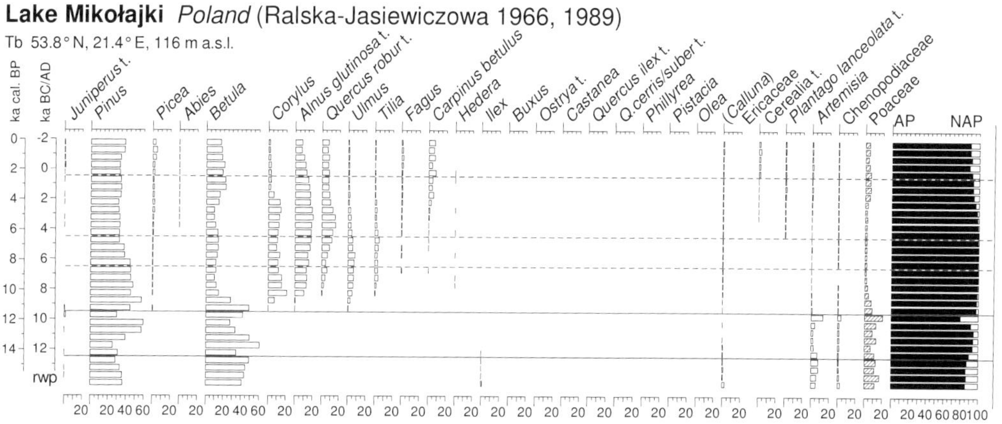

Löddigsee *Germany* (Jahns 2007)

Tb 53.4° N, 11.9° E, 43 m a.s.l.

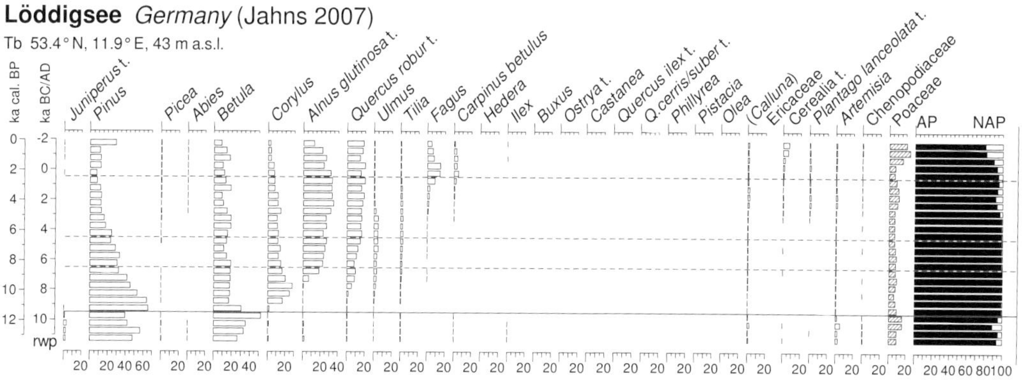

Biskupinskie Lake *Poland* (Noryskiewicz 1995)

Tb 52.8° N, 17.7° E, 80 m a.s.l.

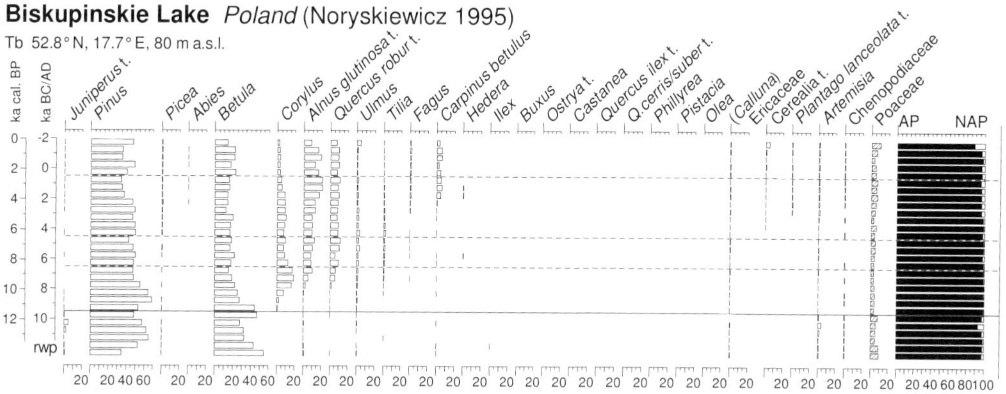

Bledowo Lake *Poland* (Binka et al. 1988, 1991)

Tb 52.6° N, 20.7° E, 78 m a.s.l.

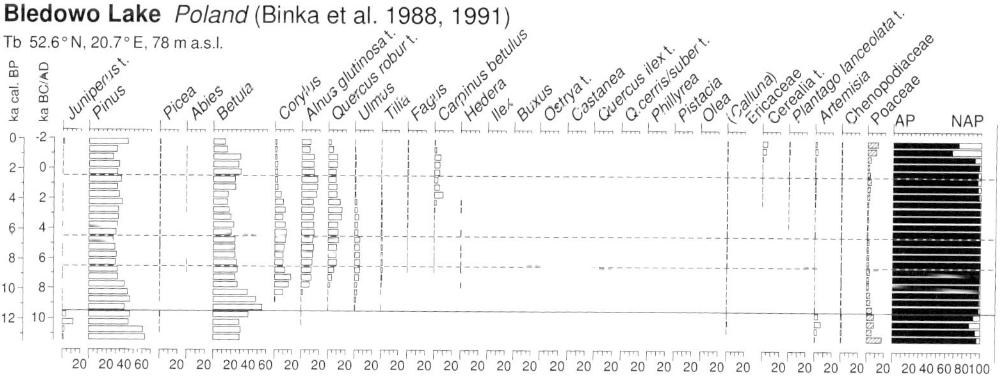

Luttersee *Germany* (Chen 1988)
Tb 51.6° N, 10.2° E, 164 m a.s.l.

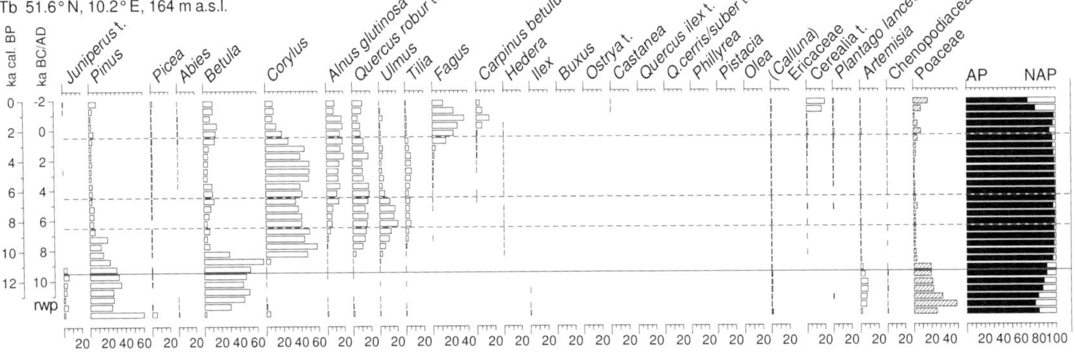

Hrabanovská Černava *Czech Republic* (Petr 2005)
Tb 50.2° N, 14.8° E, 184 m a.s.l.

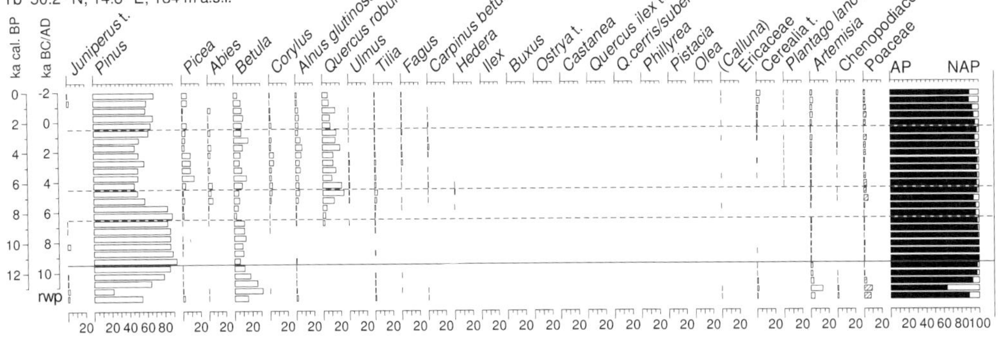

Borkovická blata *Czech Republic* (Jankovská 1980)
Tb 49.2° N, 14.9° E, 415 m a.s.l.

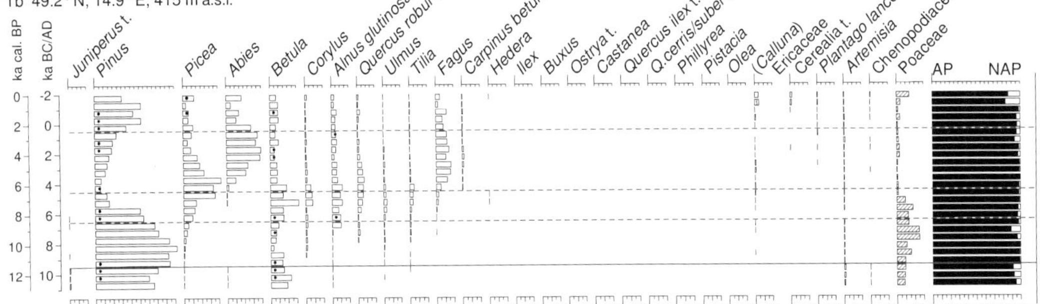

Huzenbacher See *Germany* (Rösch and Tserendorj 2011a, b)
Tb 48.6° N, 8.3° E, 747 m a.s.l.

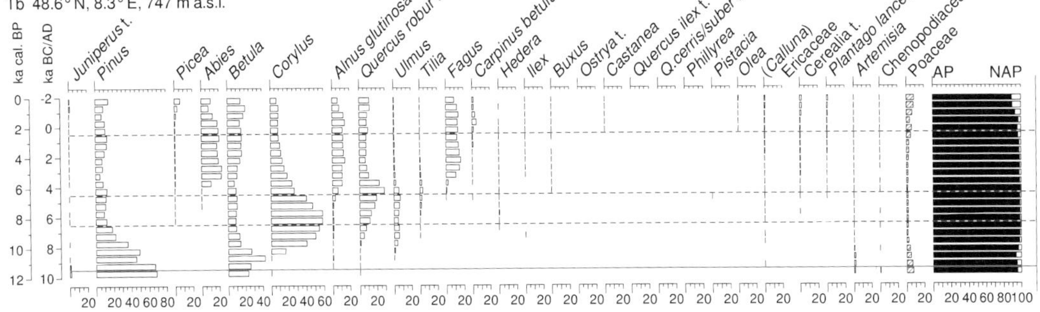

Durchenbergried *Germany* (Rösch 1986)

Tb 47.8° N, 9.0° E, 432 m a.s.l.

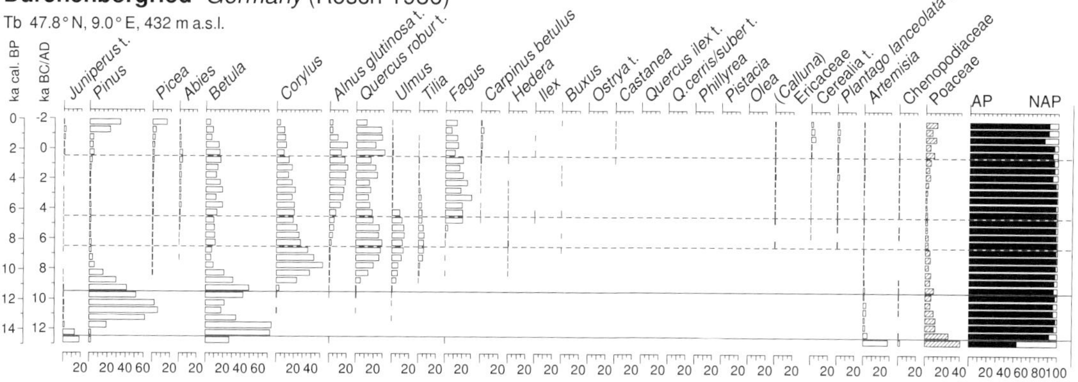

Trumer Moos *Austria* (Krisai 1975)

Tb 47.5° N, 13.1° E, 500 m a.s.l.

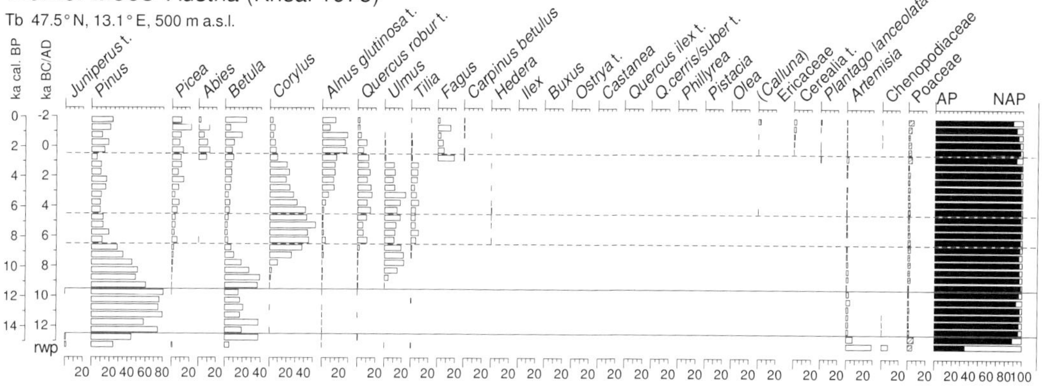

Rotsee *Switzerland* (Lotter 1988)

Tb 47.1° N, 8.3° E, 419 m a.s.l.

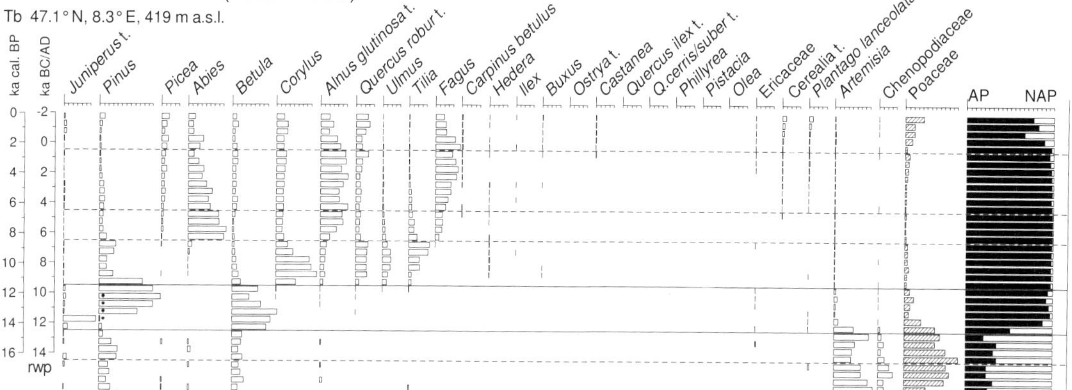

Fig. 4.3-8b. Selected sites in temperate central Europe (Tb). Late-glacial and Holocene pollen records are presented as histograms at a 500-year resolution. Some important sites are treated in other chapters, for example Lake Gościąż in chapter 2. Dots stand for findings of macrofossils. At several sites there are some macrofossil analyses but they concentrated on the Late-glacial and Early Holocene. In these cases we do not show any dots because their lack (earlier or later) could be misunderstood as absence. Some cores reached down to the Weichselian till and show reworked pollen in the bottom sample (rwp) such as *Carya*, *Liquidambar* and *Bruckenthalia* type at Skånsø in Denmark or *Abies* and *Picea* at Rotsee.

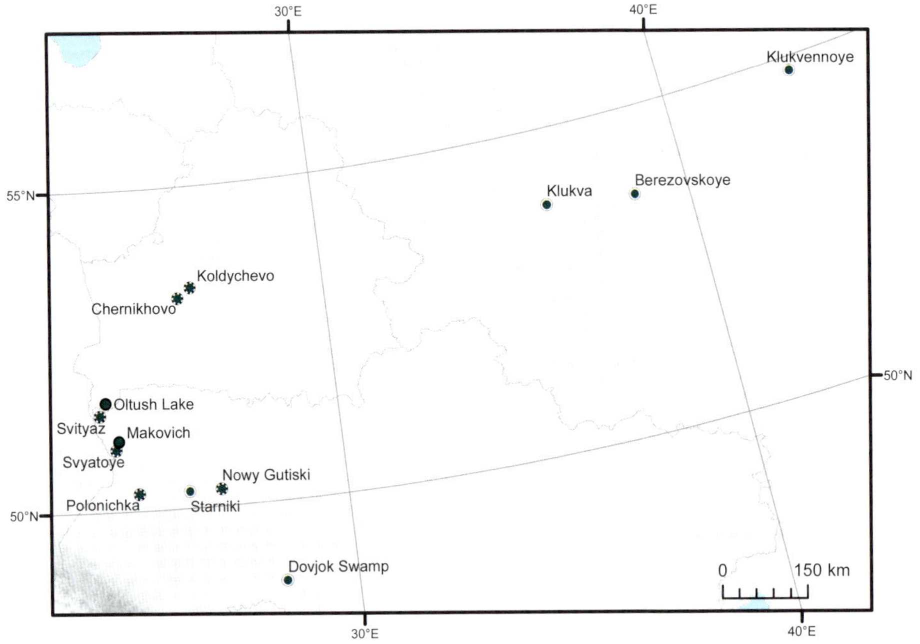

Fig. 4.3-9a. Map of selected sites in the temperate eastern European (Tc) region. Dots with black borders indicate sites shown in the histograms (Fig. 4.3-9b), white borders indicate sites used for the maps showing the history of selected taxa (see chapter 4.4) and dashed borders indicate sites used for both.

Tc, Temperate eastern European region

The region consists of the north-eastern part of the European vegetation zone of temperate deciduous forests where *Fagus* is missing (except for some single occurrences near the western limit). The region to the south-east had a different vegetation development and will be treated separately as the south-eastern European region (Tg). The northern, diffuse border of Tc runs through parts of Belarus and Russia and is the limit to the hemiboreal region (Bd). The western and south-western borders largely coincide with the eastern border of Poland and with the foothills of the Carpathians. In the south-east the Tc region is in contact with the forest-steppe and steppe of the Pontic region Po. The eastern European region of deciduous forests tapers eastwards as does the hemiboreal mixed deciduous-conifer forest, forming the so-called 'eastern European mixed-forest wedge' between the taiga in the north and the steppe in the south. Plant geographically the Tc region corresponds to the southern half of the Sarmatian floristic province of Meusel et al. (1965). The northern half of the Sarmatian floristic province, however, is considered in terms of its vegetation history as a separate region, namely the hemiboreal region (Bd), because of the important contribution of *Picea*. With respect to soil types, large parts of the region show chernozems, but peat deposits are important regionally, for example in the lowlands of the Pripyat River (Polesie).

The region Tc was outside the maximum ice-extent during the LGM. Lakes are therefore rare and mires may have hiatuses. Cryptic refugia for woody taxa may have been possible, for example in sheltered places or along rivers, but they are difficult to prove unless plant macrofossils or charcoal are available. Chernikovo in Belarus may be one of the few sites (of those studied so far) that goes back to approximatley 18 000 cal yr BP (Zernitskaya, 1985; Matveev et al., 1993).

Late-glacial and Holocene: At Staroje Lake (Zernitskaya et al., 2015) and at Svityaz (Schwörer et al., 2022), stomata of *Pinus* were found from as early as the Bølling-Allerød interstadial. At both sites, the Haploxylon *P. cembra* (section *P. strobus*) was also recorded but in very small quantities (hardly visible in the histogram). At Svityaz, *Larix* pollen and stomata were recorded for the Allerød, while

Picea pollen and stomata were found at Staroje Lake for the Younger Dryas. This can be taken as proof of local presence. Forest establishment during the Late-glacial was dominated by *Pinus sylvestris* and *Betula*, and both retained substantial shares in the forests during the Holocene (Fig. 4.3-9b). In early studies the lack of juniper may be due to the fact that the pollen grain had not yet been recognized. During the Early and Mid Holocene, deciduous trees expanded, especially *Tilia* and *Ulmus*, but also *Quercus* and *Corylus*, although a distinct phase dominated by *Corylus* is missing. As is the case in the eastern parts of central Europe, *Corylus* hardly plays a role in the east of region Tc. The similar fading-out eastwards of the *Fagus*-records mirrors the eastern limit of the modern distribution of this taxon which is sensitive to frost (especially to late frosts in spring). During the Early Holocene, a very slight expansion of *Picea* is only visible in the north (Chernikovo), reflecting the modern distribution limit.

Human impact: Some records of Cerealia are available, but the pollen type may have been overlooked at some sites (but was recorded in the south-east at Dovjok Swamp by Kremenetski, 1995 and in the north-west at Svityaz by Schwörer et al., 2022).

Svityaz *Ukraine* (Schwörer et al. 2022)

Tc 51.5° N, 23.8° E, 157 m a.s.l.

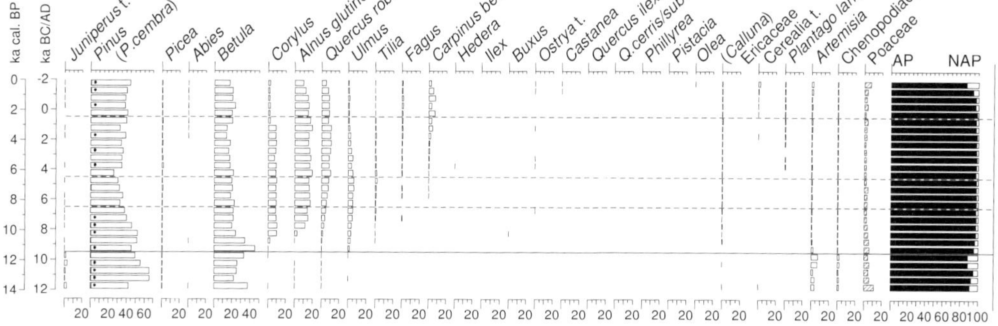

Oltush Lake *Belarus* (Yelocheva Y K 1988 unpubl. in EPD)

Tc 51.7° N, 24.0° E, 158 m a.s.l.

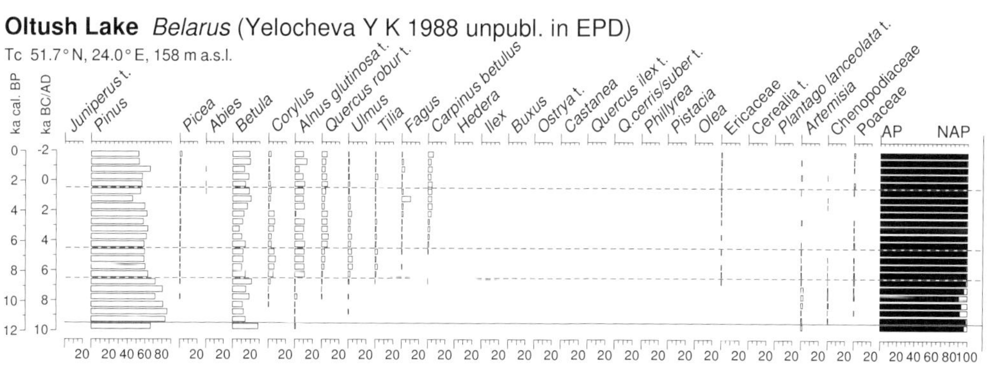

Makovich *Ukraine* (Artushenko 1957, 1959)

Tc 51.0° N, 24.2° E, 192 m a.s.l.

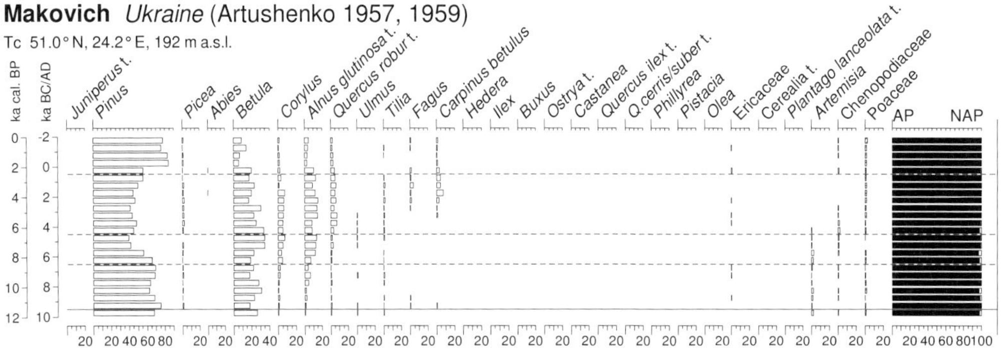

Koldychevo *Belarus* (Bogdel 1984; Elovicheva and Bogdel 1985)

Tc 53.3° N, 26.1°, 185 m a.s.l.

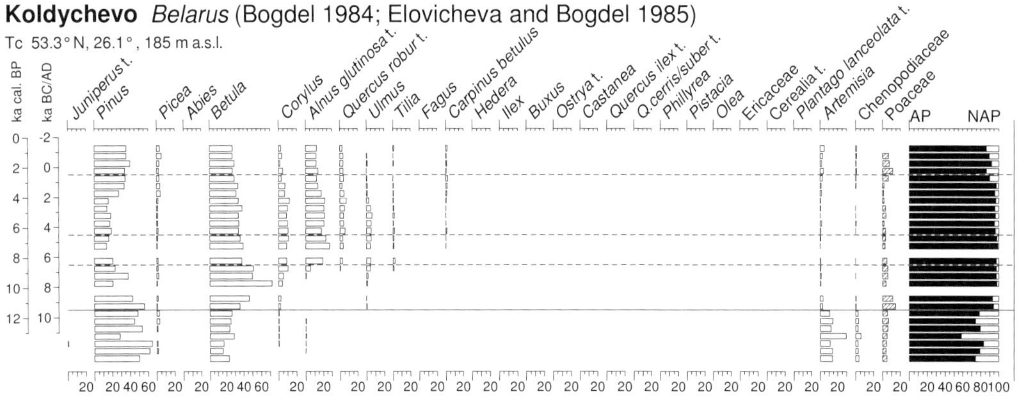

Chernikhovo *Belarus* (Zernitskaya 1985; Matveev et al. 1993)

Tc 53.4° N, 26.4° E, 168 m a.s.l.

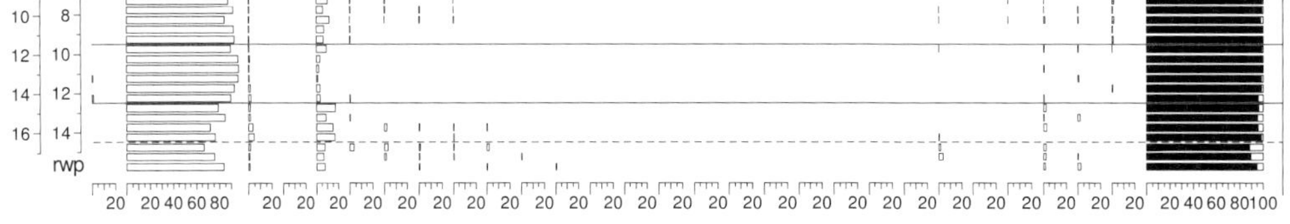

Nowy Gutiski *Ukraine* (Klimanov and Bezusko 1981; Bezusko et al. 1988)

Tc 50.3° N, 26.8° E, 210 m a.s.l.

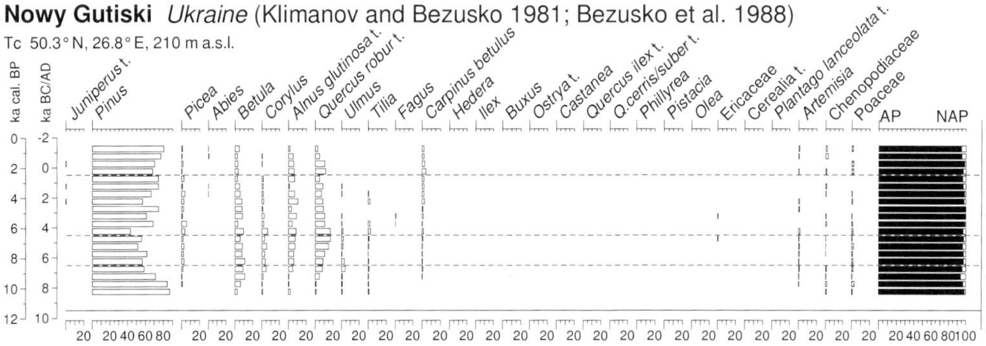

Polonichka *Ukraine* (Bezusko 1977; Artushenko et al. 1982)

Tc 50.3° N, 24.8° E, 200 m a.s.l.

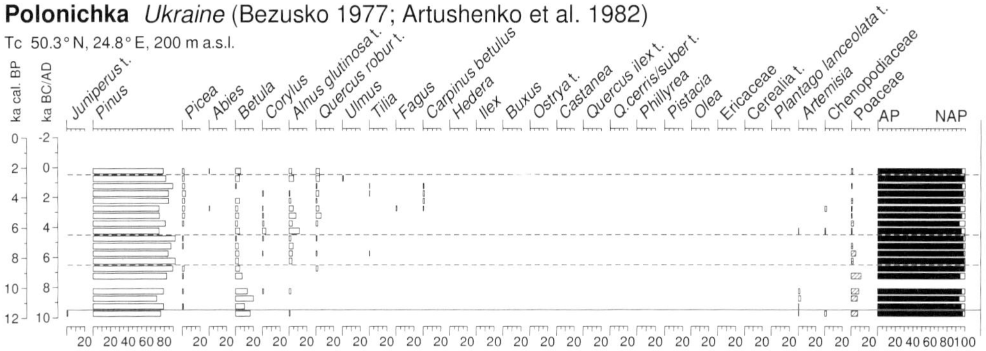

Svyatoye *Ukraine* (Artushenko 1957)

Tc 51.1° N, 24.3° E, 183 m a.s.l.

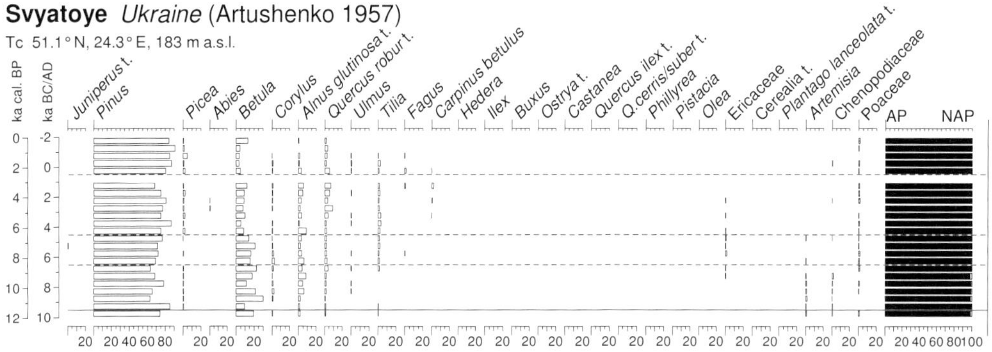

Fig. 4.3-9b. Selected sites in the temperate eastern European (Tc) region arranged from north-west to south-east as histograms at a 500-year resolution. A general problem is the absence or scarcity of *Juniperus* pollen, which we suspect to be partly 'analyst related'.

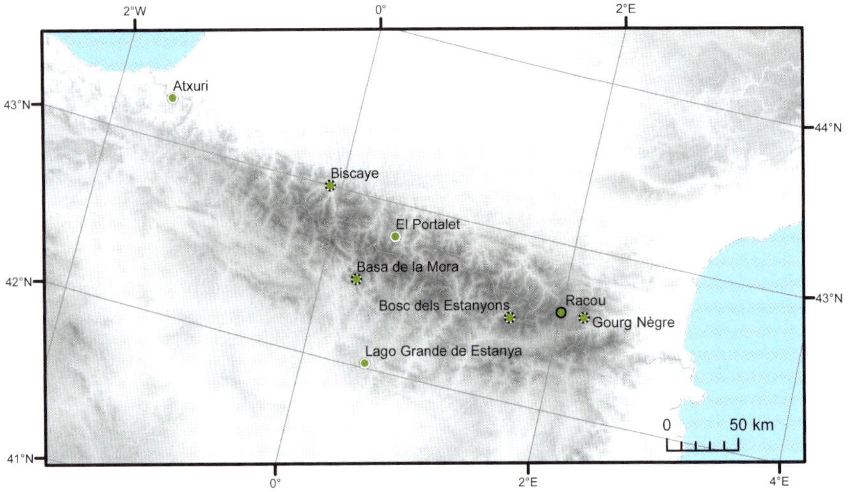

Fig. 4.3-10a. Map of selected sites in the Pyrenees (Td) region. Dots with black borders indicate sites shown in the histograms (Fig. 4.3-10b), white borders indicate sites used for the maps showing the history of selected taxa (see chapter 4.4) and dashed borders indicate sites used for both.

Td, The Pyrenees

This mountain chain, reaching elevations up to 3404 m asl, has a vegetation and forest development that justifies it being distinguished as a separate region in vegetation history. The region corresponds to the Pyrenean floristic province of Meusel et al. (1965) that these authors assign to the macaronesian-mediterranean floristic region. The extension from the Atlantic to the Mediterranean, the elevational gradients, and the contrast of southern and northern slopes produce very heterogeneous conditions in both climate and vegetation.

The maximum ice extent in the Pyrenees was before 30 000 yr BP (García-Ruiz et al., 2003). During the LGM the Pyrenees were also partly covered by glaciers which melted early in the Late-glacial (Andrieu et al., 1988; García-Ruiz et al., 2003).

Late-glacial and Holocene sediment that covers the end of the NAP-dominated pleniglacial was reached at a few sites – from the littoral zone near the Atlantic at La Moura (Oldfield, 1964; Reille, 1993a) to mid-elevations in the western to central Pyrenees at Lourdes and Biscaye (Reille and Andrieu, 1995) and even at high elevations in the east at La Borde (1660 m asl; Reille, 1990b,c). Near the bottom of the cores reworked pollen of thermophilous taxa and high percentages of long-distance transported pine pollen were sometimes found. Both of these decrease in the Oldest Dryas which is

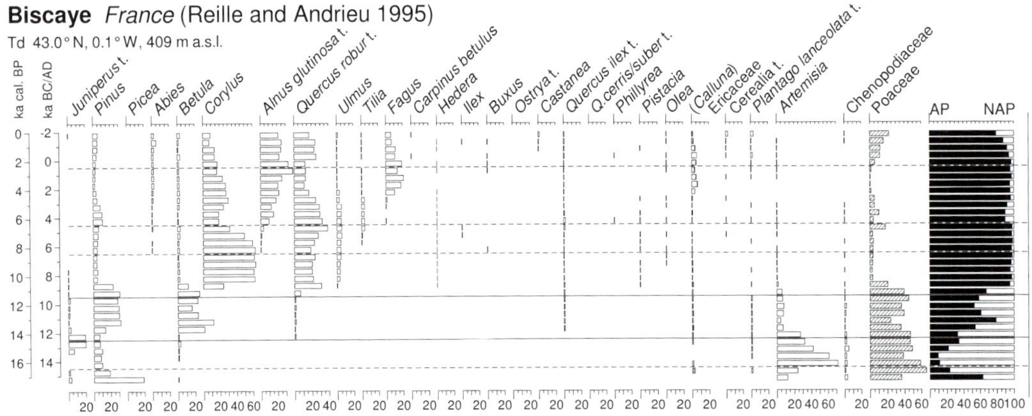

Basa de la Mora *Spain* (Pérez-Sanz et al. 2013)

Td 42.5° N, 0.3° E, 1914 m a.s.l.

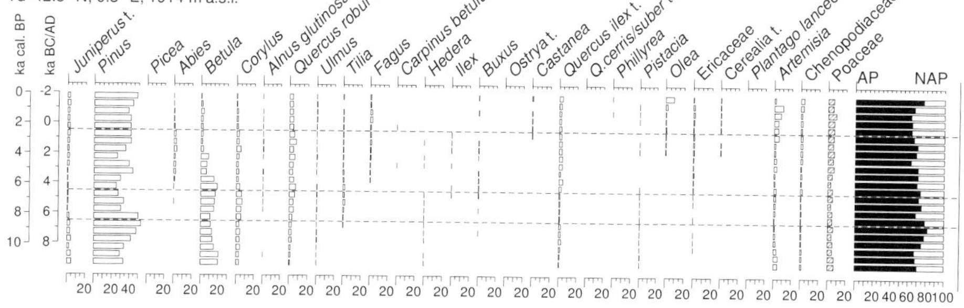

Bosc des Estanyons *Andorra* (Miras et al. 2007)

Td 42.5° N, 1.6° E, 2180 m a.s.l.

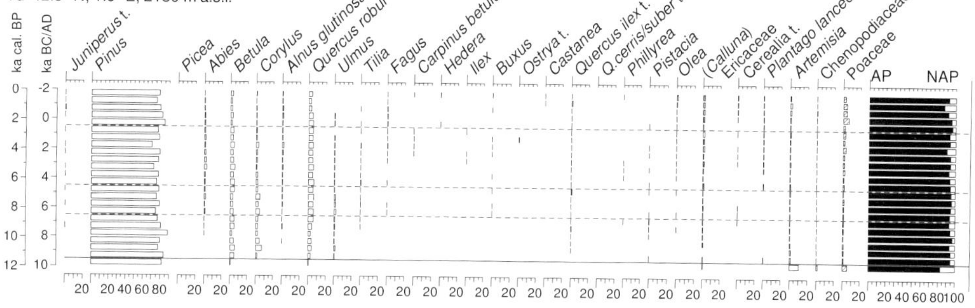

Racou *France* (Guiter et al. 2005)

Td 42.6° N, 2.0° E, 2000 m a.s.l.

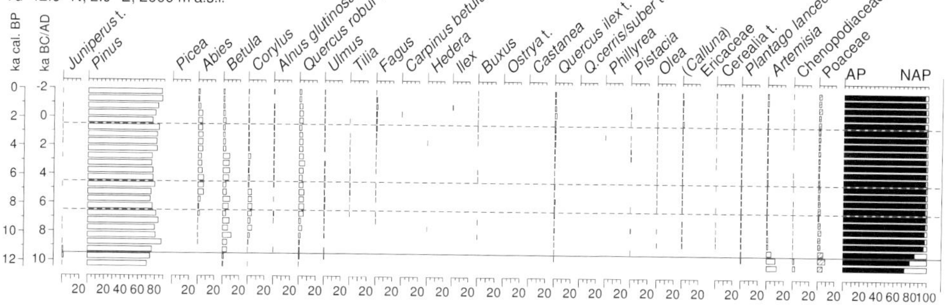

Gourg Nègre *France* (Reille and Lowe 1993)

Td 42.6° N, 2.2° E, 2080 m a.s.l.

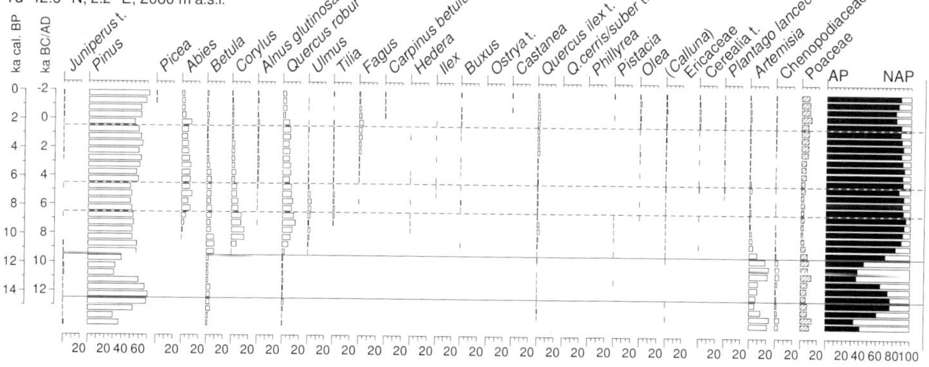

Fig. 4.3-10b. Selected sites in the Pyrenees (Td) region from west to east as histograms at a 500-year resolution. The Bølling-Allerød was reached at a few sites. During the Holocene even at high-elevation sites, Mediterranean pollen such as *Quercus ilex* type, *Olea*, *Pistacia* is regularly recorded. *Picea* is absent (except for recent plantations).

rich in NAP, especially in *Artemisia* and Chenopodiaceae. The Bølling-Allerød interstadial is characterized by woodlands, often initiated by *Juniperus*, but then becoming rich in *Betula* and *Pinus*: the local presence of the latter, at the latest during the Allerød, is documented by findings of stomata even at an elevation of 2000 m asl (Gourg Nègre; Reille, 1990c). The Younger Dryas is well recorded at all elevations as a minimum of *Pinus* and a new maximum of *Artemisia* and other NAP. This cooler and drier period is sometimes structured, as for example at Racou (2000 m asl; Guiter et al., 2005).

In the Early Holocene deciduous *Quercus* expanded rapidly and always distinctly before *Corylus* (Reille and Andrieu, 1991; Reille, 1993b). Traces of *Ulmus, Tilia* and *Fraxinus* follow (often in this sequence). *Abies* develops at mid-elevations. During the Mid Holocene *Fagus* occurs and expands. Hardly anything is known about the history of *Pinus uncinata*, which is so conspicuous in the modern subalpine belt, except that its occurrence on bogs at low to middle elevation is not relict (Reille, 1991b). Sites on the southern (and at some sites on the northern) slopes of the Pyrenees show long Holocene pollen records of *Quercus ilex* type, *Olea, Pistacia* and other Mediterranean trees. In contrast to the more eastern mountains of the Alps and the Carpathians where *Picea abies* is important, this tree never reached the Pyrenees. Since the second half of the 19th century spruce has been planted at some sites.

Human impact can be traced from the Neolithic (Miras et al., 2007; J.S. Carrión et al., 2010a).

Overviews are presented for the French side of the Pyrenees by Jalut et al. (1996) subdividing into western, Ariège, central and eastern Pyrenees and for the Spanish side by Carrión (2012) and J.S. Carrión et al. (2010a).

Transects: Jalut et al. (2000).

Fire history: Gil-Romera et al. (2014), J.S. Carrión et al. (2010a)

Te, The Alps

These high mountains in the south of the central European region, which reach elevations of up to 4810 m asl, show a vegetation history that is different from that of the surrounding lowlands, just as the Pyrenees and the Carpathians do. Today the montane and the subalpine vegetation is dominated by conifer forests. Floristically, the region corresponds to the Alpine sub-region of the central European region of Meusel et al. (1965). On the basis of floristic composition and vegetation a subdivision is usually made

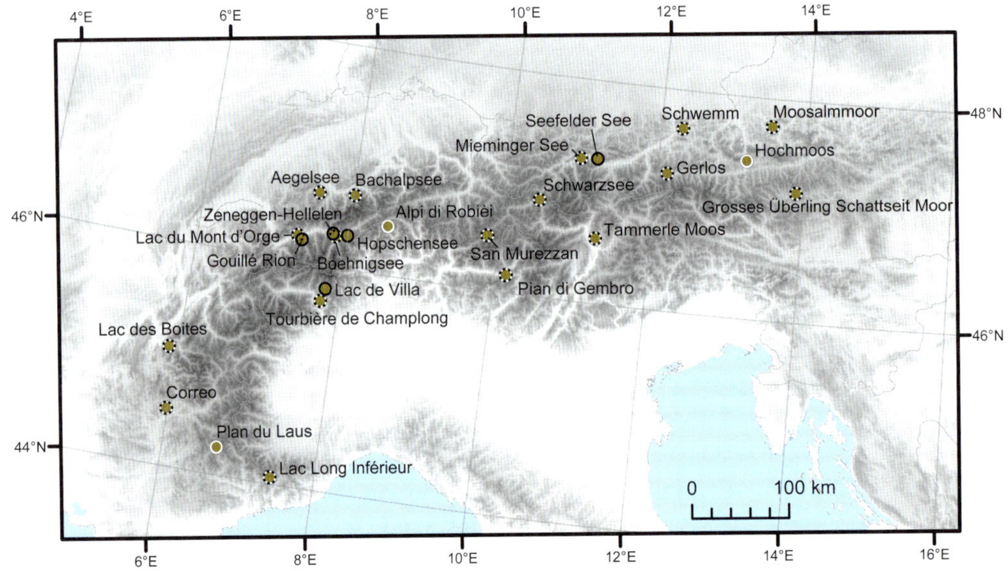

Fig. 4.3-11a. Map of selected sites for the Alps (Te) region. Dots with black borders indicate sites shown in the histograms (Fig. 4.3-11b), white borders indicate sites used for the maps showing the history of selected taxa (see chapter 4.4) and dashed borders indicate sites used for both.

into the western Alps (from the Riviera to Savoy) and the central and eastern Alps that are subdivided into the northern and southern outer Alps (Randalpen in German) with high precipitation and the central Alps with their longitudinal valleys with low precipitation. For an overview of the modern vegetation see Ozenda (1983) or Ellenberg (2009).

During the LGM the Alps were heavily glaciated with nunataks along and between the northern and southern outlet-glaciers. The south-west (Alpes Maritimes, Seealpen) and the outermost east of the Alps remained unglaciated. It should come as no surprise that the sites in these unglaciated areas show especially long vegetation records before the Bølling (de Beaulieu, 1977; Draxler, 1977; Wahlmüller, 1985; Nakagawa, 1998; Krisai et al., 1991). The nunataks played an important role in the phylogeographic pattern of many high-alpine plants (Gugerli and Holderegger, 2001; Stehlik, 2003; Tribsch and Schönswetter, 2003). The disintegration of the network of ice-streams had already started in the early Late-glacial, in the Oldest Dryas, when NAP dominated over AP. During the Bølling and Allerød interstadial the glaciers had retreated into the mountains, whereas during the Younger Dryas limited re-advances are well recorded as terminal moraines in the higher valleys (for example, the Egesen of the eastern Alps).

During the Late-glacial the establishment of shrub or trees *Juniperus*, *Betula* and *Pinus (sylvestris)* played a major role at lower to intermediate elevations, as indicated by stomata records of *Juniperus* and *Pinus* at Zeneggen-Hellelen, Lac de Mont d'Orge and Boehnigsee (and during the Holocene at higher elevations such as at Bachalpsee or Gouillé Rion). At many but not all sites the Bølling is characterized by high *Betula*-values and the Allerød by high *Pinus*-values. This is comparable to sites in southern parts of central Europe. The Younger Dryas is frequently recorded by an increase in *Artemisia* and other NAPs. The increase in *Quercus* synchronous with an increase in *Artemisia* at Lac Long Inféricur above 2000 m asl in the southern Alps may seem contradictory but is explained by de Beaulieu (1977) as follows. The site was above the timberline during the Late-glacial and Early Holocene, and the cool, dry climate of the Younger Dryas pushed the pine forests downwards, which resulted in an increase of *Quercus* in percentage terms (long-distance transported and less filtered by decreasing pine-forests at lower elevations). Timberline-depressions during the Younger Dryas are recorded throughout the Alps. At some sites conifer stomata are recorded for the Allerød and the earliest Early Holocene, but not for the Younger Dryas, for example *Juniperus* and *Pinus* at Zeneggen-Hellelen (Welten, 1982b; Ammann et al., 2014a).

For the Holocene development major differences exist between the (continental) central and the marginal (more humid) Alps. In the latter, the deciduous broadleaved trees of central Europe (*Corylus, Ulmus, Tilia*, later also *Fagus*) were important, whereas in the central Alps they were generally rare (even at low elevations, for example at Mont d'Orge). In these central Alps and their longitudinal valleys, *Quercus pubescens* may have occurred (pollen type of *Quercus robur*) at lower elevations, while at lower and intermediate elevations *Pinus sylvestris* and at high elevations *Pinus cembra* and *Larix decidua* were dominant. The latter two trees did occur in the less continental marginal Alps but were never as important as in the dry valleys of the central Alps.

A contrast between east and west is obvious among the conifers: *Abies alba* spread early from the northern Apennines into the south and south-west (de Beaulieu, 1977) and became very prominent at many sites including Lake Allos at 2227 m asl (Chapra and Dobson, 1981) and Lake Vens at 2330 m asl (Brisset et al., 2015). In the eastern Alps, *Abies* established late and, at most sites, only played a minor role, in spite of the fact that there were also refugia in the south-east (Greece, north-western Balkan, Dinarides). *Picea abies,* in contrast, shows an early spread into the eastern Alps where it becomes particularly abundant at many sites. In southern and south-central alpine sites such as Pian di Gembro or San Murezzan, spruce and fir established and expand at about the same time (Pini, 2002; Gobet et al., 2003). The later and slower range expansion of spruce westwards was mapped not only by pollen and macrofossils but also by phylogenetics: a decreasing within-population gene diversity (h_s) and a decreasing genetic differentiation (G_{ST}) towards the western Alps reflect successive founder effects (Fig. 4.3-11b; Gugerli et al., 2001a; van der Knaap et al., 2005; Figs. 2 and 4 in Tollefsrud et al., 2008).

During the Mid Holocene the forest limit and the tree limit reached higher elevations than today (see section 4.3.4 and chapter 4.9).

Lac Long Inférieur *France* (de Beaulieu 1977)

Te 41.4° N, 7.5° E, 2039 m a.s.l.

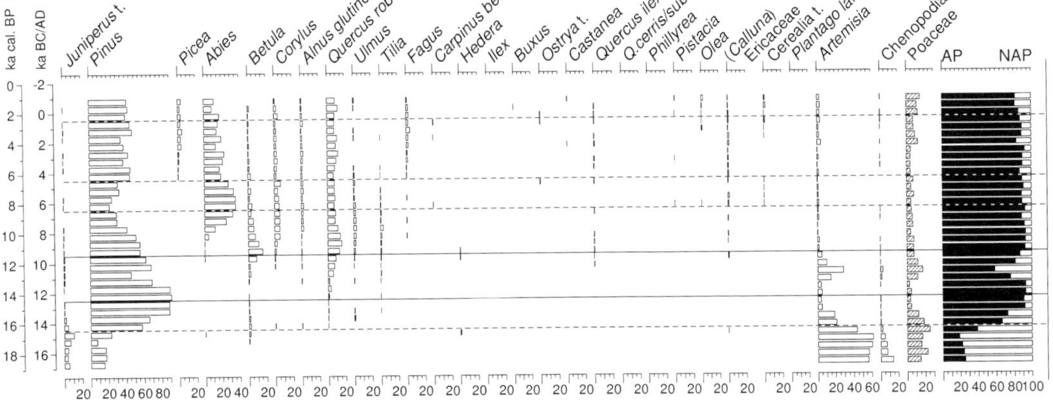

Correo *France* (Nakagawa 1998)

Te 44.5° N, 6.0° E, 1090 m a.s.l.

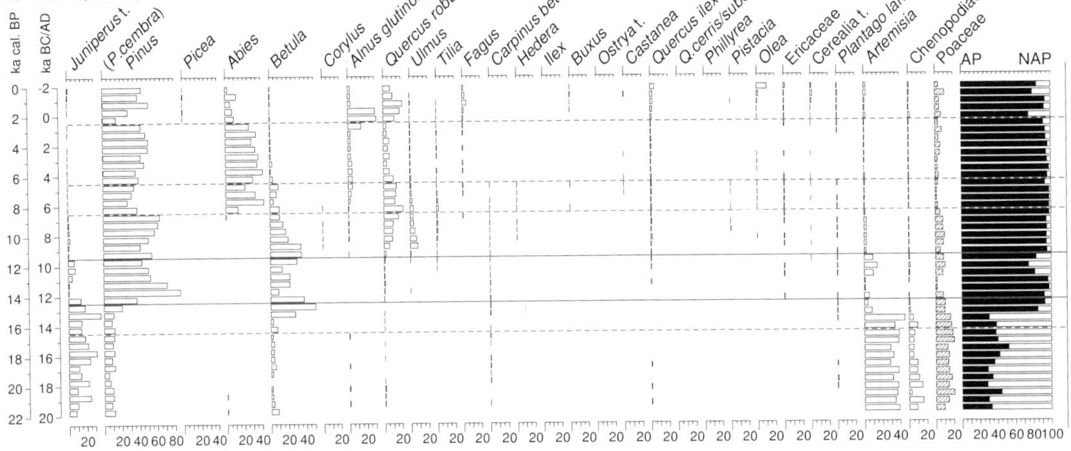

Lac des Boites *France* (Nakagawa 1998)

Te 45.1° N, 5.9° E, 1560 m a.s.l.

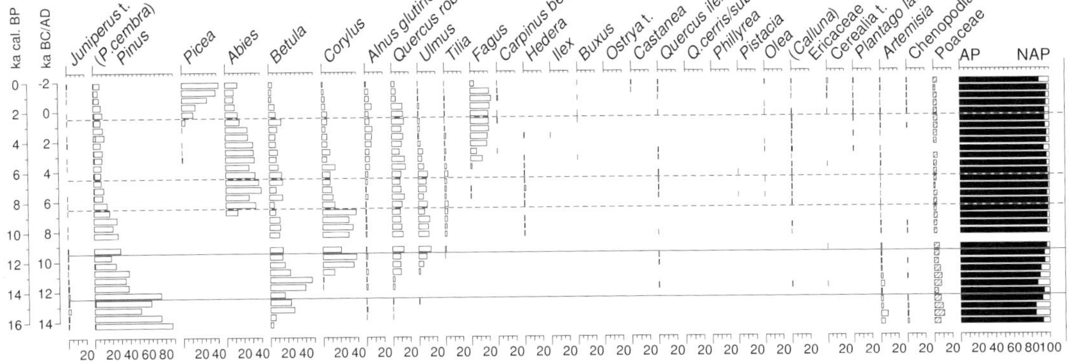

Lac de Villa *Italy* (Brugiapaglia 1996)

Te 45.7° N, 7.8° E, 820 m a.s.l.

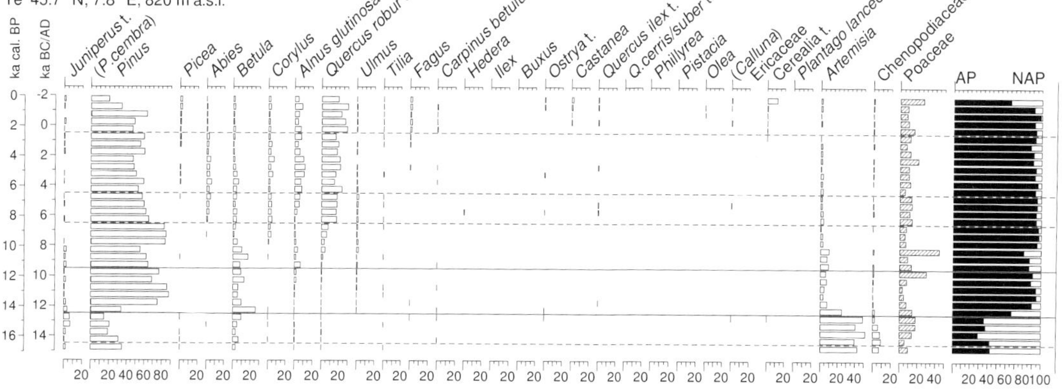

Tourbière de Champlong *Italy* (Brugiapaglia 1996)

Te 45.8° N, 7.8° E, 2320 m a.s.l.

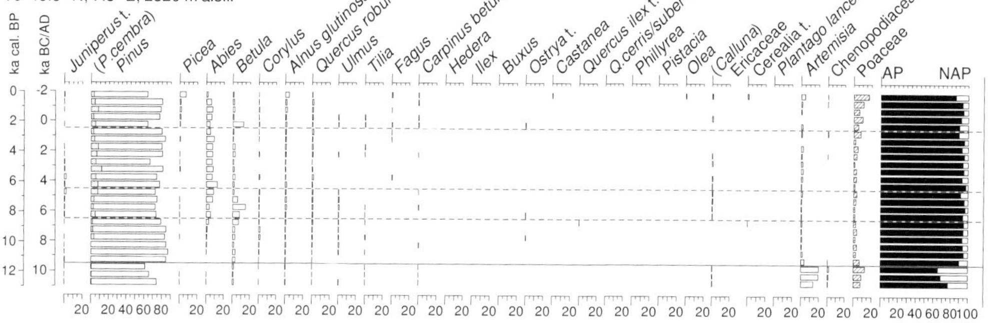

Lac du Mont d'Orge *Switzerland* (Welten 1982a)

Te 46.2° N, 7.3° E, 640 m a.s.l.

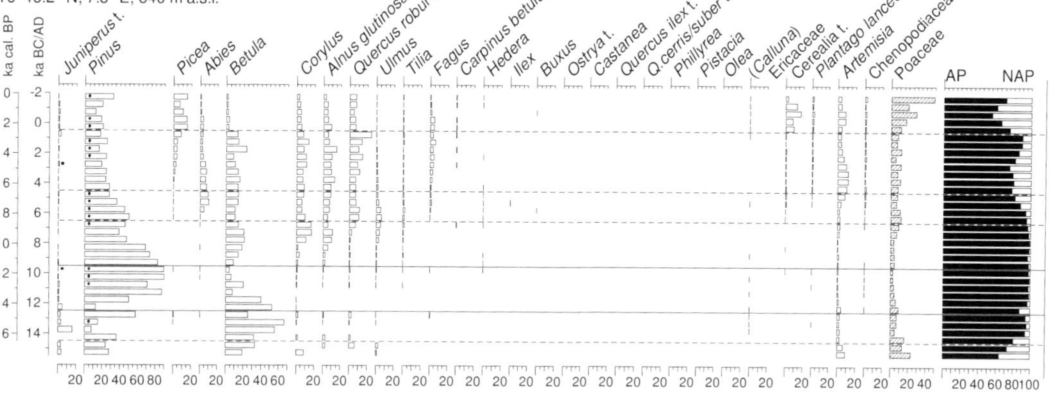

Gouillé Rion *Switzerland* (Tinner et al. 1996)

Te 46.2° N, 7.4° E, 2343 m a.s.l.

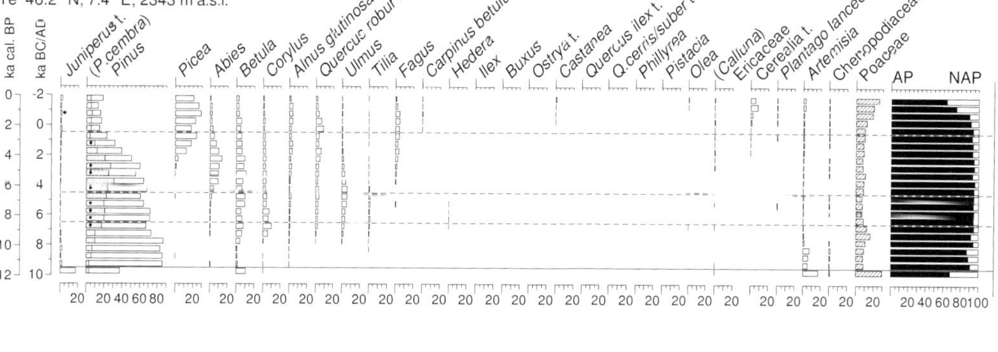

Aegelsee *Switzerland* (Wegmüller and Lotter 1990)

Te 46.6° N, 7.5° E, 989 m a.s.l.

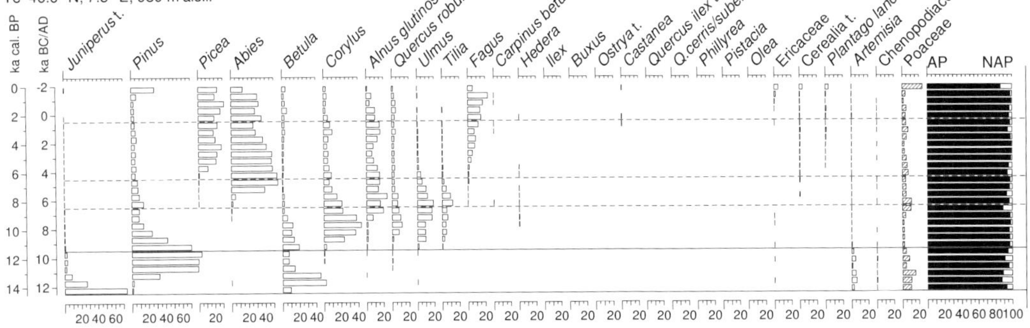

Bachalpsee *Switzerland* (Lotter et al. 2006)

Te 46.7° N, 8.0° E, 2265 m a.s.l.

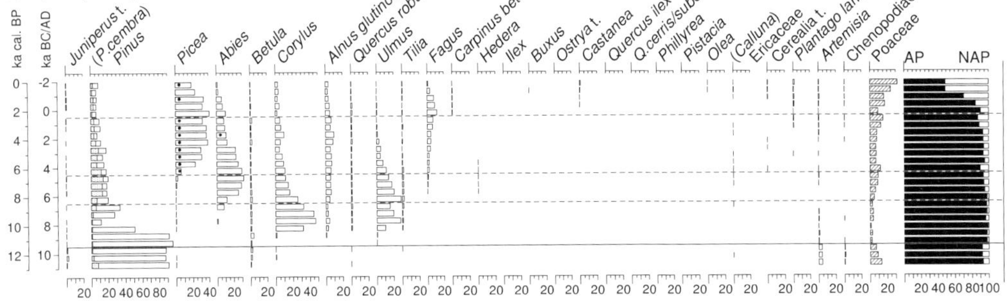

Zeneggen-Hellelen AB *Switzerland* (Welten 1982b)

Te 46.3° N, 7.8° E, 1520 m a.s.l.

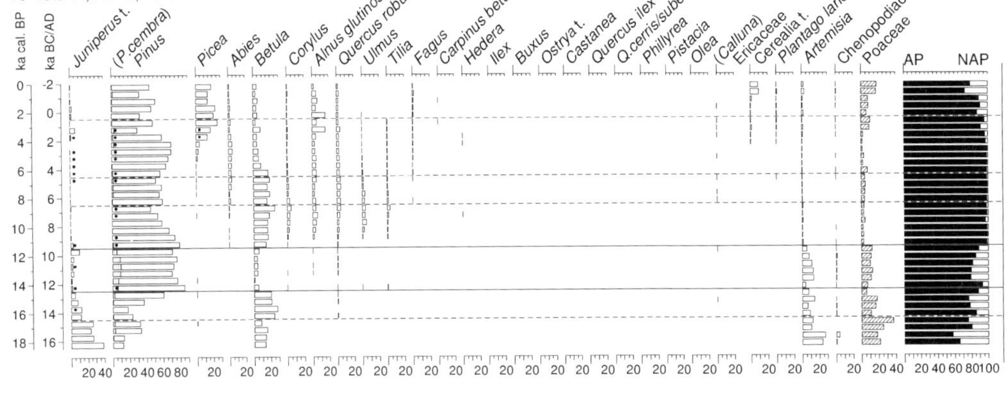

Boehnigsee *Switzerland* (Markgraf 1969)

Te 46.3° N, 7.8° E, 2095 m a.s.l.

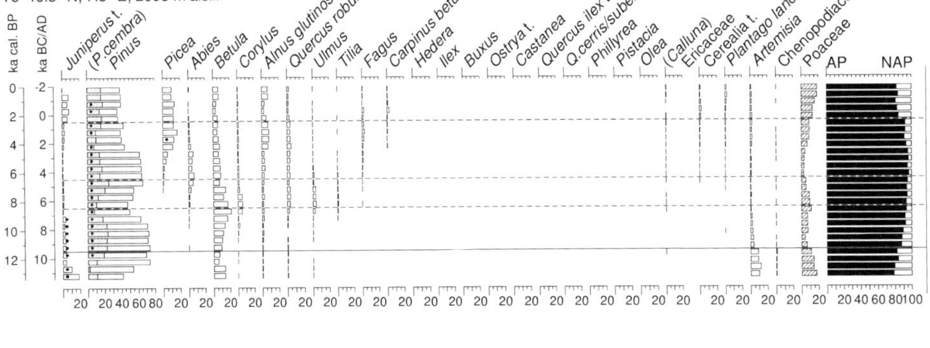

Hopschensee *Switzerland* (Welten 1982b)

Te 46.3° N, 8.0° E, 2017 m a.s.l.

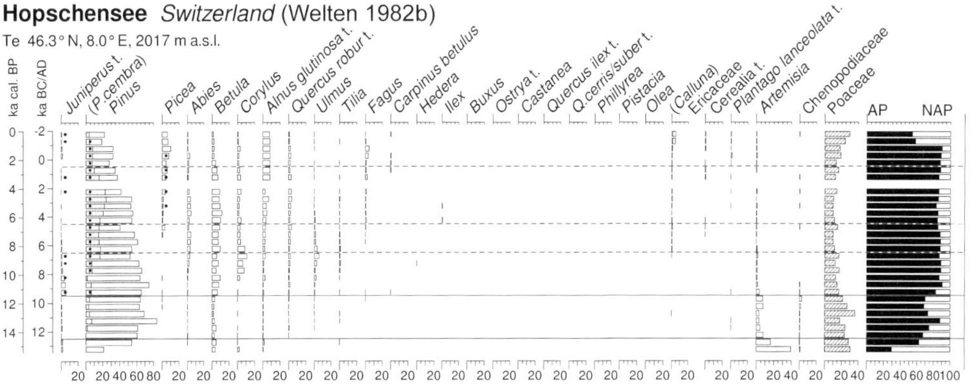

San Murezzan *Switzerland* (Gobet et al. 2003, 2005)

Te 46.5° N, 9.9° E, 1768 m a.s.l.

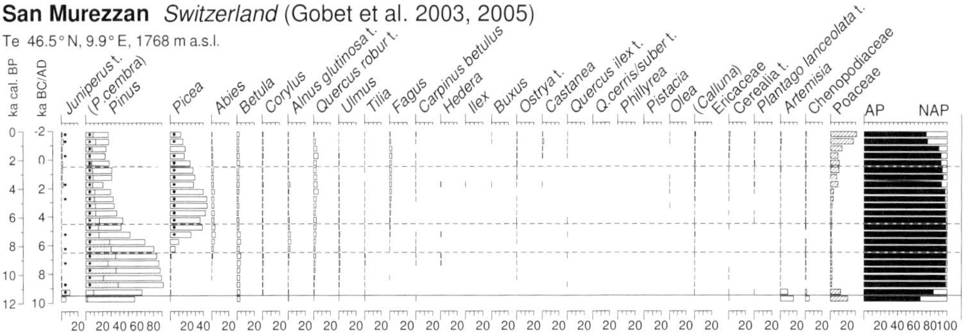

Pian di Gembro *Italy* (Pini 2002)

Te 46.2° N, 10.3° E, 1350 m a.s.l.

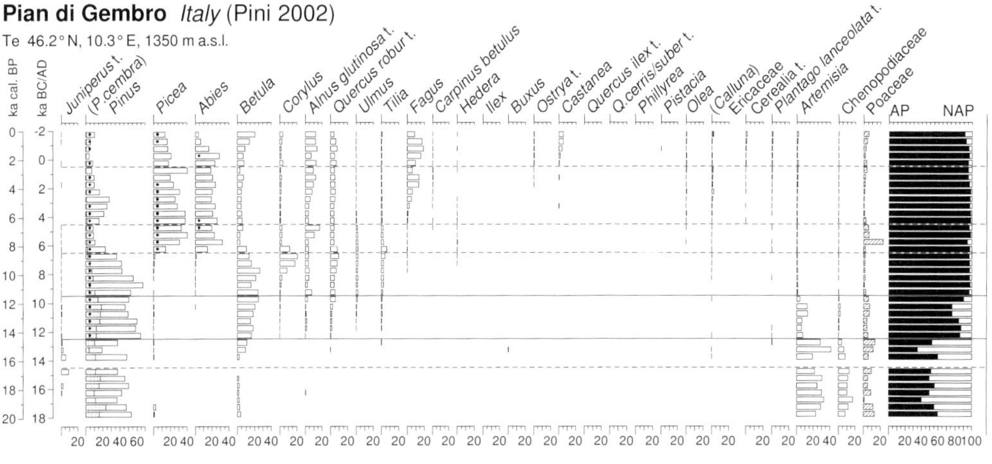

Schwarzsee, Reschenscheideck *Austria* (Welten 1982a)

Te 46.9° N, 10.5° E, 1721 m a.s.l.

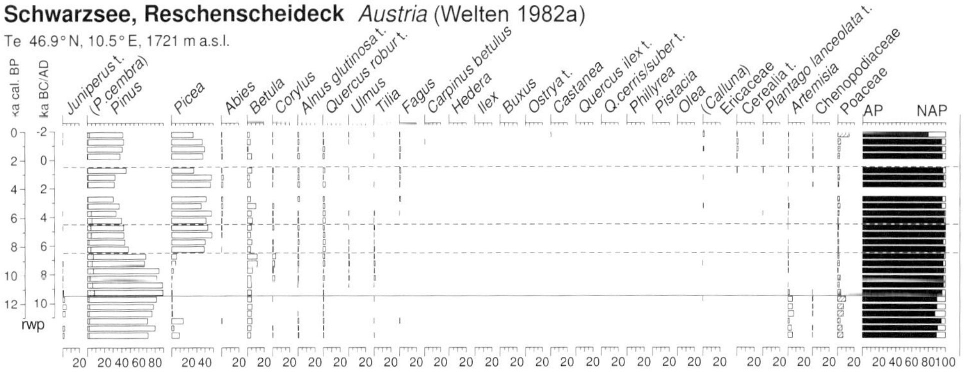

Mieminger See *Austria* (Wahlmüller 1985)

Te 47.3° N, 11.0° E, 800 m a.s.l.

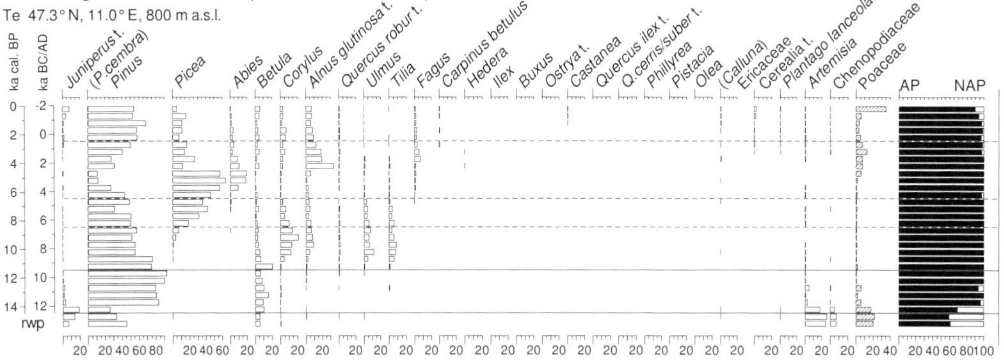

Tammerle Moos *Austria* (Wahlmüller 1990)

Te 46.6° N, 11.3° E, 1400 m a.s.l.

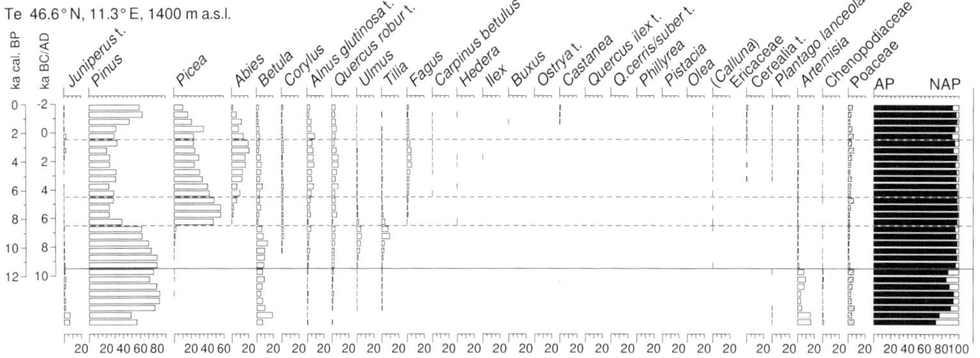

Gerlos *Austria* (Bortenschlager 1984)

Te 47.2° N, 12.1° E, 1590 m a.s.l.

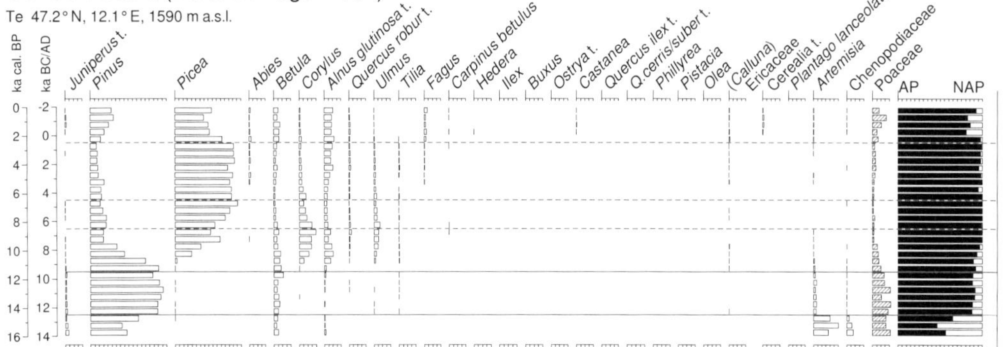

Schwemm *Austria* (Oeggl 1988)

Te 47.7° N, 12.3° E, 664 m a.s.l.

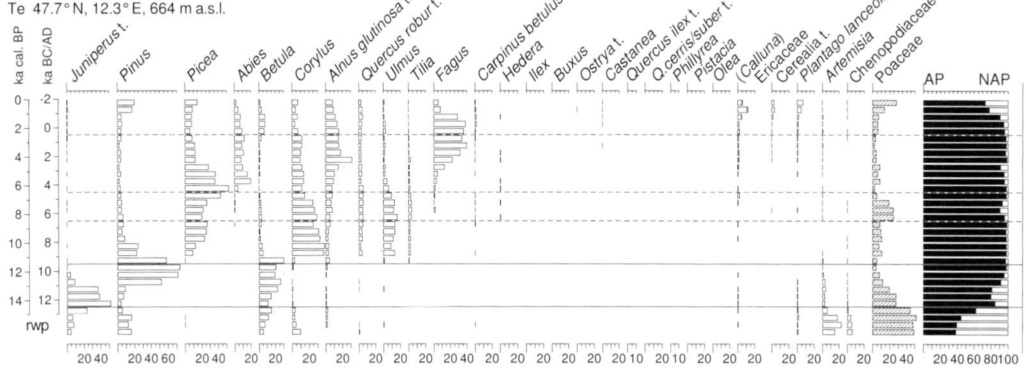

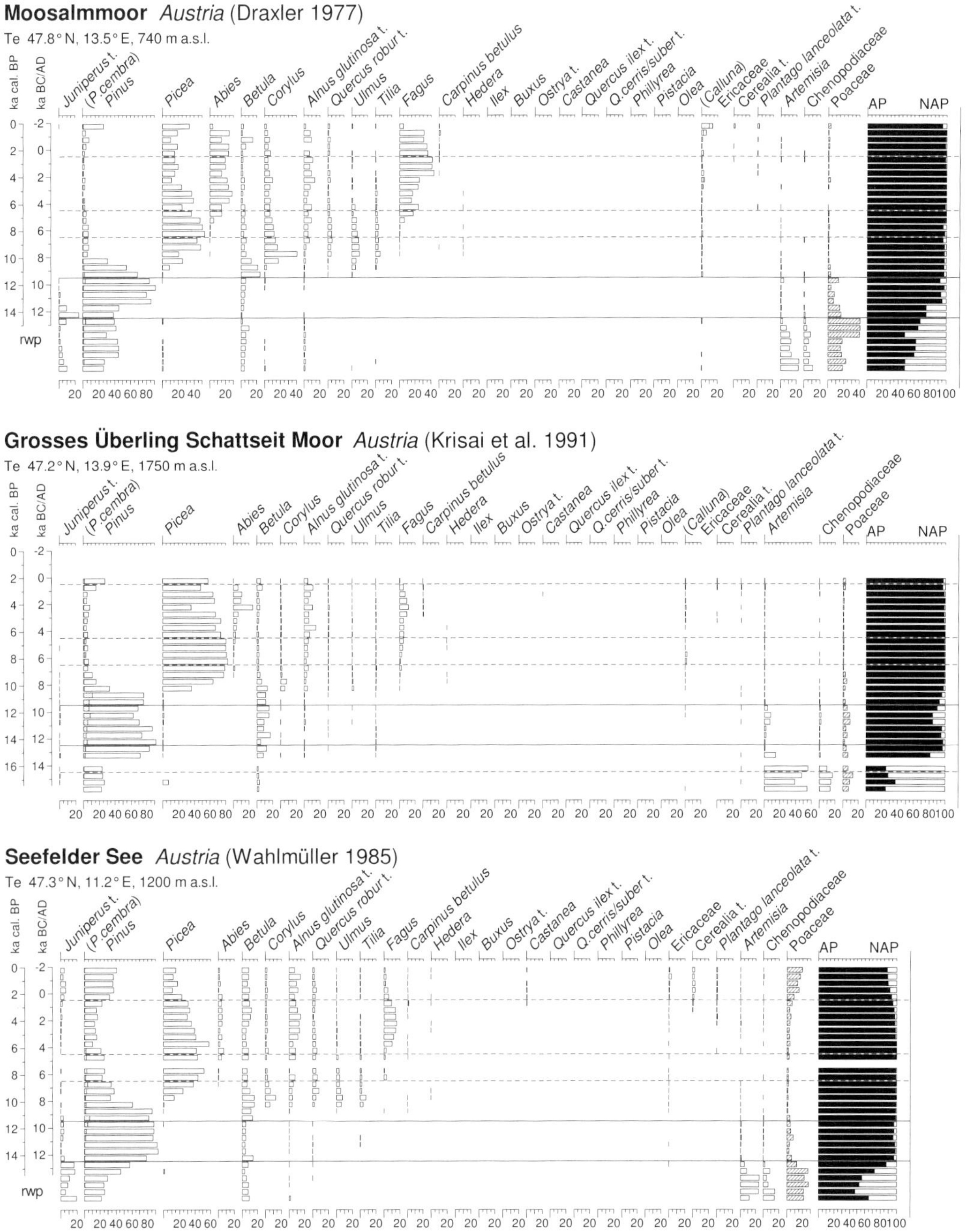

Fig. 4.3-11b. Selected sites in the European Alps (Te). Late-glacial and Holocene pollen records are presented as histograms at a 500-year resolution. The shaded parts in the histograms are *Pinus cembra* within *Pinus* and *Calluna* within Ericaceae. Not surprisingly, sites at lower elevation often show an earlier start and then more thermophilous taxa (for example *Quercus ilex* at Lac de Villa). Usually, the forest composition of the montane and subalpine belts is well reflected (*Quercus-Fagus-Abies-Picea*). Even easily dispersed pollen such as *Pinus cembra* indicates their elevational distribution by showing distinctly higher values at higher elevations. Dots indicate conifer stomata or macrofossils. In an attempt to capture the elevational differences, some of the simplified diagrams are arranged in contrasting pairs: Lac de Villa (820 m asl) and Tourbière de Champlon (2320 m asl); Lac de Mont d'Orge (640 m asl) and Gouillé Rion (2343 m asl); Zeneggen-Hellelen (1520 m asl) and Boehnigsee (2095 m asl); Aegelsee (989 m asl) and Bachalpsee (2265 m asl). In the eastern Alps the three sites Schwemm (664 m asl), Moosalmmoor (740 m asl) and Grosses Ueberlinger Schattseit-Moor (1750 m asl) may be compared as an elevational transect. The temporal resolution of 500 years is naturally not high enough to capture the eight cool phases during the Holocene (Zoller, 1960, 1967; Patzelt and Bortenschlager, 1973; Patzelt, 1977; Haas et al., 1998).

When considering the elevational gradient we find that:

- Often sites at lower elevations show more thermophilous taxa (even *Quercus ilex* at Lac de Villa) than at higher elevations, which is not surprising.
- The forest composition of the montane and subalpine belts is well illustrated in the pollen records (*Quercus-Fagus-Abies-Picea*).
- Even easily dispersed pollen like that of *Pinus cembra* indicates the elevational distribution of the taxon by showing distinctly higher values at higher elevations.
- Human impact (as estimated from increasing cultural indicators and NAP during the Mid and Late Holocene) is often earlier and stronger at lower elevations.

After many Palaeolithic and Mesolithic traces, the **human settlements** go back to 7500–6000 cal yr BP. Certainly, the most unusual finding of the Neolithic period is the Tyrolean Ice man Ötzi (Oeggl, 2009). From the Bronze Age onwards, transhumance and summer-farming above the timberline became important (Tinner et al., 1996; Tinner and Theurillat, 2003; Schwörer et al., 2014a). Forests were reduced and their composition changed.

Overviews: For the Alps in general, de Beaulieu (1977); for the western Alps, de Beaulieu et al. (1994), Nakagawa (1998) and Brisset et al. (2015); for the Central Alps, Welten (1982b) and Burga and Perret (1998); for the Eastern Alps, Kral (1974), Bortenschlager (1984), Wahlmüller (1993), Bortenschlager et al. (1996), Culiberg and Šercelj (1996) and Kofler and Oeggl (2001).

Transects: For the eastern and the entire Alps, Kral (1974, 1979); for the central Alps, Welten (1982b) and for the Younger Dryas, Ammann et al. (1993). For time-altitude diagrams see Welten (1952), Schneider (1985b).

Pollen maps: de Beaulieu (1977) and Burga and Perret (1998).

Data-model comparisons are presented by Heiri et al. (2006), Henne et al. (2011) and Schwörer et al. (2014b).

Climate reconstructions: Lotter et al. (1992b), Livingstone et al. (1999) and Heiri et al. (2003, 2004, 2014, 2015).

Phylogeography: Stehlik et al. (2002), Stehlik (2003), Tribsch and Schönswetter (2003), Tribsch (2004) and Thiel-Egenter et al. (2011).

Tf, The Carpathians

These mountains that on maps of Europe have a distinctive crescent-like shape reach elevations of 2663 m asl in the High Tatra in the north-west of their range and 2544 m asl in the south. The neighbouring regions of central Europe (Tb), eastern Europe (Tc) and south-eastern Europe (Tg) all belong to the zone of deciduous forests. The lower Bihar mountains, enclosed by the Carpathians, can be included in this Tf region. In contrast to the Pyrenees and the Alps the elevations are lower and, therefore, the modern vegetation of the high-montane and subalpine forests dominated by conifers (with *Picea abies, Larix decidua, Pinus cembra* and *P. mugo*) is less extensive; the summergreen forests of oaks, beech or beech and fir occupy vast areas in the colline and lower montane belt. Plant geographically, the region corresponds to the Carpathian subregion of Meusel et al. (1965) which the authors include partly in the central European and partly in the Pontic/south-Siberian floristic regions.

In comparison with the Alps the glaciation of the Carpathians was not extensive: local glaciers developed in the northern Carpathians in the High Tatra and Czernahora and in the southern Carpathians (Transilvanian Alps). The ice had probably already retreated into the cirques during the early Late-glacial.

Late-glacial and Holocene: in the eastern Carpathians the crater lake of St Anne (946 m asl) provided a record of the vegetation between c. 26 ka and 8 ka BP (Magyari et al., 2006, 2009b, 2014). Two lines of evidence suggest nearby refugia of tree taxa: (1) during the LGM c. 23–19 ka BP pollen of coniferous and deciduous trees occurred nearly continuously and at higher frequencies, possibly indicating woodlands in lower montane zones; (2) at the elevation of the lake, macrofossils of *Betula nana* and *B. pubescens* were found as early as 15 200 cal yr BP, *Pinus sylvestris* at 14 750 cal yr BP and *Larix decidua* at 12 920 cal yr BP, suggesting short migration routes. A third line of evidence for refugia comes from phylogeographical studies, see e.g. Grivet and Petit (2003) or Höhn et al. (2009).

Forests were established during the Bølling as recorded at several sites: Avrig (Tanţău et al., 2006), Luci (Tanţău et al., 2003a), Steregoiu (Björkman et al., 2003), Preluca Tiganului (Björkman et al., 2003); further north at Tarnawa Wyżna afforestation oc-

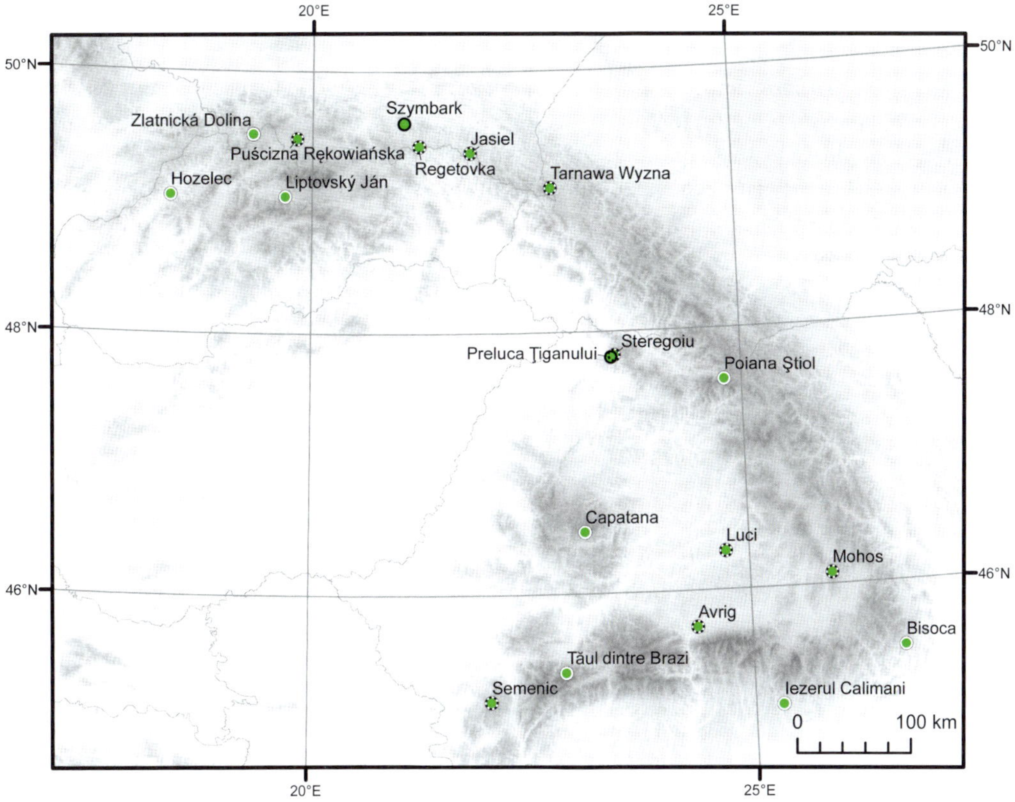

Fig. 4.3-12a. Map of selected sites in the Carpathians (Tf) region. Dots with black borders indicate sites shown in the histograms (Fig. 4.3-12b), white borders indicate sites used for the maps showing the history of selected taxa (see chapter 4.4) and dashed borders indicate sites used for both.

curred during the early Allerød (Ralska-Jasiewiczowa, 1972, 1989). These forests were characterized by conifers, namely *Pinus sylvestris, P. mugo, P. cembra, Picea abies* and *Larix* (macrofossils at Steregoiu, Preluca Tiganului, pollen at Tarnawa Wyżna); at some sites *Juniperus* was frequent (Tarnawa Wyżna, Steregoiu). *Betula* was only subdominant. In some Romanian sites *Ulmus* and *Picea* played a role as early as the Allerød (macroremains of *Picea* in Preluca Tiganului, Steregoiu). The Younger Dryas at these mid-elevation sites was often recorded as a depression in *Picea* and at some sites a depression in *Ulmus* pollen: at the low-elevation site Turbuta (275 m asl) *Ulmus* and *Picea* survived during the Younger Dryas (Feurdean et al., 2007a). At all sites vegetation became more open (higher abundances of *Artemisia* and other NAP) during this cooler and drier phase.

At the onset of the Holocene *Ulmus* re-expanded in the eastern and southern Carpathians and spread in the western Carpathians (Poland, Slovakia). *Quercus, Tilia, Fraxinus* and *Alnus* followed. *Corylus* expanded at some sites either more or less simultaneously with *Ulmus* (for example at Semenic; Rösch and Fischer 2000) or distinctly later (as at Avrig, Luci and Preluca Tiganului). During the second half of the Holocene the abundance of *Carpinus betulus* at many sites is remarkable: at some sites its maximum was distinctly before the increase of *Fagus* (Semenic, Avrig, Mohos), at other sites more or less in parallel (for example at Jasiel, Tarnawa Wyżna and Steregoiu, see Fig. 4.3-12b). The second half of the Holocene is characterized by the abundance of *Fagus sylvatica* and at some but not all sites of *Abies alba*. At higher elevations *Picea abies* and at some sites *Alnus viridis* were the dominant or subdominant taxa (*Picea* stomata at Iezerul Calimani at 1650 m asl; Fărcaş et al., 1999). Maps for the distribution and pollen abundance of

Pinus, *Picea* and *Quercus* are provided by Feurdean et al. (2011).

Human impact: activities in human settlements involving pasturage and/or agriculture can be recognized from at least the Neolithic (depending on elevation; Ralska-Jasiewiczowa, 1972a,b, 1989; Szczepanek, 1989; Wacnik, 1995; Feurdean et al., 2013a; Tanţău et al., 2014).

Overviews are provided by Ralska-Jasiewiczowa and Latałowa (1996), Rybníčková and Rybníček (1996) and Feurdean et al. (2007b, 2010, 2011, 2012, 2013a). A treasure trove of literature for the Carpathian region including both the mountains and the plains is the review by Buczkó et al. (2009).

Isopollen maps for 13 woody taxa in the western Carpathians are provided by Obidowicz et al. (2013).

Climate reconstructions: Feurdean et al. (2014), Tanţău et al. (2014) and Ruskál et al. (2020).

Phylogeography: Ronikier (2011).

Puścizna Rękowiańska *Poland* (Obidowicz 1989, 1990)

Tf 49.5° N, 19.8° E, 656 m a.s.l.

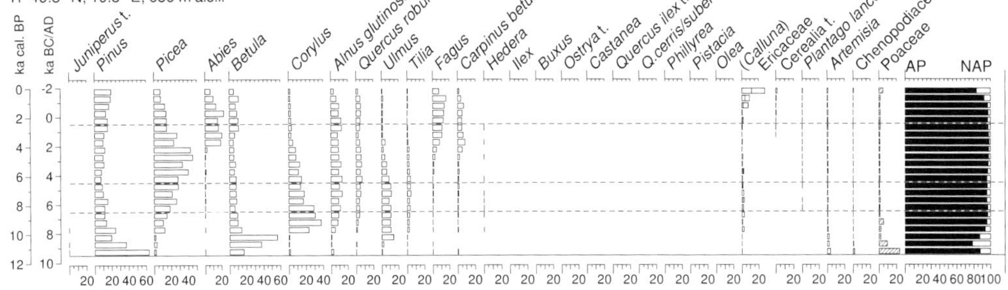

Szymbark *Poland* (Gil et al. 1974)

Tf 49.6° N, 21.1° E, 465 m a.s.l.

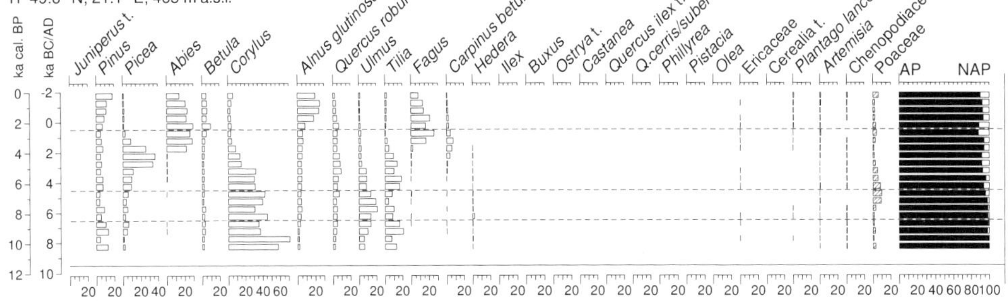

Regetovka *Slovakia* (Wacnik 1995)

Tf 49.4° N, 21.3° E, 515 m a.s.l.

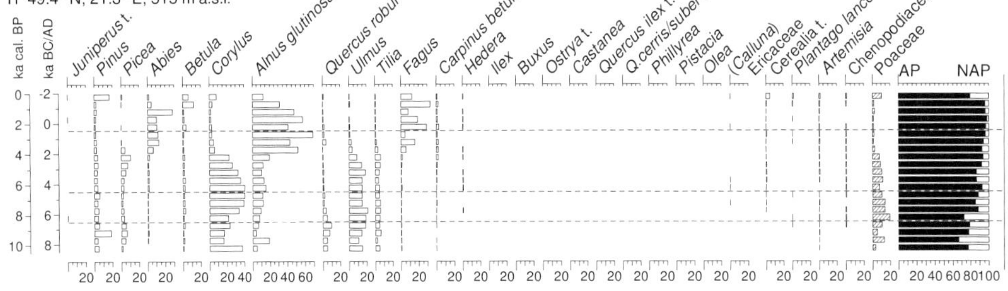

Jasiel *Poland* (Szczepanek 1987)

Tf 49.4° N, 21.9° E, 680 m a.s.l.

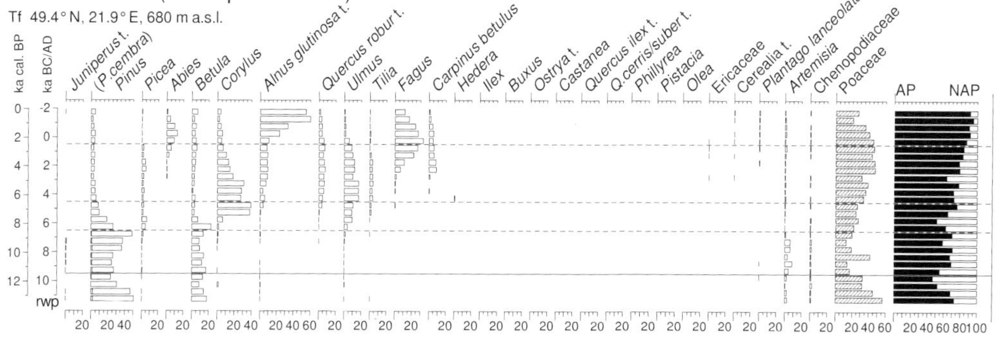

Tarnawa Wyzna *Poland* (Ralska-Jasiewiczowa 1972, 1980, 1989)

Tf 49.1° N, 22.8° E, 670 m a.s.l.

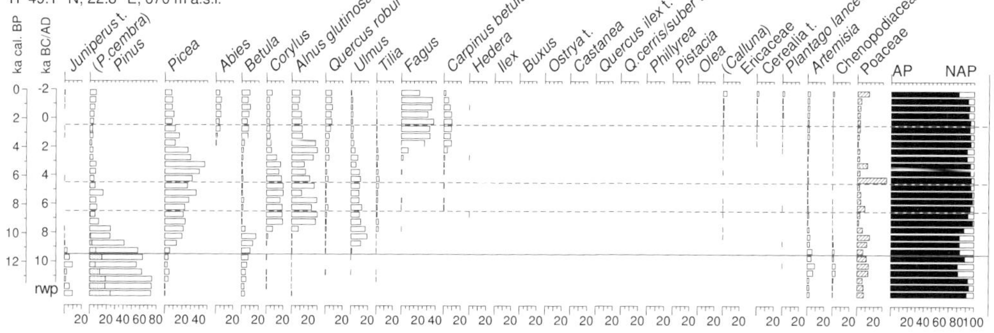

Preluca Tiganului *Romania* (Feurdean and Bennike 2004; Feurdean 2005)

Tf 47.8° N, 23.5° E, 740 m a.s.l.

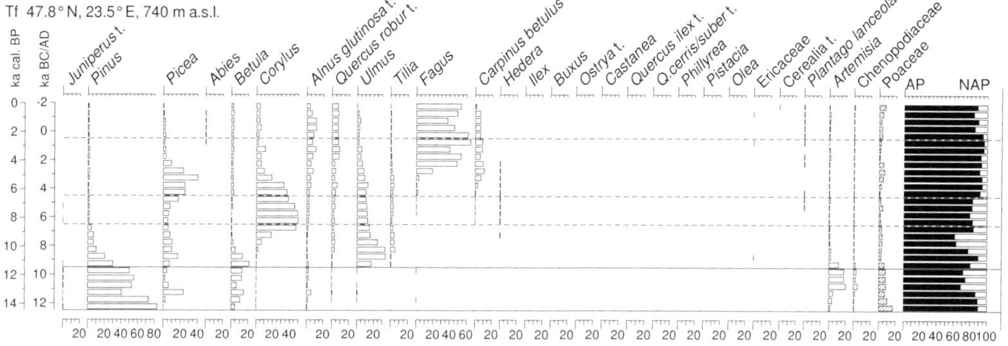

Steregoiu *Romania* (Björkman et al. 2003; Feurdean and Bennike 2004)

Tf 47.8° N, 23.5° E, 790 m a.s.l.

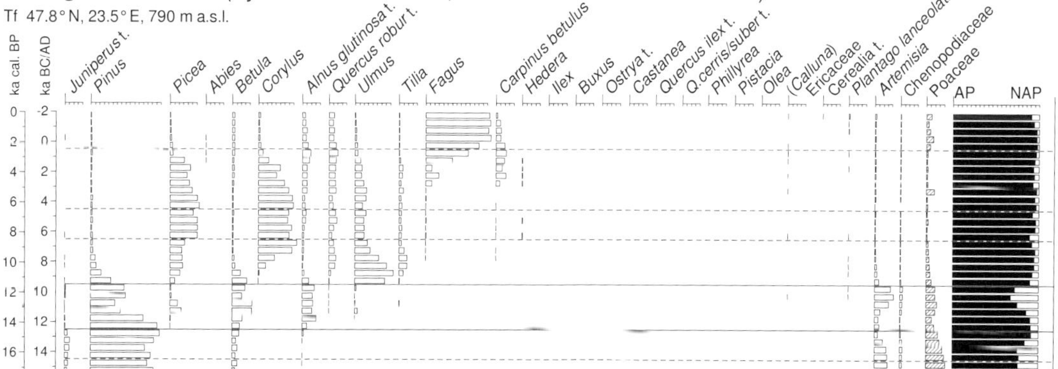

Luci *Romania* (Tantau et al. 2014)

Tf 46.3° N, 24.7° E, 1080 m a.s.l.

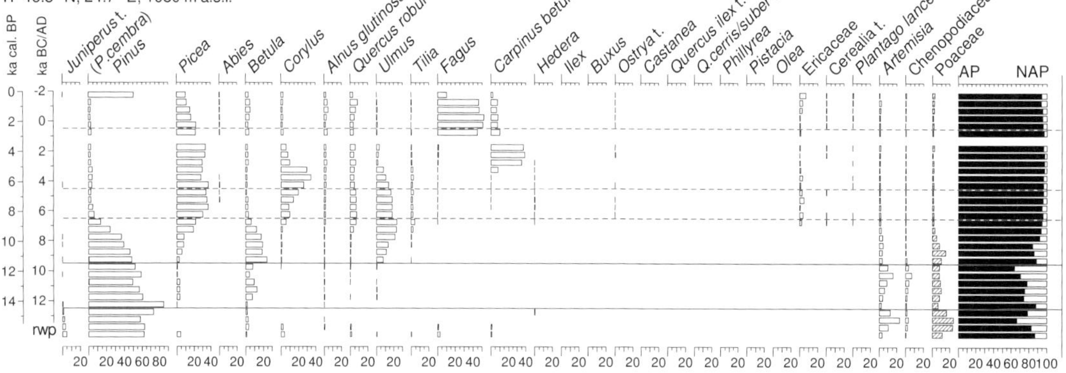

Mohos 1 *Romania* (Tanţău 2003; Tanţău et al. 2003)

Tf 46.1° N, 25.9° E, 1050 m a.s.l.

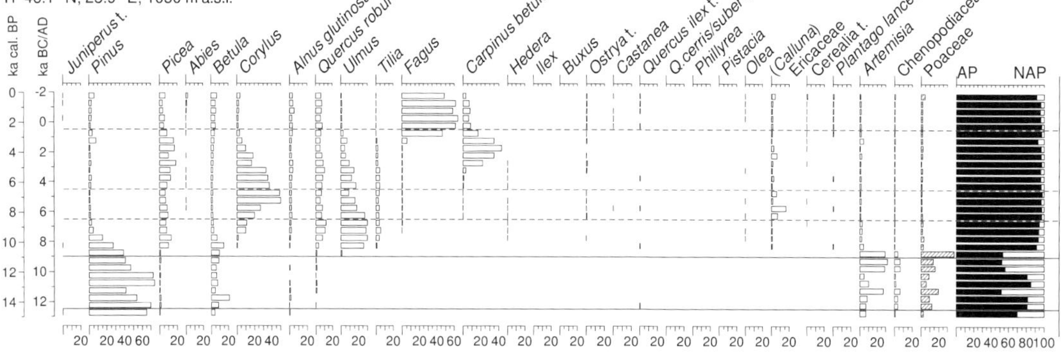

Avrig *Romania* (Tanţău 2003; Tanţău et al. 2006)

Tf 45.7° N, 24.4° E, 400 m a.s.l.

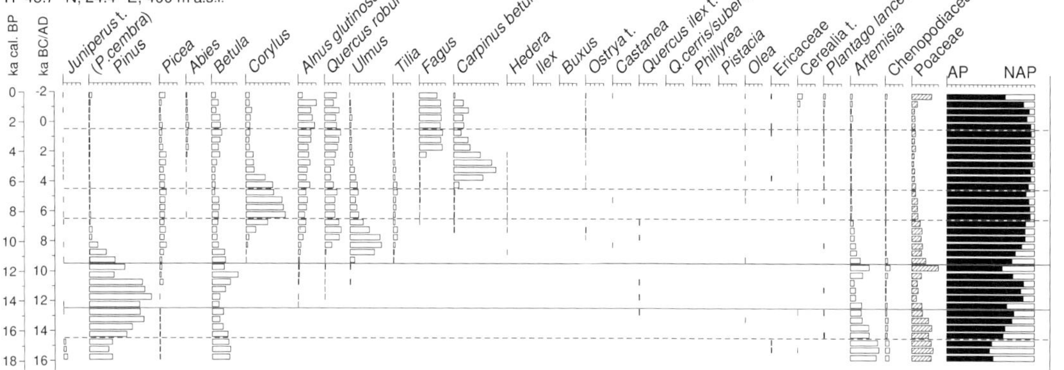

Semenic *Romania* (Rösch and Fischer 2000)

Tf 45.2° N, 22.1° E, 1400 m a.s.l.

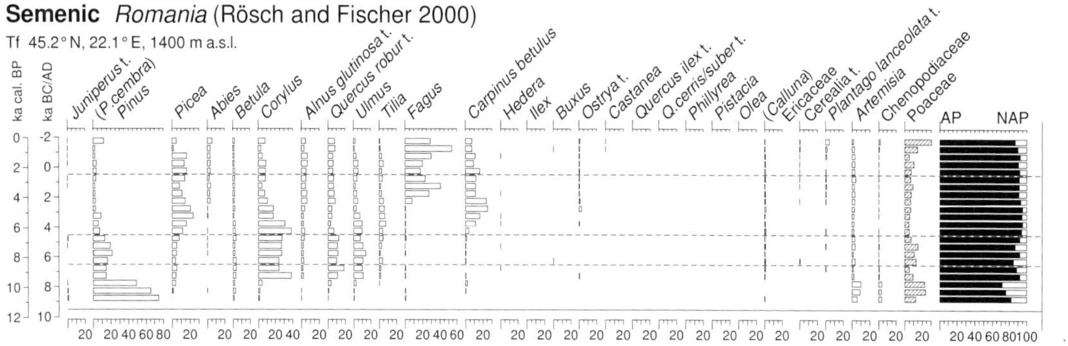

← **Fig. 4.3-12b.** Selected sites in the Carpathians (Tf) region. Late-glacial and Holocene pollen records are presented as histograms at a 500-year resolution, arranged from noth-west to south-east. At Preluca Tiganului and Steregoiu, macrofossils of *Betula* cf. *alba*, *Pinus* and *Picea* and at Preluca Tiganului also *Larix* were recorded in the Late-glacial before the Younger Dryas, suggesting nearby glacial refugia (Wohlfarth et al., 2001; Feurdean and Bennike, 2004; Feurdean et al., 2007b; Tanțău et al., 2014). For the Holocene, the strong and early development of *Carpinus* (before *Fagus*) is obvious in the pollen records (see also Fig. 4.4.3-9 and the chapter on *Carpinus*).

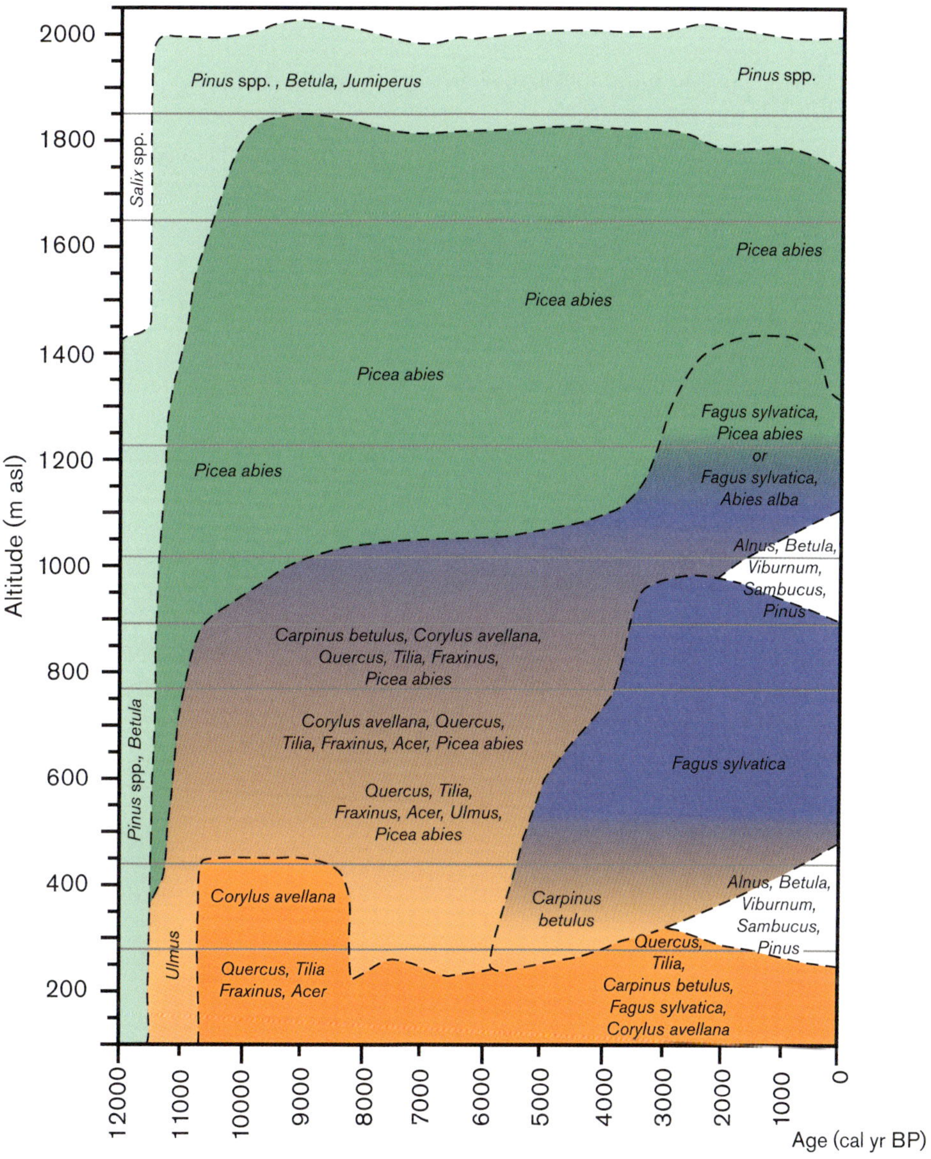

Fig. 4.3-12c. Altitude-time diagram for the Romanian Carpathians. Main forest constituents of the last 12 000 years at nine sites from <300 m asl (forest steppe) to the subalpine belt near 2000 m asl dominated today by *Pinus mugo* ssp. *mugo* with some scattered *Pinus cembra*. Modified from Feurdean et al. (2013a).

Tg, Temperate south-eastern European region

This region includes the south-eastern part of the temperate zone of deciduous forests. The limits are: towards the north, the Pannonian lowlands and the Carpathians, towards the south-west, a line parallel to the Adriatic coast (leaving a narrow strip of Mediterranean vegetation near the coast), towards the east, the coast of the Black Sea (in Bulgaria and Turkey), and towards the south along a diffuse line across the Balkan Peninsula (through Albania and central Greece to the Marmara Sea). Within these limits are landscapes with a variety of modern vegetation types and differing forest histories but, in order to avoid too much splitting, we discuss this region as one unit. Of special importance are the Dinaric Alps (up to 2571 m asl), the Balkan mountains, the Rodope mountains (up to 2925 m asl), the Greek mountains such as the Pindus mountain range, the Boras mountains or the complex of Mount Olympus (up to 2917 m asl). In these mountains the modern vegetation is characterized at low to middle elevations by *Fagus sylvatica* and *Abies alba*, at higher elevations by coniferous forests with *Picea abies, Pinus nigra* and *Pinus sylvestris*, and locally with endemic *Picea omorika*, *Pinus peuce* and *P. heldreichii*. In the subalpine belt *Pinus mugo,* in its prostrate form, is widespread. In contrast to the mountains the lowlands south of the Carpathians and the Bulgarian lowlands (between the Balkan mountains and the Rhodope mountains) have a modern vegetation dominated by *Quercus pubescens, Q. frainetto, Q. dalechampii* and *Q. cerris.*

This complexity in modern vegetation makes it clear that this region cannot form a discrete biogeographical unit. In the plant-geographical subdivision of Meusel et al. (1965) the northern part belongs to the Pannonian and Danubian floristic province (part of the Pontic–south-Siberian flora region), while the southern parts (Illyria, Albania and the Balkan parts of Tg) belong to the Macedonian–Thracian floristic province (which is part of the submediterranean sub-province of the Macaronesian–Mediterranean floristic province).

Geomorphological traces of the last glaciation are found at the highest elevations of the mountains but they are strongly restricted. Outside the local glaciations the origin of the basins is either tectonic subsidence (Ioannina, Xinias, Tenaghi Philippon), sinkholes in limestone (often combined with subsidence, such as Nisi Fen; Christanis, 1994), coastal lagoons or palaeochannels of rivers (oxbow lakes).

During the LGM no major areas were ice-covered but local glaciers occurred at higher elevations on the mountains providing potential coring sites in a group of glacial lakes (for example Rila and Pirin mountains, Boras and Pindus mountains). Quite a number of marine cores from the Black Sea have also been studied palynologically (Shopov et al., 1992; Atanassova, 1995; Filipova-Marinova, 2003) but because of their very different (quite vast) pollen catchment area we do not include the deep-sea cores here with the terrestrial sites from lakes and mires in the histograms of figure 4.3-13b. The coastal cores, however, provide the history of sea-level changes that is important for archaeology (Filipova-Marinova, 2003; Peev, 2009). Two very important terrestrial sites reach back to about 22 000 cal yr BP: Ioannina in the west (Bottema, 1974; Tzedakis, 1993; Tzedakis et al., 2002; Lawson et al., 2004) and Tenaghi Philippon in the east (Wijmstra, 1969; Tzedakis et al., 2001, 2003; Tzedakis, 2005). The record of tree taxa during the full glacial and early Late-glacial is very different at these two sites (Fig. 4.3-13b). Some tree species survived the last glacial period in refugia at intermediate elevations with orographic precipitation, that is between a lowland that was too dry and mountains that were too cold (Beug, 1975a; Denèfle et al., 2000; Bennett and Provan, 2008; Bhagwat and Willis, 2008; Birks, 2015).

During the early Late-glacial, steppes dominated by Chenopodiaceae and *Artemisia* were widespread. Sites such as Tenaghi Philippon, Xinias and Nisi Fen (Wijmstra, 1969; Bottema, 1974; Digerfeldt et al., 2000; Lawson et al., 2005) can be compared with marine cores from the Aegean Sea (Rossignol-Strick, 1995; Kotthoff et al., 2008; bisaccate pollen types are excluded from the diagrams in Kotthoff et al., 2008). The onset of the Bølling in western and central Greece (Ioannina: Bottema, 1974; Lawson et al., 2004; Nisi Fen: Lawson et al., 2005) is marked by open forests of *Pinus* and *Quercus*. The two sites are at the same elevation of c. 475 m asl, so temperatures may have been similar; but precipitation is (and most probably was) higher in western Greece than further east, thus Ioannina is less continental than Nisi Fen. This may explain the higher Late-glacial abundances of deciduous *Quercus* at Ioannina (moisture availability). On the Drama plain of north-eastern Greece (Tenaghi

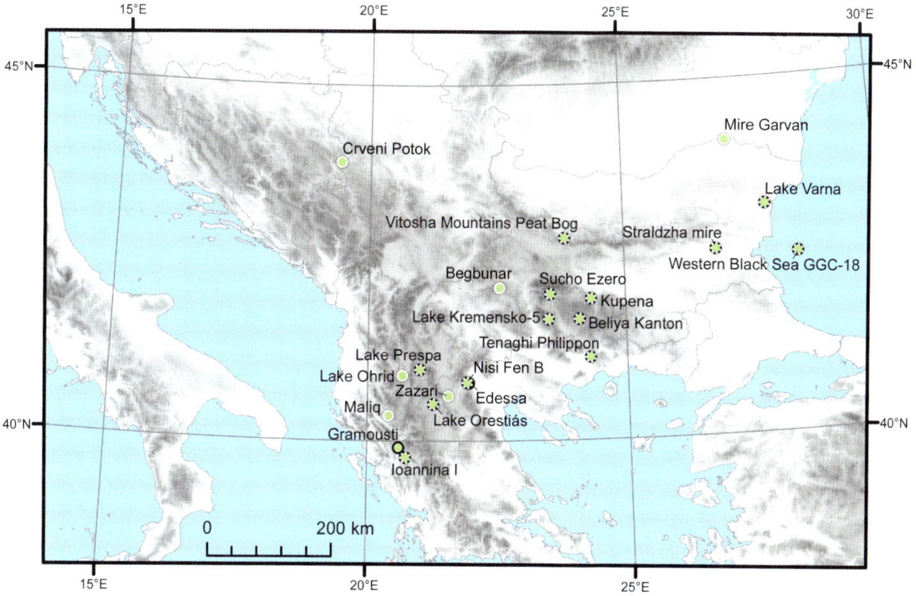

Fig. 4.3-13a. Map of selected sites in the temperate south-eastern European (Tg) region. Dots with black borders indicate sites shown in the histograms (Fig. 4.3-13b), white borders indicate sites used for the maps showing the history of selected taxa (see chapter 4.4) and dashed borders indicate sites used for both.

Philippon, only 40 m asl) deciduous *Quercus* dominated the Bølling and Allerød chronozones. At this site *Tilia, Ulmus, Fraxinus* and *Pistacia* were already present during the Late-glacial interstadial.

Also in the montane belt, at many sites the establishment of forest started during the Late-glacial interstadial (Bølling-Allerød), such as around Lake Prespa (Panagiotopoulos et al., 2013) and in the southern Bulgarian mountains where the dominant genus was *Pinus*. According to macrofossil records at Kremensko-5 and Besbog (2124-2250 m asl) in the Pirin mountains this included not only species of the section Diploxylon but also *P. peuce*, a species of the section Haploxylon. *Juniperus* and *Betula pendula* are also recorded as macrofossils from the Bølling-Allerød (Stefanova et al., 2006a). The reason for the early and rapid development of tree taxa may be the proximity of refugial areas. At lower elevations deciduous taxa occurred such as *Quercus, Ulmus, Tilia, Fraxinus, Carpinus betulus, C. orientalis* and *Corylus* together with *Abies* and possibly also *Fagus*.

The Younger Dryas is strongly expressed in the west at the intermediate elevation of Lake Ohrid and Lake Prespa (Lézine et al., 2010; Panagiotopoulos et al., 2013) and at high elevations near the timberline in the Rila mountains (Tonkov et al., 2011) and the Pirin mountains (Stefanova et al., 2006a). At the sites Besbog-3 and Kremensko-5 the YD is also well reflected in the plant macrofossil record. Sediment sequences at lowland sites do not usually reach the YD, except for Straldzha (Thracian Plain), where the YD is marked by the absence of a pollen signal and where semi-desert conditions lasted from about 17 900 to 10 300 cal yr BP (15 900–8300 BC). Connor et al. (2013) explained: "The timing and duration of Late-glacial and Early-Holocene aridity in the region is strongly linked to summer insolation (Wright et al., 2003; Stefanova et al., 2006b; Tzedakis, 2007; Leng et al., 2013). The Straldzha record suggests that the summer insolation maximum overrode the effect of the North Atlantic circulation in the continental lowlands of the Balkans. The Bølling-Allerød interstadial and Younger Dryas stadial, both temperature events that are clearly registered in the Rila and Pirin Mountains (Atanassova and Stefanova, 2003; Stefanova et al., 2006a; Tonkov et al., 2008, 2011), had a minor impact on the vegetation of the Rhodope Mountains (Huttunen et al., 1992; Bozilova et al., 2011) and no substantial effect on the vegetation of the Thracian Plain ... aridity, not temperature, seems to have been the key factor limiting

Ioannina *Greece* (Lawson et al. 2004)

Tg 39.8° N, 20.9° E, 473 m a.s.l.

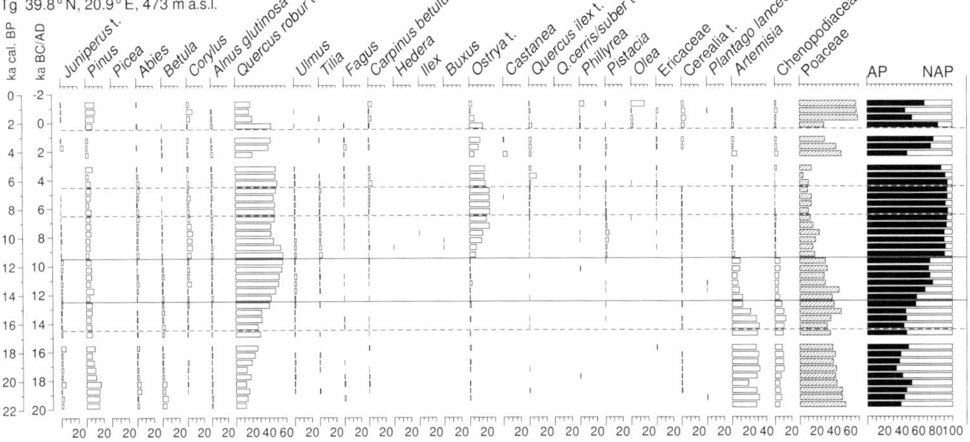

Gramousti *Greece* (Willis 1992)

Tg 39.9° N, 20.6° E, 385 m a.s.l.

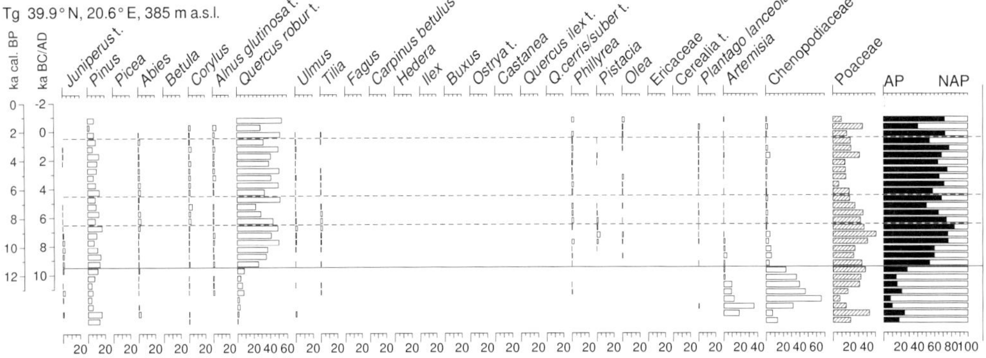

Nisi Fen *Greece* (Lawson et al. 2005)

Tg 40.1° N, 21.9° E, 475 m a.s.l.

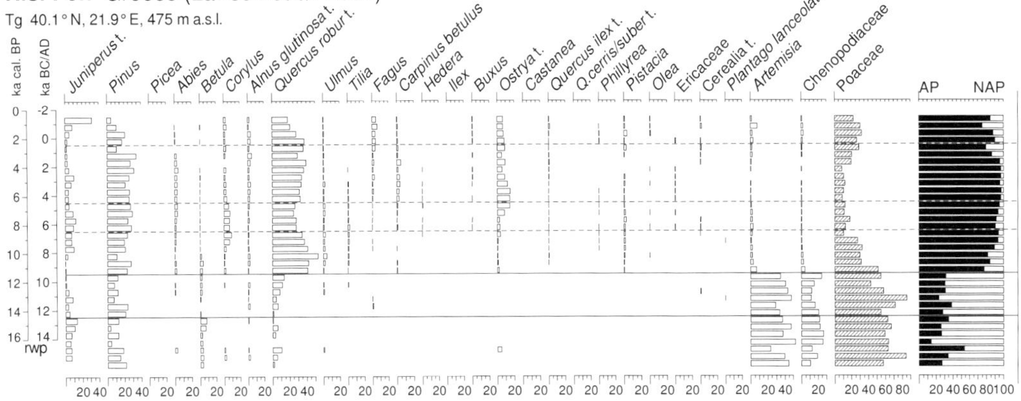

Lake Orestiás *Greece* (Kouli and Dermitzakis 2010)

Tg 40.5° N, 21.3° E, 630 m a.s.l.

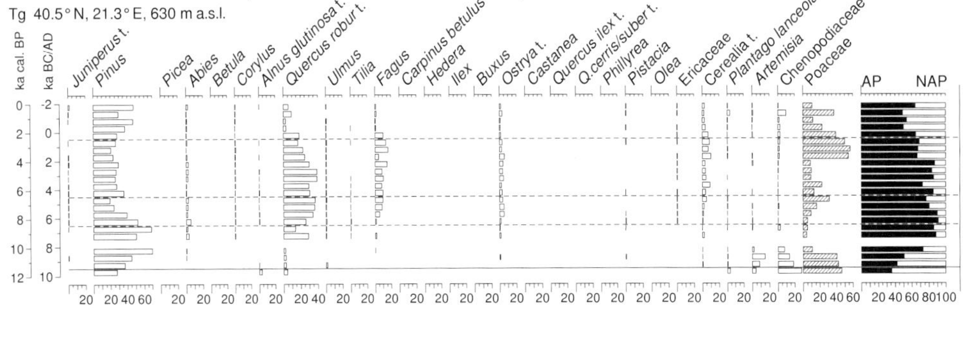

Lake Prespa *N. Macedonia* (Panagiotopoulos et al. 2013, 2014)

Tg 41.0° N, 21.0° E, 849 m a.s.l.

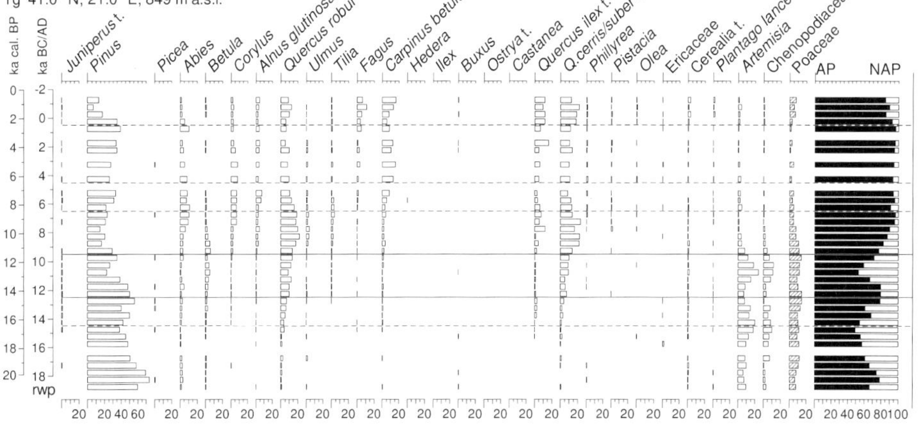

Edessa *Greece* (Bottema 1974)

Tg 40.8° N, 22.0° E, 350 m a.s.l.

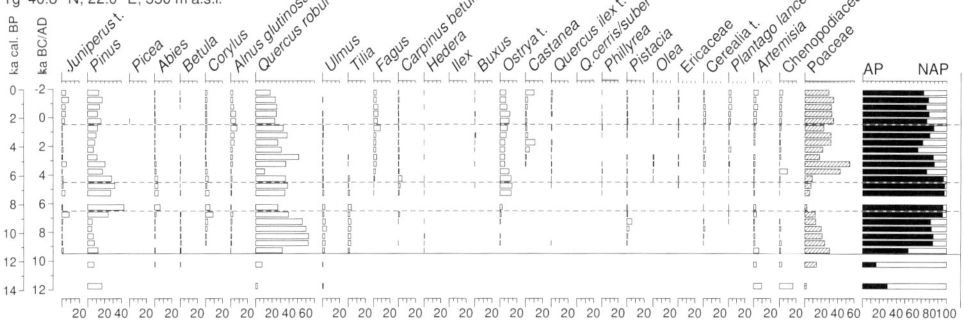

Lake Kremensko-5 *Bulgaria* (Stefanova et al. 2006)

Tg 41.7° N, 23.5° E, 2124 m a.s.l.

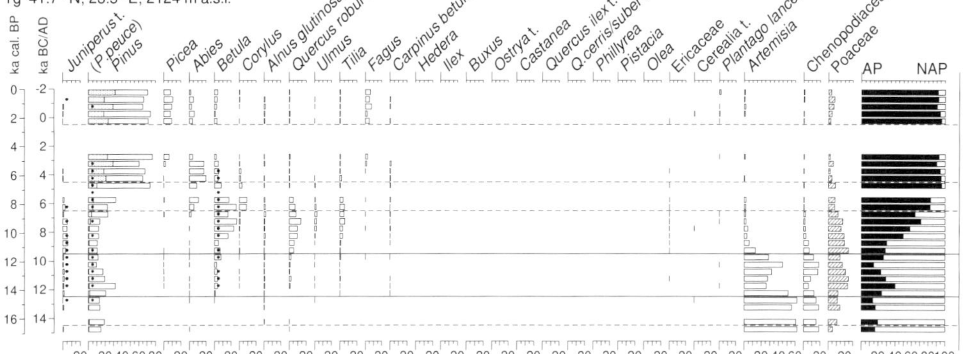

Sucho Ezero II *Bulgaria* (Bozilova and Smit 1979)

Tg 42.1° N, 23.5° E, 1900 m a.s.l.

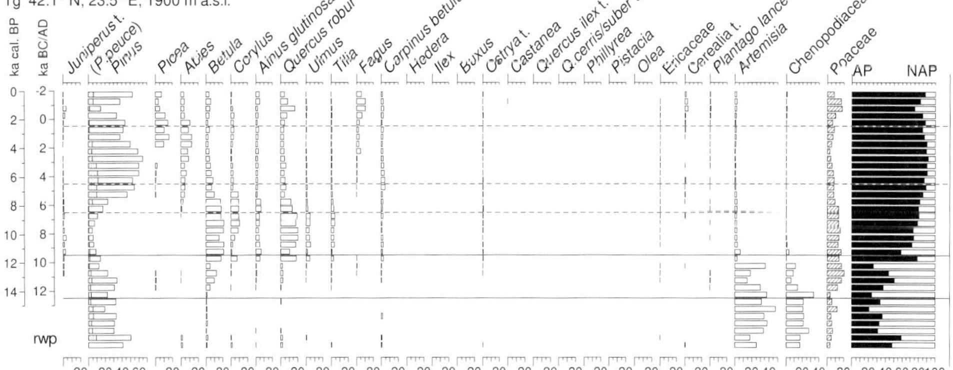

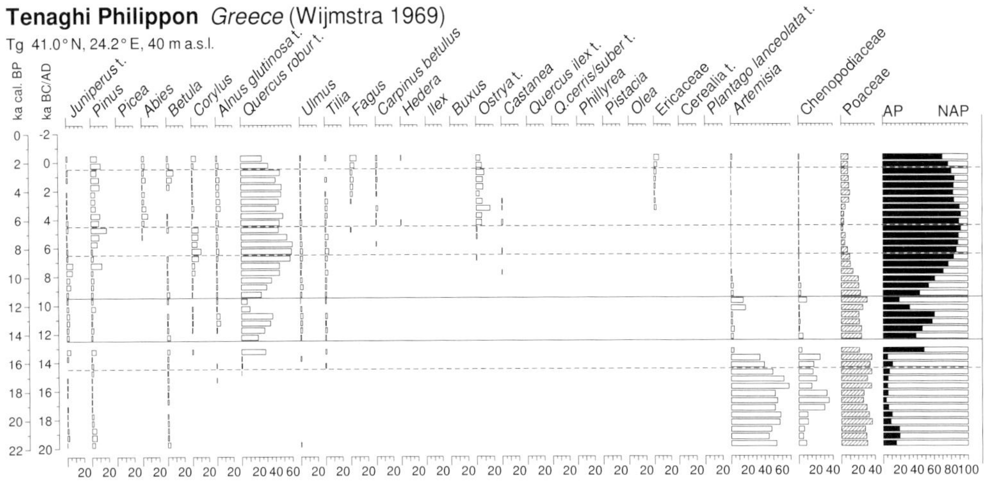
Tenaghi Philippon Greece (Wijmstra 1969)
Tg 41.0° N, 24.2° E, 40 m a.s.l.
ka cal. BP
ka BC/AD
Juniperus t.
Pinus
Picea
Abies
Betula
Corylus
Alnus glutinosa t.
Quercus robur t.
Ulmus
Tilia
Fagus
Carpinus betulus
Hedera
Ilex
Buxus
Ostrya t.
Castanea
Quercus ilex t.
Q.cerris/suber t.
Phillyrea
Pistacia
Olea
Ericaceae
Cerealia t.
Plantago lanceolata t.
Artemisia
Chenopodiaceae
Poaceae
AP
NAP

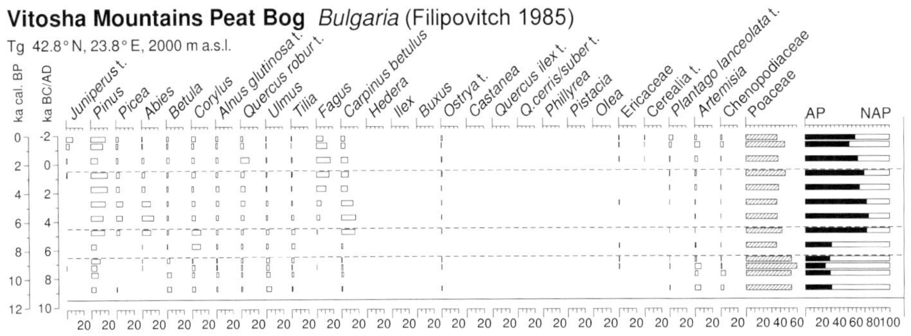
Vitosha Mountains Peat Bog Bulgaria (Filipovitch 1985)
Tg 42.8° N, 23.8° E, 2000 m a.s.l.
ka cal. BP
ka BC/AD
AP
NAP

Beliya Kanton Bulgaria (Lazarova et al. 2011)
Tg 41.7° N, 24.1° E, 1547 m a.s.l.
ka cal. BP
ka BC/AD
AP
NAP

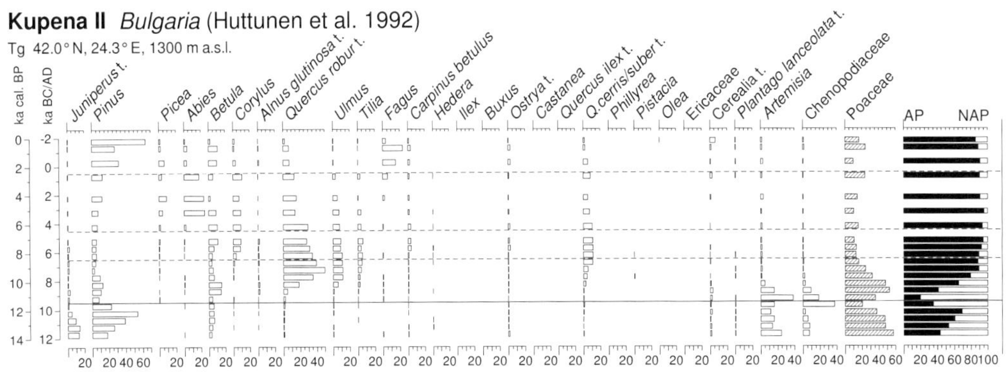
Kupena II Bulgaria (Huttunen et al. 1992)
Tg 42.0° N, 24.3° E, 1300 m a.s.l.
ka cal. BP
ka BC/AD
AP
NAP

Straldzha Mire *Bulgaria* (Connor et al. 2013)

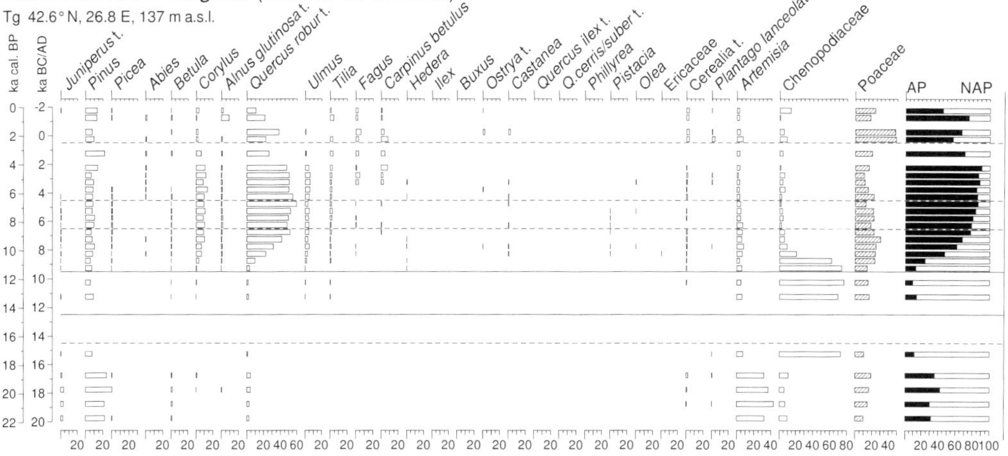

Lake Varna: Popeljanovo *Bulgaria* (Bozilova and Beug 1994)

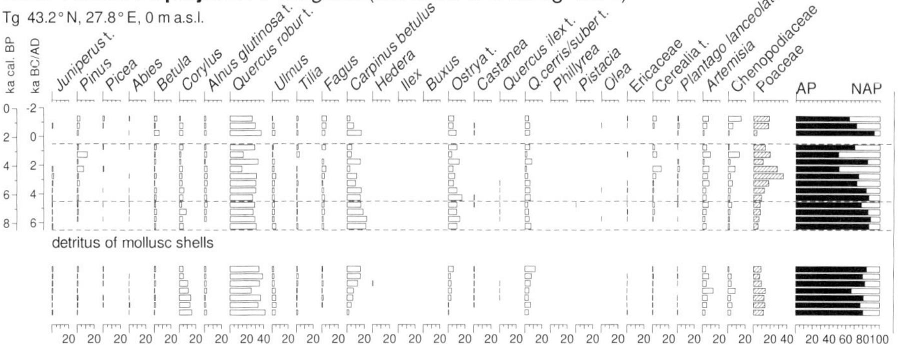

Western Black Sea: GGC-18 *Bulgaria* (Filipova-Marinova et al. 2013)

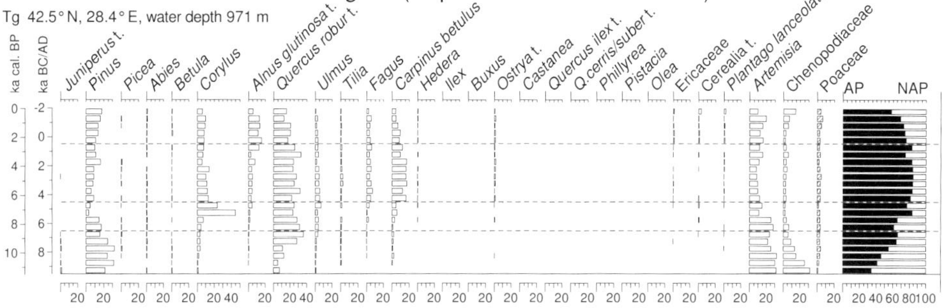

Fig. 4.3-13b. Selected sites in the cool-temperate to submediterranean south-eastern European (Tg) region. Late-glacial and Holocene pollen records are presented as histograms at a 500-year resolution. For the mountain sites of Sucho Ezero II and Kremensko-5 the portion of the Haploxylon type of *Pinus peuce* is indicated in *Pinus*. The dots in the histogram of Kremensko-5 show where macrofossils were found in the Late-glacial (and Holocene) sections, an indication for nearby glacial refugia of *Juniperus*, *Pinus* and *Betula*. At several sites the long, even Late-glacial records of the Cerealia type illustrate the problems around the identification of some pollen types in the Poaceae (e.g. *Glyceria*, *Hordeum*, *Lygeum spartum*, see Beug, 2004). The gradient from south to north intermingles on modern vegetation maps with the one from west to east, therefore the site of Tenaghi Philippon is important in several respects: not only for its long Late Pleistocene record (Wijmstra, 1969; Tzedakis et al., 2001, 2003) but also for its transitional situation between the eastern Mediterranean Mc region and the south-eastern European Tg region.

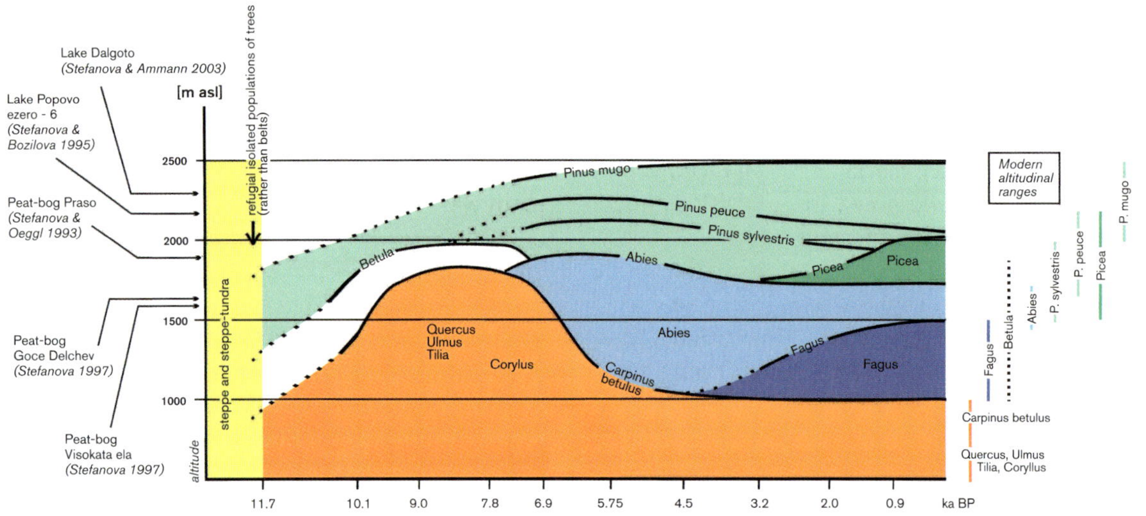

Fig. 4.3-13c. Elevation-time diagram of forest vegetation changes during the Late Glacial and Holocene in the Pirin Mountains, southern Bulgaria. Modified from Stefanova and Ammann (2003).

lowland forest development during the Straldzha semidesert phase."

At the sites of central Greece (Gramousti: Willis, 1992) the YD-signal is weak (and at several sites the YD was not reached, exceptions are Ioannina and Nisi Fen). Lawson et al. (2004, 2005) discuss the possible causes: orographic precipitation at Ioannina and Nisi Fen may have counteracted the effect from a cold North Atlantic. Further east, at Tenaghi Philippon the signal is stronger and the pollen zone Y3 shows a very sharp *Quercus* depression. We consider the original chronology by Wijmstra (1969) to be better dated than the correlation by Rossignol-Strick (1995), where the pollen zone X5 would correspond to the Younger Dryas. The YD-signal of Tenaghi Philippon is comparable to that in marine cores.

At most sites the onset of the Holocene is the sharpest change recorded. During the first half of the Holocene forest development is characterized at low and medium elevations by a strong development of *Quercus* (*Q. robur* type and sometimes *Q. cerris* type) and only small and late contributions of *Corylus* (Fig. 4.3-13c). Later, during the Mid Holocene, the strong dominance of the deciduous, mesic-thermophilous trees reached higher elevations than today (Stefanova and Ammann, 2003). *Picea abies* and *Abies alba* were present at the latest during the Early Holocene, but both species reached only higher abundances (together with *Fagus*) during the second half of the Holocene. Therefore, the forest development of south-eastern Europe shows some similarities to that of central Europe and northern Italy, but *Betula* and *Corylus* play a minor role and *Carpinus* a major and early one. Moreover, the chronology of the forest periods is somewhat different (see Figs. 4.3-8b and 4.3-13b), possibly due to the proximity of tree refugia (if compared to central Europe, but not northern Italy).

At many places traces of **human impact** go back to the Early to Mid Holocene. Sites near the Black Sea play a role in the discussions concerning 'Noah's Flood' (Ryan and Pitman, 1998), as discussed in special issues of Quaternary International (vol. 167-168 from 2007 and vol. 197 from 2009) and the book by Yanko-Hombach et al. (2007). Under long-lasting human impact the original deciduous woodlands at low to middle elevations became deciduous brushwood, termed shiblyak, in which *Paliurus spina-christi, Juniperus oxycedrus, J. communis* and *Carpinus orientalis* are important. Its deciduous nature makes it different from evergreen Mediterranean maquis. For human impact in general see Filipova (1985), Bozilova and Filipova (1986), Lazarova (1995), Bozilova and Tonkov (1998), Marinova et al. (2012) and Filipova-Marinova et al. (2013b).

Overviews: Bozilova et al. (1996) for 9 type regions in Bulgaria.

Climate reconstruction: Bordon et al. (2009), Filipova-Marinova et al. (2013b) and Feurdean et al. (2014).

Th, Northern Italian region

This region includes northern Italy, Italian Switzerland (Ticino, southern Grisons), southern Slovenia and north-western Croatia and is characterized by broadleaved summergreen forests with submediterranean (or supramediterranean) species such as *Quercus pubescens*, *Q. cerris*, *Castanea sativa*, *Ostrya carpinifolia* and *Fraxinus ornus*. Temperate deciduous trees such as *Fraxinus excelsior*, *Tilia*, *Ulmus*, *Acer* and *Quercus petraea* are often co-dominant or dominant. In the warmest areas, where winter frost is rare, evergreen Mediterranean broadleaved species (e.g. *Quercus ilex*, *Laurus nobilis*, *Cistus salviifolius*) form northernmost stands. *Fagus sylvatica*, which today characterizes the central European region (Tb), is abundant in the mountain belt above c. 1000 m asl, whereas *Quercus robur* and *Carpinus betulus* are important in the flatlands of the Po plain. Today's vegetation composition in the Th region resembles that of the Tg region and of the supramediterranean belt above c. 500–1000 m asl in the central Mediterranean region (Mb). The Late-glacial and Holocene vegetation dynamics are markedly different from those of adjacent areas (e.g. Mb, Pa, Tb, Te or Tg), leading to a clear delimitation of this European vegetation history region.

During the last glaciation most of the southern Alps and minor parts of the northern Apennines were covered by glaciers, whereas the Po plain was ice-free. Ice-free regions were forested (e.g. *Larix decidua*, *Picea abies*, tree *Betula*) or covered by steppe (Ravazzi et al., 2006; Kaltenrieder et al., 2009). LGM refugial areas of important temperate taxa such as *Corylus avellana*, deciduous *Quercus*, *Tilia*, *Ulmus*, *Fraxinus excelsior*, *Carpinus*, *Abies alba* and *Fagus sylvatica* were located in this region (Willis and McElwain, 2002; Kaltenrieder et al., 2009). Hilly areas in the lowlands provided favourable microclimatic conditions during coldest and driest periods (Gubler et al. 2018, chapter 4.11). Deglaciation of ice-covered areas occurred early at the onset of the Late-glacial (c. 19 000 cal yr BP; see Origlio) and afforestation of formerly ice-covered lowland areas started before the Bølling warming after Heinrich Event 1 at around 16 500–16 000 cal yr BP when temperatures increased by ca. 1–2°C (e.g. Origlio; Vescovi et al., 2007; Samartin et al., 2012; Bolland et al. 2020). Important trees that formed forests prior to the Bølling were *Pinus cembra*, *Betula*,

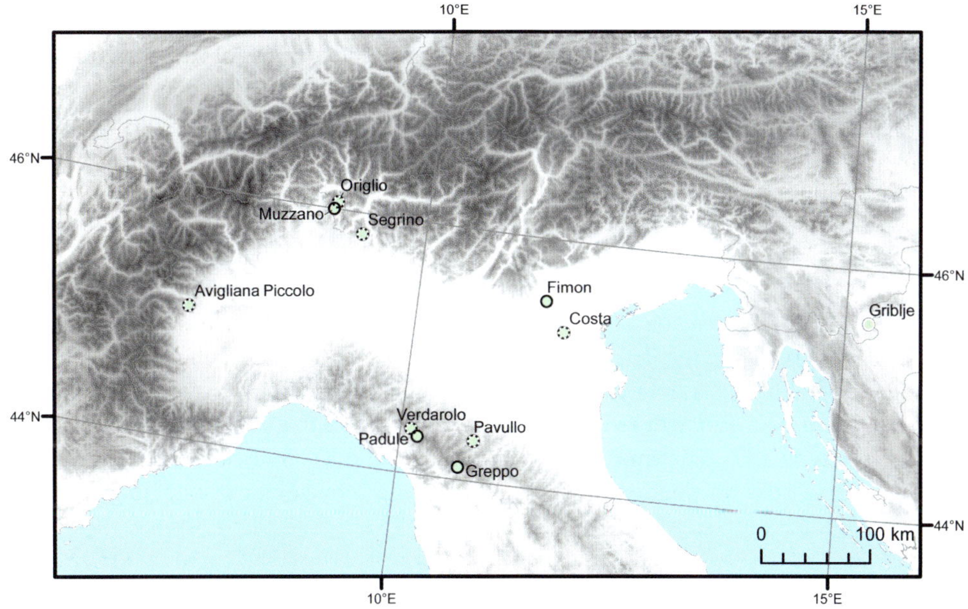

Fig. 4.3-14a. Map of selected sites in the northern Italian (Th) region. Dots with black borders indicate sites shown in the histograms (Fig. 4.3-14b), white borders indicate sites used for the maps showing the history of selected taxa (see chapter 4.4) and dashed borders indicate sites used for both.

Greppo *Italy* (Vescovi et al. 2010)

Th 44.1° N, 10.7° E, 1442 m a.s.l.

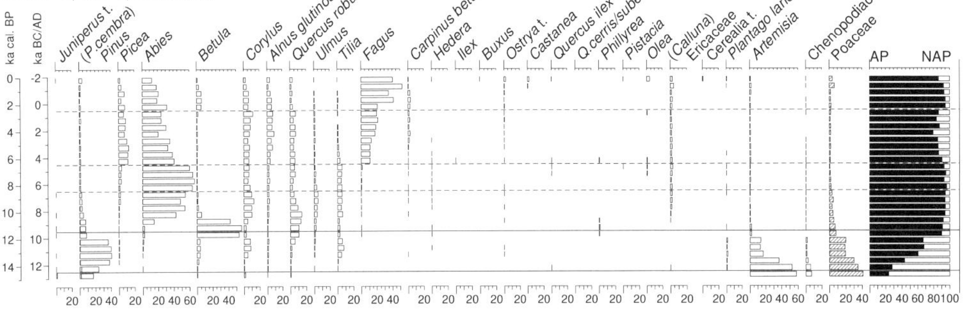

Padule *Italy* (Watson 1996)

Th 44.3° N, 10.2° E, 1187 m a.s.l.

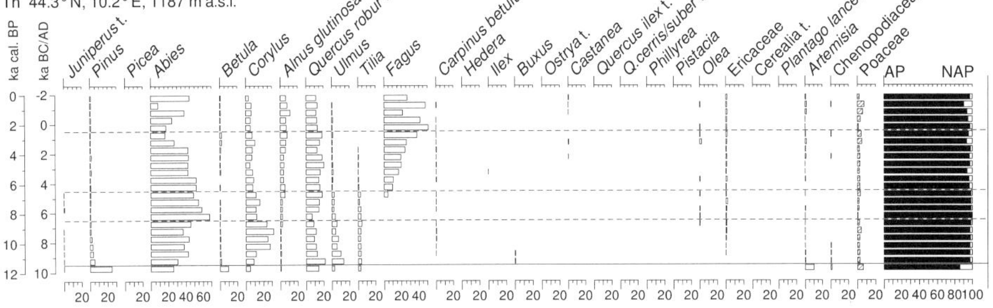

Pavullo *Italy* (Vescovi et al. 2010)

Th 44.3° N, 10.8° E, 675 m a.s.l.

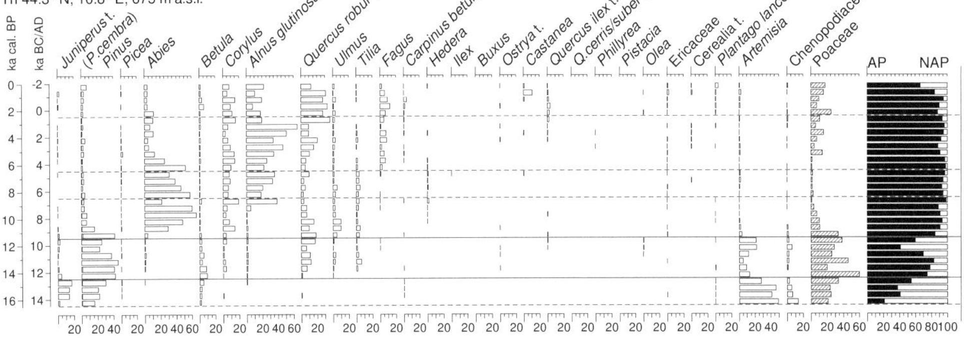

Verdarolo *Italy* (Morales-Molino et al. 2021)

Th 44.4° N, 10.1° E, 1390 m a.s.l.

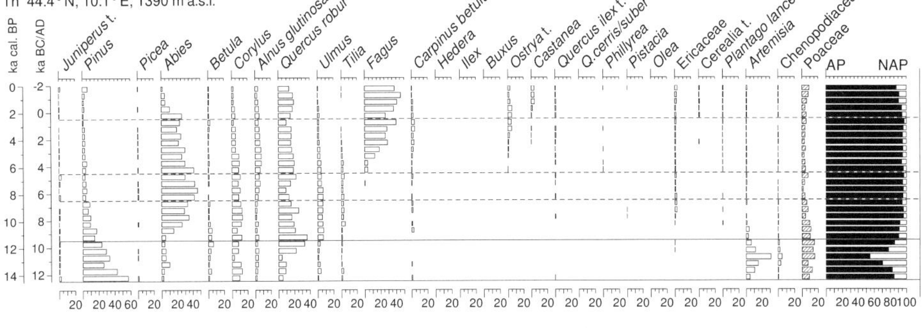

Avigliana Piccolo *Italy* (Finsinger and Tinner 2006)

Th 45.1° N, 7.4° E, 356 m a.s.l.

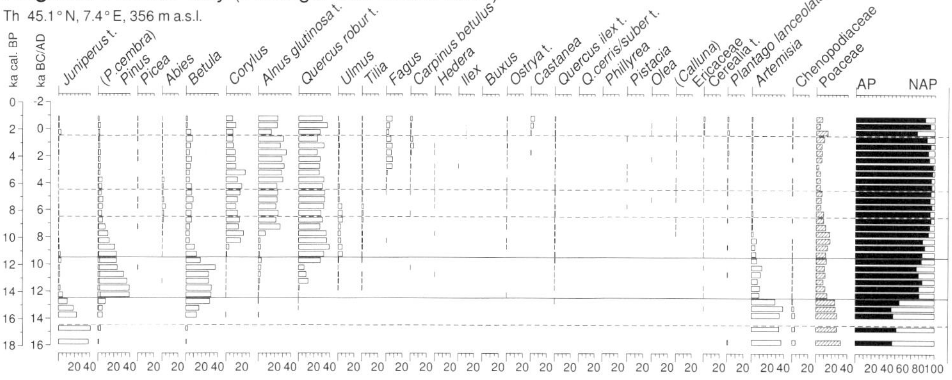

Costa *Italy* (Kaltenrieder et al. 2009, 2010)

Th 45.3° N, 11.7° E, 7 m a.s.l.

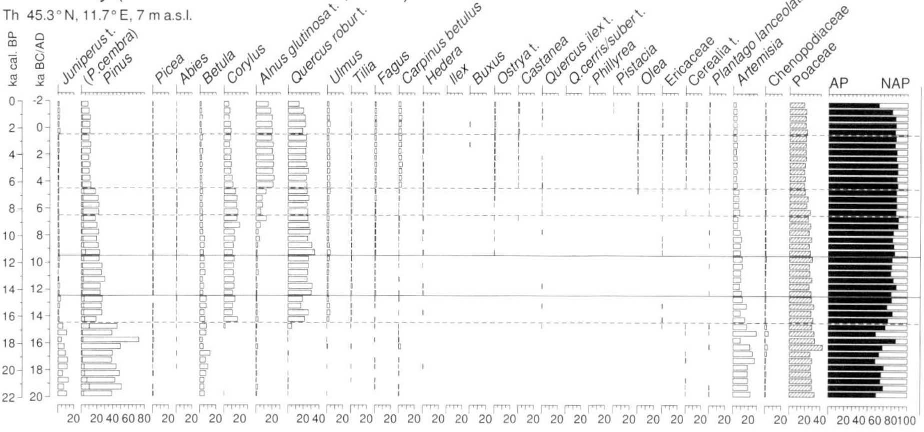

Fimon *Italy* (Valsecchi et al. 2008)

Th 45.5° N, 11.5° E, 23 m a.s.l.

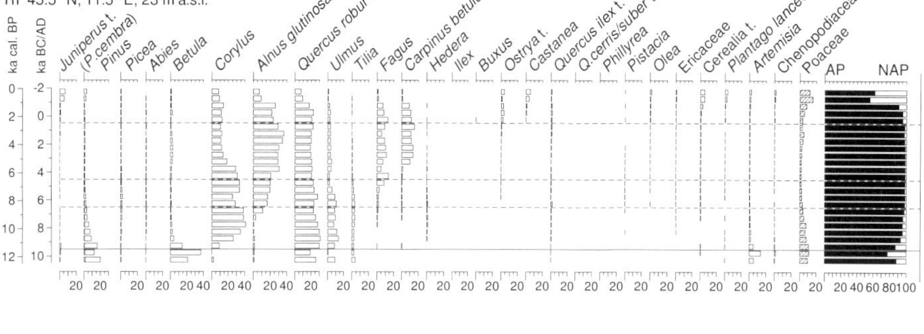

Segrino *Italy* (Gobet et al. 2000)

Th 45.8° N, 9.3° E, 374 m a.s.l.

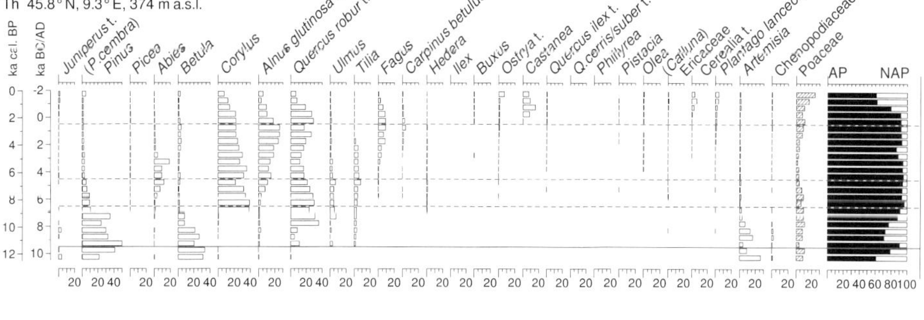

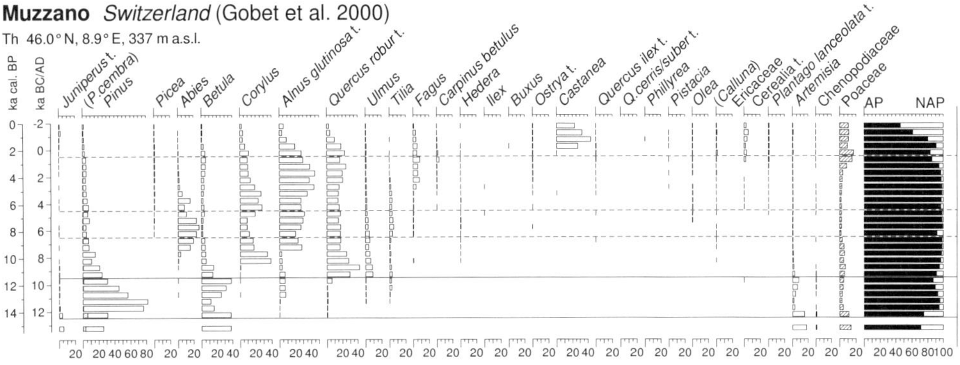

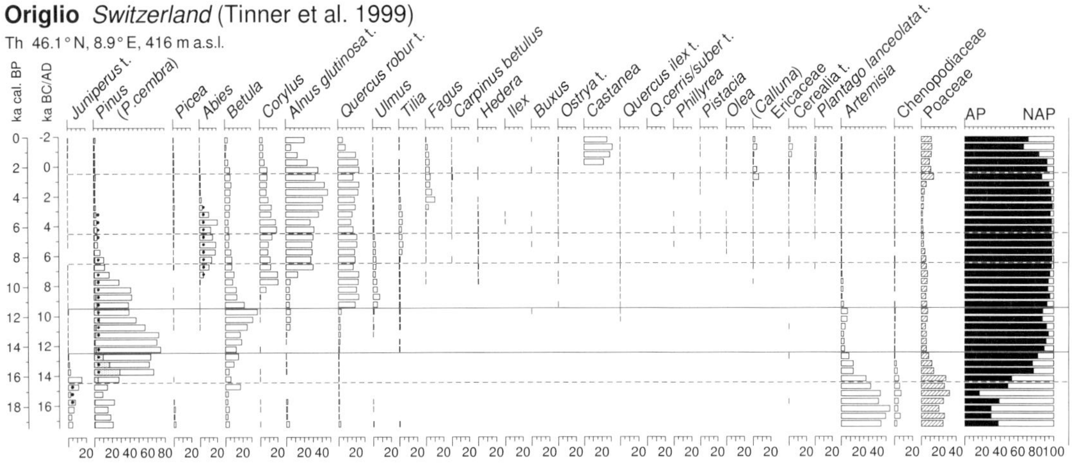

Fig. 4.3-14b. Selected sites in the temperate to submediterranean northern Italian (Th) region. Late-glacial and Holocene pollen records are presented as histograms at a 500-year resolution. It covers the southern forelands of the Alps, the northern Appenines and the Po plain down to the Mediterranean Sea, roughly corresponding to the catchment of the Po River with the exception of the eastern edge. The most unusual postglacial vegetation-history feature is the prominence of *Abies alba* in submediterranean and temperate mixed forests at c. 200–1500 m asl, while on the Po plain at c. 0–200 m (e.g. Costa, Fimon) deciduous *Quercus* pollen is more prominent. At sites above 200 m *Abies alba* pollen is replaced by *Fagus sylvatica*, deciduous *Quercus*, *Ostrya* and *Castanea sativa* during the Late Holocene, when human impact increased. Mediterranean taxa such as *Quercus ilex* and *Olea* show regular pollen occurrences during the Mid and Late Holocene at most lowland sites (e.g. Fimon, Costa, Segrino, Origlio).

Larix decidua, Juniperus communis, Pinus mugo, P. sylvestris, and *Picea abies* (Ravazzi et al., 2006; Vescovi et al., 2007) when the treeline in the region was at about 800 m asl. At the refugial site Costa *Quercus, Ulmus, Tilia* and *Corylus* also expanded at around 16 500–16 000 cal yr BP). The treeline ascended by c. 800–1000 m in response to the Bølling climate warming (Vescovi et al., 2007). Temperate taxa (e.g. *Quercus, Tilia, Ulmus, Abies alba*) already expanded during the Late-glacial from the refugial areas in the region (see Costa, Avigliana Piccolo, Origlio, Pavullo). Younger Dryas impacts on vegetation were prominent only above c. 500 m asl, suggesting that lowland vegetation was buffered against the cooling (c. 3–4°C). Nevertheless, temperate trees such as *Quercus* or *Ulmus* declined in the lowlands. Early Holocene vegetation composition was similar to that in the central European region with *Quercus, Corylus, Ulmus, Tilia, Acer*, and *Fraxinus* forming deciduous broadleaved forests, whereas *Pinus cembra, Larix decidua, Juniperus communis, Pinus mugo, P. sylvestris* and *Picea abies* migrated uphill to form the subalpine belt and timberline in the Alps (region Te). In the Northern Apennines only the highest peaks allowed boreal *Picea abies* (see Greppo) to survive locally when the climate warmed at the onset of the Holocene (Vescovi et al., 2010a). In contrast, temperate *Abies alba* became important in the lowlands and the mountains during the Early Holocene, forming species-rich forest communities together with broadleaved trees (e.g. *Tilia, Ulmus, Quercus, Acer* at Pavullo, Greppo, Verdarolo, Segrino, Muzzano, Origlio). These forests persisted until the first intense **human impact** during the Neolithic (Martinelli et al., 2017). *Fagus sylvatica* expanded during the second half of the Holocene and especially after the demise of the *Abies alba* forests (e.g. Greppo, Padule, Pavullo, Verdarolo, Segrino, Muzzano, Origlio), probably because it was more resistant to human impact such as fire and browsing (Tinner et al., 1999; Vescovi et al., 2010b). Since the Roman period about 2000 years ago *Castanea sativa* became important for fruit and timber production at sites below 1000 m asl (e.g. Origlio, Muzzano, Segrino). Human impact and specifically fire have strongly altered the vegetation composition over the past millennia. The dominance of a few disturbance adapted tree taxa in the forests (*Castanea sativa* and *Quercus* in the lowlands, *Fagus sylvatica* in the mountain belt) that we see today is primarily a result of land use and human disturbance (Gobet et al., 2000; Valsecchi et al., 2010; chapter 4.10). The past millennia have led to a strong deforestation of the area for agricultural purposes, where fire was often used not only to gain open land but also to maintain grasslands and to keep fields open. Open grasslands and *Quercus* dominance were, however, natural on the Po plain (see Costa), where braided river systems contributed to increased natural variability and disturbance (Kaltenrieder et al., 2010). Today, open lands are widespread in the lowlands but as a consequence of economic changes during the 20th and 21st century they are increasingly replaced by secondary forests in the mountains.

Climate reconstructions: Samartin et al., (2012, 2016, 2017) and Finsinger et al. (2019).

Data-model comparisons: Keller et al. (2002), Wick and Möhl (2006), Henne et al. (2015) and Ruosch et al. (2016).

Ma, Western Mediterranean region

This region comprises of the major part of the Iberian Peninsula and southern France, that is, the western part of the Mediterranean evergreen broadleaved vegetation zone at low elevations and the numerous mountain chains embedded therein. The evergreen broadleaved vegetation consists of sclerophyllous and laurophyllous trees and shrubs as well as palms. Mediterranean conifers (pines and junipers) can co-occur. Plant geographically, the very rich flora belongs to the western Mediterranean group of provinces. The region is strongly subdivided by its geomorphology: lowlands and hilly landscapes alternate with plateaux and mountains, and this exerts a strong influence on the climate. The climate shows a steep humidity gradient from the oceanic west to the much drier east: even semi-arid to arid areas are found in the south-east (substantial parts of the Almería, Murcía and Alicante provinces in south-eastern Spain) and in the central Ebro basin (in the rainshadow of the Pyrenees and the Iberian Range). This humidity gradient is visible in the modern flora and vegetation. In the oceanic west (central and southern Portugal, Extremadura in south-western Spain) and south (Andalusia, southern Spain) many species of Genisteae and *Erica* grow with evergreen *Quercus* species (such as *Quercus suber* and south-west Atlantic broom-heath shrublands). Even a few woody taxa of the laurel forests such as *Rhododendron ponticum* ssp. *baeticum, Prunus lusitanica* ssp. *lusitanica* and *Morella (Myrica) faya* occur in particularly mild and moist locations. These latter taxa are absent in the subcontinental centre of the Iberian Peninsula or in the east. Low elevation regions were never glaciated and harboured refugia for many taxa where orographic precipitation offered enough humidity during glacial periods.

Below c. 1000 m asl, the modern vegetation consists of thermomediterranean and mesomediterranean vegetation (Sainz Ollero et al., 2010). However, the vast range of latitudes (c. 36°–43° N) makes it difficult to assign fixed elevational limits to the vegetation belts. Mesomediterranean vegetation is mostly dominated by *Quercus ilex* ssp. *rotundifolia* (or *Q. rotundifolia*) woodlands, often mixed with other *Quercus* species (*Q. suber, Q. faginea* ssp. *broteroi*), and an understorey mainly consisting of evergreen shrubs (Ericaceae, Cistaceae). *Quercus suber* woodlands are also widespread on siliceous bedrock and mild and relatively moist set-

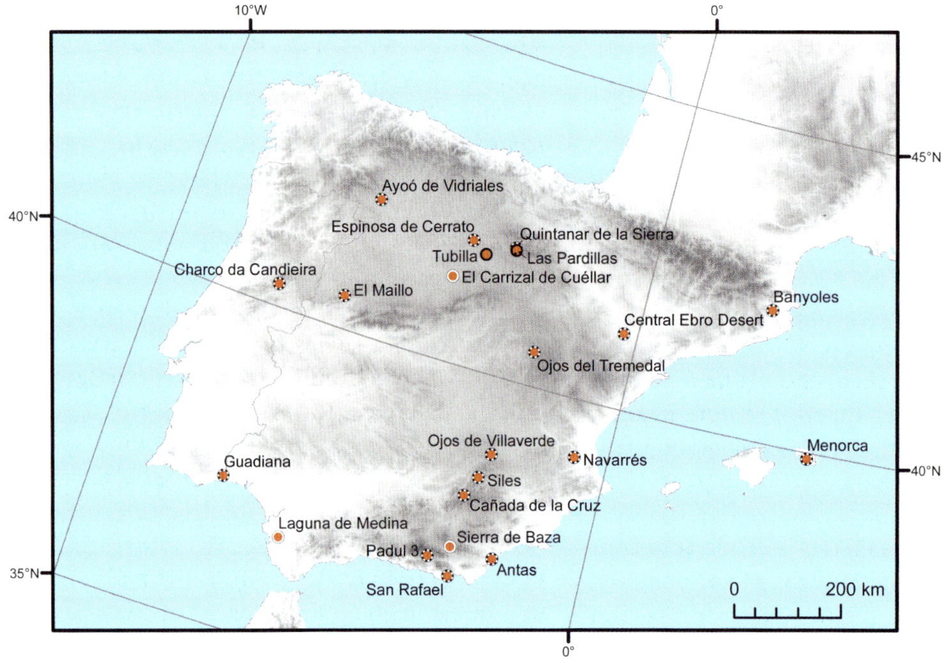

Fig. 4.3-15a. Map of selected sites in the western Mediterranean (Ma) region. Dots with black borders indicate sites shown in the histograms (Fig. 4.3-15b), white borders indicate sites used for the maps showing the history of selected taxa (see chapter 4.4) and dashed borders indicate sites used for both.

tings of the western façade of Iberia and Catalonia (north-eastern Spain), both in the mesomediterranean and thermomediterranean belts. Additionally, the thermomediterranean vegetation includes a wide array of evergreen communities like mixed evergreen forests (for example in Serra da Arrábida, Portugal, with tree *Pistacia lentiscus*, *Phillyrea latifolia*, *Quercus coccifera*, *Ceratonia siliqua* or *Viburnum tinus*), maquis, juniper (*Juniperus phoenicea* ssp. *turbinata*, *J. oxycedrus* ssp. *macrocarpa*) shrublands on coastal dunes, xerothermic scrub (*Ziziphus lotus*, *Maytenus senegalensis* ssp. *europaea*, *Periploca angustifolia*) or xerophilous pinewoods (*Pinus halepensis*, *P. pinea*). Supramediterranean and oromediterranean vegetation units are widespread at higher elevation in the inner plateau (Meseta Central) and cover many hills and several mountain chains that are characterized by a complex physiography and a high diversity of bedrocks and soils. The supramediterranean vegetation is dominated by summergreen *Quercus* forests. Nevertheless, conifers are also very important in the supramediterranean belt of the Iberian Peninsula: Pine- (*Pinus nigra*, *P. pinaster*, *P. sylvestris*) and juniper-dominated (*Juniperus thurifera*, *J. communis* ssp. *hemisphaerica*) woods are widespread on the inland Iberian mountains and high plateaux. The oromediterranean belt is also rich in coniferous forests (mostly pinewoods) including, in the very south, the endemic *Abies pinsapo* that also occurs in the supramediterranean belt.

In southern France the vegetation of the lowermost mesomediterranean belt grows up to ca. 500 m asl and is dominated by evergreen broadleaved (e.g. *Quercus ilex*, *Laurus nobilis*) and deciduous species (e.g. *Quercus pubescens*), often converted to maquis and garrigue by land use (Ozenda 1985, Beaudouin et al., 2007). In the supramediterranean belt at ca. 500–1000 m asl deciduous species (e.g. *Quercus pubescens*) are more important, sometimes together with evergreen broadleaved trees or shrubs such as *Buxus sempervirens*. In the oromediterranean (or montane) belt at ca. 1000–1600 m asl *Abies alba* and *Fagus sylvatica* are important. Finally, conifers dominate in the subalpine or altimediterranean belt, which in this book is assigned to the southernmost Alps

(Te). The treeline is located at about 2200–2400 m asl (Ozenda, 1985; Beaudouin et al., 2007).

Vegetation maps can be found not only in Bohn et al. (2003) and Sainz Ollero et al. (2010) but also in Rivas-Martínez (1987).

A note on special types of sediment: we do not include marine cores in the histograms since they are often not comparable to terrestrial records because of their low sedimentation rates and much larger pollen source area (see map in Roucoux et al., 2005). Marine records are, however, important for tracking vegetation changes on millennial timescales (Sánchez Goñi et al., 2002, 2008). In situations with a particularly high sedimentation rate allowing high-resolution analyses, marine cores can indicate synchroneities with terrestrial ones, as Fletcher and Sánchez Goñi (2008) showed by comparing the pollen records from the Alborán Sea and Padul (Pons and Reille, 1988). Pollen and charcoal records from caves (in sediment or animal droppings) are not included in this general overview because of the usual occurrence of gaps in the sequences and their unusual pollen sources, but they can be very important as records for specific periods or taxa (Carrión et al., 1999; Ochando et al., 2020).

Last Glacial Maximum: the Galician-Duero mountains and the Cantabrian Range in the north (see Ta; Jiménez-Sánchez et al., 2013; Serrano et al., 2013), the Pyrenees (treated in Td) and the Sierra Nevada (part of the Baetic ranges) in the south (Gómez-Ortiz et al., 2015), as well as several ranges of the Iberian Central System and the northern section of the Iberian Range were glaciated during the LGM (Palacios et al., 2011; Carrasco et al., 2013). For the Little Ice Age, see González-Trueba et al. (2008). Outside of the glaciated areas, the origin of the basins varies: tectonic, karstic, fluviatile, oxbow, estuary, deflation and combinations thereof. Updated maps showing the distribution of palynological study sites are provided by J.S. Carrión et al. (2010a), Postigo-Mijarra et al. (2010) and the magnificent compilation by Carrión (2012).

The Pleistocene refugia of a number of thermophilous, Mediterranean and temperate trees are documented by pollen, plant macrofossils and charcoal along the western and southern coasts and also in the Baetic ranges. For instance, on the **Atlantic coast of Portugal**, García-Amorena et al. (2007) found wood of the *Pinus sylvestris-nigra* group, partly *in situ*, dated between c. 34 and 20 uncal ka BP (c. 39–24 cal ka BP). Further, in Buraca Grande, a cave at 80 m asl in Estremadura (central Portugal), Figueiral and Terral (2002) recorded charcoal of *Pinus*, *Quercus*, Fabaceae, *Arbutus unedo*, *Buxus sempervirens*, *Olea europaea*, *Pistacia lentiscus* and *Rhamnus/Phillyrea* during the last pleniglacial. On the **Mediterranean coast of southern Iberia**, *Olea, Phillyrea, Pistacia* and *Ilex* pollen was found in Aurignacian layers (c. 32 cal ka BP) of Cueva Bajondillo (Andalusia), while charred wood of *Pinus, Quercus, Cneorum tricoccon* and *Phillyrea* was identified in Upper Paleolithic layers of Nerja Cave (Andalusia; Badal, 1996; Cortés-Sánchez et al., 2008) and charcoal fragments of several pine species, *Olea, Arbutus unedo, Pistacia lentiscus,* deciduous and evergreen oaks were found in the Mousterian and Upper Palaeolithic layers of Gorham's Cave (Gibraltar; Carrión et al., 2008). These finds are also of interest in an anthropological context because this is where Neanderthal populations survived the longest in Europe (Finlayson et al., 2006). In the **arid south-east,** two caves, the inland site Cueva Negra del Estrecho del Río Quípar (780 m asl) and the coastal site Sima de las Palomas del Cabezo Gordo (80 m asl), provide evidence of the Pleistocene persistence of temperate, Mediterranean and Ibero-North African plant taxa including, for instance, *Betula*, *Fraxinus*, *Olea, Phillyrea, Pistacia*, evergreen and deciduous *Quercus*, *Maytenus* or *Periploca* (Carrión et al., 2003b). In the **Baetic ranges**, the elevational transect in the Sierra de Segura including the sites Cañada de la Cruz (1595 m asl), Siles (1320 m asl) and Villaverde (870 m asl) suggests the close proximity of glacial refugia of *Acer, Taxus, Arbutus, Buxus, Salix, Ulmus, Phillyrea, Pistacia,* and *Olea* (Carrión et al., 2001a,b; Carrión, 2002). Maps with the location of Pleistocene refugia can be found in Carrión et al. (2008) and González-Sampériz et al. (2010) for the Iberian Peninsula, and an overview for *Olea* is presented by Y. Carrión et al. (2010d).

Late-glacial and Holocene: in the today's **thermomediterranean belt**, the Late-glacial vegetation, already during the Oldest Dryas, consisted of species-rich mosaics of steppe elements (high values of *Artemisia*, Chenopodiaceae) and woody taxa. The latter include *Olea* and *Quercus ilex* type in San Rafael (south-eastern Spain; Pantaleón-Cano et al., 2003) and *Buxus* on Menorca (Yll et al., 1997), in addition to the ubiquitous *Pinus* and *Juniperus*. During the Late-glacial interstadial (Bølling-Allerød), *Pinus*, *Juniperus* and, at some sites, *Quercus* spread. The Younger Dryas is not always a prominent feature here but is often still visible as a decline in *Quercus ilex* type, *Q. robur* type and/or *Pinus* and an increase in *Artemisia* and Chenopodiaceae (San Rafael, Guadiana estuary).

The onset of the Holocene in the thermomediterranean belt is sometimes characterized by a generalized expansion of woodland (see AP values in the histograms; Fig. 4.3-15b). Woodland composition was variable but usually dominated by evergreen taxa such as *Q. ilex* type (Guadiana, Antas), *Pinus* (Navarrés) and *Buxus-Pinus-Juniperus* (Menorca). However, steppe communities rich in xerophytes persisted during the Early Holocene in certain sites such as San Rafael (Pantaleón-Cano et al., 2003) and the Guadiana estuary (Fletcher et al., 2007). A Mid Holocene 'mesophytic optimum' starting at c. 9000–8000 cal yr BP followed (see Fig. 4.6-8 for a summary; Aranbarri et al., 2014), characterized by reduced amounts of xerophytes and the spread of evergreen broadleaved trees and shrubs such as *Q. ilex* type, *Pistacia* and *Olea* (San Rafael, Antas, Guadiana), in some cases accompanied by deciduous *Quercus* (San Rafael, Guadiana) and *Q. suber* type (Guadiana). On Menorca, the high abundances of *Buxus* and *Corylus* are a remarkable feature of the Mid Holocene. The inland location of Navarrés and the hilly surrounding terrain would contribute to explaining the persistence of pine forests until their disruption by human-induced fires at c. 6000 cal yr BP, which triggered the spread of oak woodlands, maquis and *Pinus pinaster* (Carrión and van Geel, 1999; J.S. Carrión et al., 2010b, Connor et al., 2021). This period ended at c. 6000–5000 cal yr BP, when xerophytes began re-expanding, particularly on the Iberian south-east (J.S. Carrión et al., 2010b). The cause(s) of this 'xerophytization' is/are still under discussion because of the overall paucity of pollen-independent hydroclimatic reconstructions on the Iberian Peninsula, but a Late-Holocene trend towards drier conditions (see Fig. 4.6-8; Aranbarri et al., 2014; Walzak et al., 2015; Thatcher et al., 2020) coupled with increasing human activities would probably be the likeliest scenario (as discussed in J.S. Carrión et al., 2010b). In south-western Iberia, the Guadiana estuary pollen record shows an expansion of heathlands during the past millennia related to increasing human activities.

In the today's **mesomediterranean belt,** the Late-glacial record of Padul shows that both deciduous and evergreen oaks dominated the landscape of southern Iberia during the Bølling-Allerød even at c. 800 m asl. The early record of moderate abundances of *Pistacia* is also remarkable. In the Iberian north-east, the Banyoles pollen record shows a picture more similar to temperate sites, with *Betula*, *Corylus* and deciduous *Quercus*, resulting from the close proximity to the Pyrenees. The Younger Dryas is indicated by a reduction in *Quercus*, particularly *Q. ilex* type, and an increase in *Artemisia*, Chenopodiaceae, *Pinus* and *Juniperus* (Pons and Reille, 1988). The Holocene vegetation history of southern Iberia at mid-elevations seems to have been characterized by the dominance of diverse oak woodlands with notable abundances of evergreen broadleaved shrubs (*Pistacia*, *Phillyrea*, Ericaceae). Northwards and inland, the Villaverde record shows an Early Holocene dominance of pinewoods, which were mostly replaced with mesophilous trees, mainly deciduous oaks, at c. 6000 cal yr BP. At c. 5000 cal yr BP, *Q. ilex* started to spread alongside other evergreen shrubs (*Phillyrea*, *Pistacia*, Ericaceae). The past c. 2500 years show an increase in conifers, pines and junipers, and steppic taxa (*Artemisia*, Chenopodiaceae) following the collapse of oaks. Junipers and pines were also the dominant trees in the central Ebro Desert during the Early and Mid Holocene, probably because of their competitiveness in dry and continental settings. During the past c. 5000 years, pinewoods were partly replaced by steppic communities.

In the today's **supra- and oromediterranean belts**, the Late-glacial is well recorded at many sites. In the south, Carihuela Cave (1020 m asl) and Siles (1320 m asl) show high pollen percentages of *Pinus*, *Juniperus*, *Artemisia* and Poaceae, along with minor but steady values of deciduous and evergreen *Quercus* as well as temperate and Mediterranean trees and shrubs (Carrión et al., 1998; Carrión, 2002b; Fernández et al., 2007). Northwards, Charco da Candieira (1409 m asl) records the dominance of *Pinus* and *Betula* with some *Quercus* during the Bølling-Allerød in an oceanic setting. Slight increases in *Pinus* and *Betula* are the main signature of the Late-glacial interstadial around Ayóo de Vidriales (780 m asl), with moderate abundances of heathers. In contrast, *Pinus* dominated during the Bølling-Allerød interstadial at the more continental and higher site of Quintanar de la Sierra (1470 m asl), expanding over the preceding *Juniperus* and *Hippophaë* shrubland (Peñalba, 1994; Peñalba et al., 1997). The Younger Dryas is usually quite pronounced and seen as a decrease in trees and an increase in *Artemisia* and Poaceae, such as at Quintanar de la Sierra (Peñalba, 1994; Peñalba et al., 1997) and Ayóo de Vidriales (Morales-Molino and García-Antón, 2014). However, no sign of a Younger Dryas is observed in other sites, for instance at Siles (Carrión, 2002b). On the Northern Iberian Plateau, *Pinus* and *Betula* decreased during the Younger Dryas at Ayoó de Vidriales (780 m asl) but expanded rapidly at the onset of the Holocene (Morales-Molino and García-Antón, 2014). Other supramediterranean sites of the Northern Plateau ('*Meseta Norte*') reveal the persistence of pine forests throughout the Holocene on inland dunes (Franco-Múgica et al., 2005) and dry continental settings on limestone (Espinosa de Cerrato, Tubilla del Lago; Franco Múgica et al., 2001; Morales-Molino et al., 2017). The latter pinewoods on limestone went extinct in historical times due to increasing land use and the use of fire, which in turn assisted the spread of oaks. Pinewoods were also dominant throughout the Holocene in the supramediterranean belt of the Segura mountains, although a remarkable increase in mesophilous deciduous broadleaved trees occurred during the Mid Holocene mesophytic optimum c. 7500–5000 cal yr BP (Siles; Carrión, 2002b). Westwards, the increasing oceanic influence led to an increasingly reduced relevance of pines in the Holocene landscapes, outcompeted by birches and deciduous oaks (El Maíllo, Charco da Candieira). At these western Iberian sites, heathlands were the most favoured vegetation type under rising land use during the past millennia (for example Ayóo de Vidriales, El Maíllo, Charco da Candieira). In south-eastern France, the pollen record from the Palavasian Lagoon shows that mixed deciduous-evergreen woodlands dominated by deciduous oaks, covered the region around the study site for the past c. 7500 years (Azuara et al., 2018). *Fagus* spread during the Mid Holocene, which was interpreted by the authors as indicative of moister conditions, while evergreen oaks expanded during the past c. 5000 years (Azuara et al., 2018). Iberian oromediterranean sites show a hegemonic dominance of pines throughout the Holocene, in some case with *Betula* co-dominating the mountain forests (Ojos del Tremedal, Las Pardillas, Quintanar de la Sierra). A certain amount of heathland spread occurred during the past millennia, probably related to anthropogenic clearance of the pine forests. In the north-western part of the Iberian range (Las Pardillas, Quintanar de la Sierra) the role of moisture-loving *Fagus* during the Late Holocene is remarkable, although small refugial populations may have persisted in the north-west of the Iberian Peninsula and in the Cantabrian Range during the LGM (Ramil-Rego et al., 2000; López-Merino et al., 2008). Weak or moderate human impact and favourable climatic conditions (cool, humid) likely fostered the Late Holocene development of *Fagus*, but later enhanced land use reduced the beech populations.

Human impact: the age of the onset of human impact on the vegetation of the Iberian Peninsula is highly variable across this territory, from the Neolithic to Roman Times and even Middle Ages, as nicely summarized in J.S. Carrión et al. (2010b). Regional overviews are also available, for example López-Sáez et al. (2014) for the Spanish Central System or Azuara et al. (2015) for south-eastern France. Other important publications on this topic are Yll et al. (1997), Zapata et al. (2004), Riera et al. (2006) and Connor et al. (2019).

Overviews: for the entire Iberian Peninsula, Carrión et al. (2010a), Carrión (2012) and Sadori (2013); for northern Iberia, Moreno et al. (2012); and for western Iberia, Mateus and Queiroz (1993).

Transects: Jalut et al. (2000) and Gil-Romera et al. (2009). For southern France de Beaulieu et al. (1984) provide an early overview.

Pollen maps: For the Late-glacial and Early Holocene in north-west Iberia: Muñoz Sobrino et al. (2007) and Carrión, et al. (2010a).

Climate reconstructions: Reed et al. (2001) and Carrión (2002b).

Phylogeography: Fineschi et al. (2000), Lumaret et al. (2002, 2004), Hampe et al. (2003), Petit et al. (2005), Magri et al. (2006), Gómez and Lunt (2007), Rodríguez-Sánchez et al. (2010) and Liepelt et al. (2010).

San Rafael *Spain* (Pantaleón-Cano et al. 2003)

Ma 36.8° N, 2.6° W, 0 m a.s.l.

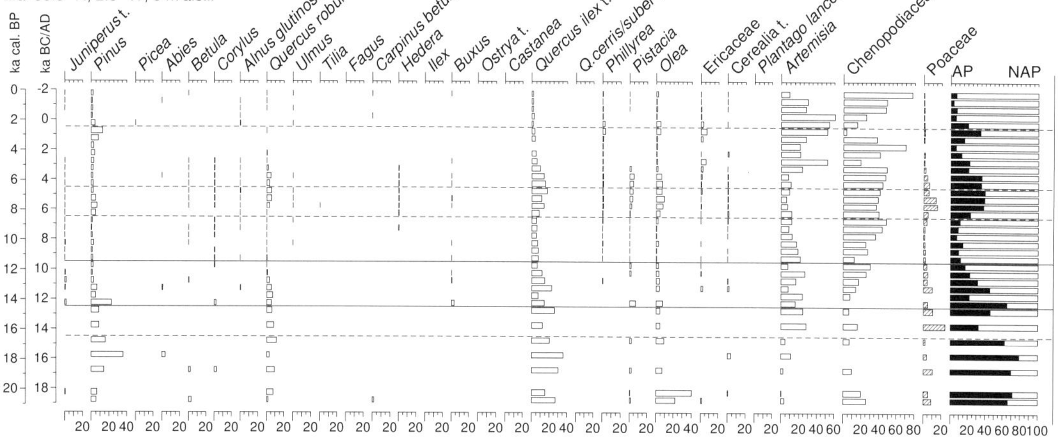

Guadiana *Spain* (Fletcher et al. 2007)

Ma 37.1° N, 7.2° W, 1 m a.s.l.

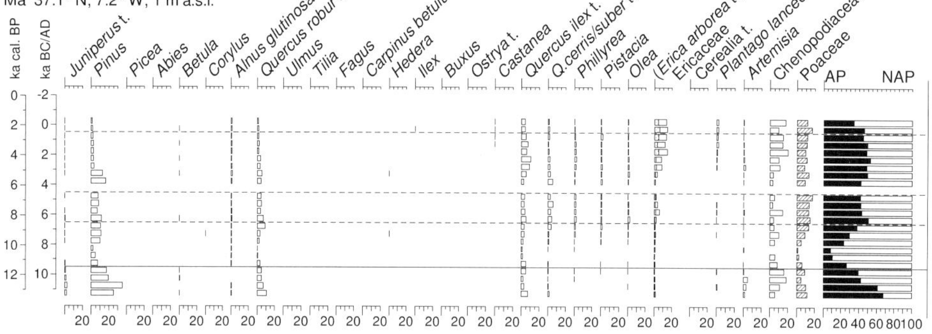

Antas *Spain* (Pantaleón-Cano et al. 2003)

Ma 37.2° N, 1.8° W, 0 m a.s.l.

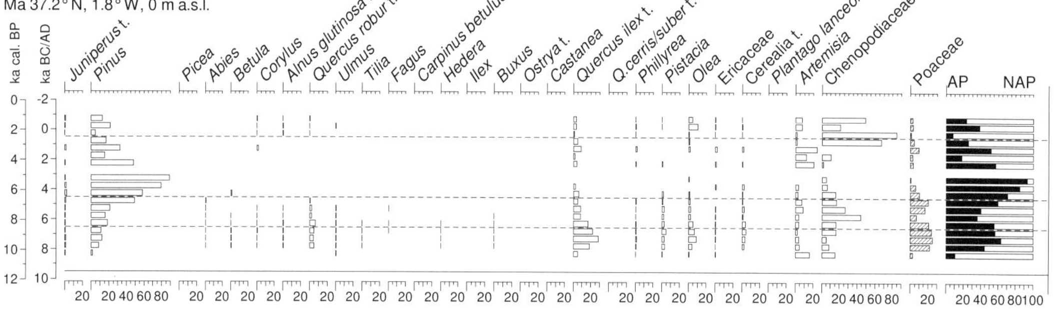

Navarrés *Spain* (Carrión and van Geel 1999)

Ma 39.1° N, 0.7° W, 225 m a.s.l.

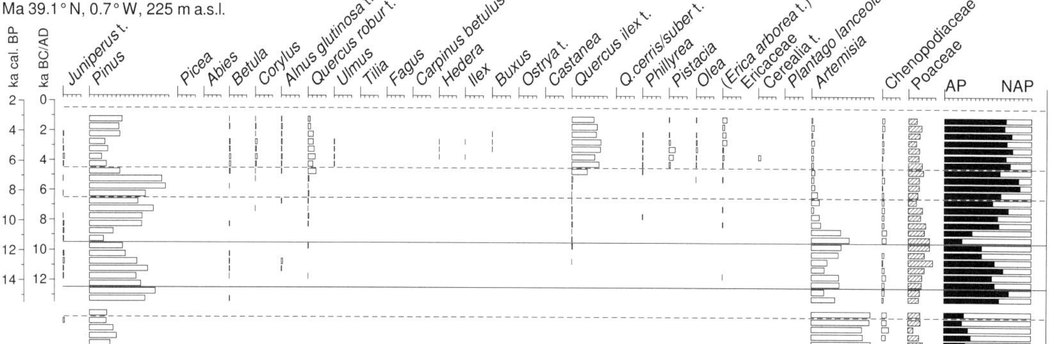

Menorca [4 sites] *Spain: Balearic Islands* (Yll et al. 1997)

Ma 39.9° N, 4.5° E, 12-47 m a.s.l.

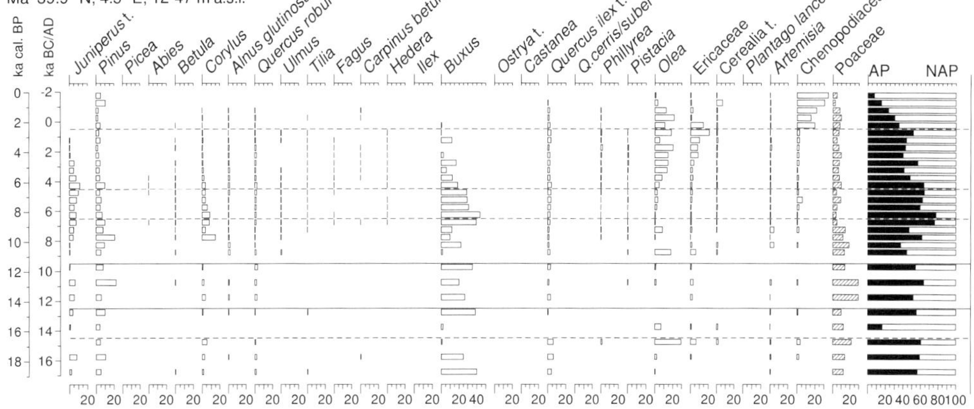

Padul 3 *Spain* (Pons and Reille 1988)

Ma 37.0° N, 3.1° W, 785 m a.s.l.

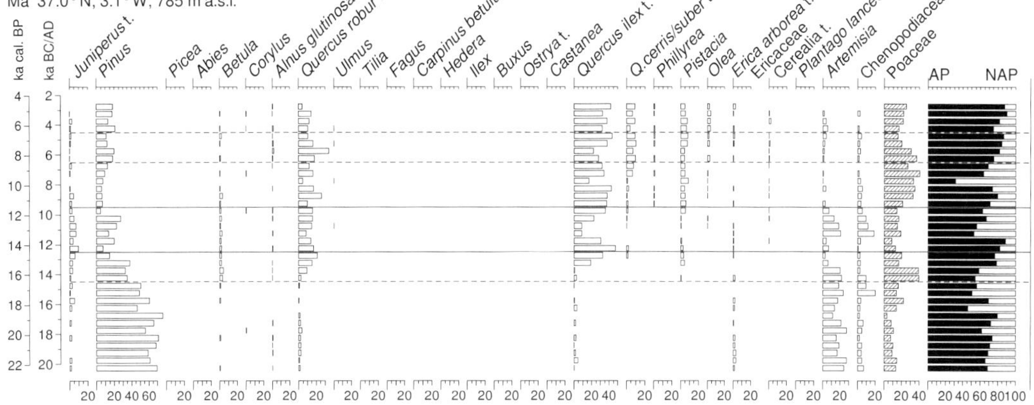

Villaverde *Spain* (Carrión et al. 2001b)

Ma 38.5° N, 2.2° W, 870 m a.s.l.

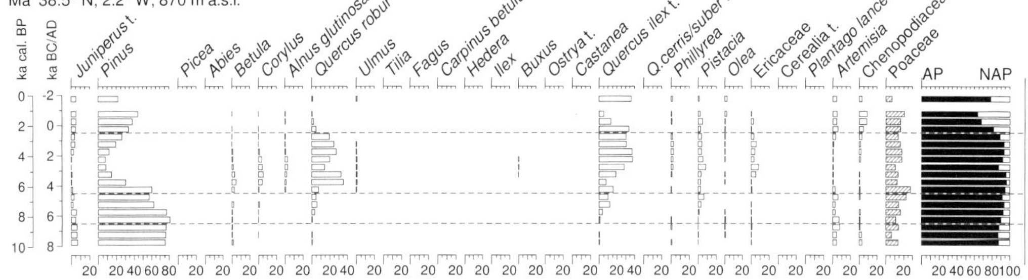

Central Ebro Desert [4 sites] *Spain* (Davis and Stevenson 2007)

Ma 41.2° N, 0.3° E, 150-357 m a.s.l.

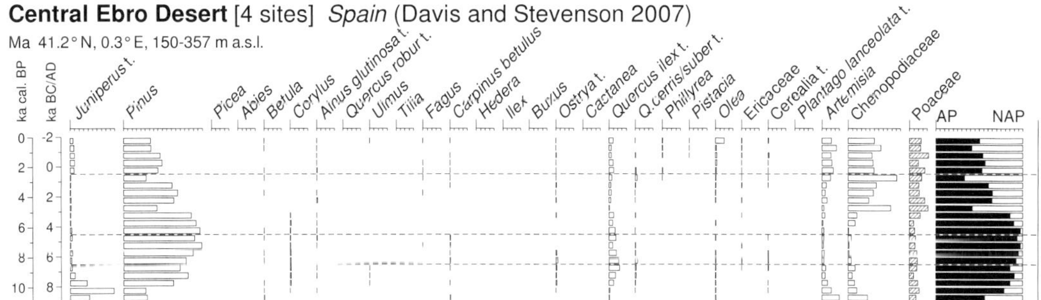

Banyoles *Spain* (Pérez-Obiol and Julià 1994)

Ma 42.1° N, 17.3° E, 173 m a.s.l.

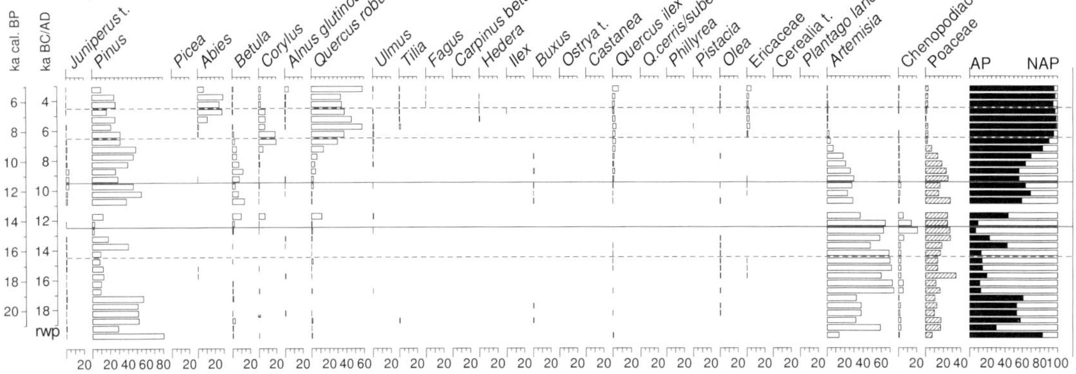

Siles *Spain* (Carrión 2002)

Ma 38.4° N, 2.5° W, 1320 m a.s.l.

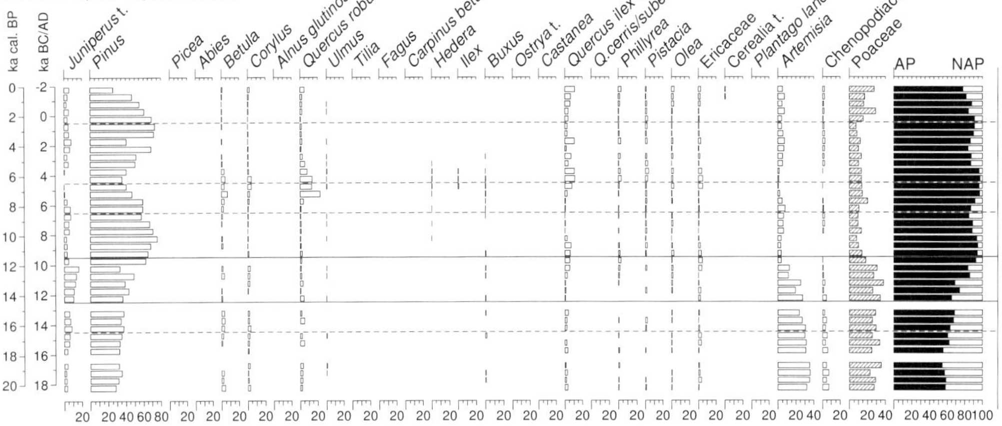

Charco da Candieira *Portugal* (van der Knaap and van Leeuwen 1995, 1997)

Ma 40.3° N, 7.6° W, 1409 m a.s.l.

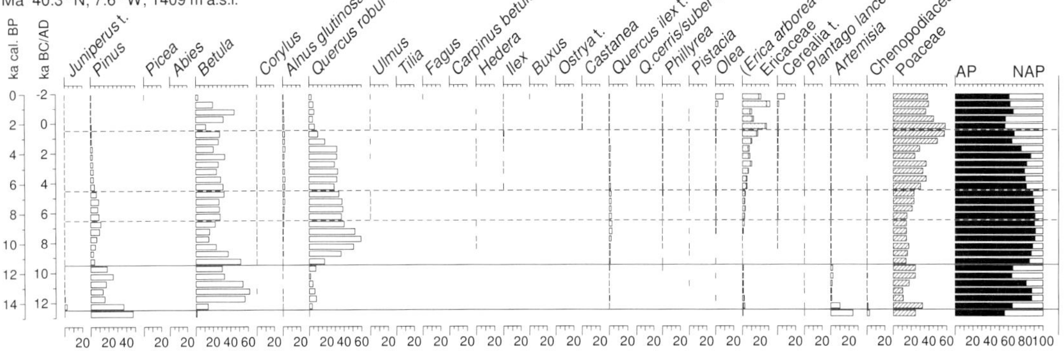

El Maillo *Spain* (Morales-Molino et al. 2013)

Ma 40.5° N, 6.2° E, 1100 m a.s.l.

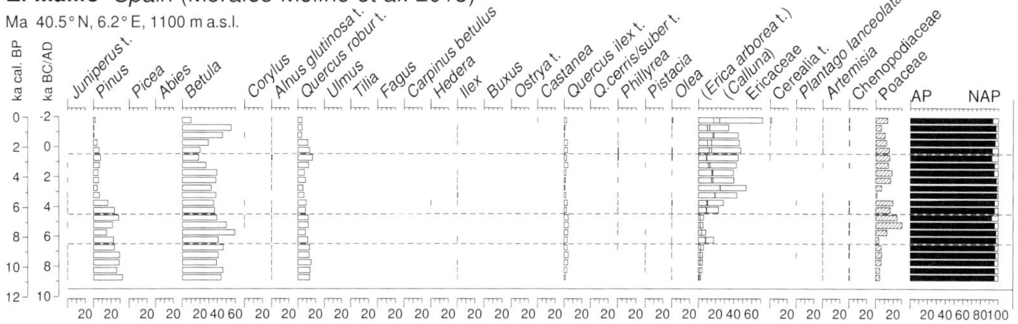

Tubilla *Spain* (Morales-Molino et al. 2017)

Ma 41.8° N, 3.6° W, 900 m a.s.l.

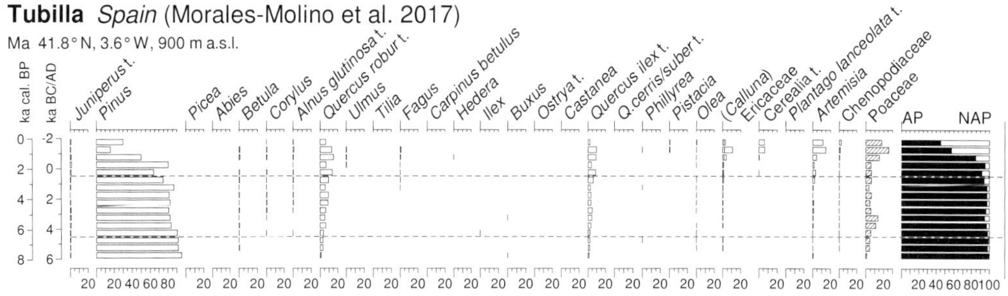

Espinosa de Cerrato *Spain* (Franco-Múgica et al. 2001)

Ma 41.9° N, 3.9° W, 885 m a.s.l.

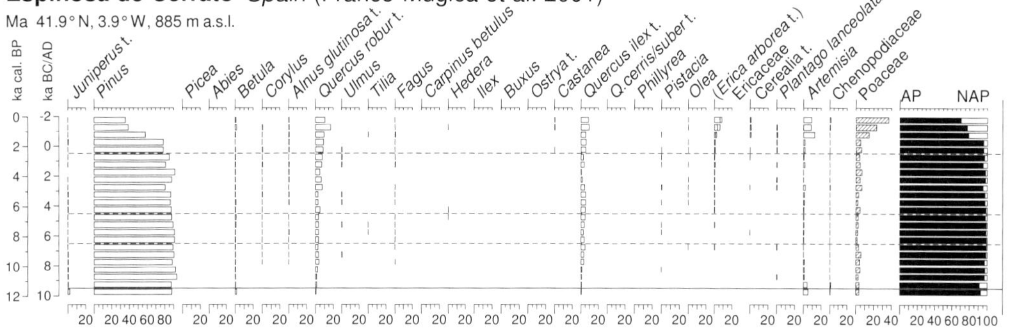

Ayoó de Vidriales *Spain* (Morales-Molino and García-Antón 2014)

Ma 42.1° N, 6.1° W, 780 m a.s.l.

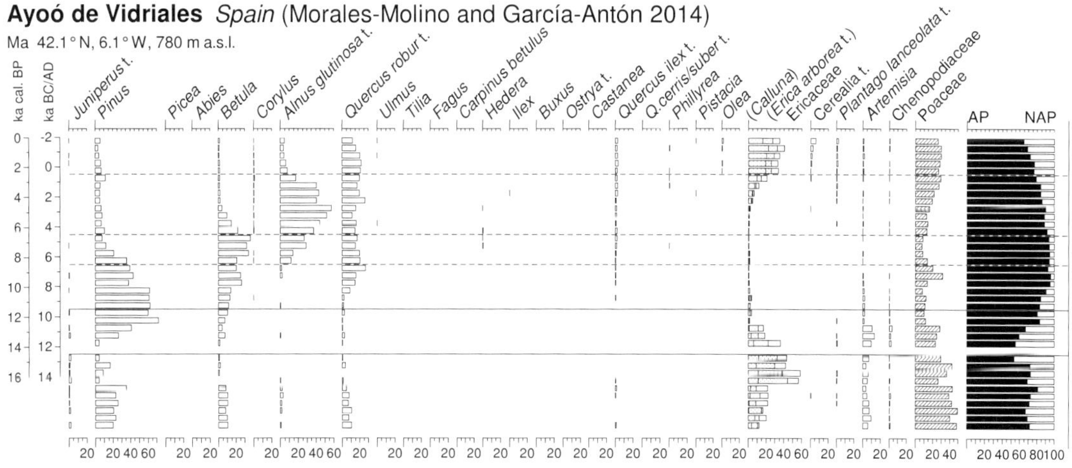

Cañada de la Cruz *Spain* (Carrión et al. 2001b)

Ma 38.1° N, 2.7° W, 1595 m a.s.l.

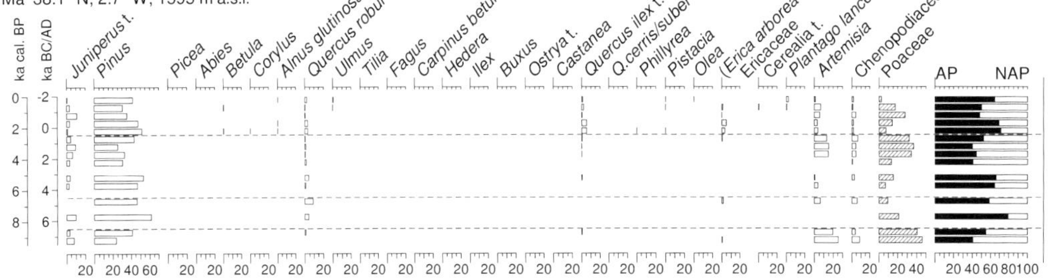

Ojos del Tremedal *Spain* (Stevenson 2000)

Ma 40.5° N, 2.0° W, 1650 m a.s.l.

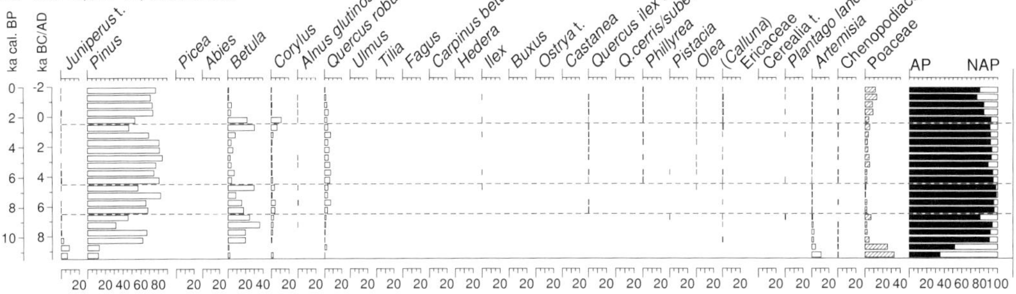

Las Pardillas *Spain* (Sánchez-Goñi and Hannon 1999)

Ma 42.0° N, 3.0° W, 1870 m a.s.l.

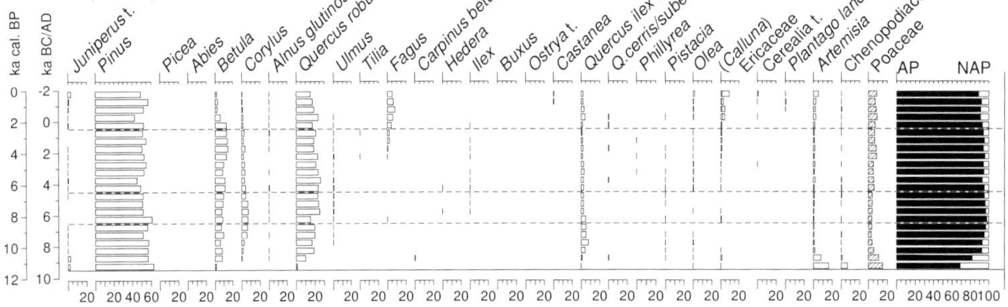

Quintanar de la Sierra *Spain* (Peñalba 1994, Peñalba et al. 1997)

Ma 42.0° N, 3.0° W, 1470 m a.s.l.

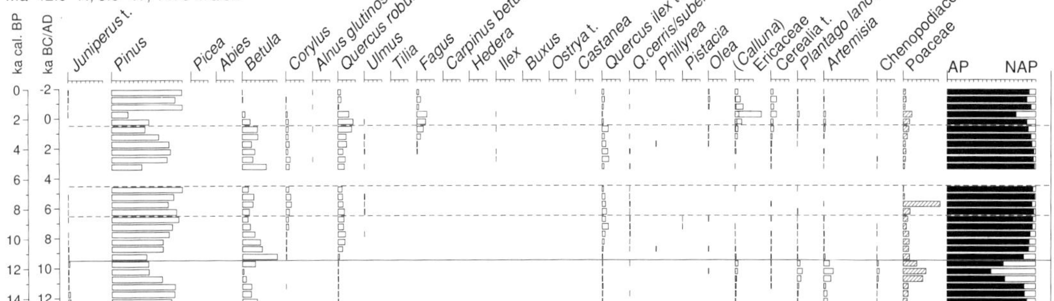

← **Fig. 4.3-15b.** Selected sites in the western Mediterranean (Ma) region. Lateglacial and Holocene pollen records are presented as histograms at a 500-year resolution. Because of the physiographic complexity of the western Mediterranean region, in particular of the Iberian Peninsula, the selected sites are arranged according to elevational belts – even localities not too far apart can be quite different in their vegetation history, such as San Rafael and Antas. For the four belts thermomediterranean, mesomediterranean, supramediterranean and oromediterranean, elevations are not given in m asl because the elevational transitions vary strongly between the latitudes 36° and 43° N. Two histograms are made by combinations of sites. Central Ebro Desert: this diagram is the average of four diagrams – Hoya del Castillo, Laguna Guallar, Salada Pequeña and Laguna Salada Chiprana. We took the title "Central Ebro Desert" from Davis and Stevenson (2007). Menorca: this diagram is the average of four diagrams – Cala'n Porter, Cala Galdana, Hort Timoner and Son Bou. Because we simplify the coordinates for the histograms, two sites in the oromediterranean belt seem to be at the same spot, but the detailed coordinates show the spatial difference: Quintanar de la Sierra: 42° 01' 49"N, 03° 01' 14" W, 1470 m asl., and Las Pardillas: 42° 02' 36"N, 03° 02' 45"W, 1870 m asl.

Mb, Central Mediterranean region

This region includes the central part of the Mediterranean biome comprising Corsica, Sardinia, Sicily, Malta, a substantial part of the Italian Peninsula (with a northernmost extension to Liguria in northern Italy) as well as a narrow band along the eastern Adriatic Sea in the Balkans. That Istria and Dalmatia belong to the central Mediterranean region (Mb) is explained by the history of some important Mediterranean floristic units, which today have an uninterrupted spatial range (e.g. *Fraxinus ornus* and *Ostrya carpinifolia*; Horvat et al., 1974, Mayer, 1984). Significant mountain chains with pronounced vegetation belts occur in this region, reaching up to 2900 m in central Italy, 2700 m in Corsica, 1830 m in Sardinia, 3320 m in Sicily, 2260 m in southern Italy and 2200 m in Dalmatia. Lowland evergreen oak forests (e.g. *Quercus ilex, Q. suber*) and tall evergreen maquis (e.g. *Ceratonia siliqua, Olea europaea, Pistacia lentiscus, Myrtus communis, Chamerops humilis*) occur in the southern part of the region, forming the thermomediterranean vegetation belt of southern Italy, Sicily, Sardinia and a multitude of southern islets. On the southernmost islets (e.g. Pantelleria, Lampedusa) coastal maquis forms an inframediterranean semi-arid belt below the thermomediterranean belt (c. 0–200 m asl) with, for example, *Juniperus turbinata* and *Periploca angustifolia* (Calò et al., 2013) a vegetation type which also occurs in the neighbouring Mediterranean areas of northern Africa. In the lowlands of the northern part of the region, for example Corsica (with the exception of a very narrow band at the coast), western central Italy, Istria and Dalmatia, as well as at intermediate elevations (c. >500 m asl) in the southern part of the region, evergreen broadleaved taxa (e.g. *Quercus ilex, Laurus nobilis*) are mixed with deciduous broadleaved species (e.g. *Quercus pubescens, Q. petraea, Q. cerris, Castanea sativa, Fraxinus ornus, Ostrya carpinifolia*) forming the mesomediterranean vegetation belt. Higher up (c. >500 m in the north, >1000 m asl in the south), evergreen broadleaved oaks disappear and broadleaved deciduous species prevail (e.g. *Quercus pubescens, Q. petraea, Q. cerris, Castanea sativa, Fraxinus ornus, Ostrya carpinifolia*) in the supramediterranean belt. In contrast to the western Mediterranean region (Ma) *Fagus sylvatica* and *Abies alba* form mountain or oromediterranean (Reisigl et al., 1992) beech and fir belts in almost all areas of the central Mediterranean region. For instance, *Fagus sylvatica* has its southernmost European occurrence in Sicily, where it forms dense, dark stands together with evergreen *Ilex aquifolium* and very exceptionally with the endemic *Abies ne-*

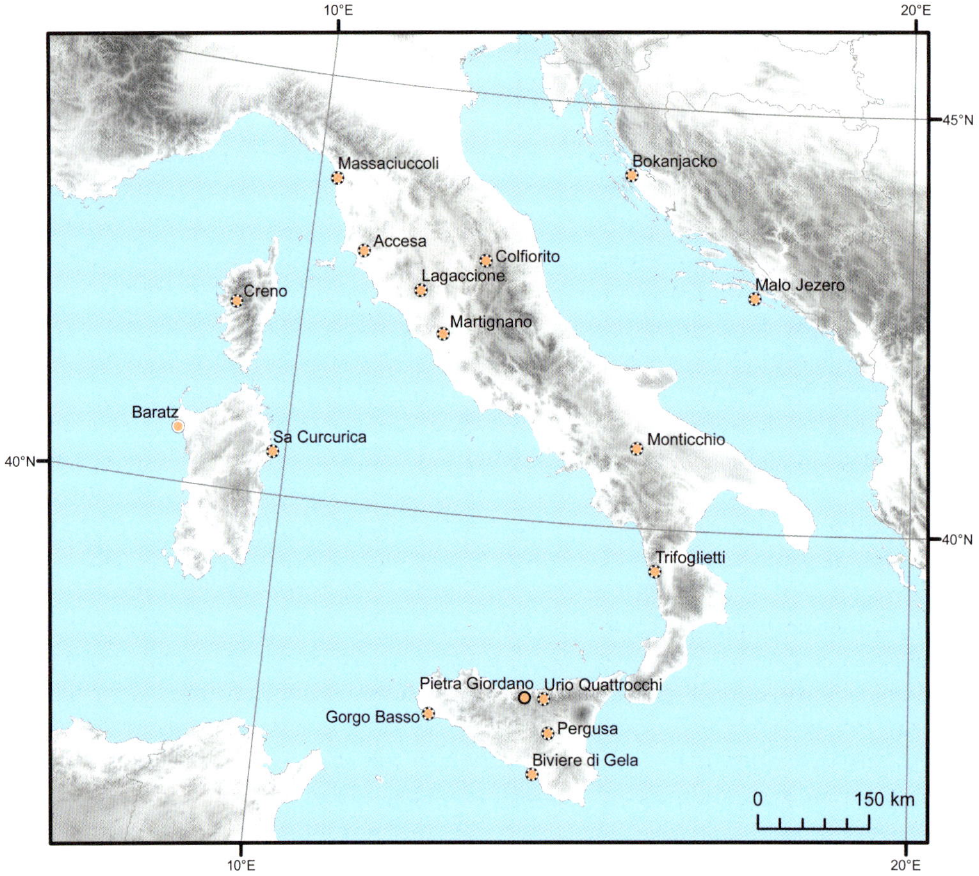

Fig. 4.3-16a. Map for selected sites in the central Mediterranean (Mb) region. Dots with black borders indicate sites shown in the histograms (Fig. 4.3-16b), white borders indicate sites used for the maps showing the history of selected taxa (see chapter 4.4) and dashed borders indicate sites used for both.

brodensis, a fir species closely related to *A. alba*. This mountain or oromediterranean vegetation can expand into the altimediterranean or subalpine belt to build the upper treeline where mountain peaks are sufficiently high. Above the timberline (c. >2000 m asl) alpine meadows or thorny cushions are dominant, for instance in Sicily with endemic *Astragalus siculus* and *A. nebrodensis*.

Local mountain glaciers expanded in Corsica and northern and central Italy during the last ice age. Most of them vanished during the Late-glacial. The Late-glacial afforestation after the ice age often started with the expansion of *Juniperus*, *Pinus* and deciduous broadleaved *Quercus* species, which most likely survived the last ice age in several refugia in this region. Other temperate trees surviving the ice age in this region were *Fagus sylvatica, Acer, Fraxinus, Abies alba, Tilia, Ostrya, Alnus* and *Ulmus* (Kelly and Huntley, 1991; Watts et al., 1996a,b; Magri and Sadori, 1999; Allen et al., 2000, Willis and McElwain, 2002). The presence of endemic temperate tree species such as *Abies nebrodensis, Zelkova sicula* or endemic shrubs such as *Astragalus siculus* and *A. nebrodensis* underscores the refugial character of the region. Subtropical Mediterranean trees and shrubs (e.g. *Quercus ilex, Laurus nobilis, Olea europaea, Phillyrea, Pistacia, Ceratonia, Chamaerops humilis*) also probably survived in sheltered habitats of the region, for instance in Corsica, Sardinia, Sicily and the southern Italian Peninsula. Sporadic ice-age finds of pollen of some of these trees and shrubs occur regularly or are abundant towards the end of the Late-glacial or at latest at the onset of the Holocene (Kelly and Huntley, 1991;

Watts et al., 1996b; Magri, 1999; Magri and Sadori, 1999; Allen et al., 2000).

The Late-glacial development of the evergreen broadleaved **thermomediterranean** vegetation is still largely unknown. Records covering more than 10 000 years show that in coastal Sicily afforestation was delayed until the Mid Holocene (c. 8000 -7000 cal yr BP), when *Quercus ilex-Olea europaea* forests expanded into drought-tolerant Early Holocene *Pistacia*-dominated maquis (e.g. Gorgo Basso or Lago Preola; Tinner et al., 2009; Calò et al., 2012), likely in response to increasing moisture availability. The vegetation history of Malta shows similarities to that of Sicily, with a strong Holocene dominance of drought-tolerant *Pistacia* maquis, however, *Quercus ilex* and *Olea europaea* stands were far less abundant than in Sicily (Djamali et al., 2013). In Sardinia, Early Holocene vegetation was dominated by *Pinus, O. europaea, Q. ilex, Erica arborea* and *E. scoparia* forests or shrublands (Beffa et al., 2016; Kalis and Schoch, 2019). Around 8000 cal yr BP, during the Holocene Thermal Maximum (HTM, Samartin et al. 2017; Fischer et al. 2018), drought- and fire- tolerant *E. arborea* and *E. scoparia* woodlands dominated (see Sa Curcurica). This special thermomediterranean vegetation type was gradually replaced by evergreen oak vegetation (*Q. ilex, Q. suber*) after c. 5500–2800 cal yr BP, when summer temperatures declined and oceanity increased, so that Sardinian vegetation composition became more similar to that in thermomediterranean Sicily and mainland southern Italy (Di Rita and Magri, 2009; Noti et al., 2009; Tinner et al., 2009; Calò et al., 2012; Beffa et al., 2016; Di Rita et al., 2018; Pedrotta et al., 2021). Very similar Holocene vegetation trajectories as in Sardinia occurred in coastal Corsica (Revelles et al., 2019) and on the islet of Cavallo south of Corsica (Poher et al., 2017). Finally, thermomediterranean evergreen broadleaved forests were replaced by open vegetation, mainly arable fields, orchards, grasslands and maquis, during the past 3000 years (e.g. Di Rita and Magri, 2009; Noti et al., 2009; Tinner et al., 2009; Calò et al., 2012; Beffa et al., 2016; Di Rita et al., 2018; Pedrotta et al., 2021). First conversions of natural vegetation into cultural landscapes started earlier. For instance, high pollen values of *Ficus*, together with Cerealia type and other cultural indicators, suggest conspicuous **agricultural activities** since the early Neolithic 7500-7000 years ago at the Sicilian site Gorgo Basso (Tinner et al., 2009).

The ice age and Late-glacial vegetation history of the today's **mesomediterranean** belt is better known. Of particular importance for the history of this vegetation belt is the Monticchio record in southern Italy, which spans more than 120 000 years (Watts et al., 1996a,b; Allen and Huntley, 2000, 2009; Allen et al. 2000). In southern Italy at c. 500–1000 m asl deciduous *Quercus* gradually invaded pine parklands and the open steppic tundra after c. 16 000 cal yr BP (e.g. Monticchio, Allen et al. 2000, 2002), probably in response to early climate warming prior to the Bølling-Allerød interstadial (Samartin et al. 2012; Bolland et al. 2020, see adjacent Th region). Open deciduous mixed-oak forests (with *Ulmus, Tilia, Fagus*) established at around 14 200 cal yr BP as a consequence of further warming during the Bølling-Allerød interstadial. Forests became closed during the Early Holocene (11 700–8200 cal yr BP), when *Quercus ilex, Phillyrea, Abies alba* and *Ostrya* expanded into the deciduous mixed-oak-forests, forming mesomediterranean communities. Similar Late-glacial and Early Holocene forest dynamics occurred in the mesomediterranean belt in central Italy (Magri, 1999; Magri and Sadori, 1999). At some sites (e.g. Martignano, Vico and Lagaccione; Kelly and Huntley, 1991; Magri, 1999; Magri and Sadori, 1999) evergreen broadleaved *Quercus ilex* had already established during the Late-glacial/Holocene transition. Some southern and central Italian mesomediterranean communities are thus at least 12 000 years old, given that vegetation composition remained more or less stable throughout the Holocene. However, mass expansions of *Quercus ilex* usually followed after 10 000 cal yr BP during the HTM, for example at Lagaccione and Accesa around 8500–8000 cal yr BP (Magri, 1999; Drescher-Schneider et al., 2007b). In Sicily at elevations >500 m asl forests expanded only at the onset of the Holocene around 11 700 years ago (Pergusa; Sadori and Narcisi, 2001). There, Holocene mesomediterranean forests were dominated by deciduous and evergreen oaks, while other taxa were less abundant. On the Dalmatian coast (e.g. Malo Jezero; Jahns and van den Bogaard, 1998; Jahns, 2002), Early-Holocene vegetation was dominated by deciduous *Quercus* forests. Evergreen broadleaved vegetation expanded into the deciduous oak forests at 7000–6000 cal yr BP, leading to today's mesomediterranean coastal vegetation belt. Species-rich mesomediterranean forests in Italy were composed of deciduous *Quercus, Tilia,*

Ulmus, Carpinus, Ostrya, Corylus, Alnus, Betula and *Castanea* and by evergreen *Quercus ilex, Olea europaea, Phillyrea, Abies alba, Pinus* and *Juniperus* (e.g. Lagaccione, Monticchio; Accesa, Massaciuccoli; Magri, 1999; Allen et al., 2002; Colombaroli et al., 2007; Drescher-Schneider et al., 2007b). Towards the Late Holocene, mesomediterranean forests became impoverished and disturbance-sensitive trees such as *Abies alba, Ulmus* and *Tilia* were lost or strongly reduced (Colombaroli et al., 2007). This change was accompanied by the expansion of open land for arable and pastoral **farming** that intensified during the past 3000–2000 years (Kelly and Huntley, 1991; Sadori and Narcisi, 2001; Drescher-Schneider et al., 2007b).

The today's **supramediterranean** and **oromediterranean** belts above the mesomediterranean belt were unforested during the Late-glacial (e.g. Creno, Colfiorito, Trifoglietti; Reille et al., 1997, 1999; Brugiapaglia and de Beaulieu, 1995; de Beaulieu et al., 2017) when steppe prevailed (e.g. *Artemisia*, Chenopodiaceae). Mixed deciduous forests with conifers (e.g. *Pinus, Abies alba*) and deciduous trees (e.g. *Fagus sylvatica, Quercus*) expanded rapidly at the onset of the Holocene c. 11 700 years ago (e.g. Creno, Trifoglietti; Reille et al., 1997, 1999; de Beaulieu et al. 2017). During the Early Holocene, deciduous species such as *Quercus cerris* and/or *Q. pubescens* were important in the supramediterranean belt (e.g. Quattrocchi; Bisculm et al., 2012). The continuous record of *Fagus sylvatica* at the supramediterranean site Quattrocchi suggests that at its southernmost range limit in Sicily the species has been abundant in the oromediterranean vegetation above that site since at least 10 000 cal yr BP. Supra and oromediterranean forests persisted during the Mid Holocene (e.g. Creno, Pietra Giordano, Trifoglietti; Reille et al., 1997, 1999; Tinner et al., 2016; de Beaulieu et al., 2017). Finally some supramediterranean and oromediterranean forests were markedly opened up during the Late Holocene to provide **agricultural** and pastoral space at some sites (e.g. Quattrocchi, Pietra Giordano; Bisculm et al., 2012; Tinner et al., 2016), while at others (e.g. Creno; Reille et al., 1997, 1999) the forests remained rather closed.

Climate reconstructions: Giraudi et al. (2011), Magny et al. (2012), Curry et al. (2016) and Samartin et al. (2017).

Data-model comparisons: Henne et al. (2013, 2015), Tinner et al. (2013) and Ruosch et al. (2016).

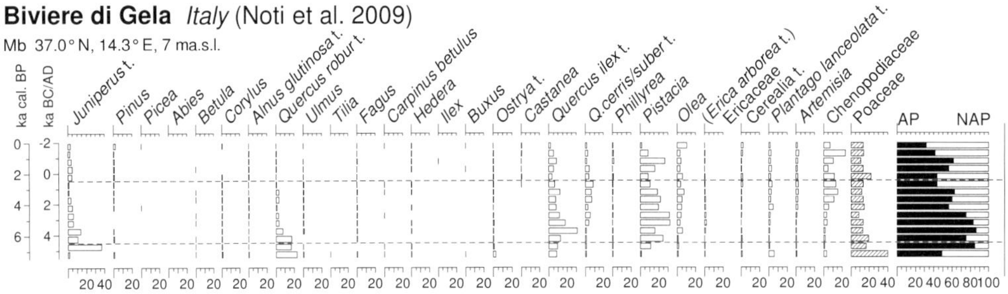

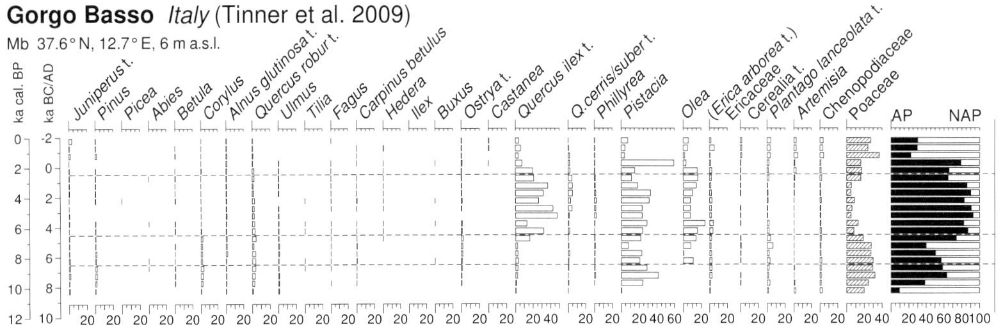

Sa Curcurica *Italy* (Beffa et al. 2016)

Mb 40.5° N, 9.8° E, 0 m a.s.l.

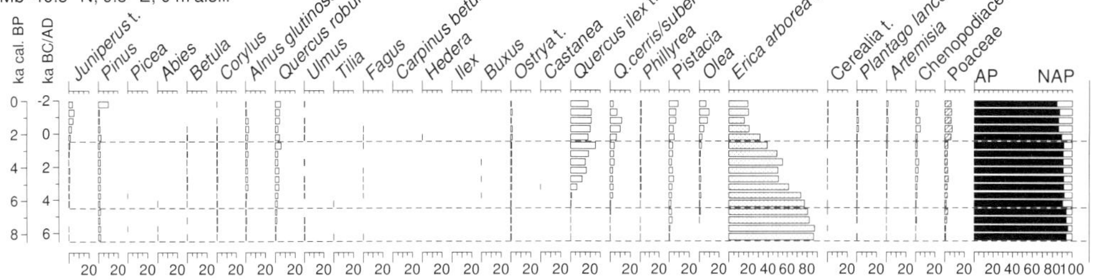

Pergusa *Italy* (Sadori and Narcisi 2001)

Mb 37.5° N, 14.3° E, 670 m a.s.l.

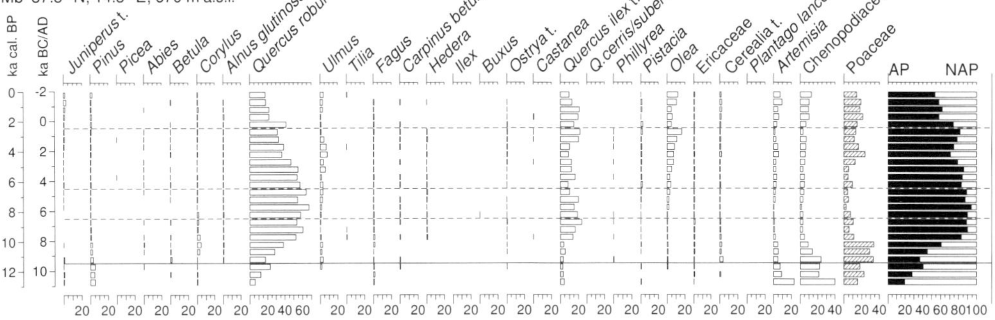

Monticchio *Italy* (Watts et al. 1996; Allen et al. 2002)

Mb 40.9° N, 15.6° E, 656 m a.s.l.

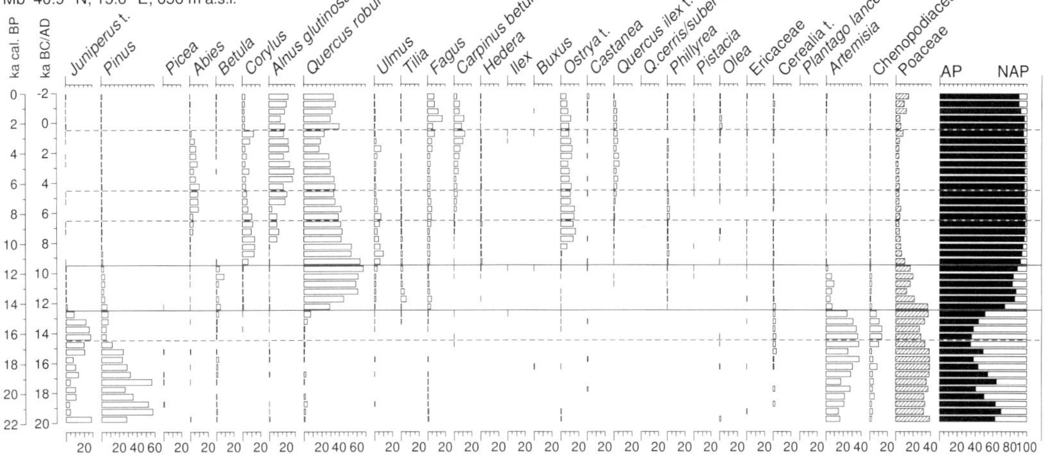

Martignano *Italy* (Kelly and Huntley 1991)

Mb 42.1° N, 12.3° E, 204 m a.s.l.

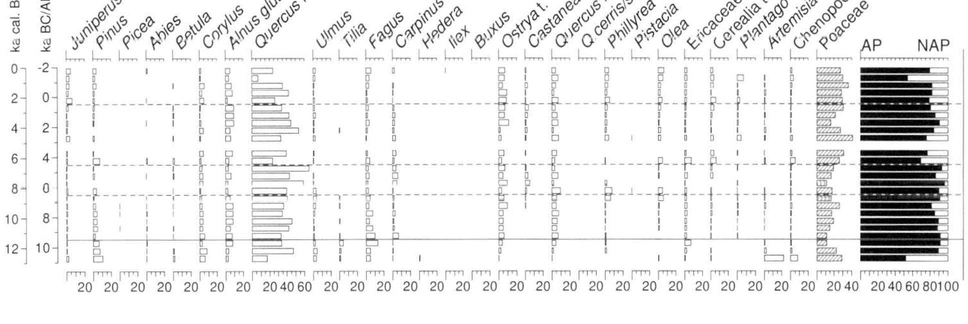

Lagaccione *Italy* (Magri 1999)

Mb 42.6° N, 11.9° E, 355 m a.s.l.

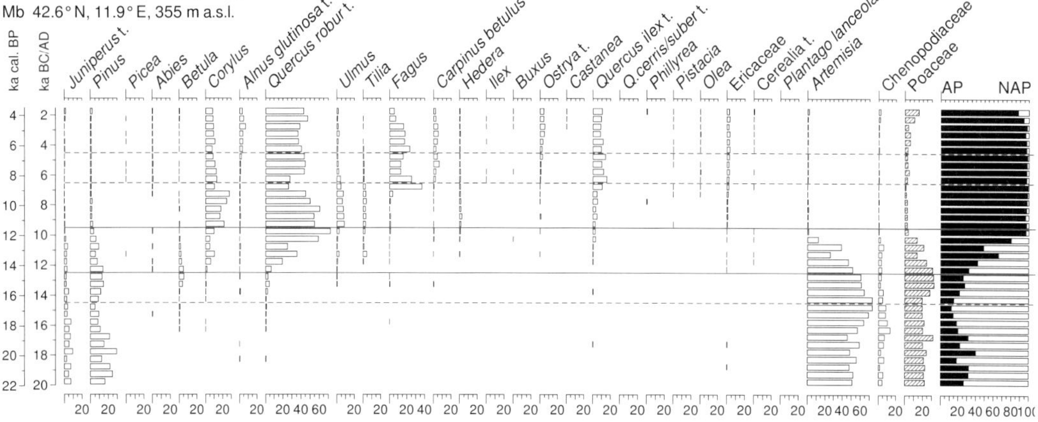

Malo Jezero *Croatia* (Jahns and Bogaard 1998; Jahns 2002)

Mb 42.8° N, 17.4° E, 20 m a.s.l.

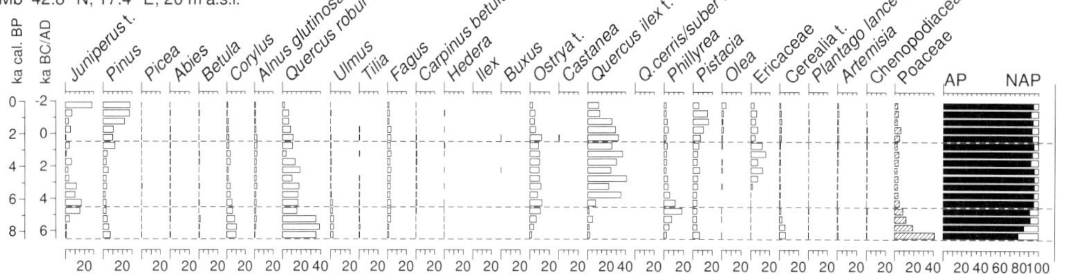

Accesa *Italy* (Drescher-Schneider et al. 2007)

Mb 43.0° N, 10.9° E, 157 m a.s.l.

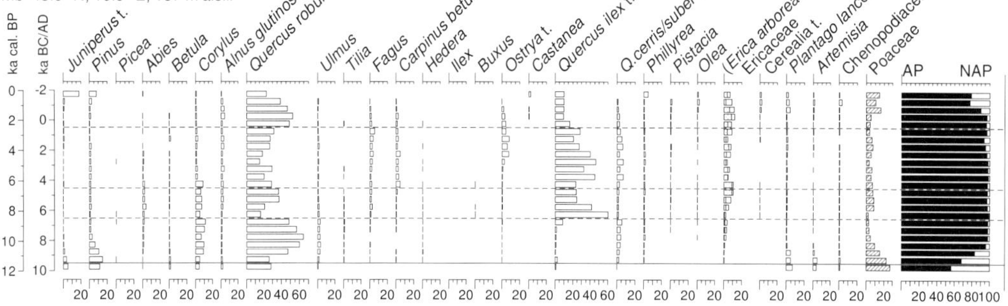

Massaciuccoli *Italy* (Colombaroli et al. 2007)

Mb 43.8° N, 10.3° E, 0 m a.s.l.

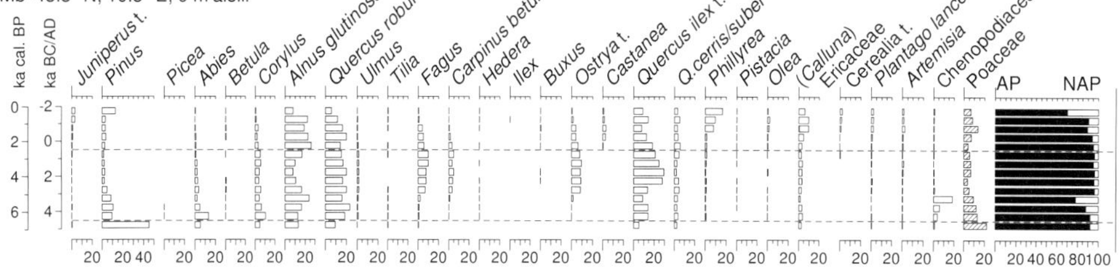

Bokanjacko *Croatia* (Grüger 1996)

Mb 44.2° N, 15.2° E, 25 m a.s.l.

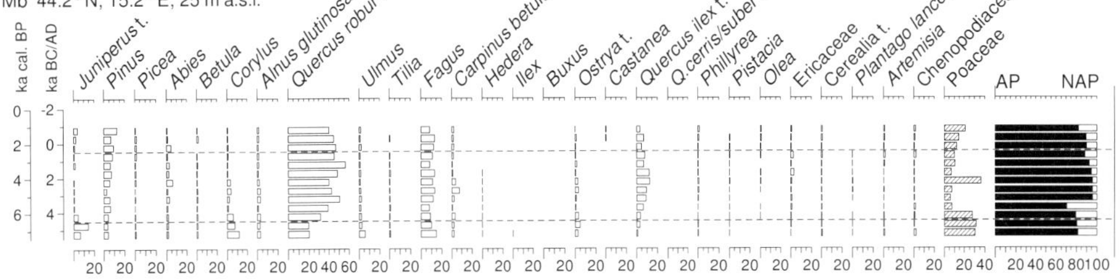

Urio Quattrocchi *Italy* (Bisculm et al. 2012)

Mb 37.9° N, 14.4° E, 1044 m a.s.l.

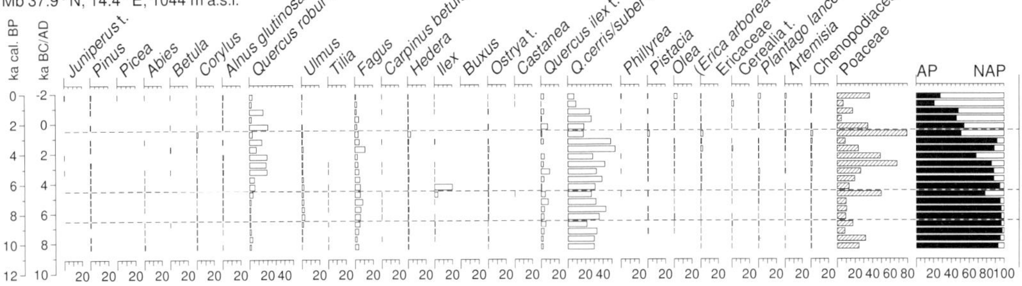

Colfiorito *Italy* (Brugiapaglia and de Beaulieu 1995)

Mb 43.0° N, 12.9° E, 752 m a.s.l.

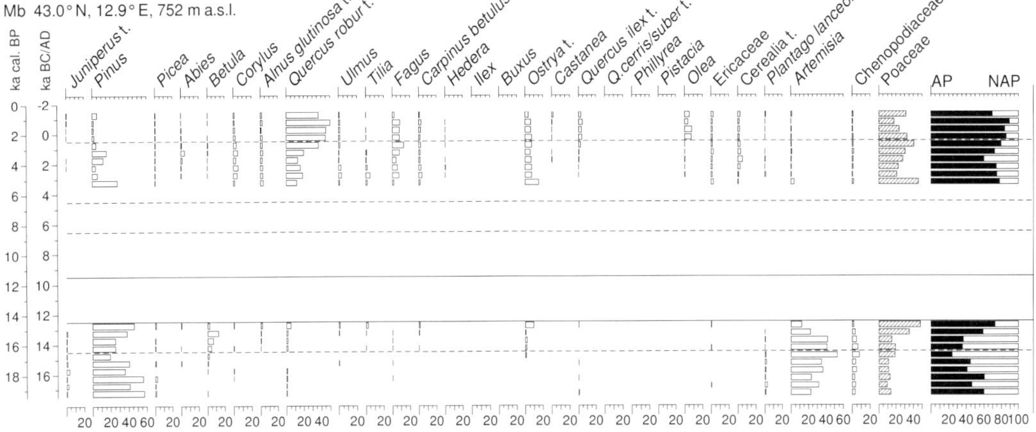

Pietra Giordano *Italy* (Tinner et al. 2016)

Mb 37.9° N, 14.1° E, 1323 m a.s.l.

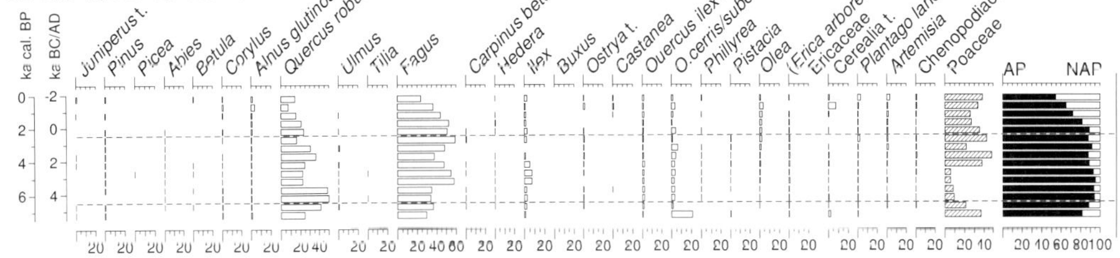

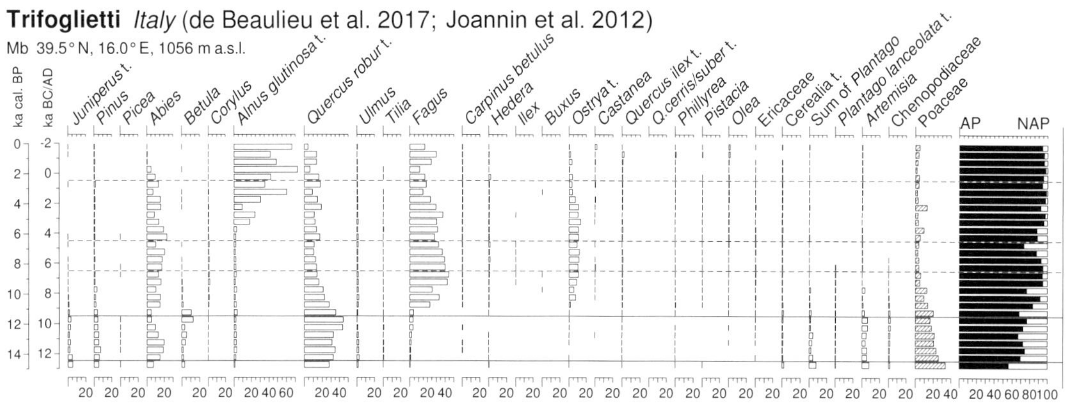

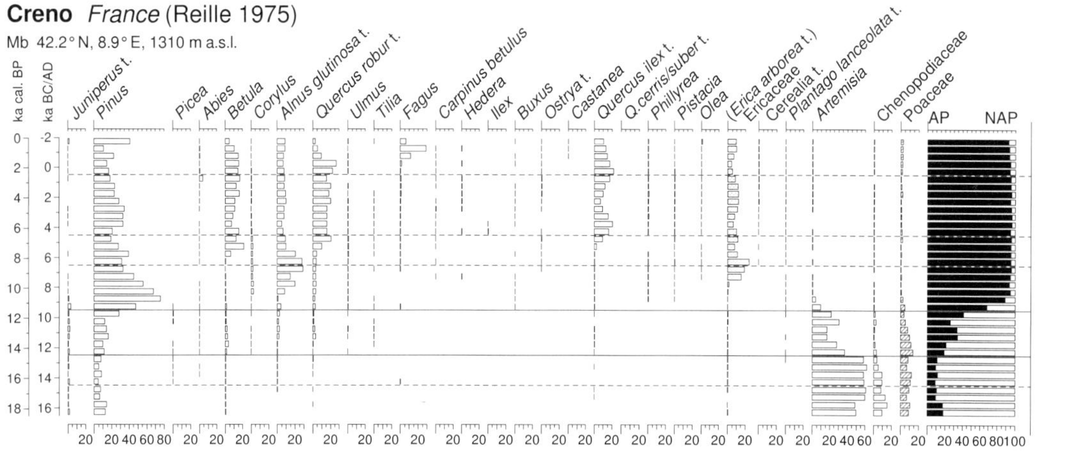

Fig. 4.3-16b. Selected sites in the central Mediterranean (Mb) region. Late-glacial and Holocene pollen records are presented as histograms at a 500-year resolution. Because of the topographic complexity of the Italian Peninsula the selected sites are arranged according to elevational belts, as even localities not too far apart can be quite different in their vegetation history. For the four vegetation belts thermomediterranean, mesomediterranean, supramediterranean and oromediterranean, elevations cannot be defined in m asl because the elevational positions vary strongly between the latitudes 36° and 45° N. In the warmest thermomediterranean belt (e.g. Biviere di Gela, Gorgo Basso, Sa Curcurica), pollen of evergreen broadleaved trees such as *Quercus ilex*, *Olea*, *Pistacia* and *Erica* dominates throughout the Holocene. In the mesomediterranean belt (e.g. Pergusa, Monticchio, Lagaccione, Accesa, Martignano), pollen of both evergreen (e.g. *Quercus ilex*, *Olea*, *Erica*) and deciduous broadleaved (e.g. *Ostrya*, *Quercus*) taxa is important and conifers such as *Abies alba* (e.g. Massaciuccoli, Accesa) or *Pinus* (e.g. Malo Jezero, Bockanjacko) can play a major role. In the supramediterranean belt, pollen of deciduous broadleaved trees (e.g. *Quercus*, *Ostrya*) is dominant (e.g. Quattrocchi, Colfiorito), occasionally with conifers (e.g. *Pinus* at Colfiorito). In the oromediterranean belt, pollen of mesophilous *Fagus* co-dominates with deciduous oaks, conifers (e.g. *Abies*, *Pinus*) and evergreen broadleaved taxa (e.g. *Ilex*, *Erica*) at the sites Pietra Giordano, Trifoglietti and Creno.

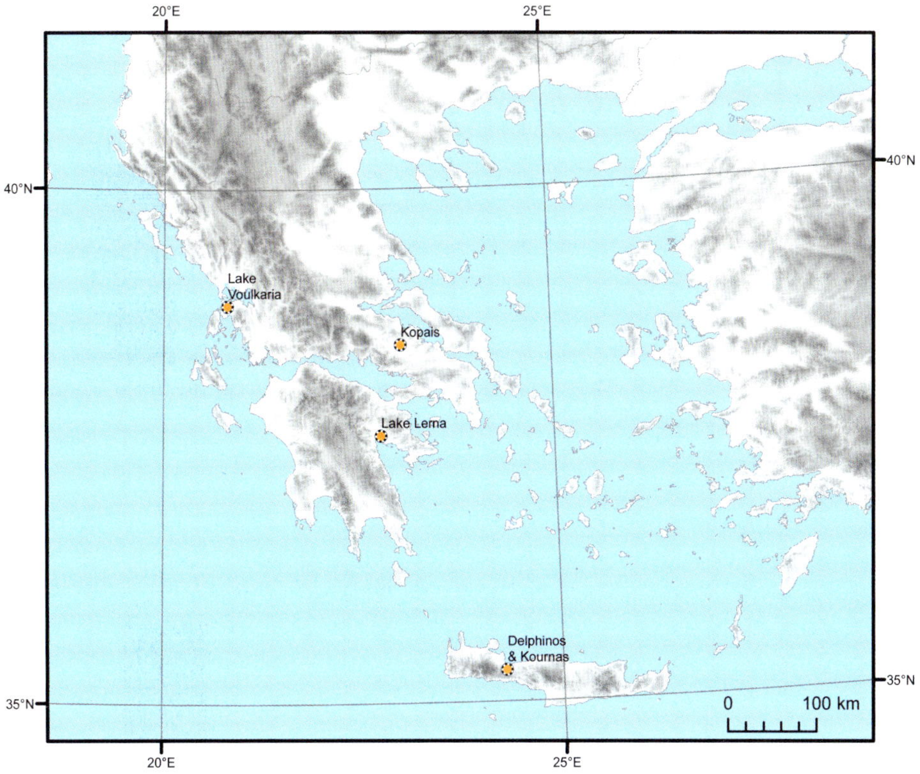

Fig. 4.3-17a. Map of selected sites in the eastern Mediterranean (Mc) region. Dots with black borders indicate sites shown in the histograms (Fig. 4.3-17b), white borders indicate sites used for the maps showing the history of selected taxa (see chapter 4.4) and dashed borders indicate sites used for both.

Mc, Eastern Mediterranean region

The eastern Mediterranean region Mc is a rather narrow strip along the coasts that is less broad than along the western and central Mediterranean coasts. The potential natural vegetation is evergreen broadleaved (Oleo-Ceratonion and Quercion ilicis; see map by Glavač in Horvat et al., 1974). Today, frost periods are absent, but very short cold snaps may occur. Seasonality of precipitation is strong with winter rains and summer droughts. Plant geographically, the region Mc corresponds to the western Hellenic, Aegean, western Anatolian and southern Anatolian floristic provinces of Meusel et al. (1965), or to the east Mediterranean floristic province of Takhtajan (1986).

Being outside of any glaciations the origin of the basins is either tectonic subsidence (e.g. Kopais), sinkholes in limestone (Lake Voulkaria), coastal lagoons, palaeochannels of rivers (oxbow lakes) or a combination of some of these. At Kopais a 60 m core reaches back to MIS 5e (Tzedakis, 1999) and a 120 m core back to MIS 11 (Okuda et al., 2001). Pollen preservation is often poor as the map in Bottema and Sarpaki (2003) shows: out of 13 sites on Crete, seven did not contain pollen.

The early **Late-glacial** was reached at only few sites, for example Kopais (Turner and Greig, 1975) but there it can easily be compared with marine cores (Rossignol-Strick, 1995; Kotthoff et al., 2008): NAP and especially Chenopodiaceae dominate the early Late-glacial, *Quercus* and *Juniperus* are prominent during the Bølling-Allerød, but decline during the Younger Dryas, when Poaceae show a peak. This means that the Younger Dryas is recorded here in the same way as it is at some sites in northern Italy, as a *Quercus*-depression. Deciduous oaks dominated the Early **Holocene** in the basin of Kopais and around Lake Voulkaria (Greig and Turner, 1974; Turner and Greig, 1975; Jahns, 2005). Traces of evergreen taxa, such as *Pistacia, Phillyrea* and *Quercus ilex* type (also named *Q. coccifera* type) are found at low abundances from the Early Holocene. They increase at some of the sites, usually *Pistacia* earlier than *Phillyrea* (*Pistacia lentiscus* is evergreen, *P. terebinthus* is summergreen; their pollen cannot usually be distinguished). *Ostrya* type (including *Carpinus orientalis*) became more frequent around 8000 cal yr BP. Between 8500 and 6000 cal yr BP the summergreen oaks decrease (dramatically at some sites). The emblematic genus *Olea* is sporadically found as early as 9000 cal yr BP at Voulkaria: it forms a continuous curve even before the beginning of the Minoan culture on Crete (Bottema and Sarpaki, 2003). Around 6000–5500 cal yr BP *Olea* increased in all records, sometimes together with *Phillyrea* and evergreen oaks. Because the rise in *Quercus ilex* was usually distinctly earlier than the one in *Olea* several authors discuss a scenario whereby a climate change to higher aridity caused an opening up of the vegetation, with less deciduous and more sclerophyllous taxa. Such a 'Mediterranization' was, therefore, perhaps not caused by but was followed by stronger human impact (Greig and Turner, 1974; Jahns, 2005). However, for Crete, Bottema and Sarpaki (2003) concluded that from c. 8500 cal yr BP onwards no major climate changes occurred but the Minoan culture especially (c. 5600–3350 cal yr BP / 3650–1400 BC) and also its followers had a strong impact on vegetation, very likely promoting *Olea* as a fruit tree (Jouffroy-Bapicot et al., 2021).

The second genus giving the name to the alliance Oleo-Ceratonion, *Ceratonia siliqua*, is virtually absent from pollen diagrams, probably because of its low pollen production related to entomophily (insect pollination). Some single grains were found at Kournas. The two other 'edible trees' *Juglans* and *Castanea* are recorded only rarely or are lacking. *Vitis* is regularly found at low to medium values but the pollen of *Vitis sylvestris* and *V. vinifera* cannot be separated (but the seeds can be distinguished). Cultivation and pasturing lasted for several millenia and resulted at many places in the development of evergreen scrubland called (in the east) phrygana. In areas of deciduous vegetation the so-called shiblyak developed. Several woody genera never reached Crete: *Fagus, Carpinus, Ostrya, Corylus, Betula*. In contrast, Crete is one of the regions where *Cupressus sempervirens* is native and still shows its broad, non-columnar form (Bagnoli et al., 2009; Caudullo and de Rigo, 2016a). Moreover, some plants such as the native palm *Phoenix theophrasti* and the endemic deciduous tree *Zelkova abelicea* are unique to Europe.

Climate reconstructions: Roberts et al. (2011).

Human impact: Allen (1990), Jahns (1993, 2005), Rackham and Moody (1996) and Bottema and Sarpaki (2003).

Phylogeography: Lumaret et al. (2002).

Delphinos & Kournas *Greece* (Bottema and Sarpaki 2003)

Mc 35.3° N, 24.5° E, 0-20 m a.s.l.

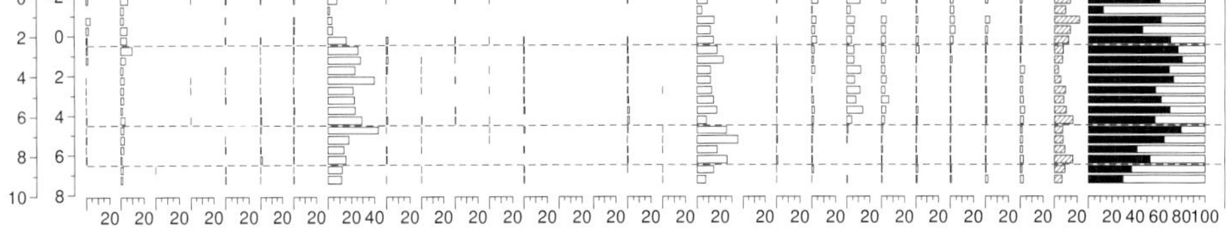

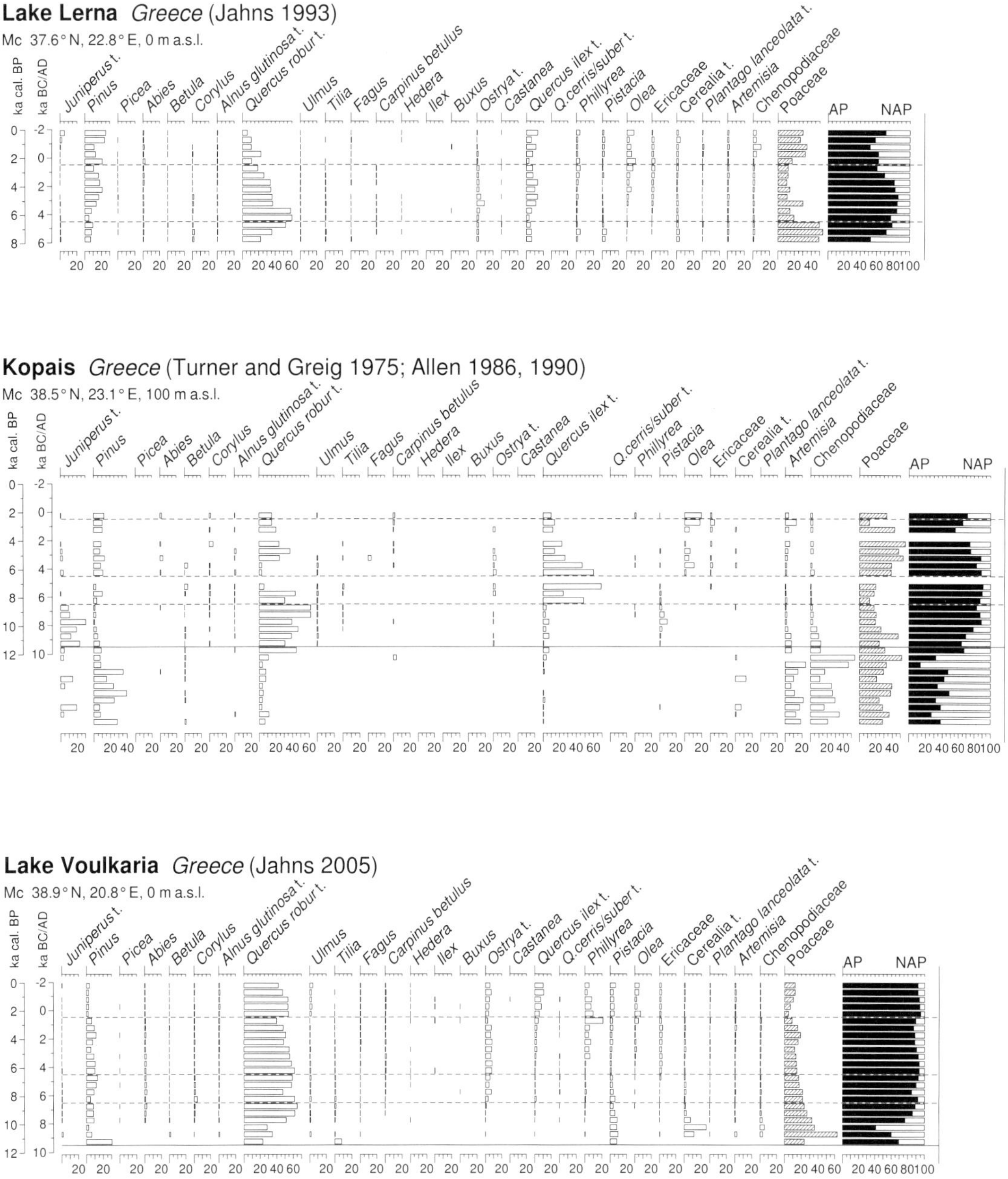

Fig. 4.3-17b. Selected sites in the eastern Mediterranean (Mc) region. Late-glacial and Holocene pollen records are presented as histograms at a 500-year resolution. The histogram from Crete combines the records from Delphinos up to 3000 cal yr BP with the ones from Kournas after 3000 cal yr BP (Bottema and Sarpaki, 2003).

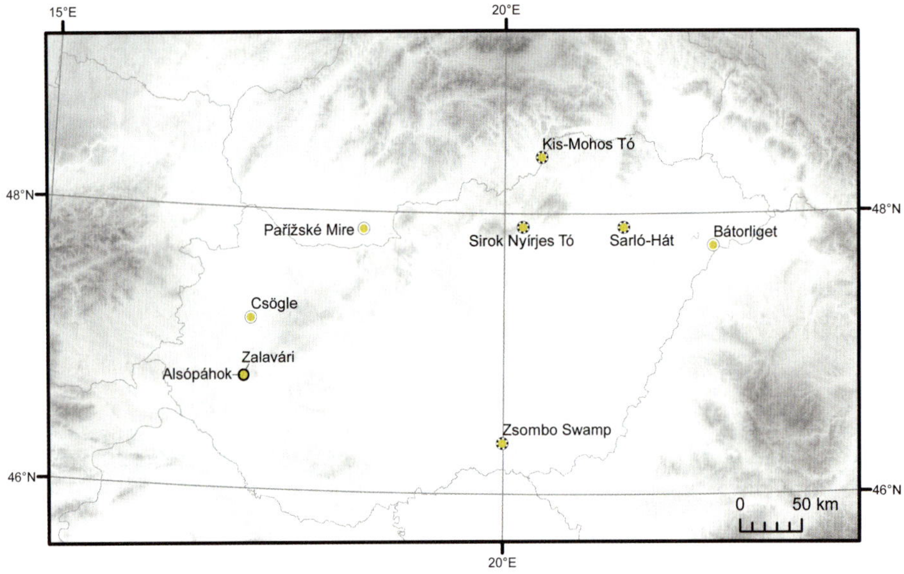

Fig. 4.3-18a. Map of selected sites in the Pannonian (Pa) region. Dots with black borders indicate sites shown in the histograms (Fig. 4.3-18b), white borders indicate sites used for the maps showing the history of selected taxa (see chapter 4.4) and dashed borders indicate sites used for both.

Pa, The Pannonian region

Pannonia or the Carpathian Basin (together with parts of the much smaller basin of Vienna) is plant geographically the westernmost province of the south-Siberian/Pontic/Pannonic floristic region *sensu* Walter and Straka (1970) and Walter (1974). The largest part of the Pannonian lowland is the Great Hungarian Plain (Nagy Magyar Alföld) surrounded by the arch of the Carpathians, the Balkan mountains and the Dinarides. The Pannonian climate is continental, summer-warm and dry: the mean annual precipitation is below 700 mm (in parts below 600 mm) and especially the period May to July is dry. The natural potential vegetation would be a mosaic of *Quercus pubescens* woods (showing the link to submediterranean vegetation) and various steppic types, some of which are controlled by edaphic conditions, for example on loess, on shallow soils over rock, on serpentine, or on salty soils. An overview of the modern vegetation with connections to vegetation history is offered by Fekete et al. (2014).

The region was never glaciated. The dominant soil types are brown forest soils and chernozems (Eckmeier et al., 2007). The natural archives are usually oxbow lakes (in palaeochannels, see Appendix 1 in Magyari et al., 2010), deflation basins (alkaline lakes, 'natron lakes') or landslide hollows. Exceptions are Sirok, a crater lake (Gardner, 2002) and Lake Balaton, a remnant of a bay of the Tethys Sea with three sub-basins and some hiatuses in the sediment series (Nagy-Bodor et al., 1995; Cserny and Nagy-Bodor, 2000). Oxbow lakes in palaeochannels and mires can be affected by hiatuses in their sediments or peat deposits. In the histograms of figure 4.3-18b we do not include the important site Nagy Mohos because it is only 400 m away from Kis-Mohos-tó. Moreover the core from the northern sub-basin has a hiatus between 14 600 cal yr BP and 9535 cal yr BP (Juhász, 2002). The core from the southern sub-basin includes a section dated 22 ka BP to 8.8 ka BP (Magyari et al., 1999, 2000) and a Mid-Holocene record of hydrological changes and human impact (Magyari et al., 2001): the latter can be compared to Sarló-hát. For the site Bátorliget see Sümegi and Gulyás (2004). As an overview of additional sites, Appendix S1 in Magyari et al. (2010) may help.

Refugia: at some sites the cores reach down to sediments of the Upper Weichselian (Fehér-tó to c. 28 cal ka BP; Sümegi et al., 1999, 2013 or Nagy Mohos to c. 21.7 cal ka BP; Magyari et al., 1999). In addition records of macro-charcoal from 31 sites of palaeo-soils in loess or archaeological sequences dated between c. 36.8 and 20.3 cal ka BP document the glacial survival of *Pinus sylvestris, P. cembra, Picea, Larix, Juniperus, Salix, Betula*, and (rare) *Carpinus* (Willis et al., 2000; Willis and van Andel, 2004). Rudner and Sümegi (2001) provide an overview of 21 Upper Weichselian sites including loess and lake deposits, soils, and archaeological sites. Some of the best continuous records are found either at the north-eastern end of the Great Plain for example at Bátorliget (Sümegi and Gulyás, 2004), Csardo-Nyríes-tó (with analyses of macrofossils and charcoal) and Báb-tava (Harrington, 1995; Willis et al., 1995; Magyari et al., 2008), at the north-west corner of the Great Plain (the Devecser Plain with Mezőlak; Sümegi, 2007) or in the northern Hungarian mountains (Bükk hills, Mátra hills; Willis et al., 1997; Magyari et al., 1999, 2001; Juhász, 2002; Gardner, 2002). The two sites Hajós-Kaszálók and Vörös-moscár in the southern centre of the Great Hungarian Plain are, unfortunately, very problematical because of hiatuses and sterile layers (Jakab et al., 2004).

Late-glacial and Holocene: the Oldest Dryas is characterized not only by high abundances of *Artemisia* and Poaceae but also by *Pinus* pollen. The local presence of pine is, at some sites, confirmed by records of charred wood of *P. cembra* (at Nagy-Mohos; Juhász, 2002).

The Late-glacial interstadial Bølling-Allerød is not well recorded because at Nagy Mohos there is a major hiatus (Bølling to Younger Dryas).

The Younger Dryas (or its end) was reached at Alsópáhok, Zalavári and Zsombo swamp but it is usually weakly expressed as a slight decrease in *Pinus* and a slight increase in *Artemisia* and Poaceae. This is consistent with the concept of a dry climate during the Younger Dryas in the eastern Mediterranean (Rossignol-Strick et al., 1992; Wick et al., 2003; Connor et al., 2013).

In the Early Holocene the rise in *Corylus* can be either before that of *Quercus* (Zalavár, Nagy Mohos) or more or less at the same time (Sarló-hát). During the Mid Holocene a major shift from 'mixed oak forest' rich in *Ulmus* to a dominance of *Fagus* and *Carpinus* occurred: the rise in *Fagus* can be before that of *Carpinus* (Alsópáhok, Zalavár, Nagy Mohos) or after it (Sirok Nyríes Tó, Sarló-hát, Csarod-Nyríes-tó, Báb-tava). At Sirok Nyríes Tó, Gardner (2002) found four cycles (between c. 6900 and 5200 cal yr BP) that involved late-successional (shade-tolerant) *Quercus* and *Carpinus* alternating with early-successional (light-demanding) *Corylus*. The cycles were possibly controlled by coppicing and/or pollarding during the late Neolithic and Copper Age and led to a dominance of *Carpinus*. A similar but younger shift in forest composition (c. 4000–1000 cal yr BP) is also discussed for the two neighbouring sites Csardo-Nyríes-tó and Báb-tava by Magyari et al. (2008): a strong long-lasting (second) elm-decline and a *Corylus*-decline are accompanied by a *Carpinus*-rise in two steps. The strongest rates of change occurred about 3700 cal yr BP when anthropogenic indicators also became more frequent.

Whereas the survival of dry grasslands and forest-steppe are discussed for central Europe as temporal and spatial 'bottlenecks' (Gradmann, 1933; Pokorný et al., 2015), their persistence through the Holocene is confirmed for the Great Hungarian Plain by either using indicator taxa for steppe or pollen percentages of woody taxa, where AP-values <70% indicate forest steppe and <50% steppe (both in surface and down-core samples (Magyari et al., 2010).

Human impact: studies are collected in the two volumes 19 and 20 of Varia Archaeologica Hungarica edited by Gál et al. (2005) and Zatykó et al. (2007).

Overviews: Buczkó et al. (2009).

Climate reconstructions: Feurdean et al. (2014).

Alsópáhok *Hungary* (Juhász 2002; Zatykó et al. 2007)

Pa 46.8° N, 17.2° E, 126 m a.s.l.

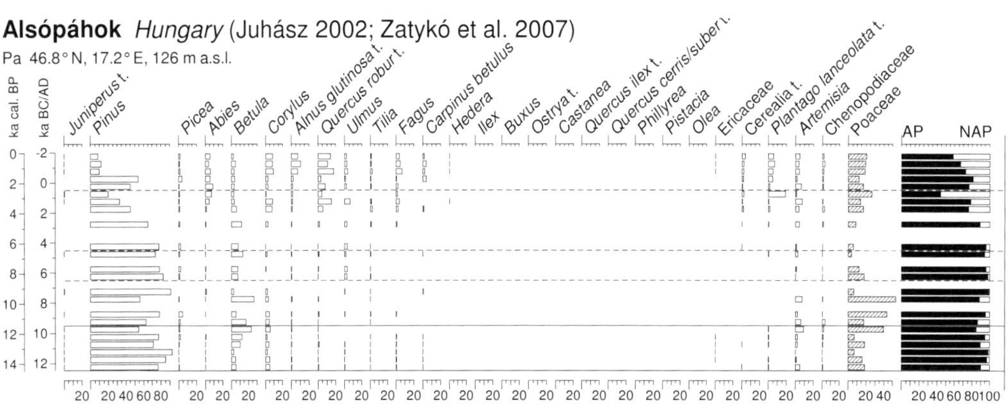

Zalavári *Hungary* (Juhász 2002)

Pa 46.8° N, 17.2° E. 188 m a.s.l.

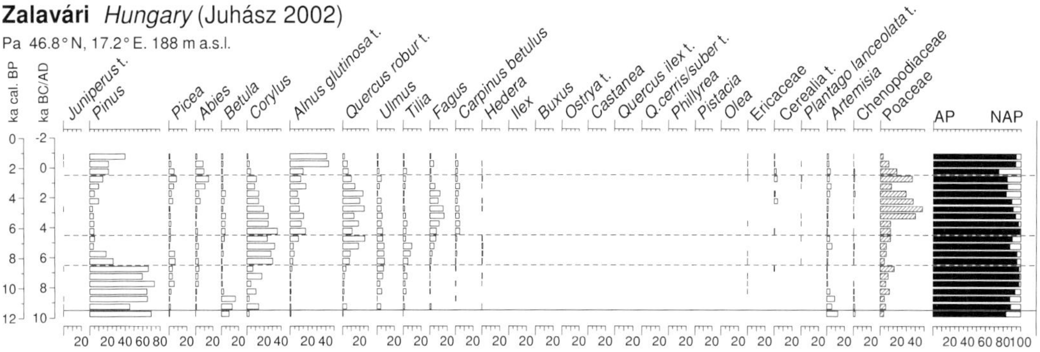

Zsombo Swamp *Hungary* (Nagy-Bodor 1995)

Pa 46.4° N, 20.0° E, 92 m a.s.l.

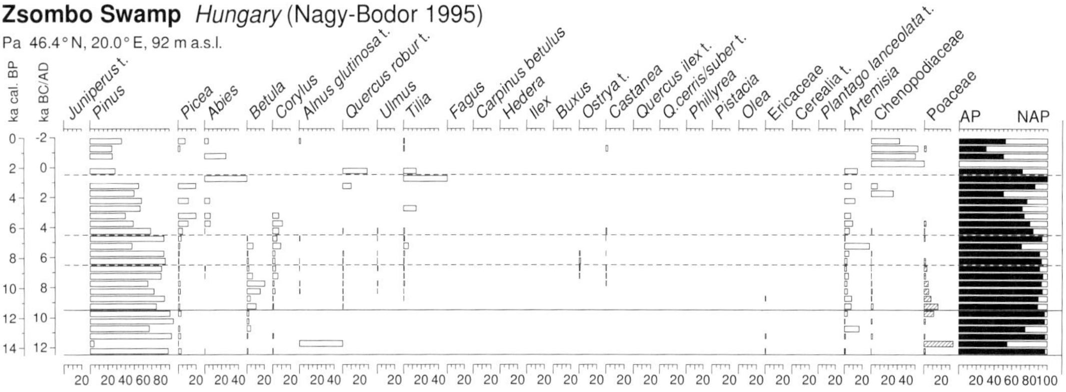

Sirok Nyírjes Tó *Hungary* (Gardner 2002)

Pa 47.9° N, 20.2° E, 200 m a.s.l.

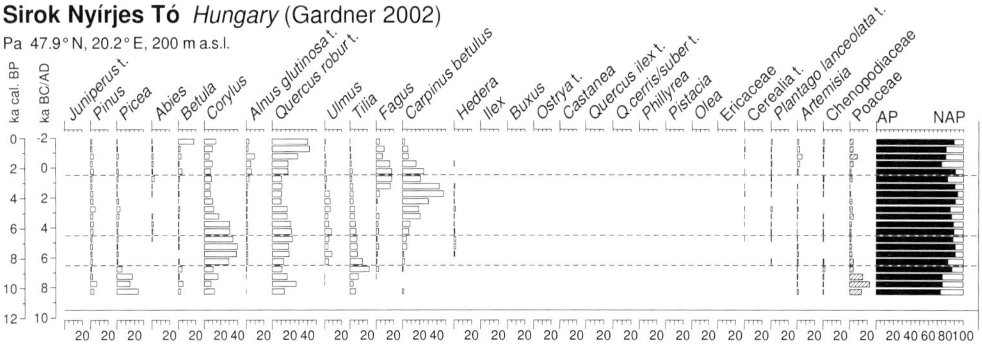

Kis-Mohos Tó *Hungary* (Willis et al. 1997)

Pa 48.4° N, 20.4° E, 310 m a.s.l.

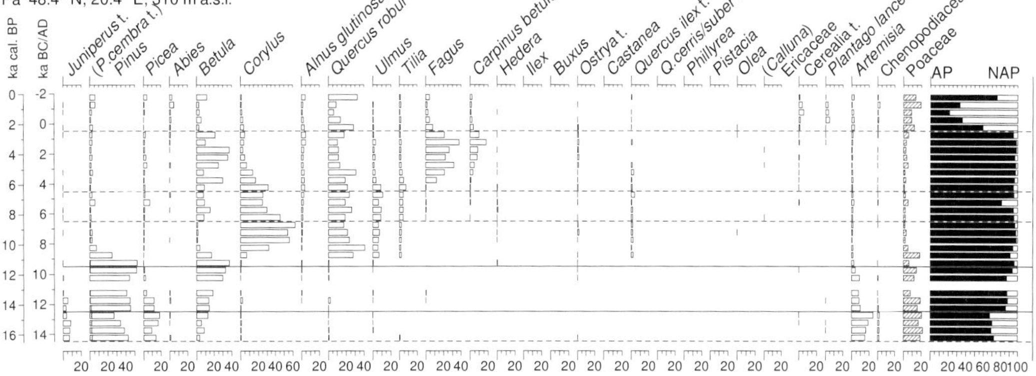

Sarló-Hát *Hungary* (Magyari et al. 2010)

Pa 47.9° N, 21.3° E, 86 m a.s.l.

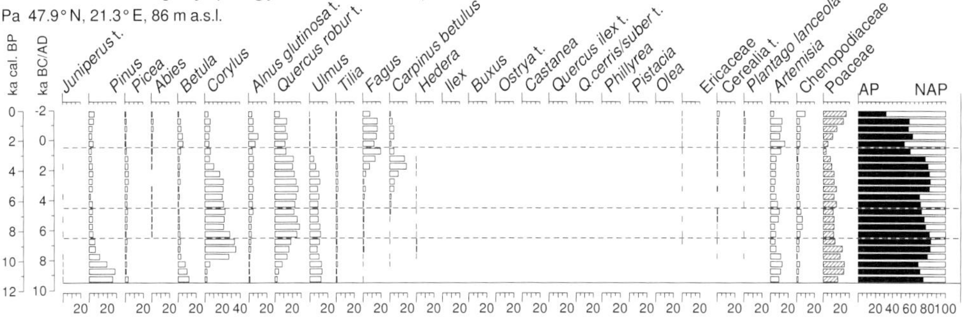

Fig. 4.3-18b. Selected sites in the Pannonia (Pa) region. Late-glacial and Holocene pollen records are presented as histograms at a 500-year resolution. At least four different landscape histories are visible: the Balaton region plus the Devecser Plain (Alsópáhok, Zalavári), the northern Hungarian hills (Bükk hills, near Kelemér and Mátra hills, with Kis-Mohos Tó and Sirok Nyríes Tó), the north-eastern lowland (with Sarló-hát) and the vast southern plain (with Zsombo Swamp).

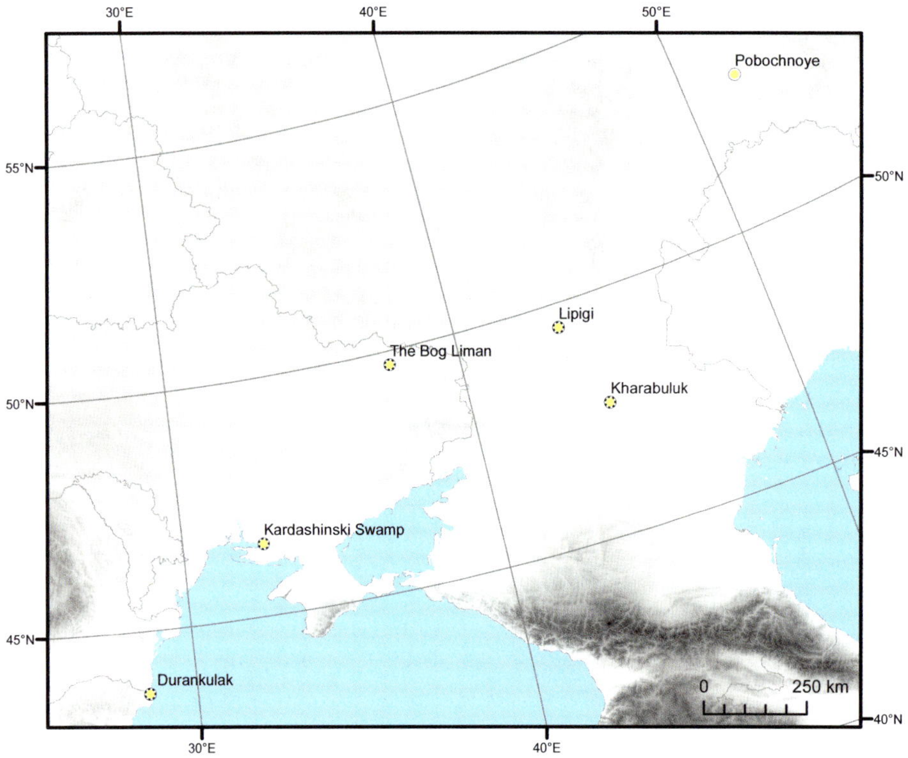

Fig. 4.3-19a. Map of selected sites in the Pontic (Po) region. Dots with black borders indicate sites shown in the histograms (Fig. 4.3-19b), white borders indicate sites used for the maps showing the history of selected taxa (see chapter 4.4) and dashed borders indicate sites used for both.

Po, Pontic region

This region comprises the landscapes adjacent to the north-western and northern shores of the Black Sea, that is the Dobrudscha in the south, the Ukrainian lowlands with the northern part of the Crimean Peninsula and the Don basin and areas adjacent to the southern Ural mountains. To the south-west the region borders on the south-eastern European region (Tg) and to the north and north-west on the eastern European region of deciduous trees (Tc). The southern limit is formed by the Jaila mountains on the Crimean Peninsula and the Great Caucasus and to the south-east by the semi-desert of the Caspian lowlands. The vegetation of the region today, with its dry climate, forms a vast steppic zone in the interior of the continent while broad transitional zones connect it in the north and north-west through forest-steppe to the temperate deciduous forests and, in the south-east, through steppe rich in *Artemisia* to the semi-desert. In Bohn et al. (2003), map 16 and accompanying descriptions are useful. Plant geographically, the region largely coincides with the Pontic floristic province of Meusel et al. (1965), whereas Takhtajan (1986) does not consider it as a separate region but includes it in the east European floristic province.

The Pontic region was not glaciated during the LGM but was affected by changes in the coastlines of the Black Sea (Fig. 1-2). Globally low sea levels during the LGM disconnected the Black Sea from the Mediterranean and caused a regression into the southern basin that became a large freshwater lake (recorded in molluscs). The paper by Ryan et al. (1997) and the popular book by Ryan and Pitman (1998) rekindled the discussion as to whether the reconnection of the Black Sea to the Mediterranean Sea was abrupt, gradual or oscillating, what role the early melt-water input or tectonics played and whether Neolithic migrations were accelerated because of flooded sea shores (Aksu et al., 2002; Chepalyga, 2006; Hiscott et al., 2007; Mudie et al., 2007; Ryan, 2007).

Late-glacial and Holocene: so far the Late-glacial vegetation history is hardly known from terrestrial sites in the Pontic region because most corings only reached the end of the Younger Dryas. An exception is the well-dated site Pobochnoye in the Buzuluk forest (Fig. 4.3-19a; Kremenetski et al., 1999c), where the transition from the Younger Dryas to the Holocene is marked by a sharp rise in *Pinus* and a decline of Chenopodiaceae. A second rise in *Pinus* with sharp declines of *Artemisia* and Chenopodiaceae occurs at Pobochnoye as late as c. 6500 cal yr BP. In contrast to the terrestrial cores the marine cores may cover the Late-glacial and even the Upper Pleistocene (Mudie et al., 2007; Filipova-Marinova et al., 2013a). Whether woody taxa formed cryptic refugia during the LGM, for example along rivers, has until now been very difficult to evaluate.

The Holocene vegetation development is characterized by generally low values of tree pollen, among which especially *Pinus* and summergreen *Quercus* are prominent while at some sites *Tilia* played a role (Fig. 4.3.2-26). The pollen type *Ostrya* including *Ostrya carpinifolia* and *Carpinus orientalis* is only recorded at the two south-western sites Durankulak and Shabla-Ezeretz. The sum of NAP increases from west to east, mirroring the decreasing importance of trees towards the true steppe in the east. Among the herbs, the high values of Poaceae, Chenopodiaceae and *Artemisia* are indicators of steppe or forest-steppe. In the southernmost parts of the Pontic region a change from forest-steppe during the first half of the Holocene to steppe during the second half occurred. In this area, with human settlements for more than 7000 years, this change may be caused either by climate or by human impact or by both. **Human impact** has been widespread since the Neolithic; for the western Black Sea coast see Filipova (1985), Bozilova and Filipova (1986), Bozilova and Tonkov (1998) and Marinova (2003); for Razdorskoye with more than ten archaeological layers see Kremenetski (1997a,b); for the northern Black Sea coast where early farming started in the Kalamitian stage of the Black Sea, 7.1–4.0 cal ka BP see Streif (1990). Further east in the eastern European forest-steppe Shumilovskikh et al. (2019) provide a comparison of an off-site record (oxbow-lake) with archaeological sites for the last 2500 years.

Climate reconstruction: Feurdean et al. (2014).

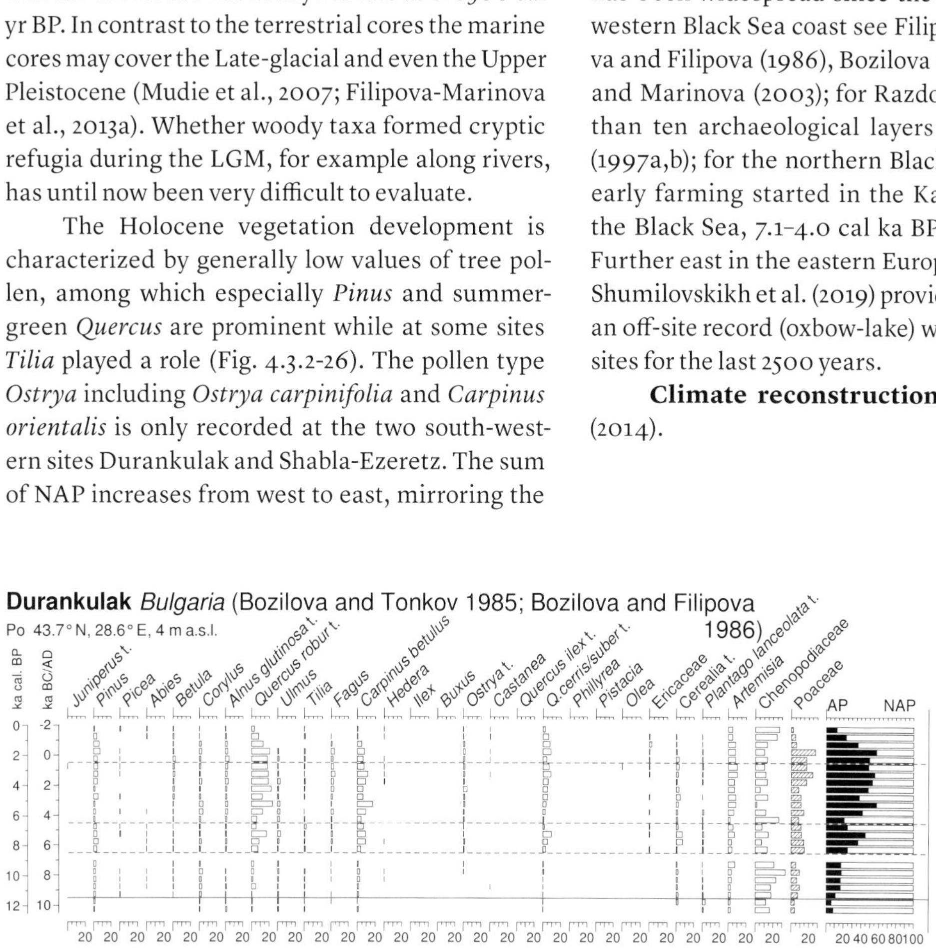

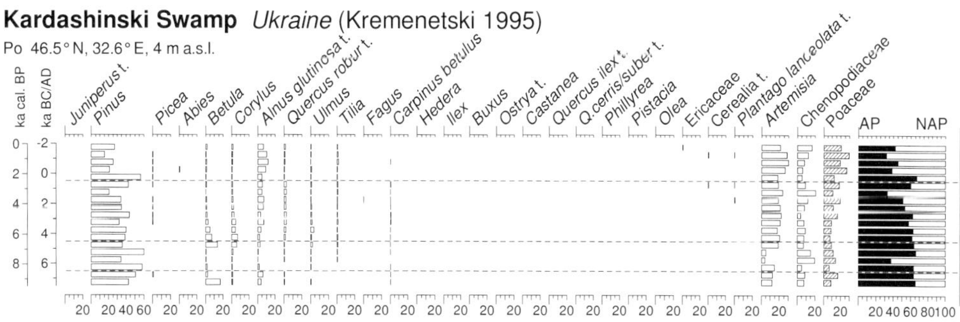

The Bog Liman *Ukraine* (Bezusko 1973)
Po 49.7° N, 37.7° E, 150 m a.s.l.

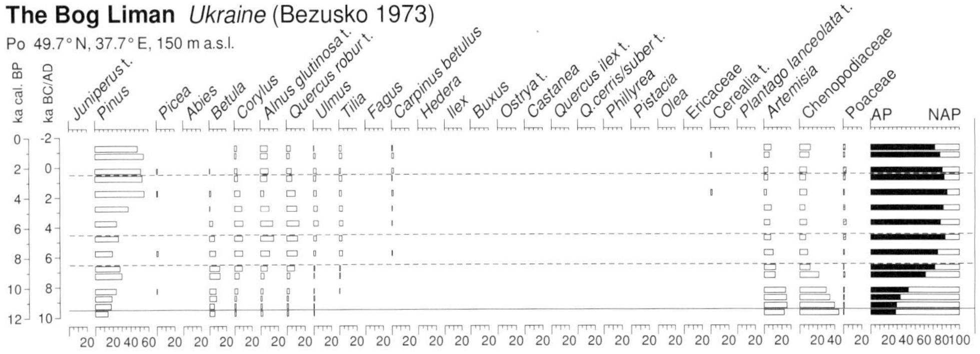

Lipigi *Russia* (Kremenetski et al. 1999a)
49.5° N, 43.4° E, 90 m a.s.l.

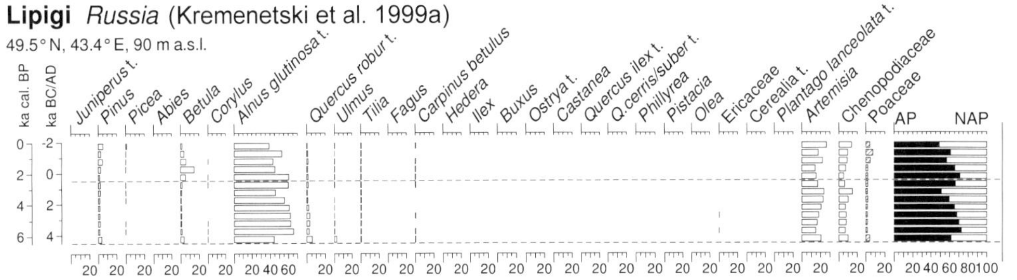

Kharabuluk *Russia* (Kremenetski et al. 1999a)
Po 47.7° N, 44.3° E, 70 m a.s.l.

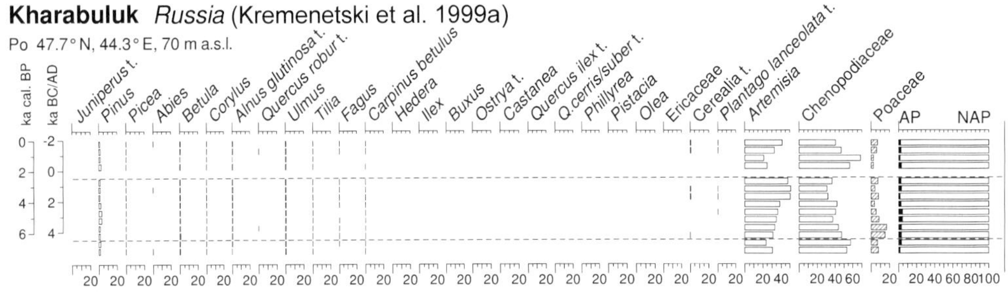

Fig. 4.3-19b. Selected sites in the Pontic (Po) region. Late-glacial and Holocene pollen records are presented as histograms at a 500-year resolution arranged from west to east. The number of potential coring sites is not high (outside any glaciation) and mostly restricted to coastal sites (e.g. Limans near the Black Sea) or swamps on river terraces. As can be expected in steppic environments the portion of the NAP is generally high and it increases towards the east. Tree pollen either comes from *Pinus*, a high pollen producer able to grow on sandy soils, or from taxa of riparian woods such as *Quercus* or *Alnus*.

Box 9

Tree species migration models

Heike Lischke

Swiss Federal Institute for Forest, Snow and Landscape Research WSL, Switzerland

Dynamic models simulating tree migration combine environmental influences with movement in space. They act on a grid with environmental conditions specific for each grid-cell (Fig. B9-1a). The KissMig model (Nobis and Normand, 2014) combines general environmental suitability for a single species with a transition rate for the species from cell to cell which is estimated from external data. Other approaches such as Forest Landscape Models (He et al., 2017; Shifley et al., 2017) or the dynamic global vegetation model LPJ-GM (Lehsten et al., 2019) additionally simulate in each grid-cell the interactions between individuals of different species – taking into account the vertical and horizontal heterogeneity in a forest stand – resulting in forest dynamics and explicit seed production and dispersal. These models differ strongly in their resolution, ranging from 1 m (iLand: Seidl et al., 2012), 20 m (LandClim: Schumacher et al., 2004), 100–1000 m (TreeMig: Lischke et al., 2006) to 1000 m (LPJ-GM). The local dynamics of each species are given by the environment-driven life-cycles of individual or groups of similar trees (Fig. B9-1c). These consist of survival and germination of seeds in the seed bank, sapling development, tree growth, mortality (increased by disturbances) and seed production from a certain height or age of maturity. The trees compete for light, that is taller trees shade smaller ones and reduce their growth and survival, in some models also for soil water. The intra- and inter-specific competition in combination with the life-cycles and heterogeneity results in community dynamics, typically forest succession.

Seed dispersal within and between grid-cells is described by simplified 'dispersal kernels', indicating the proportion of seeds landing within a given distance z from the mother tree (Fig. B9-1d). In the TreeMig model, the kernel is the sum of a steeply declining function for short distance and a flatter but longer function for the rare long-distance transport. The seeds transported into a sink-cell from the trees of this species in all other (source) cells enter the seedbank in the sink-cell. Pollen dispersal can be included analogously, of course, with different dispersal kernels.

The dispersal needs to be calculated with grid-cells not much larger than the length of the dispersal kernel (i.e. ≤1 km). To keep computational load manageable, in some approaches the local dynamics are calculated only for a subset of the grid-cells (Nabel et al., 2014; extrapolation), for example on a regular mesh, and instead of calculating seed dispersal from each cell to each other, the much more efficient Fast Fourier Transform Convolution is used (Lehsten et al., 2019).

The models' local dynamics have been parameterized from literature and current forest compositions. For the dispersal kernels rough estimates based on seed weights and dispersal vectors (e.g. wind speeds) were used. A better parameterization would require long-term spatio-temporal data of forest compositions which could perhaps come from tree pollen records.

Applications with the TreeMig model, for example in the Valais from 14 000 to 6000 cal yr BP, driven by immigration derived from pollen data show a vivid pattern of species compositions resulting from different waves of movement, and indicate that including migration and succession on top of species' climate dependence is crucial (Lischke, 2004, 2005). Simulations for the future climate change in the boreal zone exhibit a very strong migrational lag lasting for several centuries and a strong reduction of migration speed due to competing species (Epstein et al., 2007; Goetz et al., 2011). In similar simulations in Switzerland, the migrational lag is less pronounced in the shift of the tree species up the mountain slopes, but strong for some more drought-adapted species from the submediterranean region due to the barrier function of the mountains and the comparably long distances the species have to cover (Lischke, 2020).

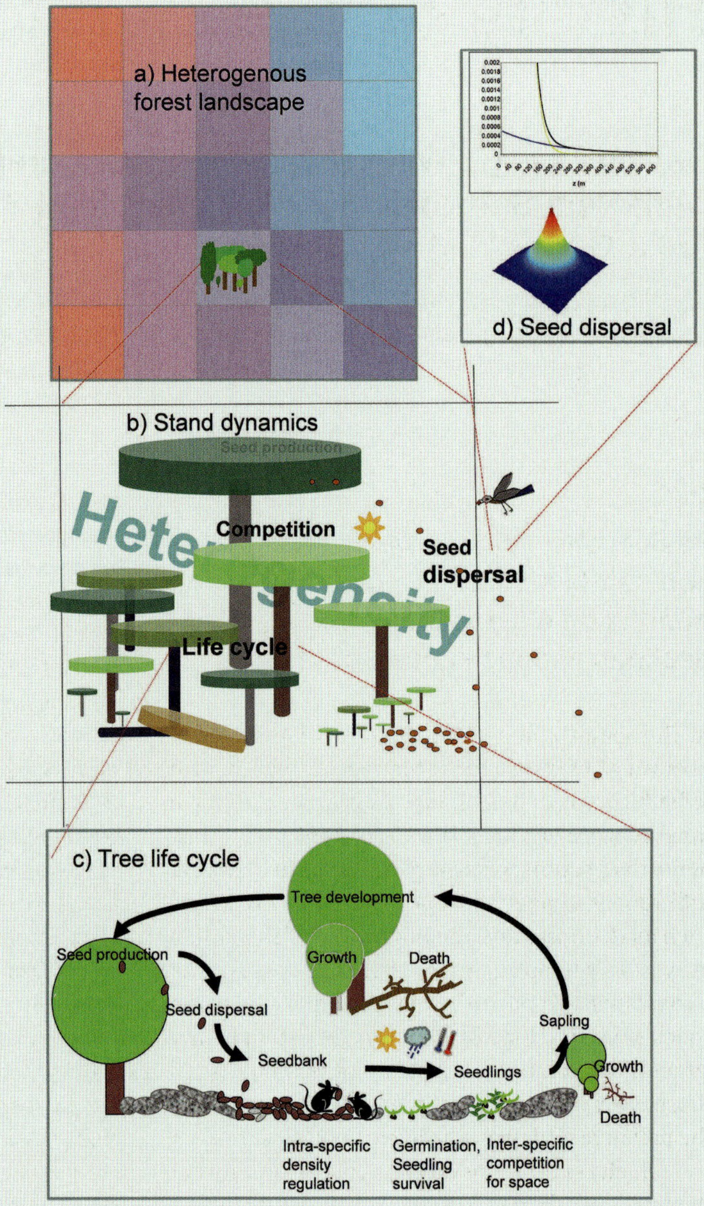

Fig. B9-1.

Concept of a Tree Species Migration Model: (a) The model acts on a grid with environmental conditions specific for each grid cell; (b) For each grid cell the interactions between individuals of different species are simulated, taking forest heterogeneity into account; (c) The local dynamics of each species are given by the environment-driven life-cycles of individual or groups of similar trees; (d) Seed dispersal within and between grid cells is described by simplified 'dispersal kernels', indicating the proportion of seeds landing in a given distance z from the mother tree.

4.4

History of selected taxa

Gerhard Lang[1], Willy Tinner[1,2], César Morales-Molino[1,2], Christoph Schwörer[1,2], Lieveke van Vugt[1,2], Erika Gobet [1,2], Brigitta Ammann[1,2]

[1] Institute of Plant Sciences, University of Bern, Bern, Switzerland
[2] Oeschger Centre for Climate Change Research, University of Bern, Bern, Switzerland

4.4.1

Basics: absence, presence, population establishment, spatial expansions, population expansions, population declines and extinctions of taxa

In European pollen records, the frequencies of taxa vary between and within regions with regard to both timing and abundances. If curves of single taxa are parallelized in space (e.g. from west to east or north to south) according to age, often temporal differences appear in important data properties such as the onset and the maximum of the pollen curves. It is tempting to interpret the onset of the curve as arrival and the increase as expansion of the species, as was done by von Post (1924, 1929) and later on Bertsch (1935, 1940) and others, though in those early days, it was impossible to synchronize the pollen records using radiocarbon dating.

Such assumptions may appear convincing but they require a thorough check. The assumption that the (usually exponential) increase of pollen before the first peak (the so-called rational limit) reflects population expansions has been widely accepted. Much more debated was, and is, the onset of the regular pollen occurrence (the so-called empirical limit). Does it reflect the immigration of a taxon or was the taxon present before in the area, though at lower population densities? Rudolph (1930) preferred this latter interpretation and assumed that at the latest during the Early Holocene all taxa had already reached central Europe. In his view, vegetation history (and as a part of it the 'central European basic succession' or 'Grundsukzession') reflected (subsequent) expansions of taxa in response to changing climate and environment instead of migrational processes. Later on, Welten (1944) corroborated Rudolph's interpretation. On the basis of spatial distribution densities of trees, Welten (1944) showed that the presence of 70 individuals of a rare tree species in a forested lake catchment with a radius of 2300 m may result in the detection of one pollen grain for every 10 000 grains counted. Seventy individuals in such a small area around a site are many, if compared to detection uncertainties of ecological field data. Welten's estimation thus shows that past occupancy might be seriously underestimated when using pollen or other (less abundant) plant remains (Fig. 4.11-4). The detection frequency (0.01%) suggested by Welten corresponds to the sedimentary situation before the onset of the empirical limit, with one pollen grain in every 20th sample for pollen sums of 500, showing that the arrival or even the initial population establishment of a species (with >50 individuals) can occur far before the onset of the empirical limit in pollen diagrams. Consequently, Welten (1944) proposed to avoid the use of pollen values for determining the arrival or local presence of a taxon and concluded that migrational reconstructions are illusory ('Scheinwanderungen'). Firbas (1949) largely took up the view of Rudolph and Welten remaining critical with respect to arrival and migration reconstructions. After discussing all relevant dispersal processes which could have induced fast migration rates at or before the onset of the Holocene (e.g. animals, wind, streams, ice floes), he partially rejected the early migrational reconstruction attempts of Bertsch (1935, 1940) to emphasize that the assessment of the relevance of migrational processes for the vegetation history of Europe primarily depends on the unequivocal evidence of the arrival time of a species. Similarly, Godwin (1956) suggested that >50% of the British flora was already established by the end of the Late-glacial. Even for species that expanded late in north-western Europe (e.g. *Fagus sylvatica*) he excluded low migrational rates as a cause (p. 208). Instead, he emphasized the presence of previously established small foci that allowed the species to expand quickly once the climatic conditions became more favorable.

Watts (1973) and later Birks (1986b) refined the relationships between the pollen-stratigraphic evidence and the pollen-inferred arrival and expansion of arboreal plants. Lang (1992) corroborated this approach by including macrofossil evidence. Macrofossils (or stomata as a proxy for macrofossils) provide better evidence of local presence than pollen, but given the low numbers preserved in the sediment, they have significantly less power as evidence of the absence of a taxon at a site. Since it is impossible to prove the spatial absence of a taxon by pollen or macrofossils (Birks, 2003, Hicks, 2006; Tinner and Lotter, 2006), the reconstruction of the presence and thus of the immigration of a taxon remains obscure (Seppä, 2007). Indeed, reconstructing immigration and migrations by pollen or macrofossils assumes away the challenges of imperfect detection (see Kéry, 2011). We thus modify the original definition by Lang (1994) to consider this unsolved caveat, avoiding the term immigration. The terms 'apparent regional' and 'apparent local presence' take into account detection uncertainty (Kéry, 2011).

On the basis of pollen and macrofossil records seven (dynamic) phases might be distinguished (Fig. 4.4.1-1). 1) No pollen and macrofossils in the sediment record: taxon might be absent. 2) Occasional pollen grains (after absolute pollen limit), no closed pollen curve, no macrofossils: apparent regional presence of the taxon, range close to the site or alternatively, the taxon might be present locally but with low numbers (cryptic presence). 3) Continuous pollen record (after empirical limit) and/or first macrofossils: apparent local presence of the taxon, local population establishment. 4) Increase of pollen abundances (after rational limit) to reach a first peak and/or regular occurrence of macrofossils: population (mass) expansion of a taxon. 5) Secondary pollen and macrofossil peaks, often smaller than the first peak: population oscillations of the taxon. 6) Decrease of pollen and/or macrofossil abundances: decline of the population of the taxon. 7) Disappearance of pollen and macrofossils in the sediment record: population collapse, potential local extinction.

With regard to pollen, this scheme may apply to wind-pollinated plants of Europe with average or moderate pollen production and dispersal (e.g. *Abies, Picea, Ulmus, Quercus, Fagus, Carpinus*, Table 4.3-1). Single occasional pollen finds (absolute limit) of insect-pollinated taxa (e.g. *Buxus, Hedera, Ilex*) may already indicate the local presence of the plant, while for strong pollen producers (e.g. *Pinus, Betula*) care should be taken when inferring the local presence only on the basis of pollen. For this latter group the (latest) local presence becomes likely when reaching the rational pollen limit.

The conventional interpretation that pollen and macrofossil abundances primarily reflect population dynamics (and not productivity changes such as increased flowering or organ production) is supported by recent interdisciplinary studies comparing dynamic modelling outputs, pollen and macrofossil records (e.g. Lotter and Kienast, 1992; Keller et al., 2002; Heiri et al., 2006; Henne et al., 2011; Schwörer et al., 2014b). Productivity changes may indeed occur at annual scales, but are supposed to become irrelevant at decadal to millennial scales, also because the long-term dynamics may exceed the life span of the individual organisms.

Another problem when tracking the spatial occurrence or expansion on the basis of palaeo evidence (e.g. pollen, stomata, macrofossils) is taxonomic resolution. Pollen types may represent species, group of species, subgenera, genera or even families. Species-rich taxa such as *Quercus robur* type cannot provide details about the history of the involved species if macrofossil evidence is lacking or if the today's spatial species distribution does not deliver additional information.

The simplest way to illustrate the spatial occurrence of a taxon in time is to depict its local pollen abundance in space for a certain time period (Fig. 4.4.1-2) by using dots with different sizes or different colours for different pollen abundances (e.g. Giesecke et al., 2014; Ruosch et al., 2016; Brewer et al., 2017). Such a procedure has the advantage of being easily producible and accessible, but the disadvantage of being unrelated to ecological

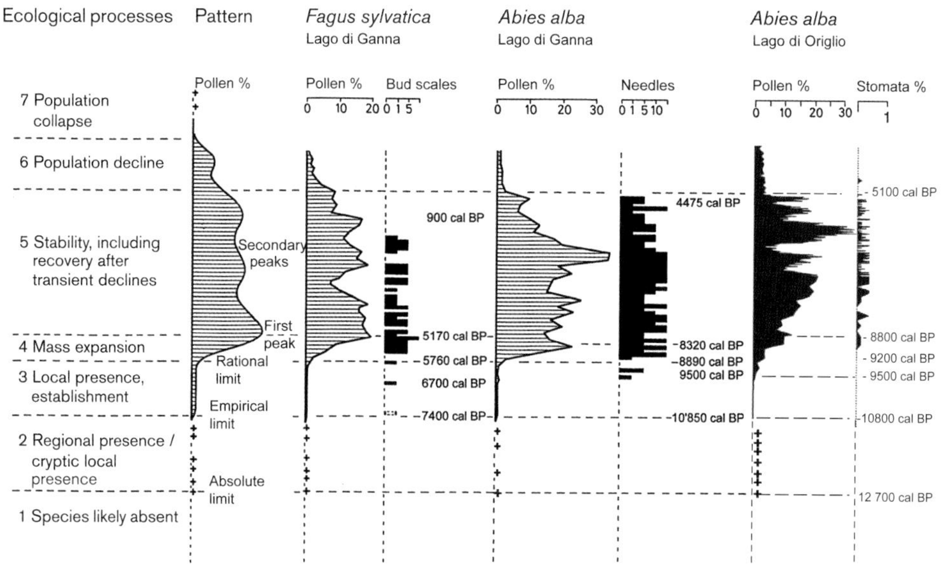

Fig. 4.4.1-1. Biostratigraphic patterns in pollen, macrofossil and stomata records and inferred ecological processes. The phases, processes and schematic pollen sequence follow Watts (1973) and Birks (1986a) and the curves of beech (*Fagus sylvatica* L.) and silver fir (*Abies alba* Mill.) follow Schneider and Tobolski (1985) and Tinner et al. (1999). Ages are in calibrated years before present (cal yr BP) where 'present' is AD 1950.

processes, such as the establishment, expansion, decline or collapse of populations. Also, because no attention is paid to the preceding and following temporal pattern of pollen abundances (e.g. closed curves v. single spikes or clusters of pollen, samples at the very base of stratigraphies), such illustrations are susceptible to outliers which might be unrelated to species or vegetation conditions. For instance, reworked material might be present at the base of biostratigraphies that may derive from older, allochthonous material. Another comparable approach is to map combined palaeobotanical evidence from pollen, macrofossils and soil charcoal to show the number of samples with past occurrences of a taxon for a certain period (Krebs et al., 2019; Fig. 4.4.1-3). On the basis of such combined multiproxy evidence, probabilistic approaches have been used to map past occurrences of plants for a certain time or period (Krebs et al., 2019; Fig. 4.4.1-4).

Past abundances, local population establishment (empirical limit) and population expansion (rational limit) of taxa are usually visualized by isopollen (see Fig. 4.4.1-5) or isochrone maps (see Fig. 4.4.2-2). Isopollen maps show sites that are connected with lines of the same pollen percentage values (on the basis of a homogenous pollen sum including AP and NAP) for a certain time or period. Such pollen maps were developed first by Szafer (1935) for Poland, and later by Firbas (1949) for central Europe and other authors for other regions (e.g. Birks, 1975; Birks and Saarnisto, 1975), as well as Huntley and Birks (1983) for the entirety of Europe. Isopollen maps have the great advantage of being relatively objective and circumventing interpretations. On the other hand, they have the disadvantage that one taxon results in several maps – one for each time interval – and that they are difficult to follow for non-specialists.

Maps with isochrones show lines which connect study sites with the same or similar findings for the same time: for instance, a line which designates locations where *Fagus* reaches the empirical limit at 3000 cal yr BP. Isochrones are thus useful to illustrate spatial expansions of taxa through time. Because the selected threshold of pollen is connected to interpretation insecurities, isochrone maps are less objective than isopollen maps, though they have the great advantage of illustrating long periods with one single map (e.g. Late-glacial and Holocene

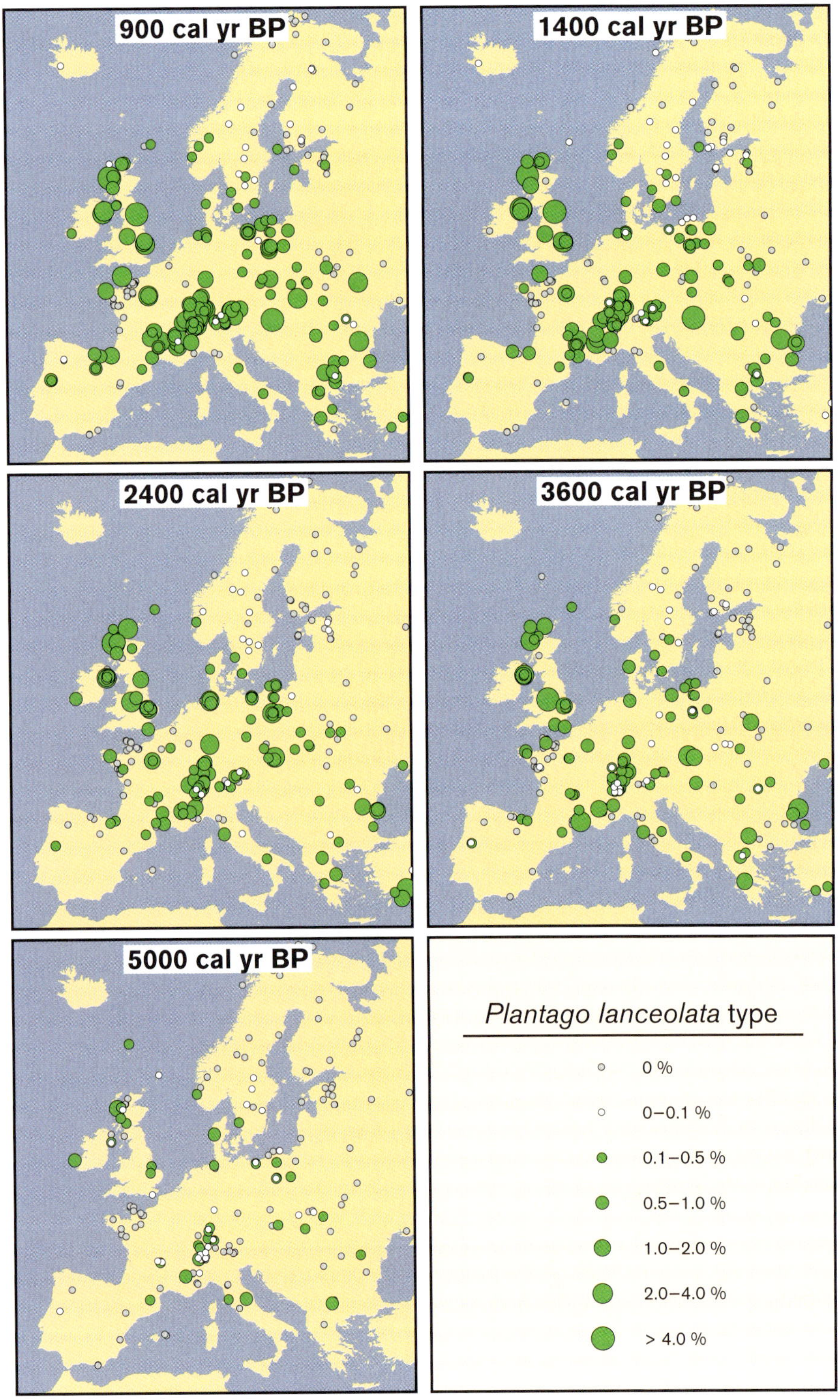

Fig. 4.4.1-2. Maps showing the presence and abundance of *Plantago lanceolata* type pollen during five time periods, chosen to represent the medieval forest clearance (900 cal yr BP), the Migration period (1400 cal yr BP), Iron Age (2400 cal yr BP), Bronze Age (3600 cal yr BP) and Neolithic (5000 cal yr BP). Modified from Giesecke et al. (2014).

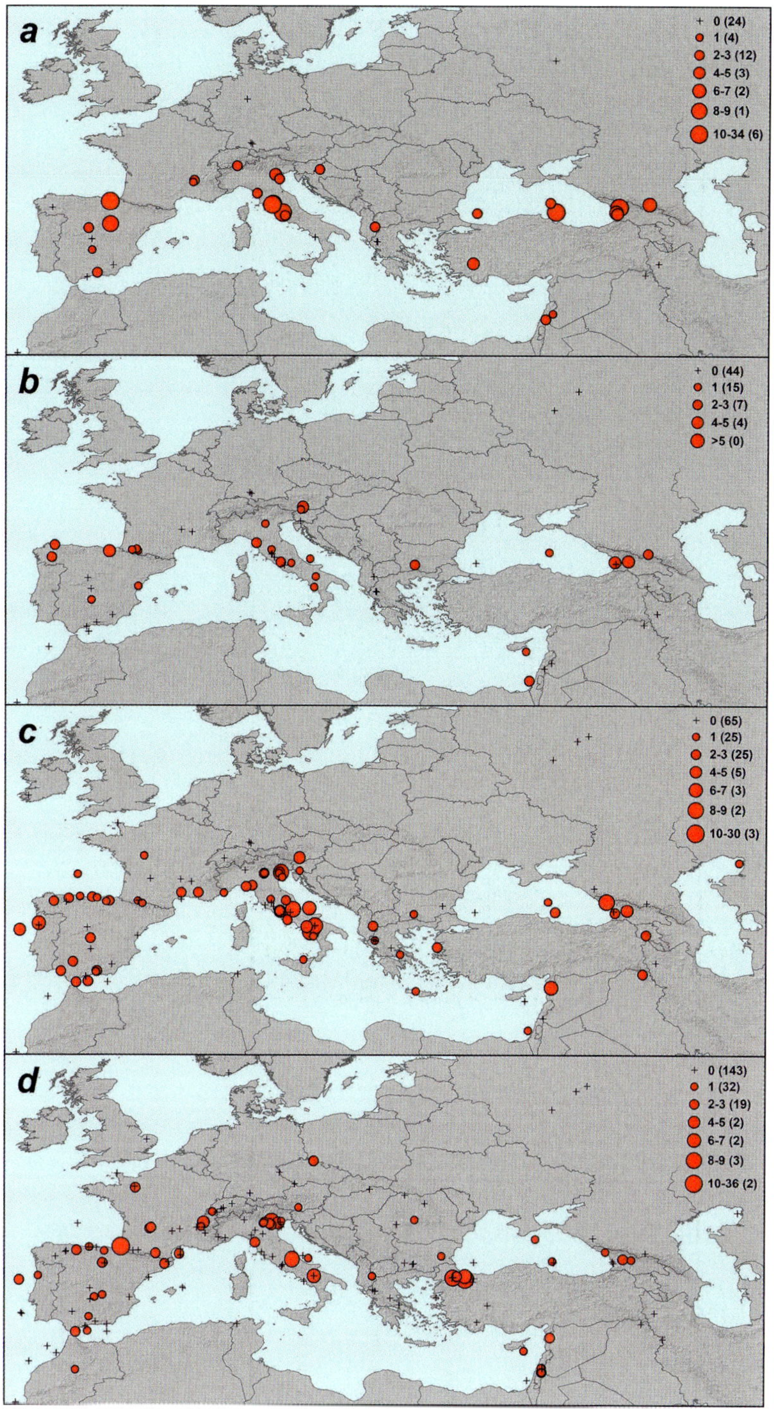

Fig. 4.4.1-3. Maps of sweet chestnut (*Castanea* spp.) occurrence for selected periods. Palaeobotanical sites covering four periods a) 191 000–110 000 cal yr BP, b) 110 000–71 000 cal yr BP, c) 71 000–29 000 cal yr BP and d) 29 000–18 200 cal yr BP, classified according to the number of positive samples. For every class the number of sites is given in brackets. Dot sizes reflect number of samples with palaeobotanical evidence (e.g. pollen, macrofossils, charcoal) for *Castanea* at the sites. Modified from Krebs et al. (2019).

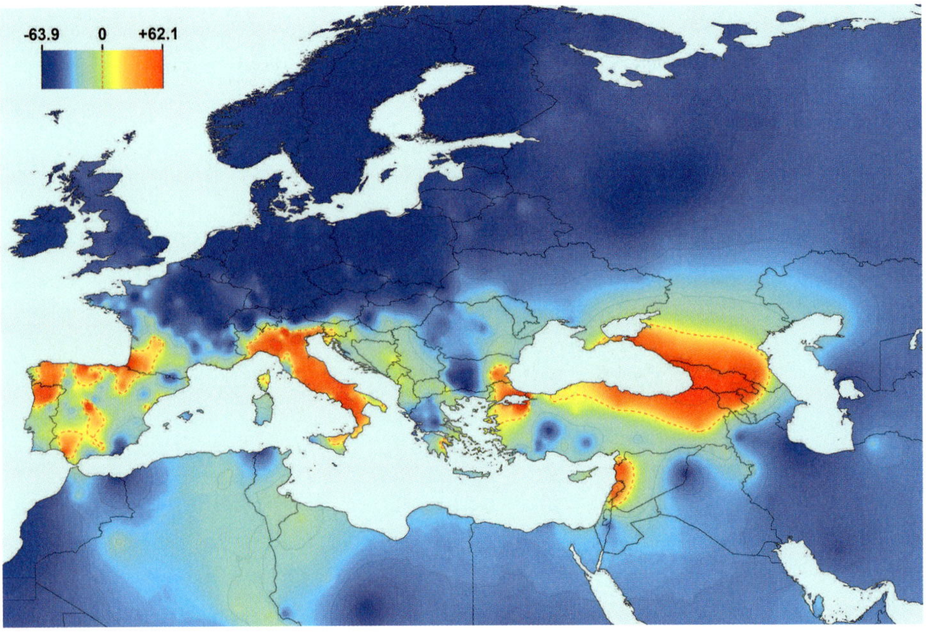

Fig. 4.4.1-4. Map of sweet chestnut (*Castanea* spp.) refugium probability within Europe, obtained by using ordinary kriging as a method of spatial interpolation, starting from positive and negative Index of Refugium Probability (IRP) values calculated for 2827 palaeobotanical sites. Modified from Krebs et al. (2019).

history). Isochrone maps can be drawn for selected pollen percentage values (e.g. Giesecke and Brewer, 2018) or for pollen patterns that are considered to reflect ecological processes (e.g. empirical limit, rational limit). Such an ecological event can be compared with today's distribution of the taxon if the range is shown on the same map (Lang, 1992), leading to a modified isochrone map.

The maps in the following chapter show the empirical, absolute or rational pollen limit of a given taxon at selected sites, as well as isochrones in 2000-year steps. The isochrones were calculated using the nearest neighbour spatial interpolation technique in ArcGIS 10.8, which is based on the weighted values of neighbouring datapoints. The maps are complemented by today's distribution of the species (Caudullo et al., 2017) and by a bottom bar diagram showing the number of sites that share similar chronologies with regard to the empirical, absolute or rational limit (as visualized by the same colour). Together they show spatial and temporal aggregations of change with regard to the apparent local presence or the population establishment of a taxon, ultimately they thus allow the reconstruction of the timing of the expansion of a taxon at the continental scale. The empirical limit is rather conservative with respect to false absences (Birks, 2014), while the absolute limit is susceptible to false presences. Therefore, the latter might be used only for taxa with rare pollen occurrence. False presences might be relevant for high pollen producers (e.g. *Pinus sylvestris*): this issue can be assessed by considering complementary proxies such as stomata and macrofossils or by choosing the rational limit. Here we define the empirical limit as continuous presence in at least five adjacent samples spanning at least 1000 years, without interruption of more than one sample. The rational limit we define as rapid initial increase in the pollen curve. These definitions are at the basis of the isochrone maps presented in this chapter. Other authors sometimes use pollen threshold values (e.g. 1%) instead of the empirical limit to reconstruct first local population establishments, mainly because the procedure is easier for numerical data analysis. Using thresholds may decrease the likelihood of false presence (for continuous records), but has the disadvantage to increase the probability of false absence, which may mean

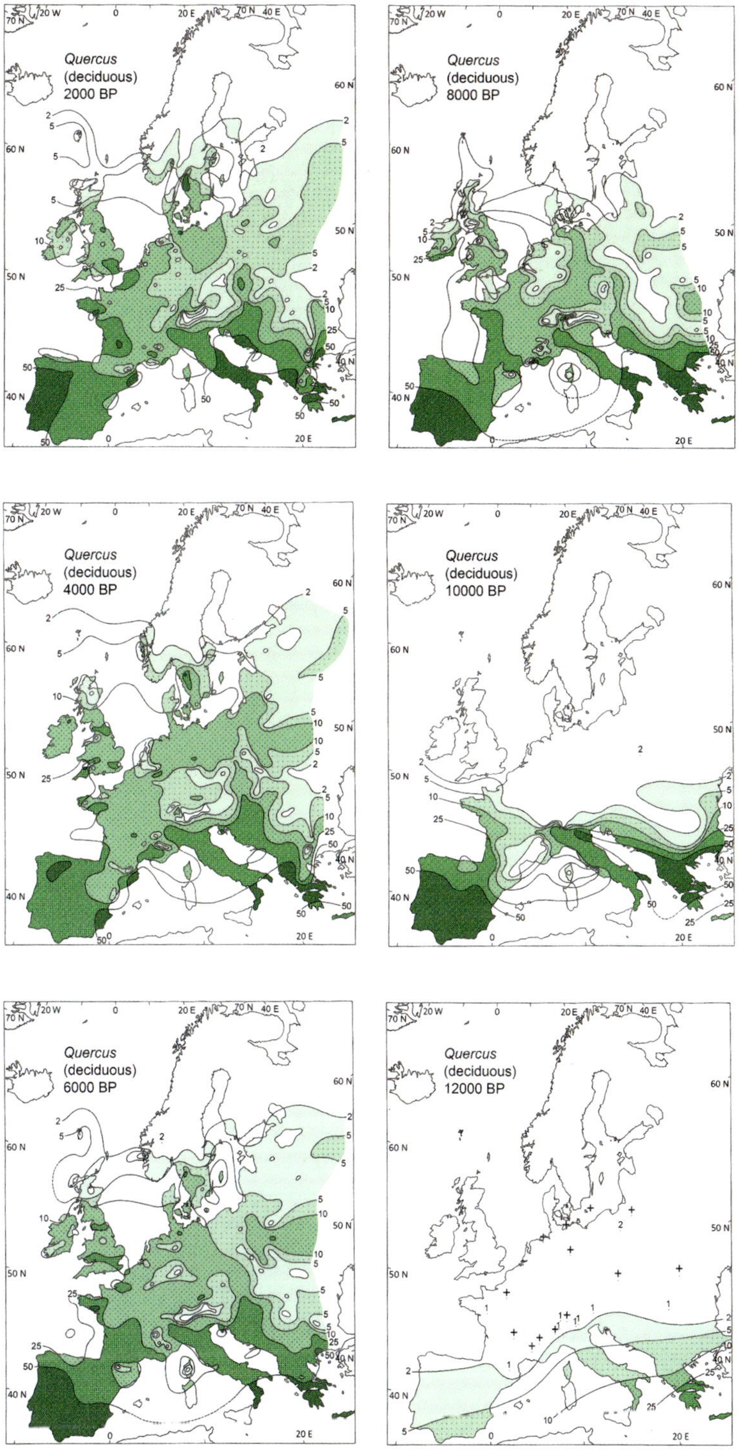

Fig. 4.4.1-5. Isopollen maps of oak (*Quercus* spp.) pollen percentages across Europe for 12 000, 10 000, 8 000, 6 000, 4 000 and 2 000 radiocarbon years before present (BP). Such maps summarize the spatial dynamics of the population history of oak during the past 12 000 years, with an emphasis on changing abundance patterns in space and time. Modified from Huntley and Birks (1983) and from Birks and Tinner (2016).

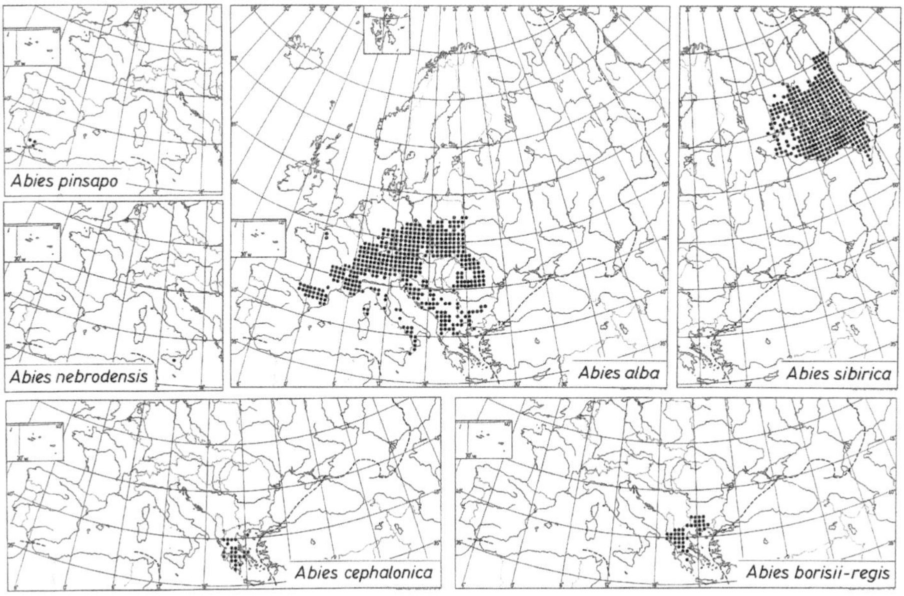

Fig. 4.4.2-1. Distribution of the six *Abies* species in Europe. After Jalas and Suominen (1973).

missing the first establishment of populations. The following discussion of selected arboreal taxa presents first the needle-leaved, then the broadleaved deciduous, and the evergreen broadleaved arboreal plants. Within these groups the plants are presented in alphabetical order of their Latin scientific names. Some examples of comparisons are included, for example *Fagus/Carpinus* or *Corylus/Quercus*: many more comparisons are possible. Our isochrone maps start at 20 000 cal yr BP to track the LGM refugia of subtropical, temperate and boreal species as well as their subsequent spreads during the Late-glacial and Holocene. LGM refugia are considered to be particularly important because during the LGM (23 000–19 000 cal yr BP) ranges of subtropical, temperate and boreal species contracted markedly for the last time (chapter 4.11).

4.4.2

Conifers

Abies

The genus *Abies* consists of about 40 species of which six species occur in Europe and another six in the adjacent northern Africa and Asia Minor (Fig.4.4.2-1; see also Schütt, 1991). Except for *A. sibirica*, which belongs to the section *Centrales*, all other species belong to the section *Marginales*. The taxa in this group are closely related and, therefore, often a combination of criteria are needed for the identification of cones, buds and needles. Processes of introgression are also known.

Only a few studies of Late-glacial and Holocene history concern species other than *A. alba* and we therefore concentrate on this species. For the species of the Mediterranean and Asia Minor it is difficult to estimate whether – after surviving the Last Glacial Maximum (LGM) – a major expansion of the range occurred. It is thus highly probable that outside the ranges of these endemic species one has to reckon with only *A. alba*, although the identification of the species on the basis of pollen morphology is not possible.

Abies

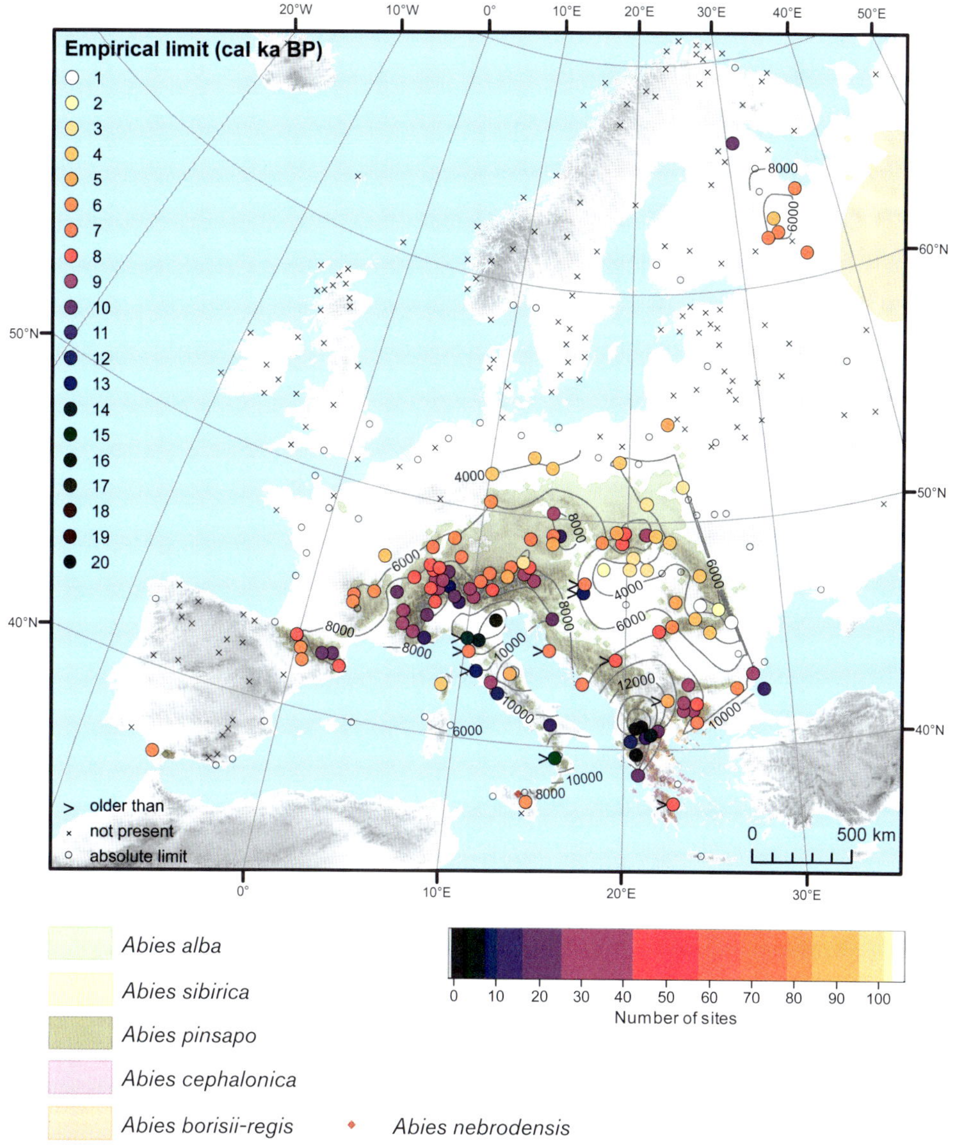

Fig. 4.4.2-2. Map showing the empirical limits of ***Abies*** and the current distribution in Europe. The coloured dots indicate the empirical limit in cal ka BP at selected sites, crosses refer to sites where the taxon was absent and small circles where it reached the absolute limit. Grey lines represent isochrones of empirical limit in 2000-year steps, based on nearest neighbour interpolation. Bar diagram shows stacked number of sites where the empirical limit was reached during the same 1000-year time interval, with colours according to the legend in the figure.

Today, *Abies alba* (silver fir) is restricted to central and southern Europe (Fig. 4.4.2-1) and, primarily in the montane forest belts, from the Pyrenees to the Balkans. North of the Carpathians, *A. alba* is also found outside the mountains in the plains of southern Poland. *Abies alba* is often associated with *Fagus sylvatica* and *Picea abies*, sometimes also with *Pinus sylvestris*, *Quercus robur* or *Carpinus betulus*.

Macrofossils of *Abies* such as seed-wings (cones disintegrate when ripe), seeds or the flat needles with an indented tip are easy to identify to the genus but difficult to the species. This is also true of the stomata and the pollen grains. Identification to species on the basis of wood anatomy is not usually possible (Schweingruber, 1990), except in some rare cases, where biometric measurements supported by statistics can be achieved (García Esteban et al., 2007, 2009). The pollen is 115–190 μm in size, vesiculate, and with a thick proximal crest. In pollen diagrams the empirical limit (onset of the continuous curve) can usually be taken as the time of apparent local presence: this is often more or less quickly followed by the rational limit (rapid increase) that can be interpreted as population expansion.

At an early stage, glacial **refugia** during the LGM (Fig. 4.4.2-2) were recognized at mid-elevations of Mediterranean mountains where orographic rain counterbalanced the dry glacial climate, for example in southern Italy (Grüger, 1977; Schneider, 1985a), northern Italy (Euganean Hills: Kaltenrieder et al., 2009), northern Apennines and possibly some small, sheltered sites in the unglaciated southern Alpine forelands (Hofstetter et al., 2006) as well as on the southern Balkan Peninsula or Greece (see Bottema, 1974; Tzedakis et al., 2002), where our map (Fig. 4.4.2-2) suggests several LGM refugia (black dots). At low elevations, records for the full glacial were also found, especially as charred wood identified to genus or species at Palaeolithic sites, for example in Dolní Věstonice (Kneblovà, 1954) and other sites in the Czech Republic (Svoboda and Svobodová, 1985; Opravil, 1994; Willis and van Andel, 2004). A full-glacial refugium of *Abies alba* in the Pyrenees was long considered improbable (Reille, 1988; Reille and Lowe, 1993) but charcoal records from the cave of Altamira (Spain) and from the French sites of Gazel and Belvis provided dates of the LGM or very early Late-glacial (c. 20 800–18 800 cal yr BP: Vernet, 1980; Uzquiano, 1992; Pérez-Obiol and Julià, 1994; see also Terhürne-Berson et al., 2004; Liepelt et al., 2009).

The **phylogeography** of *Abies alba* was originally based on isozymes and/or allozymes (Konnert and Bergmann, 1995) and already gave support to the hypothesis of multiple refugia – five refugia including an isolated one in the Pyrenees, the latter confirmed by Fady (1999). A substantial range expansion occurred from only three of these putative refugia. Later, organelle-DNA markers were developed: as for other gymnosperms the chloroplast DNA (cpDNA) in *Abies* is purely paternally inherited, whereas mitochondrial DNA (mtDNA) is maternally inherited (Vendramin and Ziegenhagen, 1997). In contrast, the variants of isozymes/allozymes are biparentally inherited. One set of haplotypes in mtDNA with five alleles primarily differentiates three groups in the seven Mediterranean species of *Abies* and two groups within *A. alba*. In the latter two, limits in the eastern Carpathians and in the north-western Balkans were mapped by Ziegenhagen et al. (2005) and Liepelt et al. (2009) and coincide with the ones shown by Konnert and Bergmann (1995). Liepelt et al. (2009) present a synthesis concentrating on *A. alba* and combine several techniques that provide maps of allelic richness, of gene diversity (estimated with different parameters: Widmer and Lexer, 2001) and overlay maps of allozyme data with the distribution of mt DNA marker variants. This synthesis emphazises the two contrasting concepts of the 'effective refugia' that contributed significantly to the recolonization of central Europe (southern Balkan, northern and/or western Balkan, northern Apennine) versus the 'isolated refugia' that did not contribute to the modern range further north (Pyrenees, Calabria). Putative refugia in the French-Italian Maritime Alps and the Massif Central were refuted on the basis of palynological and genetic evidence (Muller et al., 2007; Liepelt et al., 2009).

In addition, Liepelt et al. (2009) discuss aspects important for our understanding of evolution during the Quaternary such as suture and introgression zones (based on, for example, Breitenbach-Dorfer et al., 1997; Gömöry et al., 2004) and the differences in population genetics of the rear and the leading edge during species range expansion. The rear edge usually has high levels of genetic diversity while the leading edge is a region of secondary contacts. Both edges are, therefore, important in conservation. Introgression zones were found in the north-western Carpathians and in Croatia and Slovenia (Konnert and Bergmann, 1995; Liepelt et al., 2002, 2009).

The phylogeography of the Mediterranean species of *Abies* can be found in Ziegenhagen et al. (2005), García Esteban et al. (2007, 2009) and Liepelt et al. (2010).

Range expansions and contractions: range expansions started from the effective refugia, that is from Greece; from the central and eastern Balkans to the north-east and north-west; from Croatia and Slovenia to the north and north-east; from southern Italy to the north; and from northern Italy (continuous pollen records at 17 000–14 000 cal BP, green dots in Fig. 4.4.2-2) westwards and eastwards around the Alps into central Europe (via Massif Central, Jura Mountains, Vosges, Black Forest).

Regional spreading maps can be found in de Beaulieu (1977) and van der Knaap et al. (2005) for the entire Alps, in Muller et al. (2007) for the south-western Alps and in Wegmüller (1977) for the western and central Alps.

Rates of spreading were earlier estimated to be over 700 m yr^{-1}, but by allocating recolonisation from effective refugia (based on genetic studies) Cheddadi et al. (2014) could show that values ≤250 m yr^{-1} are more realistic and compatible with models of tree dispersal. Our own estimates on the basis of Fig. 4.4.2-2 are 190–220 m yr^{-1} (Table 4.11-1).

Human impact played a major role in reducing the spatial distribution of *Abies alba* after about 6000 cal yr BP (4000 BC) as demonstrated by comparing the Early- to Mid-Holocene distribution of *Abies* findings at low elevations and near the seashore with independent climate records, providing a case of a large discrepancy between the potential and the realized niche of a species (see Tinner et al., 2013). The high sensitivity of the species to fire and to ungulate browsing may be the major cause. In addition, the long and strong trunks of fir were highly valued for large constructions such as the roofs of naves, bridges or castles. In many central European sites with macrofossil and/or pollen records, an *Abies* decline is correlated with the Roman period (Küster, 1994). By combining the southern European palaeo-data of Tinner et al. (2013) with the dynamic global vegetation model LPX-Bern, Ruosch et al. (2016) conclude that under future warmer-than-present climate conditions "*Abies alba* may ensure important ecosystem services including stand and slope stability, infrastructure protection, and carbon sequestration...".

Juniperus

Farjon (2005) in his thorough treatment of the genus *Juniperus,* in the family Cupressaceae, comes to the following conclusion, "In this monograph, 52 species are accepted, some with subspecies or varieties, making it the most diverse genus in the family" (see also Mao et al., 2010). In Europe, one widespread species is the northern hemispheric, circumpolar *J. communis* that has about five varieties (often considered subspecies). Their amplitudes with respect to climate and soils are very broad. In contrast, the southern *J. oxycedrus, J. phoenicea, J. sabina* the south-western *J. thurifera* and the south-eastern *J. foetidissima* and *J. excelsa* have more restricted ranges. Their pollen morphology is characterized by an inaperturate but usually ruptured (split) grain with small gemmae. Beug (2004) proposes criteria to distinguish the section *Sabina* (*J. sabina, J. foetidissima, J. excelsa*). The *J. communis* type also includes *Cupressus sempervirens* as well as taxa not occurring in the Quaternary of Europe (*Taxodium, Thuja, Libocedrus*). Among the macrofossils, needles of *J. communis* can be identified, characterized particulary by the pattern of stomata on their abaxial side. Furthermore, subfossil needles allow the identification of juniper subspecies such as e.g. *Juniperus communis* ssp. *nana* (Weber, 1980; Tinner et al., 1996). For stomata on pollen slides see Fig. 2.3-5. *Juniperus* is anemophilous and zoochorous.

The Pleistocene **refugia** for juniper must have been numerous, as most long records show. For example, at La Grande Pile (France, Vosges) where the pollen record covers the period from the late Riss glaciation to the Early Holocene, distinct juniper peaks are recorded at the onsets of the Eemian interglacial and the Würmian interstadials St. Germain I (Brørup) and St. Germain II (Odderade), as well as at the onset of the Bølling interstadial (see Fig. 1 in de Beaulieu and Reille, 1992b). In northern Germany, the Brørup, Odderade and Oerel interstadials contain substantial juniper occurrences (Behre and Lade, 1986; see chapter 3). At Lago Grande di Monticchio, juniper values are high during the Saalian glaciation, but also during the

Juniperus

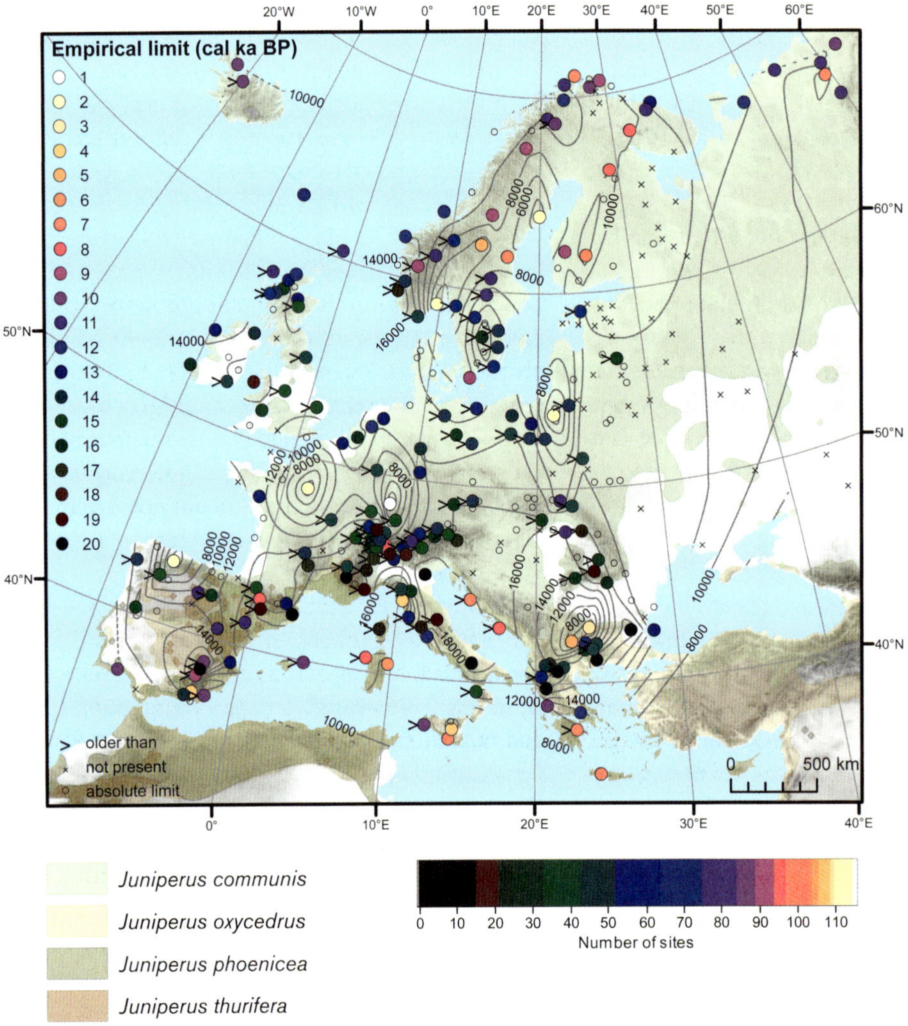

Fig. 4.4.2-3. Map showing the empirical limits of ***Juniperus*** and the current distribution in Europe. The coloured dots indicate the empirical limit in cal ka BP at selected sites, crosses refer to sites where the taxon was supposedly absent and small circles where it reached the absolute limit. Grey lines represent isochrones of empirical limit in 2000-year steps, based on nearest neighbour interpolation. Bar diagram shows stacked number of sites where the empirical limit was reached during the same 1000-year time interval, with colours according to the legend in the figure. The identification of juniper pollen is difficult and became standard only during the past decades, thus crosses in the map may also result from detection difficulties.

cool phases of MIS 4 and MIS 2 (Watts et al., 1996a). Similarly, at Kopais K93 (Greece) *Juniperus* pollen was found in varying amounts from the Eemian to the Holocene (Tzedakis, 1999). In eastern Europe, charcoal of *Juniperus* was identified at the important archaeological site Dolní Věstonice (Czech Republic; Opravil, 1994; Musil, 2003; Willis and van Andel, 2004) and in loess deposits in Hungary (Willis et al., 2000) at dates between 37–20 cal ka BP. Our new map (Fig. 4.4.2-3) shows different LGM refugia in the Iberian, Italian and Balkan peninsulas (black dots) as far north as c. 45°N.

In terms of **phylogeography**, van der Merwe et al. (2000) proposes three routes of immigration into Britain (see also Thomas et al., 2007). For central Europe, Michalczyk et al. (2010) conclude "... there are no underlying phylogeographic signals or any other meaningful geographic genetic structures. Consistent with a scenario of periglacial survival there is no correlation between genetic and geographic distances...". Again, this is an argument for numerous refugia for juniper in Europe (Fig. 4.4.2-3). For *J. thurifera*, in contrast, Terrab et al. (2008) show quite a different pattern in south-western Europe. The Iberian Peninsula in particular is rich in genetically distinct populations: they may be the sources for the populations in the south-western Alps and on Corsica. The Spanish populations are quite distinct from the ones in Morocco which means that "... the Strait of Gibraltar acted as an efficient barrier against gene flow..." (Terrab et al., 2008).

Range expansions and contractions: similar to the situation at La Grande Pile, the Late-glacial at many sites shows a 'juniper jump' sensu Denton et al. (2006) at the onset of the Bølling (Fig. 4.4.2-3, green dots). Subsequently, this heliophilous taxon was shaded out by tree birches and pines. Later, secondary juniper peaks are recorded either during the Older Dryas and/or the Younger Dryas and especially during the earliest Early Holocene (Preboreal). De Beaulieu (1977) provided an overview of these double or triple peaks of juniper in western Europe (his Fig. 25). In general the 'jumps' were early in the south and later in the north. This 'behaviour' demonstrates that *Juniperus* cf. *communis* can be considered a protocratic tree. That view is corroborated by the fact that enhanced soil fertility under *Juniperus communis* in arctic ecosystems was found through symbiosis with nitrogen fixers (both ectomycorrhiza and arbuscular mycorrhiza, a form of endomycorrhiza) and a high capability to extract phosphorus (P) from the soil (De Luca and Zackrisson, 2007). Pollen productivity is probably low in the prostrate form *J. communis* ssp. *nana* but high in the tall form *J. communis* ssp. *communis* (Birks, 1973b; Iversen, 1954; Gaillard, 1984a). Prior to the Bølling, macrofossil remains of needles indicate the presence of the subspecies *J. nana* (Weber, 1980). Therefore, the 'juniper jump' is often interpreted as a shift from dwarf forms to tree forms and hence from low to high pollen productivity. This could explain the extraordinary velocity of the increase (e.g. at Gerzensee from 25% to 85% in 40 years: Ammann et al., 2013a). Immigration or population build-up would have needed more time (because, for instance, the age at first flowering is 10–20 years). However, on the basis of our novel map (Fig. 4.4.2-3) we estimate the spreading rates of *Juniperus* at 85–220 m yr^{-1} (Table 4.11-1).

During the first part of the Holocene juniper played a minor role in most vegetation types because it is a light-demanding genus: exceptions were near timberlines, in open woodlands, scrub and heaths, near shores and on poor soils or rocks. During the Late Holocene, however, it had a remarkable come-back – as Schwartz et al. (2005) show for the Vosges, Thomas et al. (2007) for the British Isles and Gaillard et al. (1992) with a convincing quantification (Fig. 2) for southern Sweden – as grazing and other anthropogenic disturbances opened up the woodlands and favoured its expansion. The double pulse of two periods with high frequencies (Late-glacial/Early Holocene v. prehistoric human impact) were already well visualized in the pollen maps by Huntley and Birks (1983) and with additional sites by Brewer et al. (2017) and are also apparent in our map (Fig. 4.4.2-3).

Larix

Today, in the northern hemisphere, the summergreen genus *Larix* consists of ten species whereof two species occur in Europe (Fig. 4.3.3-5). Of these, *Larix decidua* Mill. (synonym *L. europaea* Lam. ex DC) occurs in the Alps and the Sudetan mountains, and as the subspecies *L. decidua* ssp. *carpathica* in the Carpathians and *L. decidua* ssp. *polonica* in the contiguous northern lowlands in Poland (the status as a subspecies is discussed in Flora Europaea and

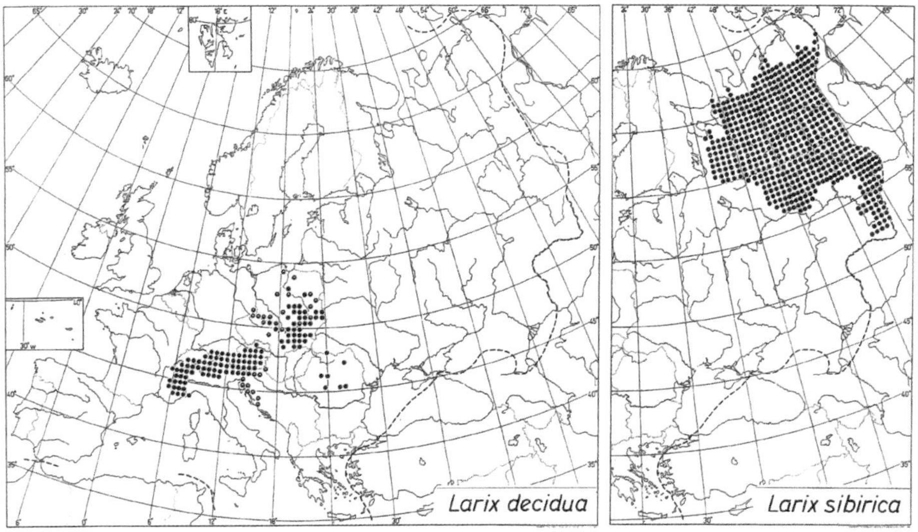

Fig. 4.4.2-4. Distribution of the two species of *Larix* in Europe. After Jalas and Suominen (1973).

in Farjon, 1990). The western Siberian larch sensu lato, *L. sibirica* Ledeb., has a large range reaching across the Ural mountains into north-eastern Europe near Lake Onego (for taxonomy and nomenclature see Farjon, 1990). In eastern Siberia it is replaced by *L. dahurica* Turcz. ex Trautv. (synonym *L. gmelinii* (Rupr.) Rupr. and *L. sibirica* Maxim non Lebed.). Overall, the European range of the genus *Larix* resembles that of *Pinus cembra* and *P. sibirica* (stone pines). Several genetic-phylogeographic studies exist for the entire genus, for example, Semerikov and Lascoux (2003) and Gros-Louis et al. (2005).

In the Alps, *L. decidua* is primarily widespread in the subalpine forests of Swiss stone pine and larch (Larici-Pinetum cembrae) near the timberline. In contrast to *P. cembra*, larch is not as restricted to the central Alps but also reaches the marginal Alps and the montane belt. In the Sudetan and Carpathian mountains, *L. decidua* ssp. *carpathica* occurs in subalpine forests, which are dominated by *Picea* and thus termed Larici-Picetum (Mayer, 1984). *Larix decidua* ssp. *polonica* differs considerably from the other taxa in so far as it is restricted to elevations between 150 and 600 m asl, to acidic soils, and to forests of oak and pine (including *Quercus petraea, Pinus sylvestris, Abies alba, Fagus sylvatica, Carpinus betulus*). *Larix sibirica* Lebed. (Farjon 1990: *L. russica*) belongs to the woody taxa reaching, in the far north, to the open forest-tundra, forming the polar treeline. In these situations the species usually occurs in monospecific tree-communities on carbonate soils (Walter, 1974). *Larix* is anemophilous and anemochorous.

Pollen grains of *Larix* are – in contrast to those of the Pinoideae (*Abies, Picea, Pinus*) – inaperturate and without sacci/vesiculae. For a long time they were hardly recorded in pollen-analytical studies. Good photographs can be found in Moore et al. (1991). As was shown by Firbas (1949), pollen productivity is low and, therefore, even low percentages may show a local presence.

Plant macrofossils such as cones, bud scales, seeds and needles can be identified to the species; stomata to the genus. The high annual production and deposition of needles – and thus of stomata – helps to counterbalance the low pollen productivity and dispersal and to improve the distinction between local presence and absence of the genus (Trautmann, 1953; Sweeney, 2004; Ammann et al., 2014). Concerning wood anatomy, the separation of *Larix* and *Picea* is not always easy (Schweingruber, 1978). This affects the taxonomic resolution of charcoal analyses, for example in Palaeolithic sites as synthesized by Willis and van Andel (2004).

The Early Weichselian interstadials such as Brørup and Odderade (= St. Germain I and II) provided macroremains or pollen of *Larix* cf. *decidua* (Behre and Lade, 1986; de Beaulieu and Reille,

Larix

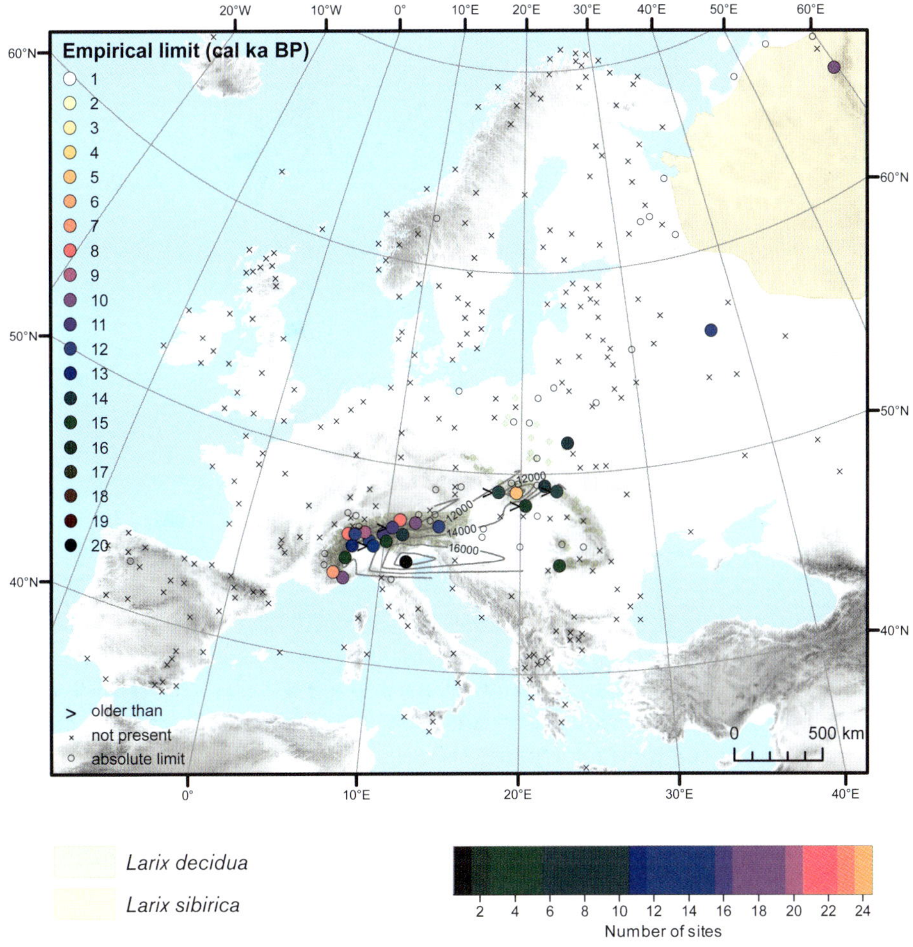

Fig. 4.4.2-5. Map showing the empirical limits of ***Larix*** and the current distribution in Europe. The coloured dots indicate the empirical limit in cal ka BP at selected sites, crosses refer to sites where the taxon was absent and small circles where it reached the absolute limit. Grey lines represent isochrones of empirical limit in 2000-year steps, based on nearest neighbour interpolation. Bar diagram shows stacked number of sites where the empirical limit was reached during the same 1000-year time interval, with colours according to the legend in the figure.

1992b; Litt, 1994). Potential full-glacial **refugia** are indicated by early findings of macroremains in the lowlands of Polish Galicia and Ukraine (Szafer and Jaron, 1935; Klimaszewski et al., 1950), by subsequent findings of wood-charcoal and pollen in central Bohemia, near the city of Prague (200 m asl, 40 000–30 000 years old: Jankovská and Pokorný, 2008, 2015) and southern Moravia at Bulhary (180 m asl, c. 25 000 cal yr BP: Rybníčková and Rybníček, 1992, 2014). In the inter-montane basins of the western Carpathians in Slovakia, *Larix* cf. *decidua* pollen and macrofossils (e.g. complete cones) are dated from 30 000 to 18 000 cal yr BP (Jankovská and Pokorný, 2008, 2015). At the mammoth-site of Niederweningen (Switzerland, about 45 000 years old) pollen of larch was found (Drescher-Schneider et al., 2007a; Furrer et al., 2007). South of the Alps, full-glacial records are documented for the Colli Euganei (Lago della Costa: Kaltenrieder et al., 2009). Indeed, Lago della Costa shows the oldest empirical limit in our map (Fig. 4.4.2-5, black dot), which is corroborated by conspicuous *Larix* needle finds (Kaltenrieder et al., 2009). Several **phylogeographic** studies cover all Eurasian taxa (Semerikov and Lascoux, 2003; Gros-Louis et al., 2005; Khatab et al., 2008). For Europe, Wagner et al. (2015a) used both nuclear clusters and mitochondrial DNA analyses to show the strong separation of types in the Alps, central Europe (Sudetan mountains and Polish lowlands) and the south-eastern Carpathians. Within the Alps, a clear west–east gradient was found. The nuclear clusters suggest multiple LGM-refugia, namely four regions in the southern Alps, two in central Europe and one in the south-eastern Carpathians. It is also of interest that translocations could be filtered out, that is "...intentional or accidental anthropogenic movements of living organisms from one place to another". This is especially important for a species that was planted in vast areas (see below and Wagner et al., 2015a).

Range expansions and contractions: the history of European larch since 130 ka is discussed and illustrated with 27 maps in Wagner et al. (2015b). Late-glacial findings of pollen (green and blue dots in Fig. 4.4.2-5) and macroremains (seeds and needles) are known in the **eastern range** of *Larix* from the Polish Carpathians and the adjacent northern lowlands (Koperowa, 1970; Harmata, 1987, 1995), from several lowland sites in Slovakia (Rybníček and Rybníčková, 2002) and from Preluca Tiganului in the northern Romanian Carpathians (Feurdean and Bennike, 2004). In rugged lowland sandstone landscapes of inner Bohemia, Late-glacial (YD) and Early Holocene (Preboreal) pollen finds indicate the local survival of larch populations until about 8000 cal yr BP, when broadleaved forests fully expanded and outcompeted them (Pokorný et al., 2017). It seems that during the Holocene, the tree withdrew from the lowlands to the Carpathian mountains and only left small populations as *L. decidua* ssp. *polonica* outside the Carpathians (where it was later largely replaced by summergreen broadleaved trees). Studies on population genetics could possibly clarify such a hypothesis.

Early Late-glacial findings in the **Alpine range** of *Larix* as pollen, needles and stomata in the foot-hills of the southern Alps indicate occurrences between 16 500 and 14 600 cal yr BP both in north-western Italy (e.g. Lago Piccolo di Avigliana, 353m asl: Finsinger et al., 2006) and north-eastern Italy (e.g. Palughetto at 1040 m asl and Pian di Gembro at 1350 m asl: Pini, 2002; Vescovi et al., 2007).

After the decay of the Alpine Würmian ice, *Larix* was able to invade the Alpine valleys relatively rapidly, in the same way as *Pinus cembra*. Later, it became outcompeted by other tree taxa and escaped to higher elevations. During the Younger Dryas a slight shift towards lower elevations is recorded as summarized in Wagner et al. (2015b). Since the Early Holocene at the latest, larch formed extensive forests in the subalpine forest belt as documented by numerous findings of macroremains in the Austrian and Swiss Alps (see chapter 4.9 on timberlines). Larch spread remarkably late into the western Alps (Fig. 4.4.2-5), which were not heavily glaciated during the LGM. It only established during the first half of the Holocene, being possibly inhibited by higher summer temperatures and more frequent droughts than today (Huntley and Birks, 1983). A delay for edaphic reasons is less probable – in the western Alps carbonates predominate which probably still had no raw humus at that time. However, on the basis of our new data compilation (Fig. 4.4.2-5) we estimate spreading rates of 90–180 m yr^{-1} (Table 4.11-1), which is only moderately lower than for other conifers with a wide European range, e.g. *Picea abies* (135–240 m yr^{-1}) or *Abies alba* (190–220 m yr^{-1}). Early **human impact** was recorded in the Upper Engadine (south-eastern Alps) where grazing of domestic animals was favoured by opening the larch forests (pastured woods, 'Lärchenweide' or 'Lärchenwiesen'), especially from the Iron Age to the High Middle Ages (e.g. Kleiber, 1974; Zoller and Brombacher, 1984; Gobet et al. 2003). Late human impact occurred in the form of plantations from the sixteenth century to a peak in the nineteenth century, when a so-called 'Lärchenmanie' or 'manie de mélèze' developed in forestry: the wide ecological tolerances of *Larix* made plantations possible even in the lowlands. Of interest were its rapid wood accumulation and the qualities of the wood, primarily its resistance to decay and its beauty for furniture and sculptures.

In brief, *Larix decidua*, similar to *Pinus cembra*, belongs to the woody taxa which show relatively minor horizontal range shifts but major vertical shifts during the Late-glacial and Holocene.

Larix sibirica has a patchy record in the European part of northern Russia (boreorussian region): for example at Mezhgornoe (pre-Polar Ural, 550 m asl; Fig. 4.3-5) where, today, it forms the alpine timberline, pollen and stomata are recorded as single findings at least since 10 600 cal yr BP and continuously after about 8300 cal yr BP (Kultti et al., 2003). For Siberia, the macroremains document occurrences during the full glacial, Late-glacial and Holocene times (Binney et al., 2009).

Picea abies

The genus *Picea* is represented in Europe today by two species. *P. abies*, with the two subspecies *P. abies* ssp. *abies* and *P. abies* ssp. *obovata* (Fig. 4.4.2-6), is widespread in northern Europe and in various mountain ranges of central Europe. The other, *P. omorika,* is endemic to a small area in the Dinaric Alps (but see chapter 3.3).

Picea has vesiculate (saccate) pollen grains that can be identified to species on the basis of size-statistics: *P. omorika* has smaller grains with sacci (seen from the side) mainly between 45 and 55 µm, whereas *P. abies* has larger grains with sacci 65–70 µm, (Müller and Weber, 1904; Firbas, 1927; Erdtman, 1943). But the curves of the size statistics overlap to some degree so the identification of single grains is not always possible (Andersen, 1961). A north–south divergence in the European spruce became obvious on the size and shape statistics of the pollen grains at a relatively early stage (Birks, 1978).

The distinction of the two species is possible if macroremains of seeds and cones are available: based on the cones even the distinction between the two subspecies *P. abies* ssp. *abies* and *P. abies* ssp. *obovata* is possible. Needles and stomata allow the identification of the genus. In wood anatomy the distinction between *Picea* and *Larix* can sometimes be difficult (Schweingruber, 1990). Late-glacial and Holocene findings of *Picea* outside the modern range of *P. omorika* can be allocated with a high probability to *P. abies*. All species are anemophilous and anemochorous.

Various percentage thresholds, ranging between 0.5% and 10%, to infer the apparent presence of *Picea* near the site have been discussed. Estimations of such **thresholds** or presence can be derived in the following ways. (1) Pollen influx can be compared with modern vegetation. In northern Finland, Hicks (2001, 2006) found that at values <25 grains $cm^{-2} yr^{-1}$ the spruce forest was more than 10 km away, with values >50–100 grains $cm^{-2} yr^{-1}$ spruce was present nearby but only sparsely, whereas values >300 grains $cm^{-2} yr^{-1}$ indicated spruce forest around the site. (2) Comparison of pollen with macrofossil records may indicate lo-

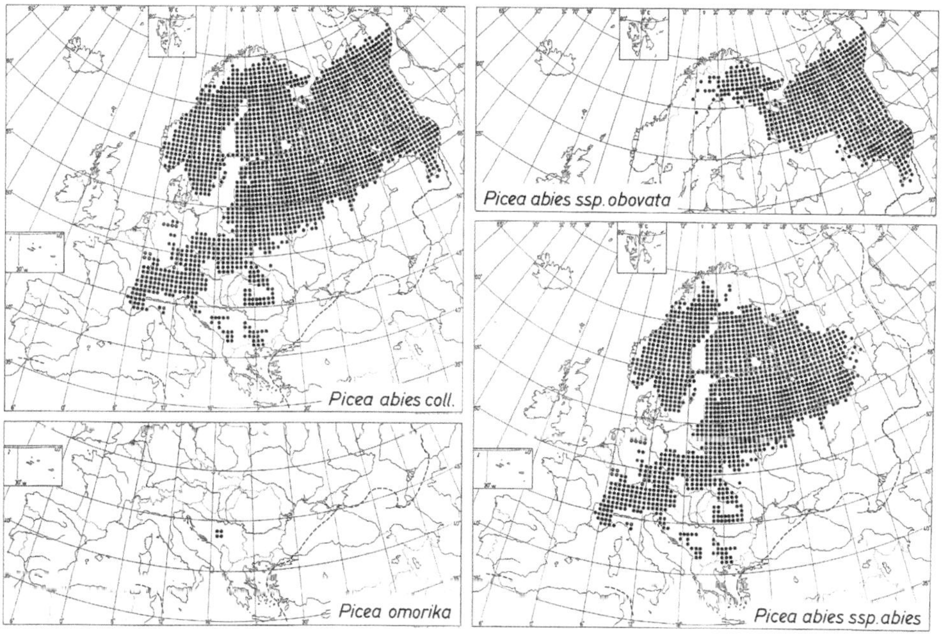

Fig. 4.4.2-6. Distribution of the two species of *Picea* in Europe, as well as the two subspecies of *Picea abies*. After Jalas and Suominen (1973).

Picea abies

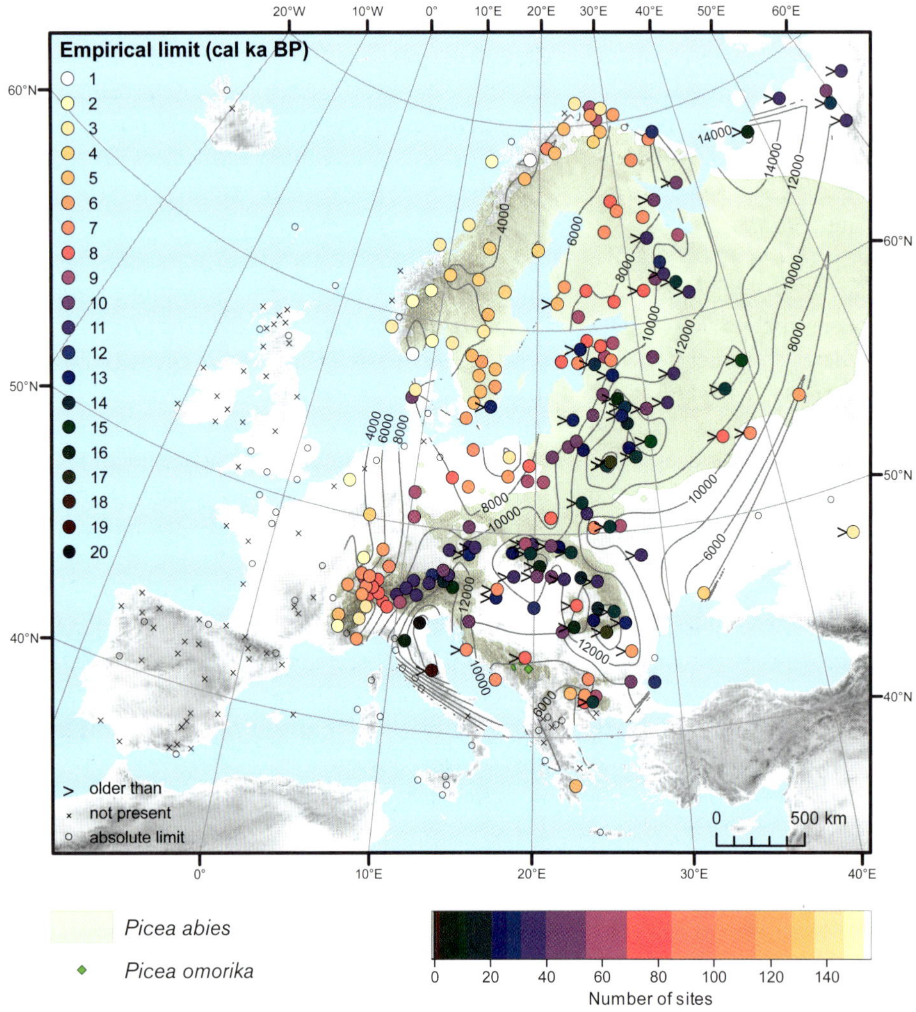

Fig. 4.4.2-7. Map showing the empirical limits of ***Picea*** and the current distribution in Europe. The coloured dots indicate the empirical limit in cal ka BP at selected sites, crosses refer to sites where the taxon was absent and small circles where it reached the absolute limit. Grey lines represent isochrones of empirical limit in 2000-year steps, based on nearest neighbour interpolation. Bar diagram shows stacked number of sites where the empirical limit was reached during the same 1000-year time interval, with colours according to the legend in the figure.

cal presence. In order to resolve the presence/absence problem, so as to avoid the misinterpretation of long-distance transported or reworked pollen, Terhürne-Berson (2005) reviewed and mapped records of both pollen and macroremains. From the early Polish literature several studies are reported by Latałowa and van der Knaap (2006) where 0.5% *Picea* pollen was accompanied by needles or seeds. At the Swedish site Klotjärnen no spruce macroremains were found before the onset of the continuous curve of *Picea* pollen but appeared for the first time when pollen values reached 0.5% (Giesecke, 2005b). Thresholds of 1%, 3%, 5%, and 10% were used for the spatial synthesis presented by Giesecke and Bennett (2004). A threshold of 2% was selected by Latałowa and van der Knaap (2006) after careful discussion. This threshold is unable to track small outlier populations. These authors also show that values between 1% and 8% do not change the general pattern of the range expansion of the taxon but are relevant for its timing. Ravazzi (2002) used a threshold of 4% for his review of the Late Quaternary in southern Europe. Here, we prefer empirical limits because they are more sensitive to first establishments of populations (Fig. 4.4.2-7), reducing the false absence problem.

Picea omorika (Serbian spruce) is restricted today to a number of local occurrences along the Drina river north of Višegrad, near the border between Bosnia and Serbia. The species seems to be a weak competitor and is confined to extreme north-exposed and steep habitats on calcareous soils at elevations between 800 and 1600 m asl (Schmidt-Vogt, 1977; Mayer, 1984). The Serbian spruce forests (Picetum omorikae) also contain other conifers such as *Picea abies, Abies alba, Pinus nigra* and *P. sylvestris*. *Picea omorika* also occurs in riparian forests with *P. abies* and *Alnus glutinosa*. According to Gigov (1956), *P. omorika* was more widespread during the Holocene than today. *Picea omorika* or *P. omorikoides* was widespread in Europe during the Tertiary, and also in Quaternary interglacials and interstadials (see chapter 3). Holocene records of *Picea omorika* are very rare. On the basis of the macrofossil- and pollen-inferred postglacial history of the species, Finsinger et al. (2017b) reconstructed good post-fire recoveries. The authors therefore suggest that prescribed burning may be beneficial to reduce the vulnerability of the threatened endemic species.

Picea abies (Norway spruce) has an extremely wide geographical distribution covering nearly the whole boreal zone of Eurasia. Within this area the polymorphic species can be subdivided into a number of forms, the taxonomic status of which can be discussed as varieties, subspecies or even species (Schmidt-Vogt, 1977). In Europe, two main subspecies are recognized, namely *P. abies* ssp. *abies* (European spruce) occupying the larger part of the area of *P. abies*, and *P. abies* ssp. *obovata* (Siberian spruce) widespread in Siberia but reaching westwards into Sweden (Fig. 4.4.2-6). In the area of overlap, hybrids with many mixed characters occur that make identification difficult (Wright, 1955; Bobrov, 1972), but possible (Popov, 2003, 2012, 2015). Tsuda et al. (2016) provide genetic results (nuclear SSR and mitochondrial DNA) of both taxa from western Europe to Lake Baikal.

The ecological amplitude of *P. abies* is very wide. Due to the great ecological and economic importance, especially of *P. abies* ssp. *abies*, a vast literature exists in forestry research. The main distribution, where pure or mixed spruce forests occur, consists of the boreal and hemiboreal zones and the montane and subalpine belts of the central and southern European mountains. These areas at high latitudes or elevations are relatively cool with long and pronounced winters (5–7 months with a mean minimum temperature below 0°C), moderately warm summers (July means of 12–17°C) and precipitation in winter as snow. In northern Europe, annual mean temperatures are 5–6°C and precipitation ranges between 450 and 600 mm. In the central European mountains, higher mean annual temperatures and precipitation of over 850 mm per year (Schmidt-Vogt, 1977) prevail. A strong tolerance against frost and late-frosts is characteristic. Edaphically, spruce does not demand high nutrient availability but it does require quite high water availability. As a drought-sensitive species, it does not form the forest limit towards the steppe (except at locally wet places: Schmidt-Vogt, 1977; Walter and Breckle, 1986; Ravazzi, 2002).

In terms of plant sociology, the European spruce forests (alliance Vaccinio-Piceion) can be subdivided into the Nordic lowland spruce forests (sub-alliance Piceion septentrionale or Linnaeae borealis-Piceion) rich in boreal or subarctic species and the central and south-eastern European mountain forests (Eu-Vaccinio-Piceion, Jahn, 1977). In both sub-alliances, four groups can be distinguished according to water and nutrient availability: spruce forests rich in *Vaccinium* (acidic and nutrient poor soils); spruce forests rich in herbs (on calcareous soils, nutrient richer); spruce forests rich in tallherbs (humid, nutrient richer); and spruce forests rich in *Sphagnum* (very humid).

Glacial **refugia**: for southern Europe, Ravazzi (2002) provides a substantial review of the Late Quaternary based on 163 sites. He writes: " ... *Picea abies* distribution experiences a complex history during the last glacial period. The maximum expansion is not of interglacial age, and the maximum contraction does not coincide with full glacial conditions. ... the spruce range expanded in southern Europe during the middle pleniglacial interstadials ... Dry maxima probably occurred during the alpine deglaciation, between the LGM and the onset of the lateglacial interstadial". Ravazzi also demonstrates the difference of spatially close refugia in respect of their importance for the spatial expansion: while from the refugium in the northern Apennines hardly any expansion was recorded, the refugia in the eastern pre-Alps were most efficient throughout the Alps from east to west (Fig. 15 in Ravazzi, 2002; Ravazzi et al., 2006). For central Italy, figure 8 in Follieri et al. (1998) provides an overview. Such interpretations are in agreement with our new data compilation (Fig. 4.4.2-7) which suggests LGM stands in northern and central Italy (black and brown dots, 20 000–19 000 cal yr BP).

Phylogeography: variation in morphological traits combined with allozymic variation also suggest a clear geographical pattern, that is gradients along longitude and latitude (Lagercrantz and Ryman, 1990). Based on the comparison of pollen and macroremains, Terhürne-Berson (2005) delimited six putative refugial areas: the plains of north-western Russia; the eastern Alpine forelands (Austria, Czech Republic); the south-eastern Alpine forelands (Slovenia, Italy); the north-western Carpathians (Poland, Slovakia); the Hungarian and Danubian plains; and the Moldavian uplands. Combining records of pollen, macrofossils and mitochondrial and chloroplast DNA, Tollefsrud et al. (2015) come to the conclusion that *P. abies* ssp. *abies* and *P. abies* ssp. *obovata* "... are genetically distinct and occupied separate LGM refugia, Norway spruce on the East European Plain and Siberian spruce in southern Siberia, where they were already widespread during the lateglacial ...".

The geographical north–south divergence became obvious with various genetic studies. (1) Chloroplast microsatellites displayed two regions called Sarmatic-Baltic and Alpine-central European (Vendramin et al., 2000). (2) Mitochondrial genomes show the same split with a rather homogenous Lineage A in the north and a very heterogeneous Lineage B in the south (Sperisen et al., 2001). They conclude a single putative refugium in north-western Russia and several distinct refugia, namely south-western Carpathians, Dinarides and a complex Alpine to north- and east-Carpathian group. The interesting contact zone of two types of *Picea abies* within the eastern Carpathians is situated south of a contact zone with *Abies alba*. (3) Nucleotide diversity also indicates three spatial groups: the Baltic-Nordic-Russian populations; the Romanian populations; and the Alpine populations (Heuertz et al., 2006b).

Range expansions and contractions: already during the Allerød, *Picea* expanded vigorously from east to west in both north-eastern and south-eastern areas (Fig. 4.4.2-7). During the Younger Dryas it contracted in the area north of 60° N but expanded in the region stretching from eastern Estonia to central Belarus (53–60° N: Latałowa and van der Knaap, 2006). These authors also show twelve north-eastern sites with a distinct increase in *Picea* during the Younger Dryas, provide European maps for the periods from pre-Allerød to the Early Holocene (Preboreal) with five percentage classes and a discussion as to whether this spruce expansion, during a cooler period, could be due to an increase of the cold-tolerant *Picea abies* ssp. *obovata*. After a careful discussion of various percentages as thresholds to estimate the local presence they chose a 2% threshold and show maps for Europe in time slices of 1000 years (including increases, decreases and no change). These maps are then used by Tollefsrud et al. (2008) as background for the mapped distributions of the mitochondrial DNA-variants. By doing so, the advantages of the two methods are combined: the timescales from the fossil records with tracing the expansion routes (marked by mtDNA) from the various refugia.

For the Holocene, **regional maps** not only show more spatial detail but also some processes of range expansion. For **Fennoscandia** and adjacent areas, Tollefsrud et al. (2009) combined nuclear and mitochondrial markers and demonstrated spatial groups within Lineage A and a putative southern and a northern migration route. Giesecke and Bennett (2004) collected and mapped pollen data from more than 330 sites and showed that the biogeographic dynamics of spruce developed in two steps: in the first phase small outpost populations arrived during the Early Holocene; in a second phase a front-like spread at high population densities occurred during the Mid to Late Holocene. Superimposed on the map of Giesecke and Bennett (2004), a transect of five sites with modern and Holocene pollen influx of *Picea* provided Seppä et al. (2009a) the possibility to estimate biomass (as t per ha). On the next finer spatial scale, based on records from 18 small forest hollows in southern Scandinavia (not from lakes or mires), the process of stand invasion and a spreading rate of c. 200 m yr^{-1} could be deduced (Bradshaw and Lindbladh, 2005), which is in excellent agreement with our own estimates of 135–240 m yr^{-1} at the European scale (see Table 4.11-1).

Two types of enigmatic findings of spruce in Scandinavia are still very much discussed.

(1) Wood and cones found at Mt. Åreskutan in nunatak situations at 1360 m asl in the central Scandes dated between 12 945 and 9650 cal yr BP were tentatively explained as a result of immigration from the Norwegian west coast (Kullman, 2001, 2002) but different lines of evidence show that this hypothesis cannot be supported (Birks et al., 2005). Ice-free islands of Late-glacial age in the Scandes are likely if, instead of the 'maximum model', the 'minimum model' with a thinner, multi-domed ice sheet is accepted (McCarroll and Nesje, 1993; Dahl et al., 1997). Wood records from Storsnasen and its neighbourhood dated between 8985 and 5560 cal yr BP (Kullman, 2001) can therefore be compared to the first step of small outlying populations (Giesecke and Bennett, 2004; Giesecke, 2005c).

(2) The findings of ancient DNA of spruce in lake sediment 22 000–17 700 cal yr BP at Andøya by Parducci et al. (2012) was questioned for a number of reasons by Birks et al. (2012a) and Vorren et al. (2013); see also Alsos et al. (2021).

In the **Polish and Slovakian Carpathians**, *Picea* was present during the Allerød, decreased during the Younger Dryas and then dominated many forests until the expansion of *Fagus* after c. 5800 cal yr BP (Ralska-Jasiewiczowa and Latałowa, 1996).

In the **north-western Romanian Carpathians**, the earliest macroremains of *Picea* were dated to c. 13 900 cal yr BP at Preluca Tiganului (Wohlfarth et al., 2001) and 13 400 cal yr BP at Steregoiu (Björkman et al., 2002). Throughout the eastern and southern Carpathians and the Apuseni Mountains, spruce expanded very rapidly, as mapped by Feurdean et al. (2011). On the phylogeographic maps based on mitochondrial DNA the northern and north-eastern Carpathians resemble the Alpine arch, whereas the south-eastern Carpathians harbour some unique haplotypes (Tollefsrud et al., 2008). In the southern Carpathians, ancient DNA analyses showed that Early-Holocene populations possibly had a higher genetic variation than modern populations, probably due to bottleneck effects (Magyari et al., 2011). Spruce established relatively late (8.2–7.4 cal kyr BP) in the mountains of the **southern Balkan Peninsula** (Pirin, Rila, Rhodopes). These mountains were, therefore, not a refugial area (Huttunen et al., 1992; Stefanova and Bozilova, 1995; Bozilova and Tonkov, 2000; Stefanova and Ammann, 2003). The mitochondrial DNA of modern populations indicate relationships to the south-eastern Carpathians and the Dinaric Alps.

For **southern Europe** and the **Alps**, figure 4.4.2-7 visualizes the spread from east to west. Ravazzi (2002) presents a spatially and temporally refined Alpine synthesis for spruce including 163 selected sites for the time period since the Eemian. By analysing mitochondrial (maternally inherited) markers in modern spruce, Gugerli et al. (2001a) could show a strong gradient from moderately to highly variable populations in the eastern Alps to monomorphic populations in the western Alps, a pattern best explained by founder effects. For **Italy**, Magri et al. (2015) provide maps presenting the populations of the south-eastern Alps and the northern Apennines.

Human impact: as Giesecke (2004) summarized for Fennoscandia: "Many palaeoecological investigations from Fennoscandia indicate that the rise of *Picea abies* pollen is associated with disturbances in the form of fire or human impact (e.g. Huttunen, 1980; Bradshaw and Hannon, 1992; Almquist-Jacobson, 1994; Segerström, 1997; Hörnberg et al., 1999). Results obtained in this study do not corroborate this hypothesis. Instead they indicate that fires were more frequent after the expansion of *Picea abies*. Forest fires have a positive effect on seedling establishment and growth, as they release nutrients that are fixed in the forest litter (DeLuca et al., 2002; Wardle et al., 2003). However, frequent fires result in a domination of *Pinus sylvestris*, while long periods without fires favour *Picea abies* (Bradshaw and Zackrisson, 1990). As a shade tolerant tree *Picea abies* is not in need of large forest openings. Therefore the successful establishment of *Picea abies* might have been assisted by disturbance on a local scale and a secondary expansion may be well explained by human impact, but different mechanisms have to be identified to explain the general pattern of spread." Similar conclusions were reached for the Alps, specifically in regard to the promotion of *Picea abies* by anthropogenic fire and grazing (see chapters 4.9 and 4.10).

In central and northern Europe, spruce has a long history of plantation outside its natural range. The wood is used in many ways from the big construction of stave churches in Norway to violin tops by the masters of Cremona (Stoel and Borman, 2008). For the top ('belly') of a violin, spruce wood was used, whereas the bottom ('back'), the ribs and the neck were made of maple wood. The density differences between early and late wood of both wood types were smaller in the classical violins than it is in the modern ones, which may modify vibrations and sound radiation (Stoel and Borman, 2008). The Cremonese masters used spruce wood from the southern Alps, for example, the area of Paneveggio, more than 200 km north-east of Cremona. Burckle and Grissino-Mayer (2003) propose an interesting hypothesis relating tree rings, the Maunder Miniumum and the sound qualities of string instruments.

Pinus Diploxylon

In Europe the subgenus Diploxylon consists of nine to ten species. Six to seven of these have relatively widespread but disjunct ranges in southern Europe.

Pinus nigra forming a group of subspecies or varieties (Fig. 4.4.2-10), *P. brutia* (Fig. 4.4.2-9) and *P. halepensis, P. heldreichii, P. pinaster* and *P. pinea* (all Fig. 4.4.2-9). The group of *P. mugo* sensu lato with the species (or subspecies) *P. uncinata, P. mugo* sensu stricto and *P. rotundata* are restricted to the mountains of central and southern Europe from the Pyrenees to the Carpathians (Fig. 4.4.2-8). The Mediterranean *Pinus* species form specific communities in certain elevational belts, probably partly naturally, partly anthropogenically: Pinetum halepensis, Pinetum pineae, Pinetum pinastris (only in the west), Pinetum leucodermis or heldreichii (in the Balkans only) and Pinetum brutiae (only in the east: Mayer, 1984). The erect trees of *P. uncinata (Pinus mugo* ssp. *uncinata)* dominate in the Pyrenees and western Alps in the upper montane and subalpine belts up to the timberline but also reach lower elevations with special soil conditions. The prostrate *P. mugo (Pinus mugo* ssp. *mugo)* dominates in the eastern Alps forming a vast subalpine belt, especially on limestone. The closely related and morphologically diverse *P. rotundata (Pinus mugo* ssp. *rotundata)* grows on bogs in the mountains from the Black Forest to the northern Carpathians, partly in erect and partly in procumbent forms. The ecology and biogeography of the Mediterranean pines are reviewed in, for example, Barbéro et al. (1998).

In contrast to the Mediterranean pine species, ***P. sylvestris*** is the only species of the subgenus Diploxylon that covers a large and, for a good part, contiguous range in Europe (Fig. 4.4.2-8): from southern Spain and the Balkans to the Arctic Sea, and eastwards beyond the Ural mountains to Siberia, Manchuria and Mongolia. The very wide range of ecological tolerances can be estimated in several ways: (1) the vast geographical distribution, (2) the wide range of climatic and edaphic indicator values according to Ellenberg et al. (1992) and Hill et al. (1999) and (3) the genetic and systematic heterogeneity: the species can be subdivided over this enormous spatial range into a large number of varieties or ecological types. According to Flora Europaea more than 150 variants (Tutin et al., 2010)

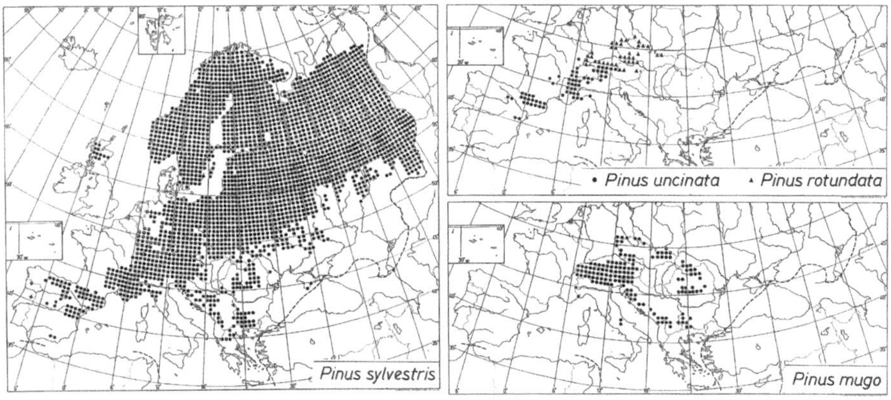

Fig. 4.4.2-8. Current distribution of *Pinus sylvestris*, *Pinus uncinata* and *Pinus mugo*. After Jalas and Suominen (1973).

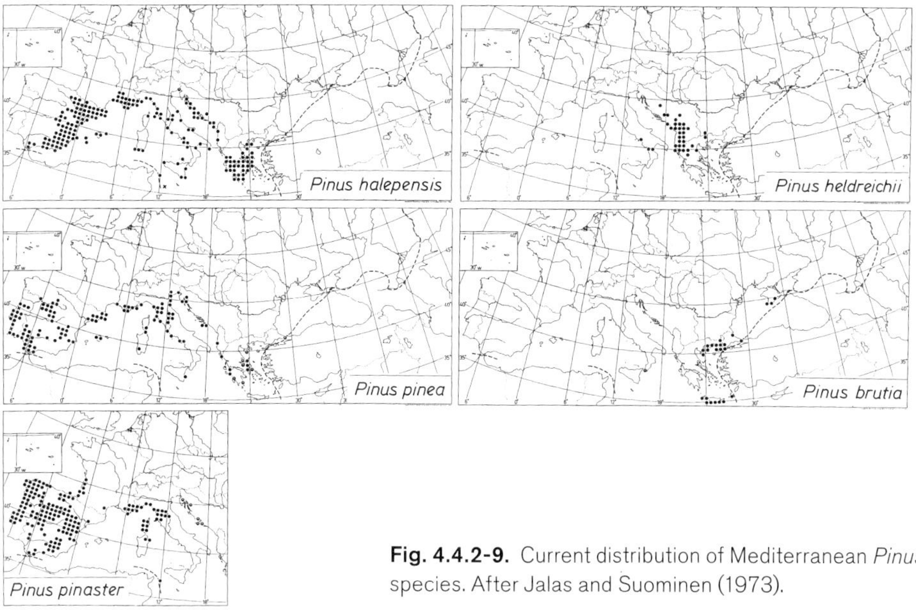

Fig. 4.4.2-9. Current distribution of Mediterranean *Pinus* species. After Jalas and Suominen (1973).

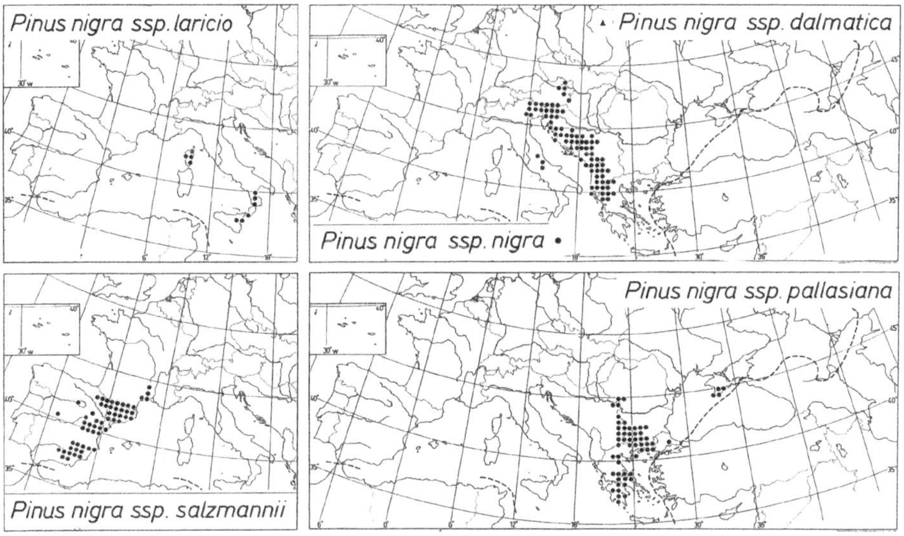

Fig. 4.4.2-10. Current distribution of the four European subspecies of *Pinus nigra*. After Jalas and Suominen (1973).

are known, partly recognized as subspecies. *Pinus sylvestris,* in its southern and central European range, is often restricted to nutrient poor, partly dry (sandy), partly wet (boggy) localities. On better soils, it cannot compete with deciduous broadleaved species and temperate conifers: this is visualized by the Ellenberg diagram for central Europe with the axes humidity and soil-pH, where *P. sylvestris* is restricted to three of the four corners. In contrast to central Europe, in the northern part of its range, where deciduous broadleaved trees and temperate conifers are either reduced or lacking because of cold climatic conditons, pine occupies the nutrient poorer and spruce the nutrient richer soils.

Pollen morphologically the numerous species within the Diploxylon-type can only partly be separated and, even then, with a low accuracy (Fig. 2.3-4). Early attempts to differentiate some taxa can

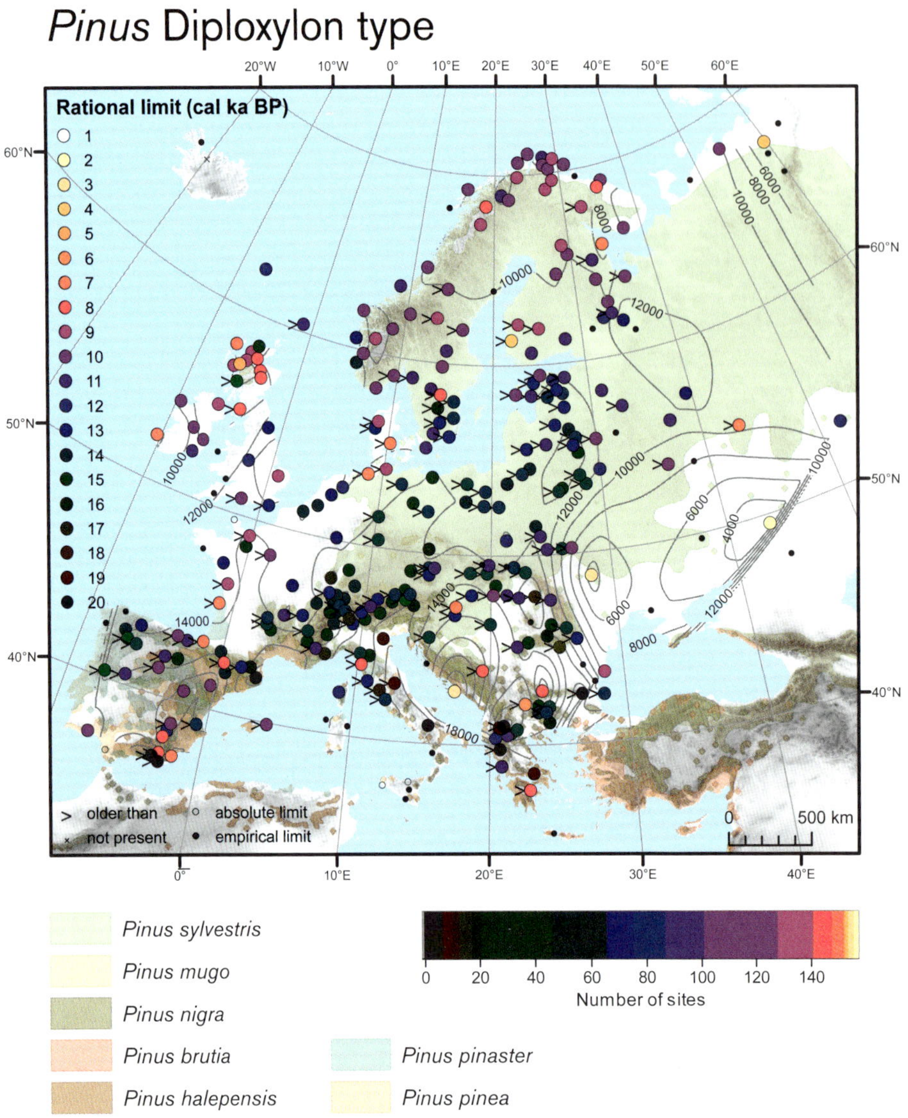

Fig. 4.4.2-11. Map showing the rational limits of ***Pinus* Diploxylon** type and the current distribution in Europe. The coloured dots indicate the rational limit in cal ka BP at selected sites, crosses refer to sites where the taxon was absent, small circles where it reached the absolute limit and small black dots where it reached the empirical limit. Grey lines represent isochrones of rational limit in 2000-year steps, based on nearest neighbour interpolation. Bar diagram shows stacked number of sites where the rational limit was reached during the same 1000-year time interval, with colours according to the legend in the figure.

be found in Hörmann (1929), Welten (1944), Lang (1952a) and Aytug (1964). The fine morphological differences between *P. sylvestris, P. mugo* sensu lato and *P. nigra* described by Klaus (1975b) have been applied to subfossil and fossil material (e.g. Schmidt, 1981) but are not always convincing. On Corsica, where today out of the subgenus Diploxylon, *P. laricio* (*P. nigra* ssp. *laricio*), *P. pinaster* and *P. halepensis* occur, Reille (1990a) presents two separate curves for *P. laricio* versus *P. pinaster/halepensis*. For the Iberian Peninsula the highland taxa *P. nigra* and *P. sylvestris* can be separated from *P. halepensis, P. pinea* and *P. pinaster* (Desprat et al., 2015).

The presence/absence problem is difficult for *Pinus* Diploxylon types if based solely on pollen because pollen production and long-distance transport are high. There is hardly any pollen spectrum in Europe, whatever its age, that does not contain pine pollen. Lisitsyna et al. (2011) provide an overview of percentage values used as thresholds in modern and fossil samples of 15–50% or 5–55%, respectively, thus including the highest percentage values of all taxa considered. However, by comparing pollen percentages with records of stomata of *Pinus sylvestris* at two sites in the Scottish Highlands, Froyd (2005) demonstrated the local presence of pine when the pollen values were as low as 0.4%. Macrofossils are therefore especially important in pine records: needles, cones or cone scales, seeds, seed wings, bark, wood (charcoal) and more can be identified (e.g. Franco-Múgica et al., 2001; Heikkilä et al., 2009). Needles can be particularly distinctive. Even isolated cuticles and stomata can be helpful for the identification to species level: for instance, on the Iberian Peninsula, García-Álvarez et al. (2009a) provide criteria to identify *P. nigra, P. pinaster, P. pinea* and *P. halepensis*. García-Álvarez et al. (2009b) present the differences between *P. sylvestris* and *P. uncinata*. In terms of wood anatomy within the Diploxylon type, three Mediterranean species are unique, namely *P. pinaster, P. pinea* and *P. halepensis* (including *P. brutia*) and two groups of species can be identified: 1) *P. sylvestris, P. mugo*, and *P. nigra* and 2) *P. pinaster* and *P. heldreichii*.

Glacial **refugia** of the **Mediterranean pine species** are generally not precisely known but in Mousterian layers in Gorham's Cave near Gibraltar, charcoal was identified as *Pinus pinea/pinaster* and *P. nigra/sylvestris*, occurring together with pollen of *Olea, Pistacia* and *Corylus* (Carrión et al., 2008). Iberian inland and highland sites also provided evidence of the presence of pines during the early-, mid-, and late-pleniglacial (last c. 70 ka) as reviewed by Carrión et al. (2003b) and González-Sampériz et al. (2010). Trunks of *P. sylvestris/nigra* dated to the last 35 ka were identified on the Atlantic coast of Portugal north of the River Tagus (García-Amorena et al., 2007). With a refined pollen morphology, Desprat et al. (2015) show that the dominant pine of the last glacial period in central and south-western Iberia was *P. nigra*. *Pinus pinaster/pinea* is also well documented as Palaeolithic charcoal from western Portugal (from Lower Aurignacian to final Magdalenian in Cabeço de Porto Marinho; Figueiral, 1995). In all long records of France, Italy and Greece the pine pollen of the Diploxylon type were not separated to estimate the role of the Mediterranean species, here charcoal analyses may help.

For *P. sylvestris*, the Pleistocene records in the three southern peninsulas are numerous, for example charcoal in Palaeolithic caves of the Cantabrian mountains (Uzquiano, 1992, 2008) or of the Portuguese Serra de Sicó (Figueiral and Terral, 2002). Every long record, either from the French Massif Central, central and southern Italy or Greece, contains a pine curve going through all cold stages (see chapter 3 and Bottema, 1979). However, also outside the southern peninsulas in sheltered habitats in central and eastern Europe, refugia of pines (both *P. sylvestris* and *P.cembra*) were demonstrated through charcoal findings, for example in famous Palaeolithic sites such as Willendorf (Haesaerts et al., 1996) or Dolní Vestoniče (Opravil, 1994; Damblon et al., 1996), and put into a broader context by Willis and van Andel (2004). At some sites, charcoal of both *Pinus* Haploxylon and Diploxylon types was found. The problem for the identification of refugia there remains that most radiocarbon dates in eastern Europe are older than 24 ka cal BP, which means that the LGM, the coldest period from 23 to 19 cal ka BP is not well documented (Tzedakis et al., 2013). A map of European refugia of *P. sylvestris* based on pollen and macroremains is provided by Cheddadi et al. (2006). However, our map (Fig. 4.4.2-11) suggests the presence of LGM refugia of *Pinus* Diploxylon on all three European peninsulas (large black dots for rational limit).

Phylogeography: many of the Mediterranean pine species show a west-east gradient, but interesting differences exist. At the one end of the spectrum, *Pinus pinea* is "Genetically depauperate but widespread..." (Vendramin et al., 2008). Chloroplast DNA showed only four haplotypes, one of them reaching from Portugal to Lebanon, one specific to central Spain and two in the Near East. In contrast to *P. pinea*, the west–east gradient in *P. halepensis* is so strong that it even led to subspecies or species, namely *P. brutia*, as shown by Bucci et al. (1998). Spatial gradients can also be important on smaller scales as illustrated for *P. pinaster* within the western Mediterranean by Vendramin et al. (1998) and Burban and Petit (2003) or within the Iberian Peninsula by Salvador et al. (2000). Burban and Petit (2003) indicate the strong difference between the high diversity of the paternally inherited chloroplast-DNA (through pollen dispersal) and the low diversity of the maternally inherited mitochondrial DNA (through restricted seed dispersal). The latter with a Moroccan, a western and an eastern (non-overlapping) mitotype are compared by Magri et al. (2007) to *Quercus suber*, for which a spatial separation during the Tertiary is discussed. In *P. sylvestris* all studies show a high genetic diversity in the Iberian Peninsula (Sinclair et al., 1999; Soranzo et al., 2000; Cheddadi et al., 2006; Pyhäjärvi et al., 2008), but also general genetic diversity gradients west–east and south–north (Naydenov et al., 2007; Pyhäjärvi et al., 2008).

Range expansions and contractions: the Late-glacial distributions and dynamics of the Mediterranean pine species are very complex. For the Iberian Peninsula, with its great physiographic diversity, a few general trends can be extracted as reviewed by J.S. Carrión et al. (2010) who found at least as many surprises as expected trends. Among the firm trends is, for example, the observation that *P. nigra/sylvestris* were important in all elevational belts, whereas *P. pinea* was confined to low elevations, especially near the coasts (see Fig. 4.4.2-9). The spreading distances of the Mediterranean species after the LGM were not very large.

Pinus sylvestris, in contrast, expanded from several refugia in the south, east and south-east of Europe more or less across the continent. Feurdean and Bennike (2004) and Feurdean et al. (2007b) report macroremains of *Pinus* sp. in the Romanian Carpathians from as early as the Bølling. Several life-history traits make *P. sylvestris* a strong pioneer: it can already flower when it is only 10–25 years old, its seeds are transported by wind and birds over several kilometres and it can grow on raw soils. As a truly protocratic pioneer, *P. sylvestris* spread across Europe at high speed. In pollen diagrams we often observe hardly any time-lag between the first local occurrence and the population expansion or between the empirical and rational limits (at least within the sampling resolution). Pine already dominated the vegetation of southern central Europe during the Allerød. This dominance then lasted longer in the north, while in southern central Europe, during the Early Holocene, the spreading deciduous, rather mesophilous taxa *Corylus, Ulmus, Quercus, Fraxinus* and *Tilia* displaced *Pinus* into three corners of the central European Ellenberg-diagram. The decline of pine in the north of Europe is later caused by competition with *Picea*. The history of pine in Scotland is rather complex as summarized in Willis et al. (1998): the two centres of Early-Holocene occurrences in the north-west and south-east mapped by Birks (1989; in uncalibrated radiocarbon years BP) may be reflected to some degree by the western/north-western and eastern clusters based on monoterpene and isozyme studies (Forrest, 1980, 1982; Kinloch et al., 1986). Interestingly the earliest pollen records in the north-west coincide, more or less, with the highest genetic diversity of modern populations. The origins of Scottish populations from both central Europe and Scandinavia seem possible. In addition, potential small western refugia are discussed by Bennett (1984) and Birks (1989) with three alternative hypotheses: glacial survival, jump-dispersal from the south or arrival from small scattered populations in the south-west or Ireland. In the British Isles today only the northern Scottish populations of pine are natural: in southern areas and in Ireland it became extinct in the Mid Holocene. Extinction also occurred in areas of northern France, Belgium and Denmark. In Scandinavia tree trunks in peat were already described and identified as *P. sylvestris* at the end of the 19th century with their absence or presence interpreted in terms of climate, as Blytt (1876) put it for Norway "during alternate rainy and dry periods". For Sweden, Sernander (1909) presented early results. The two authors then formulated what became called 'the Blytt-Sernander system', a sequence of dry continental and wet oceanic periods during the Holocene (see chapter 4.1). In Finland, after a thermal maximum during the Early and Mid Holocene (on the basis of radiocarbon-dated stumps and macrofossils north of the present latitudinal and elevational limit of pine) the decline in pine was interpreted as a climatic cooling (Hyvärinen, 1975b, 1976; Eronen, 1979; Eronen and Hyvärinen, 1982; Eronen and Huttunen, 1987, 1993; Kullman, 1989, see chapter 4.9).

For *P. sylvestris* in Europe, Cheddadi et al. (2006) provide overview maps from 18 to 7 uncal ka BP based on pollen and macrofossils and combine them with genetic data, while Brewer et al. (2017) present European maps on a calibrated time scale. They all show the great importance of this species throughout Europe during the Late-glacial and the Early Holocene. In agreement with local and regional studies, our new European data compilation (Fig. 4.4.2-11) shows that *Pinus* Diploxylon spread from the LGM refugia to cover most areas of western, central and eastern Europe already during the Late-glacial. Only in the formerly glaciated areas of northern Europe *Pinus* Diploxylon spread rather late during the Early Holocene.

Human impact: among the Mediterranean pine species, *P. pinea* may be the one that was most spread by humans and for a strong reason, namely its edible seeds. Already in the Upper Pleistocene the kernels were appreciated, as seen at Gorham's Cave (Metcalf, 1958; Carrión et al., 2008). This species was also recorded in charcoal of Würmian settlements in southern France (Massif des Maures and the Durance valley: Bazile-Robert, 1981). Human interest in this species may be the cause for the amazing genetic uniformity expressed by Vendramin et al. (2008) as "genetically depauperate but widespread".

In the discussion on 'native or introduced', vegetation history may clarify open questions in both directions: in northern and central Portugal, *P. pinaster* is today widely planted and often considered as introduced but charcoal analyses of fifty archaeological sites indicate that it was present at least since Neolithic times (Figueiral, 1995). In contrast, Reille (1989, 1991b, 1992a) found that *P. uncinata (Pinus mugo* ssp. *uncinata)* in the French Massif Central, in Pinet/Aude (south-eastern France), the Vosges and the Jura mountains was introduced in the 19th century as was *P. halepensis* on Corsica.

For *P. sylvestris* an interesting contrast between southern and northern Europe is reviewed by Willis et al. (1998). At many southern sites, anthropogenic interactions, especially the abandonment of fields and degradation of soils since the Neolithic, led to an increase in pine (that needs a lot of light but tolerates poor soils). In northern Europe different processes may have been responsible for the reductions of pine, either caused by climate change or human impact or their combination. So-called 'forest-beds' – tree trunks in peat found in Fennoscandia (see above) – are largely interpreted as signs of climatic change. In contrast, the records of numerous pine stumps found at many sites in Scotland seem ambigously climate-controlled, which triggered interesting discussions around the triangle of human impact, climate and bog hydrology (e.g. Smith and Pilcher, 1973; H.H. Birks, 1975; H.J.B. Birks 1977, 1989; Bridge et al., 1990; Willis et al., 1998). The radiocarbon dates were spread over most of the Mid and Late Holocene but with a strong cluster around c. 4500 cal yr BP. The geographically extensive spreading of this cluster through Scotland and Ireland makes a change towards a wetter climate plausible. On the other hand, the sharpness of the cluster also hints at human impact. Deforestation reduced evapotranspiration and this enhanced peat growth in the water-logged mires. If both processes – climatically or anthropogenically controlled – changed the hydrology then the question arises as to whether it was primarily the degree of conservation of stumps that increased rather than their original abundance.

Pinus Haploxylon

In Europe the subgenus Haploxylon contains three species, namely *Pinus cembra* (Swiss stone pine, Arolla pine) in the Alps and the Carpathians, *P. peuce* endemic to parts of the Balkans and *P. sibirica* spreading from north-eastern Europe into Siberia (see Fig. 4.4.2-12).

The main range of *P. cembra* is in the subalpine belt of the central Alps, where it grows in characteristic forests together with *Larix decidua* (Larici-Pinetum cembrae, or Vaccinio-Pinetum cembrae; Piceion; Vaccinio-Piceetea) that form the timberline (Fig. 4.9-5). In the western Alps this forest-belt is found between 1800 and 2500 m asl, whereas in the eastern Alps between 1700 and 2300 m asl (Mayer, 1984). In the Carpathians (Tatra, north-eastern and southern Carpathians), *P. cembra* only occurs sporadically between 1700 and 1900 m asl, usually together with *Picea abies*. There,

P. cembra does not form closed stands and a distinct belt of larch and Swiss stone pine, as found in the Alps, is missing. *Pinus cembra* is anemophilous and ornithochorous.

Pinus peuce resembles *P. cembra* in its habitats, but is restricted to the Balkan Peninsula, where it occurs in the mountains of Montenegro, Albania, North Macedonia and in south-western Bulgaria (Balkan range, Pirin, Rila and Rhodope mountains). Like *P. cembra*, this species also prefers elevations between 2000 and 2100 m asl and forms the timberline, often together with *Picea abies*.

Pinus sibirica belongs to the dark, evergreen taiga, together with *Abies sibirica* and *Picea abies* ssp. *obovata*. Like the other species of the subgenus it prefers higher elevations but may occur in cold lowlands if they are not too wet and paludified.

Pollen morphologically the Haploxylon type is differentiated from the Diploxylon type by larger grains and sacci/vesiculae that, in their outline, are not sharply delimited from the main body, but in top view show sharp limits of their structure and also by a characteristic pattern on the ventral side. Additional features can be found in Klaus (1975b) and Wegmüller (1977) and good photographs in Beug (2004) and Schmidt (1981). The separation of pollen grains of the Haploxylon and Diploxylon types can sometimes be difficult. The identification of species within the subgenera is not possible, but in contrast to the species within the Diploxylon type, the species within the Haploxylon type are mutually exclusive concerning their modern ranges.

In contrast to *Pinus* Diploxylon, the empirical limit is interpreted as a sign of local presence (Fig. 4.4.2-13). While higher threshold values (e.g. 1%) decrease the likelihood of false presence, they have the disadvantage to increase the probability of false absence, which may mean missing first or early population establishments.

Macrofossils, such as cones, bracts, seeds and needles seem to allow for species identification (Katz et al., 1977). In wood anatomy, the separation from *P. sylvestris/P. mugo* is possible (Schoch et al., 2004) but the interspecific differences in stomata are too small for a separation (Trautmann, 1953). All species of the subgenus have large heavy seeds

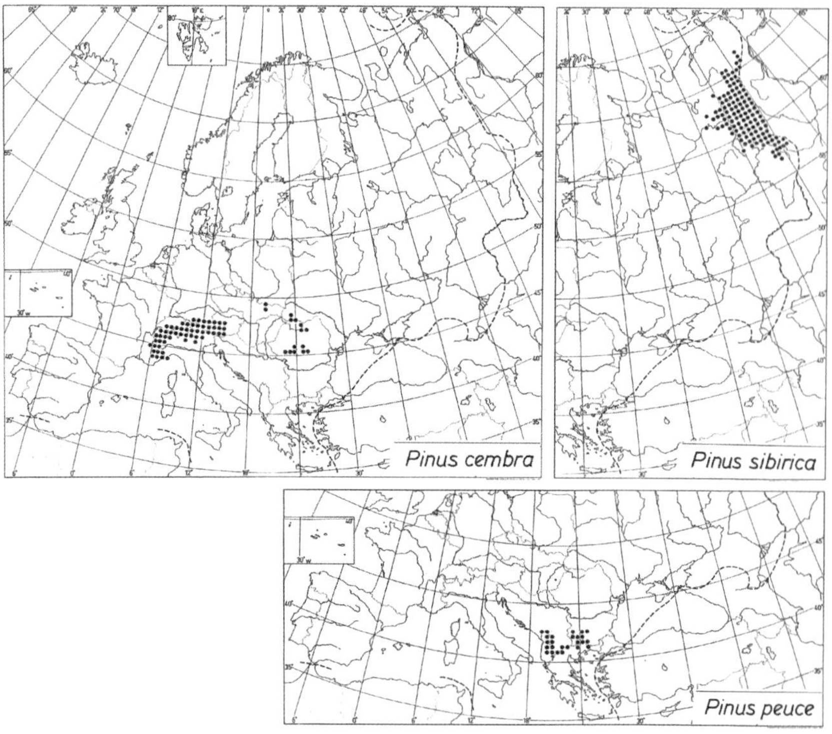

Fig. 4.4.2-12. Current distribution of *Pinus cembra*, *Pinus sibirica* and *Pinus peuce*. After Jalas and Suominen (1973).

Pinus Haploxylon type

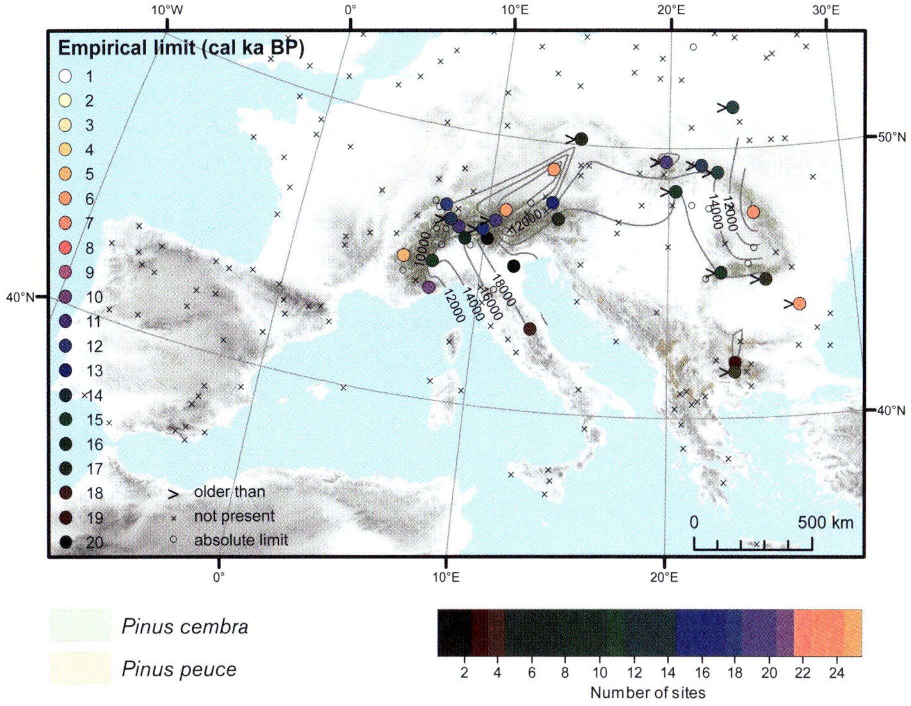

Fig. 4.4.2-13. Map showing the empirical limits of ***Pinus* Haploxylon** type and the current distribution in Europe. The coloured dots indicate the empirical limit in cal ka BP at selected sites, crosses refer to sites where the taxon was absent and small circles where it reached the absolute limit. Grey lines represent isochrones of empirical limit in 2000-year steps, based on nearest neighbour interpolation. Bar diagram shows stacked number of sites where the empirical limit was reached during the same 1000-year time interval, with colours according to the legend in the figure.

without wings (comparable to *P. pinea*, *P. edulis* or *P. monophylla*: all used as staple food by people and birds, thus ornithochorous). *Pinus cembra* is primarily dispersed by *Nucifraga caryocatactes* (Eurasian nutcracker; Bonn and Poschlod, 1998).

The full-glacial **refugia** must be outside the modern range for both *P. cembra* and *P. peuce* because their modern areas of occurrence were under ice. In contrast, the modern European part of the range of *P. sibirica* is outside the Weichselian-Valdai glaciation. Macroremains of *P. cembra* are reported from the Hungarian lowland (Scharfetter, 1938; Zolyomi, 1953) and from southern Moravia, where they are attributed to the Palaeolithic culture of Pavlov (Gravettian, see chapter 4.5.2) dated around 29 000 to 22 000 cal yr BP (Necesaný, 1951; Kneblovà, 1954). A pollen diagram from the south-Moravian site Bulhary (180 m asl) near the Palaeolithic sites of Dolní Věstonice (174 m asl) and Pavlov (205 m asl) and covering this period shows a closed pollen curve for *P.* Haploxylon of 1–10% of the total pollen sum (Rybníčková and Rybníček, 1992). Both of these famous archaeological sites provided charcoal of *P. cembra* (Opravil, 1994; Damblon, 1997; Willis and van Andel, 2004). In the western Carpathians at Šafárka (Slovakia, 600 m asl) and Jablůnka (Czech Republic, 350 m asl), *P. cembra* was identified and also dated to the last full glacial (Jankovská and Pokorný, 2008). In addition, pollen records of *P. cembra* outside of its modern range exist from the south-eastern margin of the Alps, for instance in Pokljuka (1170 m asl) and Kaznarica (550 m asl) in Slovenia (Šercelj, 1971, 1984). Wood and charcoal findings are again recorded from a famous Palaeolithic site, namely from Willendorf (Austria, 210 m asl: Haesaerts et al., 1996) at c. 30 000 to 26 000 cal yr BP. Our isochrone map (Fig. 4.4.2-13) shows LGM refugia in the unglaciated areas south of the Alps (black dots). Ages older than 17 000 cal yr BP in the unglaciated forelands of the Sudetes and the Carpathians suggest additional refugia in eastern central Europe.

Phylogeographical studies indicate that variation in the chloroplast DNA of *Pinus cembra* shows higher diversity in the Carpathian than in the Alpine populations (Höhn et al., 2009). Within the central (Swiss) Alps, a gradual decline in genetic diversity from south-east to west may indicate a common glacial refugium in the south-east (Gugerli et al., 2009). Similarities on an inter-specific level of chloroplast and mitochondrial DNA led Gugerli et al. (2001b) to the conclusion of "a relatively recent evolutionary separation of *P. cembra* and *P. sibirica* despite their currently disjunct distributions".

Range expansions and contractions: Late-glacial findings in the east – also outside but close to the modern range – include the sites in the lowlands of northern Slovakia: at Jedlova (650 m asl) and Bobrov (620 m asl) in the Oravska basin west of the High Tatra (Jankovská, 1984). These findings of macroremains together with *Larix decidua* suggest survival of the two species in the lowlands and, since the Early Holocene, their displacement to higher elevations, even to the timberlines of the Carpathians (Rybníčková and Rybníček, 1993). From the glacial refugia in the lowlands and near the foothills of the Carpathians and the south-eastern Alps, *P. cembra* was expanding rapidly after the ice retreat (Fig. 4.4.2-13), as for example illustrated by needle findings in Lago di Ganna (452 m asl, near Varese, Italy: Schneider and Tobolski, 1985) dated to the Bølling. This first Late-glacial expansion occurred before the Bølling, at latest at around 16 500 to 16 000 cal yr BP, as documented by short shoots and needles at the site Balladrum (390 m asl, near Locarno, Switzerland) and covered low to middle elevations (see chapters 4.3.2 and 4.9).

At the beginning of the Holocene the competition from temperate taxa became strong at lower elevations and *P. cembra* together with *Larix decidua* escaped to the subalpine belt, for example in the Valais (Kaltenrieder et al., 2005) and the Poschiavo valley (Switzerland; see chapters 4.3.2 and 4.9). Pollen and macrofossil investigations performed in the rugged terrain of the sandstone gorges of the inner Bohemian Basin (Czech Republic; sites Čin-Čan-Tau, 265 m asl and Vlčí důl, 310 m asl) show continuous occurrence of *P. cembra* since the Late-glacial to the 1st millenium of the Holocene and document its local extinction due to the above-mentioned competition with expanding temperate broadleaved taxa (Pokorný et al., 2017; Svoboda et al., 2018).

Early to Mid and Late Holocene records in the western Alps are documented by Ali et al. (2005) who identified and dated charcoal from 37 soil profiles and 10 travertine sequences. Some parts of the western Alps were among the last high-elevation areas colonized by *Pinus cembra* (Fig. 4.4.2-13).

Taxus baccata

Taxus is a gymnosperm genus of the temperate zone of the northern hemisphere with about seven to ten species, some of which are often considered subspecies of *T. baccata* (English yew). It is 'a conifer without cones', and therefore previously put into the order of Taxales, but is now again in the Pinales. The tree is usually dioecious (rarely monoecious). The plant is anemophilous and zoochorous. All of its parts, except the aril, are poisonous (alkaloid taxol), especially for horses, but sheep, goats, reindeer, ibex, roe and red deer can eat its needles (and cows can if they do not drink water during grazing: Uzquiano et al., 2015). Like *Abies alba*, *T. baccata* has a pectinate arrangement of needles, an adaptation of temperate conifers to competition with broadleaved species.

Today the distribution of *Taxus* in Europe consists of numerous small patches, sometimes summerized as 'widespread but uncommon', as shown in figure 4.4.2-14.

In contrast to the present, *Taxus* was a major constituent of many European forests during several interglacials (e.g. Watts, 1985; during the Eemian near Lac de Bouchet: Reille et al., 1998; near Oerle in NW-Germany: Behre and Lade, 1986; Behre, 1989). The species tends to avoid acidic soils and prefers high air humidity. Therefore, in England and Wales it occurs down to sea level, but in submediterranean areas, it occurs at mid-elevations in the mountains (Thomas and Polwart, 2003). It is a taxon with an especially limited filling of its potential range, so that there is a large difference between its potential and realized ranges sensu Svenning and Skov (2004).

Taxus baccata

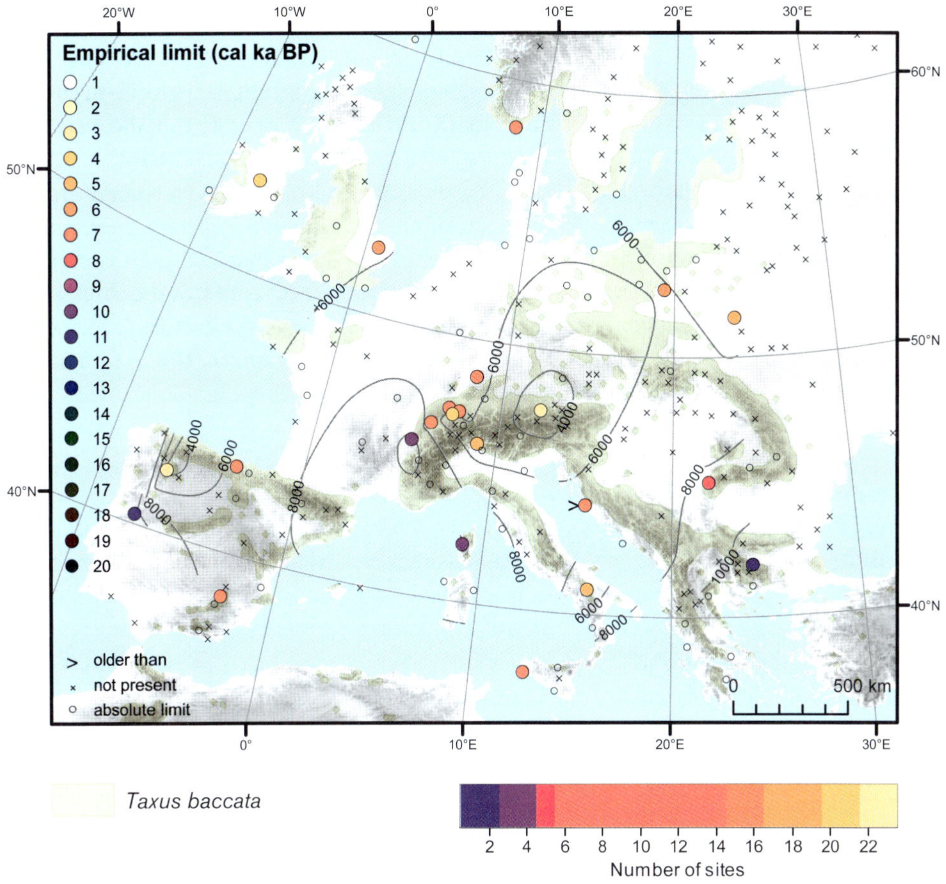

Fig. 4.4.2-14. Map showing the empirical limits of ***Taxus baccata*** and the current distribution in Europe. The coloured dots indicate the empirical limit in cal ka BP at selected sites, crosses refer to sites where the taxon was supposedly absent and small circles where it reached the absolute limit. Grey lines represent isochrones of empirical limit in 2000-year steps, based on nearest neighbour interpolation. Bar diagram shows stacked number of sites where the empirical limit was reached during the same 1000-year time interval, with colours according to the legend in the figure. Note that pollen of *Taxus baccata* is easily corroded and might not have been identified by early studies due to taxonomical difficulties.

The pollen of *Taxus* is thin-walled and is easily corroded: there are several sites where macrofossils and wood have been found but no pollen (Deforce and Bastiaens, 2007; López-Merino et al., 2010). In addition, the pollen may not have been identified in some early studies. Therefore the 'presence/absence problem' of *Taxus* may be especially serious (see Fig. 4.26 in Huntley and Birks, 1983 and Brewer et al., 2017).

The location of full-glacial **refugia** and also of Late-glacial occurrences of *Taxus* seem somewhat enigmatic (Fig. 4.4.2-14). Our isochrone map, however, shows oldest occurrences in southern Europe, where the refugia were most probably located. In such a case the absolute pollen limit (as evidence of regional presence, Fig. 4.4.1-1) might be considered in addition to the empirical limit. For instance, in northern Spain only single pollen grains were found at mid-elevation sites (Peñalba, 1989, 1994). An exception is El Portalet at 1800 m asl where a nearly continuous pollen curve during the Late-glacial indicates the presence of yew (González-Sampériz et al., 2006; Gil-Romera et al., 2014). In southern Spain at Laguna de Siles, both Late-glacial and Holocene records were found and in the Serra da Estrela, Holocene records (van der Knaap and van Leeuwen, 1995; Carrión, 2002b). For Iberia, Uzquiano et al. (2015) provide a fine overview. Other refugia might have occurred on the Balkan and Italian peninsulas as well as on the Mediterranean islands (Fig. 4.4.2-14).

Phylogeography: Its spatial genetic structure can be highly complex as shown by González-Martínez et al. (2010) for the western Mediterranean region. Aspects of conservation are important (Svenning and Magård, 1999; Thomas and Polwart, 2003; Piovesan et al., 2009; Sanz et al., 2009; Scarnati et al., 2009; Ruprecht et al., 2010; Linares, 2013; Abel Schaad et al., 2014).

Range expansions and contractions: For the Holocene in south-western Europe occurrences of *Taxus* can be amazingly sharp and short as, for example, at Laguna Negra in the Sierra de Urbión (6760–4500 cal yr BP: von Engelbrechten, 1998) and Belate (Navarra: Peñalba, 1989, 1994), or temporally more extended with high values between about 7000 and 2500 cal yr BP as reviewed for all Iberia by Uzquiano et al. (2015) based on pollen and charcoal at 44 sites. In Monticchio, southern Italy, a distinct phase of *Taxus* with *Abies* was dated to c. 4500–2570 cal yr BP by Watts et al. (1996b). On Corsica, where *Taxus* occurs today together with *Ilex aquifolium* between 850 and 1500 m asl, two phases with remarkably high pollen values were found between c. 9550 and 6200 cal yr BP by Reille et al. (1999). At some sites in the south-western Alps and the Jura mountains, *Taxus* occurred during the Mid Holocene (Atlantic: de Beaulieu et al., 1994) or in the Massif Central during the Late Holocene (de Beaulieu et al., 1985). In England, single grains of the English yew were found from 7000 cal yr BP in Caburn, East Sussex (Waller and Hamilton, 2000), or at higher frequencies from c. 6700 cal yr BP at Roudsea Wood, Cumbria (Birks, 1982). At Caburn, a population increase was found around 5500 cal yr BP, distinctly after the *Tilia*-decline and the two elm-declines. At Roudsea, a small population increase occurred shortly after the elm-decline but together with a *Tilia*-increase. At Diss Mere, several sharp peaks of *Taxus* followed the elm decline (Peglar et al., 1989). In Blekinge (southern Sweden), Berglund (1966) found *Taxus* sporadically towards the end of the Early Holocene (end of the Boreal) and then a main occurrence during the Late Holocene (Subboreal c. 3800–2300 cal yr BP). At two sites on Åland, a distinct maximum of *Taxus* was found between c. 2000 and 1200 cal yr BP (15 BC to AD 800): the maxima are not quite synchronous, but at both sites the upper end of the *Taxus* maximum coincides with the beginning of the continuous *Secale* curve, a distinct hint that the decline of yew was caused by human impact (Sarmaja-Korjonen et al., 1991). At Lake Gościąż (central Poland), the first pollen grains of yew are recorded at c. 6800 cal yr BP and higher abundances between c. 5200 and 4200 cal yr BP (Ralska-Jasiewiczowa et al., 1998) but generally the Holocene pollen values in Poland are below 1%, except in the Tuchola Forest (Krupiński et al., 2004). Also in northern Germany, the highest abundances were recorded between c. 5000 and 3000 cal yr BP (Willerding, 1968; Averdieck, 1971; Brande, 1975). For Belgium, Deforce and Bastiaens (2007) demonstrated that during the early part of the Late Holocene (Subboreal), *Taxus* grew on the coastal plain and the lower Scheldt valley in carr vegetation but it disappeared when the fens developed into bogs. Today it is restricted to the southern parts of the country where it grows on calcareous steep slopes. At the European scale a very clear Holocene pattern of spread appears with isochrones becoming younger towards the north, which points to an Early to Late Holocene spread of the species out of the three Mediterranean peninsulas where the glacial refugia were likely located (Fig. 4.4.2-14). The bar diagram shows that most of the spread occured during the Mid Holocene. On the basis of the isochrone map we estimate the spreading rates of *Taxus* to 180–270 m yr^{-1} (Table 4.11-1), a value comparable to that of other conifers (e.g. *Pinus* Diploxylon) or deciduous trees (e.g. *Quercus* or *Tilia*).

As in Britain and Ireland, the modern occurrences on steep, rocky limestone slopes may be the result of long-standing competition (shaded-out by *Fagus* and *Carpinus*) and human impact (Tittensor, 1980; Birks, 1982; O'Connell et al., 1988; Mitchell, 1990; Brande, 1994; Peñalba, 1994; Waller and Hamilton, 2000; Deforce and Bastiaens, 2007).

Human impact: because the wood of yew is both very strong and very flexible it was used by early humans in many ways, especially for bows, arrows and spears. The spear from Clacton-on-Sea, Essex, from the Lower Palaeolithic (Hoxnian, MIS 13) may be one of the oldest wooden artefacts found so far (Hazzeldine Warren, 1922; Godwin, 1975; Thomas and Polwart, 2003). The spear from Lehringen, Lower Saxony (Eem, MIS 5e) is also made of *Taxus* wood (Jacob-Friesen, 1949). Another aspect of human impact on yew populations is that its wood is an excellent fuel providing long-lasting fires. Uzquiano et al. (2015) also discuss the seeming contradiction between the content of alkaloid taxols and the use of yew as fodder for lifestock during prehistoric and historic times. In some languages interesting connections are made: Zoller (1981) mentioned that the medieval 'Iwe' and the Swabian 'Eib' mean both yew and bow. In Greek τό τόξον (the bow) and τοξικός (poisonous) sound nearly like *Taxus*; some authors consider 'toxon pharmacon' an 'etymological hybrid or distortion'. Knowledge of the special properties of the genus *Taxus* even made its way into the world of literature, such as in Shakespeare's Richard II: "Thy very beadsmen learn to bend their bows / Of double-fatal yew against thy state".

When looking at various maps it is obvious that *Taxus* is a taxon for which our knowledge of the Late-glacial and Holocene history is still very incomplete. Human impact probably worked in two opposite directions: planting (especially in graveyards) and cultivating versus over-exploitation that may have caused increases and decreases in pollen abundances (Dickson, 1993; Scheeder, 1994; Hassler-Schwarz, 1999).

4.4.3

Deciduous angiosperms

Acer

The genus *Acer* comprises c. 150 species which grow in the northern hemisphere, only occasionally it crosses the equator, such as in the mountain forests of Java and Sumatra (Schütt et al., 2002). Most *Acer* species are deciduous broadleaved, only a few are evergreen broadleaved. About 12 species grow in Europe, all of them are deciduous with the exception of *A. sempervirens*, which is native to Crete (Pfadenhauer and Klötzli, 2014). *Acer platanoides* is the most widespread European species, its natural range spans from southern Europe including the Caucasus to Fennoscandia and Russia in northern Europe. In forests, it is often a secondary species, generally found in small stands or admixed as individuals. However, under temperate continental conditions it has the potential to co-dominate with *Quercus robur* and *Tilia cordata* (Caudullo and de Rigo, 2016b). The tree is of medium size, growing 25–30 m tall and becoming c. 500 years old. The species is very tolerant to shade in juvenile and mature stages, has an intermediate drought sensitivity comparable to that of *Quercus petraea* or *Abies alba*, is rather demanding with regard to soils but tolerates cold winters, although severe winter frosts can damage it (Ellenberg, 2009). Like for all *Acer* species, the flowers are insect-pollinated, thus markedly underrepresented in pollen records.

Acer pseudoplatanus is less widespread than *A. platanoides*, but more abundant in its range, which covers central and southern Europe. *Acer pseudoplatanus* can dominate or co-dominate in temperate mountain forests, where it may occur with, for example, *Tilia cordata, T. platyphyllos, Fraxinus excelsior* and *Ulmus glabra* (Pasta et al., 2016a). Preferred habitats comprise slopes, rock debris and gorges. The species can grow rather tall, reaching 30–40 m and is long-lived with maximum ages around 600 years. The ecology of *A. pseudoplatanus* is very similar to that of *A. platanoides* with regard to its high shade tolerance as well as its intermediate drought and frost sensitivity (Ellenberg, 2009). A third maple species with a wide range is *A. campestre*. It occurs all over Europe with the exception of western Iberia, Fennoscandia and northern Russia. It is a secondary species admixed in deciduous broadleaved temperate forests and grows together with *Quercus, Ulmus, Tilia* and *Castanea sativa* (Zecchin et al., 2016). The species is markedly shorter than *A. platanoides* and *A. pseudoplatanus* (<20 m) and more drought tolerant. Southern European maple species with regional to local importance are *A. monspessulanum*, *A. opalus* (restricted to south-western Europe), *A. heldreichii* (south-eastern Europe), *A. tataricum* (south-eastern Europe), *A. hyrcanum* (south-eastern Europe), *A. sempervirens* (Crete), *A. velutinum* (Caucasus), *A. trautvetteri* (Caucasus) and *A. cappadocicum* (Caucasus; Schütt et al., 2002).

Macrofossil finds allow tracking occurrences of *Acer* species during the Pleistocene and the Holocene (e.g. Zagwijn, 1996a; Garcia-Amorena et al., 2008; Martinetto, 2009), however, the available records are rare and fragmentary. *Acer* pollen can

Acer

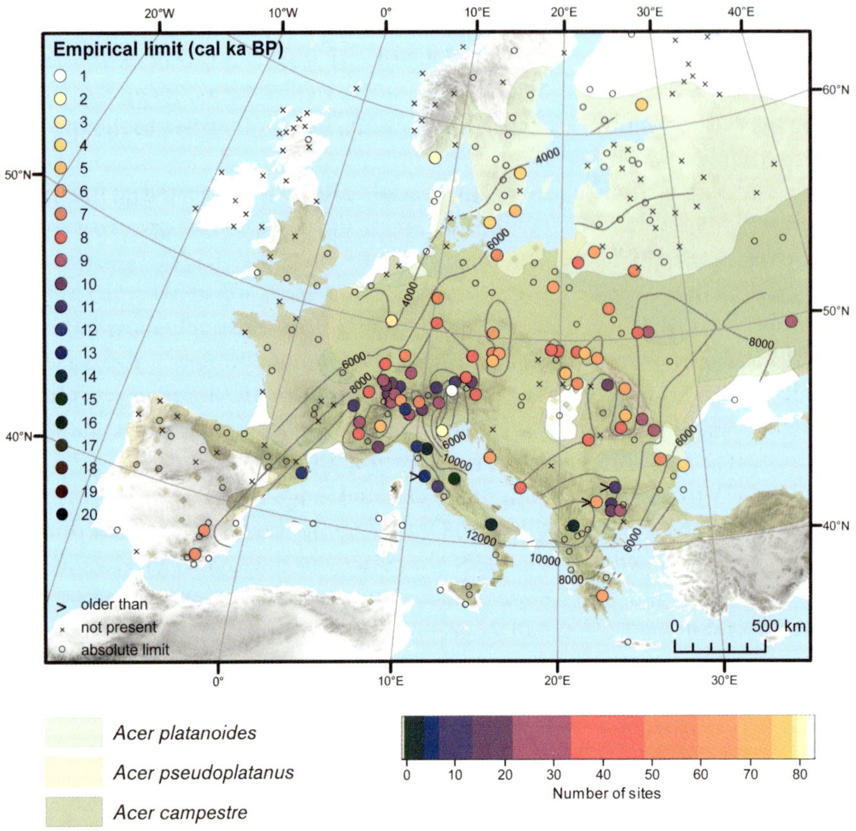

Fig. 4.4.3-1. Map showing the empirical limits of ***Acer*** and the current distribution in Europe. The coloured dots indicate the empirical limit in cal ka BP at selected sites, crosses refer to sites where the taxon was absent and small circles where it reached the absolute limit. Grey lines represent isochrones of empirical limit in 2000-year steps, based on nearest neighbour interpolation. Bar diagram shows stacked number of sites where the empirical limit was reached during the same 1000-year time interval, with colours according to the legend in the figure. Note that *Acer* is insect-pollinated, the local establishment therefore may have occurred before the empirical limit.

be resolved into four taxonomic groups, *A. campestre* type, *A. negundo* type, *A. carpinifolium* type and *A. saccharum* type (Biesboer, 1975), of which the first two are relevant for Europe (Clarke and Jones, 1980). Nevertheless, the pollen is usually determined to the genus level only.

Pollen finds of *Acer* in autochthonous organic or calcareous sediments suggest that the genus survived the Last Glacial Maximum (LGM, c. 23 000–19 000 cal yr BP) in southern European **refugia**, for example on the Italian Peninsula (Kaltenrieder et al., 2009). Given that the species is insect-pollinated the empirical limit is likely capturing early mass expansions of *Acer*, while the initial establishment of small local stands may have occurred long before (absolute limit, see chapter 4.4.1). Accordingly, the isochrone map for *Acer* (Fig. 4.4.3-1) shows rather late ages. The oldest evidence of conspicuous *Acer* populations comes from the Italian and Balkan peninsulas and is dated older than 14 000 cal yr BP (green dots in Fig. 4.4.3-1). Before the onset of the Holocene, *Acer* populations were rare and small or absent outside southern Europe, so that the range of the taxon remained small (12 000 cal yr BP isochrones) with only c. 5% of the sites reaching the empirical limit (bar diagram, Fig. 4.4.3-1).

Range expansions and contractions: During the Early Holocene the species expanded in southern Europe and central Europe south of c. 50°N. Nevertheless, at the onset of the Mid Holocene at around 8000 cal yr BP, far less than half of the today's range was covered. The spread continued during the Mid Holocene to reach Fennoscandia by 5000–4000 cal yr BP, when most of today's range was almost attained. During the Late Holocene only a minor spread to north-western and northern Europe occurred. The populations of *Acer* outside southern Europe are thus young and related to Holocene warm temperatures that allowed the maple species to spread north. Spreading rates of *Acer* were 90–260 m yr^{-1} and thus comparable with those of other species with winged diaspores (e.g. *Picea abies, Abies alba*; Table 4.11-1). The resulting postglacial dispersal distance was 2–10 km, which is in good agreement with ecological observations (Table 4.11-1). During the course of the Holocene, *Acer* abundance in the vegetation was markedly reduced through land use including burning (see chapter 4.10). Given that the wood of *Acer* is very valuable, among the best in Europe (Schütt et al., 2002), its use may additionally have reduced the occurrence of the tree.

Alnus

The genus *Alnus* Miller consists of 29 to 35 species (Bevan et al., 2019). Out of the four species of *Alnus* native to Europe, *A. cordata* is restricted to southern Italy and Corsica, *A. viridis* mainly occurs on wet slopes in mountains (with several subspecies), *A. incana* has a vast range in north-eastern Europe and Fennoscandia but is also widespread in more southern mountain areas, whereas *A. glutinosa* is very widespread throughout large parts of Europe (e.g. Fig. 4.4.3-2; McVean, 1953). All species, with exception of *Alnus viridis*, prefer wetlands and alluvial habitats. *Alnus* species live in symbiosis with *Frankia* (Actinomycetales), as well as ectomycorrhizal fungi which makes them strong pioneers. All species are anemophilous and anemochorous (and hydrochorous: Cappers, 1993a).

In pollen morphology only *A. viridis* (Chaix) DC (= *A. alnobetula* (Ehrh.) Hartig) can be separated from the other European species (Beug, 2015). For the pollen morphology of the three subgenera of *Alnus* on a global scale see Bevan et al. (2019). Macrofossils, such as wood, twigs, cones, fruits and male catkins can be partly identified to the species level.

Full-glacial **refugia** were located on the three Mediterranean peninsulas and probably, to a smaller extent, also near the southern Alps, near the Carpathians and east of the Carpathians (according to findings of Upper Palaeolithic charcoal in Mitoc Malu Galben: Damblon et al., 1996). Douda et al. (2014) provide a European map for putative refugia but wisely use several question marks at northern sites because for basal samples contamination by both reworking and long-distance transport is conceivable. Our isochrone map suggests that during the LGM the refugia of the species producing the *Alnus glutinosa* type were located on the Italian and Balkan peninsulas (black dots), possibly also on the Iberian peninsula and in France.

In a **phylogeographic** study of *A. glutinosa*, King and Ferris (1998) found 13 cpDNA haplotypes, three or four of them in various parts of south-eastern Europe, one each in southern Italy, Corsica and Iberia, a rare one in central Norway, and two types widespread in central, north-western and north-eastern Europe. Three additional haplotypes were found in north-eastern Turkey.

The **range expansions and contractions** are comprehensively presented by Douda et al. (2014) including 18 maps based on four threshold values from <0.5% to ≥10% for pollen and presence for macrofossils. Here, we prefer empirical limits because they are more sensitive to first establishments of populations (Fig. 4.4.3-2).

An interesting contrast was shown by Giesecke et al. (2011) concerning not the first establishment but the population expansion: a nearly synchronous mass expansion around the Baltic Sea from north-east Germany to southern Fennoscandia at c. 9500 cal yr BP differs strongly from the time-transgressive population expansions on the British Isles. For *A. glutinosa*, the latter is demonstrated by Bennett and Birks (1990) using 92 sites with good time control and shown on nine maps. They found that the Early Holocene scattered occurrences with no obvious geographical pattern led to a "spread and increase...patchy and erratic in space and time". The availability of suitable habitats and rare weather

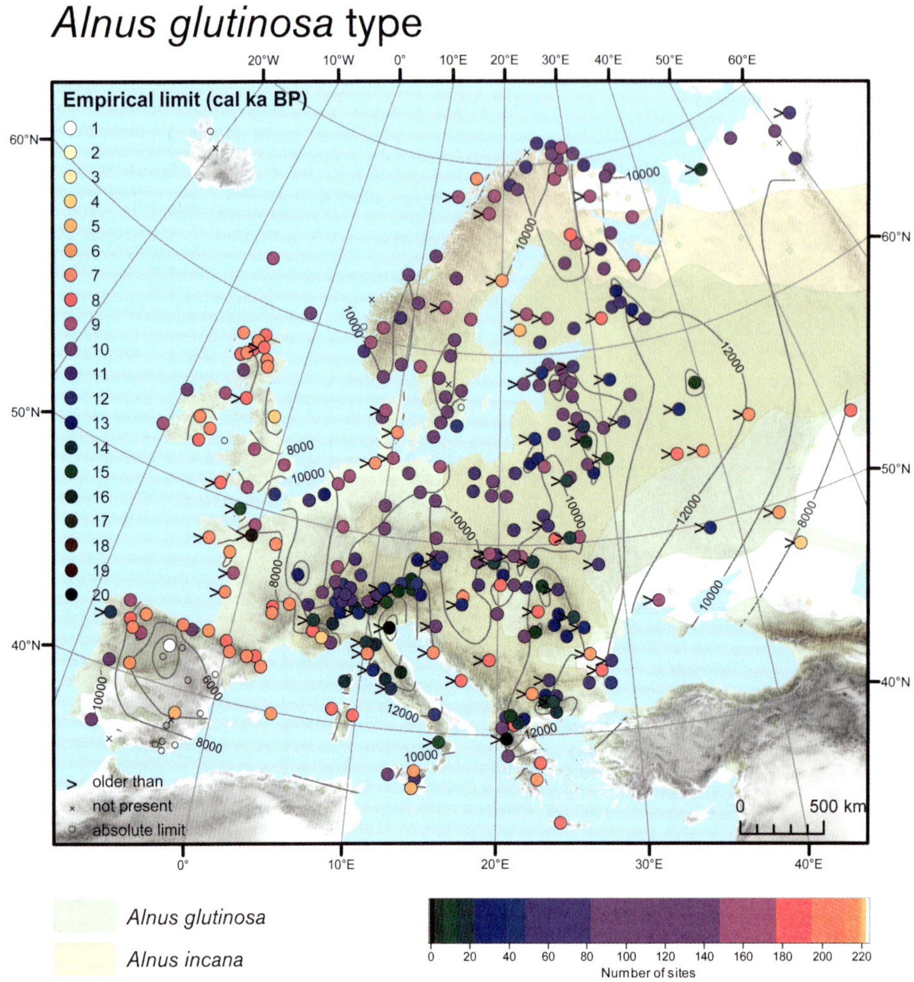

Fig. 4.4.3-2. Map showing the empirical limits of ***Alnus glutinosa*** type and the current distribution of *A. glutinosa* and *A. incana* in Europe. The coloured dots indicate the empirical limit in cal ka BP at selected sites, crosses refer to sites where the taxon was absent and small circles where it reached the absolute limit. Grey lines represent isochrones of empirical limit in 2000-year steps, based on nearest neighbour interpolation. Bar diagram shows stacked number of sites where the empirical limit was reached during the same 1000-year time interval, with colours according to the legend in the figure.

events may have controlled the abundance of this hygrophilous species. This dependency on humidity is also discussed by Pokorný et al. (2000) who found some tendency towards cyclic succession. Maps for four threshold values are provided by Giesecke and Brewer (2018). Our isochrone map (Fig. 4.4.3-2) shows that at the continental scale *Alnus glutinosa* type was almost entirely restricted to southern and eastern Europe during the Late-glacial; only c. 20% of the sites (bar diagram) show ages older than 12 000 years. During the Early Holocene the taxon expanded over wide areas of central and northern Europe and its spread culminated around 10 000 cal yr BP (bar diagram). Finally, during the Mid and Late Holocene the taxon spread to some areas in Scotland and Fennoscandia as well as in Iberia and at some sites on the Mediterranean islands. On the basis of the isochrone map we estimate the spreading rates of *Alnus glutinosa* type to 180–360 m yr^{-1} (Table 4.11-1), a value comparable to that of other deciduous trees (e.g. *Quercus* or *Tilia*).

Betula

The genus *Betula* comprises c. 40 deciduous species which are widespread in the northern hemisphere (Schütt et al., 2002). In Europe, two tree – *Betula pendula* and *B. pubescens* – and two shrub species – *B. humilis* and *B. nana* – grow (Beck et al., 2016a). The shrub species are restricted to cold environments (e.g. tundra) or wetlands (e.g. mires), while the two tree species have large ranges. North of the Alps (or c. 45°N), *B. pendula* and *B. pubescens* occur from the lowlands to high elevations, while in southern Europe birch trees are restricted to mountain ranges with cool and/or moist conditions (Morales-Molino et al., 2019). All European birch species are short-lived, light-demanding pioneers (Ellenberg, 2009; Beck et al., 2016a). *Betula pendula* and *B. pubescens* can dominate or co-dominate in boreal forests and are secondary species in temperate forests. Both trees are medium-sized and can grow up to

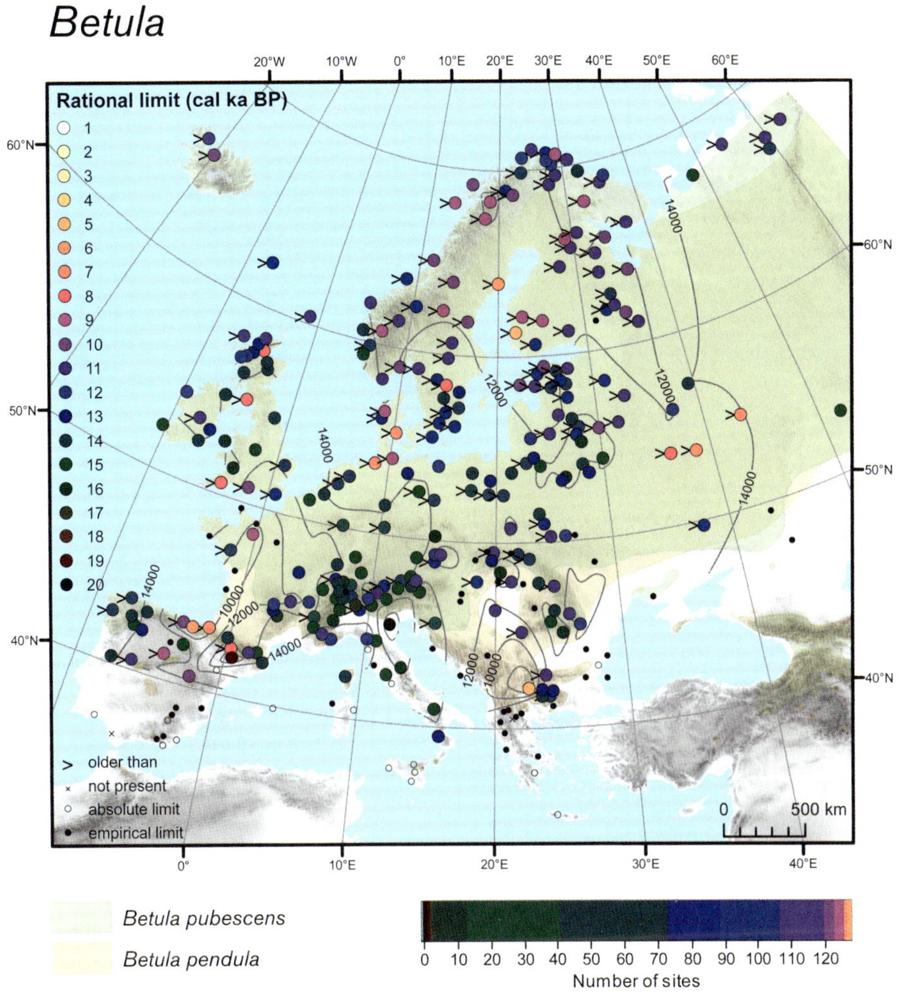

Fig. 4.4.3-3. Map showing the rational limits of ***Betula*** and the current distribution in Europe. The coloured dots indicate the rational limit in cal ka BP at selected sites, crosses refer to sites where the taxon was absent, small circles where it reached the absolute limit and small black dots where it reached the empirical limit. Grey lines represent isochrones of rational limit in 2000-year steps, based on nearest neighbour interpolation. Bar diagram shows stacked number of sites where the rational limit was reached during the same 1000-year time interval, with colours according to the legend in the figure.

c. 20–30 m and reach maximum ages of c. 120 years (Schütt et al., 2002; Beck et al., 2016a). Differences between the two *Betula* tree species comprise adaptations to cold climates and soils. *Betula pubescens* tolerates poorly drained soils, while *B. pendula* grows best on well-drained soils. *Betula pubescens* can tolerate colder conditions than *B. pendula*. Its tolerance to cold temperatures at the northern treeline makes *B. pubescens* the tree species with northernmost positions in Europe (Beck et al., 2016a).

The **phylogeographic** analysis of chloroplast haplotypes of extant *B. pendula* and *B. pubescens* shows that most genetic variation can be attributed to variation within populations. However, there is also considerable variation among populations, which generates distinct southern, western-central and eastern European genetic patterns (Maliouchenko et al., 2007). Specifically, *B. pendula* shows genetic differences in the three southern European peninsulas, which are partly mirrored in Europe north of 45°N. In addition, eastern European populations have their own distinct properties. Similarly, *B. pubescens* shows considerable genetic variability in space. Such genetic differences may have originated during the contraction and dispersal phases of the glacial–interglacial cycles. For instance, chloroplast haplotypes which are unique to Iberia and Italy suggest that some of the southern European populations are old and did not contribute much to the recolonization of northern Europe during the postglacial (Maliouchenko et al., 2007).

Macrofossil analyses allow the identification of the different *Betula* species (e.g. *B. nana*: Lotter, 1999; Bjune, 2005), however, sedimentary records displaying the occurrence of *Betula* at the species level are rather rare. Pollen analysis allows the distinction of tree and shrub *Betula* (Birks, 1968; Blackmore et al., 2003; Clegg et al., 2005), but unfortunately the pollen separation is only rarely done. Our isochrone map for *Betula* uses the rational limit, because the empirical limit is reached almost at all sites from the beginning of the records. For most records in northern Europe this applies also for the rational limit (Fig. 4.4.3-3), probably because *Betula* was already present locally when the natural archives established. *Betula* pollen is overrepresented in sedimentary records, thus the rational limit may reflect rather initial population expansions.

Macrofossil evidence from sites with autochthonous organic or calcareous sediments shows that during the Last Glacial Maximum (LGM) mixed *Betula* forests occurred in **refugia** south of the Alps. For instance, in northern Italy, tree *Betula* dominated together with *Larix decidua* during the LGM at 23 000–19 000 cal yr BP (Kaltenrieder et al., 2009). It is likely that tree *Betula* survived the LGM also north of the Alps at latitudes between 45 and 50°N (Birks and Willis, 2008; Fig. 4.4.3-3), as suggested by charcoal remains of *Betula* found in soils in association with other boreal trees (e.g. *Pinus cembra, P. sylvestris*) dated at 20 000–15 000 cal yr BP (Willis and van Andel, 2004). The available macrofossil and pollen evidence suggests that before the onset of the Bølling-Allerød interstadial at around 14 700 cal yr BP, *Betula nana* was an important taxon of the steppic tundra and tundra north of the Alps (e.g. Lotter, 1999; Rey et al., 2017, 2020). At the latest during the Early Holocene, *B. nana* was confined to its modern main range in northern Europe. For the interpretation of the *Betula* isochrone map (Fig. 4.4.3-3) we therefore assume that after c. 14 500 cal yr BP south of c. 55°N the pollen derives mostly from tree *Betula*. By 14 000 cal yr BP, *Betula* had spread over large areas of southern, western, central and eastern Europe, likely reflecting Late-glacial afforestation dynamics in response to the Bølling-Allerød warming. The today's range of *Betula* was attained during the Early Holocene, when northernmost positions were reached in Fennoscandia, northern Russia and Iceland (Fig. 4.4.3-3). Spreading rates of *Betula* were 230–990 m yr^{-1} and thus faster than those of other trees with comparable wind-dispersed diaspores (Table 4.11-1). Due to the young age of first flowering, the resulting postglacial dispersal distance was 2–30 km, which is comparable with that of other trees (Table 4.11-1). During the Mid and Late Holocene, the abundance of *Betula* markedly increased due to land-use practices including burning and other disturbance (see chapter 4.10). The flexible yet strong bark of tree birches was used in prehistorical and traditional cultures, from baskets to canoes.

Carpinus betulus

The genus *Carpinus* in the family Betulaceae (or also understood as a more narrow family: Corylaceae) consists of 25–30 species distributed in the Holarctic. In Europe, two species occur. The first, *Carpinus betulus* (European or common hornbeam), has a

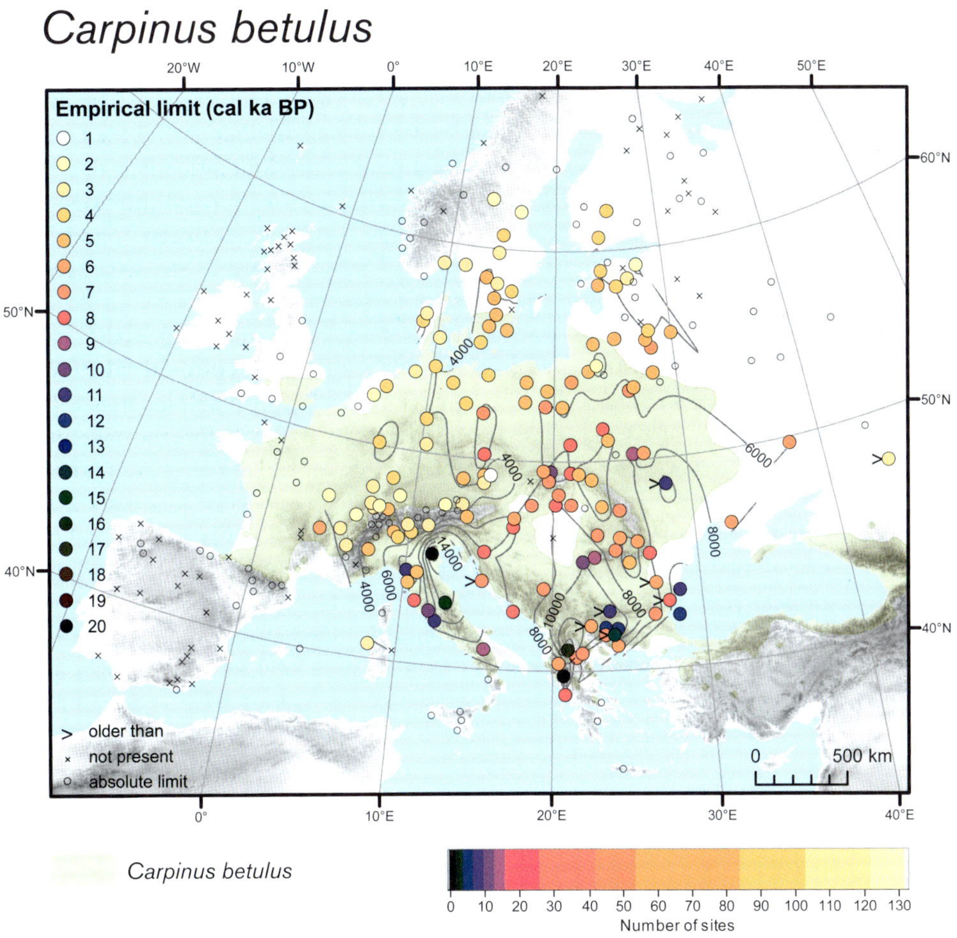

Fig. 4.4.3-4. Map showing the empirical limits of ***Carpinus betulus*** and the current distribution in Europe. The coloured dots indicate the empirical limit in cal ka BP at selected sites, crosses refer to sites where the taxon was absent and small circles where it reached the absolute limit. Grey lines represent isochrones of empirical limit in 2000-year steps, based on nearest neighbour interpolation. Bar diagram shows stacked number of sites where the empirical limit was reached during the same 1000-year time interval, with colours according to the legend in the figure.

wide range especially in central and eastern Europe, where the species reaches several hundred kilometres further east (Fig. 4.4.3-4) than *Fagus sylvatica* (Fig. 4.4.3-8). The second European species is *C. orientalis* (eastern or oriental hornbeam) which, in Europe, is restricted to the Apennine and Balkan peninsulas but also occurs in Asia Minor and the Caucasus. Morphologically the pollen and the macrofossils of the two species can be distinguished – the pollen of *C. orientalis* belongs to the *Ostrya* type.

In Europe today, *C. betulus* usually occurs with *Quercus robur* in oak-hornbeam forests summarized in plant sociology as the alliance Carpinion betuli. Because of the competition with beech forests of the alliance Fagion, the hornbeam forests are restricted in western and central Europe to areas with summer-warm climates: *C. betulus* can tolerate wet soils and can therefore occur in riparian forests (in contrast to beech). East of the range limit of beech, *C. betulus* forms mixed forests with *Tilia cordata* and *Acer platanoides* that include herb taxa similar to those of the *Fagus* forests in the west ("beech forests without beech": Ellenberg, 2009). In southern and south-eastern Europe, *C. betulus* can be found at low elevations in thermophilous oak-forests of the Quercion pubescenti-petraeae alliance and at higher elevations, up to about 1300 m asl, with *Fagus*. Undoubtedly hornbeam was fostered by humans at many places because it recovers strongly when coppiced (Rackham, 1980). *Carpinus* is anemophilous and anemochorous.

Macroremains such as wood, leaves, bud scales and fruits allow the identification of the genus, sometimes also of the species. Also the pollen grains with a diameter >40µm and 4–5 pores (stephanoporate) can, in Europe, usually be identified as *C. betulus*. Pollen productivity of *Carpinus* is higher than that of *Fagus*, but lower than *Pinus, Betula, Corylus* and *Quercus*. Usually the threshold value of 1% of total pollen is taken as an indication of local presence and values of ≥10% for a major element of a community (Lisitsyna et al., 2011). However, for our map (Fig. 4.4.3-4) we prefer the empirical limit as indicator for the establishment of first local populations because it reduces the probability of false absence. In peat, the pollen can be corroded, but it is usually well preserved in lake sediments.

For Lower, Middle and Upper Pleistocene records in Iberia, 18 sites (partly marine) are mapped by Postigo Mijarra et al. (2008) who summarize: "There is abundant fossil evidence of *Carpinus*, a taxon that was widely distributed throughout the Peninsula during the Pleistocene, and which suffered a drastic population decrease in the Holocene, possibly related to competition with other species reinforced by climate." Today the genus is nearly absent from Iberia.

On the Italian Peninsula in the south, at Lago Grande di Monticchio, *Carpinus* pollen was found at quite high values in the Early and Middle Weichselian deposits, with lower and more sporadic records during colder and drier parts of the last glacial, making **refugia** nearby very probable. Also at the central Italian sites of Valle di Castglione, Lagaccione and Lago di Vico, the early Weichselian interstadials of St. Germain I and St. Germain II are rich in *Carpinus*. The tree was then absent in the early pleniglacial and in the late pleniglacial (EPG and LPG sensu Tzedakis et al., 2013), but sporadically present in between in the middle pleniglacial at the Lazio Complex (Follieri et al., 1988, 1998 with a synthesis in Fig. 11; Magri, 1999; Magri and Sadori, 1999). In northern Italy at Lago di Fimon, *Carpinus betulus* is frequent not only during the Eemian but also in an early Würm interstadial that probably correlates to St. Germain I /Brørup (Pini et al., 2010). In the nearby Euganean Hills, substantial partly continuous records of *C. betulus* are found for the LGM and the Late-glacial (Kaltenrieder et al., 2009), suggesting that the site acted as a LGM refugium (Fig. 4.4.3-4). In France, substantial records of hornbeam were found in many long records. In the Vosges at La Grande Pile *C. betulus* shows high values during the Eemian, smaller but consistent values during St. Germain I and II and only single grains during the Pleniwürm (Woillard, 1978; de Beaulieu and Reille, 1992b). In the long records of Les Echets near Lyon and in the Massif Central at Lac de Bouchet and Praclaux, *C. betulus* was very abundant during the interglacials of Landos and Ribains (Eemian) but also quite abundant during several interstadials, especially during Bouchet 1 (de Beaulieu and Reille, 1984, Reille and de Beaulieu, 1995; Reille et al., 1998). In Hungary, a single record of charcoal of *Carpinus* sp. in a loess area was dated to 26 700 uncal yr BP (c. 31 000 cal yr BP), that is to the middle pleniglacial (Sümegi and Hertelendi, 1998; Willis et al., 2000; Willis and van Andel, 2004). Also in the longest Greek records of Ioannina, Kopais and Tenaghi Philippon, the early Weichselian interstadials show decreasing peaks of *Carpinus*, as visualized in figure 3 of Tzedakis et al. (2001), a figure that also includes Valle di Castiglione and Lac de Bouchet/Praclaux (Follieri et al., 1988; Reille and de Beaulieu, 1990). This figure 3 (going back in time to MIS 11c) and figure 7 in Follieri et al. (1998) show the occurrence of hornbeam during the early Weichselian interstadial as 'a fading echo from the Eemian'. During the early pleniglacial (MIS 4), and especially the late pleniglacial (MIS 3), the range of *Carpinus* must have been very small, consisting of isolated patches. However, our isochrone map suggests that during the LGM the tree survived on the Balkan and Italian peninsulas (black dots), from where it spread to other areas during the Late-glacial (Fig. 4.4.3-4).

Range expansions and contractions: Late-glacial records are rare and occurrences are restricted to the Italian and Balkan peninsulas (Fig. 4.4.3-4). At Italian sites, *C. betulus* is scanty or occurs at low frequencies (see Fig. 4.3-14). The same is true for the Carpathians and Balkans (Fig. 4.3-12 and 4.3-13), but at some sites Late-glacial occurrences can be nearly continuous and substantial, such as at Lake Gales and Lake Brazi in the Retezat

Mts, south-western Carpathians (Magyari et al., 2012b). Many of the findings in the Retezat Mountains for the Late-glacial and Early Holocene are identified as *C. orientalis*, just as the Late-glacial record at Bátorliget in northern Hungary (Willis et al., 1995). The regular early records of *C. betulus* in the south-western Carpathians indicate nearby refugia, a fact also confirmed by the **phylogeographic** findings of Grivet and Petit (2003) who show highest diversity of chloroplast haplotypes in the Carpathians. During the Early Holocene, north of c. 45°N *Carpinus* was confined to eastern Europe. Finally during the Mid and Late Holocene the tree reached its current range limits (Fig. 4.4.3-4).

For the Holocene an interesting discussion developed around the immigration and population expansion of *C. betulus* and its comparison to *Fagus sylvatica* and is covered in the entry for *Fagus* below.

A surprise was the detection of *Carpinus* (and *Quercus*) on Tenerife, Canary Islands, at La Laguna (560 m asl) with high pollen values between c. 4700 and 2200 cal yr BP, but then declining rapidly to extinction in the Late Holocene (de Nascimento et al., 2009). Our isochrone map suggests spreading rates of 220–310 m yr^{-1}, resulting in dispersal distances of 4–9 km per event (Table 4.11-1). Other isochrone maps for four abundance classes including the Late-glacial and the Holocene are provided by Giesecke and Brewer (2018).

In conclusion we see that in most parts of Europe *Carpinus betulus* was a late-comer in the secular succession of the Holocene, that is a mesocratic species, but in south-eastern Europe it was a protocratic species – as it was in several interglacials.

Castanea sativa

The genus *Castanea* consists of about 10–12 species in the northern hemisphere. *Castanea sativa* (sweet chestnut) is the only species native to Europe. It prefers soils developed on siliceous rocks and a climate with mild winters and warm, humid summers. It is thus considered as a warm-temperate or submediterranean species. In mountain areas, it grows up to c. 700 m asl, at favourable spots, even up to 1000 m asl. Chestnut is zoochorous and its flowering biology is rather special. Morphologically chestnut is monoecious but, because of self-sterility, it is functionally dioecious. Wild forms produce high amounts of pollen, whereas cultivars produce very little (Rudow and Conedera, 2001), which can even lead to seemingly paradoxical situations where less pollen means more cultivation. Pollination is both anemophilous and entomophilous (the latter becomes important under very humid conditions when pollen becomes sticky). The pollen morphology can be problematical because the thin-walled grains are easily corroded and separating them from those of *Lotus, Hypericum* and *Sedum* amongst others can be difficult (Sánchez-Goñi, 1993; van der Knaap and van Leeuwen, 1995; Bottema, 2000). In terms of wood anatomy the distinction of *Castanea* versus *Quercus* needs special care (Schoch et al, 2004).

The potential full-glacial **refugia** in Europe are presented by Krebs et al. (2004) who evaluate 1471 sites and develop an index for a refugium probability (IRP; see map on Fig. 4.4.1-3 and 4.4.1-4). High probabilities are found not only around the eastern and southern shores of the Black Sea but also in at least three regions of the Apennine Peninsula and in the Cantabrian mountains. For Iberia, 12 (and 2 marine) Pleistocene records were mapped by Postigo Mijarra et al. (2008), including the cave of Lezetxiki of Mousterian age (in the Spanish Basque country: Sánchez-Goñi, 1988). Central Italian, central Adriatic (Lowe et al., 1996) and north Italian cores also provided sporadic chestnut records of full glacial and Late-glacial age (Paganelli and Miola, 1991; Kaltenrieder et al., 2009). Niche modelling is largely consistent with these refugial regions (Roces-Díaz et al., 2018). The **phylogeography** of *Castanea,* based on chloroplast DNA of modern populations, does not strongly separate spatial groups, probably because of the strong human impact on the genus (Fineschi et al., 2000).

Range expansions and contractions (Fig. 4.4.3-5): Late-glacial and Early Holocene records are usually sporadic and only in rare cases continuous (e.g. Garda area: Paganelli and Miola, 1991). For the Early to Mid Holocene, the maps for 9 cal ka BP and 7.5 cal ka BP by Krebs et al. (2004) show more sites and higher values, but the percentage curves are still discontinuous (see also Bottema, 2000), that is, the empirical limit shown in Fig. 4.4.3-5 is not yet reached. In western Asia, for example, north-western Syria (Ghab Valley) the occurrence of *Castanea* pollen was also discontinuous (Yasuda et al., 2000). In northern Italy at Lago della Costa in the Colli Euganei, high and consistent pollen percentages of

Castanea sativa

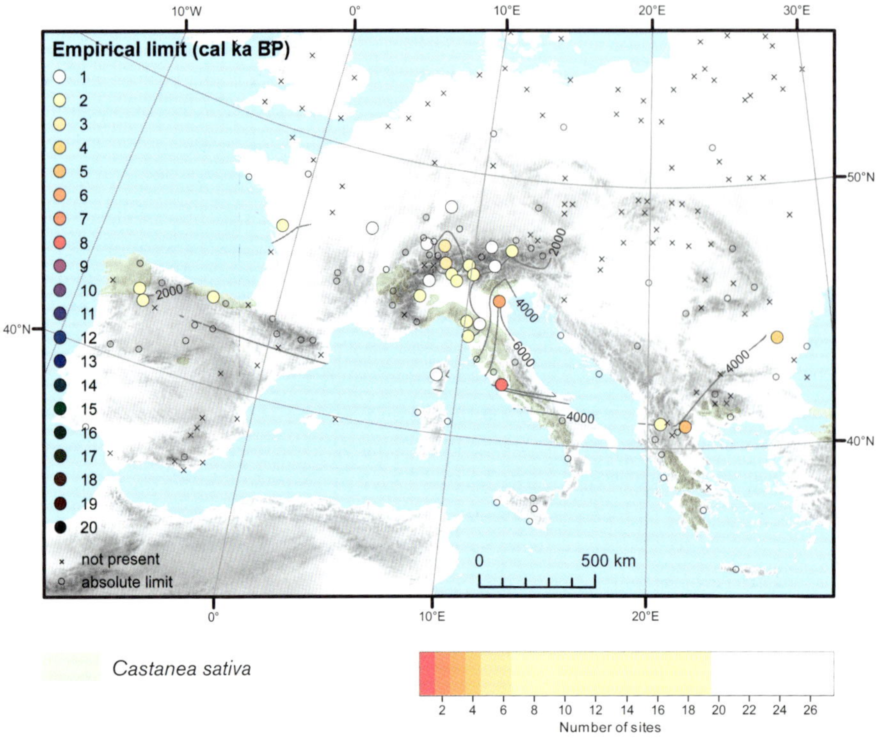

Fig. 4.4.3-5. Map showing the empirical limits of ***Castanea sativa*** and the current distribution in Europe. The coloured dots indicate the empirical limit in cal ka BP at selected sites, crosses refer to sites where the taxon was absent and small circles where it reached the absolute limit. Grey lines represent isochrones of empirical limit in 2000-year steps, based on nearest neighbour interpolation. Bar diagram shows stacked number of sites where the empirical limit was reached during the same 1000-year time interval, with colours according to the legend in the figure.

Juglans, Castanea and *Olea* together with other cultural indicators such as cereals and high charcoal values have been recorded from as early as the Neolithic (Kaltenrieder et al., 2010). However, the oldest empirical limit suggesting the establishment of local stands comes from central Italy and is dated around 8000 cal yr BP (Fig. 4.4.3-5). For the Late Holocene, Conedera et al. (2004) state: "The first unambiguous pollen data showing evident indications of the chestnut tree spreading due to human activity are found in several regions in the Anatolian Peninsula, north-eastern Greece and south-eastern Bulgaria dating back to around 3700 BP (2100–2010 BC, Van Zeist and Bottema, 1991)". Conedera et al. (2004), who also summarizes the classical Greek and Roman literature on agriculture and arboriculture, write "...that its use as food was not the primary driving force behind the introduction of chestnut into Europe by the Romans". But "Pliny ... deals above all with the agricultural aspects of chestnut coppicing, emphasizing the supremacy of the chestnut in the production of poles to support vines". And "An epigram of Martial in which chestnut is used as a symbol of inferiority in aristocratic Roman circles...". In contrast, walnut was considered 'luxury food' north of the Alps (Bakels and Jacomet, 2003). Conedera et al. (2004) summarize: "Chestnut cultivation for fruit production conquered the Roman world slowly. At the latest by the end of the pre-Christian era there were varieties of the chestnut of good quality". During the Middle Ages more varieties and qualities of chestnuts were developed that were spread by grafting. The use of the timber and the fruits coexisted, just as with *Juglans* and *Olea*. A symposium volume about ecology, cultivars and pathogens of chestnut was edited by Heiniger (2001).

Corylus

The genus *Corylus* belongs to the family Betulaceae. In the Holarctic it consists of about 15 species of which three occur in Europe (Fig. 4.4.3-6). *Corylus avellana* (common hazel) is widespread throughout nearly the whole of Europe except in the far north (today, see below); *C. maxima* is restricted to the north-eastern coast of the Adriatic Sea and northern Greece; and *C. colurna* has a wide range on the Balkan Penisula and in Anatolia. Erdogan and Mehlenbacher (2000) studied the genetics of the species in the northern hemisphere as well as cultivars.

In most parts of its range *C. avellana* is a shrub in the understorey, but in favourable habitats its stems can reach up to 6 m or more (e.g. in Sognefjord, Norway). In cultural landscapes it is frequent at the forest edge and in hedges. In coppiced forests it is also anthropogenically enhanced. In the far west of Europe, such as in Ireland, western Scotland (e.g. Isle of Skye: Birks, 1973b) and south-western Norway, on very steep slopes without grazing animals it can form shrub vegetation that looks natural. *Corylus maxima* is also a shrub and occurs in submediterranean deciduous forests. Today it is often cultivated for its large fruits. *Corylus colurna* is a tree in deciduous forests of south-eastern Europe, but is also cultivated, especially on the Turkish Black Sea coast, and now also planted in central European cities. *Corylus* is anemophilous and zoochorous.

Macroremains of *Corylus* such as wood, leaves or bud scales allow identification only to the genus, but the fruits allow distinction of the species.

The morphology of the triporate pollen grain can be difficult to separate from that of *Myrica gale* which has a very different autecology, but some good descriptions exist: Erdtman et al. (1963), Edwards (1981), Moore et al. (1991) and Blackmore et al. (2003). Because of this, in the potential distributional range of *Myrica gale* near the Atlantic coast, the *Corylus* pollen-type may include *Myrica gale* and is therefore often cautiously called 'Coryloid'. In large parts of the modern range of hazel there is a high probability that the pollen represents *Corylus avellana*. Only in south-eastern Europe can *C. colurna* and *C. maxima* be contributing to the pollen curve, because the pollen morphology does not allow their separation (Caramiello et al., 1989). Pollen productivity is high (Broström et al., 2008): a threshold value for local presence is, therefore, hard to fix. Huntley and Birks (1983) propose 2%, whereas Lisitsyna et al. (2011) propose 2–10%. In addition, these authors discuss that the thresholds were usually percentages of the total terrestrial pollen but in some studies the arboreal pollen sum (AP) was used instead. In our map we use the empirical limit to avoid inferring false absences of first stand establishments. The resulting spatial patterns attributes pre-Bølling occurences only to the three peninsulas of southern Europe, which is ecologically meaningful for a temperate species (Fig. 4.4.3-7). We need to be careful when using old literature because *Corylus* was often excluded from the pollen sum (which explains values >100%). This exclusion is not recommended because it already means an interpretation ('hazel only as understorey') and it prevents comparisons across Europe.

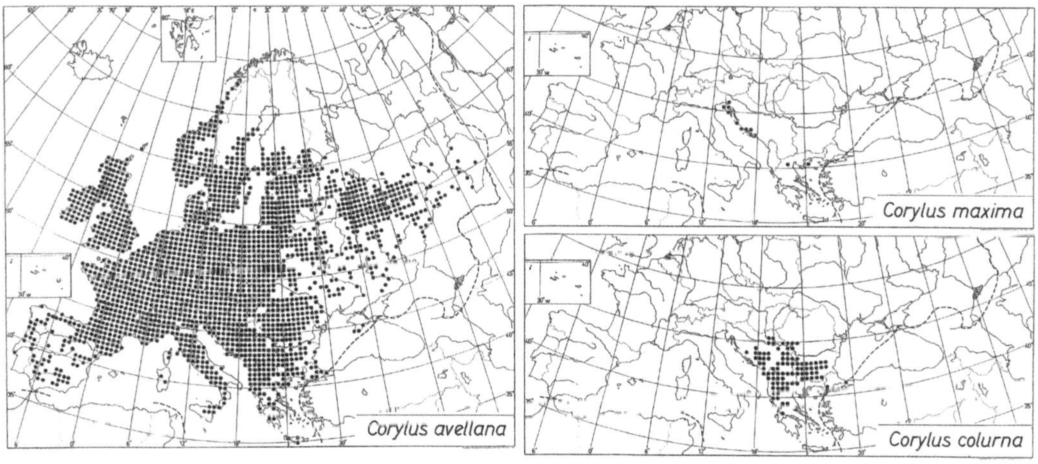

Fig. 4.4.3-6. Current distribution of the three species of *Corylus* in Europe. After Jalas and Suominen (1976).

Glacial **refugia** of *Corylus* on the Iberian Peninsula are recorded by pollen and charcoal both in the north and in the south. Northern examples are from the coast (Area Longa: Gómez-Orellana et al., 2007), from the Cantabrian Range (El Esquilleu: Uzquiano, 2008) and from Sierra de Neila (Quintanar de la Sierra: Peñalba et al., 1997; Ruiz Zapata et al., 2002). Southern examples are Siles (Carrión, 2002b) or sporadic findings in the cave Cueva Negra (from *Homo neanderthalensis* to the Holocene: Carrión et al., 2003b). González-Sampériz et al. (2010) provide an overview of more than 100 terrestrial sites (including caves) and more than 10 marine cores. In the Massif Central in France not only the early parts of several interglacials (Praclaux = Holstein; Landos; Ribains = Eemian) but also numerous interstadials are rich in *Corylus* (see the synthesis in Fig. 7 in Reille et al., 2000). Further north at Les Echets and La Grande Pile, *Corylus* was important during the Eemian and, to a lesser degree, also dur-

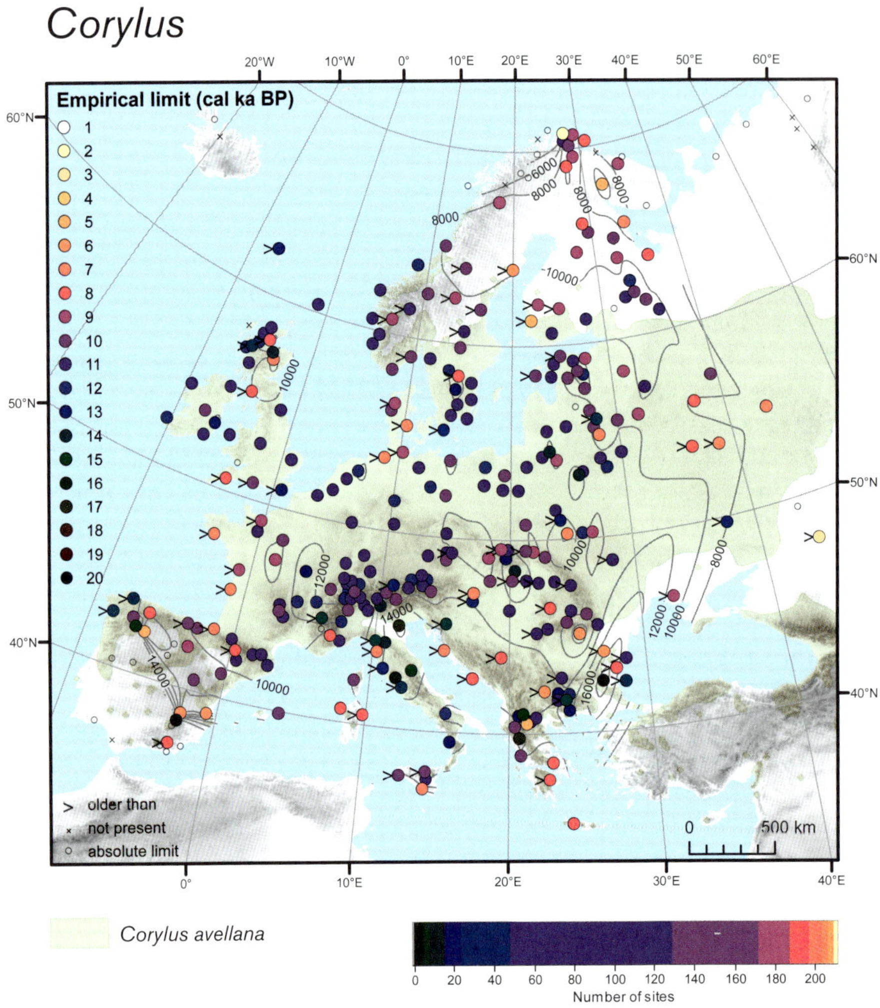

Fig. 4.4.3-7. Map showing the empirical limits of ***Corylus*** and the current distribution of *Corylus avellana* in Europe. The coloured dots indicate the empirical limit in cal ka BP at selected sites, crosses refer to sites where the taxon was absent and small circles where it reached the absolute limit. Grey lines represent isochrones of empirical limit in 2000-year steps, based on nearest neighbour interpolation. Bar diagram shows stacked number of sites where the empirical limit was reached during the same 1000-year time interval, with colours according to the legend in the figure.

ing the interstadials St. Germain I (Brørup) and St. Germain II (Odderade: Woillard, 1978; de Beaulieu and Reille, 1984, 1992). For the long records from central Italy, Follieri et al. (1998) provide an overview in their figure 11. Also in northern Italy (Lago di Fimon: Pini et al., 2010), *Corylus* was very prominent during the last interglacial, fading away during the early to middle Würm interstadials. In the nearby Euganean hills at Lago della Costa it was sporadically found almost throughout the whole period, from c. 32 to 18 cal ka BP (Kaltenrieder et al., 2009). In southern Italy, *Corylus* is recorded from the Saalian glaciation through the Eemian and sporadically during the Würmian (at Monticchio: Allen and Huntley, 2000, 2009). In the long Greek record of Ioannina, *Corylus* is also recorded more or less throughout the last 130 000 years (Tzedakis et al., 2002). Our isochrone map suggests LGM refugia on the Iberian and Balkan peninsulas (black dots), likely also on the Italian peninsula (at Lago della Costa, 17 000 cal yr BP).

The **phylogeography** of *C. avellana* shows the highest genetic diversity in the Balkans and in Italy but hardly any haplotypes in common with central Europe. This may exclude the two southern peninsulas as sources for the spread to central Europe (Palmé and Vendramin, 2002). On a north–south transect from central Sweden to central Europe, Persson et al. (2004) subdivide the allozyme diversity into five groups that probably reflect both bottlenecks and genetic drift.

Range expansions and contractions: Late-glacial pollen records are known from most of the long cores mentioned above. In addition charcoal was found at sites in the Pyrenees (e.g. at El Portalet: González-Sampériz et al., 2006). In the north-eastern part of the Iberian System, the Sierra de Neila at Quintanar de la Sierra, *Corylus* pollen was also sporadically found at low values throughout the Late-glacial (Peñalba et al., 1997; Ruiz Zapata et al., 2002). At many Iberian and French sites *Corylus* is absent in the Late-glacial sections. For Magdalenian sites in south-western France, see Vernet (1980).

For Italy and the Alps, the map by Finsinger et al. (2006) visualizes the timing during the Late-glacial versus Holocene and the relationship between *Corylus* and *Quercus* south and north of the Alps (Fig. 4.4.3-15). During the Late-glacial, *Corylus* was present at some sites on the Iberian, Italian and Balkan peninsulas (Fig. 4.4.3-7). Other isochrone maps for four abundance classes including the Late-glacial and the Holocene are provided by Giesecke and Brewer (2018).

Holocene records of *Corylus* in Mediterranean areas and the Balkan Peninsula are usually with low values, with somewhat higher values during the Early to Mid Holocene at some sites. In contrast, in the Carpathians and parts of Pannonia the Early to Mid Holocene records are substantial (see histograms in chapter 4.3). For Italy, Magri et al. (2015) present maps for the Holocene. At the continental scale, *Corylus* spread mainly during the Early Holocene, when about 80% of the European sites were populated. Although local establishment may have occasionally occurred already at 14 000–12 000 cal yr BP north of the refugial southern European peninsulas and in some instances during the Mid or Late Holocene (Fig. 4.4.3-7).

For the Holocene history of *Corylus* in central and north-western Europe at least three phenomena are remarkable. **(1)** The spatial expansion through most of western, central, eastern and partly northern Europe was extremely rapid (see Giesecke and Brewer, 2018). **(2)** A 'tail' before the population expansion is nearly absent, which means that establishment and population expansion at the sites studied are scarcely separated in time (or by less than c. 500 years). **(3)** During the Early to Mid Holocene (c. 9–7 cal ka BP; Boreal and first part of the Atlantic), *C. avellana* reached further north than today as documented by macrofossils in Sweden. This early evidence of a Holocene Thermal Maximum (HTM: Andersson, 1902), has now been refined in space and time with a data-model comparison by Seppä et al. (2015). The decline of *Corylus* in northernmost Europe after the HTM is also illustrated by the boreo-fennoscandian histograms of figure 4.3-4b (somewhat less distinct than in the records of *Ulmus*). The aspects (1) and (2), that is, the spatial expansion and population expansion, are to a certain degree connected by processes involved in population dynamics. The zoochorous spreading rates were 450–900 m yr^{-1} and thus among the highest inferred from our isochrone maps (Table 4.11-1). The resulting seed dispersal may involve relatively small 'jumps' (5–9 km), if compared to for example *Fagus* (5–34 km) or *Quercus* (3–32 km). Thus, the quick spread across the continent mainly depended on the early age of first flowering and fruiting (Table 4.11-1). In addition, transport by central European rivers flowing northwards and by Mesolithic people were already proposed as spreading agencies by Firbas (1949) and later by Huntley (1993) and Tallantire (2002). The very fast population expansion across central, north-western and northern Europe is discussed in detail in Giesecke et al. (2011). The amazing synchroneity around 10.55 cal ka BP across large parts of Europe may be best explained by a climate shift towards warmer and dry summer conditions at the end of the Preboreal oscillation (Tinner and Lotter, 2001). Tallantire (2002) adds the interesting hypothesis that *Corylus* was present during the earliest Holocene over large regions of Europe as a non-flowering shrub that then responded to a climate shift with high pollen production. Non-climatic causes for the rapid population expansion (e.g. soil development, fire frequency, human assistance) have been discussed and evaluated as less important but possibly facilitating the rapid spread (Firbas, 1949; Smith, 1970; Huntley, 1993; Tallantire, 2002; Finsinger et al., 2006; Giesecke et al., 2011). Human impact included at least two aspects, namely the nuts as a staple food and coppicing for poles and firewood.

Fagus

Today the genus is represented in Europe by only two species – *Fagus sylvatica* (present over large areas of Europe) and *Fagus orientalis* (restricted to a small area in the far south-east). The ranges of the two species overlap on the Balkan Peninsula: the taxa *F. moesiaca* and *F. taurica* are usually considered as intermediate forms, whereby *F. moesiaca* is closer to *F. sylvatica* and *F. taurica* closer to *F. orientalis* (Meusel et al., 1965; Jalas and Suominen, 1976). Denk et al. (2002) come to the conclusion that *F. moesiaca* and *F. orientalis* should be treated as synonymous to *F. sylvatica*. For the numerous types of European beech forests, the maps 10, 11 and 12 in Bohn et al. (2003) provide an overview.

Pollen morphology does not allow the separation of the two European species or subspecies, let alone the intermediate forms (Gortemaker, 1986). SEM pictures show some minor differences between the surfaces of the exines of *F. sylvatica* and *F. orientalis*, but they probably do not permit a safe identification (Fig. 2.3-4). Palynological as well as wood anatomical records, therefore, only refer to the genus. Considering the modern ranges of the two species or subspecies we can – except for the extreme south-east – exclude *F. orientalis*, at least for the Late-glacial and the Holocene. *Fagus* is anemophilous and zoochorous.

In contrast to the pollen, the morphology of the macroremains offers the possibility to separate the species or subspecies, by the leaves (especially with the analysis of the cuticles), bud scales, cupulae and fruits (Czeczott, 1932; Wagenitz, 1981). As the critical limit for an apparent local presence we may use the onset of the closed pollen curve (=empirical limit; Fig. 4.4.3-8). Thresholds of 1 or 2% *Fagus* pollen were sometimes used to distinguish absence from presence. These are rather conservative limits (in regard to false presence), because a number of sites are known where wood or other macroremains were found and dated at levels thousands of years before the pollen curve became continous or the beech values even reached 2% (e.g. as charcoal in caves in Luxembourg; Pernaud, 2001). Thus using the empirical limit seems more indicative, as it reduces the probability of inferring false absences.

Today *Fagus orientalis* (oriental beech) has a distributional range that reaches beyond Europe including the north of Turkey and the Caucasus and, within Europe, northern Greece and eastern Bulgaria – and if we include *F. moesiaca* and *F. taurica* also Albania, southern Serbia, the Balkan mountain ranges and the Crimean Peninsula (see map in Horvat et al., 1974; Fig. 4.4.3-8). *Fagus orientalis* and its two closely related taxa prefer – at least in the western part of their range – lower, warmer and drier habitats than *F. sylvatica*: they can form distinct beech communities, namely Fagion moesiacum. Their history is hardly known. At the Bulgarian Black Sea coast at Durankulak a nearly continuous pollen curve of *Fagus* is recorded from the onset of the Holocene which could possibly originate from populations of *F. orientalis* (Bozilova and Tonkov, 1998).

Fagus sylvatica (European beech) has a wide distributional range. It is often considered a characteristic member of the central European 'Geoelement' (a group of taxa with similar distributions), but its range goes far beyond central Europe. In southern Europe it does not occur in the lowlands but is found in the mountains of northern Spain,

Corsica, Italy and the Balkan Peninsula. The north-eastern range limit runs from the western Black Sea, across western Ukraine and Poland to the southern shores of the Baltic Sea (the isolated populations on the Crimean Peninsula are considered by Meusel et al. (1965) to belong to *F. orientalis*). In the north-east, the range limit has many gaps. In Scandinavia, *F. sylvatica* settled in southern Sweden and southern Norway (Fægri, 1960; Bjune et al., 2013). In the west, the species occurs in the south of England and Wales, but today it may occasionally grow also throughout nearly the whole of the British Isles (Perring and Walters, 1976). Inside the range a number of rather continental areas exist where beech does not occur, such as the Po Plain, the Pannonian basin or the Romanian lowlands near the Danube river (Wallachia). In the continental climate of the central Alps, beech is rare or absent.

The **ecological amplitude** of *F. sylvatica* is broad. Climatically it prefers humid summers and mild winters: it is very sensitive to late frosts. Edaphically, it is a strong competitor on intermediate sites of both calcareous and siliceous soils. Constantly wet, groundwater-controlled and oxygen-poor soils as well as extremely dry, shallow soils are avoided. In southern and south-eastern

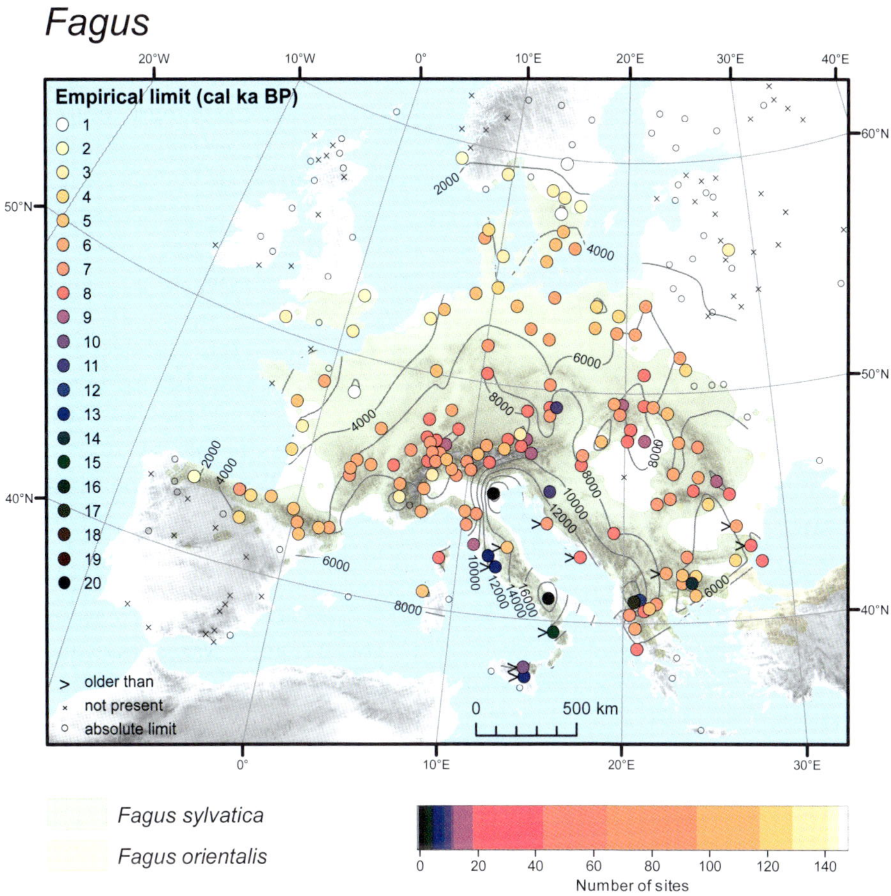

Fig. 4.4.3-8. Map showing the empirical limits of ***Fagus*** and the current distribution in Europe. The coloured dots indicate the empirical limit in cal ka BP at selected sites, crosses refer to sites where the taxon was absent and small circles where it reached the absolute limit. Grey lines represent isochrones of empirical limit in 2000-year steps, based on nearest neighbour interpolation. Bar diagram shows stacked number of sites where the empirical limit was reached during the same 1000-year time interval, with colours according to the legend in the figure.

Europe it is a mountain tree, forming a montane to subalpine belt (often with *Abies alba*) above the summergreen oak forests of the lowlands. In southern central Europe, it also prefers submontane to montane elevations, again often with *Abies alba*. The lower, warmer and drier regions are left to *Quercus* and *Carpinus*. In oceanic and suboceanic parts of its range, beech also grows at low elevations, either as the dominant species or together with *Quercus* and *Pinus*. In plant sociology beech forests are usually grouped into five major groups: the four species-rich alliances of the south, namely (1) Scillo-Fagion on the Iberian Peninsula, (2) the Fagion austro-italicum on the Apennine Peninsula, (3) the Fagion carpathicum in the Carpathians, (4) the Fagion illyricum on the Balkan Peninsula and as (5) the species-poor beech forests of the north (Fagion medio-europaeum). Beech also often plays quite an important role outside the communities dominated by *Fagus sylvatica* (Fagetalia), for example, in oak-birch forests (Quercetalia robori-petraeae). Overviews can be found in Dierschke (1989), Géhu and Julve (1989), Matuszkiewicz (1989), Müller (1989) and Pignatti (1989).

For the interglacial and glacial history of beech see chapter 3.9.1 and figure 2 in Magri et al. (2006) or figure 4 in Tzedakis et al. (2001).

Glacial **refugia** of *Fagus sylvatica* were multiple, as indicated by both palaeobotanical and genetic evidence (see Magri et al., 2006 and chapter 3.9.1).

(1) For Iberia, Postigo Mijarra et al. (2008) provide a map of Pleistocene *Fagus*-records. In the southern Pyrenees and the Cantabrian range three different, partly unique types occurred according to isozyme studies map in Magri et al. (2006).

(2) A genetically different small population was found in the lowlands of southern France. In addition, dated charcoal in south-western France (Landes de

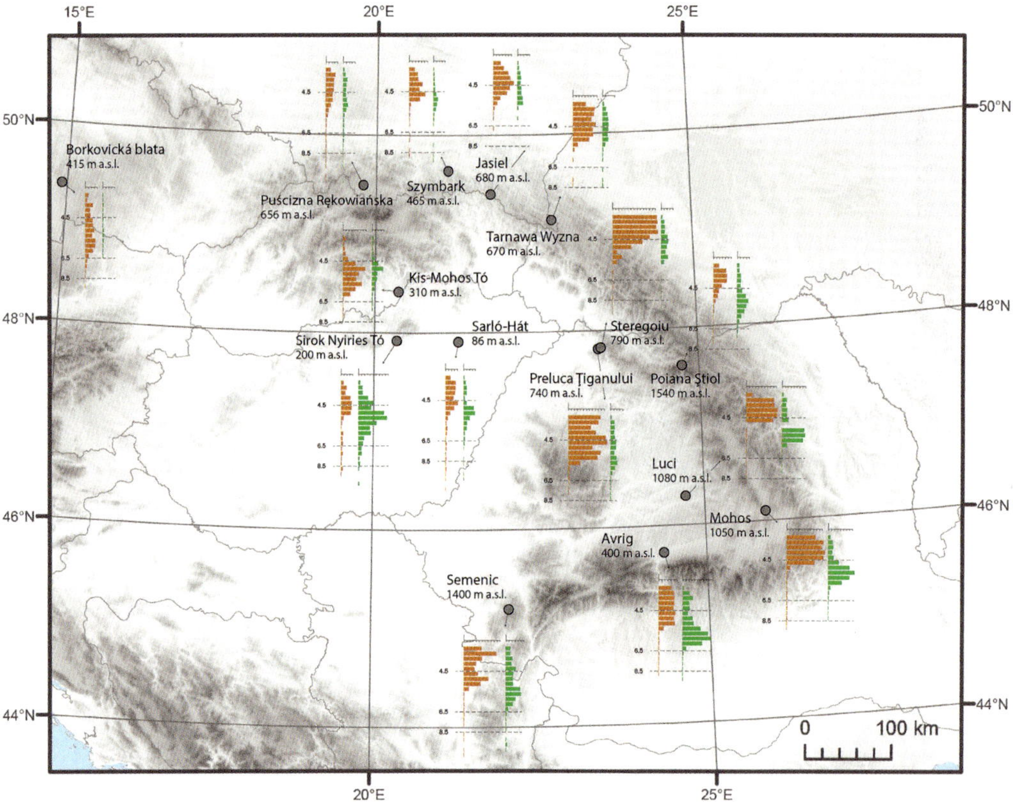

Fig. 4.4.3-9. The Holocene frequencies of *Fagus* (orange) and *Carpinus* (green) in the Carpathian mountains and some adjacent regions as percentages in 500-yr-histograms. Some of these sites can be found in Magyari et al. (2010). See Fig. 4.4.3-4 for the empirical limits of *Carpinus* and Fig. 4.4.3-9 for *Fagus*.

Gascogne) indicates survival of beech even during Heinrich-Event 1, one of the coldest and driest periods of the Oldest Dryas (de Lafontaine et al., 2014).

(3) In the western pre-Alps in France.

(4) In the south-eastern pre-Alps and hills of the Po Plain in Italy, Slovenia and possibly southern Moravia and southern Bohemia.

(5) In Istria/Croatia, Dinaric Alps, and the Pannonian mountains (Hazler et al., 1997).

(6) In the southern Balkans, at least three genetically distinct populations near palynological sites that go back only to the Mid Holocene.

(7) In southern and central Italy; for central Italy figure 4 in Follieri et al. (1998) provides an overview.

(8) In southernmost Italy with a genetically unique type.

Again, as is the case with other tree taxa, mid-elevations of mountain ranges may have provided suitable sites – not too dry (as below) and not too cold (as above) as suggested by Beug (1975a; see Fig. 4.11-3). Near the Cantabrian Range, at some sites in the Sierra de Neila in the north-western Iberian Range, beech pollen was recorded for the late Pleistocene, but that occurrence was followed by millennia without records, then sporadic records during the Late-glacial and continuous records during the Holocene (Ruiz Zapata et al., 2002; López-Merino et al., 2008). However, our isochrone map suggests that *Fagus sylvatica* survived the LGM on the Italian Peninsula (black dots around 20 000 cal yr BP) and probably also on the Balkan Peninsula (oldest empirical limit around 17 000 cal yr BP; Fig. 4.4.3-8).

Phylogeography: the allozyme variation was studied by Gömöry et al. (2007) and the phylogeny of the entire genus by Renner et al. (2016). Further studies indicate a high genetic diversity in south-eastern Europe, especially in the Dinaric Alps, the Pannonian mountains (Hazler et al., 1997) and in Greece (Hatziskakis et al., 2009; Papageorgiou et

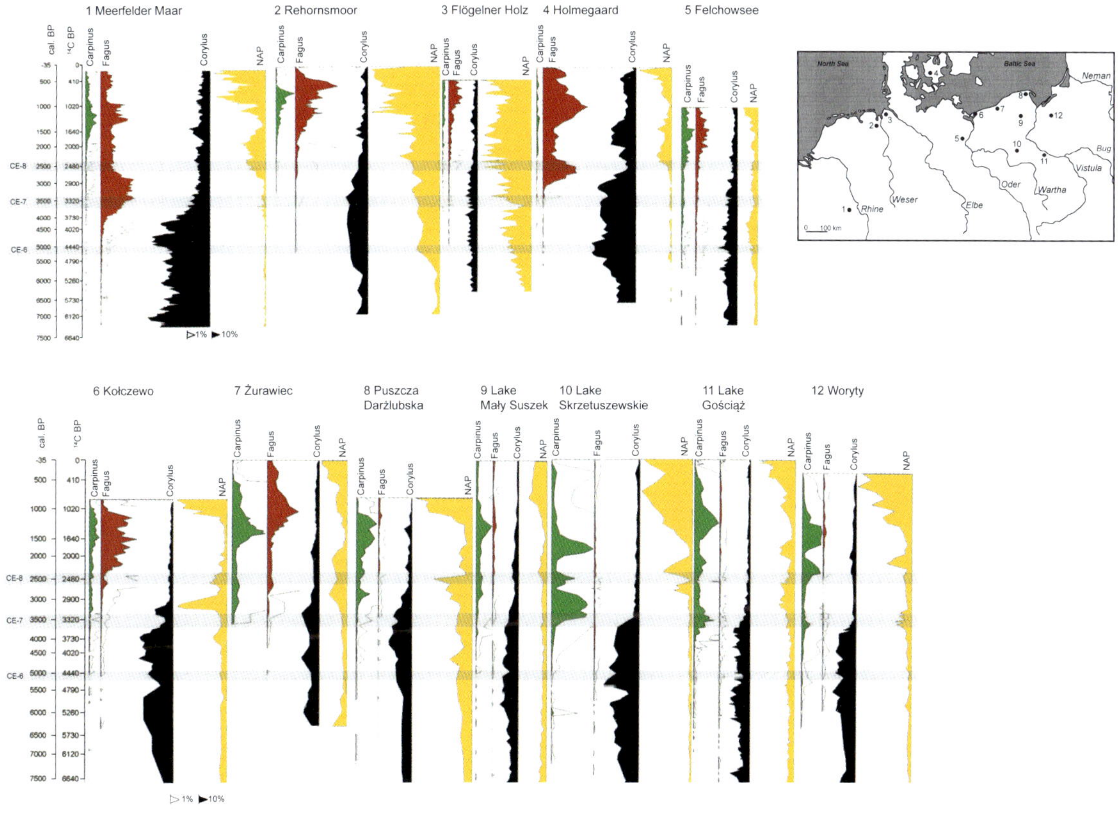

Fig. 4.4.3-10. Transect of central European pollen diagrams, sorted from west to east, demonstrating the decreasing importance of *Fagus sylvatica* and the increasing importance of *Carpinus betulus*. Modified after Ralska-Jasiewiczowa et al. (2003).

al., 2014), but genetic diversity today is also high on the Crimean Peninsula, in southern Italy and the Pyrenees (Demesure et al., 1996; Comps et al., 2001; Magri et al., 2006). For the maps of Magri et al. (2006) an innovative approach was applied that connects time, space and population genetics. Onto the maps of the palaeorecords, the genetic groups of the modern phylogeography of the beech populations were added in colours and symbols according to the seven most dominant chloroplast haplotypes and according to the isozyme groups. Whether this reasonable approach of overlay of palynological and (modern) genetic maps is correct should be tested when results of ancient DNA become available.

Range expansions and contractions: During the Late-glacial and the Early Holocene spatial expansions were rather slow and population expansions were also slow and/or late (Fig. 4.4.3-8). For instance, Ruiz-Alonso et al. (2019) present a pollen and charcoal record at Gesaleta (Western Pyrenees) since the beginning of the Holocene.

The spatial expansion of *Fagus sylvatica* reached maximum values during the Mid Holocene, which started at 8200 cal yr BP (Fig 4.4.3-8), when about two-thirds of the European range were occupied. Only a minor part of the range was filled during the Late Holocene. According to Magri et al. (2006) and Magri (2008) the populations of the numerous refugia in north-western Iberia, southern and central Italy (map in Magri et al., 2015) and the Balkans hardly migrated northwards. In contrast, the populations from the south-eastern Alpine foothills, Slovenia, the north-western Dinaric Alps and possibly southern Slovenia and southern Moravia seem to be the source of this Mid Holocene spread to central and northern Europe. The highest genetic diversity is located in the Balkans – but apparently these were not the populations that invaded the Carpathians (Gömöry et al., 1999, 2007, 2010; Brus, 2010).

An interesting discussion developed around the immigration and population expansion of *Fagus sylvatica* and its comparison to *Carpinus betulus*. Today, both species have rather similar ranges in Europe, except for the eastern-most range limit of *Carpinus*. Their Ellenberg-values for humidity and pH of soils are identical and their values for temperature and light are similar.

At numerous sites, *Fagus* pollen shows a long 'tail' of low percentages after the empirical limit, the mass expansion occurs much later. As an example, at the northern German site of Carlshof, about 3300 years were estimated between minor local records and the mass expansion. The contrast between the 'long tail' of *Fagus* and the 'short tails' of *Corylus* and *Quercus* is quite strong. Seed dispersal cannot be the reason for the difference because fruits of all three genera are heavy, zoochorous and distributed by the same animal species (Tinner and Lotter, 2006). One set of arguments concerns climate change: for central Europe, the summers of the Early and Mid Holocene were relatively dry (high insolation), so not very suitable for the humidity-loving beech (Tinner and Lotter, 2001, 2006). Giesecke et al. (2007) used a bioclimatic model for the whole of Europe (STASH) driven by three different climate models (atmospheric general circulation models) for 6000 cal yr BP and found a clear mismatch with the pollen-based distribution of *Fagus* in Europe. No single force could explain the distribution at that time, but drought during the growing season was probably a major determinant. As an indicator of intrinsic biological factors, the doubling times in pollen accumulation rates may be useful. For 10 sites south and north of the Alps and further areas in central Europe, Tinner and Lotter (2006) compared the empirical (continuous) limit with the mass expansion as indicators of local presence and population increase, respectively. Both pollen patterns were concentrated during cool and humid events of the type of the 8.2 event (CE-1 to CE-8 according to Haas et al., 1998), thus confirming a climatic control for southern and central Europe.

The other set of arguments to explain late population increases concerns **human impact**. For northern Italy, Valsecchi et al. (2008) found increases in *Fagus* pollen synchronous with increases in pollen of anthropogenic indicators and charcoal, but in addition, favourable climatic conditions (cool and humid) were a determining factor. Valsecchi et al. (2008) come to the conclusion "...that the expansion of *F. sylvatica* populations has been triggered by more than one single factor alone". Also, for north-western Iberia both climatic and anthropogenic causes of beech increases are discussed (e.g. López-Merino et al., 2008; Muñoz Sobrino et al., 2009; Pèlachs et al., 2009). For northern central Europe a west-east transect (Fig. 4.4.3-10) demonstrates (1) long, thin 'tails' of beech pollen, (2) mass expansions for several sites during the cool phases

CE-7 and CE-8, (3) increases of *Fagus* contemporaneous with increases in anthropogenic indicators and (4) in the west-east direction the decreasing importance of *Fagus* and the increasing importance of *Carpinus*. At Carlshof, population expansion was obviously fostered by anthropogenic disturbance – by forest opening and fires (Bradley et al., 2013). In southern Scandinavia, at the northern edge of the range, records from 18 small forest hollows indicate the importance of human impact (pollen indicators, charcoal) on the spread and the population increase of beech (Bradshaw and Lindbladh, 2005). For Draved Forest in Denmark a comparison of the *Fagus* and *Tilia* pollen records of the last 1500 years with climate-driven forest simulation models indicates a strong mismatch and hence the importance of human impact, which can be upscaled to north-western Europe (Bradshaw, 2008). In contrast to these early histories of establishment and expansion, are the numerous records of later reductions of beech: increased human impact (often climate-induced) led to phases of major population decreases. On the pre-Alpine foreland, the five pre-Roman Fagus peaks and declines at Lobsigensee between 6900 and 2570 cal yr BP (Ammann, 1988, 1989), can easily be compared with Thayingen-Egelsee (Troels-Smith, 1981) and Schleinsee (Müller, 1962). Between 6500 and 4600 cal yr BP (including the Neolithic to Early Bronze Age), four declines accompanied by increased fire and agricultural impacts at Burgäschisee and Moossee (Swiss Plateau) could be precisely dated thanks to varved sediments (Rey et al., 2019a).

A new form of human impact is anthropogenic climate change. Based on the results of Magri et al. (2006), Hu et al. (2009) combine the phylogeography of *Fagus* with a climate scenario under double CO_2 and show that mostly only one of the genotypes may move northwards, whereas several southern types may become extinct.

In south-eastern Europe (Fig. 4.4.3-9), *Carpinus* developed after a long Early Holocene 'tail' of low values into a population increase as early as 6500 cal yr BP and distinctly before the population expansion of *Fagus*, as seen at Sirok in Hungary (Gardner, 2002) or examples from the southern Carpathians at Semenik (Rösch and Fischer, 2000), Avrig (Tanţău et al., 2006), Mohos (Tanţău et al., 2003b), Luci (Tanţău et al., 2014) and Lake Saint Ana (Magyari et al., 2009b). In the central and northern Carpathians, the maxima of *Carpinus* and *Fagus* were then more or less synchronous (see Figs. 4.4.3-8 and 4.4.3-9). In central Europe, the initial spread of *Carpinus* is recorded at many sites in the Mid Holocene, but population expansion was delayed and very often triggered by human impact (Küster, 1997). The anthropogenic openings in the forests made it possible for hornbeam to build up larger populations. Additionally, the taxon was favoured by humans as firewood (also for smelting ores) and for coppicing (although coppicing reduces pollen production; see Gardner, 2002). At many, but not all, of these central European sites, the population expansion of *Carpinus* was later than that of *Fagus*. One of several reasons for similarities of the Holocene distribution histories of *Carpinus* and *Fagus* may be the shared glacial refugia in southern Europe, for example, around Ioannina or the Colli Euganei (Tzedakis et al., 2002; Kaltenrieder et al., 2009; Brewer et al., 2017). The high genetic diversity recorded in chloroplast haplotypes and PCR-RFLP for *Fagus* (Magri et al., 2006) and in cpDNA for *Carpinus* (Grivet and Petit, 2003) may reflect the south-eastern refugia. However, factors controlling migration rates such as maximum jumps of seeds are rather dissimilar, with *Fagus* reaching much higher maximum dispersal estimates (34 km) than *Carpinus* (9 km; see Table 4.11-1). The difference is compensated by earlier ages of first flowering, for *Carpinus* it is about 20 years, whereas for *Fagus* it is about 40–50 years, resulting in similar spreading rates for *Carpinus* (220–310 m yr^{-1}) and *Fagus* (135–425 m yr^{-1}; Table 4.11-1). Dissimilarities in the Holocene spread of the two trees may have derived from the different (climatic) niches and from the different seed dispersal mechanisms involved (i.e. animals vs wind, Table 4.11-1).

Regional syntheses: a taxon as important in temperate deciduous forests as beech is naturally treated not only by many site-specific publications but also by regional syntheses, such as Fægri (1960) for Norway; Zoller and Kleiber (1967), Wegmüller (1977), Kral (1979), and Burga and Perret (1998) for the Alps; Birks (1989) for the British Isles; Ralska-Jasiewiczowa (1983) for Poland; de Beaulieu and Reille (1984) for southern France; Rybníčková and Rybníček (1988) for the Czech Republic and Slovakia; and Pott (2000) for central Europe – to name just a few. Early overviews are found in Firbas (1949) or Godwin (1975). For the whole of Europe, see Huntley and Birks (1983), Huntley (1988), Pott (1997, 2000) and especially Magri et al. (2006) and Magri (2008).

Fraxinus excelsior and *F. angustifolia*

The genus *Fraxinus* comprises c. 65 species, which, with the exception of southernmost populations on Java, grow in the northern hemisphere (Schütt et al., 2002). *Fraxinus* species are mostly deciduous broadleaved trees or shrubs, only a few subtropical species are evergreen. Three deciduous *Fraxinus* species grow in Europe: *Fraxinus excelsior, F. angustifolia* and *F. ornus*. The latter species occurs in eastern Iberia, Italy and the Balkans (Caudullo and De Rigo, 2016c) and has its own characteristic pollen type, which can be easily differentiated from the other two *Fraxinus* species (Punt et al., 1991), thus it is not discussed here. *Fraxinus excelsior* (European ash) is the most widespread among the European *Fraxinus* species: its natural range spans from southern Europe including the Caucasus to Fennoscandia and Russia in northern Europe (Beck et al., 2016b). In the temperate forests it is a dominant or secondary species, generally found in stands or admixed as individ-

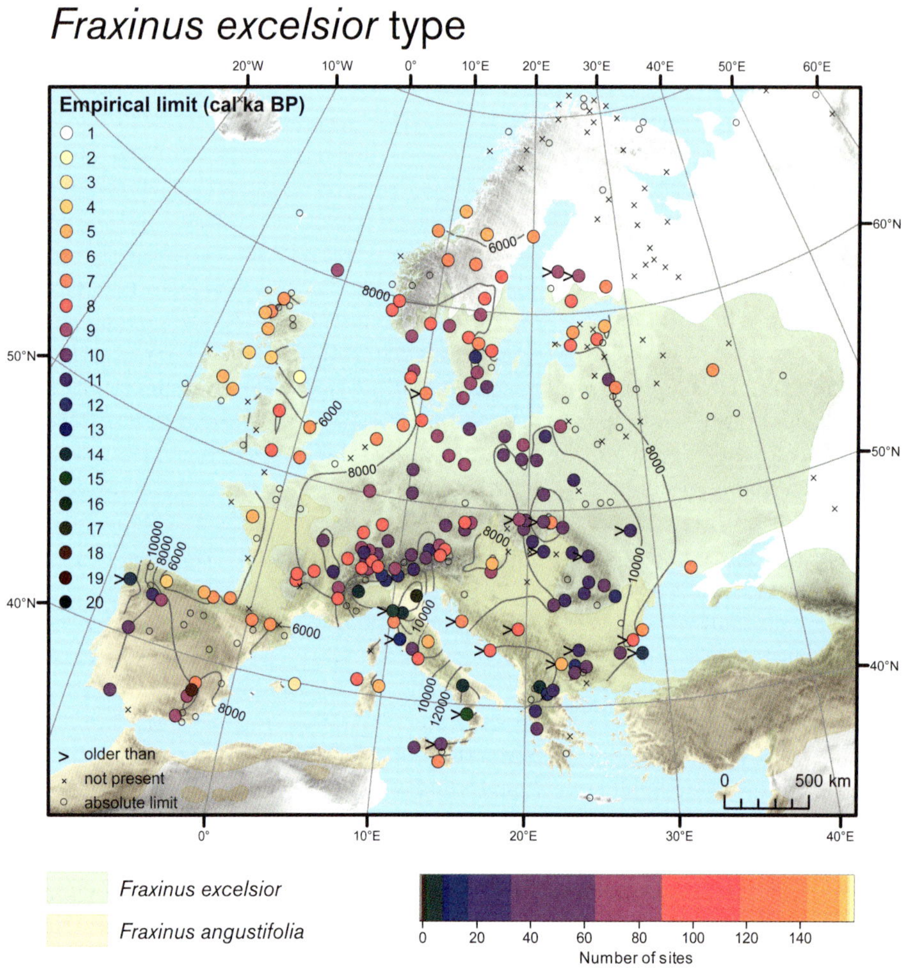

Fig. 4.4.3-11. Map showing the empirical limits of ***Fraxinus excelsior*** type and the current distribution of *F. excelsior* and *F. angustifolia* in Europe. The coloured dots indicate the empirical limit in cal ka BP at selected sites, crosse refer to sites where the taxon was absent and small circles where it reached the absolute limit. Grey lines represent isochrones of empirical limit in 2000-year steps, based on nearest neighbour interpolation. Bar diagram shows stacked number of sites where the empirical limit was reached during the same 1000-year time interval, with colours according to the legend in the figure.

uals. It is often associated with other temperate species such as *Acer pseudoplatanus*, *A. platanoides*, *Tilia platyphyllos*, *Quercus robur*, *Q. petraea*, *Ulmus minor*, *U. glabra*, *Abies alba*, *Fagus sylvatica* and *Castanea sativa* (Schütt et al., 2006; Beck et al., 2016b). The tree is rather tall, growing up to 40–45 m high and becoming c. 300 years old. It is very tolerant to shade in its juvenile stage, but rather light demanding as a mature tree. The species has an intermediate drought sensitivity and tolerates seasonal waterlogging, so that it is also found along rivers. It prefers rich soils and tolerates rather cold winters, but is very sensitive to late frost and rather sensitive to severe winter frost (Ellenberg, 2009).

Fraxinus angustifolia (narrow-leaved ash) is less widespread than *F. excelsior*. The tree is restricted to southern Europe with the exception of a north-western branch spreading to eastern France and western Switzerland as well as a north-eastern branch spreading to Hungary, northern Romania, eastern Austria and the Czech Republic. *Fraxinus angustifolia* is a dominant or secondary species in temperate forests and similarly to *F. excelsior* it is often associated with *Acer*, *Ulmus*, *Quercus* and *Tilia* species (Caudullo and Houston Durrant, 2016). *Fraxinus angustifolia* grows on moist and well-drained soils, can tolerate dry climates (>400 mm annual precipitation) but is susceptible to frost damages. Similarly to *F. excelsior*, it is also often found along rivers. The northern part of the range overlaps with that of *F. excelsior*, with which it can hybridize, generating plants with intermediate phenotypes (Caudullo and Houston Durrant, 2016). The species can grow rather tall, reaching 35–45 m and is light-demanding. The flowers of *F. excelsior* and *F. angustifolia* are wind-pollinated, thus their pollen is well represented in sedimentary pollen records.

Phylogeography: The analysis of chloroplast haplotypes of extant *Fraxinus excelsior* and *F. angustifolia* shows that there is considerable variation among populations, which generates distinct western, central and eastern European genetic patterns (Heuertz et al. 2006a). *Fraxinus excelsior* shows genetic differences in the three southern European peninsulas, which are well mirrored in Europe north of 45°N. More thermophilous *F. angustifolia* shows very similar genetic patterns in the three southern European peninsulas. Such genetic differences may have originated during the contraction and dispersal phases of the glacial-interglacial cycles. On the other hand, the wide-ranging haplotype sharing between *F. angustifolia* and *F. excelsior* suggests hybridisation in common glacial refugia in the three peninsulas or during postglacial range expansions (Heuertz et al. 2006a).

Macrofossil analyses allow the identification of the different *Fraxinus* species (e.g. *F. excelsior*; Vassio and Martinetto, 2012), however, sedimentary records displaying the occurrence of *Fraxinus* at the species level are extremely rare. Pollen analyses do not allow the distinction of closely related *F. excelsior* and *F. angustifolia* (Punt et al., 1991). For the interpretation of our isochrone map (Fig. 4.4.3-11) we assume that north of c. 45°N the pollen derives mostly from *F. excelsior*, south of c. 40°N it originates from *F. angustifolia*, while between 40 and 45°N it may reflect both species. Pollen finds of *F. excelsior* type in autochthonous organic or calcareous sediments suggest that the genus survived the Last Glacial Maximum (LGM, c. 23 000–19 000 cal yr BP) in southern European **refugia**, for example, on the Italian Peninsula (Kaltenrieder et al., 2009). Given that the two species are wind-pollinated, the mapped empirical limit is likely capturing the initial population growth of *F. excelsior* and/or *F. angustifolia* after the LGM. Such earliest evidence of local population increases is dated at around 18 000 cal yr BP in the Iberian Peninsula, around 17 000 cal yr BP in the Italian Peninsula and around 16 000 cal yr BP in the Balkan Peninsula (Fig. 4.4.3-11). We thus assume that populations of both *Fraxinus* species were expanding locally already during the Oldest Dryas (19 000–14 700 cal yr BP) as far north as c. 45°N. By c. 14 000 cal yr BP, at the onset of the Bølling-Allerød interstadial, *F. excelsior* and *F. angustifolia* were spreading out of their refugia in northern, central and southern Italy, while on the Balkan and Iberian peninsulas such early **regional spreads** are not documented in our isochrones (Fig. 4.4.3-11). However, such spreads occurred at the latest at 12 000 cal yr BP on the Iberian and the Balkan peninsulas. As observed for other temperate taxa (see e.g. *Acer*; Fig. 4.4.3-1), *F. excelsior* and *F. angustifolia* populations were rare and small or absent outside southern Europe before the onset of the Holocene, so that the range of the taxon remained small (12 000 cal yr BP isochrones) with only c. 10% of the mapped sites reaching the empirical limit (Fig. 4.4.3-11, bar diagram).

During the Early Holocene (11 700–8200 cal yr BP) the species spread rapidly in southern, central and northern Europe to reach latitudes around c. 60°N. At the onset of the Mid Holocene around 8000 cal yr BP about 70% of the today's range was already covered (Fig. 4.4.3-11). The spread of *F. excelsior* continued during the Mid Holocene, mainly in north-western and eastern Europe. By 5000 cal yr BP, westernmost and northernmost Holocene locations were reached, although locally the empirical limit was reached as late as 2000 years ago at some sites. Hence, the populations of *F. excelsior* outside southern Europe are young and related to Holocene warming, which allowed the ash species to spread north of 45°N. Spreading rates of *F. excelsior* type were 135–460 m yr^{-1} and thus faster than those of other trees with comparable winged diaspores such as *Acer, Picea abies* and *Abies alba,* but slower than those of *Ulmus* (Table 4.11-1). The resulting postglacial dispersal distance was 3–18 km (Table 4.11-1).

Human impact: During the Late Holocene, the abundance of *F. excelsior* and *F. angustifolia* was markedly reduced through land use including burning (see chapter 4.10). Given that the wood of *Fraxinus* is very valuable, among the best of Europe (Schütt et al., 2002), its use may additionally have reduced the occurrence of the tree. However, the leaves are palatable to livestock and therefore both species were used as fodder trees. This widespread usage may have contributed to maintaining *Fraxinus* trees in open landscapes. Moreover, the production of edible flakes called manna from the phloem sap of *F. angustifolia* and *F. ornus* may have reduced human eradication of the tree species in southern Europe (Caudullo and Houston Durrant, 2016).

Juglans regia

The genus *Juglans* consists of about 20 species, most of them in the Americas, at least five in Eurasia, and only *J. regia* in southern Europe (Aradhya et al., 2007). *Juglans regia* (walnut) grows in the part of riparian forests that is not flooded and on slopes: it prefers calcareous soils up to the lower montane belt. It is a zoochorous, monoecious, anemophilous species with a relatively high pollen productivity (Filipova-Marinova et al., 2010) and medium to good dispersal. Pollen morphologically *Juglans* is unique and therefore not problematical, even when the grains are corroded (Beug, 2004). As macrofossils, the nuts are remarkable and the wood is anatomically distinct.

Pleistocene records in Iberia have been mapped by Postigo Mijarra et al. (2008) for the Lower, Middle and Upper Pleistocene. Full-glacial records are documented in some caves such as Carihuela (south-eastern Spain: Carrión and Sánchez-Gómez, 1992), at Lascaux (Leroi-Gourhan, 1980), or at Lezetxiki (Spanish Basque country, cave of Mousteria age: Sánchez-Goñi, 1988). An overview of many sporadic findings in western Europe is presented by Renault-Miskovsky et al. (1984), an Iberian map of Pleistocene sites by Postigo Mijarra et al. (2008) and Pleistocene records by González-Sampériz et al. (2010). Walnut is, however, missing from many potential **refugia** sites as reviewed by Tzedakis et al. (2013).

Phylogeography: Some of the genetic studies concentrate on cultivars (e.g. Pollegioni et al., 2011; Ciarmiello et al., 2013), others on the wide distribution range in the central Asian mountains (e.g. Pollegioni et al., 2014). Interestingly, on some of the mountains of Kyrgyzstan (Fergana range) walnut was absent before c. 2000 cal yr BP (Beer et al., 2008), but plays a role in landscape genetics (Pollegioni et al., 2014).

Range expansions and contractions: Late-glacial records are known sporadically and in low abundances from the Iberian Peninsula, for example from Upper Palaeolithic layers in Gorham's Cave, Gibraltar (Carrión et al., 2008), from Solutrean at El Pirulejo (Córdoba) and from Magdalenian and Azilian sites (e.g. Arenaz: Isturiz and Sánchez-Goñi, 1990; or the cave La Riera: Leroi-Gourhan, 1986). However, our isochrone map (Fig. 4.4.3-12) suggests that during the Late-glacial local stands of *Juglans regia* established on the Balkan Peninsula (empirical limit reached).

Early Holocene records are scarce but substantial, both in the west and the east, for example in Basa de la Mora (Spanish Pyrenees, 11 300–8800 cal yr BP: Pérez-Sanz et al., 2013), eastern Morocco (Ballouche, 1991), or in Alykes Lagoon on the Greek island of Zakynthos (Avramidis et al., 2013). However, our map (Fig. 4.4.3-12) suggests that during the Early Holocene first local population increases of *Juglans regia* occurred on the Italian Peninsula (empirical limit reached).

Mid-Holocene records of walnut pollen are discontinuous but widespread from west to east: in Iberia at El Sabinar at 1117 m asl (Carrión et al., 2004), Siles at 1320 m asl (Carrión, 2002b) or Serra da Estrela (van den Brink and Janssen, 1985); in Italy at Lago di Nemi (Lowe et al., 1996) and at Lago della Costa in the Colli Euganei where high and consistent pollen percentages of *Juglans, Castanea* and *Olea* together with high charcoal values are recorded already in Neolithic sediments (Kaltenrieder et al., 2010); and at Bulgarian sites near the Danube river (Lazarova, 1995). Unusually high and consistent abundances were found on the Greek island of Zakynthos (Avramidis et al., 2013).

For the Late Holocene, Bottema (2000) presents an impressive figure showing 17 Greek and 17 Turkish sites where, from 3000 uncal yr BP (c. 3200 cal yr BP) onwards, the records are abundant and often continuous, a finding which also applies to other areas of southern Europe, such as northern Italy, southern France and neighbouring western Switzerland (Fig. 4.4.3-12). This may indicate that some natural populations existed that were anthropogenically enhanced or replaced by cultivated forms. Also, north of the Alps at Lake Constance, *Juglans* with small, thick-walled fruits were found in Neolithic and Bronze Age settlements and Isenberg (1986) provides an overview with a map discussing records in central Europe before and after the Roman period. Towards the end of the La Tène period, traces of *Juglans* were recorded both near Lake Constance by Rösch (1992) and at Logné in the Massif

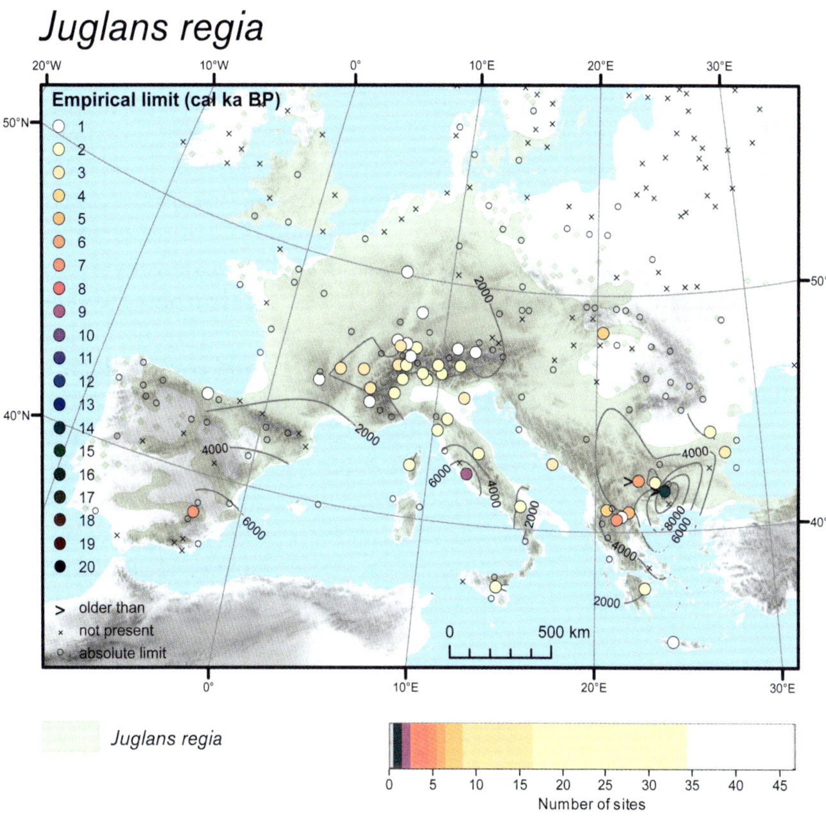

Fig. 4.4.3-12. Map showing the empirical limits of ***Juglans regia*** and its current distribution in Europe. The coloured dots indicate the empirical limit in cal ka BP at selected sites, crosses refer to sites where the taxon was absent and small circles where it reached the absolute limit. Grey lines represent isochrones of empirical limit in 2000-year steps, based on nearest neighbour interpolation. Bar diagram shows stacked number of sites where the empirical limit was reached during the same 1000-year time interval, with colours according to the legend in the figure.

Armorican of western France by Barbier and Visset (1997). All over Europe both spatial expansion and population expansions are conspicuous in Greek and Roman times – or as Isenberg (1986) made the pun: "Ex oriente nux". Archaeobotanical connections to other plant taxa are also of interest, for example, 'edible Rosaceae' or *Vitis vinifera* (Bakels and Jacomet, 2003). As Zohary et al. (2012) summarize: "Greek, Roman, and Jewish literary sources indicate that the walnut was already widely grown in Classical times in the Mediterranean Basin and south-west Asia. *J. regia* cultivation started probably before Roman times, particularly in south-west Asia". The species was appreciated for its strong and beautiful wood, its nuts and in some areas for the medical use of the exocarp. In central Europe, continuous curves and high pollen values of *Juglans* in pollen diagrams can largely be understood as an indication of the Roman invasion. However, our isochrone map shows that north of the Alps the empirical limit was only exceptionally reached before 1000 cal yr BP. The etymology of Juglans is based on 'Jovis glans', the acorn of Jupiter. As Bertsch (1951, 1953) summarized, this name was created by Cicero, but often it meant either chestnut (in Greek) or walnut (in Latin). Both fruits were considered of Anatolian origin and their species name was therefore 'pontic' or 'persian' or 'regia' (that is, coming from regions under the Persian king).

Ostrya carpinifolia and *Carpinus orientalis*

The genus *Ostrya* comprises about ten deciduous species, which grow in the northern hemisphere (Schütt et al., 2002). The closely related genus *Carpinus* has c. 30 deciduous species, which, with the exception of the North American *Carpinus caroliniana*, all grow in Eurasia (Schütt et al., 2002). The two species *Ostrya carpinifolia* and *Carpinus orientalis* produce the same pollen type (*Ostrya* type). *Carpinus betulus* has its own pollen type, thus it is not discussed here (see Fig. 4.4.3-4). *Ostrya carpinifolia* and *Carpinus orientalis* share very similar ranges and environmental niches. Both species grow on the Italian and Balkan peninsulas (Pasta et al., 2016b; Sikkerna and Caudullo, 2016) in submediterranean forests together with with *Fraxinus ornus, Quercus pubescens* and other oaks (e.g. *Q. cerris, Q. frainetto, Q. petraea, Q. congesta*). The two species also occur in the Caucasus, Anatolia, Syria and Lebanon (Pasta et al., 2016b; Sikkerna and Caudullo, 2016). *Ostrya carpinifolia* and *Carpinus orientalis* can be dominant or co-dominant under moderate to high disturbance, while under low disturbance the forests rapidly develop towards submediterranean mixed oak forests (Pasta et al., 2016b; Sikkerna and Caudullo, 2016). Both species are rather shade-tolerant (Pignatti, 2005), so that to some degree they can persist in the understorey.

Ostrya carpinifolia (European hop-hornbeam) is a small to medium-sized tree which can reach up to 25 m. After cutting it often grows as a shrub c. 3–6 m tall. The species is competitive under warm-temperate and rather moist environments, with optima around 13.5°C mean annual temperature and 1300 mm mean annual precipitation (Schütt et al., 2006). *Carpinus orientalis* (Oriental hornbeam) is smaller than *Ostrya carpinifolia,* it usually grows up to 5 m height, only exceptionally to 15 m. The species is also slightly less thermophilous and more continental than *Ostrya carpinifolia* (Pignatti, 2005). Both species are wind-pollinated and thus well represented in sedimentary pollen records.

Macrofossil analyses allow the identification of the species (e.g. *Carpinus orientalis*; Sadori et al., 2010), however, sedimentary records displaying the occurrence of *Ostrya carpinifolia* or *Carpinus orientalis* at the species level are extremely rare. Pollen analyses do not allow the distinction of the two species (e.g. Moore et al., 1991). For the interpretation of our isochrone map (Fig. 4.4.3-13), we assume that the pollen derives from both species, with a tendency to *Ostrya carpinifolia* in Italy, where the species is more abundant in the forests today than *Carpinus orientalis* (Pasta et al., 2016b; Sikkerna and Caudullo, 2016). Two different *Carpinus orientalis* chloroplast haplotypes occur in the Italian and Balkan peninsulas, which can be explained by the ice age survival of the species in isolated **refugia** in the two Mediterranean peninsulas (Grivet and Petit, 2003). Pollen finds of *Ostrya* type suggest that the taxon survived the Last Glacial Maximum (LGM, c. 23 000–19 000 cal yr BP) in southern Europe on the Balkan and Italian peninsulas. Given that the two species are wind-pollinated, the mapped empirical limit is likely capturing the initial population growth of *Ostrya carpinifolia* and/or *Carpinus orientalis* after the LGM. Such earliest evidence of local population increases is dated at around 19 000 cal yr BP in the Balkan Peninsula and around 17 000 cal yr BP in the Italian Peninsula (Fig. 4.4.3-13). We thus assume that populations of the taxon were expanding locally already during the Oldest Dryas (19 000–14 700 cal yr BP) between c. 40 and 45°N. By c. 12 000–10 000 cal yr BP, during the Early Holocene, *Ostrya carpinifolia* and/or *Carpinus orientalis* were **spreading** out of their refugia in central and southern Italy as well as the central and southern Balkans (Fig. 4.4.3-13). At around 8000 cal yr BP, at the onset of the Mid Holocene, most of today's range was already covered. Later, during the Mid Holocene, the taxon spread to northern Italy and the northern Balkans. The spreading of the taxon continued during the Late Holocene, with the youngest population expansions occurring on the southern slope of the Alps, where the empirical limit was attained as late as 2000 years ago at some sites. It is likely that this late expansion reflects *Ostrya carpinifolia*, given that today *Carpinus orientalis* is very rare in northern Italy (Sikkerna and Caudullo, 2016). In contrast to other temperate species, *Ostrya carpinifolia* and *Carpinus orientalis* remained confined to southern Europe. Spreading rates of *Ostrya* type were 30–110 m yr^{-1} and thus slower than those of other trees with wind-dispersed diaspores such as *Carpinus betulus*, *Acer*, *Picea abies* or *Abies alba* (Table 4.11-1).

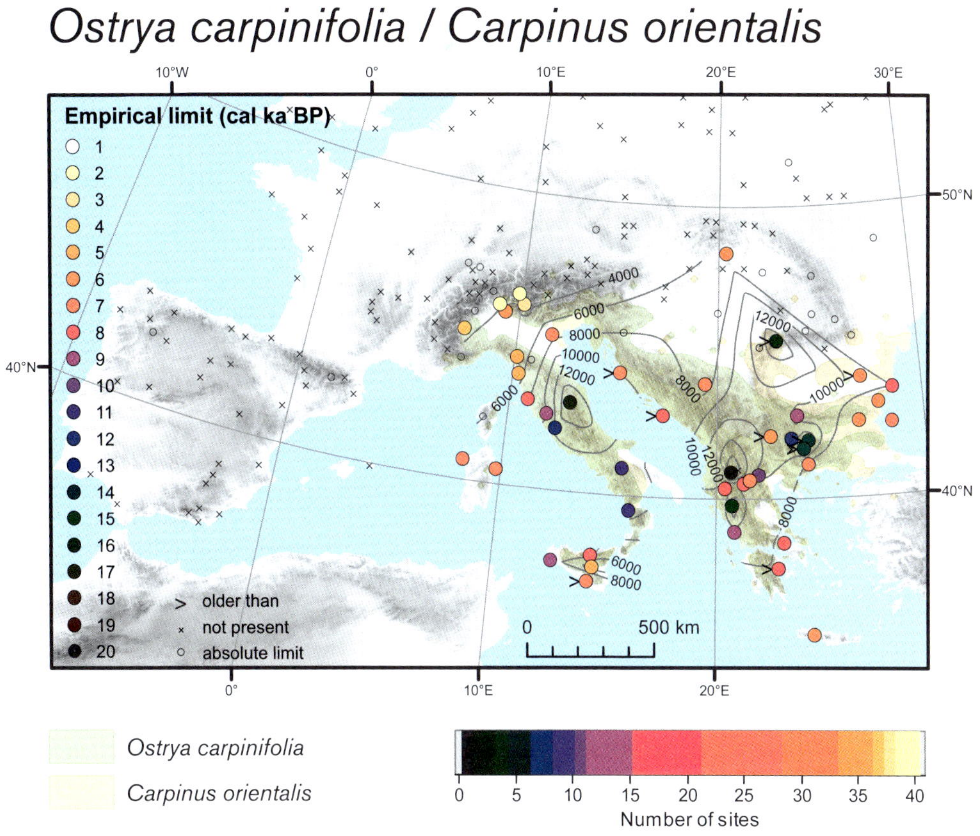

Fig. 4.4.3-13. Map showing the empirical limits of ***Ostrya carpinifolia / Carpinus orientalis*** and their current distribution in Europe. The coloured dots indicate the empirical limit in cal ka BP at selected sites, crosses refer to sites where the taxon was absent and small circles where it reached the absolute limit. Grey lines represent isochrones of empirical limit in 2000-year steps, based on nearest neighbour interpolation. Bar diagram shows stacked number of sites where the empirical limit was reached during the same 1000-year time interval, with colours according to the legend in the figure.

Human impact: During the Late Holocene the abundance of *Ostrya* type was increased through land use, particularly for wood cutting (Gobet et al., 2000) and firewood. Both species are strong resprouters and can be managed in coppiced stands for wood production. The leaves of *Carpinus orientalis* are palatable to livestock and therefore the species is used as fodder tree (Sikkerna and Caudullo, 2016).

Deciduous *Quercus*

Out of the c. 500 oak species in the world (Manos et al., 1999), more than 20 species of deciduous oaks exist in Europe, most of them in southern Europe (Flora Europaea; for selected species see Fig. 4.4.3-14).

In terms of pollen morphology, the distinction is usually only made between deciduous and evergreen taxa. *Quercus suber* is evergreen but its pollen is closer to the *Q. robur–Q. pubescens* type – it is therefore of interest that Beug (2004) additionally defines a *Quercus cerris* type that also includes *Q. crenata* and *Q. suber*. Benthem et al. (1984) also described the *Q. cerris* type. *Quercus crenata* is often considered to be a hybrid between *Q. suber* and *Q. cerris*. Extensive introgressions among oak species are widespread.

As glacial **refugia** the intermediate elevations of the mountains in the three southern peninsulas are undisputed (Huntley and Birks, 1983; Brewer et al., 2002). For the Iberian Peninsula, not only southern but also central and northern full glacial sites of surviving deciduous oaks are known and Carrión et al. (2008) review Pleistocene records of *Quercus* from Gibraltar to central and northern Spain. In Gorham's Cave (Gibraltar), the Middle Palaeolithic Mousterian Level IV (c. 39 000–28 000 cal yr BP) as well as the Upper Palaeolithic Level III (c. 22 500–19 600 cal yr BP) contain *Quercus* as both charcoal and pollen. In the cave sediments, pollen of both deciduous and evergreen oaks were recorded: in the coprolites, *Q. suber* was identified. Records from 39 Iberian sites and ten marine cores are reviewed by Carrión et al. (2008) – from the south-west (Doñana Park: Yll et al., 2004) and south-east (Bajondillo: López-Sáez et al., 2007; Nerja Cave: Badal, 1998; San Rafael: Pantaléon-Cano et al., 2003) to the north-west (e.g. Area Longa: Gómez-Orellana et al., 2007) and north-east (e.g. Abric Romaní near Barcelona: Burjachs and Julià, 1994). The high spatial differentiation of oak refugia in Iberia is also reflected by the high diversity in phylogeography (Olalde et al., 2002; Petit et al., 2002a).

For the Apennine Peninsula, Follieri et al. (1998) provide in their Fig. 10 an overview of the substantial occurrences of deciduous oaks in central Italy, from the Eemian to the Holocene. For southern Italy, Lago Grande di Monticchio provided a rich *Quercus* record covering the last 120 000 years with a minimum only around the LGM and the subsequent Oldest Dryas between 25 000 and 15 000 cal yr BP (Watts et al., 1996a; Allen et al., 1999). Also in northern Italy, climatically favourable sites such as the Colli Euganei were refugia for oak (Kaltenrieder et al., 2009). Again, multiple refugia are likely reflected in the phylogeography (Fineschi et al., 2002).

In Greece, a continuous presence of deciduous oak was found at mid elevations: Ioannina at 470 m asl in both long records, namely Ioannina 249 including the last 423 ka (Tzedakis, 1993) and Ioannina 284 including the last 133 ka (Tzedakis et al., 2002). Even during the coldest and driest periods of the early pleniglacial (68–59 ka) and the late pleniglacial (26–19 ka) pollen percentages were between 10% and 20%. At Tenaghi Philippon, at only 10 m asl but more than a degree further north than Ioannina, the oak record is strong during the early Weichselian (from Amersfoort to Odderade), is lacking during most of the pleniglacial but returns during the Late-glacial interstadial (Wijmstra, 1969). Our new isochrone map summarizes these local and regional findings very well, specifically it suggests the presence of LGM refugia on the Iberian, Italian and Balkan peninsulas (black dots, Fig. 4.4.3-14).

In **phylogeography**, a major effort was made in an EU-supported project 'FOREST' to analyse 32 chloroplast variants in more than 2600 living oak populations of eight species (*Q. robur, Q. petraea, Q. pubescens, Q. frainetto, Q. faginea, Q. pyrenaica, Q. canariensis, Q. macranthera*). Regional papers are offered by Bordács et al. (2002), Cottrell et al. (2002), Csaikl et al. (2002a,b), Fineschi et al. (2002), Jensen et al. (2002b), König et al. (2002b), Olalde et al. (2002) and Petit et al. (2002b). A first synthesis by Petit et al. (2002a) provides a synoptic map over the the six cpDNA lineages across Europe, demonstrates the strong sharing of cpDNA variants and discusses diversity measures. A second synthesis

Quercus sp. (deciduous)

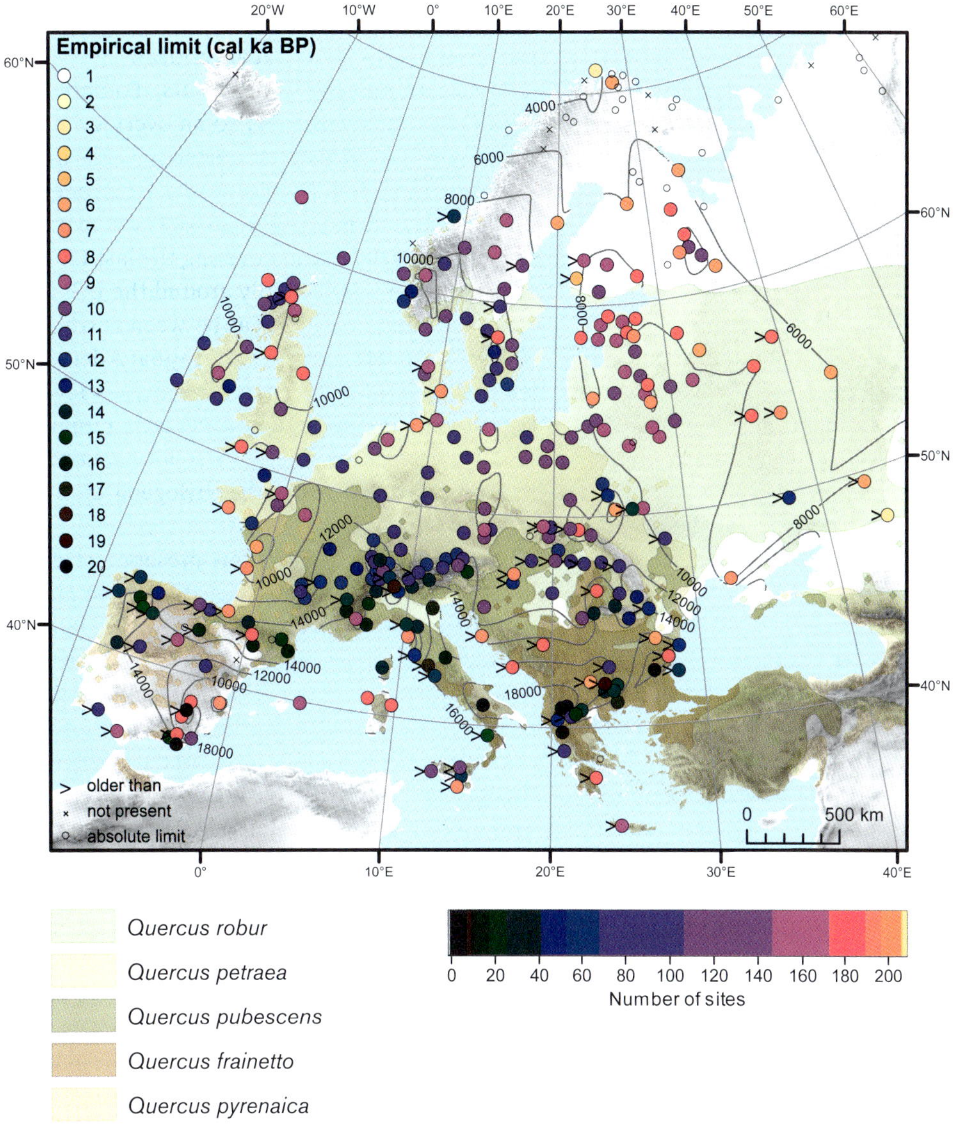

Fig. 4.4.3-14. Map showing the empirical limits of deciduous species of ***Quercus*** and the current distribution of five species in Europe. The coloured dots indicate the empirical limit in cal ka BP at selected sites, crosses refer to sites where the taxon was absent and small circles where it reached the absolute limit. Grey lines represent isochrones of empirical limit in 2000-year steps, based on nearest neighbour interpolation. Bar diagram shows stacked number of sites where the empirical limit was reached during the same 1000-year time interval, with colours according to the legend in the figure.

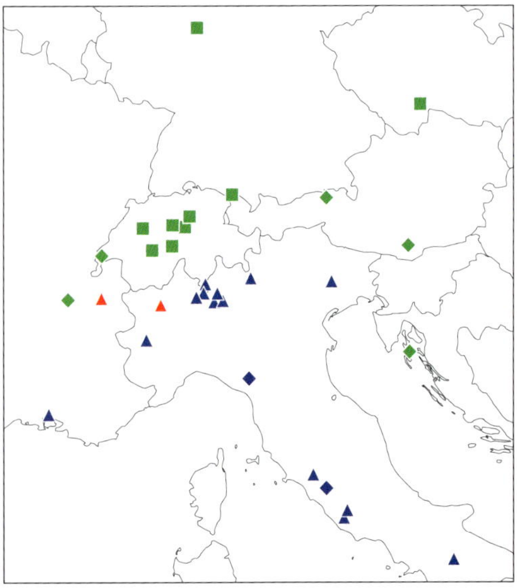

Quercus expansion before *Corylus* during the Holocene (▲) or Late-glacial (▲); *Corylus* expansion before *Quercus* during the Holocene (■); synchronous expansion during the Holocene (◆) or Late-glacial (◆)

Fig. 4.4.3-15. Map showing the spatial pattern of *Corylus* and *Quercus* population expansion, indiscating a clear north–south division, with *Quercus* expanding first in the south and *Corylus* expanding first north of the Alps. Modified after Finsinger et al. (2006).

by Petit et al. (2002c) provides maps of the six lineages and – in conjunction with palaeoecological records and their chronology, as reviewed by Brewer et al. (2002) based on the EPD – maps the supposed spreading routes of five lineages (*Q. macranthera* is restricted to the Caucasus).

Range expansions and contractions: deciduous oak is recorded for the Late-glacial in the Iberian Peninsula at a vast number of sites (Carrión et al., 2008). Examples from the south are: the core MD95-2043 in the Alborán Sea, that already shows sharply rising values of both *Quercus* types after Heinrich Event 1; at San Rafael (0 m asl) both types are frequent during the Late-glacial; and at Padul (785 m asl) a sharp increase occurs at the onset of the Bølling, (Pons and Reille, 1988; Yll et al., 1994; Pantaléon-Cano et al., 2003; Fletcher and Sánchez Goñi, 2008). A western example is Charco da Candieira in the Sierra da Estrela with about 10% oak pollen during the Allerød and a distinct *Quercus*-depression during the Younger Dryas (van der Knaap and van Leeuwen, 1997). In the north, charcoal records indicate local presence, for example at Altamira (Uzquiano, 1992). On the Apennine Peninsula, not only at southern sites such as Monticchio but also at north Italian sites such as Lago Piccolo di Avigliana or Lago della Costa, populations of *Q. robur* expanded as early as the Bølling or before (Fig. 4.4.3-15). Indeed first deciduous oak stands had likely established as early as 18 000 cal BP in the formerly glaciated Ticino area (brown dot Fig. 4.4.3-14). Again the Younger Dryas is recorded by an oak-depression (Huntley et al., 1999; Finsinger et al., 2011).

In Greece at several lowland sites, populations of *Quercus* expanded during the Late-glacial interstadial and decreased during the Younger Dryas, from west to east: Gramousti (185 m asl; Willis, 1992), Kopais (10 m asl; Turner and Greig, 1975) and Tenaghi Philippon (40 m asl; Wijmstra, 1969). At the mid-elevation site of Ioannina (473 m asl), in the west, the deciduous oaks also expanded during the Late-glacial interstadial but the contraction during the Younger Dryas was very limited (Lawson et al., 2004); in the east at the same elevation at Nisi Fen (475 m asl; Lawson et al., 2005) and the nearby site of Khimaditis (Bottema, 1974) the contraction of oak is nearly absent. Lawson et al. (2005) discuss this in terms of precipitation that decreases from west to east and increases with elevation; this may be a reflection of the fact that the Younger Dryas in Mediterranean latitudes was more a dry than a cool period (Wright et al., 2003; Connor et al., 2013). Our isochrone map (Fig. 4.4.3-14) shows that the pre-Bølling stands of deciduous *Quercus* were almost entirely restricted to the three peninsulas of southern Europe, with only about 10% of the European sites reaching the empirical limit. At the end of the Late-glacial this number had increased to c. 20%, with again almost all local initial population expansions occuring in southern Europe (Fig. 4.4.3-14).

During the Early Holocene, deciduous oaks expanded rapidly through most of southern and central Europe (Fig. 4.4.3-14) as mapped previously also by Huntley and Birks (1983), Brewer et al. (2002) (both on a timescale of uncalibrated radiocarbon years) or as reviewed and numerically explored by Giesecke et al. (2017). An early pollen and macrofossil record in south-western Slovakia is dated to 11 200 cal yr BP (Jamrichová et al., 2014). For Italy, twelve maps covering the Holocene expressing oak occurrences as pollen-percentage classes are presented by Magri et al. (2015). For Bulgaria, syntheses can be found in Bozilova et al. (1996) and Connor et al. (2013). For Romania, including the southern Carpathians, Feurdean et al. (2011) provide a Holocene synthesis based on 14 well-dated sites. Isochrone maps for four abundance classes including the Late-glacial and the Holocene are provided by Giesecke and Brewer (2018). Our isochrone map shows that at most of the non-southern European sites deciduous oak stands established during the Early Holocene (Fig. 4.4.3-14). During the Mid Holocene the taxon was also spreading in north-eastern Europe, covering the entire today's range.

Human impact on oak populations is recorded in many regions, for example as multiple uses for the last 2000 years at a site in the Czech Republic (Jamrichová et al., 2013). In general the impact is based on at least three ways of use: the high quality of the wood for construction was already appreciated for the lake dwellings of the Neolithic and Bronze Age, the bark was used for tanning and the seeds for feeding animals, especially pigs ("the best ham is growing on oaks": ten Cate, 1972).

It will be a major challenge to trace the history of the numerous southern oak species and analyses of plant macrofossils and/or aDNA will be necessary to identify the taxa.

Tilia

Depending upon how the species are delimited, the genus *Tilia* comprises up to 60 species. Five of these, belonging to two sections, occur in Europe. The section *Anastraea* includes *T. cordata, T. platyphyllos, T. rubra* and *T. dasystyla*; the section *Astrophilyra, T. tomentosa.*

Tilia cordata Mill., the small-leaved linden, has by far the largest distribution range (see map Fig. 4.4.3-16), covering approximately 6.5 million square kilometres. The range of *T. cordata* is very similar to that of *Acer platanoides*, also including a separate occurrence in the Caucasus. *Tilia cordata* shows an ecological behaviour of continental character, probably due to its relatively late frondescence (10–14 days later than *T. platyphyllos*) and some drought resistance. In central Europe it is a characteristic species of the Carpinion alliance (in the Braun-Blanquet system). Along the northern limit of its range it often occurs in conifer forests of the boreal zone. It is also characteristic around the Alps in mixed deciduous forests of the 'foehn'-valleys, which are especially warm but also humid. The chloroplast DNA variation of *T. cordata* Mill. was studied by Fineschi et al. (2003).

Tilia platyphyllos Scop., the large-leaved linden, is sensitive to late frosts and not very resistent to drought. So, in contrast to *T. cordata*, the species shows a suboceanic character, preferring warm-humid habitats, particularly on calcareous soils. In central Europe it is a characteristic species of the Tilio-Acerion alliance (in the Braun-Blanquet system). Within its range, the species is polymorphous and split into 3 or 4 subspecies. The ssp. *grandifolia* is restricted to northern Europe, ssp. *cordifolia* to northern, eastern and central Europe, ssp. *platyphyllos* to central and southern Europe and ssp. *pseudorubra* exclusively to southern Europe. In areas where *T. platyphyllos* and *T. cordata* occur together the two species frequently hybridise (*T.* x *vulgaris*).

Tilia tomentosa Moench., the silver linden, occurs in south-eastern Europe, mainly in lower mountain areas and the lowlands of the Balkan Peninsula, but also in western Anatolia. The species is drought tolerant and restricted to areas with a warm continental climate, preferring thermophilous deciduous oak forests (Quercetalia pubescentis order in the Braun-Blanquet system; Horvat et al., 1974). During the last interglacial (Eemian) the northern limit of *T. tomentosa* was more to the north than it is at present, as far north as the lowlands of northern Germany, Poland and Russia (Frenzel, 1968; Meusel et al., 1978).

Tilia rubra DC., the red linden, has a disjunct distribution with scattered occurrences on the Balkan (ssp. *rubra*) and Crimean peninsulas, the Causasus and the northern parts of Anatolia (ssp. *caucasica*). The species is taxonomically close to *T. platyphyllos*. In the vegetation monograph by Horvat et al. (1974) red linden is not mentioned (probably because some authors include *T. rubra* as well as *T. dasystyla* in *T. platyphyllos*), whereas Mayer and Aksoy (1986) refer to the species as a member of the north Anatolian deciduous forest.

Tilia dasystyla Stev., the Caucasian linden occurs in the Caucasus region and northern Iran.

Pollen morphologically the identification is easy at the genus level, but not at the species level. The differentiation between *T. cordata* and *T. platy-*

Tilia

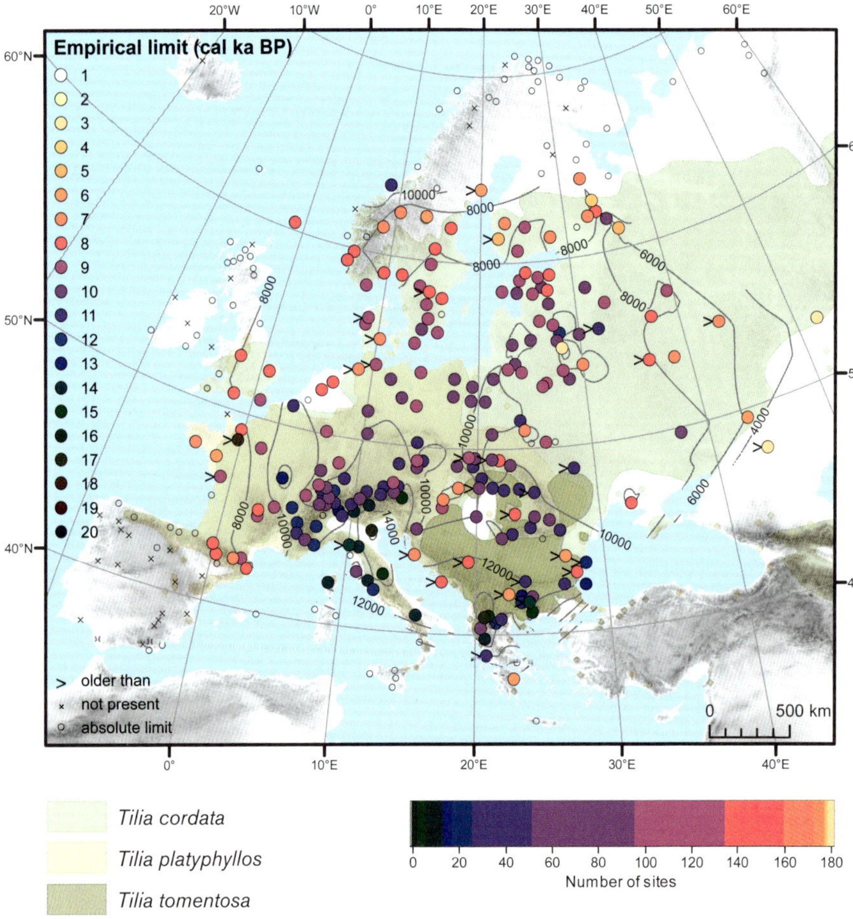

Fig. 4.4.3-16. Map showing the empirical limits of ***Tilia*** in Europe. The coloured dots indicate the empirical limit in cal ka BP at selected sites, crosses refer to sites where the taxon was absent and small circles where it reached the absolute limit. Grey lines represent isochrones of empirical limit in 2000-year steps, based on nearest neighbour interpolation. Bar diagram shows stacked number of sites where the empirical limit was reached during the same 1000-year time interval, with colours according to the legend in the figure.

phyllos has been described by several authors (Andrew, 1971; Stockmarr, 1974; Guggenheim, 1975; Christensen and Blackmore, 1988; Moore et al., 1991) but, according to Beug (2004), only about 75% of modern material could be identified in the two species.

Pollen grains of *Tilia* are exceptionally resistant to oxydation. Therefore, in glacial and Holocene sediments reworked pollen of Tertiary or interglacial age is sometimes preserved: however, these grains are usually corroded and identifiable as 'secondary pollen grains'. Even in cave deposits can *Tilia* pollen be over-represented because of its high resistance to oxydation.

All *Tilia* species are entomophilous and anemochorous. Pollen production and dispersal in *Tilia* is relatively low as is generally the case with entomophilous plants. Thus, the genus is under-represented in pollen spectra, probably by between two to four times (Andersen, 1970; Fægri and Iversen, 1989). Because of the low pollen production of these insect-pollinated species, low threshold values are proposed for the estimation of presence or absence (Lisitsyna et al., 2011; see also chapter 2.7). In our

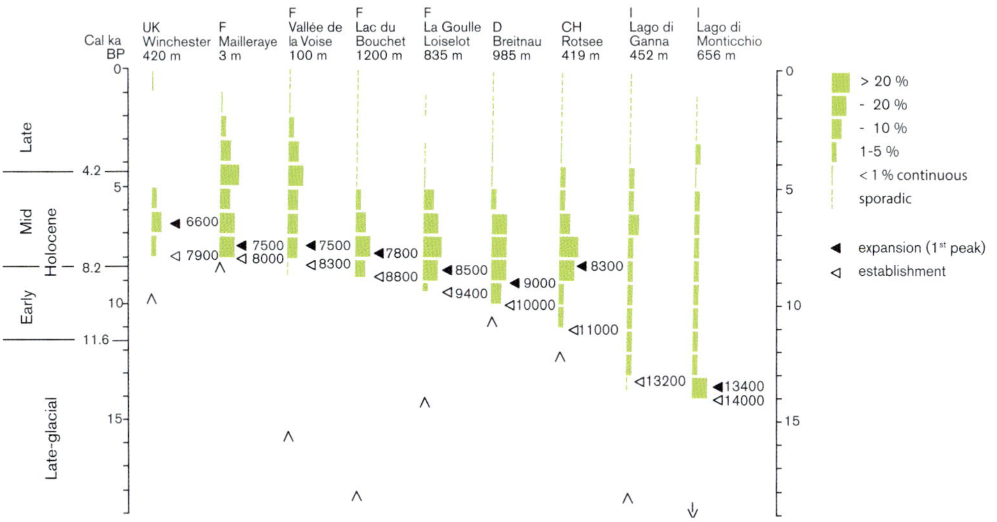

Fig. 4.4.3-17a. Expansion of ***Tilia*** in Europe, documented in pollen records along a transect from the Italian Peninsula to the British Isles.

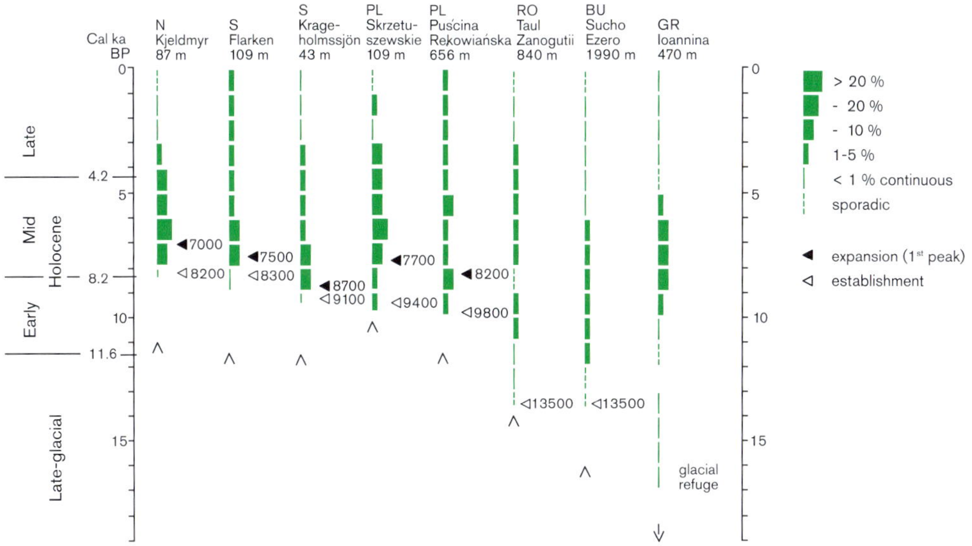

Fig. 4.4.3-17b. Expansion of ***Tilia*** in Europe, documented in pollen records along a transect from the Balkan Peninsula to southern Scandinavia. Modified after Lang (2003).

isochrone map we use the empirical limit which in this case may indicate the timing for the population expansions more than the establishment of individual trees or small stands (see chapter 4.4.1).

Macrofossils concern mostly fruits (nuts, where determination to species is often possible). Rarely, flowers with anthers and bud scales are also recorded, but their identification to species level is not always possible. The wood can be identified to the generic level. An overview of sites is given in Lang (2003).

Glacial **refugia** were probably located in the southern Balkan Peninsula as indicated by records from the west coast of the Black Sea (Filipova-Marinova, 2003), from Ioannina and Tenaghi Philippon in Greece (Bottema, 1974) and Lake Ohrid in Albania (Lézine et al., 2010). Full-glacial records were also found throughout Italy: in the south (Monticchio: Watts et al., 1996a, 2000), in the centre (Follieri et al., 1998, with a synthesis in Fig. 12) and in the north (Lago della Costa in the Euganean hills: Kaltenrieder et al., 2009; Lago di Fimon: Pini et al.,

2010). The chloroplast DNA variation of *T. cordata* Mill. was studied by Fineschi et al. (2003).

Range expansions and contractions: Late-glacial records in Bulgaria stretch from sea level (Filipova-Marinova et al., 2013b) to the mountains (e.g. Sucho Ezero or Kremensko-5; Bozilova and Smith, 1979; Stefanova et al., 2006a,b). In North Macedonia (e.g. Lake Prespa: Panagiotopoulos et al., 2013), *Tilia* remains present through the Late-glacial and only shows a depression during the Younger Dryas. In Slovenia (e.g. Lake Bled: Andrič et al., 2009) and at several sites in Italy, the *Tilia* pollen curves become continuous during the Allerød and are rare throughout the Younger Dryas, but are confirmed by some macrofossils in Lago di Ganna (near Varese: Schneider and Tobolski, 1985). This local and regional evidence is well summarized by our isochrone map suggesting Oldest Dryas population expansions in the Italian and Balkan peninsulas (17 000–15 000 cal yr BP, Fig. 4.4.3-16). Before the onset of the Holocene, *Tilia* further spread at other sites in southern Europe and eastern France (12 000 cal yr BP isochrone).

At numerous sites on both sides of the Alps the onset of the Early Holocene is marked by a rapid increase in *Tilia*: some sites north of the Alps include macroremains, for example, Rotsee near Lucerne (Lotter, 1988) or Federsee north of Lake Constance (Bertsch, 1931). The spatial expansion and the population expansion during the Early to Mid Holocene were previously illustrated in the maps of Brewer et al. (2017) and in those from Lang (2003), who also calculated spreading rates of between 190 m yr^{-1} and 820 m yr^{-1}, which are somewhat similar to the ones calculated by Huntley and Birks (1983) of 300–500 m yr^{-1}. A particularly low rate of only 50 m yr^{-1} was estimated by Birks (1986a, 1989) for England near the north-western limit of the distribution range, indicating a gradual retardation of spreading very likely caused by climatic conditions (additional information provided by Pigott and Huntley, 1978, 1980, 1981). Our new isochrone map (Fig. 4.4.3-16) suggests that *Tilia* populations expanded at many sites north of the Alps (or c. 45°N) during the Early Holocene, when *Tilia* spread at c. 60% of the European sites (bar diagram, 11 000–9000 cal yr BP). Subsequently, during the Mid and Late Holocene, the spreading of *Tilia* was substantially reduced. On the basis of our isochrone map, we estimate the spreading rates of *Tilia* to 220–570 m yr^{-1}. The resulting disperal distance is 2–19 km (Table 4.11-1). Lang (2003) noted that the interval between establishment and end of mass expansion (first peak, see Fig. 4.4.1-1) differs normally from site to site. In the case of *Tilia*, a couple of selected and simplified curves demonstrate that these differences range from 400 years (Vallé de la Voise) to more than 2500 years (Rotsee), with an average of approximately 1000 years (Fig. 4.4.3-17). The interval between establishment and first culmination of expansion was interpreted as the spreading time needed for competition with preceeding tree taxa in the Late-glacial and Early Holocene forest development. This assessment of Lang (2003) is in agreement with dynamic modelling results, which show that after the establishment of a species (empirical limit) several centuries to about a millennium are needed to reach a first biomass peak (e.g. Lotter and Kienast, 1992; Keller et al. 2002). The long duration of the expansion process is explained by intra- and interspecific competition. The *Tilia* decline during the Late Holocene can be attributed to climatic reasons and/or human disturbance (e.g. Late Holocene cooling, increasing fire incidence; see chapters 4.10 and 4.11).

To clarify the history of the two widespread species *Tilia cordata* and *T. platyphyllos* and also of the rare ones (*T. tomentosa*, *T. rubra*, *T. dasystyla*) additional studies, including the analysis of plant macrofossils and/or aDNA, in southern and south-eastern Europe are needed.

Ulmus/Zelkova

The genus *Ulmus* consists of 20–45 species of which three are important for Europe: *U. glabra* (= *U. montana* = *U. scabra*; wych elm) extends furthest north (Moskal-del Hoyo et al., 2018), while *U. minor* (= *U. carpinifolia* = *U. campestris*; field elm) reaches from the Iberian Peninsula to the Caspian Sea, and *U. laevis* (= *U. effusa*; European white elm) from central France to the Ural mountains (60°E). Hybridizations are frequent. *Zelkova* was widespread until the Pliocene but, today in Europe, is represented by just two island endemics: *Z. sicula* on Sicily and *Z. abelicea* on Crete. In terms of pollen morphology the two genera cannot be safely separated (stephanoporate, rugulate to suprareticulate: Beug, 2004). Within the genus *Ulmus*, Stockmarr (1970, 1974) proposed differentiations between *U. glabra* and *U. laevis*, which

Tolonen (1980) later supplemented (thickness of annuli) and applied to a southern Finnish site. Here *U. laevis* seemed to dominate during the Early Holocene and *U. glabra* became more important from the Mid Holocene (Atlantic period) onwards. *Ulmus* and *Zelkova* are anemophilous and anemochorous.

As macrofossils, both the fruits and the wood are anatomically distinguishable at the genus level.

Full glacial **refugia**: during the Eemian *Ulmus/Zelkova* was frequent at many sites, for example, Les Echets, La Grande Pile, Lago di Fimon, Valle di Castiglione, Lago Grande di Monticchio and Ioannina (de Beaulieu and Reille, 1984a,b, 1992a; Follieri et al., 1988; Tzedakis et al., 2002; Allen and Huntley, 2009; Pini et al., 2010, see Chapter 3). In some of these long records the presence of *Ulmus/Zelkova* is more or less continuous through either just the intermediate stages such as St. Germain I and II of the early Würm (e.g. Fimon) or throughout the Würmian (e.g. Ioannina). For Italy, an overview figure for *Ulmus* is presented by Follieri et al. (1998) and for *Zelkova* by Follieri et al. (1986). The isochrone map (Fig. 4.4.3-18) suggests LGM refugia on the Balkan Peninsula (black dot), probably also on the Italian Peninsula (empirical limit reached at c. 17 000 cal yr BP).

Range expansions and contractions: Late-glacial records of *Ulmus* are confined to southern Europe, specifically the Balkan and Italian peninsulas (see Fig. 4.4.3-18). The sharp onset and the very rapid spatial expansion of the genus to Europe north of 45°N during the Early Holocene are remarkable. Giesecke et al. (2017) estimate a spread of at least 1000 m yr^{-1}, probably second only to *Corylus* in its velocity. On the basis of our isochrone map we come to similar estimates of 150–800 m yr^{-1} for *Ulmus*, only marginally less than for *Corylus* (450–900 m yr^{-1}) and *Betula* (230–990 m yr^{-1}; Table 4.11-1). At many Holocene sites north of the Alps the population expansion of *Ulmus* occurred just after that of *Corylus* and before that of *Quercus*. For millennia, elm was an important part of 'mixed oak forest' communities, which probably have no analogues today. During the Mid Holocene, a number of population fluctuations are visible in many pollen diagrams, including the hotly debated **'elm decline'**. For the whole of Europe, Huntley and Birks (1983) provide not only maps with a comprehensive text but also 'difference-maps' at steps of 500 radiocarbon years. A concise summary of patterns, processes and hypotheses can be found in Birks (1986a). For central and north-western Europe, we give here just some examples from south to north. On the Swiss Plateau the first rapid elm decrease was before the mass expansion of *Fagus* with a second decrease accompanying the expansion of *Fagus* (c. 8000 and 7000 cal yr BP, respectively). The third and fourth declines occurred during early Neolithic times and were followed by a Bronze Age, a Roman and a medieval decline (Ammann, 1989). For northern Germany, Chen (1988) and Beug (1992) offer a detailed picture based on Luttersee, and Beug (2016) provides evidence of four elm declines based on a synthesis of ten sites: of these declines the first three are related to human impact (but not necessarily synchronous at the various sites) while the last decline occurs alongside the rapid increase of *Fagus*. For Poland, the isopollen maps of Ralska-Jasiewiczowa (1983) indicate especially sharp elm declines in areas of fertile soils. In Denmark, at Hassing Huse Mose, Andersen and Rasmussen (1993) found four elm declines dated at 6480, 6080, 5820 and 5360 cal yr BP (5820 cal yr BP or 3870 BC corresponds to the onset of the Early Neolithic and 5360 cal yr BP or 3410 BC to the onset of the Funnel Beaker culture). For Denmark and northern Germany, the quantitative reconstruction of vegetation openness was, inter alia, related to the elm decline by Nielsen et al. (2012). For The Netherlands, Zolitschka (2013) summarizes: "There are two or three land-occupation phases in the Neolithic and Bronze Ages, that show but low values of terrestrial herbs. They are negatively correlated with *Ulmus*". In Ireland, as early as 1956, Mitchell (1956) presented several sites where, after the marked elm decline around 5800 cal yr BP, three additional declines followed, the youngest at AD 500. Smith and Goddard (1991) presented Sluggan Bog and put Mitchell's finding into "...a pollen zone scheme for the non-specialist". In northern Ireland, two elm declines dated to 5810–6065 cal yr BP and 4920–4795 cal yr BP were correlated with increased erosional input (He et al., 2014). In western Ireland the sharp elm decline occurs after the first cereal grains are recorded (Molloy and O'Connell, 1987, 1991; O'Connell et al., 1988). Groenman van Wateringe (1983) summarized 13 Irish sites and showed that agricultural activities had started before the strong elm decline. In eastern Scotland, multiple elm declines were dated to c. 5975, 5680 and 5120 cal yr BP, the stronger ones correlat-

Ulmus

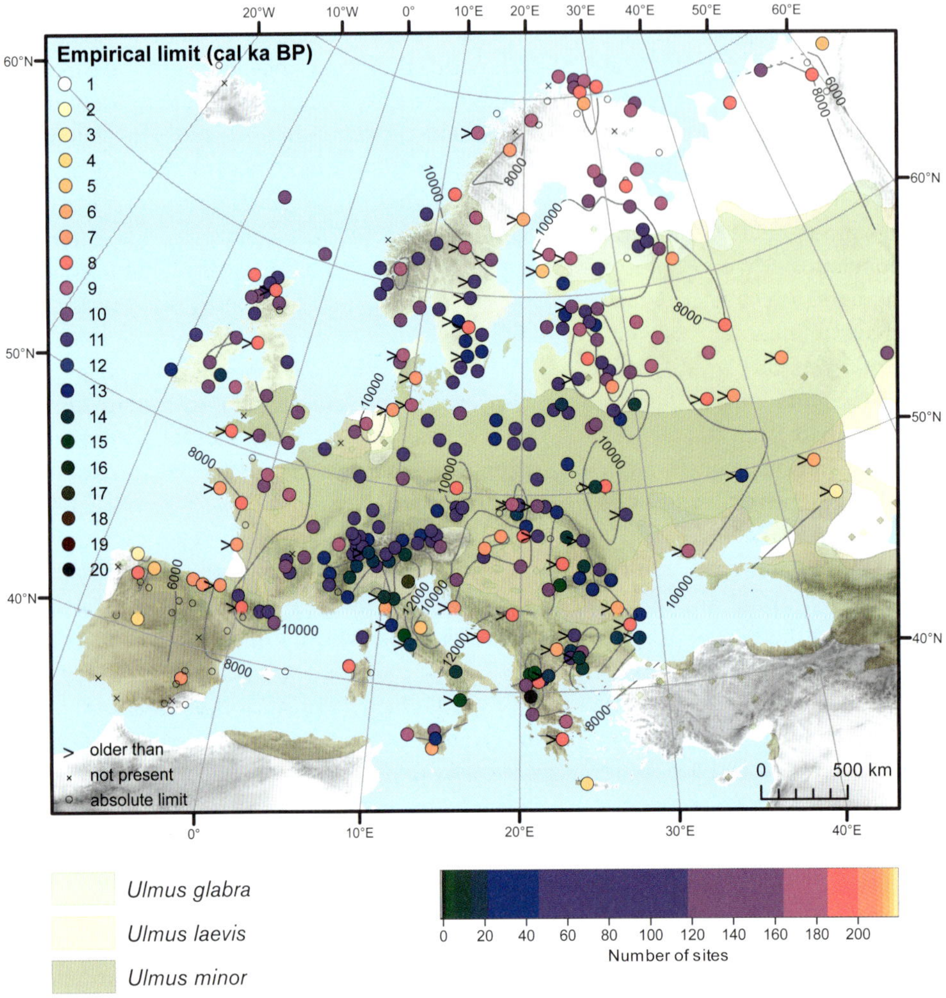

Fig. 4.4.3-18. Map showing the empirical limits of ***Ulmus*** in Europe. The coloured dots indicate the empirical limit in cal ka BP at selected sites, crosses refer to sites where the taxon was absent and small circles where it reached the absolute limit. Grey lines represent isochrones of empirical limit in 2000-year steps, based on nearest neighbour interpolation. Bar diagram shows stacked number of sites where the empirical limit was reached during the same 1000-year time interval, with colours according to the legend in the figure.

ed to erosional disturbances (Kaplan et al., 2011). In East Anglia, at Hockham, the first elm decrease was at about 6850 cal yr BP (Bennett, 1983a) after which the so-called Mid-Holocene 'elm-decline' occurred (c. 5100 cal yr BP), the studies of which are best summarized by Peglar (1993a, 1993b). "Much has been written about this event and several hypotheses proposed to explain it. These include climatic change (e.g. Iversen, 1941; Tauber, 1965; Frenzel, 1966); human activity (e.g. Iversen, 1949, 1960, 1973; Troels-Smith, 1960; Göransson, 1986); soil deterioration (Sturlodottir and Turner, 1985); and disease (e.g. Watts, 1961; Iversen, 1973; Rackham, 1980; Groenman-van Wateringe, 1983; Molloy and O'Connell, 1991). Other authors have advocated a combination of factors being responsible for the decline with different importance in different regions (e.g. Smith, 1961; Rackham, 1980; Göransson, 1986; Girling, 1988), including hydrological changes (Digerfeldt, 1997). Summaries of these various hypotheses are given by, among others, Ten Hove (1968), Iversen (1973), Rackham (1980), Huntley and Birks (1983), and Birks (1986a)." In the laminated section of the sediment from Diss Mere, Peglar (1993b) was able to show that the conspicuous decline happened within six years. She concludes: "The extreme rapidity of the elm decline strongly suggests that factors such as climatic change, felling by people, or soil deterioration were not responsible for the elm population crash" and "The form, magnitude, and rate of decrease of the *Ulmus* decline at Diss Mere are all consistent with the hypothesis of a pathogenic attack". She also successfully compares the dynamics of the elm decline at Diss Mere with the pollen declines of *Tsuga canadensis* and *Castanea dentata* in New Hampshire (Allison et al., 1986). A second argument in favour of the disease-hypothesis, in addition to the rapidity of the decline, is the fact that at a few sites the vector of the ascomycete fungus *Ceratocystis ulmi*, the bark beetles *Scolytus scolytus* and *S. multistriatus* (Rackham, 1980) were found as fossil elytra (Girling and Greig, 1985; Caspers and Freund, 1997). At West Heath Spa, Hampstead, London "...the remains of *Scolytus* at the site were discovered approximately 15 cm below the elm decline horizon" as stated by Girling (1988). *Scolytus* was recorded in the Neolithic settlement of Thayingen by its galleries in the bark (Guyan, 1955). The spores of the fungus most probably do not fossilize (personal communication by Bas van Geel, who thinks that ancient DNA may help). Girling (1988) explains: "In common with other Scolytidae, the larvae feed by excavating galleries between the bark and the sap wood, the radial pattern of these channels varying between species. *S. scolytus* almost always attacks dead wood, breeding in living trees only in unusual conditions. ... The fungus is introduced into the larval galleries. ... *C. ulmi* also travels via the root systems of suckered elms and such attacks are rapidly fatal to the clones". Rackham (1980) and Groenman-van Wateringe (1983) also explained that "... elms which shoot out vigorously after pollarding are especially prone to infection". It is difficult to resist the conclusion that the infected stone axes used by Neolithic settlers for girdling or pollarding may have sped up the process. It is clear, however, as Rackham (1980) and Tian et al. (2016) calculated, that leaf-foddering and pollarding alone could not have had such a strong effect on the elm populations considering the high number of trees to be shredded to make that effect and the low number of humans at that time. Rasmussen (1989, 1990) indicated that the remains of leaves and twigs in an archaeological context (mainly dung) include not only *Ulmus*, but also *Tilia, Fraxinus, Betula, Quercus, Acer* and more. A pathogen attack affecting *Ulmus*, but not the other genera, would explain why specifically elm declined so sharply.

Concerning the climatic aspects, Digerfeldt (1997), working in southern Sweden, found climatically controlled palaeohydrological changes that interacted with human impacts and possibly enhanced the susceptibility of *Ulmus* to the pathogen.

In conclusion, it may all depend on the definition of the 'the' in 'the elm decline' – and we may need to use the plural (Rackham, 1980). Throughout Europe the declines dated approximately between 6480 cal yr BP and 5360 cal yr BP seem to have the highest amplitude, the greatest rapidity, and the best acceptance, but before and after there were several minor elm declines. Moreover, in central and southern Europe the *Ulmus* declines were often associated with excessive anthropogenic burning and declines of other fire-sensitive trees such as *Tilia* and *Abies* (see chapter 4.10). The degree of elm recovery after declines is also far from uniform (Birks, 1986a). It may depend on soil fertility, on competition, on coppicing or on the elm species involved (e.g. a shift from *U. glabra* to *U. minor* being more abundant, as discussed by Bennett 1986a). In a comparison of Mid-Holocene elm declines with recent ones, Flynn and Mitchell (2019) show a high similarity of the dynamics at Diss Mere (Peglar, 1993b) with the Irish site Brackenstown between 1975 and 1978: this recent pandemic was caused by a different Ascomycete, *Ophiostoma novo-ulmi*, the minor decline in 1918 by the less virulent *O. ulmi* (Flynn and Mitchell, 2019).

4.4.4

Evergreen angiosperms

Buxus sempervirens

In the cosmopolitan family of the Buxaceae the genus *Buxus* consists of more than 40 species that are exclusive to the Old World. Two monoecious species occur in Europe: (1) *B. sempervirens* (common boxwood) is a subatlantic to submediterranean species with a discontinuous range with a north-eastern limit following a line from southern England over the Adriatic coast to North Macedonia (see Fig. 4.4.4-1) and (2) *B. balearica* (Balearic boxwood) which is restricted to the western Mediterranean, namely the southern coast of Spain, the Balearic islands and Sardinia (Rosselló et al., 2007).

Buxus sempervirens was long understood as a thermophilous shrub or small tree, but Di Domenico et al. (2012) summarized several physiological aspects of the species that makes it not only shade tolerant but also to some extent cold tolerant. Today it occurs in submediterranean forest communities, but in the north-western part of its range it is primarily found in the undergrowth of the Aceri monspessulani-Quercetum petraeae, further south in the Buxo-Quercetum pubescentis and in submontane beech forests (Carici-Fagetum, Buxo-Fagetum). In the south-east of its range *B. sempervirens* occurs in

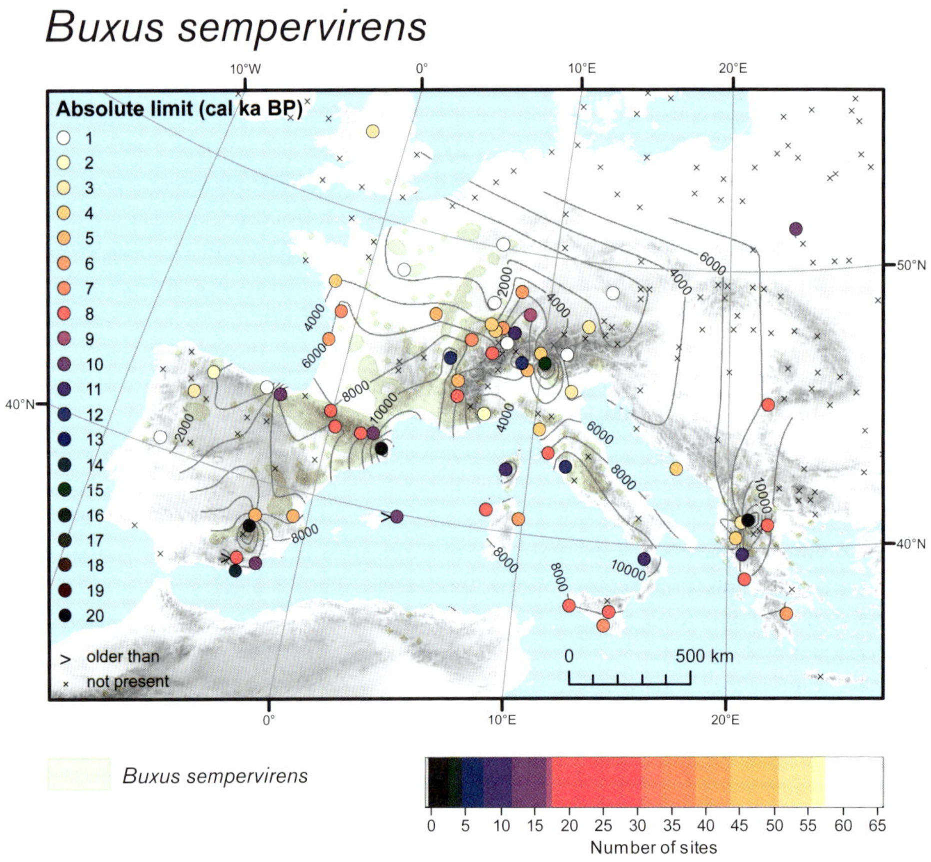

Fig. 4.4.4-1. Map showing the absolute limits of ***Buxus sempervirens*** and its current distribution in Europe. The coloured dots indicate the absolute limit in cal ka BP at selected sites, crosses refer to sites where the taxon was absent. Grey lines represent isochrones of absolute limit in 2000-year steps, based on nearest neighbour interpolation. Bar diagram shows stacked number of sites where the absolute limit was reached during the same 1000-year time interval, with colours according to the legend in the figure.

the Ostryo-Carpinion orientalis and in the Juglando-Platanetum orientalis, as well as in the *Pinus nigra* forests of North Macedonia and Greece (Horvat et al., 1974; Mayer, 1984; Oberdorfer, 1992). *Buxus sempervirens* is entomophilous and zoochorous.

Pollen morphology: the periporate pollen grains with vermiculate-rugulate or reticulate sculpture and small pores with diffuse limits and no annuli were first found in Hoxnian (Holstein interglacial) deposits in the British Isles (Histon Road: Hollingworth et al., 1950; Gort: Jessen et al., 1959) and only later were they found in Holocene deposits. Brückner (1993) provides criteria for the pollen morphology and taxonomy of 33 Eurasiatic taxa of the genus. The pollen productivity and dispersal of this entomogamous taxon are low and therefore even single findings may be interpreted as indicating a local occurrence, we therefore chose the absolute pollen limit for our isochrone map (Fig. 4.4.4-1).

Plant macrofossils such as wood, charcoal, twigs, leaves, fruits and seeds allow the identification of the genus. No indications for the distinction of the species have been found, but *B. sempervirens* seems most probable outside the western Mediterranean.

Macroremains were found from interglacial to archaeological (Upper Magdalenian, Neolithic, Bronze Age, Roman Age) deposits (see Godwin, 1956, 1975; Schweingruber, 1978; Vernet, 1980).

The full-glacial **refugia** of boxwood were most probably both in the Mediterranean region (e.g. Buraca Grande, Estremadura, Portugal, Upper Palaeolithic: Figueiral and Terral (2002) or submediterranean region (e.g. Canecaude, south-western France: Vernet, 1980) and in more temperate areas such as the Colli Euganei in northern Italy, dated between 30 and 23 cal ka BP (Kaltenrieder et al., 2009). This example from Italy illustrates the LGM bottleneck (see chapter 4.11). Several temperate and mediterranean species experienced strong range contractions during the LGM at 23 000–19 000 cal yr BP, if compared to previous periods of the full glacial. Our isochrone map (Fig. 4.4.4-1) suggests the presence of LGM refugia on the Balkan and Iberian peninsulas (black dots). Phylogenetics distinguish two major clusters, namely the European group including *B. sempervirens*, *B. balearica* and *B. colchica*, and an Asiatic group with at least seven species (van Laere et al., 2011; Thammina et al., 2016).

Range expansions and contractions: Late-glacial occurrences are known from north-eastern Spain (Menéndez Amor and Florschütz, 1963), Corsica (Reille, 1975), Calabria (Grüger, 1977) and north-western Greece (Bottema, 1974) but Late-glacial records were also found in the foothills of the south-western Alps (Clerc, 1988), the central southern Alps and Italy (Schneider, 1978; see Fig. 4.4.4-1). However, north of the Alps the record from the Oldest Dryas at Bibersee (Swiss Plateau) is in reworked material just above the till, together with several mesophilous tree taxa (van Leeuwen unpubl.; Di Domenico et al., 2012). The range expansion occurred in the first millennia of the Holocene – in the Pyrenees from east to west from the Early to Mid Holocene; in the French Massif Central, the Jura mountains and the Vosges until the end of the Early Holocene; in Italy and the Balkans during the Early Holocene, and in the lowlands to the north until the Mid Holocene. Late-glacial and Holocene maps are provided by Di Domenico et al. (2012). Remarkable is the so far singular finding about 300 km north of the modern northern limit in northern England (Oldfield and Stratham, 1963). Until the Mid Holocene, *Buxus* had apparently accomplished its spreading northwards. On the basis of our isochrone map we estimate its spreading rates to 50–190 m yr^{-1} (Table 4.11-1). Summarizing reviews of the Late-glacial and Holocene range history can be found in Lang (1970, 1992), Wegmüller (1984) and Di Domenico et al. (2012).

Buxus balearica played a major role at some sites on the Balearic Islands during the Early to Mid Holocene but it decreased rapidly around 5000–4000 cal yr BP. Yll et al. (1997) discuss the various climatic and anthropogenic reasons for its decline.

The wood of *Buxus* is highly appreciated for its visual and acoustic qualities and therefore used in sculpture and for making flutes and recorders.

Hedera

The genus *Hedera*, within the largely tropical family Araliaceae, consists of several Asian and "... approximately 12 species of ivies distributed in Eurasia, Macaronesia, and northern Africa" (Valcàrcel et al., 2003). *Hedera helix* sensu lato (common ivy), is widespread in Europe including the British Isles, but lacking from central and northern Fennoscandia and has an eastern range limit running from the eastern Baltic along the western border of Russia to the Black Sea (see Fig. 4.4.4-2).

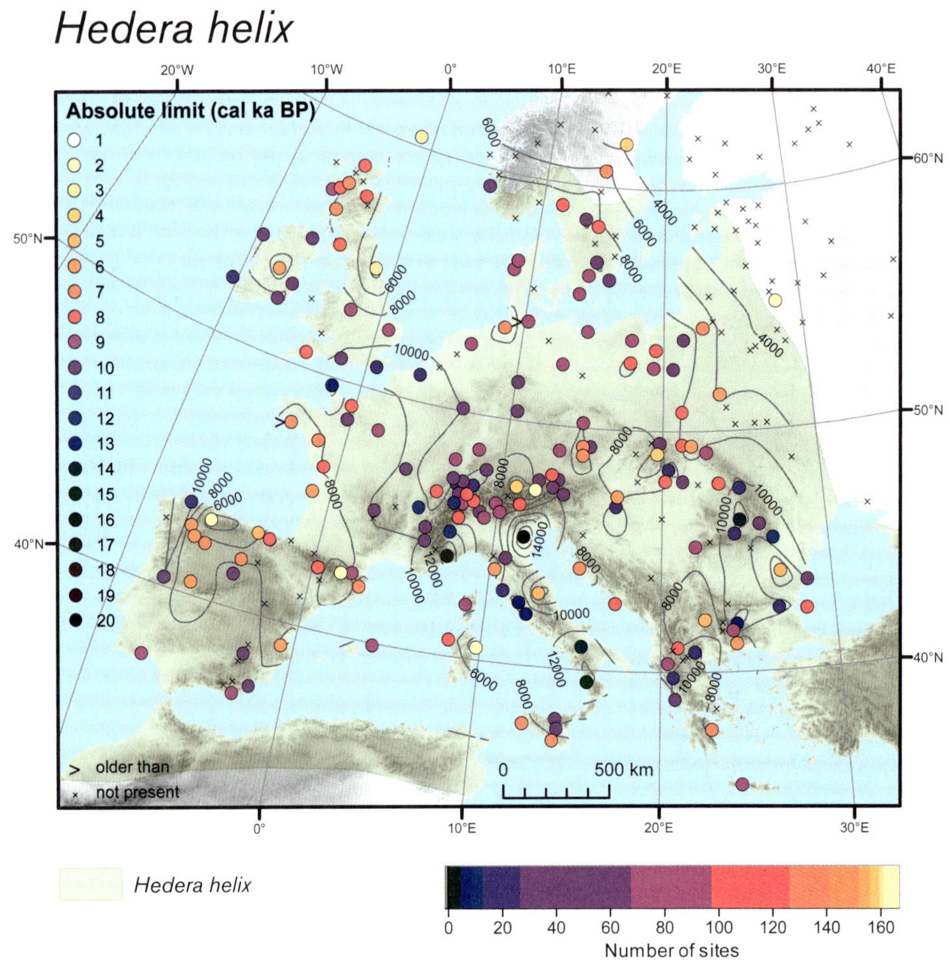

Fig. 4.4.4-2. Map showing the absolute limits of ***Hedera helix*** and its current distribution in Europe. The coloured dots indicate the absolute limit in cal ka BP at selected sites, crosses refer to sites where the taxon was absent. Grey lines represent isochrones of absolute limit in 2000-year steps, based on nearest neighbour interpolation. Bar diagram shows stacked number of sites where the absolute limit was reached during the same 1000 year-time interval, with colours according to the legend in the figure.

Hedera helix sensu lato is an evergreen liana widespread in summergreen forests. It prefers habitats with mild winters and high air humidity. Only its *forma arborea* (*forma arborescens*) reaches maturity with flowering and pollen production, otherwise it remains sterile when creeping near the ground, where it spreads only vegetatively. At such sites it is 'palynologically silent'. It flowers in autumn, is entomophilous and ornithochorous. The pollen of *Hedera* only permits identification to the genus. The very characteristic tricolporate pollen grains with an irregular reticulum that becomes finer towards the colpi were first recognized and described by Iversen (1941) but then found at numerous sites. Because the entomogamous genus does not produce large amounts of pollen, even single findings may indicate local presence (Bottema, 2001). We therefore use the absolute limit for our isochrone map (Fig. 4.4.4-2). This, however, may primarily concern horizontal pollen dispersal but not necessarily vertical dispersal, since single grains can be found at high elevation sites far outside the habitat of the plant. *Hedera* attracted a special interest in palaeoecology – together with *Ilex* and *Viscum* – when Iversen (1944) introduced the climatic envelope of modern limiting summer and winter temperatures to reconstruct temperatures for the Holocene. For *Hedera*, the limits were a minimum mean temperature of 13–15°C for the warmest month and of -1.5°C for the coldest month. These values are valid for the pollen-producing *forma arborea*. The modern range of *Hedera*, including the vegetative forms, extends a bit further eastwards into areas with colder winter temperatures and coincides largely with the European part of the Cf- and Cs-climates according to the Köppen climate classification. Generally the taxon is bound to a warm-temperate and humid climate without a drought season and only moderately cold winters. A more quantitative treatment can be found in Kühl et al. (2002) who introduced the probability density function (pdf) method for climate reconstruction (see Fig. 2.7-6).

Plant macrofossils such as wood, charcoal and seeds can be identified. In archaeological contexts it has been shown that *Hedera* has been used as leaf fodder since the Neolithic (Iversen, 1941; Rasmussen, 1989, 1990).

The full-glacial **refugia** are thought to be in the Mediterranean region: for example, at Monticchio, the record of ivy is strong during the early Weichselian (Allen and Huntley, 2000). The isochrone map (Fig. 4.4.4-2) suggests the presence of LGM refugia of *Hedera helix* on the Italian Peninsula (black dot) and all occurrences older than 15 000 cal yr BP (green dots reflecting Oldest Dryas) are restricted to the Italian Peninsula.

Phylogeography: Cytotaxonomic as well as chloroplast and nuclear evidence supports the view of Mediterranean refugia. High levels of diversity in chloroplast-DNA in the modern Mediterranean populations may reflect refugial areas, whereas the low cpDNA diversity north of the Pyrenees, Alps and Carpathians may indicate areas recolonized during the Early Holocene from a few refugia (Valcàrcel et al., 2003). In Europe the species-complex consists of diploid species (2n = 48, including *H. helix, H. azorica, H. canariensis, H. maroccana*), a tetraploid species (2n = 96, *H. hibernica*), hexaploid species (2n = 144, *H. maderensis* and *H. cypria*) and an octoploid species (2n = 192, *H. colchica*, see e.g. Vargas et al. 1999). Originally four chlorotypes (i.e. haplotypes of chloroplast DNA) were found, and this polymorphism does not coincide with the levels of ploidy. As the dendrogram by Vargas et al. (1999) and the dendrogram and map by Valcàrcel et al. (2003) demonstrate, a single chlorotype may occur in some diploids (*H. azorica, H. canariensis*) as well as in some polyploids (*H. colchica, H. maderensis*), while multiple chlorotypes again occur in both diploids (*H. helix* ssp. *helix, H. helix* ssp. *rhizomatiphera*) and polyploids (*H. hibernica, H. cypria*). Later, Grivet and Petit (2002) found 13 chloroplast haplotypes and their map similarly illustrates a high genetic diversity in the southern peninsulas, but also a west-east differentiation.

Range expansions and contractions: Late-glacial records are reported from north-eastern Spain (Menéndez Amor and Florschütz, 1963), from the margin of the southern Alps (Grüger, 1968) and from the Balkan Peninsula (Huttunen et al., 1992). Our isochrone map suggests that after 15 000 cal yr BP, likely in connection with climate warming of the Bølling-Allerød interstadial, *Hedera helix* spread in the Italian and Balkan peninsulas, and at c. 12 000 cal yr BP also in France and the Alps (Fig. 4.4.4-2). A further expansion north of the Pyrenees, the Alps and the Carpathians took place during the first millennia of the Holocene (similar to that of *Corylus*). The limits of the modern range were more or less reached by the end of the Early Holocene (c. 8000 cal yr BP), with exception of northern Europe, where the species expanded until c. 4000–3000 years ago. Long-distance dispersal and fast migration velocities are probable because the berries are only ripe in spring and may therefore be transported over long distances by birds migrating northwards. Indeed, our isochrone map suggests spreading rates of 185–650 m yr^{-1}, which is faster than for example *Quercus, Tilia* or *Fagus* (Table 4.11-1).

The later Holocene history of the genus is also of interest. According to the pollen records the frequencies of *H. helix* sensu lato increased up until the end of the Early Holocene and during the Mid Holocene (c. 8200–5000 cal yr BP), but has decreased during the Late Holocene, with the exception of the west of the British Isles (Lang, 1954; Stalling, 1987; Godwin, 1975). The lack of *Hedera* pollen on the European continent after c. 5000 cal yr BP was interpreted as an indication of decreasing winter temperatures and increasing continentality (Iversen, 1944; Godwin, 1975). Eastern parts of the modern range of *Hedera*, where only vegetative reproduction occurs, may therefore be considered a relict from the Mid Holocene. However, during the past 5000 years the species also strongly declined in warm southern Europe and the declines were often associated with human impact and burning (see chapter 4.10).

Ilex aquifolium

The genus *Ilex* in the partly evergreen, partly summergreen tropical-subtropical family of the Aquifoliaceae consists of at least 440 species out of which only two are widespread in Europe. *Ilex aquifolium* (English holly) occurs on the British Isles, in western and parts of central Europe and the Mediterranean area (see Fig. 4.4.4-3). *Ilex colchica* (colchic or Black Sea holly) has a range similar to *Hedera colchica* in the southern and eastern Black Sea area. More restricted are the ranges of *I. imerethica* and *I. stenocarpa* in the Caucasus. In Macaronesia (Canary Islands, Madeira, Azores), the genus is represented by *I. canariensis* and *I. perado* (*I. platyphylla*). These eastern (Caucasian) and western (Macaronesian) endemic groups are considered to be Tertiary relicts. For large parts of Europe, only *I. aquifolium* needs to be considered for the Late-glacial and the Holocene. The plant is entomophilous and ornithochorous.

Its pollen morphology is unique with three colpi and dimorphous clavae (big and small: Fægri and Iversen, 1989; Beug, 2015). The evergreen shrub or low tree of *I. aquifolium* is a characteristic member of the Atlantic–Mediterranean and subatlantic–submediterranean flora (Geoelement). Its eastern range limit is often used as an example of a climatically controlled limit because it coincides largely with the 0°C January isotherm, or – even better – with the isoline reflecting at least 345 days during a year with a minimum above 0°C (Hintikka, 1963).

Iversen (1944) could show that *I. aquifolium* is restricted to areas with January mean temperatures above -0.5 °C. Kühl et al. (2002) provide a gridded map overlaying January temperatures with the range of *I. aquifolium* and its pdf-curves for January and July temperatures: it is even more restricted to mild winter temperatures than *Hedera helix* (see Fig. 2.7-6).

Plant macrofossils (wood, charcoal, leaves, seeds) allow the identification to the genus as do the tricolpate pollen grains with their intectate-clavate sculpture. The first subfossil findings may be those of Fægri (1940) in south-western Norway.

The pollen productivity and pollen dispersal of this entomophilous (entomogamous) plant are low, probably even lower than for *Hedera* and *Viscum*. Therefore even single grains can be interpreted as representing a local occurrence. Accordingly, we use the absolute limit in our isochrone map (Fig. 4.4.4-3). To assess the first establishment of the plant needs high pollen sums.

Just as for *Buxus* and *Hedera* the full-glacial **refugia** of *Ilex aquifolium* are thought to be in the Mediterranean area. Single grains of *Ilex* were found in full glacial sediments on the Iberian Peninsula, for example on the Atlantic coast of Galicia (Area Longa, >45 ka BP: Gómez-Orellana et al., 2007) and in Mousterian deposits in Gorham's Cave (Carrión et al., 2008), in Italy at Lago Grande di Monticchio during the Odderade interstadial (between 64 and 74 ka BP: Allen and Huntley, 2000), at Cánolo Nuovo during the Hengelo interstadial (Calabria: Grüger, 1977) and at Lago di Vico (>38 ka BP: Magri and Sadori, 1999) and in Greece at Ioannina between 25.6 and 24.3 ka BP (Lawson et al., 2004). As in the case of *Buxus sempervirens* at Costa in Italy, the disappearance of *Ilex aquifolium* at Ioannina after c. 24 000 cal yr BP until c. 10 000 cal yr BP illustrates the LGM bottleneck, which likely led to strong or maximum range contractions of temperate and Mediterranean species (see chapter 4.11).

A west–east axis is also reflected in the **phylogeography**. Studies on chloroplast DNA by Rendell and Ennos (2003) provide support for two major refugial areas, one on the Iberian and one on the Apennine (possibly also Balkan) Peninsula. The two chloroplast haplotypes on the British Isles may reflect either two arrivals from the south-west and south-east or the fact that garden plants escaped and behaved like native populations.

Range expansions and contractions: Late-glacial findings are rare, but single pollen grains are known from southern Italy (Calabria: Grüger, 1977; Basilicata: e.g. Watts et al., 1996b). In agreement, our isochrone map (Fig. 4.4.4-3) suggests that *Ilex aquifolium* stands re-established in southern Italy during the Oldest Dryas (green dot).

During the Early Holocene findings are scattered, but in La Moura (south-western France) Oldfield (1964) found this characteristic grain just after the end of the Younger Dryas. Single grains are also recorded in the British Isles (Godwin, 1956) and in south-western Norway (Sollesnes and Fægri, 1951).

Ilex aquifolium

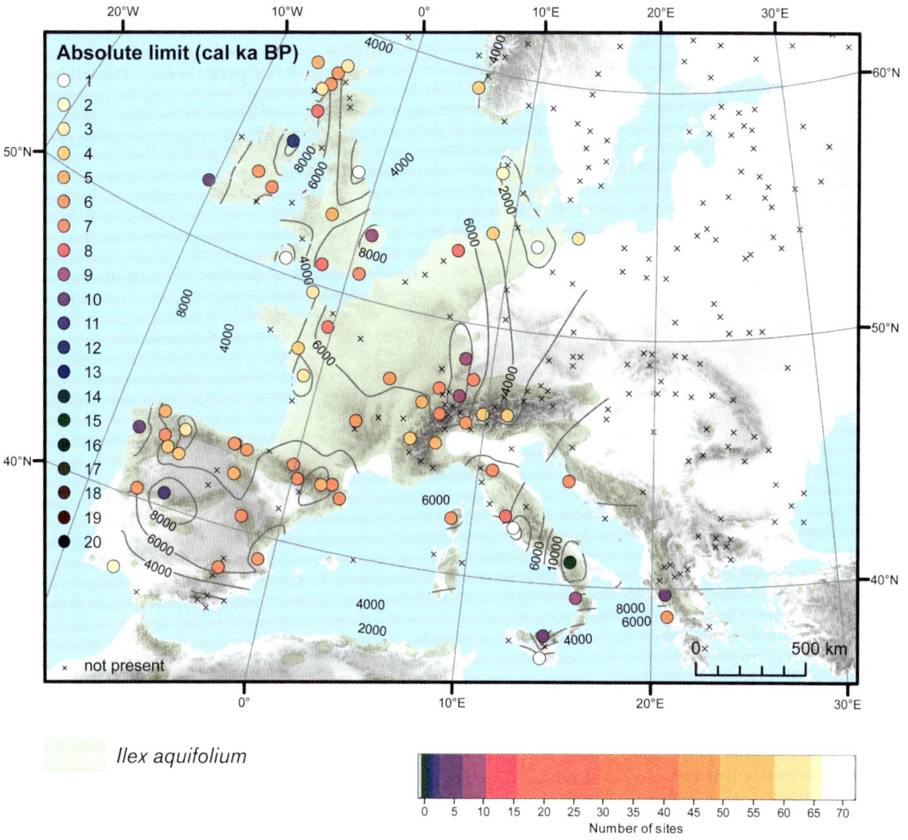

Fig. 4.4.4-3. Map showing the absolute limits of ***Ilex aquifolium*** and its current distribution in Europe. The coloured dots indicate the absolute limit in cal ka BP at selected sites, crosses refer to sites where the taxon was absent. Grey lines represent isochrones of absolute limit in 2000-year steps, based on nearest neighbour interpolation. Bar diagram shows stacked number of sites where the absolute limit was reached during the same 1000-year time interval, with colours according to the legend in the figure.

During the Holocene in Spain, continuous records of *Ilex* are characteristic for the period rich in mesic trees, for example at El Sabinar (6.6–4.8 cal ka BP: Carrión et al., 2004). Single early findings are also recorded near its modern eastern limit, for example in northern Switzerland (Rösch, 1983) or even outside the modern range in southern Moravia (Rybníček and Rybníčková, 1968; Rybnícková, 1974). The other early palynological records are not only spread through the Mid Holocene as for *Hedera* but also through the early part of the Late Holocene. In agreement with such local evidence our new isochrone map (Fig. 4.4.4-3) illustrates the Holocene species dynamics at the continental scale. During the Early Holocene, *Ilex aquifolium* spread on the Italian, Iberian and Balkan peninsulas and also in southern central Europe and the British Isles, with Ireland preceding England at the Late-glacial/Holocene transition. During the Mid and Late Holocene, the species finally spread in the Alps and to the north reaching Scotland and southern Scandinavia as well as to the east, attaining northern Germany (Fig. 4.4.4-3).

Similarly to the berries of *Hedera* and *Viscum*, the fruits of *I. aquifolium* (a drupe containing 2–4 pyrenes) are dispearsed by birds. This endozoochorous dispersal can lead to major jumps in seed dispersal and thus high velocities of migration. On the basis of our isochrone map we estimate the spreading rates of *Ilex aquifolium* to 110–340 m yr^{-1} (Table 4.11-1).

Interesting for climate history are the findings outside the modern range: in (1) southern Norway during the Late Holocene (Hafsten, 1956; Danielsen, 1970); (2) the Czech Republic during the Early Holocene (Rybníček and Rybníčková, 1968; Rybnícková, 1974); (3) Hungary during the Mid Holocene (Jarai-Komlodi, 1968) and (4) on the Adriatic coast during the Early Holocene (Beug, 1961, 1962). This means that during the Early to Mid Holocene holly could slightly transgress its modern eastern limit, possibly because of a shift in the 0°C or -0.5°C January isotherm.

Olea europaea

The genus *Olea* consists of approximately 40 taxa out of which *O. europaea* (olive) is the only one growing around the Mediterranean Sea and the only one that was domesticated for its fruits which are rich in oil (Zohary et al., 2012). The wild and domesticated forms are often understood as varieties, *Olea europaea* var. *sylvestris* and *O. europaea* var. *europaea*, respectively. Wild forms are also called oleaster, but it is difficult to distinguish original wild forms from feral (escaped) forms. Its flowers are hermaphrodite or polygamous: it is an anemophilous, zoochorous species. All forms prefer calcareous soils. Oleasters and most cultivars are self-incompatible (Villemur et al., 1984). A large number (hundreds) of cultivated forms exist. "Under cultivation, the reproductive biology of the tree changed. While wild olives reproduce by sexual reproduction from seed, domestic varieties are maintained by vegetative propagation and are, in fact, clones" (Zohary et al., 2012).

Pollen morphologically *Olea* identification can be problematic. If grains are corroded the separation from some Brassicaceae needs special care (Beug, 2015). The pollen morphology of *Olea* does not allow the distinction between wild and cultivated forms. In contrast, the separation may be possible on charred wood either by special morphometric characters in the wood anatomy or through elemental analysis (Terral, 1996; Terral and Arnold-Simard, 1996).

Wood charcoal of middle and late pleniglacial age was found at sites in both the western (Buraca Grande in western Portugal, Zafarraya and Gorham's Cave in southern Spain) and eastern Mediterranean (Santorini and Klisoura Cave) as reviewed and mapped by Y. Carrión et al. (2010d, based on Friedrich, 1978; Figueiral and Terral, 2002; Terral et al., 2005; Carrión et al., 2008; Ntinou, 2010). The frequencies of the records are higher in MIS 3 than in MIS 2. At some of the sites, olive remains (wood or endocarps) were found together with other Mediterranean taxa such as *Pistacia lentiscus*. Full-glacial pollen records with high abundances are recorded for the west (e.g. Carihuela Cave: Carrión and Sánchez-Gómez, 1992), as well as for the east (e.g. Island of Lesbos, Greece, near the Turkish coast: Margari et al., 2009). Our isochrone map (Fig. 4.4.4-4) suggests the presence of LGM **refugia** on the Iberian Peninsula (black and brown dots).

The indication of widespread full-glacial refugia indicated by macrofossils and pollen of olive is supported by **phylogeography** (Lumaret et al., 2000, 2004; Lumaret and Ouazzani, 2001; Breton et al., 2006, 2009; Besnard et al., 2009, 2013; Diez et al., 2015). Native oleasters were present both in the western and eastern Mediterranean: Breton et al. (2006) suggest seven refugial areas while Besnard et al. (2002a) suggest at least five refugial areas.

Range expansions: Late-glacial records of olive pollen are scarce but San Rafael (Almería, Spain: Pantaléon-Cano et al., 2003) and Balearic sites (Yll et al., 1995) are remarkable examples. Outside Europe, in the Ghab Valley (north-western Syria: Yasuda et al., 2000) scattered records are found in the Late-glacial, but they become continuous and rapidly increase during the late Younger Dryas, that is, in the pre-Pottery Neolithic A. Additionally, our isochrone map suggests the establishment of Late-glacial stands on the southern Italian Peninsula (Fig. 4.4.4-4).

Early Holocene records of wood charcoal were found in both western (Mesolithic sites on Iberia: see overview in Y. Carrión et al., 2010d), and eastern sites (e.g. pre-Pottery Neolithic B: Willcox, 1996). The isochrone map suggests that during the Early Holocene, *Olea europaea* spread on the Iberian and Italian peninsulas as well as in Sicily and Sardinia (empirical limit reached; Fig. 4.4.4-4). At many sites pollen records are sporadic during the Early Holocene but more continuous from the Mid Holocene onwards, for example at several Iberian sites (Y. Carrión et al., 2010d) and in Italy (Laghi Albano and Nemi, Lago

Olea europaea

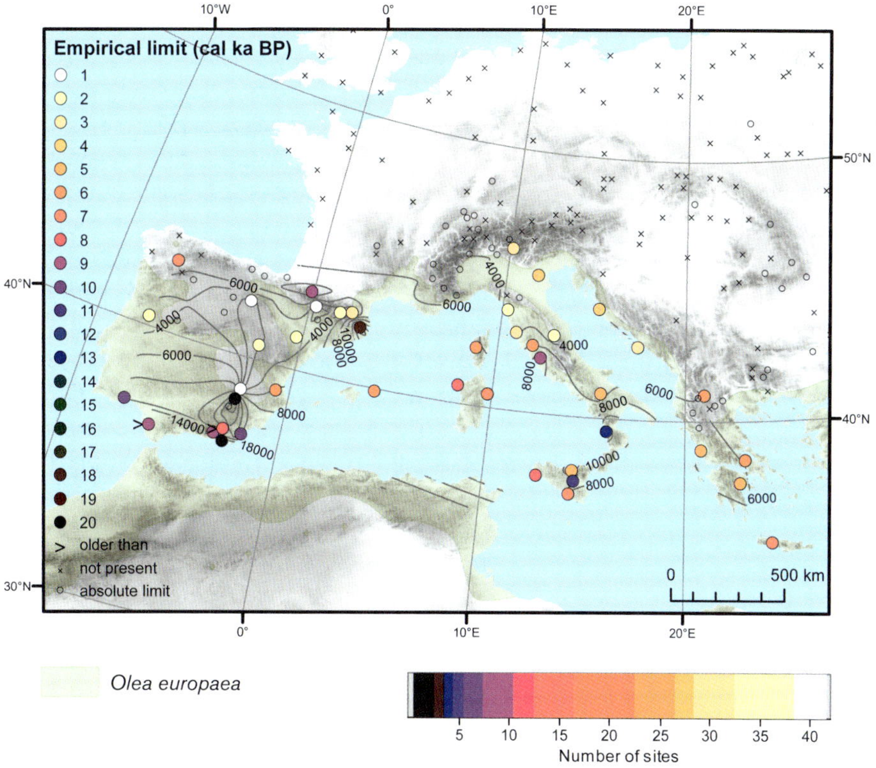

Fig. 4.4.4-4. Map showing the empirical limits of ***Olea europaea*** in Europe. The coloured dots indicate the empirical limit in cal ka BP at selected sites, crosses refer to sites where the taxon was absent and small circles where it reached the absolute limit. Grey lines represent isochrones of empirical limit in 2000-year steps, based on nearest neighbour interpolation. Bar diagram shows stacked number of sites where the empirical limit was reached during the same 1000-year time interval, with colours according to the legend in the figure.

di Vico: Lowe et al., 1996; Magri and Sadori, 1999; Lago Grande di Monticchio: Allen et al., 2002). In Greece, *Olea* generally spread only during the Mid Holocene. Oldest evidence includes a pollen record from the island of Zakynthos in the Ionian Sea where *Olea* reached its empirical limit at the transition from the Early to the Mid Holocene (c. 8200 cal yr BP; Avramidis et al., 2013). The maps based on EPD (Brewer et al., 2017) agree with our isochrone map showing expansions after c. 6000 cal yr BP. In contrast to natural Late-glacial and Early Holocene spreads of *Olea europaea* in thermomediterranean communities, such rather late expansions were probably associated with farming activities in the mesomediterranean vegetation belt (Mercuri et al., 2013). On the basis of our isochrone map we estimate the spreading rates of *Olea europaea* to 45–175 $m\ yr^{-1}$ (Table 4.11-1).

Human impact: as Zohary et al. (2012) state: "It is likely that olives were collected from the wild long before their domestication". Domestication is well documented in eastern Mediterranean regions such as Israel and Jordan after 6800 cal yr BP, even outside the natural range of *Olea* but later, during the Near Eastern and Greek Bronze Age (including Minoan Crete: Bottema and Sarpaki, 2003), records become more frequent (Blitzer, 1993; Kaniewski et al., 2012; Zohary et al., 2012; Valamoti et al., 2018). Langgut et al. (2019) use three criteria for dating the onset of horticulture in pollen diagrams: (1) a fairly rapid and sudden increase in *Olea* pollen,

(2) this increase unaccompanied by increases of sclerophyll taxa (which would rather indicate a climatic cause) and (3) this increase of *Olea* combined with archaeological data indicating local or regional increases of human populations. These authors also provide useful maps and histograms summarizing the early olive cultivation in the southern Levant, a second centre around the Aegean Sea (Crete) and then the spread westwards. However, the indication of cultivated *Olea europaea* in pollen records also requires its association with cultural indicators such as cereals or adventive weeds (Jouffroy-Bapicot et al. 2021, see chapter 4.5). Domestication in western Mediterranean regions is less well known but is indicated first by wood-anatomical differences of Neolithic charcoal in eastern Spain (Cova de Cendres, from Magdalenian to Bronze Age: Terral and Arnold-Simard, 1996) and second by morphometric differentiation of extant and archaeological endocarps (Terral et al., 2004). The separation into western and eastern types is in accordance with the phylogeography based on mitochondrial and chloroplast DNA (Besnard et al., 2002a, 2002b). In addition to local western domestication, van Zeist (1980) suggested imports from the east to Iberia from as early as the Phoenician. In addition, archaeobotany provides a number of archaeologically interesting contributions, for example, about crop processing (Margaritis and Jones, 2008a, 2008b).

As can be seen from the histograms in chapter 4.3.2, the olive records reflect the human impact at many Mediterranean lowland sites; the decorative wood, delicious fruits and precious oil are appreciated over millennia. Olive twigs also have an emblematic connotation in Greek, Roman, Jewish, Christian, Islamic and modern culture: the twigs are symbols of peace from Noah to Picasso. Despite its clear association with human impact, it is clear that before the onset of the Neolithic, *Olea europaea* was a natural component of European thermomediterranean vegetation, forming forests with *Quercus ilex*, *Q. coccifera* and other evergreen broadleaved trees (Figures 4.3-15, 4.3-16).

Phillyrea

Phillyrea (Oleaceae) comprises two species of hermaphrodite or androdioecious evergreen shrubs or small trees, *Phillyrea angustifolia* L. and *Phillyrea latifolia* L., which are distributed throughout most of the Mediterranean region and Asia Minor (Andrés, 2011). A third taxon named *Phillyrea media* L. (= *P. latifolia* ssp. *media* (L.) P. Fourn.) is now considered a synonym of *P. latifolia* because of the lack of consistency and notorious variability of its diagnostic features (Andrés, 2011). The distribution range of *P. angustifolia* embraces the western and central Mediterranean region, whereas *P. latifolia* has a circum-Mediterranean distribution range that extends over southern Europe, north-western Africa and south-western Asia (de Bolòs and Vigo, 1995; Andrés, 2011). Regarding their ecological requirements, both *P. angustifolia* and *P. latifolia* show no preference concerning soils, are light demanding and drought tolerant, and require mild winters to thrive (Costa et al., 2005; Rameau et al., 2008; Andrés, 2011). *Phillyrea* species are relevant components of evergreen broadleaved woodlands (*Quercus ilex*, *Q. suber*), xero-thermophilous pinewoods (*Pinus pinaster*, *P. halepensis*, *P. pinea*), and evergreen and semi-deciduous shrublands (for example maquis, pseudomaquis: Quézel and Médail, 2003; Costa et al., 2005; Rameau et al., 2008; Andrés, 2011; Čarni et al., 2018). *Phillyrea* species occur more rarely in the understorey of thermophilous deciduous oak woods (*Quercus pyrenaica*, *Q. canariensis*: Quézel and Médail, 2003; Costa et al., 2005).

The pollen of *Phillyrea* is very rounded both in equatorial (elliptic or circular) and polar (usually circular) views (P 15.5–25 µm, E 19.5–28.0 µm, P/E 0.78–1.02; Punt et al., 1991). Regarding the apertures, *Phillyrea* pollen grains are tri- or tetrazonocolporate with long or rather long, usually narrow and sunken colpi (Punt et al., 1991). However, the pores are poorly defined (Punt et al., 1991), which makes some authors describe this pollen type as just colpate (see Beug, 2004). The exine is rather thin to thick, with the sexine thicker than the nexine, with sexine 1 of short and distinct collumellae of similar thickness to sexine 2, and a semitectum of capita more or less circular in cross-section (Punt et al., 1991). The ornamentation is finely reticulate with thin and simplicolumellate muri narrower than the width of the lumina (Punt et al., 1991). The lumi-

Phillyrea

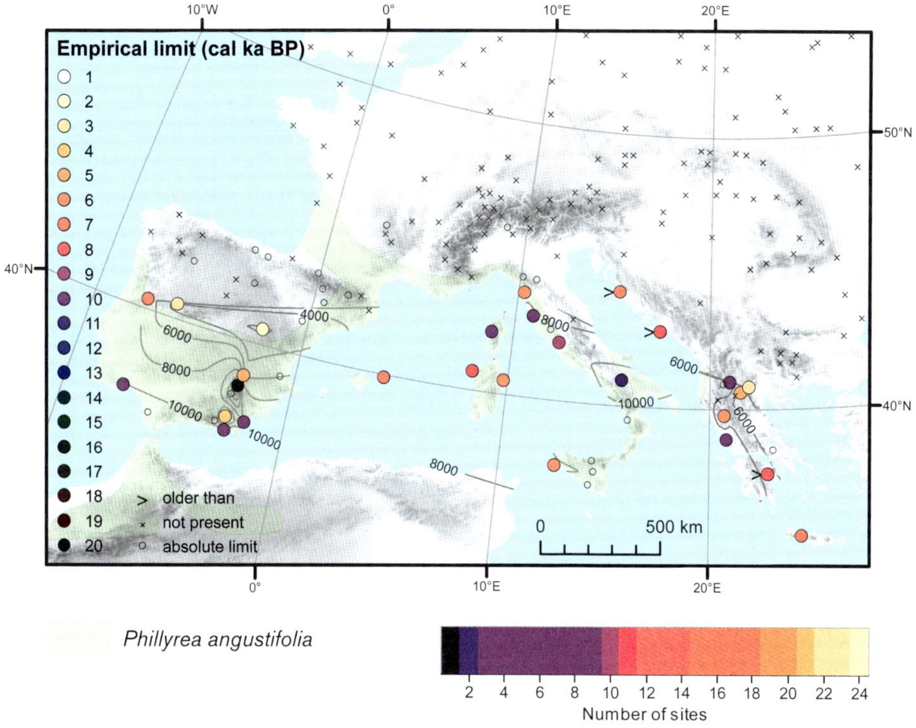

Fig. 4.4.4-5. Map showing the empirical limits of ***Phillyrea*** and the current distribution of *Phillyrea angustifolia* in Europe. The coloured dots indicate the empirical limit in cal ka BP at selected sites, crosses refer to sites where the taxon was absent and small circles where it reached the absolute limit. Grey lines represent isochrones of empirical limit in 2000-year steps, based on nearest neighbour interpolation. Bar diagram shows stacked number of sites where the empirical limit was reached during the same 1000-year time interval, with colours according to the legend in the figure.

na is irregular, angular and homogeneous in size (1–2(2.5) µm) throughout (that is, not decreasing in size towards the colpi: Punt et al., 1991). In summary, *Phillyrea* shows a distinct pollen morphology but similar to other Oleaceae, so pollen analysts must pay particular attention to identify it correctly.

Both species of *Phillyrea* are anemophilous and their seeds are dispersed by animals (Herrera et al., 1994; Traveset, 1994; Perea et al., 2013).

According to Schweingruber (1990), it is not possible to distinguish the two species of *Phillyrea* on the basis of their wood anatomy. In transverse section, *Phillyrea* wood is diffuse-porous, with the pores and vascular tracheids in dendritic groups surrounded by vasicentric paratracheal parenchyma, and 1–3-seriate marginal bands of parenchyma. In tangential section, the most conspicuous features are the 1–2(3)-seriate rays, ten cells high on average (up to 15 cells) with larger marginal cells. Finally, a radial section features heterogeneous rays with 1–2(3) rows of square and upright marginal cells. Ray-vessel pits are numerous and quite small, while perforation plates are simple and vessels have spiral thickenings. Ray and parenchymatic cells contain small crystals. It is important to note that the similar wood anatomy of *Phillyrea* and *Rhamnus* explains the abundance of *Rhamnus/Phillyrea* identifications in anthracological studies (Badal et al., 1994; Figueiral and Terral, 2002; Schroedter and Nelle, 2015; Connolly et al., 2019; Aranbarri et al., 2020).

The European palynological record suggests that *Phillyrea* LGM **refugia** were located in the warmest areas of the Mediterranean peninsulas (Tzedakis et al., 2013; Manzano et al., 2017). Indeed, *Phillyrea* pollen is recorded since the Last Glacial Maximum in the Cazorla-Segura Mountains of south-eastern Spain (Carrión, 2002b; Figures 4.3-15, 4.4.4-5). During the Early Holocene, *Phillyrea* also expanded at many other Mediterranean sites, reaching its northern range limits during the Mid Holocene (Fig. 4.4.4-5). As previously indicated, uncertainties in the identification of *Phillyrea* charcoal fragments limit the usefulness of the anthracological record to detect the occurrence of *Phillyrea* glacial refugia.

The available palaeobotanical data do not suggest any relevant change in the Holocene distribution of *Phillyrea* but show a Mid-Holocene establishment or, more probably, spread in most of the major Mediterranean islands, that is the Balearic Islands, Sardinia, Sicily and Crete (Figure 4.4.4-5). In general, *Phillyrea* is consistently present although usually at low abundances for most of the Holocene in thermomediterranean and mesomediterranean settings throughout Mediterranean Europe (Figures 4.3-15, 4.3-16, 4.3-17). Towards the cooler supramediterranean and oromediterranean vegetation belts, *Phillyrea* abundances largely decrease (actually many records are discontinuous; Figures 4.3-15, 4.3-16) and almost disappear in submediterranean records (see Figures 4.3-13, 4.3-14). A remarkable exception is the supramediterranean site Siles, which shows a rather continuous presence of *Phillyrea* pollen since the LGM and particularly during the Late-glacial (Carrión, 2002b; Figure 4.3-15). *Phillyrea* was usually part of evergreen-dominated or mixed deciduous-evergreen Mediterranean woodlands (*Quercus ilex* type, *Q. cerris/suber* type, *Juniperus*, *Pistacia*, *Olea*, *Q. robur* type) such as in Delphinos and Kournas, Lake Lerna, Biviere di Gela, Gorgo Basso, Sa Curcurica, San Rafael, Accesa, Massaciuccoli, Padul and Villaverde (Figures 4.3-15, 4.3-16, 4.3-17). *Phillyrea* was particularly abundant in some central and eastern Mediterranean sites such as Malo Jezero and Lake Voulkaria (Figures 4.3-16, 4.3-17), in woodlands dominated by deciduous trees and shrubs (for example *Quercus robur* type, *Ostrya* type, *Fagus*) but with notable abundances of evergreen elements (*Phillyrea*, *Pistacia*, *Olea*, *Quercus ilex* type) particularly at Malo Jezero.

Pistacia

Pistacia is a genus of Anacardiaceae consisting of about ten species of trees and shrubs broadly distributed in five disjoint areas, the main one comprising the Mediterranean region and western Asia (Yi et al., 2008; Al-Saghir and Porter, 2012). Three species, *Pistacia lentiscus* L., *P. terebinthus* L. and *P. atlantica* Desf., and a putative hybrid, *P. saportae* Burnat, are native to Mediterranean Europe (Yi et al., 2008; Al-Saghir and Porter, 2012). *Pistacia lentiscus* is an evergreen shrub or small tree distributed throughout the Mediterranean region and on the Canary Islands, where it grows on all sorts of soils in areas with mild winters (Güemes and Sánchez-Gómez, 2015). It is quite light demanding and drought tolerant (Rameau et al., 2008). In the Mediterranean region, *P. lentiscus* is a typical component of evergreen oak woodlands (e.g. *Quercus ilex*, *Q. suber*), xerophilous Mediterranean pinewoods (*Pinus halepensis*), mixed thermophilous evergreen forests, mixed *Q. ilex–P. halepensis* woodlands, juniper (*Juniperus phocenicea* ssp. *turbinata*, *J. oxycedrus* ssp. *macrocarpa*) woodlands on coastal sand dunes and scrub (maquis and garrigue), always in warm settings (Quézel and Médail, 2003; Costa et al., 2005; Rameau et al., 2008). *Pistacia lentiscus* gets rarer as seasonality increases and winters get harsher, but it is still present in some thermophilous deciduous oak woodlands (e.g. *Quercus pyrenaica* in southern Iberia: Costa et al., 2005). The deciduous *P. terebinthus,* a shrub or a small tree, is also quite widespread in the Mediterranean region, where one can find it in mesomediterranean and submediterranean shrublands, evergreen and deciduous woodlands, and on scree slopes, where it can be particularly abundant. Deciduous *Pistacia terebinthus* is also light demanding and drought tolerant but more resistant to frost than evergreen *P. lentiscus* (Rameau et al., 2008), so it may thrive at higher elevation and under more continental climates such as in inland Iberia and northern Greece. *Pistacia terebinthus* is relatively frequent in the understorey of submediterranean deciduous (*Quercus faginea*, *Q. trojana*) and evergreen (*Q. ilex*) oak woodlands and pinewoods (*Pinus nigra*, *P. halepensis*; Costa et al., 2005; Rameau et al., 2008). Likewise, *P. terebinthus* is not a rare component of maquis communities (Costa et al., 2005; Rameau et al., 2008). However, *P. terebinthus* may become particularly abundant, even dominant or co-dominant,

Pistacia

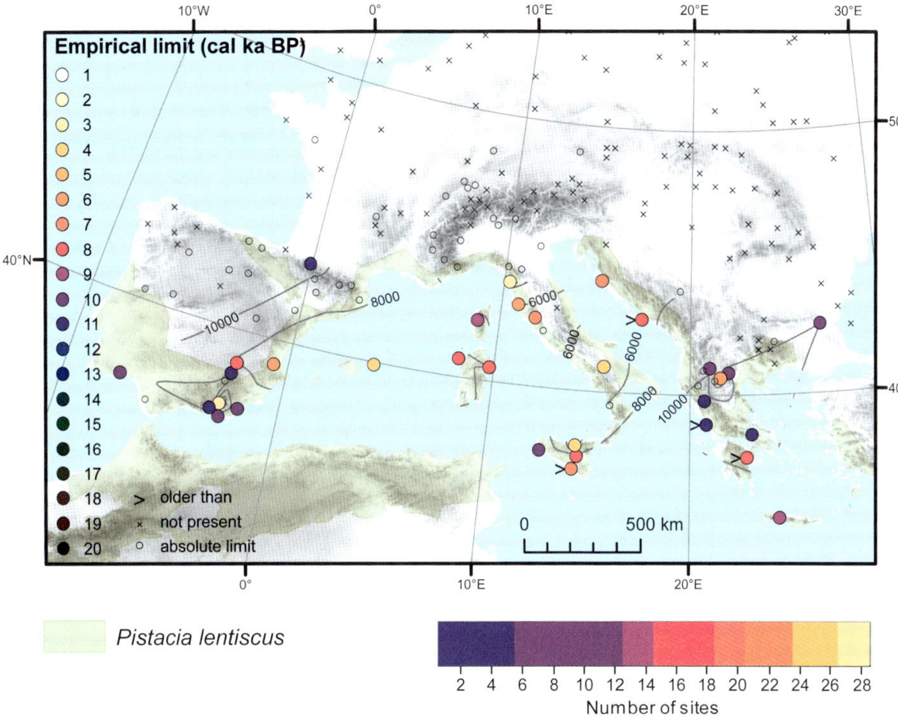

Fig. 4.4.4-6. Map showing the empirical limits of ***Pistacia*** and the current distribution of *Pistacia lentiscus* in Europe. The coloured dots indicate the empirical limit in cal ka BP at selected sites, crosses refer to sites where the taxon was absent and small circles where it reached the absolute limit. Grey lines represent isochrones of empirical limit in 2000-year steps, based on nearest neighbour interpolation. Bar diagram shows stacked number of sites where the empirical limit was reached during the same 1000-year time interval, with colours according to the legend in the figure.

on rocky habitats, steep slopes, scree and even cliffs (Costa et al., 2005; Rameau et al., 2008). In such habitats, it is not rare to find *P. terebinthus* in highly diverse mixed Mediterranean deciduous stands (*Sorbus aria, S. torminalis, Amelanchier ovalis, Acer opalus* ssp. *granatense*, *A. monspessulanum*, *Fraxinus ornus*, *Q. faginea*), for instance in the Baetic ranges of south-eastern Spain (Costa et al., 2005). *Pistacia palaestina* Boiss. has been cited for the islands of Crete and Chios in Greece (Zakynthinos and Rouskas, 1996), but the latest taxonomical proposals frame this taxon within *P. terebinthus* (Yi et al., 2008; Al-Saghir and Porter, 2012). Finally, *P. atlantica* is also a deciduous shrub or small tree with a wide distribution from the Canary Islands to Pakistan, growing in rather dry and usually warm habitats (Al-Saghir and Porter, 2012). In Europe, *P. atlantica* has a very narrow distribution restricted to several isolated populations in Greece (Quézel and Médail, 2003; Al-Saghir and Porter, 2012). This species is particularly resistant to drought, so it is a relevant component of the thermophilous maquis, mixed evergreen thermophilous woodlands and wooded steppes (Quézel and Médail, 2003).

Pistacia pollen is spheroidal (P/E 0.83–0.96), usually with an oval outline in equatorial view, reticulate to micro-reticulate (lumina 0.8–1.4(1.8) µm) and 28–37 µm wide (Beug, 2004). The pollen grains may be zonoporate (4–5 pores, 1–2 slightly outside the equator) or periporate (6–8 pores: Beug, 2004). The pores are oval or slit-like, with an irregular outline. The pore contour is often slightly protruding, like an operculum. The exine is 1.5 μm thick, columellate (Beug, 2004).

Both *P. lentiscus* and *P. terebinthus* are wind-pollinated (Herrera, 1988) and their seeds are dispersed by birds and mammals (Herrera, 1984; Perea et al., 2013).

Regarding the wood anatomy of *Pistacia*, this description is based on Schweingruber (1990). *Pistacia lentiscus* wood is mostly ring-porous in transverse section, usually with a single and loosely packed ring of pores in the early wood while in the late wood the pores are often arranged in radial rows and clusters. Parenchyma is sparse and mainly paratracheal. The tangential section features mostly 2–3-seriate rays which are up to 25 cells high while the uniseriate ones are 3–10 cells high. The marginal cells of the rays are often conspicuously enlarged and resin canals are present but at a variable density. Finally, the radial section shows heterogeneous rays, often with one marginal row of square or upright cells. These marginal cells of the rays contain large, prismatic and solitary crystals. Vessel-ray pits are large, perforation plates are simple and the vessels feature distinct spiral thickenings. The wood of *P. terebinthus* and *P. atlantica* is very similar but the rays are usually broader (3–5-seriate). According to Schweingruber (1990), *P. lentiscus* could be distinguished from *P. terebinthus* and *P. atlantica* on the basis of the arrangement of the pores, the shape of the rays, the resin canals and the frequency of crystals, but that would be difficult. Indeed, charcoal identifications to species level are not rare in the literature (Badal et al., 1994; Aranbarri et al., 2020).

Available palaeobotanical evidence, even if limited, indicates that *Pistacia* survived the last pleniglacial in small glacial **refugia** located close to the coast in the warmest areas of the Mediterranean region (Figueiral and Terral, 2002; Carrión et al., 2003b; Tzedakis et al., 2013; Manzano et al., 2017). More specifically, pollen and charcoal data from caves in south-eastern Spain show an almost continuous presence of *Pistacia* (in some charcoal fragments identified to species level, mainly *Pistacia lentiscus*) during MIS 3 and MIS 2, including the pleniglacial (Aura Tortosa et al., 2002; Carrión et al., 2003b, 2008; Vidal-Matutano et al., 2015; Manzano et al., 2017). A number of deciduous and evergreen woody taxa are also present in these pollen and charcoal assemblages. A similar pattern has been reported from Greece and the Dalmatian coast (Willis, 1994; Eastwood, 2004; Tzedakis et al., 2013).

Not much **phylogeographical** data are available regarding *Pistacia*, but Martínez-López et al. (2020) highlighted the importance of long-distance dispersal events mediated by migrating frugivorous birds in shaping the genetic structure to present-day *P. lentiscus* populations.

Range expansions: According to the available evidence (Fig. 4.4.4-6), the distribution range of *Pistacia* in Europe has been quite stable throughout the Holocene and *Pistacia* species have only been relevant in Mediterranean and submediterranean vegetation. *Pistacia* reaches the empirical limit, indicative of local presence, at the onset of the Holocene or during the Early Holocene over most of southern Europe (southern Iberia, Sicily, southern Balkans, Greece). The northern range limit in Italy and on the Dalmatian coast was attained during the Mid and Late Holocene. However, the relative paucity of well-dated and sufficiently long pollen records in the warmer areas of the Mediterranean region makes one take these observations with caution.

Concerning the postglacial dynamics of *Pistacia*, there are significant regional differences regarding its abundance and cohabiting taxa that are worth noting. In the western Mediterranean, only thermomediterranean and southern mesomediterranean sites show a consistently relevant presence of *Pistacia* whereas supramediterranean, oromediterranean and northern mesomediterranean sites mostly record scattered occurrences indicative of regional occurrence. South-eastern Iberian thermomediterranean sites such as San Rafael and Antas (Fig. 4.3-15) show a continuous presence of *Pistacia* since the onset of the Holocene and a remarkable spread during the Mid Holocene alongside other evergreen broadleaved trees and shrubs (e.g. *Olea* and *Quercus ilex* type) at the expense of steppe or semi-desert communities. Finally, the local extirpation of *Pistacia* occurred at c. 5000 cal yr BP. Other western Mediterranean thermomediterranean sites show a continuous occurrence of *Pistacia* at low abundances throughout the Holocene without remarkable changes (see Guadiana, Navarrés or Menorca; Fig. 4.3-15). Southern west Mediterranean mesomediterranean sites such as Padul and Villaverde show a continuous presence of *Pistacia* at variable abundance over the entire Holocene in mixed oak forests (with pines at Villaverde; Fig. 4.3-15). *Pistacia* is also much better represented at warm and coastal than at cooler and inland sites in the central Mediterranean. Island thermomediterranean sites show striking abundances of *Pistacia*, particularly in Sicily (Biviere di Gela, Gorgo Basso) where *P. lentiscus* seems to have dominated the coastal vegetation together with *Olea europaea* and *Q. ilex* until the past 1000–2000 years (Fig. 4.3-16). In Sardinia, the Sa Curcurica record suggests that *Pistacia* was a prominent component of the understorey of *Erica arborea*, *Q. ilex* and *Q. suber* woodlands but never dominated (Fig. 4.3-16). Mesomediterranean, supramediterranean and oromediterranean Italian pollen sites record continuous but low *Pistacia* pollen percentages, suggestive of the regional presence of *Pistacia* at low abundances. Interestingly, at Malo Jezero on the Dalmatian coast (Croatia; Fig. 4.3-16) *Pistacia* has moderate abundances for most of the Holocene, increasing remarkably from c. 2500 cal yr BP onwards. Here, *Pistacia* grew in open habitats at the beginning of the record and since c. 6500 cal yr BP with *Q. ilex* and *Ostrya*. Finally, the eastern Mediterranean sites show a continuous presence of *Pistacia* but never dominance or co-dominance (see Delphinos and Kournas, Lake Lerna or Lake Voulkaria; Fig. 4.3-17). Nevertheless, it is worth noting that the Lake Voulkaria pollen record largely resembles Malo Jezero with respect to the moderate abundance of *Pistacia* throughout most of the Holocene, particularly during the Early Holocene (Figures 4.3-16, 4.3-17).

Evergreen oaks: *Quercus ilex, Q. coccifera, Q. suber*

In Europe, the three main evergreen oaks are *Quercus ilex, Q. coccifera* and *Q. suber*. In addition *Q. alnifolia* is endemic to Cyprus. *Quercus suber* has a western and central Mediterranean distribution range: as a distinctly acidophilous species it is widespread on the western half of the Iberian Peninsula where soils over siliceous rocks are common, but restricted to rare siliceous outcrops in eastern Spain and further east, while it is abundant in Sardinia and rather common in several areas of Sicily, Corsica and Italy. *Quercus suber* can hybridize with *Q. ilex* (e.g. Elena-Rossello et al., 1992). *Quercus ilex* is a major component of the Mediterranean forest and maquis vegetation. It is widespread on the Iberian Peninsula except for Cantabria and Catalonia (*Q. ilex* ssp. *rotundifolia* or *Q. rotundifolia*). In southern France and Italy, *Q. ilex* occurs primarily near the coasts but also up to mid elevations. In the more continental eastern Mediterranean it is restricted to the coastal vegetation, for instance a narrow strip at the eastern Adriatic coast but more widespread in Greece. It is name-giving to the Quercion ilicis of the central and northern Mediterranean and important in the Oleo-Ceratonion of the southern Mediterranean vegetation zones (map 14 in Horvat et al., 1974; Bohn et al., 2003). For the ecological requirements of *Q. ilex* the symposium volume edited by Romane and Terradas (1992) may be consulted. *Quercus coccifera* and the related *Q. calliprinos* are distributed in the western and eastern half of the Mediterranean region, respectively. Taxa with coriaceous leaves persisting over winter and replaced in spring are called semi-evergreen (e.g. *Q. crenata, Q. faginea. Q. macrolepis, Q. trojana*).

By pollen morphology the evergreen species *Q. ilex* and *Q. coccifera* are usually distinguished together from deciduous taxa as *Q. ilex* type or *Q. ilex/coccifera* type. The pollen of the semi-evergreen taxa is separated by some but not all authors as *Q. cerris* type, *Q. crenata* type, *Q. suber* type or *Q. pseudosuber* type. Descriptions and keys can be found in Planchais (1962), Smit (1973), Colombo et al. (1983), Benthem et al. (1984), Reille (1992b) and Beug (2015). Some authors distinguish only two pollen-morphological types, namely 'deciduous versus evergreen', thus evergreen *Q. suber* would likely be included in 'deciduous oaks', because its pollen type is morphologically closer to that of deciduous oaks than to the *Q. ilex* type. An 'empty column' for *Q. suber* in the histograms of chapter 4.3.2 does not necessarily mean the absence of cork oak – it may be included under '*Q. robur* type' (for Iberian to Apennine peninsulas). In cases of poor pollen preservation, even the distinction between summergreen and evergreen oaks is not always possible (e.g. at Kopais, Greece; Turner and Greig, 1975).

In **phylogeography**, *Q. suber* is strongly separated from other taxa when allozymes are considered (Toumi and Lumaret, 2001). The distribution of its chloroplast haplotypes shows distinct longitudinal bands (Lumaret et al., 2005) that Magri et al. (2007) explain by the break-up of microplates during the Oligocene and Miocene: this spatial pattern may therefore be of Tertiary origin. However, multiple Quaternary refugia for *Q. suber, Q. ilex* and *Q. coccif-*

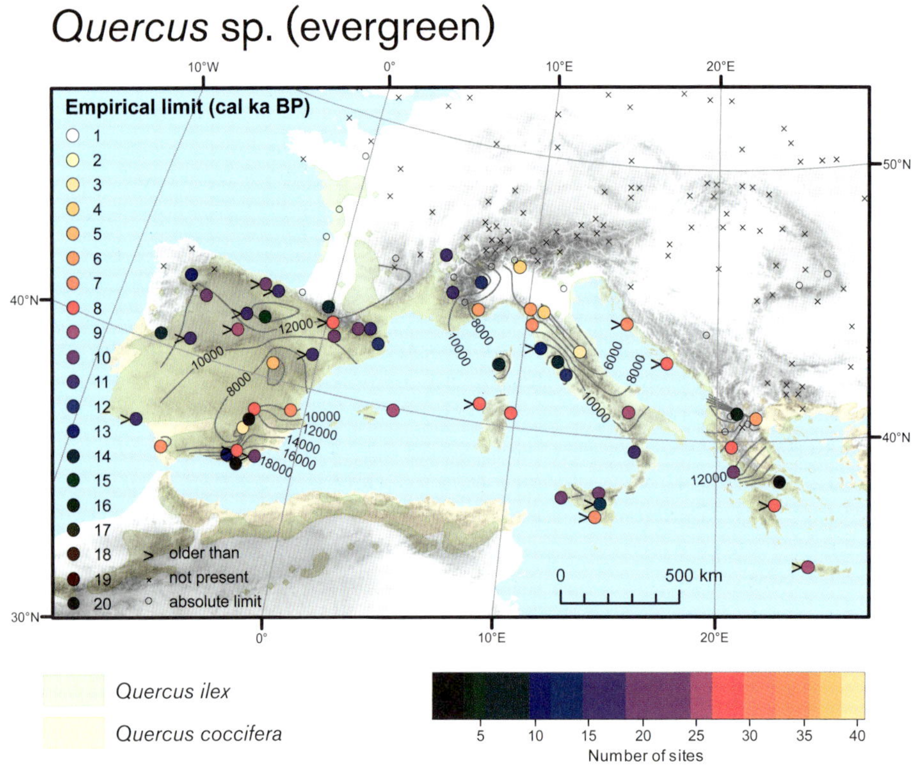

Fig. 4.4.4-7. Map showing the empirical limits of ***Quercus ilex*** type and the current distribution of *Quercus ilex* and *Quercus coccifera* in Europe. The coloured dots indicate the empirical limit in cal ka BP at selected sites, crosses refer to sites where the taxon was absent and small circles where it reached the absolute limit. Grey lines represent isochrones of empirical limit in 2000-year steps, based on nearest neighbour interpolation. Bar diagram shows stacked number of sites where the empirical limit was reached during the same 1000-year time interval, with colours according to the legend in the figure.

era (López de Heredia et al., 2007) are also indicated by molecular and palaeoecological evidence. *Quercus ilex* shows a high diversity of chloroplast DNA with the most ancestral lineages in the south-east (Crete and Greece) and more recent and numerous haplotypes in central and western Mediterranean regions (Lumaret et al., 2002; Petit et al. (2005).

The pleniglacial **refugia** are not known for all Mediterranean oak species. For *Q. suber*, records between 36 000 and 31 000 cal yr BP were found at Navarrés (Valencia, Spain: Carrión and van Geel, 1999). A review of 39 terrestrial (partly archaeological) and nine marine records on the Iberian Peninsula for *Q. ilex* is compared with phylogeographic maps by López de Heredia et al. (2007). At Padul, a nearly continuous record of *Q. ilex* is well dated between c. 34 and 16 cal ka BP (Pons and Reille, 1988). For the Apennine Peninsula, Follieri et al. (1998), in their Fig. 3, provide an overview of the substantial occurrences of *Q. ilex* type after the Eemian. Also at Ioannina (north-western Greece, 470 m asl: Tzedakis et al., 2002; Lawson et al., 2004), after the high values of the Eemian, evergreen oaks are recorded as minor occurrences throughout the full and late-glacial periods until the rise in the Holocene. However, our isochrone map suggests the presence of LGM refugia of evergreen oaks on the Balkan and Iberian peninsulas (black dots; Fig. 4.4.4-7).

Range expansions and contractions: during the Late-glacial high pollen percentages of *Q. ilex* and low but nearly continuous values of *Q. suber* are recorded at Padul (Pons and Reille, 1988) and in at least three marine cores, as reviewed by Carrión et al. (2000). Charcoal of both *Q. ilex* and *Q. suber* of final Magdalenian ages was found in Cabeço de Porto Marinho and Buraca Grande (Portuguese Estremaduro: Figueiral, 1995; Figueiral and Terral, 2002). In the histograms of chapter 4.2.2, early Late-glacial records of the *Q. ilex* type with high values are shown for thermomediterranean sites such as San Rafael or Menorca (Fig. 4.3-15). High and continuous values are then found during the Late-glacial interstadial. Low and scattered values are recorded at higher elevations. In the Italian crater lakes, evergreen oak pollen is only sporadically found during the Late-glacial (e.g. Follieri et al., 1988, 1993; Magri, 1999; Magri and Sadori, 1999; Allen et al., 2002). The same is true for Greek sites, where evergreen oaks are either missing or sporadic during the Late-glacial (e.g. Ioannina; Lawson et al., 2004). However, our new isochrone map (Fig. 4.4.4-7) suggests that during the Late-glacial evergreen oak expanded on the Iberian, Italian and Balkan peninsulas as well as on Sicily and Corsica (green dots). In some cases evergreen oaks (most likely *Q. ilex*) reached northermost positions in Italy and northern Iberia already at the end of the Late-glacial c. 12 000 years ago (blue dots).

At numerous sites the pollen record of the *Q. ilex* type either starts or becomes continuous during the Early Holocene. Maximum pollen percentages of the *Q. ilex* type were found on the Iberian Peninsula at sites in the thermomediterranean belt of the south-east, for example at San Rafael (or Antas, Almería) between c. 20 000 and 5080 cal yr BP (or 7200 cal yr BP at Antas: Pantaléon-Cano et al., 2003). The Mid Holocene is characterized at these Spanish sites by a sharp decline of evergreen and deciduous oaks which is interpreted in terms of an aridification between c. 6800 and 4500 cal yr BP (sharpest between c. 5700 and 5200 cal yr BP), a phenomenon also discussed by Pantaléon-Cano et al. (2003) for other regions within and beyond the Mediterranean region (e.g. Yll et al., 1997). For many areas the interaction between drier climates and human impact has been discussed. For the Apennine Peninsula, Magri et al. (2015) present a fine set of Holocene maps that show (1) the presence of evergreen oaks already at the beginning of the Holocene, (2) maximum values south of the Po Plain around 6 ka cal BP and (3) high values (>30%) at some coastal sites.

In the eastern Mediterranean region the pollen values of the evergreen oaks are generally low, but increase with human impact (e.g. Lake Voulkaria: Jahns, 2005). Similar increases with stronger human impact are also documented in southern Italy, such as near Naples for 800 BC–AD 800 by Grüger and Thulin (1998). However, declines of evergreen oaks are also documented together with increasing charcoal, for example, at Lago Battaglia on the peninsula of Gargano by Caroli and Caldara (2007). Again the alternative explanations or their interactions are discussed: human impact and/or aridification. For the north-west Italian site of Lago Massaciuccoli, statistical analyses of a core contiguously sampled for pollen and charcoal led to the unexpected conclusion on anthropogenic burning "...that *Quercus ilex*, which is currently considered a fire-adapted tree species, was also disadvantaged by fire..." (Colombaroli, 2007). Moreover, due to its low flammability *Quercus ilex*, together with *Olea europaea*, may have significantly reduced fire incidence under natural conditions (see chapter 4.5.8). For the western coast of Corsica, charcoal stratigraphy in soils (pedanthracology) indicated a shift from *Pinus nigra* ssp. *laricio* with an understorey of Ericaceae (*Erica arborea, Arbutus unedo*) to primarily *P. pinaster* and *Q. ilex* between c. 2000 and 1000 cal yr BP (Carcaillet et al., 1997). The authors propose the 'allelopathic inhibition hypothesis' that increasing anthropogenic fires destroyed the thick litter of pines and Ericaceae that earlier had allelopathic effects unfavourable to the germination of *Q. ilex*. In a synthesis for the western Mediterranean, Reille and Pons (1992) show that *Q. ilex* was present during the Late-glacial interstadial and the Early Holocene but more frequent in semi-arid areas (e.g. Padul: Pons and Reille, 1988) than in sub-humid to humid areas (southern and south-eastern France and eastern Pyrenees). Reille and Pons (1992) summarize: "... in the northern part of the western Mediterranean Basin and in the mountains of North Africa the establishment of the *Quercus ilex* and *Quercus suber* forests occurred after 6000 BP, at times varying from one region to another, and always as a result of human impact on nature". This view is in strong contrast with the natural spread of *Q. ilex* recorded at the European scale before the onset of the Neolithic (Fig. 4.4.4-7), it might be thus reconsidered with the new evidence that was gained during the past 30 years of research (see chapters 4.3.2 and 4.5.8). A much younger change in holm oak occurred in the opposite direction after the Second World War because of "... the utilisation of fuel oil as cheap fuel, which caused coppices to be left uncut, numerous *Quercus ilex* forests began to evolve floristically into *Quercus pubescens* forests ..." (Reille and Pons, 1992). On the basis of our isochrone map we can estimate the spreading rates of evergreen oaks to 100–160 m yr^{-1} (Table 4.11-1).

Viscum album

The genus *Viscum*, in the largely hemi-parasitic family of the Viscaceae (formerly Loranthaceae), consists of about 100 species in the Old World, most of them in Africa and Madagascar, but also in southern Asia and Australia (Zuber, 2004). Only two species have expanded their range into Europe. *Viscum album* (mistletoe) can be subdivided into four subspecies (see Fig. 4.4.4-8) and also occurs through Eurasia to Japan. *Viscum cruciatum* is restricted to the European south-west (Portugal, southern Spain). Among the other hemi-parasitic Loranthaceae, *Loranthus europaeus* (oak mistletoe) occurs in eastern central Europe, south-eastern Europe and the Apennine Peninsula. In the Mediterranean area, the related taxon *Arceutobium oxycedri*, living on juniper, occurs. This species was recorded for the last pleniglacial in Padul, southern Spain (Pons and Reille, 1988).

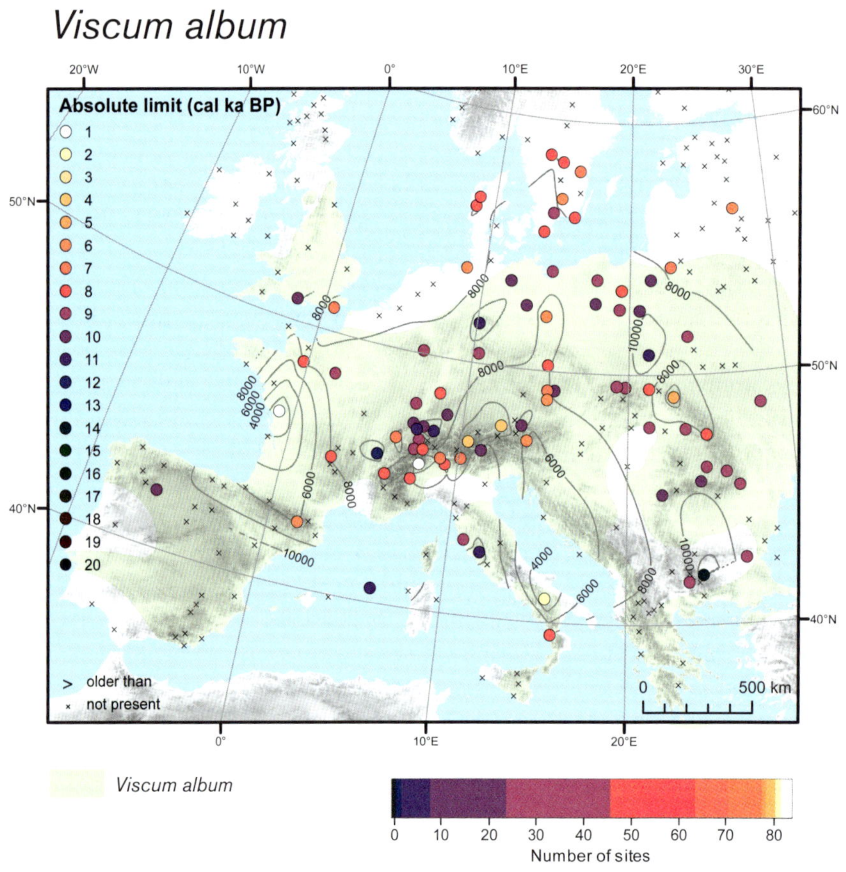

Fig. 4.4.4-8. Map showing the absolute limits of ***Viscum album*** and its current distribution in Europe. The coloured dots indicate the absolute limit in cal ka BP at selected sites, crosses refer to sites where the taxon was absent. Grey lines represent isochrones of absolute limit in 2000-year steps, based on nearest neighbour interpolation. Bar diagram shows stacked number of sites where the absolute limit was reached during the same 1000-year time interval, with colours according to the legend in the figure.

Viscum album is dioecious and the subspecies can only be identifed by their berries, thus male plants cannot be distinguished morphologically. It is entomophilous and ornithochorous. The subspecies are host-specific (see Zuber and Widmer, 2000; Zuber, 2004).

- *V. album* ssp. *album* on dicotyledonous trees (often on *Tilia* or *Populus*, but also on *Quercus, Ulmus* and *Fraxinus*; never on *Fagus*).
- *V. album* ssp. *abietis* on *Abies*.
- *V. album* ssp. *austriacum* on *Pinus* spp. and rarely on *Picea* or *Larix*. In spite of the large range of *Pinus sylvestris*, this mistletoe occurs only in central Europe and very rarely in southern Europe. In the Alps it may reach an elevation of c. 800 m asl.
- *V. album* ssp. *creticum* (endemic to Crete) on *Pinus halepensis* ssp. *brutia*.

Iversen (1944) showed convincingly that *Viscum album* is restricted to areas with rather warm summers where July mean temperatures are not below 15.5°C in oceanic regions and 18–19°C in continental regions. However, January mean temperatures as low as -8°C are tolerated. Kühl et al. (2002) include the species in their development of quantification of climate reconstruction by the probability density function method for indicator taxa (see Fig. 2.7-6).

Plant macroremains such as wood and twigs, that are occasionally found in Neolithic settlements, permit the identification of the species (Schweingruber, 1978). The pollen grains are (like *Ilex*) tricolpate with an intectate-clavate sculpture, but with a narrow cylindric shape and less dense clavae. Iversen (1941) provided the first records. Pollen morphological differences between the subspecies are not known but would be of interest. Pollen productivity of this entomophilous-amphiphilous taxon is low: therefore curves in pollen diagrams are usually not continuous and even single grains may indicate local presence. We therefore use the absolute limit in our isochrone map (Fig. 4.4.4-8).

The full-glacial **refugia** may have been in south-western Europe (MIS 2 according to González-Sampériz et al., 2010) and south-eastern Europe, (e.g. at Kupena according to Huttunen et al., 1992).

Phylogeographic studies of the subspecies (except *V. album* ssp. *creticum*) are provided by Zuber and Widmer (2009) on chloroplast haplotypes. Their maps show the highest haplotype diversity for *V. album* ssp. *album* in the east, south-east and south, for *V. album* ssp. *abietis* in the south-east and east and for *V. album* ssp. *austriacum* in the south of Europe. This largely confirms the view of palaeoecologists that the refugia were in south-eastern Europe. Asia Minor has special haplotypes in all three subspecies.

Range expansions and contractions: Late-glacial occurrences of *Viscum* in southern Europe are often erratic, but occasionally present on all the southern peninsulas: in Spain at, for example, Cova de les Cendres (Badal, 2006), in Italy at Lago di Accesa (Allerød: Drescher-Schneider et al., 2007b) and in Bulgaria at Kupena (late Allerød: Huttunen et al., 1992). In central Europe, at many sites, single grains are recorded in the bottom samples of early Late-glacial age or just above the till. Together with single grains of *Tilia* or *Ulmus* or even Tertiary spores, they obviously indicate reworked material (see figures for Nussbaumersee, Bibersee, Lac des Boites, Löddigsee, Nagy-Mohos in chapter 4.3). However, our isochrone map (Fig. 4.4.4-8) suggests the Late-glacial presence of *Viscum album* in the Balkan peninsula (green dot). Towards the end of the Late-glacial the species had expanded to the Rhône valley in southern France (blue dot).

During the first three millennia of the Holocene the expansion of the range into central and western Europe took place rapidly (Fig. 4.4.4-8). We may give two examples. (1) In the central and northern Alps, Welten (1982b) recorded *Viscum* (without reworked material) already during the Early Holocene. (2) In the Iberian Central Range at El Maíllo at 1100 m asl, nearly regular Holocene occurrences were found until c. 7500 cal yr BP (Morales-Molino et al., 2013). The rapid expansion during the Early Holocene is comparable to that of *Hedera*.

Because the pollen grains of the subspecies cannot be distinguished only suppositions are possible about the host plants.

Remakable are findings of *V. album* north of its modern northern range limit (see Fig. 4.4.4-8), namely in northern England (Godwin, 1956), Denmark (Iversen, 1944; Fredskild, 1975; and others), southern Norway (Hafsten, 1956; Danielsen, 1970) and southern and central Sweden (Florin, 1957; Berglund, 1966, Digerfeldt, 1972; Gaillard, 1984b; and others). These findings suggest – just as the records of *Corylus* and waterplants – higher summer temperatures, especially around the Early to Mid Holocene. After c. 5750 cal yr BP the records of *Viscum album* throughout its range become less frequent. This decline was, as already mentioned by Iversen (1944), somewhat later than for *Hedera*. In addition to climatic causes for this decrease of *V. album* an additional cause may have been the immigration and expansion of *Fagus sylvatica*, which is not a potential host.

4.4.5

Aquatic macrophytes

Some aquatic taxa show interesting changes in their biogeographical ranges during the Late-glacial and the Holocene, especially in northern Europe. These changes were not only expansions, as most tree taxa experienced, but also reductions. At an early stage, Iversen and Olsen (1943) discussed the relationships between taxa and water chemistry. Iversen (1954) gave three reasons why water plants are useful climate indicators:

- they have short life cycles (especially when compared to trees)
- their seeds are rapidly dispersed by water-fowl (promoting rapid spread), and
- they have no need for developed soils.

These contrasts in life-history traits between water plants and trees were used for the comparison of summer-temperature reconstructions based on pollen and aquatic macrofossils in northern Europe, with potential lags found in the estimation of the onset of the Holocene Thermal Maximum (HTM) as recorded by trees (Väliranta et al., 2015; see Fig. 4.4.5-7). The common northern distribution limits of water plants in northern Finland compiled by Väliranta et al. (2015; see circles on distribution maps in their supplementary figure) were then applied to macrofossil records of Late-glacial and Early Holocene sites in Denmark and southern Sweden (Schenk et al., 2020; see Table 4.4.5-1).

The study of water plants is a field where the interaction between pollen and plant macrofossil analyses is very fruitful.

In many hydrophilous taxa, pollination occurs under water (hydrogamous). Protection of the pollen against desiccation is therefore not needed. Their very thin wall is considered as endexine (intine), only built mainly of cellulose and pectines and without ectexine built of sporopollenin: it is thus scarcely or never preserved (Takahashi, 1995). For such taxa the macroremains are the only subfossil traces (e.g. *Ceratophyllum, Najas*).

Many of the taxa produce a range of macroremains (e.g. fruits, seeds, spines or hairs on leaves) that can be identified to genus or even species (*Najas, Potamogeton*). For the identification of *Potamogeton* fruits early studies include Jessen (1955), Aalto (1970) and Cappers (1993b) beside the manuals for macrofossils (see Birks, 2013).

Some anemophilous taxa produce large amounts of pollen some of which may be identified to species (*Myriophyllum alterniflorum* v. *M. spicatum* and *M. verticillatum*).

In small lakes, vegetative remains, especially leaves, get less widely transported from their source than pollen and may help to reconstruct the spatial pattern of the aquatic vegetation and possibly lake-level changes (e.g. Hannon and Gaillard, 1997; Zhao et al., 2006)

Also of special interest is whether the water plants take up bicarbonates as a carbon-source for photosynthesis or not. If not, their macroremains may serve for radiocarbon dating (see chapter on methods Table 2.4-2). For the role of water plants as indicators of lake developments, see chapter 4.6. For the occurrence of various taxa under various pH-conditions see Table 4.6-4.

We concentrate in the following on the examples of six species. For *Nuphar* and *Nymphaea* maps can be found in Brewer et al. (2017), in their supplementary information.

In the following six maps, some of the records at some sites are not directly dated by radiocarbon but just by palynological correlation.

Ceratophyllum submersum

The perennial and rootless genus *Ceratophyllum* (family Ceratophyllaceae) occurs in Europe as two species, *C. demersum* and *C. submersum*, both of which are dependent on nutrient-rich water. While today *C. demersum* reaches to nearly 70°N in Norway, *C. submersum* prefers summer-warm habitats and, along with *Nymphaea* and *Potamogeton* communities, reaches to almost 59°N in Estonia and Sweden (Fig.4.4.5-2).

The pollen is not preserved but the spines of the leaves (especially numerous in *C. demersum*)

Table 4.4.5-1. Fennoscandian Holocene summer temperatures based on various biological records.

Site	Climate or vegetation type	Location	Warmest period (cal yr BP)	Biotic record	July temperature warmer than today	References
Barheivatn	Oceanic northern boreal, just above birch tree-line	69°42'N 19°51'E	8500–4500	Pollen + macros	c. 2°C	Bjune et al. (2004)
Dalmutladdo	Continental northern boreal, birch	69°10'N 20°43'E				
Tsuolbmajavri	Ecotone from boreal forest to oro-arctic tundra	68°41'N 22°05' E 526 m	8200–6700	Pollen	1.4–1.7°C	Seppä and Birks (2001)
Tsuolbmajavri			Warmest c. 6200	Diatoms	< 1°C	Korhola et al. (2000)
Vuoskkujárvi	Subarctic to high boreal: mountain birch	68°20'N 19°06'E 348 m	c. 8000–4000	Pollen	c. 1°C	Bigler et al. (2002)
			c. 7000–3000	Diatoms	c. 1.5°C	
			Holocene	Chironomids	c. 1.2°C	
Trettetjørn	Oceanic	60°43'N 7°00'E 810 m	8000–4000 with 2 cool phases 6600, 5000	Pollen + macros	1.9°C	Bjune et al. (2005)
Vestre Øykjamyrtjørn	Highly oceanic	59°4'N 6°00'E 570 m	8000–4000	Pollen + macros	1.4°C	Bjune et al. (2005)
Finnish sites	Southern to central Finland	North to 63.9°N	c. (9000) 8000 to 2600?	Pollen + macros	3.9°C	Eriksson (1966), Vuorela and Aalto (1982)

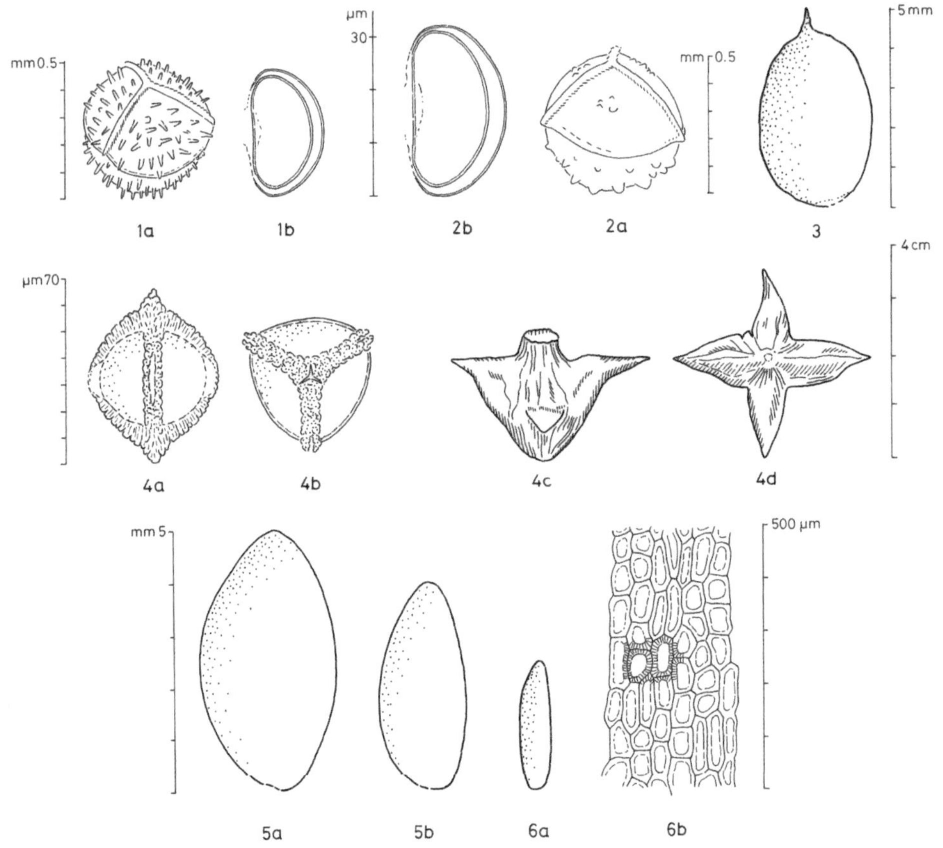

Fig. 4.4.5-1. Spores, pollen grains and fruits of some water plants: 1a: *Isoëtes echinospora*, macrospore, 1b: microspore. 2a: *Isoëtes lacustris*, macrospore, 2b: microspore. 3: *Ceratophyllum submersum*, fruit. 4a: *Trapa natans*, pollen grain in equatorial view, 4b: in polar view, 4c: fruit in side view, 4d: fruit seen from above. 5a: *Najas marina* ssp. *marina*, fruit, 5b: *Najas marina* ssp. *intermedia*, fruit. 6a: *Najas flexilis*, fruit, 6b: epidermis of the seed. 1a, 2a after Watts (1959); 1b, 2b after Lang (1952b) 3 after Backman (1943); 4a, 4b after Erdtmann (1943) and Moore et al. (1991); 4c, 4d after Katz et al. (1965); 5a, 5b, 6a after Lang (1967); 6b after Backman (1950).

can be found on pollen slides, though they are hardly identifiable to species. The fruits are very distinctive for the species (see Fig. 4.4.5-1).

In addition to subfossil records from within its present-day range, Backman (1943) listed about 20 records, mostly from Finland (Fig. 4.4.5-2). The northernmost of these records is near 66°N. Most of the findings are from gyttja samples from the period between 9000 cal yr BP and 2600 cal yr BP. The expansion of the northern range limit by seven degrees of latitude (nearly 800 km) during this period suggests temperatures higher than today (see Table 4.4.5-1). However, some sites may also have been lost through the infilling of the lakes.

Isoëtes echinospora and *I. lacustris*

These fascinating 'living fossils' belong to the division of the Lycopodiophyta and are some of the earliest heterosporous plants. Their megaspores are trilete, while the microspores are monolete. Their photosynthesis follows the CAM path, but the advantage that the name-giving Crassulaceae on dry habitats use, namely economising water, cannot be important for these water plants. However, Keeley (1981, 1998, 2014) explains that carbon fixation during the night is strong for all aquatic *Isoëtes*, because CO_2 concentrations in the water are higher during the night-time. In Crassulaceae (and some other taxa of

Ceratophyllum submersum

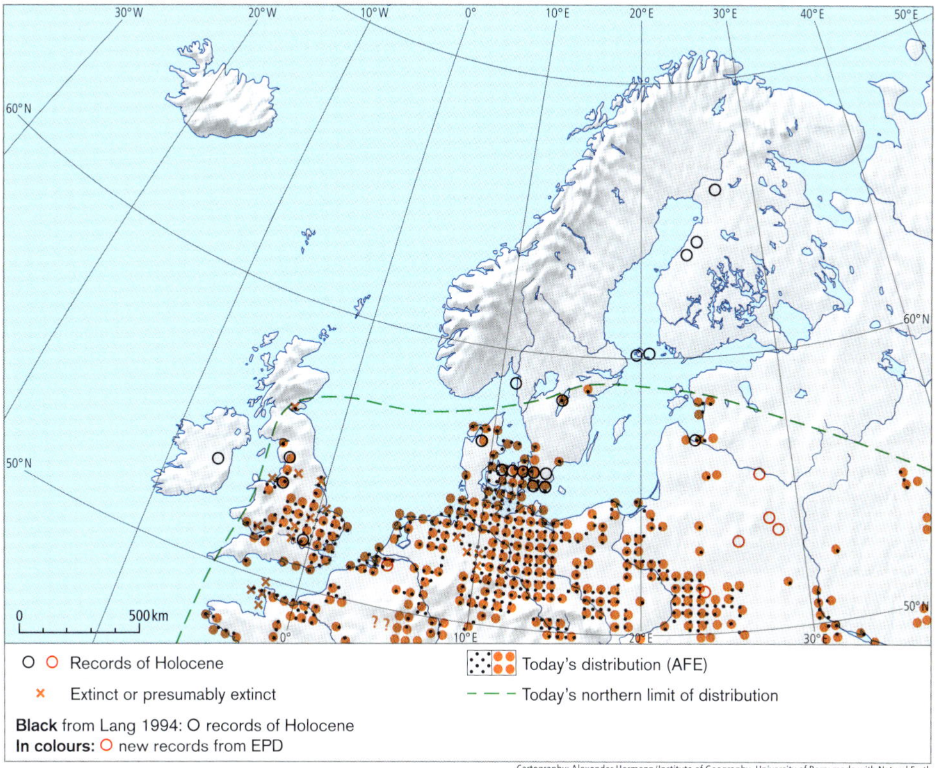

Cartography: Alexander Hermann/Institute of Geography, University of Bern; made with Natural Earth

Fig. 4.4.5-2. *Ceratophyllum submersum.* Subfossil records of fruits at selected sites from the Holocene (circles) and modern range (dots). Modern range according to Atlas of Flora Europaea (AFE). Subfossil records after Backman (1943), Hultén (1971), Godwin (1975) and sites in the European Pollen Database (EPD).

dry habitats) the advantage of nocturnal carbon fixation is different: they close their stomata during the day to conserve water. The two species of northern and central Europe, *Isoëtes echinospora* and *I. lacustris,* are perennial, submerged 'water ferns' that prefer nutrient-poor lakes with no, or very low, dissolved carbonates (Table 4.6-4). Both settle near the shore at a water depth of down to c. 2 m. They form the community of Isoëtetum echinosporae (order of Littorelletea) but also occur in other communities of oligotrophic and dystrophic lakes (see Table 4.6-4). In the few lakes of the southern Black Forest which contain *Isoëtes, I. echinospora* prefers the shallower parts, influenced by seasonal lake-level changes and ice, whereas *I. lacustris* occurs at greater depths. In addition to a few isolated occurrences in the siliceous mountains of southern central Europe both species have a similar main range, including the British Isles and Fennoscandia (Fig. 4.4.5-3a and 4.4.5-3b).

The trilete megaspores can be identified to species based on their morphology, and the monolete microspores on their size difference (e.g. Oberdorfer, 1931; Birks, 1973a): *I. echinospora* 18–31 µm, mean 24 µm; *I. lacustris* 27–41 µm, mean 34 µm (Fig. 4.4.5-1). Unfortunately, the distinction has not been made at all sites.

Most parts of the modern ranges of both species were under the ice during the Last Glacial Maximum (LGM), thus their occurrences must be of Late-glacial or Holocene age. From the periglacial area, microspores of *I. echinospora* of Younger Dryas age were found in the small lakes of the Eifel in Germany (Straka, 1975) but not in the large, deep Maar lakes. As figure 4.4.5-3a illustrates, Late-glacial *I. echinospora* have been recorded outside of

Isoëtes echinospora

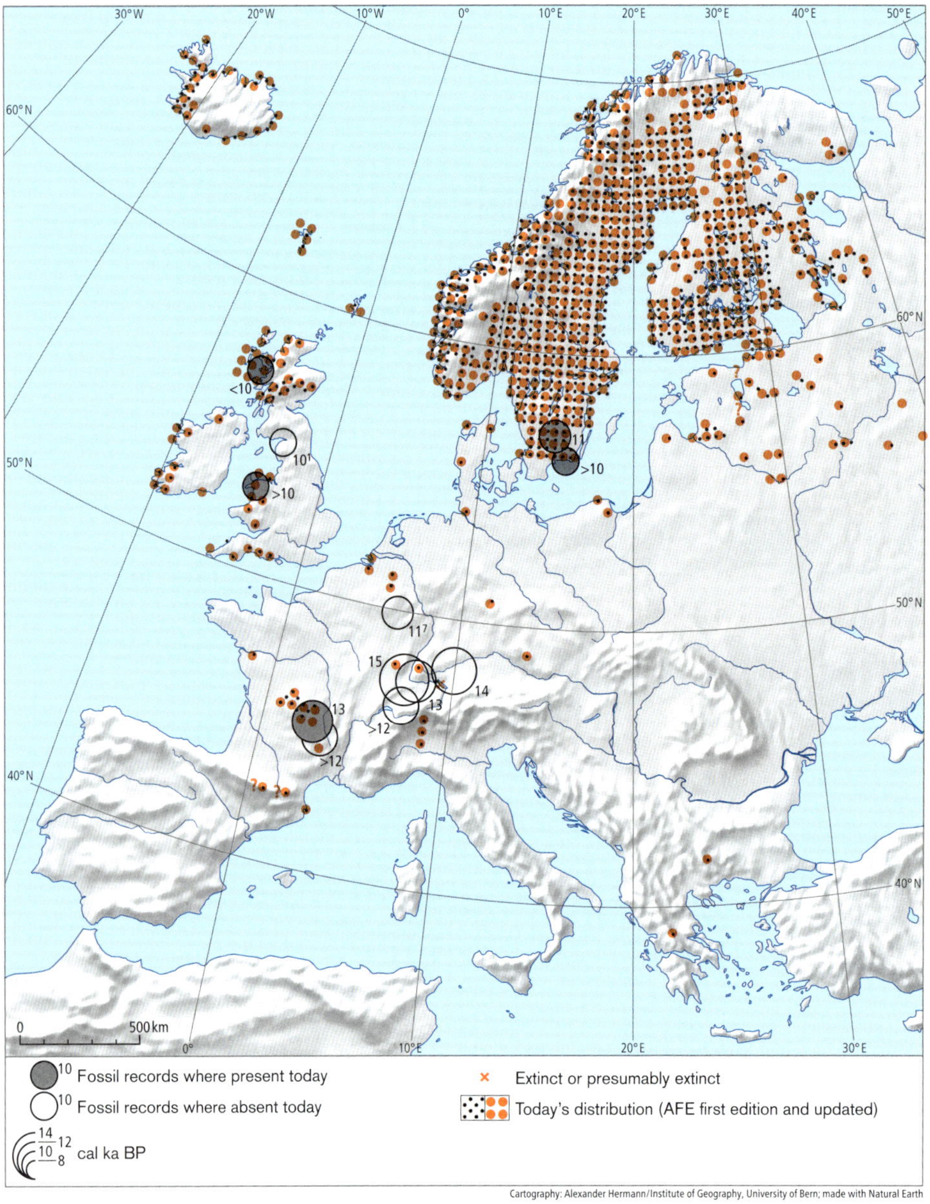

Fig. 4.4.5-3a. ***Isoëtes echinospora.*** Oldest subfossil records of spores after the Last Glacial Maximum (LGM) and modern range. Modern range according to Atlas of Flora Europaea. Subfossil records according to Oberdorfer (1931), Göttlich (1957), Godwin (1960), Lang and Trautmann (1961), Seddon (1962), Berglund (1966), Walker (1966), Welten (1967), Digerfeldt (1972), Birks (1973b), Lang (1975), Straka (1975), Schloss (1979) and sites in the European Pollen Database (EPD).

Isoëtes lacustris

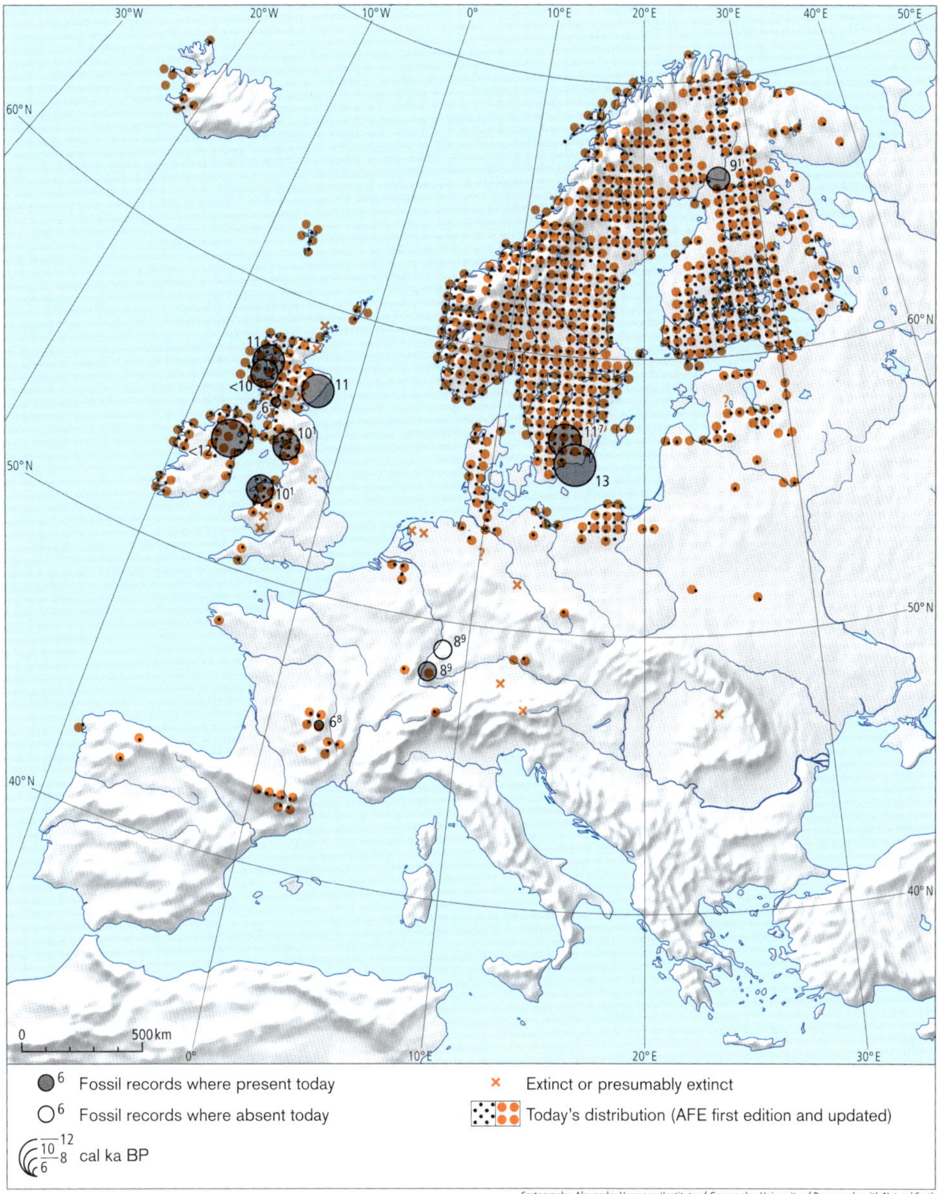

Cartography: Alexander Hermann/Institute of Geography, University of Bern; made with Natural Earth

Fig. 4.4.5-3b. ***Isoëtes lacustris.*** Oldest subfossil records of spores after the Last Glacial Maximum (LGM) and modern range. Modern range according to Atlas of Flora Europaea (AFE). Subfossil records according to Oberdorfer (1931), Jessen and Farrington (1938), Lang (1955, 1958, 1975), Göttlich (1957), Godwin (1960), Lang and Trautmann (1961), Seddon (1962), Berglund (1966), Walker (1966), Welten (1967), Vasari and Vasari (1968), Digerfeldt (1972), Birks (1973b), Reynaud (1974), Straka (1975), Schloss (1979), Stewart et al. (1984) and sites in the European Pollen Database (EPD).

the glaciated areas and beyond the Younger Dryas/ Loch Lomond ice limit. In most of these lakes, large populations developed during the Allerød.

Late-glacial records of *I. lacustris*, however, have so far only been found at the margin of the Fennoscandian Ice, and the development of large populations seems to be restricted to the Holocene.

At least 13 sites are known so far where both species were recorded and their co-occurrence can be

- a succession from early *I. echinospora* to later *I. lacustris*, for example at Loch Meodal on the Isle of Skye at about 9800 cal yr BP (H.J.B. Birks, personal communication)
- as a long-lasting co-occurrence with a shift in dominance from *I. echinospora* to *I. lacustris*, for example at Lake Solsø (Odgaard, 1988)
- as a long-lasting co-occurrence where both species have high abundances, for example at Hoydalar on the on the Faroe Islands (Jóhansen, 1975)
- as a long-lasting co-occurrence of both species with a constant dominance of *I. lacustris*, for example at Dallican Water (Bennett et al., 1992), Lake Trummen (Digerfeldt, 1972), Skånsø (Odgaard, 1988) or Färjesjön (Berglund, 1966).

The hypothesis of Lang (1992) that the glacial refugia of *I. echinospora* were in the west and south-west of Europe, whereas those of *I. lacustris* were only in the west, needs to be checked with additional studies. The population decline and even disappearance of both species in northern and central Europe may be caused by both infilling and eutrophication of the lakes.

Najas flexilis

The genus *Najas* (family Najadaceae) consists of 35–40 species in temperate and tropical areas worldwide. These species are annual, submersed and mainly dispersed by water fowl.

Najas flexilis is an amphi-Atlantic species with a large range in eastern North America and some scattered occurrences in the Pacific Northwest and northern and central Europe. For Europe, Backman (1948) mentioned 47 modern records (British Isles, Fennoscandia, central Europe, western Russia; see Fig. 4.4.5-4).

This monoecious, tiny and inconspicuous plant with narrow leaves inhabits shallow, oligo- to mesotrophic but base-rich lakes and ponds. In terms of plant-sociology it is a rare constituent of the Najadetum intermedia (Lang, 1967, 1990; see below). In Fennoscandia, *Najas flexilis* usually grows together with various *Chara* species, sometimes with *Isoëtes lacustris* (Luther, 1945).

Because the pollen grains of this hydrogamous (hydrophilous) genus have no ectexine, they are not preserved in lake sediments. However, the narrow cylindrical, yellowish to brownish fruits, 2–3 mm in length and with a single seed are frequent macroremains. After the decay of the thin fruit-wall the hard seed, with its isodiametric to elongated sclerenchym cells, can be identified (Fig. 4.4.5-1).

Early findings were published especially in northern Europe (Sandegren, 1920). Backman (1948) could list 214 sites, of which 143 were in Finland and with the southernmost site near Lake Garda (Italy). Additional sites have been subsequently discovered. Overall, the findings were far more numerous in the past than they are now. The majority of the sites, both in northern Europe and in the Alps, are inside the limits of the last glaciation. Just a few sites are recorded from the former periglacial area (southern England, central Germany, western Russia). Because many of the early findings are not well dated, the temporal indications remain quite vague. It seems that the establishment and population expansion occurred in the southern parts of the former Fennoscandian ice during the Early Holocene, but farther north, in contrast, only during the Mid Holocene. Already in the later part of the Mid Holocene, abundances had started to decrease and this process continued through the Late Holocene resulting in a drastically reduced range. Three mechanisms have been suggested as triggers for this reduction: (1) filling-in of the lakes, (2) decreasing (summer) temperatures and (3) changing water chemistry. The filling-in process of lakes certainly played a role at many sites, but it cannot be the principal reason, because other water plants such as *Isoëtes* would also have been affected, which is not often the case. Decreasing temperatures are not very probable, at least for the early part, because the population decline starts distinctly before the 'Subatlantic climate deterioration'. There remains the explanation that nutrients were reduced as well as carbonates (acidification).

This is in line with the local extinction of *Najas flexilis* in the Black Forest after the Mid Holocene, together with other trophically demanding taxa such as *Ceratophyllum demersum* and *Potamogeton pusillus* (Lang, 1955).

Najas marina sensu lato

The annual, dioecious and submerged *Najas marina* sensu lato is a cosmopolitan species with its main range in continental, temperate Eurasia. The need to split this taxon, which is very variable in Europe, into two ecologically different taxa is indisputable, but their taxonomic rank and the accompanying nomenclature are under debate. We base our division on Hess et al. (1976) and Oberdorfer (1990) who assume a species rank for the two following taxa.

(1) *Najas marina* L. sensu stricto, (*N. major* All., *N. marina* L. ssp. *marina*, *N. marina* L. var. *communis* Rendle).

This taxon is a large (often more than a metre tall) submerged plant of meso- to eutrophic lakes. In southern central Europe it is a characteristic species of the Potamogetono-Najadetum marinae (Oberdorfer, 1992). Within the range of *N. marina* sensu lato it has a more southern distribution, from northern to southern Germany. It is absent in northern Europe and on the south coast of the Baltic Sea (Doll and Pankow, 1989). The fruits are from 4 to >5 mm long and easy to identify (Fig. 4.4.5-1). A more detailed description with figures is given by Jacomet (1986).

(2) *Najas intermedia* Wolfgang (*N. marina* L. ssp. *intermedia* (Wolfg.) Casper, *N. marina* L. var. *intermedia* (Wolfg.) Aschers.

Najas flexilis

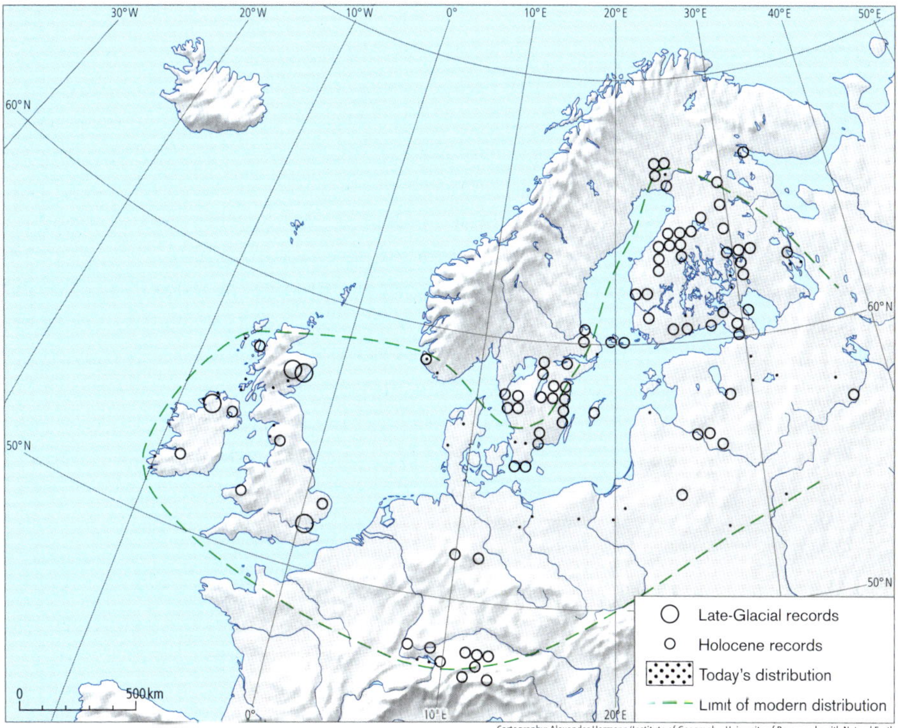

Fig. 4.4.5-4. ***Najas flexilis.*** Late-glacial and Holocene subfossil records of fruits and modern range (dots). The dashed line indicates the range limit today (connection of the extreme single occurrences). Subfossil records of seeds according to Backman (1948), Hultén (1971) and Godwin (1975). Modern range according to Backman (1948), Hultén (1971), Perring and Walters (1976) and Atlas of Flora Europaea.

Najas marina

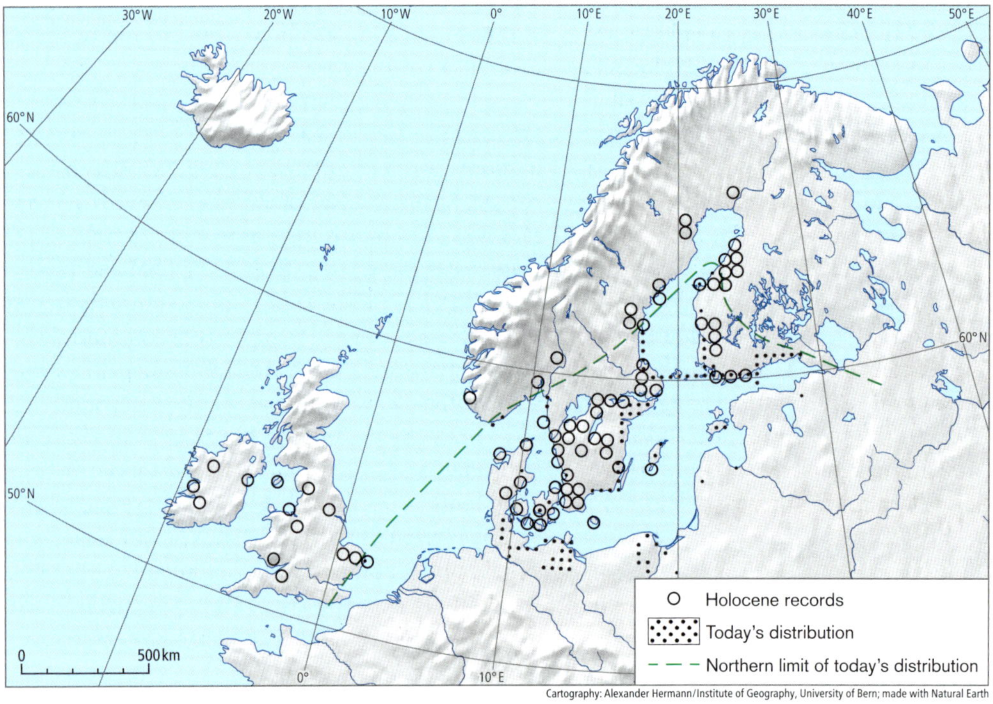

Fig. 4.4.5-5. ***Najas marina.*** Holocene subfossil records of fruits (circles) and modern range (dots). The dashed line indicates the northernmost occurrences today. Subfossil records of seeds according to Backman (1941), Hultén (1971) and Godwin (1975) . Modern rage according to Hultén (1971), Perring and Walters and (1976) Atlas of Flora Europaea.

This is a much smaller (at most 30 cm tall) submerged plant of base-rich oligo- to mesotrophic lakes. In southern central Europe it is a charcteristic species of the Najadetum intermediae, together with *Chara* species. Compared to the range of *N. marina* sensu lato it shows a more northern distribution. In northern Europe and on the British Isles only this taxon occurs, whereas towards the south, its range reaches lakes north of the Alps (Lake Constance, some lakes of the Bavarian, Swiss, and Carinthian Alpine foreland), where it overlaps with the range of *N. marina* sensu stricto. From the area around the Baltic Sea, *N. marina* L. var. *brevifolia* Rendle has been described, which is confined to the brackish coasts and may be considered as a variety of *N. intermedia*. The range of *N. intermedia* is certainly not yet sufficiently well known.

The fruits are usually only 3–4 mm long, and are both shorter and narrower than those of *N. marina* sensu stricto (Fig. 4.4.5-1). A more detailed description can be found in Jacomet (1986) and photographs in Godwin (1975).

The nomenclature is somewhat difficult if the proposal of Viinikka (1976) is included: the more northern European taxon would be called *N. marina* ssp. *marina*, and the central to southern European taxon *N. marina* ssp. *major*.

Unfortunately the numerous findings of *Najas marina* sensu lato were only separated into the two taxa in more recent papers, for example, Lang (1967), Jacomet (1986) and Jacomet et al. (1989), despite the fact that Neuweiler (1905) had already shown the differences between the fruits. Therefore the history of the two taxa needs to be studied further in central Europe. However, for northern Europe we can assume that during the Holocene only *Najas intermedia* occurred. From a comparison between the modern range and the subfossil record in northern Europe, including the British Isles (Fig. 4.4.5-5), we may expect both trophic and thermal factors as causes for the range shifts. This is because first, the records from the Early to Mid Holocene are nearly all from freshwater sites, in more or less base-rich, oligo- to mesotrophic waters (similar to today in northern German inland sites and in the piedmont lakes of the Alps) but, later in the Late Holocene, *N. intermedia* was restricted to coastal brackish sites and became locally extinct in freshwater lakes. This suggests that the increasing acidification of the Fennoscandian lakes made them unsuitable for the species and restricted its range to the more base-rich coastal sites. Second, the subfossil records reached further north (to about 66°30'N) than they do today (to about 63°30'N). This difference is smaller than was recorded for *Ceratophyllum submersum* and *Trapa natans*, but is still c. 3 degrees of latitude (ca. 330 km). Higher temperatures during the maximum northern expansion of the range are, therefore, probable (see Table 4.4.5-1).

Trapa natans

The genus *Trapa* (family Trapaceae) occurs in the Old World as three to eleven species, depending on the taxonomy and nomenclature. The very variable *Trapa natans* inhabits suitable sites in large parts of southern Eurasia, but shows a patchy distribution – as many ornithochorous water plants do. Today its range in Europe is restricted to central and southern Europe and western Russia, with higher frequencies in more continental areas. This annual water plant is rooted, forms rosettes of swimming leaves and prefers high summer temperatures and nutrient-rich waters. According to Oberdorfer (1992) and Horvat et al. (1974) the species is found abundantly in central and south-eastern Europe in the communities of Trapetum natantis and Nymphoidetum peltatae, but also in other communities of the Nymphaeion.

Both the pollen and the fruits can be easily identified and are usually well preserved in lake sediments (Fig. 4.4.5-1). The fruits are morphologically extremely variable (Gams, 1958), especially in having two or four thorns formed out of the calyx. In addition to the one-seeded nut (2–4 cm large) the tips of the calyx-thorns (c. 0.5 cm) with bristles fossilize and may also be found in large numbers (e.g. Vuorela and Aalto, 1982). The starch-rich seeds are edible and taste like those of *Castanea sativa* – which

Trapa natans

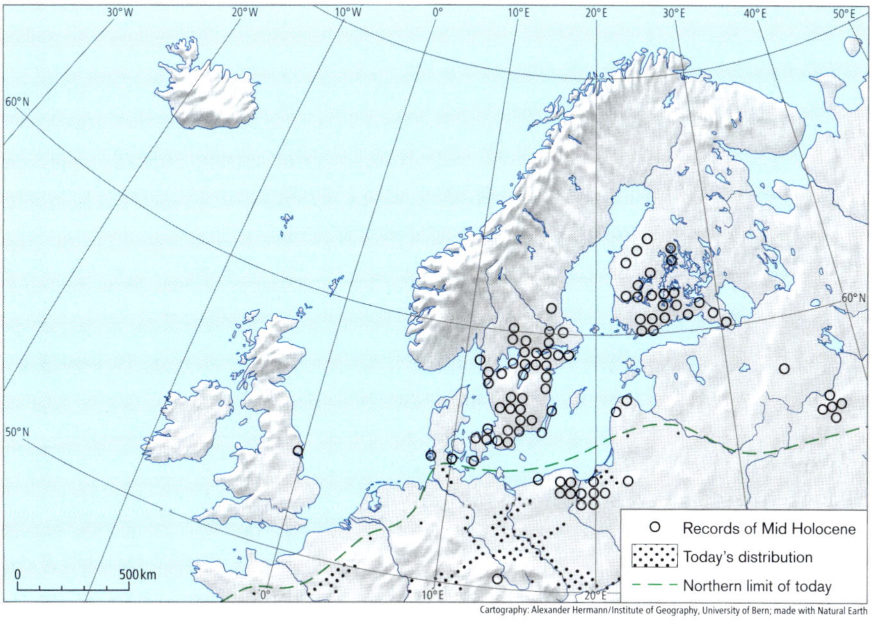

Cartography: Alexander Hermann/Institute of Geography, University of Bern; made with Natural Earth

Fig. 4.4.5-6. *Trapa natans*. Mid Holocene subfossil records (circles) and modern range (dots). The dashed line indicates the northern most single occurrences today. Subfossil records of seeds and pollen grains according to Szafer (1966), Hultén (1971) and Flenley et al. (1975). Modern range according to Gams in Hegi (1926), Szafer (1966) and Haeupler and Schönfelder (1988).

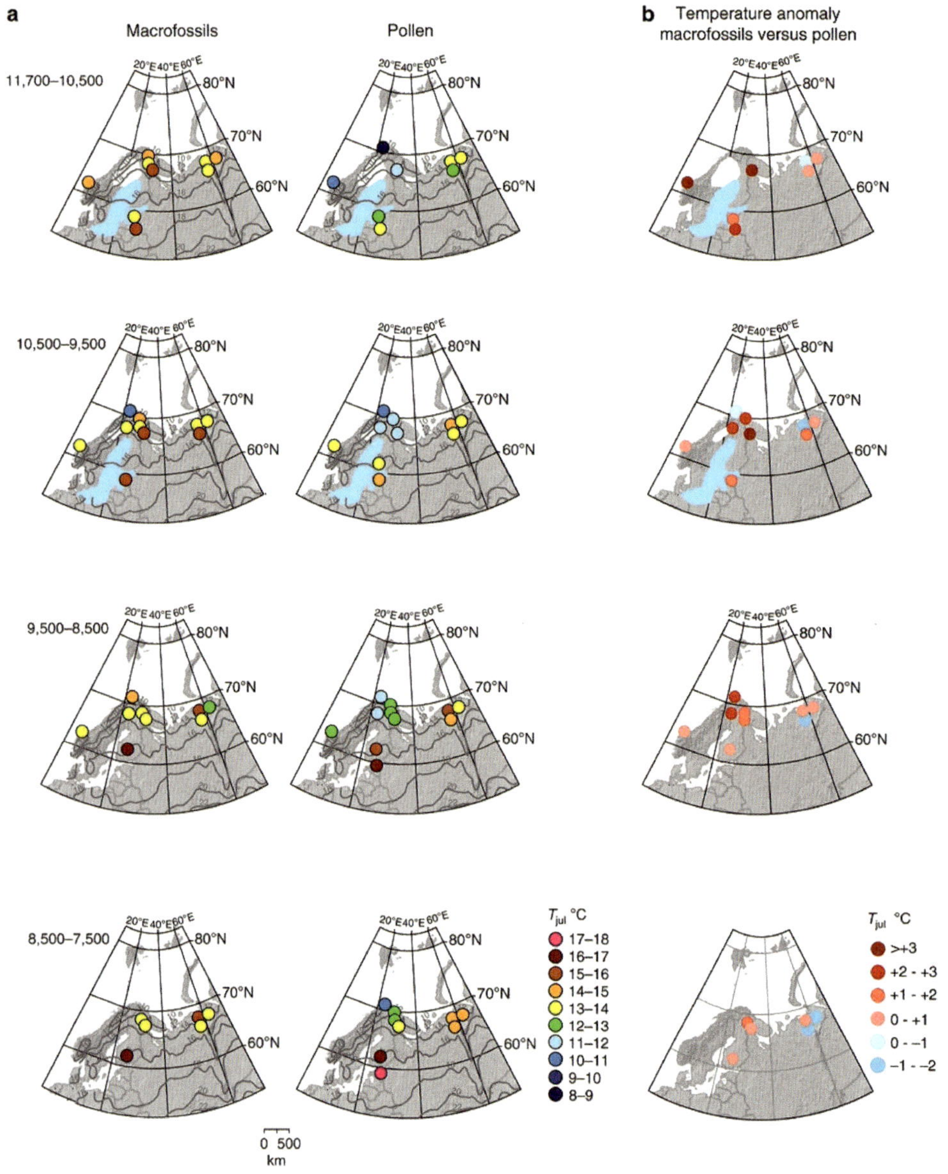

Fig. 4.4.5-7. A comparison of macrofossil- and pollen-based mean T_{Jul} in northern Europe for four early postglacial time windows. (a) Spatio-temporal pattern of pollen- and macrofossil-based mean July temperatures (T_{JulM} and T_{JulP}) for different time windows. In addition, essential information about the ice-margin position and postglacial stages of the Baltic Sea are shown (adapted and modified after Wohlfahrt et al., 2008). Modern T_{Jul} isolines (in grey) are also shown. (b) Temperature anomalies between T_{JulM} and T_{JulP} reconstructions. Anomalies are shown as positive (red) when the macrofossil-based temperatures are higher than the pollen-based temperatures and negative (blue) for the reverse. From Väliranta et al. (2015) under a Creative Commons license.

explains the name 'water-chestnut' in recipies from the Far East.

The Holocene findings of *Trapa natans* in northern Europe already attracted the interest of Quaternary scientists a hundred years ago because this species is completely lacking from the region today (Fig.4.4.5-6). Early reports of numerous records and careful discussions were provided by Lindberg (1910, 1916), Malmström (1920), Gams (1927) and Samuelsson (1934), and more recently by Digerfeldt (1972) and Eriksson (1966); the latter compiled 70 sites in Finland alone. Most of the records are based on 'stone-nuts': pollen findings are more rare. Among them, the one in north-east England (Flenley et al., 1975) is remarkable because, so far, it is unique for the British Isles.

The time period represented by the records of *Trapa natans* in northern Europe is long, namely from about 11.5 to 2.5 cal ka BP: in southern Finland, abundances increased during the second half of this period, that is, during the Mid Holocene. The study by Vuorela and Aalto (1982) may serve as an example. They found pollen and numerous stone-nuts of *Trapa natans* in gyttja of Mid Holocene age – interestingly together with remains of anthropochorous species such as *Polygonum aviculare*, *Stellaria media* and *Sonchus arvensis* (sometimes termed 'weeds'). This co-occurrence, plus the fact that many *Trapa* nuts were cracked open and were found near a Neolithic settlement, points to the possible use of nuts for human consumption.

This archaeological link indicates one possible cause for the wider range of *Trapa natans* in the Mid Holocene, namely (intentional) dispersal by prehistoric people. Human impact may have determined the northern range limit either by increasing the velocity at which it moved northwards or by enhancing its abundance. In any case a plant, whether cultivated or not, cannot grow beyond its climatic limit. There is little doubt that summer temperatures higher than today are the main cause for the northern range limit during the hypsithermal (Table 4.4.5-1). According to Gams (1926) *Trapa natans* needs water temperatures above 20°C for the development of flowers. A modern estimation of the temperature requirements of *Trapa natans* can be found in Kühl et al. (2002) and their modified Figure in figures 2.7-5 and 2.7-6. Figure 4.4.5-6 shows the subfossil Mid-Holocene range limit, with the northernmost record reaching nearly 64°N, whereas today it is at about 56°30'N. This difference of about 7.4 degrees latitude or almost 840 km corresponds to the values found for *Ceratophyllum submersum*, for which human use can be excluded.

Open questions – some answers

Water plants can be important for climate reconstructions due to their short life cycles, their long-distance dispersal (by waterfowl) and their independence from mature soils (Iversen, 1954). Here we summarize some examples.

(a) For the end of the Holocene Thermal Maximum (HTM), we may take a range shift for *Trapa natans* of 7.4 degrees latitude and a lapse rate of 0.53°C per degree latitude, which results in a cooling of 3.9°C. This estimate is much higher than what we expect according to new quantitative studies that usually show Mid-Holocene summer temperatures only ≤2°C warmer than today (see references mentioned in Table 4.4.5-1). What might be the cause of such a discrepancy? Why does the 'indicator-species approach' with *Trapa natans* suggest an almost double value for the Late-Holocene cooling?

1. The difference may be real if one takes into account the bioclimatic differences between the northern boreal and arctic sites of the modern studies summarized in Table 4.4.5-1, compared to the hemiboreal sites of the Holocene records of *Trapa natans* and *Ceratophyllum submersum*. Differences are possible in a north–south direction or in a west–east direction (a stronger hypsithermal warming deeper in the continent, as modelled by Davis et al., 2003).
2. The difference may be an artefact caused either by the lapse rate applied (which is dependent on humidity) for estimating the shift of the northern range limit over 7.4° latitude, or by inherent uncertainties in the various proxies (e.g. the prediction error being nearly as large as the temperature change, or the differences between various proxies (see discussion in Brooks and Birks, 2001; Seppä and Birks, 2001; Bigler et al., 2002; Korhola et al., 2002; Bjune et al., 2004, 2005).

(b) For the onset of the Holocene Thermal Maximum (HTM) in Fennoscandia a refined study is offered by Väliranta et al. (2015) by comparing reconstructed July temperatures based on pollen (T_{JulP}) with the ones based on aquatic macrofossils (T_{JulM}; see Fig. 4.4.5-7). They applied this comparison to 13 sites: from the arctic tundra (Kharinei: Salonen et al., 2011), the arctic treeline (Tumbulovaty: Kultti et al., 2004; Väliranta et al., 2011), the subarctic (three sites including Jansvatnet: Birks et al., 2012b), through various boreal sites to three boreo-nemoral sites. The comparison was elaborated and illustrated for four time windows of the Early to Mid Holocene (11 700 to 7500 cal yr BP). Conclusions are "In all regions, ... the largest positive T_{jul} anomalies occur in the north and the difference becomes smaller towards the mid-Holocene", and "The most striking difference between macrofossil- and pollen-based reconstructions is exhibited in subarctic and northern boreal Fennoscandia, with macrofossils inferring 2–4°C higher temperatures until 8500 cal yr BP. ... Several aquatic species suggest a T_{JulM} of at least ca. 13–14°C". Figure 4.4.5-7 from Väliranta et al. (2015) is both informative and challenging.

If the synchronous trend holds true, its cause may be orbitally controlled insolation changes. If minor cool oscillations can be demonstrated, either solar activity or ocean and atmospheric circulation may be considered (Stuiver and Braziunas, 1993a,b; Linderholm et al., 2009).

4.4.6

Glacial plants

Under 'glacial plants' we here include taxa that had wide spatial distributions in Europe during glacial periods, but restricted distributions during the Holocene (and earlier interglacials). Among these taxa there are many with modern ranges split into various small areas far apart, that is, distances not to be crossed by seed dispersal and thus disjunct ranges. The species are often called 'glacial relicts', which also applies when not all of these isolated occurrences lie within their glacial ranges, but reached some of the regions during the Late-glacial or the Holocene.

The five selected examples belong to very different biogeographical groups (geoelement, geographisches Florenelement of Walter and Straka, 1970), and also to different 'growth forms' according to Raunkiaer (1905, 1934) and to different 'plant functional types' developed after Woodward (1987) and Woodward and Williams (1987) by Prentice et al. (1992a) and Cramer (1997) (see Table 4.4.6-1).

More examples were already treated by Tralau (1963) with useful maps concerning the arctic species *Cornus suecica, Diapensia lapponica, Ranunculus hyperboreus, Salix polaris* and the arctic-alpine species *Arctostaphylos alpina, Dryas octopetala, Salix reticulata, Selaginella selaginoides* and *Thalictrum alpinum*. A review within a larger context can be found in H.H. Birks (2008a) and H.J.B. Birks and Willis (2008).

Like in other contexts the study of plant macrofossils is especially fruitful if applied to glacial plants because of the high taxonomical resolution (see sections on *Betula nana, Salix herbacea* and Table 4.4.6-2).

Betula nana

In the holarctic genus *Betula* (c. 50 species) the species *B. nana* belongs to the section *Nanae* that consists of only a few low-growing shrubby species. The modern circumpolar distribution of *B. nana* consists of *B. nana* ssp. *nana* (Suk.) Hult. occurring from Greenland to Europe and western Asia, and *B. nana* ssp. *exilis* (Suk.) Hult. occurring in North America and eastern and central Asia (Kuparinen et al., 2007; Nash et al., 2016). In Europe, besides the arctic and boreal regions, disjunct occurrences are recorded in the French Massif Central, the Alps and the Carpathians (see distribution map from Atlas Flora Europaea in Fig. 4.4.6-1).

In the main range in the area of dwarf shrub tundra and the adjacent forest-tundra near the polar timberline, *B. nana* forms mostly tree-less shrub communities, often together with low-growing *Salix* species on minerogenic soils. Further south in Fennoscandia the species can be the undergrowth in open birch woods, at the edges of mires it also goes into conifer woodlands or onto bogs or oligotrophic fens (Oxycocco-Empetrion hermaphroditi, northern German bog alliance). Several single sites in central and south-eastern Europe are restricted to open or sparsely wooded bogs with *Pinus sylvestris* and *P. rotundata* (Sphagnion magellanici, central European bog alliance). It seems not yet clear to what degree the dwarf birches on minerogenic soils in the arctic and the ones on peat in lower latitudes are ecotypes. They both form ectomycorrhizal (EMC)

Table 4.4.6-1. Some characteristics of five selected glacial taxa.

Species	Biogeographical range	Growth form	Fossil record
Betula nana	Arctic-subarctic	Chamaephyte	Macrofossils
Ephedra spp.	Pontic-central Asian	Evergreen chamaephyte or phanerophyte	Pollen
Hippophaë rhamnoides	Pontic-central Asian	Deciduous phanerophyte (with N_2-fixers)	Pollen, stellate hairs and macrofossils
Koenigia islandica	Arctic	Therophyte (= annual)	Pollen, macrofossils
Salix herbacea	Arctic-alpine	Chamaephyte	Macrofossils

Betula nana

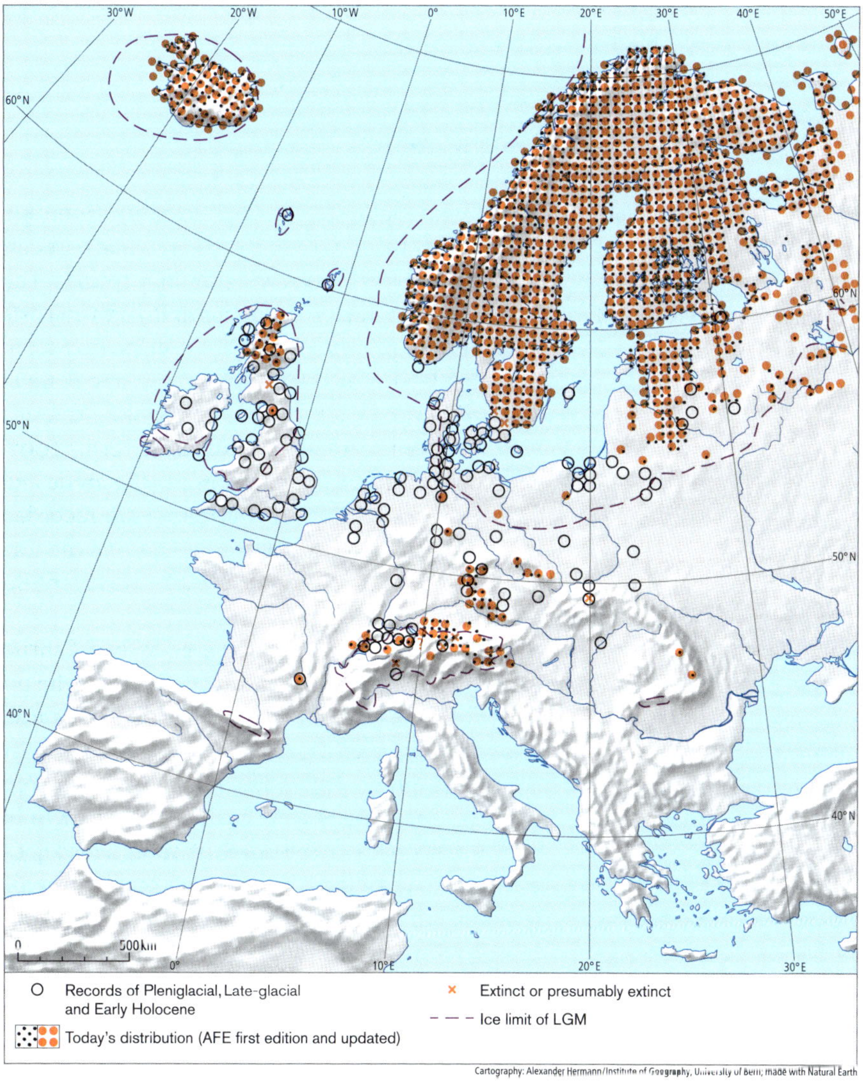

Cartography: Alexander Hermann/Institute of Geography, University of Bern; made with Natural Earth

Fig. 4.4.6-1. *Betula nana*. Late-Quaternary records of macrofossils (selected sites) and modern distribution according to the Atlas of Flora Europaea (AFE; Jalas and Suominen, 1976).

Table 4.4.6-2. Vegetation development on raw soils during the early Late-glacial in the northern Alpine foreland (Swiss Plateau and south-west German Alpine foreland). Ch: Chamaephytes (dwarf and creeping shrubs, cushion plants). H: Hemicryptophytes (perennial herbs). Pm: Macrophanerophytes (trees). Pn: Nanophanerophytes (shrubs). After Lang (1952a,b), Weber (1977, 1979, 1980), Gaillard (1984c) and Ammann (1989).

Local pollen zone	Vegetation	Characteristic species	Duration (years)	Chronozone
5. *Betula alba*	Birch wood	Pm *Betula pendula* Pm *Betula pubescens*	~ 500	Bølling
	Afforestation c. 14 700 cal yr BP (onset of Bølling)			
4. *Juniperus Hippophaë*	Shrub vegetation	Pn *Hippophaë rhamnoides* Pn *Juniperus communis*	100–200	
3. *Betula nana Artemisia*	Dwarf shrub vegetation	Pn *Betula nana* Ch *Selaginella selaginoides* H *Carduus defloratus* and more from 2 (see below)	500–1000	
2. *Artemisia Helianthemum* Cyperaceae	Pioneer meadows	Ch *Arabis alpina* Ch *Armeria alpina* Ch *Artemisia* spec. Ch *Dryas octopetala* Ch *Gypsophila repens* Ch *Helianthemum alpestre* Ch *Salix herbacea* Ch *Salix reticulata* Ch *Salix retusa* Ch *Saxifraga aizoides* Ch *Saxifraga oppositifolia* H *Polygonum viviparum* H *Potentilla aurea/crantzii* H *Thalictrum alpinum*	1000–1500	Oldest Dryas
1. *Artemisia-Pinus*	Raw soils with hardly any vegetation	Pollen of *Pinus* may be either long-distance transport or reworked (with Tertiary pollen and spore types)	?	

associations, the diversity of which shows opposing effects in experimental treatments: with long-term warming diversity increased, whereas with N and P fertilization diversity decreased (Deslippe et al., 2011). This may become important with global warming.

Among the macrofossils of *B. nana*, the narrowly winged fruits, the female catkin scales and the small leaves are so characteristic that they were already identified in the 19th century (e.g. Nathorst, 1870; Andersson, 1888). They belong to the most often found plant remains of the 'Dryas floras'. Drawings and photographs can be found in BOX 1 by Hilary Birks and, for example, Tralau (1963), Katz et al. (1965), Godwin (1975) and Birks (2013). In cases where leaves are numerous and well preserved they could be used to estimate stomatal density that shows a relationship to the atmospheric CO_2 concentration (Chang et al., 1997; Ruddiman et al., 2016), more often analysed on *Salix herbacea* (Beerling et al., 1995; Nguetsop et al., 2004). The variability of the female catkin scales and the fruits was studied by Białobrzeska and Truchanowiczowna (1960). As van Dinter and Birks (1996) demonstrate, even on wingless *Betula* fruits, *B. nana* can be separated from fruits of tree birches. The pollen morphological separation is possible (Eneroth, 1951; Terasmae, 1951; Wenner, 1953; Andersen, 1961; Birks, 1968; Gaillard, 1983; Usinger, 1978a, 1978b, 1985). For recent developments and applications of *Betula nana* pollen determination see Clegg et al. (2005) and Rey et al. (2020). Genetic studies on Svalbard showed low genetic diversity within samples and intermediate within populations (Alsos and Engelskjøn, 2002). On Iceland, introgressions between dwarf birch and tree birches are frequent as shown, for example, by Maliouchenko et al. (2007) or in high resolution by Thórsson et al. (2010). Relict southern populations, such as the ones in Poland, show a high genetic diversity (Dąbrowska et al., 2006; Jadwiszczak et al., 2012).

Numerous records of macrofossils indicate that during the end of the pleniglacial and through the Late-glacial (but before the forest development), *B. nana* had a large range throughout central and north-western Europe (Fig. 4.4.6-1), keeping in mind that the density of sites investigated is not even, with a certain concentration seen in north-western Europe where tundra of Weichselian age is recorded. Further east, where during the final Weichselian glaciation steppic tundra was dominant, the records are more scarce, but at Lake St. Anna in the south-eastern Carpathians, Magyari et al. (2014) found a seed of *B.nana* with an age of 15 150 cal yr BP. There are no records south of a line from the French Massif Central (Lang and Trautmann, 1961; Vasari and Vasari, 1985) over the southern Alps (Schneider and Tobolski, 1985) and into the northern Hungarian Plain (uncertain record of *B. nana* or *B. humilis*; Zolyomi, 1953). This means that the species was not growing in the dry glacial steppes but, at higher elevations with orographic rain, conditions may have been more favourable in southern mountains (e.g. Lake St. Anna at 946 m asl).

Late-glacial and Holocene: in general the fossil records indicate that the modern scattered occurrences of dwarf birch south of the arctic range are relicts from the last ice age. This is also valid if *B. nana* changed its habitat from steppe tundra to bog surfaces during the Late-glacial and Early Holocene, because earlier bogs hardly existed. Only in rare cases could the species be found since the onset of the Holocene at sites where it still grows today, such as Melbecker Moor in Lüneburg Heath (Overbeck and Schneider, 1938), Radauer Born in the Harz Mountains (Beug, 1957), and in Chanaleilles in the Massif Central (Vasari and Vasari, 1985).

High frequencies and abundances of dwarf birch were found for some centuries during the Late-glacial in at least three areas of major ice retreats.

(1) In the circum-Alpine sites a characteristic sequence in vegetation development starting in the Oldest Dryas was recorded: after an early period of raw soils nearly devoid of plants, a species-rich pioneer phase, a dwarf shrub period rich in *B. nana* (but also with some tree birches) followed that lasted probably 500–1000 years (Table 4.4.6-2). A shorter scrub-phase followed in which *Juniperus*, *Hippophaë* and increasing tree birches dominated (*B. pendula* and *B. pubescens*). This latter phase corresponds to the onset of the Bølling interstadial as is evidenced by many sites with increasing $\delta^{18}O$ values and rapid changes in the Coleoptera and Chironomid fauna (e.g. Siegenthaler and Eicher, 1979; Eicher et al., 1981; Elias and Wilkinson, 1983; Gaillard and Lemdahl, 1994; Brooks and Heiri, 2013). Whether the earlier increase in *B. nana* is a phenomenon of succession or an indicator of a warming climate is less clear. The structure inside the Oldest Dryas as shown by Vescovi et al. (2007) and a 'shoulder' in the oxygen-isotope curves before the Bølling-jump in several ice cores indicates some early slight climatic warming of 1-2°C at around 16 500-16 000 cal yr BP as inferred from chironomid records south and north of the Alps (Samartin et al. 2012, Bolland et al. 2020, Rey et al. 2020; chapter 4.3, e.g. Th and Mb regions).

(2) Also in the realm of the Fennoscandian ice sheet, periods of dominant or nearly dominant *B. nana* are recorded – according to the more northern sites later than nearer to the Alps – until the Younger Dryas or even later (Schütrumpf, 1936; Iversen, 1954; Berglund, 1966).

(3) In the North Atlantic, Jóhansen (1985) could show that after the last glaciation, *B. nana* developed on the Shetland Isles first around 11 500 cal yr BP, on the Faroe Islands around 10 850 cal yr BP and on Iceland possibly only around 8600 cal yr BP.

Ephedra

The genus *Ephedra*, usually allocated to the Gnetacae inside the Cycadophyta (in the Gymnosperms) consists of several dozen species. All – with the exception of a few winding plants – are low shrubs with reduced leaves growing in the vast dry regions north and south of the Tropics of Cancer and Capricorn in Eurasia, North and South America, but not in Australia. Bolinder et al. (2016) provide morphologies and evolutionary trees since the Cretaceous. Ickert-Bond and Wojchiechowski (2004) demonstrate a clear phylogeographical difference between the Old World versus the New World taxa based on nuclear and chloroplast DNA. According to Stapf (1889) the genus can be subdivided into sections (Table 4.4.6-3):

(1) Section *Pseudobaccatae* (with yellow to red berry-like seeds) contains, in the Old World, the series *Leptocladae* (thin twigs), *Pachycladae* (thick twigs) and *Scandentes* (climbing) and, in North and South America, *Antisyphiliticae*.

(2) Section *Alatae* (with winged dry seeds) contains the Old World Series of *Tropidolepsis* (with hard covering scales) and the New World Series of *Habrolepides* (tender covering scales).

(3) The small section *Asarca* with a few species in south-western North America.

In Europe and adjacent Africa and Asia today, only *Leptocladae* occur with *E. distachya* and *E. major* (= *E. nebrodensis*, including *E. procera*), as well as *Scandentes* with *E. fragilis* (= *E. altissima*, including *E. fragilis* var. *campylopoda*).

All *Ephedra* species are plants of semi-deserts or steppes. In the diversity centre of Central Asia, *Ephedra* species including *E. distachya* are important constituents of the semi-deserts in the mountains with hot summers and extremely cold winters, usually associated with *Artemisia* and Chenopodiaceae. Some species climb up to great elevations, for example in the Andes up to 4700 m asl or in the Himalayas up to 5400 m asl (*E. intermedia*). In Siberia they nearly reach the polar circle and the 'cold pole' (Gams, 1952).

The species occurring today in southern and south-eastern Europe and northern Africa show various affinities to communities.

E. distachya (Fig. 4.4.6-2), the species reaching furthest north in Europe today, occurs in south-eastern Europe on *Stipa*-steppe (Walter, 1974), in the southern and central Alps in 'Felsensteppe' (Ephedro-Artemisietum, partly as ssp. *helvetica*; Braun-Blanquet, 1961) and on Atlantic and Mediterranean coasts in communities of sandy dunes.

E. major (Fig. 4.4.6-2), a shrub of 1–2 m height, grows on Teneriffe (Canary Islands) and in Mediterranean dry scrub and semi-desert communities with mild winters.

E. fragilis (Fig. 4.4.6-2), a climber on shrubs and low trees, grows in maquis and other shrub communities in the Mediterranean region.

Pollen morphologically *Ephedra* forms the special group of the Polyplicatae (Fig. 2.3-3). According to Welten (1957) for Europe and McAndrews et al. (1973) for North America, three groups can be distinguished by the number and form of the pseudocolpi (Table 4.4.6-3):

1. The *E. distachya* type is found in numerous species of the section *Pseudobaccatae* and is clearly identical to the north American *E. nevadensis* type.
2. The *E. fragilis* type sensu stricto includes only the species of the series *Scandentes* in the Old World.
3. The *E. strobilacea* type, which by some authors is called *E. fragilis* type sensu lato when the *Scandentes* are included, contains the species of the section *Alatae* and is is clearly identical with the North American *E. torreyana* type. Usually the identification of the three groups is not problematic, but the distinction of the species is difficult and very uncertain. In addition, Beug (1956) described various size classes depending on incomplete formation of the tetrads.

Macroremains: as for many taxa of dry habitats the record of Ephedra is extremely small or non-existent from the usual wet coring sites; pack-rat middens in North America are an exception.

The first records of *Ephedra* were reported by Brorson-Christensen (1949) but interpreted as reworked pre-Quaternary pollen. Later, more findings in southern Scandinavia, central Europe and the Alps disproved this interpretation (Iversen, 1951, 1954; Lang, 1951; Zagwijn, 1952; Hafsten, 1953; Müller, 1953; Krog, 1954; Welten, 1955). Since then, numerous additional records have been published covering nearly all of Europe northwards to Scotland (Birks, 1973b), the Lofoten Islands (Moe, 1983) and Finnish Lapland (Sorsa, 1965): at various frequencies all three pollen types were found (Fig 4.4.6-3). Chronologically, the findings reach from the Middle Pleistocene to the present with a clear maximum during the Late-glacial and rather rare single grains from the Holocene. As a summary, long European records of *Ephedra* include:

- Alborán Sea MD95-2043: *E. distachya* type, nearly continuous through MIS 3–MIS 2, especially high during Heinrich Event (HE)5–1 (Fletcher and Sánchez Goñi, 2008).

- Padul: *Ephedra* type not indicated, sporadic from Middle Würm, high in Oldest and Younger Dryas (Pons and Reille, 1988).
- La Grande Pile: both *Ephedra* types, in various cold periods before and after the Eemian (Woillard, 1978).
- Les Echets: *Ephedra* type not specified, in various cold periods after the Eemian (de Beaulieu and Reille, 1984a,b).
- Lago della Costa: both *Ephedra* types, between c. 33 and 16 cal ka BP, quite frequent but not continuous (Kaltenrieder et al., 2009).
- Lago di Fimon: both *Ephedra* types, from Termination II, from late Eemian to various phases of the Würm (Pini et al., 2010).
- Monticchio: both *Ephedra* types, high *E. distachya* type in Saalian, then sporadic, high in Melisey (Allen and Huntley, 2009).
- Ioannina: *Ephedra* type not specified, during full and Late-glacial (Lawson et al., 2004).
- Kopais: both *Ephedra* types, in various cold periods before and after the Eemian (K93, 60 m long: Tzedakis, 1999; core from 1984, 120 m long: Okuda et al., 2001).

For the *E. distachya* type (the most frequently recorded type, but rarely above 1–2% of the total pollen sum) most authors tend to assume local presence near the sites. Even southern Scandinavia was not excluded from this interpretation, where the island of Öland, famous for its steppic flora, was considered a potential pollen source (Iversen, 1951, 1954, 1973 – but see also Königsson, 1968). These ideas were supported by the study of Welten (1957) about modern pollen deposition within the range of *Ephedra distachya* ssp. *helvetica* in the central Swiss Alps: even at short distances of 50–100 m away from abundant *Ephedra* patches the pollen could be below 1%. Therefore, pleniglacial and Late-glacial occurrences near the ice margin were thought to be possible: this may be in accordance with the records of steppic animals.

On the other hand, more and more findings of single grains from Mid-Holocene deposits call for caution: grains of the *E. distachya* type were recorded in southern Norway (Hafsten, 1953, 1956), southern Sweden (Königsson, 1968) and even Finnish Lapland (Sorsa, 1965), and grains of the *E. fragilis* type in southern Norway and southern Sweden (Küttel, 1985). Long-distance transport over large distances was then indicated by the record of Sahara-dust together with the *E. fragilis* type in modern firn of Austrian glaciers (Bortenschlager, 1965, 1967, 1970; Fritz, 1976). At the southern fringe of the Alps at Lago di Ganna, Drescher-Schneider (1993) found both the *E. distachya* and the *E. fragilis* type from the early Late-glacial to the Late Holocene together with pollen grains originating in Africa, namely of *Calligonum* (Polygonaceae), *Moltkiopsis ciliata* (Boraginaceae) and *Lygeum spartum* (Poaceae). Two more records of *Ephedra* also indicate possibilities of extreme long-distance transport of pollen: (1) records from Tristan da Cunha and Gough Island in the middle of the South Atlantic 4000 km away from the closest population in South America (Hafsten, 1961) and (2) records in the Great Lakes region in North America c. 1400 km away from the nearest occurrence in south-western North America (Maher, 1964).

What conclusions can be drawn from these partly contradictory observations and newer studies about the pleniglacial and Late-glacial ranges of *Ephedra*? Because no macroremains of the genus have so far been found in Europe, interpretations rely only on pollen records – just as for *Artemisia*, for which, however, the pollen values are much higher (see chapter 3 on the pleniglacial). For the very low values of *Ephedra* the distinction between local presence and long-distance transport remains problematic: pollen analysis touches here its methodological limits. The assignment to species for the *E. distachya* type is presumably *E. distachya* because of its ecology and distribution (see Table 4.4.6-3). Welten (1957) also showed the high correspondence in pollen morphology (fossil and modern). Regarding the modern fragmented range (Fig. 4.4.6-3), its occurrence in southern Europe and westwards to the Atlantic coast during the pleniglacial and late Würm seems most probable, but whether it expanded north of a line from Brittany to the Alps and the Carpathians seems doubtful. For western Scotland, Birks (1973b) assumed long-distance transport because he also found grains of *Ambrosia*, a genus of the New World. Fredskild (1975) came to the same conclusion on finding the *E. distachya* type together with *Ambrosia* in Late-glacial sediments in Langeland, Denmark. The Finnish records may also be caused by long-distance transport. As a potential source, *E. monosperma* on Öland was assumed (e.g. Tynni, 1959; Vasari, 1962; Sorsa, 1965), but also here long-distance transport cannot be excluded, for in-

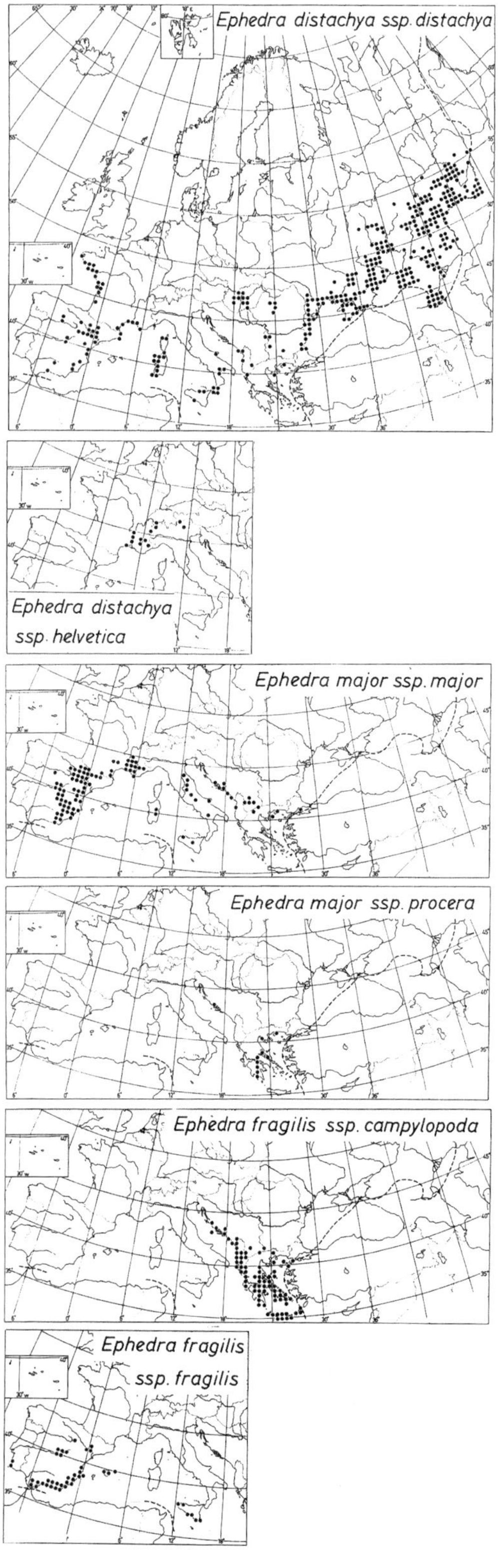

Fig. 4.4.6-2. Current distribution of ***Ephedra*** in Europe. After Jalas and Suominen (1973).

Table 4.4.6-3. Pollen types of the genus ***Ephedra***. Pollen morphology, species and their modern range (after Stapf, 1889; Welten, 1957).

1. *E. distachya* type: 5 – 8 'intercolpi', pseudocolpi branched: section *Pseudobaccatae*	
Series *Leptocladae*	Old World
E. distachya (inc. *E. helvetica*)	Atlantic/Mediterranean/ Pontic/South Siberian
E. major (E. nebrodensis, E. procera)	Mediterranean-Anatolian/ central Asian
E. equisetina	Turanian
E. gerardiana	Central Asian
E. monosperma	Turanian/central/east Asian
Series *Pachycladae*	Old World
E. intermedia	Turanian/central Asian
E. pachyclada	Turanian
E. sarcocarpa	Turanian
Series *Antisyphilitica*	New World
E. antisyphilitica	SW North American
E. nevadensis	SW North American
2. *E. fragilis* type: 8 – 12 (-17) 'intercolpi', pseudocolpi not branched: section *Pseudobaccatae*	
Series *Scandentes*	Old World
E. fragilis (E. altissima)	Macaronesian-Mediterra-nean
E. alte	North Saharan/East Medi-terranean
E. foliata	Turanian
3. *E. strobilacea* type: 14 – 20 'intercolpi', pseudocolpi not branched: section *Alatae*	
Series *Tropidolepides*	Old World
E. alata	North Saharan
E. przewalskii	Central Asian
E. strobilacea	Turanian
Series *Habrolepides*	New World
E. torreyana	SW North American
E. trifurca	SW North American

Ephedra distachya

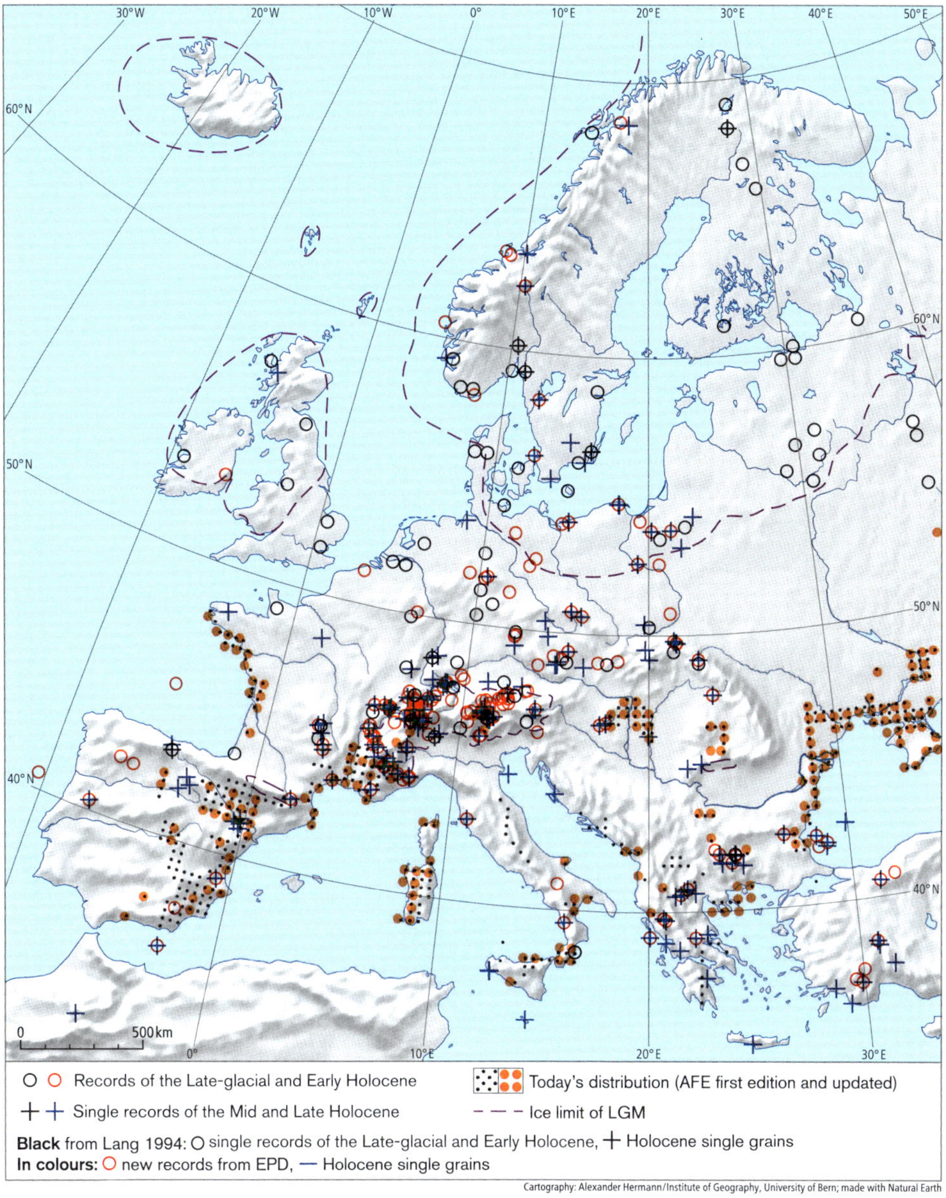

Fig. 4.4.6-3a. Finds of ***Ephedra distachya* type** (*Leptocladae*) pollen from the Late-glacial and Early Holocene (circles) and single grains (+) from the Mid and Late Holocene. Current distribution according to the Atlas of Flora Europaea (AFE; Jalas and Suominen, 1973).

Ephedra fragilis

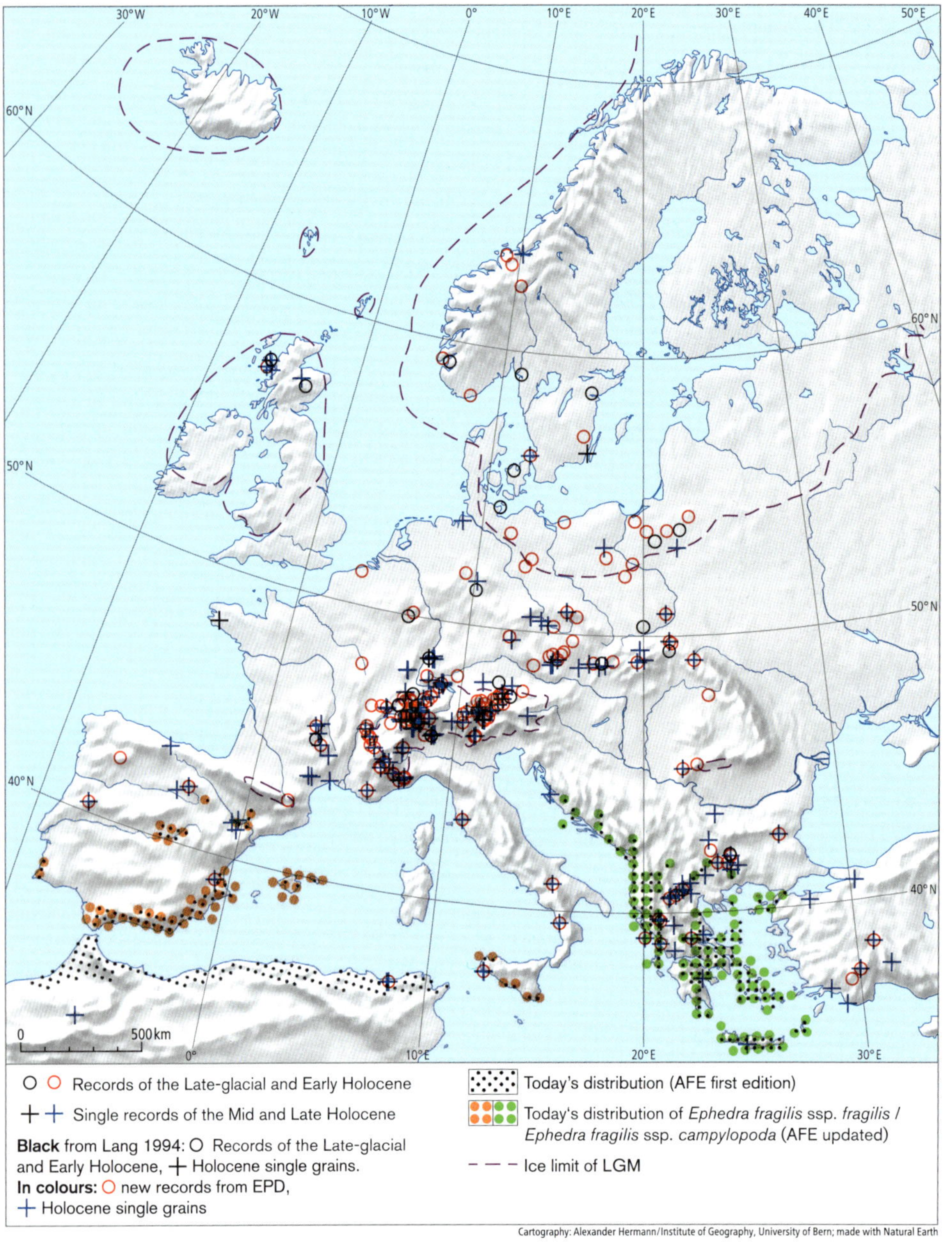

Cartography: Alexander Hermann/Institute of Geography, University of Bern; made with Natural Earth

Fig. 4.4.6-3b. Finds of ***Ephedra fragilis* type** (*Scandentes*) pollen from the Late-glacial and Early Holocene (circles) and single grains from the Mid and Late Holocene (+). The current distribution of the two subspecies ssp. *fragilis* and ssp. *campylopoda* are indicated with two different colours in their modern distribution; they cannot be separated as pollen types. The range in North Africa is roughly copied from Stapf (1889). EPD: European Pollen Database, AFE: Atlas of Flora Europaea.

stance from regions east of the Black Sea. Also for the central European periglacial regions, sporadic occurrences of the *E. distachya* type are now usually considered as long-distance transport. Around the northern limits of the Alps, Late-glacial *E. distachya* may be possible but macrofossils would help clarification. The pleniglacial and Late-glacial occurrence of *E. fragilis* or even *E. strobilacea* or *E. alata,* however, can be excluded for areas north of the Alps. Single grains may be transported over long distances from the south and south-east. In southern Europe, in contrast, these species have been more widespread during glacial periods than today. To sum up, we can take the genus *Ephedra* as an example with restricted and refugial areas during warm and wet periods (interglacials), comparable to arctic-alpine species (Birks and Willis, 2008), and in contrast to temperate or Mediterranean tree species. One could simplify their ecological requirements by saying that the arctic alpines like the cold, and *Ephedra* likes dry conditions: both groups were therefore wider spread under glacial climates and survive in 'interglacial refugia' today (chapter 4.11).

A vast literature exists about *Ephedra* as a pharmaceutical plant.

Hippophaë rhamnoides

The genus *Hippophaë* in the family of Elaeagnaceae is confined to the Palaearctic. It consists of seven species out of which only *H. rhamnoides* (sea buckthorn) occurs in Europe (Swenson and Bartish, 2003). According to the monographic studies of Rousi (1971) the species contains nine subspecies but only four of them occur in Europe. *Hippophaë rhamnoides* is a woody pioneer on sandy to gravelly raw soils with relatively high pH. Secondary anthropogenic habitats are road sides or gravel pits.

The ranges of the four subspecies coincide largely with the four chloroplast haplotypes A–D of Bartish et al. (2006).

1. Ssp. *rhamnoides* (= ssp. *maritima* v. Soest) occurs along the coasts of the Atlantic Ocean in France and the English Channel, along the North Sea, Baltic Sea, Bothnian Bay, but also inland in Scotland and Norway, where its northernmost site is at 68°50'N; in southern Norway a site at 1100 m asl is known. In plant sociology: Hippophaëo-Salicetum arenariae / Salicion arenariae / Prunetalia. The cpDNA haplotype is type A.

2. Ssp. *fluviatilis* v. Soest is widespread in the Alps (up to 1900 m asl) and pre-Alps, but also in the Pyrenees, the Upper Rhine valley and the Apennines. In plant sociology: Salici-Hippophaëtum / Berberidion /Prunetalia, but also in Cytiso-Pinetum and Erico-Pinetum / Erico-Pinion / Erico-Pinetalia. The cpDNA haplotype is type C.

3. Ssp. *carpathica* Rousi with a range in the Carpathians, along the Danube and its tributaries (Bavaria, Austria, Hungary, former Yugoslavia) to the mouth of the Danube into the Black Sea. In plant sociology like ssp. *fluviatilis*: Salici-Hippophaëtum / Berberidion /Prunetalia, but also in Cytiso-Pinetum and Erico-Pinetum / Erico-Pinion / Erico-Pinetalia. The cpDNA haplotypes are types A and C.

4. Ssp. *caucasica* Rousi occurring in the Caucasus, in Armenia and eastern Turkey and at the Bulgarian Black Sea coast. The cpDNA haplotype is type B.

The ranges of the closely related ssp. *fluviatilis* and ssp. *carpathica* are in contact on the north-eastern edge of the Alps, where intermediate forms (and the two non-monomorphic populations with cpDNA haplotypes A and C) occur. Most populations studied by Bartish et al. (2006) are monomorphic concerning cpDNA haplotypes, but over 60% are polymorphic in the nuclear DNA *Chsi* (intron of chalcone synthase): the spatial distribution of the latter is more complex than the one of cpDNA haplotypes, but still with clear east–west and south–north gradients.

Hippophaë rhamnoides is a summergreen, thorny shrub growing up to 6–10 m. As the other members of Elaeagnaceae it lives in symbioses with *Frankia* (Actinomycetales/ Actinobacteria) forming nodules on the roots. This fixing of nitrogen from the atmosphere makes *Hippophaë* a strong pioneer on minerogenic raw soils poor in organic matter. It is therefore often found on sandy or gravelly ground along rivers or sea shores.

Sea buckthorn is dioecious, anemophilous and zoochorous. Pollen grains are tricolporate and psilate/scabrate and were recognized as subfossils already by von Post (1924). No significant differences between the subspecies recognized by Rousi (1971) were found by Sorsa (1971). Pollen productivity seems to be low, so that Firbas (1934) concluded that 5% *Hippophaë* pollen means a substantial

Hippophaë rhamnoides

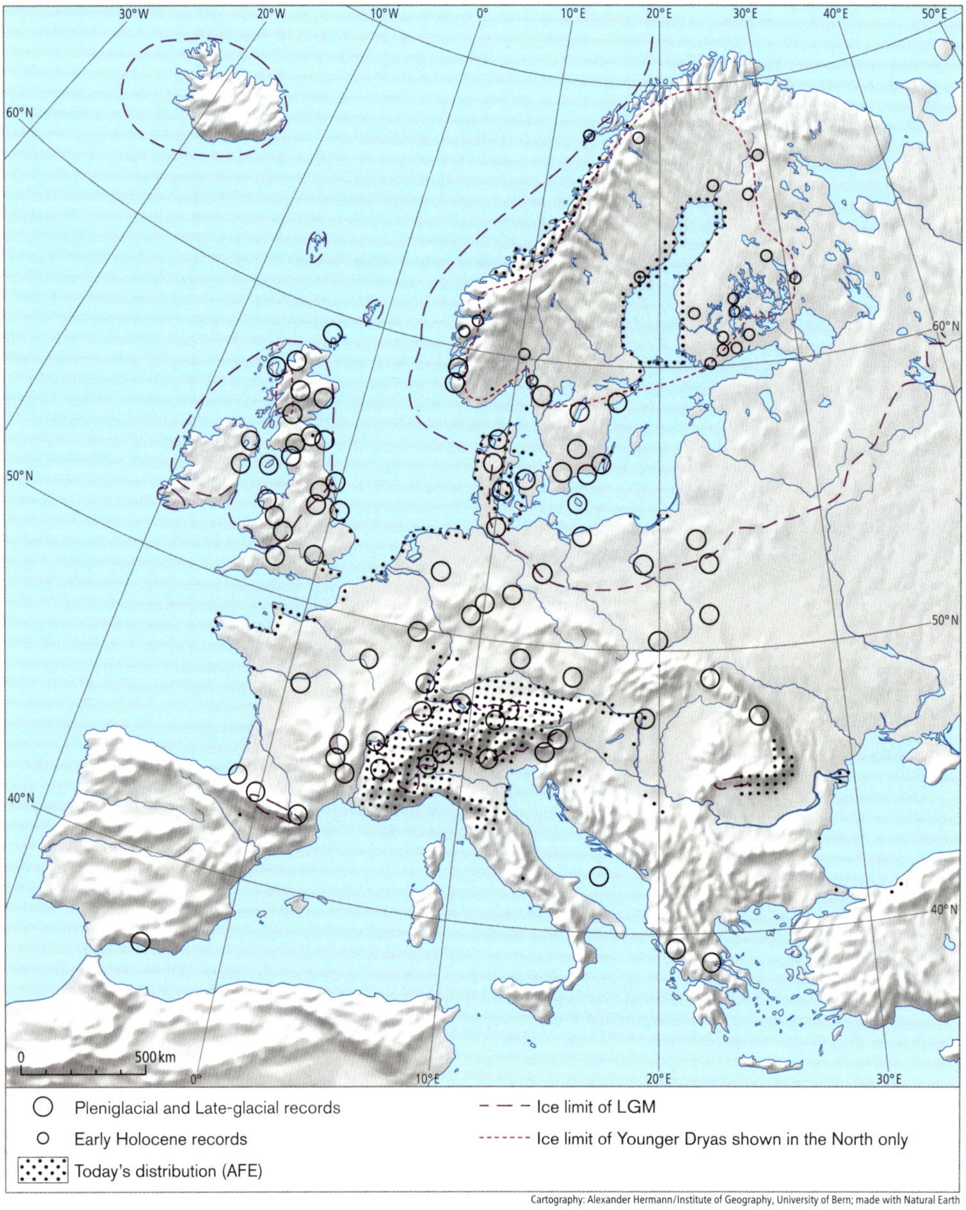

Fig. 4.4.6-4a. Finds of ***Hippophaë rhamnoides*** pollen from the pleniglacial and Late-glacial, from the Early Holocene and its current distribution. The extent of the ice sheets during the Last Glacial Maximum (LGM) is indicated for all of Europe, the one during the Younger Dryas only in Fennoscandia. AFE: Atlas of Flora Europaea.

Hippophaë rhamnoides

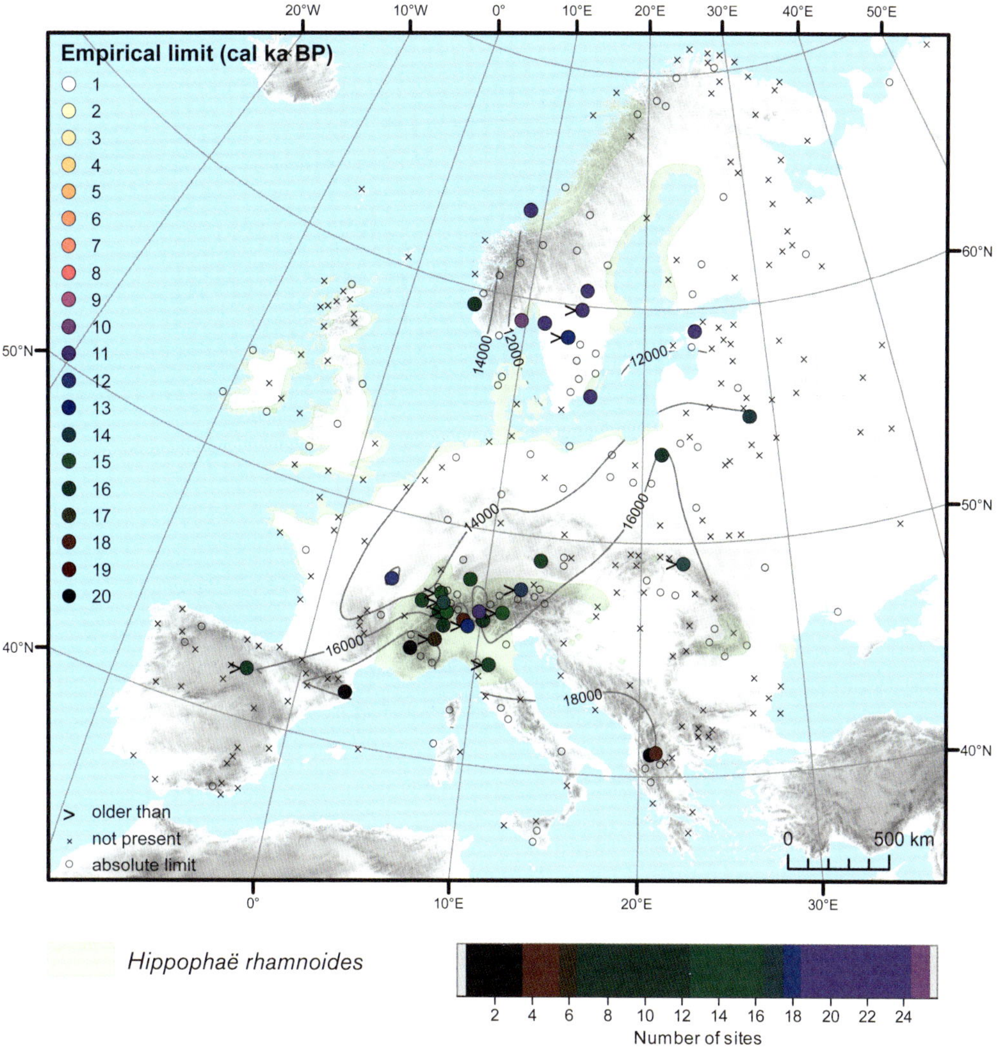

Fig. 4.4.6-4b. Empirical limit of ***Hippophaë rhamnoides*** and its current distribution in Europe. The coloured dots indicate the empirical limit in cal ka BP at selected sites, crosses refer to sites where the taxon was absent and small circles where it reached the absolute limit. Grey lines represent isochrones of empirical limit in 2000-year steps, based on nearest neighbour interpolation. Bar diagram shows stacked number of sites where the empirical limit was reached during the same 1000-year time interval, with colours according to the legend in the figure.

coverage in the neighbourhood. Macrofossils: the bud scales and the narrow lanceolate leaves are covered more or less densely with hairs of shield or stellate form that reduce transpiration: these hairs can be found in subfossil samples. The conspicuous orange fruits are berries with usually a single seed. The berries are very rich in vitamin C and are traditionally used for juice and jams. The fruits are primarily dispersed by birds: after passing through their intestines still 95–100% of the seeds are able to germinate (e.g. Darmer, 1952). In contrast to the pollen grains, the seeds show significant differences between the subspecies: lengths of seeds vary between 2.8 and 5.3 mm; widths between 1.4 and 2.7 mm (figures in Florschütz, 1958; Katz et al., 1965; Beijerinck, 1976). It is remarkable that *Hippophaë* can also propagate vegetatively by root-suckers.

Because of the distinct morphology of pollen and macroremains, a large number of records have been reported, for example by Gams (1943), Sandegren (1943), Firbas (1949a), Godwin (1956, 1975), Hafsten (1956, 1966), Berglund (1966) and Reynaud (1976).

During the last pleniglacial and Late-glacial sea buckthorn was widespread in Europe. Uzquiano (1992) identified its wood at Altamira in a layer of Lower Magdalenian. During the ice retreat, it played a major role on newly deglaciated terrain (comparable to modern *Sheperdia canadensis* near North American glaciers). We provide in figures 4.4.6-4a and 4.4.6-4b two maps the comparison of which is of interest. Specifically, the isochrone map (Fig. 4.4.6-4b) suggests LGM occurrences on the three southern European peninsulas (black and brown dots).

Range expansion and contraction: during the Late-glacial a peak of populations of *Hippophaë* is often found (usually together with *Juniperus*) between the treeless 'steppe-tundra' and the birch-pine forest. This *Juniperus-Hippophaë* phase dominates in southern central Europe in the first part of the Bølling, and on the British Isles and in northern Europe in the Allerød and the Younger Dryas. But this dominance of the two shrubs is found only inside the former glaciated area where raw soils inside the maximal moraines occurred. The maps by Brewer et al. (2017) are very useful for *Hippophaë* because they just concern one species (in contrast to e.g. *Salix* with numerous species of various autecologies). Our isochrone map (Fig. 4.4.6-4b) suggests that much in contrast to temperate species *Hippophaë rhamnoides* expanded from southern Europe to central Europe and northern Europe already during the Late-glacial, establishing in the formerly glaciated period as soon as the northern ice cap melted.

During the Early Holocene, after the retreat of the inland ice from the middle-Swedish and south-Finnish end-moraines, *Hippophaë* proceeded to northern Fennoscandia. With the Holocene forest expansions from south to north and from east to west, the heliophilous shrub was weak in competition and its vast distributional range of the Late-glacial became split up into isolated areas in the southern central and south-eastern Europe and coastal and riverine occurrences.

Koenigia islandica

The genus *Koenigia* in the Polygonaceae family contains – depending on species concepts – up to ten species (Melchior, 1964). They have arctic-subarctic or antarctic ranges (*K. fuegiana* on Tierra del Fuego). The only species in Europe is *K. islandica* that has a large northern hemispheric to circumpolar distribution (North America, Greenland, northern Eurasia). Among the therophytes it is the one that reaches furthest north. A taxon usually considered a variety was ascertained as a separate species *K. hadacii* by Löve and Löve (1975). It occurs only on Svalbard and its chromosome number is 2n = 14, while in *K. islandica* it is 2n = 28. The range of *K. islandica* in Europe includes Svalbard, Bear Island, Jan Mayen, Iceland, the Faroe Islands, the northern part of the Kola Peninsula and the Scandinavian fjell region from the North Cape to the latitude of Bergen (Fig. 4.4.6-5). On the British Isles, the inconspicuous plant was discovered only in 1951 at some sites on Skye and Mull (Inner Hebrides, Scotland) together with *Diapensia lapponica* among others (Lousley, 1951). It does not occur in the Alps today. Phylogenetic studies can be found in Fan et al. (2013) and Schuster et al. (2015).

Koenigia islandica is a tiny, prostrate, annual, anemophilous plant. It is a pioneer on open, wet ground rich in clay and/or silt in snowbeds, along ponds or brooks (Gjaerevoll, 1956; Hadač, 1971; Böcher et al., 1978; Feilberg et al., 1984). On Skye, the occurrence of the species is described in alpine-subalpine gravel communities (Androsacion

Koenigia islandica

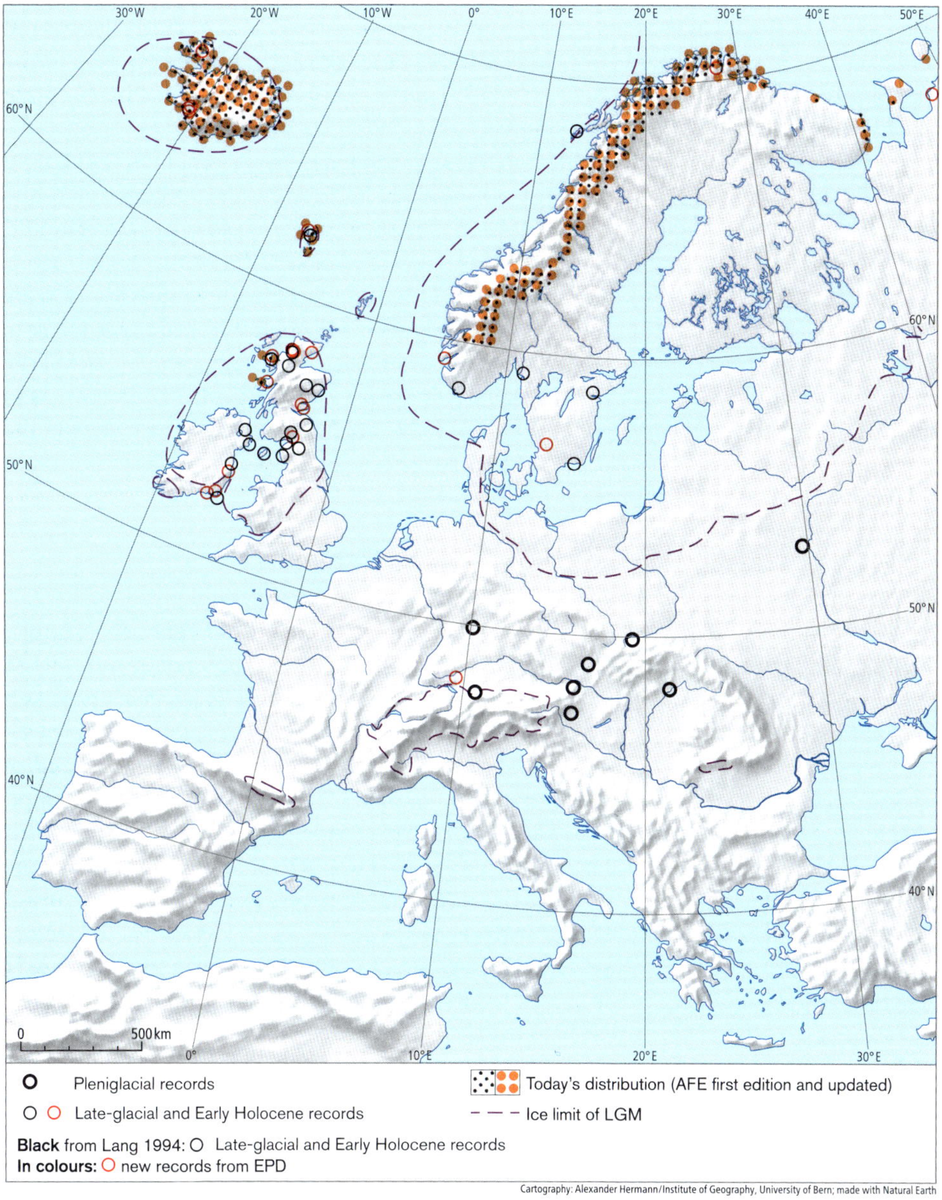

Fig. 4.4.6-5. Finds of ***Koenigia islandica*** pollen from the pleniglacial and from the Late-glacial and Early Holocene, its current distribution and the maximal extent of the ice sheets during the Last Glacial Maximum (LGM). EPD: European Pollen Database, AFE: Atlas of Flora Europaea.

alpinae, Androsacietalia, Thlaspietea rotundifolii) and spring communities (Cardamino-Montion, Montio-Cardaminetalia, Montio-Cardaminetea) by Ratcliffe (1959) and Birks (1973b). A low-growing plant like *Koenigia* is naturally highly dependent on soil temperatures. For the modern Scottish sites, Conolly and Dahl (1970) made a plausible case that the species is restricted to regions where the maximum temperatures in summer do not exceed 21°C. In more continental regions this value may be higher. Studying the modern pollen deposition, Birks (1973a) found that pollen transport through the air is very weak but precipitation easily washes the pollen out. Pollen deposition in lakes may therefore be primarily by inwash together with minerogenic matter: the palynological record may thus be from within the hydrological catchment.

The pollen morphology of *K. islandica* is very characteristic, namely stephanocolpate-echinate with numerous, very short colpi (Figs in Berglund, 1966; Godwin, 1975; Moore et al., 1991; Hedberg, 1997; Beug, 2004). Pollen with similar morphology occur in some other taxa of Polygonaceae but because of their plant geography they may not be relevant here (Hedberg, 1946). Also the fruits are characteristic: pear-shaped nutlets of maximum 1.5 mm in length, with longish epidermis cells that show protruding cell walls (Dickson, 1970).

The subfossil records of *K. islandica* are, so far, mainly based on pollen records with the exception of Whitrig Bog in the lowlands of south-eastern Scotland. From this site came not only the first pollen record by R. Andrew at a time when the modern Scottish sites were not yet known (in Godwin, 1956) but also later macrofossils found by A.P. Conolly (in Godwin, 1975).

Records from the last pleniglacial are known from the northern and eastern surroundings of the Alps and from adjacent periglacial regions to the north and the east (Koperowa, 1962; Frenzel, 1964, 1987; Koperowa and Środoń, 1965; Járai-Komlódi, 1966a,b). Late-glacial and some Early Holocene records are known from Svalbard (Środoń, 1960), Andøya (Vorren et al., 2009), more frequently from the British Isles (Godwin, 1975; Coxon, 1986), from Scandinavia (Hafsten, 1958, 1963; Berglund, 1966; Florin, 1969; Danielsen, 1970) and the Faroe Islands (Jóhansen, 1968). Most of these records are from the Younger Dryas, just a few are from the Early Holocene.

In conclusion, the following history of the range becomes clear: during the last pleniglacial, *K. islandica* was widespread in the periglacial regions of central and eastern Europe south of the Fennoscandian ice shield. After the melting of the continental ice this pioneer plant was forced back into the newly deglaciated areas in the north and northwest, but could not expand into the then ice-free Alpine region. The modern isolated populations in Scotland are obviously relicts from the Late-glacial. *Koenigia islandica* is therefore a good example of a species that, during the last glacial-interglacial cycle, experienced a very strong displacement from south to north combined with a strong range reduction: its modern range is an example of an 'interglacial relict' (see chapter 4.11 on refugia).

Salix herbacea

In the species-rich genus *Salix, S. herbacea* belongs to the group of the low-growing species with modern main ranges beyond the polar and alpine treelines. Today's distribution of the northern hemispheric/amphi-Atlantic *S. herbacea* is in the arctic-alpine of Europe. It reaches from Iceland, via the British Isles to north-western Fennoscandia, but also to the Pyrenees, the Alps, the Carpathians and some isolated mountains in Spain, Italy and the Balkans (Fig. 4.4.6-6). Since Carolus Linnaeus, *S. herbacea* has been considered 'the smallest tree on Earth'. The woody stems and branches remain mostly under ground, while the leaves and flowers reach only a few centimetres above the soil surface. Its life form is transitional between a chamaephyte (survival-buds above ground) and hemicryptophyte (survival-buds just below the soil surface). *Salix herbacea* is a species of habitats where the duration of the snow cover is long, where it is thus protected from frost. In the Alps it occurs primarily in hollows above the timberline in snow-beds (Salicion herbaceae; Salicetalia herbaceae; Salicetea herbaceae) but also in alpine gravel and meadow communities. It prefers soils on siliceous bedrock, but may occur on calcareous rock. In the main range of the arctic it occurs in many different communities in various habitats that range from dry to humid (Beerling, 1998).

The macrofossil records are primarily based on the very characteristic round, crenate-serrate

Salix herbacea

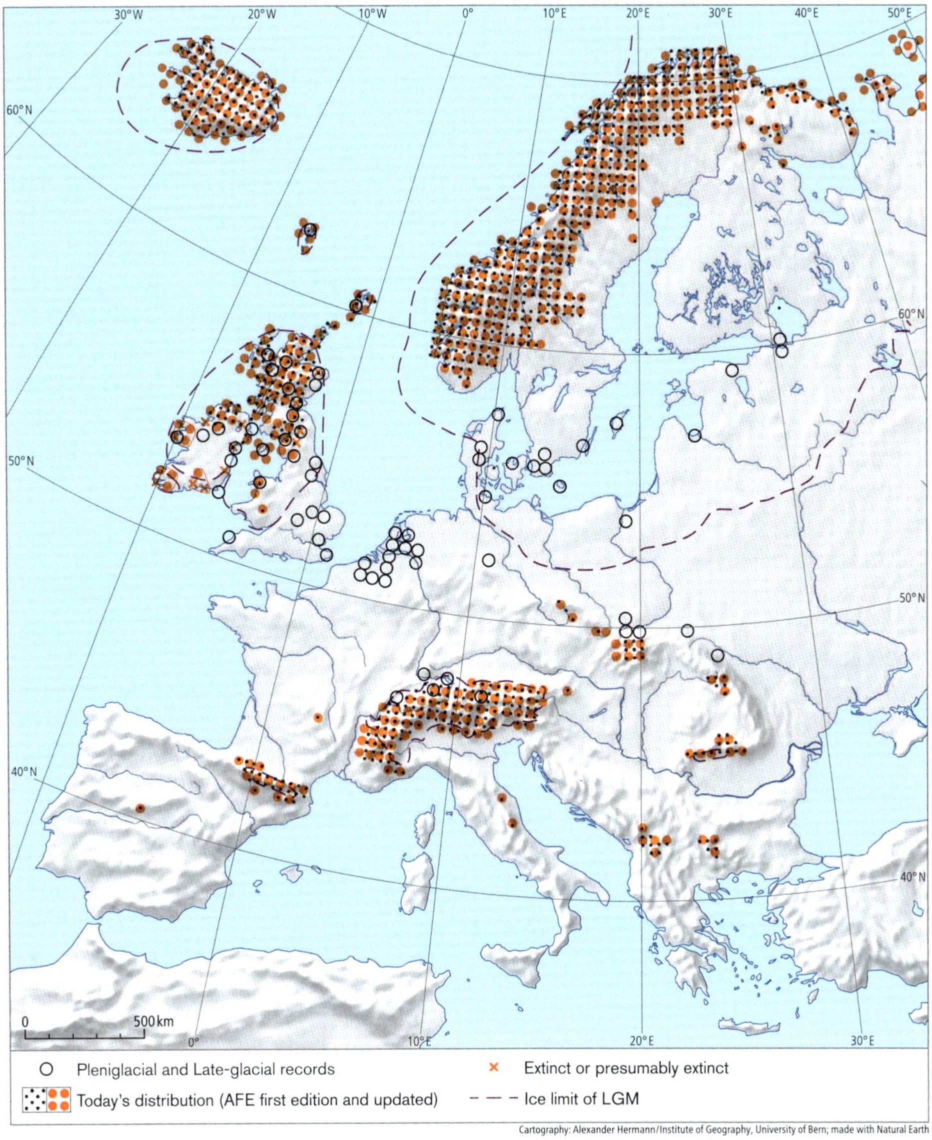

Fig. 4.4.6-6. Finds of ***Salix herbacea*** macrofossils from the pleniglacial and Late-glacial, its current distribution and the maximal extent of the ice sheets during the Last Glacial Maximum (LGM). AFE: Atlas of Flora Europaea.

leaves where the veins reach into the 'teeth'. In *S. polaris*, fragments of leaves may look similar but the leaf margins are entire and the veins do not reach the leaf margins. Pictures can be found in Godwin (1975) and Birks (2013). Such leaves were identified in the early records from southern Sweden (Nathorst, 1870; Andersson, 1888). In pollen morphology, the identification to species is more difficult (Straka, 1952; Erdtman et al., 1963).

As figure 4.4.6-6 summarizes, the pleniglacial and Late-glacial records show large gaps in the European lowlands between the glaciated areas. For pre-Weichselian glacials, Tralau (1963) did not find such large central European gaps and he discussed whether the separation into a northern and an Alpine-Carpathian range only developed in the last glacial period. On the other hand, it seems possible that the larger number of fossil records in the north-west of Europe indicates the development of oceanic tundra communities, as was sometimes discussed for *Betula nana*. In a multidisciplinary approach, Alsos et al. (2009) combine fossil records with phylogeography and species distribution modelling. Based on amplified fragment length polymorphism (AFLP) they show in impressive figures a very deep and probably very old split between Canada/western Greenland and Europe/ eastern Greenland/Iceland/Jan Mayen. Lesser but still clear differentiations were found for Svalbard/ Scotland–southern Scandinavia/northern Scandinavia and north-western Russia versus the Pyrenees and versus the Alps and the Carpathians. Not surprisingly the distribution modelling under climate warming suggests greatest losses of diversity in the southern populations.

Salix herbacea also plays an important role in reconstructing changes of atmospheric CO_2 because its subfossil leaves are robust, with thick cuticles often well preserved and thus used for studies of stomatal density (number of stomata per unit leaf ares) or stomatal index (number of stomata proportional to the sum of stomata plus epidermal cells). Especially changes in atmospheric CO_2 during the rapid transitions into and out of the Younger Dryas are of interest (Beerling and Chaloner, 1992; Beerling et al., 1995; Rundgren and Björck, 2003), but also changes during the Holocene (Rundgren and Beerling, 1999).

4.5

Anthropogenic changes to the vegetation

Karl-Ernst Behre[1], W.O. (Pim) van der Knaap[2,3], Gerhard Lang[2], César Morales-Molino[2,3], Willy Tinner[2,3]

[1] Lower Saxony Institute for Historical Coastal Research, Wilhelmshaven, Germany
[2] Institute of Plant Sciences, University of Bern, Bern, Switzerland
[3] Oeschger Centre for Climate Change Research, University of Bern, Bern, Switzerland

4.5.1

Fundamentals

Archaeobotany and palaeoethnobotany

The strong increase in human activities during the course of the Holocene led to a growing acceleration of their impact on the vegetation, which resulted in the open and completely remodelled landscape of today. Due to the close interrelationship between vegetation and human settlements, archaeological excavations provide important and mostly well-dated plant material especially for the time before written sources. It is primarily macro-remains, such as fruits and seeds but also vegetative particles, particularly wood, that are unearthed. Most of these are carbonized but waterlogged sites with fresh material also exist.

In contrast to pollen, these remains can often be determined to species level and thus they complement the pollen evidence. Taken together they enable the reconstruction of the prehistoric environment of the settlements and the nature of the human interference: clearing, grazing, and the various types of arable farming as well as the impact of mining, salt extraction and glass making on the forests.

The domestication and history of cultivated plants is the main focus for the investigation of macro-remains, but the associated weeds are also important because they can be used, for example, to illuminate field rotation or different ways of harvesting. The origin of weed species coming as apophytes from natural habitats in the region or as archaeophytes, introduced by farming or trade from distant areas can also be shown through the investigation of plant macro-remains from archaeological excavations.

The pioneer in this field was Oswald Heer, who, in 1865, published the seminal paper "*Die Pflanzen der Pfahlbauten*", where he described the well preserved plant material from the waterlogged pile dwellings along the shores of Swiss lakes that had been discovered in the preceding dry winters. During the following decades botanical investigations of prehistoric settlements slowly spread from Switzerland to other parts of central Europe and beyond. By and by these studies developed into a discipline in its own right. In the 1950s the term palaeoethnobotany was introduced by Hans Helbaek to define the study of archaeological plant remains. This term is derived from ethnobotany, which was defined as the study of the interrelations of primitive humans and plants. Nowadays the more convenient term archaeobotany is mostly employed. It may, however, seem less appropriate than palaeoethnobotany because, by definition, it does not focus on human-plant interactions. On the other hand, almost all plant remains preserved in archaeological contexts arrived there directly or indirectly through the activity of ancient people (Willerding, 1978; Pearsall, 1989).

Another related term is palaeobotany, which is nowadays confined to palaeontology in connection with geology.

There are several surveys for Europe in addition to monographs: Renfrew (1973), Körber-Grohne (1987), van Zeist et al. (1991), Jacomet and Kreuz (1999), Zohary et al. (2012), Knörzer (2007), Cappers and Neef (2012) and Neef et al. (2012). A special journal in this field *Vegetation History and Archaeobotany* exists, founded in 1992, as well as a database: Archaeobotany (2021).

The reflection of human activities in pollen diagrams

With the arrival of Neolithic farming cultures, the forests that prevailed in Europe were opened up both by clearings for fields and by forest grazing. This can be seen in the pollen diagrams, where the tree pollen (AP) decreases while the non-arboreal pollen (NAP) increases. The ratio between AP and NAP in the pollen diagrams provides good information about the activities of people in the surrounding area. However, changes in forest density or forest composition may also have natural causes. Therefore, independent criteria that show human activities are necessary and, with respect to the pollen diagrams, these are called 'anthropogenic indicators' (Table 4.5.1-1; Behre, 1981, 1986).

These anthropogenic indicators include a wide range of species which differ considerably in their value for this purpose. To make their use easier, two groups of pollen-producing plants are distinguished: primary and secondary anthropogenic indicators. Primary anthropogenic indicators are defined as all those species which are cultivated in fields, orchards and gardens. They do not, however, include forest trees or meadow plants even if they have been planted or sown. The most important primary anthropogenic indicators are the cereals (Ce in Table 4.5.1-1), of which the pollen of *Secale* and *Zea* can be easily identified, while it is more difficult to identify the pollen of the other cereals, for which the microscope method of phase contrast has to be applied (Beug, 2015). In most parts of Europe the separation of Cerealia pollen from that of wild grasses can usually be done on the basis of size and on the structure of the pore. However, an absolute distinction of pollen from wild grasses and pollen from cereals is impossible. There is a small overlap of both types that must not be neglected and, therefore, it is necessary to use the name 'Cerealia type', especially where early, single finds are concerned (Behre, 2007b). When working in the Near East, the identification of Cerealia pollen is much more difficult, if not impossible. Here the Cerealia pollen type includes cultivated cereals, the wild ancestors of the cereals which occur there and several other wild grasses. Although less pronounced, a similar problem also occurs in the Mediterranean, where occasionally wild grasses may produce the Cerealia pollen type.

When looking at the values of Cerealia in pollen diagrams one has always to keep in mind that *Secale*, being an allogamous species, has high pollen productivity and a good dispersal capacity while the autogamous cereals, *Triticum*, *Hordeum* and *Avena* produce much pollen which remains in the hulls and is, therefore, only poorly dispersed. Because of this, there is a strong difference between their appearance in the pollen diagrams and their real presence in the surrounding area.

As can be seen in Table 4.5.1-1, even in central Europe, which has been intensively investigated, only a few of the cultivated plants are frequently encountered in pollen diagrams or can be identified by their pollen. The letter 'O' denotes cultivated plants other than cereals such as the food plants *Fagopyrum* and *Vicia faba* as well as plants used for both oil and fibres such as *Linum usitatissimum* and *Cannabis sativa/Humulus*, which are recorded in many pollen diagrams. A great number of species can, however, not be identified at all or belong to genera or families that include wild plants so that their appearance has less significance. Moreover, pulses and most of the fruit trees in particular are underrepresented in the pollen record as they are insect-pollinated.

Fortunately other useful trees such as *Castanea*, *Juglans*, *Olea* and *Ficus* are well represented in pollen diagrams: they are particularly important in southern and south-central Europe and, in many areas, indicate the cultivation of nuts and fruits. Their pollen, however, cannot be distinguished from those of their wild forms. Furthermore, *Juglans* and *Castanea* have become forest trees in large

Table 4.5.1-1. The principal anthropogenic indicators in pollen diagrams and their occurrences in the various farming contexts. The frequency of occurrence is considered without reference to variation in pollen production and dispersal. Circles and crosses show their different frequencies in the different vegetation units. Ce: Cereals; O: Other easily detectable cultivated plants; Tr: Fruit and nut trees; Ad: Adventive species (hemerochores); Ap: Apophytes; Fa: Apophytic plant families. Life forms: T: Therophytes; H: Hemicryptophytes; G: Geophytes; Ch: Chamaephytes; P: Phanerophytes. From Behre (1981), updated.

		Life forms	Winter crops	Spring crops/root crops	Fallow land	Moist meadows and pastures	Dry pastures/heathlands	Grazed woodland	Footpath and ruderal habitats	Natural habitats
Primary indicators										
Ce	*Secale cereale*	T	◯	+	+					
	Hordeum type	T	○	◯	+					
	Triticum type	T	○	◯	+					
	Avena type	T		◯	+					
	Zea mays	T		◯						
O	*Fagopyrum*	T		◯	+					
	Linum usitatissimum type	T		◯	+					
	Vicia faba	T		○	+					
	Cannabis/Humulus	T		○	+					
Tr	*Castanea sativa*	P								
	Juglans regia	P								
	Olea europaea	P								+
	Ficus carica	P								+
Secondary indicators										
Ad	*Centaurea cyanus*	T	◯	+	+					
	Agrostemma githago	T	◯	+	+	○				
	Papaver type	T	◯	○	○				○	+
	Anagallis type	T	○	○	○				○	+
	Scleranthus annuus	T	○	+	+		+			
	Spergula arvensis	T	○	○	○		+		+	
	Fallopia convolvulus type	T	○	○	○				○	
	Polygonum aviculare	T	○	○	○	○		○	○	○
	Persicaria	T	○	○	○			+	○	○
	Mercurialis annua	T		○		+				◯
	Plantago lanceolata	H			○	◯	+	+	+	
	Plantago major, P.media	H		+	○	○	○	○	◯	
	Ambrosia	T		○	○				◯	+
Ap	*Rumex acetosella*	H	○	○	○	+	○	○	+	○
	Rumex acetosa	H			○	◯		○	+	+
	Trifolium repens type	H	+	○	○	○	+	○	○	○
	Succisa pratensis	H				○	○			○
	Urtica	H		○	○			○	◯	○
	Rumex obtusifolius	H		○		○		○	◯	○
	Artemisia (vulgaris)	H			○			○	○	+
	Melampyrum (pratense)	T					○	◯		○
	Pteridium aquilinum	G			+	○	○	◯		○
	Polypodium vulgare	Ch					○	○		○
	Calluna vulgaris	Ch					◯	○		○
	Juniperus type	P					○	○		+
Fa	Poaceae		○	○	◯	◯	○	○	○	○
	Cyperaceae				○	◯	○	○		◯
	Caryophyllaceae		○	○	○	○	+	+	○	○
	Brassicaceae		○	○	○	○	+	○	○	○
	Ranunculaceae		+	+	○	◯		○	○	○
	Apiaceae				○	○	+	○	○	○
	Chenopodiaceae			○	○				○	+
	Campanulaceae	H			○	+	○	○		+
	Asteraceae		○	○	○	○	○	○	○	○
	Cichorioideae		○	○	○	○	○	○	○	○

parts of southern Europe as well as in France but are not harvested for food purposes there. Also in southern Switzerland *Castanea* spreads into the forests and grows spontaneously in a natural way.

Apart from the primary anthropogenic indicators, there are a number of other species that can be used to trace different human activities in the past. These are the secondary anthropogenic indicators, which are not intentionally grown by people but are favoured in various ways or unintentionally introduced by people and their economy. They comprise weeds, ruderal and other species as well as components of the meadows and pastures, briefly: all wild and synanthropic species that show a positive response to any form of human impact. These species are generally recognized as falling into one of two categories: apophytes (Ap) and adventive (Ad) species. The apophytes are native plants which are favoured, directly or indirectly, by cultural activity and colonize newly established biotopes. Adventive species, on the other hand, are synanthropic plants that are not native in the respective area but were unintentionally introduced by people from another region. There are additional names for the latter that underline their form of transport: by humans (anthropochores) or generally by culture including animals (hemerochores). With respect to their time of appearance these plants have been divided into archaeophytes that appeared before AD 1500 and neophytes, which came later, in most cases from the New World or east Asia.

Most of the secondary anthropogenic indicators occur in a range of farming contexts and plant communities but have their core area in one or the other as is shown in a simplified way in Table 4.5.1-1. Because of this, the evaluation of their pollen curves has to be done with care. The differentiation into the various groups takes cognizance only of the farming practice and not of other factors such as soil fertility and humidity. The ways of farming changed considerably from the Neolithic to modern times and dramatically within the last century. The information presented in Table 4.5.1-1 is, therefore, based primarily on modern plant communities and on the occurrence of these species before the introduction of the chemical control of weeds.

The most important secondary anthropogenic indicator is *Plantago lanceolata*; its spread is strictly related to farming and it is not only an important indicator of a cattle-based economy and ruderal habitats but also of old arable farming in that it was present in fields before the introduction of the mouldboard plough, which extirpated perennial weeds. It is common all over Europe apart from northern Scandinavia and Finland. Its reliability as an indicator in the Mediterranean region is, however, limited because other taxa, for example *P. lagopus*, are non-indicator plants but grow on dry and sandy ground and produce similar pollen. Likewise, there are also other secondary indicators in the Mediterranean that cannot clearly be separated from related non-indicator taxa. Given the occurrence of native wild plants which produce pollen types that cannot be distinguished from those of crops (e.g. cereals, *Olea*, *Juglans*, *Castanea*, *Vitis*) and adventives (e.g. *Plantago lanceolata*), probabilistic approaches combining both, primary and secondary indicators, have been developed for human impact reconstructions in the Mediterranean realm (e.g. the land use probability index LUP, Deza-Araujo et al., 2022).

In addition to herbs, shrubs or dwarf shrubs such as *Juniperus* or *Calluna* are also important indicators of pastures. Several families also have to be regarded as secondary anthropogenic indicators (Fa in Table 4.5.1-1), because many of their members, whose pollen cannot be separated, show human activity: the most important is Poaceae.

Several of the apophytes in Table 4.5.1-1 had been quite common during the Late-glacial, when natural open vegetation prevailed in most parts of Europe. In particular *Artemisia*, *Juniperus,* Chenopodiaceae and the grasses had high pollen frequencies but do not indicate any human activity at that time.

When considering the huge region of Europe as a whole, there are several species which are apophytes in one area and aliens in another, in accordance with their climate and soil requirements. Therefore Table 4.5.1-1 is mainly valid for central Europe and the immediately surrounding areas and has to be modified for other regions as has already been done for northern Norway (Vorren, 1986) and France (Brun, 2011).

Another approach to the reconstruction of past land-use was taken by Gaillard et al. (1994). Here pollen datasets of modern land use were compared with fossil pollen samples from different periods by means of canonical correspondence analysis. This method has been applied to southern Sweden. The main problem is, however, to find suitable modern analogues of former farming practices.

Anthropogenic indicators, upland herb pollen and wild grasses in selected pollen diagrams

Histograms represent half millennia. Cultural periods (right) are Mesolithic period (Mes), Neolithic period (Neol), Bronze Age (Bron), Pre-Roman Iron Age (Iron), Roman period (Rom), Migration period (Migr), Medieval period (MA), and Modern times (Mod). Percentages are calculated on an upland pollen sum that excludes wild grasses; these grasses are calculated on an upland pollen sum that includes grasses. NAP - non-arboreal pollen (upland herb pollen); Poaceae - grasses; Anthropogenic indicators include the herbaceous crop plants.

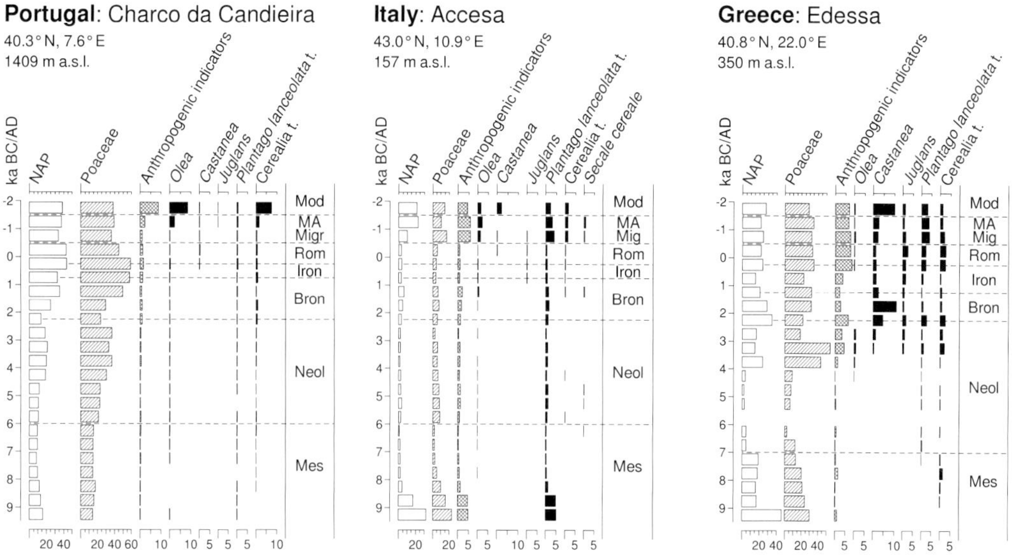

Fig. 4.5.1-1. Charco da Candieira: van der Knaap and van Leeuwen (1995, 1997); anthropogenic indicators include both the herbaceous and woody crop plants. Accesa: Drescher-Schneider et al. (2007b). Edessa: Bottema (1974).

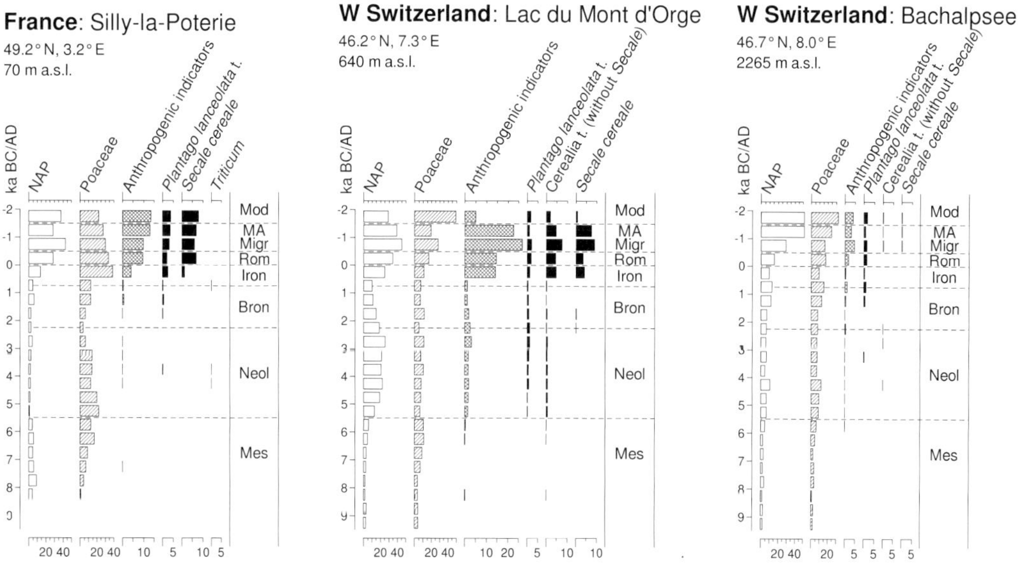

Fig. 4.5.1-2. Silly-la-Poterie: van Zeist and van der Spoel-Walvius (1980). Lac du Mont d'Orge: Welten (1982b). Bachalpsee: Lotter et al. (2006).

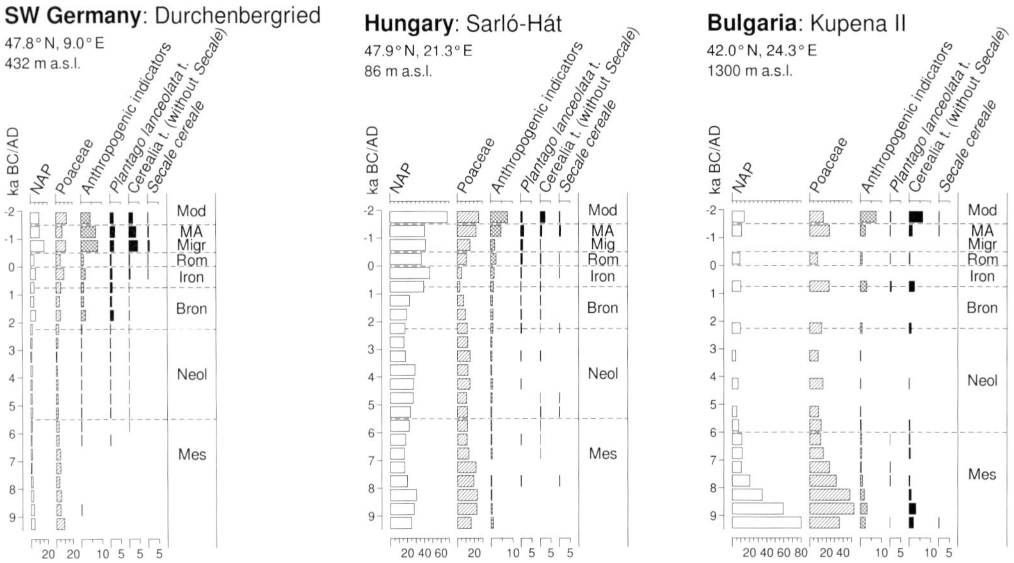

Fig. 4.5.1-3. Durchenbergried: Rösch (1986). Sarló-Hát: Magyari et al. (2010). Kupena II: Huttunen et al. (1992).

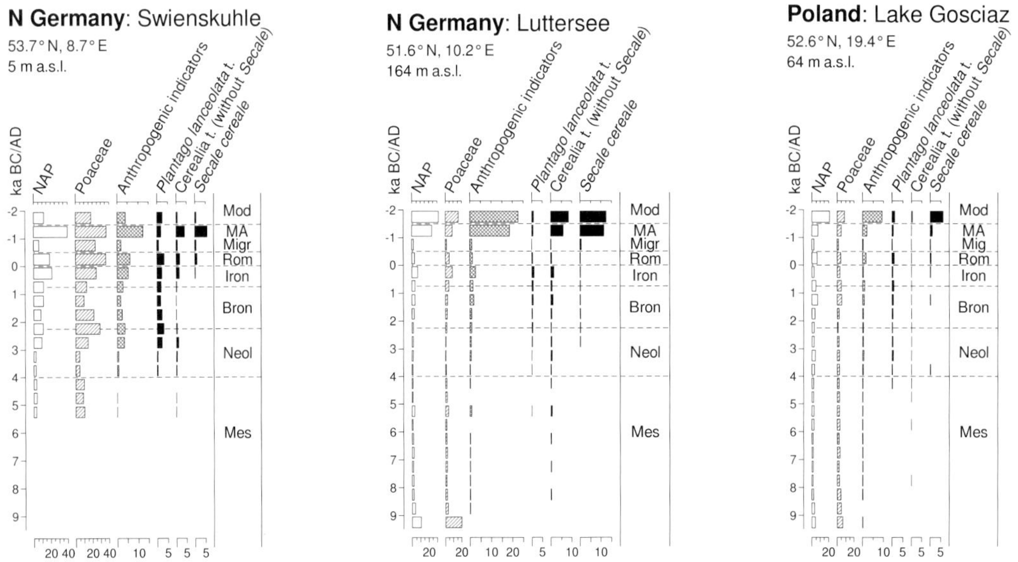

Fig. 4.5.1-4. Swienskuhle: Behre and Kučan (1994). Luttersee: Chen (1988). Lake Gosciaz: Ralska-Jasiewiczowa et al. (1998).

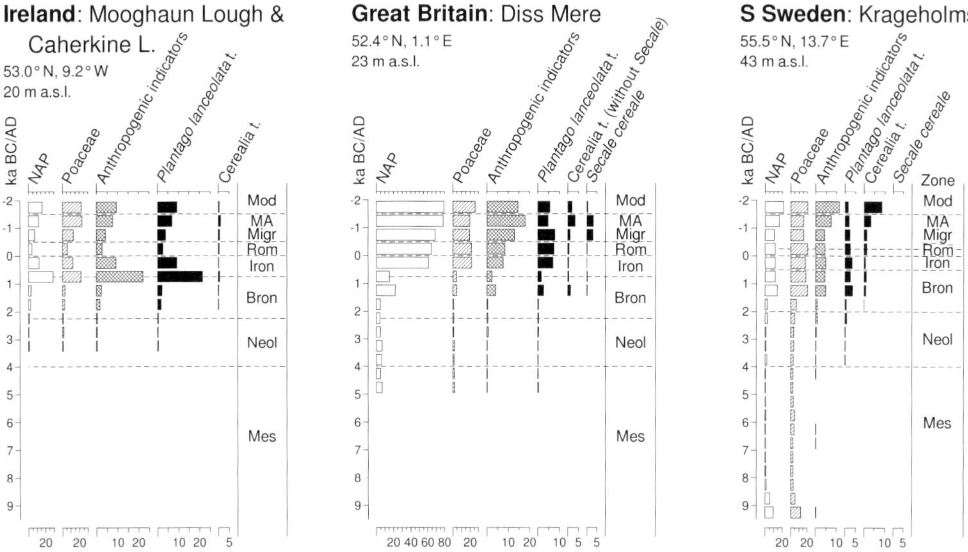

Fig. 4.5.1-5. Mooghaun Lough and Caherkine Lough combined: O'Connell et al. (2001). Diss Mere: Peglar (1993a). Krageholmssjön: Gaillard (1984b); Gaillard and Digerfeldt (1991).

Fig. 4.5.1-6. Loch Cleat: Williams (1977); Birks and Williams (1983). Hobbister: Farrel (2015). Årstrand/Lofoten: Nilssen (1988).

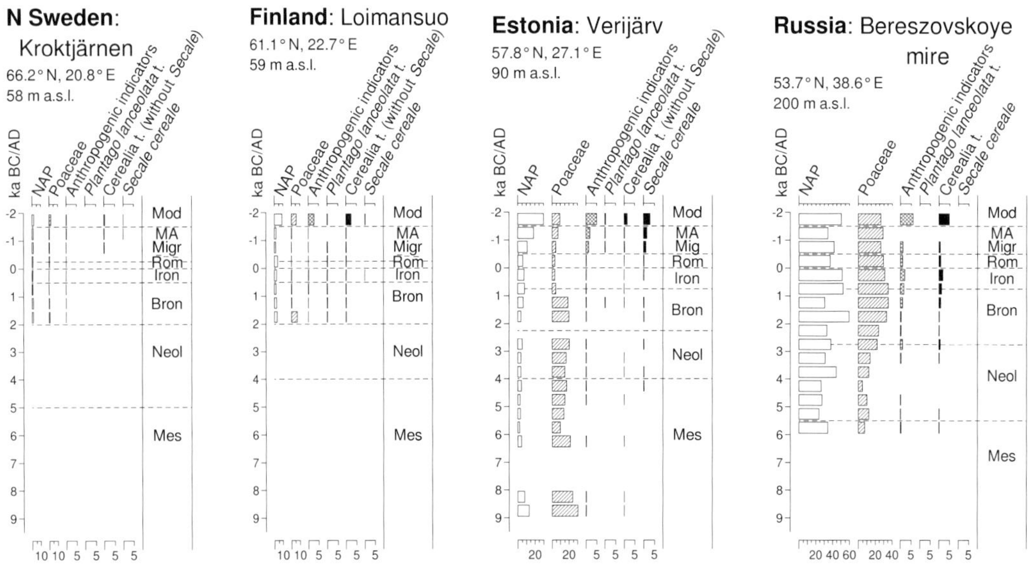

4.5.1-7. Kroktjärnen: Segerström (1990). Loimansuo: Vuorela (1975). Verijärv: Niinemets and Saarse (2006). Bereszovskoye mire: Novenko et al. (2012).

4.5.2

People and environment during Palaeolithic and Mesolithic times

The first humans who entered Europe were *Homo erectus*. They came from Africa and the first remains to be found were in Spain and dated to around 780 000 years BP. *Homo sapiens neanderhtalensis* evolved from *H. erectus* and occupied large parts of Europe from the Eemian until the Pleni-Weichselian. Between 50 000 and 40 000 years BP another human, the modern *H. sapiens sapiens*, who had also descended from *H. erectus* and originated in Africa, entered Europe from the south-east via the Levant. For about 5000 years both lived side by side, but around 38 000 years BP, *H. neanderthalensis* became extinct and *H. sapiens* remained the only human species (Terberger, 2019).

In archaeological terms, this took place during the **Palaeolithic**, which lasted until c. 11 600 years BP when it was followed by the Mesolithic. The very long Palaeolithic period is subdivided into three stages and, within these, several archaeological cultures, defined by their characteristic stone artefacts, have been distinguished: Early Palaeolithic (Acheuléen and Clactonien), Middle Palaeolithic (Mousterian 200 000–40 000 years BP) and Late Palaeolithic (Aurignacian 38 000–32 000, Gravettian 28 000–2 1000, Solutrean 22 000–18 000, Magdalenian 18 000–12 000 years BP). These names show that many Palaeolithic sites are known from France and it was there where research on these old cultures started. Most of these large culture groups are divided into regional ones. Well known for the last part of the Palaeolithic is the Pentknife point complex in Britain, the Azilien and Federmesser groups in France and Germany and the Bromme culture in southern Scandinavia and the surrounding area. The last Palaeolithic culture in north Germany and south Scandinavia was the Ahrensburg culture which, like the Azilien, ended in the west around 11 600 BP (=9650 BC).

The subsequent Mesolithic period was of a different duration. Beginning around 9650 BC, it lasted until the onset of farming cultures, that is between c. 11 000 cal yr BP (9000 BC) in the Levant and 6000 cal yr BP (4000 BC) in northern central and north Europe (Fig. 4.5.3-4). Because of the very short duration of this period in the Near East, the term Mesolithic is not used there and this period is included in the last part of the Late Palaeolithic, called Epipalaeolithic.

Important cultures from the earliest part of the **Mesolithic** are the Sauveterrien in south-west

Europe, the Beuronien from the Paris Basin to the Carpathians and the Maglemosian north of that. Major cultures that existed in the north at the end of the Mesolithic until 4000 BC were the Swifterbant culture in the The Netherlands, the Ertebölle culture in Denmark and the surrounding areas and the Neman-Zedmar culture from north Poland to the east (Terberger, 2019).

During the Early Palaeolithic, population density was very low, probably around 100 000 people in the whole of Europe, but increased step by step. People had to follow the environmental changes caused by the alternation of warm and cold periods and there was no human impact on the landscape at that time. The humans were foragers: fishers, hunters and gatherers and lived in dense forests, open woodland and grass steppe but preferred to live along the sea coast or near lakes and rivers where fish and sea mammals were available. In mountainous areas they lived in caves and rock shelters (abris in French) and in such situations, particularly in late Palaeolithic times, they produced rock paintings with detailed hunting sceneries like the famous ones at Altamira in northern Spain or Lascaux in the Dordogne in France.

During the Palaeolithic a great variety of large animals were still available for the hunters such as *Hippopotamus* and *Trogontherium* (giant beaver) in the waterways and, elsewhere, others such as aurochs, wild horse and the impressive elephants: first the southern mammoth, *Mammutus meridionalis*, which was followed by the forest elephant, *Palaeoloxodon antiquus*, in the warm stages and then the mammoth, *Mammutus primigenius* in the cold steppe (von Koenigswald, 2007). Most of the large mammals, however, became extinct at the end of the last glaciation and there is still a vivid debate about the reasons for this more or less synchronous extinction. There are strong arguments that this was caused by overkill by the growing number of humans with increasingly better weapons but climate change also has support as the main reason for this extinction.

Following the excavation of the unique site of Schöningen, north Germany, it has become clear that the evolution of well-organized group hunting started very early. Here a hunting and butchering camp was excavated that contained several thousand bones of large mammals, including more than 20 skeletons of wild horse, dated to around 300 000 years BP. Even at that time *Homo erectus* used skilfully produced spears of which eight complete ones were unearthed together with many lithic tools (Thieme, 2007).

Many sites of reindeer hunters which date to the end of the Palaeolithic have been excavated in northern central Europe and Scandinavia. In the open grassland of the Late-glacial the hunters followed the herds of wild reindeer that migrated further and further to the north as afforestation spread from the south at the beginning of the Mesolithic. Because these hunters and foragers did not yet have permanent settlements, their impact on the vegetation was still very small and is not visible in the pollen diagrams. Though absolute proof that fire was used in conjunction with hunting is difficult to achieve, it must still be regarded as a possibility. Real forest fires were only likely in coniferous woods. There are numerous sites in the sandy areas of the Netherlands and north-west Germany where there is a thin layer of charcoal of Allerød age which has been overblown by a cover of sand during the subsequent Younger Dryas period, the so-called Usselo horizon. Uncertainty remains as to whether the fires were started by Palaeolithic hunters or naturally by thunderstorms. Similar activities may also have taken place during the Mesolithic and there are indications, for example from Star Carr in England (Mellars and Dark, 1998) and Zutphen in the Netherlands (Bos et al., 2006), that active burning of reed swamp areas has been used as a hunting technique. At several European localities, Mesolithic woodland disturbance, often associated with fire, has been considered to have an anthropogenic origin (Simmons and Innes, 1985), although natural processes cannot be fully excluded.

The vegetation of the Mesolithic from the Early Holocene until the Mid Holocene is characterized by total forest cover in most parts of Europe, with exception of the forest steppes in the south-east and above or north of the treeline. In the dense woodlands the provision of an adequate food supply by hunting and gathering alone was difficult and, therefore, the Mesolithic settlements were concentrated on coastal areas, at lake margins and along rivers, where fishing was also possible.

A special activity was the intensive gathering of molluscs, which resulted in shell middens of varying size along the coasts. Best known are the middens in north Denmark, where hundreds of shell mounds have been recorded, the largest being several hundred metres long and several metres thick: here they belong to the Ertebölle culture, dated to the late Mesolithic. Also in northern Spain and Portugal there are several localities with large heaps of shells which date from the Mesolithic: most important is the Muge site close to Lisbon. Other concentrations of shell mounds have been found in Brittany and western Scotland but they are missing from the Mediterranean (Bailey, 2008). In many coastal areas, however, they were later submerged and have been hidden by the sea since the Mesolithic.

Remains of gathered plant food have been found in many Mesolithic excavations. This comprises not only seeds and fruits (most commonly hazel nuts), but also the vegetative parts of a number of plants, such as leaves, roots and tubers (Kubiak-Martens, 2002).

4.5.3

History of cultivated plants (all dates given in calibrated years BC)

Domestication

Wild plants have several ways of dispersing their seeds in order to survive and to find new places in which to grow. They are transported by wind and water, inside and outside of animals, and must be well adapted for this process. This is in sharp contrast to cultivated plants where the seeds should not readily escape from the parent plant but keep together so that collection is easy. Domesticated species have lost the mechanisms of dispersal and possess new characteristics. This can be shown in the archaeobotanical material.

During the Mesolithic period, hunters and gatherers collected fruits and seeds when they became ripe. Eventually they started to grow some of the important species. However, just sowing seeds in a field does not produce a domesticated plant; there was still a long way to go. In at least ten Near East sites the archaeobotanical evidence suggests pre-domestic cultivation. The oldest cereal remains showed that, as a first step covering several centuries, an increase in grain size took place, probably through the selection of the seeds (Fuller, 2007). The next step, during which the cereals lost their natural way of dispersal, was much more radical (Fuller et al., 2010).

The ears of the wild cereals, as well as of other grasses, have a central axis, called a rachis, which is articulated and consists of internodes. In wild cereals the axis is brittle, which means that the individual spikelets disarticulate at maturity and disperse (Fig. 4.5.3-1). In contrast to this, domesticated cereals have a tough rachis and the spikelets keep together. This form is unable to disperse and offers distinct advantages for human use. It needs the protection of and dissemination by humans and can easily be harvested by them.

This tough rachis is the result of a mutation. When such a mutation occasionally occurred in the wild it was lost. However, if it occurred in a field of the earliest farmers, this was very useful for them, because the farmer could harvest the whole spike without loss of the grain, which was unavoidable with a brittle spike. So, probably, an unconscious form of selection happened. As a consequence, grains with this mutation were enriched in the seed and the cereal population unintentionally changed

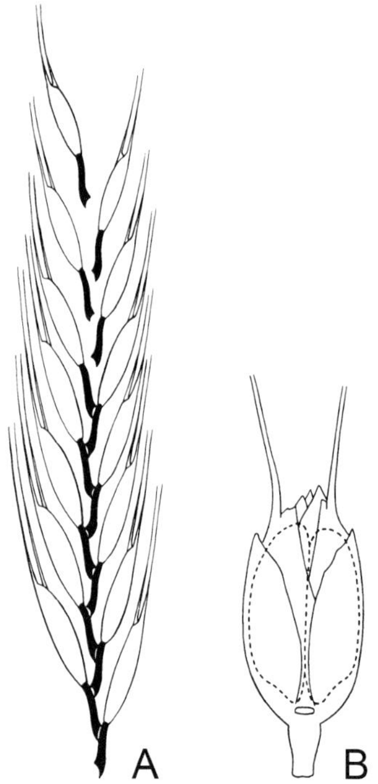

Fig. 4.5.3-1. A: Schematized ear (spike) of wild emmer wheat *(Triticum dicoccoides)*; the ear disarticulates at maturity. B: Emmer wheat spikelets (two grains enclosed by bracts). From van Zeist (1992).

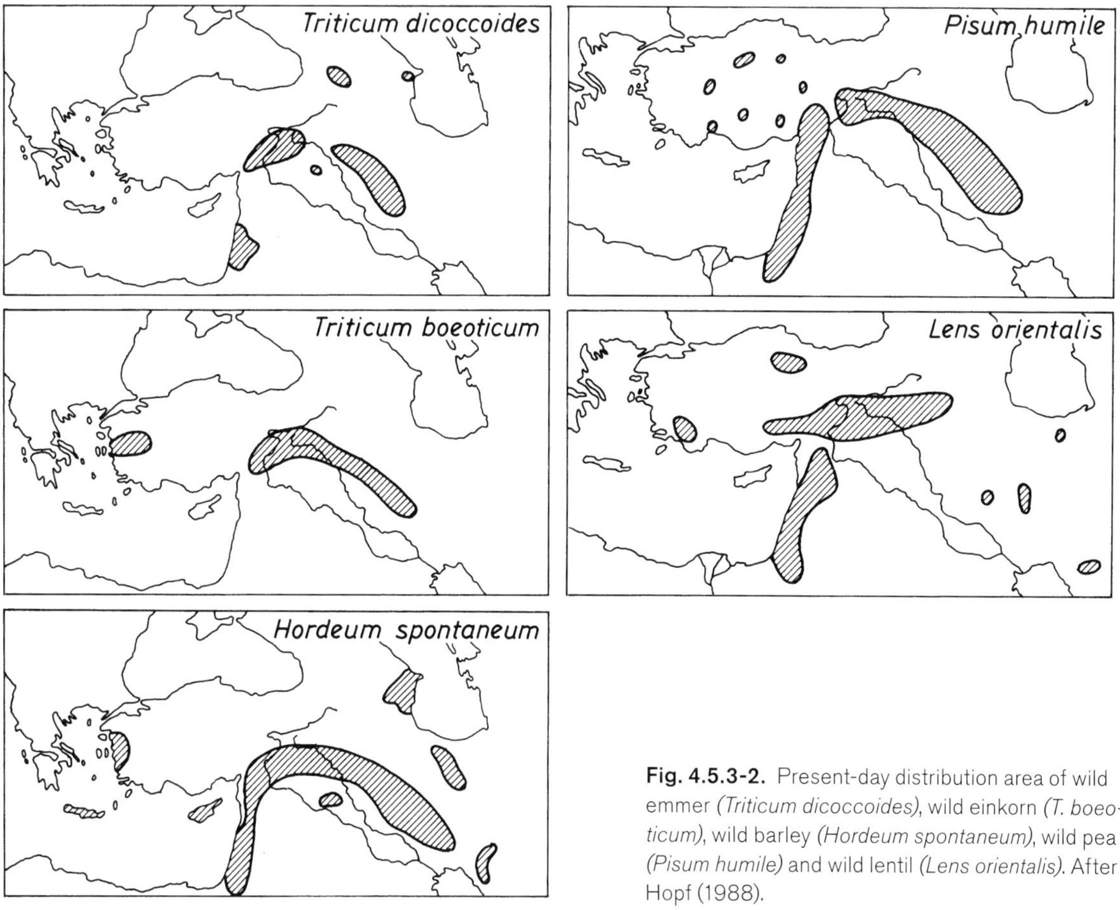

Fig. 4.5.3-2. Present-day distribution area of wild emmer *(Triticum dicoccoides)*, wild einkorn *(T. boeoticum)*, wild barley *(Hordeum spontaneum)*, wild pea *(Pisum humile)* and wild lentil *(Lens orientalis)*. After Hopf (1988).

to non-shattering plants, a process which, however, probably took more than 1000 years. This is the most important characteristic of a domesticated cereal: others are the loss or reduction of hairs, barbs and awns and the loss of germination inhibitions.

The loss of an efficient seed-dispersal mechanism can also be observed in other crop plants. In wild field pea and lentil the pods open at maturity and the seeds are flung away, but in the domesticated forms the pods remain closed. The same is also true for flax (van Zeist, 1992).

For thousands of years people gathered the seeds and fruits of the wild ancestors of cultivated plants such as cereals, legumes and others. The distribution area of many of these species, for example, *Triticum dicoccoides, T. boeoticum, Hordeum spontaneum, Pisum humile* and *Lens orientalis* is situated in the Near East and covered an area from south Palestine across southern Turkey to the Zagros Mountains in Iran (Fig. 4.5.3-2). In this area, the first domestication of plants took place around 8500 BC as can be shown by the archaeobotanical evidence from the sites of earliest agriculture, which have only been found here. In consequence, this wide arc has been called the Fertile Crescent.

The first plants that were domesticated include einkorn, *Triticum monococcum* ssp. *monococcum* (diploid), originating from its ssp. *boeoticum;* emmer, *T. dicoccum,* same as *T. turgidum* ssp. *dicoccum* (tetraploid) from *T. dicoccoides*, same as *T. turgidum* ssp. *dicoccoides;* barley, *Hordeum vulgare,* from *H. spontaneum;* lentil, *Lens culinaris,* from *L. orientalis;* pea, *Pisum sativum,* from *P. humile;* and flax (linseed), *Linum usitatissimum,* from *L. bienne.* They compose the Neolithic crop assemblage, which was also the initiator of food production in Europe and on other continents (Zohary et al., 2012; Nesbitt, 2002; Fuller, 2007; Asouti and Fuller, 2012; Willcox et al., 2012).

The Neolithic spread of agriculture from the Near East to Europe

The origin of agriculture and the earliest cultivated plants have been documented at several important sites in the Fertile Crescent, among others from Beidha in Jordan, Jericho in Palestine, Tell Aswad and Abu Hureyra in Syria, Cayönü in Turkey or Jarmo and Ali Kosh in Iran (Zohary et al., 2012). It is the area where the transformation of human foragers into agriculturally based societies took place for the first time. This process has been called the 'Neolithic Revolution' although it lasted for much longer than a modern revolution. By the end people not only had arable farming but herded domesticated animals and lived in fixed houses in permanent settlements and used pottery.

The chronology of the earliest Neolithic distinguishes between the period before and after the invention of pottery. The first appearance of domesticated plants is recorded in the second part of the pre-pottery Neolithic (PPN B), which is around 8500 BC (Nesbitt, 2002).

The so-called 'Neolithic package', which contained not only the Neolithic crop assemblage mentioned above but also domesticated sheep, goats, cattle and pigs as well as fixed rectangular buildings and pottery, spread from the Near East via Anatolia to Europe.

It is still a matter of debate as to whether invading tribes introduced the Neolithic culture into the Mesolithic regions and replaced or absorbed the local people or whether the indigenous Mesolithic populations adopted the Neolithic way of economy and living. Probably both processes were involved but to different degrees (Cunliffe, 2008). In several cases the Neolithic layers at an archaeological site directly overlap the Mesolithic strata and in addition to the bones of domesticated animals there are the remains of wild ones, thus showing the transition between the different economies.

The start of the Neolithic way of life led to a considerable change in the vegetation around the settlements due to the clearing of forest for fields, and because of grazing by domesticated animals, this opening of the forests is visible in the pollen diagrams (Figs. 4.5.1-1 to 4.5.1-7).

The first farmers arrived in Cyprus before 8000 BC (Lucas et al., 2012) and by around 7000 BC domesticated plants, together with the other constituents of the Neolithic package, had crossed the Aegean Sea and can be traced in the famous Franchthi Cave in southern Greece. The Neolithic economy soon advanced to northern Greece as recorded for example in Nea Nikomedia from around 6300 BC. Quite a number of sites with early farming are known from the adjacent Balkans. The first are situated in Bulgaria, where the most important settlement Karanovo starts at 6200 BC and by around 6000 BC several Neolithic cultures were established further north: the Starčevo culture along the Danube in Serbia, northwest of it the Körös culture

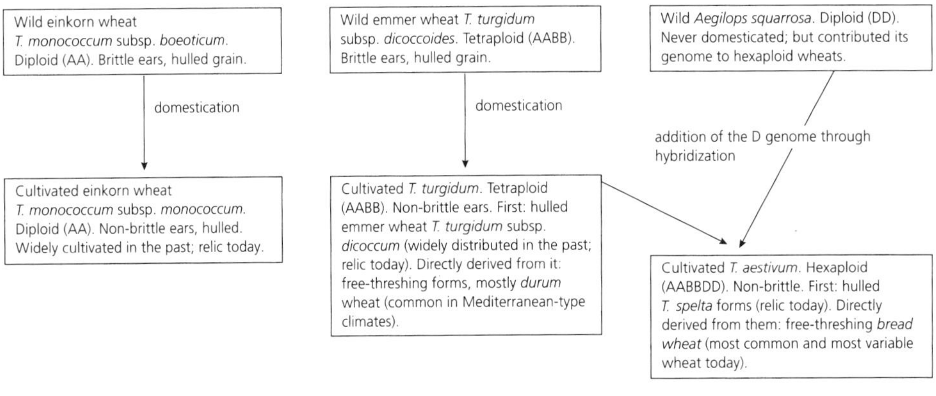

Fig. 4.5.3-3. The wild progenitors of the main cultivated wheats and their principal derivatives under domestication. The common names, still used in archaeobotany, are mentioned in the text. From Zohary and Hopf (2000).

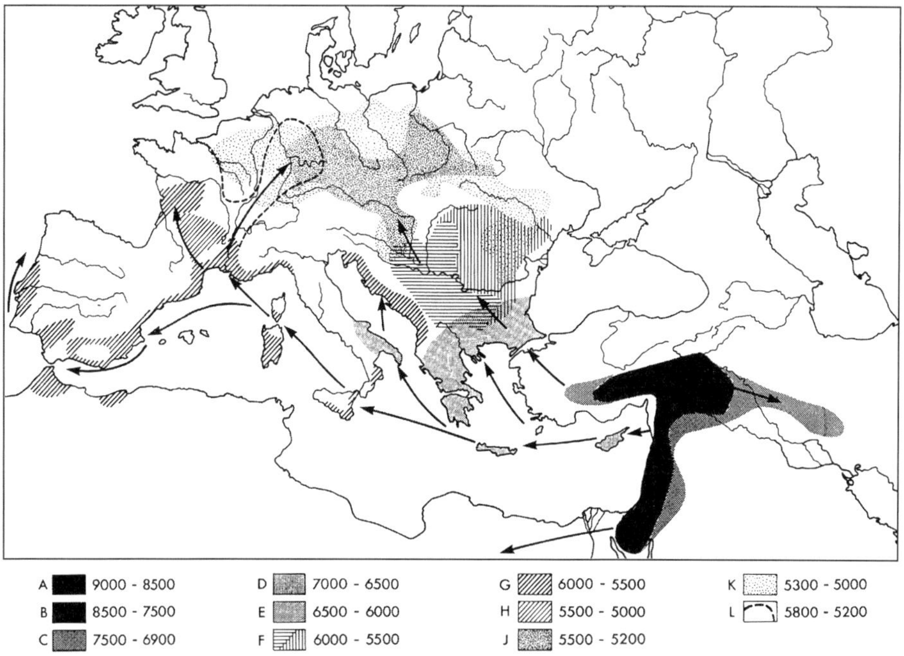

Fig. 4.5.3-4. The spread of the Neolithic economy from its origin in the Near East to Europe up until 5000 BC. A and B: Pre-pottery Neolithic (PPN); C: late PPNB and early pottery; D: first expansion to Europe (eastern Greece) and across the sea (Cyprus, Crete); E: west and north Greece, Bulgaria, southern Italy; F: Starčevo-Körös-Criş; G: Impressa and Cardial; H: Roucadourien; J: oldest linearbandkeramik (LBK; linear pottery culture); K: old to youngest LBK; L: La Hoguette (north and east border). All years are in cal BC. From Lüning (2000), slightly modified.

in Hungary and to the north the Criş culture in the Carpathian Basin in Romania (Fig. 4.5.3-4).

In addition to this continental direction of the spread of agriculture there was a maritime spread westwards into the Mediterranean basin. By 6500 BC the Neolithic package had already reached southern Italy, as documented by the Impressa-pottery. From here neolithisation proceeded northwards very quickly moving around the Adriatic Sea and along the west coast of Italy. Further to the west, from southern France across Spain to southern Portugal, by 5500 BC all coastal areas were occupied by Neolithic communities, which are characterized by the Cardial pottery, named after the imprints of *Cardium* shells (Lüning, 2000). From the very beginning, the farmers grew all the members of the Neolithic crop assemblage. Because the ancestors of these crops do not belong to the native vegetation of this area there is no doubt at all that they were introduced as part of the Neolithic package.

The inland areas of the northern Mediterranean for some time remained Mesolithic: here the Neolithic economy advanced firstly along the rivers, in particular along the Rhone to the north where it met the same way of life that had come there from the southeast (Jacomet and Kreuz, 1999; Cunliffe, 2008; Müller, 2009).

Starting around 5500 BC and originating in the Körös culture there was a fast expansion of the Neolithic way of life and economy along the Upper Danube which then spread across the loess areas in central Europe. This was the Linearbandkeramik (LBK), a particularly homogeneous culture, named after the characteristic incised decoration on their pottery. In the space of only 300 years it occupied large parts of the central European loess area from southern Poland across Czechia and lower Austria to the middle Rhine and then expanded considerably in a second phase from 5300 to 5000 BC to the west as far as the Paris basin. The LBK farmers brought with them the technology for constructing well-developed longhouses and the complete Neolithic package with domesticated plants and animals. In addition to this, the poppy, *Papaver somniferum*, which was lacking in the south-eastern Neolithic, was domesticated in the western part of the LBK area. It is still debated whether this culture spread through its adoption by indigenous hunter-gatherers or whether it was introduced by newly immigrating people, but in this case most archaeologists support the latter, in particular because the archaeological evidence in the loess region does not reveal a gradual transition from the hunter-gatherer way of life to entirely sedentary farming communities as in the Balkans or Greece (Bakels, 2009).

After this fast expansion, Neolithic cultures stopped at the northern boundaries of the loess areas. Beyond, Mesolithic cultures still prevailed for almost a millennium, with little contact with their advanced neighbours. In northern central Europe a big step took place around 4100 BC, when the Neolithic way of life spread, together with the Trichterbecher (TRB = funnel beaker) culture, into northern Germany, the northern part of the Netherlands and southern Scandinavia. By about the same time the remaining inland areas of Spain and France as well as most parts of Great Britain had also adopted the Neolithic economy (Lüning, 2000; Cunliffe, 2008; Müller, 2009).

Younger crop plants

The free threshing (naked) wheat species bread wheat, *Triticum aestivum* (hexaploid) and hard wheat, *T. turgidum* with its ssp. *durum* (tetraploid) which dominate today have a different history. *Triticum aestivum* does not belong to the first Neolithic crop assemblage. This naked wheat was derived directly from the hexaploid hulled wheat *T. spelta,* which only came into existence after the domestication of the tetraploid *T. dicoccum*. After its Neolithic cultivation, *T. dicoccum* expanded into the Middle East and, between 7000 and 6000 BC, reached the area of a diploid wild grass, *Aegilops squarrosa*. There both fused to the hexaploid glume spelt wheat, *T. spelt*a, from which *T. aestivum* quickly developed. After that, bread wheat also spread from the Near East on the route via the Balkans to central Europe, where it arrived in the late LBK period (Zohary et al., 2012).

The parent species *T. spelta* has also been cultivated in southern Russia and Transcaucasia but, according to archaeobotanical and genetical investigations, separately from it, another form of spelt wheat developed in central Europe by an introgression of genes from emmer into bread wheat. This European spelt came into existence around 2300 BC and is still cultivated on a small scale in southern Germany (Akeret, 2005).

Very soon after the domestication of the tetraploid hulled emmer, *T. dicoccum*, a free-threshing form evolved from it in the Near East. This naked wheat, *T. turgidum* especially its var. *durum*, spread on a western Mediterranean route via France and Spain and reached central Europe in the 5th millennium and eventually the TRB culture after 4000 BC. It met the hexaploid naked wheat in parts of south Germany (Maier, 1996; Kirleis and Fischer, 2014).

Today, the hexaploid bread wheat is the most common in the world, the tetraploid hard wheat is, however, also widespread in rather dry and warm regions: in Europe it is confined to the Mediterranean. For both species, many thousand cultivars have been developed. The glume wheats nowadays have only relic character.

The millets *Panicum miliaceum* and *Setaria italica* are of central Asian origin. They have been domesticated in east Asia and entered Europe from the east. Broomcorn millet, *Panicum miliaceum,* is the more important and this had already arrived in Poland in LBK times. Younger Neolithic records are from Romania and Czechia. During the Bronze Age, it spread to the west and also became an important crop plant in central Europe, France and Italy. Foxtail millet, *Setaria italica*, was very rare in the west during the Bronze Age and became a minor crop in southern Europe from the Roman period onwards.

Oats, *Avena sativa,* are regarded as a secondary crop and derived from the weedy *A. fatua*. In this case, domestication took place in Europe. The first records of *A. sativa* are from the last millennium BC, but in the 1st century AD oats became established as a major crop plant (van Zeist et al., 1991).

Rye, *Secale cereale*, has a more complicated history. Its ancestor is *Secale montanum*, which was already domesticated in the early Neolithic in

Table 4.5.3-1. Cereals and other useful plants of Europe. Circles: frequency of cultivation (smallest one: weed or collected). Mean values are indicated; there are often considerable differences between the various regions of Europe. The areas of the wild ancestors are indicated. From Lang (1994), extended.

	Origin	Neolithic	Bronze Age	Pre-Roman Iron Age	Roman Period	Middle Ages	Older Modern Times	Younger Modern Times
Carbohydrate plants								
Triticum aestivum/durum	W.As	○	○	○	○	○	○	◯
Triticum dicoccum	W.As	◯	○	○	○	◦	◦	∘
Triticum monococcum	W.As	○	○	◦	◦	◦	◦	∘
Triticum spelta	W.As	◦	○	○	○	○	○	◦
Hordeum vulgare	W.As	○	○	○	◦	◦	◦	○
Avena sativa	W.As		∘	∘	○	○	○	○
Panicum miliaceum	C.As	◦	○	○	◦	◦	◦	◦
Setaria italica	As	◦	◦	◦	◦	◦	◦	◦
Secale cereale	W.As			∘	○	◯	◯	○
Zea mays	C.Am						◦	◯
Fagopyrum esculentum	E.As					◦	○	◦
Solanum tuberosum	S.Am						○	◯
Protein plants								
Pisum sativum	W.As	◯	○	○	○	○	○	○
Lens culinaris	W.As	○	○	○	◦	◦	◦	◦
Vicia faba	W.As		◦	○	○	○	◦	◦
Phaseolus vulgaris	C.+S.Am						◦	○
Cicer arietinum	W.As	∘	∘	◦	◦	○	○	○
Oil and fibre plants								
Linum usitatissimum	Eur,W.As	◦	○	○	○	○	○	◦
Cannabis sativa	C.As			◦	◦	○	○	◦
Papaver somniferum	Med	○	○	○	○	○	◦	◦
Camelina sativa	E.Med	∘	◦	◦	○	◦	◦	∘
Brassica rapa	Eur	∘	∘	∘	∘	○	○	◦
Vegetables								
Cucumis sativus	S.As			∘	∘	◦	○	○
Cucumis melo	S.As,Afr			∘	◦	◦	○	○
Lycopersicon esculentum	C.+S.Am						◦	○
Beta vulgaris	Med,Atl			∘	∘	∘	◦	◯
Dye plants								
Reseda luteola	Med	∘		∘	◦	○	◦	∘
Isatis tinctoria	W.As			◦	◦	○	◦	
Rubia tinctorum	SW.As			◦	◦	○	○	∘
Fruit trees and nuts								
Olea europaea	Med	∘	◦	◯	◯	◯	◯	◯
Juglans regia	E.M,W.As		∘	◦	○	○	○	○
Castanea sativa	W.As		∘	◦	○	○	○	○
Ficus carica	Med	∘	∘	◦	◯	◯	◯	◯
Citrus div. spec.	SE.As			◦	○	○	◯	◯
Malus domestica	Eur	∘	∘	◦	○	○	◯	◯
Prunus avium	Eur	∘		∘	○	○	○	○
Prunus domestica	W.As	∘	∘	∘	◦	○	○	○
Vitis vinifera	Eur	∘	∘	◦	◯	◯	◯	◯

Anatolia. From there *Secale cereale* likely migrated to central Europe as a weed among other cereals and single grains of domesticated but weed rye have been recorded from there and from the route of its migration across the Balkans since the early Neolithic. The status of rye changed from a weed to a crop plant during the early Iron Age; its competitiveness with respect to the other cereals was probably enhanced by the climatic deterioration at that time. The beginning of rye being a crop in its own right during the pre-Roman Iron Age and the Roman period presumably took place independently in different areas of Europe (Behre, 1992).

Horse or faba bean, *Vicia faba*, is a pulse that entered Europe at the end of the Neolithic and spread fast as a main crop in the Bronze Age in the Mediterranean and central Europe. Its ancestors are unknown but it had already been domesticated in the early Neolithic in the Near East.

Another latecomer is gold of pleasure, *Camelina sativa*, an oil plant that played a major role in central and eastern Europe from the Bronze Age up until modern times but nowadays has become almost extinct. In a weedy form (*C. sativa* ssp. *linicola*) it came from the eastern steppes and infested the flax fields: only later was it picked up as an oil crop. An important modern oil plant is the rapeseed, *Brassica napus*, which has only been grown from the 17th century onwards. It is a tetraploid species and probably originated by hybridization between the diploid *B. rapa* and the diploid cabbage, *B. oleracea*, both of which are native and have been used over large areas of Europe since classical antiquity.

The last important crop plant that arrived in Europe before those coming from America was buckwheat, *Fagopyrum esculentum*. It is not wheat, but belongs to the Polygonaceae: however, it is used in the same way as cereals. The home of buckwheat is China and it was only introduced into Europe in the high Middle Ages in Germany and soon became an important crop plant on poor soils, particularly in north Germany, the Netherlands and France. Nowadays it is mainly grown in eastern Europe (Körber-Grohne, 1987; Zohary et al., 2012; Bakels, 2009).

Fruits and nuts

European fruits and nuts, trees or shrubs, have different histories: some were domesticated here from native parents, others have been introduced.

Apples, *Malus domestica*, and pears, *Pyrus communis*, have a similar history. The distribution area of their wild forms extends throughout Europe and their fruits were collected long before their domestication. Wild apples and pears were found in many Neolithic to Iron Age sites but the time of their domestication is not known: probably it took place at different times in different areas. Written sources from classical Greece show that grafting was in use in that period, which is a clear indication of domestication. In Roman times grafting became common all over the Empire, but not yet outside of it.

The sweet cherry, *Prunus avium*, is also native all over Europe, while the sour cherry, *P. cerasus*, which is tetraploid, has its ancestor in easternmost Europe and adjacent Asia, from where it was brought to other parts of Europe. Both cherries must have been collected at a very early date, since their cultivation is first mentioned in classical sources from Greece and Rome. Plums, *Prunus domestica*, are hexaploid and have a different origin. They do not have relatives in most parts of Europe and probably developed in the Caucasus region by crossings of a wild form of *P. domestica* with *P. cerasifera*. In central and more distant parts of Europe wild forms still appeared in the Bronze Age. Plum cultivation depends almost entirely on grafting and the earliest written sources of plum planting and grafting are from Roman times. *Prunus amygdalus* (almond), on the other hand, can easily be planted as seeds. Its wild form is native in the Near East and southeast Europe. Its fruits were also collected very early and probably already cultivated in the Bronze Age. Quince, *Cydonia oblonga*, is native in Iran and the Caucasus region and also reached the Mediterranean in classical times. Similarly pomegranate, *Punica granatum*, a minor crop in southern Europe, came from this area but is also native in the Balkans.

Other related fruit species, peach, *P. persica*, and apricot, *P. armeniaca*, are of east Asian origin and were brought to Europe during the Greek period via Iran and Armenia as is also evident by their names. Citrus fruits also have their origin in southeast Asia and India. Only the citron itself, *Citrus medica*, arrived in Europe as early as Greek times, most of the other cultivars did not appear until the Islamic period but spread rapidly through the Mediterranean basin (Zohary et al., 2012).

Three classical fruits characterize the Mediterranean region: olive, *Olea europaea,* fig, *Ficus carica* and the grape vine, *Vitis vinifera.* The wild forms of all of them are native in the region and their fruits were collected from the wild long before they were domesticated. Archaeobotanical investigations, as well as some old depictions, show that all three were cultivated at least in the Bronze Age and, in the Levant, already at the end of the Neolithic. At that time, large scale horticulture started in the Mediterranean basin. *Vitis* has the least temperature demands of these three: the area of its wild form *V. sylvestris* includes central Europe, where it climbed the trees along the Danube and the Rhine but today has almost become extinct there due to forest clearing. In the Gallo-Roman and Roman period domesticated grapes were cultivated all over the Roman Empire and in the high Middle Ages vine growing even extended to the Baltic countries but, at the beginning of the Little Ice Age, these vineyards were destroyed by frost.

Large fruit trees are represented in the Mediterranean by walnut, *Juglans regia*, and chestnut, *Castanea sativa*. For a long time, the region of the Balkans and further east has been regarded as their natural distribution area but new pollen data show that both were already present further west in southern Europe at least since the Neolithic, e.g. from 4200 BC in the Po valley (Kaltenrieder et al., 2006; see Figures 4.4.3-5 and 4.4.3-15). Although old archaeobotanical finds are still lacking, their nuts will have been collected from the very earliest times. Charred remains of nuts from both trees are only found in the later Iron Age, but in Roman times these classical fruit trees became very common and their cultivation spread all over the Roman Empire, as is documented by a steep rise in their pollen curves around the birth of Christ.

Cultivated plants from the New World

Soon after the discovery of America a large number of cultivated plants were brought from the New World to Europe, where some of them gained enormous importance. Maize, *Zea mays*, was brought to Europe in connection with the second voyage of Columbus. It spread very quickly through the Mediterranean and, by the end of the 16th century, it was already grown as far east as Turkey and Syria. Later, in the 19th and 20th centuries, it became a very important fodder plant north of the Alps as well. Of similar importance is the potato, *Solanum tuberosum*, which also came to Europe through Columbus. However, it spread much more slowly because it was a short-day plant and had to be adapted to the longer days in more northern Europe. At the same time, people were, at first, hesitant about eating it because it is poisonous unless cooked. Only in the 19th century did it become the widespread crop plant that it is today.

Other important useful plants from America are tomato, *Lycopersicon esculentum*, tobacco, *Nicotiana tabacum*, the *Phaseolus* beans, the squashes and pumpkins of the genus *Cucurbita* and at a comparatively late date the sunflower, *Helianthus annuus.*

4.5.4

People and environment during the Neolithic

Neolithic chronology

The Neolithic is defined as the first prehistoric culture practicing arable and pastoral farming, connected to fixed houses and permanent villages which also means the start of human impact on the vegetation and landscape. This is the economic change from foraging to farming, called the Neolithic revolution, as described earlier.

In accordance with the spread of agriculture from the Near East (Fig .4.5.3-4), the Neolithic started in Europe at different times in different regions and so there is no fixed date for its beginning: its onset is not synchronous but metachronous (Table 4.5.4-1).

In Europe, the Neolithic started around 7000 BC when the package of domesticated plants and animals together with pottery and permanent settlements arrived in southern Greece. It spread across the Balkans, where the Karanovo, Starčevo and, further to the north, the Körös and Kriş cultures developed. In parallel with this the Neolithic spread across the Mediterranean Sea to the west where it is expressed by the Impressa culture in Italy and the Cardial cultures further to the west. A very fast ex-

Table 4.5.4-1. Comparison of archaeological and historical periods in Europe.

Britain + Ireland	Northern Europe	Northern C. Europe	Southern C. Europe	Eastern Europe	Southeastern Europe	Greece	Italy + Iberia	France	BC/AD
Modern period	Modern period	Modern period	Modern period	Modern period	Modern period	Modern period	Modern period	Modern period	+2000
High + Late Middle Ages	Middle Ages	High + Late Middle Ages	High + Late Middle Ages	High + Late Middle Ages	High + Late Middle Ages	High + Late Middle Ages	High + Late Middle Ages	High + Late Middle Ages	
Early MA + Migration period	Viking period	Early MA + Migration period	Early MA + Migration period	Early MA + Migration period	Early MA + Migration period	Early MA + Migration period	Early MA + Migration period	Early MA + Migration period	+1000
	Germanic Iron Age								
Roman period	Roman Iron Age	Roman period	Roman period	Roman period	Roman period	Roman period	Roman period	Roman period	0
Pre-Roman Iron Age	Pre-Roman Iron Age	Pre-Roman Iron Age	Pre-Roman Iron Age	Pre-Roman Iron Age	Pre-Roman Iron Age	Pre-Roman Iron Age	Pre-Roman Iron Age	Pre-Roman Iron Age	-1000
Bronze Age	Bronze Age	Bronze Age	Bronze Age	Bronze Age	Bronze Age	Bronze Age	Bronze Age	Bronze Age	-2000
					Chalcolithic	Chalcolithic			
Neolithic	Neolithic	Neolithic	Neolithic	Neolithic	Neolithic	Neolithic	Neolithic	Neolithic	-3000
									-4000
									-5000
Mesolithic	Mesolithic	Mesolithic	Mesolithic	Mesolithic	Mesolithic		Mesolithic	Mesolithic	-6000
						Mesolithic			-7000

pansion took place from the Danube basin to southern central Europe with the culture of the Linearbandkeramik (LBK). It arrived at the Rhine as early as 5500 BC and soon spread further into France. In the east, it reached western Ukraine and Hungary and, around 5000 BC, occupied most of the European loess belt with its rich soils. Large and important settlements from this period are Sittard in the Netherlands, Langweiler in Germany and Bylany in Czechia, but there are many others (Lüning, 2000; Bakels, 2009).

The next Neolithic cultures north of the Alps after 5000 BC were Cerny, Rössen, Großgartach and Stichbandkeramik, followed by Michelsberg in the west and Lengyel and Tripolye in the east. Only around 4000 BC was the Mesolithic replaced in Europe in its remaining areas north of the uplands, from the The Netherlands to Poland, by the Trichterbecher (TRB, funnel beaker) culture and, in Britain, by the Windmill Hill culture. At the end of the Neolithic around 2800 BC two large groups formed: the bell beaker (Glockenbecher) culture that extended from Spain and Britain to eastern Germany and is characterized by typical pottery drinking vessels and, in the east far into Russia, by the corded ware (Schnurkeramik) culture, called after the ornamentation from the impressions of cords on their pottery. Both culture groups continued into the subsequent Bronze Age when the corded ware flourished (Müller, 2009).

A fascinating feature of the Neolithic is the megalithic tomb (Thorpe, 2004). The first were erected along the west Iberian coast and in northwest France around 4700 BC where especially the huge menhirs, often in straight alignments, and the dolmens of Carnac in Brittany are most impressive. Soon after, these features were also built in western Britain and Northern Ireland and several centuries later, around 3600 BC, the well-known passage graves became common in the The Netherlands, northern Germany, Denmark and southern Sweden, in most cases belonging to the TRB culture. It is noteworthy that these monuments were erected before the pyramids in Egypt.

Already in the Neolithic, at around 7000 BC, copper had been discovered in Anatolia and from 4500 BC onwards copper mining started in the ore rich areas of the Balkans. From there copper metallurgy spread to the north-west. It was known in the TRB culture but was not important. The famous Neolithic Iceman or Ötzi, who died 3100 BC, already possessed a copper axe (Sperl, 1992). This period was the Copper Age, situated in time before the Bronze Age, that is before people found out that adding tin to copper formed the harder bronze. This transition phase to the metal ages is, in most parts of Europe, included in the Neolithic. In the Balkans, however, where a considerable industry of copper smelting had developed, this period between the Neolithic proper and the Bronze Age became a separate phase in the chronology and was called Chalcolithic or Eneolithic (Bogucki and Crabtree, 2004; Cunliffe, 2008; Müller, 2009).

Neolithic land use and human impact on vegetation

A consequence of farming that started with the Neolithic Revolution was the first visible impact on the vegetation, caused by clearing forests for fields and by forest grazing. One of the best known cultures of the early Neolithic is the Linearbandkeramik (LBK), which occupied the extensive loess areas. Many settlements with large longhouses have been excavated. Despite the good state of archaeological research there are only a few suitable pollen diagrams from around these sites, because the loess region is a very dry one and peat or lake deposits where pollen is well preserved are very rare. The settlers also had problems finding a suitable water supply and had to dig wells down to 12 m deep and more in order to reach groundwater: several of these wells have been excavated in recent years (Weiner, 1998).

There are only a few pollen diagrams from close to LBK settlements, for example Wange in Belgium (Bakels, 2009), Sersheim (Smettan, 1986), Wetterau (Stobbe, 1996) and Luttersee (Beug, 1992) all in Germany or Pleszow in Poland (Wasylikowa et al., 1985). They show that these settlements had been created in forests in which, in most cases, linden was the dominant tree because of the good soils. Cereals, as well as other anthropogenic indicators, clearly reflect the LBK occupation phase with *Plantago lanceolata* being very important. The opening of the forests was, however, rather limited, probably mainly for small fields, while the areas grazed by cattle must have been much further away: a main characteristic of the large farmhouses is the lack of byres. The settlements, with their fields, existed like islands in the extensive forests. Although the forest composition shows responses due to human activities, they differ between the various pollen diagrams and there is still a strong need for future investigations.

Archaeological excavations have provided quite a lot of archaeobotanical remains and have shown that *Triticum dicoccum, T. monococcum, Pisum, Lens culinaris* and *Linum usitatissimum* were most important while *Hordeum vulgare* occurred much less in the early Neolithic. *Papaver* was an additional crop plant in the west while the first farmers on the Balkans also grew naked wheat, *Triticum aestivum/durum* as well as *Cicer arietinum, Vicia ervilia* and *V. sativa/Lathyrus cicera* (not separated) (Kreuz et al., 2005; Kreuz, 2012). A new glume wheat similar to *T. timopheevii* appeared in the early Neolithic and dissapeared already in the late Bronze Age (Jones et al., 2000; Toulemonde et al., 2015).

Quite different from the LBK settlements on dry ground were the lake-shore villages. The first of these was discovered in Switzerland near Zürich during the dry summer of 1854. Later it turned out that they existed in many lakes around the Alps. Following the development of special lake-dwelling research in the 20th century they were found not only in Switzerland and the adjoining Germany and Austria, but also in Slovenia, Italy and France. Altogether, about 1000 of these sites are now known. The oldest ones date from the early Neolithic shortly after 5000 BC and the last ones ceased to be used around 800 BC: the flourishing period of the lake-shore settlements was the Bronze Age (Menotti, 2004; Suter and Schlichtherle, 2009).

These lake-shore settlements are also unique for vegetation history because they are waterlogged and provide large amounts of uncarbonized plant remains with excellent preservation. It was Oswald Heer who first discovered the new dimension of this material and, already in 1865, published the pioneer paper: *Die Pflanzen der Pfahlbauten*. Important Neolithic lake dwellings with large-scale archaeobotanical investigations are, among others, those from the lower part of Lake Zurich (Jacomet, 1985) or Lake Constance (Maier, 2001). The inhabitants of the lake-shore villages were farmers but they also hunted and fished. In the Neolithic sites, wild animals still played an important role alongside the domestic ones. The main crops were *Triticum durum/turgidum, T. dicoccum, T. monococcum, Hordeum vulgare, Linum usitatissimum* and *Papaver somniferum*. The first of these is a naked wheat that has only been identified recently and is very important because it indicates close contact to south-west Europe where it had been introduced. In addition to crop plants, collected seeds and fruits were important and here even wild apples could be found in the waterlogged layers (Jacomet, 2004).

In the large region of northern central Europe and beyond, the Neolithic started around 4000 BC and there is good evidence for human impact during the TRB culture. In the pollen diagrams the onset of the Neolithic is often marked by a steep fall in elm, the so-called elm decline (see chapter 4.4.3). This remarkable feature is often believed to be caused by human interference that promoted the elm disease which is spread by the elm bark beetle (Peglar and Birks, 1993; Clark and Edwards, 2004).

During the first phase after the elm decline there are only very small quantities of anthropogenic indicators, including single Cerealia-type pollen grains and there are no distinct signs of forest openings. Troels-Smith (1953, 1984) interpreted this as a special type of farming – the leaf-fodder husbandry – an interpretation that is widely accepted. In this economy the clearings are thought to have been very small, just large enough for the settlements and the small fields. It is assumed that the cattle were not driven into the forest, but kept in enclosures at the settlements (houses at that time did not yet have byres) and fed with the foliage of pollarded trees, mainly from *Ulmus,* but also from *Fraxinus, Tilia* and other woody species. At that time meadows for cutting hay did not yet exist. Such leaf-fodder has been in use at a broad scale from the Neolithic until modern times and is still collected in several remote areas of Europe.

Several centuries later, this leaf-fodder economy was followed by quite a different way of farming, the so-called *landnam* phase, first described by Iversen (1941). This *landnam* phase was, in fact, not a phase but lasted several centuries and, in places, as shown in Fig. 4.5.4-1 for up to 1000 years or more and extended into the early Bronze Age. In the areas around the settlements human impact was much stronger than ever before and this is reflected in drastic changes in the pollen diagrams. The main crop plants were naked and hulled barley, emmer, einkorn and naked wheat (Kirleis et al., 2012), but the distinctive characteristic of this culture was the great importance of pastoral farming. Now cattle were no longer kept in enclosures but the animals grazed in the forest and also did so during the winter time and this had severe consequences for the woodlands. Not only twigs with leaves but also bark and especially the sprouts of young trees and shrubs were eaten by the animals. This intensive forest grazing resulted in the opening up of the forests around the settlements and the formation of a green carpet of grasses and other herbs on the ground.

The *landnam* phase was first described from pollen diagrams from the young moraine areas in Denmark, after which it was recognized in diagrams from a large region north of the central European uplands and as far as western Ireland, where it is very distinct but without the typical *Tilia*-fall, because linden did not reach Ireland in the Holocene (O'Connell and Molloy, 2001). Its strong impact on the vegetation is, however, best documented in pollen diagrams from poor sandy areas in the old moraine districts. An impressive example is given in Fig. 4.5.4-1 from north-west Germany. Its course is as follows: phase A registers the preceding leaf-fodder economy which ends with a strong decrease in *Quercus* together with a *Tilia* decline, which is a special regional criterion for the onset of the *landnam* phase. Phase B shows a very strong increase in primary as well as secondary anthropogenic indicators. In particular the curve of Poaceae really explodes and reaches up to 248% (percentages based on total tree pollen). This shows the opening up of the forest in the surrounding area. Most interesting is the course of *Calluna* which indicates the degradation of the sandy soil by grazing. Its curve decreases again at the end of the occupation phase and so confirms that the patches of heathland were caused and maintained by human activities alone. In phase C, after the *landnam* phase, the area was abandoned and at first the pioneer tree *Betula* occupied the ground. Finally, in phase D the natural woodland regenerated, now with prevailing oak but only a little *Tilia*, because this tree had lost its competitiveness due to

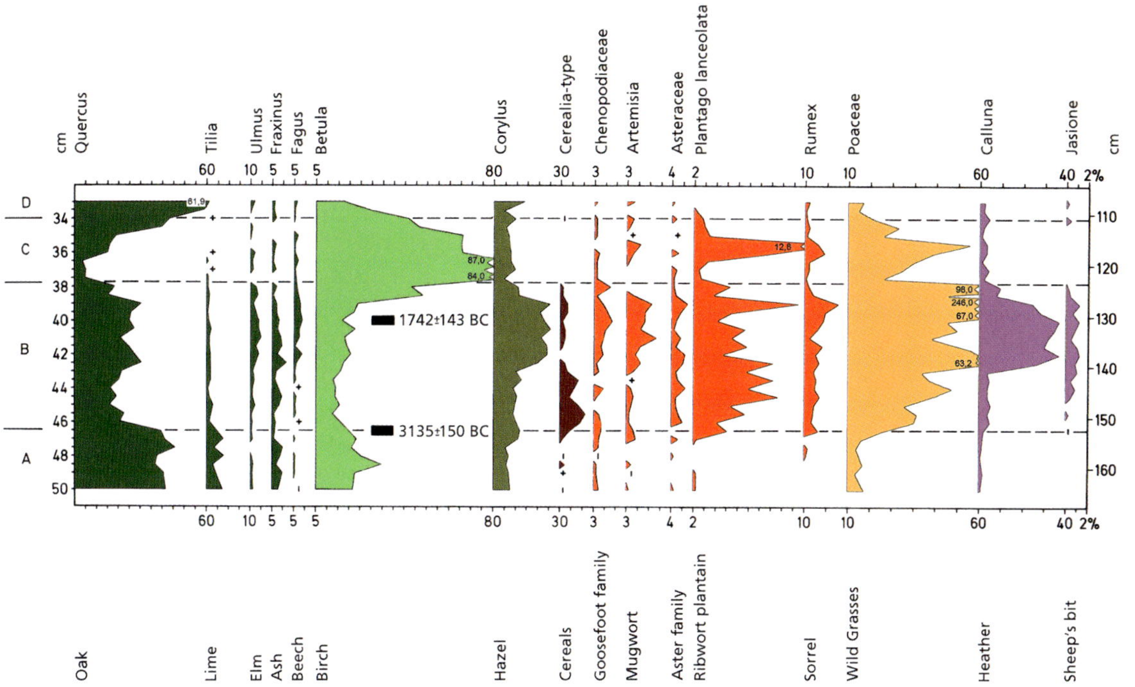

Fig. 4.5.4-1. Landnam phase of the Neolithic funnel beaker (TRB) culture on Pleistocene ground in northern Germany. Only relevant curves; ∑AP = 100%. A primeval high forest with leaf-fodder husbandry; B *landnam* phase – pastoral and arable farming; C succession of natural reforestation; D secondary high forest. From Behre (2008b).

the preceding climate deterioration (Behre, 1988). Charcoal and pollen records from various areas of Europe show that besides grazing, Neolithic burning also contributed to changes in forest structure and composition, creating open spaces for arable and pastoral farming, disfavouring disturbance-sensitive and promoting disturbance-resistant species (see chapter 4.10).

A special source for connecting prehistoric cultures with former economies is provided by pollen spectra from dated burial mounds that are available particularly from the Netherlands. It was van Zeist (1967) who, at an early stage, published spectra from Neolithic grave monuments that showed two distinct types of land use. He suggested that in the Netherlands the people of the TRB (funnel beaker) and bell beaker (Glockenbecher) cultures practiced the leaf-fodder economy, whereas the people of the protruding-foot beaker culture used the *landnam* system. However, Casparie and Groenman-van Waateringe (1980) presented different results from the same region. Their interpretation was that the differences in the barrow pollen spectra mainly indicated differences in soil fertility and hydrological situations.

4.5.5

People and environment in the Bronze Age

Bronze Age chronology

The lower and upper boundary of this period as well as the subdivision into Early, Middle and Late Bronze Age are very variable in Europe. The Bronze Age follows the copper time, the Chalcolithic in the south-east of Europe, after which bronze was discovered as a hard alloy of copper and tin. The Bronze Age started around 2200 BC in the Balkans and Mediterranean regions. The spread of bronze-making technology to the north took several centuries and it eventually reached southern Scandinavia around 1900 BC. The upper boundary of the Bronze Age marks the introduction of iron and this was delayed in the same way, resulting in the end of the Bronze Age being 1200 BC in the Balkans and 500 BC in southern Scandinavia.

Within this period, a large number of different cultural groups were formed and only the most important ones will be mentioned here. Well known is the high civilization of the Minoans on Crete that flourished with monumental palaces in the Early Bronze Age: at the same time its counterpart on mainland Greece was the Helladic culture. Probably in connection with the eruption of the volcano Thera on the island of Santorini at around 1450 BC, which caused huge devastation in the Aegean, the region was next occupied by the Mycenaean culture. Although the Bronze Age has been called the first Golden Era of Europe, the other large culture groups did not have as high standards as in Greece. In the Early Bronze Age, the Aunjetitz culture extended over central and eastern Europe while, in Britain, the Wessex culture had developed. During the Middle Bronze Age a large region north of the Alps was occupied by the Hügelgräberkultur, which had descended from the Auntjetitzer/ Únětice culture, and is characterized by a huge number of tumuli (Hügel). The Late Bronze Age was dominated in most parts of Europe by the urnfield culture, subdivided into many regional or local groups (Hänsel, 1998; Bogucki and Crabtree, 2004). In the north, in southern Scandinavia and around the southern Baltic, the term Nordic Bronze Age is used for the whole period and is subdivided into six subperiods (Montelius I–VI), the last of which ends around 500 BC, that is in the Iron Age further south.

Bronze Age land use and human impact on vegetation

During the Bronze Age, population density increased considerably and, correspondingly, clearings for fields as well as forest grazing opened up the woodland around the settlements, as can be seen in many European pollen diagrams. On sandy soils in the northern region, patches of *Calluna* heathland developed which is recorded not only in pollen diagrams but also often by the presence of heath sods that were used to build the numerous tumuli.

In the upland areas of Britain with high rainfall and low summer temperatures, as well as on flat plateaux with poor drainage, human activities involving clearing forests resulted in changes to the water regime, which is connected with the spread of extensive blanket bogs. This started primarily in the Bronze Age and extended into the Iron Age and was facilitated by the climate deterioration at that time (Moore, 1982).

In large parts of the Alps mountain pasture started in the Bronze Age, exceptionally already during the Neolithic, which resulted in forest openings and, in several areas, in the lowering of the tree line (see chapters 4.3.2 and 4.9).

The Bronze Age was when pile-dwellings flourished, as mentioned above, in connection with the Neolithic, when several of them started. Now lake-shore settlements became common everywhere along the lakes around the Alps and the large amount of well-preserved plant-remains enables the reconstruction of these special environments (Jacomet and Karg, 1996). There was a great demand for timber to construct the houses, platforms and fences and this led to the first known systematic forest management in the areas close to these settlements. The different ways of selecting and cutting timber can also be demonstrated by dendroarchaeology (Billamboz, 2009).

Human impact upon the landscape and vegetation during the Bronze Age increased not only because of the growing population but also because of considerable changes in both arable and pastoral farming. A very important innovation was the construction of the ard, the first plough that was drawn by oxen. Up until then, only hand hooks had been used for ploughing. These were small and pulled by men. The use of the ard led to a major expansion of fields and thus raised crop production considerably. At the end of the Bronze Age early field systems are known, mainly from Britain, where this new way of agriculture was even called an 'agricultural revolution'. In the urnfield culture ploughing was now probably routinely applied to the same plots of land. When ploughing on slopes soil was eroded downhill so lynchets were banked up, which formed kinds of terraces, which are preserved in many places in England (Fowler, 1983; 1984). A similar way of farming on more level ground is represented by the so-called Celtic fields, a prehistoric parcelling system that, in the Iron Age, spread across large parts of northern Europe and will be discussed in chapter 4.5.6.

The crop plants of the Neolithic continued to be grown including *Triticum dicoccum, T. monococcum,*

T. durum/aestivum, Hordeum vulgare, Lens culinaris, Pisum sativum and *Linum usitatissimum*. In addition to those already cultivated in the Early Bronze Age, there are some newcomers such as *T. spelta* and *Panicum miliaceum*. The latter had already arrived in the east during the Neolithic but spread to most other parts of Europe during the Bronze Age. Later, *Vicia faba, Camelina sativa* and *Vicia ervilia* were grown and, in the west, *Setaria italica*. At the very end of the Bronze Age, the first grains of *Avena sativa* are recorded (Behre, 1998; Bakels, 2009). In the Mediterranean, broad-scale horticulture with *Olea europaea, Vitis vinifera, Ficus carica* and others started in the Bronze Age. Progress was also made in harvesting tools in that sickles made of bronze came into use around 1500 BC and replaced flint as a cutting tool.

An important change also took place in animal husbandry. Up until the end of the Early Bronze Age, houses had no byres and cattle spent the whole year outdoors which, in the *landnam* phase, led to severe forest damage in winter time, as described in the preceding chapter. Probably as a consequence of a deterioration in the climate, byres to stall at least cattle during the winter were constructed, as can be seen from the ground plans of the excavated houses (Arnoldussen and Fokkens, 2008; Kristiansen, 1998). This means that the most damaging forest browsing in winter time ceased and was replaced by the collection of leaf fodder which was stored in special barns. So forest damage did not keep pace with the strong increasing population and its cattle.

4.5.6

Pre-Roman Iron Age, Roman Period and the Migration Period

Chronology

The term Iron Age has been used with different meanings. In most parts of Europe, it means the Pre-Roman Iron Age only, which is clearly separated from the subsequent Roman period, however, in Scandinavia the term Iron Age includes both the Pre-Roman as well as the Roman period and extends until the Viking period (see Table 4.5.4-1).

In most European regions, the beginning of the Iron Age is set around 800 BC but, in accordance with the spread of iron technology it is somewhat earlier in south-east Europe, where this technology was already established in Greece around 1200 BC, and later in northern Europe, where it becomes common at around 500 BC. As is the case with the Bronze Age the main terminology of the subdivisions within the Iron Age is tied to the various culture groups. However, the earliest of these cultures did not come into existence exactly at the defined time boundary but rather the culture adopted the new technology at this point. So the important Lusation culture in eastern central Europe already came into existence around 1300 BC, in the Bronze Age, and continued into the Iron Age until 500 BC. The same is true for the famous Hallstatt culture that spread across large parts of the northern Alpine foreland from eastern France to Hungary. The culture began in the Bronze Age, around 1250 BC, and lasted until 500 BC in the Iron Age when it was followed over a still larger region by the La Tène period. At this time, large parts of Europe from France almost to the Black Sea were inhabited by the Celts and north of them by Germanic tribes.

Another important cultural group was the nomad Scythian horse people who, in the 6th century BC, advanced from their Asian homeland to the west and even expelled the Cimmerians from their settlements in what today is Hungary.

The Mediterranean region was dominated in the east by the Greeks, who, from around 750 BC, established quite a number of separate city states with a high civilization from where the coasts around the Mediterranean Sea and the Black Sea were colonized. This colonization was accompanied by the spread of the new farming technology and cultivated plants that came from the east and the south. From this time, written sources also exist that complement the archaeological evidence. Further to the west the Etruscans also played a major role until they were followed by the Romans.

The boundary between the Pre-Roman Iron Age and the Roman period is also shifting. For most of central and western Europe the Pre-Roman Iron Age ends with the Roman conquest that occurred during the last two centuries BC and the first century AD. For the area outside the former Roman Empire, the term Roman period is also used, although their inhabitants had an identity of their own as exhibited by the far spread Germanic tribes or the Przeworsk and Wielbark cultures in the area of modern Poland. Most characteristic of these barbarians, as they were called by the Romans, is the lack of cities in the areas they occupied (Bogucki and Crabtree, 2004; von Schnurbein, 2009).

The decline of the Roman Empire took place gradually and historians define the end of this period as being between the incursion of the Huns in AD 375 and the demolition of Rome in AD 476. The definition of the start of the Migration period is correspondingly between these dates while the end of the Migration period is also variable and is set between AD 600 and 800 for the different parts of Europe. Some authors include the Migration period into the following Early Middle Ages.

Land use and impact on vegetation from the Pre-Roman Iron Age to the Migration period

Due to the growth of population and the introduction of new farming techniques, human impact on the landscape and the vegetation increased considerably in the course of the Pre-Roman Iron Age. Starting in the east, the Mediterranean region was totally transformed from the 8th century BC onwards. The rise of Ericales, Poaceae and other herbs in the pollen diagrams indicates increasing deforestation that now also occurred in mountainous areas. Both arable and pastoral farming were considerably improved and the advanced civilization produced its characteristic environment, culminating in nicely arranged gardens. Large orchards for fruit and nuts changed the appearance of the landscape around cities and deforestation increased strongly. Over large areas the cultural landscape was completed.

The number of cultivated plants increased: apart from olives, figs and grapes, which had been cultivated in the Bronze Age, now apple, pear, plum and cherry were also planted. New species reached Greece in classical times before the birth of Christ from the east, including hemp (*Cannabis sativa*), quince (*Cydonia oblonga*), peach (*Prunus persica*) and citron (*Citrus medica*) while apricot (*Prunus armeniaca*) and pistachio (*Pistacia vera*) only arrived in Roman times (Zohary et al., 2012). During the Greek Iron Age, a very sophisticated horticulture that culminated in the invention of grafting was developed, enabling the farmers to multiply the types of fruit and nuts and so improve them considerably. Charcoal and pollen records show that in southern and central Europe burning for agricultural purposes often peaked during the Iron Age, for example to keep the fields open (see chapters 4.5.8 and 4.10).

The development of the cultural landscape in the northern part of Europe was totally different from that in the south but there were some comparable events. On the Pleistocene deposits, the first ecological devastation that deserves this expression took place in the Pre-Roman Iron Age. Around 500 BC a new way of farming which had developed in Britain during the Late Bronze Age and is called Celtic fields, spread to continental Europe (Brongers, 1976; Zimmermann, 1976). The wide distribution of this field system of rectangular plots has only recently become apparent through the employment of airborne laser altimetry (Kooistra and Maas, 2008; Arnold, 2011). On the continent, many of these Celtic fields have been discovered in an area stretching from the Netherlands in the west to Estonia in the east. In Denmark they are called Oldtidsagre. These field systems usually cover around 100 hectares and their common feature is the cellular structure of the system. Detailed investigations have shown that the plots were enclosed by walls and served as fields that were regularly manured by the accumulation of soil from the surrounding area as well as with dung. This large-scale collection of soil material from the area around resulted in severe destruction of the forests and, in particular, of their soils. Because a rotation system was used, only a part of a Celtic field was concurrently in use and the whole field was very large. This way of land cultivation ended in the 1st century AD (Behre, 2000).

Other human activities led to changes in forest composition. The regeneration in forest clearings or pastures favoured the expansion of beech, which now had a better chance to spread than it had previously because many of the potentially competitive tree species were removed. In pollen diagrams from lowland Poland, however, it is the *Carpinus* curve which often registers this settlement phases because it was primarily the hornbeam stands which were cleared and this resulted in an increase of other tree pollen types (Ralska-Jasiewiczowa and Latałowa, 1996).

During the Roman period, human impact increased all over Europe in accordance with the rapidly growing population and so the natural forests were substantially destroyed. The way of farming differed very much, however, inside and outside the border of the Roman Empire. In the Roman world agriculture flourished. The new cities in the Roman provinces as well as the military bases created a big demand for food that had to be produced. Roman villas were built everywhere and were comparable to large modern farms. The techniques of advanced Roman horticulture, together with wine and fruit trees and even nut trees such as *Juglans* and *Castanea,* were brought to the different provinces. The Roman border was like a fire-wall and there was only limited agrarian exchange with the barbarians to the north. The forests outside the Roman region were, however, not as dense and impassable as described by Roman officers and authors. There is good evidence regarding the agriculture of the Germanic tribes who now exchanged the primitive ard for the more effective mouldboard plough. Around the birth of Christ new crops, in particular rye and oats which had just previously been developed from weedy forms in these areas, spread and were grown intensively, while emmer and naked barley disappeared. Considerable progress was made in animal farming. There is now clear evidence of mown meadows which were widely created during this time; further south the first hay meadows are mentioned from the Pre-Roman Iron Age (Knörzer, 1975).

After the decline of the Roman Empire, during the Migration period, widespread displacement of settlements occurred. Many agricultural areas were abandoned, resulting in a significant regeneration of woodland. In some parts of continental Europe an almost complete interruption of habitation took place. These processes are clearly reflected in the pollen diagrams as well as in the rare archaeological evidence.

4.5.7

Middle Ages and the Modern period

Chronology

In different parts of Europe, the start of the (Early) medieval period is dated to different times. In central Europe, a date of around AD 700 is regarded as the onset of the Early Middle Ages and so it is defined in this book. It is the time when the Carolingians, who ruled over large parts of Europe, rose to power, a time which was followed by the Ottonian dynasty. AD 1000 is set as the beginning of the High Middle Ages. In Scandinavia, however, the Viking Age that lasts until AD 1050, is not included in the Middle Ages and, in eastern Europe, the boundary Early / High Middle Ages is often assigned to AD 1100. The Late Middle Ages are defined as lasting from AD 1250 to 1500, the start of the Modern times.

Medieval and modern land use and the transformation of vegetation and landscape

At the onset of the medieval period, large areas of Europe were occupied by regenerated or still primeval woodland but during the Middle Ages most of these forests were, step by step, destroyed. In accordance with the rapid rise in population, many new settlements were founded during the Early Middle Ages and these continued to expand later: their names indicate that this was done in forests which were subsequently cleared. In the High Middle Ages cities were also established in great numbers outside the former Roman Empire. In addition to the requirements of the rural areas the cities required huge amounts of timber, firewood and especially cereals to feed their citizens. New fields became necessary for this production, so forests were continuously cleared and the remaining woodland was over-exploited not only to obtain wood but even more by forest grazing and the cutting of both leaf fodder and bedding for the animals. A consequence of overgrazing, connected with the degradation of the soil and burning, was the strong expansion of heathland over large areas of Atlantic Europe as well as the formation of *maquis* and *garrigue* vegetation in southern Europe (chapter 4.5.8).

The upcoming mining industry required large quantities of timber and, in particular, charcoal and the glassworks ruined the forests in a special way since great volumes of wood were burnt only to gain the potash that was necessary for glass production.

Together with the increasing population, there was an increasing demand for food and this led to the development of regulated field systems. The most important one was the three-field system which was first initiated in northern France in

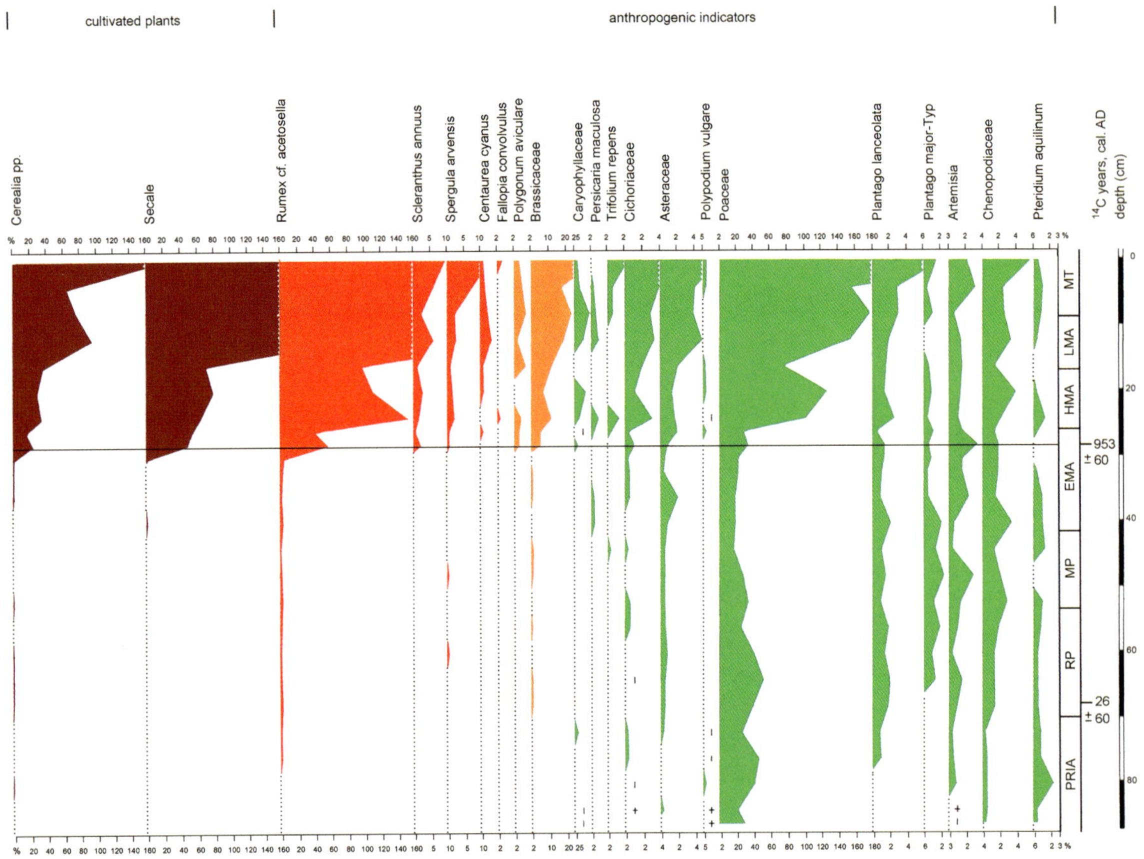

Fig. 4.5.7-1. Example of the change in anthropogenic indicators due to the introduction of a new agricultural form: the plaggen practice in the Middle Ages with permanent rye cultivation. Detail from an original pollen diagram from Dunum, northwest Germany. Dark red = cereals; light red = weeds of winter grain crops, yellow and green = other anthropogenic indicators. PRIA = Pre Roman Iron Age; RP = Roman period; MP = Migration period; EMA = Early Middle Ages; HMA = High Middle Ages; LMA = Late Middle Ages; MT = Modern Times. From Behre (2014).

the Early Middle Ages and, from the 11th century, spread across large parts of central Europe where it prevailed until the early 19th century. Under this system the arable land of a village was divided into three fields: one was sown with winter crops, the second with summer crops and the third was left fallow. In rotation the planting on the individual fields changed every year.

Special agrarian techniques were developed, mainly to compensate for the lack of nutrients in poor soils. On the poor old morainic soils, the practice of plaggen was employed over a large area stretching from Belgium to Denmark. This was triggered by the permanent cultivation of winter rye. Heather or sometimes also grass sods were stripped from the heathland, composted and enriched with dung if available and applied to the fields. This technique was employed on a one-field system. It started around AD 1000 and increased steadily during the Middle Ages and Modern times until it ceased in the 19th century due to the introduction of artificial fertilizers (Fig. 4.5.7-1). In the meantime the plaggen soils of the fields had been raised up by more than 1 m due to the accumulated material. Huge areas were necessary to fulfil the demand of plaggen since 20 to 40 times the size of the fields were required for plaggen cutting. In the course of the sod cutting the ground was regularly broken and the soil was exposed to wind action which resulted in severe sand drifts. When carried out for many centuries, this system resulted in an extreme destruction of the vegetation as well as of the landscape as

Fig. 4.5.7-2. Agricultural regions of Europe during the first decades of the 20th century. 1: Arctic-subarctic region with reindeer herding. 2: Boreal barley region (barley exceeds oat; additional cultivation of potatoes, cabbage and fodder turnips). 3: Oat region (oat exceeds barley and other cereals). 4: Rye region (rye exceeds other cereals). 5: Western wheat region (wheat exceeds other cereals). 6: Sub-mediterranean wheat-maize region. 7: Ukrainian steppe region with wheat. 8: Mediterranean olive region. 9: Subtropical barley region. After Troll (1925).

a whole (Pape, 1970; Behre, 1976; Groenman-van Waateringe, 1992; Spek, 2004).

Another system with strong impact on the natural forests was the slash-and-burn culture, a form of shifting agriculture which today is common in tropical rain forests but which was in use as a well-developed rotational farming culture in the Nordic countries. It was probably practiced from as early as 2000 years ago, but only became important in the Middle Ages and was particularly important from the 16th century until the 19th century. It was common in Finland and expanded from there into Sweden. Plots preferably of spruce forest were cut and, once dry, the timber was burnt and seed was sown into the still warm ash. Rye was the most common crop but occasionally barley and oats and sometimes even turnips and potatoes were grown. Once abandoned, the fields provided grazing for the cattle. After the first cultivation period, forest regeneration only reached a birch-alder stage before it was burnt again. In Sweden a rotation of 20–30 years was operated, while in Finland the cycle was longer but decreased as population rose. Around the villages, the area was devastated with different phases of forest regeneration (Vuorela, 1986; Hamilton, 1997).

Other forms of slash-and-burn culture have been practiced in several parts of Europe. A well-known example is the sophisticated *Hauberg-wirtschaft* in Siegerland, west Germany. This was a rotational system with cycles of 18–22 years. After slashing and burning, rye, oats and buckwheat were grown for 1–2 years, which was followed by grazing and finally coppicing of the trees that had sprouted (Pott, 1986).

A comparable burning method was practiced with the cultivation of buckwheat on extremely poor raised bogs that were drained and burnt in order to concentrate the small amount of nutrients. Buckwheat could be grown for six years but then the fields were exhausted and a fallow period of 30 years had to follow before the next cultivation phase. This rotational system was applied between the 16th century and around AD 1900 in large areas of the northern Netherlands and north-west Germany and resulted in the large-scale destruction of the raised bogs there.

The destruction of the forests continued into Modern times and now also included the mountainous areas. Timber became rare and in many places and the authorities started to protect and manage forests in order to have timber but also to provide hunting facilities for the privileged. Systematic management of the remaining forests only started in the 17th century. From around AD 1800 and continuing during the 19th century large-scale reforestation took place and in many regions of Europe heaths and waste land were afforested. Due to the introduction of artificial fertilizers, it was no longer necessary to cut sods for manure and there was less need for extensive grazing because, now, wool and meat were imported from overseas. Starting in Britain, firewood was replaced by coal and later by petrol and gas.

The systematically planted forests, however, were in many areas quite different from the preceding natural forests. Now conifers were favoured instead of native deciduous trees and some of them, such as *Pseudotsuga menziesii, Pinus strobus,* and others, were introduced from America, *Larix kaempferi* (= *leptolepis*) from Japan.

With respect to cultivated plants, rye became the dominant crop in large parts of central and eastern Europe while wheat dominated in the west and south and barley and oats in the north and northwest (see Fig. 4.5.7-2). There were, however, only a few newcomers during the medieval period: already in the Early Middle Ages the cultivation of various citrus fruits had started in the south while north of the Alps, in the Late Middle Ages, buckwheat arrived in the west, coming from its Asian origin. In Modern times, vegetables including turnips and other food plants became increasingly more important. After the discovery of America, a large number of cultivated plants were introduced: quite a number of them, particularly potatoes, maize, tomatoes and sunflowers, now belong to the most important crops in Europe.

4.5.8

History of some anthropogenic vegetation types

Dry grasslands and steppe

Large areas in the hilly and montane elevational belt of Europe were, or still are, covered by arid and semi-arid grasslands. As the pollen diagrams from many regions show, they occupy areas that had previously been woodland. Thousands of years of exploitation, mainly by grazing, produced these wide open areas. In many areas the process started in the Bronze Age but increased strongly in the Middle Ages. An important consequence of this deforestation has been the change in the local climate near the soil surface. After deforestation, solar radiation was no longer intercepted at the level of the woodland canopy but reached to the ground surface. This greatly increased the range of temperature, humidity and wind speed at this level so that the ground, which was previously fairly moist throughout the year, dried out at certain times and also became much warmer in summer and colder in winter than previously and the local climate took on many of the features of the general climate of continental steppes and Mediterranean karst

(Ellenberg, 1988). New plant communities, adapted to these factors, formed these dry grasslands.

The calcareous grasslands with a low nutrient content in the uplands of central Europe, France and Britain, were extensive and formed large areas of rough grazing (Ellenberg, 1988; Küster, 1993). Their typical species-rich Festuco-Brometea have to be differentiated from the species-poor Nardetalia communities on acidic soils, where *Nardus stricta* is the characteristic species though both were formed and maintained by long-term grazing. These extensive, grazed, dry grasslands reached their greatest expansion in the 18th and 19th centuries. Since then, vast areas of these montane pastures have become occupied by shrub communities or are afforested.

For a long time the vegetation history of central Europe and the surrounding areas was dominated by the question of whether this region had been completely forested before human impact started in the Neolithic or whether open landscapes with steppe vegetation also existed at that time. Gradmann argued in his *Steppenheidetheorie* (1933) that the first farmers, who settled in the dry loess areas, moved into a semi-open landscape and defended it against the spreading forest. This was supported by the existence of chernozem soils and the occurrence of steppe species in the modern vegetation. Numerous pollen diagrams in particular from the driest areas in central Europe with less than 500 mm precipitation east of the Harz mountains (Müller, 1953; Litt, 1992), in the Kujawy region in central Poland (Ralska-Jasiewiczowa et al., 1998) and the modern steppe on the island of Öland in the Baltic (Königsson, 1968), however, indicate closed forests until they were opened up by the first farmers.

To the south-east, in the dry lowlands of Bohemia (Pokorný et al., 2015) and Moravia (Rybničková and Rybniček, 1972), there are low pollen quantities of steppe elements such as *Artemisia* in the Mid Holocene pollen diagrams but no indications of forest-steppe. In the inner Carpathian lowlands of Romania, the Early Holocene forest-steppe was occupied by coniferous and deciduous broadleaf forests in the Mid Holocene but the woodland was not fully closed: in the late Neolithic, deforestation started and led to a predominance of grassland from about AD 1 (Feurdean et al., 2007a, 2015).

Contiguous with the region of the modern forest steppes east of the Carpathians there is unambiguous evidence for areas with natural forest-steppe existing during the Mid Holocene and later, as there is in southern Dobrudza in north-east Bulgaria (Bozilova and Tonkov, 1998). Further east, the Ukrainian forest-steppe and steppe are registered as natural in many pollen diagrams with *Artemisia* and Chenopodiaceae as dominant taxa (Kremenetski, 1995, 1999).

A special case is represented by the Great Alföld – the Hungarian puszta – which is often mistakenly referred to as a wide natural steppe. In the Mid Holocene it was, however, occupied by deciduous forests with some open areas with steppe oak woods. In addition to alluvial forests there were also grasslands on the edaphically constrained floodplains which were subject to repeated disastrous floods from the Tisza river. This is confirmed not only by the pollen data but by the large numbers of charcoal fragments from Neolithic settlements in the area (Moskal-del Hoyo, 2013). From the Bronze Age onwards, extensive grazing of the woodlands led to the development of a vast secondary steppe landscape. Its main expansion took place through clearing and grazing from the 15th century onwards and particularly during the Osmanic Empire in the 16th and 17th centuries (Lóczy, 1988; Magyari et al., 2010, 2012a). Now a relict steppe area at Hortobágy is protected and artificially preserved against incoming forest.

Maquis and garrigue

Natural or semi-natural tall forests are very rare in the Mediterranean region. Besides areas of human settlements and intensive agriculture, two vegetation types dominate. Evergreen sclerophyllous shrubby woodlands usually 3-4 m tall called maquis (French), macchia (Italian), mancha (Spanish) or xeromuvi (Greek) and open shrub and dwarf shrub communities with plants shorter than 1 m called garrigue (or garigue, French), gariga (Italian), tomillar (Spanish) or phrygana (Greek).

The maquis is composed of plant communities which in phytosociology are usually assigned to the Quercion ilicis alliance for the mesomediterranean and to the Oleo-Ceratonion for the thermomediterranean vegetation belt (Reisigl et al., 1992; Pignatti, 1997). These denominations derive from the plant-sociological assumption that, if human pressure ceased, such communities would become

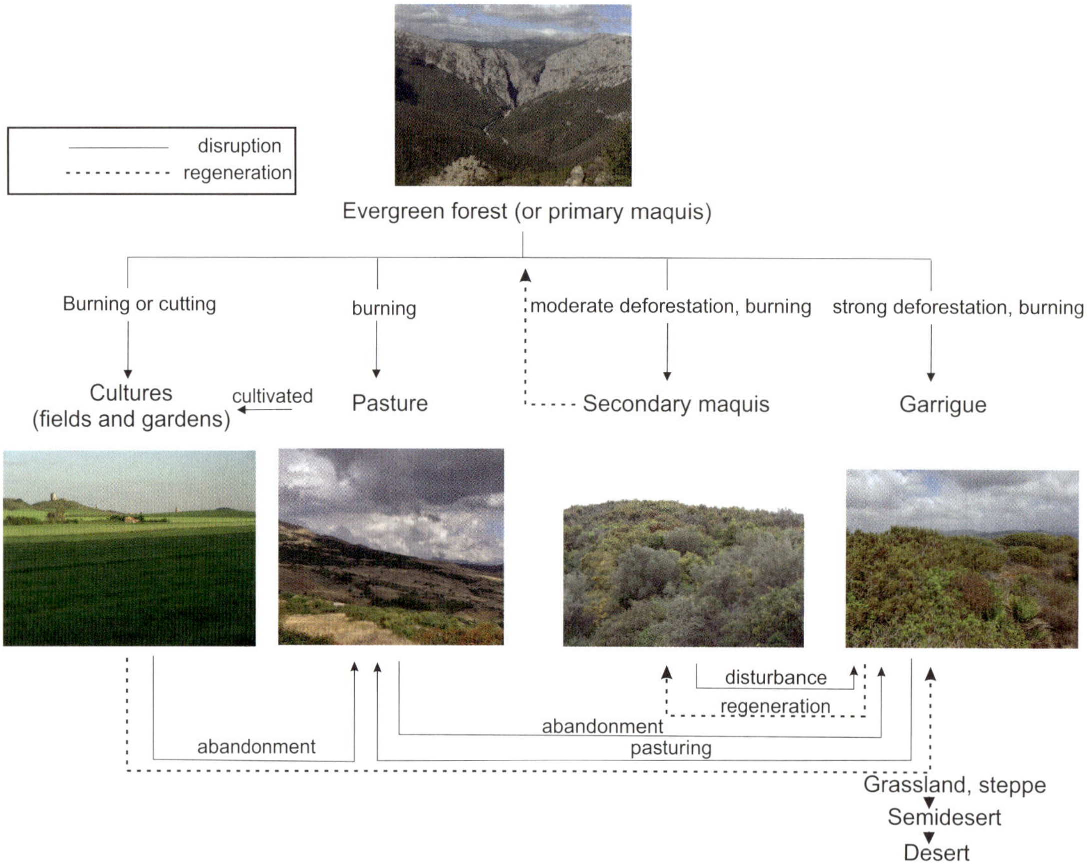

Fig. 4.5.8-1. Mediterranean vegetation types and their primary drivers following Chiappini (1988). Evergreen forest is the original natural vegetation type, the picture shows dense *Quercus ilex* forests at Sa Gorropu in Sardinia (Photo C. Morales-Molino). In the area, *Quercus ilex* forms stands together with other Mediterranean species such as *Olea europaea, Arbutus unedo* and *Pistacia lentiscus*. Through moderate burning, forests are converted into secondary maquis as at Sithonia in Greece (Photo C. Morales-Molino). Garrigue develops with frequent burning as at Capo Caccia in Sardinia, where the European fan palm *Chamaerops humilis* is abundant (Photo W. Tinner). Through burning and cutting the forests can also be transformed into pastures as in Sicily (Photo C. Schwörer) or fields as in Castilla (Photo C. Morales-Molino).

forests or woodlands dominated by *Quercus ilex* or *Olea europaea* and *Ceratonia siliqua*, respectively. Over centuries wood was burned and/or cut every 10–20 years for various agricultural purposes (Chiappini, 1988). Usually the trees and shrubs are therefore young, and often shorter than 6 m (Fig. 4.5.8-1). The garrigue is composed of herbaceous and shrub species <1 m and is frequently used as pasture, especially by sheep and goats (Fig. 4.5.8-1). In phytosociology, plant communities of the garrigue belong to the Ononido-Rosmarinetea on limestone, to the Cisto-Lavanduletea on siliceous rocks and to the therophytic (annual or ephemeral) Thero-Brachypodietea.

It has been hypothesized that maquis and garrigue vegetation types are anthropogenic stages of evergreen forest degradation (Fig. 4.5.8-1), primarily shaped by different intensities of human impact by periodic burning, wood cutting and pasturing (e.g. Chiappini, 1988; Reisigl et al., 1992; Allen, 2001). Less certain is from which types of primary vegetation these secondary types developed. Early pollen studies such as Beug (1962, 1975b), Brande (1973), Bottema (1979, 1980), Pons (1980), van Zeist and

Bottema (1982), Reille (1990a) and Reille and Pons (1990) suggested different vegetation histories in the thermomediterranean (Oleo-Ceratonion) and mesomediterranean (Quercion ilicis) belts. Modern studies have corroborated this notion.

In the **western Mediterranean** (Fig 4.3-15), some thermomediterranean sites such as San Rafael at 0 m asl (Pantaleón-Cano et al., 2003) were forested since the Late-glacial, that is, long before significant human impact started. Later, agricultural and pastoral activities during the Neolithic and subsequent cultural periods affected evergreen broadleaved *Quercus ilex*- and *Pinus*-dominated forests. In dry south-eastern thermomediterranean Iberia, dense forests never established during the Holocene, although woody species reached maximum abundances during the Mid Holocene around 7000–6000 years ago, probably in response to increased moisture availability, before being reduced by land use (Carrión et al., 2002, 2010c,d; Pantaleón-Cano et al., 2003). On the Balearic Islands, special vegetation types occurred in the coastal belt (for example *Buxus*-dominated forests in Minorca) that were gradually replaced by maquis (Ericaceae, *Pistacia*, *Cistus*, *Olea*), when human impact increased (Burjachs et al., 2017; Servera-Vives et al., 2018).

In the cooler and moister mesomediterranean environments of Iberia (Fig. 4.3-15), rather closed mixed forests with evergreen oaks, pines and deciduous trees occurred since the Late-glacial or at latest since the Early Holocene (Pons and Reille, 1988; Carrión et al., 2001a,b, 2003a; Carrión, 2002b; Morales-Molino et al., 2017; Ramos-Román et al., 2018). It is during the last 5000–4000 years that evergreen shrublands with a remarkable proportion of Ericaceae, especially on acidic bedrock, or evergreen *Quercus ilex*-dominated oak woods with a diverse understorey of evergreen shrubs spread (e.g. Villaverde: *Erica, Pistacia, Phillyrea, Myrtus*; Carrión et al., 2001). Replaced forest communities comprised pine, deciduous oaks and other temperate deciduous trees (Carrión et al., 2001a,b, 2003a; Morales-Molino and García-Antón, 2014; Ramos-Román et al., 2018). The main drivers of this vegetation shift were likely enhanced human activity and fire occurrence, possibly under an increasingly drier climate (Carrión et al., 2001a,b, 2003b; Morales-Molino and García-Antón, 2014; Ramos-Román et al., 2018).

Similar patterns emerge from the **central Mediterranean** (Fig. 4.3-16). The warmer coastal thermomediterranean sites (e.g. Lago Preola, Gorgo Basso and Biviere di Gela in Sicily) became densely forested only at the onset of the Mid Holocene, at around 8000–7000 cal yr BP (Noti et al., 2009; Tinner et al., 2009; Calò et al., 2012). Before that time, natural maquis, dominated by evergreen *Pistacia lentiscus* and many herbaceous species prevailed, probably because of Early-Holocene climatic conditions too dry for forest growth. After 8000–7000 cal yr BP, Sicilian natural or quasi-natural Mid-Holocene thermomediterranean forests were dominated by *Quercus ilex* and *Olea europaea*. Forests were substituted by maquis and garrigue at latest during the Greek colonization period around 700 BC, when human fires massively disrupted coastal forest vegetation (Tinner et al., 2009). Similar to the Balearic Islands, a special vegetation history is recorded in the thermomediterranean belt of Sardinia, where *Erica arborea* and *Erica scoparia* shrublands and woodlands dominated during the Early and Mid Holocene to be gradually substituted by *Quercus ilex* forests and managed open land during the Late Holocene (e.g. Sa Curcurica; Beffa et al., 2016), probably in response to climate change (moister summers, drier winters), declining fire occurrence, and human impact.

At the cooler and moister mesomediterranean sites in Sicily, Corsica, Italy and Dalmatia (Fig. 4.3-16), forests had established during the Late-glacial (e.g. Monticchio, Martignano, Lagaccione) or Early Holocene (e.g. Pergusa in Sadori and Narcisi, 2001; Sadori et al., 2008 and Accesa in Drescher-Schneider et al., 2007b; Colombaroli et al., 2008, 2009). If, compared to the warmer Sicilian thermomediterranean sites afforestation occurred earlier, it was probably because moisture availability was higher during the Late-glacial and Early Holocene. Most of these forests were initially dominated by deciduous species (e.g. oaks, see Pergusa, Monticchio, Martignano, Lagaccione, Malo Jezero, Accesa). Subsequently, evergreen broadleaved species expanded during the Early and Mid Holocene (e.g. Pergusa, Martignano, Lagaccione, Accesa, Malo Jezero), possibly in response to declining seasonality with less pronounced frost occurrence and reduced winter precipitation (see Colombaroli et al., 2009; Calò et al., 2012; Tinner et al., 2013; Curry et al., 2016 for palaeoclimatic inferences). However, in Corsica Holocene vegetation dynamics were different, resembling those in thermomediterranean Sardinia, with a gradual Mid- to Late-Holocene expansion of

Q. ilex into *Erica arborea* woodlands (Reille, 1992a). Some coastal sites (e.g. Fango or Saleccia) show a marginal relevance of deciduous species (Reille, 1992a), suggesting that a narrow band of thermomediterranean vegetation occurred at the lowest elevations on Corsica during the Holocene. During the Late Holocene, mesomediterranean forests were disrupted and maquis, garrigue and grasslands expanded in the central Mediterranean. In these open communities, light-demanding shrubs such as *Cistus, Pistacia, Phillyrea, Erica* and *Juniperus* (e.g. Reille, 1992a; Jahns and Bogaard, 1998, Colombaroli et al., 2007, 2009) expanded massively, most likely as a consequence of human impact and burning. This process intensified during the past 3000 to 2000 years, when modern maquis and garrigue landscapes established.

Similarly to the central Mediterranean, thermomediterranean sites in the **eastern Mediterranean** (Fig. 4.3-17; e.g. Kournas, Lerna) reached maximum arboreal cover c. 7000 to 2500 years ago (Jahns, 1993; Bottema and Sarpaki, 2003). Thermomediterranean sites in the eastern Mediterranean show a more pronounced co-dominance of deciduous oaks than at western and central Mediterranean sites, where evergreen broadleaved species clearly dominated during the Holocene. At the cooler and moister mesomediterranean sites (Fig. 4.3-17), evergreen broadleaved species were co-dominant (e.g. Kopais; Turner and Greig, 1975; Allen, 1986, 1990) or at least of secondary importance (Voulkaria; Jahns, 2005). Comparable to the western and central Mediterranean, the onset of forest vegetation disruption and maquis formation started during the Late Holocene, more specifically between the Late Bronze Age (e.g. Lerna) and Antiquity (e.g. Delphinos and Kournas). Remarkably, some sites (e.g. Voulkaria) remained forested until the present. Mesomediterranean and thermomediterranean vegetation belts spread less north than in the western and central Mediterranean during the Holocene. This spatial gradient that encompasses c. 400–500 km of latitudinal difference (e.g. mesomediterranean central Italy c. 44° N vs. submediterranean northern Greece c. 40°N) is probably the result of increasing continentality to the east, which through increased winter frost may significantly affect the northern distribution limit of evergreen broadleaved subtropicals such as *Pistacia lentiscus, Olea europaea* and *Quercus ilex* (Nardini et al., 1998; Larcher, 2000).

Early studies had put forward the hypothesis that evergreen broadleaved ***Quercus ilex*** had expanded into the deciduous forests of the mesomediterranean belt due to increased human impact during the Mid and Late Holocene (see discussions in Reille, 1992a; Carcaillet et al., 1997; Colombaroli et al., 2009). This hypothesis can be rejected for all sites where *Quercus ilex* expanded prior to Neolithic human impact (e.g. at Pergusa, Laggacione, Martignano, Accesa), but may partly apply for other sites, where the expansion occurred during the Late Holocene. However, pollen-independent climatic proxies and modelling results point to complex linkages between the *Q. ilex* expansion and climate change, specifically moisture availability and frost occurrence (Colombaroli et al., 2009; Calò et al., 2012; Tinner et al., 2013; Beffa et al., 2016; Curry et al., 2016; Pedrotta et al., 2021). Moreover, in the mesomediterranean environments of northern (and thus rather cool) regions such as southern France, Corsica, Liguria or Croatia, *Quercus ilex* may have benefitted from both climatic change (e.g. decreasing winter frost occurrence and/or drier winters) and human impact during the Late Holocene. Independent from this question, which is important to assess the natural conditions in the absence of human impact, many records from the western and central Mediterranean show that **human disturbance**, including deliberate burning, decisively contributed to the expansion of maquis and garrigue during the Late Holocene in both the thermomediterranean and mesomediterranean belts (Colombaroli et al., 2007; Tinner et al., 2009; Carrión et al., 2010b). Comparisons between charcoal and pollen series show that frequent fire promoted disturbance-adapted arboreal taxa such as *Chamaerops humilis, Pistacia lentiscus, Erica arborea, E. scoparia, Phillyrea, Cistus, Juniperus, Pinus* and many herbs (e.g. Colombaroli et al., 2007; Tinner et al., 2009; Calò et al., 2012; Beffa et al., 2016).

We conclude that in most areas of the Mediterranean, today's maquis and garrigue have developed in response to human impact through the millennia, in particular burning for pastoral and arable purposes. Some of the maquis and garrigue communities have Bronze Age and Iron Age legacies and are therefore in their nature comparable to the northwestern European *Calluna vulgaris* heathlands. Where mean annual precipitation was below c. 350–400 mm (limit of tree growth in the Mediterranean;

Tinner et al., 2009), maquis and garrigue occurred naturally together with isolated tree stands. Occasionally, natural maquis may have also occurred on very shallow and rocky soils where precipitation exceeded the minimum requirements for tree growth. We infer that in the absence of human impact most maquis and garrigue would be replaced within decades by forests in both the thermomediterranean and mesomediterranean belts.

Thermomediterranean forests would be naturally dominated by woody species such as *Quercus ilex, Q. coccifera, Q. suber, Olea europaea, Erica arborea, E. scoparia* and *Pistacia lentiscus*, depending on local (e.g. edaphic and disturbance) conditions. This assessment is in best agreement with dynamic vegetation modelling evidence (Henne et al., 2013, 2015). Interestingly, if restored, such thermomediterranean evergreen oak-olive forests would help mitigate projected climate-driven increases in wildfire risk, given that they are far less flammable than the garrigue communities or pine and eucalypt plantations (Henne et al., 2015). However, palaeo-validated models suggest that under moderate global warming conditions (c. +3°C locally) such forests would be rapidly (c. 50-150 years) replaced by thermomediterranean or inframediterranean maquis and garrigue, reinstalling vegetation conditions similar to those of the Early Holocene (Henne et al., 2015).

Mesomediterranean forests would be characterized by the mixed occurrence of Mediterranean evergreen broadleaved (e.g. *Quercus ilex*) and temperate or submediterranean taxa (e.g. *Quercus petraea, Q. pubescens, Ostrya, Corylus, Ulmus, Tilia, Abies*; see Monticchio or Accesa). We assume that under natural conditions temperate species would concentrate on deep soils with enough moisture, whereas Mediterranean evergreen broadleaved species would be likely restricted to shallow soils, for example on rocky sites or hilltops. Again these results are in best agreement with dynamic vegetation models for current climatic conditions under natural fire and browsing disturbance regimes (Henne et al., 2015). Such palaeo-validated models suggest that under moderate global warming conditions (c. + 3°C locally), mesomediterranean vegetation would rapidly (ca. 50-150 years) develop into thermomediterranean communities, with a persistent loss of temperate and submediterranean species (Henne et al., 2015).

Atlantic heathland

Vast areas of dwarf shrub heath have long played a prominent role in the western European coastal regions from Norway to the Netherlands and further south from Brittany to Portugal as well as across the British Isles (Fig. 4.5.8-2). This typical vegetation formation has a specific name in the different regions: heath or Heide in the Germanic languages or bruyère, brughiera, lande and brezal in languages of Celtic and Roman origin. The Atlantic heathlands demand an oceanic, humid climate with mild winters. They generally occupy acidic soils of low fertility and are characterized by the dominance of a few Ericaceae species, mainly *Calluna vulgaris*, mixed with some *Empetrum, Vaccinium* and grass species, while further to the south Fabaceae and Genistoideae are also frequent. Some of the Ericaceae are mostly restricted to or even occur only on the Iberian Peninsula *(Erica lusitanica, E. mackaiana, E. umbellata* etc.) while others reach from the Iberian Peninsula to Ireland and Britain (*Erica ciliaris, Erica australis, E. vagans, Daboecia cantabrica*) or even to west Norway and the Faroe islands (*E. cinerea*). In the majority of heathlands, trees or shrubs are lacking while, in others, shrubs such as *Juniperus communis, Ulex europaeus* or *Cytisus scoparius* are encountered. Quite a number of Genistoideae shrubs, for example *Genista tridentata, Genista florida* as well as the dwarf shrubs *Ulex minor* or *Argyrolobium zanonii* only occur on the Iberian Peninsula (Gimingham 1972).

The aspect of the large dwarf shrub heaths led to the impression that it was the natural vegetation and there has been a long-lasting debate as to whether the Atlantic heathlands were natural or created by people. Pollen analysis, pedology and the modern development of the heathlands following the abandonment of grazing and litter accumulation showed that almost all of these heathlands were created by human activities. In the Atlantic regions, there are very few natural heathland communities, on moving dunes, above the treeline and in harsh climate regions such as the extreme north-west.

The evidence from north-west Scotland shows that there were no forests in the Outer Hebrides; the postglacial vegetation started with species rich grasslands and very small areas of scattered birch and hazel scrub. By 6500 BC *Calluna* heaths expanded as a result of acidification. Further to the south-east most parts of the Isle of Skye were al-

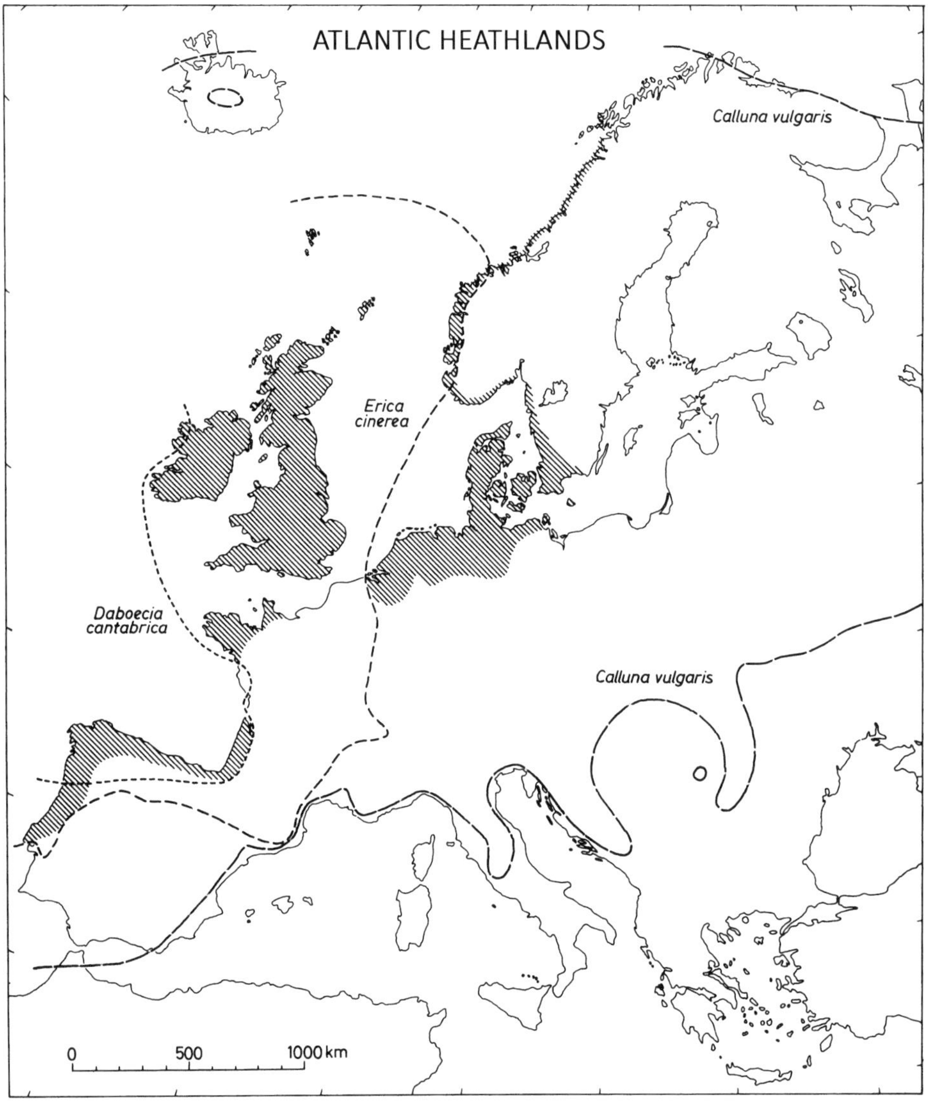

Fig. 4.5.8-2. Map of the lowland heath region of western Europe, with eastern distribution boundaries of some Ericaceae (after Gimingham, 1972; Kaland, 1986).

ready well forested in the Mid Holocene with *Betula*, *Alnus* and *Corylus* as well as some *Ulmus* and *Quercus,* but since about 4000 BC there has been extensive forest destruction by people that caused the spread of heath and bogs (Birks and Williams, 1983; Birks, 1991b).

Also on the Orkney, Shetland and the Faroe Islands, heath started to replace grassland vegetation while small amounts of birch and hazel scrub existed between 6000 and 5000 BC and so preceded the human activities that later caused the strong expansion of heathland (Bunting, 1996; Johansen, 1996). Numerous pollen diagrams from other parts of Britain showed the anthropogenic origin of all heath and grassland except the high-montane type. Investigations from several areas on sandy ground showed that the lowland heaths of southern England were created by human manipulation of the vegetation, including fire, and that their maintenance required, and still requires, active intervention. Here the first traces of heathlands date to the Late Neolithic, while more extensive areas developed from

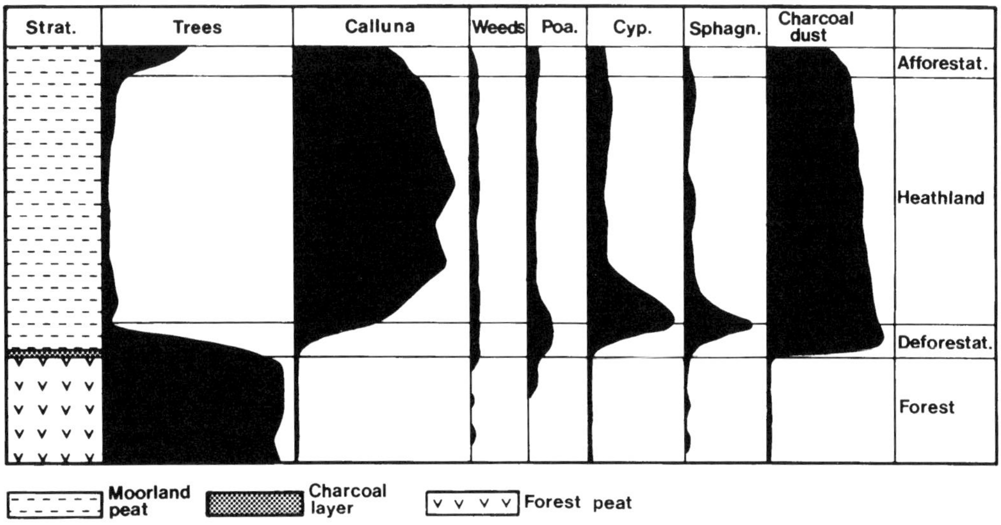

Fig. 4.5.8-3. Simplified pollen diagram from a small bog in the Norwegian coastal heath showing heath formation. Poa. = Poaceae; Cyp. = Cyperaceae; Sphagn. = *Sphagnum.* From Kaland (1986).

the Late Bronze Age. The maximum expansion of heathlands started in the High Middle Ages and extensive areas existed in southern England until the late 19th century (Groves et al., 2012).

A special case is represented by the Norwegian coastal heath which has been intensively studied by Kaland (1986) and Prøsch-Danielsen and Simonsen (2000). The first investigations, carried out by Fægri (1940), had led to the assumption that this coastal heathland was caused by climate but, later he conceded that human impact was an additional factor of heath formation. In the meantime, numerous pollen diagrams were presented, most of them situated close to prehistoric or historical farms, which had been archaeologically dated. Some profiles showed an abrupt change from forest to heathland with forest peat at the base, followed by a considerable layer of charcoal and a sequence of moorland peat (Fig. 4.5.8-3); others show more different and stepwise changes in deforestation.

The pollen diagrams illustrate the strong decline of the forest and the contemporaneous expansion of heathland, together with the appearance of anthropogenic indicators combined with a high influx of charred particles that indicate regular burning to maintain the heathland. It is now without doubt that the formation of the Norwegian coastal heathland was caused not by changes in climate as previously assumed but by human activities with the help of fire. Figure 4.5.8-4 shows that the final establishment of these heaths was a continuous process with three pronounced clearing periods lasting from the Neolithic until the Middle Ages, but was mainly completed by the end of the Bronze Age. At only a very few number of sites could small patches of pre-Neolithic heath be traced. The rapid regeneration of the forest after the decline in the use of the heath for pasture in recent times clearly shows that the heathland is completely dependent on continuous land use through a combination of grazing and heath burning.

The factors behind the formation of the large heathlands in Belgium to Denmark had, for a long time, been subject to controversial debate regarding their natural versus anthropogenic origin. This ended with the evidence from numerous pollen diagrams which showed that these heathlands were of anthropogenic origin. They were formed by the destruction of the preceding forests by people and their cattle and also through soil exhaustion by arable farming. When this ends, woodland is re-established, the continuous survival of heaths being dependent upon permanent grazing. In some areas there were additional management practices such as the use of fire or the regular removal of humus from the uppermost layers.

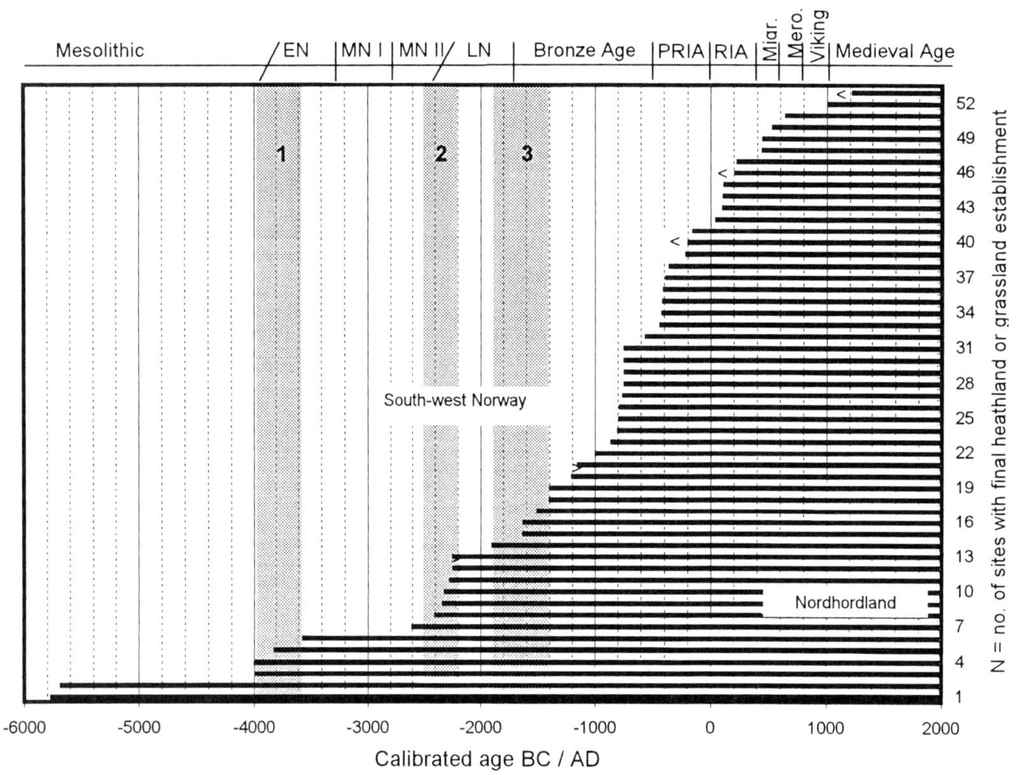

Fig. 4.5.8-4. Schematic representation of the dates of the final establishment of heathland or grassland and permanent infields in the south-western Norwegian coastal heath region. Main forest clearance periods (1–3) are marked. The archaeological periods are dated here as follows: EN, Early Neolithic 3900/3800–3300 BC; MN I, Middle Neolithic I 3300–3800 BC; MN II, Middle Neolithic II 2800–2400/2300 BC; LN, Late Neolithic 2400/2300–1700 BC; Bronze Age 1700–500 BC; PRIA, Pre-Roman Iron Age 500–0 BC; RIA, Roman Iron Age AD 1–400; Migr, Migration period AD 400–600; Mero, Merovingian Age AD 600–800; Viking Age AD 800–1030; Middle Ages AD 1030–1537. Top = south; bottom = north. From Prøsch-Danielsen and Simonsen (2000), slightly modified.

Apart from a few records of small local disturbances, the earliest records of small heaths relate to areas with Neolithic habitation. In several places, the characteristic heath-podzol has been found immediately below burial mounds and a pollen diagram from Skarrild, west Jutland, from the soil below a Neolithic single grave, shows the development from a natural alder-birch forest to a *Calluna* heath (Odgaard, 1985). Sometimes marked *Calluna* peaks can also be found from peat profiles near to Neolithic settlements and it can be shown that the heath disappeared again after the occupation phase (Behre and Kučan, 1994). Numerous records for Neolithic heaths are provided by the pollen spectra from within and beneath Neolithic graves in the Netherlands (van Zeist, 1967; Casparie and Groenman-van Waateringe, 1980) as well as in Denmark (Andersen, 1998).

During the Bronze Age the formation of *Calluna* heaths increased and many burial mounds were erected using heath sods. A further extension took place during the Iron Age and Roman period but the large-scale expansion only began in the Middle Ages and by AD 1800 most parts of the Pleistocene soils of the Netherlands, north-west Germany and Jutland were covered by extensive treeless heaths which were used for grazing and sod cutting. With the decrease in this traditional land use, natural reforestation took place accompanied by large-scale plantations mainly with conifers not native to these areas (Behre, 1988, 2000; Odgaard, 1994).

Along the west coast of France there were also extensive heaths and grasslands, the Lande Bretonne, Landes de Gascogne and especially Les Landes in the south-west. In Brittany there are still today large areas of open vegetation with *Calluna vulgaris,* several *Erica* species, *Ulex europaeus, Pteridium aquilinum* and others as well as scrubland (bocage). Several pollen diagrams show that before humans took up farming there the interior of Brittany was nearly completely covered with forest. The clearing of this woodland started in the Neolithic and increased in the Gallo-Roman period, culminating in the Middle Ages and Modern Times and resulted in a completely open landscape. At the beginning of the 19th century, large areas became reforested by plantations of *Pinus pinaster* (van Zeist, 1964; Joly and Visset, 2009).

The Landes further south underwent a similar process. Originally they were covered mainly by deciduous oak forests. Clearing and grazing started in the Neolithic and increased stepwise until the wide open Landes were formed in the Middle Ages and Modern Times. Here, as early as the Pre-Roman Iron Age, *Pinus pinaster* spread naturally and replaced the preceding oak forests. It has been used intensively as firewood for the production of sea salt. In the last two centuries it has been planted and forests have returned to large parts of the former Landes (Marambat, 1992).

The Atlantic heathlands in the lowlands along the coasts of the Iberian Peninsula have a similar history. Here, it was mainly pine that constituted the primary forests. The landscape opened up considerably during the Late Bronze Age and the degradation of the vegetation reached its first maximum in the Roman era when the area became treeless and formed a transition to the steppe-like garrigues of the hinterland with high amounts of Ericaceae, *Pteridium* and *Genista/Ulex*. This period was generally followed by a regeneration phase with increasing values of pine, oak or hazel while heathland diminished. During the Middle Ages and Modern Times anthropogenic impact was even stronger than in the Roman period and, together with intensive cultivation, heathland or garrigue spread again (Mateus, 1992; Janssen, 1994).

Grazed and managed forests

In prehistoric times, forests were the main source of food for domestic animals: they also provided litter for their bedding and wood and timber for many purposes. When animal husbandry started in the Neolithic, there were not yet meadows, and cattle had to graze in the forest. This led to openings in the forest around the settlements and therewith to the formation of a herb carpet on the ground.

A reliable record of forest grazing is difficult to find because the vegetation units of small open patches can only be detected in pollen diagrams from small mires close to settlements (Fig. 4.5.4-1). These patches are represented by large amounts of grass pollen and the occurrence of pollen from species of woodland clearings such as *Epilobium, Digitalis, Urtica, Senecio* type, as well as from *Calluna, Juniperus* and *Vaccinium* (Berglund et al., 1986; Pott, 1986).

Forest grazing was practiced since the beginning of livestock farming but planned management first started in the Middle Ages and has continued until Modern Times, transferring the woodland into typical grazed forests (*Hudewälder,* Pott and Hüppe, 1991). The forest composition changed in the following way. Mantle and fringe communities developed at woodland edges and parts of the forests were characterized by huge single oaks which served as mast trees for the pigs' food. In Roman Times people differentiated between '*silvae glandiferae*' (forests where pigs were fattened upon acorns and beechnuts) and '*silvae vulgaris pascue*' (the normal pasture forests; Ellenberg, 1988).

For modern viewers these open old forests with gnarled oaks give the impression of primeval woodland but they have been transformed by people for various purposes for at least several centuries. These impressions became the basis of a hypothesis, which was put forward by Vera (1997, 2000) and has been heavily discussed in recent years. Vera denies the existence of closed forests as climax or late-successional vegetation in pre-Neolithic times in lowland temperate Europe. In his hypothesis, grazing by large herbivores had a decisive influence on the original vegetation and led to a mosaic of small grasslands with groups of trees. This view of a park landscape is, however, not in accordance with the wealth of palynological and other data and has, therefore, been rejected by palaeoecologists (Bradshaw, 2002;

Mitchell, 2005): it is, however, still alive in discussions concerning forest management.

Another exploitation of the forests was the gathering of leaf fodder for winter feeding through the practice of pollarding and shredding, that is cutting off the branches of the trees. *Ulmus, Fraxinus* and *Acer* and, in Scandinavia, also *Betula* (Austad, 1985) were mainly used for this purpose, which is still in use in some remote areas in the Alps and southern Europe. In famine years almost all tree and shrub species were taken (Brockmann-Jerosch, 1936).

The great demand for small-diameter wood for posts as well as for tanbark and charcoal was solved by two other methods of forest management. 1) By coppicing, where trees are cut to near ground level every few years and then regrow from the stool. This method, carried out at defined intervals, is still in use in large areas of Europe, especially in France. 2) Lopping, whereby trees are cut 2.5 m above ground level and grow again from the permanent base, called bolling, to produce successive crops of wood (Rackham, 1976, 1980).

Meadows and pastures

In addition to fields, the modern cultural landscape is characterized by the presence of meadows and pastures. These would not exist without human activity: only in the saltmarshes at the coasts and in mountainous areas above the treeline or in the forest-steppes and steppes of south-eastern Europe is there more or less extensive natural grassland although small patches also occur around bogs or lakes and along rivers. While the pastures often developed as a result of cattle grazing, the meadows had to be established artificially. The occasional mowing of pastureland does not yet form a meadow but as soon as a meadow is mown regularly once or twice a year, its species composition changes and the characteristic meadow communities develop.

Without the scythe and grazing animals, there would be neither meadow nor intensive pasture in most parts of Europe. Both, however, are mainly composed of plant species that come from the native flora. On moist soils there are poor litter meadows, which are cut once a year for obtaining straw and belong to the communities of the Molinion, while on the better soils that are fertilized and mown twice a year, the communities belong to the Calthion. Most fodder meadows, which have to be manured regularly, have Arrhenatherion communities, characterized by the false oat grass, *Arrhenatherum elatius* and these can be mown several times a year (Ellenberg, 1988; Pott, 1995).

These anthropogenic plant communities developed gradually over the course of prehistory. They are difficult to identify by pollen analysis because wild grass pollen cannot be determined to species level and may also come from grazed open forests. Macro-remains are better sources, in particular since there are keys for uncarbonized fruitlets of Poaceae (Körber-Grohne, 1964, 1991). Table 4.5.8-1 shows the record of grassland species since the Neolithic period. It gives hints for the development of grassland communities but since almost all of these species are part of the natural vegetation, sparse occurrences are not yet proof of human-made meadows and pastures. Syntheses by Greig (1988), Knörzer (1975, 1996), Körber-Grohne (1990) and Willerding (2007) show that in Neolithic times there were not yet grassland communities in the modern sense and certainly not hay meadows. These probably came into existence in the course of the Late Bronze Age and the pre-Roman Iron Age, when in most parts of central and northern Europe, the stall feeding of cattle in winter became important following the climatic deterioration of the last millennium BC.

Of great importance for the creation of meadows were iron scythes of which the first ones are known from the Alpine region from around the second century BC. Later, in Roman Times, the finds become numerous and are from several parts of Europe (Capelle, 2005). The Roman author Columella (transl. 1972) described in detail how to lay out meadows for cutting hay and pointed out that meadows were of prominent importance for Roman farming.

In prehistoric times there was often no clear distinction between meadows and pastures nor between the latter and grazed forests. Outside the areas of the former Roman Empire, a separation of meadows and pastures is first mentioned from the 9th century AD (Poschlod, 2015). However, the highly productive fodder meadows of today only spread during the later Middle Ages and the first records of the diagnostic *Arrhenatherum elatius* are from this period (Table 4.5.8-2).

A special method for the creation and extension of meadows was practiced over large parts of Fennoscandia. Regular flooding of wetland

Table 4.5.8-1. Grassland species of central and north-west Europe. Records of macro-remains from selected species in the various cultural epochs. NL: Neolithic, BA: Bronze Age, IA: Pre-Roman Iron Age, RP: Roman period, MA: Middle Ages. After Greig (1988), Körber-Grohne (1990) and Knörzer (2007). From Lang (1994), updated.

	NL	BA	IA	RP	MA
Grasses					
Anthoxanthum odoratum	O	.	O	O	O
Bromus hordeaceus	O	O	O	O	O
Festuca rubra/ovina	O	O	O	O	O
Lolium perenne	O	.	O	O	O
Molinia caerulea	O	.	.	O	O
Phleum pratense	O	O.	O	O	.
Poa pratensis	O	O	O	O	O
Poa trivialis	O	O	O	O	O
Holcus lanatus	.	O	O	O	O
Cynosurus cristatus	.	.	O	O	O
Dactylis glomerata	.	.	O	O	O
Helictotrichon pratensis	.	.	O	.	.
Briza media	.	.	.	O	.
Bromus erectus	.	.	.	O	O
Festuca pratensis	.	.	.	O	O
Holcus mollis	.	.	.	O	.
Alopecurus pratensis	.	.	.	.	O
Arrhenatherum elatius	.	.	.	.	O
Trisetum flavescens	.		.	.	O
Other herbs					
Achillea millefolium	O	.	O	O	O
Cerastium holosteoides	O	O	O	O	O
Daucus carota	O	O	O	O	O
Heracleum sphondylium	O	.	.	O	O
Knautia arvensis	O	.	O	O	O
Leontodon cf. *autumnalis*	O	.	.	O	O
Leucanthemum vulgare	O	.	O	O	O
Lychnis flos-cuculi	O	O	O	O	O
Pastinaca sativa	O	O	.	O	O
Prunella vulgaris	O	O	O	O	O
Ranunculus acris	O	O	O	O	O
Rumex cf. *acetosa*	O	O	.	O	O
Silaum silaus	O	.	.	O	O
Stellaria graminea	O	O	O	O	O
Trifolium dubium	O	.	O	O	.
Trifolium repens	O	O	O	O	O
Galium mollugo	.	O	O	O	O
Lathyrus pratensis	.	O	.	.	.
Linum catharticum	.	O	O	O	O
Medicago lupulina	.	O	O	O	O
Pimpinella saxifraga	.	O	.	O	.
Ranunculus bulbosus	.	O	.	O	O
Scabiosa columbaria	.	O	.	O	O
Caltha palustris	.	.	O	O	O
Centaurea jacea	.	.	O	O	.
Hypochaeris radicata	.	.	O	O	O
Pimpinella major	.	.	O	.	O
Succisa pratensis		.	O	.	.
Trifolium campestre	.	.	O	O	O
Trifolium pratense	.	.	O	O	O
Veronica chamaedrys	.	.	O	O	.
Anthriscus sylvestris	.	.	.	O	O
Carum carvi	.	.	.	O	O
Galium verum		.	.	O	O
Salvia pratensis	.	.	.	O	.
Senecio jacobaea	.	.	.	O	O
Tragopogon pratensis	.	.	.	O	.
Bellis perennis	.	.	.	.	O

favoured the biomass production of tall sedges and hampered the growth of mosses, thus making possible sustained hay-making. This technique was applied from the 16th century until about 1950 (Vasari and Väänänen, 1986).

Communities of field weeds and ruderals

When arable farming started at the beginning of the Neolithic, numerous species were introduced that accompanied the cultivated plants. In most cases, these were annual species which became combined as various communities of field weeds and ruderals, depending on edaphic factors and the different types of agriculture. There are different views with regard to their structure and classification, which reflect the traditional or the actual way of arable farming (Oberdorfer, 1983; Ellenberg, 1988; Hüppe and Hofmeister, 1990; Pott, 1995). The traditional classification is mainly based on the type of cultivation and separates the weed communities of winter crops (Secalietea) from those of summer crops and annual ruderal communities (Chenopodietea) while the classification of modern plant communities is mainly based on the base content in the fields and separates the winter and summer crop communities only on a lower rank as alliances: Sperguletalia arvensis on acidic soils versus Papaveretalia on base-rich soils. The latter system takes into account the modern methods of permanent crop rotation, artificial fertilising and so forth and is less usable for periods before 1950.

In connection with a rural way of life, and from the Neolithic onwards, spaces inside and outside of the settlement areas came into existence where organic waste and rubbish (latin *rudus*, gen. *ruderis*) were deposited. These uncultivated ruderal places, generally rich in nitrogen, were mostly occupied by perennial weeds which formed several ruderal communities. This perennial weed vegetation is generally summarized in the Artemisietea group. Within this class the Artemisietalia on cool, damp soils and the Onopordetalia in warm and dry habitats, show a strong anthropogenous character.

The majority of the annual field weeds and the annual and perennial ruderal plants have been imported from the south-east of Europe and its marginal and neighbouring areas, the dry steppe regions, from where the cereal species also came. Because the soils were always kept open to grow the cultivated plants, annual aliens were able to hold their ground against their strong competitors, which were mainly perennial species from the native forest flora.

The species promoted by human activities are generally recognised as falling into one or other of two following categories: anthropoochores or apophytes. **Anthropochores** are not native in the respective area, but were mostly unintentionally introduced by people. Most of these weeds had been introduced before the end of the Middle Ages and are called archaeophytes; a large number arrived later and are called neophytes. The second category consists of the **apophytes**, which are native taxa that are favoured by cultural activity and spread directly or indirectly to colonise newly established biotopes (Table 4.5.8-2). The distinction as to whether a species belongs to the apophytes or archaeophytes is sometimes difficult. The fossil records are important for deciding this but such records are sparse from the pre-Neolithic period compared with the great number of investigated culture layers from the Neolithic onwards. At a finer scale, transitions may also be recognised. In individual parts of a larger area, some native apophytes may show themselves as anthropochores by penetrating into new areas, which have been changed by human activity. For further information on this latter topic, the classical study by Linkola (1916) relating to Finland, can still be recommended.

The weed flora only attained its modern dimensions gradually and underwent major changes in connection with the different agricultural practices, such as the kind of tillage, sowing, crop rotation and fallowing. The record of fossil macro-remains indicates a continuous increase in the number of archaeophytes from the Neolithic to the Middle Ages, while written sources complement the evidence for the gradual spread of neophytes in modern times. From the Neolithic until the Roman period, arable agriculture was probably done as ley farming with several years of field crops followed by longer fallow periods but also with permanent cultivation without crop rotation and fallow phases (Willerding, 1986, 1988; Rösch, 1998; Jacomet et al., 2016).

The agricultural implements used in conjunction with the cultivation system also have a role in determining the weed flora of a cultivated area.

Table 4.5.8-2. Apophytes and Archaeophytes in the weed and ruderal vegetation of central Europe. Evidence of selected species by macro-remains and pollen records in the different cultural epochs. Apo: Apophytes, already recorded from pre-Neolithic times; the following are neophytes, first recorded in NL: Neolithic, BA: Bronze Age, IA: pre-Roman Iron Age, RP: Roman period, MA: Middle Ages. Community: centre of occurrence in modern plant communities (classes, -etea; orders, -etalia). Agropyretea (couch-grass communities), Agrostietea (pioneer communities of trampled and flooded places), Arrhenatheretalia (manured meadows and pastures), Artemisietalia (mugwort and burdock communities), Bidentetea (bur-marigold mud banks), Calystigietalia (valley and river bank communities), Polygono-Chenopodietalia (arable and garden weeds of rich soil), Festuco-Brometea (poor and dry calcareous swards), Glechometalia (ground-ivy forb communities), Molinietalia (moist meadows), Onopordetalia (perennial dry ruderal communities), Polygono-Poetea (annual trampled communities), Secalietea (weed communities of winter crops), Sedo-Scleranthetea (open sandy swards), Sisymbrietalia (annual ruderal communities). After Willerding (1986) and Knörzer (2007), with additions. Plant-sociological assignment after Oberdorfer (1990). From Lang (1994) with additions.

	Apo	NL	BA	IA	RP	MA	Community
Biennials and perennials							
Aegopodium podagraria	?	.	.	O	O	O	Glechometalia
Carduus nutans	O	.	.	O	O	O	Onopordetalia
Cirsium arvense	O	O	O	O	O	O	Artemisietalia
Euphorbia cyparissias	O	.	O	.	O	O	Festuco-Brometea
Lamium album	O	O	.	O	O	O	Artemisietalia
Linaria vulgaris	O	O	.	O	O	O	Onopordetalia
Malva sylvestris	O	O	O	O	O	O	Onopordetalia
Mentha arvensis	O	O	O	O	O	O	Molinietalia
Plantago lanceolata	?	O	O	O	O	O	Arrhenatheretalia
Plantago major	O	O	O	O	O	O	Agrostietea
Plantago media	O	.	O	O	O	O	Arrhenatheretalia
Potentilla anserina	O	.	.	O	O	O	Agrostietea
Potentilla argentea	?	O	.	.	O	O	Sedo-Scleranthetea
Ranunculus repens	O	O	O	O	O	O	Agrostietea
Rumex acetosella	O	O	O	O	O	O	Sedo-Scleranthetea
Stellaria graminea	O	O	O	O	O	O	Arrhenatheretalia
Urtica dioica	O	O	O	O	O	O	Calystegietalia
Alliaria petiolata	.	O	.	.	O	O	Glechometalia
Arctium lappa	.	O	O	.	O	O	Artemisietalia
Arctium minus	.	O	O	O	O	O	Artemisietalia
Artemisia vulgaris	.	O	.	.	.	O	Artemisietalia
Carduus crispus	.	O	.	O	O	O	Calystegietalia
Chenopodium bonus-henricus	.	O	O	O	O	O	Artemisietalia
Cirsium vulgare	.	O	.	O	O	O	Onopordetalia
Echium vulgare	.	O	O	O	O	O	Sedo-Sclerantetea
Lapsana communis	.	O	O	O	O	O	Glechometalia
Malva moschata	.	O	.	.	.	.	Arrhenatheretalia
Melilotus officinalis	.	O	.	.	.	O	Onopordetalia
Nepeta cataria	.	O	.	O	O	O	Onopordetalia
Picris hieracioides	.	O	O	O	O	O	Onopordetalia
Saponaria officinalis	.	O	O	.	O	O	Agropyretea
Verbena officinalis	.	O	O	O	O	O	Agrostietea
Veronica serpyllifolia	.	O	.	O	O	.	Agrostietea
Ballota nigra	.	.	O	O	O	O	Artemisietalia
Barbarea vulgaris	.	.	.	O	O	O	Agrostietea
Dipsacus fullonum	.	.	.	O	O	O	Artemisietalia
Cichorium intybus	.	.	.	.	O	O	Agrostietea
Potentilla reptans	.	.	.	.	O	O	Agrostietea
Reseda lutea	.	.	.	.	O	O	Onopordetalia
Bunias orientalis	.	.	.	.	.	O	Artemisietalia
Cardaria draba	.	.	.	.	.	O	Agropyretea
Leonurus cardiaca	.	.	.	.	.	O	Artemisietalia
Marrubium vulgare	.	.	.	.	.	O	Onopordetalia
Muscari comosum	.	.	.	.	.	O	Sedo-Scleranthetea
Annuals							
Atriplex patula	O	O	O	O	O	O	Polygono-Chenopodietalia
Bidens cernua	O	O	O	O	O	O	Bidentetea
Centaurea cyanus	O	O	O	O	.	O	Secalietea
Cerastium arvense	O	.	.	O	.	O	Agropyretea
Chenopodium album	O	O	O	O	O	O	Polygono-Chenopodietalia
Euphorbia peplus	O	.	.	.	O	O	Polygono-Chenopodietalia

Table 4.5.8-2 continued

Fallopia convolvulus	O	O	O	O	O	O	Secalietea
Galeopsis tetrahit	O	O	O	O	O	O	Secalietea
Persicaria hydropiper	O	O	O	O	O	O	Bidentetea
Persicaria lapathifolia	O	O	O	O	O	O	Bidentetea
Persicaria maculosa	O	O	O	O	O	O	Polygono-Chenopodietalia
Persicaria minor	O	O	O	O	O	O	Bidentetea
Persicaria mitis	O	O	.	O	O	O	Bidentetea
Polygonum aviculare	O	O	O	O	O	O	Polygono-Poetea
Rumex maritimus	O	.	.	O	O	O	Bidentetea
Stellaria media	O	O	O	O	O	O	Polygono-Chenopodietalia
Viola arvensis	O	O	O	O	O	O	Secalietea
Aethusa cynapium	.	O	O	O	O	O	Secalietea
Agrostemma githago	.	O	O	O	O	O	Secalietea
Anagallis arvensis	.	O	O	O	O	O	Polygono-Chenopodietalia
Aphanes arvensis	.	O	.	O	O	O	Secalietea
Arabidopsis thaliana	.	O	.	.	.	.	Sedo-Scleranthetea
Arenaria serpyllifolia	.	O	O	O	O	O	Secalietea
Avena fatua	.	O	O	O	O	O	Secalietea
Bromus arvensis	.	O	O	O	O	O	Polygono-Chenopodietalia
Bromus secalinus	.	O	O	O	O	O	Secalietea
Bromus sterilis	.	O	O	O	O	O	Sisymbrietalia
Capsella bursa-pastoris	.	O	O	O	O	O	Polygono-Chenopodietalia
Chenopodium ficifolium	.	O	O	O	O	O	Polygono-Chenopodietalia
Chenopodium polyspermum	.	O	O	O	O	O	Polygono-Chenopodietalia
Convolvulus arvensis	.	O	O	O	O	O	Agropyretea
Descurainia sophia	.	O	.	O	O	O	Sisymbrietalia
Digitaria ischaemum	.	O	O	O	O	O	Polygono-Chenopodietalia
Echinochloa crus-galli	.	O	O	O	O	O	Polygono-Chenopodietalia
Fumaria officinalis	.	O	O	O	O	O	Polygono-Chenopodietalia
Galium aparine	.	O	O	O	O	O	Calystegietalia
Galium spurium	.	O	O	O	O	O	Secalietea
Geranium dissectum	.	O	O	.	O	O	Polygono-Chenopodietalia
Hordeum murinum	.	O	.	O	.	O	Sisymbrietalia
Hyoscyamus niger	.	O	O	O	O	O	Onopordetalia
Lamium purpureum	.	O	O	O	O	O	Polygono-Chenopodietalia
Lithospermum arvense	.	O	O	O	O	O	Secalietea
Lolium temulentum	.	O	O	O	O	O	Secalietea
Malva pusilla	.	O	O	O	O	O	Polygono-Poetea
Matricaria recutita	.	O	.	.	O	O	Secalietea
Myosotis arvensis	.	O	O	O	O	O	Secalietea
Papaver argemone	.	O	O	O	O	O	Secalietea
Papaver rhoeas	.	O	O	O	O	O	Secalietea
Raphanus raphanistrum	.	O	O	O	O	O	Secalietea
Reseda luteola	.	O	.	.	O	O	Onopordetalia
Scleranthus annuus	.	O	O	O	O	O	Secalietea
Setaria pumila	.	O		O	O	O	Polygono-Chenopodietalia
Setaria viridis	.	O	O	O	O	O	Polygono-Chenopodietalia
Silene latifolia	.	O	O	O	O	O	Artemisietalia
Sinapis arvensis	.	O	O	O	O	O	Secalietea
Sisymbrium officinale	.	O	O	O	O	O	Sisymbrietalia
Solanum nigrum	.	O	O	O	O	O	Polygono-Chenopodietalia
Sonchus oleraceus	.	O	O	O	O	O	Polygono-Chenopodietalia
Spergula arvensis	.	O	O	O	O	O	Polygono-Chenopodietalia
Thlaspi arvense	.	O	O	O	O	O	Polygono-Chenopodietalia
Torilis japonica	.	O	O	O	O	O	Glechometalia
Trifolium arvense	.	O	O	O	O	O	Sedo-Scleranthetea
Urtica urens	.	O	O	O	O	O	Sisymbrietalia
Valerianella dentata	.	O	O	O	O	O	Sedo-Scleranthetea
Valerianella locusta	.	O	O	O	O	O	Sedo-Scleranthetea
Veronica hederifolia	.	O	O	O	O	O	Polygono-Chenopodietalia
Vicia hirsuta	.	O	O	O	O	O	Secalietea
Vicia tetrasperma	.	O	O	O	O	O	Secalietea
Adonis aestivalis	.	.	O	.	.	O	Secalietea
Anchusa arvensis	.	.	O	.	O	O	Polygono-Chenopodietalia
Anthemis arvensis	.	.	O	O	O	O	Secalietea
Bromus tectorum	.	.	O	O	O	O	Sisymbrietalia
Conringia orientalis	.	.	O	.	.	O	Secalietea
Datura stramonium	.	.	O	.	.	O	Polygono-Chenopodietalia
Erodium cicutarium	.	.	O	O	O	O	Sedo-Scleranthetea
Euphorbia helioscopia	.	.	O	O	O	O	Polygono-Chenopodietalia
Euphorbia platyphyllos	.	.	O	.	.	O	Polygono-Chenopodietalia
Geranium columbinum	.	.	O	.	.	.	Polygono-Chenopodietalia
Geranium pusillum	.	.	O	.	O	O	Polygono-Chenopodietalia
Heliotropium europaeum	.	.	O	.	.	O	Polygono-Chenopodietalia
Mercurialis annua	.	.	O	.	O	O	Polygono-Chenopodietalia
Orlaya grandiflora	.	.	O	.	O	O	Secalietea
Rapistrum perenne	.	.	O	.	.	.	Secalietea

Anthemis cotula	.	.	.	O	O	O	Secalietea
Digitaria sanguinalis	.	.	.	O	O	O	Polygono-Chenopodietalia
Euphorbia exigua	.	.	.	O	O	O	Secalietea
Fumaria vaillantii	.	.	.	O	.	O	Secalietea
Lamium amplexicaule	.	.	.	O	O	O	Polygono-Chenopodietalia
Malva neglecta	.	.	.	O	.	O	Sisymbrietalia
Odontites vulgaris	.	.	.	O	O	O	Arrhenatherietalia
Odontites verna	.	.	.	O	O	O	Secalietea
Plantago arenaria	.	.	.	O	.	.	Sisymbrietalia
Poa annua	.	.	.	O	O	O	Polygono-Poetea
Ranunculus arvensis	.	.	.	O	O	O	Secalietea
Tripleurospermum perforatum	.	.	.	O	O	O	Polygono-Chenopodietalia
Vicia villosa	.	.	.	O	O	O	Secalietea
Anchusa officinalis	.	.	.	.	O	O	Onopordetalia
Barbarea intermedia	.	.	.	.	O	.	Polygono-Chenopodietalia
Legousia speculum-veneris	.	.	.	.	O	O	Secalietea
Lepidium campestre	.	.	.	.	O	O	Polygono-Chenopodietalia
Lepidium ruderale	.	.	.	.	O	O	Polygono-Poetea
Neslia paniculata	.	.	.	.	O	O	Secalietea
Nigella arvensis	.	.	.	.	O	O	Secalietea
Portulaca oleracea	.	.	.	.	O	O	Polygono-Poetea
Silene noctiflora	.	.	.	.	O	O	Secalietea
Xanthium strumarium	.	.	.	.	O	O	Sisymbrietalia
Cerinthe minor	.	.	.	.	.	O	Onopordetalia
Cynoglossum officinale	.	.	.	.	.	O	Onopordetalia
Elsholtzia ciliata	.	.	.	.	.	O	Sisymbrietalia

Table 4.5.8-2 continued

Here it is the resistance of a species to particular cultivation method, rather than their general ecological characteristics, which is important. The mattock and the ard rip apart the soil but do not destroy perennating organs such as roots and rhizomes so the spread of grasses, plantain and other perennial species is favoured, especially with extended fallow periods. The coming of the mouldboard plough, first recorded in central Europe in the third century BC, resulted in a completely new biotope in which perennial plants were by and large destroyed, to be replaced by annual therophyte species. This is very important for the occurrence of the most important of anthropogenic indicators, *Plantago lanceolata*, which nowadays signifies undisturbed grassland but also indicates arable farming in prehistoric times (Behre, 1981).

With the introduction of the rotating three-field system and the one-field system with the practice of plaggen (see chapter 4.5.7) new communities of weed vegetation developed. A good example is the formation of the Secalietea by the permanent winter rye cultivation on the plaggen soils around the southern North Sea (Behre, 1976). After the introduction of artificial fertilizers in the middle of the 19th century fallow was not necessary any more and the continuous ploughing of the fields led to the elimination of most of the perennial field weeds. Modern cultivation methods with cleaning of the seed material and the employment of herbicides and insecticides eventually resulted in a severe reduction in the weed flora as a whole (Holzner and Immonen, 1982).

Looking at the ruderal or weedy plant communities from the Neolithic until the Modern period, it is seen that these changed slowly through time due both to the immigration of new species and through alterations in agricultural practices. This has been demonstrated in several areas (Knörzer, 1971b; Wasylikowa, 1981; Jones, 1988).

In order to reconstruct former agricultural practices with the help of weed communities, the archaeobotanical data have to be compared with the plant associations as defined before the rigid changes in agriculture from about 1940 onwards (Tüxen, 1937) rather than with the contemporary composition and classification of plant communities.

Neophytes

The neophytes constitute the youngest members of the European flora. This term marks the species that have been introduced intentionally or unintentionally after the discovery of America in 1492. They differ considerably from the archaeophytes, the majority of which were introduced from the Near East in connection with arable farming. The neophytes originate from many regions of the world: most of them came from North and South America but a large number also came from the Far East and quite a number from Africa, Australia and elsewhere. Many of them are fully naturalized and look like native plants such as *Solidago canadensis* north of the Alps or *Opuntia ficus-indica* south of the Alps (Table 4.5.8-3).

There are many pathways of introduction and modes of dispersal. Apart from cultivated food plants, spices and other useful species, which have been treated in chapter 4.5.3, many species were deliberately introduced by gardeners, foresters or aquarists. Sooner or later they escaped from their primary artificial habitat or were planted out into the wild. Many more species were introduced unintentionally by commercial traffic with ships, railways and other forms of transport and were able to spread on their own into suitable habitats where they became naturalized. This is the case with most of the annual neophytes.

It is noteworthy that the number of neophytes which immigrated into natural or near-natural communities of the European vegetation is low. The European forests, mires and even the alpine vegetation are obviously saturated and offer only a few ecological niches for new species. Nevertheless, there are some species that have succeeded in becoming established in such habitats such as *Acorus calamus* or *Sarracenia pupurea* in wet environments.

On the other hand the newcomers had good chances in more or less open or disturbed areas such as waste places in and around settlements and along roads and railways. These semi-natural or artificial sites became the homes of immigrants such as *Conyza canadensis, Juncus tenuis* or *Oenothera biennis*, which created and enriched ruderal communities in the classes Polygono-Poetea and Stellarietea. Other preferred habitats for neophytes are riverbank communities.

Water bodies form special cases since these can host several aquatics often planted out by aquarists and spread by streams as well as by migrant birds. Examples are *Azolla filiculoides, Sagittaria latifolia* or *Elodea canadensis*, which first appeared in 1836 in Ireland but were intentionally put into the waters at the Berlin botanical garden in 1859 from where it rapidly spread across Europe (Kowarik, 2010).

An important group of neophytes is represented by the 'invasive species', which often spread very rapidly and cause significant and unwanted ecological, economic or human-health impacts (Ruiz and Carlton, 2003). Some of them, even woody species such as *Prunus serotina* or *Rhododendron ponticum* (mainly in Britain and Ireland), had initially been planted intentionally but naturalized and got out of control. Well known are the weeds *Galinsoga parviflora* and the tall, poisonous biannual *Heracleum mantegazzianum*. Riverbank areas became occupied by the perennials *Solidago canadensis* and *S. gigantea, Fallopia japonica* and *F. sachalinensis* as well as by American *Aster* species and so on. There are severe conflicts between neophytes and nature conservation that result in various ways of combat.

The frequency of introduction of neophytes is increasing enormously, mainly due to the strong growth in trade and traffic (Fig. 4.5.8-5). More than half of the present neophytes only arrived in Europe after AD 1900 and the speed of arrival is increasing (Kowarik, 2010).

Apart from fruit trees quite a number of neophytic trees have been introduced because of their ability to grow on poor habitats or because of their high yields of wood and timber. Particularly since the middle of the 19th century *Pseudotsuga menziesii, Pinus strobus, Picea sitchensis* and *Larix kaempferi* have become permanent constituents of the forests over large parts of Europe. South of the Alps, *Eucalyptus* species are now common, and in France and surrounding areas, the Roman archaeophyte *Castanea sativa* has become a forest tree. Fully naturalized are *Rhododendron ponticum* (chiefly in Britain and Ireland) and *Prunus serotina*, both already mentioned and, to the east, *Robinia pseudoacacia* as well as in the south *Acacia dealbata* and others. Since the 1970s even a neophytic palm, *Trachycarpus fortunei*, has started to naturalize and, due to global warming, has become an invasive tree in the forests of southern Switzerland and neighbouring areas (Walther et al., 2007).

Table 4.5.8-3. Selection of primarily central European neophytes, their geographical origin and time of introduction. Cult: cultivated, Erg: Ergasiophygophytes (escaped), Xen: Xenophytes (adventives). Compiled after several authors.

	Origin	Type	Ecology
Trees and shrubs			
Acacia dealbata	Australia (1792)	Cult	forests
Acer negundo	N America (1688)	Cult	parks
Ailanthus altissima	E Asia (ca. 1740)	Cult	parks and roads
Amorpha fruticosa	N America (1724)	Cult	riverbanks
Buddleja davidii	E Asia (19th C.)	Cult	parks
Eucalyptus div. spec.	Australia (19th C)	Cult	forests
Fraxinus pennsylvanica	N America (18th C)	Cult	parks
Ginkgo biloba	E Asia (18th C)	Cult	parks and roads
Gleditsia triacanthos	N America (18th C)	Cult	parks
Larix kaempferi	E Asia (19th C)	Cult	forests
Mahonia aquifolium	N America (1823)	Cult	parks
Paulownia tomentosa	E Asia (1834)	Cult	parks
Picea sitchensis	N America (19th C)	Cult	forests
Pinus strobus	N America (1705)	Cult	forests
Prunus serotina	N America (1623)	Cult	parks
Pseudotsuga menziesii	N America (1827)	Cult	forests
Quercus rubra	N America (1724)	Cult	forests
Rhododendron ponticum	Caucasus (1763)	Cult	parks
Robinia pseudoacacia	N America (1629)	Cult	parks
Symphoricarpos albus	N America (1817)	Cult	parks
Thuja occidentalis	N America (1532)	Cult	parks
Thuja orientalis	E Asia (1690)	Cult	parks
Trachycarpos fortunei	E Asia (1830)	Cult	parks
Tsuga canadensis	N America (18th C)	Cult	parks
Biennials and perennials			
Acorus calamus	S Asia (16th C)	Erg	lakeshores
Artemisia verlotiorum	E Asia (19th C)	Erg	riverbanks
Aster lanceolatus	N America (19th C)	Erg	riverbanks
Aster novae-angliae	N America	Erg	riverbanks
Aster tradescantii	N America (19th C)	Erg	riverbanks
Carpobrotus edulis	S Africa	Erg	sand
Erigeron annuus	N America	Erg	riverbanks
Fallopia japonica	E Asia (1823)	Erg	riverbanks
Fallopia sachalinensis	E Asia (1863)	Erg	riverbanks
Fragaria indica	SE Asia	Erg	scrub
Helianthus tuberosus	N America	Erg	riverbanks
Heracleum mantegazzianum	Caucasus (ca. 1890)	Erg	riverbanks
Juncus tenuis	N America (1824)	Xen	pathways
Kalmia angustifolia	N America	Erg	fens
Lupinus polyphyllus	N America	Erg	clearings
Mimulus guttatus	N America (ca. 1850)	Erg	riverbanks
Opuntia ficus-indica	M America	Erg	ruderal
Phytolacca americana	N America	Erg	ruderal
Rudbeckia laciniata	N America (17th C)	Erg	riverbanks
Sarracenia purpurea	N America	Erg	fens
Solidago canadensis	N America (1632)	Erg	riverbanks
Solidago gigantea	N America (1758)	Erg	ruderal
Spartina alterniflora	N America (1872)	Erg	sea shores
Veronica filiformis	W Asia (ca. 1930)	Erg	grassland
Annuals			
Ambrosia artemisiifolia	N America	Xen	ruderal
Amaranthus albus	N America (18th C)	Xen	ruderal
Amaranthus retroflexus	N America (18th C)	Xen	ruderal
Bidens connata	N America (ca. 1865)	Xen	lake shores
Bidens frondosa	N America (1896)	Xen	lake shores
Conyza canadensis	N America (17th C)	Erg	ruderal
Cuscuta gronovii	N America	Xen	lake shores

Galinsoga ciliata	M/S America (1850)	Erg	fields
Galinsoga parviflora	S America (ca. 1800)	Erg	fields
Impatiens balfourii	Himalaya	Erg	shrubs
Impatiens glandulifera	Himalaya (ca. 1900)	Erg	riverbanks
Impatiens parviflora	E Asia (1837)	Erg	Forests
Iva xanthiifolia	N America	Xen	Ruderal
Lepidum densiflorum	N America (1879)	Xen	Ruderal
Lepidum virginicum	M/N America	Xen	ruderal
Matricaria discoidea	NE Asia (1850)	Erg	pathways
Oxalis eurpaea	N America (19th C)	Xen	fields
Oxalis dillenii	N America (1961)	Xen	fields
Veronica persica	W Asia (1805)	Erg	fields
Xanthium saccharatum	N America (19th C)	Erg	riverbanks
Aquatic plants			
Azolla filiculoides	M America	Erg	lakes
Crassula helmsi	Australia (1911)	Xen	lakes
Elodea canadensis	N America (1836)	Xen	lakes
Elodea nuttallii	N America (1941)	Xen	lakes
Hydrocotyle ranunculoides	N America	Xen	ditches
Lagarisiphon major	S Africa	Erg	lakes
Myriophyllum aquaticum	S America (19th C)	Erg	lakes
Sagittaria latifolia	N America	Erg	lake shores

Table 4.5.8-3 continued

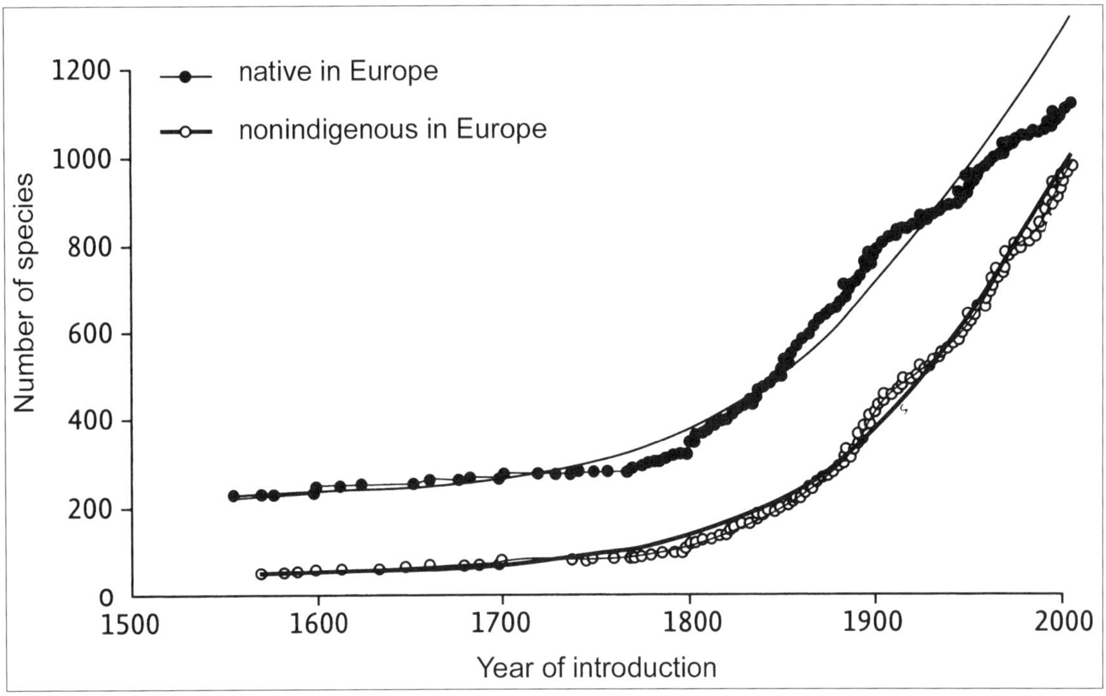

Fig. 4.5.8-5. Neophytes. Increase of non-native species in Europe after AD 1500. Upper curve: species native in parts of Europe but occurring in at least one country as non-native. Lower curve: non-indigenous species all over Europe. From Lambdon et al. (2008).

4.6

Development of European lakes

Gerhard Lang[1], Brigitta Ammann[1,2]

[1] Institute of Plant Sciences, University of Bern, Bern, Switzerland
[2] Oeschger Centre for Climate Change Research, University of Bern, Bern, Switzerland

4.6.1

Introduction

Alongside the few 'exotic' environmental archives, such as coprolites (Kelso and Solomon, 2006; Yll et al., 2006), soils (Andersen, 1986) or ice cores (Azuara et al., 2015; Brugger et al., 2018a) the greatest natural archives for biological remains on the continents are the sediments in lakes and the peat of mires. For physical and chemical records ice cores and stalagmites also play important roles.

The challenges, opportunities and risks with using lakes and mires as natural archives for tracing environmental history arise from the fact that both the local and the regional signals are simultaneously embedded in the sediment or the peat. Lakes and mires 'write down' their own histories. If we are able to read these 'autobiographies' and if we can connect them to the regional histories we may understand the relevant processes behind biotic changes at several scales. The precondition is that in the biotic record we can separate local events from regional ones.

For the study of vegetation history an estimate of the sedimentation rate is relevant, because it controls the temporal resolution that can be achieved and thus the sampling strategy for a specific question.

For selected aquatic taxa see chapter 4.4.6.

4.6.2

Types of lakes and their vegetation

Lakes are more than pollen traps. They are model ecosystems where abiotic (e.g. light, oxygen, temperature) and biotic processes (e.g. food web, nutrient cycling) interact. Lakes also integrate environmental changes within the lake with changes in the catchment by incorporating micro- and macro-fossils of both aquatic and terrestrial origin.

The origin and age of lakes

Lakes (and ponds) are waterfilled basins in the landscapes and may have various origins. Following early classifications by Davis (1882) and Forel (1901), Hutchinson (1957), in his first volume, wrote a chapter on the origin of lakes in the form of a global account, which was organized into 76 types. Table 4.6-1 summarizes those of Hutchinson's types that are relevant for Europe. The approximate ages given in this table do not necessarily concern the age of the basin formation but the age of the water filling, which then determines the sedimentation process and thus potentially the length of the environmental record. Löffler (2004) provides a thorough and global treatment of the topic.

Many lakes have an origin controlled by a combination of agencies: wind action along coasts (forming tombolos or drowning estuaries by dune formation), valleys (and their glaciers) may follow tectonic lines (faults), glacial erosion causing over-

Geo-morphic agencies		Type No. and description	Examples (mainly from Europe)
Tectonic activities	01.	Relict lakes cut off from the sea by epeirogenic uplift	Caspian Sea, Aral Sea (from Thetys)
	04.	Upwarping all around basin	Lake Victoria, central Africa
	06.	In syncline dammed by an anticline	Fählensee (CH)
	09.	Tectonic graben or multiple grabens, rift-lakes	Lake Baikal (1642 m deep), Dead Sea, Lake Tanganyika (1670 m deep)
Volcanic activity	10.	Crater in cinder cone	
	11.	Maar lakes	Eifel (DE), Auvergne (FR)
	12.	Explosion craters	Auvergne (FR)
	13.	Collapsed calderas	Crater Lake (Oregon USA)
	14.	Conca (plural conche), shallow, after several eruptions	Conca di Bolsena (IT), Lago di Bracciano (IT), Mono Lake (California USA)
	18. -19.	Dammed by lava	Auvergne (FR)
Landslide, rockfalls	20.-22.		Lac de Chaillexon Doubs (FR/CH)
Glacial activity	23.-25.	Lakes held by ice or by moraines in contact with existing ice, proglacial lakes	
		Glacial rock basins	
	26.	On ice-scoured old surfaces	Numerous in Fennoscandia, GB, CA
	27.	Cirque lakes, near snowline, in accumulation area of firn, often over-deepend	Innumerable examples in Alps and other mountains
	28a.	In valley rock basins below snow line, often in chains (paternoster lakes)	
	28b.	Fjord lakes, in overdeepened valleys	Nordfjord lakes; Hornindalsvatn (514 m deep, NO)
	28c.	Piedmont lakes, in formerly glaciated areas, often overdeepend (transitional to type 30)	Windermere in the English Lake District (UK); north and south of the Alps: Lago di Garda (IT), Lago di Como (IT), Lago Maggiore (IT); Zürichsee (CH); Ammersee (AU); Starnbergersee = Würmsee (DE)
	28d.	Large rock basins, formerly under continental ice	Great Laurentian Lakes (CA, USA), Great Slave Lake (CA)
	29.	Ice-cauldron lakes (pot-holes, gryte, Gletschertopf)	Loch Rannoch (Scotland), many places in the Alps, e.g. Maloja
		By morainic and outwash dams	
	30.	Lakes dammed by terminal or recessional moraines (transitions to 28c)	Numerous, e.g. Chiemsee (DE), Zürichsee (CH)
	31.	Lakes dammed by outwash filling a valley below the former glacier	
	32a.	Lakes formed in lateral valley by terminal moraine of main valley	Lac de Clairvaux (Jura FR)
	32b.	Lakes formed in lateral valley by lateral moraine of main valley	Lac de Barterand (Jura FR)
	33.	Lakes between recessional moraines, often with outwash and dead-ice	Lille Le (Dalsland SE)
		Drift basins	
	34.-38.	Various kettleholes in till, pitted outwash plains, in valleys as a chain of lakes (similar to type 40)	Numerous in Fennoscandia, parts of N DE, e.g. Schleswig-Holstein
	39.	Lakes in kettles between eskers or drumlins	Schleinsee (DE); Greifensee (CH)
	40.	Tunnel valleys: plunge pools in ground moraines; alignement of lakes determined by subglacial valleys (formed under high hydrostatic pressure)	Jelser Seen (DE)

Peri-glacial	**41.-42.**	**Thermokarst lakes, thaw lakes,** or collapsed pingos, in permafrost areas	
Solution lakes	43a.	Doline lakes	
	43b.	Uvala lakes, fusion of several dolines	Muttersee (CH)
	44.	Polje, tectono-karstic lakes	
	45.	Lakes in caves, by solution and deposition of calcareous sinter	
	46.	Lakes formed by subsidence after solution of soluble salts (NaCl or gypsum)	Mansfelder See (Saxony DE)
Fluviatile action	48.	Plunge pool below waterfall now dry	
		Fluviatile dams	
	49.	A lateral tributary dammed a main stream	Lake Pepin (Minnesota-Wisconsin USA)
	50.	Fluviatile deposits of deltas divided an original lake into two	Derwentwater and Bassenthwaite in English Lake District; Brienzersee (CH); Thunersee (CH)
	52.	Lateral lakes when levees of main stream back-water up a tributary	Numerous lakes on the lower Danube
	53.	Deltaic levee lakes	Étang de Vaccarès on Rhone Delta (FR)
		Lakes of mature flood plains	
	55.	Oxbow or isolated loops of meanders	Along the Wistula River (PL); billabongs (AU)
Aeolian activity	60.	Basins dammed by wind-blown sand	
	62.	Lakes between well-oriented dunes	Cherry County (Nebraska USA)
	63.	Deflation basins, lunettes	
At shorelines	64.	Maritime coastal lakes in drowned estuaries	France; Portugal; Greece; Baltic coast; Limans (BG, UA)
	65.	Lakes enclosed by two tombolos or spits joining an island to the mainland	Stagno di Orbetello (IT)
	66.-68.	Spits cut off marginal basins from larger lakes	
Meteoritic impacts	75.-76.		Kaali (EE), Lake Bosumtwi (GH), Lake El'gygytgyn (Siberia, RU)

Table 4.6-1. The origin of lakes. Examples from Hutchinson (1975, 156-163). Here is a selection out of 76 types. Many lakes originated from combinations of agencies.

deepening as, for example, in Lago di Como, where the surface of the bedrock is 212 m below sea level. Valleys can be blocked by landslides, rockfalls, moraines or lava flows. Landslides and rockfalls were frequent after the ice retreated from slopes. In endorheic systems the lake can be a remnant of the sea after tectonic or glacio-tetonic movements (e.g. Caspian Sea, Lake Ladoga, Neusiedlersee in Hungary and Austria, Balaton in Hungary) or – within continents – big deflation basins (e.g. Lake Eyre in Australia, possibly Lake Chad).

In terms of numbers, the lakes of glacial origin are the most abundant but lakes of other origin may be more important for two reasons: spatially the rarer lakes outside former glaciated areas may improve the spatial coverage of environmental history – timewise some of the non-glacial lakes may be much older than those in glaciated areas and thus contain very long records (for example volcanic lakes such as Maars in Germany, volcanic lakes in the French Massif Central, or meteoritic impact craters).

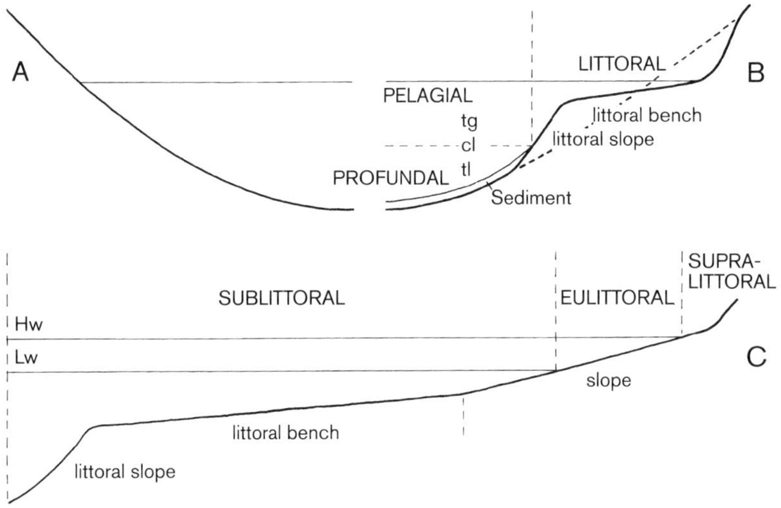

Fig. 4.6-1. Shore morphology and biotopes of larger lakes. A. Lake basin just after its formation. B. Gradual changes in morphology and biotopes, with the compensation level (cl) between the trophogenic layer (tg) and the tropholytic layer (tl). C. Subdivisions of the littoral zone. Hw = high water level, Lw = low water level.

Circulation types of extra-tropical lakes based on Hutchinson and Löffler (1956) and Lang (1994).

Type of mixing	Characteristics	Examples in Europe
Amictic lakes	Permanent ice-cover, no over-turns	Polar and high-alpine sites
Meromictic lakes	Only the upper layer (mixolimnion) circulates, but not the lower part (monimolimnion); limit: chemocline	Längsee (A), Lago di Cadagno (CH), Ulmener Maar (D), Valkiajärvi (SF)
Holomictic lakes	The whole water body circulates (epilimnion, metalimnion, hypolimnion)	
Cold-monomictic	Full over-turn in summer, rest of the year stagnation under ice	Märjelen See at Aletsch glacier (CH)
Dimictic	Full over-turn both in spring and autumn, stagnation in summer and in winter	Widespread in the temperate zone
Warm-monomictic	Full over-turn in winter, stagnation during the rest of the year	Lakes in the English Lake District (UK), Lago di Como, Lago di Garda (I), Lake Lucerne (CH)

Table 4.6-2. Lake types according to their behaviour in annual deep-mixing, also called 'over-turn' or 'turn-over' or 'circulation' (the latter not to be confused with circulation caused by rivers, the Coriolis effect or wind). The original system by Hutchinson and Löffler (1956) includes tropical lakes. A revised classification was proposed by Lewis (1983) including shallow lakes.

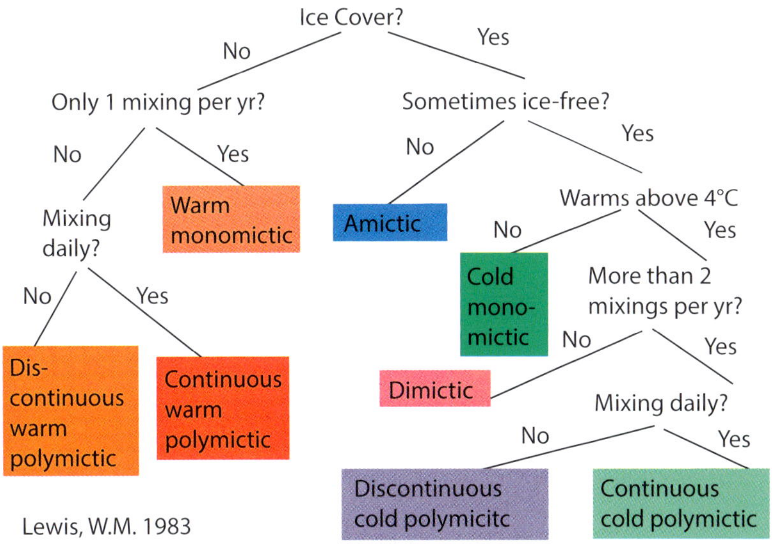

Fig. 4.6-2. Key to identifying the circulation type of a lake. Modified from Lewis (1983).

The morphometry, habitats and circulation types of lakes

Basins do not keep the same shape after becoming a lake. Wind, waves and horizontal circulation may form sublittoral terraces, of which the land-side can be erosional and the lake-side can be aggradational (for example carbonates precipitated where water plants are photosynthetically active; see Fig. 4.6-1).

The free open water is called the pelagial, which contrasts with the lake bottom. The deepest part of this bottom is often without vegetation and is called the profundal. In the zones at the shores, the eulittoral lies between the lowest and the highest seasonal water-levels. Below the eulittoral a sequence of vegetation zones may be observed either termed sublittoral or infralittoral. Such zonations cannot be applied to shallow ponds.

The nutrient budget of a lake is not only dependent on input from surface waters or from the atmosphere but also on internal recycling as a result of the mixing of the water body. The oxygen concentration at different water depths and in different seasons also depends on the mixing. Two physical peculiarities of water play an important role in the way in which the water is mixed:

The density anomaly: the highest density of water at 4°C results in bottom water of that temperature (cooler and warmer water is lighter). This means that lakes do not freeze to the ground as long as any water is liquid, which is of primary importance to aquatic life.

The mixing of a waterbody by wind occurs only when the water is of equal temperature throughout its depth (homothermic conditions). In most lakes there are, therefore, periods of circulation (overturn, turn-over) and periods of stagnation. Table 4.6-2 provides an overview of lake types concerning their deep-circulation (not including tropical lakes).

In temperate Europe the dimictic lake type with three layers of water during periods of stagnation is widespread. During the warm season the upper layer is more or less of equal temperature, rich in oxygen and light (epilimnion) and below is the metalimnion or thermocline, where steep gradients of temperature and oxygen are observed. Below it, to the bottom, is the hypolimnion: cool water, often 4°C, dark, poorer in oxygen, with often higher nutrient concentrations. Dimictic lakes turn over in spring and in autumn and exhibit both a summer stagnation and a winter stagnation (the latter often under ice, coldest at the top).

When being used for studying vegetation history the circulation behaviour of a lake plays a role through the processes connected to the re-suspension and re-sedimentation of littoral plant remains and pollen, that can result in sediment focusing (Davis, 1973; Davis and Ford, 1982). For shallow lakes such as Lake Balaton (average depth of 3.5 m) storms may stir up sediment to a depth of 10 cm; therefore, the pollen composition of samples must be considered as an average and high-resolution sampling will not improve the temporal resolution (Zólyomi, 1995; Cserny and Nagy-Bodor, 2000; O'Sullivan and Reynolds, 2004).

Ecological types of lakes

From the early days of limnology, the classification of lake types based on their trophic and ecological conditions was a matter of concern (Forel, 1901; and later, Hutchinson, 1975; Ruttner, 1962). It seems that many concepts of ecology (food webs, energy flow and nutrient cycling) were first developed in aquatic ecology, rather than terrestrial ecology – and the reason could be that lakes offer outer limits to the ecosystem: an outer limit to the biota, that is the shore, and an outer limit to the input of nutrients, dissolved or suspended matter, that is the hydrological catchment (Odum and Odum, 1953). In contrast, terrestrial ecosystems such as forests, steppes or alpine meadows do not have such sharp outer limits.

The central criterion for most classifications of lakes is the **trophic state** (nutrient status; see Fig. 4.6-3). Whereas in terrestrial ecosystems (except in the tropics) nitrogen (as nitrates or ammonium) may often be the limiting factor, in aquatic systems it is usually phosphate. Table 4.6-3 summarizes the four most important lake types and the criteria for their classification. However, the seasonal fluctuations in different components (for example Si, N) are also important. Many textbooks on limnology are available for extended data and their relationships (Dokulil et al., 2001; Wetzel, 2001; Schwoerbel and Brendelberger, 2005). Many aspects are treated in O'Sullivan and Reynolds (2004) and especially macrophytes by Pokorný and Květ (2004). In today's lakes, aspects of human impact are highly relevant (Phillips et al., 2008; Hering et al., 2015).

In terms of the **alkalinity** of the lake water the dystrophic lakes have the lowest pH, the oligotrophic lakes poor in CaCO3 an intermediate pH, and both oligotrophic lakes rich in CaCO3 and eutrophic lakes have a pH of 7 or higher (Tables 4.6-3, 4.6-4, 4.6-5 and Fig. 4.6-4). On the pH gradient the distribution of the taxa is not the same as on the nutrient gradient, as the comparison of Table 4.6-4 with Table 4.6-5 demonstrates: low pH can be found either in oligotrophic lakes with low carbonate concentrations or in dystrophic lakes. As Table 4.6-4 shows in several species aggregates, the different subspecies occur in different pH ranges (for example *Sparganium erectum, Potamogeton natans*). In cases where macrofossils of waterplants can be identified to species or even subspecies, they can serve as indicators for past pH in the lake water. However, if the identification of macrofossils is not possible to subspecies, then the reconstruction of the pH becomes imprecise, ranging over 2–3 classes.

Floristic lake types

The above-mentioned lake types, based on productivity, are neither homogenous inside their type, nor are the limits between the types sharp. Therefore, refinements are possible, especially for certain regions rich in lakes. One approach uses the sediment type (depending on the geologiocal setting of the lakes (Lundqvist, 1927); another approach uses the bottom fauna of the profundal (Jónasson, 2004). For botanists the species composition of macrophytes in the various littoral belts is a meaningful indicator (e.g. in the English Lake District; Poland; Scandinavia). In figure 4.6-4 and Table 4.6-5 a classical and remarkable example from Finland is summarized (Maristo, 1941). Later the changes in 113 lakes over a 40 year period were presented by Rintanen (1996), who also lists more recent literature on Finnish lake types. It may be of interest to compare the map of Maristo's lake types with the map of mire types in Finland (Eurola et al., 1984 or Atlas of Finland) and to think about reasons for similarities and dissimilatities between the two maps. Another comparison of interest is the one of figure 4.6-4 on pH and taxa with Table 4.6-5. The latter includes in the 11 lake types both pH gradients and nutrient gradients combined (carbonate-rich and carbonate-poor lakes occur in both oligotrophic and eutrophic lakes, that means that the gradients of pH and nutrients do not always run in parallel).

4.6.3

The developmental history of lakes

Every lake goes through an evolution from its early formation as a basin to (potentially) its disappearance through infilling processes. Wetzel (1983) and other limnologists use the term "ontogeny of lakes". An essential process is the sedimentation of inorganic and organic matter that records abiotic and biotic changes both in the lake and in its catchment: it results in a reduction of the water volume (the reduction may be unimportant in very deep lakes). In addition to sediment type, the trophic state, lake level, temperature, and other characteristics may also change over time. These endogenic and exogenic processes are linked but are discussed separately in the following sections.

Infilling or terrestrialization of lakes

In the numerous glacigenic lakes of Europe the bottom sediment is usually a clay or a silt overlying the till. The sediment is poor in organic remains and is sometimes laminated or even varved, if the lake is still in contact with the glacier. In large and deep lakes inorganic material may even prevail throughout the Holocene, at least near the inlets. In smaller basins, with smaller catchments, the inorganic deposits may give way to organic sediments, either detritus gyttja built up by algae and organic remains (if the water is poor in bicarbonate) or carbonate-gyttja (even lake marl, if carbonates are above 40–60%). This transition from minerogenic to organic (or carbonatic) sediment may occur when the vegetation

Table 4.6-3. Ecological lake types and their main characteristics in terms of nutrient availability, pH, morphology and so forth, following Ellenberg (1963) and Schmidt (1974).

Lake Type / Characteristic	Oligotrophic, poor in $CaCO_3$	Oligotrophic, rich in $CaCO_3$	Eutrophic	Dystrophic
Nutrients, especially P	low ($P<0.05$ mg/l)	low ($P<0.05$ mg/l)	high ($P>0.1$ mg/l)	very low ($P<0.01$ mg/l)
Productivity	low	low	high	low
Oxygen near bottom	high	high	low	very low
pH	5-7	> 7.5	> 7	< 5
Shore morphology	steep to gentle	steep, or with sublittoral marl-terrace	gentle	gentle
Water depth	great to shallow	great	great to shallow	shallow
Secchi depth	great	great	low to intermediate	very low
Water colour	yellowish	blue	greenish	yellow to brown
Sediment type	sand, silt, clay, gyttja	silt, calcareous gyttja, lake-marl	gyttja	dy
Sedimentation rate	organic low, inorganic often high	generally low, but often high in littoral	high	low to intermediate
Phytoplankton in pelagial	Desmidiaceae	Chlorophyceae, diatoms	Cyanophyceae, diatoms	Desmidiaceae, diatoms
Macrophytes in sublittoral	*Nitella, Isoëtes, Lobelia*	*Chara, Najas*	*Potamogeton, Zannichellia*	Usually absent
Macrophytes in eulittoral	*Carex rostrata*	*Phragmites, Littorella*	*Phragmites, Typha*	*Sphagnum*
Occurrence in Europe	siliceous mountains, sandy areas	piedmont of the Alps	widely distributed	mire areas

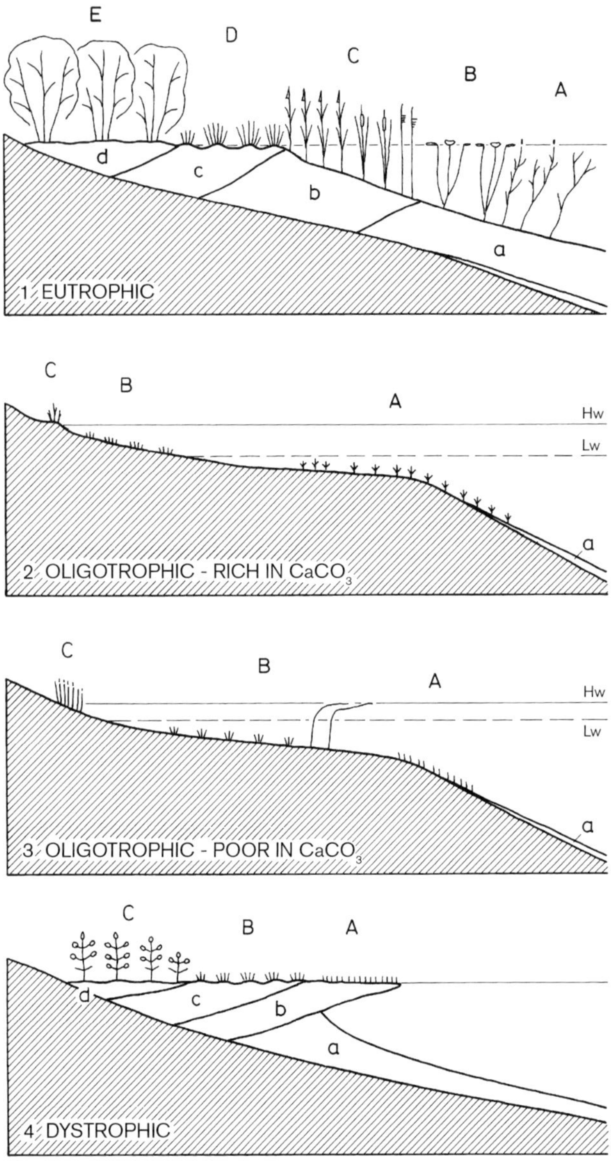

Fig. 4.6-3. Ecological lake types with their shore vegetation and sediments. (1) Eutrophic lakes: today's vegetation, A: Potamogetonion, B: Nymphaeion, C: Phragmition, D: Magnocaricion, E: Alnion glutinosae, sometimes replaced by Molinietalia. Sediments and peat, a: gyttja, at its base with clay; b: *Phragmites*-peat; c: peat of tall *Carex* spp.; d: carr peat. (2) Oligotrophic-calcareous lakes: Today's vegetation, A: *Chara* and *Najas* (Charetalia and Potamogetonion p.p.), B: Eleocharition acicularis and Deschampsion litoralis, C: the surf zone or breaker zone with nitrophilous vegetation. Sediments, a: calcareous gyttja and lake marl, at the base with clay. (3) Oligotrophic lakes poor in calcareous components: today's vegetation, A: Nitelletalia, B: Isoëtion lacustris and Lobelion, C: *Carex rostrata,* Caricion nigrae. Sediments, a: gyttja with clay at the base. (4) Dystrophic lakes: today's vegetation, A: *Carex limosa*, Rhynchosporion albae, B: *Eriophorum vaginatum,* Sphagnion magellanici, C: *Pinus sylvestris* or *P. rotundata,* Dicrano-Pinion. Sediments and peat, a: dy; b: *Cuspidata*-peat; c: *Eriophorum*-peat; d: *Pinus*-carr. Hw = high water level, Lw = low water level. Examples (1) and (4) refer to large lakes, examples (2) and (3) to small lakes.

Table 4.6-4. Water plants and their pH ranges; the range classes 1 (very acid, pH 2.5–5.5) and 5 (alkaline, 6.5– >8.5) do not occur among the genera considered. Extracted from Landolt et al. (2010) Flora Indicativa: Ecological Indicator Values and Biological Attributes of the Flora of Switzerland and the Alps.

pH ranges, in three groups (out of five)	4: neutral to alkaline	3: weakly acid to neutral	2: acid	Very large variation
Narrow variation	5.5–>8.5	4.4–7.5	3.5–6.5	
Large variation	4.5–>8.5	3.5– >8.5	2.5–7.5	
Species				
Hippuris vulgaris	+			
Ceratophyllum demersum	+			
C. submersum	+			
Potamogeton gramineus	+			
P. lucens	+			
P. perfoliatus	+			
P. pusillus (2 subspecies)	+			
P. compressus (4 subspecies)	+			
P. compressus ssp. *obtusifolius*		+		
P. pectinatus aggregate	+			
ssp. *pectinatus*	+			
ssp. *helveticus*	+			
ssp. *filiformis*		+		
P. praelongus		+		
P. crispus		+		
P. natans aggregate				X
ssp. *nodosus*	+			
ssp. *natans*		+		
ssp. *polygonifolius*			+	
P. alpinus			+	
Nymphaea alba (2 subspecies)		+		
Nuphar lutea ssp. *lutea*	+			
ssp. *pumila*		+		
Sparganium emersum		+		
S. natans		+		
S. erectum aggregate				X
S. erectum ssp. *erectum*	+			
ssp. *oocarpum*		+		
ssp. *microcarpum*			+	
ssp. *neglectum*			+	
S. hyperboreum			+	
S. angustifolium			+	
Lemna trisulca	+			
L. minuta		+		
L. minor aggregate		+		
ssp. *gibba*	+			
ssp. *minor*		+		
ssp. *turionifera*		+		
Myriophyllum spicatum	+			
M. verticillatum		+		
M. heterophyllum		+		
M. alterniflorum			+	
Utricularia vulgaris (2 subspecies)		+		
U. minor (2 subspecies)		+		
U. intermedia, aggregate				X
ssp. *intermedia*	+			
ssp. *stygis*		+		
ssp. *ochroleuca*			+	

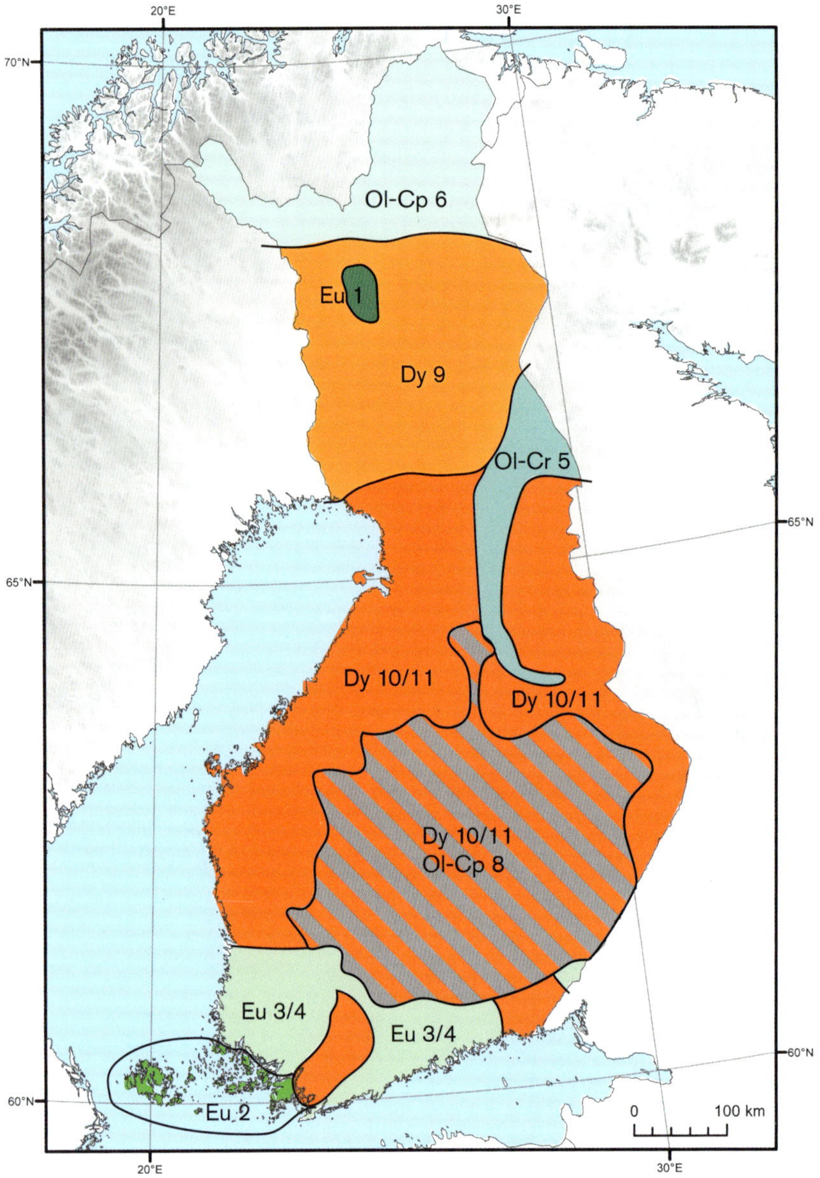

Fig. 4.6-4. Main distribution of floristic lake types in Finland, see Table 4.6-5. After Maristo (1941).

→ **Table 4.6-5.** Floristic lake types in Finland based on the frequencies of their macrophytes. 1–4 Eutrophic lakes (EU): 1 Stratiotes-type; 2 *Potamogeton*-type; 3 *Typha/Alisma*-type; 4 *Schoenoplectus lacustris*-type; 5 Oligotrophic calcareous type (OL CR): *Potamogeton filiformis/Chara*-type; 6–8 Oligotrophic poor in carbonates (OL CP): 6 *Carex*-type; 7 *Lobelia*-type; 8 *Phragmites*-type; 9–11 Dystrophic types (DY): 9 Elodeïd-type; 10 *Equisetum/Phragmites*-type; 11 *Equisetum*-type. O for frequency >50% of the sites. + for frequency <50% of the sites. ★ indicator taxa, ♦ unique and strong indicator taxa. Simplified from Maristo (1941); life-forms according to Du Rietz (1930) and Mäkirinta (1978).

	Eu 1	Eu 2	Eu 3	Eu 4	Ol Cr 5	Ol Cp 6	Ol Cp 7	Ol Cp 8	Dy 9	Dy 10	Dy 11
Elodeïds											
Potamogeton zosterifolius	O	.	+	.	.	.	.	.	+	.	.
Potamogeton alpinus	O	.	.	+	+	+	+	.	+	+	+
Myriophyllum spicatum	O	.	.	.	O	+	.	.	.	.	.
Potamogeton praelongus	O	O	.	.	O	+	+	.	O	+	.
Potamogeton perfoliatus	O	O	O	O	O	O	+	O	O	O	.
Potamogeton pusillus	+	O	+	.	+	O	+	+	O	+	.
Potamogeton obtusifolius		O	O	+	.	.	.	.	+	.	.
Potamogeton filiformis	+	.	.	.	♦	.	.	.	.	.	.
Ranunculus peltatus	.	.	+	.	O	O	O	O	O	+	+
Potamogeton gramineus	+	.	+	+	O	O	+	+	O	+	.
Myriophyllum alterniflorum	.	.	+	+	O	O	O	O	O	O	.
Nymphaeids											
Potamogeton natans	.	O	O	O	O	.	O	O	O	O	+
Nymphaea candida	.	O	O	O	+	.	+	+	O	O	O
Nuphar lutea	O	O	O	O	O	+	O	O	O	O	O
Nuphar pumila	O	.	.	+	.	O	.	+	O	+	+
Sparganium friesii	.	+	O	+	+	.	O	O	O	O	+
Sparganium angustifolium	+	.	.	.	O	O	+	.	+	+	+
Ceratophyllids											
Stratiotes aloides	♦	.	.	.	.	.	.	.	.	.	.
Lemna trisulca	.	O	.	.	.	.	.	.	+	.	.
Ceratophyllum demersum	+	O	+	.	.	.	.	.	.	.	.
Utricularia vulgaris	.	O	+	+	+	.	+	+	O	+	+
Lemnids	.										
Spirodela polyrhiza		O	+	.	.	.	.	.	.	.	.
Lemna minor	+	O	O	.	.	.	.	.	.	+	.
Hydrocharis morsus-ranae	.	.	+	.	.	.	.	.	.	.	.
Charids											
Chara fragilis	.	O	+	.	.	.	.	+	.	+	.
Chara strigosa	+	.	.	.	♦	.	.	.	.	.	.
Nitella flexilis	.	+	+	.	+	+	+	+	O	+	.
Isoëtids											
Eleocharis acicularis	.	.	O	O	+	O	+	O	O	+	+
Subularia aquatica	.	.	+	.	O	O	O	O	O	+	+
Lobelia dortmanna	.	.	+	.	+	.	♦	★	++	★	★
Isoëtes echinospora	.	.	.	+	O	O	+	+	O	+	+
Isoëtes lacustris	.	.	+	.	O	O	O	O	O	O	+
Helophytes											
Typha latifolia	.	O	O	.	.	.	.	.	.	.	.
Typha angustifolia	.	.	O	+	.	.	.	.	.	.	.
Equisetum fluviatile	O	O	O	O	O	O	O	O	O	O	O
Phragmites australis	O	O	O	O	O	.	O	O	O	O	O
Alisma plantago-aquatica	.	O	O	O	.	.	+	O	.	+	+
Schoenoplectus lacustris	+	+	O	O	+	.	O	O	O	+	+

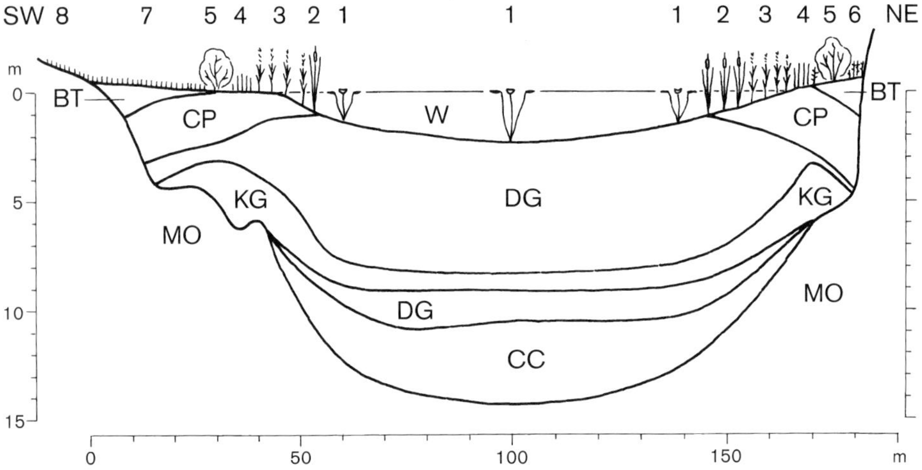

Fig. 4.6-5. Stratigraphy and today's vegetation at the south-eastern Buchensee (430 m asl) near Lake Constance, south-west Germany, an example of a dead-ice kettlehole in the Würmian till of the Rhine glacier. Terrestrialisation is advanced and has left a small eutrophic calcareous lake. MO = ground-moraine (till); CC = clay rich in carbonates and clay-gyttja (older part of the Late-glacial); DG = detritus gyttja (lower layer Allerød to Mid Holocene, in its lower part with Laacher See tephra; upper layer Late Holocene); KG = calcareous gyttja (lake marl, Late Holocene); CP = *Carex*-peat; BT = alder-carr peat. Today's vegetation, 1: Nymphaeetum albae (floating leaves). 2: Typhetum angustifoliae. 3: Cladietum marisci. 4: Caricetum lasiocarpae. 5: Salicetum pentandrae-cinereae. 6: Valeriano-Filipenduletum, replacing the Alnetum glutinosae. 7: Angelico-Cirsietum oleracei, replacing the Alnetum glutinosae. 8: Arrhenathereteum elatioris, replacing the Fagion. Elevational scale 5× exaggerated. After Lang (1973).

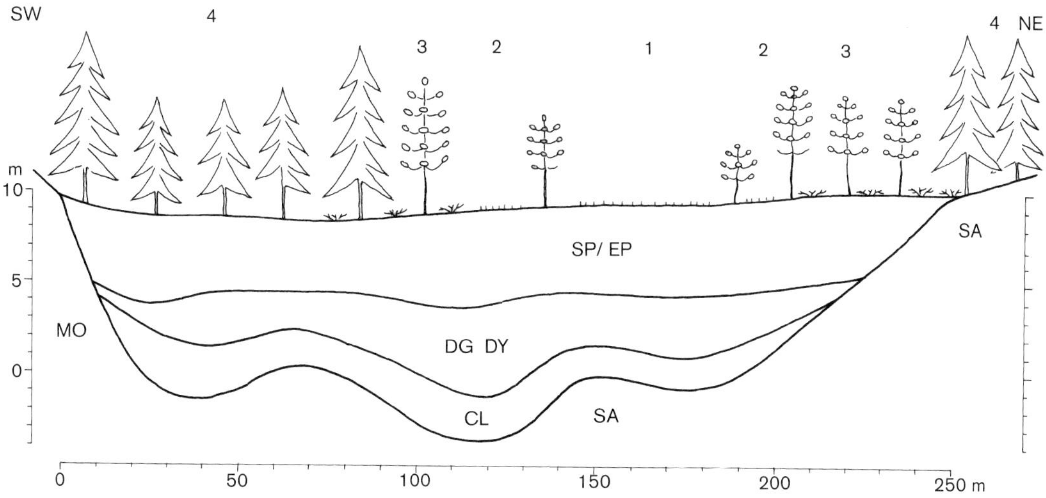

Fig. 4.6-6. Stratigraphy and today's vegetation at Rotmeer (965 m asl) in the southern Black Forest, south-west Germany. Example of a lake dammed by a terminal moraine of the Würmian glaciation. The lake developed, via a dystrophic stage into a floating mire that today shows slightly ombrotrophic features with *Pinus rotundata* – an early stage of an ombrogenous bog. MO = ground-moraine (till) and SA = sand free of carbonates. CL = clay free of carbonates and some clay-gyttja (upper part with Laacher See tephra). DG detritus gyttja, Early to Mid Holocene. DY = dy, Mid Holocene. SP = *Sphagnum* peat, lower part *S. Cuspidata*, upper part *S. Cymbifolia* and EP = *Eriophorum*-peat, Late Holocene. 1: Caricetum limosae and Sphagnetum magellanici. 2: Sphagnetum magellanici and Eriophoretum vaginati. 3: *Pinus mugo*-Sphagnetum and Vaccinio-uliginosi-Pinetum rotundatae. 4: Bazzanio-Picetum. Modified from von der Goltz et al. (1976), Lang (2005) and Lotter and Hölzer (1994).

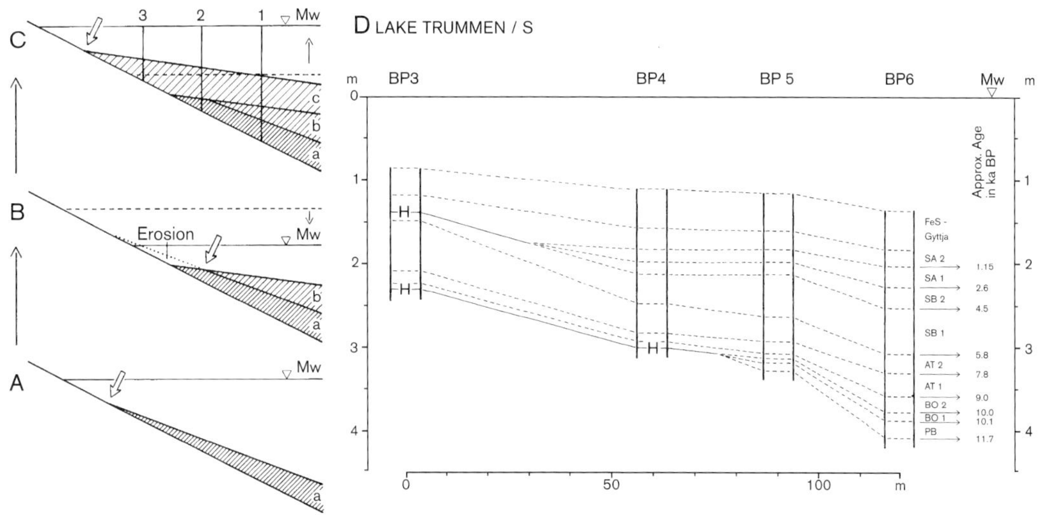

Fig. 4.6-7. Stratigraphic effects of lake-level changes. Left: three schematic time slices (A–C) distinguished by bio-stratigraphy (pollen analysis) or lithostratigraphy (sedimentology). Oblique white arrows indicate the upper limit of sedimentation. Profile 1 contains the full sediment sequence a-b-c, whereas profiles 2 and 3 are incomplete. An upper part of the sediment a deposited during period A was eroded during period B with a lower lake level. During period C, with a much higher lake level, the sediments a and b are covered by sediment c. Right (D): Lake Trummen in southern Sweden as an example. Profiles BP5 and BP6 are complete; profile BP4 shows a hiatus before the younger Boreal, i.e. before 10 ka BP; profile BP3 shows this hiatus and in addition a hiatus between the older Subboreal and the younger Subatlantic, i.e. 4.5–1.15 ka BP. Modified from Digerfeldt (1972).

in the surrounding area becomes denser, and is most pronounced with afforestation, when erosive input from the catchment decreases and higher litter production enhances pedogenesis and nutrient cycling. This transition is time-transgressive, as is afforestation, that is earlier in southern Europe (Late-glacial), later in northern Europe (Holocene). Sedimentation rates can vary enormously, whereby inorganic sediments may grow rapidly at a centimetre or decimetre per year, but organic matter may increase by a millimetre or less per year.

Near the shore some special sediments may develop in relation to the aquatic macrophytes, namely:
1. The organic component may consist of the detritus and macrofossils of the plants forming the vegetation belts: coarse detritus gyttja towards the centre of the lake, then *Phragmites*- and *Carex*-peat, then *Alnus*-carr peat at the shore.
2. The inorganic component can be silt or sand and – in situations where the water contains enough bicarbonates – carbonatic gyttja or lake marl. The photosynthesis of algae and macrophytes takes up CO_2 and therefore pulls the balance towards carbonate precipitation:

$$2\,HCO_3^- + Ca^{2+} \leftrightarrow CaCO_3 + H_2O + CO_2\uparrow$$

We, therefore, often observe calcareous incrustations on stems and leaves of higher water plants or Characeae.

If the erosive input can be excluded these carbonates may be used for the following.
1. The measurement of the ratio of oxygen and carbon isotopes that may indicate past temperatures and productivity (Siegenthaler and Eicher, 1979; von Grafenstein et al., 1992, 2013; Last and Smol, 2001; Bernasconi and McKenzie, 2013).
2. The morphotypes of the carbonate concretions may indicate water-depth and its changes and, therefore, lake-level fluctuations (see below; Magny, 1998, 2007).

The vegetation belts around lakes are a classic example of succession, here a 'hydroseries', sensu Yanko-Hombach and Yanina (2019), that is an autogenous change of communities caused by the sequence of plant associations themselves, where other influences from outside the plant communities are not involved. Note that in the example given in figure 4.6-5 the lines between the sediment (and peat) types are not isochrones. In the centre of the lake gyttja may still be deposited while near the shore, peat is already forming. Another word of caution is needed because we may be tempted to replace time by space in the discussion of succession. Because some processes are slow and because we cannot watch for millennia we tend to extrapolate from the spatial pattern of vegetation belts onto the sequence of communities over time. This may be true in some cases but certainly not in all. Grosse-Brauckmann (1974) and Rasmussen et al. (2014a) show that ecological conditions may have changed so much in the past that today's vegetation belts do not reflect past communities.

In dystrophic lakes too, the processes in the profundal and littoral do change over time: humic acids may precipitate as dy, while near the shore, floating mats of *Sphagnum* may develop and progress towards the lake centre. The sequence in the deposits may then go from dy to *Sphangnum cuspidatum* peat, to *Eriophorum* peat and to *Pinus*-carr peat (an example is given in Fig. 4.6-6).

In rare cases eutrophic lakes can show floating mats (quaking mires, Schwingrasen) formed by species of the Magnocaricion. Oligotrophic lakes generally fill in more slowly because of their low productivity. They may also develop floating carpets that grow concentrically and may eventually coalesce, thus forming an oligotrophic quaking mire.

Changes in the trophic state through the development of a lake are often recorded in the macrofossils of waterplants that have preferences for oligo-, meso- or eutrophic conditions. At least two directions of changes have been recorded:
1. Oligotrophication. In areas originally rich in phosphates and nitrates in the surrounding soils, but leached through time, the lake water may become poor in nutrients.
2. Eutrophication. This is frequently caused by human impact (agriculture in the catchment; see chapter 4.5), but also occurs naturally either when more fertile soils developed in the catchment or when the basins became more shallow.

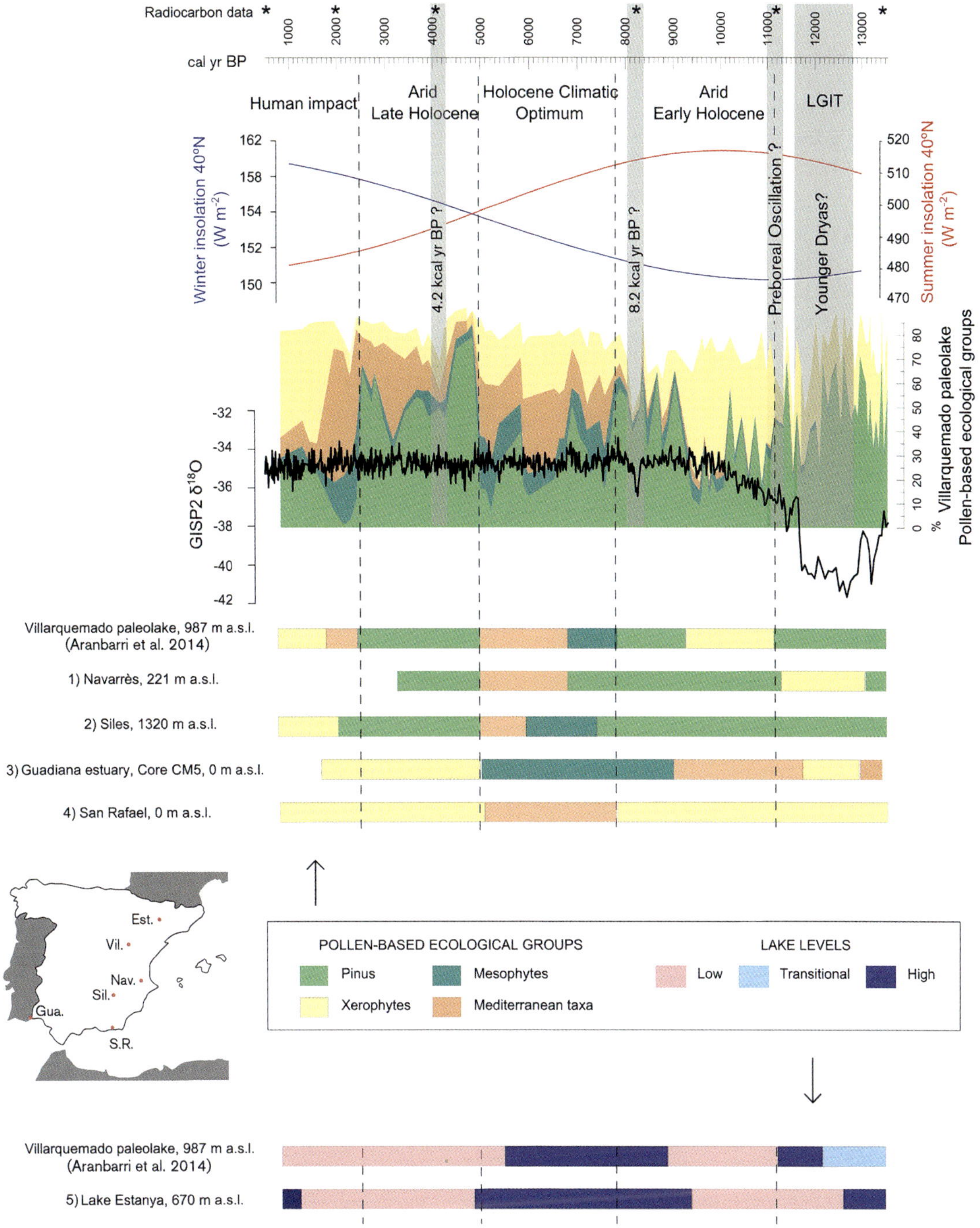

Fig. 4.6-8. Villarquemado with its pollen-based ecological groups through the Late-glacial and Holocene compared to summer and winter insolation (calculated for 40°; Hammer et al., 2001), GISP2 isotope values (Stuiver et al., 1995) and four Iberian sites: 1) Navarrès (Carrión and van Geel, 1999); 2) Siles (Carrión, 2002b); 3) Guadiana estuary Core CM5 (Fletcher et al., 2007); 4) San Rafael (Pantaléon-Cano et al., 2003). Lake-level changes at Villarquemado are inferred from sedimentological/geochemical data and the hygro-hydrophyte pollen assemblages and compared with lake-levels at Lake Estanya (Morellón et al., 2009). Modified from Aranbarri et al. (2014).

Potential disturbances of stratigraphic sequences in lakes

Site selection involves several spatial scales. For a chosen landscape Berglund (1986) discusses the criteria for the selection of sampling sites according to the research questions (biological, limnological or archaeological). At the local scale the selection of profundal cores versus littoral cores depends on both the research question and the coring equipment available – the centre of a lake is often preferable but possibly too deep for the available corer.

Disturbances in the stratigraphy of the sediment can be caused by slumps, resuspension and redeposition resulting in sediment focusing (Davis, 1968) and hiatuses, although the latter can be 'a blessing in disguise' (see below).

Lakes of certain origin are more prone to disturbance than others, for example thaw lakes in permafrost areas can be very difficult because of the instability of the shores, as visualized by pictures of 'drunken forests'.

Lake-level changes

Records of changing lake levels are of major relevance for our understanding of environmental, climatic and archaeological history (Jones and Jordan, 2007; Magny, 2007 – both reprinted in 2013). Criteria for lower or higher lake levels come from at least six possible indicators of shifting shore lines. In all six approaches the refinement depends on the number of cores taken on a littoral-profundal transect.

a) Hiatuses, their onset and end: the 'Digerfeldt-approach' (Digerfeldt, 1972, 1986; see Fig. 4.6-7). Sedimentologists have discussed in detail the so called 'sedimentation limit' that depends primarily on wind fetch, wave action and aquatic plants.

b) Sand layers or coarse organic matter, their onset and end: (Gaillard, 1984, 1985; Gaillard and Digerfeldt, 1991; Hannon and Gaillard, 1997; Gaillard and Birks, 2007). Dearing (1997) also indicated the importance of several sedimentary indicators including intermittent complete mixing (ICM), peripheral wave action (PWA), random redistribution (RR) and sliding and slumping (SS).

For approaches (a) and (b) the onset may be caused by either no-deposition and/or by erosion.

c) Aquatic plant-species are adapted to certain water depths: from submerged to floating-leaved to littoral taxa (see Fig 4.6-3; Hannon and Gaillard, 1997). Changes in plant macrofossils in cores arranged as transects from the shore to the centre of the lake may, therefore, be interpreted as periods of lowering or rise of the lake level. For lowering the difference from the effect of the natural infilling process needs to be evaluated (see Fig. 4.6-7 for an example based on Digerfeldt, 1972, 1986) Combinations of approaches (a), (b) and (c) are quite conclusive (Väliranta et al., 2005).

d) In carbonatic lakes: morphotypes of carbonate concretions can reflect the former water depth as demonstrated by Magny (1992, 1993, 1995, 1999, 2001) and Magny and Ruffaldi (1995).

e) Algae and aquatic invertebrates can also, in some cases, be used for the reconstruction of lake levels because many organisms show a clear relationship between their habitat and water depth. This approach resembles the one using plant macrofossils (Hannon and Gaillard, 1997; Gaillard and Birks, 2007). The palaeolimnological approach looks into the ratio of profundal to littoral taxa (P/L), where a shift from littoral to profundal taxa (in a single core) indicates a rising lake-level. Examples based on biostratigraphies of diatoms and Cladocera were presented by Hyvärinen and Alhonen (1994) and Hyvärinen and Mäkelä (1996), who also carefully discuss possible confounding factors.

f) Drowned tree stumps are not only macrofossils but also megafossils (also somewhat related to approach b). For example, the results of Hyvärinen and Alhonen (1994) and Hyvärinen and Mäkelä (1996) were linked to submerged trunks of Scots pine and their dendrochronology by Eronen et al. (1999). In areas where a dendrochronological master curve for the relevant taxon exists, such megafossils contain their own calendar.

Closed lakes (that is non-overflowing, arheic or endorheic lakes) are ideal for the reconstruction of lake-levels because they function more or less like a 'long-term rain-gauge'. Lakes with an outlet, however, are more difficult for at least two reasons:

1. Increased precipitation over the catchment may simply mean more outflow from the lake (as in seasonal fluctuations as for example with snow-melt).

2. Outlets can be bedrock thresholds or sedimentary complexes that are often controlled by erosional phenomena. In addition, activities in the tributar-

ies to the outflow may cause damming or eroding of the outlet at times. In karstic areas lake-levels may also be affected by changing infiltration processes.

Lake-level changes have been used to reconstruct water-budgets of the past. At least for closed lakes the precipitation P over the catchment minus the potential evapotranspiration pE of the catchment (P–pE) control lake-level changes. A remarkable study in a large closed basin is provided by Aranbarri et al. (2014) (see Fig. 4.6-8).

Additional questions arise when lake-level changes are interpreted as records of climatic changes. Gaillard (1985) pointed to a certain synchroneity of lake-level changes in both southern Sweden and central Europe. Synchroneity over a wider space points to a common cause and climatic change is an obvious explanation (changing P–pE). Gaillard (1985) was careful in the conclusions because many of the central European records were from open lakes. Factors that can complicate the climate inferred from lake-level changes are summarized in a critical review by Dearing (1997). Two types of comparisons of lake-level changes with other records that integrate annual precipitation and temperatures are (1) with Holocene glacier fluctuations (Holzhauser et al., 2005) or (2) tree-lines (Hyvärinen and Alhonen, 1994; Hyvärinen and Mäkelä, 1996). Links between lake-level changes and climatic parameters are discussed by Harrison et al. (1991), Prentice et al. (1992b), Guiot et al. (1993), Harrison and Digerfeldt (1993), Harrison et al. (1993) and Yu and Harrison (1995a), and possible connections to solar and orbital forcing by Magny (1993, 1994, 1999, 2004) and Magny et al. (2001, 2003, 2011). Databases on lake levels are discussed and used, for example, by Yu and Harrison (1995b), Qin et al. (1998), or in GLSDB (Global Lake Status Data Base).

Because lake shores are great archives of Neolithic, Bronze Age and Iron Age settlements, lake-level changes have strongly interacted with these prehistoric settlements (Magny, 1992; Jacomet, 1998; Arbogast et al., 2006; Magny et al., 2006; Bleicher et al., 2018).

4.7

Development of mires

Hans Joosten[1,2], Lukas Guth[1], Gerhard Lang[3], Brigitta Ammann[3,4]

[1] Institute of Botany and Landscape Ecology, Greifswald University, Greifswald, Germany
[2] Greifswald Mire Centre, Greifswald, Germany
[3] Institute of Plant Sciences, University of Bern, Bern, Switzerland
[4] Oeschger Centre for Climate Change Research, University of Bern, Bern, Switzerland

4.7.1

Introduction

Mires are ecosystems in which – under permanently water-saturated, oxygen-poor soil conditions – dead plants do not completely decay. The semi-decomposed plant material piles up as layers of peat that, over time, may reach several metres thickness storing palaeoenvironmental information in a stratigraphical way.

Terminologically and conceptually a differentiation is made between 'land where peat is accumulating' (a '*mire*') and 'land where peat is present' (a '*peatland*'). In a mire, intricate relationships exist between vegetation, water and accumulating peat. A peatland is defined more widely. In a drained peatland, peat accumulation does not take place anymore and the peat no longer has a genetic relationship with the vegetation above.

Peatlands that have been drained, cultivated or cut-over for peat extraction will no longer bear their original vegetation and may even have lost most of their peat. However, they may still contain important palaeoecological information. Draved Mose raised bog in Denmark, for example, originally covered c. 500 ha, but Andersen (1966a), Aaby and Tauber (1975) and Aaby (1976) had only a remnant of 3.5 ha to work with. Similarly Dupont (1986) combined biostratigraphies with isotope stratigraphies from two narrow peat sections in a remnant of what formerly must have been the largest (c. 3,000 km^2!) contiguous raised bog complex of mainland Europe: the Bourtangerveen on the border of the Netherlands and Germany (Joosten et al., 2017a). The Wachseldorn Untermoos in Switzerland is also such a lonely peat remnant but Heeb and Welten (1972) found some of the earliest Cyperaceae-peat there, dated by radiocarbon to the beginning of the Bølling. Whereas the remaining peat layers of a drained and degraded peatland may thus still constitute a palaeo-archive, only a *mire* continues to accumulate new layers of peat, which will also, in the future, provide information on vegetation and environmental development of recent times and times that are still to come (Greiser and Joosten, 2018).

In general, peatlands are linked to humid climates, as their distribution over Europe reflects (Table 4.7-1). Much peatland (~10% or more of the area) is found in countries of northern and western Europe (except for Svalbard, Iceland, the Faroes and the Azores, which have smaller proportions), with the highest value (27%) being found in Finland. In central Europe, values are mostly below 5%, except in the more oceanic Netherlands, where – similar to Denmark – peatlands formerly covered very extensive areas but have largely disappeared as a result of centuries of drainage-based land use. In eastern Europe, peatland occurrence is again larger, except for Ukraine with its southern tendencies. In southern and south-eastern Europe, the peatland area remains below 1% of each country's area, except in the mountainous Republic of North Macedonia and in Romania with its extensive Danube delta.

The proportion of mires related to the total peatland area varies more strongly. A high proportion, however, does not necessarily mean that the country has been cautious about destroying mires. It also reflects that drained peatlands have the ten-

Table 4.7-1. Estimated peatland and mire areas of biogeographical Europe per country (after Tanneberger et al., 2017). Unless otherwise indicated, the data refer to areas with a minimum peat thickness of 30 cm. The country areas are total surface areas, including land and inland water bodies but excluding polar regions and uninhabited islands. Superscripted numbers refer to notes located below the table.

Country	country area	peatland area		mire area	
	km²	km²	% of country area	km²	% of peatland area
Northern Europe					
Estonia	45,227	9,150	20.23	3,250	35.52
Finland	337,010	90,000[1]	26.71	35,000[1]	38.89
Iceland	103,000	5,777[2]	5.61	2,112[2]	36.56
Latvia	64,562	7,514	11.64	3,165	42.12
Lithuania	65,300	6,460[4]	9.89	1,781	27.57
Norway	323,787	44,700[1]	13.81	37,700[1]	84.34
Svalbard	62,422	3,000[1]	4.81	3,000[1]	100
Sweden	450,295	66,450	14.76	52,300	78.71
Western Europe					
Azores	2,333	160	6.86	80	50.00
Faroe Islands	1,393	17.6	1.26	17.6	100
Ireland	69,825	14,664.7[3]	21.00	2,692.7[3]	18.36
United Kingdom	242,495	26,838.3	11.07	8,750	32.60
Central Europe					
Andorra	468	5	1.07	5	100
Austria	83,871	1,200	1.43	175	14.58
Belgium	30,528	247.8	0.81	60	24.21
Czech Republic	78,866	285.4	0.36	150	52.56
Denmark	43,094	2,029	4.71	137	6.75
France	551,500	2,875	0.52	875	30.43
Germany	357,137	12,800	3.58	250	1.95
Hungary	93,026	300	0.32	75[1]	25.00
Liechtenstein	160	2.6	1.63	1	38.46
Luxembourg	2,586	3.5	0.14	0.1	2.86
Netherlands	37,354	2,733.4	7.32	150	5.49
Poland	311,888	14,950	4.79	2,390	15.99
Slovakia	49,036	60	0.12	30.6	11.77
Switzerland	41,285	280[6]	0.68	120	42.86
Eastern Europe					
Belarus	207,600	25,605	12.33	8,630	33.70
Russian Federation (European part)	~ 4,000,000	~ 235,000/ ~ 680,000[8]	~ 6.0/ ~ 17.0[8]	> 150,000[9]	65–75[10]
Ukraine	603,500	10,000[1]	1.66	6,395[1]	63.05
South and south-eastern Europe					
Albania	28,748	44	0.15	3.5	7.95
Armenia	29,743	47	0.16	37	78.72
Azerbaijan	86,600	2.7	< 0.1	2.7	100

Country	country area km²	peatland area		mire area	
		km²	% of country area	km²	% of peatland area
Bosnia and Herzegovina	51,209	180	0.35	162.5	90.00
Bulgaria	110,900	208	0.19	54	25.96
Croatia	56,594	33.1	0.06	0.3	0.76
Cyprus	9,251	< 0.1	< 0.1	< 0.1	100
Georgia	69,700	170	0.24	165	97.06
Greece	131,957	103	0.08	45	13.70
Italy	301,339	750	0.25	120	16.00
Republic of North Macedonia	25,713	281	1.09	13	4.63
Republic of Moldova	33,846	10	< 0.1	1	10.00
Montenegro	13,812	75	0.54	50	66.67
Portugal	89,879	271	0.29	10	3.69
Romania	238,391	7,690[5]	3.23	1,427.7	18.57
Serbia	88,361	100	0.11	12.5	12.50
Slovenia	20,273	83.9	0.41	4.3	5.07
Spain	505,992	350	0.07	200	57.14
Turkey	783,562	220[1]	< 0.1	30[7]	13.64
Total		~ 593,727		~ 320,000	~ 54

Table 4.7-1 continued

[1] >0 cm of peat; [2] >0 cm of organic soil with >12% organic carbon content; [3] ≥45 cm of peat if undrained and ≥30 cm of peat if drained; [4] ≥30 cm of peat if undrained and ≥20 cm of peat if drained; [5] >20 cm of organic soil with 20% organic matter content; [6] >10 cm of organic soil with 20% organic matter content; [7] >20 cm of peat and >1 ha; [8] peatlands (>30 cm peat) together with paludified shallow peatlands (<30 cm peat); [9] very rough estimate because statistics for agriculture and forestry do not distinguish between peatlands, shallow peatlands and paludified mineral lands within drained lands; [10] rather rough estimates – in some regions more than two-thirds of the peatland area may be drained, whereas in other regions almost the entire peatland area is undisturbed.

dency to turn into mineral soils by oxidation of the peat. To what extent that happens depends on the thickness of the original peat layers, the duration and intensity of drainage-based land use (consider arable agriculture versus low-intensity grazing) and the climate-determined rate of peat loss (high in warm, low in cool and humid climates). A high proportion of mires in a country may thus indicate both a good and a bad conservation status.

4.7.2

Mire classification

Mire classification, that is the grouping of peatlands according to their similarities, must have originated within the everyday categorisation of land. Present-day peatland names and terms reflect the properties that were considered to be relevant, such as the instability and wetness of the soil, the shape and environmental setting, the presence of a special soil substance or the characteristic colour. Use-oriented characteristics included the suitability for providing construction materials, fodder or fuel. Several of these properties were strongly correlated, which enabled one property to be used as an indicator for others. The other side of the coin was that with the development of separate scientific disciplines the same terms started to be used for different concepts. Joosten et al. (2017b) list, for example, 22 different meanings of the term 'bog' (or its linguistic equivalent) used in Europe.

This enormous diversity has, in the course of time, led to numerous approaches to peatland classification, reflecting a multitude of environmental factors, local and regional conditions, interests and traditions. Whereas with scientific progress typologies tended to become more diverse and complex, simple systems persisted by tradition and practical use. As a consequence the peatland and palaeoecological literature abounds in types, terms and concepts, with various degrees of (in)consistency. For example, Dau (1823) distinguished "Hochmoore" ('high mires') with a convex surface and "Wiesen- oder Sumpf-Moore" ('meadow or swamp mires') with a horizontal surface, as well as "Holzmoore" (wood mires) with woody peat and "Meermoore" ('sea mires') along the sea. This typology thus uses a landscape morphologic criterion to separate the first two, a plant physiognomic criterion to typify the third and a topographic criterion to define the fourth class. Similarly the common classification 'eu-', 'meso-', 'oligo-' and 'ombrotrophic' (Eurola et al., 1984) refers in the first three terms to the amount of available nutrients and in the latter to their source.

The meaning of the terms may also have changed over time, for example that of the commonly used 'topogenous' and 'soligenous'. Lennart von Post (1926) defined topogenous mires as having originated "from topographic conditions only" and included groundwater fed valley mires and spring mires in this concept. Soligenous mires, in his view, were merely those where peat formation was not only "induced and continued" by direct precipitation but also by "meteoric water flowing in from the surrounding terrain". In 1948, Hugo Sjörs redefined the terms, with topogenous referring to mires with an approximately horizontal water table and stagnant water and soligenous to those with a sloping water table and moving water. This redefinition led, amongst others, to a typological switch of spring-fed mires from topogenous to soligenous. Later the term soligenous has also been limited to denote mires fed by artesian water (Masing, 1975; Boch and Mazing, 1979; Wołejko, 2000). In current European mire science the concepts 'topogenous' and 'soligenous' are thus used in (at least) three different ways. The approach of Sjörs (1948) is the most common one, but the recent pan-European EUNIS habitat classification takes again the original von Post (1926) definition (Joosten et al., 2017a).

Many of the almost 20 well-established approaches to peatland classification in Europe (Joosten et al., 2017a) relate to how the peatland looks, how it can be used or where it is situated but provide little information on how the peatland is (or has been) functioning.

Variables that are commonly used to classify mire sites include water-table level (and its fluctuations), origin of the water, pH and base saturation of water and soil, nutrient availability, vegetation physiognomy, productivity and floristic composition (see 4.7.3), peat stratigraphy, or a combination of these variables (see 4.7.4). All these properties may apply to spots, that is they are 'non- dimensional' (although some properties obviously occupy horizontal [vegetation] or vertical space [water level, peat stratigraphy]). Such variables are suitable for classifying local mire spots and for describing assemblages (for example mosaics) of different spots in a larger mire. Most palaeoecological reconstructions depart from such spatially restricted point sampling (corings).

Also the classification and denomination of entire mire bodies ('massifs') may use such 'non-dimensional' properties when one or a few of these properties dominate the mire, for example a eutrophic fen, a string-flark fen or a palsa mire. In addition to 'point' properties, however, mire classification also has to consider the properties of the entire mire, that is of the mire 'massif'.

The most classic division on the scale of the mire massif is that between a 'bog': a convex peatland that receives its water and nutrients solely from atmospheric precipitation (i.e. 'ombrotrophic'), and a 'fen': a flat or concave peatland in a depression that also receives water that has been in contact with the mineral soil or bedrock (i.e. 'geotrophic'='minerotrophic'). Because of their water supply, bogs are in essence acid and nutrient poor, whereas fens may receive water of various quality. Some fens receive groundwater that is also acid and nutrient poor. Such 'transitional mires' function from their landscape positioning and water supply like a fen, but their

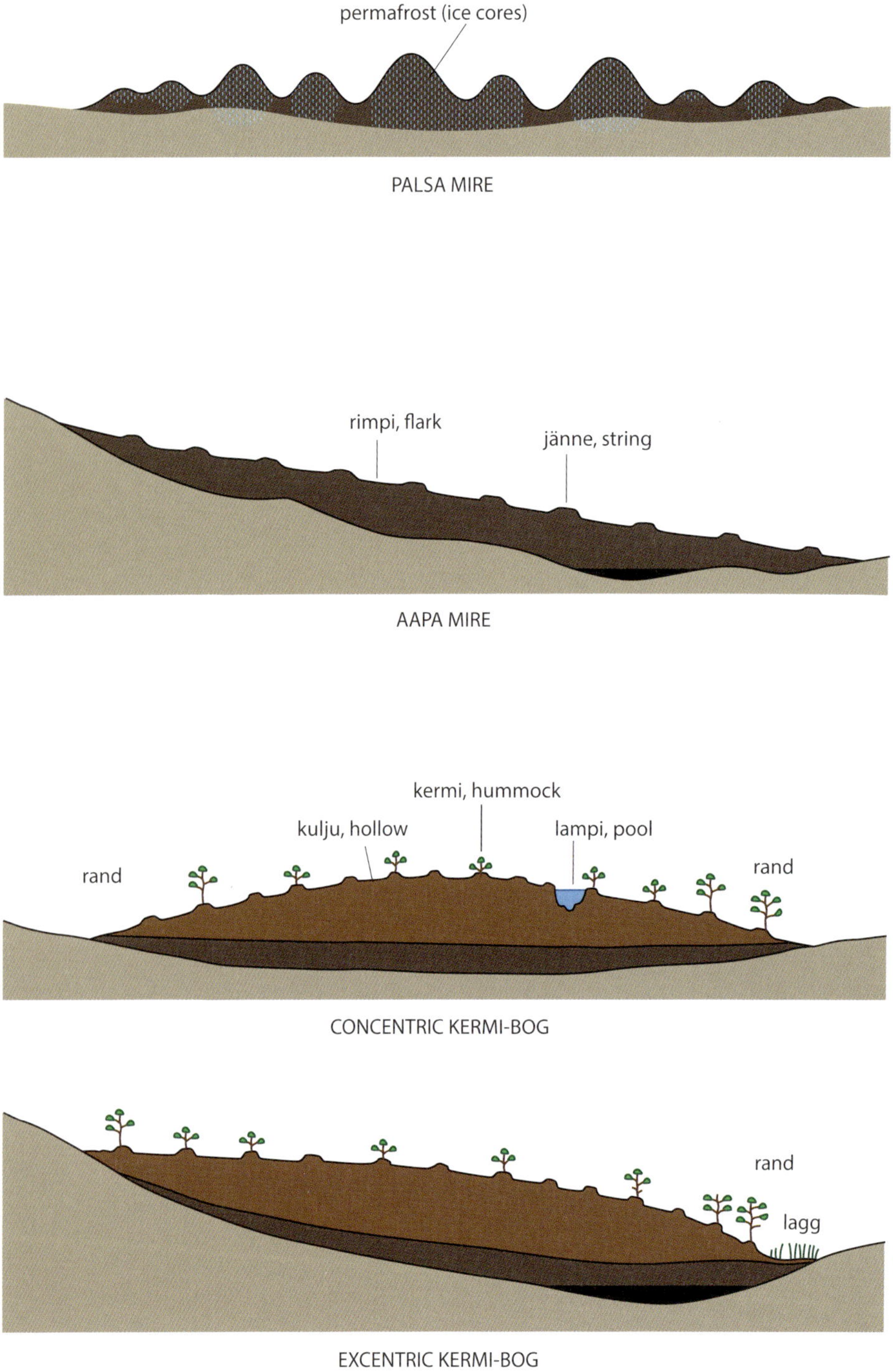

Fig. 4.7-1. Hydromorphic mire types. Stippled blue: permafrost; blue: water; black: lake sediment, dark brown: minerogenous peats; light brown: ombrogenous peats (after Ruuhijärvi, 1983; Eurola et al., 1984; Joosten et al., 2017a).

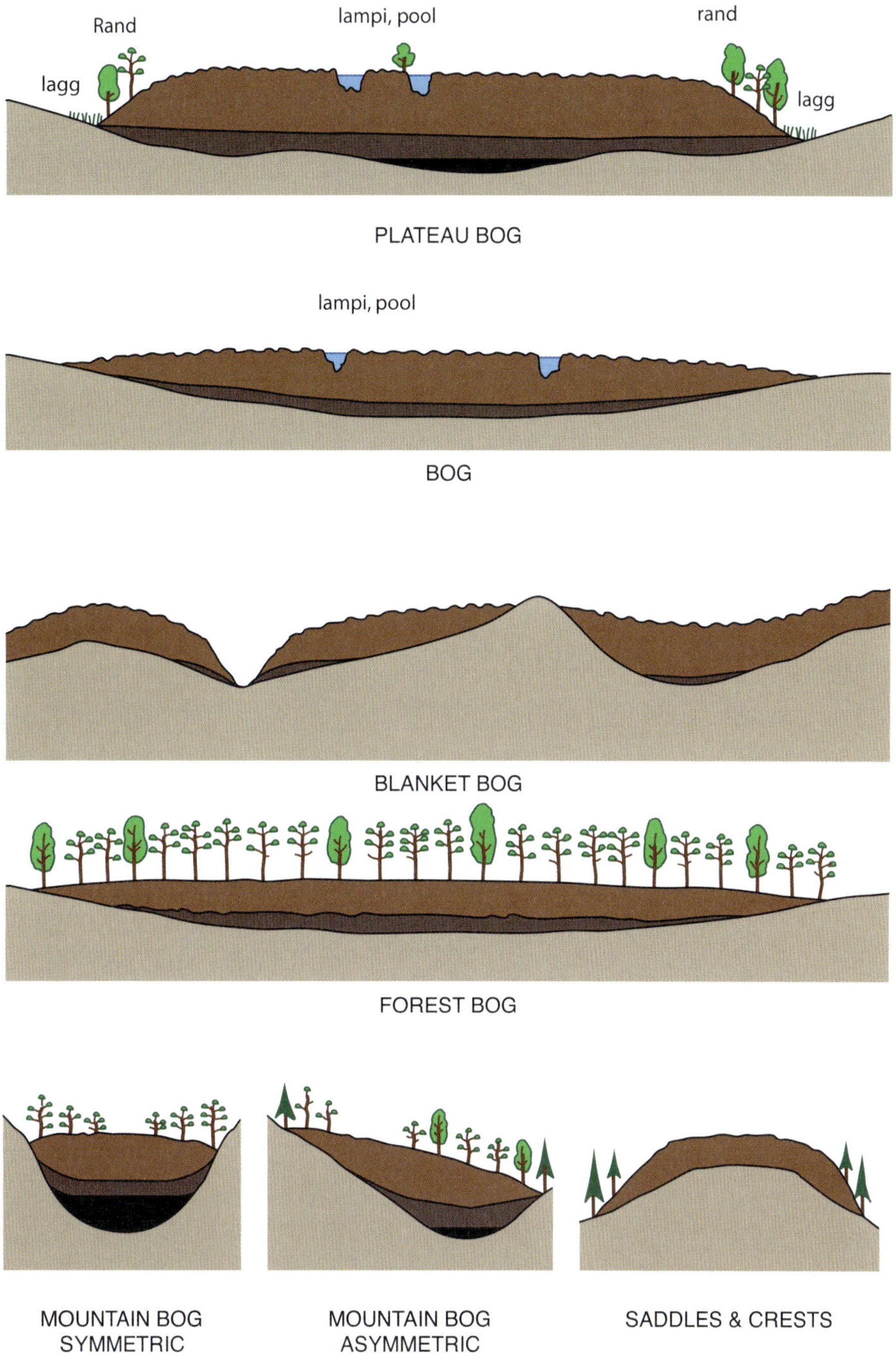

Fig. 4.7-1. continued

vegetation and hydrochemistry is similar to that of a bog. Within Europe, the majority of peatlands are 'fens' (including transitional mires), but their proportion strongly differs regionally (Tanneberger et al., 2021).

The distinction between ombrotrophy (bogs) and minerotrophy (fens) is indicated by the presence of 'mineral soil water indicators' (consider "Mineralbodenwasserzeigergrenze"; Du Rietz, 1954). No plant species is strictly ombrotrophic: bogs, as merely rainwater-fed mires, can floristically only be identified by the absence of typical fen species. These fen indicator species may differ regionally, depending on climatic and biogeographical conditions (Rydin and Jeglum, 2008), and must thus, in palaeoenvironmental contexts, be interpreted with critical awareness.

Another classic differentiation is that between 'terrestrialization' (peat formation by filling in open water: 'wet becomes dry') and 'paludification' (peat formation starting directly over a formerly dry mineral soil: 'dry becomes wet'; Charman, 2002). Whereas terrestrialization is often presented as the paradigm of peatland initiation, on an area basis, paludification is the more important peatland initiation mechanism. A special case is peat formation directly on wet ground that has emerged from the sea in areas of land uplift due to isostatic rebound, such as the coasts of Finland (Aario, 1932; Huikari, 1956).

These classical criteria have given rise to various classification approaches of mire massifs, with increasing functional inclusiveness:

- The'topographic' classification looks at the location, topography or land-setting in which the mires are situated. Examples include 'valley mire' (in a valley), 'saddle mire' (draped like a saddle between mountains), 'floodplain mire', 'basin mire' and 'oxbow mire'.
- The 'hydromorphic classification' characterizes mire massifs on the basis of their overall shape (microrelief, 'Grossform'), the types and arrangement of microrelief elements (microrelief, 'Kleinform'), and the proportion of ombro- and minerotrophic components (Fig. 4.7-1).
- The 'hydrogeomorphic' classification integrates factors external to the peatland, such as the topographic location of the peatland within the surrounding landscape, the associated source of water and the flow and strength of water movement.
- The 'hydrogenetic' classification focusses on the hydrological processes that drive peat formation and peatland development, especially on the feedback mechanisms between i) water flow and fluctuations, ii) vegetation and iii) peat formation, and on the role of peatland development in landscape hydrology (Table 4.7-2; Joosten et al., 2017a).

In its simplest form, a mire massif is surrounded on all sides by mineral soil or water. The origin of such a massif may lie in a single point from which the massif expanded, or in a larger homogeneous area where peat accumulation started everywhere simultaneously (see Fig. 4.7-9). Individual mire massifs that developed in separate depressions may expand and grow together to form a mire complex, that is one united mire body, which develops further as a single entity.

4.7.3

Floristic and ecological mire types

The number of peat-forming plant species, that is species that leave a substantial volume of recognisable macrofossils sedentarily (i.e. on the spot where the plant material has been produced) in Europe is less than 200. Peat-forming species must produce decay-resistant tissues and be able to reach mass occurrence while having largely anoxic conditions in the root zone. Among the vascular plants these are various Cyperaceae taxa (e.g. *Carex*, *Eriophorum*, *Cladium*), some Poaceae species (notably *Phragmites* but also *Molinia* and *Glyceria*), trees such as *Alnus*, *Betula*, *Picea* and *Pinus*, (dwarf) shrubs such as Ericaceae, *Myrica* and some *Salix* species and herbs such as *Menyanthes*, *Narthecium*, *Scheuchzeria*, *Thelypteris* and *Equisetum fluviatile*.

Also, among the mosses, only a few genera contribute substantially to peat formation, including *Aulacomnium*, *Calliergon*, *Calliergonella*, *Drepanocladus/Warnstorfia*, *Meesia*, *Paludella*, *Polytrichum*, *Scorpidium*, *Tomentypnum* and – most importantly – the peatmosses (*Sphagnum*). These taxa provide the close-knit matrix of vegetation and surficial peat in which many niches for other peatland species are generated (Joosten et al., 2017a). An overview of modern peat-producing vegetation types is presented in Table 4.7-3.

In addition to macroscopic remains, peat contains a large variety of remnants of plants, animals

Table 4.7-2. The main groups of hydrogenetic mire types (after Joosten et al., 2017a).

Main groups	Main hydrogenetic mire types	Palaeoecological implications
Mire with a horizontal water table and without water flow or with water moving alternately in both directions along its slope → **Horizontal mire**	Mire developing in or over an open water body → **terrestrialization mire**	Peat accumulation stops when the peat has completely filled up the pre-existing anoxic space. Peat formation only continues in the case of *allogenic* water level rise, by for example climate change, sea level rise, subsoil sealing, deforestation and interventions in the drainage system (by beavers or humans).
	Mire developing as the result of a rising water-table → **water rise mire**	
	Mire developing by regular flooding by rivers (seasonal), lakes (wind) or seas (lunar tides) → **floodwater mire**	
Mire with an inclining water level and water flowing in one direction along its slope(s) → **Inclining mire**	Uppermost peat compact and water mainly flowing *over* the peat body. May have rather steep slopes → **surface flow mire**	Peat accumulation may continue uninterruptedly over many millennia as peat formation *autogenically* raises the water level in the peatland (and often also in the catchment). In surface-flow and acrotelm mires, water flow is in the top layer of the mire, enabling the development of regular surface patterning perpendicular to the water flow (for example aapa mires and con- and eccentric raised bogs). Percolation mires do not have such surficial flow and do not develop such patterning.
	Uppermost peat/vegetation with a conspicuous and effective gradient in porosity. Water mainly flowing through the uppermost part of the coherent peatmoss vegetation/peat body → **acrotelm mire**	
	Uppermost and deeper peat porous with water flowing *through* a major part of the peat body → **percolation mire**	

and microbiota that have been living on the spot. Moreover, peat may contain an even larger diversity of materials that have been transported into the mire through the air (pollen and spores, dust) or by water (as in the case of floodwater mires). These 'fossils' provide information on landscape development much further away and outside the mire.

Classification approaches that link floristic and environmental variables (Ellenberg et al., 1992) are especially suitable for indication and interpretation. The 'ecological mire type' concept of Succow (1988; Fig. 4.7-2), for example, combines pH, nutrient availability (C/N) and floristic composition, whereas the GEST-approach (Greenhouse gas Emission Site Type; Couwenberg et al., 2011) allows the quantification of greenhouse gas fluxes from the prevailing peatland vegetation type. The detailed reconstruction of palaeoenvironmental conditions by way of such indicator approaches ('uniformitarianism'; Birks and Birks, 1980) is, however, only valid when no changes in the species spectrum and climate have occurred (which could have modified the ecological amplitudes of species by changing competition; Kotowski et al., 1998; Koska et al., 2001) and if (meaningful) genetic changes have been absent.

Table 4.7-3. The peat forming phytosociological vegetation types of Europe (from Joosten et al., 2017a).

Class	Order	Alliance	Characterisation
Oxycocco-Sphagnetea Br.-Bl. et Tx. ex Westhoff et al. 1946	Sphagnetalia medii Kästner et Flössner 1933	Oxycocco microcarpi-Empetrion hermaphroditi Nordhagen ex Du Rietz 1954 nom. conserv. propos.	Bog hummock and wooded bog vegetation; boreal non-oceanic regions
		Sphagnion medii Kästner et Flössner 1933	Bog lawn and carpet vegetation; nemoral-boreonemoral non-oceanic regions
	Erico-Ledetalia palustris Tx.1937	Oxycocco-Ericion tetralicis Nordhagen ex Tx.1937	Bog and poor fen hummock (partly tree-covered) and lawn; nemoral-boreal oceanic regions
		Ericion tetralicis Schwickerath 1933	Oceanic dwarf shrub mire vegetation, transitional to heathland; nemoral-boreal oceanic regions
Scheuchzerio palustris-Caricetea fuscae Tx. 1937	Scheuchzerietalia palustris Nordhagen ex Tx. 1937	Scheuchzerion palustris Nordhagen ex Tx.	Mainly carpet of bog and poor fen; wide distribution
	Caricetalia davallianae Br.-Bl. 1950 nom. conserv. propos.	Caricion davallianae Klika 1934	Small sedge rich fen; nemoral-boreal zones
		Caricion atrofuscae-saxatilis Nordhagen 1943	Small sedge extremely rich fen; alpine and upper boreal zones
		Caricion viridulo-trinervis Julve ex Hájek et Mucina in Theurillat et al. 2015	Small sedge rich fen of halophytic dune slack in Atlantic seaboards
	Caricetaliae fuscae Koch 1926	Drepanocladion exannulati Krajina 1933	Arctic-alpine fen/tundra with thin or no peat
		Caricion fuscae Koch 1926 nom. conserv. propos.	Mire margin, small sedge poor / intermediate fen
		Sphagno-Caricion canescentis Passarge (1964) 1978 nom. conserv. propos.	*Sphagnum*-dominated, poor fen; nemoral-boreal zones
		Festucion frigidae (Rivas-Mart. et al. 2002	Poor spring and fen vegetation; south-west Europe
		Anagallido tenellae-Juncion bulbosi Br.-Bl. 1967	Rich/intermediate oceanic fen; Iberia
		Caricion intricatae Quezel 1953	Rich/intermediate fen; Mediterranean (Corsica)
		Narthecion scardici Horvat ex Lakusic 1968	Rich/intermediate fen; Mediterranean (Balkan)
	Sphagno warnstorfii-Tomentypnetalia Lapshina 2010	Sphagno warnstorfii-Tomentypnion nitentis Dahl 1956	Mire margin/low hummock, intermediate-rich fen; mainly boreal/subalpine and alpine zones
		Caricion stantis Matveyeva 1994	Mainly rich fen; arctic zone
		Stygio-Caricion limosae Nordhagen 1936	Carpet and mud bottom of rich fen; wide distribution
		Saxifrago-Tomentypnion Lapshina 2010	Brownmoss moderately rich fen; boreal continental region

Class	Order	Alliance	Characterisation
Phragmito-Magno-caricetea Klika in Klika & Novak 1941	Phragmitetalia Koch 1926	Phragmition communis Koch 1926	Reed and tall sedges of freshwater wetlands; nemoral-southern boreal (submontane) zones
		Typhion laxmannii Nedelcu 1968	Upper littoral reed fen; continental sections
	Nasturtio-Glycerietalia Pignatti 1953	Glycerio-Sparganion Br.Bl. & Sissingh 1942	Stream reed beds in nemoral zone
	Magnocaricetalia Pignatti 1953	Magnocaricion elatae Koch 1926	Mesotrophic tall sedge fen; nemoral zone
		Magnocaricion gracilis Gehu 1961	Eutrophic tall sedge fen; nemoral zone
	Bolboschoenetalia maritimi Hejny in Holub et al.1967	Scirpion maritime Dahl et Hadač 1941	Maritime (brackish)-water reed beds; mainly nemoral oceanic regions
Littorelletea uniflorae Br.-Bl. et Tx. ex Westhoff et al. 1946	Littorelletalia uniflorae Koch ex Tx. 1937	Sphagno-Utricularion T. Mülller et Görs 1960	Pools in poor fen, wide distribution
		Scorpidio-Utricularion minoris Pietsch 1965	Pools in rich fen, wide distribution
Montio-Cardaminetea Br.-Bl. et Tx. ex Klika et Hadač 1944	Montio-Cardaminetalia Pawlowski et al. 1928	Mniobryo-Epilobion hornemanni Nordhagen 1943	Soft, cold water, boreal, alpine and arctic northern Europe
		Cardamino-Montion Br.-Bl. 1926	Soft, cold water, subalpine – alpine belts in central-western Europe
		Swertio perennis-Anisothecion squarrosi Hadac 1983	Cold water springs in montane belt of central Europe
		Epilobio nutantis-Montion Zechmeister 1994	Cold water springs in submontane and montane belts of western Europe
		Cratoneurion commutati Koch 1928	Cold water springs with calcareous water in Europe
	Cardamino-Chrysosplenietalia Hinterlang 1992	Caricion remotae Kästner 1941	Wooded poor springs; mainly in submontane-montane regions in southern Europe
Alnetea glutinosae Br.-Bl. et Tx ex Westhoff et al. 1946	Alnetalia glutinosae Tx. 1937	Alnion glutinosae Malcuit 1929	Wooded regularly flooded fen/swob; mainly nemoral-submontane/southern boreal regions
	Salici pentandrae-Betuletalia pubescentis Clausnitzer in Dengler et al. 2004	Salici pentandrae-Betulion pubescentis Clausnitzer in Dengler et al. 2004	Wooded birch/bay willow rich fen; nemoral boreal outside the most oceanic regions
	Sphagno-Betuletalia pubescentis Scamini et Passarge 1959	Betulion pubescentis Lohmeyer et Tx. ex Oberd. 1957	Wooded birch poor fen; nemoral-boreal regions
Frangulетea Doing ex Westhoff et Den Held 1969	Salicetalia auritae Doing 1962	Salicion cinereae T. Müller et Görs ex Passarge 1961	Fen scrub/woodland; mainly nemoral in oceanic regions
		Alno incanae-Salicion pentandrae K.-Lund 1981	Fen scrub/woodland; mainly poor-/intermediate, boreal regions

Table **4.7.-3** continued

Class	Order	Alliance	Characterisation
Vaccinio-Piceetea Br.-Bl. in Br.-Bl. et al. 1939	Ledo palustris-Laricetalia gmelinii Ermakov in Ermakov et Alsynbayev 2004	Empetro-Piceion obovatae Morozova et al. 2008	Wooded mires with long-frozen soil/permafrost; northern boreal continental regions
	Vaccinio uliginosi-Pinetalia sylvestris Passarge 1968	Vaccinio uliginosi-Pinion sylvestris Passarge 1968	Open pine woods in poor fen/bog vegetation; boreal-nemoral non-oceanic regions
	Eriophoro-Piceetalia abietis Passarge 1968	Eriophoro-Piceion abietis Passarge 1968	Spruce woods in poor fens; boreal-nemoral non-oceanic regions
	Calamagrostio purpureae-Picetalia obovatae Lapshina 2010	Calamagrostio canescentis-Piceion abietes Solomeshch in Willner et al. 2015	Herb and grass dominated spruce woodlands; boreal non-oceanic regions
Molinio-Arrhenatheretea Tx. 1937	Molinietalia caeruleae Koch 1926	Molinion caerulea W. Koch 1926	Anthropogenic grasslands/fens; mainly nemoral oceanic regions
		Calthion palustris Tx. 1937	Herb-rich anthropogenic grass-lands/fens; mainly nemoral region
Juncetea maritimi Br.-Bl. in Br.-Bl. et al. 1952	Juncetalia maritimi Br.-Bl. ex Horvatic 1934	Juncion maritimi Br.-Bl. ex Horvatic 1934	Salt marshes transitional to fens, thin–no peat; Mediterranean and nemoral Atlantic coast
Festuco-Puccinellietea	Scorzonero-Juncetalia gerardii Vicherek 1973	Juncion gerardii Wendelberger 1943	Saline grasslands/fens with thin–no peat, from Atlantic coast to continental Russia

Table **4.7.-3** continued

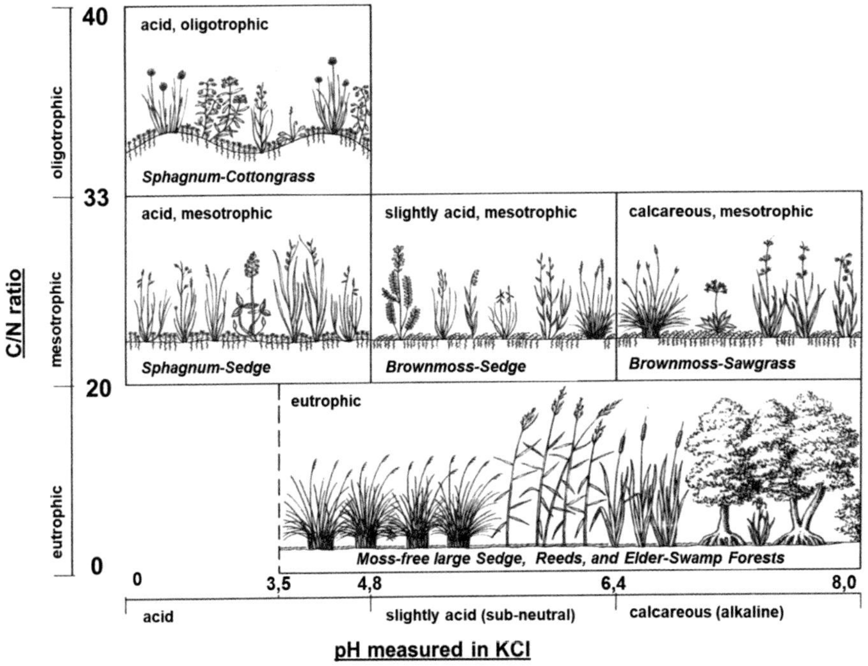

Fig. 4.7-2. The Succow (1988) system of ecological mire types.

Table 4.7-4. Main occurrences of ombrotrophic (grey) and minerotrophic (white) hydromorphic mire massif types in the various mire regions of Europe. C=characteristic type, x=additional (after Moen et al., 2017).

Mire region / **Mire type**	Percolation bog	Rim raised bog	Sloping fen	String-flark mire	Palsa mire	Polygon mire	Blanket bog	Atlantic raised bog	Typical raised bog	Wooded raised bog	Plane bog	Percolation fen	Flat fen
Colchis mire region	C												x
Northern fen region		x	C	C			x				x	x	x
Palsa mire region				x	C								x
Arctic seepage and polygon mire region					x	C							x
Atlantic bog region							C	C			x		x
Typical raised bog region									C	C	x		x
Continental fen and bog region										C	x	x	x
Nemoral-submeridional fen region											x	x	C
Southern European marsh region													C
Central and southern European mountain compound region			x								x	x	C

4.7.4

Mire regionality

Mire ecosystems differ over Europe. This regional diversity of mire types and mire vegetation is linked to differences in climate. In the south, flat fens prevail. Northwards these are replaced by percolation fens, raised bogs, aapa mires, palsa mires, and furthest north by polygon mires. Further there are marked differences from the western, oceanic blanket bogs to the eastern groundwater-fed mires on the continental steppes. The change in mire types from the lowlands to the high mountains, for example in the Alps and the Scandinavian Mountains, somewhat reflects the latitudinal zonation.

Moen et al. (2017) distinguish within biogeographical Europe ten mire regions (Fig. 4.7-3), mainly based on the distribution of hydromorphic mire types (Table 4.7-4), but also on the distribution of mire features, sites and complexes as well as mire flora and vegetation. Differences within the regions are related to bedrock, relief, soil, water conditions and other local environmental factors.

4.7.5

Development of mires

4.7.5.1

Initiation of peat growth

The timing of the first peat formation in a terrestrializing lake depends on the regional climate (effective moisture controlling lake levels) and on local factors such as basin topography, sediment accumulation rate and the hydrosere (the vegetation succession, as often reflected in vegetation zonation), depending on water depth and nutrient status.

Peat initiation by paludification is brought about by changes in climate (more precipitation or less evapotranspiration), sea-level rise or rising groundwater due to anthropogenic reasons. An important example of the latter is deforestation, which reduces evapotranspiration (Moore, 1973, 1975, 1984) and enhances groundwater recharge (Kim and Jackson, 2012). Paludification may also result from mechanisms that seal off the ground, such as by humus colloids (as in the case of kettle-hole mires; Gaudig et al., 2006), or water-repellent charcoal (Fig. 4.7-4). The situation, that fire leads to more water in the landscape, may seem paradoxical but examples are given by Tallis (1975), Tuittila et al. (2007) and Väliranta et al. (2007).

Dates of peat initiation can, therefore, be used for reconstructing changes in climatic humidity and in land use, if both processes can be disentangled.

Peat initiation in Europe had already started in the Late-glacial, especially during the Allerød chronozone, but the onset of peat formation was more frequent during the Early to Mid Holocene (Succow and Joosten, 2001). Whereas these early mires were fens formed by terrestrialization or paludification, during the Holocene, bogs could establish and expand (even directly over mineral soil: G: "wurzelechtes Hochmoor", Overbeck, 1975), especially when the climate was wetter and cooler. In some regions, damming of rivulets for running water mills has in the last millennium led to a peak in the formation of new water-rise mires (Succow and Joosten, 2001).

With due caution, climate development can be reconstructed from data on peat initiation. Korhola (1995) studied peat initiation in southern Finland. He excluded the 23 sites where peat growth had started by terrestrialization, because that process may be dominated by local lake-water depth or topography. The oldest peat dated from about 10 700

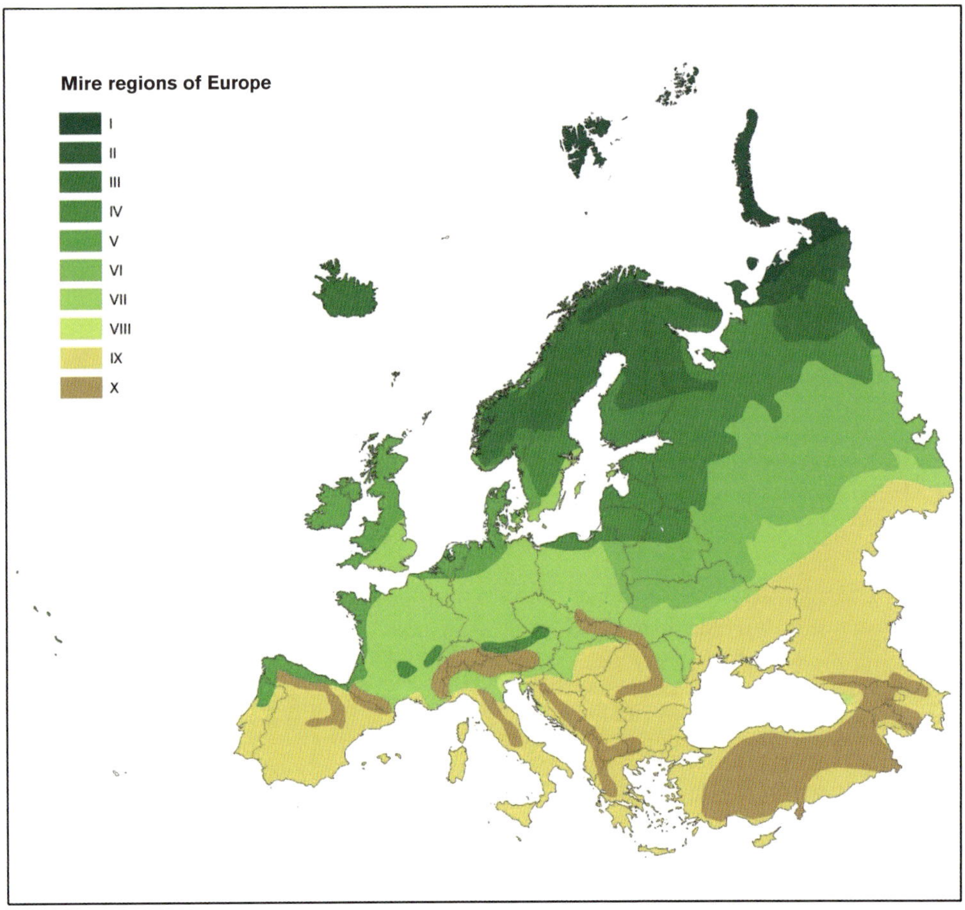

Fig. 4.7-3. Mire regions of Europe (from Tanneberger et al., 2021). I. Arctic seepage and polygon mire region, II. Palsa mire region, III. Northern fen region, IV. Typical raised bog region, V. Atlantic bog region, VI. Continental fen and bog region, VII. Nemoral-submeridional fen region, VIII. Colchis mire region, IX. Southern European marsh region, X. Central and southern European mountain compound region.

Table 4.7-5. Typical mire massif types (Joosten et al., 2017a; Moen et al., 2017).

1. Tundra seepage and polygon mires: thin peat layers built up mainly by brownmosses and rootlets over permafrost. Polygon mires have ice-wedge based polygonal structures and occur in Europe on Svalbard and in north-east Russia.

2. Palsa mires: fens with mounds with an ice core, usually of sedge peat but, when higher, also with bog vegetation and *Sphagnum* peat at the surface. The permafrost nucleus grows by hygroscopically attracting water from the surroundings. Because of the expanding ice core and exposed situation, the surface eventually dries out and breaks open, the ice core melts and the palsa mound erodes and collapses. In the northern part of their range more extensive flat palsas occur ('peat plateaux', sometimes with polygonal structures and then called 'polygonal peat plateau mires'), while in the southern part the palsa mounds are higher and more isolated. Palsa mires occur in northern Russia and northern Scandinavia (where they extend southwards in the Scandinavian Mountains) and in central Iceland.

3. String-flark (or aapa) mires: often large fens of sedge peat with raised strings, sometimes made-up of *Sphagnum* peat, and very wet, often water-filled flarks. The strings and flarks develop perpendicular to the slope of the mire. Aapa mires occur in the northern and central boreal zone.

4. Sloping fens: fens with a smooth and markedly sloping surface on mountain slopes. They are common in the (highly) oceanic areas of Iceland, Ireland, Great Britain and west Norway and in low alpine vegetation zones.

5. Typical raised bogs: bogs with a dome or shield shape. They occur in various types: *Concentric* raised bogs (kermi bogs) have a concentric surface pattern consisting of elongated hummocks (kermi) and hollows or pools. They are common in the boreonemoral/southern boreal parts of Finland and Sweden, the eastern parts of the Baltic countries and further east in Russia. *Eccentric* raised bogs with an eccentric pattern develop when the underground is more sloping and are more common further north in Fennoscandia. *Plateau* raised bogs have a raised flat and open expanse ('plateau') with irregularly distributed surface features, a steep marginal forested slope and a marked 'lagg' (marginal fen strip). They are the dominant raised bog type in southernmost Finland and the western parts of the Baltic countries, but also occur mixed with other raised bog types in other parts of Europe. *Wooded* raised bogs have an (open) forest cover of *Pinus sylvestris*. They occur in areas with less precipitation, mainly in eastern Europe. Also montane and submontane raised bogs often have *Pinus* (*sylvestris, rotundata/uncinata* or *mugo*) growing on them. They occur scattered in the mountains and foothills of central (partly also eastern and southern) Europe.

6. Rim raised bogs: sharply domed bogs in the markedly oceanic parts of the middle boreal zone. They are common in Norway in areas with long-lasting snow cover and high summer humidity.

7. Atlantic raised bogs: bogs with rather flat domes, which often coalesce with other domes to form a bog complex. They lack a marginal forest and often also an explicit lagg. They are common in the central Irish lowlands, and further occur along the British coasts (scattered), in Bretagne (France), northern Spain, Belgium, the Netherlands, northwest Germany, Denmark and along the coast of Norway.

8. Blanket bogs: treeless bogs that cover the underlying, often undulating mineral subsoil like a blanket. They occur in the most oceanic (coastal and mountainous) areas of Europe: in Ireland, Britain, western Norway, and in the westernmost parts of France and northern Spain, where abundant precipitation allows ombrotrophic mire vegetation to cover the entire terrain and even hills in the landscape. Blanket bogs have often originated and developed as a result of human activities such as land clearing, burning and livestock grazing. If the blanket bog massifs are close to flat, they may be difficult to separate from plateau raised bogs and plane bogs.

9. Percolation bogs: bogs with a domed, open and featureless, monotonous expanse dominated by *Sphagnum*, slightly decomposed *Sphagnum* peat over large depths, extremely high peat accumulation rates (4 mm per year) and a remarkably large mire oscillation capacity. Percolation bogs are currently restricted to the warm-temperate, humid Colchis region in Georgia at the eastern Black Sea coast.

10. Plane (or flat) bogs are not distinctly domed, they are flat or weakly sloping without an explicit marginal slope or lagg. They are treeless and hummocks and hollows are irregularly distributed. Plane bogs are young ombrotrophic mire massifs or occur in areas where peat accumulation is very slow, for example, in upland (including alpine) areas. The plane bog is thus a heterogeneous mire type.

11. Percolation fens are slightly sloping, often sedge–brownmoss dominated mires without surface patterning. They are bound to landscapes where groundwater supply is large and evenly distributed over the year and produce weakly decomposed, highly permeable and elastic peat. They prevail in more continental areas and are in more oceanic areas often rapidly succeeded by bogs.

12. Flat fens and spring mires: minerotrophic mires of various sizes and nutrient states, in depressions, valleys and on slopes where groundwater is present. They are azonally common over large parts of Europe together with other mire types. However, at low elevations in southern and eastern Europe they are often the only mire types occurring.

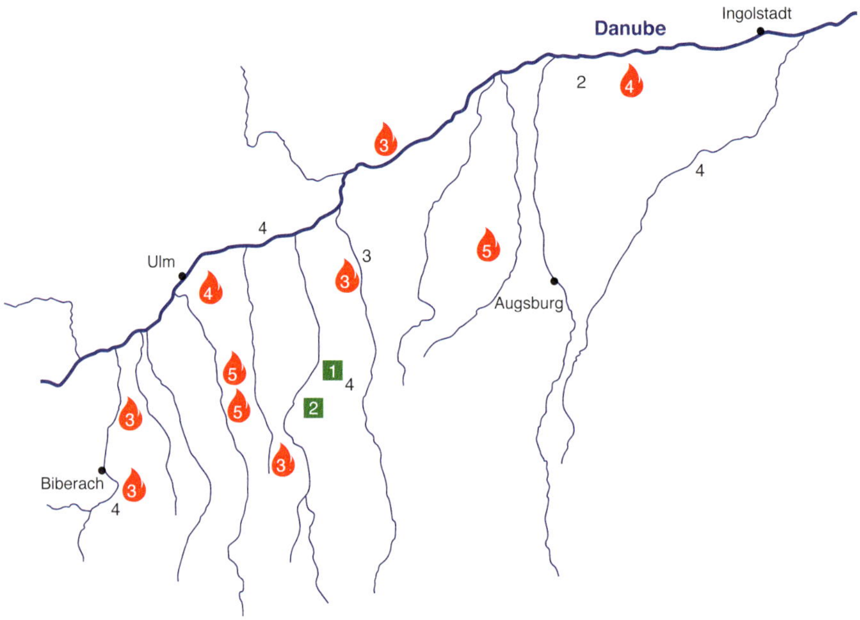

Fig. 4.7-4. Peat initiation and its approximate age on river terraces in southwest Germany: 1: Oldest Dryas. 2: Younger Dryas. 3: Subboreal 5.8-2.6 cal ka BP. 4: subatlantic 2.6-0 cal ka BP. 5: Medieval times. Red flame: charcoal recorded at the onset of peat formation. Green square: no charcoal found. Bare numbers: presence of charcoal not tested. Modified from Frenzel (1983).

cal yr BP, whereas the periods 8000–7300 cal yr BP (11 sites) and 4300–3000 cal yr BP (7 sites) were found to be phases of intensified peat initiation. The regional consistency and synchroneity of these changes was corroborated further inland by Tolonen (1968, 1987) and by high lake levels in southern Sweden (Digerfeldt, 1972, 1974, 1986; Gaillard and Digerfeldt, 1991, Harrison and Digerfeldt, 1993). At Lakkasuo, however, initial paludification was dated at 6500 cal yr BP, that is in the 'quiet' period (Tuittila et al., 2007). An example of interacting processes was the start of peat formation during the drier period 7300 to 4300 cal yr BP (5 sites) on sites that are sometimes underlain by charcoal (Korhola, 1995). In the southern Pennines, northern England, Tallis (1991a) found three distinct phases of peat initiation: around 10 000 cal yr BP, 8300–7800 cal yr BP and c. 5700 cal yr BP. He distinguished between water-collecting sites (with early peat initiation) and water-shedding sites (from which peat subsequently spread). Comparing his studies with 57 sites from northern England, he concluded that Early Holocene peat initiation was generally confined to the uplands, except in water-collecting basins. The phase of peat initiation around 8300–7800 cal yr BP was climatically controlled, whereas the younger phases are related to human impact (perturbation during the Mesolithic and deliberate forest clearance during the Late Neolithic and Bronze Age).

In a global synthesis, Morris et al. (2018) noticed changes in precipitation to be significant for peat initiation in some areas but mean growing-season temperature to be more important in deglaciated areas in Europe, North America and Patagonia. A mean growing-season temperature above 0°C is regarded as a threshold for widespread peat formation.

In conclusion, some mires in Europe started to form shortly after the ice retreat, whereas the majority started to form between c. 8000 and 3000 yr BP with an emphasis on the second half of this period (especially during the Mid to Late Holocene) and with a certain delay towards the north (Table 4.7-6)

4.7.5.2

Pathways of mire development with emphasis on the fen–bog transition (FBT)

Whereas mire development is classically displayed as the progressive succession of well-defined phases, in reality much more multidirectional variety exists. Detailed reviews of succession in mires are presented by Overbeck (1975), Rybníček (1984), Zobel (1988), Succow and Joosten (2001), Charman (2002) and Rydin and Jeglum (2008). Rybníček (1973) carefully compared modern plant communities in vegetation zones (that is in space) with sequences of lake deposits and peat (in time) (Fig. 4.7-5).

Among these pathways (Fig. 4.7-6) the transition from minerogenous to ombrogenous peat – the fen-bog transition (FBT) – is of special interest because of its possible palaeoecological and palaeoclimatological inferences. In this transition the moss genus *Sphagnum* plays an important role. *Sphagnum* has the prominent capacity to improve its own niche by acidifying its surroundings as an 'ecosystem engineer' (Skene, 1915; Clymo, 1963; van Breemen, 1995).

The palaeoclimatological interpretation of the FBT has to consider that the end of minerogenic influence ('Mineralbodenwassergrenze', see section 4.7.2) may be hard to determine by indicator taxa or chemical peat analysis, certainly in the case of an acid, nutrient-poor water supply from the mineral catchment. The onset of *Sphagnum* in peat stratigraphy primarily indicates a change in local hydrology, which does not necessarily have to be climatically controlled but may also result from other allogenic and autogenous processes, such as natural succession (Overbeck, 1975; Frenzel, 1983; Wassen and Joosten, 1996).

Moreover, the best ecological information is linked to the species rather than to the genus (Fig. 4.7-7) and various *Sphagnum* species, such as *Sphagnum teres* and *S. squarrosum*, are not indicative of acid, nutrient-poor (ombrotrophic) conditions (Tolonen, 1968; Daniels and Eddy, 1990; Michaelis, 2019).

In northern Europe the transition from fen to bog occurred primarily during the Mid Holocene and the first part of the Late Holocene (Table 4.7-6).

In north-west Germany the fen-to-bog transition in some mires had already started during the Early Holocene (Overbeck, 1975; Table 4.7-6), whereas in eastern Germany the transition took place from the Mid Holocene onwards (Succow and Lange, 1984). In the British Isles, important contributions to the FBT question came from the classical sites of Bolton Fell (Barber, 1981; Barber et al., 1994) and Walton Moss (Barber et al., 1998; Hughes et al., 2000) in northern England and from comparing the development of Tregaron Bog (Wales) with that of 13 other sites in Wales, Scotland, Northern Ireland and Ireland (Hughes and Barber, 2003). The latter study concludes "All that is required for ombrotrophic conditions to develop is the separation of the mire growing surface from the supply of surface or subsurface waters, so that meteoritic supply becomes the sole source of water and nutrients. The separation can occur as a consequence of vertical accumulation of peat mass, but equally a lowering of the bog water table would also promote isolation and possibly speed up the 'forward' rate of succession to ombrotrophy". The timing of the FBT in their dataset shows a wide range from c. 10 000 to 4300 cal yr BP, whereas at least two distinctive pathways to ombrotrophy were found: 1) via the 'wet-pioneer oligotrophic community' and 2) via the 'dry-pioneer oligotrophic community'. The first one is usually linked to high effective precipitation, the latter to low precipitation prevailing. A clear exception is the record of Tregaron Bog, where the FBT may have been affected by the proximity of the Teifi river channel, which has shifted and incised on a number of occasions during the Early to Mid Holocene (Godwin and Mitchell, 1938; Hughes and Barber, 2003). These studies warn against simple interpretations of changes in peat stratigraphy in terms of climate change.

A similar caution may also apply to the interpretation of changing composition of *Sphagnum* species. Since *Sphagnum* mosses grow clonally, they tend to 'move around' in the carpet, so that analysis of peat profiles retrieved with small diameter corers may overestimate species turnover as the clones may enter and exit the narrow core as they grow upwards. In an elegant combination of experiments and permanent plots (time scale 1–10 yrs), remapping (40–60 yrs) and peat stratigraphical studies (100–2000 yrs), Rydin and Barber (2001) showed the long-term co-existence of *Sphagnum* species, illustrating net competitive equivalence among species in spite of small niche differences.

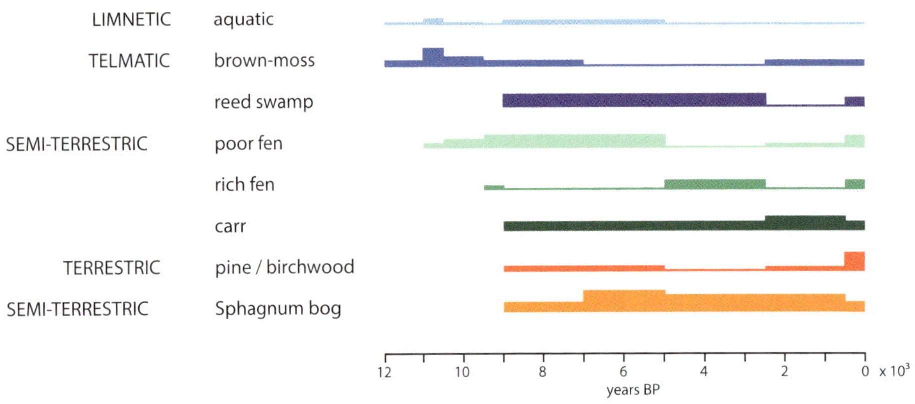

Fig. 4.7-5. The frequency (arbitrary scale) of different wetland communities in central Europe during the Late- and postglacial periods. Time scale in calibrated kilo-years BP (not linear). Adapted from Rybníček (1973) by Tallis (1983).

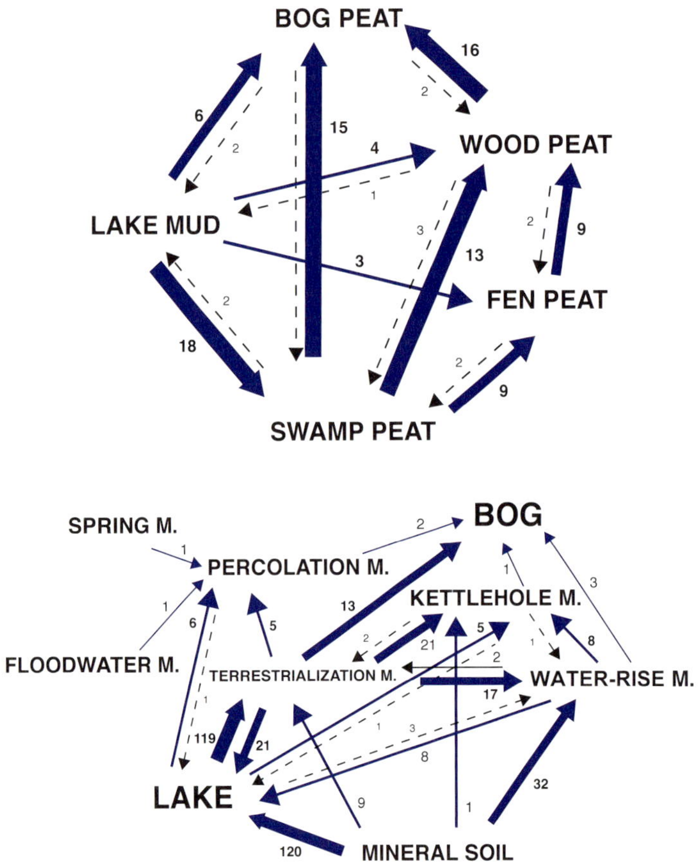

Fig. 4.7-6. a) Stratigraphic sequences in peat profiles in Britain (Tallis, 1983, based on data from Walker, 1970). The numbers refer to the number of observed cases of that particular transition. b) Reconstructed succession of hydrogenetic mire types in north-east Germany (modified after Succow and Joosten, 2001). Solid arrows show pathways of various frequency, dotted arrows rare reversals of the normal hydroseral sequence.

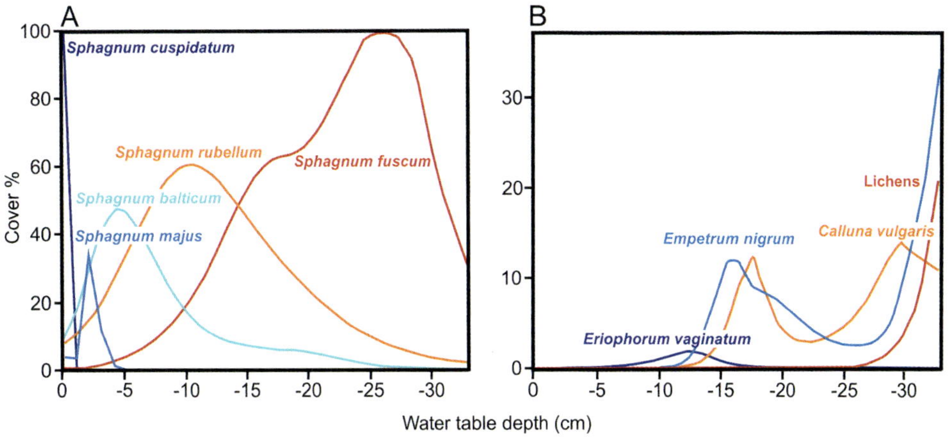

Fig. 4.7-7. Relationship between *Sphagnum* species (A), selected vascular plant species (B) and water-table depth below surface in Kontolanrahka bog, south Finland (after Väliranta et al., 2007).

	Early				**Mid**				**Late Holocene**			
cal ka BP	11.6	10	8.9			5.7			2.6			
uncal ka BP	10	9	8	7	6	5	4	3	2	1	0	
Finnish north Karelia		-		6	18	6		21	33	9	6	-
North-west Germany		-	3	16	16	16		18	13	13	5	-

Table 4.7-6. Approximate age of the fen-to-bog transition. The numbers represent the percentage of bogs per millennium. Northern Karelia: 33 bogs near the northern limit of the typical raised bog region (Tolonen, 1967). North-west Germany: 38 bogs near the southern limit of the Atlantic bog region (Overbeck, 1975).

4.7.5.3

Horizontal and vertical peat growth

The progression of mire growth can be very different among mire types. A raised bog is virtually a huge drop of water (Bakker, 1992), whose vertical growth cannot keep up with its horizontal expansion (Aartolahti, 1965). The increasingly flatter shape results in a more humid centre with less plant productivity, which again slows down peat accumulation and height increase compared to the younger and more marginal parts of the bog (Fig. 4.7-8; Joosten, 1993).

Belyea and Baird (2006) present five conceptual models of raised bog growth, which differ in the mode of peatland initiation and the relationship between rates of height growth and lateral expansion (Fig. 4.7-9). The bog grows in height until it reaches either a hydrological (12.65 m, A, B) or a biological limit (8 m, C, D, E). In all cases, lateral expansion is limited by physiographic constraints, such as two stream channels. These five models are idealized representations, but probably cover the range of possibilities that occur in nature.

Korhola (1994) showed how direction and rate of lateral expansion of five raised bogs in southern Finland were largely determined by the gradient of the underlying terrain and by small-scale topographic features posing from time-to-time barriers to expansion. The periods of more rapid expansion were not synchronous between the mires and phases of horizontal expansion did not always correspond

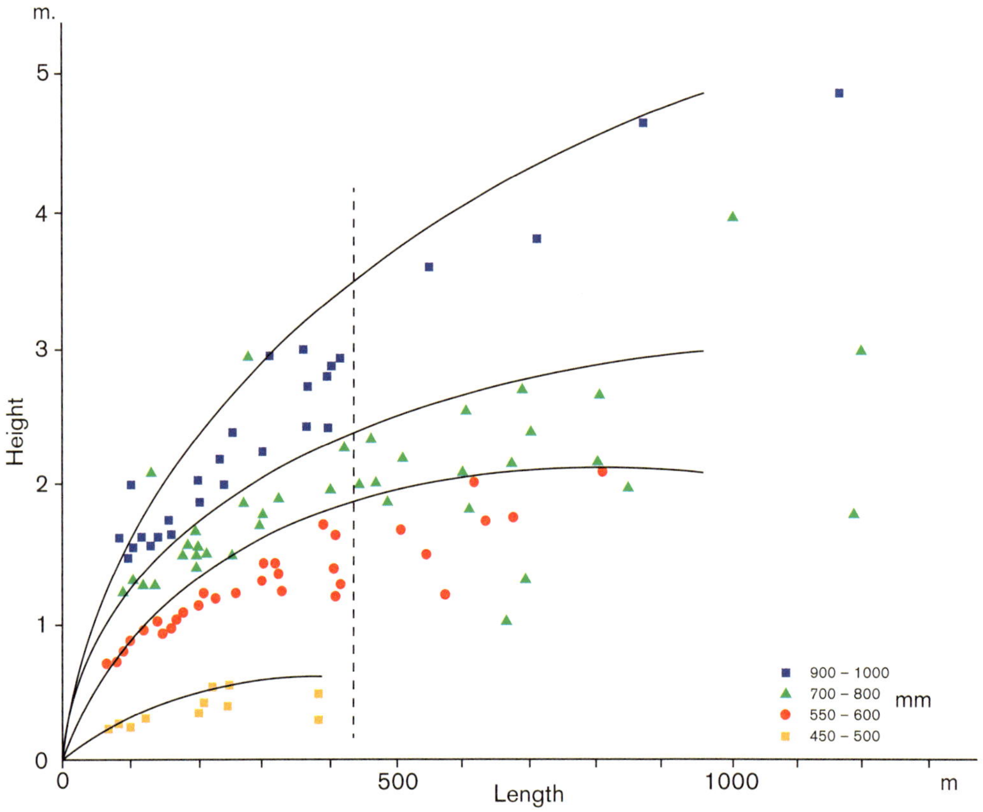

Fig. 4.7-8. The relationship between horizontal (x-axis) and vertical peat growth (y-axis) in metres under different amounts of (recent) annual precipitation (Tallis, 1983 based on data from Granlund, 1932).

with major phases of vertical growth. The period between c. 6000 and 4000 yr BP emerged, however, as a period with less expansion, thereby pointing to a possible drier climate.

The only distinguishable synchronous feature was the pronounced decline of lateral growth in recent millennia because all areas topographically susceptible to paludification had already been overtaken by mire vegetation. The study clearly shows that the development of bogs can be influenced by many internal and external factors and that it is hazardous to assume any constancy in deposition and expansion rates without knowing the exact environmental settings.

Similarly complex in their development may be mires that grow uphill. In the peatlands of the Harz mountains, Germany, horizontal and vertical peat growth did not only depend on the macroclimate but also on various local geomorphologic and hydrologic conditions, resulting in different histories for each mire (Henrion, 1982; Beug et al., 1999; Fig. 4.7-10).

In fens, the growth mechanism and therefore the age distribution of the peat layers is dependent on the hydrogenetic mire type (Joosten et al., 2017a). In terrestrialization mires, where peat formation takes place in or over 'open' water, peat may accumulate on the bottom of the water body ('bottom-up') or in/under a floating mat ('top-down'). Peat layers in the former type generally have horizontal isochrones, whereas isochrones (and stratigraphy!) in peatlands originated by floating mat terrestrialization are clearly concave or more complex, depending on whether the floating mat developed simultaneously over the entire basin or whether ingrowth started from the margins of the basin and developed centripetally (Gaudig et al., 2006).

In water-rise mires, vertical peat formation follows the rising groundwater table. Therefore the age of the peat at a certain height is (in undrained

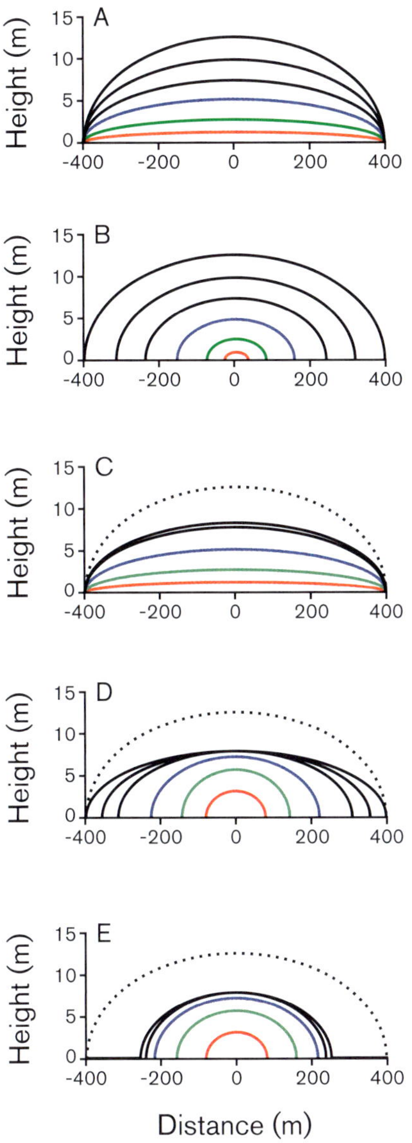

Fig. 4.7-9. Five models of bog development showing profiles at 1000, 2500, 5000, 7500, 10 000 and 12 650 years after initiation. The dotted profile in models C–E is the same as the final profile in models A and B. In models A and B, the bog grows in height at a constant rate until it reaches a maximum height of 12.65 m, whereas in models C–E, the bog grows in height at a decreasing rate toward an asymptotic maximum of 8 m. In models A and C, the bog initiates simultaneously across the entire interfluvial plain; in models B, D and E, the bog initiates at a central locus and subsequently expands. Model E is a special case in which the bog grows in lateral extent at a rate determined by height growth (modified from Belyea and Baird, 2006).

conditions) everywhere the same (horizontal isochrones), and horizontal expansion depends on the steepness of the basin and the speed of groundwater rise. Percolation mires and surface-flow mires grow up with an inclined surface, which leads to slightly tilted peat isochrones. Horizontal expansion upslope may stop when the mire reaches the upper level of the groundwater supply. However, because peat formation raises the drainage base, percolation mires are able to raise the groundwater level in their catchment and to effectuate an upslope movement of the groundwater discharge points, thus facilitating upslope horizontal expansion (Michaelis, 2002). Spring mires start from the point of groundwater discharge and develop a dome-like shape similar to raised bogs.

4.7.5.4

Peatland patterning and stratigraphy

Peatland ecosystems can show striking surface patterns, which have prompted research since the early days of peatland science. The pattern of hummocks and hollows on living bogs and observations of their differential growth led Sernander (1909) and von Post and Sernander (1910) to the hypothesis of cyclic succession: the hummocks would become drier and drier during their vertical growth and would eventually stop growing, whereas the hollows would steadily grow and eventually become hummocks. The idea was later elaborated in 'space-for-time-substitution' by Osvald (1923, 1925, 1949, 1970), studying contemporary plant communities and relief patterns in bogs. The impressive concept of 'cyclic bog regeneration' was considered to "stand among the most beautiful and well-grounded achievements of bog science" (Kulczyński, 1949) and remained the central paradigm of bog growth and development for a long time (Tansley, 1953; Moore and Bellamy, 1976), until it was subverted by high-resolution palaeoecological studies (Walker, 1961; Walker and Walker, 1961; Casparie, 1969, 1972; Klinger, 1976; Moore, 1977; Barber, 1981; Svensson, 1988). These studies showed that hummocks can remain hummocks over millennia and that stratigraphic changes between drier and wetter phases may reflect changes in climatic conditions. Such climatic control, however, does not preclude the possibility of occasional 'cyclic regeneration', as was shown by Tolonen (1966, 1967, 1968) in Finland.

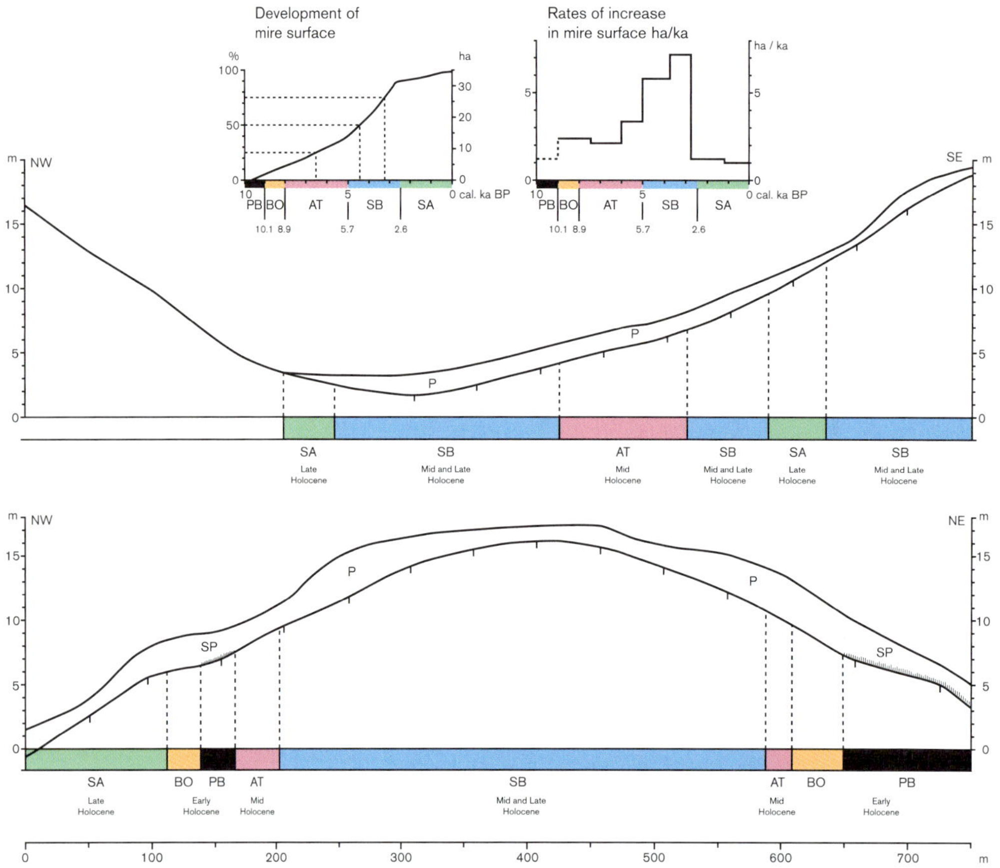

Fig. 4.7-10. Development of the mire at Magdbett in the Harz Mountains, Germany. The mountain peatland started from paludification nuclei on both sides of the pass during the Early Holocene (in black), grew uphill across the pass and uphill on both sides and thus became a 'saddle mire'. Top: areal increase over time, in % and hectares (left) and as rate in ha per thousand years (right). Bottom: longitudinal transect over the pass; middle: transversal transect over the centre of the pass. Modified from Henrion (1982) and Beug et al. (1999). PB: Preboreal, BO: Boreal, AT: Atlantic, SB: Subboreal, SA: Subatlantic chronozones.

Horizontally much more extensive changes within bog-peat profiles that must reflect large-scale environmental dynamics are the so-called 'recurrence surfaces'. Weber (1910) described in northwest Germany a widespread sharp transition from strongly humified blackish to slightly humified yellowish-brown *Sphagnum* peat. This transition, the so-called 'Grenzhorizont' (or later: Schwarz-Weiss-Kontakt SWK) was originally considered to be synchronous within and between bogs. The strongly humified peat was interpreted to have developed under drier conditions (the 'Stillstandskomplex', retardation phase) in the Mid to Late Holocene; the slightly humified peat under wet conditions in the Late Holocene. Later, numerous radiocarbon dates suggested the diachronous (time-transgressive) nature of the Grenzhorizont (Averdieck, 1957; Overbeck et al., 1957; Overbeck, 1975). Then, van Geel et al. (1996) and van Geel and Renssen (1998) reported a sudden and sharp rise in ^{14}C-levels at the relevant time (making traditional radiocarbon dates difficult to interpret) and suggested that a decline in solar activity around 850 BC had triggered an abrupt and widespread climate shift to cooler and wetter conditions (Speranza et al., 2002; Mauquoy et al., 2004).

The Grenzhorizont is the most prominent but not the only recurrence surface. In Sweden, Granlund (1932) and Nilsson (1935) described up to five abrupt transitions from strongly to slightly humified peat, with the numbers varying among different

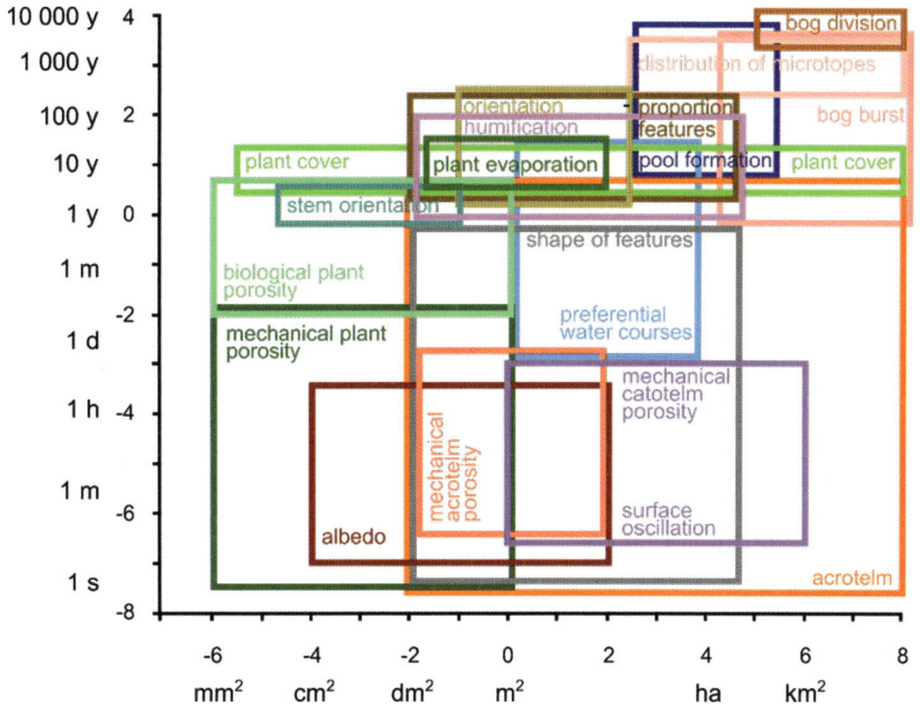

Fig. 4.7-11. Tentative overview of hydrological self-regulation mechanisms of a living raised bog in space and time. On the horizontal axis (in 10^x m²) the surface over which the mechanisms work, on the vertical axis the reaction time with which they react to changes (in 10^x y, with more familiar units on the secondary axes). Note the logarithmic scale of the axes (from Couwenberg, 2021).

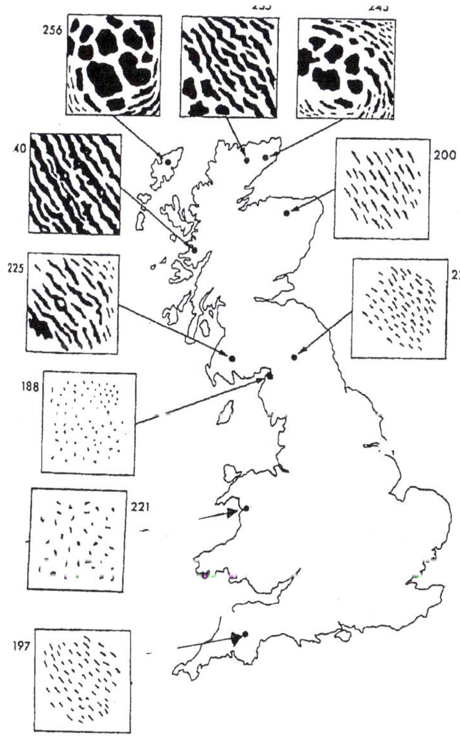

Fig. 4-7-12. Distribution and general pattern of wet structures on bogs in Great Britain. The insets are drawn at the same scale. Hollows and pools are in black, ridges in white. The number near each square indicates the average annual number of rainy days for each site. After Lindsay et al. (1985).

mires. Overbeck et al. (1957) noted three major recurrence surfaces in seven bog profiles in north-west Germany, whereas Nilsson (1964) found seven surfaces in Ageröds Mosse (south Sweden). In Finland, Aartolahti (1965) found thin, highly humified peat layers, which were not synchronous, not even within the same bog. In Rusland Moss (Cumbria, UK), Dickinson (1975) found two continuous recurrence surfaces across the raised bog, but also discontinuous stratigraphic changes.

The reason for this stratigraphic complexity is that raised bogs are self-organizing ecosystems, which, over time, develop a wide range of self-regulation mechanisms that operate on all scales in time and space in a complex, iterative way (Fig. 4.7-11; Couwenberg, 2021). Both climate change and autonomous bog development (see section 4.7.6.3) may lead to hydrological changes that – because of the feedback mechanisms – work out in locally strongly differentiated ways, instigating horizontal (surface pattern) and vertical (stratigraphy, humification) changes. When deriving palaeoenvironmental information from small sections of a bog (a core, a two-dimensional peat face), one must be aware of these complexities, which may, for example, result in a wetter climate leading to locally drier conditions (Couwenberg and Joosten, 2005). Broad-scale processes, such as climate change, which are synchronous over a larger area, can therefore only be reliably derived from bogs by 'stacking' many small scale observations from a larger area (Charman et al., 2006), and preferably testing them against independently derived climate information from other proxies and types of archives, for example stable isotopes from speleothems (Charman, 2001; McDermott, 2014) or chironomids from lake sediments (Langdon et al., 2004).

Possibly the most fascinating phenomona of mires are the regular patterns they display on various scales of observation, varying from hummock–hollow complexes observable in the field, to string-flark patterns that most impressively show on aerial photographs, up to patterns, so huge that they can only be identified convincingly on satellite imagery (Wright and Coffin, 1992; Kirpotin et al., 2021). The driving forces behind these patterns are mainly external cryogeological processes (for example, polygon mires in continuous permafrost areas) and internal biological self-organisation (aapa mires, patterned bogs in boreal and temperate areas) (Seppälä and Koutaniemi, 1985; Couwenberg and Joosten, 2005). Both drivers are not always strictly separable and interferences are common in the subarctic and northern boreal zones (palsa, ribbed palsa and peat plateau mires; Joosten, 2019).

In sloping peatlands (also slightly sloping ones) with water flow as a driving force, the development of regular, oriented patterns can be explained by self-organisation based on the difference in hydraulic conductivity between hummocks (lower) and hollows (higher) and their preference for lower and higher water table levels, respectively (Couwenberg, 2021). In flat peatlands the concentration of nutrient availability in more luxurious and, therefore, more transpiring vegetation may be the positive feedback mechanism driving patterning (Rietkerk et al., 2004; Eppinga et al., 2009, 2010).

The location, size and abundance of bog hollows is controlled by both climate and surface relief (Glaser, 1998; Couwenberg and Joosten, 2005). Pool-pattern in Britain has been related to the average annual number of rainy days (Fig. 4-7-12; Lindsay et al., 1985, 1988), but Belyea (2007) rejected this relationship in favour of stronger topographic control.

Palaeoecological research has shown that once a pool has originated a regressive development may set in, making the pools deeper and wider (Foster and Wright, 1990; Fig. 4.7-13).

Forster and Wright (1990) found three main types of pool morphometries: (a) steep sided with a flat bottom, (b) with undulating bottoms and (c) two basins separated by a ridge. The pool age and also the time span of the hiatus between peat and gyttja decreases from the centre towards the margin.

The hiatus between the top of the peat (under the pool) and the bottom of the gyttja is attributed to peat decomposition, which may be accelerated by oxygen input from photosynthesis of algae and by methanogenesis in the underlying peat, leading to the breaking up of the peat (Foster et al., 1983; Foster and Fritz, 1987). When the pool gets larger in surface and deeper, algal growth results in the accumulation of gyttja, thus burying the decomposed peat (Foster et al., 1983).

Further reading about mires and peatlands: Overbeck (1975), Moore (1984), Göttlich (1990), Lindsay (1995), Dierssen and Dierssen (2001), Succow and Joosten (2001), Charman (2002), Steiner (2005), Rydin and Jeglum (2008) and Joosten et al. (2017c).

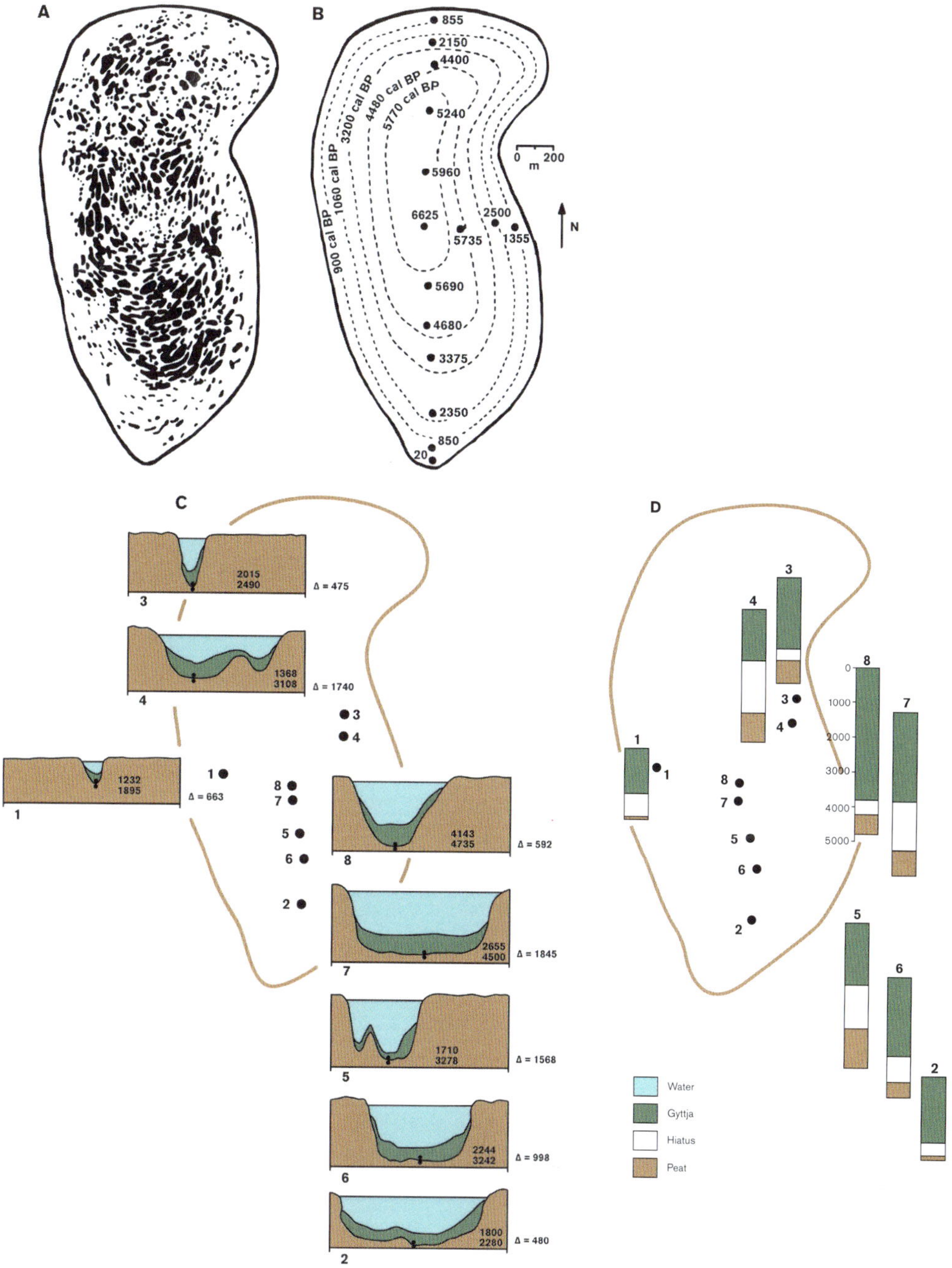

Fig. 4.7-13. The development of the pool system in Hammarmosse, central Sweden. (A) Distribution and size of the open-water pools (black), (B) basal peat dates and interpolated isochrones for bog expansion, (C) location and ages of the pools studied and (D) stratigraphy of each pool on a common time scale (uncal yr BP). In the cross sections of the pools the water is blue, the gyttja green and the peat beige; black dots indicate the position of the ^{14}C samples of the top of the peat and the base of the gyttja, dates are in the right corner. The delta value indicates the duration of the hiatus, which is longest near the centre where the decomposition of the peat started. All dates are presented as calibrated years BP, except for those in D. Modified after Foster and Wright (1990).

4.7.6

The conservation of peatland archives

Peatlands are important archives of palaeoenvironmental, archaeological and historical information (Overbeck, 1975; Birks and Birks, 1980; Godwin, 1981; Barber, 1993; Gorham et al., 2001; Charman, 2002; Zaccone et al., 2007; Wieder et al., 2009; Jeffers et al., 2015b), including information relevant for conservation biology (Birks, 1996a, 2012b) and for peatland conservation and restoration (Joosten, 1995; McCarroll et al., 2016, 2017).

Peatland archives contain a wide collection of information carriers, varying from plant macro- and microfossils, animal remains, archaeological artefacts, minerals, charcoal, volcanic ash, biochemicals, stable isotopes and all kinds of other natural or anthropogenic matter that has ended up in the peatland (Birks and Birks, 1980; Berglund, 1986). The stratigraphy of the deposits facilitates chronology (higher is younger) whereas the waterlogged conditions enable the preservation of organic matter. The peatland archive thus forms a multi-proxy record of the development of the peatland itself and its wide surroundings (Godwin, 1981; Frenzel, 1983; Barber, 1993; Chambers et al., 2012), which is an important peatland ecosystem service (Millennium Ecosystem Assessment, 2005, Davies and Bunting, 2010; Birks, 2012b; Haines-Young and Potschin, 2013; Gearey and Fyfe, 2016) and an aspect of ecosystem diversity (Joosten, 2001). Holocene climate and environmental change were first reconstructed from the stratigraphical arrangement of peat types (Rennie, 1807; von Chamisso, 1824; Dau, 1829; Steenstrup, 1842), but since then many sophisticated methods have been developed to 'decipher' the information recorded in the peatland archive. It may be expected that analytical methods and information processing will continue to progress rapidly, whereas the questions asked about the past will also change.

Peatland conservation thus faces the challenge to conserve sites that will satisfy future demands for information about the past, without knowing the questions to be addressed nor the methodologies that will be available. This implies that a comparative evaluation (ranking) of peatlands with respect to their archive value must be based on (easily assessable) criteria that capture the widest possible potential information content.

Greiser and Joosten (2018) assessed the archive value of 49 individual peatlands in a region in eastern Germany based on generalized and objective criteria, independent of any research question, in order to evaluate peatland archive value in nature conservation practice. Because future demands for information are unknown, they plead for a precautionary approach implying the conservation of as many archives as reasonably possible (Joosten and Clarke, 2002). This principle is supported by the consideration that extracting palaeoecological information from a peatland often implies the partial destruction of the archive. Whereas the development and application of non-destructive research methods are increasing (Grosjean et al., 2014; Dommain et al., 2015), palaeoecology is still – like archaeology (Reich, 1987) – in essence a 'destructive science'.

In addition to their significance for biodiversity, in many parts of the world peatlands are the predominant wetland type for cultural heritage, notably through their capacity to preserve archaeological remains and the palaeoenvironmental record under waterlogged conditions. This has been recognized by the European Archaeological Council's 2001 Strategy and Statement of Intent for the Heritage Management of Wetlands, which argued that there is much common ground in the conservation of biodiversity and the cultural heritage of peatlands (Ramsar Convention, 2004; Coles, 2004): both these values are safe as long as the peats remain water-saturated.

The Paris Agreement and the IPCC 1.5° report (IPCC, 2018) have made clear that the demands of climate policy run fully parallel with securing the palaeo-record of peatlands. To limit global climate change to 1.5–2.0°C above pre-industrial levels we have to conserve all remaining pristine peatlands, and we have to rewet and restore all drained and degraded ones within the next 30–50 years. This means that if we reach the climate targets the future of the peatland archive will also be protected. The only precondition is that during restoration existing archaeological and palaeoecological values remain – where possible – undamaged. Possible conflicts can be minimized by prior inventory (consider Coles, 1995; Coles et al., 2001; Greiser and Joosten, 2018), by involving specialists in management planning, by regular monitoring (Thom et al., 2019), and by providing information to the executing personnel about valuable sites and how they should be considered during restoration work (Joosten, 2021).

4.8

Soil development and vegetation dynamics

Gerhard Lang[1], Brigitta Ammann[1,2]

[1] Institute of Plant Sciences, University of Bern, Bern, Switzerland
[2] Oeschger Centre for Climate Change Research, University of Bern, Bern, Switzerland

4.8.1

How can soil development be traced?

Pedology has a number of ways to derive the history of soil types, for example chronosequences but what methods does palaeoecology offer?

1. Indicator taxa: Iversen (1954, 1958), Gaillard (1984), Kuneš et al. (2011) and others used indicator taxa recorded as pollen and/or plant macrofossils interpreting a shift from pioneer taxa to late-successional taxa as largely indicating a development from raw soils to various types of mature soils. Conclusions on soil moisture can be drawn based on macroremains of indicator taxa (Birks, 2015).

2. Geochemistry of the sediment: independent of biostratigraphies certain elements of the sediment (Willis et al., 1997; Braun et al., 2005) or stable isotopes such as $\delta^{13}C$ (Hammarlund et al., 1997) may indicate a shift in soil type.

3. Nitrogen-fixing taxa: *Hippophaë* and *Alnus* played a role, but see also Hu et al. (2001), Kuneš et al. (2011), Ammann et al. (2013) and figure 4.8-1.

4.8.2

How can spatial differences across Europe be explained?

There are just a few studies in Europe that relate vegetation history to soil development during the Late-glacial and the Holocene. Soil development depends strongly on the initial conditions including bedrock geology, topography, drainage and especially whether the region was glaciated or not.

1. In formerly glaciated areas. For northern and north-western Europe, Iversen (1954, 1958) designed a first overview for Denmark that then found its counterparts in Ireland, Shetland and the Faroe Islands (Birks and Peglar, 1979; Birks 1986a). This illustrated the development from minerogenic raw skeletal (oligocratic and protocratic) soils to brown earth (mesocratic) by the accumulation of organic material and to podsols (oligocratic and telocratic) by leaching. With increasing organic matter the water-retaining capacity, the nutrient content, and the concentration and diversity of soil organisms also increase. Even in the cryocratic and protocratic soil types, differences in grain-size distributions (from clay to sand) induced regionally and locally varying water-holding capacities, for example, *Artemisia* sites on sand and gravel and *Rumex* sites on clay-rich tills (Pennington, 1986). In Denmark, in the eastern regions that were covered by Fennoscandian ice during the Last Glacial Maxiumum (LGM) influenced by the Baltic, the clayey and calcareous sediments recorded a development of tree-birch woodland during the Allerød. In contrast, the western and northern regions (outside the last glaciation) with their sandy, non-calcareous substrate remained treeless throughout the entire Late-glacial (Mortensen et al., 2014). For the Allerød in northeast Germany, Theuerkauf and Joosten (2009) demonstrate the abundance of pine on sand and the abundance of birch on boulder clay. In the southern Alps, Gobet et al. (2000) show how the contrasting types of bedrock (siliceous v. limestone) favour the

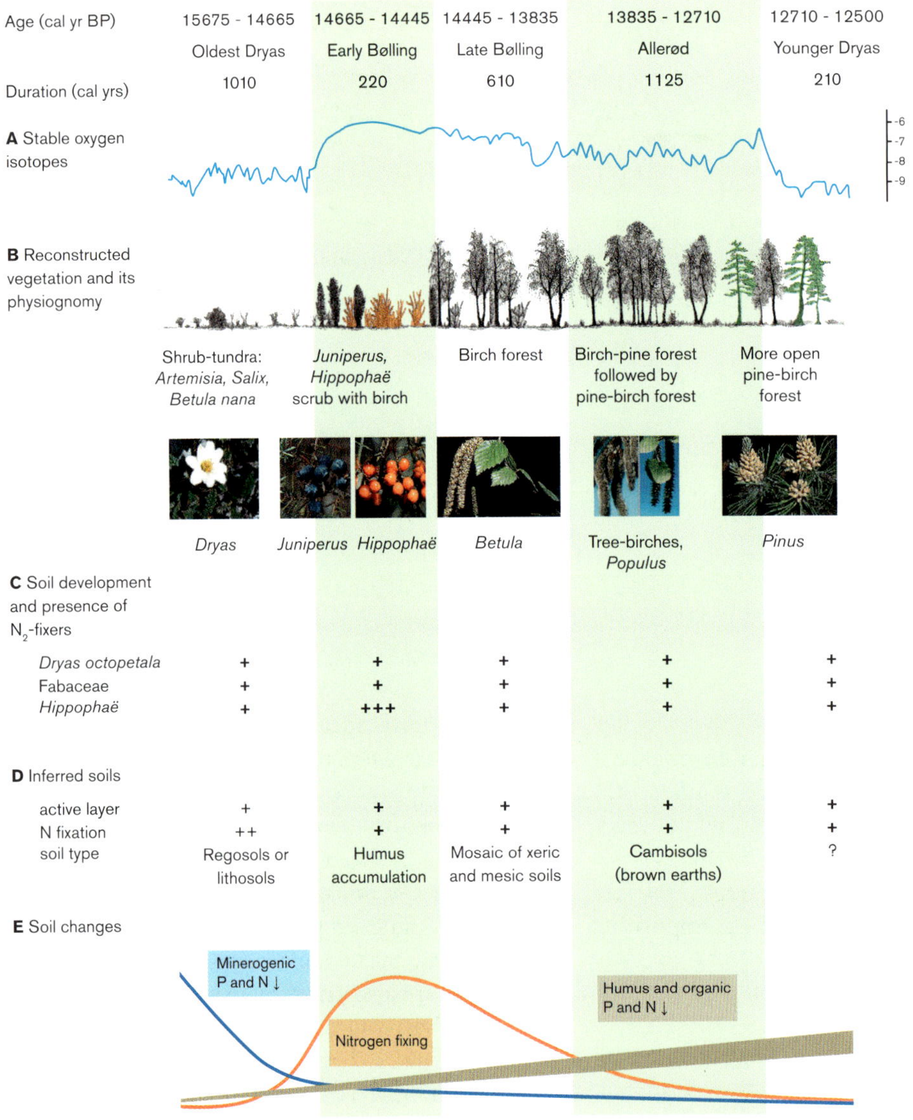

Fig. 4.8-1. Soil development after deglaciation, the example at Gerzensee (Switzerland): a primary succession of plant taxa (recorded as pollen) indicates various symbionts through the early Late-glacial. In A the $\delta^{18}O$-jump stands for the transition from the Oldest Dryas to the Bølling. B shows the Late-glacial vegetation dynamics, C the importance of nitrogen-fixing plants, D the soil types and E the corresponding nutrient dynamics. Modified after Ammann et al. (2013a) and Birks et al. (2016a).

strong or weak development of *Castanea sativa* during the Holocene.

2. Outside the area of the Last Glacial Maximum glaciations. In contrast to the central and north European trajectory where (after deglaciation and primary raw soils) brown earths developed into podsols, an example was found by Willis et al. (1997) in north-eastern Hungary (outside any glaciation) where the development went in the opposite direction. Here, Late-glacial podsols under a pine and spruce forest turned into an Early Holocene brown earth under deciduous forest (Fig. 4.8-2). The changes are recorded in both the pollen and geochemistry of the lake sediment (Braun et al., 2005). The shift in the vegetation occurred in around 100 years, whereas the shift in soil types took about 1000 years.

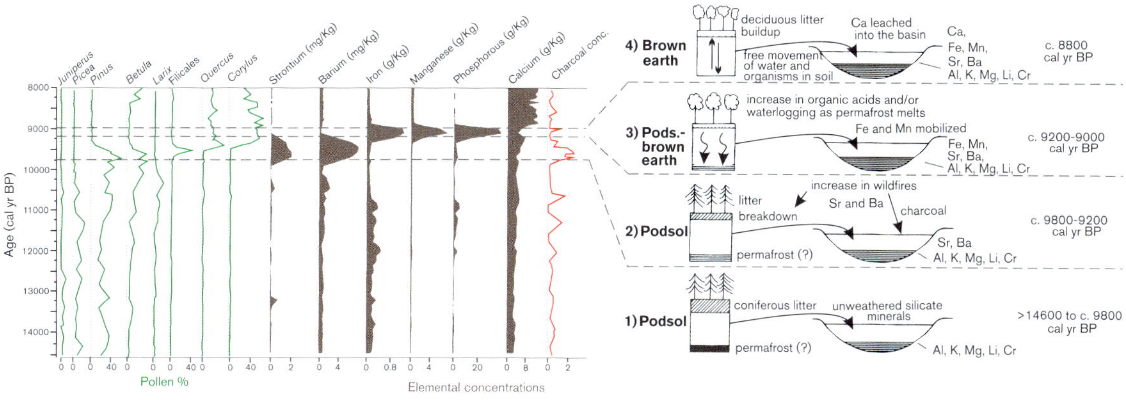

Fig. 4.8-2. Soil development at Kis-Mohos Tó (Hungary) outside the glaciations: pollen and geochemistry indicate a soil development from podsols to brown earths. Modified after Willis et al. (1997).

4.8.3

What interactions with soil development have occurred?

Climate has a strong influence on soil development because besides abiotic weathering, temperature, humidity and their seasonal distribution to a large degree also control both the accumulation of organic matter (through vegetation composition and density, litter-types, thickness and decomposition rates) and leaching processes.

Vegetation is linked to soil development by at least three feedback-loops.

1. Via the litter quality: firstly conifer needles produce a lower pH than leaves of deciduous trees and secondly various factors, such as temperature and humidity, control litter decomposition (Willis et al., 1995, 1997; Sjögersten and Wookey, 2004; Kuneš et al., 2011).

2. Via nitrogen-fixers in the vegetation such as *Alnus, Hippophaë* and Fabaceae (Smolander and Sarsa, 1990; DeLuca and Zackrisson, 2007; Kuneš et al., 2011; Ammann et al., 2013a, see "interactions" below).

3. Vegetation (or the absence of it or its openness) may control erosion that becomes visible as sand or silt deposits in the lake sediments or soil profiles.

Ecologically important interactions

1. Soil types. A species needing a certain soil type may not spread before such soils are available, (see Henne et al., 2011 for *Picea*). Species that arrive in a certain region then form its species pool which, qualitatively and quantitatively, determines the potential species composition of communities. These may then influence possible species interactions (Jeffers et al., 2011a) including competition, facilitation and symbioses/mutualisms. Soil types are also the basis for the so-called 'calcicole riddle' (Ewald, 2003) as explained in chapter 5.7.

2. Symbiotic associations between phanerogams and soil fungi. Among the latter, several types of mycorrhiza are especially important. For boreal and temperate forests Read (1993a,b) suggested that often, with the succession from pioneers or grasslands or ruderal communities to various stages of shrub or tree communities, not only the soil types but also the associated fungal symbionts may change from primarily arbuscular mycorrhiza (AM, or AMF, formerly called vesicular-arbuscular mycorrhiza VA, a form of endomycorrhiza) to more and more ectomycorrhiza (ECM) and later increasingly to ericoid mycorrhiza (ERM). This temporal sequence and the qualitative and quantitative composition of the groups are dependent on both the availability of host plants and on the dispersal mechanisms of the mycobiont propagules (Cázares et al., 2005). Read (1993a,b) underlined that "there are inevitably stag-

es during which cohorts of species with different types of mycorrhiza coexist". In a review including primary successions studied in boreal, temperate and subtropical forests Dickie et al. (2013) come to the conclusion that "...Mycorrhizal types do not demonstrate consistent shifts through ecosystem development. Rather, most mycorrhizal types can dominate at any stage of ecosystem development... In contrast to coordinated shifts in soil fertility, plant traits and ecological processes throughout ecosystem development, shifts in fungal communities within and among mycorrhizal types are weak or idiosyncratic."

In addition, some plant taxa can form more than one type of symbiosis with mycorrhiza, so called dual mycorrhiza, for example *Salix* (Dhillion, 1994), or *Quercus* (Dickie et al., 2001). In recently deglaciated terrain not only AM but also ECM were found, for example on *Salix polaris* (Fujiyoshi et al., 2011), *S. herbacea* (Mühlmann and Peintner, 2008) and *S. helvetica* (Welc et al., 2012, 2014). Dhillion (1994) studied nine *Salix* species in Norway where some of the species showed dual mycorrhiza. AM was often recorded earlier and ECM later. On a temporally and spatially refined scale, AM was often found on juvenile plants and young roots, whereas ECM was primarily found on mature plants and older roots.

In addition to the 'classical' mycorrhiza-groups (AM, ECM, ERM), there are several unrelated genera of fungi united under the term 'dark septate endophytes (DSE)' that do not form intracellular coils (Jumpponen, 2001). As Jumpponen and Trappe (1998) summarize "They have been reported for nearly 600 plant species representing about 320 genera and 100 families... a group that functionally and ecologically overlaps with soil fungi, saprotrophic rhizoplane-inhabiting fungi, obligately and facultatively pathogenic fungi and mycorrhizal fungi..." The abbreviation UTS in the literature stands for fungi of uncertain trophic status. Future work may improve our knowledge of the secular successions (as forms of 'long-term primary successions') if we can integrate studies of pollen and plant macrofossils to trace successions with studies on mycorrhiza in space and time. Both phanerogams and fungi need strong criteria for the presence or absence of taxa (Richard, 1987; Pardoe, 2001; Cázares et al., 2005; Jumpponen et al., 2012). It is of interest that taxa regularly found as pollen or macrofossils in the sediments of the Oldest Dryas such as *Salix* (Dhillion, 1994), *Helianthemum* (Read et al., 1977) and *Juniperus* (Thomas et al., 2007) are associated with various types of mycorrhiza, while nodulated legumes have symbiotic *Rhizobium* (bacteria), and *Dryas* and *Hippophaë* have *Frankia* (actinomycetes). For the British flora, useful check-lists are found in Harley and Harley (1987a,b). In conclusion, mycorrhiza is a vast field of research that has its own journals (*Mycorrhiza*, and parts in *Plant Signaling & Behavior*) and that contributes to the research field informally called www = wood wide web (with numerous contributions on the 'ordinary' web, to close the circle).

4.8.4

Human impact

This is indicated in at least two ways:

1. Erosion after deforestation is recorded as layers of sand or silt in the sediments or soils, sometimes accompanied by reworked (even Tertiary) pollen.

2. When forest soils are found above modern treelines – these soils are a witness to local high-elevation forests before the lowering of the treeline, for example by transhumance since the Bronze Age in the Alps as derived from soil biosequences and palaeoecological evidence (Tinner et al., 1996; see chapter 4.9).

4.9

Treeline and timberline dynamics

Willy Tinner[1,2], Gerhard Lang[1], Christoph Schwörer[1,2]

[1] Institute of Plant Sciences, University of Bern, Bern, Switzerland
[2] Oeschger Centre for Climate Change Research, University of Bern, Bern, Switzerland

4.9.1

Polar treeline and timberline

Today's polar treeline runs along the northernmost part of Fennoscandia and further along the northern edge of the Kola Peninsula and the northern Russian mainland (Fig. 4.9-1). In continental northern Russia the limit of tree growth (>2–3 m growth height) lies only slightly to the north of the Arctic Circle (66°32' N) and is formed by *Picea abies* ssp. *obovata*, while in oceanic Fennoscandia it goes beyond 71° N and consists of *Betula pubescens* ssp. *tortuosa*. South of the polar treeline, dwarf shrub tundra, open birch stands and isolated birch and pine forests form vegetation mosaics. These parklands build the forest tundra and are bordered to the south by closed *Pinus sylvestris* forests that form the polar timberline (or forest limit) between 69° and 70° N (Fig. 4.9-1). The polar treeline and timberline represent heat-deficiency limits, whereby, as a rough rule, the minimum for the existence of conifer trees (>2–3 m) is assumed to be 30 days per year with a temperature average above 10°C (e.g. Walter and Breckle, 1986). The global polar treeline has also been associated with growing seasons of 90–106 days and seasonal mean temperatures between 5.1 and 6.9°C, similar to other treeline positions in the world (6.4 ± 0.7°C; Körner, 2012).

Clear indications of previously more northerly treeline and timberline positions are given by the long-known finds of subfossil *Pinus* trunks in lakes and bogs north of the current distribution limit of *Pinus sylvestris* (Fig. 4.9-1; Juul, 1925). Early pollen analytical studies also investigated the postglacial vegetation history of polar treeline and timberline (Auer, 1927; Aario, 1940, 1943), whereby it was shown that the vegetational interpretation of percentage pollen diagrams presents considerable difficulties and needs to be supplemented by absolute pollen occurrences (concentrations in these early studies). Pollen-percentage distortions were clearly recognized by the investigations of modern pollen precipitation along a north-south transect from the treeless tundra to the spruce forest area (Aario, 1940). The analysis of absolute pollen values was fostered by the development of influx (or pollen accumulation rate) calculations (Welten, 1944). After further developments by Davis (1963, 1966) absolute pollen analysis was successfully used in investigations at the Fennoscandian treeline by Hyvärinen (1975a, 1976, 1985) and the results were combined with radiocarbon dating of subfossil woods (Eronen, 1979; Eronen & Hyvärinen, 1982). For instance, sediment records were retrieved from four lakes situated in different vegetation types between Lake Inari and the Varanger Peninsula and percentage and absolute pollen diagrams were calculated (Fig. 4.9-2): Akuvaara (pine forest), Suovalampi (birch pine forest), Bruvatnet (open birch stands) and Domsvatnet (tundra). The pollen concentration and pollen influx values in the Holocene sequence of the diagrams show a progressive development with a culmination between 9000 and 4000 cal yr BP, probably due to environmental and biotic processes, among the most important ones being higher summer temperatures during the Holocene Thermal Maximum (HTM, c. 10 000–5000 cal yr BP). Specifically, tundra was followed by birch woods at Domsvatnet and at Suovalampi, which were followed

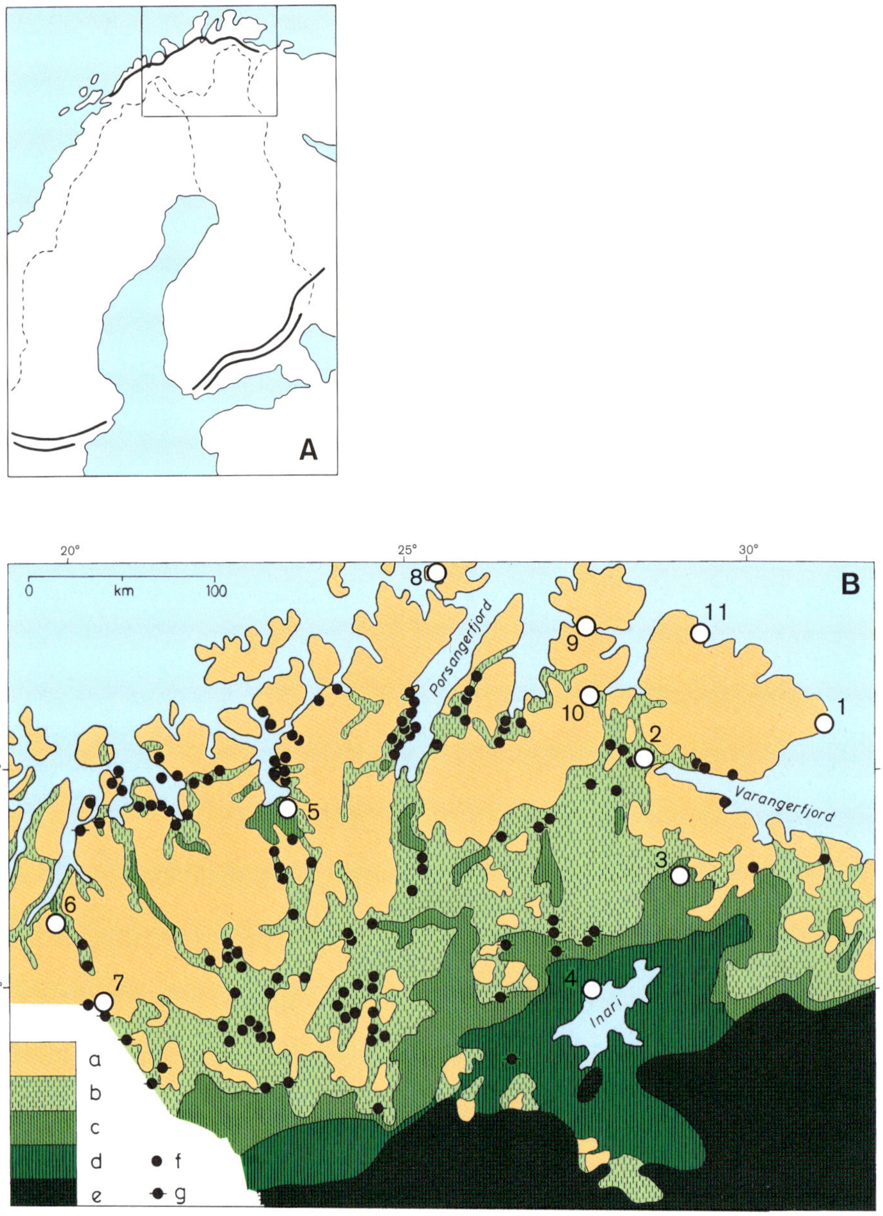

Fig. 4.9-1. Today's vegetation in northern Fennoscandia (a-e), subfossil *Pinus* wood finds from the Holocene (f, g) and pollen analysis sites (1-11). a: Treeless tundra. b: Open birch wood (*Betula pubescens* ssp. *tortuosa*). c: Open birch-pine forest. d: Closed pine forest. e: Pine and spruce forest. f: Subfossil *Pinus* trunks, undated (Juul, 1925). g: Subfossil *Pinus* trunks, radiocarbon dated (Eronen, 1979). 1: Domsvatnet (Hyvärinen, 1976). 2: Bruvatnet (Hyvärinen, 1975a). 3: Suovalampi (Hyvärinen, 1975a). 4: Akuvaara (Hyvärinen, 1975a). 5: Alta (Hyvärinen, 1985). 6: Rättuvarri (Eronen and Hyvärinen, 1982). 7: Mukkavaara (Eronen and Hyvärinen, 1982). 8: Liten Cap'pesjav'ri (Huntley et al., 2013). 9: Hopseidet (Seppä, 1998). 10: Ifjord (Seppä, 1998). 11: Over Gunnarsfjorden (Huntley et al., 2013). Modified from Hyvärinen (1985). The small sketch (A) shows the position of the map section (B). The thick lines in sketch A indicate the ice margins of the Younger Dryas: in the north, Tromsø-Lyngen moraines; in the south, Salpausselkä moraines. According to Hyvärinen (1975a).

in turn by pine forests at the latter site. After these successions, a regressive period followed extending to the present day. For instance, birch woods were replaced by tundra at Domsvatnet, while at Bruvatnet and Suovalampi pine forests declined. Conversely, vegetation remained stable at Akuvaara, which is still today located in the pine forest (see Figs. 4.9-1 and 4.9-2).

These early findings proved that after the rapid melting of the inland ice at the onset of the Early Holocene, the northern limits of the open birch, birch-pine and closed pine forests shifted to the north during the period between 10 000 and 4000 cal yr BP, with a maximum northern expansion at 9000 to 6000 cal yr BP. In the vicinity of the fjords, *Pinus sylvestris* reached the Arctic Ocean everywhere during the HTM, as numerous wood finds unambiguously show (Fig. 4.9-1). The spread of *Pinus sylvestris* apparently took place relatively early from the north-east (Varangerfjord) and was completed at the outermost northern boundary between 9500 and 8000 cal yr BP. These early exhaustive studies suggest that the shift of the species limit in response to the HTM may have been in the range of some dozen to about 100 kilometres north compared to today (Eronen & Hyvärinen, 1982, Huntley et al., 2013).

Subsequent pollen, macrofossil and megafossil studies corroborated these notions. At the onset of the Holocene, coastal areas in Scandinavia were already ice-free, providing evidence for earliest afforestation. Macrofossil records show that the treeline reached the coastal Kråkenes area in south-western Norway (62° N) at c. 10 900 cal yr BP, c. 650 years after the onset of the Early Holocene warming (Birks, 2003, 2015). The treeline shifted north keeping pace with temperature rise, reaching cool north-western Norway c. 1450 years after the onset of the Holocene (Birks, 2015), most likely in response to further Early Holocene warming. Climatic cooling after 5000 cal yr BP resulted in a southward retreat of the polar treeline, creating the ecotone area of today's forest tundra with its mosaic of tundra, open birch woods and isolated islands scattered with birch and pine forests. In detail, this development was overlaid by smaller fluctuations that followed the finer climatic oscillations at decadal to centennial scales.

Vegetation responses to warming after the Little Ice Age included the expansion of treelines. Since about AD 1900, *Pinus sylvestris* has invaded the birch forest area, probably in response to summer and winter temperature increases (Erkkamo, 1956; Heino, 1978; Hustich, 1947, 1948, 1958). In the past decades birch and pines have re-occupied former Mid-Holocene positions in response to global warming (Kullman, 2013). This tree and forest expansion process may last for a few centuries until new equilibrium conditions are reached with climate, as observed in response to rapid and drastic climate warming at the onset of the Holocene (Birks, 2015).

4.9.2

Alpine treeline and timberline

The upper limit of tree growth is only reached in the mountains of Europe that are of sufficient height. As at the polar treeline, the alpine treeline is defined as the upper limit of individual trees (>2–3 m growth height; Ellenberg, 2009) and the timberline (or forest limit) as the upper limit of more or less closed forests. Regarding the elevational vegetation belts, the upper limit of tree growth is usually defined as the dividing line between the subalpine and the alpine vegetation belts, so that the subalpine belt represents the uppermost (potential) forest level. The transition area between the forest and treeline is often described as an ecotone, sometimes also 'kampfzone' meaning 'battle zone' to indicate the ecologically extreme situation for tree growth (Ellenberg, 2009). In the Alps, the treeline ecotone usually comprises at least 100 m of elevational difference, usually 200–300 m (Ozenda, 1985). Whether such an ecotone (or biome transition zone) between subalpine forests and treeless alpine communities would exist without human influence has long been debated (e.g. Ellenberg, 1963, 2009; Holtmeier, 1985, 1989; Mayer, 1970; Körner, 2012). A comparison with North America shows that under natural conditions, treeline and timberline can be separated, forming transitional zones as observed in Europe at the polar and alpine treelines (Arno and Hammerly, 1993). The alpine tree and timberlines, like the polar lines, are mainly the result of decreasing temperatures with increasing elevation, which result in a strong decline of plant tissue formation (Körner, 2012). For trees, conditions become more critical than for shrubs and herbs, because their taller size decouples them from more favour-

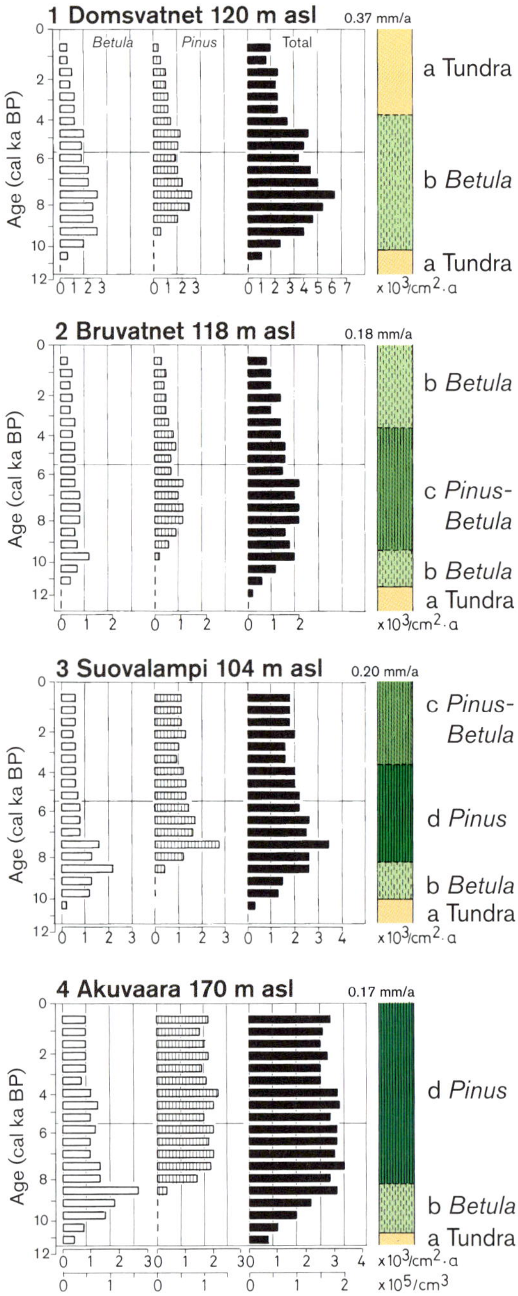

Fig. 4.9-2. Pollen diagrams (absolute diagrams) from the area of the polar tree and timberline in northern Fennoscandia (Fig. 4.9.-1). Simplified and presented as histograms with estimated mean values for periods of 500 years each and a uniform concentration scale (100 000 pollen grains cm^{-3}). Top right: sedimentation rate (mm yr^{-1}). The influx scale (1000 pollen grains cm^{-2} yr^{-1}) varies between sites (according to the sedimentation rates). Right margin: vegetation interpretation. According to diagrams by Hyvärinen (1975a, 1976), modified.

able microclimatic conditions close to the ground. Additional stress factors comprise winter frost drought (Tranquillini, 1979), which makes it difficult to extract water from frozen soil, leading to the desiccation of organs and eventually the death of plants. In the Alps, the climatic threshold value associated with the treeline varies with species if the July mean temperatures are considered. For instance *Picea abies* treelines are located where July mean temperatures reach 9.5°C, while the threshold for *Pinus cembra* treelines are July mean temperatures of 7.5°C (Landolt, 1992). These differences result from variability in solar radiation that permits continental species such as *Pinus cembra* to reach higher elevations under drier and sunnier conditions. Generally, in the Alps, the climatic threshold for tree growth is about 100 days per year with a temperature average of more than 5°C (Ellenberg, 2009), which corresponds well to values observed at global polar treelines (90–106 days and seasonal mean temperatures between 5.1 and 6.9°C) and to the globally recognized limits for tree growth (growing season mean temperature 6.4 ± 0.7°C; Körner, 2012).

4.9.3

Northern Europe

In the Scandinavian mountains (Scandes), which are located almost entirely in the boreal zone (see chapter 4.3 and Fig. 4.9-1), only three woody species occur in the treeline ecotone. At the upper treeline it is *Betula pubescens* ssp. *tortuosa*, which forms the subalpine Fjell birch bush or forest. At the upper timberline *Pinus sylvestris* and *Picea abies* co-occur. In southern Norway, the upper timberline reaches almost 1000 m asl, while *Betula* forms the treeline at about 1200 m asl. In northern Scandinavia the alpine and the polar timberline merge (Figs. 4.9-1 and 4.9-3). In an early phase, reconstructions of upper timberline and treeline changes were mainly based on the investigation of radiocarbon-dated subfossil wood above the current distribution boundaries of the respective tree species (G. Lundqvist, 1959; J. Lundqvist, 1969; Karlén, 1976, 1982; Kullman, 1980, 1987a,b, 1988, 1989). Surprisingly, new megafossil finds suggest maximum Holocene tree species occurences up to 700 m higher than today (corrected for isostatic land uplift) at c. 11 000–7000 cal yr BP, with highest positions reached around 9000

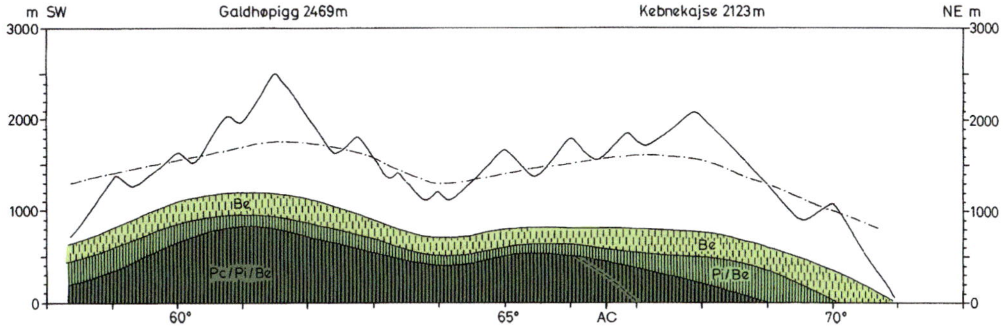

Fig. 4.9-3. Simplified vegetation profile of a longitudinal section through the Scandinavian mountains (Scandes), from southern Norway (left) to the North Cape (right). The dominant tree species are indicated in the elevation levels. Be: *Betula pubescens* ssp. *tortuosa* (fjell birch bush and forest). Pc: *Picea abies* (spruce forest), reaches less far north at the western slope (dotted line) than at the eastern slope. Pi: *Pinus sylvestris*. AC: Arctic Circle. Dashed dotted: snow line. Elevation about 150 times exaggerated. After Mayer (1984), modified.

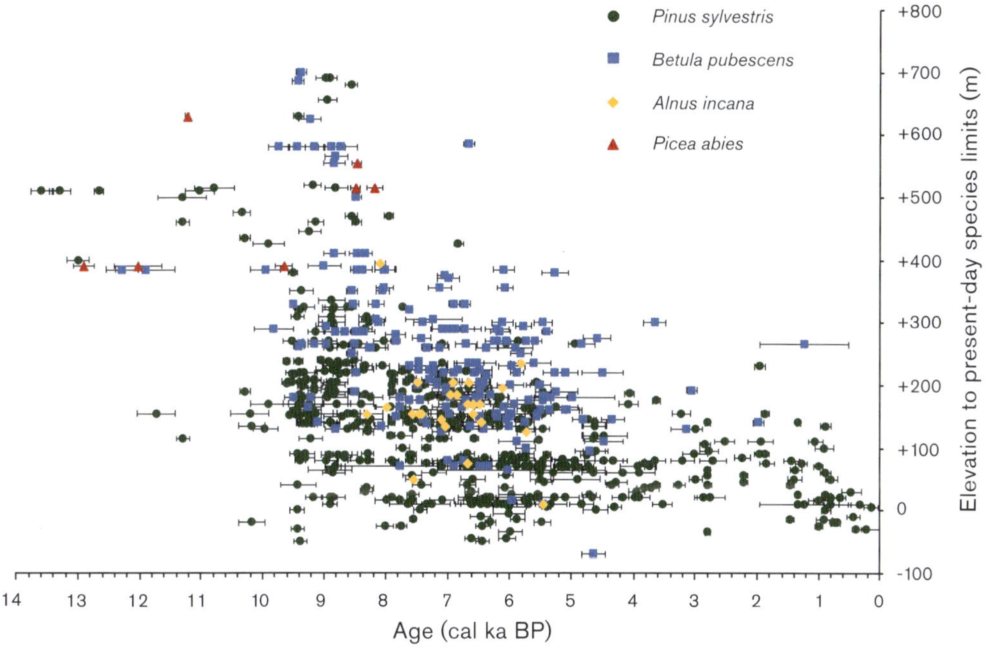

Fig. 4.9-4. Elevation of radiocarbon-dated subfossil Holocene wood finds in the Scandes in relation to the present upper limit of the individual species. Compiled with data from Lundqvist (1969), Kullman (1980, 1987a, 1988, 1992a, 1995, 2002, 2004, 2013, 2017), Kullman and Kjällgren (2000), Kullman and Öberg (2015) and Paus and Haugland (2017).

cal yr BP for *Pinus sylvestris* and *Betula pubescens* (Kullman, 1992a, 1995, 2002, 2004, 2013; Kullman and Kjällgren, 2000; Kullman and Öberg, 2015; Fig. 4.9-4).

Recent macrofossil studies along elevational transects in the Scandes corroborate that highest treeline positions c. 100–400 m higher than today were reached c. 10 000–5000 cal yr BP (Barnekow, 1999; Bergman et al., 2005; Eide et al., 2006). However, maximum treeline positions in the Scandes reached only elevations c. 170–210 m higher than today during the HTM according to Paus and Haugland (2017). These marked differences regarding the reconstruction of highest tree or timberline positions might be explained by local peculiarities and/or the approaches involved. Pollen (percentages, concentrations, influx) and macrofossil (percentages, concentration, influx) analyses are approaches which quantitatively reconstruct communities and past biomass abundances, they thus primarily record timberline (or forest line) dynamics. Megafossils reflect the occurrence of single individuals, they may thus also record scattered tree individuals growing at or even above the treeline. Such individuals represent the species limit, where growth is possible under extremely favourable microclimatic settings (Tinner, 2007), as also commonly observed today above the treeline (Körner, 2012; Carcaillet and Blarquez, 2019). Megafossil-inferred very high treeline positions for the period before 10 000 cal yr BP are particularly difficult to explain (see Bergman et al., 2005), when compared with the macrofossil and pollen reconstructions as well as with plant-independent palaeoclimatic evidence (e.g. Coope et al., 1998; Brooks and Birks, 2001; Velle et al., 2005a,b; Shala et al., 2017) suggesting summer temperatures comparable to or below those of the past 2000 years, during the period 11 600–10 000 cal yr BP.

Similar investigations as from the Scandes are available from the mountains of Scotland. In the Cairngorm Mountains, which rise up to 1300 m, the current treeline is typically around 490–650 m asl (Edwards et al., 2019), whereas the climatic upper limit of *Pinus sylvestris* lies between 600 and 700 m asl (potential treeline). Subfossil trunks have been found in bogs up to nearly 800 m asl, about 150 m higher than today. According to radiocarbon data, these trees were growing between 8500 and 4000 cal yr BP (Pears 1968, 1972, 1988; Dubois and Ferguson, 1985, 1988).

4.9.4

Central and southern Europe

In the central and southern European high mountain ranges (Pyrenees, Alps, Carpathians, Dinarides), subalpine forest communities are more diverse than in the Nordic mountains (Ellenberg, 2009; Körner, 2012). This is particularly the case in the Alps, where the moist and oceanic conditions in the northern and southern Alps are very different from those in the dry and continental central Alps (or inner Alps), creating a strong environmental gradient, which promotes diversity. In the central Alps, summit heights of over 4800 m asl are reached, contributing to a pronounced mass elevation effect (Fig. 4.9-5), and conifers predominate or co-dominate from the colline to the subalpine belt. The upper treeline is formed by *Larix decidua* and *Pinus cembra*, together with small trees (e.g. *Betula pendula, Sorbus aucuparia*) and shrubs (e.g. *Pinus mugo* ssp. *mugo* on carbonatic and *Alnus viridis* on siliceous bedrock). *Pinus mugo* and *Alnus viridis* form thickets of crooked wood where disturbance (avalanche, fire) is high. The timberline (or forest limit) can reach 2200–2400 m above sea level, but is usually lowered by several hundred metres by land use, while the treeline, which is hardly influenced by humans can reach up to 2500 m asl (Landolt, 1992; Leonelli et al., 2011). In the northern Alps, the upper forest limit is usually formed by *Picea abies*, while in the southern Alps *Larix decidua* is also abundant. In the north-western Alps upright *Pinus mugo* ssp. *uncinata* is found in subalpine forests. Today's forest limits usually reach 1800–2000 m asl in the northern and southern Alps, although they are often located at lower elevations as a consequence of widespread summer farming activities (Mayer, 1984; Ozenda, 1985). The treeline in the outer northern and southern Alps can reach 2100 m asl (Landolt, 1992; Ellenberg, 2009).

In the Alps, the reconstruction of timberline changes is largely based on pollen, stomata and macrofossil analyses. In the mountains, percentage pollen analyses alone are insufficient to track timberline changes, as the pollen rain does not only originate from the local vegetation, but also from lower and more distant locations due to regional and long-distance transport. This has been shown by the investigations of Jochimsen (1986) on modern pollen in the subalpine and alpine belts of the

Austrian central Alps, as well as Markgraf (1980) in the Swiss northern Alps. Several studies therefore include pollen concentrations and influx as well as macrofossils (percentages, concentration, influx) to refine reconstructions of timberline fluctuations (Lang and Tobolski, 1985; Tinner et al., 1996; Wick and Tinner, 1997; Wick et al., 2003b; Lotter et al., 2006; Schwörer et al., 2014a). After developing calibration datasets to compare pollen and vegetation composition, Welten (1950, 1952) produced the form of the 'elevation-time diagram' to show the development of the different vegetation belts from the lowlands to high mountain peaks, including the timberline. This original spatio-temporal illustration of vegetation dynamics found its way into several classic studies (e.g. Mayer, 1966; Wegmüller, 1977; Kral, 1979; Schneider, 1985b; Burga, 1987, 1988; Zoller, 1987) and textbooks (e.g. Fægri and Iversen, 1989; Tallis, 1991b). Recently this approach of Welten (1952) has been applied to macrofossils (Heiri et al., 2014; Fig. 4.9-6) and to other mountain ranges elsewhere in Europe (e.g. Feurdean et al., 2013b; Stefanova and Ammann, 2003; see Fig. 4.3-12c).

The post-LGM treeline dynamics of the Alps faithfully tracked climate change and in particular summer temperatures (Tinner and Vescovi, 2007; Fig. 4.9-6). During the LGM when summer temperatures were c. 10°C colder than today (Samartin et al., 2016; Finsinger et al., 2019) treelines in the unglaciated Alpine forelands were located c. 2000 m below their today's occurrence (Tinner and Vescovi, 2007; Finsinger et al., 2019) at c. 300–500 m asl (see also Fig. 4.4.7-3). Subalpine forests dominated by *Larix decidua* and tree *Betula*, for instance, were growing on the Po Plain at 7 m asl during the LGM (23 000–19 000 cal yr BP; Kaltenrieder et al., 2009). In response to summer climate warming of c. 2°C after Heinrich event 1 around 16 700–16 000 cal yr BP (Samartin et al., 2012), the *Pinus cembra*-*P. sylvestris*-*Betula* (likely *B. pendula*)-dominated treeline rose from c. 300–500 to c. 900 m asl in Ticino south of the Alps (Tinner et al., 1999; Hofstetter et al., 2006; Vescovi et al., 2007), while north of the Alps tree *Betula* expanded into the tundra (Rey et al., 2017), forming a forest tundra ecotone below c. 600 m asl (Rey et al., 2017; Fig. 4.9-6). At 14 700–14 500 cal yr BP, in response to a further increase of summer temperature of c. 3–4°C at the onset of the Bølling interstadial (Lotter et al., 2012), *Betula*-*Pinus sylvestris* forests expanded also in the lowlands north

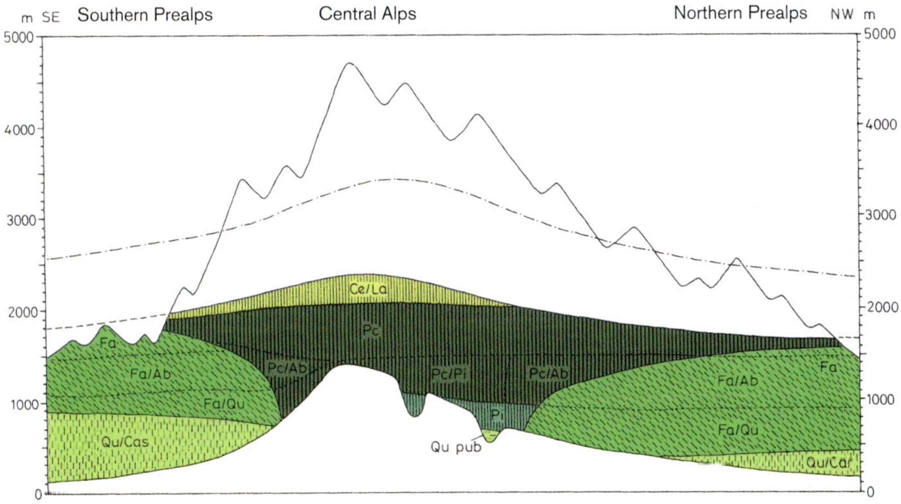

Fig. 4.9-5. Simplified vegetation profile of a cross-section through the Alps, from Ticino (left) to the Swiss Jura (right). The elevation levels indicate the (anthropogenically modified) predominant tree species in central locations. Ab: *Abies alba*. Car: *Carpinus betulus*. Cas: *Castanea sativa*. Ce: *Pinus cembra*. Fa: *Fagus sylvatica*. La: *Larix decidua*. Pc: *Picea abies*. Pi: *Pinus sylvestris*. Qu: *Quercus robur* and *Quercus petraea*. Qu pub: *Quercus pubescens*. Dashed dotted line: snowline. Elevation about 15 times exaggerated. After Ellenberg (2009), modified.

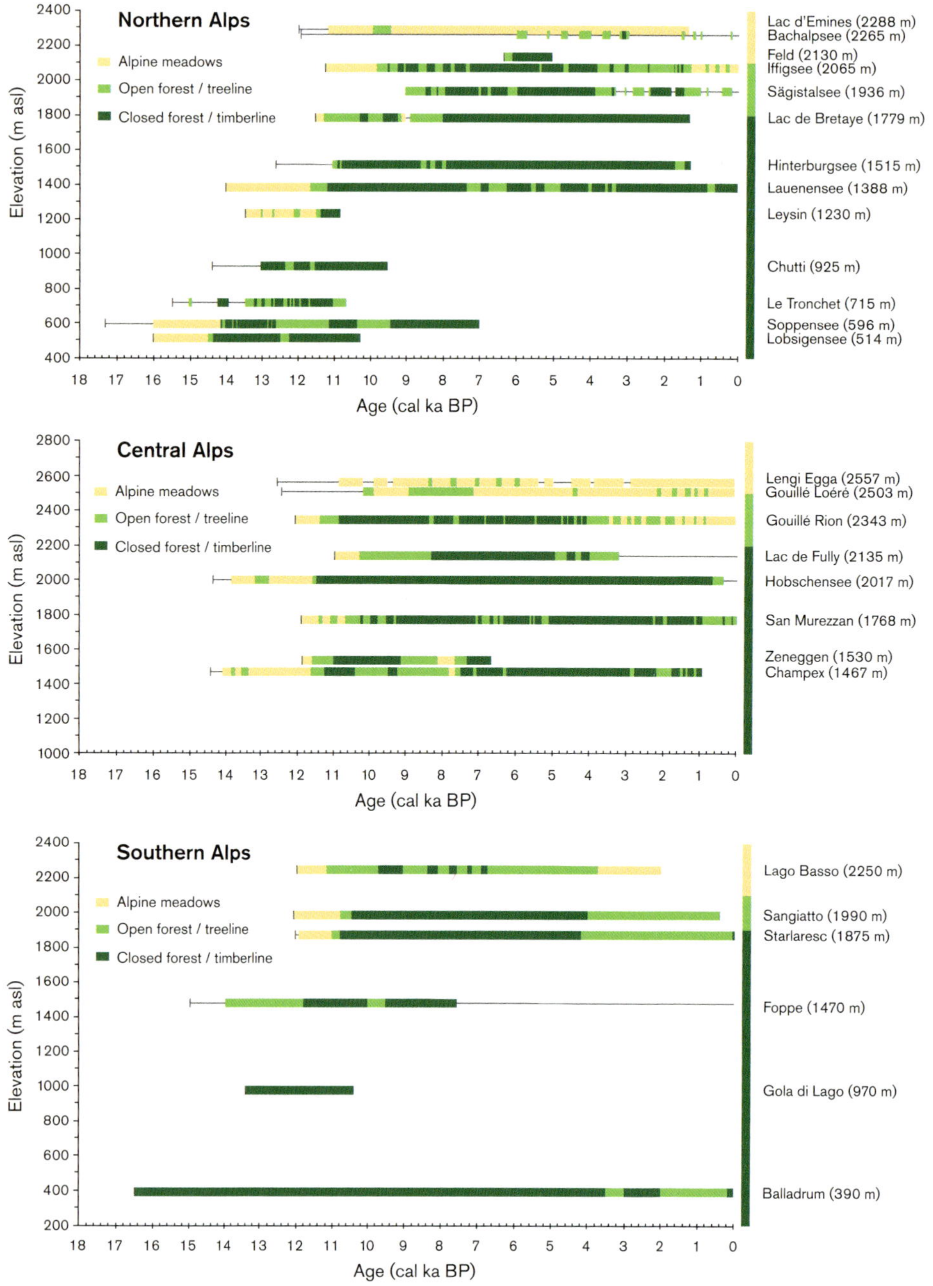

Fig. 4.9-6. Elevation-time diagram of the Late-glacial and Holocene local vegetation development based on plant macrofossil records in the Swiss Northern, Central and Southern Alps. Grey lines indicate the length of the record and absence of macrofossils. Horizontal yellow bars represent macrofossils indicative of alpine meadows, light-green bars individual macrofossils from tree species, suggesting treeline conditions or open forests, whereas dark-green bars represent numerous tree macrofossils indicating closed forest conditions below the forest limit. Vertical bars indicate present-day vegetation. Data from Welten (1982b), Wick and Tinner (1997), Tobolski and Ammann (2000), Gobet et al. (2003, 2005), Tinner and Theurillat (2003), Hofstetter et al. (2006), Finsinger and Tinner (2007), Heiri et al. (2014), Thöle et al. (2016), Vescovi et al. (2018), as well as unpublished data. Modified and expanded from Heiri et al. (2014).

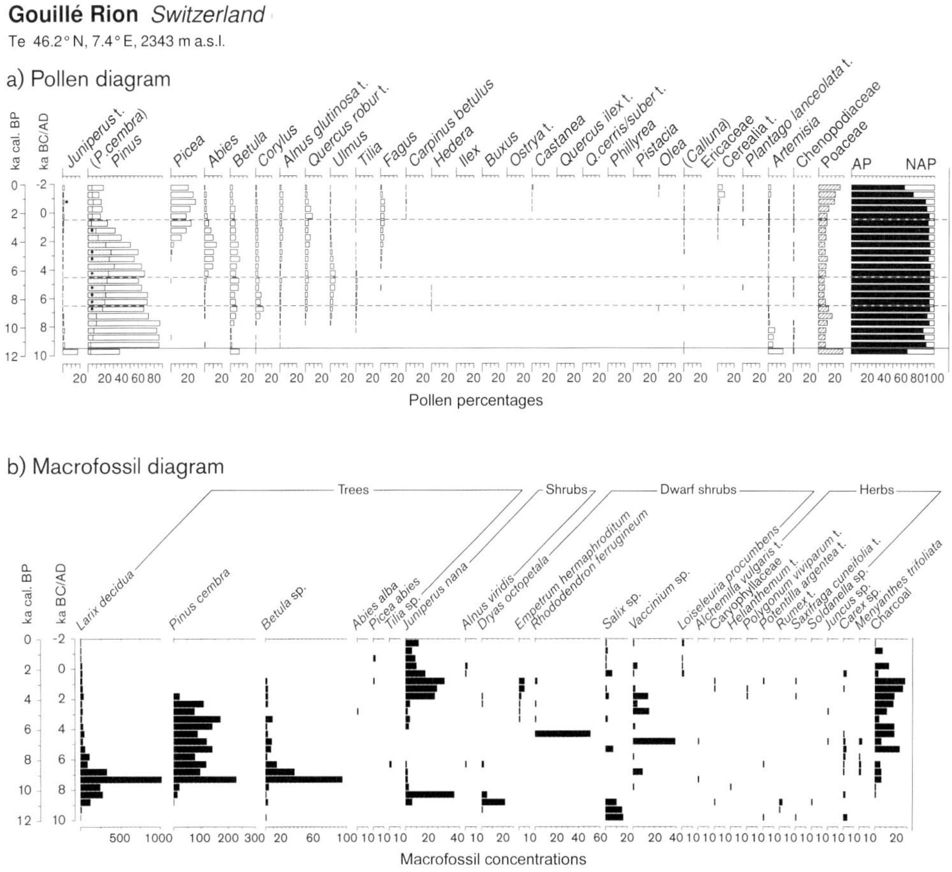

Fig. 4.9-7. Pollen and macrofossil diagrams from the alpine timberline in the Swiss Central Alps: Gouillé Rion (2343 m asl). Pollen diagram (above) according to Tinner et al. (1996), simplified and presented as histogram with mean values for periods of 500 years each. Macrofossil diagram (below) according to Tinner et al. (1996) and Tinner and Kaltenrieder (2005), combined from two profiles and presented as a simplified histogram.

of the Alps (Ammann et al., 2013a,b), resulting in a gradual upward shift of treeline to c. 1300–1400 m asl (Heiri et al., 2014; Fig. 4.9-6). Similar climate-response processes led to treeline positions of c. 1800 m asl in the southern Alps (Tinner and Vescovi, 2007; Vescovi et al., 2007). The treeline was only temporarily lowered during the Younger Dryas cooling (onset c. 12 800 cal yr BP) before it reached today's elevations at the onset of the Holocene around 11 500–11 000 cal yr BP (e.g. Tinner and Theurillat, 2003; Gobet et al., 2005; Heiri et al., 2014).

At the site Gouillé Rion which is located close to today's treeline (2343 m asl) in the Valais (Central Alps), alpine plant communities (e.g. with *Salix herbacea*) established at 11 600–11 500 cal yr BP, when temperatures increased by c. 3–4°C within less than a century according to oxygen isotope records (Tinner and Kaltenrieder, 2005; Figs. 4.9-7 and 4.9-8). Subsequently *Larix decidua* trees reached the site at c. 11 350 cal yr BP, probably in response to further warming by 1–2°C. Given that the treeline during the Younger Dryas was located at c. 1500 m asl in the central Alps (Gobet et al., 2005), this finding implies that the treeline moved 800 m upslope within c. 300 years in response to climate warming of c. 4–6°C at the onset of the Holocene (Tinner and Kaltenrieder, 2005). At this site forests dominated by *L. decidua* persisted until 9600 cal yr BP, when *Pinus cembra* became important. The dominance of *L. decidua* for two millennia is mainly explained by marked continentality, in particular dry summer conditions. Heliophilous and drought-adapted species such as *Dryas octopetala* and *Juniperus*

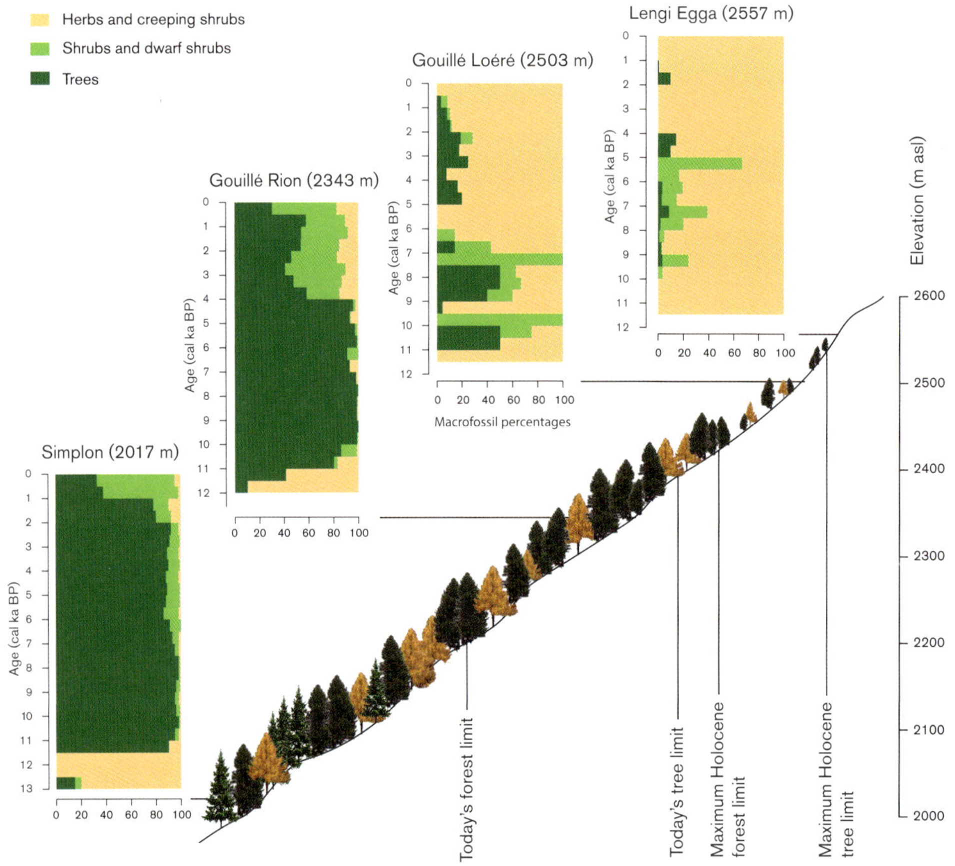

Fig. 4.9-8. Virtual transect along an elevational gradient from closed forests to treeline in the Swiss central Alps (Valais), showing local vegetation dynamics since the Late-glacial at different sites based on macrofossil percentage data. The macrofossil percentage diagrams show the rapid afforestation at the beginning of the Holocene, highest treeline positions during the Holocene Thermal Maximum and a decrease of forest elevation and return to alpine meadows at higher elevation during the Late Holocene and recent past. The vegetation transect represents reconstructed mountain forests dominated by *Pinus cembra* and *Larix decidua* for the period 10 000–5000 cal yr BP. Today, Gouillé Loéré and Lengi Egga are situated in the alpine vegetation belt above the treeline, dominated by herbs (e.g. *Carex curvula*) and creeping shrubs (e.g. *Salix herbacea*). Simplon and Gouillé Rion are located in the treeline ecotone, characterized by a mixture of trees, shrubs (e.g. *Juniperus nana, Rhododendron ferrugineum*) and herbs. Modified from Tinner and Theurillat (2003) and Tinner (2007).

nana could persist in the open *L. decidua* treeline forests but were out-competed when the shade-tolerant *P. cembra* expanded forming dark forests (Fig. 4.9-7). This example shows that besides temperature, moisture availability was decisive with regard to treeline vegetation composition and structure. Similar results pointing to moisture availability and seasonality as important factors controlling treeline vegetation composition during the Early Holocene were obtained from the northern Swiss Alps (Schwörer et al., 2014a). Macrofossil and pollen data show that at the site Iffigsee at 2065 m asl at today's treeline, *L. decidua* forests dominated during the Early Holocene to be gradually replaced by *P. cembra* at 8000–4000 cal yr BP. Dynamic vegetation models considering the climatic course of the Early Holocene suggest that this Early Holocene *L. decidua* dominance at the treeline was closely related to reduced moisture availability resulting from low precipitation and/or shallow soils (Schwörer et al., 2014b; Fig. 4.9-9). Earlier macrofossil and pollen investigations in the northern Swiss Alps (Wegmüller and Lotter, 1990) suggested delayed afforestations at 1800 m asl with *P. cembra* and *L. decidua* expanding only at 9000 cal yr BP, followed

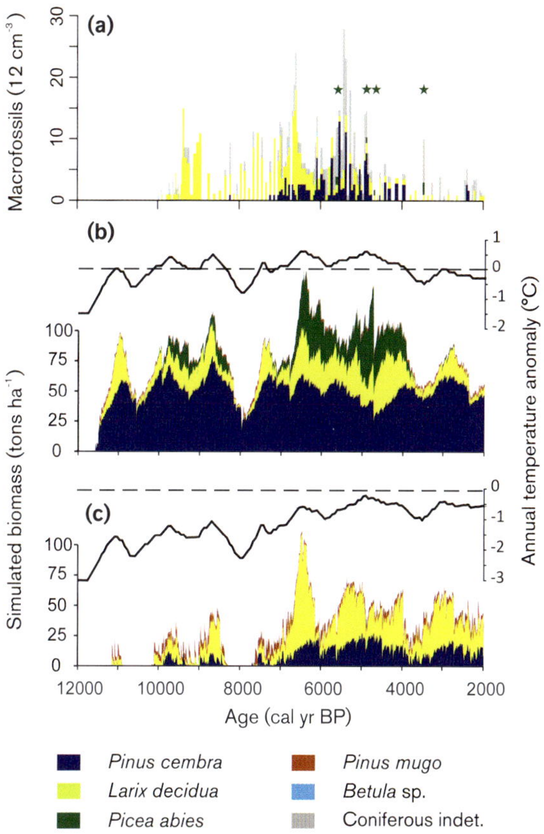

Fig. 4.9-9. Macrofossil record from Iffigsee compared to dynamic LANDCLIM simulations of local forest dynamics under different climate and soil scenarios. (a) Tree macrofossil concentrations (number of macrofossils cm^{-3} of sediment) at Iffigsee. The green stars highlight samples with spruce macrofossils. (b) Top: chironomid-inferred temperature input used in the model. Bottom: forest biomass simulated within a 50-m radius of Iffigsee with temperatures according to the original Hinterburgsee July temperature reconstruction (Heiri et al., 2003, 2004), present soil and precipitation. (c) Same as (b) but with monthly temperatures corrected for orbitally forced insolation changes, with reduced soil water availability (bucket size = 4 cm) and precipitation (70% of observed modern values). Modified from Schwörer et al. (2014b).

by a spread of *Picea abies* at c. 6000 cal yr BP. At the site Iffigsee, *L. decidua* expanded at around 10 000 cal yr BP which is still a delay of c. 1350 years when compared to afforestation dynamics in the central Alps. This difference might be explained by assuming stronger seasonality in the Early Holocene in the northern than in the central Alps, as a result of different atmospheric circulation patterns due to the remaining ice masses in the north of Europe (Schwörer et al., 2014a). The central and southern Alps were sheltered from possible cold air excursions from the far north by the barrier effect of the Alps and seasonality might not have been as pronounced as in the northern Alps due to the proximity of the Mediterranean Sea. Locally, edaphic conditions may have delayed or impeded the establishment of forests below the regional treeline (Finsinger and Tinner, 2007; Ponel et al., 2011).

In the Alps, treelines and timberlines reached their highest positions during the period 10 000–6000 cal yr BP (Tinner and Theurillat, 2003; Schwörer et al., 2014a,b; Figs. 4.9-6 and 4.9-8). For instance *Larix decidua* trees (macrofossils) reached the site Bachalpsee in the northern Swiss Alps, which is today above the treeline (2265 m asl), only at around 8000 cal yr BP (Lotter et al., 2006). For the Austrian Alps, Bortenschlager (1972, 1977) assumed uppermost Holocene treeline positions c. 200 m higher than today. In the Swiss central Alps the Holocene treeline reached elevations of c. 2530 m asl, that is c. 180 m higher than today (Tinner and Theurillat, 2003; Fig. 4.9-8). This upper limit of tree growth is documented by the presence of a macrofossil-inferred alpine herbaceous community at the site Lengi Egga (2557 m asl) throughout the Holocene. Finds of *Salix herbacea, Alchemilla vulgaris, Cerastium alpinum, Cirsium spinosissimum, Helianthemum, Leontodon, Potentilla aurea, Saxifraga stellaris, Saxifraga aizoides, Silene exscapa*, and *Taraxacum* were abundant, whereas trees and shrubs were very rare or absent (Tinner and Theurillat, 2003; Fig. 4.9-8). Early- and Mid-Holocene treeline oscillations were confined to c. 100–200 m.

From about 6000 cal yr BP, and especially after 3000 cal yr BP, the timberline began to sink for several hundreds of metres, to reach elevations around 2000–2200 m asl in the central Alps and 1800–2000 m asl in the northern and southern Alps. This timberline recession was primarily caused by land-use activities that included burning to gain land for summer farming (Gobet et al., 2003; Wick et al., 2003b, Schwörer et al., 2014a, 2015) and the Late-Holocene (summer) cooling. During this period, *Picea abies* invaded the subalpine forests in the western Alps, whereas in the eastern Alps it had reached modern timberline positions already during the Early Holocene, probably because of the closer proximity of the species' refugia (Ravazzi, 2002) and more favourable climatic conditions (Henne et al., 2011). The use of fire for deforestation strongly promoted *Alnus viridis* thickets (Gobet et al., 2003) and together with excessive browsing reinforced the expansion of *Picea abies* (Rey et al., 2013; Schwörer et al., 2014a, 2015) and tall herb (e.g. *Epilobium angustifolium*) communities. Interestingly, human impact also caused the disappearance of special forest communities such as (macrofossil-inferred) *Pinus cembra-Abies alba* timberline forests (Wick et al., 2003b; Lotter et al., 2006) and species-rich subalpine *Pinus cembra-Abies alba-Larix decidua-Picea abies-Acer-Ulmus* forests (Thöle et al., 2016).

In the northern Romanian Carpathians, the timberline reached positions higher than 1500–1800 m asl by 10 200 cal yr BP. Uppermost timberline and treeline positions in response to warm summer conditions were attained at 8500–4900 cal yr BP (Feurdean et al., 2016), when treeline and timberline were c. 100–200 m higher than today. As observed in the Alps, both increasing human impact and climate cooling contributed to the treeline descent of the Late Holocene in the northern Romanian Carpathians. Macrofossil records from the southern Romanian Carpathians suggest afforestation of high elevations during the Bølling and Allerød interstadial, when the treeline was at about 1700–1800 m asl (Magyari et al., 2012b; Orbán et al., 2018), comparable to the situation in the southern Alps. The treeline and timberline ascended only moderately during the Early Holocene reaching c. 1900 m asl (e.g. site Lia, 1910 m asl; Vincze et al., 2017). Highest treeline and forest positions of c. 2000 m and 1800 m respectively, were attained at 8000–3000 cal yr BP (Orbán et al., 2018). During the HTM, the treeline and timberline were thus only c. 100 m higher than today (treeline at c. 1900 m; Magyari ct al., 2012b). Subalpine forests included *Pinus cembra, P. mugo, Picea abies* and partly *Larix decidua* and *Abies alba*. The main determinant of treeline dynamics was climate during the Early and Mid Holocene, while human impact, mainly burning and grazing, became important after 4200 to 2600 cal yr BP (Orbán et al., 2018). The moderate descent of treeline positions after 3000 cal yr BP was thus a combined effect of both climate cooling and human impact (Vincze et al., 2017).

First reconstructions of treeline ecotone dynamics from the Pyrenees reveal that the timberline reached c. 2250 m asl, exceeding today's uppermost limit by c. 300–400 m at the end of the HTM (Leunda et al., 2019). Forest vegetation was composed of *Pinus uncinata* and tree *Betula*. After 4200 cal yr BP alpine vegetation dominated by *Dryas octopetala* expanded, while the treeline descended, likely as a result of both climate cooling and livestock impact (Leunda et al., 2019). Surprisingly, the treeline ascended during the past 2000 years, temporarily regaining maximum positions of c. 2250 m, whereas the timberline remained low at c. 1900 m, likely as a result of pastoral activities.

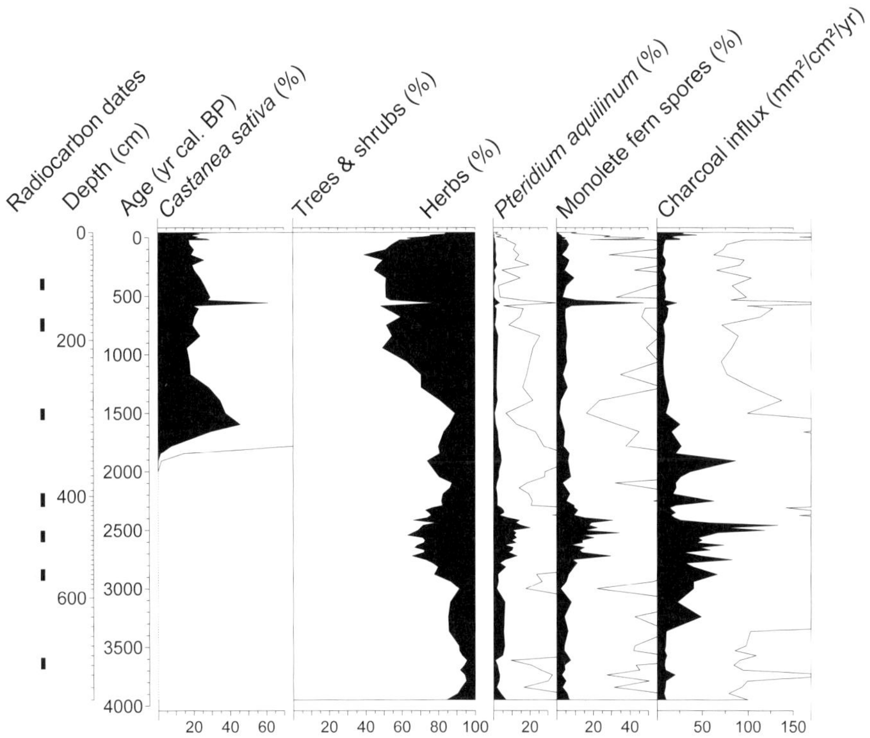

Fig. 4.10.2-3. Strongly simplified pollen percentage and charcoal influx diagram from Lago di Origlio showing the linkage between the human use of fire and the introduction and spread of *Castanea sativa*. Calibration efforts suggest that natural background fire frequency was at about 1–3 mm^2 cm^{-2} yr^{-1}, corresponding to c. 4 fires 1000 km^{-2} yr^{-1} or 60 burnt hectares 1000 km^{-2} yr^{-1} (for more details on conversion of charcoal influx into fire numbers or area burnt see Tinner et al., 1999). Before the introduction of *Castanea sativa* (sweet chestnut) lowermost charcoal influx was at c. 30 mm^2 cm^{-2} yr^{-1} corresponding to c. 80 fires 1000 km^{-2} yr^{-1} and highest charcoal influx at over 120 mm^2 cm^{-2} yr^{-1} corresponding to c. 180 fires 1000 km^{-2} yr^{-1} or c. 3000 burnt hectares 1000 km^{-2} yr^{-1}. Fires were used to open the forest for arable and pastoral farming during the Bronze and Iron Ages (4200–2200 cal yr BP, 2250–200 BC). Around 2000 years ago, intense burning was used to prepare the ground for the sweet chestnut economy. Highly fire-adapted plants such as ferns including *Pteridium aquilinum* benefitted strongly from fire disturbance. After the introduction of *Castanea sativa* during Roman times charcoal influx declined markedly to c. 10 mm^2 cm^{-2} yr^{-1}, which corresponds to c. 10–15 fires 1000 km^{-2} yr^{-1}. Low fire frequencies in the densely settled area were probably the result of first fire-fighting measures to protect the sweet chestnut groves. *Castanea sativa* is very fire resistant (see e.g. strong positive response to fires c. 600 years ago and Table 4.10.2-1), but fire endangers fruit and timber production. After c. AD 1950 fire frequencies increased again, which reflects the loss of economic importance of sweet chestnut groves. The course of vegetation and fire history over the past 4000 years is characteristic for the Insubrian region (Ticino, Piedmont, Lombardy) and expands to other areas of northern Italy including the Apennines. Empty lines show 10x exaggerations. Modified from Tinner et al. (1999).

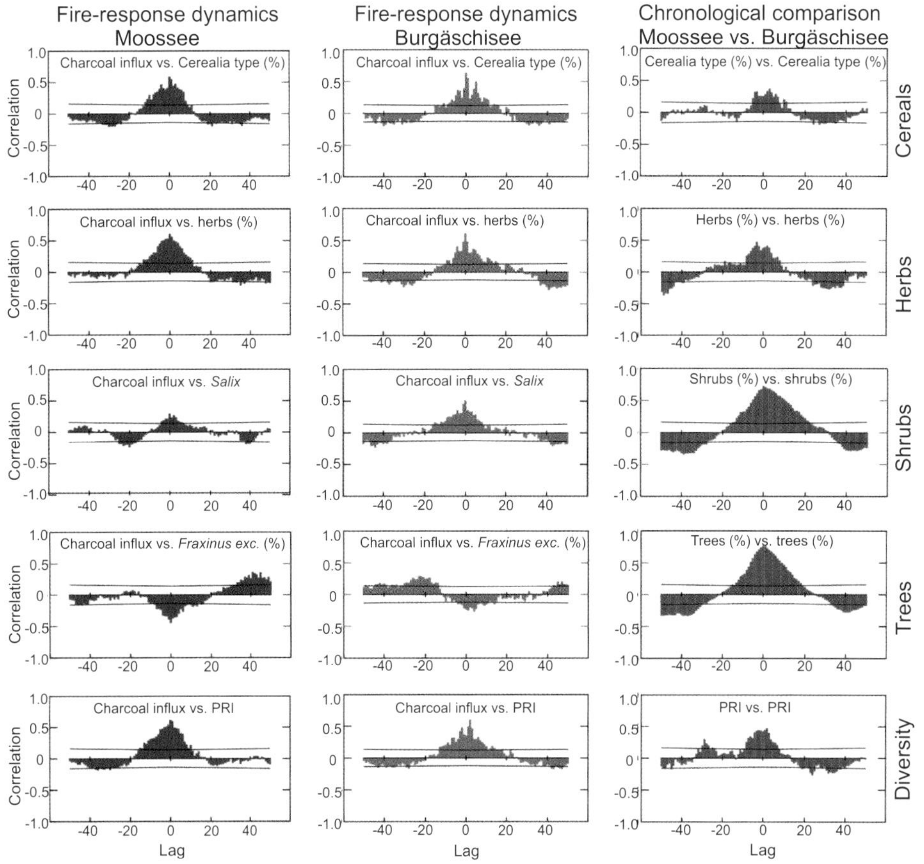

Fig. 4.10.2-4. Cross-correlograms from Moossee and Burgäschisee, western Swiss Plateau for the period 6500–4150 cal yr BP (4550–2200 BC). Microscopic charcoal influx is the independent variable and percentages of selected pollen types and palynological richness (PRI) are dependent variables in order to assess the fire response dynamics at the two sites. Influx is compared with percentages to guarantee independence of variables, that is to avoid spurious correlation effects due to chronological, depositional or percentage distortions, for example, at the same site. 1 lag = 11.1 ± 0.2 cal years (Moossee) and 8.7 ± 0.6 cal years (Burgäschisee). Cross-correlograms comparing the two sites (i.e. Moossee vs. Burgäschisee for Cerealia type) are to assess synchrony, leads and lags between the sites. The two sites were compared on the basis of the two original and independent high-precision varve and wiggle-matching chronologies (for more details see Rey et al. 2019a), without any tuning. For this comparison between the sites 1 lag = 11.1 ± 0.2 cal years (resolution of Burgäschisee was adjusted to that of Moossee). The solid horizontal lines mark the significance level for the correlation coefficients ($p = 0.05$). The distance between the two small lakes (31 and 21 ha, respectively) is c. 25 km, the pollen catchments are clearly distinct, while the microscopic charcoal catchments overlap. The cross-correlation analyses show that most fires were related to farming and crop production (Cerealia type maximum at lag 0 at both sites). Slash-and-burn activities caused the spread of herbs and shrubs and the decline of trees. Diversity increased during periods of increased fire activity, at Moossee with the fires (lag 0), at Burgäschisee c. 20 years after the fires (lag 2). These results contrast with the results from Lago di Origlio, south of the Alps, where fire activity was higher and diversity declined. Together this suggests that diversity benefitted from intermediate disturbance levels. Most surprisingly, cross-correlation analyses between the sites show that declines of forests (tree %) and expansion of shrublands (shrub %) were coeval at the two sites (maxima at lag 0). Land use and diversity dynamics were more complex, with one site leading or lagging the other by c. 30 years (maximum correlations at positive and negative lags 3 for Cerealia type and herb %) or c. 20 years (maximum correlations at positive and negative lags 2 for PRI). Modified from Rey et al. (2019a).

burning activities can be facilitated by ring-barking or girdling of trees and shrubs, a practice which was likely used in Europe since prehistory (Innes et al., 2013).

In **southern Europe**, the long-term use of fire resulted in the creation of widespread grassland, heathland, maquis and garrigue communities, dominated by fire-prone shrubs and herbs, while native Mediterranean forests of, for example, *Quercus ilex, Q. suber, Q. pubescens, Q. cerris* or wild *Olea europaea* were gradually confined to remote or infertile areas (e.g. rocky slopes, mountains; Fig. 4.5.8-1). In some areas slash-and-burn drove fire-sensitive species and communities to local extinction (e.g. Tinner et al., 1999, 2009; Keller et al., 2002; Colombaroli et al., 2007; Kaltenrieder et al., 2010; Valsecchi et al., 2010; Vescovi et al., 2010a; Henne et al., 2013, 2015, Martinelli et al., 2017), markedly affecting the course of vegetation history (Fig. 4.10.2-1). Fire sensitive trees that were strongly reduced or eliminated locally by excessive human fire disturbance included *Abies alba, Tilia cordata, T. platyphyllos, Ulmus glabra, U. minor, Fraxinus excelsior, Acer campestre, A. pseudoplatanus* and *Hedera helix* (Fig. 4.10.2-2; Tinner et al., 2000). A pioneer study applied novel ancient DNA (aDNA) approaches to subfossil needles of *Abies alba* collected from lake sediments (Origlio, Switzerland) and dating back from 7200 to 5800 cal yr BP (Schmid et al., 2017). This procedure allowed the reconstruction of genetic variation of extinct lowland *Abies alba* populations before, during and after stand-replacing fires at 6500–6200 cal yr BP. The genetic variation present before fire disruption was recovered afterwards, most likely by in situ recolonization from scattered trees or tree stands in the Origlio area that escaped fires. This main finding contradicts an alternative hypothesis of Wick and Möhl (2006), who suggested that the extinction of southern European lowland *Abies alba* populations c. 5000 years ago was possibly caused by reduced genetic variation that impeded adaptation to changing environments (Schmid et al., 2017).

If human fires were extremely frequent and intense, fire-resistant tree species such as *Pinus nigra, Quercus pubescens* or *Q. ilex* were also replaced (Colombaroli et al., 2007; Tinner et al., 2009; Bisculm et al., 2012; Morales-Molino et al., 2017), in some cases by flammable shrubs such as *Erica, Cistus*, Fabaceae and Lamiaceae, thus increasing fire susceptibility (e.g. Carrión et al., 2010c). Fires were used to create monospecific forests for fruit production (e.g. *Castanea sativa*; Fig. 4.10.2-3) but were supressed when the orchards needed to be protected from fire damages (Morales-Molino et al., 2015). However, fire-adapted arboreal communities existed in southern Europe before the advent of agriculture. For instance, during the Early and Mid Holocene, frequent droughts and summer heat likely contributed to maintain flammable natural maquis or woodlands dominated by *Pistacia lentiscus* in coastal Sicily or *Erica arborea* and *E. scoparia* in coastal Sardinia (Tinner et al., 2009; Beffa et al., 2016). Other taxa advantaged by frequent fires in Southern Europe were *Chamaerops humilis, Phillyrea, Cistus, Pteridium aquilinum* and *Calluna vulgaris*.

Natural fire activity inferred from charcoal influx for before the onset of the Neolithic was low in **temperate Europe** (e.g. Tinner et al., 2005a). Subsequently, the impact of slash-and-burn activities became important (e.g. Clark et al., 1989) but was less pronounced than in southern Europe, likely because of lower temperatures and higher moisture. Ecological patterns similar to those reconstructed for the Mediterranean emerged in response to anthropogenic fires, specifically the pronounced promotion of strong resprouters such as *Corylus avellana, Calluna vulgaris, Alnus, Betula* and *Salix* and herbaceous species (Fig. 4.10.2-4; Clark et al., 1989; Odgaard, 1992; Tinner et al., 2005a; Bradshaw et al., 2010; Rösch et al., 2014; Jamrichová et al., 2017; Rey et al., 2019a). The increased use of fire for land use during the Late Holocene reduced fire-sensitive late-successional trees or lianas such as *Tilia, Ulmus, Fraxinus, Acer, Abies alba* and *Hedera helix* (Tinner et al., 2005a; Bradshaw et al., 2010; Rey et al., 2017, 2019a; Novenko et al., 2019). Fire incidence also reduced relatively fire-resistant trees such as *Fagus, Carpinus* and *Quercus* (e.g. Bradley et al., 2013; Gałka et al., 2015): in contrast to fire-sensitive trees, however, they were able to recover after repeated fire disturbance (e.g. Rey et al., 2019a). Where fire impact was particularly strong and associated with other recurrent disturbances such as browsing, this ultimately led to monospecific, disturbance-adapted forests (dominated by e.g. *Fagus, Quercus*), shrublands, heathlands and grasslands that are characteristic of modern temperate European landscapes (Birks and Tinner, 2016; Jamrichová et al., 2017).

Pollen type	Palaeoecologically derived fire sensitivity (definition see below)		Comparison with ecological studies * (only for medium fire frequencies)	
	Medium fire frequency	High fire frequency	Species concerned	Agreement
Abies	1	1	*Abies alba*	+
Acer	2	2	*Acer campestre* *Acer pseudoplatanus*	+ ++
Alnus glutinosa type	5	3	*Alnus glutinosa*	+
Betula	5	4	*Betula pendula*	++
Calluna	4-5	6	*Calluna vulgaris*	+
Castanea	5	?	*Castanea sativa*	++
Corylus	5	3-4	*Corylus avellana*	-
Fagus	3	3	*Fagus sylvatica*	++
Fraxinus excelsior type	2	2	*Fraxinus excelsior*	++
Hedera	1	1	*Hedera helix*	++
Ilex	1	1	*Ilex aquifolium*	++
Quercus (deciduous)	4	3	*Quercus pubescens* *Quercus petraea*	+/- +/-
Salix	5	3	*Salix caprea*	+
Sambucus nigra type	5	5	*Sambucus nigra*	?
Tilia	2	2	*Tilia cordata* *Tilia platyphyllos*	++
Ulmus	2	2	*Ulmus glabra*	+
Vitis	2	2?	*Vitis sylvestris*	+
Monolete fern spores	5	6	Fern species	?
Pteridium aquilinum	6	6	*Pteridium aquilinum*	++
Most upland herbs (e.g. Poaceae, Asteraceae Rosaceae, *Anemone nemorosa*, *Humulus* type, *Potentilla* type, *Trifolium repens* type)	5	5	A multitude of species	++

Fire frequency classes:
Low fire frequency (response weak, not shown): < c. 10 fires 1000 km^{-2} $year^{-1}$
Medium fire frequency: >c. 10 < c. 100 fires 1000 km^{-2} $year^{-1}$
High fire frequency: >100 fires 1000 km^{-2} $year^{-1}$

Fire sensitivity classes:
1 = Fire-intolerant: fire rapidly leads to local extinction. Usually no regeneration
2 = Fire-damaged: fire slowly leads to local extinction
3 = Fire-sensitive: fire leads to considerable decreases, usually no local extinction
4 = Fire-indifferent: fire frequency usually does not influence presence
5 = Fire-enhanced: fire favours abundance
6 = High abundance largely depends on fire

Agreement with ecological literature:
- disagreement
\+ agreement in one source
++ agreement in more than one source
? data not available

Table 4.10.2-1. Attribution of fire-sensitivity values for important plant taxa, derived from palaeoecological data from Lago di Origlio and Lago di Muzzano (modified from Tinner et al. 2000).

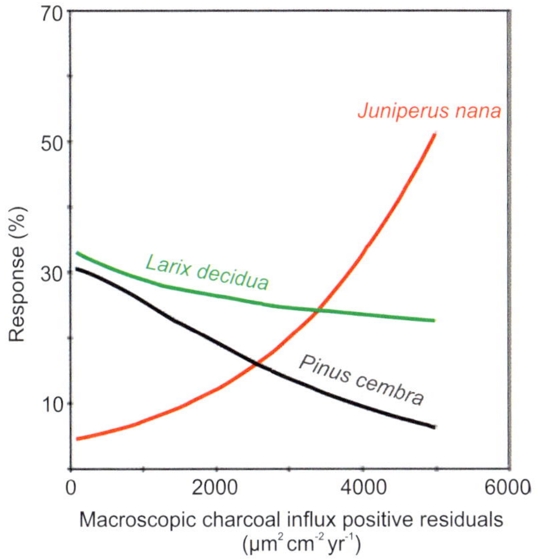

Fig. 4.10.2-5. Species response curves to macroscopic charcoal influx positive residuals as a proxy for fire activity. Species responses were obtained using a generalized additive model (GAM). The analyses reveal the rather high fire sensitivity of *Pinus cembra*, a moderate fire sensitivity of *Larix decidua* and the high fire resistance of *Juniperus nana*. Modified from Colombaroli et al. (2010).

Prehistorical and historical slash-and-burn cultivation altered natural fire regimes in **boreal Europe**, for example in southern Finland (Alenius et al., 2013; Clear et al., 2013). While the use of fire to open forest for small scale cultivation may have started as early as in the Neolithic around 7000 cal yr BP in the boreal forests of southern Finland, large-scale slash-and-burn activities connected to intense arable farming arose only during the past c. 2000–1500 years (Alenius et al., 2013; Clear et al., 2013). Fire disturbance connected to agricultural activities caused disruptions of boreal forests dominated by *Pinus, Betula, Picea* and *Alnus* and the mass expansion of herbs and the heliophilous shrub or small tree *Juniperus* (Alenius et al., 2013). Expansions of the resprout ers *Alnus* and *Betula* as well as herbaceous species after fires are documented in Russian boreal forests, where low fire activity was strongly connected to high abundance of fire-sensitive *Picea* throughout the Holocene (Kuosmanen et al., 2014). Given that in Russia *Picea* was present since the Early Holocene the authors conclude that low fire incidence favoured spruce, rather than the species reduced fire occurrence, an alternative hypothesis developed in Scandinavia, where spruce expanded late (Ohlson et al., 2006; 2011). Although fires may have initially facilitated the expansion of *Picea* populations (Kuosmanen et al., 2014; Ohlson et al., 2017), in the south of Scandinavia the tree was itself replaced by *Fagus sylvatica* after local Late-Holocene fire disturbance (Ohlson et al., 2017). In hemiboreal and temperate southern Sweden the use of fire for agricultural purposes was of secondary importance for vegetation (Bradshaw et al., 2010). Fire disturbance in southern Sweden was, however, positively connected to increases of fire-adapted trees or shrubs such as *Betula*, *Pinus* and *Calluna* (Table 4.10.2-1). Similarly to southern, western and central Europe, extreme uses of fire for agricultural purposes led to drastic changes of vegetation structure and composition in northern Europe. Among the most striking cases is the creation of widespread *Calluna vulgaris* heathlands by regular intentional burning in formerly densely forested coastal Norway c. 4500–2000 years ago (Kaland, 1986; 2014). Anthropogenic fire-dependent *Calluna* heathlands of Neolithic or younger age were not confined to boreal Norway, they were also established in temperate Atlantic Europe (e.g. Behre, 1976, 1988; Odgaard, 1992), ranging from Scotland in the north to Portugal in the south (Kaland, 2014; see chapter 4.5.8).

Ecologically closely related to the boreal forests of northern Europe are the **subalpine forests** of the temperate and Mediterranean ranges. Postglacial fire-response patterns of vegetation are similar to those observed in other biomes of Europe, specifically there is little evidence of alteration of vegetation by natural fire regimes before the onset of agriculture. High-resolution time-series analyses from the Alps show that with the advent of agriculture, anthropogenic burning strongly promoted resprouters and herbaceous species at the cost of late-successional trees, in particular conifer species (Gobet et al., 2003; Rey et al., 2013; Schwörer et al., 2015). Cross-correlation analyses show that excessive prehistoric anthropogenic fire disturbance was connected to mass expansions of fire-resistant *Alnus viridis* as well as ferns and herbs (e.g. monolete spores, *Botrychium, Selaginella selaginoides, Epilopium* type, *Thalictrum, Aconitum, Aster* type, *Urtica*, Caryophyllaceae, Cichorioideae) in the Upper Engadine (Gobet et al., 2003). In agreement, in the lower Engadine, prehistoric fire promoted *Alnus viridis* thickets, *Calluna vulgaris* and *Corylus avel-*

lana shrublands, *Pteridium aquilinum* stands and meadow communities (Dietre et al., 2017). In the Bernese Oberland, Neolithic anthropogenic fires promoted *A. viridis* thickets and herbs such as *Artemisia* and Poaceae, while during the Bronze Age fire contributed to a further expansion of *Alnus viridis* and herbs (Rey et al., 2013; Schwörer et al., 2015; Thöle et al., 2016). Fires were associated with arable and pastoral farming as revealed by significant cross-correlations with, for example, *Plantago lanceolata*, Cerealia type and *Sporormiella* (Rey et al. 2013; Schwörer et al., 2015). However, during times with most intense pastoral farming (the past 1000 years in both the Engadine and Bernese Oberland) and very high fire incidence, *A. viridis* became less abundant and subalpine meadows expanded (Gobet et al., 2003; Rey et al., 2013; Schwörer et al., 2014a). Indeed, the combination of frequent fires, trampling and intense grazing may lead to collapses of *A. viridis* populations (Genries et al., 2009b). However, close links between prehistoric fire disturbance and the expansions of *A. viridis* and *Juniperus nana* are reported from other Swiss areas such as the north-east Grisons and the Valais (Fig. 4.10.2-5; Finsinger and Tinner, 2007; Colombaroli et al., 2010; Röpke et al., 2011). Similarly, in the western Italian Alps, *A. viridis* and *Juniperus* expanded in response to fire (Pini et al., 2017).

In the Alps, prehistoric burning led to massive forest disruptions (Kothieringer et al., 2015), a lowering of the timberline by several hundred metres (Colombaroli et al., 2010; Schwörer et al., 2014a; Dietre et al., 2020) and changes in forest structure and composition (Gobet et al., 2003; Stähli et al., 2006; Genries et al., 2009a,b; Rey et al., 2013; Schwörer et al., 2014a, 2015; Pini et al., 2017). *Pinus cembra* and *Abies alba* declined in response to Neolithic and Bronze Age fires, which over the millennia, together with grazing, led to the today's dominance of *Picea abies* in the subalpine forests (Wick et al., 2003b; Schwörer et al., 2015; Thöle et al., 2016). Fire-sensitive *Picea abies* only exceptionally directly benefitted from fire, given that only in some time series the species increased with burning (Schwörer et al., 2015), while in others it was unaffected (e.g. Gobet et al., 2003) or even declined (Rey et al., 2013). Although prehistorical fire reduced both *Pinus cembra* and *Larix decidua* over the centuries and millennia, continued moderate burning and grazing advantaged the latter, resulting in the formation of larch meadows, a special vegetation type of the Engadine (Gobet et al., 2003). Similarly, a very long persistence and dominance of *Larix decidua* in the subalpine vegetation has been attributed to fire and possibly other disturbances in the western Italian Alps (Blarquez et al., 2010). Changes in the subalpine belt also included the disappearance of temperate trees such as *Ulmus, Tilia* and *Acer* that together with conifer species likely formed a transitional belt in the Alps comparable to the nemoboreal vegetation type in Fennoscandia, before fire and browsing confined them to lower, temperate elevations (Thöle et al., 2016; Dietre et al., 2017).

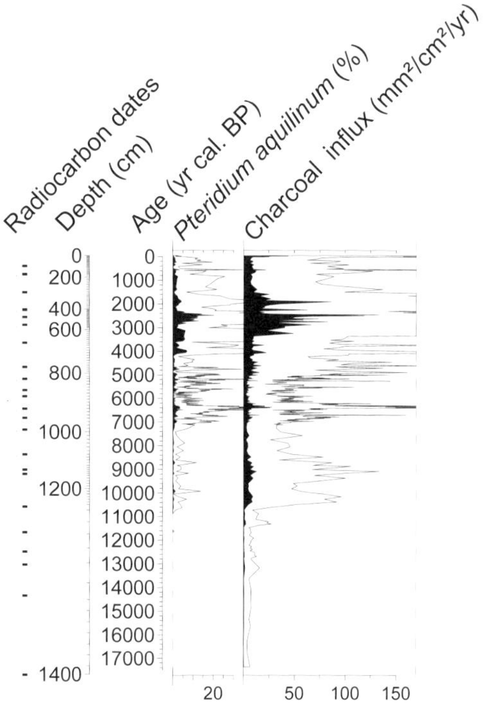

Fig. 4.10.2-6.
Spore percentage and microscopic charcoal influx diagram. Postglacial fire-dependence of *Pteridium aquilinum* at Lago di Origlio, Ticino, southern Switzerland. Only if microscopic charcoal influx is > c. 5 mm^2 cm^{-2} yr^{-1} corresponding to c. 8 fires 1000 km^{-2} yr^{-1} or 120 burnt hectares 1000 km^{-2} yr^{-1} does *Pteridium aquilinum* become relevant in moist, warm-temperate Insubrian ecosystems. Modified from Tinner et al. (1999).

Evidence for beneficial effects of Holocene fire activity on *Picea* comes from the **Balkans** for the species *Picea omorika* in the Dinaric Alps (Finsinger et al., 2017b). However, records from the Romanian Carpathians suggest that, like the situation in Fennoscandia, the occurrence of *Picea abies* was connected to low fire incidence in subalpine forests (Feurdean et al., 2016, 2017). Correspondingly to what observed in southern Scandinavia, *Picea abies* was in some cases replaced by *Fagus sylvatica* after fire disturbance (Feurdean et al., 2016, 2017). When compared to *Picea abies, Fagus sylvatica* is more fire-tolerant. Although its resprouting capacity is only moderate and the resulting shoots tend to die back, *Fagus sylvatica* is a very efficient post-fire (non-serotinous) masting seeder (Ascoli et al., 2013, 2015; Maringer et al., 2016, 2020). In the Romanian Carpathians grazing and burning for land use contributed to the lowering of the treeline and the expansion of shrublands and meadows during the Late Holocene (Geanta et al., 2014; Vincze et al., 2017). Ecological response patterns of subalpine ecosystems were similar to those observed in the Alps, with fires reducing important fire-sensitive trees such as *Abies alba* and *Pinus cembra* and favouring shrubs such as *Alnus viridis* and *Corylus avellana* as well as herbaceous species (Geanta et al., 2014). In the subalpine forests of the Spanish **Pyrenees** anthropogenic fires increased during the Late Holocene (Garcés-Pastor et al., 2017; Leunda et al., 2020). Burning caused the decline of fire-sensitive species such as *Abies alba*, whereas resprouters such as *Corylus avellana* and Ericaceae expanded (Garcés-Pastor et al., 2017; Leunda et al., 2020). However, the available records suggest that fire impacts on subalpine vegetation were less pronounced than in the Alps or the Carpathians (Leunda et al., 2020).

Fire-ecological studies show that the vegetation of Europe was profoundly altered by fire disturbance during the past 10 000 years. Some fire-related changes in **European vegetation** occurred before the onset of agriculture, such as the expansion of fire-adapted *Pteridium aquilinium* in the southern Alps c. 10 800 cal yr BP (Fig. 4.10.2-6; Tinner et al., 1999). This case is particularly interesting because the species is known to create so-called fire-spirals, that is to promote fire by producing a highly flammable layer of dried fronds every autumn (FEIS, 2020). However, most fire-related Holocene vegetation shifts occurred in response to excessive anthropogenic burning for arable and pastoral farming and included the conversion of forests into parklands, shrublands, grasslands or meadows. In the forests, fire-sensitive species were strongly reduced, while fire-resistant species expanded (Table 4.10.2-1; Tinner et al., 1999, 2000, 2005a; Gobet et al., 2003; Colombaroli et al., 2007, 2010; Rey et al., 2013; Geanta et al., 2014; Schwörer et al., 2014a). Such fire-related vegetational shifts led to an impoverishment and homogenization of European forest vegetation, ultimately leading to the establishment of monospecific forests. Surprisingly, the loss of fire-sensitive forest species also occurred in vegetation types in which fire is almost absent today, such as the beech forests of central Europe (Clark et al., 1989; Tinner et al., 2005a; Bobek et al., 2019; Rey et al., 2019a). Therefore this prehistoric fire legacy cannot be inferred from observations but is only accessible through paleoecological records. Fire-ecological studies also show that some vegetation types were less flammable than others (e.g. olive-stone-oak forests in southern Europe; mixed silver fir or mixed beech forests in southern and central Europe; spruce forests in northern Europe and the Alps). The example of the evergreen olive–oak forests from thermomediterranean Europe suggests that low flammability can be unrelated to climatic conditions (e.g. dry and hot summer season in the Mediterranean). Here low Holocene flammability can be explained by the traits of the plants (e.g. inflammable cork or litter) or communities (e.g. soil and air moisture-retaining evergreen Mediterranean forests). During the Mid and Late Holocene some of these forests were replaced by highly flammable (e.g. maquis, pine or eucalypt plantations) or fire-adapted vegetation types (e.g. sweet chestnut forests, deciduous oak forests, beech forests, shrublands, heathlands, grasslands).

From Holocene studies of the boreal and hemiboreal it emerges that low fire activity might be decisive for maintaining *Picea* populations under future global change conditions (Clear et al., 2013). Conversely, it has been hypothesized that maintaining moderate fire disturbance may contribute to the preservation of species-rich European ecosystems which were artificially created by humans over the past millennia (e.g. subalpine meadows, *Calluna* heathlands, *Castanea sativa* forests, *Ostrya carpinifolia* shrublands, Mediterranean maquis and garrigues; see Colombaroli et al., 2010; Colombaroli and Tinner, 2013). However, the available long-term fire-ecological studies also clearly show that excessive fire activity in the future may further disrupt, transform and impoverish the remnant forests in Europe if no management measures are taken to confine excessive anthropogenic burning. Thus, finding plant community-resolved fire-management strategies will be crucial for maintaining forest ecosystems, biodiversity and the related provisioning services.

4.10.3

Grazing and browsing

Grazing and browsing can strongly affect vegetation structure and composition. For instance, in central European ranges, browsing strongly influences the survival rate of palatable seedlings such as *Abies alba* or *Taxus baccata* (Ellenberg, 2009). Moreover, excessive human-induced grazing and browsing can promote less palatable species (e.g. *Picea abies*) at the expense of palatable ones (e.g. *Abies alba*; Ellenberg, 2009). Past herbivory has been conventionally deduced from pollen-based reconstructions of pastureland communities (Behre, 1981, 1988). More recently, new proxies such as spores of coprophilous fungi (e.g. van Geel and Aptroot, 2006; Fyfe et al., 2008; Bisculm et al., 2012; Rey et al., 2013; Eddudóttir et al., 2020) and sedimentary aDNA (e.g. Bajard et al., 2017; Giguet-Covex et al., 2019; Edwards, 2020) of herbivores allow new pollen-independent and/or animal-specific insights (Fig. 4.10.3-1).

Effects of browsing and grazing may include impoverishment of forest vegetation. For instance, the tree flora of the British Isles was diverse until c. 1000 years ago. Subsequently heavy browsing and grazing led to the creation of *Quercus* woodland with an associated flora of high conservation value (Bradshaw et al., 2015). Similarly, grazing likely contributed to the stability of *Quercus* woodlands in central Europe throughout the Holocene (Petr et al., 2020). Other effects of grazing herbivores were soil erosion in Iceland (Eddudóttir et al., 2020), disruption of peatlands, with substantial declines of *Sphagnum* and *Betula* in Spain (Morales-Molino et al., 2019), as well as the expansion of non-palatable *Picea abies* and nitrophilous *Urtica dioica* in the Alps (Schwörer et al., 2015; Fig. 4.10.3-1).

During the Holocene, herbivory was often associated with burning of vegetation (e.g. Carrión et al., 2010c; Connor et al., 2013; Rey et al., 2013; Kothieringer et al., 2015; Schwörer et al., 2015; Morales-Molino et al., 2017; Conedera et al., 2017; Hajkova et al., 2018), suggesting that since the Neolitihic, vegetation was deliberately burned to facilitate grazing. High-resolution continuous time-series allow the numerical exploration of leads and lags between vegetation, fire and grazing proxies to better address the question of whether fire was intentionally used to create grasslands and meadows for domestic animals (Rey et al., 2013). Time-series analyses show that during the Bronze Age, *Sporormiella* abundances increased significantly after fire disturbance in the Bernese Alps. The *Sporormiella*-inferred increase of grazing activity lasted for c. 28–56 years after burning to then gradually decline (Rey et al., 2013). Before the Bronze Age, during

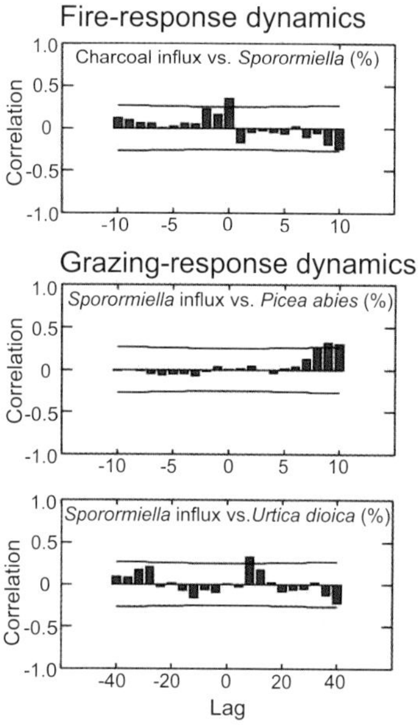

Fig. 4.10.3-1. Cross-correlograms showing the linkage between fire and grazing disturbance as well as grazing responses of *Picea abies* and *Urtica dioica* at Iffigsee, Bernese Alps, Switzerland for the period 6960–5160 cal yr BP (5010–3210 BC). The independent variable is given first, the dependent variable second. Microscopic charcoal influx is a proxy for regional fire activity and *Sporormiella* percentages and influx for grazing activities. Influx are number of particles (charcoal) or spores (*Sporormiella*) cm^{-2} $year^{-1}$. Influx is compared with percentages to guarantee independence of variables, that is, to avoid spurious correlation effects due to chronological, depositional or percentage distortions for example. 1 lag = 29 ± 7.5 years. The solid horizontal lines mark the significance level for the correlation coefficient ($p = 0.05$). The cross-correlation analyses show that fire and grazing were closely linked, without any leads or lags (maximum correlation at lag 0). *Urtica dioica* and *Picea abies* responded positively to grazing disturbance after c. 60 (lag 2) and 300 years (around lag 10). Modified from Schwörer et al. (2015).

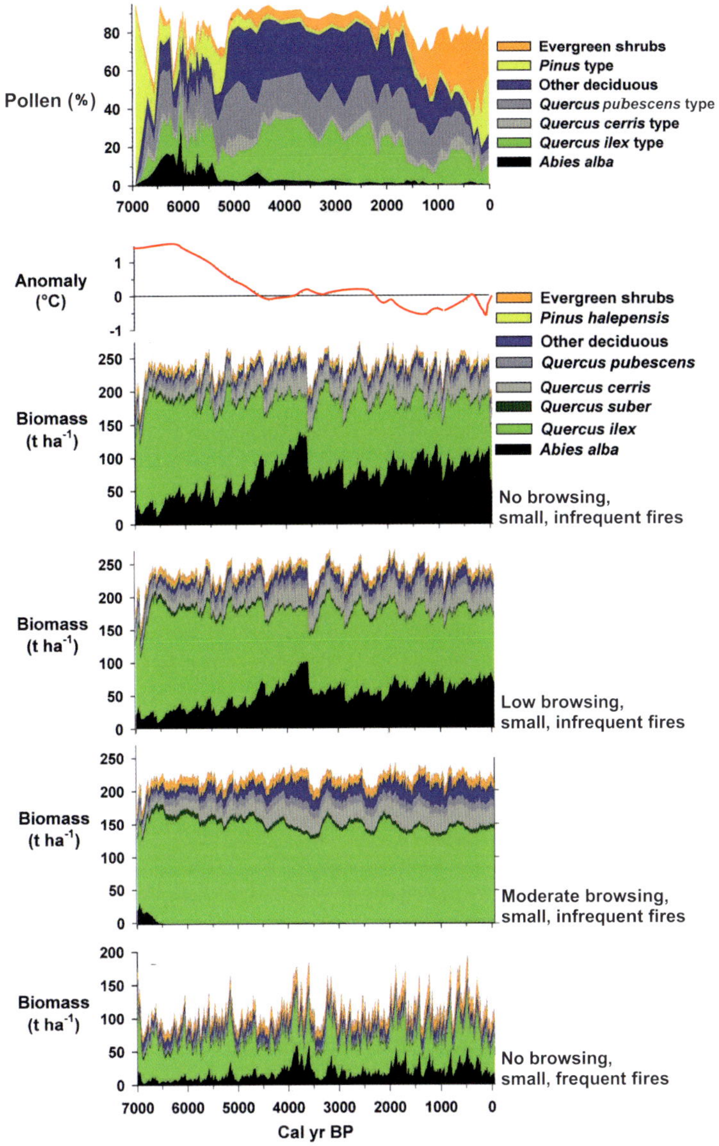

Fig. 4.10.3-2. Comparison of pollen percentage data from Lago di Massaciuccoli in Tuscany (Italy) with LandClim dynamic modelling outcomes in tonnes of biomass of vegetation around the lake. Lago di Massaciuccoli is a Mediterranean lagoon with a size of c. 700 ha (7 km²). Simulations included 4 km² of representative landscapes around the lake, 4 km² uphill and 4 km² in the dunes at the sea shore. Here only the uphill simulations are shown, in the coastal dunes *Quercus ilex* is dominant in all scenarios and *Abies alba* almost lacking. The numeric climatic scenario for the simulations (summer temperature trends and variability) comes from chironomid records from the nearby Tuscan-Emilian Apennines (c. 60 km distance). The model suggests that with no browsing and small, infrequent fires *Abies alba* would have expanded during the Late Holocene in response to decreasing summer temperatures and the resulting increasing summer moisture availability. With low browsing and small, infrequent fires (corresponding to natural conditions, with herbivores controlled by carnivores) *Abies alba* would have been less abundant but the Late-Holocene expansion would have occurred as well. If browsing becomes moderate, *Abies alba* collapses within c. 500 years and *Quercus ilex* expands. The palaeo record shows that the final collapse of *Abies alba* at the Mediterranean Sea shore was strongly linked to human disturbance, mainly fire and agricultural activities including grazing. Modified from Henne et al. (2013).

the Neolithic, *Sporormiella*-inferred herbivory increased only together with fires, suggesting that the beneficial effect of burning on grazing was shorter (maximum of c. 15 years; Schwörer et al., 2015). This difference might be explained by the increasing (pollen-inferred) creation of open land, which with progressing time, may have demanded less ignition to maintain good pastureland.

Sensitivity studies on the basis of dynamic, process-based vegetation models (e.g. LandClim) allow backcasts or retrodictions of past browsing activities that can be compared with the palaeo evidence (e.g. Henne et al., 2013; Fig. 4.10.3-2). If climate reconstructions and other disturbance agents are included in such simulations, they offer the unique possibility of disentangling and identifying the true forcing factors of long-term vegetation change (e.g. climate v. browsing) under controlled conditions that reflect the current state of knowledge (e.g. in regard to life traits of species). Such a procedure can be considered equivalent to experiments in the past with the ultimate goal of improving projections of future vegetation change under global-change conditions.

4.10.4

Frost, drought, flood, waterlogging, avalanches, wind-throw, volcanic eruptions, landslides, rockfalls and pathogens

Frost and drought are weather extremes that may strongly affect vegetation composition (Ellenberg, 2009). Plant species have different tolerances to frost and drought. For instance, *Fagus sylvatica, Castanea sativa, Fraxinus excelsior* and *Abies alba* are sensitive to late spring frost, while *Pinus sylvestris, P. nigra, Betula pendula* and *B. pubescens* are not (Ellenberg, 2009). Similarly, *Picea abies, Fagus sylvatica, Alnus* sp. and *Populus nigra* are very sensitive to drought, while *Pinus* sp., *Quercus* sp., *Ulmus* sp., *Tilia* sp., *Acer* sp., *Carpinus betulus, Castanea sativa, Ostrya carpinifolia, Corylus avellana* and *Abies alba* are less affected (Ellenberg, 2009). Accordingly, it has been hypothesized that late spring frost and/or drought inhibited the expansion of *Fagus sylvatica* in Europe during the Early and Mid Holocene, when climate was more continental (Tinner and Lotter, 2001, 2006; Giesecke et al., 2007). Instead, recurrent summer drought during the summer insolation maximum around 10 000 years ago may have caused high abundances of *Corylus avellana* in Europe during the Early Holocene (Tinner and Lotter, 2001; Finsinger et al., 2006). During the Late Holocene, drought may have induced collapses of drought-sensitive *Alnus* in boreal and temperate Europe (Stivrins et al., 2017; Latałowa et al., 2019). In the Alps, Early- and Mid-Holocene summer drought likely determined the vegetation composition at the treeline (Tinner and Kaltenrieder, 2005; Heiri et al., 2006; Schwörer et al., 2014b). Similarly, Early- and Mid-Holocene summer drought affected southern European vegetation, delaying afforestation and promoting natural maquis and other drought-adapted vegetation types during the generally dry Early Holocene (e.g. Tinner et al., 2009; Connor et al., 2013; Beffa et al., 2016; Curry et al., 2016; Budsky et al., 2019; Morellon et al., 2018).

While clear links exist between Holocene vegetation clearances for land use and increased flood (e.g. Vannière et al., 2013; Joannin et al., 2014; Brisset et al., 2017) and avalanche activity (Fouinat et al., 2018), only little is known about the effects of wind-throw, waterlogging, flood and avalanche disturbance on long-term vegetation composition, structure and dynamics (Bradshaw and Hannon, 2004). Plant and animal remains found underneath volcanic deposits document the composition of communities destroyed by the Laacher See eruption c. 12 900 years ago, however, it is unlikely that this disturbance had long-lasting effects on extra-local vegetation (Litt et al., 2003). Landslides may have advantaged disturbance-adapted shrubs such as *Corylus avellana* (Simova et al., 2019). It has been hypothesized that wetlands and increasing landscape heterogeneity, which originated from landslides, may have influenced the local course of vegetation history, by promoting suitable habitats for species that were regionally rare or absent (Simova et al., 2019). Lake deposits in the Alps document the impact of a huge rockfall on local vegetation around 9600 cal yr BP in the Kander valley in the Bernese Alps (Tinner et al., 2005b). Early-Holocene forests were severely disturbed by the rockfall event, even at several km distance. The existing mixed *Ulmus-Tilia-Corylus* forest collapsed almost entirely in response to the rockfall, forest fires strongly increased and ferns and grasses expanded. Subse-

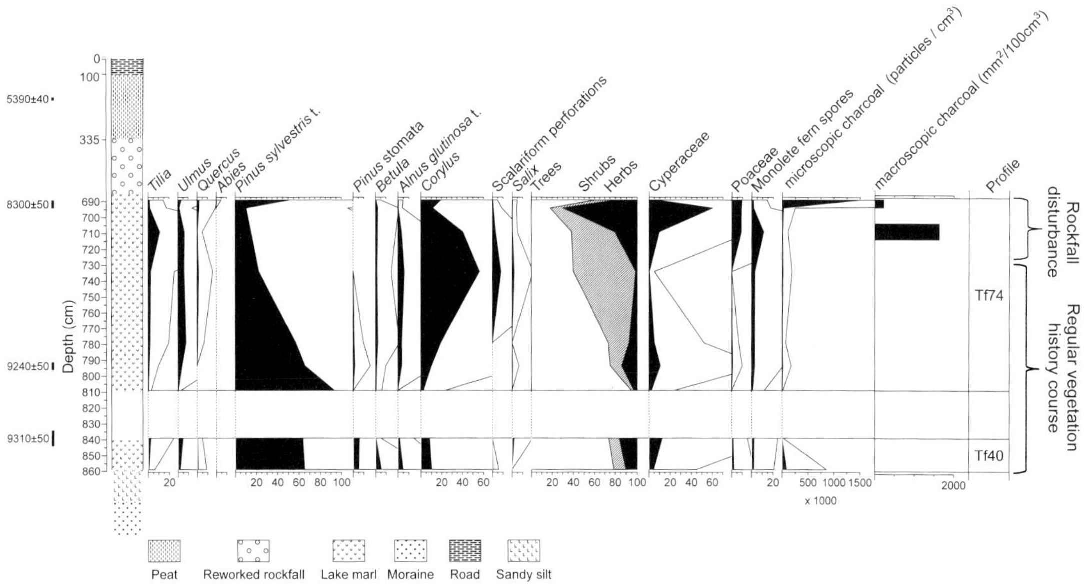

Fig. 4.10.4-1. Pollen percentage and charcoal concentration diagram of Tellenfeld, Bernese Alps, Switzerland. The former lake of Tellenfeld was destroyed by reworked rockfall material (see lithology), likely deposited by a huge debris flow covering an area of c. 2 km². It likely resulted from the outburst of a rockfall lake uphill in the Kander valley. The rockfall of the Kander valley occurred c. 10 km uphill of Tellenfeld and comprised ca. 0.8 km³ of material, destroying >5 km² of vegetation. On top of the reworked rockfall material a mire developed (peat): the available radiocarbon date shows that no rockfall material has reached this outermost downhill location of rockfall debris during the past 6000 years, unambiguously dating the main event to before 6000 cal yr BP. In the sediments just below the reworked rockfall material a drastic anomaly in Holocene vegetation dynamics occurred at around 9600 cal yr BP. Local fires (macroscopic charcoal peak) drastically increased and disturbance-adapted ferns (monolete spores) expanded. This change was followed by the almost complete collapse of forests (arboreal pollen <40%) around the site. Subsequently, an anomalous expansion of *Pinus sylvestris* forests (when compared to adjacent sites) occurred, before the catastrophic deposition of reworked rockfall material finally destroyed the lake around 9100 cal yr BP. The reconstructed Early-Holocene vegetation anomaly resembles observed environmental and vegetation patterns after rockfalls with fires (spark ignition), rapid expansions of opportunistic ferns and grasses as well as the establishment of late-successional *Pinus sylvestris* forests on boulders. Modified from Tinner et al. (2005b).

quently, *Pinus sylvestris* became very abundant, as often observed after such catastrophic events (Fig. 4.10.4-1; Tinner et al., 2005b).

Pathogens may infest plants leading to population collapses, as impressively shown by the disappearance of the formerly very important North American *Castanea dentata* forests during the 20th century (Rigling and Prospero, 2018). This example illustrates that pathogen attacks are usually species or genus specific. Therefore, in the palaeo record, pathogen attacks may be discerned from less species-specific disturbances such as fire, drought, frost, browsing or grazing (e.g. Tinner et al., 2000). More elegantly, pathogen infestations might be reconstructed by analysing remains of the pathogens that are preserved in the natural archives, however, such direct evidence remains very rare (Waller, 2013). Comparison with the dynamics of modern pathogen infestations (e.g. Flynn and Mitchell, 2019) led to the conclusion that the steep fall in *Ulmus* pollen values observed during the Mid Holocene was probably a result of a combination of prehistoric human activities and a tree pathogen (Peglar, 1993b; Grosvenor et al., 2017; for more details on *Ulmus* see 4.4.3). The Mid-Holocene elm decline was very rapid with pollen values halving within 5 years (Birks and Tinner, 2016). Such rapid responses are characteristic for pathogen attacks and similar to those observed after fire disturbance (e.g. Colombaroli et al., 2007; Birks and Tinner, 2016). Pathogens such as bark beetles may have affected *Picea abies* in the past as they do in the present, however, the available evidence remains elusive (van der Knaap et al., 2020).

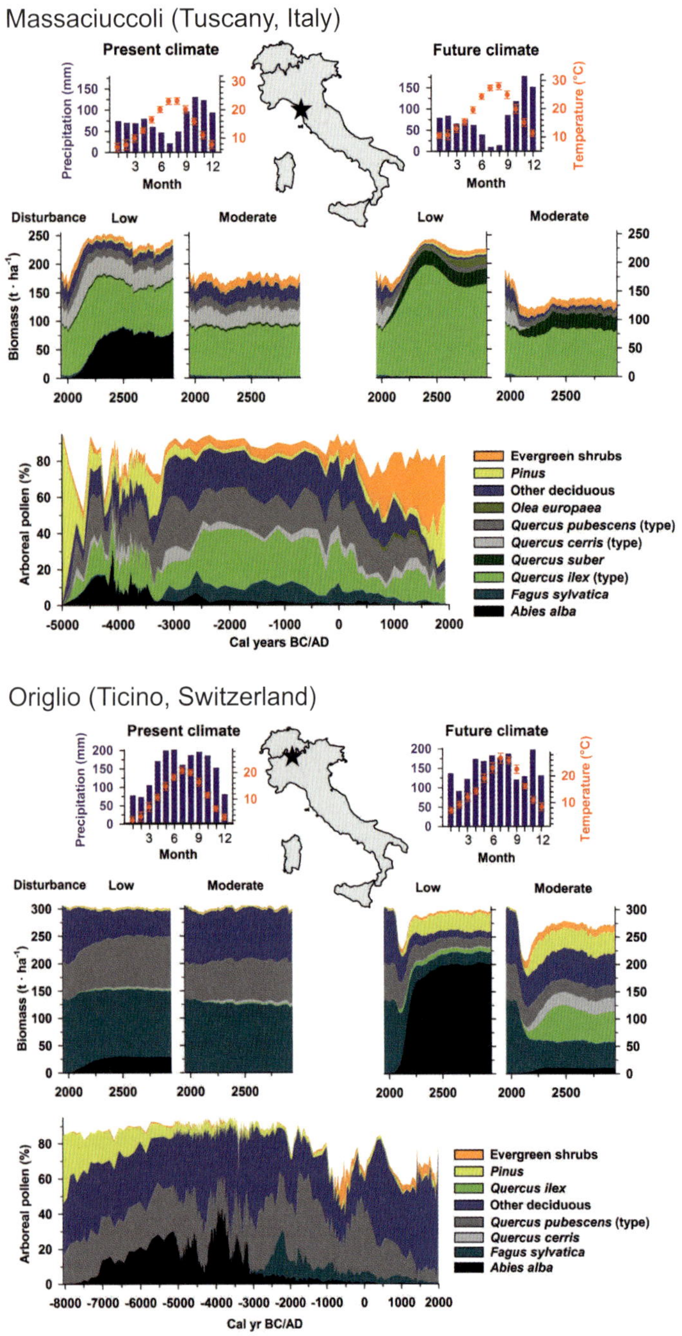

Fig. 4.10.5-1. Holocene pollen data from the two sites Lago di Massaciuccoli (Tuscany, Italy) and Lago di Origlio (Ticino, Switzerland) and simulations of future vegetation dynamics around the sites under different disturbance scenarios. Under current climatic conditions and reduced fire and browsing disturbance, extinct Mediterranean *Abies alba-Quercus ilex* (Massaciuccoli) as well as mixed warm-temperate *Abies alba*-deciduous broadleaved forests reappear after c. 100–200 years from now (AD 2100–2200). If instead the disturbance regime remains moderate under present climatic conditions these extinct communities are unable to recover at either site. Under future climates c. 4°C warmer, forests at Massaciuccoli become exclusively dominated by evergreen broadleaved species (e.g. *Quercus ilex, Q. suber*), regardless of the disturbance regime, which strongly determines biomass availability. An ecological surprise emerges at Lago di Origlio under future climates and reduced disturbance, with *Abies alba* becoming the dominant species under subtropical humid conditions after c. 100–200 years around AD 2100–2200. If instead future climate disturbance remains moderate, biomass strongly declines at Massaciuccoli and *Quercus ilex* expands at Origlio, forming unprecedented mixed evergreen-deciduous broadleaved forests in which *Abies alba* is far less abundant. On the basis of the palaeoecological record these simulations seem realistic (see comparison with Mid Holocene c. 8000–6000 years ago when summer temperatures were higher and disturbance lower than today). On the other hand, on the basis of current species distribution and neglecting Holocene disturbance legacies, such dynamic simulations would appear highly unlikely and thus be rejected. Modified from Henne et al. (2015).

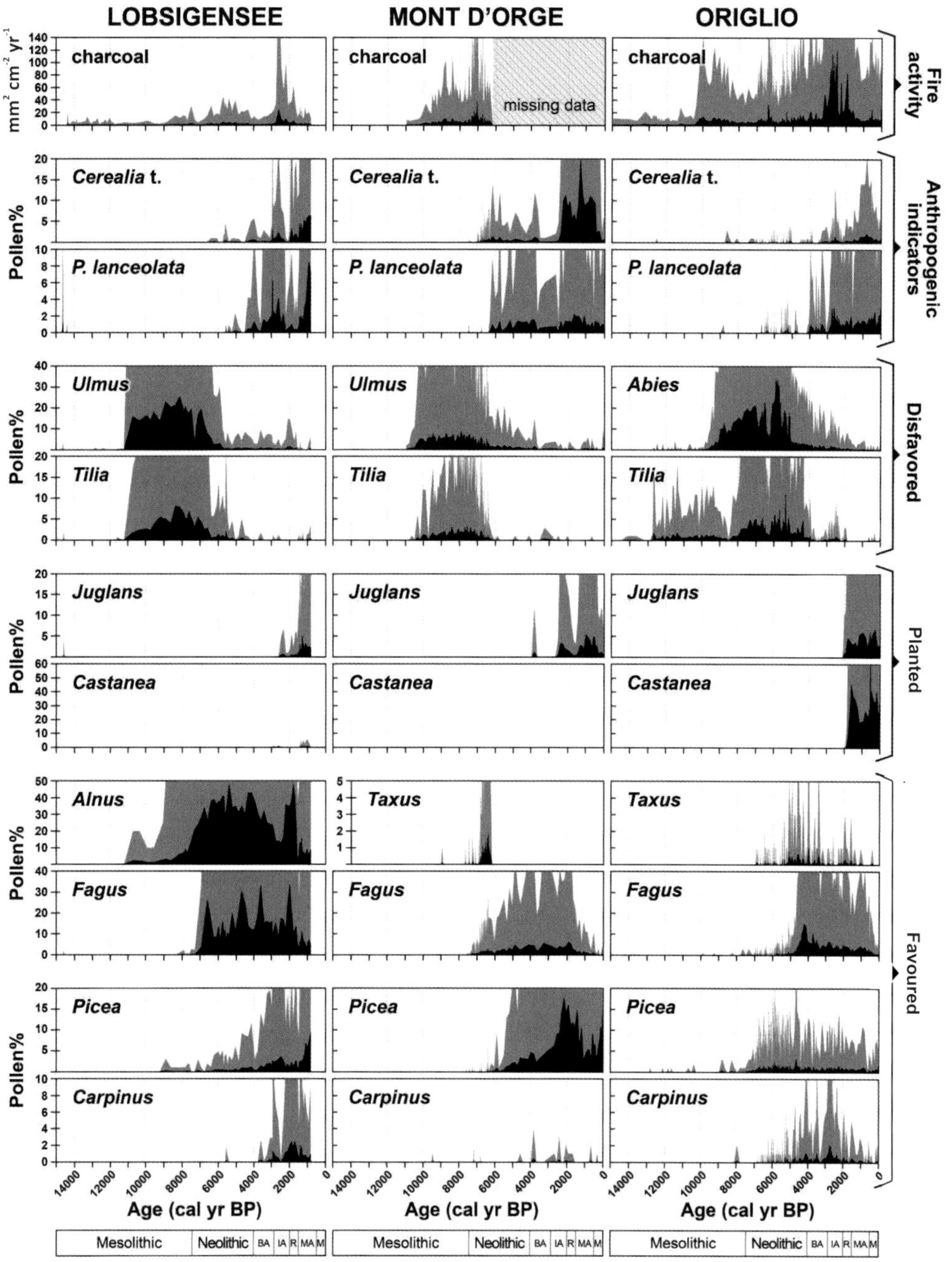

Fig. 4.10.5-2. Selected pollen types in percentages and charcoal influx from three study sites in Switzerland: Lobsigensee (Swiss Plateau), Lac du Mont d'Orge (Valais) and Lago di Origlio (Ticino). The close link between anthropogenic indicators (Cerealia type, *Plantago lanceolata*) suggests that after c. 5000 cal yr BP most fires were ignited by humans and closely related to land use (slash-and-burn). This caused a strong reduction of some tree taxa (e.g. *Ulmus, Tilia, Abies*), while others were strongly favoured (e.g. *Fagus, Alnus, Picea, Taxus, Carpinus*). Over the millennia, human impact and fires caused competitional advantages or disadvantages, substantially altering the realized environmental niches of the involved plant species. Modified from Conedera et al. (2017).

4.10.5

Summary

Established proxies such as macroscopic and microscopic charcoal or spores from dung fungi allow better assessments of fire, grazing and browsing disturbance. New proxies such as sedimentary aDNA and environmental DNA (eDNA) allow novel insights into species and population dynamics that were so far difficult or impossible to reconstruct by optical approaches. Moreover, dynamic disturbance impact modelling provides the unique possibility to conduct studies under controlled conditions, for instance to assess past and future tipping points or sensitivities as well as the spatial organisation of species and communities (e.g. Henne et al., 2013; Tinner et al., 2013; Figures 4.10.3-2, 4.10.5-1).

Natural disturbance agents affected vegetation before the advent of agriculture, for example during previous interglacials or the pre-Neolithic Holocene. However, natural interglacial disturbance variability usually did not cause major community or ecosystem changes, neither in regard to structure (e.g. forest cover), nor in regard to composition (e.g. broad-scale establishment of fire-adapted vegetation). Rather stable Holocene environmental conditions allowed late-successional species to dominate for centuries to millennia under natural disturbance regimes. The available evidence suggests that in some cases disturbance may have contributed to rapid vegetation adjustments to climate change, for example, *Fagus sylvatica* that replaced *Picea abies* in southern Scandinavia possibly in response to decreased Late Holocene seasonality and increased fire disturbance (Ohlson et al., 2017). Similarly, in temperate central and southern Europe, *Fagus sylvatica* was advantaged by both a step-wise climatic change towards more oceanic and humid conditions during the Late Holocene as well as excessive human disturbance that caused the exclusion of its main disturbance-sensitive competitors *Abies alba, Tilia, Fraxinus, Acer,* and *Ulmus* (Tinner and Lotter, 2006; Giesecke et al., 2007; Tinner et al., 2013). In the thermomediterranean vegetation natural fire disturbance probably contributed to maintain fire and drought-adapted communities (e.g. *Erica arborea, E. scoparia, Pistacia lentiscus* stands in coastal Sardinia or Sicily) when climatic conditions were very dry and warm.

With the introduction of farming, anthropogenic fire, browsing and grazing strongly affected European vegetation structure and composi-

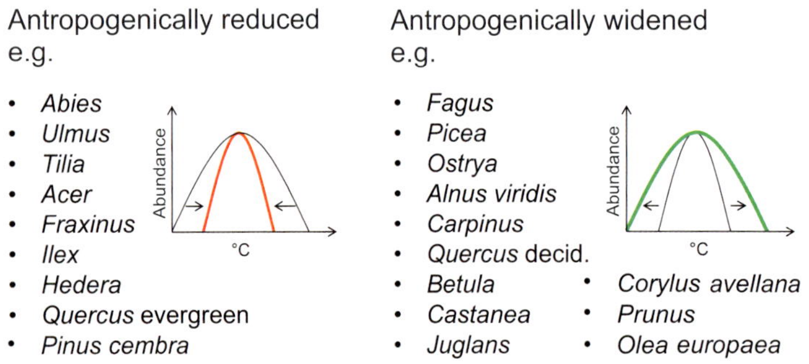

Fig. 4.10.5-3. Legacy of past anthropogenic disturbance on extant realized climatic niches of European plant taxa. Most changes occurred during prehistory or antiquity and have thus a millennial-long history. Reduced and widened realized climatic niches are not only important to understand today's distribution of plant species in Europe, they may also control future vegetation responses to climatic change. Assuming a dynamic equilibrium with today's climatic conditions, for example for species distribution modelling (SDM), results in an overestimation of the future climatic response capacity or resilience of species with widened niches. In contrast, the future climatic response capacity or resilience of species with anthropogenically reduced niches is markedly underestimated. Dynamic, process-based modelling relying on ecophysiological traits of the species involved can be used to overcome future effects of anthropogenic legacy issues (see Fig. 4.10.5-1).

tion. Disturbance-adapted ecosystems such as heathlands, maquis, garrigue, grasslands, fields, parklands, and monospecific woodlands became widespread in Europe, causing huge forest losses and a marked impoverishment of forest vegetation (Tinner and Ammann, 2005). Some formerly widespread forest communities collapsed completely, for instance the species-rich, warm-loving mesomediterranean and submediterranean mixed silver fir forests of southern Europe. Such local extinctions over wide environmental gradients imply that excessive disturbance such as anthropogenic burning or browsing affected the realized niche of species, creating millennial long legacies that cannot be recognized without palaeoecological analyses. Monospecific forests of, for example, *Picea abies, Quercus* sp., *Fagus sylvatica, Pinus* sp. and *Castanea sativa* that are widespread in Europe have been created by humans over the millennia (Fig. 4.10.5-2). Land use may have stretched the realized niches of these favoured species to their fundamental niche limits, while the realized niches of disfavoured species may have been strongly reduced (Fig. 4.10.5-3). Because today's occurrence of some important European tree species is co-controlled by past land use, in some cases reaching back to the Neolithic period (Fig. 4.10.5-2), correlative models that disregard land use and that assume that the today's ranges of the species are in equilibrium with today's climate (species distribution models, niche models, bioclimatic models) may fail to anticipate future species responses to climate change (Tinner et al., 2013; Ruosch et al., 2016). Instead, using palaeo-validated dynamic, process-based models that include land use and disturbance may help to produce better simulations and scenarios of species and ecosystem responses to future disturbance and climate change (Fig. 4.10.5-1; Schwörer et al., 2014b; Henne et al., 2015). Disturbances such as drought and late frost may have markedly influenced vegetation composition in Europe: however, in the absence of high-frequency reconstructions for drought and frost this issue remains speculative. Similarly the effects of other disturbance agents are poorly resolved, among them pathogens seem of particular importance given that in some cases there is evidence of high relevance for some tree species (e.g. *Ulmus* decline).

Records from previous interglacials unambiguously show that despite recurrent disturbance impacts, closed forests prevailed in Europe under natural conditions (e.g. Wegmüller, 1992; Tzedakis et al., 1997; Reille et al., 1998; Svenning, 2002; Kühl and Gobet, 2010). This does not mean that disturbance agents such as fire or megaherbivores had no relevance for natural forest structure or composition (Sandom et al., 2014). The density of past natural forests, which in part predates the appearance of *Homo sapiens* in Europe, can be assessed by comparison of fossil arboreal pollen abundances with modern pollen-vegetation relations (e.g. Welten, 1950; Wright et al., 1967; Beer et al., 2007; Canellas-Bolta et al., 2009; Lambert et al., 2017). From such calibrations it emerges that forests covered most areas of Europe during the past interglacials, when modern humans were absent and the European interglacial megafauna still rich (e.g. forest elephant, hippopotamus; see chapter 3). Only during glacial periods, when conditions were particularly cold and dry as well as after the onset of wide-scale agricultural activities during the past 5000 years, did parklands and grasslands become important at the continental scale. A regional exception is the natural steppe and forest steppe forming the pontic-pannonic vegetation type of south-eastern Europe that likely remained unforested during the entire Holocene (Kremenetski et al., 1999; Gobet et al., 2017; Makhortykh and de Capitani, 2017). Locally, drought disturbance may have contributed to the survival of steppic species in the forested European environments (e.g. Hejcman et al., 2013; Pokorny et al., 2015).

The available evidence suggests that disturbances were significant agents of Holocene European vegetation change. Unfortunately, with the exception of fire, grazing and browsing, only a few studies are currently available that unambiguously document the effects of past disturbance on European vegetation history. Future disturbance-focused research strategies will require very high temporal resolutions and new proxies (e.g. aDNA for insect infestations) that in combination with established proxies may deliver new insights into long-term disturbance ecology. Such new insights are greatly required for validating dynamic, process-based models to gain better and more realistic assessments of future vegetation dynamics under global-change conditions (Fig. 4.10.5-1).

4.11

Causes of glacial-interglacial vegetation dynamics

Willy Tinner[1,2], Brigitta Ammann[1,2], Gerhard Lang[1]

[1] Institute of Plant Sciences, University of Bern, Bern, Switzerland
[2] Oeschger Centre for Climate Change Research, University of Bern, Bern, Switzerland

4.11.1

Glacial and interglacial stages

Only a few examples of arboreal plants are discussed in this book and even less attention is brought to herbaceous species. Refining the existing substantial gaps will need further research efforts, for example with regard to taxonomic resolution of herbaceous species that may come with new techniques, such as aDNA approaches. Nevertheless, our overviews in chapters 2, 3 and 4 already show an extremely varied picture of the spread of floristic elements during the Holocene and also during other interglacials. In the end, each plant species has its own history, including the herb flora. It is also clear that similar ranges of the species today do not mean similar range histories (Lang, 1971). Roughly, with regard to the glacial–interglacial cycles, the conditions that controlled the occurrence of plant species over such long periods can be divided into five groups (see Fig. 4.11-1; Birks and Tinner, 2016).

First, during the **cryocratic** ('cold-ruled') stage, which corresponds to glacial conditions, arctic and alpine species thrived locally in ice-free areas including on nunataks. Most plants occurring north of c. 45°N were herbs or dwarf shrubs. Temperate and boreal taxa survived in the southern European peninsulas south of c. 45°N, occasionally boreal trees may also have survived in microrefugia north of c. 45°N. Mediterranean tree stands were confined to favourable areas in southern Europe south of c. 40°N, where they co-occurred with temperate and boreal species. North of the Alps (or c. 45°N), soils were exposed to permafrost and solifluction.

Second, during the subsequent **protocratic** ('initial-ruled') stage at the onset of the interglacials, some species, particularly boreal trees, spread when temperatures were still lower than average Holocene temperatures. Examples of protocratic species are *Betula pendula, B. pubescens, Pinus sylvestris, Larix decidua, Pinus cembra, Picea abies, Sorbus aucuparia* and *Juniperus communis* depending on the location in Europe north of c. 45°N. In southern Europe, temperate deciduous broadleaved trees (e.g. *Quercus, Ulmus, Ostrya, Carpinus*) formed open forests, in which subtropical evergreen broadleaved trees (e.g. *Olea europaea, Quercus ilex, Pistacia lentiscus*) were present but not abundant. North of the Alps, soils were shallow, with reduced water holding capacities.

Third, during the subsequent **mesocratic** ('mid-ruled') stage, biomass peaked towards the end, under maximum interglacial temperatures and rather high moisture availability, similar to or slightly higher than those of the Holocene Thermal Maximum (HTM). Temperate deciduous broadleaved taxa such *Corylus, Quercus, Ulmus, Tilia, Fraxinus* and *Acer* expanded to reach their maximum European range limits. They built conspicuous populations in southern, western, central, eastern and parts of northern Europe within the first millennia of the mesocratic interglacial stages, forming temperate forests. In southern Europe, Mediterranean and warm-temperate conifers (e.g. *Pinus halepensis, P. pinea, P. pinaster, Abies alba,*

A. cephalonica) expanded into the evergreen and deciduous broadleaved forests (e.g. Tzedakis and Bennett, 1995; Allen et al., 2002; Tzedakis, 2007; Allen and Huntley, 2009). During this stage, forests were closed, and herbs and shrubs confined to locally disturbed areas (from e.g. fire, floods, drought, frost). North of the Alps, soils developed such as brown earths that replaced raw soils such as regosols, now offering conditions like those south of the Alps (or c. 45°N).

Fourth, during the **oligocratic** ('poorness-ruled') stage, coniferous trees expanded north of c. 45°N and species such as *Abies alba, Picea abies, Carpinus betulus* and *Fagus sylvatica* became important, depending on the interglacial of interest. Peatlands and heaths expanded (Fig. 4.11-1). These vegetation shifts were likely caused by declining temperatures and increasing humidity. Here, the behaviour of *Carpinus betulus* is of interest: in some early and middle Pleistocene interglacials it belonged to the mesocratic taxa in the temperate biome. Similarly, some taxa behaved as mesocratic in southern Europe (e.g. *Abies*), while north of c. 45°N they had a more oligocratic character. For the Holocene, on the basis of macrofossil finds, we know that the same species had such complex behaviours (e.g. *Abies alba*), likely reflecting their broad climatic niches and high competitiveness under very dissimilar climates (from nemoboreal to mesomediterranean). The example of *Pinus*, however, with its subtropical and boreal species with very different climatic niches shows that several species may also have caused an apparently dissimilar behaviour of the same genus along European spatiotemporal climatic gradients. In southern Europe, mesophilous taxa such as *Fagus*, *Alnus* and *Abies* gradually replaced evergreen broadleaved trees (e.g. *Quercus ilex*) during this fourth oligocratic stage, while continental broadleaved deciduous trees (e.g. *Tilia, Ulmus, Quercus*) remained important (Birks and Tinner, 2016). Towards the end, soils became increasingly acidified, especially north of the Alps, likely in response to raised precipitation, cooling and the resulting vegetation shifts (e.g. *Picea abies* or *Calluna vulgaris* expansions: Tibbett et al., 2019).

Finally, during the fifth **telocratic** ('end ruled') stage, further climatic cooling severely affected vegetation composition all over Europe. In western, central and eastern Europe, deciduous broadleaved taxa such as *Tilia*, *Ulmus* and *Quercus* declined together with frost-sensitive, evergreen broadleaved *Ilex* and *Hedera*. In southern Europe forest cover declined. Some frost-resistant temperate taxa such as deciduous *Quercus* remained important, while frost-sensitive temperate conifers such as *Abies* collapsed and steppic vegetation expanded. Subsequently, after the telocratic stage a new cryocratic phase began, which became increasingly dry (Fig. 4.11-1).

According to most authors (e.g. Birks and Birks, 2004) interglacials cover all four non-cryocratic phases, while according to others, the interglacials are restricted to the warm phases similar or warmer than the Holocene, that is, the mesocratic and oligocratic stages (e.g. Tzedakis, 2007). If compared in more detail to the Holocene, this latter interglacial definition and subdivision may include an early, mid and late (interglacial or Holocene) phase, while the protocratic and telocratic stages would correspond to Late-glacial conditions with pre-temperate or post-temperate climates in western, central and eastern Europe as well as pre-mediterranean and post-mediterranean conditions in southern Europe (Tzedakis, 2007). A subdivision of Mediterranean mesocratic and oligocratic interglacials into three periods (Tzedakis, 2007), with Mediterranean forest followed by deciduous forest, and succeeded by montane coniferous forest, may apply to submediterranean and in part to mesomediterranean conditions, but be challenging for the thermomediterranean vegetation type, in which evergreen broadleaved species may have dominated throughout the interglacials (see chapter 4.3.2). Regardless of which stages belong to interglacials, they can be distinguished from interstadials by the mesocratic and oligocratic phase, with European vegetation dynamics in response to summer temperatures similar or higher than those of the Holocene (Jessen and Milthers, 1928). Although in glacial-interglacial classifications much attention is paid to the dynamics during the interglacial period, the duration of interglacials was substantially shorter than that of glacials, so that the Quaternary is mostly characterized by cold to cool, cryocratic to protocratic and telocratic conditions (chapter 3, Fig. 3.3-1; Birks and Willis, 2008).

Under natural conditions, regionally varying dynamics of species and communities in time and space were likely caused by four main factors and their various ways of interaction: (1) location

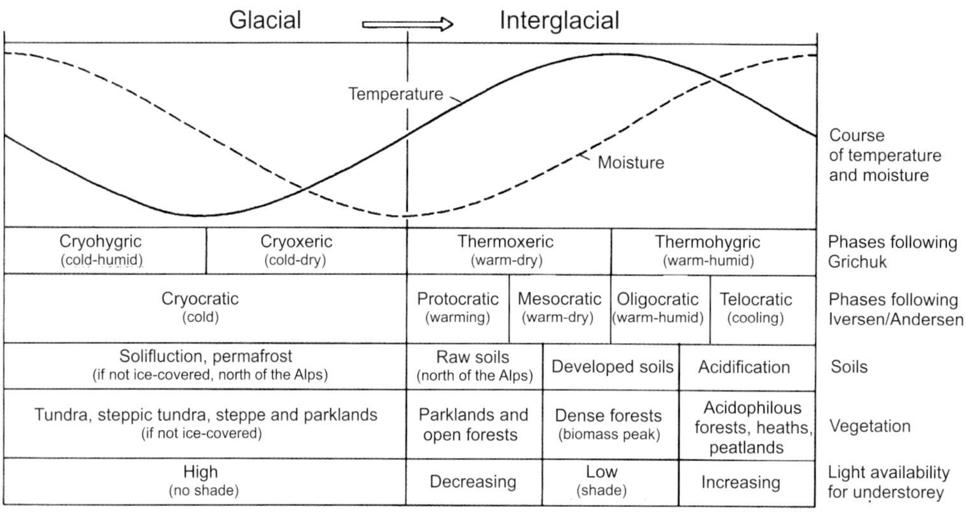

Fig. 4.11-1. Schematic illustration of the Glacial-Interglacial cycle. Following Iversen, 1958; Grichuk, 1964; Andersen, 1966a; Birks and Birks, 2004.

of refugia, (2) spatial spread, survival, extinction and speciation of plants, (3) biotic processes such as competition for resources according to the different niches of species and (4) varying environmental conditions, in particular climate, soils and disturbance. Later on, during the current interglacial, human impact emerged as a consequence of production economies involving farming, metallurgy and fossil-fuel emissions.

4.11.2

Refugia

Quaternary refugia are areas or regions during glacial-interglacial cycles in which the species reach their maximum spatial range contraction (Stewart et al. 2010). Following this definition boreal, temperate and Mediterranean species are in their refugia during cold glacial periods, while arctic-alpine species are in their refugia during warm interglacial periods (Fig. 4.11-1). Unambiguous evidence exists for the presence of refugia for temperate species during the Last Glacial Maximum (LGM) and other glacials in the southern peninsulas of Iberia, Italy, and the Balkans. All southern refugia have mountain complexes that offered mid-elevation conditions with enough moisture supply (Beug 1967, 1975a; Bennett et al. 1991; Willis 1994; Feurdean et al. 2007b; Birks and Willis 2008; Fig. 4.11-2). In addition moist localities in lowlands, for example, valley bottoms, likely also acted as refugia of temperate taxa (Beug 1975a; Kuneš et al. 2008a; Kaltenrieder et al., 2009; Samartin et al., 2016; Gubler et al., 2018; Fig. 4.11-3). The refugia in the southern European peninsulas were likely of small spatial extent and confined to sheltered positions offering enough temperature and moisture for temperate and Mediterranean tree species, as shown by the low to very low abundances of such taxa in pollen records that cover the entire LGM (see chapter 4.3; e.g. Monticchio, Lagaccione or Costa in Italy) and other full glacial periods. In the maps showing the initial establishment of plant populations (usually empirical pollen limit) and their spread throughout Europe, we assume that those sites with pollen evidence dated at 20 000–18 000 cal yr BP (black to brown colours) acted as LGM refugia (see e.g. Fig. 4.4.3-1). Phylogeographic maps provide genetic information that can be coupled to the fossil survival evidence in southern Europe (e.g. for *Fagus*: Magri et al. 2006; see chapter 4.4.3 for angiosperm genera), however, so far, comprehensive palaeogenetic evidence that can be directly linked to the survival of temperate or Mediterranean trees in their ice-age refugia is missing. Although palaeogenetic evidence regarding the survival of boreal trees such as *Picea* in northern Scandinavia outside the LGM ice sheet is a matter of debate (Birks et al., 2012a; Parducci et al., 2012; Vorren et al., 2013), it shows

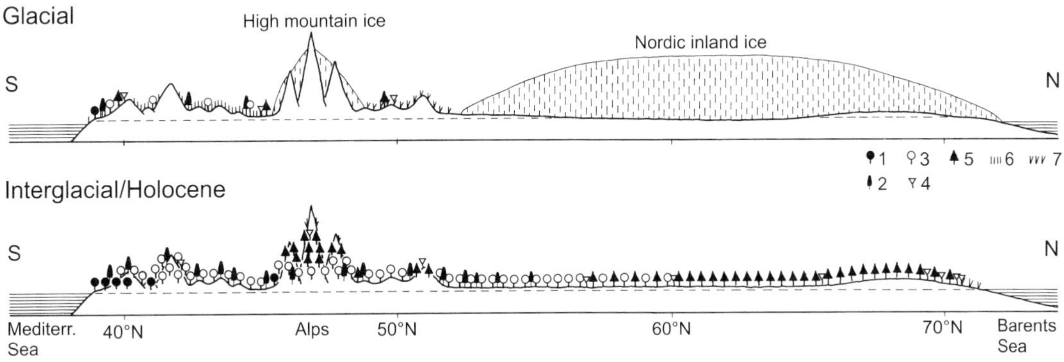

Fig. 4.11-2. Vegetation composition during the glacial–interglacial cycles. The schematic vegetation conditions during the Middle and Young Pleistocene glacials (top) and interglacials including the Holocene (bottom) are shown along a south–north transect in Europe. Plant functional types: 1 = Mediterranean evergreen broadleaved trees (e.g. *Quercus ilex, Olea europaea*), 2 = Mediterranean and temperate needleleaved trees (e.g. *Pinus halepensis, Abies alba*), 3 = Temperate deciduous broadleaved trees (e.g. *Quercus robur, Tilia cordata*), 4 = Boreal deciduous broadleaved trees (e.g. *Betula pendula*), 5 = Boreal needleleaved trees (e.g. *Picea abies, Pinus sylvestris*), 6 = Herbs and grasses of the steppe; 7 = Herbs and grasses of the steppe tundra and tundra. Elevation c. 100x exaggerated. During the glacials, Mediterranean and temperate trees were confined to small areas south of c. 45°N (Alps) where moisture and temperature was high enough to permit their growth. Boreal trees also occurred north of the Alps with northernmost positions around c. 50°N. During the protocratic phase of the interglacials, trees and other plants expanded rapidly out of their refugia, tracking the melting of the ice. At the onset of the mesocratic phase migrational lags had decayed and forest vegetation reached a dynamic equilibrium with climate, probably within centuries to at most a few millennia, depending on the distance from the refugia. Following van der Hammen et al. (1971), Birks and Willis (2008) and Tzedakis et al. (2013).

the potential of palaeogenetic approaches, if, for example, they are rigorously applied to identify genetic dynamics of temperate and Mediterranean trees during the LGM (Tzedakis et al., 2013). Similar controversies are also known from the Alps, with regard to the LGM survival of trees on nunataks above today's timberline elevations (Carcaillet and Blarquez, 2017; Finsinger et al., 2019).

While there is consensus that Mediterranean subtropical evergreen broadleaved trees, shrubs and palms (e.g. *Quercus ilex, Laurus nobilis, Olea europaea, Pistacia lentiscus, Chamaerops humilis*) survived the LGM only south of c. 40°N and that boreal trees survived as far north as c. 50°N south of the ice sheets, the glacial occurrence of temperate trees north of 45°N is debated (e.g. Birks and Willis, 2008; Tzedakis et al., 2013). Similar discussions are also known for so-called 'intermediate' Late-glacial refugia, which may have allowed temperate species to spread to areas north of the Alps before the onset of the Holocene (Giesecke, 2016; Robin et al., 2016). The existence of so-called 'cryptic' glacial refugia of temperate trees north of the Alps has been hypothesized on the basis of ecological analogies (Birks and Willis, 2008) and macroscopic charcoal finds that were taxonomically determined through wood anatomy (or anthracology: Willis and van Andel 2004). This approach usually allows the identification to the genera although *Picea* and *Larix* are sometimes hard to separate (Schweingruber, 1978, Schweingruber et al., 2006). Some of the macroscopic charcoal finds were radiocarbon-dated and generally provided radiocarbon ages older than the LGM (i.e. older than 23 000–19 000 cal yr BP). In their seminal work, Willis and van Andel (2004) unambiguously demonstrate the contrast between the middle pleniglacial (59–24 ka BP, MIS 3), with a highly diverse woody flora including temperate trees north of 45°N and the LGM, when boreal trees such as *Pinus sylvestris, P. cembra, Juniperus, Picea* and *Salix* were still growing north of 45°N, but temperate trees were confined to latitudes south of c. 45°N (Willis and van Andel, 2004). Such anthracological studies confirm the available pollen-inferred evidence (chapter 4.4) that the spatial contraction of the ranges of the temperate trees reached its apex during the LGM, likely confining them to their spatially isolated refugia in southern Europe (e.g. Pons and Reille, 1988; Allen et al., 1999; Magri, 1999; Kaltenrieder et al., 2009). Further analyses on the basis of life-history traits, habitat preferences, present geographical distributions and genetic studies showed that taxa with large and heavy fruits or seeds were limited to southern Europe, while wind-dispersed, habitat-generalists which reproduce vegetatively may have also survived in eastern and northern Europe (Bhagwat and Willis, 2008).

In contrast to the rather well-studied LGM-refugia of Mediterranean, temperate and boreal taxa, less is known about the refugia of arctic-alpine taxa during the interglacials. The difficulties lie mostly in the rather low taxonomic resolution of the pollen morphology. This gap can be partly filled by plant macrofossil analyses which reach higher taxonomic resolutions. For instance, of 70 arctic-alpine species recorded at LGM and Late-glacial sites, only seven species were also found as pollen (Birks and Willis, 2008). Another difficulty comes from the position of the refugia at the top of the mountains or in the far north. However, because we can use the Holocene as an analogue for other interglacials we know that isolated populations can occur at high elevations that act as refugia, when compared to the wide ranges during the much longer cold periods of the Quaternary.

The finding that temperate species likely survived the ice-age maxima (e.g. LGM) exclusively in southern Europe has important implications: present-day populations of temperate trees outside southern Europe are young and thus likely to contribute less to genetic diversity (Petit et al., 2003; Tzedakis et al., 2013). This refugial biodiversity legacy of the glacial–interglacial cycles is underscored by the occurrence of temperate species and genera that are restricted to latitudes south of c. 45°N (e.g. *Fraxinus ornus, Quercus cerris, Q. pyrenaica, Q. faginea, Q. trojana, Abies nebrodensis, A. cephalonica, A. pinsapo, Ostrya carpinifolia, Celtis australis, Zelkova sicula, Z. abelicea*). Taken together, the available palaeoecological evidence implies that, at the continental scale, southern European temperate ecosystems can be considered as long-term sanctuaries of European biodiversity, which has important implications for biodiversity and nature conservation.

4.11.3

Spread of species

Can the temporal sequence of taxa during the interglacials be explained by different dispersal velocities (or migration rates, e.g. in m yr^{-1}) and taking into account the location of refugia? This was one of the first questions posed by early macrofossil and pollen analysts at the onset of the 20th century (e.g. C.A. Weber, 1914; H.A. Weber 1918; Firbas, 1923; Rudolph and Firbas, 1924; Stark, 1925; Bertsch, 1935). Indeed, in general, in Europe, pollen-, stomata- and macrofossil-inferred abundances of boreal and temperate species usually increase first in the south, then in the centre, the west and the east, and last in the north.

One way to tentatively assess this question is to estimate the dispersal velocities (m yr^{-1}) of various taxa in the palaeo-record and to compare the resulting average dispersal events (km) with measured dispersal values to check if the reconstructed palaeo-record values are plausible (Clark, 1998). This procedure requires the assumption that sedimentary evidence allows the reconstruction of the initial establishment of local populations (e.g. empirical pollen limit) to infer differences in ages and distances between the various first local presences of plant taxa. However, population growth models show that the lag between the arrival of first individuals of a species and the time when the population density reaches the size required for registration in pollen records can encompass several centuries to millennia (MacDonald, 1993; Fig. 4.11-4). Population densities needed for being registered in the sediment records may vary according to the species (e.g. pollen or biomass productivity), competition, environmental conditions and the size of the archive. Welten (1944), who on the basis of annually laminated (or varved) sediments invented pollen influx (grains cm^{-2} yr^{-1}), showed that the presence of c. 70 individuals of a tree species in a forested lake catchment with a radius of 2300 m may result in the detection of one grain for every 10 000 pollen grains counted and thus remain below the empirical limit. Macrofossil and stomata records have lower numbers, so that the registration of species usually occurs even later than in pollen records (see chapter 4.4.1 and Fig. 4.4.1-1). Given that the time between the arrival of first individuals and the establishment of populations with enough individuals to be tracked in the fossil record can vary according to the initial rates of population growth (MacDonald, 1993; Fig. 4.11-4) the following estimation of spreading rates is only tentative and relies on the assumption that on average this (hidden) initial population growth phase is (temporally) comparable for the same taxon at the European scale.

If the establishment of *Fagus* populations with a few individuals needed about 1000 years for the c. 350 km between the Bavarian Forest and the lowland near the Harz mountains, a velocity of c. 350 m yr^{-1} would result. The potential dispersal velocity (V) can be considered as directly proportional to the dispersal distances of diaspores (Djd), and as indirectly proportional to the maturation age or first flowering (Aff), V = Djd / Aff . Differences between the palaeo-inferred and measured dispersal distances may point to other variables or processes controlling the spread of species. If modern dispersal distances of diaspores are clearly longer than those calculated for the interglacials, additional factors such as climate may have limited the spatial expansion. Conversely, if modern dispersal distances are distinctly shorter than those estimated for interglacials, additional unknown refugia may have been involved. Finally, if the modern and the past dispersal distances are similar, then range shifts may have been largely controlled by average dispersal, although other factors cannot be excluded (e.g. soils, climate). Such assumptions are challenged by the occurrence of extreme dispersal events (e.g. through storms or floods, so called 'jump dispersals'; Davis and Sugita, 1997) which can by far exceed the measured average dispersal capacity of species, thus making rapid spread possible, despite short average dispersal distances (Clark, 1998). Another difficulty is that our knowledge of the dispersal distances of diaspores is still incomplete and often depends on random observa-

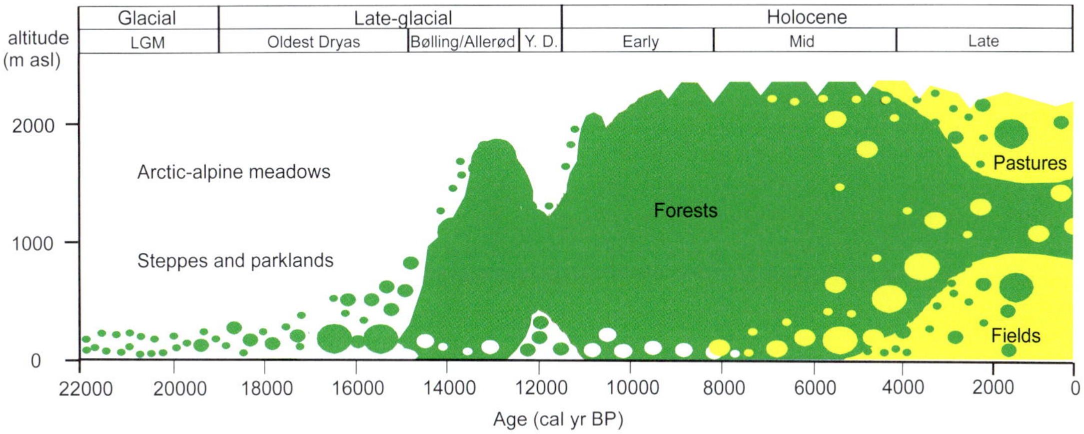

Fig. 4.11-3. Schematic illustration showing the development of vegetation belts for northern Italy. During the Last Glacial Maximum (LGM), fragmented forests and scattered tree stands existed in the lowlands, mostly dominated by boreal species, with rare occurrences of temperate species in sheltered habitats. Forests expanded during the Late-glacial in response to warming, the treeline reached c. 800 m asl at 16 000 cal yr BP and c. 1800 m at 14 000–13 000 cal yr BP. At low elevations, steppic communities and parklands persisted on the Po Plain, possibly because of dry conditions and river disturbance. Further warming at the onset of the Holocene allowed a further expansion of forests, which reached modern treeline elevations above 2000 m asl. Some habitats on the Po Plain remained naturally unforested. Holocene treeline oscillations were related to climate change, as for example, at c. 8200 cal yr BP, at the transition from the Early to the Mid Holocene. After 8000 cal yr BP, human impact disrupted forests to create pastures and fields. During the Late Holocene after c. 5000 cal yr BP, human impact also caused a substantial lowering of the upper timberline. Following Beug (1975a), Tinner and Vescovi (2007) and Kaltenrieder et al. (2009, 2010).

Table 4.11-1. Spreading rates and dispersal distances of some important arboreal taxa. The spreading rates are derived from the isochrone maps in chapter 4.4. The empirical limit is used for the delimitation of isochrones with the exception of *Betula* and *Pinus* Diploxylon, where the rational limit is used (*) and *Buxus sempervirens*, *Hedera helix* and *Ilex aquifolium*, where the absolute limit is used. The dispersal distances observed today are reviewed by Bonn and Poschlod (1998).

Taxon	**Age of first flowering (years)**		**Spreading rates, average to maximum** (m yr^{-1})	**Dispersal distance and vector**		
	Free standing	In closed communities		Postglacial (km)	Maximum observed today (km)	Anemo/ Hydro/ Zoochor
Protocratic taxa						
Betula (tree)*	10–15	20–30	230–990	2–30	3	A
Juniperus communis	10–20	-	85–220	1–4		Z
Pinus Diploxylon*	10–30	30–50	120–490	1–25	2	A
Pinus Haploxylon			45–105			Z
Larix			90–180			A
Alnus	12–20	40	180–360	2–14		A/H
Mesocratic taxa						
Acer	20–30	40	90–260	2–10	4	A
Buxus sempervirens			50–190			Z
Corylus avellana	10	-	450–900	5-9	10	Z/H
Fraxinus excelsior t.	25	40	135–460	3–18		A
Hedera helix			185–650			Z
Ilex aquifolium			110–340			Z
Olea europaea			45–175			Z
Phillyrea			110–160			Z
Pistacia			90–130			Z
Quercus (deciduous)	30–60	50–80	100–400	3–32	20–30	Z
Quercus (evergreen)			100–160			Z
Tilia	10–30	-	220–570	2–19		A
Ulmus	30–40	-	150–800	5-32		A
Oligocratic taxa						
Abies	30–60	60–80	190–220	6–18	8	A
Picea	30–40	60–70	135–240	4–17		A
Carpinus betulus	20	30	220–310	4-9		A
Fagus	40–50	50–80	135–425	5–34	10	Z
Taxus baccata			180-270			Z

tions (for overviews, see Bonn and Poschlod, 1998). Nevertheless, if maximum dispersal distances calculated for the Holocene are compared with dispersal distances observed today (Table 4.11-1) similar orders of magnitude emerge for some taxa (e.g. *Quercus, Fagus, Corylus*), suggesting that dispersal may indeed have influenced the postglacial spread of some species in Europe. Interestingly, differences between reconstructed and observed dispersal capacity are reduced for protocratic (or boreal) taxa, if refugia 45–50°N are taken into account (Feurdean et al., 2013b).

Reconstructed dispersal rates may change over time as shown by three examples. (1) Birks (1989) showed for *Tilia* on the British Isles that the dispersal rates on the northern and north-western range limits sank dramatically after c. 7000 yr BP and ended around 5500 yr BP, so that Ireland and Scotland were not reached. The interpretation is that before c. 7000 yr BP dispersal factors were important for the taxon, whereas after 7000 yr BP climatic constraints may have disadvantaged its further spread. (2) An opposite dynamic was suggested by Huntley and Birks (1983) for *Carpinus*: slow dispersal rates until about 2000 yr BP, high rates afterwards. (3) Un-reduced dispersal rates until the present were mapped as isochrones for *Fagus* in southern England, suggesting that the tree has not yet completed its spread to the north, where the climatic limits would be located (Birks, 1989).

The patterns observed may result from an interplay between varying environmental conditions at decadal to millennial scales and the dispersal capacity of species. Much of the interpretations of early studies relied on the assumption that the postglacial or Holocene climate was stable. Thus the start of the spread of species was set at the end of the LGM (c. 19 000 years ago) or at the end of the Late-glacial (c. 12 000 years ago). However, the available palaeoclimatic evidence suggests that the postglacial was characterized by numerous climatic shifts. Strong temperature oscillations occurred during the period 19 000–11 000 cal yr BP, while precipitation and moisture regimes experienced marked changes also during the Holocene (e.g. Holzhauser et al., 2005; Nichols et al., 2009; Giraudi et al., 2011; Magny et al., 2012; Zanchetta et al., 2012; Regattieri et al., 2014; Holmes et al., 2016). A simplified view on postglacial plant spread assuming stable postglacial environmental conditions also included the non-consideration of factors other than climate influencing the spread of plants such as disturbance (e.g. frost, fire, wind) and human impact (Tinner and Lotter, 2006).

Another way to evaluate the hypothesis that the postglacial continental spread of plants is only controlled by their dispersal ability is to consider the involved dispersal vectors and mechanisms. Specifically, the hypothesis can be falsified if species surviving in the same refugia, owing similar maturation ages and dispersal vectors (e.g. jays and nutcrackers for *Quercus* and *Fagus* or wind for *Acer, Tilia, Ulmus, Picea* and *Abies*) displayed very different (recurring) timings of population expansions over Europe during the interglacials (Tinner and Lotter, 2006). Such comparisons imply that the mesocratic behaviour of deciduous *Quercus* (early spread) and the telocratic behaviour of *Fagus* (late spread) cannot be explained by the dispersal mechanisms or spread efficiency of the involved species, given that they survived the LGM together in refugia south of the Alps (e.g. Willis and McElwain, 2002; Kaltenrieder et al., 2009) and that both trees use the same animal species for their spread (Tinner and Lotter, 2006). This finding is underscored by the very similar spreading rates of the two trees during the postglacial (Table 4.11-1), which are 100–400 m yr^{-1} for *Quercus* and 135–425 m yr^{-1} for *Fagus*, if directly inferred from European isochrones. Similar conclusions about the absence of major migrational lags (or disequilibria with climate) were reached for *Picea abies* and *Abies alba*, which in some regions of Europe expanded only late during the Holocene, despite their efficient wind dispersal (winged diaspores: Giesecke et al., 2017). Taken together, inferring species-specific dispersal capacities and velocities from the palaeo-record remains a challenge (Giesecke, 2007), specifically because it is impossible to track the first arrival of individuals (while it may be possible for the establishment of initial, small populations, see 4.3.1) and because other factors such as competition, climate, disturbance or humans may have delayed or accelerated the spread of species. Nevertheless, the palaeo-evidence provides a good empirical approximation to how rapidly plant populations can move under varying environmental conditions.

Taxon-specific spreading rates in m yr^{-1} might be used to estimate when a potential migrational lag may have decayed at the continental scale, once the conditions were favourable for the spread of the species. Assuming a start of the spread at about 45°N the hypothetical time needed to arrive to 60°N for northernmost positions of temperate trees would be c. 1800 years for *Corylus*, 2000 years for *Ulmus*, 3800 years for *Fagus*, 4000 years for *Quercus*, 4700 years for *Tilia*, 6900 years for *Picea* and 7500 years for *Abies*, if the fastest velocities of spread inferred from the palaeo-records could be constantly reached (Table 4.11-1). This estimation shows that if the onset of the Holocene is set as the starting point for the spread of tree populations, migrational lags of temperate species may have all decayed millennia ago, even at the northernmost suitable positions. Even if tentative, this information is useful to assess future response of species that may move north in response to global warming.

Species-specific rates of postglacial plant spread (dispersal velocity, migration rates) are relevant for the estimation of whether biota may respond fast enough with their range limits to present and future climatic changes. Such estimations require integration of palaeo-data with models (e.g. Lischke, 2006; Cunze et al., 2013). Several model types are available (see box 9 by Heike Lischke, chapter 5.2.3.3 with Fig. 5.1). The inferred high migration rates of plants are referred to as Reid's paradox (Clark, 1998; Clark et al., 1998), which can be solved by assuming that rare dispersal events reach distances far beyond the observed average distances. In general, early-successional trees spread faster than mid- and late-successional trees (Table 4.11-1). This difference is the result of different maturation ages and of different ecological requirements that, in part, include the ability of most pioneers to grow on raw soils, while late-successional species, particularly mesophilous ones, may prefer developed soils.

Mediterranean evergreen broadleaved species are today in similar northernmost positions, as the temperate species were during the LGM (northern range limits at c. 45°N). Model estimates suggest that, climate-wise, such species may be able to grow as far north as Scandinavia within this century (e.g. Overpeck et al., 2003) if the spread is not delayed by dispersal. The palaeo-record shows that continental spreads can be very fast but still require at least some centuries to overcome more than 1500 km of distance under natural conditions. However, today the vegetation is fragmented and this may affect the spread of species. On the other hand, gardens and parks harbour Mediterranean and other subtropical species far north of 45°N, which may facilitate their rapid spread (Walther et al., 2002, 2007).

The dispersal and spreading capacity of trees and other plants is very difficult to observe (Giesecke, 2007) but crucial for their survival if environmental conditions are changing, as they were during the glacial–interglacial cycles or as they will be in the near future under global-change conditions. If all Quaternary glacial–interglacial cycles are taken together, it results that temperate forests of Europe north of c. 45°N may have lost c. 60% of their original arboreal diversity, while for Europe south of c. 45°S the loss was more moderate but may still have reached about 30%. Although it is unclear how many species went extinct because the taxonomic resolution of pollen analysis does not reach the species level, lost Pliocene European arboreal genera such as *Magnolia, Liquidambar, Parthenocissus, Taxodium, Actinidia, Sequoia, Sciadopitys, Nyssa, Pterocarya, Carya, Eucommia, Tsuga* and *Phellodendron* (see chapter 3; Willis and McElwain, 2002) may have comprised many different species, as suggested by the macrofossil record (Mai, 1995). This finding emphasizes the risk of biodiversity losses during times of strong environmental change if rapid spread in and out of environmentally suited refugia is reduced or impeded by, for instance, insufficient dispersal ability, physical barriers or fragmented habitats.

While there is abundant evidence of loss of European taxa during the Quaternary, speciation is difficult to capture on the basis of microscopy approaches, probably because the pollen and spore morphotypes are conservative and may not change over millions of years (see identical pollen morphology of European and American genera such as *Tilia, Castanea, Fagus, Acer, Fraxinus, Quercus, Alnus, Picea, Abies, Pinus*). Here novel approaches such as aDNA may provide new insights. On the basis of the fossil record, it has been hypothesized that the regular reorganization of species ranges in response to the Quaternary glacial–interglacial cycles diminishes or even impedes the opportunity for speciation, given that the selection processes are likely to be lost if the plant communities are completely reorganized every 20 000 to 100 000 years (Bartlein and Prentice, 1989; Bennett, 1990; MacDonald, 1993). However, the isolation of populations of temperate and Mediterranean species in small refugial patches usually lasted about 100 000 years: such a length of time may have provided occasions for macro-evolutionary changes (Huntley and Webb, 1989; MacDonald, 1993), which may have contributed to the high species richness observed in the Mediterranean refugial areas of Europe.

4.11.4

Competition

The assessment of past competition dynamics (see chapter 5.2.4) on the basis of fossil records or process-based, dynamic vegetation models requires the assumption that the traits and environmental niches of species remained stable over the period of interest, which for the Quaternary comprises over 2 million years and thus evolutionary times. Elegant numerical comparisons between American (*Fagus grandifolia*) and European (*Fagus sylvatica*) beech emphasize that the climatic niche of these related tree species remained stable over millions of years (Huntley et al., 1989). Indeed, both species have similar traits and almost identical realized climatic niches with unimodal distributions and optima at -1 to -4°C (January) and 18°C (July). As a result, North American climatic response surfaces can accurately predict the spatial occurrence of beech in Europe and vice versa, despite their spatial isolation since c. 50 million years ago (Huntley et al., 1989). Similar results were reached for organisms with short generation times (Lotter et al., 1999), implying that environmental niches and important traits may stay stable over millions of years despite macroevolutionary shifts and speciation.

The time between the arrival of first reproductive individuals and the time when the population

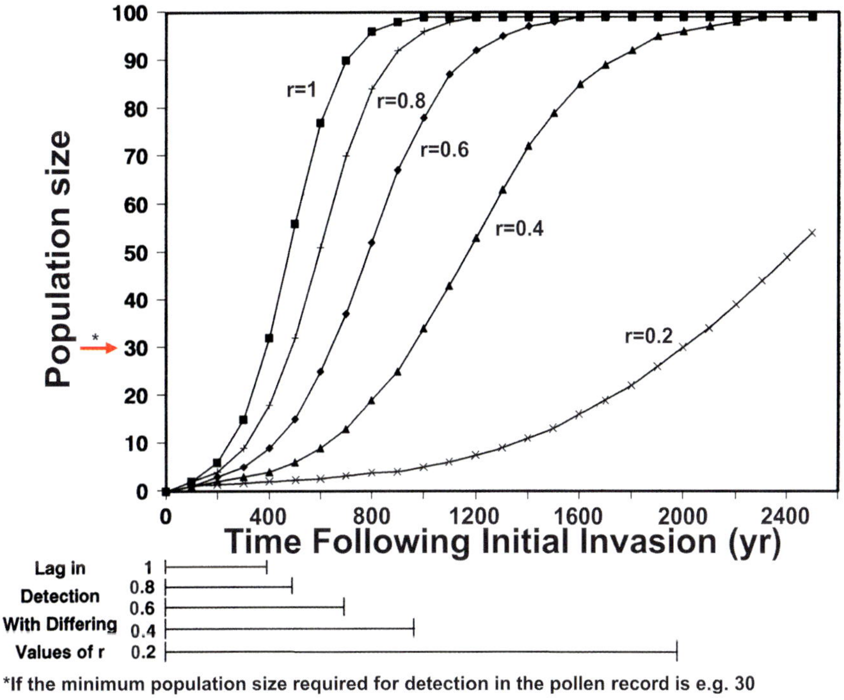

Fig. 4.11-4. Impact of different rates of population growth (r) on the detection of plant taxa in the fossil pollen record. The lag in detection is controlled by both, the size of the population required for detection (e.g. depending on pollen productivity, environmental conditions, size of archive) and the rates of population growth. The relationship between r and the lag time is non-linear and follows the power function $t_{lag}=a\,r^{-1}$, where t is time, a is a constant related to the population density required for detection of the species in the pollen record and r the rate of population growth. Modified from MacDonald (1993).

density reaches the size to appear in pollen records can encompass several centuries to millennia (Fig. 4.11-4; MacDonald, 1993). In pollen diagrams, this period often corresponds to the phase between the absolute and empirical limit for most taxa. Once a small population attains a certain size, the pollen signal reaches its empirical limit and the population becomes recognizable in the pollen record (see chapter 4.4.1; Fig. 4.4.1-1). The establishment of an initial population of few individuals (i.e. the apparent local presence) and its mass expansion corresponding to the rational pollen limit are also usually separated by several to many generations. In pollen diagrams these two events (empirical limit, rational limit) are reflected with changing clarity, depending on the taxon, the sampling resolution and the quality of the chronology. Best records are from annually laminated (or varved) sediments (Fig. 2.4-1) because they provide sufficient time control and resolution to estimate reliable pollen influx values. Reliable pollen influx (pollen grains cm^{-2} yr^{-1}) is superior to pollen concentration or percentage calculation to assess invasion and competition effects, because it is unaffected by sedimentary processes (e.g. changing sedimentation rates that affect pollen concentration) or internal biases (variables influencing each others that affect pollen percentages: MacDonald, 1993). In reality, however, pollen influx is very rarely unbiased by sedimentary processes, so that it is wise to also consider pollen percentages and concentrations.

In usual pollen diagrams with samples every few centimetres hardly any or very small temporal differences between apparent local arrival (empirical limit) and expansion (rational limit) are visible for protocratic (boreal) taxa: this difference becomes clearer in mesocratic (temperate) taxa and can be up to 2000 years for oligocratic (temperate) taxa (see taxa with long preceding tails in chapter 4.3, e.g. Fig. 4.3.2-8). Such contrasts may reflect competition according to the different ecological requirements or niches of species. Protocratic taxa are heliophilous boreal pioneers entering steppe or tundra (without tall trees), able to grow on raw soils. Therefore, after the initial population establishment, the mass expansion was rapid, with hardly any competition (e.g. for light) by other species. Among the mesocratic (temperate) taxa some are heliophilous, some are half-shade trees. First individuals were in competition with less competitive light-demanding trees, thus the delay between establishment of first small populations and mass expansion increased but usually remained confined to 200–300 years. Oligocratic (temperate) taxa are half-shade or shade-tolerant and moisture demanding trees: they were in competition with the relatively stable communities of mesocratic trees, therefore they needed more time between the establishment of first populations and the mass-expansion, usually 500 to 1000 years, exceptionally up to 2000 years. These explanations may sound simple, but the processes behind are complex, as modern ecology and ecological modelling tell us.

Competition effects can be explored through the use of dynamic models. Such models rely on the integration of life traits, preferences and tolerances or niches of species, as derived from ecological observations and ecophysiological experiments. Because the knowledge about the niches of the species is limited and because of the simplification of reality inherent to models (e.g. just a few important environmental niches can be captured), such attempts remain tentative and require validation by empirical data in time, as delivered, for example, from palaeoecology. Dynamic models that integrate species-specific traits or have been adjusted to include species-specific traits include ForClim, LandClim, LPX-Bern and LPJ-Guess (Lotter and Kienast, 1992; Heiri et al., 2006; Miller et al., 2008; Henne et al., 2011; Ruosch et al., 2016). Experiments aiming at disclosing past competition dynamics include simulations of past vegetation trajectories as well as more static sensitivity studies. In their seminal study, Lotter and Kienast (1992) showed that dynamic models are able to realistically simulate the course of vegetation history if the timing of the species invasion is set according to the palaeo-record. Specifically, species-specific population growth rates were similar in both the simulations and the palaeo-record: the competing species usually reached their maximum population sizes after a few centuries to one millennium. This early simulation of competition dynamics confirmed the notion that, despite the complex relationship between plant population size and pollen abundance, pollen records are well suited to infer past population and competition dynamics (e.g. Watts, 1973; Godwin, 1975; Bennett, 1983b, 1988; MacDonald, 1993; Lotter, 1999). Later, dynamic modelling efforts including comparisons with macrofossil records (e.g. Heiri et al., 2006), confirmed the close link between pollen and macrofossil records and stochastically simulated plant population and competition dynamics. Most importantly, these results together with detailed influx studies on varved sediments (Lotter, 1999) could rule out doubts that other causes, such as extremely slow advances (over centuries) of invading populations with a sharply defined front, may mimic population increase patterns in the pollen records and thus blur the reconstructed competition dynamics (MacDonald, 1993). Taken together the combined fossil and modelling evidence suggests that competition may have delayed population size increases by some decades to centuries at most and thus be of secondary importance with regard to the millennial-long pattern of vegetation change observed in the glacial–interglacial cycles.

4.11.5

Climate and other environmental factors

The notion that climate and climate changes controlled the vegetation development of the Late-glacial and the Holocene was at the centre of early research, which is illustrated by palaeobotanically derived climate concepts such as the Boreal or Atlantic periods in the Blytt-Sernander system (Blytt, 1876; Sernander, 1909) or the Older or Younger Dryas terms (Jessen and Jonassen, 1935; Table 4.1-2). The importance of climate change for postglacial vegetation dynamics was emphasized throughout the 20th century (e.g. Rudolph, 1928, 1930; Firbas, 1939, 1949; Godwin, 1956; Prentice, 1986b, Webb, 1986, Huntley, 1990b). On the basis of biogeographic observations, specifically the distribution of today's communities in dependence to climate, early authors had hypothesized that the postglacial vegetation sequence ('Grundsukzession', basic succession) observed in temperate Europe was the consequence of similar climatic gradients in the past. Thus the shift from Late-glacial steppic conditions to Late-glacial *Betula-Pinus* communities as well as the Early Holocene establishment of mixed hazel-oak forests and the Mid to Late Holocene establishment of *Fagus-Abies* communities would have been driven by climate shifts (e.g. Rudolph, 1930; von Post, 1946). This hypothesis does not exclude competition effects but assumes that migrational lags had stopped by the time of the climatic shift, so that the plant species were able to respond rapidly to climate change. The corresponding dynamic equilibrium hypothesis (Huntley et al., 1989; Huntley, 1996; Tinner and Lotter, 2006) implies that temperature, moisture and other climatic properties (e.g. seasonality and resulting disturbance such as frost or drought) released step-wise population expansions across the continent that can be evaluated by multiproxy approaches (MacDonald, 1993).

Multiproxy high-resolution studies show that rapid warming such as at the onset of Bølling (termination 1a) and at the transition to the Holocene had profound effects on plant populations (e.g. Ammann, 2000; Birks and Ammann, 2000; Ammann et al., 2013a,b), likely releasing the spread of boreal (c. 14 500 cal yr BP) and temperate species (c. 11 500 cal yr BP) in Europe north of the Alps (Fig. 4.8-1). Similarly, high-resolution studies on annually laminated sediments show that the empirical limit of *Fagus sylvatica* and the subsequent exponential expansion of the species at c. 8200 cal yr BP is synchronous over wide areas of southern central Europe (Tinner and Lotter, 2001, 2006; Fig. 4.11-5). The abrupt response of (ornithochorous) *Fagus sylvatica* to cooling and increasing moisture availability after 8200 cal yr BP implies that individual beech trees had reached the sites before the onset of the climate change, but were too rare to release an unambiguous pollen signal (Welten, 1944; MacDonald, 1993; Tinner and Lotter, 2001; Hicks, 2006). Other examples of fast-spreading (ornithochorous) genera able to respond rapidly to climatic change are *Hedera, Ilex* and *Viscum* (Iversen, 1944; Kühl et al., 2002). Climatically controlled population expansions and declines have been inferred for *Corylus avellana* in Fennoscandia (see Fig. 4.3-4). During the Early to Mid Holocene, the species expanded beyond its modern range limit and it retreated during the Late Holocene, thus tracking the Holocene Thermal Maximum in both pollen and macrofossil evidence (Andersson, 1902; Seppä et al., 2015).

Taken together the available evidence suggests that European tree species were able to rapidly track postglacial climatic changes (Giesecke et al., 2017; Giesecke and Brewer, 2018). This finding implies that plants, despite their sessile life, can move rapidly over long distances thanks to the very efficient dispersal by wind, animals and water of

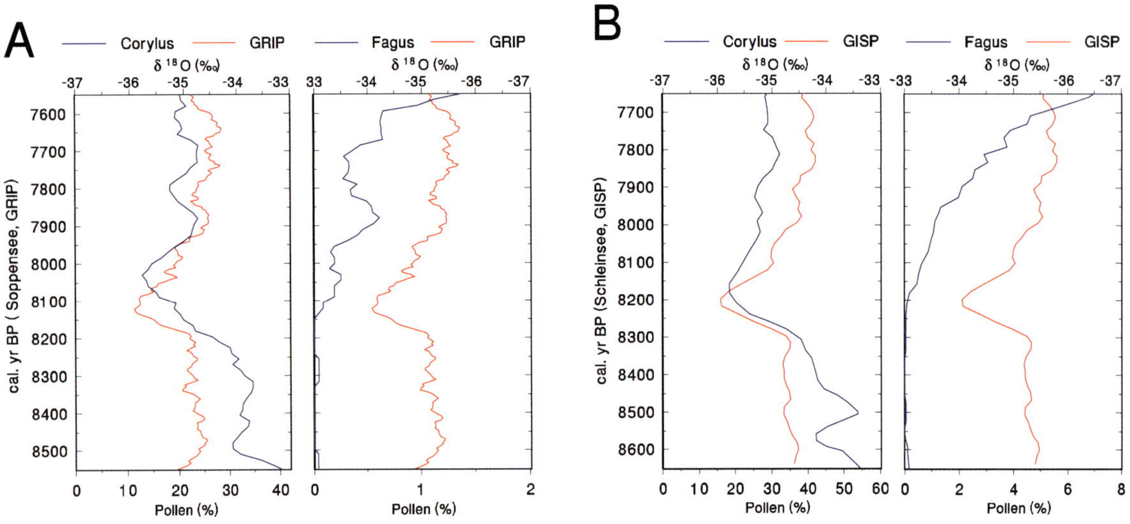

Fig. 4.11-5. Synchronous invasions of *Fagus sylvatica* in Switzerland (Soppensee) and southern Germany (Schleinsee) c. 8200 years ago. The sites are representative of a distance spanning c. 500 km, from eastern France to Bavaria. Soppensee and Schleinsee have very precise varve (annual lamination) chronologies. Both oxygen isotope records from Greenland (GRIP, GISP) are presented according to their original chronologies. Similar oxygen isotope shifts were recorded in lake sediment records of southern central Europe but do not reach the precision and resolution of the Greenland records (von Grafenstein et al., 1998). The initial increase of *Fagus* pollen occurred at 8175 ± 45 cal yr BP (if the chronological uncertainties of both pollen records are averaged), when climate cooling of c. 1.5–2°C reached its apex and moisture availability increased, marking the end of the Early Holocene and the onset of the Mid Holocene. The invasion of *Fagus* was associated with a marked decline of *Corylus* and expansions of *Abies, Pinus, Betula* and *Tilia*. The rapid and synchronous invasion of *Fagus sylvatica* at distant sites demonstrates that the species was already present in southern central Europe before 8200 cal yr BP (absolute limit reached). However, the pollen records allowed tracking the climate-driven invasion of the tree species only once a threshold for recognition (the empirical limit) was attained (see Fig. 4.4.1-1). Lines show locally weighted linear least squares regression (LOWESS) smoothed pollen and oxygen isotope values. Modified from Tinner and Lotter (2001).

seeds and fruits. Specifically, if the dynamic equilibrium hypothesis is true, it would imply that plant species may move even more rapidly than currently inferred from the palaeo-record (see chapter 4.11.2 and Table 4.11-1). Regardless of such open questions, it is undisputed that the glacial–interglacial cycles and their main ecosystem consequences such as biome expansions and contractions, involving increasing and decreasing biomass or biodiversity, are governed by climate change resulting from external (e.g. orbital parameters) and internal (e.g. ocean circulation, volcanism) Earth-system drivers.

Other environmental drivers such as soil development, permafrost occurrence, atmospheric CO_2 concentration, or disturbance (e.g. frost, drought, fire, wind, flood, landslides) likely played important although secondary roles for glacial–interglacial vegetation dynamics. Permafrost occurrence during the glacials may have locally controlled the occurrence of plant species (Birks and Willis, 2008) directly through reduction of plant growth (Körner, 1999, 2012) or indirectly, given that together with low CO_2 it may have exacerbated water stress on plants (Tzedakis et al., 2013). During the cryocratic stage, under dry conditions, increased erosion, low CO_2 and soil properties such as reduced water storage capacity may have become decisive for plant growth, particularly for mesophilous species (Henne et al., 2011; Schwörer et al., 2014b). Moreover, the occurrence of plant species, specifically evergreen broadleaved taxa may have been controlled by frost disturbance under both glacial and interglacial conditions (see chapter 4.10). Similarly, frequent drought disturbance may have limited moisture-demanding plants during dry periods, such as the cryocratic, protocratic or mesocratic stages.

4.11.6

Human impact

The archaeobotanical European record unambiguously shows that species were transported intentionally or unintentionally by prehistoric people into new areas. Such taxa could only survive without special care (gardening) if they were still inside their species-specific realized climatic niches or boundaries. As shown in chapter 4.5.3, early agricultural groups and/or techniques expanded in the Early Holocene mainly from the Near East (Table 4.5.3-1; Fig. 4.5.3-2), which caused a pattern of change along a south–north gradient. This pattern of change spatially resembles the succession of important plant taxa across the European continent. Thus, because of the high similarity of development in space and time between some important tree taxa and societal change, it could be suggested that human activities facilitated not only the spread of crops and weeds but also influenced Holocene tree and shrub trajectories. As a general concept such an assumption is certainly exaggerated, for spreads of trees and shrubs occurred also during interglacials long before any human impact on vegetation. However, caution is needed to assess if any differences between the Holocene and previous interglacials might have been caused by human impact. Even if anthropogenic factors are not plausible as the primary cause of Holocene tree spread, they may have influenced (intentionally or unintentionally) the spread and abundance of certain tree and shrub species (see chapter 5.2.4).

The very rapid spread of *Corylus* over large areas of Europe at the beginning of the Holocene has been associated with Mesolithic people interested in the attractive and nutritious nuts (see chapters 4.4.3 and 4.5.2). Such a link cannot be excluded, indeed hazelnuts were a primary constituent of Mesolithic diet (Regnell, 2012; Bishop et al., 2015) and Mesolithic activities may have affected the occurrence of plant species (Kuneš et al., 2008b). However, a thorough evaluation of the competing hypotheses suggests that climate, specifically a pronounced continentality with hot and dry summers, was likely the main driver of the Early Holocene spread of *Corylus* across Europe, similar to previous interglacials, when human impact was absent or negligible (Finsinger et al., 2006).

It has been hypothesized that the Late Holocene spread of *Fagus sylvatica* may have been anthropogenically advantaged in the northern parts of central Europe and southern Scandinavia. Already Iversen (1973: p. 108) assumed that the northern spread of beech was partly due to human impact. Opening of forests by prehistoric people for land use created gaps which likely favoured the establishment of seedlings of invasive woody taxa. For both *Fagus* and *Carpinus*, Ralska-Jasiewiczowa et al. (2003) provide a careful discussion showing similarities and dissimilarities on an extensive west–east transect in northern lowlands of central Europe: *Fagus* is strong in the west, *Carpinus* in the east, reflecting their climatic requirements and tolerances (see Figs 4.4.3-7 and 4.4.3-9). Indeed, also in northern central Europe the spread of *Fagus* was connected to (step-wise) climatic shifts towards cooler and moister conditions (Tinner and Lotter, 2006). It is thus likely that a combination of both human impact and climate change contributed to the spread of *Fagus sylvatica* in northern central Europe and Scandinavia. Extreme cases of human influence on ranges of tree species concern European fruit trees such as *Castanea sativa*, which during the Roman and early medieval times was introduced at the continental scale for fruit and timber production where climate was suitable (Conedera et al., 2004; chapter 4.4.3).

The contrary to what is observed for species favoured by humans likely happened to the tallest tree species of Europe, *Abies alba*. Our European isochrone map (Fig. 4.4.2-2) impressively shows that the spread of the species came to a sudden end c. 6000–4000 years ago. Given the close link of *Abies alba* population collapses and prehistoric human disturbance c. 6000–4000 years ago (see chapter 4.10) it is likely that the further Late Holocene spread of the species was inhibited by early land use of Neolithic and Bronze Age cultures (Zagwijn, 1992; Tinner and Lotter, 2006; Tinner et al., 2013). In agreement, the European range of *Abies* was larger during previous interglacials, spreading much further north than during the Holocene (Zagwijn, 1996a). Most tree and shrub species were probably in between such exemplary cases of anthropogenically advantaged and disadvantaged plants and it is likely that their ranges were less affected than their abundances (see chapters 4.10.2 and 4.10.3 for e.g. *Picea abies, Alnus viridis, Larix decidua, Tilia, Acer, Fraxinus, Hedera, Quercus ilex* or *Olea europaea*). Increased human impact including regular burning ultimately caused the widespread deforestation of Europe (e.g. Fig. 4.11-3) and the spread of extensive pastures, fields, heaths, maquis and settlements (Birks and Tinner, 2016). Prehistoric and historic deforestation and creation of open land was particularly intense in the lowlands of Mediterranean Europe, where today unplanted forest is extremely rare. Almost all temperate forests of Europe have been extensively managed by selective land use over many millennia, during the past centuries this process also affected remote boreal forests that were increasingly used for industrial purposes.

4.11.7

Summary

The strong climatic and environmental changes of the Quaternary forced rapid species responses, which comprised survival in refugia, subsequent continental spreads and contractions as well as extinctions. The resulting vegetation history patterns can be considered as a result of very complex interactions between dispersal, population dynamics and competition as well as climate and other environmental change (e.g. frost, drought, soils, CO_2 concentration). Moreover, during the past millennia, human impact became increasingly important for European vegetation: forests were disrupted, the ranges and niches of species changed, open land was created and artificial forest ecosystems introduced for timber or fruit production, promoting open land biodiversity, but reducing forest ecosystem evenness and richness (Colombaroli and Tinner, 2013). Some of these changes have very long legacies so that their anthropogenic nature can only be revealed by palaeoecological approaches (e.g. monospecific *Castanea sativa* forests south of the Alps, monospecific deciduous *Quercus* and *Fagus sylvatica* forests north of the Alps).

The traits and environmental niches of species can explain recurrent glacial and interglacial dynamics of plants at the population, community and ecosystem levels, specifically their spread, abundance and survival over periods of strong environmental change. Neutral processes such as species dispersal, recruitment or mortality (Jackson and Blois, 2015; Mitchell et al., 2019; see chapter 5.2.4) may have determined the local and regional absence or presence of species, and thus influenced the course of Quaternary vegetation history, but altogether they appear of secondary importance. The fossil record also suggests that interactions of species with their abiotic environment (e.g. responses to climate change) were more important than biotic interactions (e.g. competition for resources, facilitation, predation, herbivory or parasitism). This implies that the glacial and interglacial changes in vegetation structure and composition, including the shifts from one successional stage to the other (cryocratic, protocratic, mesocratic, oligocratic and telocratic phases), are best explained by the prevalence of Grinnellian niche processes (environmental interactions) over Eltonian niche processes (biotic interactions) and neutral processes (Jackson and Blois, 2015; see chapter 5.2.4).

Palaeoecological studies reveal that plant species responded very rapidly to climate change (e.g. Ammann, 2000; Birks and Ammann, 2000; Ammann et al., 2013a,b; Tinner and Lotter, 2001; Tinner and Kaltenrieder, 2005). The plant species responses resulted in community and ecosystem changes if the magnitude of the climatic change exceeded important thresholds which were limiting or enabling plant growth (e.g. expansion of forest into tundra and vice versa). In some cases, strong climate change in temperatures (>2–5°C) and/or precipitation (>200–500 mm annual means) occurred very rapidly, within a few decades to centuries. Given that some of the most important organisms involved have life-history processes that exceed by far the duration of observations and experiments, the reconstruction of past plant responses to rapid environmental change is essential to assess future vegetation responses under global warming conditions. Carefully focussed future palaeoecological research and cooperation with stochastic, dynamic, process-based vegetation modelling (e.g. Heiri et al., 2006; Henne et al., 2011, 2013, 2015; Ruosch et al., 2016) provides the numeric basis for refined scenarios of plant species responses to future climate and land use changes (see chapters 4.10.5 and 5.2.3.3).

Southern European temperate communities are generally richer in species than those north of the Alps. However, if compared to North America and eastern Asia, the European arboreal flora is impoverished, mainly as an effect of physical barriers such as mountain ranges running west–east and the Mediterranean Sea, which impeded a further spread to the south during times of rapid and strong environmental change, as caused by glacial–interglacial climatic shifts (Birks and Tinner, 2016). Only some European tree species were able to survive the strong climatic oscillations of glacial–interglacial cycles, while others likely underwent extinction. The latter finding highlights the high risk of species losses during periods of rapid climate change. Fragmented habitats as a consequence of human impact may exacerbate the difficulties of plant species to naturally spread in and out of their current ranges.

5.

Vegetation history, evolution and modern ecology

H. John B. Birks[1,2]

[1] Department of Biological Sciences and Bjerknes Centre for Climate Research, University of Bergen, Bergen, Norway
[2] Environmental Change Research Centre, University College London, London, United Kingdom

5.1

Introduction

Studies on floristic and vegetational history in the late 18th century and the 19th century were initially based on plant megafossil remains (e.g. tree stumps) preserved in peat and lakes (see chapter 6). Plant macrofossils (e.g. seeds, fruits, leaves) began to be studied systematically in the late 19th century and contributed greatly to early ideas on floristic and vegetational history. The addition of pollen analysis to the techniques of Quaternary botany following von Post's seminal lecture in Kristiania (now Oslo) (von Post, 1916a, 1967) allowed the subject to develop rapidly. It was primarily used as a geological tool for correlation, reconstructing forest and climate history, and relative dating of archaeological material, sea-level changes and peat-stratigraphical changes (HJB Birks, 2008b; Birks and Berglund, 2018). As in Birks (2019), I use 'Quaternary botany' to include the study of plant megafossils, plant macrofossils and plant microfossils preserved in peat and in lake and ocean sediments. It is currently expanding to include ancient-DNA and other biomarkers.

The predominantly geological slant of Quaternary botany began to change to a more botanical and ecological slant in the 1920s and 1930s with innovative studies by, for example, Aario, Bertsch, Firbas, Godwin, Rudolph, Szafer and Welten (Birks and Berglund, 2018; see chapter 6). Perhaps the major event in the development of ecological Quaternary botany was the 1933 Baltic Course on pollen analysis led by von Post. Nearly all the participants were young geologists from the Baltic countries, except for two botanists – Knut Fægri from Norway and Johannes 'Johs' Iversen from Denmark. They had recently defended their doctoral theses on dynamic and functional plant ecology, respectively. Despite very different personalities, they became life-long friends. During the 1933 course they planned to write a small monograph about pollen analysis, not as a geological relative dating tool, but on how it could become an ecological tool for studying long-term vegetation dynamics and human impact on vegetation, and for providing a factual basis for plant geography (Birks and Berglund, 2018). Their monograph appeared in 1950 (Fægri and Iversen, 1950) and it quickly led to a major shift towards more ecologically based research in Quaternary botany (Birks and Berglund, 2018; Birks, 2019).

The historical development of Quaternary botany as a whole can be roughly grouped into three main phases – (1) an essential pioneer phase (from about 1825 to 1950), (2) a building phase (1951–1973) and (3) a mature phase (1974–present) (see chapter 2). Three main working approaches (Fægri, 1974; Ball, 1975; Birks, 1985) can be recognized within these phases: (1) a descriptive exploratory approach (stage I sensu Fægri, 1974) with the essential detection, description, and archiving of basic patterns - an approach that naturally dominated the pioneer phase; (2) a narrative approach (stage II sensu Fægri, 1974) with plausible but largely untestable inductively based explanations for the patterns detected in the descriptive approach within the pioneer and building phases; and (3) an analytical approach (stage III sensu Fægri, 1974) within the mature phase in which falsifiable hypotheses about the processes behind the observed patterns are proposed, evaluated, tested and rejected (see chapter 2.5).

The aim of this chapter is to discuss selected contributions that Quaternary botany has made to evolution, modern ecology and plant geography in Europe. These are (1) biotic responses over evolutionary and ecological times to environmental change (e.g. extinction, speciation, adaptation, persistence, stasis, spreading); (2) nature of the plant 'community'; (3) pollen richness and diversity; and (4) disturbance and the 'intermediate-disturbance' hypothesis. It will also briefly outline some applications of Quaternary botany to topics such as conservation and restoration ecology, the evaluation of native status and naturalness, palaeolimnology, Quaternary palaeoclimatology and megaherbivore demise. I conclude by outlining possible future directions and challenges in Quaternary botany.

Although the emphasis of the chapter (and of the book as a whole) is Europe, my discussion, where appropriate, includes relevant examples form outside Europe, particularly North America. This chapter draws heavily on Birks (2019). Other relevant reviews include Birks (1986a, 1993); Bennett (1997); Jackson and Williams (2004); Birks (2008b, 2014); Jackson and Blois (2015); and Birks and Berglund (2018).

I avoid using numerous abbreviations to simplify reading. Those that are widely used are yr(s) = year(s); kyr = thousand years; Myr = million years; cal = calibrated; BP = before present with present defined as 1950 CE; LGM = last glacial maximum; and MIS = marine isotope stage.

Table 5-1. Numbers of total and partial (species) extinctions and regional exterminations of taxa in Late Pliocene sequences in Belarus and Early or Middle Pleistocene interglacials in Poland along with their assumed correlations with Marine Isotope Stages (MIS) and interglacial stages in western and central Europe (Lindner et al., 2013). Based on Stachowicz-Rybka (2015a) and data from Velichkevich and Granoszewski (1996); Velichkevich and Lesiak (1996); Velichkevich and Mamakowa (2003); Velichkevich and Zastawniak (2003); Velichkevich et al. (2004); Stachowicz-Rybka (2011, 2015a,b,c); and Drzymulska (2018).

	Approx age (kyr)	**MIS**	**Western & central Europe stage**	**Polish stage**	**Total extinctions**	**Species extinctions**	**Exterminations**	**Number taxa identified**
Middle Pleistocene	370–425	11	Holsteinian	Mazovian	1	5	4	198
	480–620	13-15	Cromerian IV	Ferdynandovian 2	0	17	4	174
Early Pleistocene	710–760	17	Cromerian III	Domuratovian and Ferdynandovian I (Podlasian p.p.)	0	12	3	155
	790–815	19/20	Cromerian II/ Glacial A1	Augustovian (Podlasian p.p.)	0	13	3	147
	865–1000	21	Cromerian I	Podlasian p.p.	0	18	4	162
Late Pliocene				Kholmech	1	78	24	170

p.p. = in part

5.2

Biotic responses to environmental change in the Quaternary

5.2.1

Introduction

The Quaternary (last 2.6 Myr) is a period of very marked and widespread climatic and associated environmental change (Oldfield, 2005; Bradley, 2015), with multiple glacial–interglacial oscillations driven by variations in orbital insolation as a result of periodic fluctuations in Earth's orbit (Bradley, 2015). Glacial-stage conditions account for 75–80% of the Quaternary, whereas the remaining 20–25% consists of shorter interglacial stages (10–20 kyr) during which conditions were similar to, or slightly milder than, the present day (Jackson and Overpeck, 2000; Tzedakis et al., 2012, 2017). During major parts of the glacial stages much of Europe north of 40°N was covered by large terrestrial ice-sheets and widespread permafrost with temperatures possibly 10–25°C lower than present. High aridity and temperatures 2–5°C lower than today were characteristic of more southerly areas (Birks and Willis, 2008). Global atmospheric CO_2 concentrations were as low as 180 ppmv (parts per million on a volume basis) during glacial stages rising to pre-industrial levels of 280 ppmv in interglacial stages (Birks and Willis, 2008; Bradley, 2015). Given these marked changes between the alternating glacial and interglacial stages, an obvious question is how did biota respond to these changes?

There are five major types of biotic response to Quaternary environmental changes – extinction; adaptation and evolution; persistence; stasis; and distributional range shifts (dispersal and range expansion or contraction) (Bennett, 1997; Nogués-Bravo et al., 2018). Given the rapid, frequent and marked alternations in climate during the Quaternary with its short interglacial stages and long glacial stages, many extinctions, much adaptation and speciation, and major range-shifts should be expected. Such Quaternary biotic responses involve both evolutionary time (extinction, adaptation, genetic differentiation, stasis) and ecological time (range-shifts, persistence). The bulk of available Quaternary botanical data and hence this chapter relate to ecological time. However, I first discuss what is known about biotic responses in evolutionary time within the Quaternary.

5.2.2

Evolutionary time

Biotic responses within evolutionary time comprise extinctions, adaptation and speciation, and stasis. What is the evidence for plant extinction and speciation in the Quaternary?

5.2.2.1

Extinction and extermination (extirpation)

The term extinction is increasingly used in ecology, biogeography, conservation science and management. Despite this widespread use, the various kinds of extinction are rarely, if ever, distinguished (Dasgupta et al., 2019). Following Birks (2019) I use total extinction for the loss of a genus, family or higher category; partial extinction for the loss of a member (e.g. species) of a group (e.g. genus) incidental to the evolution of the phyletic line (= species extinction); and regional extinction or extermination for the loss of a taxon in one area although it continues to occur in other areas, so-called extirpations. All three types of extinction have occurred throughout the history of life on Earth (Bennett, 1997). Did they occur in the Quaternary and are they common in the last 2.6 Myr?

What is known about total or partial (species) extinctions in the Quaternary (prior to human activities) is based almost entirely on detailed plant-macrofossil studies of Late Pliocene sequences and Early and Middle Pleistocene interglacials in Poland and Belarus (Birks, 2019; see Table 5.1).

Late Pliocene sequences in south-eastern Belarus (Velichkevich and Zastawniak, 2003) yielded macrofossils of 170 taxa in 82 genera and 43 families (Table 5.1). Most (137) are identified to species level. Of these 137, 78 (48%) are partial (species) extinctions although one reappears in the Middle Pleistocene where it subsequently suffered total extinction (see below). There is a preponderance of extinct taxa in aquatic families such as Potamogetonaceae

(14 extinction species out of 16 taxa), Cyperaceae (11 out of 24), Salvinaceae (4 out of 4) and Haloragaceae (3 out of 4). In contrast only three tree or shrub taxa underwent partial extinction.

Stachowicz-Rybka (2011, 2015a,b,c) provides comparable detailed macrofossil data from six Early Pleistocene interglacials in Poland (Table 5.1). The rich macrofossil assemblages contain several extinct species as well as regionally exterminated taxa such as *Azolla filiculoides* and *Dulichium arundinaceum*. The numbers of extinct species are almost the same in each of the Early Pleistocene interglacials (MIS 21, (20)/19, 17, 15/13) and belong almost entirely to genera that today grow in open-water or telmatic conditions. The rich macrofossil floras from the Middle Pleistocene Mazovian interglacial (MIS 11, ≡ Holsteinian) have been critically re-examined (Velichkevich and Mamakowa, 1999, 2003; Velichkevich et al., 2004). Five extinct species are present in the 198 taxa identified (Table 5.1) along with four exterminated and one totally extinct taxon (*Aracites interglacialis*). As in the Late Pliocene and Early Pleistocene sequences, these extinctions are largely confined to aquatic or wet-ground taxa. Eemian interglacial sequences in Poland (Kupryjanowicz, 2008) contain one extinct species and several regionally exterminated aquatic and terrestrial taxa. Vistulian (≡ Weichselian) assemblages contain three extinct species, all aquatics, and several regionally exterminated 'arctic-alpine' taxa that occur today further north in Fennoscandia (Velichkevich and Mamakowa, 1999)

At present there appears to be only one totally extinct plant taxon in the entire Quaternary at a time when human intervention can be discounted. This is *Aracites interglacialis* (Mamakowa and Velichkevich, 1993) whose distinctive seeds disappear near the close of the Masovian stage in Poland, Belarus, Russia, Ukraine, Finnish Lapland and the English Midlands. Its taxonomic relationships are unclear as it has been assigned to six different families! It is generally thought today to belong to the Araceae (Velichkevich and Zastawniak, 2006). Its ecology is not known but judging by its associated fossils, it grew on oligotrophic mires.

It is unclear why the species extinctions documented by detailed macrofossil analyses are largely of aquatic taxa (Birks, 2019). A possible hypothesis is that pre-Weichselian (Vistulian) aquatic and telmatic floras were simply richer than today. Due to a series of environmental bottlenecks associated with rapid environmental changes – such as climatic shifts caused by alternating glacial and interglacial stages, rapid interstadial climatic change and associated 'cryptic biotic transitions' (Cooper et al., 2015) – species extinctions were frequent, leaving an impoverished temperate aquatic flora with species that today have broad ecological tolerances and widespread geographical range (Birks, 2019). The long-term history of the aquatic flora may thus be similar to that of European Late Cenozoic trees and shrubs (van der Hammen et al., 1971; Lang, 1994) with the progressive species extinction of many aquatic taxa, leaving a depauperate aquatic flora comprising a small number of taxa with wide ecological tolerances and good dispersal abilities, and hence widespread ranges. In western Europe, Middle or Late Pleistocene interglacials show a few species extinctions in the genera *Potomogeton* and *Najas* and many regional exterminations of mainly aquatic or wetland taxa (e.g. *Brasenia schreberi, Euryale ferox, Nymphoides aquatica, Salvinia natans*) (Birks, 2019).

In the Late Quaternary there appears to have been only one total extinction of a flowering plant (*Paschalocous disperta*, Easter Island Palm; Dransfield et al., 1984), but this is almost certainly due to human activity. There is one well-documented natural species extinction (*Picea critchfieldii*) at about 15 000 yrs ago in the south-eastern United States (Jackson and Weng, 1999).

In contrast to the rather few known examples of total (2) or species extinctions (c. 70) of plants through the Quaternary (Nogués-Bravo et al., 2016), there are very many exterminations (regional extinctions) or extirpations as a result of changes in geographical ranges. Extermination is a process that covers a wide range of spatial scales ranging from continents, biogeographical regions, countries, islands, to individual catchments of lakes studied palaeoecologically. These exterminations include tree and shrub genera (van der Hammen et al., 1971; Lang, 1994; Svenning, 2003; see chapter 3.3 and Table 3.3.1); arctic and alpine plants at the end of glacial stages (Tralau, 1963; Godwin, 1975; Lang, 1994; Birks, 2008a; Birks and Willis, 2008); and a range of plants in the Holocene (Birks, 2019). Possible explanations for these exterminations are diverse and include climate change, habitat loss and, in the Holocene, human activity (see Bennett, 1995, 1997; Marchant et al., 2009; Birks, 2019).

5.2.2.2 Adaptation and speciation

Besides extinction and exterminations in response to environmental changes within evolutionary time, plants may adapt to changes in their environment. This involves a shift in a population towards a phenotype that best matches the changing environment. Davis et al. (2005) propose that such adaptive evolution would aid long-term persistence and may permit the exploitation of new niches in response to changing environments. Despite great advances in phylogenetics using DNA of extant taxa and the elucidation of many evolutionary trees, there is, as far as I know, little or no evidence for plant speciation during the Quaternary. It most probably occurred to create, for example, the many neoendemic taxa in the southern Alps, the South African Cape and western Australia, but there appear to be no data at present. Possible explanations for the lack of speciation are (1) the fossil data, particularly pollen, are incomplete and pulses of genetic differentiation and resulting speciation may go unrecorded and (2) geographic isolation in cold dry stages may simply have been too short for speciation events.

Available Quaternary botanical data provide no evidence for speciation even though it probably occurred to create the many microspecies and neoendemics found today, nor any evidence for adaptation although this may have occurred but is currently undetectable. There is weak evidence for total and some partial extinctions and much evidence for regional extinctions or exterminations.

Given the current Quaternary data, stasis appears therefore to be the major response of plants in evolutionary terms during the Quaternary with its multiple glacial and interglacial stages (Bennett, 1990, 1997). Species appear to have persisted unchanged (at least morphologically) over these glacial–interglacial oscillations. Stasis appears to have existed despite considerable environmental change. The apparent evolutionary stability of species through these oscillations is unexpected as speciation and extinction would be expected to be the major evolutionary responses to global Milankovitch oscillations (Bennett, 1990, 1997).

The Quaternary record thus provides the 'missing tier' in Gould's (1985) three-tier hierarchy of dominant processes controlling evolutionary patterns seen in the fossil record (Bennett, 1990; 1997; Table 5.2).

Bennett (1997) argues that Milankovitch cycles and associated climatic instability are the main causes of stasis within the Quaternary. Over ecological time with relatively stable climate, adaptation and microevolution by natural selection may occur. As climate changes in response to Milankovitch cycles, different biotic assemblages may form. Any adaptations accumulated in ecological time are likely to be lost (exaptation) under the changing climate. Gene pools are thus stirred by recombining previously separated populations. Orbital forcing of climate may undo much of any evolutionary change accumulated at a microevolutionary level in ecological time (Table 5.2). Mass extinctions undo any lineage trends resulting from peripatric speciation occurring in the 1–20 Myr time-scale (Bennett, 1990, 1997).

Stasis appears to be common throughout the history of life on Earth (Bennett, 1990, 1997) even though its commonness is not predicted from modern genetic studies (Benton and Harper, 2009). A striking example of evolutionary stasis is *Osmunda claytoniania*. Palaeobotanical evidence suggests this fern has not changed even at the level of fossilized nuclei and chromosomes for at least 180 Myr (Bomfleur et al., 2014). There are many hypotheses about the causes of stasis (Bennett, 1997; Gould, 2002; Futuyma, 2005). In the context of Quater-

Table 5-2. Temporal hierarchy of dominant processes controlling evolutionary patterns seen in the fossil record (from Bennett, 1990, 1997).

Tier	Periodicity	Cause	Evolutionary process
First	-	Natural selection	Microevolutionary change within species; 'ecological moments'
Second	20–100 kyr	Milankovitch orbital forcing	Disruption of assemblages; loss of any accumulated adaptive microevolutionary changes
Third	-	Geographic isolation	Allopatric speciation; 'macroevolution'
Fourth	~26 Myr	Mass extinctions	Loss of species and higher taxa

nary botany, the hypothesis of habitat tracking (Eldredge, 1989) is strongly supported by the predominant biotic response in ecological time, namely range shifts.

Ecological forecasts using species distribution models predict that future global warming may cause widespread regional exterminations and even species or total extinctions (e.g. Thuiller et al., 2005; Botkin et al., 2007; Dasgupta et al., 2019). There is a clear discrepancy between such pessimistic and worrying forecasts and the Quaternary plant record in terms of species or total extinctions (Nogués-Bravo et al., 2016). This is called the 'Quaternary conundrum' (Botkin et al., 2007) and understanding it is critically important in developing realistic forecasts of the effects of future climate change on plant diversity.

5.2.3

Ecological time

5.2.3.1

Introduction

Biotic responses within ecological time comprise persistence ('toleration'), possibly with a shift in habitat, extinction and range shifts (Bennett, 1997; Dawson et al., 2011; Nogués-Bravo et al., 2018).

5.2.3.2

Persistence

As discussed above, there is little evidence for true or partial (species) extinctions in evolutionary and hence in ecological time, but much evidence for exterminations and hence range shifts ('migration') over large geographical areas. Such shifts are discussed later. Is there any Quaternary botanical evidence for persistence in relation to environmental change?

Table 5-3. Estimated rates of spread (m year^{-1}) of trees during the Holocene in Britain and Ireland (Birks, 1989) and on the European mainland based on Huntley and Birks (1983), Feurdean et al. (2013b) and Giesecke and Brewer (2018), and on the pollen maps of Brewer et al. (2017). For additional taxa and traits see Table 4.11-1.

	Britain & Ireland (Birks 1989)	**Europe (Huntley and Birks 1983)**	**Europe (Feurdean et al. 2013b)**	**Europe (Brewer et al. 2017)**	**Europe (Giesecke and Brewer 2018)**
Abies	-	40–300	170	200	250
Alnus	500-600	50–2000	100	200–500	600
Betula	250	>2000	260	?	800
Carpinus	-	50–1000	60	400	500
Corylus	500	1500	na	1000	1000
Fagus	100–200	200–300	90	300	400
Fraxinus	50-200	200–500	na	200	-
Picea	-	80–500	70	250	500
Pinus	<100–700	1500	195	?	600
Quercus	50-500	150–500	165	500	500
Tilia	50–500	300–500	na	500	500
Ulmus	100–550	500–1000	na	1000	1000

- = not estimated or considered; ? = not estimable; na = not available

Some trees, mainly conifers, and clonal alpine cushion or mat plants may have survived in situ for centuries and even millennia (some are over 1500 yrs old) and presumably have tolerated a range of environmental changes including climatic change at decadal, centennial or even millennial scales (Birks, 2019). Some taxa have persisted over at least the last 21 kyr within or near their last glacial maximum (LGM) refugia (Bennett et al., 1991; Magri et al., 2006; Birks and Willis, 2008) or over much of the Holocene at or near their Mid-Holocene range-limit where they persist vegetatively (e.g. Pigott and Huntley, 1978, 1980, 1981). Several of the habitats where such long-lived taxa grow today are extreme habitats such as very dry shallow soils (often on limestone), vertical rock-faces, rock outcrops, or in deep ravines. It is possible that some taxa may have shifted habitat over short distances (1.5–10 km) to sites of different aspect, slope, topography or elevation in response to environmental changes (Dawson et al., 2011), as is occurring today in alpine areas in response to current environmental changes (e.g. Steinbauer et al., 2018).

As stasis is the overriding evolutionary process for plants in the Quaternary and as extinction, persistence and local habitat shifts are rare in ecological time, range-shifts or spread are the principal ecological processes that drive the long-term changes recorded in interglacial-stage pollen diagrams. Range-shifts and vegetational (and hence palynological) change can be a response to exogenous, abiotic factors, to endogenous, biotic factors and to neutral processes (Jackson and Blois, 2015). An additional factor in the Holocene is the role of humans in influencing vegetation and palynological change (Birks, 1986a). These drivers of change are discussed in chapters 5.2.4, 4.5 and 4.10.

5.2.3.3

Range-shifts and tree-spreading rates

From the earliest days of pollen analysis (e.g. von Post, 1916a,b, 1924, 1967; Fries, 1967) spatial patterns in pollen-stratigraphical data have been explored by mapping pollen values of different taxa at selected time intervals. Von Post's motto was "Think horizontally, work vertically" (Edwards et al., 2017) and he applied this idea throughout his career. Maps of pollen values (e.g. Szafer, 1935; Firbas, 1949; Huntley and Birks, 1983; Brewer et al., 2017) and of the times of particular pollen-stratigraphical events (e.g. Birks, 1989; Lang, 1994; Giesecke and Brewer, 2018) reveal surprisingly complex spatial and temporal patterns of pollen changes and hence of taxon spread and range expansion or contraction (see chapter 4.4). Some tree taxa seemingly spread at very fast rates from a range of presumed refugial source areas (e.g. Giesecke et al., 2017; Giesecke and Brewer, 2018). Pollen maps raise many questions about the interpretation of pollen-stratigraphical data in terms of local presence, invasion and range expansion, about how trees spread and expanded over large areas so quickly, about what ecological processes may have caused major pollen-stratigraphical changes, and whether tree pollen assemblages existed in the past that appear to have no modern analogues today (Birks, 2019).

Estimated rates of spread of trees during the Holocene for the European mainland and for Britain and Ireland are summarized in Table 5.3; new

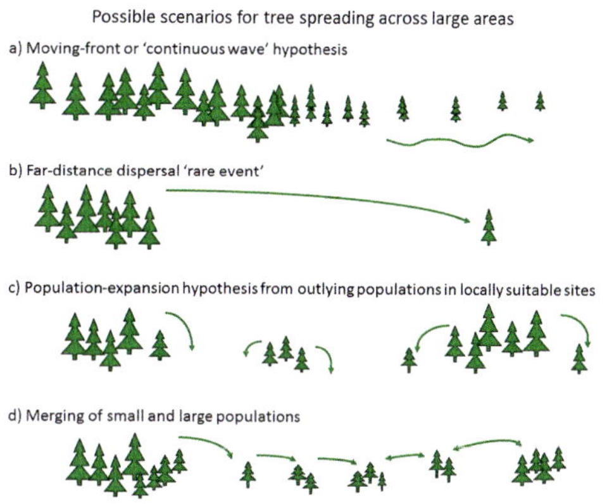

Fig. 5-1. Possible scenarios for tree spreading across a large area in an interglacial stage.
a) The moving-front or continuous wave hypothesis where trees 'march' across the landscape. b) Rare far-distance dispersal events form small outlying populations. c) Populations expand from the outlying populations into locally favourable sites or enclaves. d) Merging of large and small populations. Small scattered populations expand (as in c) and are a source for dispersal events, as are the large populations. Scenarios b, c and d combined are likely to be most important. Based, in part, on Davis (1987) and Giesecke (2013).

estimates on the basis of the isochrone maps of chapter 4.4 are presented in Table 4.11-1.

The spreading rates based on the pollen maps of Brewer et al. (2017) are rough estimates based on simple visual inspection of their maps at 1 kyr intervals, whereas the Giesecke and Brewer (2018) estimates are based on isochrone maps. The estimates from Feurdean et al. (2013b) attempt to take account of known microrefugia (Birks and Willis, 2008) by considering LGM tree presences outside the main southern macrorefugia in the Iberian, Apennine and Balkan peninsulas. The ranges of tree-spreading rates in mainland Europe are between 250 and 1000 m yr^{-1} (Feurdean et al., 2013b; Brewer et al., 2017; Giesecke and Brewer, 2018). The earlier rates of 40–2000 m yr^{-1} suggested by Huntley and Birks (1983) are probably overestimates of tree-spreading rates.

Despite the key role that tree-spreading and range-shifts play in interglacial vegetation history, the mechanisms that enable trees to move at such fast rates and of the processes that allow trees to invade and expand into existing forest vegetation are poorly understood (Birks, 2019). Reid (1899) estimated that for *Quercus* trees to spread through Britain (965 km) without any external assistance would take about 1 Myr. This conundrum of how trees spread so rapidly is called 'Reid's Paradox' (Clark et al., 1998).

Pollen maps give the impression that trees spread as a continuously moving front across the landscape (Fig. 5.1a) and invade and occupy all suitable sites. An alternative hypothesis proposes that taxa spread by far-distance chance dispersal of propagules (Fig. 5.1b) from LGM refugia into favourable habitats or into forest openings to form small outlying populations outside the main range of the taxon. These outlying populations mature and, in turn, act as seed parents for local expansion and further establishment in new localities (Fig. 5.1c, 5.1d). The result is a discontinuous expansion of the range-limit that may appear as a continuous range-expansion at the broad spatial scale of pollen maps (Birks, 1989, 2019).

There is ever-increasing genetic and palaeobotanical evidence for tree populations in microrefugia further north than is assumed from pollen data (Birks, 2019; see also chapter 4.11). Such populations may have comprised trees in locally favourable habitats but were too small to be detected by pollen analysis (Birks and Willis, 2008). These northern outlying populations as well as the larger populations in macrorefugia further south (Birks and Willis, 2008) expanded in the Early Holocene. Without knowledge of these microrefugia, their expansion can be easily interpreted as rapid spread of more southerly populations, resulting in overly high estimated rates of spreading (Table 5.3; Giesecke, 2013; Birks, 2019).

The available but limited data from dispersal biology, tree phylogeography and genetics, and increases in pollen-accumulation rates (as a proxy for population increases) are consistent with the outlying population hypothesis (Fig. 5.1b–d) which depends on dispersal from LGM microrefugia and macrorefugia (Birks, 2019). This hypothesis was first presented by Rudolph (1930) and explicitly decouples tree spread from tree-population expansion (Giesecke and Brewer, 2018). Pollen-stratigraphical data can often unambiguously record population expansion but are very poor at identifying outlying population arrival and local geographical spread (Watts, 1973; Bennett, 1983b, 1985, 1988). Such decoupling of dispersal and expansion (Giesecke and Brewer, 2018) highlights what has been termed 'migration lags' in tree-spreading (Davis, 1978; Bennett, 1983b; Birks, 1986a). This concept is ambiguous: it can result from 'dispersal limitation' or 'expansion limitation' or both. All can result in delayed expansion of a pollen taxon, which can be interpreted as a migration lag (Birks, 2019). Detecting dispersal limitation and expansion limitation from pollen-stratigraphical data is difficult as it requires not only high-resolution palynological and macrofossil data but also an independent palaeoclimate reconstruction to establish if the regional climate was suitable for the taxon in question (Birks, 2019). Birks and Birks (2008a) demonstrate that tree *Betula* may have exhibited a dispersal limitation of about 700 cal yrs in the Early Holocene at Kråkenes in western Norway. Expansion limitation may be reflected by the long discontinuous tail of low pollen values prior to the main expansion and rise in pollen values of *Picea abies* and *Pinus* in northern Europe (Birks, 2019). Such a tail may, of course, result from far-distance extra-regional pollen dispersal. Here other botanical data such as plant macrofossils or conifer stomata can help distinguish between local presence of small populations or extra-regional pollen transport (Froyd, 2005; Ammann et al., 2014; see chapter 4.11.2).

5.2.4

Drivers of interglacial vegetational and palynological change

Pollen diagrams record not only the expansion or contraction of taxa but also changes in the relative or absolute frequencies of the different taxa present in the pollen-source area of the site under investigation. What are the drivers of such changes? There are three major hypotheses concerning these drivers: 1) environmental forcing, 2) biotic interactions and 3) neutral processes (Fig. 5.2). In the case of the Holocene, a fourth hypothesis is human impact (Fig. 5.2; see chapters 4.5, 4.10 and 4.11).

The hypothesis of environmental forcing (Fig. 5.2) proposes that pollen-stratigraphical changes are a response to external (i.e. exogenous) abiotic environmental factors such as climate change (von Post, 1946). Vegetation composition is assumed to be structured primarily by whether the fundamental or Grinnelian niches of its potential members overlap with the prevailing environmental conditions and hence by the physiological and demographic responses to taxa to the external abiotic environment (Jackson and Overpeck, 2000; Jackson and Blois, 2015). These external factors can result in two major types of biotic change (Williams et al., 2011a) – extrinsic change externally driven by abrupt climate change; and intrinsic change resulting from internal thresholds, tipping points and other non-linear biotic responses to gradual climate change. This hypothesis assumes that plant distributions, abundances, vegetation and hence pollen assemblages are in equilibrium over the spatial and temporal scales of interest.

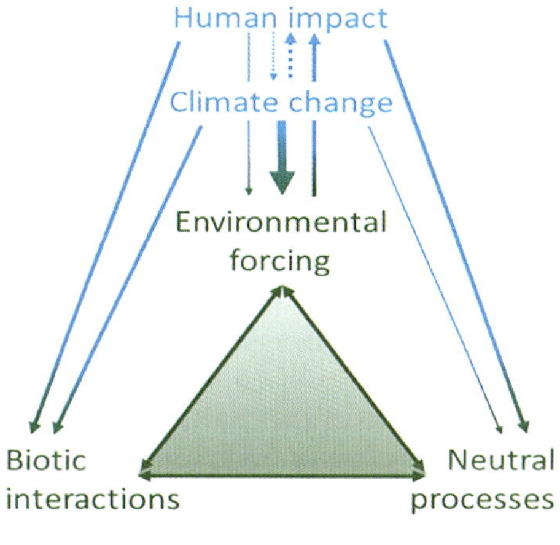

Fig. 5-2. The three major hypotheses and underlying processes for observed changes in Quaternary interglacial plant-assemblage composition and abundance. In the case of the Holocene, a fourth hypothesis is human impact which can influence the processes underlying the three major hypotheses especially biotic interactions. In the real world, all the underlying processes may interact. Based on Jackson and Blois (2015) and modified from Birks (2019).

The hypothesis of biotic interactions (Fig. 5.2) proposes that pollen-stratigraphical changes are a response to internal (i.e. endogenous) biotic factors ('traits' - see Birks, 2020) such as growth rates, longevity and competition; to ecological properties or 'traits' such as shade tolerance; and to disturbances such as fire or storms (Iversen, 1960, 1973; Birks, 1986a; Lacourse, 2009). This hypothesis is based on the Eltonian niche concept and considers vegetation to be structured primarily by interactions between taxa (Jackson and Blois, 2015).

The hypothesis of neutral processes (Fig. 5.2) proposes that pollen-stratigraphical changes are structured by random processes, particularly propagule dispersal, recruitment and mortality, and proximity of glacial refugia. Such neutral processes are non-equilibrial and may be influenced by historical legacies of, for example, past survival or demographic and dispersal events (Jackson and Blois, 2015).

The fourth hypothesis of human impact (Fig. 5.2) is only relevant to Holocene pollen-stratigraphical changes, especially in Europe but also elsewhere (Birks and Berglund, 2018; Birks, 2019). It proposes that the natural rates and directions of change that occurred in response to environmental forcing, biotic interactions and neutral processes in the first 4–5 kyr of the Holocene in Europe may have been disrupted and changed by the appearance and ever increasing importance of *Homo sapiens* and the spread and expansion of agriculture.

Do Quaternary botanical data permit the testing of these four major hypotheses (Fig. 5.2) over long time periods? There is much support for the environmental forcing hypothesis (Fig. 5.2). The strong spatio-temporal broadscale coherence of pollen assemblages over broad areas and the generally close correspondence with independent palaeoclimate reconstructions and with palaeoclimate simulations from general circulation models indicate that exogenous climate change is the major driver to explain such broadscale coherence (e.g. Birks, 2008b; Jackson and Blois, 2015; Birks, 2019)

Whilst the role of exogenous climate change in driving vegetation change and thus pollen assemblages at the Late-glacial/Holocene transition at about 11 700 cal yr BP is very clear and undisputed (e.g. Giesecke et al., 2011; Seddon et al., 2015), the role of climate change in driving pollen-stratigraphical changes in Europe within the Holocene is less clear (see chapter 4.11.4). Seddon et al. (2015) suggest that the underlying dynamics of palynological change are different at different times, with extrinsic linear responses (Williams et al., 2011a) in the Late-glacial and Early Holocene to broadscale climate forcing, but changing to intrinsic non-linear site-specific responses (Williams et al., 2011a) throughout much of the Holocene. Many Holocene pollen-stratigraphical changes in Europe are not unambiguously explicable in terms of our current understanding of past climate change and of interactions between climate variables during the Holocene. Jackson and Blois (2015) propose that "species and populations are not billiard balls knocked around the landscape by changing climate; climate change drives community change via processes that involve individuals, populations, and communities".

At the finer spatial scale of small lakes or hollows within forest stands (e.g. Iversen, 1973; Andersen, 1978; Jacobson and Bradshaw, 1981; Bradshaw, 2013), observed changes in pollen composition and abundance may represent the results of endogenous biotic interactions between individuals (e.g. Lacourse, 2009) and the impacts of disturbance and, in the Holocene, human activity (Fig. 5.2; Jackson and Blois, 2015). These drivers may themselves be influenced by changing climate. Jackson and Blois (2015) emphasize that "[a]lthough these interactions play out locally and often briefly, even the broadest geographic patterns are spatial aggregations of local interactions, and all temporal changes, regardless of scale, are ultimate outcomes of population processes and interactions. Paleoecological dynamics ... cannot be understood fully without considering population processes and species interactions."

The neutral-process hypothesis (Fig. 5.2) derives from Hubbell's (2001) neutral theory of biodiversity and biogeography. It assumes that taxon-niche properties are not relevant to vegetation composition or structure as taxa are ecologically equivalent and thus that this composition and structure result entirely from neutral ecological drift (Jackson and Blois, 2015). Properties of such vegetation include virtually unlimited membership; no equilibrium; and are influenced by historical legacies of past dispersal and demographic events (Hubbell, 2001; Jackson and Blois, 2015; Birks, 2019). Such neutral ecological-drift processes are effectively random with respect to Grinnelian and/or Eltonian taxon-niche properties, resulting in vegetation composition and structure that is unpredictable from either taxon interactions or environmental forcing (Jackson and Blois, 2015). The potential impact of historical legacies or contingencies ('ecological memory'; Ogle et al., 2015; Benito et al., 2020) is being increasingly recognized (see Flessa and Jackson, 2005b) over a range of timescales (100–1000 yr) including Holocene timescales. For example, Davis (1976, 1981) interprets spatial and temporal patterns of Holocene forest development in eastern North America as largely resulting from random ('neutral') processes of propagule dispersal and location of LGM refugia (see Lindbladh, 1999; Svenning and Skov, 2007; Jackson et al., 2009a; Svenning et al., 2015 for further examples). The role of historical legacies may also be important over longer Quaternary interglacial–glacial timescales (e.g. Tzedakis and Bennett, 1995; Herzschuh et al., 2016; Herzschuh, 2019). Such historical legacies represent neutral processes and can arise by a range of interactions across different temporal and spatial scales (Jackson and Blois, 2015). These include climate variation and biotic responses at different temporal scales due to the strong impact of historical processes such as dispersal, recruitment, mortality and survival. Although such processes may involve aspects of Eltonian and/or Grinnellian niche concepts, the processes are essentially neutral because future outcomes cannot be predicted solely from niche properties and the environment (Jackson et al., 2009a; Birks, 2019).

The fourth major hypothesis (Fig. 5.2) for driving changes in Holocene plant assemblages is human impact in the so-called *Homo sapiens* phases (Birks, 1986a), especially in Europe but also elsewhere (Birks, 2019). The role of humans influencing rates and directions of causative ecological processes and observed palynological patterns is extensively reviewed by, for example, Iversen (1973), Birks (1986a), Behre (1986, 1988), Huntley and Webb (1988), and Birks and Berglund (2018) (see chapters 4.11.5, 4.5 and 4.10). Human activity resulted in local and regional forest destruction and the creation of clearings, cultivated areas, heaths, meadows and pastures, and other open habitats. Such activity may have disrupted and changed the natural rates and directions of change that occurred in response to environmental forcings, biotic interactions and neutral processes in the first 4–5 kyr of the Holocene in Europe. It is extremely difficult to identify and distinguish between natural and human-induced changes as climate changed, biota interacted and neutral processes continued at the same time as human impacts started and expanded. The challenge is to isolate unambiguously the effects of human activity from the other potential drivers of change (e.g. Berglund, 2003; Henne et al., 2013).

With palaeoecology increasingly involving many sedimentary proxies (e.g. Flessa and Jackson, 2005b; Birks and Birks, 2006; Birks and Berglund, 2018), it is often now possible to reconstruct the past environment, especially climate, independently of the Quaternary botanical record. As a result, there is increasing evidence for interactions between the main natural drivers of change (Fig. 5.2) plus human impact (e.g. Vescovi et al., 2007; Valsecchi et al., 2008; Tinner et al., 2013; Rey et al., 2019a). Neutral processes can interact with environmental forcing (Jackson et al., 2009a) as can biotic interactions (Fig. 5.2). A small change in temperature or moisture can shift the competitive balance in favour of one taxon, leading to its increase at the expense of the other taxa (e.g. Booth et al., 2012; Clifford and Booth, 2015; Jackson and Blois, 2015). Birks (2019) discusses how complex interactions between climate, tree growth and herbivory may have caused the abrupt Mid-Holocene decline of *Tsuga canadensis* in eastern North America (Marsicek et al., 2013).

An obvious question to ask is whether a given vegetation type is a result of environmental forcing, biotic interactions, neutral processes, or human impact (Birks, 2019). Jackson and Blois (2015) argue that this is a 'futile' question as plant assemblages are a result of many interacting factors (Levins and Lewontin, 1980). Environmental change, biotic interactions including human impact, and neutral processes can all be important drivers of inter-glacial tree and forest dynamics and community composition and structure depending on the spatial and temporal scales. No attempt is made here to evaluate the relative importance of the four major drivers of vegetation dynamics (Fig. 5.2) as their relative importance is almost certainly a complex function of spatial and temporal scales, age, geography, topography, biogeography and change (but see chapter 4.11 on causes of glacial-interglacial vegetation dynamics). Moreover, interactions are likely to be the rule rather than the exception. Jackson and Blois (2015) emphasize that endogenous biotic interactions and neutral processes should be considered more explicitly in the interpretation of all pollen-stratigraphical data, just as human impact is commonly considered in detail in the interpretation of such data in Europe. There is thus an increasing need to consider multiple causal hypotheses in the interpretation of Quaternary botanical data.

Much ecological theory and methodology assume that ecological systems are in equilibrium in space and time. Thus environmental variation is assumed to be stationary or biotic responses to environmental change are thought to be stable over broad spatial and temporal scales (Rapacciuolo et al., 2019). Quaternary botanical data suggest, however, that ecological systems may rarely approach such assumed equilibrium (e.g. Svenning and Sandel, 2013; Birks, 2019). Ecological and biogeographical methods such as species-distribution modelling and species-area relationships can readily produce misleading predictions (e.g. Smith, 2010; Yates et al., 2018) under non-equilibrium conditions. Rapacciuolo et al. (2019) discuss three new approaches that incorporate non-equilibrium concepts into predictive ecology and hence can be incorporated into Quaternary palaeoecology (see Gaüzère et al., 2018; Chesson and Yang, 2019; Newman, 2019; Sandel, 2019 for details).

5.3

Nature of the plant 'community'

5.3.1

Introduction

The term 'community' has generated much debate, controversy and confusion in modern ecology, particularly plant ecology (e.g. Vellend, 2010, 2016; Stroud et al., 2015; Wilson et al., 2019). The debates centre on what is a 'community' and how are 'communities' structured and organized. Ecologists use the term for the total living biotic component of an ecosystem including plants, animals, fungi and microbes. Plant ecologists commonly use 'community' to refer to the botanical component of an ecosystem.

How 'communities' are structured centres on three main hypotheses (see Fig. 5.2 and chapter 5.2.4; Jackson and Blois, 2015). Clements (1905) developed his holistic or 'organismal community' concept and viewed it as a super-organism or discrete unit with sharp boundaries in space and time (McIntosh, 1998). Gleason (1917, 1926, 1939) proposed his individualistic or continuum concept of the 'community' in which the abundance of species populations varies gradually along complex environmental gradients but individually and not necessarily in the same way as other species' populations. The Clementsian concept relates most closely to the environmental-forcing hypothesis whereas the Gleasonian concept relates most closely to the biotic-interaction hypothesis (Fig. 5.2). Hubbell (2001) proposed a third model, the neutral theory of biodiversity (Fig. 5.2) in which all species are functionally equivalent, and the abundance of a species population changes as a result of stochastic processes such as random recruitment and death. Composition is thus a balance between speciation or dispersal that increases richness and random extinctions that decrease richness (Jackson and Blois, 2015). Do Quaternary botanical data provide support for any of these models?

5.3.2.

Quaternary botany and the 'community' concept

Quaternary botanical data suggest that plants react individually to their environment such as climate, soils, biota, topography and other factors and not as collective units such as 'communities' or floristic elements (e.g. Fægri, 1947, 1963; Iversen, 1954, 1960; Andersen, 1961; Cushing, 1963, 1965). These data thus support the individualistic concept of Gleason as well as Whittaker's (1967) concept of locally ordered complexity of present-day vegetation in contrast to the Clementsian concept of 'communities' as distinct integrated units in space and time.

The hypothesis that tree taxa spread independently (e.g. Davis, 1976, 1981; Birks, 1989) does not exclude the idea that taxa such as forest herbs, shrubs and ferns spread and expanded with the trees (Walker, 1982; Jackson and Overpeck, 2000). The fundamental niches of many forest taxa that co-exist today may be broadly similar on several environmental dimensions. Some niche variables may link to vegetation physiognomy and many forest plants may require a cool, moist and shaded microclimate. It is not surprising therefore to find pollen or spores of forest shrubs, herbs and ferns expand in the past at the same time or very soon after the expansion of the tree taxa that they grow under (e.g. Birks and Mathewes, 1978). The Gleasonian concept recognizes that individualistic distributions of different taxa can give rise to seemingly discrete units as well as to a continuum.

Overall, Quaternary botanical studies suggest that at the broad temporal and spatial scales of such studies, the Late Quaternary world is quasi-deterministic (Jackson, 2000) with strong Gleasonian aspects (Jackson and Overpeck, 2000) and some Clementsian ideas (e.g. Wilson et al., 2019). There is increasing evidence for the role of long-term neutral processes such as historical legacies in vegetation history (e.g. Davis, 1976, 1981; Tzedakis and Bennett, 1995; Jackson and Blois, 2015; Svenning et al., 2015; Herzschuh et al., 2016; Herzschuh, 2019; see also chapter 4.11). As Webb (1954) notes "the pattern of variation shown by the distribution of species among quadrats of the Earth's surface chosen at random hovers in a tantalizing manner between the continuous and the discontinuous." The same comment applies, in part, to the temporal patterns of variation shown by Quaternary botanical data.

5.3.3

Do pollen assemblages reflect plant 'communities'?

Despite major advances in the ecological interpretation of pollen-stratigraphical data, fossil pollen assemblages are often equated with plant 'communities' and quantitative methods designed for analysing modern 'community' data are applied directly to fossil pollen-assemblage data (e.g. Blois et al., 2013a,b, 2014; Lyons et al., 2016; Eroglu et al., 2018).

Birks (2019) discusses the question of whether pollen assemblages reflect plant 'communities' in relation to what is currently known about modern pollen deposition, representation and pollen-source area at different spatial scales ranging from medium-sized lakes or mires (5–100 ha) to smaller basins (<5 ha) and very small hollows (20–50 cm diameter). Pollen assemblages are a rich source of information about floristic and vegetation composition and dynamics (Birks, 1993; Davis, 2000). They reflect the vegetation composition of the pollen-source area of the site under study (Davis, 2000). This source area will commonly support several vegetation types or 'communities'. Pollen assemblages thus do not reflect a 'community' but a landscape mosaic of different vegetation types (Cushing, 1963). The one exception to this is pollen assemblages from very small hollows (e.g. Bradshaw, 1988, 2013), which can provide high spatial resolution of past vegetation at the stand-scale. The use of quantitative models developed for modern 'community' data to analyse fossil pollen-assemblage data from medium-sized lakes or mires runs the danger of producing misleading or even erroneous results and interpretations in terms of vegetation structure, stability, or dynamics (Birks, 2019).

5.3.4

No-analogue pollen assemblages

5.3.4.1

Introduction

No-analogue assemblages are fossil pollen assemblages that do not resemble modern pollen assemblages and hence lack modern analogues. Iversen's (1954; see also Iversen, 1973) classic reconstruction based on detailed pollen analysis of the Danish Late-glacial was one of the first studies to suggest that the Late-glacial flora, vegetation, landscape and environment had no widespread modern analogue. This idea of the unique nature of the Late-glacial flora, vegetation and environment and the lack of any widespread modern analogues is supported by results from later studies in Europe and the Americas (e.g. Cushing, 1963, 1965; Birks, 1976, 1981a; Jackson and Williams, 2004; Gonzales et al., 2009b; Correa-Metrio et al., 2012).

Jackson and Williams (2004) review no-analogue fossil assemblages and how to identify them. They also discuss the problems of false-negative matches which suggest that no modern analogous assemblage exists when, in reality, an analogue exists, and of false-positive matches which suggest that a modern analogue exists when, in reality, it does not.

5.3.4.2

How do no-analogue pollen assemblages arise?

Jackson and Williams (2004) propose three alternative hypotheses to explain no-analogue pollen assemblages in light of Birks' (2003) critique of the interpretation of Late-glacial pollen assemblages.

A 'deep no-analogue' hypothesis where the pollen assemblage reliably reflects groups of plant taxa growing together in the same pollen-source area but which do not occur together today (e.g. *Armeria maritima, Astragalus alpinus, Koenigia islandica, Helianthemum, Polemonium caeruleum* in the north-west European Late-glacial) (e.g. Godwin, 1975).

A 'shallow no-analogue' hypothesis where fossil pollen assemblages reflect anomalous pollen-source areas or landscapes in which individual vegetation types resemble modern types in composition but with a spatial arrangement (density, extent, mosaic structure) unlike any today (e.g. the Danish Late-glacial as envisaged by Iversen (1954, 1973) or the central Minnesotan Late-glacial as reconstructed by Birks (1976)).

A 'no no-analogue' hypothesis where the fossil pollen assemblages reflect anomalously steep gradients in regional vegetation and climate, resulting in the close proximity of different vegetation types that today are widely separated (leading to long-distance pollen dispersal and the occurrence of, for example, pollen of deciduous trees and prairie and desert taxa in *Picea*-dominated pollen assemblages in the central Minnesotan Late-glacial) (e.g. Cushing, 1963; Birks, 1976).

When Late-glacial pollen and plant-macrofossil assemblages are studied together in detail, the pollen assemblages often appear to have no convincing modern analogues whereas the macrofossil assemblages often closely resemble the composition of modern vegetation types (e.g. Baker, 1965; Birks, 1976, 1984, 2003; Birks and Mathewes, 1978; Jackson et al., 2000). This provides support for the 'no no-analogue' hypothesis. If it is correct, far-distant transported extra-regional pollen may represent a larger component of Late-glacial pollen assemblages than is generally envisaged (Fægri, 1981).

These three hypotheses are presented by Jackson and Williams (2004) in relation to false-negative matches. They also present three hypotheses to explain false-positive matches when fossil pollen assemblages resemble modern assemblages but other evidence (e.g. plant macrofossils) suggest that the past vegetation as a whole lacks modern analogues despite there being positive fossil and modern pollen matches. These hypotheses are (1) the 'deep no-analogue' hypothesis; (2) the 'shallow no-analogue' hypothesis; and (3) the hypothesis of 'low pollen production' of certain taxa (e.g. deciduous trees) relative to other taxa (e.g. conifers) in response to the lower CO_2 atmospheric concentrations during the LGM.

No-analogue pollen assemblages and by inference past vegetation have been widely interpreted as a response to unusual no-analogue ecological and biogeographical factors (Jackson and Williams, 2004). These factors include lags in spreading or population expansion, soil development, climate, geographical barriers, megafauna, location of LGM refugia and a combination of factors (Cushing, 1963; Birks, 2019).

Jackson and Williams (2004) present an alternative and more general hypothesis for the occurrence of no-analogue pollen assemblages and vegetation types. It is based on the existence of unique environmental conditions and gradients and on the theory of realized environmental space (RES) and potential niches developed by Jackson (2000) and Jackson and Overpeck (2000). Individual taxa are assumed to have unique responses to the multivariate environmental and biotic factors that define the dimensions of the hypervolume that contains the realized and potential niches of the taxa. Taxa can occur together if their fundamental niches overlap and if this overlap lies within the RES at a particular time. If the RES changes, the potential niches of the taxa may no longer lie within the RES and the composition of the vegetation assemblage may change. Past pollen and hence vegetation assemblages with modern analogues may occur in areas and at times when the RES matches environmental conditions somewhere on Earth today (Jackson and Williams, 2004). Past pollen and vegetation assemblages lacking modern analogues would have occurred in areas and at times when the RES had no modern counterpart, as in the LGM and Late-glacial. Taxa with potential niches that overlap with the past RES would form no-analogue vegetation types and hence no-analogue pollen assemblages (Birks, 2019). The RES and hence the potential niche of a taxon can be generalized along the lines of the realized niches of the taxon to include biotic variables (e.g. mega-herbivory) that may prevent individuals of that taxon from occurring in parts of its fundamental niche and thus limit its potential niche. If biotic variables such as mega-herbivory disappear, the RES will change and the potential niche of the taxon will expand.

5.3.4.3

Future no-analogue systems

The occurrence of no-analogue pollen assemblages and, by inference, no-analogue vegetation assemblages in the past has important ecological, evolutionary, conservation and management implications that are now being considered in ecological theory (e.g. Ackerly, 2003; Jackson and Blois, 2015; Radeloff et al., 2015), dynamic modelling (e.g. Reu et al., 2014), conservation science (e.g. Williams et al., 2013a,b; Hobbs et al., 2014; Jackson, 2016; Ordonez et al., 2016) and species-distribution modelling (e.g. Veloz et al., 2012; Catullo et al., 2015). The idea of 'no-analogue' ecosystems ('novel ecosystems') developing in the (very) near future in response to the rapid and on-going changes in climate, land use, nitrogen deposition, biological invasions, biotic interactions and disturbance regimes is now receiving serious attention in many aspects of modern ecology and biogeography, such as restoration ecology; management; species, ecosystem, and landscape conservation; predictive ecology and biogeography; ecological theory; species modelling; and ecosystem dynamics (see Birks, 2019 for details). Hobbs et al. (2013, 2014) provide a thorough overview of future novel ecosystems in the 'new ecological world order'.

5.3.4.4

Ecological attributes and traits in Quaternary botany

There has been a great upsurge of interest and activity in modern plant ecological research in trait-based ecology (Birks, 2020). This recent upsurge in ecology proposes "Rebuilding community ecology from functional traits" to "clean up the mess" of community ecology (McGill et al., 2006). Given the considerable research efforts on traits in contemporary ecology, Birks (2020) considers the question are these trends in plant ecology impacting Quaternary botany? He reviews the use of ecological attributes (= ecological characteristics) and ecological traits in Quaternary botany and shows that their use over the last 120 years has been and remains an integral part of the subject, just as ecological attributes and traits have a long history in ecology (Raunkiær, 1907, 1934; Iversen, 1936). Ecological attributes are features of a taxon such as range size, climatic tolerances, habitat types and ecological indicator values that describe the overall ecology of the taxon (Hill et al., 2004, 2007). Ecological traits have many definitions (Birks, 2020). The one that Birks (2020) adopts is that a trait is "a well-defined, measurable property of organisms, usually measured at the individual level and used comparatively across species. A functional trait is one that strongly influences organismal performance" (McGill et al., 2006). Traits are always attributes, but many attributes (e.g. geographical distribution) are not traits because they may depend on, for example, historical events (Mark Hill, pers. comm.).

Examples of the use of ecological attributes in Quaternary botany include Andersson (1902, 1903); Iversen (1944); Wasylikowa (1978); Kühl (2003); Kuneš et al. (2011); Mortensen et al. (2011); Reitalu et al. (2015); and Felde and Birks (2019). In contrast, ecological traits have rarely been used in Quaternary botany (e.g. Iversen, 1960; Bhagwat and Willis, 2008; Lacourse, 2009; Jabłońska et al., 2019; Reitalu et al., 2019; Blaus et al., 2020). As Birks (2020) discusses, the major limitations in using attributes or traits in Quaternary botany is the taxonomic level at which many Quaternary plant fossils, particularly pollen, can be identified to (see Reitalu et al., 2015). Plant macrofossils have the advantage over pollen in that fossil leaves, seeds and fruits can often be identified to species level (Birks, 2014) thereby allowing an effective use of attributes such as indicator values with macrofossil assemblages (e.g. Wasylikowa, 1978; Felde and Birks, 2019) and potentially of traits (Felde, Birks and ter Braak, in prep.).

5.4

Pollen richness and diversity

5.4.1

Introduction

There is considerable interest at present amongst ecologists and biogeographers in biodiversity, in particular biodiversity changes in the last 50–200 years, the so-called Anthropocene, and in the future (e.g. Dynesius and Jansson, 2000; Kreft and Jetz, 2007; Ellis et al., 2012; Schimel et al., 2013; McGill et al., 2015). The term biodiversity can mean many different aspects of biological organization. Here it refers to taxon richness and diversity. Birks et al. (2016b) discuss the distinction between richness and diversity in a palynological context.

McGill et al. (2015) discuss 15 measures of diversity trends within the Anthropocene. These are

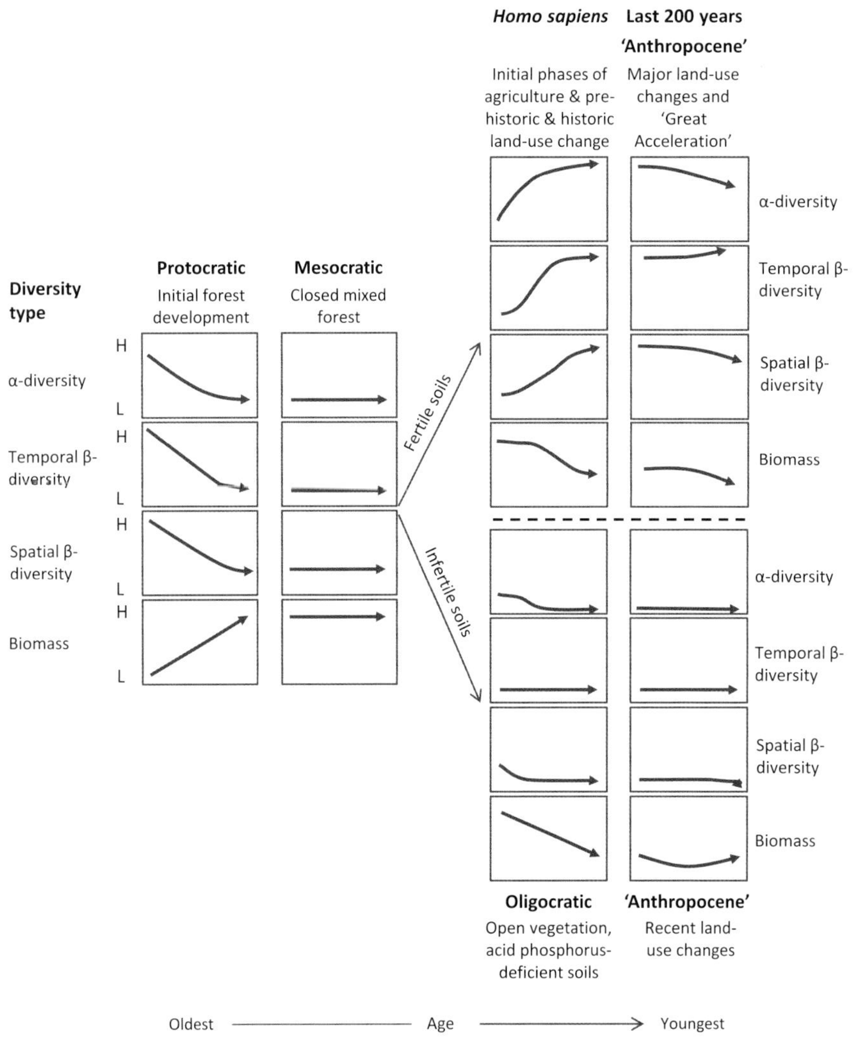

Fig. 5-3. Schematic summaries of the general major trends in the four types of diversity proposed by McGill et al. (2015) at the meta-community scale of organization (sensu McGill et al., 2015; Mittelbach and McGill, 2019) within the protocratic, mesocratic, *Homo sapiens* and oligocratic phases of the Holocene and the last 200 yrs ('Anthropocene' and 'Great Acceleration') based on terrestrial palynological data from northern and central Europe. The vertical axis for all four diversity types runs from low (L) to high (H). The horizontal axis for the four phases and the last 200 years reflects time from oldest (left) to youngest (right). The trends are schematic trajectories of diversity change with the arrows representing the end-point in the different phases. Redrawn from Birks et al. (2016a).

α-diversity (richness); spatial β-diversity (spatial compositional change); temporal β-diversity (temporal compositional change or turnover); and biomass (a rough estimate of abundance) considered at four spatial scales (global; biogeographical; meta-community (sensu Mittelbach and McGill, 2019); local). Are the Anthropocene trends presented by McGill et al. (2015) simply a continuation of trends in the last 15 000 yrs?

Pollen assemblages are the commonest and most abundant terrestrial fossil group in the Late Quaternary and they mainly reflect broad-scale patterns of past flora and vegetation (e.g. Felde et al., 2014a,b) and, by inference, floristic and landscape richness and diversity at the meta-community or regional scale (Meltsov et al., 2013; Matthias et al., 2015; Birks et al., 2016b; Felde et al., 2016, 2018) and at the biogeographical scale (Reitalu et al., 2019). Palynological richness (α-diversity) can be estimated by rarefaction analysis (Birks and Line, 1992). Interpretation of such estimates has, until recently, been controversial (see Odgaard, 1999, 2001; Goring et al., 2013; Birks et al., 2016b). Recent studies show that modern pollen richness (α-diversity) is, in part, a reflection of contemporary floristic richness (Felde et al., 2016; Reitalu et al., 2019) and, in part, a reflection of landscape mosaic structure (spatial β-diversity; Matthias et al., 2015; Meltsov et al., 2013). Spatial β-diversity of pollen assemblages can be estimated by diversity partitioning for a particular time (Birks et al., 2016b; Felde et al., 2016). Compositional turnover (temporal β-diversity) can be quantified as a rate-of-change (Birks, 2012a) or turnover (Birks, 2007c) metric. Biomass changes can be roughly estimated from fossil pollen-accumulation rates calibrated to catchment biomass by modern accumulation rates (Seppä et al., 2009a,b).

5.4.2

Holocene richness and diversity trends in northern and central Europe

Birks et al. (2016c) use Holocene palynological data (0–11 500 cal yr BP) from over 50 sites in northern and central Europe to estimate temporal trends at the meta-community or regional scale in the four types of diversity distinguished by McGill et al. (2015) (Fig. 5.3). General trends are estimated only for the four phases of the Holocene (protocratic, mesocratic (see Fig. 3.3.2), oligocratic, *Homo sapiens*; Birks, 1986a). These phases do not have defined upper or lower boundaries as they begin and end at different times at different sites (Birks, 1986a, 2019). Figure 5.3 also shows general trends for the last 200 yrs (Anthropocene) following Steffen et al.'s (2007) concept of the 'Great Acceleration' beginning about 1800 CE.

Richness (α-diversity) and β-diversity are initially high in the protocratic phase (Fig. 5.3) but decline as the diverse floras and complex fine-scale vegetational mosaic on recently deglaciated areas are replaced by a less fine-grained mosaic of grassland, heath and open woodland. Biomass increases through this succession. During the forested mesocratic phase (Fig. 5.3) α- and β-diversity are relatively constant and low as the landscape mosaic was probably rather simple. Biomass is maximal. In northern and western oceanic areas with nutrient-poor soils, the oligocratic phase has low α- and β-diversity and decreasing biomass in a low-mosaic landscape of heath, bog and open woodland. In contrast, in the *Homo sapiens* phase (Fig. 5.3) with human activity, forest clearance, and agriculture on fertile soils, α- and β-diversity increase and biomass decreases as complex mosaic landscapes develop. In the very recent past (Anthropocene), there is a small decrease in α-diversity, spatial β-diversity and biomass and a small increase in temporal β-diversity at sites on fertile soils. Sites on nutrient-poor soils show little change in α- or temporal β-diversity and a very small decrease in spatial β-diversity and a recent increase in biomass. Recent Anthropocene changes in diversity at sites on fertile soils are thus not a simple continuation of trends initiated in the *Homo sapiens* phase, whereas Anthropocene diversity trends at sites on infertile soils are primarily a continuation of trends that started in the oligocratic phase. Some of these recent trends identified from palynological data (Fig. 5.3) accord with the general trends suggested by McGill et al. (2015) based on various types of organisms and surveys except for trends in α-diversity with a decrease according to Birks et al. (2016c) and an increase hypothesized by McGill et al. (2015).

It is important to emphasize that the analysis by Birks et al. (2016c) is based entirely on terrestrial regional-scale palynological data from northern and central Europe whereas McGill et al. (2015) present trends for all areas, habitats and biota com-

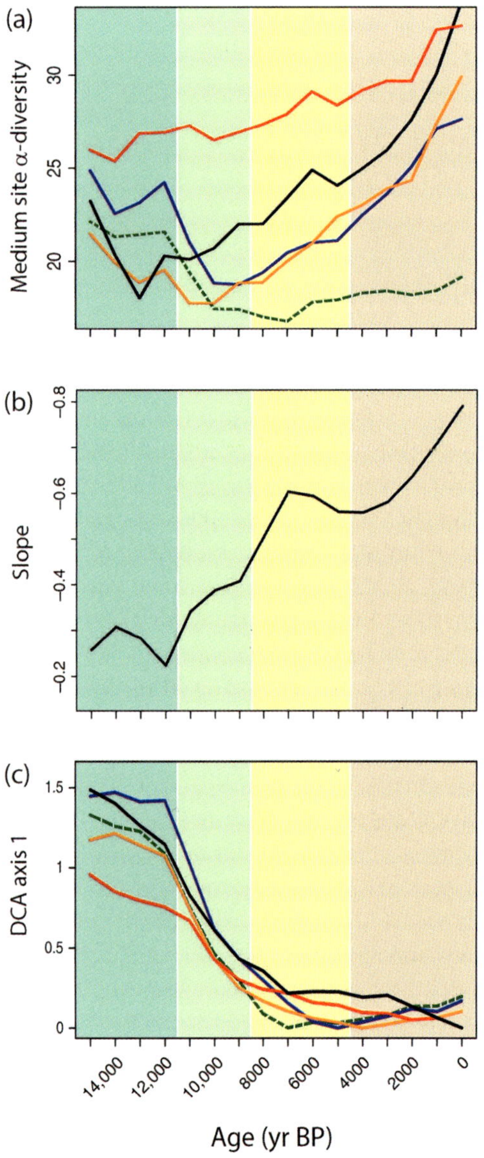

Fig. 5-4. (a) Median pollen richness (α-diversity) over all sites per geographic region. (b) Slope of the linear regression between pollen richness (α-diversity) versus latitude for 1000-year intervals. (c) Scores of regionally combined pollen samples on axis 1 of a detrended correspondence analysis (DCA) as an estimate of temporal β-diversity. The line colour coding indicates the geographical regions: green (dashed) = Boreal, blue = Temperate Oceanic; orange = Temperate Continental, red = Meridional/Submeridional, black = Alps. The temporal intervals 15–11.5, 11.5–8.5, 8.5–4.5 and 4.5–0 cal ka BP are shown by the vertical shaded bands. All plots are against age in cal yr BP. Redrawn from Giesecke et al. (2019).

bined. The diversity patterns in Figure 5.3 are likely to be different in different biogeographical regions (e.g. the Alps, the Mediterranean basin, North America, the tropics) and in different ecological systems (e.g. coastal, marine, freshwater). Some aspects of geographical variation in patterns are discussed in the next chapter.

5.4.3

Changes in palynological richness across Europe

Giesecke et al. (2019) use taxonomically harmonized pollen-stratigraphical data for the last 15 cal kyr from 749 sites across Europe and Turkey to estimate pollen richness (α-diversity) and temporal β-diversity. Given the large number of sites, they grouped the sites into five geographical regions – Boreal (north of 57°N); Temperate Oceanic (west of 11°W); Temperate Continental (east of 11°W); Meridional/Submeridional (Iberia, Mediterranean Basin, Turkey); and the Alps. The primary aim of their study is to test whether the latitudinal gradient of pollen richness changes with the northward spread of taxa in Europe after the LGM. To test this, Giesecke et al. (2019) estimate pollen richness (α-diversity) in two ways and pollen temporal β-diversity in three ways. Figure 5.4 shows the estimated median pollen richness based on sites in each geographical region. In the Late-glacial, highest median richness is in the Meridional/Submeridional region and with broadly similar values in the other four regions. At the onset of the Holocene, median richness decreases in all regions except in the Meridional/Submeridional region (Fig. 5.4a). Richness starts to increase at about 8 cal kyr BP with the order of median palynological richness being, from highest to lowest, Meridional/Submeridional, Alps, Temperate Continental, Temperate Oceanic and Boreal. The slope of the regression between pollen richness and latitude (Fig. 5.4b) increases in the first 3 cal kyr of the Holocene, reaches a peak at 7 cal kyr BP (coinciding with the period of maximum forest cover in Europe), and increases further in the last 3 kyr, presumably as a result of accelerated opening of forest cover and increased landscape mosaic structure due to human activities. Temporal β-diversity (Fig. 5.4c) is highest in all areas in the Late-glacial, decreases in the Early Holocene and remains low after about 7 cal kyr BP. The highest compositional change occurs at the onset of the Holocene in the Boreal and Temperate Oceanic regions (see Fig. 3b in Giesecke et al., 2019).

The results of Giesecke et al.'s (2019) study show that climate warming in the Early Holocene reduced pollen richness at mid- to high-latitudes due to forest development. Deforestation and the spread of agriculture in the last 5 cal kyr had a greater impact on palynological richness than the climate warming at the onset of the Holocene. This study demonstrates that non-climatic human impacts on the latitudinal pollen richness gradient have been of a larger magnitude than climate change. These findings contrast with the widespread prediction that future climate warming will cause a northward spread of species, leading to increased richness at mid- to high-latitudes and a weakening of the latitudinal floristic richness gradient (Giesecke et al., 2019).

5.4.4

Conclusions

Although richness and diversity of pollen assemblages are imperfect and incomplete reflections of past vegetation or landscape richness and diversity, they can document important changes in richness and diversity within the Late Quaternary of northern and central Europe (chapter 5.4.2) and across Europe as a whole (chapter 5.4.3). Detailed palynological studies can provide unique insights into diversity changes over a range of temporal and spatial scales and into shifting diversity patterns resulting from environmental and anthropogenic drivers in the Late Quaternary (e.g. Connor et al., 2012; Giesecke et al., 2012, 2014a; Ammann et al., 2013a; Colombaroli et al., 2013; Reitalu et al., 2015; Felde et al., 2018). There is clearly a need to explore further palynological richness and diversity patterns in different biomes and geographical areas so as to assess and understand pollen-assemblage diversity changes more extensively through time.

5.5

Disturbance and the 'intermediate-disturbance diversity' hypothesis

5.5.1

Introduction

Pollen stratigraphies typically consist of two major types of pattern – intervals where the composition and values of different taxa are relatively constant, and intervals where the composition and values of some or all taxa change (e.g. Ritchie, 1985). Changes may be abrupt (c. 20–100 cal yr) or gradual (c. 200–500 cal yr) (Williams et al., 2011a). As Fægri and Iversen (1950) point out, constancy in pollen composition and values over long periods may reflect vegetational and environmental constancy ('stability') or the lack of ecological sensitivity at sites far from critical ecotones (see Birks, 2019). A 10–11 cal kyr Holocene pollen stratigraphy from north-west Europe typically consists of about 80% of relatively constant 'stable' pollen values and about 20% of changing values, as shown by the variable-barriers partitioning method that is designed to identify and delimit stable phases, gradual transitions and abrupt changes (Birks and Gordon, 1985). Birks (2019) discusses what long intervals of relatively constant pollen values may reflect ecologically, even when the inherent binomial and multinomial counting errors in pollen counting are considered. Here I discuss causes of change in pollen stratigraphies.

5.5.2

Change and disturbance

Ecological thresholds, where a pollen stratigraphy and, by inference, vegetation type or ecosystem changes from one 'stable state' (constancy) to another, usually within a relatively short interval, have long been recognized in pollen-stratigraphical records (e.g. Smith, 1965; Watts, 1973; Ritchie, 1985; Birks, 1986a; Willis et al., 2010). Past human impact is well known to be a driver of such changes in the Holocene (Fig. 5.2), with evidence to suggest that the likelihood of an ecological threshold may increase when humans reduce ecosystem resilience (Willis et al., 2010). There are, of course, many thresholds that occur in the absence of humans and a key question is what combination of climatic and no-climatic variables can result in a switch from one 'stable state' to another, a so-called 'regime shift'. Such thresholds can include climate change (e.g. Jackson et al., 2009b); human disturbance (e.g. Rey et al., 2013, 2019a); natural disturbances such as fire, pathogens or storms (e.g. Foster et al., 1998); invasion of native biota (e.g. Watts, 1973); introduction of non-native biota (e.g. Simberloff, 2000; von Holle et al., 2003); and interactions between some or all of these drivers (e.g. Carrión et al., 2001b; Jackson et al., 2009a,b; Willis et al., 2010; Fyfe et al., 2018b).

Williams et al. (2011a), in their discussion of abrupt palynological changes, distinguish between abrupt change driven by extrinsic drivers and abrupt change driven by intrinsic drivers. Abrupt changes driven by intrinsic factors include biotic responses to millennial-scale changes during the Late-glacial (e.g. Birks and Ammann, 2000; Williams et al., 2002; Ammann et al., 2013a,b). Abrupt changes driven by intrinsic factors may result from the accumulated biotic responses to multiple drivers including climate change interacting with local biotic and abiotic processes and by stochastic processes such as disturbance or local climate extremes. The result is a 'temporal mosaic' of ecological change in which the timing and rate of change may vary among taxa within sites and among sites (Williams et al., 2011a). The Early-Holocene aridification of central North America (e.g. Grimm, 1983, 2001; Clark et al., 2001, 2002; Grimm et al., 2011) provides a striking example of such a temporal mosaic and of intrinsic abrupt changes.

In his classic study of the Big Woods area of south-central Minnesota, Grimm (1983, 1984) provides a clear example of multiple stable states and abrupt intrinsic changes driven by interactions between landscape type (hilly moraines, outwash plains, streams, lakes) and disturbance by fire. Grimm (1983, 1984) proposes that the past fire-probability pattern was primarily influenced by fire occurrences created by native Americans, the local distribution and occurrence of fire breaks, soil texture and moisture, flammability of the existing vegetation and accumulation of fuel load, and the likelihood of fire spread being determined by the areal extent and lateral continuity of the landscape patch. The contemporary composition of forest stands in this part of Minnesota (and elsewhere in southern and central parts of the state) cannot be predicted based solely on abiotic environmental factors today because historical legacies resulting from the long-term site history of disturbance by fire and succession after fire are a strong control on the landscape and its vegetation today (Berland et al., 2011; Birks, 2019).

Fire palaeoecology is a major research area in Quaternary botany (see chapter 4.10.2 and Birks (2019) for examples of detailed studies). As fire histories become increasingly detailed, interest is focusing on the interaction of fire, climate, vegetation, biomass and fuel load, topography, land use and fauna including humans (e.g. Whitlock et al., 2003; Colombaroli et al., 2009; Tinner et al., 2009; Tweiten et al., 2009, 2015; Iglesias et al., 2015) and on the role of intrinsic and extrinsic abrupt changes ('tipping points') in ecosystem dynamics (Williams et al., 2011a; Birks, 2019).

5.5.3

The 'intermediate-disturbance diversity' hypothesis

Ecologists have long recognized that disturbances, both natural and anthropogenic, play a key role in both short-term and long-term ecological dynamics (e.g. White, 1979; Pickett and White, 1987; Frelich, 2002; Brice et al., 2019; Newman, 2019). Connell (1978) presents his 'intermediate-diversity disturbance' hypothesis which proposes that there is a unimodal relationship between taxon richness and disturbance, with the richness peak at levels of intermediate disturbance (see Wilkinson (1999) for an interesting discussion on the history of this hypothesis). The robustness of the hypothesis has often been questioned (e.g. Millar and Woolfenden, 1999; Roxburgh et al., 2004; Miller et al., 2011; Fox, 2013a,b) despite many empirical demonstrations of a widespread unimodal pattern (e.g. Hiura, 1995; Mackey and Currie, 2001; Rejmánek et al., 2004; Randall Hughes et al., 2007; Catford et al., 2012; Sheil and Burslem, 2013; Huston, 2014). Kuneš et al. (2019) test this hypothesis in the long-term perspective of the last 12 cal kyr using palynological data from over 90 sites in central Europe. They estimate pollen richness by rarefaction analysis and the frequency of disturbance on the basis of the indicator values for disturbance frequency and severity compiled by Herben et al. (2016). These values for 1248 species are based on a large number of vegetation plots in central Europe from nearly 40 vegetation types. These values can be used in community and landscape ecology to estimate different aspects of the disturbance regime and its role in vegetation dynamics (e.g. Herben et al., 2016). Kuneš et al. (2019) show that disturbance frequency is highest in the early and late Holocene and lowest in the mid Holocene. The disturbance-diversity relationship changes during the Holocene from a unimodal to a monotonic increasing pattern at about 4000 cal yr BP. This change is interpreted as a result of land-use changes in the last 4000 cal yrs creating an increasingly patchy landscape mosaic that provided more habitats for disturbance-tolerant 'ruderal' taxa. Kuneš et al. (2019) propose a general and novel hypothesis that taxon invasion, and hence the structure and composition of the regional species pool, determine, to some extent, biotic responses to extrinsic factors such as disturbance. This study illustrates the value of synthesizing detailed palynological data covering broad temporal and spatial scales and of using modern indicator values to test a critical ecological hypothesis. Kuneš et al. (2019) conclude that intermediate disturbance can result in unimodal or monotonic relationships with richness depending on factors such as the regional species pool. This study helps to explain why different diversity-disturbance relationships (e.g. Mackey and Currie, 2001; Roxburgh et al., 2004; Miller et al., 2011) are found in nature, and is thus an important contribution to disturbance ecology.

5.6

Additional applications of Quaternary botany to modern ecology, biogeography and conservation biology

5.6.1

Introduction

Because of space constraints, it is not possible in this chapter to discuss or review the many additional applications of Quaternary botany to ecology, biogeography and conservation biology. Flessa and Jackson (2005b), Bradshaw and Sykes (2014), Birks and Berglund (2018) and Birks (2019) discuss several such applications.

In this section I outline briefly some important applications and give selected key references for each group of applications. They fall into four general groups in addition to the four major topics discussed above in this chapter which relate to biotic responses to environmental change (5.2), ecological theory (5.3), biodiversity research (5.4) and disturbance ecology (5.5).

5.6.2

Ecological and biogeographical aspects of interglacial and glacial stages

The nature and location of glacial-stage refugia and long-term soil development in unglaciated and glaciated areas. Key references about LGM refugia include Bennett et al. (1991), Bennett and Provan (2008), Birks and Willis (2008), Gavin et al. (2014) and Birks (2015). Useful references about long-term soil development include Iversen (1958), Andersen (1964, 1966b, 1994), Willis et al. (1997), Birks and Birks (2004), Jeffers et al. (2011a,b, 2012, 2018), Kuneš et al. (2011) and Ammann et al. (2013a).

5.6.3

Forest dynamics and management

Quaternary botanical studies provide valuable insights into temperate forest dynamics and management in a number of ways. Key references include:

Topic	References
Forest structure in temperate areas	Vera (2000); Birks (2005); Mitchell (2005); Fyfe (2018b)
Tree invasion and forest-stand dynamics	Watts (1973); Green (1982); Birks (1985, 1989); MacDonald (1993); Gillson et al. (2008)
Tree-population biology	Bennett (1983b, 1986b, 1988); MacDonald (1993)
Forest ecosystem dynamics	Henne et al. (2011, 2013); Tinner et al. (2013); Connor et al. (2018); Jeffers et al. (2018)
Vegetation stability and change	Fægri and Iversen (1950); Smith (1965); Grimm (1983); Ritchie (1985); Delcourt and Delcourt (1991); Williams et al. (2011a)
Antiquity of forest types	Jackson (2006); Jackson et al. (2014)
Extent of forest cover and landscape openness	Andersen (1970); Prentice (1986a, 1988a); Sugita (2007a,b); Gaillard et al. (2008); Williams et al. (2009, 2011b); Fyfe et al. (2013); Fang et al. (2019)

5.6.4

Conservation biology, management, restoration ecology and related topics

In the last 20–30 yrs, Quaternary botanists have addressed research questions of a more applied nature related to conservation biology (see Birks, 1996a, 2019; Willis and Birks, 2006; Birks and Berglund, 2018), including human impacts on tropical rainforests (e.g. Willis et al., 2004; Bush et al., 2011; Cole et al., 2014). Key references include:

Topic	References
Assessment of naturalness	Oldfield (1970); Edwards (1986); Birks (1996a, 2012b); Willis and Birks (2006)
Fragility and other conservation evaluation criteria	Wright (1974); Lynch (1998); Froyd and Willis (2008); Finsinger et al. (2017a)
Potential natural vegetation	Jackson (2013a)
Ecological thresholds and resilience	Froyd and Willis (2008); Virah-Sawmy et al. (2009, 2010); Willis et al. (2010, 2018); Willis and Bhagwat (2010)
Impacts of future climate change	MacDonald (2010); Cole et al. (2011); Correa-Metrio et al. (2013); Henne et al. (2013); Svenning and Sandel (2013)
Native and introduced species	Jackson (1997); Willis and Birks (2006)
Future no-analogue assemblages and novel ecosystems	Hobbs et al. (2013, 2018); Jackson (2013b)
Ecosystem services	Jeffers et al. (2015b); Saulnier-Talbot (2016)

5.6.5

Islands

Islands and their biota have long fascinated ecologists, biogeographers and evolutionary biologists over the last 175 yrs. Island palaeoecology has made considerable advances in the last 40–50 yrs in the understanding of island biota and past environments of islands (Nogué et al., 2017). Key references include:

Topic	References
Compositional stability prior to human colonization	Colinvaux and Schofield (1976); Birks (2019)
Plant extinctions and exterminations	Dransfield et al. (1984); Bennett (1995); Prebble and Dowe (2008)
Assessing native or non-native status	van Leeuwen et al. (2005, 2008); Coffey et al. (2011); Nogué et al. (2017)
Establishing baseline conditions	Burney and Burney (2007); Burney (2010); Bush et al. (2014)
Impact of vertebrate extinctions and introduced species	Froyd et al. (2014); Nogué et al. (2017)

Quaternary botanical investigations are also proving valuable in detailed studies of Late Pleistocene and Early-Holocene megafaunal extinctions. Quaternary botany clearly cannot resolve the cause(s) of such extinctions but it can investigate if and when vegetation change occurred – before, at, or after a megafaunal extinction event (e.g. Gill et al., 2009; Wang et al., 2017; Jeffers et al., 2018).

5.6.6

Multiproxy studies and links with other Quaternary research techniques

Quaternary palaeoecology is a rapidly developing research area with many new proxies and sources of biological or environmental information being developed and refined (e.g. Flessa and Jackson, 2005b; Smol, 2008; Bradley, 2015). Multiproxy studies involving a range of proxies are increasingly more common. Many show the importance of integrating various non-botanical proxies in the context of the vegetation and ecology of the site's catchment as reconstructed by pollen analysis and plant macrofossils (e.g. Ammann et al., 1985, 2013a,b; Birks et al., 2000; Lotter and Birks, 2003a,b; Birks and Birks, 2006). Examples of new proxies relevant to Quaternary botany include ancient DNA (e.g. Pedersen et al., 2016; Schmid et al., 2017; Edwards, 2020), stable-isotope analysis (e.g. Leng, 2006) and sediment geochemistry (e.g. Croudace and Rothwell, 2015).

5.7

Future directions and challenges

In the last 200 years, particularly the last 70 years, Quaternary botany, mainly pollen analysis but also plant-macrofossil and charcoal analysis, has made many important and unique contributions to ecology, evolutionary theory, biogeography, and conservation biology. These include

- demonstrating the rapid dynamics of tree spreading and the complexity of range expansion (chapter 5.2.3.3)
- the paucity of true extinctions compared to the large number of regional and local exterminations, the so-called 'Quaternary conundrum' (Botkin et al., 2007; chapters 5.2.2, 5.2.3)
- the occurrence of novel no-analogue assemblages in the past (chapter 5.3.4)
- changing patterns of richness and diversity (chapter 5.4)
- the changing relationship between diversity and intermediate disturbance (chapter 5.5.3)
- human impacts on tropical systems (Birks, 2019)
- answers to questions of direct relevance to conservation and management both on the mainland and islands (Birks, 2019; chapters 5.6.4, 5.6.5).

It has also provided many important insights into the floristic, vegetational, and environmental history of different areas (Birks and Berglund, 2018; Birks, 2019).

In their review of the 100 years of Quaternary pollen analysis, Birks and Berglund (2018) ask "Where are we today?" in light of Gill's (2013) question on her Contemplative Mammoth blog "Is pollen analysis dead? Palaeoecology in the era of Big Data". Seddon et al. (2014) present 50 priority research questions in palaeoecology, many of which involve Quaternary botany, often in conjunction with other palaeoecological techniques such as palaeolimnology, geochronology and biogeochemistry. These 50 questions fall into six themes:

1. human–environment interactions in the Anthropocene
2. biodiversity, conservation, and novel ecosystems
3. biodiversity over long timescales
4. ecosystem processes and biogeochemical cycling
5. comparing, combining and synthesizing information from multiple records
6. new developments in palaeoecology.

Many of the questions posed highlight the need for developing a common and strong agenda for ecology and palaeoecology, along the lines proposed by Flessa and Jackson (2005a,b). Attempting to answer the 50 questions of Seddon et al. (2014) and many other questions and developing a common and close agenda will challenge Quaternary botanists and palaeoecologists for years to come and will hopefully prevent the early death of pollen analysis and related techniques. Perhaps the suggestion by Gill (2013) that "the reports of my death are greatly exaggerated" is correct!

Birks (2019) presents ten suggestions that might help to increase the rigour of Quaternary botanical studies and could contribute to developing a common agenda with ecology and biogeography:

- critically consider the research question being asked
- adopt high fieldwork and laboratory standards
- greater appreciation of the past literature, especially non-electronic publications
- closer integration of theoretical ideas about, for example, pollen representation and scale in project design
- greater consideration of basic natural history, field botany and ecology in Quaternary botany
- consideration of multiple causal hypotheses and multiproxy studies
- greater need for replication and for recognition of the numerical properties of Quaternary botanical data (e.g. 'closed' percentage multivariate data; temporal ordering)
- greater attention to detailed identifications and the use of plant macrofossils to supplement pollen-stratigraphical data
- more critical numerical analysis of Quaternary botanical data
- the need to follow and read the ever-increasing ecological, biogeographical, palaeoecological, archaeological and geological literature directly relevant to Quaternary botany.

The sixth theme of the questions in Seddon et al. (2014) is new developments in palaeoecology. What new developments are likely to be rewarding in our understanding of the history and dynamics of past flora, vegetation, landscapes and ecosystems?

1. Increasing development and use of carefully designed and well archived databases of primary Quaternary botanical data (Brewer et al., 2012), such as Neotoma (Grimm et al., 2013; Williams et al., 2018), the Czech Quaternary Palynological Database (PALYCZ; Kuneš et al., 2009), and the European Pollen Database (Fyfe et al., 2009; Giesecke et al., 2014b, 2016). The recent and important uses of such databases to address ecological and biogeographical questions are well shown by, for example, Brewer et al. (2017), Giesecke et al. (2017, 2019), Giesecke and Brewer (2018), Felde et al. (2020) and Kuneš et al. (2019)

2. Increasing use of species attributes (e.g. Hill et al., 2004; Birks, 2020) including ecological indicator values (e.g. Ellenberg et al., 1992; Landolt et al., 2010) to infer past environmental conditions (e.g. Kuneš et al., 2011, 2019; Mortensen et al., 2011; Felde and Birks, 2019; see chapter 5.3.4.4). Such an approach is widely used in ecology (e.g. Schaffers and Sýkora, 2000; Diekmann, 2003). A modified version is commonly used in palaeolimnology (e.g. Birks et al., 1990; Smol, 2008). It can readily be extended and used more widely in Quaternary botanical studies as it has recently in biogeography to evaluate species-distribution models (e.g. Hellegers et al., 2019).

3. Increasing use of ancient plant DNA to supplement existing Quaternary botanical data types. The methodology of ancient DNA analysis is rapidly developing and constantly improving (e.g. Pedersen et al., 2016; Schmid et al., 2017; Clarke, 2019; Clarke et al., 2019a,b; Edwards, 2020). The hypothesis by Parducci et al. (2012) that *Pinus* and *Picea* grew on Andøya in northern Norway during the LGM may, according to Alsos et al. (2020), be incorrect as shown by much improved DNA analytical technology, replication procedures, and critical consideration of background contamination. The critique by Birks et al.

(2012a) is now validated. This is a valuable example of how science progresses by testing earlier ideas as new techniques become available and with replication.

4. Applying Quaternary botanical tools, including ancient DNA, to test long-held ecological or biogeographical hypotheses or concepts such as the equilibrium theory of island biogeography (MacArthur and Wilson, 1967), latitudinal richness gradient (Silvertown, 1985; Grubb, 1987; Haskell, 2001; Giesecke et al., 2019), potential natural vegetation (Tüxen, 1956; Jackson, 2013a), antiquity of particular vegetation types (Jackson, 2006, 2013a; Birks, 2019), system equilibrium (Birks and Birks, 2008a; Crump et al., 2019) and long-term post-disturbance succession and convergence (e.g. Burge et al., 2020).

5. Addressing unresolved and challenging ecological questions such as the fascinating calcicole riddle namely why are there so many calciphilous species in the central European flora (Pärtel, 2002; Chytrý et al., 2003; Ewald, 2003). The hypotheses presented to answer this riddle (Grubb, 1986, 1987; Ewald, 2003) all involve, in part, a historical explanation. Grubb's (1986, 1987) hypothesis is in terms of 'apparency', that is the number of suitable sites available integrated through space and time (Feeny, 1976). Most soils were rejuvenated during the cryocratic and protocratic phases in Iversen's (1958) glacial–interglacial cycle with the additions of glacial till or loess, both generally base-rich. There may thus have only been small areas on which acid-tolerant species could grow in these phases. During the Quaternary in Europe there has, Grubb suggests, been a greater 'apparency' of calcareous grassland than acidic grassland. There has also always been a substantial amount of steppe on the widespread loess of central Asia and eastern Europe disturbed by natural fire and grazing fauna. There has, as far as is known, been no comparable treeless or savannah areas on acidic soils. In contrast, along the eastern seaboard of North America there is an extensive coastal plain of mainly acidic sandy soils where very species-rich vegetation of acid-tolerant herbs occur today (Walker and Peet, 1983). Ewald's (2003) hypothesis is also historical and is related, in part, to Grubb's (1986, 1987) hypothesis. Ewald (2003) proposes that range contractions in glacial stages result in extinction of more acidophilous than calciphilous species, because acid soils were much rarer during glacial stages, It proposes that highly leached acid soils prevailed for long periods prior to the onset of Quaternary glaciations and the high 'apparency' of acid soils may thus have favoured the evolution of a predominantly acidophilous flora, namely long-term adaptation proportional to habitat area (Rozenzweig, 1995). Glacial stages reduced the total habitat for acidophilous species as periglacial processes such as denudation, solifluction, and loess deposition created young calcareous soils at the expense of mature acid soils, causing a decline in acid habitats and their flora. Iversen (1958) suggests that unleached calcareous soils in the protocratic phase may have favoured the rapid adaptive radiation of new taxa. As a result, the mesocratic flora is selectively impoverished in acidophilous species, causing an ecological disparity that cannot be simply explained by the contemporary environment (cf. Wohlgemuth and Gigon, 2003). Ewald's (2003) hypothesis rests on an asymmetry between speciation and extinction, as pre-Quaternary adaptive radiation was rapidly reversed by extinction, while the opposite process requires long time periods. This type of process is an example of 'ecological drift' (Hubbell, 2001). Ewald's (2003) analysis shows that across vegetation types in central Europe the species pool for calcareous habitats is larger today than the species pool for the more widespread acid soils. Ewald concludes that this disparity reflects the species-area relationship of the past and is thus an example of ecological drift. The relationship between local species richness and soil pH may be determined by regional species pool size, which in turn, reflects the relative abundance of soil types during the evolutionary history of the flora (see also Pärtel, 2002). Ewald proposes that acidophiles underwent a Pleistocene bottleneck whereas calciphiles did not (Peet et al., 2003). Ewald emphasizes that this hypothesis requires testing by palaeoecologists and biogeographers. Modern ecological and phylogenetic studies, ancient plant DNA and detailed Quaternary botanical studies could, with some ingenuity, be used to address the challenge of understanding why there are so many calciphilous species in the central European flora today.

Herben and Chytrý (2003) comment "[w]hile functional explanations have dominated ecology for rather a long time, interest in evolutionary and historical explanations is much more recent (see also Zobel, 2001). Such explanations have thoroughly changed our views on causation in ecology."

Over fifty years ago, the American limnologist, ecologist and palaeoecologist Ed Deevey wrote a provocative and stimulating article entitled "Coaxing history to conduct experiments" (Deevey, 1969). In this article he highlights several ecological and biogeographical topics that would benefit from a temporal perspective. Much of what Deevey wrote in 1969 is still applicable and relevant 50+ years later in 2023. It provides not only a challenging agenda but also a rich research philosophy for many future developments in Quaternary palaeoecology and botany that are closely linked to important ecological and biogeographical questions.

6.

History of the Quaternary vegetation sciences

Gerhard Lang[1], Karl-Ernst Behre[2]

[1] Institute of Plant Sciences, University of Bern, Switzerland,
[2] Lower Saxony Institute for Historical Coastal Research, Wilhelmshaven, Germany

6.1

Introduction

Under the roof of geology, palaeontology is a strong branch within which the discipline of palaeobotany has been established. This discipline encompasses the evolution and differentiation of plant families and species. The term was also initially applied to the study of the development stages of the modern flora and vegetation and is still sometimes used in that way. For better clarification, however, it should strictly be used only for pre-Quaternary times, where it mainly includes the botanical investigation of the Tertiary. Instead the term Quaternary Botany might be used (see chapter 5). The term vegetation history has now become widely accepted for the chronological development of the modern flora and vegetation.

6.2

Quaternary geology in the 19th century

To fully understand the development of vegetation history we have to go briefly into the older history of Quaternary geological research. Already in the 18th century people became aware of numerous erratic blocks in northern central Europe as well as in the surroundings of the Alps and sought an explanation for these. For north Germany it soon became evident that these strange erratics were of Scandinavian origin. One theory about how they came to the south was that they were transported by huge flood events and, on the basis of this idea, Buckland (1823) created the terms **Diluvium** (=flooding, now the Pleistocene) and **Alluvium** (=washed together, now the Holocene) which were in use for more than 100 years. One idea, presented by Sefström in 1836 (Nilsson, 1983), was that of a so-called *rullstensflod* (flood of rolled stones), which had swept down over Sweden from north to south, carrying with it great masses of cobbles and gravel. Another theory, brought up primarily by the famous Charles Lyell (1830, 1833), was that the boulders were transported by drift ice (drift hypothesis). Eventually, from around 1840 onwards, the idea of a former much larger extension of the glaciers came up for discussion and it was Otto Torell from Sweden, who definitively helped to establish this theory in a landmark lecture to the German Geological Society in Berlin in 1875.

There remained, however, a long lasting debate between monoglacialists and polyglacialists as to whether there had been only one or several glaciations in the course of the Quaternary. It was eventually Albrecht Penck and Eduard Brückner (1901-1909) who, as a result of their fundamental work, introduced the fourfold division of the glacial sequence in the Alps to which they applied the names Günz – Mindel – Riss – Würm (intentionally in alphabetical order) that finally established polyglacialism. Even earlier, in 1879, Penck had proposed a threefold division of the Ice Age in northern Germany. The three ice advances from the north were called Elster, Saale and Weichsel, while there was no evidence of a preceding one, equivalent to the Alpine Günz glaciation.

The term interglacial for the time between two glaciations had already been created by Oswald Heer (1865a). The name Quaternary for the ice age denominates the youngest period in the fourfold division of the full geological sequence, first introduced by Giovanni Arduino in 1759, whose geological time scale consisted of Primary (now ±Palaeozoic), Secondary (now Mesozoic), Tertiary (= Palaeogene and Neogene) and Quaternary (together, now Cenozoic). This concept was forgotten but became revived in the middle of the 18th century by Jules Desnoyers, Abraham Werner and others. The term Quaternary, or *Quaternaire* in its modern meaning, was first applied by Adolph von Morlot (1858). Following different linguistic branches, the French term *Quaternaire* became in English *Quaternary* but in German *Quartär* and in Swedish *kvartär* (Nilsson, 1983).

6.3

The first steps towards Quaternary vegetation history

As early as the late 17th and the following 18th century botanists engaged themselves with the geographic distribution of plants and tried to explain this distribution. Prominent scientists for this were Joseph Pitton de Tournefort (1656–1708), Carl von Linné (1707–1778) and Carl Ludwig Willdenow (1765–1812) but Alexander von Humboldt (1769–1859) has to be regarded as the real founder of plant geography or geobotany. In his two fundamental works *Essai sur la Géographie des Plants* (1805) and *Ideen zu einer Geographie der Pflanzen* (1807), as well as in *Ideen zu einer Physiognomik der Pflanzen* (1808) he already distinguished the term 'flora' (entirety of the taxa of a region) from 'vegetation' (entirety of the plant communities of a region). Although Humboldt focused mainly on the physiognomy of plants and their relationship to modern climate, he was aware that their distribution had also to be explained by historical reasons. Nevertheless, most botanists concentrated their interests mainly on the pattern of modern plant distribution.

It took some time before the botanical community fully realized that it was absolutely necessary to look back into the past to understand the modern distribution of plants (as well as of animals). Two of the first who practised these thoughts were the geologists Charles Lyell (1797–1875), known as the founder of the 'actualistic principle' and Edward Forbes (1815–1857), who in 1846 published his important essay *On the Connection between the distribution of the existing Fauna and Flora of the British Isles, and the Geological Changes which have affected their Area, especially during the Epoch of the Northern Drift.*

Even before peat bogs had been discovered as important sources for past vegetation it was Heinrich Dau (1790–1831) who placed special emphasis on them in Denmark. In 1829 he published a book on the bogs of Zealand and described the occurrence of trunks of coniferous trees that did not grow wild anymore in this country. This was the start of forest history, which was continued in Denmark by Japetus Steenstrup (1813–1897) who in 1841 published his forest history which was divided into four periods: the *Populus, Pinus, Quercus* and *Alnus* periods.

6.4

Development of vegetation history in the 19th and early 20th centuries

Decisive evidence about vegetation history was accumulated by the Swiss botanist Oswald Heer. He first devoted himself to the palaeobotany of the Tertiary and is regarded to be one of the founders of the subject. Later, however, he turned to interglacial shale coals in Switzerland (Heer, 1858) and the Holocene plant remains from the pile dwellings (Heer, 1865b). These investigations were contemporary with the fundamental work *On the Origin of Species* by Charles Darwin that came out in 1859 and which both influenced and stimulated vegetation history considerably. Somewhat later, Gabriel Nathorst, a Swedish botanist, discovered remains of glacial plants, first in Scania (1870) and later in many other places in Europe. Now it became evident that it was climate which caused the changes of vegetation in the past and this was in accordance with the final breakthrough of the Ice Age theory at this time. Also in Scotland James Geikie found plant fossils in Late-glacial and Holocene deposits (1867). In Britain it was Clement Reid who in 1899 published his groundbreaking book *The Origin of the British Flora* in which he stated "Both geographic and climatic changes had played an important part in shaping our flora as we now see it".

Changes in bog vegetation during the postglacial as shown by tree-stump layers were explained by Axel Blytt (1876) from Norway as evidence for an alternation between dry and rainy periods. This view was developed further by the Swedish botanist Rutger Sernander, who distinguished four periods that were recorded in the bogs which he called Boreal (dry), Atlantic (humid), Subboreal (dry and warm) and Subatlantic (humid and cool), terms that in a modified form are still in use today, in the Blytt-Sernander system.

It was Adolf Engler from Germany who summarized the state of the art of the whole world in his epoch-making *Versuch einer Entwicklungsgeschichte der Pflanzenwelt* (1879/1882) in which he combined palaeobotanical data with evidence from modern plant geography. This was followed by regional publications such as *Geschichte der Vegetation Schwedens* by Gunnar Andersson, *Den Danske Planteverdens Historie efter istiden* by Eugen Warming (1904), *Die Geschichte der Pflanzenwelt des norddeutschen Tieflandes seit der Tertiärzeit* by Carl A. Weber (1906) and *Die Entwicklung der deutschen Flora* by Paul Graebner (1912), while Marie Jerosch (1903) and John Briquet (1906) described the history of the Alpine Flora. New insights were opened up by the combination of linguistic and historical studies with the palaeontological and archaeological evidence in the classic work of Johannes Hoops *Waldbäume und Kulturpflanzen im germanischen Altertum*, which appeared in 1905. At the International Botanical Congress in Vienna (1905) as well as at the International Geological Congress in Stockholm (1910) special attention was directed to the history of vegetation and climate. It is remarkable that at that time there was little interest in these topics in the Mediterranean countries.

6.5

Introduction and first steps of pollen analysis

Interest in pollen grains, which are so important today, started in the first half of the 19th century, but this early interest was mainly from a taxonomical point of view. It was Carl J. Fritsche, who first described the form and structure of pollen grains in a precise way, while in 1834 H. Mohl published a book in which he described more than 1000 species and depicted them with schematic figures. He already separated them according to the number of furrows and pores as is usual right up until today. Another even more comprehensive publication came out in 1890, when the book of Hugo Fischer appeared, in which he described the pollen morphology of more than 2200 plant species. Soon after the publication of Mohl's book fossil pollen grains were also being observed. Heinrich R. Göppert (1836, 1842) described some pollen grains from the Tertiary and Christian G. Ehrenberg found 1838 spruce pollen in Swedish Cretaceous deposits. For further references see de Klerk (2018) and Birks and Berglund (2018).

Following this there was very little interest in fossil pollen grains, an interest that was only resumed by Weber from the bog research station in Bremen, Germany, who is regarded as the originator of pollen analysis with his papers of 1893 and 1896. Later he started with quantitative pollen analysis, working first with well-preserved interglacial material. In his samples he recorded the quantity of all the important tree pollen taxa and already referred to large amounts of *Taxus* pollen in an Eemian deposit, a species, which was identified and recorded again only in the 1960s. More precise percentage calculations were carried out by the Swedish botanist Gustaf Lagerheim (in Witte 1905) and later also by Carl Weber (1910, cited in H.A. Weber, 1918).

The breakthrough of pollen analysis took place in 1916, when the Swedish state geologist Lennart von Post gave a lecture in Kristiania (now Oslo), where he presented illustrated pollen percentage calculations in a diagram. With this method the Holocene changes in vegetation history could be easily pursued in time and compared across different regions. Because his publication was made in Swedish and appeared in the middle of the First World War it was scarcely perceived and only became fully noticed after the war when his student Gunnar Erdtman published two comprehensive studies written in German (1920 and 1921). Another important publication from this early period was the book of Knud Jessen (1920) on the bogs of Zealand, however, still in Danish. From then onwards pollen analysis was realised to be an outstanding method for the study of vegetation history and became very popular.

From the 1920s pollen analysis spread fast across large parts of Europe, mainly north of the Alps. Within a few years quite a number of papers appeared: from Finland (Auer, 1923), Estonia (Thomson, 1925), Latvia (Galenieks, 1925), Poland (Szafran, 1926), Czechoslovakia (Rudolph and Firbas, 1922), Germany (Stark, 1923; Bertsch, 1924), Switzerland (Spinner, 1925) and from the former USSR (Gerazimov, 1924; Kudrachov, 1924; Dokturovsky, 1925). Year after year new papers appeared in increasing numbers.

The advantages of quantitative pollen analysis in comparison with the analysis of macro-remains that had been the main method up until then were apparent. Because of the omnipresence of pollen and spores in bogs and lake deposits and the large catchment area of the pollen rain, this method was able to reconstruct vegetation changes over a wide area rather than just local developments. It was not by chance that pollen analysis was developed in northern Europe, where the forest vegetation comprises relatively few tree species which are wind pollinated and the pollen of which can easily be distinguished. In southern Europe important tree genera such as *Pinus* and *Quercus* have many species that cannot, or only uncertainly, be distinguished pollen morphologically, and the share of insect-pollinated tree species is much larger than in the north. Therefore, in the beginning, the method of pollen analysis had less chance of being employed and this was even worse in regions such as Australia, where the species-rich genus *Eucalyptus* dominates large parts of the continent.

Against this background it is easy to understand why, in its first decades, pollen analysis concentrated on central and northern Europe. The first bibliography started as "Literature on pollen statistics" and was edited from 1927 by Erdtman. By 1950 there were around 2000 citations of papers on European Quaternary deposits: only 33 of these were from the Mediterranean countries Portugal, Spain and Italy while Greece was still missing.

6.6

Progress in pollen analysis

Until the 1930s pollen analysis was rather simple and confined to a few tree pollen types. But from then on the interest also widened to include non-tree pollen and, with this step, forest history changed to vegetation history. In 1934 Firbas pointed to the relationship between non-arboreal pollen (NAP) and arboreal pollen (AP) as a measure for forest density within a certain area and the need to separate the AP was fully realized. Pollen morphology received much more attention. Looking back it is almost unbelievable that *Artemisia* for example, was for a long time known as a special pollen type but was called '*Salix* a' until 1943, when Erdtman identified it as what it really is. Other important new identifications at that time were that of *Ephedra distachia* (Brorson-Christensen, 1949) together with other *Ephedra* species by Welten (1957), of *Helianthemum* (Iversen, 1944; Welten, 1944), *Sanguisorba officinalis* and *S. minor* (van der Hammen and Iversen, 1951) and of *Centaurea cyanus* (Iversen, 1947). Progress was also made in the recognition of woody and other forest plants such as *Ilex, Hedera* and *Viscum* (Iversen, 1944) or *Juniperus* (Iversen, 1934, more exactly Bertsch, 1961).

The first, and still rather simple, pollen keys were published by Meinke (1927) and Bertsch (1942). The basic papers on pollen morphology only followed after these and came mainly from Scandinavia where, after the Erdtman volume (1943), the fundamental booklet *Pollenmorphologische Definitionen und Typen* was published by Iversen and Troels-Smith (1950). This is the basis of most of the following pollen keys, in particular for the *Textbook of Modern Pollen Analysis* by Fægri and Iversen which came out in its first edition in 1950. Parallel with it, a pollen morphology but with another terminology, was also presented in Scandinavia (Erdtman et al., 1961, 1963). This terminology was taken up in the key by Moore and Webb, which was first published in 1978. With the publication of the comprehensive *Leitfaden der Pollenbestimmung* by Beug (first edition 2004) the terminology of Iversen and Troels-Smith has become widely established, at least in central Europe. Parallel pollen atlases have been published, for example by Punt et al. 1976-2003 for north-west Europe (with family keys), Nilsson et al. (1977) for northern Europe (with electron microscope photographs) and Reille (1992b, 1995) for Europe and north Africa as well as an atlas of European moss spores by Boros et al. (1975).

In its first phases pollen analysis was often used as dating method as had been proposed by von Post, who was a geologist. With its application over larger regions it was soon realised that the sequence of tree pollen curves was not synchronous but dependent on climate and migration and therefore could not be used for supra-regional parallelization.

The number of pollen diagrams and also their quality with respect to time resolution and the number of species distinguished increased very rapidly and from about 1950 a number of regional syntheses came out that were well-founded but still without absolute dates, for example based on ^{14}C. Most important were the *Waldgeschichte Mitteleuropas* by Firbas (1949, 1952), followed by the *History of the British Flora* by Godwin (1956) that spanned the whole Quaternary and the *Forest history and palaeogeography of the USSR in the Holocene* by Neustadt (1957). They contain a lot of information and are still worth reading today. Finally a compendium of regional syntheses of palaeoecological studies of lakes and mires from all Europe was edited by Berglund et al. (1996a).

With the increase of identifiable pollen taxa it became possible to describe in detail the changes of entire vegetation units. The general form and intensity of human impact in earlier periods, as well as certain cultivation methods, can be shown by the indicator method and the ecological breadth of individual anthropogenic indicators was illustrated by Behre (1981). The indicator method had also been applied in climate history by Iversen as early as 1944, with respect to *Viscum*, *Hedera* and *Ilex*.

For examining pollen-vegetation relationships, alongside the indicator method the modern analogue approach was developed and most recently, for example within the frame of the POLLANDCAL-network, entire modern vegetation units are compared with recent pollen spectra and used to reconstruct former conditions (Gaillard et al., 2008).

An important addition to conventional relative pollen diagrams is the calculation of pollen concentration and the construction of pollen influx or accumulation rate diagrams, a method first developed in Switzerland by Max Welten (1944) and subsequently in the USA by Margaret B. Davis (e.g. 1966). In recent times numerical methods have also become important in pollen analysis (Birks and Gordon, 1985).

In order to compare the pollen diagrams and connect them with archaeological or other historical data it became necessary to subdivide them into pollen zones. Apart from local pollen zones, later defined as pollen assemblage zones (PAZ; Cushing, 1967), which are biostratigraphic units for a certain site, regional zonations were established with chronologically defined boundaries. In the beginning they were dated by connections with archaeological finds and/or with stages of the Baltic. The first of these zonations were developed almost contemporaneously for Scania (Nilsson, 1935), Denmark (Jessen, 1935) and East Prussia (Gross, 1935). The latter two became the basis of the Firbas zonation, which became prevalent in central Europe (Firbas, 1949). Zonations for other parts of Europe followed. After the introduction of radiocarbon dating the boundaries had to be revised. Many modern diagrams show only the ^{14}C time-scale alongside their local zones.

Radiocarbon dates can be obtained from organic matter such as peat and other plant remains and pure humus, but also from bones and shells (see chapter 2.4). At first bulk samples were used that required a minimum amount of suitable material. The application of the radiocarbon method started in the 1950s and from then on it was possible to correlate pollen diagrams over large distances. This method soon became standard for investigations into vegetation history and so pollen analysis lost its role for geochronological dating. Substantial progress in the radiocarbon method was made in the late 1970s, when the accelerator mass spectrometry (AMS) method was developed for which only very small samples are required and even single seeds are sufficient to obtain precise dates. The age range of the radiocarbon method is, however, limited. In its standard technique it reaches back to around 50 000 years BP but using large quantities of organic material in a high precision counter even older samples, up to 70 000 years BP, can be measured.

In the first years when this method was applied the conventional ^{14}C dates were regarded as correct but it soon became apparent that ^{14}C dates differ from actual ages given in solar years. The ^{14}C time-scale has to be calibrated with the help of very precise dendrochronological dates. During the last decades it has been possible to extend the dendrochronological record back to the Late-glacial and to calibrate ^{14}C-dates for that period as well (see chapter 2.4). Calibration is now common practice, but it has to be kept in mind that older publications may contain uncalibrated dates. A new aspect of dendrochronology, which is also important for vegetation history, is dendroclimatology that focuses on those features of tree-rings that indicate former precipitation and temperature.

Another method of absolute time control is provided by the counting of annually laminated sediments. It was Gerard de Geer who in 1912 started to count and measure varves of clay deposits in the region of the Baltic and established a 'Swedish Time Scale' (1940) from the Late-glacial to the present, which, however, has been revised twice since then. Nevertheless, it is very useful for at least relative correlations. The same method was also applied on annually laminated lake sediments and led to the first estimates of the duration of interglacials (Müller, 1974a,b). In recent times several Holocene lakes with long laminated sequences have been discovered (e.g. Lake Gościąż: Ralska-Jasiewiczowa et al., 1998) and on this basis a very precise absolute dating of the European Late-glacial could be achieved (cf. Litt et al., 2001).

For the dating of Pleistocene deposits, quite a number of geophysical and other methods have been developed which use mineral material (cf. Geyh and Schleicher, 1990; Taylor and Aitken, 1997). The most important are thermoluminescence (TL), optically stimulated luminescence (OSL), potassium-argon dating (K-Ar), uranium-thorium dating (U-Th) and amino acid dating.

At a relatively early stage, pollen analysis was also applied to Pleistocene sediments. After Weber's initial publication on Eemian deposits (1896), however, it took some time until more general interest extended into these older layers. In 1928, Szafer and Trela published a paper on an interglacial site near Poznan in Poland and in the same year the groundbreaking book on *Interglacial Fresh-water Deposits in Jutland and Northwest Germany* by Jessen and Milthers came out. From then on the number of investigations on interglacial sites increased steadily, in particular after World War II.

An exceptional area for the study of the pre-glacial Lower and Early Middle Pleistocene is East Anglia, where the Cromer Forest Bed series is exposed at many places along the coast. It was Clement Reid who discovered these very complex organic sites in the 1870s and in 1882 published *The Geology of the Country around Cromer* where he described a large number of species from this unique area on the basis of their macro-remains; the corresponding pollen diagrams were published later (West, 1980).

Another important region for the study of the Lower and Middle Pleistocene is along the lower Rhine, where Zagwijn (1960, 1963) put together pollen diagrams from various interglacial sites in the Netherlands and established the actual subdivision of these periods, as can be recognized by their Dutch names. Later, a long continental profile where the numerous Lower Pleistocene interglacials are developed in superposition, was discovered in Lieth, northern Germany (Menke, 1975) and this continuous sequence became the most informative profile for the entire Lower Pleistocene in north-central Europe. The temporal continuation for the lower Middle Pleistocene, containing five interglacial phases, is the long continental record of Gorleben, northern Germany (Müller, 1992).

In addition to many pollen profiles from single interglacials and interstadials an increasing number of long continental records from the late Pleistocene have provided detailed pollen diagrams and enabled reliable correlations. The first key profile for the Eemian and Weichselian was Le Grande Pile (Woillard, 1975 and later), followed by Les Echets (de Beaulieu and Reille, 1984a,b), both in France, Oerel (Behre and Lade, 1986) and Groß Todtshorn (Freund and Caspers, 1997) both in northern Germany, Gröbern (Litt, 1994) in eastern Germany and more recently Sokli in northern Finland (Helmens et al., 2000) and Horoski Duże in eastern Poland (Granoszewski, 2003). A synthesis of the Pleistocene vegetation history of northern Eurasia was given by Frenzel (1968) followed in 1994 by the first edition of this book on the Quaternary vegetation history of the whole of Europe written by Gerhard Lang.

Several important interglacial sites were investigated in the Alps, such as Meikirch, Sulzberg and Dürnten (Welten, 1982a) as well as Gondiswil (Wegmüller, 1992) in Switzerland, Samerberg (Grüger, 1979) in Germany, Mondsee (Klaus, 1987;

Drescher-Schneider, 2000b) in Austria or Lake Fimon (Pini et al., 2010) in Italy.

Substantial progress was made by the discovery in several countries of crater lakes, where long Pleistocene sequences could be cored. In France the volcanic region of Velay in the Massif Central yielded three important localities: Lac du Bouchet, Ribains and Praclaux and the palynological correlation of their deposits could be supported by tephra layers (de Beaulieu et al., 2001; Reille et al., 1998, 2000). Similar crater lakes with long sequences have been investigated in Italy, for example, Valle di Castiglione and Lagaccione (Follieri et al., 1998 and later) as well as Monticchio (Watts et al., 1996a,b; Allen and Huntley, 2000, 2009). They are complemented by two other long records: Leffe (Lona, 1950; Ravazzi, 2003) and Pianico Sèllere (Ravazzi, 2003) that had already been discovered long before.

Long sequences were also cored in Greece: Ioannina and the most prominent site Tenaghi Philippon (Tzedakis et al., 2001, 2006). The latter spans more than a million years and pollen analyses by Wijmstra started there as early as 1969.

In spite of the considerable number of well analysed long continental records there is still a lot to be done to correlate them with the great number of climatic stages in both the ice cores and the marine record. Currently great attention is being paid to the Weichselian refuges in southern Europe and in the Balkans as far north as Slovakia both by means of the fossil records (Willis et al., 2000; Jankovská and Pokorný, 2008) and through genetic studies (e.g. Magri et al., 2006; see chapter 4).

6.7

Peatland research

At a very early stage, bogs gained attention as sources of climate and vegetation changes. Peat cutting was common and offered an insight into peat stratigraphy at many places. There are a comparatively large number of early publications on bogs and peat growth, compiled by Gorham (1953) for the British Isles and for a wider geographical area by Overbeck (1975). The oldest treatise on bogs was published in 1658 in the Netherlands by Martin Schookius: *Tractatus de turfis, ceu cespitibus bituminosus,* followed in France by Charles Patin: *Traité des tourbes combustibles* (1663) and in Ireland by the renowned book *On the bogs and loughs in Ireland* by William King (1685), and succeeded by others in various countries. The first studies in Denmark by Dau (1829) later became the basis for the Blytt-Sernander system (see above).

As early as 1811 it was the Scotsman William Aiton (quoted in Barber 1981), who recognized the hummock-hollow complex and its possible role in the growth of bogs. This regeneration cycle theory had been forgotten and was only revived 100 years later by von Post and Sernander (1910) and Osvald (1923). It became a central topic in the discussion about peat growth but after increasing doubts from 1961 onwards, these autogenic cycles were finally refuted by Casparie (1972) and Barber (1981) and explained by local changes in hydrology.

Of great interest was, of course, the boundary horizon between the lower dark, highly humified peat and the upper pale fresh peat that is very distinct in the raised bogs of several countries. It was Weber who in 1900 called this phenomenon the '*Grenzhorizont*' and explained it by a terrestrialization hypothesis because he assumed a hiatus of 1000 years between the two layers. In addition to this clear feature Granlund (1932) described a series of four 'recurrence surfaces' in the stratigraphy of Swedish bogs. The number of these recurrence layers has increased steadily since then and their datings have varied sometimes even within a single bog. Their explanation as indicators of climatic change between dry and humid phases or as local hydrologic conditions triggered many investigations on the basis of macro-remains and on the degree of humification. We now understand that it is often a complex interrelation between climatic fluctuations, local drainage alterations and also, in younger periods, human impact.

For the characterization of peat and other unconsolidated sediments a conclusive identification system of symbolic notation was proposed by Troels-Smith (1955; see chapter 2.2.3) which, however, is not yet generally used. A first atlas on macrofossils in peat was published by Katz and Katz as early as 1933 (substantially improved as Katz et al., 1977) and detailed descriptions were produced by Grosse-Brauckmann (1972, 1974). A broad synthesis of bog research and the state of the art was presented by Overbeck (1975).

6.8

Archaeobotany

Archaeobotany, also called palaeoethnobotany, is very much related to vegetation history and often regarded as a branch of this subject. The two terms, however, are not identical but focus on different aims. While archaeobotany mainly covers the study of the plant remains found in archaeological contexts that were cultivated or utilized by people in ancient times, palaeoethnobotany, a term created by the Danish scholar Hans Helbaek, is wider and also includes the ecological aspects of human–plant relationships. This more accurately covers present-day practice and research goals. Nowadays the term archaeobotany is commonly employed because it sounds more suitable, although it is less appropriate.

The first steps on this subject were made in Egypt by Carl Kunth (1826), who studied the desiccated fruits and seeds in ancient tombs. In Europe it was Oswald Heer, mentioned already, who seriously investigated the plant remains of the Swiss pile dwellings that became visible in the extremely dry winter of 1863/64 and in 1865 published his epoch-making work *Die Pflanzen der Pfahlbauten*. He continued in 1872 with another important treatise *Über den Flachs und die Flachskultur im Altertum* and is regarded as the father of archaeobotany. It took, however, some time before archaeobotany proceeded. After the beginning of the 20th century scholars such as Neuweiler, Netolitzky, Arnaudov and Werneck started investigations in Switzerland as well as in other parts of Europe. This first period of research more or less terminated with the overview of the most important topic at that time *Geschichte unserer Kulturpflanzen* by Karl and Franz Bertsch (1947). At around 1950, archaeobotany received new impulses. Quite a number of investigations started, in the Near East in particular, to unravel the evolution and domestication of food plants and their introduction into Europe. In Europe itself interest in archaeobotany spread across the entire continent, in Denmark (e.g. Helbæk, 1950) and England (Helbæk, 1952) or in Sweden (Hjelmqvist, 1955). Soon several further centres of archaeobotany came into being, among others around van Zeist in the Netherlands, Hopf, Knörzer, Körber-Grohne, Willerding and Schultze-Motel in Germany, Jacomet in Switzerland, Jones in Britain, Tempir and Opravil in Czechoslovakia, Klichowska and Wasylikowa in Poland or Lisitsina in Russia as well as Follieri south of the Alps. From the 1970s archaeobotanical studies increased sharply in most European countries and nowadays in several countries most research positions for vegetation history are even connected with archaeological institutions.

In the 1950s and 1960s the aims of archaeobotany became broader as economic and ecological problems came into prominence. Questions concerning ancient agricultural practices, such as the cultivation of fields, sowing time, harvesting methods, crop processing and storage, gained considerably in interest and archaeobotanists realized the importance of weeds as well as of vegetative remains such as chaff. Also new genetic methods were applied and led to connections with geneticists.

For a long time archaeobotany was dominated by investigations on prehistoric material. Together with the rise of medieval archaeology during the last decades archaeobotany has also focused on these sites and has become an important addition and sometimes also a corrective to written sources from this period.

The first syntheses on archaeobotany were published by Renfrew (1973) and van Zeist et al. (1991) and a textbook appeared from Jacomet and Kreuz in 1999, while the book of Zohary and Hopf (2000) covers the plant domestication of the entire Old World. Already in 1968 Schultze-Motel started to compile all related publications and published an annual bibliography on archaeological plant remains in the journal *Die Kulturpflanze*. From 1995 this bibliography was continued by Kroll and transferred to *Vegetation History and Archaeobotany* and since 2002 it is available online on the internet. Digital plant atlases were published by Cappers et al. (2006) and Neef et al. (2012).

For many years, papers on vegetation history were published in widely scattered journals of various orientations but with the establishment of vegetation history as a discipline in its own right, special journals on this subject were increasingly offered. Already in 1954 the journal *Grana Palynologica* was issued in Sweden, renamed in 1970 as *Grana*. In 1959 *Pollen et Spores* followed in France and from the beginning included a comprehensive bibliography on pollen analysis but ceased regrettably in 1989. Since 1960 *Acta Palaeobotanica* is edited in

Poland and in the Netherlands *Review of Palaeobotany and Palynology* has been published since 1967, the latter two, however, focus mainly on papers on the pre-Quaternary. More recent is the journal *Vegetation History and Archaeobotany*, published from 1992 in Germany: it has a direct link to archaeobotany, which has become one of the most important consumers of our scientific results.

Fig. 6-1. Important European Quaternary botanists. Top left: Oswald Heer (1809–1883). Top right: Adolf Engler (1844–1930). Bottom left: Alfred Gabriel Nathorst (1850–1921). Bottom right: Carl Albert Weber (1856–1931).

Fig. 6-2. Important European Quaternary botanists. Top left: Karl Bertsch (1878–1965). Top right: Lennart von Post (1884–1951). Bottom left: Knud Jessen (1884–1971). Bottom right: Wladyslaw Szafer (1886–1970).

Fig. 6-3. Important European Quaternary botanists. Top left: Werner Lüdi (1888–1968). Top right: Helmut Gams (1893–1976). Bottom left: Fritz Overbeck (1898–1983). Bottom right: Harry Godwin (1901–1985).

Fig. 6-4. Important European Quaternary botanists. Top left: Franz Firbas (1902–1964). Top right: Mark Ilich Neustadt (1903–1985). Bottom left: Johannes Iversen (1904–1971). Bottom right: Max Welten (1904–1984).

Fig. 6-5. Important European Quaternary botanists.
Left: Knut Fægri (1909–2001). Right: Maria Follieri (1932–2012).

7.

Appendices

7.1

Tables

Table 7-1. Palaeoecological sites used for the maps generated in chapter 4.4. The number refers to the map in Fig. 4.3-2. Site names in bold are represented with a histogram in chapter 4.3.

Name	Lat	Lon	Elevation	Country	Region	Nr.	Reference
Čap'pesjav'ri	71.07	25.37	41	NOR	A	1	Huntley et al. (2013)
Lake Hopseidet	70.83	27.72	225	NOR	A	2	Seppä (1996)
Over Kobbkrokvatnet	70.70	29.29	51	NOR	A	3	Huntley et al. (2013)
Jansvatnet	70.65	23.67	53	NOR	A	4	Birks et al. (2012c)
Lake Ifjord	70.43	27.63	317	NOR	A	5	Seppä (1996)
Domsvatnet	70.32	31.03	120	NOR	A	6	Hyvärinen (1976)
Rybachiy	69.60	32.40	60	RUS	A	7	Kremenetski et al. (1997)
Yarnishnoe-3	69.10	36.10	54	RUS	A	8	Snyder et al. (2000)
Poteryanny Zub	68.80	35.30	131	RUS	A	9	Gervais et al. (2002)
Ortino	68.10	54.10	2	RUS	A	10	Kaakinen and Eronen (2000), Väliranta et al. (2003, 2011)
Bolvan I	67.43	48.43	168	RUS	A	11	Paus et al. (2003)
Lake Kharinei	67.37	62.75	110	RUS	A	12	Salonen et al. (2011)
Lake Tumbulovaty	67.10	59.70	115	RUS	A	13	Kultti et al. (2004), Väliranta et al. (2011)
Endlevatn	69.20	16.00	36	NOR	Ba	14	Vorren (1978), Vorren and Alm (1985), Vorren et al. (2013), Parducci et al. (2012)
Donvold	68.13	13.58	10	NOR	Ba	15	Nilssen (1983)
Efstadalsvatn	65.92	-21.66	123	ISL	Ba	16	Caseldine et al. (2003)
Barðalækjartjörn	65.42	19.87	413	ISL	Ba	17	Eddudóttir et al. (2016)
Blåvasstjønn	64.91	11.66	92	NOR	Ba	18	Ramfjord (1982)
Grasvatn	63.70	8.70	45	NOR	Ba	19	Paus (1982)
Saksunarvatn	62.25	-7.18	23	FRO	Ba	20	Jóhansen (1982)

Kråkenes	62.03	5.00	39	NOR	Ba	21	Birks, HH et al. (2000)
Dallican Water	60.39	-1.10	56	GBR	Ba	22	Bennett et al. (1992)
Lochan an Druim	58.47	-4.70	25	GBR	Ba	23	Birks HH (1984)
Little Loch Roag	58.13	-6.88	27	GBR	Ba	24	Birks, HJB (1973b), Birks and Madsen (1979)
Loch Sionascaig	58.06	-5.18	74	GBR	Ba	25	Pennington et al. (1972)
Coire Bog	57.85	-4.42	259	GBR	Ba	26	Birks HH (1975)
Bruvatnet	70.18	28.42	119	NOR	Bb	27	Hyvärinen (1975a)
Trollvatnet	69.88	23.47	188	NOR	Bb	28	Hyvärinen (1985)
Suovalampi	69.58	28.83	104	FIN	Bb	29	Hyvärinen (1975a)
Akuvaara	69.13	27.68	170	FIN	Bb	30	Hyvärinen (1975a)
Mukkavaara	68.92	21.00	535	FIN	Bb	31	Hyvärinen (1993)
Lake Tsuolbmajavri	68.69	22.08	526	FIN	Bb	32	Seppä and Birks (2001), Seppä and Hicks (2006)
Badsjön	68.33	18.75	800	SWE	Bb	33	Barnekow (2000)
Chuna	67.95	32.48	475	RUS	Bb	34	Solovieva et al. (2005)
Sjuodjijaure	67.40	18.10	826	SWE	Bb	35	Rosén et al. (2001)
Krugloye	66.37	37.58	140	RUS	Bb	36	Kremenetski and Patyk-Kara (1997)
Kangerjoki	66.12	29.00	288	FIN	Bb	37	Hicks (1975)
Solnechnoe	65.83	34.33	10	RUS	Bb	38	Elina (1981)
Maanselänsuo	65.62	29.60	248	FIN	Bb	39	Vasari (1965)
Zapovednoe	65.12	32.63	110	RUS	Bb	40	Elina (1981)
Vasikkasuo	64.67	27.87	270	FIN	Bb	41	Vuorela (1991) , Vuorela and Kankainen (1991)
Rugozero	64.08	32.63	140	RUS	Bb	42	Elina (1981)
Kassjön	63.92	20.02	84	SWE	Bb	43	Renberg and Segerström (1981)
Zaruckoe	63.90	36.25	20	RUS	Bb	44	Elina (1981)
Abborrtjärnen	63.88	14.45	387	SWE	Bb	45	Giesecke (2005c)
Punozerka	62.82	33.58	157	RUS	Bb	46	Elina and Filimonova (1987)
Flåfattjønna	62.33	10.40	1110	NOR	Bb	47	Paus (2010)
Styggtjärnen	62.32	13.57	715	SWE	Bb	48	Giesecke (2005c)
Gotnavolok	62.20	33.80	110	RUS	Bb	49	Elina et al. (1987), Elina and Filimonova (1996)
Ylimysneva	62.13	22.87	172	FIN	Bb	50	Huttunen (1990)

Bezdonnoe	62.03	32.77	121	RUS	Bb	51	Elina (1981), Elina and Filimonova (1996)
Aholampi	61.88	25.22	114	FIN	Bb	52	Koivula (1987)
Klotjärnen	61.82	16.53	235	SWE	Bb	53	Giesecke (2005b)
Ladoga Lake	61.56	31.34	5	RUS	Bb	54	Arslanov et al. (1996)
Brurskardtjörni	61.42	8.67	1309	NOR	Bb	55	Bjune (2005)
Siikasuo	61.30	22.07	35	FIN	Bb	56	Vuorela (1991)
Syrjälänsuo	61.22	28.12	83	FIN	Bb	57	Vuorela (unpubl.)
Trettetjörn	60.72	7.00	810	NOR	Bb	58	Bjune (2005), Bjune et al. (2005)
Holtjärnen	60.65	14.92	232	SWE	Bb	59	Giesecke (2005b)
Hirvilampi	60.63	24.25	114	FIN	Bb	60	Rankama and Vuorela (1988)
Vestre Oykjamyrtjorn	59.82	6.00	570	NOR	Bb	61	Bjune (2005), Bjune et al. (2005)
Lilla Gloppsjön	59.80	14.63	198	SWE	Bb	62	Almquist-Jacobson (1994)
Lake Llet-Ti	66.50	59.30	50	RUS	Bc	63	Väliranta et al. (2006), Klimaschewski (unpubl.)
Mezhgornoe	65.26	59.67	550	RUS	Bc	64	Kultti et al. (2003)
Onego Lake	61.70	34.90	33	RUS	Bc	65	Khomutova (1976, 1989), Khomutova and Elina (1990)
Glubokoe	61.07	36.05	50	RUS	Bc	66	Elina (1981)
Ilmen Lake	58.30	31.23	18	RUS	Bc	67	Khomutova (1989), Davydova et al. (1992)
Chistic	57.34	33.00	205	RUS	Bc	68	Surova and Chernavskaya (1997)
Lake Nero	57.18	39.45	95	RUS	Bc	69	Gunova (1972, 1975)
Lake Maardu	59.43	25.00	32	EST	Bd	70	Saarse et al. (1990a)
Liivjarve Bog	59.22	27.58	46	EST	Bd	71	Koff (1989)
Åntu Sinijarv	59.13	26.33	95	EST	Bd	72	Saarse and Liiva (1995)
Morttjern	59.06	11.62		NOR	Bd	73	Birks et al. (2012b)
Lake Jarveotsa	59.03	24.28	42	EST	Bd	74	Poska (1994)
Raigastvere Lake	58.60	26.67	52	EST	Bd	75	Pirrus et al. (1987)
Lake Flarken	58.58	13.67	109	SWE	Bd	76	Digerfeldt (1977)
Mire Pelisoo	58.47	22.38	33	EST	Bd	77	Saarse et al. (1990b)
Mire Saviku	58.40	27.23	30	EST	Bd	78	Sarv and Il'ves (1975)
Lake Ermistu	58.37	23.97	17	EST	Bd	79	Heinsalu and Veski (1991)
Dags Mosse	58.33	14.70	260	SWE	Bd	80	Göransson (1989)
Påidre	58.27	25.63	506	EST	Bd	81	Saarse et al. (1995)

Mabo Mosse	58.02	16.07	117	SWE	Bd	82	Göransson (1977)
Kirikumae	57.67	27.25	183	EST	Bd	83	Saarse et al. (1996), Saarse and Rajamäe (1997)
Kansjon	57.63	14.53	308	SWE	Bd	84	Jacobson (unpubl.)
Bergakyllen	57.17	16.15	73	SWE	Bd	85	Svensson (1989)
Lake Trummen	56.87	14.83	161	SWE	Bd	86	Digerfeldt (1972)
Novolsky	56.77	26.18	168	BLR	Bd	87	Zernitskaya et al. (1988), Matveev et al. (1993)
Kurjanovas	56.50	27.50	111	LTV	Bd	88	Heikkilä et al. (2009)
Färshesjön	56.17	15.87	14	SWE	Bd	89	Berglund (1966)
Myshetskoe-Dolgoe	56.07	37.33	198	RUS	Bd	90	Kremenetski et al. (2000), Borisova (2014)
Osvea	56.05	28.08	132	BLR	Bd	91	Zernitskaya (1991)
Mire Petrolivo	56.00	31.98	175	RUS	Bd	92	Gunova and Sirin (1995)
Ezerisch	55.85	30.00	165	BLR	Bd	93	Bogdel (1984), Elovicheva and Bogdel (1985)
Shepeta Bog	55.78	25.17	110	LTU	Bd	94	Kabailiene (1971, 1990)
Gretskoe	55.63	27.77	134	BLR	Bd	95	Bogdel (1984), Elovicheva and Bogdel (1985)
Tytuvenu Tyrelis	55.58	23.30	125	LTU	Bd	96	Kabailiene (1971, 1990)
Dolgoe	55.23	28.18	173	BLR	Bd	97	Zernitskaya (1991)
Leikishke Bog	54.53	23.53	120	LTU	Bd	98	Kabailiene (unpubl.)
Lake Hancza	54.27	22.82	229	POL	Bd	99	Lauterbach et al. (2011b)
Bezymiannoe	54.22	30.00	190	BLR	Bd	100	Bogdel (1984), Elovicheva and Bogdel (1985)
Bebrukas Lake	54.09	24.12	2	LTU	Bd	101	Kabailiene (1965, 1986), Shulija et al. (1967)
Sudoble Lake	54.03	28.10	165	BLR	Bd	102	Bogdel (1984), Elovicheva and Bogdel (1985)
Lochinskoe	53.55	28.60	164	BLR	Bd	103	Bogdel (1984), Elovicheva and Bogdel (1985)
Blomøy	60.53	4.88	36	NOR	Ta	104	Mangerud (1970)
Sandvikvatn	59.28	5.50	130	NOR	Ta	105	Paus (1982, 1988, 2013)
Loch Clair	57.56	-5.34	92	GBR	Ta	106	Pennington et al. (1972)
Abernethy Forest	57.33	-3.72	221	GBR	Ta	107	Birks, H.H. (1970); Birks and Mathewes (1978)
Loch Ashik	57.25	-5.83	30	GBR	Ta	108	Birks and Williams (1983)
Loch Cleat	57.07	-6.33	38	GBR	Ta	109	Birks and Williams (1983)

Morrone Birkwoods	57.00	-3.43	425	GBR	Ta	110	Huntley (1994), Huntley and Birks (1979)
Gallanech Beg	56.38	-5.50	70	GBR	Ta	111	Fay Davies (unpubl.), Macklin et al. (2000)
By Loch Dungeon	55.12	-4.32	396	GBR	Ta	112	Birks, HH (1975)
Sluggan Moss	54.93	-6.30	52	IRL	Ta	113	Smith and Goddard (1991)
Thorpe Bulmer	54.72	-1.30	100	GBR	Ta	114	Bartley et al. (1976), Bartley and Morgan (1990)
Cregganmore	54.25	-9.60	60	IRL	Ta	115	McKeever (1984)
Clara Bog	53.32	-7.63	59	IRL	Ta	116	Connolly (1999)
Arts Lough	52.95	-6.43	490	IRL	Ta	117	Bradshaw and McGee (1988)
King's Pool	52.81	-2.11	70	GBR	Ta	118	Bartley and Morgan (1990)
Mekelermeer	52.77	6.62	15	NLD	Ta	119	Bohncke (1991)
Llyn Gwernan	52.68	-4.87	165	GBR	Ta	120	Lowe (1981), Lowe et al. (1988), Lowe and Lowe (1989)
Hockham Mere	52.50	0.80	33	GBR	Ta	121	Godwin and Tallantire (1951), Bennet (1983)
Uddelermeer	52.25	5.76	26	NLD	Ta	122	Bohncke et al. (1988), Bohncke and Wijmstra (1988)
Ballinloghig Lake	52.20	-10.31	84	IRL	Ta	123	Barnosky (1988)
Belle Lake	52.20	-7.00	33	IRL	Ta	124	Craig (1973, 1978)
Notsel	51.55	4.77	5	NLD	Ta	125	Bohncke et al. (1987)
Llanilid	51.52	-3.45	60	GBR	Ta	126	Walker and Harkness (1990), Harkness and Walker (1991)
Vinderhoute	51.08	3.62	4	BEL	Ta	127	Verbruggen (1979), Verbruggen et al. (1996)
Pannel Bridge	50.91	0.68	3	GBR	Ta	128	Waller (1987, 1993)
Cranes Moor	50.83	-1.73	45	GBR	Ta	129	Grant et al. (2009, 2014)
Dartmoor	50.60	-4.00	603	GBR	Ta	130	Fyfe and Woodbridge (2012)
Urville-Nacqueville	49.68	-1.76	5	FRA	Ta	131	Clet-Pellerin (1991, unpubl.)
Bellengreville	49.12	-0.22	20	FRA	Ta	132	Clet-Pellerin (2005, unpubl.)
La Vie	48.55	-0.26	238	FRA	Ta	133	Barbier (1999)
Guidel	47.77	-3.50	2	FRA	Ta	134	Fernane et al. (2015)
Lutinière	46.44	-0.86	4	FRA	Ta	135	Visset (unpubl.)
Le Marais de la Perge (South)	45.38	-1.12	2	FRA	Ta	136	Diot and Tastet (1995)

Pena Vella	43.44	-7.52	700	ESP	Ta	137	Ramil-Rego et al. (1998), Iriarte-Chiapusso et al. (2016)
Puerto de Los Tornos	43.15	-3.43	920	ESP	Ta	138	Peñalba (1989)
Lago de Ajo	43.05	-6.15	1570	ESP	Ta	139	McKeever (1984), Allen et al. (1996)
Saldropo	43.05	-2.72	625	ESP	Ta	140	Peñalba (1989, 1994)
Lagoa de Lucenza	42.59	-7.11	1375	ESP	Ta	141	Leira and Santos (2002)
PRD-4	42.53	-8.52	260	ESP	Ta	142	López-Merino et al. (2012)
Laguna de la Roya	42.22	-6.77	1608	ESP	Ta	143	Allen et al. (1996)
Skogstjern	59.01	9.64	57	NOR	Tb	144	Wieckowska-Lüth et al. (2017)
Dalane	58.20	8.00	40	NOR	Tb	145	Eide et al. (2006)
Skånsö	56.51	8.83	5	DNK	Tb	146	Odgaard (1994)
Ran Viken	56.28	14.30	213	SWE	Tb	147	Digerfeldt (1973, 1974)
Lake Solsø	56.13	8.63	41	DNK	Tb	148	Odgaard (1988, 1994)
Krageholmssjön	55.50	13.73	43	SWE	Tb	149	Gaillard (1984b) , Gaillard and Digerfelt (1991)
Dallund Sø	55.48	10.28	12	DNK	Tb	150	Rasmussen (2005)
Lake Belau	54.10	10.27	39	GER	Tb	151	Dörfler et al. (2012)
Lake Racze	53.97	14.61	-9	POL	Tb	152	Latałowa and Tobolski (1989), Latałowa (1992)
Lake Mikolajki	53.77	21.42	116	POL	Tb	153	Ralska-Jasiewiczowa (1966, 1989)
Maly Suszek	53.73	17.77	115	POL	Tb	154	Miotk-Szpiganowicz (1992)
Swienskuhle	53.70	8.70	5	GER	Tb	155	Behre (1976), Behre and Kucan (1986, 1994)
Löddigsee	53.43	11.85	43	GER	Tb	156	Jahns (2007)
Strazym Lake	53.33	19.46	71	POL	Tb	157	Noryskiewicz (1987), Rozanski et al. (1985)
Biskupinskie Lake	52.78	17.73	80	POL	Tb	158	Noryskiewicz (1995)
Lake Gościąż	52.58	19.35	64	POL	Tb	159	Ralska-Jasiewiczowa et al. (1998)
Bledowo Lake	52.55	20.67	78	POL	Tb	160	Binka et al. (1988, 1991)
Schwanengrabenrinne	52.52	13.06	32	GER	Tb	161	Wolters (2002)
Grosser Treppelsee	52.15	14.45	52	GER	Tb	162	Giesecke (2001)
Luttersee	51.58	10.16	164	GER	Tb	163	Chen (1988)
Czajkow	50.78	21.28	206	POL	Tb	164	Szczepanek (1971)
Rappershausen	50.38	10.39	384	GER	Tb	165	Hahne (1992)

Hrabanoská Černava	50.20	14.80	184	CZE	Tb	166	Petr (2005), Petr and Novák (2014)
Holzmaar	50.12	6.88	336	GER	Tb	167	Litt and Stebich (1999), Litt et al. (2009)
Rasna	49.23	15.37	680	CZE	Tb	168	Rybnícek (1973), Rybnícková (1974)
Borkovická blata	49.22	14.90	415	CZE	Tb	169	Jankovská (1980)
Rachelsee	48.98	13.40	1071	GER	Tb	170	van der Knaap et al. (2020)
Červené blato	48.85	14.93	470	CZE	Tb	171	Jankovská (1980)
Huzenbacher See	48.57	8.35	747	GER	Tb	172	Rösch and Tserendorj (2011a,b)
Auneau	48.46	1.79	127	FRA	Tb	173	Hervé (unpubl.)
Altenweiher	48.01	6.99	926	FRA	Tb	174	De Valk (1981)
Trumer Moos	47.94	13.07	500	AUT	Tb	175	Krisai (1975)
Durchenbergried	47.80	9.00	432	GER	Tb	176	Rösch (1986)
Tourbière de Mur de Sologne	47.41	1.61	115	FRA	Tb	177	Planchais (1970)
Etang de la Gruère	47.24	7.05	1012	CHE	Tb	178	Fankhauser (1995), van der Knaap et al. (2000)
Champ Gazon	47.20	4.05	585	FRA	Tb	179	Jouffroy-Bapicot et al. (2013)
Burgäschisee	47.17	7.67	465	CHE	Tb	180	Rey et al. (2017)
Rotsee	47.07	8.32	419	CHE	Tb	181	Lotter (1988)
Lobsigensee	47.03	7.30	514	CHE	Tb	182	Ammann (1989)
Praz Rodet	46.57	6.17	1040	CHE	Tb	183	Mitchell (1995), van der Knaap et al. (2000)
Hières sur Amby	45.79	5.28	227	FRA	Tb	184	Clerc (1988)
Lake Aydat	45.66	2.99	837	FRA	Tb	185	Lavrieux et al. (2013)
Gourgon	45.61	3.88	1367	FRA	Tb	186	Cubizolle et al. (2014)
La Taphanel	45.27	2.68	975	FRA	Tb	187	De Beaulieu et al. (1982)
Peyre	44.96	2.72	1100	FRA	Tb	188	Surmely et al. (2009)
Les Roustières	44.71	3.09	1196	FRA	Tb	189	Ponel et al. (2016)
Klukvennoye	54.78	43.45	110	RUS	Tc	190	Novenko et al. (2018)
Klukva	53.83	36.25	210	RUS	Tc	191	Novenko et al. (2015)
Berezovskoye	53.67	38.59	210	RUS	Tc	192	Novenko et al. (2012)
Chernikhovo	53.42	26.43	168	BLR	Tc	193	Zernitskaya (1985), Matveev et al. (1993)
Koldychevo	53.27	26.07	185	BLR	Tc	194	Bogdel (1984), Elovicheva and Bogdel (1985)

Svityaz	51.51	23.84	157	UKR	Tc	195	Schwörer et al. (2022)
Makovich	50.97	24.22	192	UKR	Tc	196	Artushenko (1957, 1959)
Nowy Gutiski	50.27	26.83	210	UKR	Tc	197	Klimanov and Bezusko (1981), Artushenko et al. (1982), Bezusko et al. (1988)
Polonichka	50.27	24.75	200	UKR	Tc	198	Bezusko (1977)
Starniki	50.27	26.02	198	UKR	Tc	199	Klimanov and Bezusko (1981), Khotinskiy et al. (1994)
Dovjok Swamp	48.75	28.25	274	UKR	Tc	200	Kremenetski (1995)
Atxuri	43.25	-1.55	500	ESP	Td	201	Pérez-Díaz et al. (2015)
Biscaye	43.00	-0.10	409	FRA	Td	202	Mardones and Jalut (1983), Reille (1990c), Reille and Andrieu (1995)
Gourg Nègre	42.60	2.20	2080	FRA	Td	203	Reille and Lowe (1993)
Basa de la Mora	42.50	0.30	1914	ESP	Td	204	Pérez-Sanz et al. (2013)
Bosc dels Estanyons	42.50	1.60	2180	AND	Td	205	Miras et al. (2007)
Lago Grande de Estanya	42.03	0.53	670	ESP	Td	206	Vegas-Villarúbia et al. (2013)
Moosalmmoor	47.75	13.52	740	AUT	Te	207	Draxler (1977)
Schwemm	47.66	12.30	664	AUT	Te	208	Oeggl (1988)
Hochmoos	47.42	13.20	800	AUT	Te	209	Wahlmüller (1992, 1993)
Mieminger See	47.29	10.97	800	AUT	Te	210	Wahlmüller (1985)
Gerlos	47.24	12.14	1590	AUT	Te	211	Bortenschlager (1984)
Grosses Überling Schattseit-Moor	47.17	13.90	1750	AUT	Te	212	Krisai et al. (1991)
Schwarzsee	46.87	10.48	1721	AUT	Te	213	Welten (1982a)
Bachalpsee	46.67	8.02	2265	CHE	Te	214	Lotter et al. (2006)
Aegelsee	46.64	7.54	989	CHE	Te	215	Wegmüller and Lotter (1990)
Tammerle Moos	46.58	11.28	1400	AUT	Te	216	Wahlmüller (1990)
Lej da San Murezzan	46.49	9.85	1768	CHE	Te	217	Gobet et al. (2003, 2005)
Alpi di Robièi	46.44	8.52	1892	CHE	Te	218	Welten (1982b)
Zeneggen-Hellelen	46.28	7.84	1520	CHE	Te	219	Welten (1982b)
Lac du Mont d'Orge	46.23	7.34	640	CHE	Te	220	Welten (1982b)
Pian di Gembro	46.16	10.15	1350	ITA	Te	221	Pini (2002)
Lac de Villa	45.69	7.76	820	ITA	Te	222	Brugiapaglia (1996)
Lac des Boites	45.06	5.89	1560	FRA	Te	223	Nakagawa (1998)
Correo	44.51	5.98	1090	FRA	Te	224	Nakagawa (1998)

Plan du Laus	44.24	6.70	2121	FRA	Te	225	De Beaulieu (1977)
Lac Long Inférieur	44.06	7.45	2093	FRA	Te	226	De Beaulieu (1977)
Zlatnicka Dolina	49.52	19.28	850	SVK	Tf	227	Rybnícek and Rybnícková (1985)
Puścina Rękowiańska	49.48	19.82	656	POL	Tf	228	Obidowicz (1989, 1990, 1993)
Regetovka	49.43	21.28	515	SVK	Tf	229	Wacnik (1995)
Jasiel	49.37	21.89	680	POL	Tf	230	Szczepanek (1987, 1989)
Tarnawa Wyzna	49.10	22.83	670	POL	Tf	231	Ralska-Jasiewiczowa (1972, 1980, 1989)
Hozelec	49.05	18.30	685	SVK	Tf	232	Jankovská (1988)
Liptovsky Jan	49.04	19.68	660	SVK	Tf	233	Dolejsova (unpubl.)
Steregoiu	47.81	23.54	790	ROU	Tf	234	Björkman et al. (2003, 2002), Feurdean and Bennike (2004), Feurdean et al. (2008)
Poiana Ştiol	47.60	24.80	1540	ROM	Tf	235	Tantau et al. (2011)
Capatana	46.47	23.14	1220	ROM	Tf	236	Fărcaş et al. (2003)
Luci	46.29	24.74	1080	ROU	Tf	237	Tanţău et al. (2003a, b, 2014), Tanţău (2003)
Mohos	46.08	25.92	1050	ROU	Tf	238	Tanţău et al. (2003a, b), Tanţău (2003), Feurdean et al. (2007)
Avrig	45.72	24.38	400	ROU	Tf	239	Tanţău et al. (2006), Feurdean et al. (2007)
Bisoca	45.50	26.70	875	ROM	Tf	240	Tanţău et al. (2009), Tanţău (2003)
Taul dintre Brazi	45.40	22.90	1740	ROM	Tf	241	Buczkó et al. (2009), Magyari et al. (2009a)
Semenic	45.18	22.06	1400	ROU	Tf	242	Rösch and Fischer (2000)
Iezerul Calimani	45.10	25.30	1650	ROM	Tf	243	Fărcaş et al. (1999), Feurdean et al. (2007)
Mire Garvan	44.10	27.00	20	BGR	Tg	244	Lazarova (1995)
Crveni Potok	43.91	19.42	1090	SRB	Tg	245	Finsinger et al. (2017b)
Lake Varna	43.20	27.80	0	BGR	Tg	246	Bozilova and Beug (1994)
Vitosha Mountains Peat Bog	42.83	23.83	2000	BGR	Tg	247	Filipovitch (1985, 1988)
Straldzha mire	42.60	26.80	137	BGR	Tg	248	Connor et al. (2013)
Western Black Sea GGC-18	42.50	28.40	0	BGR	Tg	249	Filipova-Marinova et al. (2013a)
Peat-bog Begbunar	42.15	22.55	1750	BGR	Tg	250	Lazarova et al. (2009, 2015)

Sucho Ezero	42.05	23.53	1900	BGR	Tg	251	Bozilova and Smith (1979), Bozilova et al. (1986)
Kupena	41.98	24.33	1300	BGR	Tg	252	Huttunen et al. (1992)
Lake Kremensko	41.70	23.50	2124	BGR	Tg	253	Stefanova et al. (2006a)
Beliya Kanton	41.70	24.10	1547	BGR	Tg	254	Lazarova et al. (2011)
Tenaghi Philippon	41.16	24.30	40	GRC	Tg	255	Wijmstra (1969), Turner and Greig (1975), van der Wiel and Wijmstra (1987a,b)
Lake Prespa	41.00	21.00	849	MKD	Tg	256	Panagiotopoulos et al. (2013)
Lake Ohrid	40.92	20.67	693	ALB	Tg	257	Lézine et al. (2010)
Nisi Fen	40.82	21.92	475	GRC	Tg	258	Lawson et al. (2005)
Zazari	40.63	21.55	606	GRC	Tg	259	Gassner et al. (2020)
Lake Orestiás	40.51	21.26	630	GRC	Tg	260	Kouli and Dermitzakis (2010)
Maliq	40.35	20.41	818	ALB	Tg	261	Denèfle et al. (2000)
Ioannina	39.76	20.73	470	GRC	Tg	262	Bottema (1974), Tzedakis (1993), Lawson et al. (2004), Tzedakis et al. (2002)
Lago di Origlio	46.06	8.94	416	CHE	Th	263	Tinner et al. (1999)
Lago del Segrino	45.83	9.27	374	ITA	Th	264	Gobet et al. (2000)
Griblje	45.57	15.28	160	SLO	Th	265	Andrič (2007)
Lago della Costa	45.27	11.74	7	ITA	Th	266	Kaltenrieder et al. (2009, 2010)
Lago Piccolo di Avigliana	45.05	7.39	356	ITA	Th	267	Finsinger and Tinner (2006), Finsinger et al. (2006)
Lago Verdarolo	44.36	10.12	1390	ITA	Th	268	Morales-Molino et al. (2021)
Pavullo nel Frignano	44.32	10.84	675	ITA	Th	269	Vescovi et al. (2010b)
Banyoles	42.13	2.75	173	ESP	Ma	270	Pérez-Obiol and Julia (1994)
Ayoó de Vidriales	42.10	-6.10	780	ESP	Ma	271	Morales-Molino et al. (2014)
Quintanar de la Sierra	42.03	-3.02	1470	ESP	Ma	272	Peñalba (1989, 1994), Peñalba et al. (1997), Ruiz Zapata et al. (2002)
Espinosa de Cerrato	41.96	-3.94	885	ESP	Ma	273	Franco-Múgica et al. (2001)
El Carrizal de Cuéllar	41.32	-4.15	860	ESP	Ma	274	Franco-Múgica et al. (2005)
Central Ebro Desert	41.20	-0.30	150-357	ESP	Ma	275	Davis and Stevenson (2007)

Ojos del Tremedal	40.54	-2.05	1650	ESP	Ma	276	Stevenson (2000)
El Maillo	40.50	-6.20	1100	ESP	Ma	277	Morales-Molino et al. (2013)
Charco da Candieira	40.34	-7.57	1409	PRT	Ma	278	Van der Knaap and van Leeuwen (1995, 1997)
Menorca	39.90	4.10	21	ESP	Ma	279	Yll et al. (1994, 1995, 1997)
Navarrés	39.10	-0.68	225	ESP	Ma	280	Carrión and van Geel (1999)
Ojos de Villaverde	38.80	-2.37	870	ESP	Ma	281	Carrión et al. (2001b)
Siles	38.40	-2.50	1320	ESP	Ma	282	Carrión (2002b)
Cañada de la Cruz	38.06	-2.69	1595	ESP	Ma	283	Carrión et al. (2001b), Carrión (2002b)
Guadiana	37.26	-7.45	1	PRT	Ma	284	Boski et al. (2002, 2008), González-Vila et al. (2003), Fletcher et al. (2007)
Sierra de Baza	37.23	-2.70	1900	ESP	Ma	285	Carrión et al. (2007), Gil-Romera et al. (2009, 2010)
Antas	37.21	-1.82	0	ESP	Ma	286	Yll et al. (1995), Pan-taleón-Cano et al. (2003)
Padul 3	37.00	-3.10	785	ESP	Ma	287	Menéndez-Amor and Florschütz 1963, Florschütz et al. (1971), Pons and Reille (1988), Valle-Hernandez et al. (2003)
San Rafael	36.77	-2.60	0	ESP	Ma	288	Yll et al. (1994, 1995), Pantaleón-Cano et al. (2003)
Laguna de Medina	36.62	-6.05	30	ESP	Ma	289	Reed et al. (2001)
Bokanjacko	44.18	15.23	25	HRV	Mb	290	Grüger (1996)
Massaciuccoli	43.80	10.30	0	ITA	Mb	291	Colombaroli et al. (2007)
Colfiorito	43.03	12.93	752	ITA	Mb	292	Brugiapaglia and de Beaulieu (1995)
Lago dell'Accesa	42.99	10.89	157	ITA	Mb	293	Drescher-Schneider et al. (2007b)
Malo Jezero	42.80	17.40	0	HRV	Mb	294	Jahns and van den Bogaard (1998), Jahns (2002)
Lagaccione	42.60	11.90	355	ITA	Mb	295	Magri (1999)
Lac de Creno	42.20	8.90	1310	FRA	Mb	296	Reille (1975)
Lago di Martignano	42.12	12.33	204	ITA	Mb	297	Kelly and Huntley (1991)
Lago Grande di Monticchio	40.94	15.60	656	ITA	Mb	298	Watts et al. (1996b), Allen et al. (2002)

Baratz	40.68	8.23		ITA	Mb	299	Pedrotta et al. (2021)
Sa Curcurica	40.50	9.80	0	ITA	Mb	300	Beffa et al. (2016)
Trifoglietti	39.50	16.00	1048	ITA	Mb	301	de Beaulieu et al. (2017)
Urio Quattrocchi	37.90	14.40	1044	ITA	Mb	302	Bisculm et al. (2012)
Gorgo Basso	37.62	12.65	6	ITA	Mb	303	Tinner et al. (2009)
Lago di Pergusa	37.50	14.50	670	ITA	Mb	304	Sadori and Narcisi (2001)
Biviere di Gela	37.00	14.30	7	ITA	Mb	305	Noti et al. (2009)
Lake Voulkaria	38.87	20.83	0	GRC	Mc	306	Jahns (2005)
Kopais	38.48	23.07	10	GRC	Mc	307	Turner and Greig (1975), Allen (1986, 1990)
Lake Lerna	37.60	22.80	0	GRC	Mc	308	Jahns (1993)
Delphinos and Kournas	35.30	24.30	0-20	GRC	Mc	309	Bottema and Sarpaki (2003) , Jouffroy-Bapicot et al. (2021)
Kis-Mohos-tó	48.40	20.40	310	HUN	Pa	310	Willis et al. (1997, 1998)
Sarló-hát	47.90	21.30	86	HUN	Pa	311	Magyari et al. (2010)
Sirok Nyríes-tó	47.90	20.20	200	HUN	Pa	312	Gardner (2002)
Parížske Mire	47.87	18.47	123	SVK	Pa	313	Jamrichová et al. (2017)
Bátorliget	47.77	22.27	150	HUN	Pa	314	Willis et al. (1995), Sümegi and Gulyás (2004)
Csögle	47.22	17.26	134	HUN	Pa	315	Juhász (2002)
Zalavári	46.80	17.20	188	HUN	Pa	316	Juhász (2002, Zatykó et al. (2007))
Zsombo Swamp	46.36	19.99	92	HUN	Pa	317	Nagy-Bodor et al. (1995)
Pobochnoye	52.97	51.88		RUS	Po	318	Kremenetski et al. (1999a)
The Bog Liman	49.70	37.70	150	UKR	Po	319	Bezusko (1973)
Lipigi	49.50	43.40	90	RUS	Po	320	Kremenetski et al. (1999b)
Kharabuluk	47.70	44.30	70	RUS	Po	321	Kremenetski et al. (1999b)
Kardashinski Swamp	46.52	32.62	4	UKR	Po	322	Kremenetski (1995)
Lake Durankulak	43.67	28.55	4	BGR	Po	323	Bozilova and Tonkov (1985, 1998), Bozilova and Atanassova, (1990), Marinova (2003)

Table 7-2. A, Arctic sites, from northwest to southeast, in bold sites with a histogram in Fig. 4.3-2b.

Site	Lat. °N	Long. °E	Alt. (m asl)	Time span (cal ka BP)	Landscapes, remarks	References
Semmeldalen	76.7	15.3	25		Svalbard	Zelikson (1971)
Skardtjørna	78.00	13.67		9.25-0		H.H. Birks (1991a)
Jansvatnet	70.65	23.67	53	14.7-9.3	Multi-proxy	H.H. Birks et al. (2012c)
Čap'pesjav'ri	71.07	25.37	41	13.3-0	West-East transect	Huntley et al. (2013)
Over Gunnardsford	71.04	28.17	78	12.5-0		Allen et al. (2007), Huntley et al. (2013)
Over Kobbkrokvatnet	70.7	29.29	51	13.2-0		Huntley et al. (2013)
Ifjord	70.4	27.6	317	11.8-0	North - South transect	Seppä (1996)
Hopseidet	70.8	27.7	225	11.5-0		Seppä (1996)
Skaidejavri	70.05	27.87	183	10.7-0		Seppä (1996)
Rautuselkä	69.57	28.53	136	11-0		Seppä (1996)
Domsvatnet	70.3	31.0	120	11.5-0		Hyvärinen (1976)
Rybachiy	69.6	32.4	60	14-2.5		Kremenetski et al. (1997)
Poteryanny Zub	68.8	35.3	131	11.5-0	Paired sites	Gervais et al. (2002), in tundra
Yarnyshnoe-3	69.07	36.07	54	12.8-1		Snyder et al. (2000), in forest-tundra
Bolvan I Timan Ridge	67.4	48.4	168	16.5-0	West-East transect	Paus et al. (2003)
Ortino	68.1	54.1	2	11.5-0		Kaakinen and Eronen (2000), Väliranta et al. (2003, 2011)
Tumbulovaty	67.1	59.7	115	11.5-0		Kultti et al. (2004), Väliranta et al. (2011)
Kharinei	67.4	62.8	110	12-0		Salonen et al. (2011)

Table 7-3. Td, the Pyrenees, selected sites from West to East, in bold sites with a histogram in Fig. 4.3-10b.

Site	Lat. °N	Long. °W-°E	Alt. (m asl)	Time span (cal ka BP)	References
Atxuri	43.25	1.55 W	500	7.5 – 3	Peñalba (1989), Galop et al. (2004), Pérez-Díaz et al. (2015)
La Moura	43.47	1.52 W	40	>15-0	Oldfield (1964), Reille (1993b)
El Portalet	42.8	0.55 W	1802	33 – 6	González-Sampériz et al. (2006)
Estarrès	43.09	0.41 W	376	>14 - 0	Andrieu (1987), Jalut et al. (1988,1992), Andrieu et al. (1993)
Castet	43.03	0.37 W	850	11 - 0	Jalut et al. (1988), Reille (1991a)
Biscaye	43.03	0.08 W	409	>17 1	Mardones and Jalut (1983), Reille (1990a), Reille and Andrieu (1995)
Le Monge	43.03	0.02 W	394	>15 – 1.5, hiati in LG	Alimen et al. (1965)
Lourdes	43.03	0.08 W	430	>17 - 0	Alimen et al. (1965), Kolstrup (1980)

Barbazan	43.03	0.62 E	450	26 - 11	Andrieu (1991)
Basa de la Mora	42.53	0.32 E	1914	10 - 0	Pérez-Sanz et al. (2013)
Tramacastilla	42.72	0.37 E	1668	33.6 -0, hiati	Montserrat-Martí (1992)
La Bassa d'Ules	42.73	0.77 E	1600	>14 - 0	Aubert (1993), Jalut et al. (1996)
Ibon de las Ranas	42.82	0.50 E	2078	11 - 0	Montserrat-Martí (1992)
Freychinède	42.8	1.44 E	1350	>15 - 0	Jalut et al. (1982), Jalut et al. (1992), Reille (1993a)
Bosc dels Estanyons	42.48	1.63 E	2180	11.7 - 0	Miras et al. (2007)
Pinet	42.87	1.97 E	880	11.7 - 0	Keller (1929), Jalut (1974), Reille (1990b), Reille and Lowe (1993)
Laurenti	42.68	2.03 E	1860	>14 - 0	Jalut (1974), Reille (1990b)
Fournas	42.7	2.06 E	1500	5 - 0	Jalut (1974), Reille (1990b)
Bassettes	42.66	2.02 E	1900	13 – 1, hiati	Reille (1990b), Reille and Lowe (1993)
Balcère	42.59	2.06 E	1764	>15 – 1 hiatus: no Boreal	Van Campo and Jalut (1969), Reille (1990b), Reille and Lowe (1993)
Racou	42.6	2.0 E	2000	12.5 - 0	Guiter et al. (2005)
La Borde	42.53	2.08 E	1660	>14 - 5	Reille (1990a, b), Reille and Lowe (1993)
Gourg Nègre	42.63	2.22 E	2000	17–0, hiati in LG	Reille (1990b), Reille and Lowe (1993)
La Moulinasse	42.68	2.24 E	1330	>17 - 10, some hiati, 2.7 - 0	Jalut (1974), Reille (1990b), Reille and Lowe (1993)
Lake Estanya	42.03	0.7 E (S-foothills)	670	19 - 9	Vegas-Villarúbia et al. (2013)

Table 7-4. Tf, the Carpathians, including parts of Tg, the temperate south-eastern European region, sites from north-west to south-east, in bold sites with a histogram in Fig. 4.3-12b or Fig. 4.3-13b.

Site	No in Buczkó et al. (2009)	Lat. °N	Long. °E	Alt. (m asl)	Time span (cal ka BP)	Land-scapes	References
Liptovsky Jan		49.0	19.7	660	15.5-1.6		Dolejsova (1988, EPD)
Puścina Rekowiańska	P 1	49.5	19.8	656	10.2-0	Orawa-Nowy Targ	Obidowicz (1988, 1989, 1990, 1993)
Szymbark	P 8	49.6	21.1	465	8.9-2.5	Lower Beskid	Gil et al. (1972, 1974), Szczepanek (1989)
Jasiel	P 4	49.4	21.9	680	12-0		Szczepanek (1987, 1989)
Regetovka	S 11	49.4	21.3	515	8-0	Busov	Wacnik (1995)
Tarnawa Wyzna	P 6	49.1	22.8	670	13.2-0	Bieszczady Mts	Ralska-Jasiewiczowa (1972, 1987, 1989)
Preluca Tiganului	R 2	47.81	23.53	730	14.4-0	Gutaiului Mts	Wohlfarth et al. (2001), Björkman et al. (2002), Feurdean and Bennike (2004), Feurdean (2005), Feurdean et al. (2008)
Steregoiu	R 3	47.81	23.54	790	>14.7-0		Björkman et al. (2002, 2003), Feurdean and Bennike (2004), Feurdean et al. (2008)

Poiana Ştiol	R 22	47.6	24.8	1540	11-0	Rodnei Mts	Tantau et al. (2011), Tanţău and Fărcaş (2005)
Iezerul Calimani	R 4	45.1	25.3	1650	>16.75-0	Caliman Mts	Fărcaş et al. (1999, 2003, 2006), Feurdean et al. (2007)
Luci	R 21	46.3	25.7	1079	16-0	Harghitei Mts	Tanţău et al. (2003a, b, 2006), Fărcaş et al. (2006)
Sfânta Ana	R 17	46.12	25.9	950	9,000-800		Magyari et al. (2006, 2009b, 2014)
Mohoş	R 20	46.1	25.9	1050	13.89-0		Tanţău et al. (2003b, 2014), Tanţău (2006), Fărcaş et al. (2003, 2006), Feurdean et al. (2007)
Bisoca	R 19	45.5	26.7	875	ca. 12-0	Buzăului Sub-carpathians	Tanţău et al. (2003c), Tanţău (2006), Fărcaş et al. (2006)
Avrig	R 18	45.7	24.4	400	17.5-0	Făgăraş Depression	Tanţău et al. (2006), Feurdean et al. (2007)
Taul dintre Brazi	R 5	45.4	22.9	1740	15-0	Retezat Mts	Buczkó et al. (2009), Magyari et al. (2009a)
Taul Zanogutii	R 13	45.46	22.8	1840	14.8-0		Fărcaş et al. (1999, 2003, 2006), Feurdean et al. (2007a)
Semenic	R 14	45.2	22.1	1600	>7.62-0	Semenic Mts	Rösch and Fischer (2000), Fărcaş et al. (2003, 2006)
Straldzha	42.63	26.77	138	37.5-5	Riverine marsh in tectonic graben	Danube plain	Connor et al. (2013)
Mire Garvan	44.1	27.0	20	7-0	Mire at Danube River		Lazarova (1995)
Srebarna-1 - 4	44.08	27.12	20	7-0	Karstic doline near tributaries to Danube		Lazarova (1995), Lazarova and Bozilova (2001)
Arkutino-1, -2	42.4	27.7	-4	7-0	Freshwater lake behind dunes	Coast of the Black Sea	Bozilova and Beug (1992)
Varna Arsenala	43.17	27.8	0	7- 0	Liman (lagoon) in tectonic graben		Bozilova and Ivanov (1985), Bozilova and Filipova (1986), Bozilova and Beug (1994)
Varna Beloslav-Poveljanovo -1, -2	43.2 43.18	27.67 27.65	0	6.5- 0			Bozilova and Beug (1994)
Varna Core 3			-6	7.8-0			Filipova-Marinova et al. (2013b)
BlackSea GGC-18	42.776	28.677	-971	12-0	Former Tethys	Sea	Filipova-Marinova et al. (2013a)

Table 7-5. Selected sites in the Western Mediterranean, in bold sites with a histogram in Fig. 4.3-15b.

Region on map Fig. 4.2-1	Site	Lat. °N	Long. °W-°E	Alt. (m asl)	Time span (k cal yr BP)	Page Carrión et al. (2012)	References
M3, Regional picture	Laguna de Río Seco	37.03	3.33 W	3020	9.54	855	Anderson et al. (2011)
M3	Baza	37.23	2.71 W	1900	8.4-0.06	857	Carrión et al. (2007a), Gil-Romera et al. (2009, 2010)
T3	**Las Pardillas**	42.0	3.0 W	1870	11.45-1.45	405	Sánchez-Goñi and Hannon (1999)
M3 Pinus nigra	**Cañada de la Cruz**	38.06	2.69 W	1595	9.36-0.05	851	Carrión et al. (2001b), Carrión (2002)
M3	Gádor	36.9	2.9 W	1530	6.85-1.16	865	Carrión et al. (2003)
T3 conifers	**Quintanar de La Sierra**	42.03	3.02 W	1470	21.8-<3.3	381	Peñalba (1989, 1994), Peñalba et al. (1997), Ruiz Zapata et al. (2002)
T3	**Charco da Candieira***	40.34	7.58 W	1400	13,89-0.15	739	Van der Knaap and van Leeuwen (1995, 1997)
M3 Pinus nigra	**Laguna de Siles**	38.39	2.51 W	1320	20,28-0.5	835	Carrión (2002b)
T3	**El Maíllo**	40.55	6.21 W	1100	ca. 11-0		Morales-Molino et al. (2013)
M2 cave	Cueva de la Carihuela	37.44	3.44 W	1020	117-1.25	798	Carrión (1990, 1992a), Carrión et al. (1998, 1999), Fernández et al. (2007)
T3/M2	Vallefondo	42.28	6.20 W	1000	Last 800 yrs		Morales-Molino et al. (2011)
M2	**Espinosa de Cerrato**	41.96	3.94 W	885	11.05-<1.27	407	Franco-Múgica et al. (2001)
M2	**Ojos de Villaverde**	38.8	2.37 W	870	9.73-<1.16	581	Carrión et al. (2001b)
M2 Dunes, pines	El Carrizal de Cuéllar	41.32	4.15 W	860	>9.65-0	413	Franco-Múgica et al. (2005)
M2	**Padul**	37.01	3.61 W	785	54- 5.75	805	Menéndez-Amor and Florschütz (1963), Florschütz et al. (1971), Pons and Reille (1988), Valle-Hernandez et al. (2003)
T3/M2	**Ayoó da Vidriales**	42.12	6.07 W	780	19-0		Morales-Molino et al. (2014)
M2	Lago Grande de Estaña	42.03	0.53 E	670	11.38-0.16	225	Riera et al. (2004, 2006), Morellón et al. (2008, 2009), González-Sámperiz (2010)
M2	**Banyoles**	42.1	2.8 E	173	30-0	284	Pérez-Obiol and Julia (1994)
M1	Es Grau Balearic Isl	39.95	4.36 E	0	8-0	695	Carrión et al. (2012), Burjachs et al. (2017)
M1	**Navarrés**	39.1	0.7 W	225	19-0.9	614	Carrión and van Geel (1999)
M2	**Central Ebro basin***	41.2	0.3 W	150-357	10.5-0	combined	Davis and Stevenson (2007)

	Menorca Galdana	39.9	4.0 E	47	9-0	688	Yll et al. (1995)
M1, SW-Iberia contrast to Antes & San Rafael	Laguna de Medina	36.62	6.05 W	30	9-0	863	Reed et al. (2001)
M1 estuary	CM5 Beliche **Guadiana**	37.26	7.45 W	25	13-1.6	737	Boski et al. (2002, 2008), González-Vila et al. (2003), Fletcher et al. (2007)

*Four sites in the Central Ebro Desert: Hoya del Castillo, Laguna Guallar, Salada Pequeña and Laguna Salada Chiprana.

Table 7-6. Pa, Pannonia, selected sites ± from west to east, grouped in landscapes, in bold sites with a histogram in Fig. 4.3-18b.

Site	**No in Buczkó et al. (2009)**	**Lat. °N**	**Long. °E**	**Alt. (m asl)**	**Time span (k cal yr BP)**	**Origin of basin**	**Landscapes, regions**	**References**
Mezőlak	**H11**	47.35	16.5	123	15 - 0	Backswamp hollow in Bitwa creek	Devecser Plain	Ilon et al. (2006), Sümegi (2007)
Alsópáhok	**H 4**	46.75	17.20	104	11 - 0	Hévíz-Páhok subsidence	Balaton region	Juhász et al. (2007)
Zalavári	**H 5**	46.67	17.21	106	11.5- 0	Little Balaton subsidence		Juhász et al. (2007)
Csögle		47.2	17.26	134	12 - 0	Peated paleochannel	Alluvia of Marcal River (Western part of Transdanubia)	Juhász (2002)
Parížske močiare		47.88	18.47	150	11.2 – 5.5	Backswamp behind the Danube river	Little Hungarian Plain	Jamrichova et al. (2014)
Zsombó		46.4	20.0	92	14 - 0	Hollow between dunes		Nagy-Bodor et al. (1995)
Sirok Nyíres-tó	**H 19**	47.94	20.19	450	9 - 0	Landslide hollow	Mátra hills	Gardner (2002, 2005)
Kis-Mohos-tó	**H 15**	48.41	20.41	306	14.6 – 0	Landslide hollow	N-Hungarian Mid-Mountains	Willis et al. (1997, 1998)
Nagy Mohos	**H 16**	48.34	20.43	294	25.3 - 0	Valley mire, behind landslide	Putnok Hills at Kelemér	Magyari et al. (1999, 2000, 2001), Sümegi et al. (2008), Zólyomi and Járainé Komlódi (2008), Juhász (2002)
Vörös-mocsár	**H 27**	46.39	19.15	91	13 - 0	Peated river (yazoo) paleochannel of the Danube river		Jakab et al. (2004), Nagy-Bodor et al. (1995, EPD)

Sarló-hát	**H 24**	47.97	21.17	86	11.4- 0	Palaeo-channel	Nagy Magyar Alföld	Polgár, Middle Tisza Plain	Magyari et al. (2010)
Bátorliget	**H 26**	47.77	22.27	150	21 - 0	Deflation hollow		Nyírség alluvial fan	Willis et al. (1995), Sümegi and Gulyás (2004)
Nyírjes-tó, Csaroda	**H 18**	48.18	22.49	420	11 - 0	Palaeo-channel		Bereg Plain	Harrington (1995), Sümegi (1999), Juhász (2005)

Table 7-7. Po, Pontic region, selected sites, from west to east, in bold sites with a histogram in Fig. 4.3-19b.

Site	Lat. °N	Long. °E	Alt. (m asl)	Time span (k cal yr BP)	Origin of lake	References
Shabla-Ezeretz	43.83	28.85	1	ca. 7-0	Liman, separated from the Black Sea by a sand strip (brackish)	Filipova (1985), Bozilova and Filipova (1986), Filipova-Marinova and Bozilova (1990)
Durankulak	43.25	28.38	4	ca. 9-0	Liman, separated from the Black Sea by a sand strip (brackish)	Bozilova and Tonkov (1985 ,1998), Bozilova and Filipova (1986), Marinova (2003)
Dovjok swamp	48.75	28.25	274	ca. 7.5-0	Floodplain Murafa River	Kremenetski (1995)
Kardashinski swamp	46.52	32.62	4	ca. 8-0	Higher floodplain of Dnjepr River	Kremenetski (1995)
Bog Liman	49.73	37.67	150	12.5-1		Bezusko (1973)
Lipigi	49.54	43.44		6.1-0	Terrace of Don River	Kremenetski et al. (1999b)
Kharabuluk mire	47.77	44.43	70	7.5-0	Floodplain of Zelmen River	Kremenetski et al. (1999b)
Pobochnoye near Buzuluk	52.97	51.88	ca. 60	ca. 14-0	On second terrace of Samara River	Kremenetski et al. (1999a)

7.2

Acknowledgements of the authors

We would like to thank all palynologists who contributed their data to the EPD, Neotoma and Alpadaba.

Special thanks go to Bent Aaby, Vojtěch Abraham, Carole Adolf, Judy R. M. Allen, Svend Th. Andersen†, Josu Aranbarri, Britta Allgöwer, Ariane Ballmer, Ruth Beer, Hermann Behling, Claudio A. Belis, Nadine Berthel, Hans-Jürgen Beug, Katharina Bieri-Steck, Christian Bigler, Martina Bisculm, Felix Bittmann, Anne E. Bjune, Jacques-Louis de Beaulieu, Madlena Beckmann, Muriel Bendel, Giorgia Beffa, Alexander Bolland, Petra Boltshauser Kaltenrieder, Francisco Bouzas, Richard Bradshaw, Stefan Brönnimann, Sandra Brügger, Elisabetta Brugiapaglia, Harald Bugmann, Conradin Burga, Carla Buser, Christopher Carcaillet, José S. Carrión, Camilla Calò, Lanfredo Castelletti, Gina Clarke, Ben Clegg, Daniele Colombaroli†, Marco Conedera, Simon Connor, Sevil Coşgun, Hervé Cubizolle, Philippe Curdy, Brandon Curry, Kristian Dalsgaard, André-Marie Dendievel, Mara Deza Araujo, Christine Dolder, Ilse Draxler, Ruth Drescher-Schneider, Laura Dziomber, Stefan Eggenberg, Sebastian Eggenberger, Regina Egli-Gerber, Wenche Eide, Helmut Elsenbeer, Sorina Fărcaş, Vivian Felde, Basilio Ferrante, Angelica Feurdean, Walter Finsinger, Hubertus Fischer, Caitlyn Florentine, Marie-José Gaillard, Mariusz Gałka, Kathrin Ganz, Sylvia Gassner, Daniel Gavin, Peter Germann†, Bruce R. Gervais, Thomas Giesecke, Graciela Gil Romera, Adrian Gilli, Penélope González Sampériz, Eric C. Grimm†, Martin Grosjean, Eberhard Grüger, Christoph Grupp, Moritz Gubler, Jean Nicolas Haas, Albert Hafner, Maija Heikkilä, Oliver Heiri, Paul Henne, Alexander Hermann, Stefan Hersberger, Simone Hofstetter, Adam Hölzer, Priska Hubschmid, Feng Sheng Hu, Ursula Huonder, Stefanie Jacomet, Ingrid Jansen, Cathy Jenks, Richard Jones†, Fortunat Joos, Anu Kaakinen, Katarzyna Kajukało, Mihkell Kangur, Dmytro Kiosak, Pim de Klerk, Roman Knöpfli, Tiiu Koff, Nadezhda Kotova, Grzegorz Kowalewski, Radka Kozáková, Teresa Krause, Norbert Kühl, Seija Kultti, Petr Kuneš, Patrik Krebs, Robert Krisai†, Meinrad Küttel, Tommaso La Mantia, Mariusz Lamentowicz, Małgorzata Latałowa, Konrad Lauber†, Maria Leunda, Thomas Litt, André F. Lotter, Marie-France Loutre, Jürg Luterbacher, Glen M. MacDonald, Michel Magny, Donatella Magri, Jock H. McAndrews, Sergey Makhortykh, Mirosław Makohonienko, Aldo Marchetto, Elisa Martinelli, Jason McLachlan, Yannick Miras, Krystyna Milecka, Adrian Möhl, Peter Friis Møller, Scott D. Mooney, Castor Muñoz Sobrino, Atle Nesje, Ebbe H. Nielsen, Roland Noti, Michael O'Connell, Bent Odgaard, Klaus Oeggl, Christopher Paciorek, Salvatore Pasta, Aage Paus, Tiziana Pedrotta, Sylvia Peglar, Ana Pérez Sanz, Roberta Pini, Petr Pokorný, Giovanni Procacci, Claire M. Rambeau, Cesare Ravazzi, Maurice Reille, Fabian Rey, José Luis Rivera, Astrid Röpke, Manfred Rösch, Laura Sadori, Stéphanie Samartin, J. Sakari Salonen, Patrick Schläfli, Siegfried Schloss, Margit Schwikowski, Ulf Segerström, Alexander Sennikov, Heikki Seppä, Jakob Schwander, Jeffrey A. Snyder, Markus Stähli, Martina Stebich, Tony Stevenson, Werner E. Stöckli, Pál Sümegi, Willi Tanner, Ioan Tanţău, Yunuén Temoltzin-Loranca, Antoine Thévenaz, Jean-Paul Theurillat, Lena Thöle, Eva Trachsel, Mathias Trachsel, Chronis Tzedakis, Kazimierz Tobolski†, Minna Väliranta, Verushka Valsecchi, Boris Vannière, Jacqueline van Leeuwen, Ulrike van Raden, Eveline Venanzoni, Elisa Vescovi, Peter von Ballmoos, Yann Vitasse, Karl-Dag Vorren, Notburga Wahlmüller, Mike J.C. Walker, Heinz Wanner, Samuel Wegmüller†, Michael Wehrli, Lucia Wick, Jack Williams, Kathy J. Willis, Cathy Whitlock, Steffen Wolters, Herb Wright†, Gianni Zanchetta, Heinrich Zoller† and Heinz J. Zumbühl.

Acknowledgements of H. John B. Birks

I am very indebted to Brigitta Ammann for her invitation to write this chapter and to the many colleagues who have influenced my ideas about the contributions that Quaternary botany is making to modern ecology, evolutionary ideas, and biogeography. These colleagues include Brigitta Ammann, Keith Bennett, Hilary Birks, Ed Cushing†, Vivian Felde, Thomas Giesecke, Eric Grimm†, Mark Hill, Brian Huntley, Steve Jackson, Petr Kuneš, Triin Reitalu, Jim Ritchie†, Alistair Seddon, Willy Tinner, Cathy Whitlock, Jack Williams, and Herb Wright†. I am extremely grateful, as always, to Cathy Jenks for her skilful preparation of the manuscript. This chapter was prepared with support from the Norwegian Research Council project 249894 – IGNEX Interglacials and Glacials – Natural Experiments in Biodiversity Dynamics and the European Union's Horizon 2020 research and innovation programme (grant agreement 741413) – HOPE: Humans on Planet Earth – Long-term impacts of biosphere dynamics. It is a contribution to the IGNEX and HOPE projects.

Bibliography

Aaby, B., 1976. Cyclic climatic variations in climate over the past 5,500 yr reflected in raised bogs. Nature 263, 281-284.

Aaby, B., 1979. Characterization of peat and lake deposits. In: Berglund, B. (Ed.), Project Guide for IGCP, pp. 77-98.

Aaby, B., 1983. Forest development, soil genesis and human activity illustrated by pollen and hypha analysis of two neighbouring podzols in Draved Forest, Denmark. Geol. Survey Denm. 2, 5-111.

Aaby, B., 1986. Trees as anthropogenic indicarors in regional pollen diagrams from eastern Denmark. In: Behre, K.E. (Ed.), Anthropogenic Indicators in Pollen Diagrams. Balkema, Rotterdam, pp. 73-93.

Aaby, B., Berglund, B., 1986. Characterization of peat and lake deposits. In: Berglund, B.E. (Ed.), Handbook of Holocene Palaeoecology and Palaeohydrology. John Wiley & Sons, Chichester, pp. 231-246.

Aaby, B., Berglund, B.E., 2003. Characterization of peat and lake deposits. In: Berglund B.E. (Ed.) Handbook of Holocene Palaeoecology and Palaeohydrology. Blackburn Press, Caldwell NJ

Aaby, B., Tauber, H. 1975. Rates of peat formation in relation to degree of humification and local environment, as shown by studies of a raised bog in Denmark. Boreas 4, 1-17.

Aalto, M., 1970. Potamogetonaceae fruits. I. Recent and subfossil endocarps of the Fennoscandian species. Acta Bot. Fennica 88, 1-85.

Aario, L., 1932. Pflanzentopographische und paläogeographische Mooruntersuchungen in N-Satakunta. Fennia 55, 1-179.

Aario, L., 1940. Waldgrenzen und subrezente Pollenspektren in Petsamo Lappland. Ann. Acad. Sei. Fenn. A 54/8, 1-120.

Aario, L., 1941. Waldgrenzen und subrezente Pollenspektren in Petsamo, Lappland. Zeitschrift für Botanik 36, 593-595.

Aario, L., 1943. Über die Wald- und Klimaentwicklung an der lappländischen Eismeerküste in Petsamo. Ann. Bot. Soc. Zool. Bot. Fenn. Vanamo 19/1, 1-158.

Aarnes, I., Bjune, A.E., Birks, H.H., Balascio, N., Bakke, J., Blaaw, M., 2012. Vegetation responses to rapid climate changes during the last deglaciation 13 500–8000 years ago on southwest Andøya, arctic Norway. Veget. Hist. Archaeobot. 21, 17-35.

Aartolahti, T., 1965. Oberflächenformen von Hochmooren und ihre Entwicklung in Südwest-Häme und Nord-Satakunta. Fennia 93, 1-268.

Abel Schaad, D., Pulido, F., López-Sáez, J., Alba Sánchez, F., Nieto Lugilde, D., Franco Múgica, F., Pérez Díaz, S., Ruiz Zapata, M., Gil García, M., Dorado Valiño, M., 2014. Persistence of tree relicts in the Spanish Central System through the Holocene. Lazaroa 35, 107-131.

Abraham, V., Oušková, V., Kuneš, P., 2014. Present-day vegetation helps quantifying past land cover in selected regions of the Czech Republic. PLoS ONE 9(6), e100117.

Abraham, V., Hicks, S., Svobodová-Svitavská, H., Bozilova, E., Panajiotidis, S., Filipova-Marinova, M., Jensen, C. E., Tonkov, S., Pidek, I. A., Święta-Musznicka, J., Zimny, M., Kvavadze, E., Filbrandt-Czaja, A., Hättestrand, M., Karlıoğlu Kılıç, N., Kosenko, J., Nosova, M., Severova, E., Volkova, O., Hallsdóttir, M., Kalniņa, L., Noryśkiewicz, A., Noryśkiewicz, B., Pardoe, H., Christodoulou, A., Koff, T., Fontana, S. L., Alenius, T., Isaksson, E., Seppä, H., Veski, S., Pędziszewska, A., Weiser, M., Giesecke, T., 2021. Patterns in recent and Holocene pollen influxes across Europe; the Pollen Monitoring Programme Database as a tool for vegetation reconstruction, Biogeosciences 18, 4511-4534.

Ackerly, D.D., 2003. Community assembly, niche conservatism, and adaptive evolution in changing environments. Int. J. Plant Sci. 164, S165-S184.

Adam, K.D., 1951. Der Waldelefant von Lehringen – eine Jagdbeute des diluvialen Menschen. Quartär 5, 79-92.

Aiton, W., 1811. Treatise on the Origin, Qualities and Cultivation of Moss-earth, with Directions for converting it into Manure. (cited in Barber 1981).

Akeret, Ö., 2005. Plant remains from a Bell Beaker site in Switzerland, and the beginnings of *Triticum spelta* (spelt) cultivation in Europe. Veget. Hist. Archaeobot. 14, 279-286.

Aksu, A., Hiscott, R., Yaşar, D., Işler, F., Marsh, S., 2002. Seismic stratigraphy of Late Quaternary deposits from the southwestern Black Sea shelf: evidence for non-catastrophic variations in sea-level during the last ~10 000 yr. Mar. Geol. 190, 61-94.

Al-Saghir, M.G., Porter, D.M., 2012. Taxonomic revision of the genus *Pistacia* L. (Anacardiaceae). Am. J. Plant Sci. 3, 12-32.

Alenius, T., Mokkonen, T., Lahelma, A., 2013. Early farming in the northern boreal zone: Reassessing the history of land use in southeastern Finland through high-resolution pollen analysis. Geoarchaeology 28, 1-24.

Alenius, T., Haggren, G., Koivisto, S., Vanhanen, S., Sugita, S., 2017. Landscape dynamics in southern Finland during the Iron Age and the Early Modern Era - Pollen-based landscape reconstruction (LRA), macrofossil and historical data from Western Uusimaa. J. Arch. Sci.: Reports 12, 12-24.

Ali, A.A., Carcaillet, C., Talon, B., Roiron, P., Terral, J.-F., 2005. *Pinus cembra* L. (Arolla pine), a common tree in the inner French Alps since the early Holocene and above the present tree line: a synthesis based on charcoal data from soils and travertines. J. Biogeography 32, 1659-1669.

Alimen, M.-H., Florschütz, F., Menendez-Amor, J., 1965. Etude géologique et palynologique sur le Quaternaire des environs de Lourdes, in: Actes 4ème Conférence Internationale Etudes Pyrénéennes. Pau-Lourdes, pp. 7-26.

Allen, H.D., 1986. Late Quaternary of the Kopais Basin, Greece: Sedimentary and Environmental History. PhD thesis, University of Cambridge.

Allen, H.D. 2001. Mediterranean Ecogeography. Pearson Education, Harlow.

Allen, H.D., 1990. A postglacial record from the Kopais Basin, Greece. In: Bottema, S., Entjes-Nieborg, G., van Zeist, W. (Eds.), Man's Role in the Shaping of the Eastern Mediterranean. Balkema, Rotterdam, 173-182.

Allen, J.R.M., Huntley, B., 2000. Weichselian palynological records from southern Europe: correlation and chronology. Quat. Int. 73/74, 111-125.

Allen, J.R.M., Huntley, B., 2009. Last interglacial palaeovegetation, palaeoenvironments and chronology: a new record from Lago Grande di Monticchio, southern Italy. Quat. Sci. Rev. 28, 1521-1538.

Allen, J.R.M., Huntley, B., Watts, W., 1996. The vegetation and climate of northwest Iberia over the last 14 000 yr. J. Quat. Sci. 11, 125-147.

Allen, J.R.M., Brandt, U., Brauer, A., Hubberten, H.W., Huntley, B., Keller, J., Kraml, M., Mackensen, A., Mingram, J., Negendank, J.F.W., Nowaczyk, N.R., Oberhansli, H., Watts, W.A., Wulf, S., Zolitschka, B., 1999. Rapid environmental changes in southern Europe during the last glacial period. Nature 400, 740-743.

Allen, J.R.M., Watts, W., Huntley, B., 2000. Weichselian palynostratigraphy, palaeovegetation and palaeoenvironment; the record from Lago di Monticchio, southern Italy. Quat. Int 73/74, 91 - 110.

Allen, J.R.M., Watts, W.A., McGee, E., Huntley, B., 2002. Holocene environmental variability - the record from Lago Grande di Monticchio, Italy. Quat. Int. 88, 69-80.

Allen, J.R.M., Long, A.J., Ottley, C., Pearson, D.G., Huntley, B., 2007. Holocene climate variability in northernmost Europe. Quat. Sci. Rev. 26, 1432-1453.

Allison, T., Moeller, R., Davis, M., 1986. Pollen in laminated sediments provides evidence for a mid-Holocene forest pathogen outbreak. Ecology 67, 1101-1105.

Alm, T., 1993. Øvre Æråsvatn - palynostratigraphy of a 22,000 to 10,000 BP lacustrine record on Andøya, northern Norway. Boreas 22, 171-188.

Alm, T., Birks, H.H., 1991. Late Weichselian flora and vegetation of Andøya, northern Norway - macrofossil (seed and fruit) evidence from Nedre Æråsvatn. Nord. J. Bot. 11, 465-476.

Almquist-Jacobson, H., 1994. Interaction of Holocene climate, water balance, vegetation, fire, and cultural land-use in the Swedish Borderland. Lundqua 30. Lund, Sweden, pp. 1-82.

Alsos, I.G., Engelskjøn, T., 2002. Conservation genetics and population history of *Betula nana, Vaccinium uliginosum*, and *Campanula rotundifolia* in the Arctic Archipelago of Svalbard. Arct. Alp. Res. 34, 408-418.

Alsos, I.G., Alm, T., Normand, S., Brochmann, C., 2009. Past and future range shifts and loss of diversity in dwarf willow (*Salix herbacea* L.) inferred from genetics, fossils and modelling. Global Ecol. Biogeogr. 18, 223-239.

Alsos, I.G., Sjögren, P., Brown, A.G., Gielly, L., Merkel, M.K.F., Paus, A., Lammers, Y., Edwards, M.E., Alm, T., Leng, M.J., Goslar, T., Langdon, C.T., Bakke, J., van der Bilt, W.G.M., 2020. Last Glacial Maximum environmental conditions at Andøya, northern Norway; evidence for a northern ice-edge ecological "hotspot". Quat. Sci. Rev. 239, 106364.

Alsos, I.G., Lammers, Y., Kjellman, W.E., Føreid Merkel, M.K., Bender, E.M., Rouillard, A., Erlendsson, E., Benediktsson, Í.Ö., Farnsworth, W.R., Brynjólfsson, S., Gislárdóttir, G., Eddudóttir, S.D., Schomakker, A., 2021. Ancient sedimentary DNA shows rapid post-glacial colonisation of Iceland followed by relatively stable vegetation until the Norse settlement (Landnám) AD 870. Quat. Sci. Rev. 259, 106903.

Amesbury, M.J., Swindles, G.T., Bobrov, A., Charman, D.J., Lamentowicz, M., Mallon, G., Mazei, Y., Mitchell, E.A.D., Payne, R.J., Roland, T.P., Turner, E.T., Warner, B.G., 2016. Development of a new pan-European testate amoeba transfer function for reconstructing peatland palaeohydrology. Quat. Sci. Rev. 152, 132-151.

Ammann, B., 1988. Palynological evidence of prehistoric anthropogenic forest changes on the Swiss Plateau. In: Birks, H.H., Birks, H.J.B., Kaland, P.E., Moe, D. (Eds.), The Cultural Landscape - Past, Present and Future. Cambridge University Press, Cambridge, pp. 289-301.

Ammann, B., 1989. Late-Quaternary palynology at Lobsigensee - Regional vegetation history and local lake development. Diss. Bot. 137, 157.

Ammann, B., 1994. Differential flotation of saccate pollen - a nuisance and a chance. Diss. Bot. 234, 101-110.

Ammann, B., 2000. Biotic responses to rapid climatic changes: Introduction to a multidisciplinary study of the Younger Dryas and minor oscillations on an altitudinal transect in the Swiss Alps. Palaeogeogr. Palaeoclimatol. Palaeoecol. 159, 191-201.

Ammann, B., Tobolski, K., Züllig, H., Chaix, L., Hofmann, W., Elias, S.A., Wilkinson, B., Siegenthaler, U., Eicher, U., Andrée, M., Oeschger, H., 1985. Lobsigensee - Late-glacial and Holocene environments of a lake on the central Swiss Plateau. Diss. Bot. 87, 127-170.

Ammann, B., Birks, H.J.B., Drescher-Schneider, R., Juggins, S., Lang, G., Lotter, A.F., 1993. Patterns of variation in late-glacial pollen stratigraphy along a north-west-south-east transect through Switzerland - a numerical analysis. Quat. Sci. Rev. 12, 277-286.

Ammann, B., Gaillard, M.-J., Lotter, A.F., 1996. Switzerland. In: Berglund, B.E, Birks, H.J.B., Ralska-Jasiewiczowa, M., Wright, H.E. (Eds.), Palaeoecological Events During the Last 15 000 Years. Regional Syntheses of Palaeoecological Studies of Lakes and Mires in Europe. Wiley, Chichester, pp. 647-666.

Ammann, B., Birks, H.J.B., Brooks, S.J., Eicher, U., von Grafenstein, U., Hofmann, W., Lemdahl, G., Schwander, J., Tobolski, K., Wick, L., 2000. Quantification of biotic responses to rapid climatic changes around the Younger Dryas - a synthesis. Palaeogeogr. Palaeoclim., Palaeoecol. 159, 313-347.

Ammann, B., van Leeuwen, J.F.N., van der Knaap, W.O., Lischke, H., Heiri, O., Tinner, W., 2013a. Vegetation responses to rapid warming and to minor climatic fluctuations during the Late-glacial Interstadial (GI-1) at Gerzensee (Switzerland). Palaeogeogr., Palaeoclimatol., Palaeoecol. 391, 40-59.

Ammann, B., van Raden, U.J., Schwander, J., Eicher, U., Gilli, A., Bernasconi, S.M., van Leeuwen, J.F.N., Lischke, H., Brooks, S.J., Heiri, O., Novakova, K., van Hardenbroek, M., von Grafenstein, U., Belmecheri, S., van der Knaap, W.O., Magny, M., Eugster, W., Colombaroli, D., Nielsen, E., Tinner, W., Wright, H.E., 2013b. Responses to rapid warming at Termination la at Gerzensee (central Europe): Primary succession, albedo, soils, lake development, and ecological interactions. Palaeogeogr. Palaeoclimatol. Palaeoecol. 391B, 111-131.

Ammann, B., van der Knaap, W.O., Lang, G., Gaillard, M.-J., Kaltenrieder, P., Rösch, M., Finsinger, W., Wright, H.E., Tinner, W., 2014. The potential of stomata analysis in conifers to estimate presence of conifer trees: examples from the Alps. Veget. Hist. Archaeobot. 23, 249-264.

Amon, L., Veski, S., Vassiljev, J., 2014. Tree taxa immigration to the eastern Baltic region, southeastern sector of Scandinavian glaciation during the Late-glacial period (14,500-11,700 cal. B.P.). Veget. Hist. Archaeobot. 23, 207-216.

Andersen, S.T., 1961. Vegetation and its environment in Denmark in the Early Weichselian Glacial (Last Glacial). Danm. Geol. Undersøg. II 75, 1-175.

Andersen, S.T., 1964. Interglacial plant successions in the light of environmental changes, Report of the VIth International Congress on Quaternary, Warsaw 1961. Vol II: Palaeobotanical Section, Łodź, pp. 359-368.

Andersen, S.T., 1965. Interglacialer og interstadialer i Danmarks Kvartaer. Medd. Dansk Geol. Foren. 15, 486-506.

Andersen, A., 1966a. Geologi og arkaeologi i Draved Mose. Medd. Danm. Geol. Foren. 16, 255-258.

Andersen, S.T., 1966b. Interglacial vegetation succession and lake development in Denmark. The Palaeobotanist 15, 117-127.

Andersen, S.T., 1970. The relative pollen productivity and pollen representativity of North European trees and correction factors for tree pollen spectra. Danm. Geol. Unders. Ser. II 96, 1-99.

Andersen, S.T., 1973. The differential pollen productivity of trees and its significance for the interpretation of a pollen diagram from a forested region. In: Birks, H.J.B., West, R.G. (Eds.), Quaternary Plant Ecology. John Wiley & Sons, pp. 109-115.

Andersen, S.T., 1978. Local and regional vegetational development in eastern Denmark in the Holocene. Danmarks Geologiske Undersøgelse Årbog 1976, 5-27.

Andersen, S.T., 1986. Palaeoecological studies of terrestrial soils. In: Berglund, B.E. (Ed.), Handbook of Holocene Palaeoecology and Palaeohydrology. Wiley, Chichester, pp. 165-177.

Andersen, S.T., 1994. History of the terrestrial environment in the Quaternary of Denmark. Bull. Geol. Soc. Denmark 41, 219-228.

Andersen, S.T., 1998. Pollen analytical investigation of barrows from the Funnel Beaker and Single Grave Cultures in the Vroue area, West Jutland, Denmark. J. Dan. Archaeol. 12, 107-132.

Andersen, S.T., Rasmussen, K., 1993. Radiocarbon wiggle-dating of elm declines in northwest Denmark and their significance. Veget. Hist. Archaeobot. 2, 125-135.

Andersen, S.T., Aaby, B., Odgaard, B.V., 1996. Denmark. In: Berglund, B.E., Birks, H.J.B., Ralska-Jasiewiczowa, M., Wright, H.E. (Eds.), Palaeoecological Events During the Last 15 000 Years. Regional Syntheses of Palaeoecological Studies of Lakes and Mires in Europe. Wiley, Chichester, pp. 215-231.

Anderson, R., Jiménez-Moreno, G., Carrión, J.S., Pérez-Martínez, C., 2011. Postglacial history of alpine vegetation, fire, and climate from Laguna de Río Seco, Sierra Nevada, southern Spain. Quat. Sci. Rev. 30, 1615-1629.

Andersson, G., 1888. Redogörelse för senare tiders undersökningar af torfmossor, kalktuffer och sötvattensleror, särdeles med hänsyn till den skandinaviska vegetationens invandringshistoria. Bot. Notiser.

Andersson, G., 1897. Die Geschichte der Vegetation Schwedens. Bot. Jahrb. Syst. Pflanzengeschichte Pflanzengeographie 22, 433-550.

Andersson, G., 1902. Hasseln i Sverige fordom och nu. Sver. Geol. Unders. Ser. C. 3, 1-168.

Andersson, G., 1903. Klimatet i Sverige etter istiden. Nordisk Tidsskrift 1, 1-26.

Andersson, G., 1909. The climate of Sweden in the Late-Quaternary period. Sver. Geol. Unders. Årsbok 3, 1-88.

Andrée, M., Oeschger, H., Siegenthaler, U., Riesen, T., Moell, M., Ammann, B., Tobolski, K., 1986. ^{14}C dating of plant macrofossils in lake sediments. Radiocarbon 28, 411-416.

Andreev, A.A., Klimanov, V.A., 2000. Quantitative Holocene climatic reconstruction from Arctic Russia. J. Paleolimnol. 24, 81-91.

Andrén, T., Björck, S., Andrén, E., Conley, D., Zillén, L., Anjar, J., 2011. The development of the Baltic Sea Basin during the last 130 ka. In: Harff, J., Björck, S., Hoth, P. (Eds.), The Baltic Sea Basin. Springer, Heidelberg, pp. 75-97.

Andrés, C., 2011. *Phillyrea* L. In: Castroviejo, S., Aedo, C., Cirujano, S., Laínz, M., Montserrat, P., Morales, R., Muñoz, F., Navarro, C., Paiva, J., Soriano, C. (Eds.), Flora Iberica, vol. 11. Real Jardín Botánico de Madrid-CSIC, Madrid, Spain, pp. 139-143.

Andrew, R., 1971. Exine pattern in the pollen of British species of *Tilia*. New Phytol. 70, 683-686.

Andrič, M., 2007. Holocene vegetation development in Bela krajina (Slovenia) and the impact of first farmers on the landscape. The Holocene 17, 763-776.

Andrič, M., Massaferro, J., Eicher, U., Ammann, B., Leuenberger, M., Martinecic, A., Marinova, E., Brancelj, A., 2009. A multi-proxy Late-glacial palaeoenvironmental record from Lake Bled, Slovenia. Hydrobiologia 631, 121-141.

Andrieu, V., 1987. Le paléo-environnement du piémont nord-pyrénéen occidental de 27000 B.P. au postglaciaire: la séquence de l'Estarres. C. r. Acad. sci., Sér. 2, Méc. phys. chim. sci. univers sci. terre 304, 103-108.

Andrieu, V., 1991. Dynamique du paléoenvironnement de la vallée montagnarde de la Garonne (Pyrénées centrales, France) de la fin des temps glaciaires à l'actuel (PhD thesis). University of Toulouse, Toulouse, France.

Andrieu, V., Hubschman, J., Jalut, G., Herail, G., 1988. Chronologie de la déglaciation des Pyrenées françaises. Dynamique de sédimentation et contenu pollinique des paléolacs: application à l'interprétation du retrait glaciaire. Bull. AFEQ 34-35, 55-67.

Andrieu, V., Eicher, U., Reille, M., 1993. The end of the last pleniglacial in the Pyrenees (France)-Pollenanalytical, isotopic and radiometric data. Comp. Rend. Acad. Sci. 316, 245-250.

Antonsson, K., Brooks, S.J., Seppä, H., Telford, R.J., Birks, H.J.B., 2006. Quantitative palaeotemperature records inferred from fossil pollen and chironomid assemblages from Lake Gilltjärnen, northern central Sweden. J. Quat. Sci. 21, 831-841.

Appleby, P.G., 2001. Chronostratigraphic techniques in recent sediments. In: Last, W.M., Smol, J.P. (Eds.), Tracking Environmental Change Using Lake Sediments, Vol. 1: Basin Analysis, Coring, and Chronological Techniques. Kluwer Academic Publishers, pp. 171-203

Aradhya, M., Potter, D., Gao, F., Simon, C., 2007. Molecular phylogeny of *Juglans* (Juglandaceae): a biogeographic perspective. Tree Genet. Genomes 3, 363-378.

Aranbarri, J., González-Sampériz, P., Valero-Garcés, B., Moreno, A., Gil-Romera, G., Sevilla-Callejo, M., García-Prieto, E., Di Rita, F., Mata, M., Morellón, M., Magri, D., Rodríguez-Lázaro, J., Carrión, J.S., 2014. Rapid climatic changes and resilient vegetation during the Lateglacial and Holocene in a continental region of south-western Europe. Global Planet. Change 114, 50-65.

Aranbarri, J., Alcolea, M., Badal, E., Vila, S., Allué, E., Iriarte-Chiapusso, M.J., Sebastián, M., Magri, D., González-Sampériz, P., 2020. Holocene history of Aleppo pine (*Pinus halepensis* Mill.) woodlands in the Ebro Basin (NE Spain): Climate-biased or human-induced? Rev. Palaeobot. Palynol. 279, 104240.

Arbogast, R., Jacomet, S., Magny, M., Schibler, J., 2006. The significance of climate fluctuations for lake level changes and shifts in subsistence ecomomy during the late Neolithic (4300-2400 B.C.) in central Europe. Veget. Hist. Archaeobot. 15, 403-418.

Archaeobotany database, 2021. Last visited 13 December 2021. www.archaeobotany.de/database.html.

Arno, S.F., Hammerly, R.P., 1993. Timberline - Mountain and Arctic Forest Frontiers. The Mountaineers, Seattle.

Arnold, V., 2011. Celtic Fields und andere urgeschichtliche Ackersysteme in historisch alten Waldstandorten Schleswig-Holsteins aus Laserscan-Daten. Archäologisches Korrespondenzblatt 41, 439-455.

Arnoldussen, S., Fokkens, H. (eds.), 2008. Bronze Age Settlements in the Low Countries, Oxbow Books, Oxford, 200 pp.

Arslanov, K.A., Gey, N., Davydova, N., Dzhinoridze, R., Koshechkin, B., Pushenko, M.Y., Rybalko, A., Spiridonov, M., Subetto, D., Khomutova, V., 1996. Novye dannye po pleistotsenovoi i golotsenovoi istorii ozera Ladoga (New data on the Late Pleistocene and Holocene history of Lake Ladoga). Izv Ross Geogr Obshestva 128 2, 12–21.

Artushenko, A.T., 1957. History of the development of the West Ukrainian Polessye vegetation in Late Glacial and Post-Glacial times on the basis of the palynological studies. Ukr. Bot. Zhurnal 14, 12–29.

Artushenko, A.T., 1959. Allerod vegetation on the Russian plain territory in connection with the general development of the vegetation cover during the post glacial period in the East and Middle Europe. Ukr. Bot. Zhurnal 44, 772–785.

Artushenko, A.T., Arap, R.Y., Bezusko, L.G., 1982. Istoriya rastitelnosti zapadnyih oblastey Ukrainyi v chetvertichnom periode [The history of vegetation in Western Ukraine in the Quaternary]. Naukova dumka, Kiev, Ukraine.

Ascoli, D., Castagneri, D., Valsecchi, C., Conedera, M., Bovio, G., 2013. Post-fire restoration of beech stands in the Southern Alps by natural regeneration. Ecol. Eng. 54, 210-217.

Ascoli, D., Vacchiano, G., Maringer, J., Bovio, G., Conedera, M., 2015. The synchronicity of masting and intermediate severity fire effects favors beech recruitment. For. Ecol. Manage. 353, 126-135.

Asouti, E., Fuller, D. Q., 2012. From foraging to farming in the southern Levant: the development of Epipalaeolithic and Pre-pottery Neolithic plant management strategies. Veget. Hist. Archaeobot. 21, 149-162.

Atanassova, J., 1995. Palynological data of three deep water cores from the western part of the Black Sea. In: Bozilova, E., Tonkov, S. (Eds.), Advances in Holocene Palaeoecology in Bulgaria. Pensoft, Sofia., pp. 68-83.

Atanassova, J., Stefanova, I., 2003. Late-glacial vegetational history of Lake Kremensko-5 in the northern Pirin Mountains, southwestern Bulgaria. Veg. Hist. Archaeobot. 12, 1-6.

Aubert, S., 1993. Etude palynologique de la tourbière de la Bassa d'Ulès (Val d'Aran, Espagne). DEA Géographie et Aménagements, Université de Toulouse, Toulouse, France.

Auer, V., 1923. Suotutkimuksia Kuusamon ja Kuolajärven vaaraalueilta. Commun. Ex inst. quaest. Forestalium Finlandiae editae, Helsinki, 6, 1-257.

Auer, V., 1927. Untersuchungen über die Waldgrenzen und Torfböden in Lappland. Comm. ex Inst. Oliaest. Forest. Finl. 12/4, 1-52.

Aura Tortosa, J.E., Jordá Pardo, J.F., Pérez Ripoll, M., Rodrigo García, M.J., Badal García, E., Guillem Calatayud, P., 2002. The far south: the Pleistocene-Holocene transition in Nerja Cave (Andalucía, Spain). Quat. Int. 93-94, 19-30.

Austad, L., 1985. Vegetasjon i Kulturlandskapet: Lauvingstraer. Sogn og Fjordange Distrikthogskule, Skifter 1985, 2.

Austin, M.P., 1976. On non-linear species response models in ordination. Vegetatio 33, 33-41.

Austin, M.P., 1987. Models for the analysis of species' response to environmental gradients. Vegetatio 69, 35-45.

Austin, M.P., Gaywood, M.J., 1994. Current problems of environmental gradients and species response curves in relation to continuum theory. J. Veg. Sci. 5, 473-482.

Averdieck, F.-R., 1957. Zur Geschichte der Moore und Wälder Holsteins. Ein Beitrag zur Frage der Rekurrenzflächen. Nova Acta Leoplod. N.F. 19/130, 1-152.

Averdieck, F.-R., 1967. Die Vegetationsentwicklung des Eem-Interglazials und der Frühwürm-Interstadiale von Odderade/Schleswig-Holstein. Fundamenta B 2, 101-125, Köln.

Averdieck, F.-R., 1971. Zur postglazialen Geschichte der Eibe (*Taxus baccata* L.) in Nordwestdeutschland. Flora 160, 28-41.

Averdieck, F.-R., 1998. Palynologische Neuuntersuchung der Typlokalität Odderade. Meyniana 50, 25-40.

Avramidis, P., Geraga, M., Lazarova, M., Kontopoulos, N., 2013. Holocene record of environmental changes and palaeoclimatic implications in Alykes Lagoon, Zakynthos Island, western Greece, Mediterranean Sea. Quat. Int. 293, 184-195.

Aytug, B., 1964. Diagnose des pollens de *Pinus sylvestris* L. Pollen et Spores 4, 284 - 296.

Azuara, J., Combourieu-Nebout, N., Lebreton, V., Mazier, F., Müller, S., Dezileau, L., 2015. Late Holocene vegetation changes in relation with climate fluctuations and human activity in Languedoc (southern France). Clim. Past 11, 1769-1784.

Azuara, J., Lebreton, V., Peyron, O., F, M., Combourieu Nebout, N., 2018. The Holocene history of low altitude Mediterranean *Fagus sylvatica* forests in southern France. J. Veg. Sci. 29, 438-449.

Backman, A.L., 1941. *Najas marina* in Finnland während der Postglazialzeit. Acta Bot. Fenn., Helsinki, 30, 1-38.

Backman, A.L., 1943. *Ceratophyllum submersum* in Nordeuropa während der Postglazialzeit. Acta. Bot. Fenn., Helsinki, 31, 1-38.

Backman, A.L., 1948. *Najas flexilis* in Europa während der Quartärzeit. Acta Bot. Fenn., Helsinki, 43, 1-44.

Backman, A.L., 1950. *Najas tenuissima* (A. Br.) Magnus einst und jetzt. Soc. Scient. Fenn., Comm. Biol. X, 19, 1-36.

Badal, E., 1996. La végétation du Paléolithique Supérieur et de l'Epipaléolithique aux alentours de la Cueva de Nerja (Málaga, Espagne). Revue d'Archéometrie, Supplément Actes du colloque de Périgueux 199, 171-176.

Badal, E., 1998. El interés económico del pino piñonero para los habitantes de la Cueva de Nerja. In: Sanchidrián, J.L., Simón Vallejo, M.J. (Eds.), Las culturas del Pleistoceno Superior en Andalucía. Patronato Cueva de Nerja, Málaga, pp. 287-300.

Badal, E., 2006. Carbones y cenizas, ¿qué nos cuentan del pasado?. In: Carrión, J.S., Fernández, S., Fuentes, N. (Eds.), Paleoambientes y cambio climático. Agencia de Ciencia y Technologóa de la Region de Murcia, Murcia, pp. 103-116.

Badal, E., Bernabeu, J., Vernet, J.L., 1994. Vegetation changes and human action from the Neolithic to the Bronze Age (7000-4000 B.P.) in Alicante, Spain, based on charcoal analysis. Veget. Hist. Archaeobot. 3, 155-166.

Bagnoli, F., Vendramin, G., BuoanamiciI, A., Doulis, A., González-Martínez, S.C., La Porta, N., Magri, D., Raddi, P., Sebastiani, F., Fineschi, S., 2009. Is *Cupressus sempervirens* native in Italy? An answer from genetic and palaeobotanical data. Mol. Ecol. 18, 2276-2286.

Bailey, G., 2008. Mesolithic Europe: overview and new problems. In: Bailey, G., Spikins, P. (Eds.), Mesolithic Europe. Cambridge University Press, Cambridge, pp. 357-371.

Baillie, M.G.L., 1994. Dendrochronology raises questions about the nature of the AD 536 dust-veil event. Holocene 4, 212-217.

Baillie, M.G.L., 2002. Future of dendrochronology with respect to archaeology. Dendrochronologia 20, 69-85.

Bajard, M., Poulenard, J., Sabatier, P., Etienne, D., Ficetola, F., Chen, W.T., Gielly, L., Taberle, P., Develle, A.L., Rey, P.J., Moulin, B., de Beaulieu, J.L., Arnaud, F., 2017. Long-term changes in alpine pedogenetic processes: Effect of millennial agro-pastoralism activities (French-Italian Alps). Geoderma 306, 217-236.

Bakels, C., 2009. The Western European Loess Belt. Agrarian History, 5300 BC – AD 1000, Springer, Dordrecht, Heidelberg, 295 pp.

Bakels, C., Jacomet, S., 2003. Access to luxury foods in central Europe during the Roman period: the archaeobotanical evidence. World Archaeology 34, 542-557.

Baker, R.G., 1965. Late-glacial pollen and plant macrofossils from Spider Creek, southern St Louis county, Minnesota. Geol. Soc. Am. Bull. 76, 601-610.

Baker, R.G., 2013. Plant macrofossil records: Holocene of North America. In: Elias, S.A., Mock, C.J. (Eds.), *Encyclopedia of Quaternary Science,* 2nd ed., Vol. 3. Elsevier B.V. pp. 2459-2475.

Bakker, T.W.M., 1992. The shape of bogs from a hydrological point of view. Int. Peat J. 4, 47-54.

Bakker, E.S., Gill, J.L., Johnson, C.N., Vera, F.W.M., Sandom, C.J., Asner, G.P., Svenning, J.C., 2016. Combining paleo-data and modern exclosure experiments to assess the impact of megafauna extinctions on woody vegetation. Proc. Nat. Acad. Sci. USA 113, 847-855.

Ball, I.R., 1975. Nature and formulation of biogeographical hypotheses. Syst. Zool. 24, 407-730.

Ball, I.R., 1980. The Status of Historical Biogeography., XVII Congr. Int. Ornithology, pp. 1283-1288.

Ballouche, A., 1991. Probabilité de l'indigénat du noyer (*Juglans* L.) au Maroc, d'après l'analyse pollinique. In: Ballouche, A., Maley, J. (Eds.), Palaeoecology of Africa and the surrounding islands. First symposium on African Palynology. Balkema, Rabat, pp. 185-190.

Barber, K.E., 1981. Peat Stratigraphy and Climatic Change. A palaeoecological test of the theory of cyclic peat bog regeneration. Balkema, Rotterdam.

Barber, K.E., 1993. Peatlands as scientific archives of past biodiversity. Biodivers. Conserv. 2, 474-489.

Barber, K.E., Chambers, F.M., Maddy, D., Stoneman, R., Brew, J.S., 1994. A sensitive high-resolution record of late Holocene climatic change from a raised bog in northern England. Holocene 4, 198-205.

Barber, K.E., Dumayne-Peaty, L., Hughes, P.D.M., Mauquoy, D., Scaife, R., 1998. Replicability and variability of the recent macrofossil and proxy-climate record from raised bogs: field stratigraphy and macrofossil data from Bolton Fell and Walton Moss, Cumbria, England. J. Quat. Sci. 13, 515-528.

Barber, K.E., Maddy, D., Rose, N., Stevenson, A.C., Stoneman, R., Thompson, R., 2000. Replicated proxy-climate signals over the last 2000 yr from two distant UK peat bogs: new evidence for regional palaeoclimate teleconnections. Quat. Sci. Rev. 19, 481-487.

Barber, K.E., Chambers, F., Maddy, D., 2003. Holocene palaeoclimates from peat stratigraphy: macrofossil proxy climate records from three oceanic raised bogs in England and Ireland. Quat. Sci. Rev. 22, 521-539.

Barber, K.E., Chambers, F., Maddy, D., 2004. Late Holocene climatic history of northern Germany and Denmark: peat macrofossil investigations at Dosenmoor, Schleswig-Holstein, and Svanemose, Jutland. Boreas 33, 132-144.

Barbéro, M., Loisel, R., Quézel, P., Richardson, D., Romane, F., 1998. Pines of the Mediterranean Basin. In: Richardson, D. (Ed.), Ecology and Biogeography of *Pinus*. Cambridge University Press, Cambridge, pp. 153-170.

Barbier, D., 1999. Histoire de la végétation du nord-mayennais de la fin du Weichsélien à l'aube du XXIeme siècle. Mise en évidence d'un Tardiglaciaire armoricain. Interactions Homme-Milieu (PhD thesis). Université de Nantes, Nantes, France.

Barbier, D., Visset, L., 1997. Logné, a peat bog of European ecological interest in the Massif Armorican, western France: bog development, vegetation and land-use history. Veget. Hist. Archaeobot. 6, 69-77.

Barnekow, L., 1999. Holocene tree-line dynamics and inferred climatic changes in the Abisko area, northern Sweden, based on macrofossil and pollen records. Holocene 9, 253-265.

Barnekow, L., 2000. Holocene regional and local vegetation history and lake-level changes in the Torneträsk area, northern Sweden. J. Paleolimn. 23, 399-420.

Barnekow, L., Sandgren, P., 2001. Palaeoclimate and tree-line changes during the Holocene based on pollen and plant macrofossil records from six lakes at different altitudes in northern Sweden. Rev. Palaeobot. Palynology 117, 109-118.

Barnekow, L., Bragee, P., Hammarlund, D., St. Amour, N., 2008. Boreal forest dynamics in north-eastern Sweden during the last 10,000 years based on pollen analysis. Veget. Hist. Archaeobot. 17, 687-700.

Barnosky, C.W., 1988. A Late-Glacial and Post-Glacial Pollen Record from the Dingle Peninsula, County Kerry. Proc. R. Ir. Acad. [B] 88B, 23-37.

Bartish, I.V., Kadereit, J.W., Comes, H.P., 2006. Late Quaternary history of *Hippophae rhamnoides* L. (Elaeagnaceae) inferred from chalcone synthase intron (Chsi) sequences and chloroplast DNA variation. Mol. Ecol. 15, 4065-4083.

Bartlein, P.J., 2013. Timescales of Climate Change. In: Elias, S.A. (Ed.), Encyclopedia of Quaternary Science. Elsevier, Amsterdam, pp. 1873-1883.

Bartlein, P.J., Prentice, I.C., 1989. Orbital variations, climate and paleoecology. Trends Ecol. Evol. 4, 195-199.

Bartlein, P.J., Edwards, M.E., Shafer, S.L., Barker, E.D., 1995. Calibration of radiocarbon ages and the interpretation of paleoenvironmental records. Quat. Res. 44, 417-424.

Bartley, D.D., Morgan, A.V., 1990. The palynological record of the King's Pool, Stafford, England. New Phytol. 116, 177-194.

Bartley, D.D., Chambers, C., Hart-Jones, B., 1976. The Vegetational History of Parts of South and East Durham. New Phytol. 77, 437-468.

Bastin, B., Cordy, J.-M., Gewelt, M., Otte, M., 1986. Fluctuations climatiques enregistrées depuis 125000 ans dans les couches de remplissage de la grotte Scladina (Province de Namur, Belgique). Bull. A. F. E. Q. 1986(1/2), 169-177.

Baum, T., Mainberger, M., Taylor, T., Tinner, W., Hafner, A., Ebersbach, R., 2020. How many, how far? Quantitative models of Neolithic land use for six wetland sites of the northern Alpine forelands between 4300 and 3700 BC. Veget. Hist. Archaeobot. 29, 621-639.

Bazile-Robert, E., 1981. Le pin pignon (*Pinus pinea* L.) dans le Würm récent de Provence. Géobios 14, 395 - 307.

Beaudouin, C., Jouet, G.L., Suc, J.P., Berne, S., Escarguel, G., 2007. Vegetation dynamics in southern France during the last 30 ky BP in the light of marine palynology. Quat. Sci. Rev. 26, 1037-1054.

Beck, P., Caudullo, G., de Rigo, D., Tinner, W., 2016a. *Betula pendula, Betula pubescens* and other birches in Europe: distribution, habitat, usage and threats. In: San-Miguel-Ayanz, J., de Rigo, D., Caudullo, G., Houston Durrant, T., Mauri, A. (Eds.), European Atlas of Forest Tree Species. Publication Office of the European Union, Luxembourg, pp. 70-73.

Beck, P., Caudullo, G., Tinner, W., De Rigo, D., 2016b. *Fraxinus excelsior* in Europe: distribution, habitat, usage and threats. In: San-Miguel-Ayanz, J., De Rigo, D., Caudullo, G., Houston Durrant, T., Mauri, A. (Eds.), European Atlas of Forest Tree Species. Publication Office of the European Union, Luxembourg, pp. 98-99.

Becker, B., 1980. Tree-ring dating and radiocarbon calibration in south-central Europe. Radiocarbon 22, 219-226.

Becker, B., 1993. An 11,000-year German oak and pine chronology for radiocarbon calibration. Radiocarbon 35, 201-213.

Beer, R., Tinner, W., Carraro, G., Grisa, E., 2007. Pollen representation in surface samples of the *Juniperus*, *Picea* and *Juglans* forest belts of Kyrgyzstan, central Asia. Holocene 17, 599-611.

Beer, R., Kaiser, F., Schmidt, K., Ammann, B., Carraro, G., Grisa, E., Tinner, W., 2008. Vegetation history of the walnut forests in Kyrgyzstan (central Asia): natural or anthropogenic origin? Quat. Sci. Rev. 27, 621-632.

Beerling, D.J., 1998. *Salix herbacea* L. J. Ecol. 86, 872 - 895.

Beerling, D.J., Chaloner, W., 1992. Stomatal density as an indicator of atmospheric CO_2 concentration. Holocene 21, 71-78.

Beerling, D.J., Birks, H.H., Woodward, F.I., 1995. Rapid late-glacial atmospheric CO_2 changes reconstructed from the stomatal density record of fossil leaves. J. Quat. Sci. 10, 379-384.

Beffa, G., Pedrotta, T., Colombaroli, D., Henne,

P.D., van Leeuwen, J.F.N., Susstrunk, P., Kaltenrieder, P., Adolf, C., Vogel, H., Pasta, S., Anselmetti, F., Gobet, E., Tinner, W., 2016. Vegetation and fire history of coastal north-eastern Sardinia (Italy) under changing Holocene climates and land use. Veget. Hist. Archaeobot. 25, 271-289.

Behling, H., Cohen, M.C.L., Lara, R.J., 2001. Studies on Holocene mangrove ecosystem dynamics of the Bragança Peninsula in north-eastern Pará, Brazil. Palaeogeog. Palaeoclim. Palaeoecol. 167, 225-242.

Behre, K.-E., 1970. Die Flora des Helgoländer Süßwasser-"Töcks", eines Eem-Interglazials unter der Nordsee. Flora 152, 133-146.

Behre, K.-E., 1974. Die Vegetation im Spätpleistozän von Osterwanna/Niedersachsen. Geologisches Jahrbuch A 18, 3-48.

Behre, K.-E., 1976. Beginn und Form der Plaggenwirtschaft in Nordwestdeutschland nach pollenanalytischen Untersuchungen in Ostfriesland. Neue Ausgrabungen und Forschungen in Niedersachsen 10, 197-224.

Behre, K.-E., 1981. The interpretation of Anthropogenic Indicators in pollen diagrams. Pollen et Spores 23, 225-245.

Behre, K.-E. (Ed.) 1986. Anthropogenic Indicators in Pollen Diagrams. Balkema, Rotterdam, 246 pp.

Behre, K.-E., 1988. The rôle of man in European vegetation history. In: Huntley, B.W., Webb, T. (Eds.), Vegetation History. Kluwer, Dordrecht, pp. 633-672.

Behre, K.-E., 1989. Biostratigraphy of the Last Glacial Period in Europe. Quat. Sci. Rev. 8, 25-44.

Behre, K.-E., 1992. The history of rye cultivation in Europe. Veget. Hist. Archaeobot. 1, 141-156.

Behre, K.-E., 1998. Landwirtschaftliche Entwicklungslinien und die Veränderung der Kulturlandschaft in der Bronzezeit Europas. In: Hänsel, B. (Ed.), Mensch und Umwelt in der Bronzezeit Europas. Oetker-Voges Verlag, Kiel. 91-109.

Behre, K.-E., 2000. Frühe Ackersysteme, Düngemethoden und die Entstehung der nordwestdeutschen Heiden. Archäologisches Korrespondenzblatt 30, 135-151.

Behre, K.-E., 2002. Landscape development and occupation history along the southern North Sea coast. In: Wefer, G., Berger, W., Behre, K-E., Jansen, E. (Eds.), Climatic Development and History of the North Atlantic Realm. Springer, Berlin-Heidelberg, pp. 299-312.

Behre, K.-E., 2007a. A new Holocene sea-level curve for the southern North Sea. Boreas 36, 82-102.

Behre, K.-E., 2007b. Evidence for Mesolithic agriculture in and around central Europe? Veget. Hist. Archaeobot. 16, 203-219.

Behre, K.-E., 2008a. Comment on: "Mesolithic agriculture in Switzerland? A critical review of the evidence" by W. Tinner, E.H. Nielsen and A.F. Lotter. Quat. Sci. Rev. 27, 1467-1470.

Behre, K.-E., 2008b. Landschaftsgeschichte Norddeutschlands. Wachholtz, Neumünster, pp. 308.

Behre, K.-E. (Ed.), 2012. The chronological setting of the Palaeolithic sites of Schöningen. RGZM-Verlag, Mainz.

Behre, K.-E., 2014. Ostfriesland – Die Geschichte einer Landschaft und ihrer Besiedlung. Brune-Mettcker, Wilhelmshaven, 332 pp.

Behre, K.-E., Kučan, D., 1986. Die Reflektion archäologisch bekannter Siedlungen in Pollendiagrammen verschiedener Entfernung - Beispiele aus der Siedlungskammer Flögeln, Nordwestdeutschland, in: Behre, K.-E. (Ed.), Anthropogenic Indicators in Pollen Diagrams. Balkema, pp. 95–114.

Behre, K.-E., Kučan, D., 1994. Die Geschichte der Kulturlandschaft und des Ackerbaus in der Siedlungskammer Flögeln, Niedersachsen, seit der Jungsteinzeit. Probleme der Küstenforschung im südlichen Nordseegebiet 21, pp. 227.

Behre, K.-E., Lade, U., 1986. Eine Folge von Eem und 4 Weichsel-Interstadialen in Oerel/Niedersachsen und ihr Vegetationsablauf. Eiszeitalter und Gegenwart 36, 11-36.

Behre, K.-E., van der Plicht, J., 1992. Towards an absolute chronology for the last glacial period in Europe: radiocarbon dates from Oerel, northern Germany. Veget. Hist. Archaeobot. 1, 111-117.

Behre, K.-E., Brande, A., Küster, H., Rösch, M., 1996. Germany. In: Berglund, B.E., Birks, H.J.B., Ralska-Jasiewiczowa, M., Wright, H.E. (Eds.), Palaeoecological Events During the Last 15 000 Years. Regional Syntheses of Palaeoecological Studies of Lakes and Mires in Europe. Wiley, Chichester, pp. 507-551.

Behre, K.-E., Hölzer, A., Lemdahl, G., 2005. Botanical macro-remains and insects from the Eemian and Weichselian site of Oerel (northwest Germany) and their evidence for the history of climate. Veget. Hist. Archaeobot. 14, 31-53.

Beijerinck, W., 1976. Zadenatlas der Nederlandsche Flora. Bakhuys & Meesters, Wageningen.

Belyea, L.R., 2007. Climatic and topographic limits to the abundance of bog pools. Hydrol. Processes 21, 675-687.

Belyea, L.R., Baird, A., 2006. Beyond "The Limits to Peat Bog Growth": Cross-scale feedback in peatland development. Ecol. Monogr. 76, 299-322.

Benito, B.M., Gil-Romera, G., Birks, H.J.B., 2020. Ecological memory at millennial time-scales: the importance of data constraints, species longevity and niche features. Ecography 43, 1-10.

Bennett, K.D., 1983a. Devensian late-glacial and Flandrian vegetational history at Hockham mere, Norfolk, England. New Phytol. 95, 489-504.

Bennett, K.D., 1983b. Postglacial population

expansion of forest trees in Norfolk, UK. Nature 303, 164-167.

Bennett, K.D., 1984. The post-glacial history of *Pinus sylvestris* in the British Isles. Quat. Sci. Rev. 3, 133-155.

Bennett, K.D., 1985. The spread of *Fagus grandifolia* across Eastern North America during the last 18000 years. J. Biogeogr. 12, 147-164.

Bennett, K.D., 1986a. Competitive interactions among forest tree populations in Norfolk, England, during the last 10000 years. New Phytol. 103, 603-620.

Bennett, K.D., 1986b. The rate of spread and population increase of forest trees during the Postglacial. Philos. Trans R. Soc. B: Biol. Sci. 314, 523-531.

Bennett, K.D., 1988. Holocene geographic spread and population expansion of *Fagus grandifolia* in Ontario, Canada. J. Ecol. 76, 547-557.

Bennett, K.D., 1990. Milankovitch cycles and their effects on species in ecological and evolutionary time. Paleobiology 16, 11-21.

Bennett, K.D., 1995. Insularity and the Quaternary tree and shrub flora of the British Isles. In: Preece, R.C. (Ed.), Island Britain: A Quaternary Perspective. Geological Society Special Publication 96, pp. 173-180.

Bennett, K.D., 1996. Determination of the number of zones in a biostratigraphical sequence. New Phytol. 132, 155-170.

Bennett, K.D., 1997. Evolution and Ecology: The Pace of Life. Cambridge University Press, Cambridge. 241 pp.

Bennett, K.D., 2000. Psimpoll and pscomb: computer programs for data plotting and analysis. Quaternary Geology, Earth Sciences, Uppsala University, Uppsala.

Bennett, K.D., Birks, H.J.B., 1990. Postglacial history of alder (*Alnus glutinosa* (L.) Gaertn.) in the British Isles. J. Quat. Sci. 5, 123-133.

Bennett, K.D., Provan, J., 2008. What do we mean by 'refugia'? Quat. Sci. Rev. 27, 2449-2455.

Bennett, K.D., Willis, K.J., 2001. Pollen. In: Smol, J.P., Birks, H.J.B., Last, W.M. (Eds.), Tracking Environmental Change Using Lake Sediments, Vol. 3: Terrestrial, Algal, and Siliceous Indicators. Kluwer, Dordrecht, pp. 5-32.

Bennett, K.D., Fossitt, J., Sharp, M., Switsur, V., 1990. Holocene vegetational and environmental history at Loch Lang, South Uist,Western Isles, Scotland. New Phytol. 114, 281-298.

Bennett, K.D., Tzedakis, P.C., Willis, K.J., 1991. Quaternary refugia of north European trees. J. Biogeogr. 18, 103-115.

Bennett, K.D., Boreham, S., Sharp, M.J., Switsur, V.R., 1992. Holocene History of Environment, Vegetation and Human Settlement on Catta Ness, Lunnasting, Shetland. J. Ecol. 80, 241-273.

Benthem, F., van, Clarke, G., Punt, W., 1984. Fagaceae. In: Punt, W., Clarke, G. (Eds.), The Northwest European Pollen Flora. Elsevier, Amsterdam, pp. 87-110.

Benton, M.J., Harper, D.A.T., 2009. Introduction to Paleobiology and the Fossil Record. Wiley-Blackwell, New York. 592 pp.

Ber, A., 2005. Polish Pleistocene Stratigraphy - A review of interglacial stratotypes. Netherlands J. of Geosciences 84/2, 61-76.

Berglund, B.E., 1966. Late-Quaternary vegetation in eastern Blekinge, southeastern Sweden. A pollen-analytical study. II. Post-Glacial time. Opera Bot. 12, 1-190.

Berglund, B.E. (Ed.), 1986. Handbook of Holocene Palaeoecology and Palaeohydrology. John Wiley & Sons, Chichester.

Berglund, B.E. (Ed.), 1991. The Cultural Landscape During 6000 Years in Southern Sweden - the Ystad Project. Munksgaard, Copenhagen.

Berglund, B.E., 2003. Human impact and climate changes - synchronous events and a causal link? Quat. Int. 105, 7-12.

Berglund, B.E., Lagerlund, E., 1981. Eemian and Weichselian stratigraphy in South Sweden. Boreas 10, 323-362.

Berglund, B.E., Emanuelsson, U., Persson, S., Persson, T., 1986. Pollen/vegetation relationships in grazed and mowed plant communities of South Sweden, In: Behre, K.-E. (Ed.), Anthropogenic Indicators in Pollen Diagrams. Balkema, Rotterdam, pp. 37-51.

Berglund, B.E., Birks, H.J.B., Ralska-Jasiewiczowa, M., Wright, H.E., (Eds.) 1996a. Palaeoecological Events During the Last 15 000 Years. Regional Syntheses of Palaeoecological Studies of Lakes and Mires in Europe. Wiley, Chichester.

Berglund, B.E., Digerfeldt, G., Engelmark, R., Gaillard, M.-J., Karlsson, S., Miller, U., Risberg, J., 1996b. Sweden. In: Berglund, B.E., Birks, H.J.B., Ralska-Jasiewiczowa, M., Wright, H.E. (Eds.), Palaeoecological Events During the Last 15 000 Years. Regional Syntheses of Palaeoecological Studies of Lakes and Mires in Europe. Wiley, Chichester, pp. 233-280.

Bergman, J., Hammarlund, D., Hannon, G., Barnekow, L., Wohlfarth, B., 2005. Deglacial vegetation succession and Holocene tree-limit dynamics in the Scandes Mountains, west-central Sweden: stratigraphic data compared to megafossil evidence. Rev. Palaeobot. Palynol. 134, 129-151.

Berland, A., Shuman, B., Manson, S.M., 2011. Simulated importance of dispersal, disturbance, and landscape history in long-term ecosystem change in the Big Woods of Minnesota. Ecosystems 14, 398-414.

Bernasconi, S., McKenzie, J., 2013. Lake Sediments, In: Elias, S.A., Mock, C.A. (Eds.), Encyclopedia of Quaternary Science, 2nd ed. Elsevier, Amsterdam, pp. 351-359.

Bertini, A., 2010. Pliocene to Pleistocene palynoflora and vegetation in Italy: State of the art. Quat. Int. 225, 5-24.

Bertini, A., Sadori, L., 2010. Palaeobotanical and palynological records from Italy. Quaternary International 225/1, 1-4.

Bertsch, K., 1924. Paläobotanische Untersuchungen im Reichermoos. Jahresh. Verein vaterländ. Naturkunde in Württemberg 80, 1-90.

Bertsch, K., 1931. Paläobotanische Monographie des Federsees. Bibl. Bot. 103, 1-127.

Bertsch, K., 1935. Der deutsche Wald im Wechsel der Zeiten. FF Heine, Tübingen.

Bertsch, K., 1940. Geschichte des deutschen Waldes. Gustav Fischer, Jena.

Bertsch, K., 1942. Lehrbuch der Pollenanalyse. Stuttgart.

Bertsch, K., 1951. Der Nussbaum (*Juglans regia*) als einheimischer Waldbaum. Veröff. der Würtembergischen Landesstelle für Naturschutz und Landschaftspflege (in Ludwigsburg und Tübingen) 20, 65-68.

Bertsch, K., 1953. Die Walnüsse der Bodensee Pfahlbauten. Vorzeit am Bodensee 1, 1-8.

Bertsch, A., 1961. Untersuchungen an rezentem und fossilem Pollen von *Juniperus*. Flora 150, 503-513.

Bertsch, K., Bertsch, F., 1947. Geschichte unserer Kulturpflanzen. Stuttgart.

Besnard, G., Khadari, B., Baradat, P., Bervillé, A., 2002a. Combination of chloroplast and mitochondrial DNA polymorphisms to study cytoplasm genetic differentiation in the olive complex (*Olea europaea* L.). Theor. Appl. Genet. 105, 139–144.

Besnard, G., Khadari, B., Baradat, P., Bervillé, A., 2002b. *Olea europaea* (Oleaceae) phylogeography based on chloroplast DNA polymorphism. Theor. Appl. Genet. 104, 1353–1361.

Besnard, G., Rubio de Casas, R., Christin, P., Vargas, P., 2009. Phylogenetics of *Olea* (Oleaceae) based on plastid and nuclear ribisomal DNA sequences: Tertiary climatic shifts and lineage differentiation times. Annals Bot. 104, 143-160.

Besnard, G., Khadari, B., Navascués, M., Fernández-Mazuecos, M., El Bakkali, A., Arrigo, N., Baali-Cherif, D., Brunini-Bronzini de Caraffa, V., Santoni, S., Vargas, P., Savolainen, V., 2013. The complex history of the olive tree: from Late Quaternary diversification of Mediterranean lineages to primary domestication in the northern Levant. Proc. R. Soc. B 280, 20122833.

Beug, H.-J., 1956 Pollendimorphismus bei *Ephedra*. Naturwissenschaften 14, 332-33.

Beug, H.-J., 1957. Untersuchungen zur spätglazialen und frühpostglazialen Floren- und Vegetationsgeschichte einiger Mittelgebirge. Flora 145, 167-211.

Beug, H.-J., 1961. Beiträge zur postglazialen Floren- und Vegetationsgeschichte in Süddalmatien: Der See Malo Jezero auf Mljet. Teil I: Vegetationsentwicklung. Flora 150, 600-631.

Beug, H.-J., 1962. Über die ersten anthropogenen Vegetationsveränderungen in Süddalmatien an Hand eines neuen Pollendiagrammes von "Malo Jezero" auf Mljet. Veöff. Geobot. Inst. Rübel Zürich 37, 9-15.

Beug, H.-J., 1967. On the forest history of the Dalmatian coast. Rev. Palaeobot. Palynol. 2, 271-279.

Beug, H.-J., 1975a. Changes of climate and vegetation belts in the mountains of Mediterranean Europe during the Holocene. Bull. Geol. 19, 101-110.

Beug, H.-J., 1975b. Man as a factor in the vegetational history of the Balkan Peninsula. Proceedings of the first International Symposium on Balkan Flora and Vegetation. Varna, June 1973, pp 72-78.

Beug, H.-J., 1979. Vegetationsgeschichtlich-pollenanalytische Untersuchungen am Riss/Würm-Interglazial von Eurach am Starnberger See/Obb. Geologica Bavarica 80, 91-106.

Beug, H.-J., 1992. Vegetationsgeschichtliche Untersuchungen über die Besiedlung im Unteren Eichsfeld, Landkreis Göttingen, vom frühen Neolithikum bis zum Mittelalter. Neue Ausgrabungen und Forschungen in Niedersachsen 20, 261-339.

Beug, H.-J., 2004. Leitfaden der Pollenbestimmung für Mitteleuropa und angrenzende Gebiete. Verlag Dr. Friedrich Pfeil, München.

Beug, H.-J., 2015. Leitfaden der Pollenbestimmung für Mitteleuropa und angrenzende Gebiete, 2nd ed. Friedrich Pfeil, München, 542 pp.

Beug, H.-J., 2016. Die Spät- und nacheiszeitliche Vegetationsentwicklung am Nordrand der niedersächsischen und hessischen Mittelgebirge (Harz bis Weser). Verlag Dr. Friedrich Pfeil, München.

Beug, H.-J., Henrion, I., Schmüser, A., 1999. Landschaftsgeschichte im Hochharz - Die Entwicklung der Wälder und Moore seit dem Ende der letzten Eiszeit. Goslar. Papierflieger, Clausthal-Zellerfeld.

Bevan, A., Palmisano, A., Woodbridge, J., Fyfe, R., Shennan, S., 2019. The changing face of the Mediterranean – Land cover, demography and environmental change: Introduction and overview. Holocene 29, 703-707.

Bezusko, L.G., 1973. Concerning the problem of development of vegetation in the Left-Bank Forest-Steppe of the Ukraine in Holocene from data of spore-pollen investigations. Ukr. Bot. Zhurnal 30, 228–237.

Bezusko, L.G., 1977. Actualnye voprosy sovremennoyi botaniki [Concerning the problem of formation of vegetation cover of Bug-Styr district in Holocene from data of palaepalynology (Maloe Polessie)], in: Regional'naya Paleogeografiya. Naukova Dumka, Kiev, Ukraine, pp. 141–144.

Bezusko, L.G., Klimanov, V.A., Sheliag-Sosonko, Y.R., 1988. Climatic conditions in Ukraine in the Late Post-glacial and Holocene., in: Paleoklimaty Golotsena Evropeiyskoy Territorii SSSR. PANGAEA, Moscow, Russia.

Bhagwat, S.A., Willis, K.J., 2008. Species persistence in northerly glacial refugia of Europe: a matter of chance or biogeographical traits? J. Biogeogr. 35, 464-482.

Białobrzeska, M., Truchanowiczówna, J., 1960. The variability of shape of fruits and scales of the European birches (*Betula* L.) and their determination in fossil materials (in Polish). Monogr. Bot. 9, 1-85.

Bialozyt, R., Bradley, L., Bradshaw, R., 2012. Modelling the spread of *Fagus sylvatica* and *Picea abies* in southern Scandinavia during the late Holocene. J. Biogeogr. 39, 665-675.

Biesboer, D.D., 1975. Pollen morphology of the Aceraceae. Grana 15, 19-27.

Bigler, C., Larocque, I., Peglar, S.M., Birks, H.J.B., Hall, R., 2002. Quantitative multiproxy assessment of long-term patterns of Holocene environmental change from a small lake near Abisko, northern Sweden. Holocene 12, 481-496.

Bigler, C., Barnekow, L., Heinrichs, M., Hall, R., 2006. Holocene environmental history of Lake Vuolep Njakajaure (Abisko National Park, northern Sweden) reconstructed using biological proxy indicators. Veget. Hist. Archaeobot. 15, 309-320.

Billamboz, A., 2009. Jahrringuntersuchungen in der Siedlung Forschner und weiteren bronze- und eisenzeitlichen Feuchtbodensiedlungen Südwestdeutschlands. Siedlungsarchäologie im Alpenvorland 11, 399-555.

Binka, K., Nitychoruk, J., 2001. Late Saalian climate changes in Europe in the light of pollen analysis and the problem of two-step deglaciation at the oxygen isotope stage 6/5e transition. Boreas 30, 307-316.

Binka, K., Madeyska, T., Marciniak, B., Seroczynska, K., Wieckowski, K., 1988. Bledowo Lake (central Poland): History of vegetation and lake development during the last 12 kyr. Bull Acad Pol Sci 36, 147–158.

Binka, K., Ciesla, A., Lacka, B., Madeyska, T., Marciniak, B., Szeroczynska, K., Wieckowski, K., 1991. The development of Bledowo Lake (central Poland) - A palaeoecological study. Stud. Geol. Pol. 100, 3–85.

Binney, H.A., Willis, K.J., Edwards, M.E., Bhagwat, S.A., Anderson, P., Andreev, A., Blaauw, M., Damblon, F., Haesaerts, P., Kienast, F., Kremenetski, K., Krivonogov, S., Lozhkin, A., MacDonald, G.M., Novenko, E., Oksanen, P., Sapelko, T., Väliranta, M., Vazhenina, L., 2009 The distribution of late-Quaternary woody taxa in northern Eurasia: evidence from a new macrofossil database Quat. Sci. Rev. 28, 2445–2464.

Birks, H.J.B., 1968. The identification of *Betula nana* pollen. New Phytol. 67, 309-314.

Birks, H.H., 1970. Studies in the vegetational history of Scotland I. A pollen diagram from Abernethy Forest, Inverness-Shire. J. Ecol. 58, 827-846.

Birks, H.H., 1973a. Modern macrofossil assemblages in lake sediments in Minnesota. In: Birks, H.J.B., West, R.G. (Eds.), *Quaternary Plant Ecology*. Blackwell Scientific Publications, Oxford, pp. 173-189.

Birks, H.J.B., 1973b. Past and present vegetation of the Isle of Skye: A palaeoecological study. Cambridge University Press, Cambridge, UK.

Birks, H.H., 1975. Studies in the vegetational history of Scotland. IV. Pine stumps in Scottish blanket peats. Philos. Trans. R. Soc. London B 270, 181-226.

Birks, H.J.B., 1976. Late-Wisconsinan vegetational history at Wolf Creek, central Minnesota. Ecol. Monogr. 46, 395-492.

Birks, H.J.B., 1977. The Flandrian forest history of Scotland: a preliminary synthesis. In: Shotton, F. (Ed.), British Quaternary studies – recent advances. Clarendon Press, Oxford, pp. 119-135.

Birks, H.J.B., 1978. Geographic variation of *Picea abies* (L.) Karsten pollen in Europe. Grana 17, 149-160.

Birks, H.H., 1980. Plant macrofossils in Quaternary lake sediments, Advances in Limnology 15. Schweizerbart, Stuttgart.

Birks, H.J.B., 1981a. Late Wisconsin vegetational and climatic history at Kylen Lake, Northeastern Minnesota. Quat. Res. 16, 322-355.

Birks, H.J.B., 1981b. The use of pollen analysis in the reconstruction of past climates: a review. In: Wigley, T.M.L., Ingram, M.J., Farmer, G. (Eds.), Climate and History. Cambridge University Press, Cambridge, pp. 111-138.

Birks, H.J.B., 1982. Mid-Flandrian forest history of Roudsea Wood National Nature Reserve, Cumbria. New Phytol. 90, 339 354.

Birks, H.H., 1984. Late-Quaternary pollen and plant macrofossil stratigraphy at Lochan an Druim, north-west Scotland, in: Haworth, E.Y., Lund, J.W.G. (Eds.), Lake Sediments and Environmental History: Studies in Palaeolimnology and Palaeoecology in Honour of Winifred Tutin. University of Leicester Press, Leicester, pp. 377-405.

Birks, H.J.B., 1985. Recent and possible future mathematical developments in quantitative paleoecology. Palaeogeogr., Palaeoclimatol., Palaeoecol. 50, 107-147.

Birks, H.J.B., 1986a. Late-Quaternary biotic changes in terrestrial and lacustrine environments, with particular reference to north-west Europe. In: Berglund, B.E. (Ed.), Handbook of Holocene Palaeoecology and Palaeohydrology. J. Wiley & Sons, Chichester, pp. 3-65.

Birks, H.J.B., 1986b. Numerical zonation, comparison and correlation of Quaternary pollen-stratigraphical data. In: Berglund, B.E. (Ed.), Handbook of Holocene Palaeoecology and Palaeohydrology. John Wiley & Sons, Chichester, pp. 743-774.

Birks, H.J.B., 1989. Holocene isochrone maps and patterns of tree-spreading in the British Isles. J. Biogeogr. 16, 503-540.

Birks, H.H., 1991a. Holocene vegetational history and climatic change in west Spitsbergen - plant macrofossils from Skardtjørna, an arctic lake. Holocene 1, 209-218.

Birks, H.J.B., 1991b. Floristic and vegetational history of the Outer Hebrides, In: Pankhurst, R., Mullin, J. (Eds.), Flora of the Outer Hebrides, Pelagic, Exeter, pp. 32-37.

Birks, H.J.B., 1993. Quaternary paleoecology and vegetation science - current contributions and possible future developments. Rev. Palaeobot. Palynol. 79, 153-177.

Birks, H.J.B., 1995. Quantitative palaeoenvironmental reconstructions. In: Maddy, D., Brew, J.S. (Ed.), Statistical Modelling of Quaternary Science Data. Quaternary Research Association, Cambridge, pp. 161-254.

Birks, H.J.B., 1996a. Contributions of Quaternary palaeoecology to nature conservation. J. Veg. Sci. 7, 89-98.

Birks, H.J.B., 1996b. Great-Britain: Scotland. In: Berglund, B.E., Birks, H.J.B., Ralska-Jasiewiczowa, M., Wright, H.E. (Eds.), Palaeoecological Events During the Last 15 000 Years. Regional Syntheses of Palaeoecological Studies of Lakes and Mires in Europe. Wiley, Chichester, pp. 95-143.

Birks, H.J.B., 1996c. Statistical approaches to interpreting diversity patterns in the Norwegian mountain flora. Ecography 19, 332-340.

Birks, H.J.B., 1997a. Is ecology still a sick science? an appraisal of Simberloff's (1980) diagnosis of community ecology. In: Strand, R., Bristow, G.A. (Eds.), Naturvitere Filosoferer. Senter for vitenskapsteori, University of Bergen, pp. 59-76.

Birks, H.J.B., 1997b. Reconstructing environmental impacts of fire from the Holocene sedimentary record. In: Clark, J.S., Cachier, H., Goldammer, J.G., Stocks, B.J. (Eds.), Sediment records of biomass burning and global change. Springer-Verlag, Berlin, pp. 295-311.

Birks, H.H., 2000. Aquatic macrophyte vegetation development in Kråkenes Lake, western Norway, during the late-glacial and early-Holocene. J. Paleolimnol. 23, 7-19.

Birks, H.H., 2001. Plant macrofossils. In: Smol, J.P., Birks, H.J.B., Last, W.M. (Eds.), Tracking Environmental Change Using Lake Sediments, Vol. 3: Terrestrial, Algal, and Siliceous Indicators. Kluwer, Dordrecht, pp. 49-74.

Birks, H.H., 2003. The importance of plant macrofossils in the reconstruction of Lateglacial vegetation and climate: examples from Scotland, western Norway, and Minnesota, USA. Quat. Sci. Rev. 22, 453-473.

Birks, H.J.B., 2005. Mind the gap: how open were European primeval forests? Trends Ecol. Evol. 20, 154-156.

Birks, H.H., 2007a. Plant Macrofossils: Introduction. In: Elias, S.A. (Ed.), Encyclopedia of Quaternary Science. Elsevier, Amsterdam, pp. 2266-2288.

Birks, H.J.B., 2007b. Pollen Methods and Studies: Numerical Analysis Methods. In: Elias, S.A. (Ed.), Encyclopedia of Quaternary Science. Elsevier, Amsterdam, pp. 2514-2521.

Birks, H.J.B., 2007c. Estimating the amount of compositional change in late-Quaternary pollen-stratigraphical data. Veget. Hist. Archaeobot. 16, 197-202.

Birks, H.H., 2008a. The Late-Quaternary history of arctic and alpine plants. Plant Ecol. Div. 1, 135-146.

Birks, H.J.B., 2008b. Holocene climate research - progress, paradigms, and problems. In: Battarbee, R.W., Binney, H.A. (Eds.), Natural Climate Variability and Global Warming: A Holocene Perspective. Wiley-Blackwell, Oxford, pp. 7-57.

Birks, H.J.B., 2012a. Analysis of stratigraphical data, in: Birks, H.J.B., Lotter, A.F., Juggins, S., Smol, J.P. (Eds.), Tracking Environmental Change Using Lake Sediments Volume 5: Data Handling and Numerical Techniques. Springer, Dordrecht, pp. 355-378.

Birks, H.J.B., 2012b. Ecological palaeoecology and conservation biology: controversies, challenges, and compromises. Int. J. Biodiv. Sci. Ecosys. Serv. Manage. 8, 292-304.

Birks, H.H., 2013. Plant Macrofossils: Introduction. In: Elias, S.A., Mock, C.J. (Eds.), Encyclopedia of Quaternary Science, 2nd ed. Elsevier, Amsterdam, pp. 593-612.

Birks, H.J.B., 2014. Challenges in the presentation and analysis of plant macrofossil stratigraphical data. Veget. Hist. Archaeobot. 23, 309-330.

Birks, H.H., 2015. South to north: Contrasting late-glacial and early-Holocene climate changes and vegetation responses between south and north Norway. Holocene 25, 37-52.

Birks, H.H., 2017. Plant macrofossil introduction. In: Reference Module in Earth Systems and Environmental Sciences, Elsevier.

Birks, H.J.B., 2019. Contributions of Quaternary botany to modern ecology and biogeography. Plant Ecol. Div. 12, 189-385.

Birks, H.J.B., 2020. Reflections on the use of ecological attributes and traits in Quaternary botany. Front. Ecol. Evol. 8, 166.

Birks, H.H., Ammann, B., 2000. Two terrestrial records of rapid climate change during the glacial-Holocene transition (14,000-9,000 calendar years BP) from Europe. Proc. Natl. Acad. Sci. USA 97, 1390-1394.

Birks, H.J.B., Berglund, B.E., 2018. One hundred years of Quaternary pollen analysis 1916-2016. Veget. Hist. Archaeobot. 27, 271-309.

Birks, H.J.B., Birks, H.H., 1980. Quaternary Palaeoecology. Edward Arnold, London.

Birks, H.H., Birks, H.J.B., 2000. Future uses of pollen analysis must include plant macrofossils. *J. Biogeogr.* 27, 31-35.

Birks, H.H., Birks, H.J.B., 2003. Reconstructing Holocene climates from pollen and plant macrofossils. In: Mackay, A.W., Battarbee, R.W., Birks, H.J.B., Oldfield, F. (Eds.), *Global Change in the Holocene.* Edward Arnold, London, pp. 342-357.

Birks, H.J.B., Birks, H.H., 2004. The rise and fall of forests. Science 305, 484-485.

Birks, H.H., Birks, H.J.B., 2006. Multi-proxy studies in palaeolimnology. Veget. Hist. Archaeobot. 15, 235-251.

Birks, H.J.B., Birks, H.H., 2008a. Biological responses to rapid climate changes at the Younger Dryas-Holocene transition at Kråkenes, western Norway. Holocene 18, 19-30.

Birks, H.J.B., Birks, H.H., 2008b. Quaternary Palaeoecology. Blackburn Press, Caldwell.

Birks, H.H., Birks, H.J.B., 2014. To what extent did changes in July temperature influence Lateglacial vegetation patterns in NW Europe? Quat. Sci. Rev 106(15), 262-277.

Birks, H.J.B., Birks, H.H., 2016. How have ancient-DNA studies from sediments contributed to the reconstruction of Quaternary floras? *New Phytol.* 209, 499-507.

Birks, H.H., Bjune, A.E., 2010. Can we detect a west Norwegian tree line from modern samples of plant remains and pollen? Results from the DOORMAT project. Veget. Hist. Archaeobot. 19, 325-340.

Birks, H.J.B., Gordon, A.D., 1985. Numerical Methods in Quaternary Pollen Analysis. Academic Press, London. 317 pp.

Birks, H.J.B., Line, J.M., 1992. The use of rarefaction analysis for estimating palynological richness from Quaternary pollen-analytical data. Holocene 2, 1-10.

Birks, H.J.B., Lotter, A.F., 1994. The impact of the Laacher See Volcano (11000 yr B.P.) on terrestrial vegetation and diatoms. J. Paleolimnol. 11, 313-322.

Birks, H.J.B., Madsen, B.J., 1979. Flandrian vegetational history of Little Loch Roag, Isle of Lewis, Scotland. J. Ecol. 67, 825-842.

Birks, H.H., Mathewes, R.W., 1978. Studies in the vegetational history of Scotland V. Late Devensian and early Flandrian pollen and macrofossil stratigraphy at Abernethy Forest, Inverness-shire. New Phytol. 80, 455-484.

Birks, H.J.B., Peglar, S.M., 1979. Interglacial pollen spectra from Sel Ayre, Shetland. New Phytol. 83, 559-575.

Birks, H.J.B., Ransom, M. E., 1969. An Interglacial peat at Fugla Ness, Shetland. New Phytologist, 68, 777-796.

Birks, H.J.B., Saarnisto, M., 1975. Isopollen maps and principal component analysis of Finnish pollen data for 4000, 6000, and 8000 years ago. Boreas 4, 77-96.

Birks, H.J.B., Seppä, H., 2004. Pollen-based reconstructions of late-Quaternary climate in Europe - progress, problems, and pitfalls. Acta Palaeobot. 44, 317-334.

Birks, H.J.B., Seppä, H., 2010. Late-Quaternary palaeoclimatic research in Fennoscandia - A historical review. Boreas 39, 655-673.

Birks, H.J.B., Tinner, W., 2016. Past forests of Europe. In: San-Miguel-Ayanz, J., de Rigo, D., Caudullo, G., Houston Durrant, T., Mauri, A. (Eds.), European Atlas of Forest Tree Species Publication Office of the European Union Luxembourg, pp. 36-39.

Birks, H.H., van Dinter, M., 2010. Lateglacial and early Holocene vegetation and climate gradients in the Nordfjord-Alsund area,

Birks, H.J.B., Williams, W., 1983. Late-Quaternary vegetational history of the Inner Hebrides. Proc. R. Soc. Edinburgh 83B, 269-292.

Birks, H.J.B., Willis, K.J., 2008. Alpines, trees, and refugia in Europe. Plant Ecol. Divers. 1, 147-160.

Birks, H.H., Switsur, V.R., West, R.G., 1975. Studies in the vegetational history of Scotland IV. Pine stumps in Scottish blanket peats. Philos. Trans. R. Soc. Lond. B Biol. Sci. 270, 181-226.

Birks, H.H., Birks, H.J.B., Kaland, P., Moe, D., (eds.), 1988. The Cultural Landscape. Past, Present and Future. Cambridge University Press, Cambridge, 521 pp.

Birks, H.J.B., Line, J.M., Juggins, S., Stevenson, A.C., ter Braak, C.J.F., 1990. Diatoms and pH reconstruction. Philos. Trans R. Soc. B: Biol. Sci. 327, 263-278.

Birks, H.H., Battarbee, R.W., Birks, H.J.B., Bradshaw, E.G., Brooks, S.J., Duigan, C.A., Jones, V.J., Lemdahl, G., Peglar, S.M., Solem, J.O., Solhoy, I.W., Solhoy, T., Stalsberg, M.K., 2000. The development of the aquatic ecosystem at Kråkenes Lake, western

Norway, during the late-glacial and early Holocene - a synthesis. J. Paleolimn. 23, 91-114.

Birks, H.H., Peglar, S.M., Boomer, I., Flower, R.J., Ramdani, M., 2001. Palaeolimnological responses of nine North African lakes in the CASSARINA Project to recent environmental changes and human impact detected by plant macrofossil, pollen, and faunal analyses. *Aquatic Ecol.* 35, 405-430.

Birks, H.H., Larsen, E., Birks, H.J.B., 2005. Did tree-*Betula*, *Pinus* and *Picea* survive the last glaciation along the west coast of Norway? A review of the evidence, in light of Kullman (2002). J. Biogeogr. 32, 1461-1471.

Birks, H.H., Birks, H.J.B., Bjune, A.E., Felde, V.A., 2012a. XXXVI International Bog ('Moor') Excursion of the Institute of Plant Sciences, University of Bern, southern and south-eastern Norway, 8-13 September 2012, Excursion Guide. University of Bergen, Norway and Bjerknes Centre for Climate Research, Bergen, Norway.

Birks, H.H., Giesecke, T., Hewitt, G., Tzedakis, P.C., Bakke, J., Birks, H.J.B., 2012b. Comment on "Glacial Survival of Boreal Trees in Northern Scandinavia". Science 338, 742.

Birks, H.H., Jones, V., Brooks, S.J., Birks, H.J.B., Telford, R.J., Juggins, S., Peglar, S.M., 2012c. From cold to cool in northernmost Norway: Lateglacial and early Holocene multi-proxy environmental and climate reconstructions from Jansvatnet, Hammerfest. Quat. Sci. Rev. 33, 100-120.

Birks, H.J.B., Lotter, A.F., Juggins, S., Smol, J.P., (Eds.), 2012d. Tracking Environmental Change Using Lake Sediments, Vol. 5: Data Handling and Numerical Techniques. Springer, Dordrecht.

Birks, H.H., Aarnes, I., Bjune, A.E., Brooks, S.J., Bakke, J., Kühl, N., Birks, H.J.B., 2014. Lateglacial and early-Holocene climate variability reconstructed from multi-proxy records on Andøya, northern Norway. Quat. Sci. Rev. 89, 108-122.

Birks, H.J.B., Birks, H.H., Ammann, B., 2016a. The fourth dimension of vegetation. Science 354, 412-413.

Birks, H.J.B., Felde, V.A., Bjune, A.E., Seppä, H., Giesecke, T., 2016b. Does pollen-assemblage richness reflect floristic richness? A review of recent developments and future challenges. Rev. Palaeobot. Palynol. 228, 1-25.

Birks, H.J.B., Felde, V.A., Seddon, A.W.R., 2016c. Biodiversity trends within the Holocene. Holocene 26, 994-1001.

Bisculm, M., Colombaroli, D., Vescovi, E., van Leeuwen, J.F.N., Henne, P.D., Rothen, J., Procacci, G., Pasta, S., La Mantia, T., Tinner, W., 2012. Holocene vegetation and fire dynamics in the supra-mediterranean belt of the Nebrodi Mountains (Sicily, Italy). J. Quat. Sci. 27, 687-698.

Bishop, R.R., Church, M.J., Rowley-Conwy, P.A., 2015. Firewood, food and human niche construction: the potential role of Mesolithic hunter-gatherers in actively structuring Scotland's woodlands. Quat. Sci. Rev. 108, 51-75.

Bittmann, F., 1992. The Kärlich Interglacial, Middle Rhine region, Germany: vegetation history and stratigraphic position. Veget. Hist. Archaeobot. 1, 243-258.

Bittmann, F., 1996. Vegetationsgeschichtliche Untersuchungen an mittel- und jungpleistozänen Ablagerungen des Neuwieder Beckens (Mittelrhein). Jahrb. Röm.-German. Zentralmuseum Mainz, 83-190.

Bittmann, F., 2012. Die Schöninger Pollendiagramme und ihre Stellung im mitteleuropäischen Mittelpleistozän. In: Behre, K.-E. (Ed.) Die chronologische Einordnung der paläolithischen Fundstellen von Schöningen. Forschungen zur Urgeschichte aus dem Tagebau von Schöningen 1, Mainz, pp. 97-112.

Björck, S., 2008. The late Quaternary development of the Baltic Sea basin. In: von Storch, H. (Ed.), Assessment of Climate Change for the Baltic Sea Basin. Springer, Heidelberg, pp. 398-407.

Björck, S., Wohlfahrt, B., 2001. 14C Chronostratigraphic techniques in Paleolimnology. In: Last, W.M., Smol, J.P. (Eds.), Tracking Environmental Change Using Lake Sediments, Vol. 1: Basin Analysis, Coring, and Chronological Techniques. Kluwer, Dordrecht, pp. 205-245.

Björck, S., Walker, M.J.C., Cwynar, L.C., Johnsen, S., Knudsen, K.-L., Lowe, J.J., Wohlfarth, B., and INTIMATE Members, 1998. An event stratigraphy for the Last Termination in the North Atlantic region based on the Greenland ice-core record: A proposal by the INTIMATE group. J. Quat. Sci. 13, 283-292.

Björkman, L., Feurdean, A., Cinthio, K., Wohlfarth, B., Possnert, G., 2002. Lateglacial and early Holocene vegetation development in the Gutaiului Mountains, northwestern Romania. Quat. Sci. Rev. 21, 1039-1059.

Björkman, L., Feurdean, A., Wohlfarth, B., 2003. Late-Glacial and Holocene forest dynamics at Steregoiu in the Gutaiului Mountains, Northwest Romania. Rev. Palbot. Palynol. 124, 79-111.

Björse, G., Bradshaw, R.H.W., Michelson, D.B., 1996. Calibration of regional pollen data to construct maps of former forest types in southern Sweden. J. Paleolimnol. 16, 67-78.

Bjune, A.E., 2005. Holocene vegetation history and tree-line changes on a north–south transect crossing major climate gradients in southern Norway—evidence from pollen and plant macrofossils in lake sediments. Rev. Palaeobot. Palynol. 133, 249–275.

Bjune, A.E., Birks, H.J.B., 2008. Holocene vegetation dynamics and inferred climate changes at Svanavatnet, Mo i Rana, northern Norway. Boreas 37, 146-156.

Bjune, A.E., Birks, H.J.B., Seppä, H., 2004. Holocene vegetation and climate history on a continental-oceanic transect in northern Fennoscandia based on pollen and plant macrofossils. Boreas 33, 211-223.

Bjune, A.E., Bakke, J., Nesje, A., Birks, H.J.B., 2005. Holocene mean July temperature and winter precipitation in western Norway inferred from palynological and glaciological lake-sediment proxies. Holocene 15, 177-189.

Bjune, A.E., Seppä, H., Birks, H.J.B., 2009. Quantitative summer-temperature reconstructions for the last 2000 years based on pollen-stratigraphical dara from northern Fennoskandia. J. Paleolimnol. 41, 43-56.

Bjune, A.E., Birks, H.J.B., Peglar, S.M., Odland, A., 2010. Developing a modern pollen-climate calibration data set for Norway. Boreas 39, 674-688.

Bjune, A.E., Helvik, I., Birks, H.J.B., 2013. The Fagus sylvatica forests in the Larvik region, south-eastern Norway: their origin and history. Veget. Hist. Archaeobot. 22, 215-229.

Blaauw, M., Heegaard, E., 2012. Estimation of Age-Depth Relationships. In: Birks, H.J.B., Lotter, A.F., Juggins, S., Smol, J.P. (Eds.), Tracking Environmental Change Using Lake Sediments: Data Handling and Numerical Techniques, Developments in Paleoenvironmental Research. Springer Netherlands, Dordrecht, pp. 379-413.

Blaauw, M., van Geel, B., Heuvelink, G.B.M., van der Plicht, J., 2003. A numerical approach to 14-C wiggle-match dating of organic deposits: best fits and confidence intervals. In: An Investigation of Holocene sun-climate relationships using numerical C-14 wiggle-match dating of peat deposits. Ponsen & Looijen BV, Wageningen, Amsterdam, pp. 25-54.

Blaauw, M., Christen, J.A., Bennett, K.D., Reimer, P.J., 2018. Double the dates and go for Bayes — Impacts of model choice, dating density and quality on chronologies. Quat. Sci. Rev. 188, 58-66.

Blackmore, S., Steinmann, J.A.J., Hoen, P.P., Punt, W., 2003. Betulaceae and Corylaceae. Rev. Palaeobot. Palynol. 123, 71-98.

Blarquez, O., Bremond, L., Carcaillet, C., 2010. Holocene fires and a herb-dominated understorey track wetter climates in subalpine forests. J. Ecol. 98, 1358-1368.

Blarquez, O., Carcaillet, C., Elzein, T.M., Roiron, P., 2011. Needle accumulation rate model-based reconstruction of palaeo-tree biomass in the western subalpine Alps. Holocene 22, 579-587.

Blaus, A., Reitalu, T., Gerhold, P., Hiiesalu, I., Massante, J.C., Veski, S., 2020. Modern pollen-plant diversity relationships inform palaeoecological reconstructions of functional and phylogenetic diversity in calcareous fens. Front. Ecol. Evol. 8.

Bleicher, N., 2013. Summed radiocarbon probability density functions cannot prove solar forcing of central European lake-level changes. Holocene 23, 755-765.

Bleicher, N., Antolín, F., Heiri, O., Häberle, S., Hüster Plogmann, H., Jacomet, S., Pümpin, C., Rentzel, P., Schäfer, M., Schibler, J., Wiemann, P., van Hardenbroek, M., Tóth, M., Zweifel, N., 2018. Middens, currents and shorelines: Complex depositional processes of waterlogged prehistoric lakeside settlements based on the example of Zurich-Parkhaus Opéra (Switzerland). J. Archaeol. Sci. 97, 26 - 41.

Blitzer, H., 1993. Olive cultivation and olive production in Minoan Crete. In: Amouretti, M., Brun, J. (Eds.), La production du vin et l'huile en Mediterranée de l'Age du Bronze à la fin du XVIeme siècle. Aix-en-Provence, pp. 163-175.

Blois, J.L., Williams, J.W., Fitzpatrick, M.C., Ferrier, S., Veloz, S.D., He, F., Liu, Z.Y., Manion, G., Otto-Bliesner, B., 2013a. Modeling the climatic drivers of spatial patterns in vegetation composition since the Last Glacial Maximum. Ecography 36, 460-473.

Blois, J.L., Williams, J.W., Fitzpatrick, M.C., Jackson, S.T., Ferrier, S., 2013b. Space can substitute for time in predicting climate-change effects on biodiversity. Proc. Natl. Acad. Sci. USA 110, 9374-9379.

Blois, J.L., Gotelli, N.J., Behrensmeyer, A.K., Faith, J.T., Lyons, S.K., Williams, J.W., Amatangelo, K.L., Bercovici, A., Du, A., Eronen, J.T., Graves, G.R., Jud, N., Labandeira, C., Looy, C.V., McGill, B., Patterson, D., Potts, R., Riddle, B., Terry, R., Tóth, A., Villaseñor, A., Wing, S., 2014. A framework for evaluating the influence of climate, dispersal limitation, and biotic interactions using fossil pollen associations across the late Quaternary. Ecography 37, 1095-1108.

Blytt, A., 1876. Essay on the immigration of the Norwegian flora during alternate rainy and dry periods. Cammermeyer, Christiania.

Bobek, P., Svobodova-Svitavska, H., Pokorný, P., Samonil, P., Kuneš, P., Kozakova, R., Abraham, V., Klinerova, T., Svarcova, M.G., Jamrichova, E., Krauseova, E., Wild, J., 2019. Divergent fire history trajectories in Central European temperate forests revealed a pronounced influence of broadleaved trees on fire dynamics. Quat. Sci. Rev. 222.

Bobrov, E., 1972. Die introgressive Hybridisation in der Gattung *Picea*. A. Dietr. Symp. Biol. Hung. 12, 141-148.

Boch, M.S., Mazing, V.V. 1979. Боч, М.С. & Мазинг, В.В. (1979) Экосистемы болот СССР. [Mire ecosystems of the USSR] Nauka, Leningrad, 187 pp. (in Russian)

Böcher, T., Fredskild, B., Holmen, K., Jacobsen, K., 1978. Grønlands Flora, 3rd ed. P. Haase & Søn, Copenhagen.

Bogdel, I.I., 1984. Razvitie prirody Belorrussii v golotsene [Holocene environmental changes of Byelorussia] (PhD thesis). Belarusskii Universety, Minsk, Belarus.

Bogucki, P., Crabtree, P., 2004. Ancient Europe, 8000 B.C. - A.D. 1000: an Encyclopedia of the Barbarian world. Scribner, New York., 658 pp.

Bohn, U., Gollub, G., Hettwer, C., Neuhäuslová, Z., Schlüter, H., Weber, H., 2003. Map of the Natural Vegetation of Europe. Landwirtschaftsverlag, Bonn.

Bohncke, S., 1991. Palaeohydrological changes in the Netherlands during the last 13000 years. PhD thesis, Vrije University, Amsterdam, The Netherlands.

Bohncke, S., Wijmstra, L., 1988. Reconstruction of Late-Glacial lake-level fluctuations in The Netherlands based on palaeobotanical analyses, geochemical results and pollen-density data. Boreas 17, 403–425.

Bohncke, S., Vandenberghe, J., Coope, R., Reiling, R., 1987. Geomorphology and palaeoecology of the Mark valley (southern Netherlands): palaeoecology, palaeohydrology and climate during the Weichselian Late Glacial. Boreas 16, 69–85.

Bohncke, S., Wijmstra, L., Woude, J.V.D., Sohl, H., 1988. The Late-Glacial infill of three lake successions in The Netherlands: Regional vegetational history in relation to NW European vegetational developments. Boreas 17, 385–402.

Bohncke, S.J.P., Bos, J.A.A., Engels, S., Heiri, O., Kasse, C., 2008. Rapid climatic events as recorded in middle Weichselian thermokarst lake sediments. Quat. Sci. Rev. 27, 162-174.

Bolinder, K., Norbäck Ivarsson, L., Humphreys, A.M., Ickert-Bond, S.M., Han, F., Hoorn, C., Rydin, C., 2016. Pollen morphology of *Ephedra* (Gnetales) and its evolutionary implications. Grana 55, 24-51.

Bolland, A., Rey, F., Gobet, E., Tinner, W., Heiri, O., 2020. Summer temperature development 18,000-14,000 cal. BP recorded by a new chironomid record from Burgaschisee, Swiss Plateau. Quat. Sci. Rev. 243.

Bomfleur, B., McLoughlin, S., Vajda, V., 2014. Fossilized nuclei and chromosomes reveal 180 million years of genomic stasis in royal ferns. Science 343, 1376.

Bond, W.J., van Wilgen, B.W., 1996. Fire and Plants. Chapman and Hall, London.

Bonn, S., Poschlod, P., 1998. Ausbreitungsbiologie der Pflanzen Mitteleuropas: Grundlagen und kulturhistorische Aspekte. Quelle und Meyer, Wiesbaden.

Bonny, A.P., 1978. The effect of pollen recruitment processes on pollen distribution over the sediment surface of a small lake in Cumbria. J. Ecol. 66, 385-416.

Booth, R.K., Jackson, S.T., Sousa, V.A., Sullivan, M.E., Minckley, T.A., Clifford, M.J., 2012. Multi-decadal drought and amplified moisture variability drove rapid forest community change in a humid region. Ecology 93, 219-226.

Bordács, S., Popescu, F., Slade, D., Csaikl, U., Lesur, I., Borovics, A., Kézdy, P., König, A., Gömöry, D., Brewer, S., Burg, K., Petit, J., 2002. Chloroplast DNA variation of white oaks in northern Balkans and in the Carpathian Basin. For. Ecol. Managem. 156, 197-209.

Bordon, A., Peyron, O., Lézine, A., Brewer, S., Fouache, E., 2009. Pollen-inferred Late-Glacial and Holocene climate in southern Balkans (LakeMaliq). Quat. Int. 200, 19–30.

Borisova, O.K., 2014. Paleoenvironment and models of adaptations of lake settlements in the Mesolithic and Neolithic of the forest zone of eastern Europe., in: Materials of the International Conference. PANGAEA, St. Petersburg, Russia, pp. 37–40.

Boros, Á., Járai-Komlódi, M., Tóth, Z., Nilsson, S., 1975. An Atlas of Recent European Moss Spores. Scientia Publishing, Budapest.

Bortenschlager, S., 1965. Funde afrikanischer Pollen in den Alpen. Naturw. 52, 663-664.

Bortenschlager, S., 1967. Pollenanalytische Ergebnisse einer Firnprofiluntersuchung am Kesselwandferner (3240 m), Ötztal, Tirol. Grana Palynol. 7, 259-274.

Bortenschlager, S., 1970. Neue pollenanalytische Untersuchungen von Gletschereis und gletschernahen Mooren in den Ostalpen. Zeitschr. Gletscherk. Glazialgeol. 6, 107-118.

Bortenschlager, S., 1972. Der pollenanalytische Nachweis von Gletscher- und Klimaschwankungen in Mooren der Ostalpen. Ber. Deutsche Bot. Ges. 85, 113-122.

Bortenschlager, S., 1977. Ursachen und Ausmass postglazialer Waldgrenzschwankungen in den Ostalpen. In: B. Frenzel (Ed.), 1977: Dendrochronologie und postglaziale Klimaschwankungen in Europa. Erdwiss. Forschung, Wiesbaden, 13, 260-266.

Bortenschlager, S., 1984. Beiträge zur Vegetationsgeschichte Tirols I. Inneres Ötztal und unteres Inntal. Berichte des naturwissenschaftlich-medizinischen Vereins in Innsbruck 71, 19-56.

Bortenschlager, S., Oeggl, K., Wahlmüller, N., 1996. Austria. In: Berglund, B., Birks, H., Ralska-Jasiewiczowa, M., Wright, H.E. (Eds.), Palaeoecological Events During the Last 15 000 Years. Regional Syntheses of Palaeoecological Studies of Lakes and Mires in Europe. Wiley, Chichester, pp. 667-685.

Bos, J.A.A, Dickson, J.H., Coope, G.R., Jardine,

W.D., 2004. Flora, fauna and climate of Scotland during the Weichselian middle pleniglacial - palynological, macrofossil and coleopteran investigations. Palaeogeogr. Palaeoclim. Palaeoecol., 204, 65-100.

Bos, J.A.A., van Geel, B., Groenewoudt, B.J., Lauwerier, R.C.G.M., 2006. Early Holocene environmental change, the presence and disappearance of early Mesolithic habitation near Zutphen (The Netherlands). Veget. Hist. Archaeobot. 15, 27-43.

Bos, J.A.A., Helmens, K.F., Bohncke, S.J.P., Seppä, H., Birks, H.J.B., 2009. Flora, vegetation and climate at Sokli, northeastern Fennoscandia, during the Weichselian middle pleniglacial. Boreas 38, 335-348.

Boski, T., Moura, D., Veiga-Pires, C., Camacho, S., Duarte, D., Scott, D.B., Fernandes, S.G., 2002. Postglacial sea-level rise and sedimentary response in the Guadiana Estuary, Portugal/Spain border. Sediment. Geol., Coastal Environment Change During Sea-Level Highstands 150, 103-122.

Boski, T., Camacho, S., Moura, D., Fletcher, W., Wilamowski, A., Veiga-Pires, C., Correia, V., Loureiro, C., Santana, P., 2008. Chronology of the sedimentary processes during the postglacial sea level rise in two estuaries of the Algarve coast, Southern Portugal. Estuar. Coast. Shelf Sci. 77, 230-244.

Botkin, D.B., Saxe, H., Araujo, M.B., Betts, R., Bradshaw, R.H.W., Cedhagen, T., Chesson, P., Dawson, T., Etterson, J.R., Faith, D.P., Ferrier, S., Guisan, A., Hansen, A.S., Hilbert, D.W., Loehle, C., Margules, C., New, M., Sobel, M.J., Stockwell, D.R.B., 2007. Forecasting the effects of global warming on biodiversity. BioScience 57, 227-236.

Bottema, S., 1974. Late Quaternary vegetation history of northwestern Greece. PhD Thesis, University of Groningen, Groningen, 190 pp.

Bottema, S., 1979. Pollen analytical investigations in Thessaly (Greece). Palaeohistoria 21, 19-40.

Bottema, S., 1980. Pollen analytical investigations on Crete. Rev. Palaeobot. Palynol. 31, 193-217.

Bottema, S., 2000. The Holocene history of walnut, sweet-chestnut, manna-ash and plane tree in the Eastern Mediterranean. Pallas 52, 35-59.

Bottema, S., 2001. A note on the pollen representation of ivy (*Hedera helix* L.). Rev. Palaeobot. Palynol. 117, 159-166.

Bottema, S., Sarpaki, A., 2003. Environmental change in Crete: a 9000-year record of Holocene vegetation history and the effect of the Santorini eruption. Holocene 13, 733-749.

Boyd, W.E., 1986. The role of mosses in modern pollen analysis: the influence of moss morphology on pollen entrapment. Pollen et Spores 28, 243-256.

Bozilova, E., Atanassova, J., 1990. Paläoökologische Bedingungen und die Pflanzengeschichte der Umgebung von Durankulak. Durankulak 1, 197-205.

Bozilova, E., Beug, H.-J., 1992. On the Holocene history of vegetation in SE Bulgaria (Lake Arkutino, Ropotamo region). Veg. Hist. Archaeobot. 1, 19-32.

Bozilova, E., Beug, H.-J., 1994. Studies on the vegetation history of Lake Varna region, northern Black Sea coastal area of Bulgaria. Veg. Hist. Archaeobotany 3, 143-154.

Bozilova, E., Djankova, M., 1976. Vegetation development during the Eemian in the North Black Sea Region. Fitologija 4, 25-33.

Bozilova, E., Filipova, M., 1986. Paleoecological environment in northeastern Black Sea area during Neolithic, Eneolithic and Bronze periods. Studia Praehistoria 8, 160-165.

Bozilova, E., Ivanov, I., 1985. Paleoenvironment in the area of the Varna Lake during Eneolithic and Bronze Age on the basis of palynological, paleoethnobotanical and archaeological evidence. Bull Musee Nat Varna 21, 43-78.

Bozilova, E., Smit, A., 1979. Palynology of lake Sucho Ezero from South Rila Mountain (Bulgaria). Phytologia 11, 54-67.

Bozilova, E., Tonkov, S., 1985. Paleoecological studies in lake Durankulak. Annu. Sofia Univ. Fac. Biol. 76, 25-29.

Bozilova, E., Tonkov, S., 1998. Towards the vegetation and settlement history of the southern Dobrudza coastal region, north-eastern Bulgaria: a pollen diagram from Lake Durankulak. Veget. Hist. Archaeobot.7, 141-148.

Bozilova, E., Tonkov, S., 2000. Pollen from lake Sedmo Rilsko reveals southeast European postglacial vegetation in the highest mountain area of the Balkans. New Phytol. 148, 315-325.

Bozilova, E., Tonkov, S., Pavlova, D., 1986. Pollen and plant macrofossil analyses of the Lake Suho Ezero in the South Rila mountains. Godishnik Na Sofiyskiya Ouniversitet BF Kn. 80, 48-57.

Bozilova, E., Filipova, M., Filipovich, L., Tonkov, S., 1996. Bulgaria. In: Berglund, B.E., Birks, H.J.B., Ralska-Jasiewiczowa, M., Wright, H.E. (Eds.), Palaeoecological Events During the Last 15 000 years - Regional syntheses of palaeoecological studies of lakes and mires in Europe. John Wiley & Sons, Chichester, pp. 701-728.

Bozilova, E., Lazarova, M., Tonkov, S., 2011. The postglacial vegetation history of the Western Rhodopes Mountains. Biodiversity of Bulgaria. Biodiversity of Western Rhodopes (Bulgaria and Greece) II, 4, 11-19.

Bradley, R.S., 1990. Holocene paleoclimatology of the Queen Elizabeth Islands, Canadian High Arctic. Quat. Sci. Rev. 9, 365-384.

Bradley, R.S., 2015. Paleoclimatology. Reconstructing Climates of the Quaternary, 3rd ed. Academic Press, Oxford.

Bradley, L.R., Giesecke, T., Halsall, K., Bradshaw, R.H.W., 2013. Exploring the requirement for anthropogenic disturbance to assist the stand-scale expansion of *Fagus sylvatica* L. outside southern Scandinavia. Holocene 23, 579-586.

Bradshaw, R.H.W., 1988. Spatially-precise studies of forest dynamics. In: Huntley, B., Webb, T. (Eds.), Vegetation History. Kluwer, Dordrecht, pp. 725-751.

Bradshaw, R.H.W., 2002. Book review on F.M.W. Vera: Grazing Ecology and Forest History. For. Ecol. Managem. 165, 327-329.

Bradshaw, R.H.W., 2008. Detecting human impact in the pollen record using data-model comparison. Veget. Hist. Archaeobot. 17, 597-603.

Bradshaw, R.H.W., 2013. Stand-Scale Palynology. In: Elias, S.A., Mock, C.J. (Eds.), Encyclopedia of Quaternary Science, 2nd ed. Elsevier, Amsterdam, pp. 846–853.

Bradshaw, R.H.W., Hannon, G., 1992. Climatic change, human influence and disturbance regime in the control of vegetation dynamics within Fiby Forest, Sweden. J. Ecology 80, 625-632.

Bradshaw, R.H.W., Hannon, G.E., 2004. The Holocene structure of north-west European temperate forest induced from palaeoecological data. In: Honnay, O., Verheyen, K., Bossuyt, B., Hermy, M. (Eds.), Forest Biodiversity: Lessons from History for Conservation. CABI, pp. 11-26.

Bradshaw, R.H.W., Lindbladh, M., 2005. Regional spread and stand-scale establishment of *Fagus sylvatica* and *Picea abies* in Scandinavia. Ecology 86, 1679-1686.

Bradshaw, R.H.W., McGee, E., 1988. The extent and time-course of mountain blanket peat erosion in Ireland. New Phyt. 108, 219–224.

Bradshaw, R.H.W., Sykes, M.T., 2014. Ecosystem Dynamics. From the Past to the Future. Wiley-Blackwell, Chichester. 334 pp.

Bradshaw, R.H.W., Zackrisson, O., 1990. A two-thousand-year history of a northern Swedish boreal forest stand. J. Veg. Sci. 1, 519-528.

Bradshaw, R.H.W., Tolonen, K., Tolonen, M., 1997. Holocene records of fire from the boreal and temperate zones of Europe. In: Clark, J.S., Tolonen, K., Tolonen, M. (Eds.), Sediment records of biomass burning and global change. Springer, Berlin, pp. 347-365.

Bradshaw, E.G., Rasmussen, P., Odgaard, B.V., 2005. Mid- to late-Holocene land-use change and lake development at Dallund Sø, Denmark: synthesis of multiproxy data, linking land and lake. Holocene 15, 1152-1162.

Bradshaw, R.H.W., Lindbladh, M., Hannon, G.E., 2010. The role of fire in southern Scandinavian forests during the late Holocene. Int. J, Wildl. Fire 19, 1040-1049.

Bradshaw, R.H.W., Jones, C.S., Edwards, S.J., Hannon, G.E., 2015. Forest continuity and conservation value in Western Europe. Holocene 25, 194-202.

Brande, A., 1973. Untersuchungen zur postglazialen Vegetationsgeschichte im Gebiet der Neretva-Niederungen (Dalmatien, Herzegowina). Flora 162, 1-44.

Brande, A., 1975. Vegetationsgeschichtliche und pollenstratigrafische Untersuchungen zum Paläolithikum von Mauern und Meilenhofen (Fränkische Alb). Quartär 26, 74-106.

Brande, A., 1994. Eibe und Buche im Holozän Brandenburgs. Diss. Bot. 234, 225-239.

Brattbakk, I., 1986. Vegetasjonsregioner – Svalbard og Jan Mayen. Kartblad 4.1.3, Nasjonalatlas for Norge.

Brauer, A., Endres, Ch., Zolitschka, B., Negendank, J.F.W., 2000. AMS Radiocarbon and Varve Chronology from the annually laminated sediment record of Lake Meerfelder Maar, Germany. Radiocarbon 42, 355-368.

Brauer, A., Wulf, S., Mangili, C., Moscariello, A., 2007. Tephrochronological dating of varved interglacial lake deposits from Pianico-Sellere (Southern Alps, Italy) to around 400 ka. J. Quat. Sci. 22, 85-96.

Braun-Blanquet, J., 1925. Die Brachypodium ramosum-Phlomis lychnitis-Assoziation der Roterböden Südfrankreichs. In: Brockmann-Jerosch, H. (ed.), 1925. Festschrift Carl Schröter. Veröff. Geobot. Inst. Rübel Zürich 3, 304-320.

Braun-Blanquet, J., 1961. Die inneralpine Trockenvegetation. Von der Provence bis zur Steiermark. Fischer, Stuttgart. 273 pp.

Braun-Blanquet, J., 1964. Pflanzensoziologie. Grundzüge der Vegetationskunde, 3. ed. Springer, Berlin, Wien, New York.

Braun, M., Sümegi, P., Tóth, A., Willis, K.J., Szalóki, I., Margitai, Z., Somogyi, A., 2005. Reconstruction of long-term environmental changes at Kelemér-Kis-Mohos-Tó. Archaeologica Hungarica, Environmental Archaeology in North-Eastern Hungary 10, 25-38.

Breitenbach-Dorfer, M., Konnert, M., Pinsker, W., Starlinger, F., Geburek, T., 1997. The contact zone between two migration routes of silver fir, *Abies alba* (Pinaceae), revealed by allozyme studies. Plant Syst. Evol. 206, 259-272.

Breton, C., Tersac, M., Bervillé, A., 2006. Genetic diversity and gene flow between the wild olive (Oleaster, *Olea europaea* L.) and the olive: Several Plio-Pleistocene refuge zones in the Mediterranean Basin suggested by simple sequence repeats analysis. J. Biogeogr. 33, 1916-1928.

Breton, C., Terral, J., Pinatel, C., Médail, F., Bonhomme, F., Bervillé, A., 2009. The origins of the domestication of the olive tree. Compt. Rend. Biol. 332, 1059-1064.

Brewer, S., Cheddadi, R., de Beaulieu, J.-L., Reille, M., data contributors, 2002. The spread of deciduous *Quercus* throughout Europe since the last glacial period. For. Ecol. Managem. 156, 27-48.

Brewer, S., Jackson, S.T., Williams, J.W., 2012. Paleoecoinformatics: applying geohistorical data to ecological questions. Trends Ecol. Evol. 27, 104-112.

Brewer, S., Giesecke, T., Davis, B.A.S., Finsinger, W., Wolters, S., Binney, H.A., de Beaulieu, J.L., Fyfe, R., Gil-Romera, G., Kühl, N., Kuneš, P., Leydet, M., Bradshaw, R.H.W., 2017. Late-glacial and Holocene European pollen data. J. Maps 13, 921-928.

Brice, M.-H., Cazelles, K., Legendre, P., Fortin, M.-J., 2019. Disturbances amplify tree community responses to climate change in the temperate-boreal ecotone. Global Ecol. Biogegor. 28, 1668-1681.

Bridge, M.C., Haggart, B.A., Lowe, J.J., 1990. The history and palaeoclimatic significance of subfossil remains of *Pinus sylvestris* in blanket peats from Scotland. J. Ecol. 78, 77-99.

Briquet, J., 1906. Le développement des Flores dans les Alpes occidentales, avec aperçu sur les Alpes en général. Résultats scient. Congr. Int. Bot. Vienna 1905.

Brisset, E., Guiter, F., Miramont, C., Revel, M., Anthony, E., Delhon, C., Arnaud, F., Malet, E., de Beaulieu, J.-L., 2015. Lateglacial/Holocene environmental changes in the Mediterranean Alps inferred from lacustrine sediments. Quat. Sci. Rev. 110, 49-71.

Brisset, E., Guiter, F., Miramont, C., Troussier, T., Sabatier, P., Poher, Y., Cartier, R., Arnaud, F., Malet, E., Anthony, E.J., 2017. The overlooked human influence in historic and prehistoric floods in the European Alps. Geology 45, 347-350.

Brockmann-Jerosch, J., 1936. Futterlaubbäume und Speiselaubbäume. Ber. Schweiz. bot. Ges. 46, 594-612.

Broecker, W., 1992. Defining the boundaries of the Late-glacial isotope episodes. Quat. Res. 381, 135-138.

Brongers, J.A., 1976. Air photography and Celtic field research in the Netherlands. Nederlandse Oudheden 6. ROB, Amersfoort, 118 pp.

Brooks, S.J., Birks, H.J.B., 2001. Chironomid-inferred air temperatures from Lateglacial and Holocene sites in north-west Europe: progress and problems. Quat. Sci. Rev. 20, 1723-1741.

Brooks, S.J., Heiri, O., 2013. Response of chironomid assemblages to environmental change during the early Late-glacial at Gerzensee, Switzerland. Paleogeogr. Paleoclimatol. Paleoecol. 391, 90-98.

Brorson-Christensen, B., 1949. Pollenanalytisk datering af en rensdyrknogle fra Nebbe Mosse, Ö. Vemmelöv Socken. In: Althin, C.A., Brorson-Christensen, B., Berlin, H., 1949. Renfyndet från Nebbe mosse och Sveriges senglaciala bebyggelse. K. Human. Vetenskapss. Lund. Årsber. 1948-1949.

Broström, A., Gaillard, M.-J., Ihse, M., Odgaard, B.V., 1998. Pollen-landscape relationships in modern analogues of ancient cultural landscapes in southern Sweden - a first step towards quantification of vegetation openness in the past. Veget. Hist. Archaeobot. 7, 189-201.

Broström, A., Sugita, S., Gaillard, M.-J., 2004. Pollen productivity estimates for the reconstruction of past vegetation cover in the cultural landscape of southern Sweden. Holocene 14, 368-381.

Broström, A., Sugita, S., Gaillard, M.-J., Pilesjö, P., 2005. Estimating the spatial scale of pollen dispersal in the cultural landscape of southern Sweden. Holocene 15, 252-262.

Broström, A., Nielsen, A.B., Gaillard, M.-J., Hjelle, K. L., Mazier, F., Binney, H., Bunting, M.J., Fyfe, R., Duffin, K., Meltsov, V., Poska, A., Räsänen, S., Soepboer, W., von Stedingk, H., Suutari, H., Sugita, S., 2008. Pollen productivity estimates of key European plant taxa for quantitative reconstruction of past vegetation - a review. Veget. Hist. Archaeobot. 17, 461-478.

Brubaker, L.B., 1975. Post-glacial forest patterns associated with till and outwash in Northcentral Upper Michigan. Quat. Res. 5, 499-527.

Brückner, P., 1993. Pollen morphology and taxonomy of Eurasiatic species of the genus *Buxus* (Buxaceae). Grana 32, 65-78.

Brugger, S.O., Gobet, E., Schanz, F.R., Heiri, O., Schwörer, C., Sigl, M., Schwikowski, M., Tinner, W., 2018a. A quantitative comparison of microfossil extraction methods from ice cores. J. Glaciol. 64, 432-442.

Brugger, S.O., Gobet, E., Sigl, M., Osmont, D., Papina, T., Rudaya, N., Schwikowski, M., Tinner, W., 2018b. Ice records provide new insights into climatic vulnerability of central Asian forest and steppe communities. Glob. Planet. Change 169, 188-201.

Brugiapaglia, E., 1996. Dynamique de la végétation tardiglaciaire et holocène dans les Alpes Italiennes nord-occidentales (MSc thesis). Université de Marseille, Marseille, France.

Brugiapaglia, E., de Beaulieu, J.-L., 1995. Etude de la dynamique végétale tardiglaciaire et Holocène en Italie centrale: le marais de Colfiorito (Ombrie). Comptes Rendus de l'Académie des Sciences. Serie IIa 321, 617-622.

Brun, C., 2011. Anthropogenic indicators in pollen diagrams in eastern France: a critical review. Veget. Hist. Archaeobot. 20, 135-142.

Bruns, M., Levin, I., Münnich, K.O., Hubberten, H.W., Fillipakis, S., 1980. Regional sources of volcanic carbond dioxide and their influence on 14C content of present-day plant material. Radiocarbon 22, 532-536.

Brus, R., 2010. Growing evidence for the existence of glacial refugia of European beech (*Fagus sylvatica* L.) in the south-eastern Alps and north-western Dinaric Alps. Period. Biol. 112, 239-246.

Bucci, G., Anzidei, M., Madaghiele, A., Vendramin, G.G., 1998. Detection of haplotypic variation and natural hybridization in halepensis-complex pine using chloroplast simple sequence repeat (SSR) markers. Mol. Ecol. 7, 1633 - 1643.

Buckland, W., 1823. Reliquiae Diluvianae. J. Murray, London.

Buczkó, K., Magyari, E.K., Bitušík, P., Wacnik, A., 2009. Review of dated Late Quaternary palaeolimnological records in the Carpathian Region, east-central Europe. Hydrobiologia 631, 3–28.

Budsky, A., Scholz, D., Wassenburg, J.A., Mertz-Kraus, R., Spotl, C., Riechelmann, D.F.C., Gibert, L., Jochum, K.P., Andreae, M.O., 2019. Speleothem delta C-13 record suggests enhanced spring/summer drought in south-eastern Spain between 9.7 and 7.8 ka – A circum-Western Mediterranean anomaly? Holocene 29, 1113-1133.

Bunting, M.J., 1996. The development of heathland in Orkney, Scotland: pollen records from Loch of Knitchen (Rousay) and Loch of Torness (Hoy). Holocene 62, 193-212.

Bunting, M.J., Hjelle, K.L., 2010. Effect of vegetation data collection strategies on estimates of relevant source area of pollen (RSAP) and relative pollen productivity estimates (relative PPE) for non-arboreal taxa. Veget. Hist. Archaeobot. 19, 365-374.

Bunting, M.J., Middleton, D., 2005. Modelling pollen dispersal and deposition using HUMPOL software, including simulating windroses and irregular lakes. Rev. Palaeobot. Palynol. 134, 185-196.

Bunting, M.J., Gaillard, M.-J., Sugita, S., Middleton, R., Broström, A., 2004. Vegetation structure and pollen source area. Holocene 14, 651-660.

Bunting, M.J., Farrell, M., Broström, A., Hjelle, K.L., Mazier, F., Middleton, R., Nielsen, A.B., Rushton, E., Shaw, H., Twiddle, C.L., 2013. Palynological perspectives on vegetation survey: a critical step for model-based reconstruction of Quaternary land cover. Quat. Sci. Rev. 82, 41-55.

Burban, V., Petit, R., 2003. Phylogeography of maritime pine inferred with organelle markers having contrasted inheritance. Mol. Ecol. 12, 1487–1495.

Burckle, L., Grissino-Mayer, H., 2003. Stradivari, violins, tree rings, and the Maunder Minimum: a hypothesis. Dendrochronologia 21, 41 - 45.

Burga, C.A., 1987. Gletscher- und Vegetationsgeschichte der Südrätischen Alpen seit der Späteiszeit. Denkschr. Schweiz. Naturf. Ges., Basel/Boston, 101, 1-162.

Burga, C.A., 1988. Swiss vegetation history during the last 18000 years. New Phytol. 110, 581-602.

Burga, C.A., Perret, R., 1998. Vegetation und Klima der Schweiz seit dem jüngeren Eiszeitalter, 1 ed. Ott, Thun.

Burge, O.R., Bellingham, P.J., Arnst, E.A., Bonner, K.I., Burrows, L.E., Richardson, S.J., Wiser, S.K., Wood, J.R., Wilmshurst, J.M., 2020. Integrating permanent plot and palaeoecological data to determine subalpine post-fire succession, recovery and convergence over 128 years. J. Veg. Sci., 10.1111/jvs.12887.

Burjachs, F., Julià, R., 1994. Abrupt climatic changes during the last glaciation based on pollen analysis of the Abric Romani, Catalonia, Spain. Quat. Res. 42, 308-315.

Burjachs, F., Pérez-Obiol, R., Picornell-Gelabert, L., Revelles, J., Expósito, I., Yll, E.-I., 2017. Overview of environmental changes and human colonization in the Balearic Islands (Western Mediterranean) and their impacts on vegetation composition during the Holocene. J. Archaeol. Sci. Rep. 12, 845-859.

Burney, D.A., 2010. Back to the Future in the Caves of Kaua'i: a scientist's adventures in the dark. Yale University Press, New Haven. 216 pp.

Burney, D.A., Burney, L.P., 2007. Paleoecology and "inter-situ" restoration on Kaua'i, Hawaii. Front. Ecol. Environ. 5, 483-490.

Burr, G.S., 2007. Radiocarbon Dating: Causes of Temporal Variations.. In: Elias, S.A. (Ed.), Encyclopedia of Quaternary Science. Elsevier, Amsterdam, pp. 2931-2941.

Bush, M.B., Flenley, J.R., Gosling, W.D. (Eds.), 2011. Tropical Rainforest Responses to Climatic Change, 2nd ed. Springer, Berlin. 454 pp.

Bush, M.B., Restrepo, A., Collins, A.F., 2014. Galápagos history, restoration, and a shifted baseline. Restor. Ecol. 22, 296-298.

Calcote, R., 1995. Pollen source area and pollen productivity: evidence from forest hollows. J. Ecol. 83, 591-602.

Calò, C., Henne, P.D., Curry, B., Magny, M., Vescovi, E., La Mantia, T., Pasta, S., Vannière, B., Tinner, W., 2012. Spatio-temporal patterns of Holocene environmental change in southern Sicily. Palaeogeogr. Palaeoclimatol. Palaeoecol. 323, 110-122.

Calò, C., Henne, P.D., Eugster, P., van Leeuwen, J., Gilli, A., Hamann, Y., La Mantia, T., Pasta, S., Vescovi, E., Tinner, W., 2013. 1200 years of decadal-scale variability of Mediterranean vegetation and climate at Pantelleria Island, Italy. Holocene 23, 1477-1486.

Canellas-Bolta, N., Rull, V., Vigo, J., Mercade, A.,

2009. Modern pollen-vegetation relationships along an altitudinal transect in the central Pyrenees (southwestern Europe). Holocene 19, 1185-1200.

Cao, X., Tian, F., Li, F., Gaillard, M.-J., Rudaya, N., Xu, Q., Herzschuh, U., 2019. Pollen-based quantitative land-cover reconstruction for northern Asia covering the last 40 ka cal BP. Clim. Past 15, 1503-1536.

Capelle, T., 2005. Sense, in Hoops' Reallexikon der Germanischen Altertumskunde, 2nd ed. 28, 177-179.

Cappers, R.T.J., 1993a. Seed dispersal by water: a contribution to the interpretation of seed assemblages. Veget. Hist. Archaeobot. 2, 173-186.

Cappers, R.T.J., 1993b. The identification of Potamogetonaceae fruits found in The Netherlands. Acta Bot. Nederland. 42, 447-460.

Cappers, R.T.J., Neef, R., 2012. Handbook of Plant Palaeoecology. Groningen Archaeological Studies 19. Barkhuis, Groningen, 475 pp.

Cappers, R.T.J., Bekker, R.M., Jans,J.E.A., 2006. Digitale Zadenatlas van Nederland. Groningen.

Capraro, L., Asioli, A., Backman, J., Bertoldi, R., Channell, J.E.T., Massari, F., Rio, D., 2005. Climatic patterns revealed by pollen and oxygen isotope records across the Matuyama-Brunhes Boundary in the central Mediterranean (southern Italy). In: Head, M.D., Gibbard P.L. (Eds.), Early-Middle Pleistocene Transition: the Land-Ocean evidence. Geological Society London, Special Publication 247, 159-182.

Capron, E., Landais, A., Bard, E., Blunier, T., Dahl-Jensen, D., Dokken, T., Gersonde, R., Parrenin, F., Schulz, M., Vinther, B., Waelbroeck, C., 2013. Dating and synchronizing paleoclimatic records over the last interglacial. PAGES 21, 26-27.

Caramiello, R., Polini, V., Siniscalco, C., 1989. Flora Palinologica Italiana. Schede di *Corylus avellana* L. cv. Tonda Gentile delle Langhe (TGL), e Corylus maxima Miller e valutazioni statistiche. Aerobiologia 5, 73-83.

Carcaillet, C., Blarquez, O., 2017. Fire ecology of a tree glacial refugium on a nunatak with a view on Alpine glaciers. New Phytol. 216, 1281-1290.

Carcaillet, C., Blarquez, O., 2019. Glacial refugia in the south-western Alps? New Phytol. 222, 663-667.

Carcaillet, C., Barakat, H., Panaïotis, C., Loisel, R., 1997. Fire and late-Holocene expansion of *Quercus ilex* and *Pinus pinaster* on Corsica. J. Veget. Sci. 8, 85-94.

Carcaillet, C., Almquist, H., Asnong, H., Bradshaw, R.H.W., Carriòn, J.S., Gaillard, M.J., Gajewski, K., Haas, J.N., Haberle, S.G., Hadorn, P., Muller, S.D., Richard, P.J.H., Richoz, I., Rösch, M., Sanchez Goñi, M.F., von Stedingk, H., Stevenson, A.C., Talon, B., Tardy, C., Tinner, W., Tryterud, E., Wick, L., Willis, K.J., 2002. Holocene biomass burning and global dynamics of the carbon cycle. Chemosphere 49, 845-863.

Čarni, A., Matevski, V., Kostadinovski, M., Ćušterevska, R., 2018. Scrub communities along a climatic gradient in the southern Balkans: maquis, pseudomaquis and shibljak. Plant Biosys. 152, 1165-1171.

Caroli, I., Caldara, M., 2007. Vegetation history of Lago Battaglia (eastern Gargano coast, Apulia, Italy) during the middle-late Holocene. Veget. Hist. Archaeobot. 16, 317-327.

Carrasco, R., Pedraza, J., Domínguez-Villar, D., Villa, J., Willenbring, J., 2013. The plateau glacier in the Sierra de Béjar (Iberian Central System) during its maximum extent. Reconstruction and chronology. Geomorphology 196, 83-93.

Carrión, J.S., 1990. Evolución paleoambiental durante el Pleistoceno Superior en el sureste de Espana. Las secuencias polınicas de las Cuevas de la Carihuela (Granada) y Beneito (Alicante) (PhD thesis). Murcia University, Murcia, Spain.

Carrión, J.S., 1992a. Late Quaternary pollen sequence from Carihuela Cave, southeastern Spain. Rev. Palaeobot. Palynol. 71, 37-77.

Carrión, J.S., 1992b. A palaeoecological study in the western Mediterranean area. The Upper Pleistocene pollen record from Cova Beneito (Alicante, Spain). Palaeogeogr. Palaeoclim. Palaeoecol. 92, 1-14.

Carrión, J.S., 2002a. A taphonomic study of modern pollen assemblages from dung and surface sediments in arid environments of Spain. Rev. Palaeobot. Palynol. 120, 217-232.

Carrión, J.S., 2002b. Patterns and processes of Late Quaternary environmental change in a montane region of southwestern Europe. Quat. Sci. Rev. 21, 2047-2066.

Carrión, J.S., 2010. The concepts of potential natural vegetation (PNV) and other abstractions (trying to pick up fish with wet hands). J. Biogeogr. 37, 2214-2215.

Carrión, J.S. (Ed.), 2012. Paleoflora y Paleovegetación de la Península Ibérica e Islas Baleares : Plioceno-Cuaternario, Murcia.

Carrión, J.S., Munuera, M., 1997. Upper Pleistocene palaeoenvironmental change in Eastern Spain: new pollen-analytical data fro Cova Beneito (Alicante). Palaeogeography, Palaeoclimatology, Palaeoecology 128, 287-299

Carrión, J.S., Sánchez-Gómez, P., 1992. Palynological data in support of the survival of walnut (*Juglans regia* L.) in the western Mediterranean area during last glacial times. J. Biogeogr. 10, 623-630.

Carrión, J.S., van Geel, B., 1999. Fine-resolution Upper Weichselian and Holocene palynological record from Navarrés (Valencia, Spain) and a discussion about factors of Mediterranean forest succession. Rev. Palaeobot. Palynol. 106, 209-236.

Carrión, J.S., Munuera, M., Navarro, C., 1998.

The palaeoenvironment of Carihuela Cave (Granada, Spain): a reconstruction on the basis of palynological investigations of cave sediments. Rev. Palaeobot. Palynology 99, 317-340.

Carrión, J.S., Munuera, M., Navarro, C., Burjachs, F., Dupré, M., Walker, M.J., 1999. The palaeoecological potential of pollen records in caves: the case of Mediterranean Spain. Quat. Sci. Rev. 18, 1061-1073.

Carrión, J.S., Parra, I., Navarro, C., Munuera, M., 2000. Past distribution and ecology of the cork oak (*Quercus suber*) in the Iberian Peninsula: a pollen-analytical approach. Divers. Distrib. 6, 29 -44.

Carrión, J.S., Munuera, M., Dupré, M., Andrade, A., 2001a. Abrupt vegetation changes in the Segura Mountains of southern Spain throughout the Holocene. J. Ecol. 89, 783–797.

Carrión, J.S., Andrade, A., Bennett, K.D., Navarro, C., Munuera, M., 2001b. Crossing forest thresholds: inertia and collapse in a Holocene sequence from south-central Spain. Holocene 11, 635–653.

Carrión, J.S., Sánchez-Gómez, P., Mota, J.F., Yll, R., Chaín, C., 2003a. Holocene vegetation dynamics, fire and grazing in the Sierra de Gádor, southern Spain. Holocene 13, 839-849.

Carrión, J.S., Yll, E.-I., Walker, M.J., Legaz, A.J., Chaín, C., López, A., 2003b. Glacial refugia of temperate, Mediterranean and Ibero-North African flora in south-eastern Spain: new evidence from cave pollen at two Neanderthal man sites. Global Ecol. Biogeogr. 12, 119-129.

Carrión, J.S., Yll, E.-I., Willis, K.J., Sánchez, P., 2004. Holocene forest history of the eastern plateaux in the Segura Mountains (Murcia, southeastern Spain). Rev. Palaeobot. Palynol. 132, 219-236.

Carrión, J.S., Fuentes, N., González-Sampériz, P., Sánchez Quirante, L., Finlayson, J.C., Fernández, S., Andrade, A., 2007. Holocene environmental change in a montane region of southern Europe with a long history of human settlement. Quat. Sci. Rev. 26, 1455–1475.

Carrión, J.S., Finlayson, C., Férnández, S., Finlayson, G., Allué, E., López-Sáez, J., López-García, P., Gil-Romera, G., Bailey, G., González-Sampériz, P., 2008. A coastal reservoir of biodiversity for Upper Pleistocene human populations: palaeoecological investigations in Gorham's Cave (Gibraltar) in the context of the Iberian Peninsula. Quat. Sci. Rev. 27, 2118–2135.

Carrión, J.S., Fernández, S., González-Sampériz, P., Gil-Romera, G., Badal, E., Carrión-Marco, Y., López-Merino, L., López-Sáez, J., Fierro, E., Burjachs, F., 2010a. Expected trends and surprises in the Lateglacial and Holocene vegetation history of the Iberian Peninsula and Balearic Islands. Rev. Palaeobot. Palynol. 162, 458-475.

Carrión, J.S., Fernández, S., Jiménez-Moreno, G., Fauquette, S., Gil-Romera, G., González-Sampériz, P., Finlayson, C., 2010b. The historical origins of aridity and vegetation degradation in southeastern Spain. J. Arid Environ. 74, 731-736.

Carrión, Y., Kaal, J., López-Sáez, J.A., López-Merino, L., Cortizas, A.M., 2010c. Holocene vegetation changes in NW Iberia revealed by anthracological and palynological records from a colluvial soil. Holocene 20, 53-66.

Carrión, Y., Ntinou, M., Badal, E., 2010d. Olea europaea L. in the North Mediterranean Basin during the Pleniglacial and the Early–Middle Holocene. Quat. Sci. Rev. 29, 952-968.

Carrión, J.S., Fletcher, W., Fernández, S., González-Sampériz, P., López-Merino, L., Peña, L., Burjachs, F., López-Sáez, J.A., García-Antón, M., Carrión, Y., Uzquiano, P., Postigo, J.M., Barrón, E., Allué, E., Badal, E., Dupré, M., Fierro, E., Munuera, M., Rubiales, J.M., García Amorena, I., Jiménez Moreno, G., Gil Romera, G., Leroy, S., García-Martínez, M.S., Montoya, E., Yll, E., Vieira, M., Rodríguez-Ariza, M.O., Anderson, S., Peñalba, C., Gil García, M.J., Pérez Sanz, A., Albert, R.M., Díez, M.J., Morales, C., Gómez Manzaneque, F., Parra, I., Ruiz Zapata, B., Riera, S., Zapata, L., Ejarque, A., Vegas, T., Rull, V., Scott, L., Andrade, A., Pérez Díaz, A., Abel Schaad, D., Moreno, E., Hernández-Mateo, L., Sánchez Baena, J.J., Riquelme, J.A., Iglesias, R., Franco, F., Chaín, C., Figueiral, I., Grau, E., Matos, M., Jiménez Espejo, F., Arribas, A., Garrido, G., Finlayson, G., Finlayson, C., Ruiz, M., Pérez Jordá, G., Miras, Y., 2012. Paleoflora y paleovegetación de la peninsula Iberica e Islas Baleares: Plioceno-Cuaternario. Ministerio de Economía y Competitividad, Madrid, Spain.

Caseldine, C., Geirsdóttir, Á., Langdon, P., 2003. Efstadalsvatn – a multi-proxy study of a Holocene lacustrine sequence from NW Iceland. J. Paleolimnol. 30, 55–73.

Casparie, W.A., 1969. Bult- und Schlenkenbildung im Hochmoortorf: Zur Frage des Moorwachstums. Vegetatio 19, 146-180.

Casparie, W.A., 1972. Bog development in southeastern Drenthe (The Netherlands). The Hague.

Casparie, W., Groenman-van Waateringe, W., 1980. Palynological analysis of Dutch barrows. Palaeohistoria 7, 20-65.

Caspers, G., Freund, H., 1997. Vegetation und Paläoklima der Weichsel-Kaltzeit im nördlichen Mitteleuropa, Schr. Deutsche Geolog. Gesellschaft, Hannover.

Caspers, G., Freund, H., 2001. Vegetation and climate in the early- and pleni-Weichselian in northern central Europe. J. Quat. Sci. 16, 31-48.

Catford, J.A., Daehler, C.C., Murphy, H.T., Sheppard, A.W., Hardesty, B.D., Westcott, D.A.,

Rejmánek, M., Bellingham, P.J., Pergl, J., Horvitz, C.C., Hulme, P.E., 2012. The intermediate disturbance hypothesis and plant invasions: implications for species richness and management. Perspect. Plant Ecol. Evol. Syst. 14, 231-241.

Catullo, R.A., Ferrier, S., Hoffmann, A.A., 2015. Extending spatial modelling of climate change responses beyond the realized niche: estimating, and accommodating, physiological limits and adaptive evolution. Global Ecol. Biogegor. 24, 1192-1202.

Caudullo, G., de Rigo, D., 2016a. *Cupressus sempervirens* in Europe: distribution, habitat, usage and threats. In: San-Miguel-Ayanz, J., de Rigo, D., Caudullo, G., Houston Durrant, T., Mauri, A. (Eds.), European Atlas of Forest Tree Species. European Commission, Luxembourg.

Caudullo, G., de Rigo, D., 2016b. *Acer platanoides* in Europe: distribution, habitat, usage and threats. In: San-Miguel-Ayanz, J., De Rigo, D., Caudullo, G., Houston Durrant, T., Mauri, A. (Eds.), European Atlas of Forest Tree Species. Publication Office of the European Union, Luxembourg, pp. 54-55.

Caudullo, G., de Rigo, D., 2016c. *Fraxinus ornus* in Europe, distribution, habitat, usage and threats. In: San-Miguel-Ayanz, J., De Rigo, D., Caudullo, G., Houston Durrant, T., Mauri, A. (Eds.), European Atlas of Forest Tree Species. Publication Office of the European Union, Luxembourg, pp. 100-101.

Caudullo, G., Houston Durrant, T., 2016. *Fraxinus angustifolia* in Europe: distribution, habitat, usage and threats. In: San-Miguel-Ayanz, J., De Rigo, D., Caudullo, G., Houston Durrant, T., Mauri, A. (Eds.), European Atlas of Forest Tree Species. Publication Office of the European Union, Luxembourg, p. 97.

Caudullo, G., Welk, E., San-Miguel-Ayanz, J., 2017. Chorological maps for the main European woody species. Data in Brief 12, 662-666.

Cázares, E., Trappe, J., Jumpponen, A., 2005. Mycorrhiza-plant colonization patterns on a subalpine glacier forefront as a model system of promary succession. Mycorrhiza 15, 405-416.

Cepek, A.G., Erd, K., 1975. Das Holstein-Interglazial im Raum Neuruppin – ein neues pollenstratigraphisches Richtprofil und seine quartärgeologische Bedeutung. Zeitschr. Geolog. Wissenschaften 3, 1151-1178.

Chamberlain, AC., 1975. The movement of particles in plant communities. In: Monteith, J.L. (Ed.),Vegetation and the Atmosphere. Academic Press, London, pp. 155-203.

Chambers, F.M., 1996. Great Britain - Wales. In: Berglund, B.E., Birks, H.J.B., Ralska-Jasiewiczowa, M., Wright, H.E. (Eds.), Palaeoecological Events During the Last 15 000 Years. Regional Syntheses of Palaeoecological Studies of Lakes and Mires in Europe. Wiley, Chichester, pp. 77-94.

Chambers, F.M, Booth, R.K., De Vleeschouwer, F., Lamentowicz, M., Le Roux, G., Mauquoy, D., Nichols, J.E., van Geel, B., 2012. Development and refinement of proxy-climate indicators from peats. Quat. Int. 268, 21-33.

Chang, P., Ji, L., Li, H., 1997. A decadal climate variation in the tropical Atlantic Ocean from thermodynamic air-sea interactions. Nature 385, 516-518.

Chapra, S., Dobson, H., 1981. Quantification of lake trophic typologies of Naumann (surface quality) and Thienemann (oxygen) with special reference to the Great Lakes. J. Great Lakes Res. 7, 182-193.

Charman, D.J., 2001. Biostratigraphic and palaeoenvironmental applications of testate amoebae. Quat. Sci. Rev. 20, 1753-1764.

Charman, D.J., 2002. Peatland Systems and Environmental Change. John Wiley & Sons, Chichester.

Charman, D.J., Blundell, A., Chiverrell, R., Hendon, D., Langdon, P., 2006. Compilation of non-annually resolved Holocene proxy climate records: stacked Holocene peatland palaeo-water table reconstructions from Northern Britain. Quat. Sci. Rev. 25, 336-350.

Charman, D.J., Barber, K.E., Blaauw, M., Langdon, P., Mauquoy, D., Daley, T., Hughes, P., Karofeld, E., 2009. Climate drivers for peatland palaeoclimate records. Quat. Sci. Rev. 28, 1811-1819.

Cheddadi, R., Mamakowa, K., Guiot, J., de Beaulieu, J.-L., Reille, M., Andrieu, V., Granoszewski, W., Peyron, O., 1998. Was the climate of the Eemian stable? A quantitative climate reconstruction from seven European pollen records. Palaeogeogr. Palaeoclim. Palaeoecol. 143, 73-85.

Cheddadi, R., Vendramin, G.G., Litt, T., François,L., Kageyama, M., Lorentz, S., Laurent, J.-M., de Beaulieu, J.-L., Sadori, L., Jost, A., Lunt, D., 2006. Imprints of glacial refugia in the modern genetic diversity of *Pinus sylvestris*. Glob. Ecol. Biogeogr. 15, 271-282.

Cheddadi, R., Birks, H.J.B., Tarroso, P., Liepelt, S., Gömöry, D., Stefan Dullinger, S., Meier, E., Hülber, K., Maiorano, L., Laborde, H., 2014. Revisiting tree-migration rates: *Abies alba* (Mill.), a case study. Veget. Hist. Archaeobot. 23, 113-122.

Chen, S.-H., 1988. Neue Untersuchungen über die spät- und postglaziale Vegetationsgeschichte im Gebiet zwischen Harz und Leine (BRD). Flora 181, 147-177.

Chepalyga, A., 2006. The lateglacial great flood in the Ponto-Caspian Basin. In: Yanko-Hombach, V., Gilbert, A., Panin, N., Dolukhanov, P. (Eds.), The Black Sea Flood Question: Changes in Coast-line, Climate and Human Settlement. Springer, New Jersey, pp. 119-128.

Chesson, P., Yang, P.J., 2019. Populations as fluid on a landscape under global environmental change. Front. Ecol. Evol. 7, art363.

Chiappini, M. 1988. Guida alla flora pratica della Sardegna. C. Delfino, Sassari, 462 pp.

Childe, G., 1951. Social Evolution, London.

Christanis, K., 1994. The genesis of the Nissi peatland (northwestern Greece) as an example of peat and lignite deposit formation in Greece. Int. J. Coal Geol. 26, 63-77.

Christensen, P., Blackmore, S., 1988. The Northwest European pollen flora, 40: Tiliaceae. Rev. Palaeobot. Palynol. 57, 33-43.

Chytrý, M., Tichý, L., Rolecek, J., 2003. Local and regional patterns of species richness in central European vegetation types along the pH/calcium gradient. Folia Geobot. 38, 429-442.

Ciarmiello, L., Pontecorvo, G., Piccirillo, P., De Luca, A., Carillo, P., Kafantaris, I., Woodrow, P., 2013. Use of nuclear and mitochondrial single nucleotide polymorphisms to characterize English walnut (*Juglans regia* l.) genotypes. Plant Mol. Biol. Rep. 31, 1116-1130.

Clark, J.S., 1998. Why trees migrate so fast: Confronting theory with dispersal biology and the paleorecord. Am. Nat. 152, 204-224.

Clark, S.H.E., Edwards, K.J., 2004. Elm bark beetle in Holocene peat deposits and the northwest European elm decline. J. Quat. Sci. 19, 525-528.

Clark, J.S., Merkt, J., Müller, H., 1989. Post-glacial fire, vegetation, and human history on the northern alpine forelands, south-western Germany. J. Ecol. 77, 897-925.

Clark, J.S., Fastie, C., Hurtt, G., Jackson, S.T., Johnson, C., King, G.A., Lewis, M., Lynch, J., Pacala, S., Prentice, I.C., Schupp, E.W., Webb, T., Wyckoff, P., 1998. Reid's paradox of rapid plant migration - Dispersal theory and interpretation of paleoecological records. BioScience 48, 13-24.

Clark, J.S., Grimm, E.C., Lynch, J., Mueller, P.G., 2001. Effects of Holocene climate change on the C4 grassland/woodland boundary in the Northern Plains, USA. Ecology 82, 620-636.

Clark, J.S., Grimm, E.C., Donovan, J.J., Fritz, S.C., Engstrom, D.R., Almendinger, J.E., 2002. Drought cycles and landscape responses to past aridity on prairies of the northern Great Plains, USA. Ecology 83, 595-601.

Clarke, C.L., 2019. Late-Quaternary dynamics of northern plant communities: A sedimentary ancient DNA (*seda*DNA) perspective. PhD thesis, University of Southampton.

Clarke, G.C.S., Jones, M.R., 1980. Aceraceae. In: Punt, W., Clarke, G.C.S. (Eds.), The Northwest European Pollen Flora II. Elsevier, Amsterdam, pp. 181-193.

Clarke, C.L., Edwards, M.E., Brown, A.G., Gielly, L., Lammers, Y., Heintzman, P.D., Ancin-Murguzur, F.J., Bråthen, K.-A., Goslar, T., Alsos, I.G., 2019a. Holocene floristic diversity and richness in northeast Norway revealed by sedimentary ancient DNA (*seda*DNA) and pollen. Boreas 48, 299-316.

Clarke, C.L., Edwards, M.E., Gielly, L., Ehrich, D., Hughes, P.D.M., Morozova, L.M., Haflidason, H., Mangerud, J., Svendsen, J.I., Alsos, I.G., 2019b. Persistence of arctic-alpine flora during 24,000 years of environmental change in the Polar Urals. Scientific Reports 9, 19613.

Clear, J.L., Seppä, H., Kuosmanen, N., Bradshaw, R.H.W., 2013. Holocene fire frequency variability in Vesijako, Strict Nature Reserve, Finland, and its application to conservation and management. Biol. Cons. 166, 90-97.

Clegg, B.F., Tinner, W., Gavin, D.G., Hu, F.S., 2005. Morphological differentiation of *Betula* (birch) pollen in northwest North America and its palaeoecological application. Holocene 15, 229-237.

Clements, F.E., 1905. Research Methods in Ecology. University Publishing Co., Lincoln, Nebraska. 214 pp.

Clerc, J., 1988. Recherches pollenanalytiques sur la paléo-écologie Tardiglaciaire et Holocène du Bas-Dauphiné. PhD Thesis, University of St. Jérôme, Marseille, France.

Clifford, M.J., Booth, R.K., 2015. Late-Holocene drought and fire drove a widespread change in forest community composition in eastern North America. Holocene 25, 1102-1110.

Clymo, R.S., 1963. Ion exchange in *Sphagnum* and its relation to bog ecology. Ann. Bot. Lond. N.S. 27, 309-324.

Coffey, E.E.D., Froyd, C.A., Willis, K.J., 2011. When is an invasive not an invasive? Macrofossil evidence of doubtful native plant species in the Galápagos Islands. Ecology 92, 805-812.

Cohen, K.M., Finney, S.C., Gibbard, P.L., Fan, J.-C., 2013 (updated 2016). The ICS International Chronostratigraphic Chart. Episodes 36, 199-204.

COHMAP Members, 1988. Climatic changes of the last 18,000 years: observations and model simulations. Science 241, 1043-1052.

Cole, K.L., Ironside, K., Eischeid, J., Garfin, G., Duffy, P.B., Toney, C., 2011. Past and ongoing shifts in Joshua tree distribution support future modeled range contraction. Ecol. Appl. 21, 137-149.

Cole, L.E.S., Bhagwat, S.A., Willis, K.J., 2014. Recovery and resilience of tropical forests after disturbance. Nat. Commun. 5, 3906.

Coleman, A.W., 1996. Are the impacts of events in the Earth's history discernable in the current distributions of freshwater algae? Hydrobiologia 336, 137-142.

Coles, B., 1995. Archaeology and wetland

restoration. In: Wheeler, B.D., Shaw, S.C., Fojt, W.J., Robertson, R.A. (Eds.), Restoration of Temperate Wetlands. Wiley, Chichester, pp. 19-32.

Coles, B., 2004. Steps towards the heritage management of wetlands in Europe. J. Wetl. Archaeol. 4, 183-198.

Coles, B., Olivier, A., Bull, D. (Eds.), 2001. The Heritage Management of Wetlands in Europe. EAC Brussels/ WARP, Exeter, 207 pp.

Colinvaux, P.A., Schofield, E.K., 1976. Historical ecology in the Galápagos Islands. I. A Holocene pollen record from El Junco Lake, Isla San Cristobel. J. Ecol. 64, 989-1012.

Colombaroli, D., Tinner, W., 2013. Determining the long-term changes in biodiversity and provisioning services along a transect from Central Europe to the Mediterranean. Holocene 23, 1477-1486.

Colombaroli, D., Marchetto, A., Tinner, W., 2007. Long-term interactions between Mediterranean climate, vegetation and fire regime at Lago di Massaciuccoli (Tuscany, Italy). J. Ecol. 95, 755-770.

Colombaroli, D., Vannière, B., Emmanuel, C., Magny, M., Tinner, W., 2008. Fire-vegetation interactions during the Mesolithic-Neolithic transition at Lago dell'Accesa, Tuscany, Italy. Holocene, 18, 679-692.

Colombaroli, D., Tinner, W., van Leeuwen, J.F.N., Noti, R., Vescovi, E., Vannière, B., Magny, M., Schmidt, R., Bugmann, H., 2009. Response of broadleaved evergreen Mediterranean forest vegetation to fire disturbance during the Holocene: insights from the peri Adriatic region. J. Biogeogr. 36, 314-326.

Colombaroli, D., Henne, P.D., Kaltenrieder, P., Gobet, E., Tinner, W., 2010. Species responses to fire, climate and human impact at tree line in the Alps as evidenced by palaeo-environmental records and a dynamic simulation model. J. Ecol. 98, 1346-1357.

Colombaroli, D., Beckmann, M., van der Knaap, W.O., Curdy, P., Tinner, W., 2013. Changes in biodiversity and vegetation composition in the central Swiss Alps during the transition from pristine forest to first farming. Divers. Distrib. 19, 157-170.

Colombo, P., Lorenzoni, F., Grigoletto, F., 1983. Pollen grain morphology supports the taxonomical discrimination of Mediterranean Oaks (*Quercus*, Fagaceae). Plant Syst Evol 141, 273-284.

Columella, L.J.M., translated by Ahrens, K., 1972. De re rustica. Über Landwirtschaft. Akademie-Verlag, Berlin, pp. 466.

Comps, B., Gomory, D., Letouzey, J., Thiebaut, B., Petit, R.J., 2001. Diverging trends between heterozygosity and allelic richness during postglacial colonization in the European beech. Genetics 157, 389-397.

Conedera, M., Krebs, P., Tinner, W., Pradella, M., Torriani, D., 2004. The cultivation of *Castanea sativa* (Mill.) in Europe, from its origin to its diffusion on a continental scale. Veget. Hist. Archaeobot. 13, 161-179.

Conedera, M., Tinner, W., Crameri, S., Torriani, D., Herold, A., 2006. Taxon-related pollen source areas for lake basins in the southern Alps: an empirical approach. Veget. Hist. Archaeobot. 15, 263-272.

Conedera, M., Tinner, W., Neff, C., Meurer, M., Dickens, A.F., Krebs, P., 2009. Reconstructing past fire regimes: methods, applications, and relevance to fire management and conservation. Quat. Sci. Rev. 28, 555-576.

Conedera, M., Colombaroli, D., Tinner, W., Krebs, P., Whitlock, C., 2017. Insights about past forest dynamics as a tool for present and future forest management in Switzerland. For. Ecol. Manage. 388, 100-112.

Connell, J.H., 1978. Diversity in tropical rain forests and coral reefs. Science 199, 1301-1310.

Connolly, A., 1999. The Paleoecology of Clara Bog, Co. Offaly (PhD thesis). Trinity College, Dublin, Ireland.

Connolly, R., Jambrina-Enríquez, M., Herrera-Herrera, A.V., Vidal-Matutano, P., Fagoaga, A., Marquina-Blasco, R., Marin-Monfort, M.D., Ruiz-Sánchez, F.J., Laplana, C., Bailon, S., Pérez, L., Leierer, L., Hernández, C.M., Galván, B., Mallol, C., 2019. A multiproxy record of palaeoenvironmental conditions at the Middle Palaeolithic site of Abric del Pastor (Eastern Iberia). Quat. Sci. Rev. 225, 106023.

Connor, S.E., van Leeuwen, J.F.N., Rittenour, T.M., van der Knaap, W.O., Ammann, B., Björck, S., 2012. The ecological impact of oceanic island colonization - a palaeoecological perspective from the Azores. J. Biogeogr. 39, 1007-1023.

Connor, S.E., Ross, S.A., Sobotkova, A., Herries, A.I.R., Mooney, S.D., Longford, C., Iliev, I., 2013. Environmental conditions in the SE Balkans since the Last Glacial Maximum and their influence on the spread of agriculture into Europe. Quat. Sci. Rev. 68, 200-215.

Connor, S.E., Colombaroli, D., Confortini, F., Gobet, E., Ilyashuk, B.P., Ilyashuk, E.A., van Leeuwen, J.F.N., Lamentowicz, M., van der Knaap, W.O., Malysheva, E., Marchetto, A., Margalitadze, N., Mazei, Y., Mitchell, E.A.D., Payne, R.J., Ammann, B., 2018. Long-term population dynamics: theory and reality in a peatland ecosystem. J. Ecol. 106, 333-346.

Connor, S.E., Vannière, B., Colombaroli, D., Anderson, R.S., Carrion, J.S., Ejarque, A., Romera, G.G., Gonzalez-Sampcriz, P., Hoefer, D., Morales-Molino, C., Revelles, J., Schneider, H., van der Knaap, W.O., van Leeuwen, J.F.N., Woodbridge, J., 2019. Humans take control of fire-driven diversity changes in Mediterranean Iberia's vegetation during the mid-late Holocene. Holocene 29, 886-901.

Connor, S.E., Araujo, J., Boski, T., Gomes, A., Gomes, S.D., Leira, M., Freitas, M.D., Andrade, C., Morales-Molino, C., Franco-Mugica, F., Akindola, R.B., Vannière, B., 2021. Drought, fire and grazing precursors to large-scale pine forest decline. Divers. Distrib. 27, 1138-1151.

Conolly, A., Dahl, E., 1970. Maximum summer temperature in relation to the modern and Quaternary distributions of certain arctic-montane species in the British Isles. In: Walker, D., West, R.G. (Eds.), Studies in Vegetation History of the British Isles. Cambridge University Press, Cambridge, pp. 159-223.

Cook, G.T., van der Plicht, J., 2007. Radiocarbon Dating: Conventional Method. In: Elias, S.A. (Ed.), Encyclopedia of Quaternary Science. Elsevier, Amsterdam, pp. 2899-2911.

Coope, G.R., Lemdahl, G., Lowe, J.J., Walkling, A., 1998. Temperature gradients in northern Europe during the last glacial–Holocene transition (14–9 ^{14}C kyr BP) interpreted from coleopteran assemblages. J. Quat. Sci. 13, 419-433.

Cooper, A., Turney, C.S.M., Hughen, K.A., Brook, B.W., McDonald, H.G., Bradshaw, C.J.A., 2015. Abrupt warming events drove Late Pleistocene Holarctic megafaunal turnover. Science 349, 602-606.

Correa-Metrio, A., Bush, M.B., Cabrera, K.R., Sully, S., Brenner, M., Hodell, D.A., Escobar, J., Guilderson, T.P., 2012. Rapid climate change and no-analog vegetation in lowland Central America during the last 86,000 years. Quat. Sci. Rev. 38, 63-75.

Correa-Metrio, A., Bush, M.B., Lozano-García, S., Sosa-Nájera, S., 2013. Millennial-scale temperature change velocity in the continental northern neotropics. PLoS ONE 8, e81958.

Cortés-Sánchez, M., Morales-Muñiz, A., Simón-Vallejo, M., Bergadà -Zapata, M., Delgado-Huertas, A., López-García, P., López-Sáez, J., Lozano-Francisco, M., Riquelme-Cantal, J., Roselló-Izquierdo, E., Sánchez-Marco, A., Vera-Peláez, J., 2008. Palaeoenvironmental and cultural dynamics of the coast of Málaga (Andalusia, Spain) during the Upper Pleistocene and early Holocene. Quat. Sci. Rev. 27, 2176–2193.

Costa, M., Morla, C., Sainz, H. (Eds.), 2005. Los Bosques Ibéricos. Una Interpretación Geobotánica. Ed. Planeta, Barcelona, Spain.

Cottrell, J., Munro, R., Tabbener, H., Gillies, A., Forrest, G., Deans, J., Lowe, A., 2002. Distribution of chloroplast DNA variation in British oaks (*Quercus robur* and *Q. petraea*): the influence of postglacial colonisation and human management. For. Ecol. Managem. 156, 181–195.

Couwenberg, J., 2021. Self-regulation and self-organization of raised bogs. PhD thesis, University of Greifswald.

Couwenberg, J., Joosten H., 2005. Self-organization in raised bog patterning: the origin of microtope zonation and mesotope diversity. J. Ecol. 93, 1238-1248.

Couwenberg, J., Thiele, A., Tanneberger, F., Augustin, J., Bärisch, S., Dubovik, D., Liashchynskaya, N., Michaelis, D., Minke, M., Skuratovich, A., Joosten, H., 2011. Assessing greenhouse gas emissions from peatlands using vegetation as a proxy. Hydrobiologia 674, 67-89.

Coxon, P., 1986. A radiocarbon dated early Post Glacial pollen diagram from a pingo near Millstreet, County Cork. Ir. J. Earth Sci. 8, 9-20.

Craig, A.J., 1973. Studies on the ecological history of southeast Ireland, using pollen influx analysis and other methods (PhD thesis). Trinity College, Dublin, Ireland.

Craig, A.J., 1978. Pollen Percentage and Influx Analyses in South-East Ireland: A Contribution to the Ecological History of the Late-Glacial Period. J. Ecol. 66, 297–324.

Cramer, W., 1997. Using plant functional types in a global vegetation model. In: Smith, T. (Ed.), Plant Functional Types: Their Relevance to Ecosystem Properties and Global Change. Cambridge University Press, Cambridge, pp. 271-288.

Croudace, I.W., Rothwell, R.G. (Eds.), 2015. Micro-XRF Studies of Sediment Cores. Springer, Dordrecht. 656 pp.

Crump, S.E., Miller, G.H., Power, M., Sepúlveda, J., Dildar, N., Coghlan, M., Bunce, M., 2019. Arctic shrub colonization lagged peak postglacial warmth: Molecular evidence in lake sediment from Arctic Canada. Global Change Biol. 25, 4244-4256.

Csaikl, U., Burg, K., Fineschi, S., König, A., Mátyás, G., Petit, J., 2002a. Chloroplast DNA variation of white oaks in the alpine region. For. Ecol. Managem. 156, 131–145.

Csaikl, U., Glaz, I., Baliuckas, V., Petit, R., Jensen, J., 2002b. Chloroplast DNA variation of white oak in the Baltic countries and Poland. For. Ecol. Managem. 156, 211–222.

Cserny, T., Nagy-Bodor, E., 2000. Limnogeology of Lake Balaton (Hungary). In: Gierlowski-Kordesch, E.H., Kelts, K.R. (Eds.), Lake Basins Through Space and Time. AAPG Studies in Geology, Vol. 46, pp. 605-618.

Cubizolle, H., Argant, J., Fassion, F., Latour-Argant, C., Deng-Amiot, Y., Dietre, B., 2014. L'histoire de la végétation depuis la fin du tardiglaciaire et l'évolution de l'emprise humaine à partir du milieu de l'holocène dans le Massif Central oriental (France). Quat. Rev. Assoc. Fr. Pour Létude Quat. 209–236.

Cui, Q.Y., Gaillard, M.-J., Lemdahl, G., Sugita, S., Greisman, A., Jacobson, G.L., Olsson, F., 2013. The role of tree composition in Holocene fire history of the hemiboreal and southern boreal zones of southern Sweden, as revealed by the application of the Landscape Reconstruction Algorithm: Implications for biodiversity and climate-change issues. Holocene 23, 1747-1763.

Cui, Q.Y., Gaillard, M.-J., Lemdahl, G., Stenberg, L., Sugita, S., Zernova, G., 2014. Historical land-use and landscape change in southern Sweden and implications for present and future biodiversity. Ecol. Evol. 4, 3666-3570

Culiberg, M., Šercelj, A., 1995. Anthracotomical and palynological research in the paleolithic site Sandalja II (Istra, Croatia). Razprave IV. Razreda Sazu 36, 49-57.

Culiberg, M., Šercelj, A., 1996. Slovenia. In: Berglund, B.E., Birks, H.J.B., Ralska-Jasiewiczowa, M., Wright, H.E. (Eds.), Palaeoecological Events During the Last 15 000 Years. Regional Syntheses of Palaeoecological Studies of Lakes and Mires in Europe. Wiley, Chichester, pp. 687-700.

Cundill, P.R., 1986. A new design for pollen trap for modern pollen studies. J. Biogeogr. 13, 83-98.

Cundill, P.R., 1991. Comparison of moss polsters and pollen trap data: a pilot study. Grana 30, 301-308.

Cunliffe, B., 2008. Europe Between the Ocean. Yale University Press, New Haven, 480 pp.

Cunze, S., Heydel, F., Tackenberg, O., 2013. Are plant species able to keep pace with the rapidly changing climate? PLOS One 8, e67909.

Curry, B., Henne, P.D., Mesquita-Joanes, F., Marrone, F., Pieri, V., La Mantia, T., Calo, C., Tinner, W., 2016. Holocene paleoclimate inferred from salinity histories of adjacent lakes in southwestern Sicily (Italy). Quat. Sci. Rev. 150, 67-83.

Cushing, E.J., 1963. Late-Wisconsin pollen stratigraphy in east-central Minnesota. PhD thesis, University of Minnesota.

Cushing, E.J., 1965. Problems in the Quaternary phytogeography of the Great Lakes region. In: Wright, H.E., Frey, D.G. (Eds.), The Quaternary of the United States. Princeton University Press, Princeton, New Jersey, pp. 403-416.

Cushing, E.J., 1967. Late-Wisconsin pollen stratigraphy and the glacial sequence in Minnesota. In: Cushing E.J., Wright, H.E. (eds.), Quaternary Paleoecology, 59-88. New Haven and London.

Czeczott, H., 1932. Distribution of *Fagus orientalis* Lipsky. Veröff. Geobot. Inst. ETH Stiftung Rübel, Zürich 8, 362-387.

Dąbrowska, G., Dzialuk, A., Burnicka O., Ejankowski, W., Gugnacka-Fiedor W., Goc, A., 2006. Genetic diversity of postglacial relict shrub *Betula nana* revealed by RAPD analysis. Dendrobiology 55, 19-23.

Dahl, S.O., Nesje, A., Øvstedal, J., 1997. Cirque glaciers as morphological evidence for a thin Younger Dryas ice sheet in east-central southern Norway. Boreas 26, 161-180.

Damblon, F., 1997. Palaeobotanical study of representative Upper Palaeolithic sites in the central European Plain: a contribution to the Sc-004 Project. Préhistoire Européenne 11, 245-253.

Damblon, F., Haesaerts, P., van der Plicht, J., 1996. New datings and considerations on the chronology of Upper Palaeolithic sites in the Great European Plain. Préhistoire Européenne 9, 177-231.

Daniels, R., Eddy, A., 1990. Handbook of European Sphagna. Institute of Terrestrial Ecology, Natural Environment Research Council, London.

Danielsen, A., 1970. Pollen-analytical Late Quaternary studies in the Ra district of Ostfold, Southeast Norway. Univ. Bergen Aarb. Mat. Naturv. Ser. 14, 1-146.

Darmer, G., 1952. Der Sanddorn als Wild- und Kulturpflanze. Leipzig.

Dasgupta, P., Raven, P.H., McIvor, A.L. (Eds.), 2019. Biological Extinction - New Perspectives. Cambridge University Press, Cambridge. 442 pp.

Dau, J.H.C., 1823. Neues Handbuch über den Torf, dessen Natur, Entstehung und Wiedererzeugung, Nutzen im allgemeinen und für den Staat. [New handbook about peat, its nature, origin, reproduction, and general and national benefits] J.C. Heinrichsche Buchhandlung, Leipzig, 248 pp. (in German)

Dau, J.H.C., 1829. Allerunterthänigster Bericht an die Königliche Dänische Rentekammer über die Torfmoore Seelands. Copenhagen and Leipzig.

Davies, A.L., Bunting, M.J. 2010. Applications of palaeoecology in conservation. Open Ecol. J. 3, 54-67.

Davis, W., 1882. On the classification of basins. Proc. Boston Soc. Nat. Hist. 21.

Davis, M.B., 1963. On the theory of pollen analysis. Am. J. Sci. 261, 897–912.

Davis, M.B., 1966. A method for determination of absolute pollen frequency. In: Kummel, B., Raup, D.M. (Eds.), Handbook of Paleontological Techniques. Freeman, San Francisco, pp. 674-686.

Davis, M.B., 1968. Pollen grains in lake sediments: Redeposition caused by seasonal water circulation. Science 162, 796-799.

Davis, M.B., 1973. Redeposition of pollen grains in lake sediment. Limnol. Oceanogr. 18, 44-52.

Davis, M.B., 1976. Pleistocene biogeography of temperate deciduous forests. Geoscience and Man 13, 13-26.

Davis, M.B., 1978. Climatic interpretation in Quaternary sediments. In: Walker, D., Guppy, J.C. (Eds.), Biology and Quaternary Environments. Australian Academy of Science, Canberra, pp. 35-51.

Davis, M.B., 1981. Quaternary history and the

stability of forest communities. In: Shugart, H.H., West, D.C., Botkin, D.B. (Eds.), Forest Succession. Springer, New York, pp. 132-177.

Davis, M.B., 1987. Invasions of forest communities during the Holocene: beech and hemlock in the Great Lakes region. In: Gray, A.J., Crawley, M.J., Edwards, P.J. (Eds.), Colonization, Succession and Stability. Blackwell Scientific Publishing, Oxford, pp. 373-393.

Davis, M.B., 2000. Palynology after Y2K — Understanding the source area of pollen in sediments. Annu. Rev. Earth Planet. Sci. 28, 1-18.

Davis, M.B., Ford, M., 1982. Sediment focusing in Mirror Lake, New Hampshire. Limnol. Oceanogr. 27, 137-150.

Davis, B.A.S., Stevenson, A., 2007. The 8.2 ka event and early-mid Holocene forests, fires and flooding in the Central Ebro Desert, NE Spain. Quat. Sci. Rev. 26, 1695-1712.

Davis, M.B., Sugita, S., 1997. Reinterpreting the fossil pollen record of Holocene tree migration. In: Huntley, B., Cramer, W., Morgan, A.V., Prentice, I.C., Allen, J.R.M. (Eds.), Past and Future Rapid Environmental Change: The spatial and evolutionary responses of terrestrial biota. NATO ASI Series, Berlin, pp. 181-193.

Davis, B.A.S., Brewer, S., Stevenson, A., Guiot, J., 2003. The temperature of Europe during the Holocene reconstructed from pollen data. Quat. Sci. Rev. 22, 1701-1716.

Davis, M.B., Shaw, R.G., Etterson, J.R., 2005. Evolutionary responses to changing climate. Ecology 86, 1704-1714.

Davydova, N.N., Subetto, D.A., Khomutova, V.I., 1992. Lake Ilmen [in Russian], in: Istoria Ozer Vostochno-Evropeiskoi Ravniny [The Lake History of East-European Plain]. St. Petersburg, Russia, pp. 101–117.

Dawson, T.P., Jackson, S.T., House, J.I., Prentice, I.C., Mace, G.M., 2011. Beyond predictions: biodiversity conservation in a changing climate. Science 332, 53-58.

Dawson, A., Paciorek, C.J., McLachlan, J.S., Goring, S., Williams, J.W., Jackson, S.T., 2016. Quantifying pollen-vegetation relationships to reconstruct ancient forests using 19th-century forest composition and pollen data. Quat. Sci.Rev.137, 156-175.

Dawson, A., Cao, X., Caput, M., Hopla, E., Li, F., Edwards, M., Fyfe, R., Cajewski, K., Coring, S.J., Herzschuh, U., Mazier, F., Sugita, S., Williams, J.W., Xu, Q., Gaillard, M.-J., 2018. Finding the magnitude of human-induced Northern Hemisphere land-cover transformation between 6 and 0.2 ka BP. PAGES Magazine 26, 34-35.

Dawson, A., Paciorek, C.J., Goring, S.J., Jackson, S.T., McLachlan, J.S., Williams, J.W. 2019. Quantifying trends and uncertainty in prehistoric forest composition in the upper Midwestern United States. Ecology, e02856.

de Beaulieu, J.-L., 1977. Contribution pollenanalytique à l'histoire tardiglaciaire et Holocène de la végétation des Alpes Méridionales Françaices. PhD Thesis. University of Aix-Marseille, Marseille, France, 358 pp.

de Beaulieu, J.-L., Reille, M., 1984a. A long Upper Pleistocene pollen record from Les Échets, near Lyon, France. Boreas 13, 111-132.

de Beaulieu, J.-L., Reille, M., 1984b. The pollen sequence of Les Échets (France): A new element for the chronology of the Upper Pleistocene. Géogr. Phys. Quat. 38, 3-9.

de Beaulieu, J.-L., Reille, M., 1992a. The last climatic cycle at La Grande Pile (Vosges, France) a new pollen profile. Quat. Sci. Rev. 114, 431-438.

de Beaulieu, J.-L., Reille, M., 1992b: Long Pleistocene pollen sequences from the Velay Plateau (Massiv Central, France). Vegetation History and Archaeobotany 1, 233-242.

de Beaulieu, J.-L., Pons, A., Reille, M., 1982. Recherches pollenanalytiques sur l'histoire de la végétation de la bordure nord du Massif du Cantal (Massif Central, France). Pollen Spores 24, 251–300.

de Beaulieu, J.-L., Couteaux, A., Pons, A., Reille, M., Triat-Laval, H., 1984. Première approche d'une histoire postwürmienne dans le sud-est de la France. Rev. Paléobiol. spéciale, 11-24.

de Beaulieu, J.-L., Pons, A., Reille, M., 1985. Recherches pollenanalytiques sur l'histoire tardiglaciaire et holocène de la végétation des Monts Aubtac (Massif Centrale, France). Rev. Palaeobot. Palynol. 44, 37-80.

de Beaulieu, J.-L., Richard, H., Ruffaldi, P., Clerc, J., 1994. History of vegetation, climate and human action in the French Alps and the Jura over the last 15,000 years. Diss. Bot. 234, 253-275.

de Beaulieu, J.-L., Andrieu-Ponel, V., Reille, M., Grüger, E., Tzedakis, P.C., Svobodová, H., 2001. An attempt at correlation between the Velay pollen sequence and the Middle Pleistocene stratigraphy from central Europe. Quat. Sci. Rev. 20, 1593-1602.

de Beaulieu, J.-L., Brugiapaglia, E., Joannin, S., Guiter, F., Zanchetta, G., Wulf, S., Peyron, O., Bernardo, L., Didier, J., Stock, A., Rius, D., Magny, M., 2017. Lateglacial-Holocene abrupt vegetation changes at Lago Trifoglietti in Calabria, Southern Italy: The setting of ecosystems in a refugial zone. Quat. Sci. Rev. 158, 44-57.

de Bolòs, O., Vigo, J., 1995. Flora dels Països Catalans. Volum III (Pirolàcies-Compostes). Editorial Barcino, Barcelona, Spain.

de Geer, G., 1912. A geochronology of the last 12000 years. Proc. 11th Internat. Geological Congress 1910, 241-253.

de Geer, G., 1940. Geochronologica Suecica: Principles. Kgl. Svenska Vetenskap Handl. Ser. 3. 18, 1-367.

de Jong, J., 1988. Climatic variability during the past three million years as indicated by vegetational evolution in northwest Europe and with emphasis on data from The Netherlands. Philos. Trans. R. Soc. London, B 318, 603-617.

de Klerk, P., 2008. Patterns in vegetation and sedimentation during the Weichselian Late-glacial in north-eastern Germany. J. Biogeogr. 35, 1308-1322.

de Klerk, P., 2018. The roots of pollen analysis: the road to Lennart von Post. Veg. Hist. Archaeobot. 27, 393-409.

de Lafontaine, G., Amasifuen Guerra, C., Ducousso, A., Petit, R.J., 2014. Cryptic no more: soil macrofossils uncover Pleistocene forest microrefugia within a periglacial desert. New Phytol. 204, 715-729.

de Nascimento, L., Willis, K.J., Fernández-Palacios, J., Criado, C., Whittaker, R., 2009. The long-term ecology of the lost forests of La Laguna, Tenerife (Canary Islands). J. Biogeogr. 36, 499-514.

de Valk, E.J., 1981. Late Holocene and present vegetation of the Kastelberg (Vosges, France) (PhD thesis). University of Utrecht, Utrecht, The Netherlands.

Dearing, J.A., 1997. Sedimentary indicators of lake-level changes in the humid temperate zone: a critical review. J. Paleolimnol. 18, 1-14.

Deevey, E.S., 1969. Coaxing history to conduct experiments. BioScience 19, 40-43.

Deforce, K., Bastiaens, J., 2007. The Holocene history of *Taxus baccata* (Yew) in Belgium and neighbouring regions. Belg. J, Bot. 140, 222-237.

Delcourt, H.R., Delcourt, P.A., 1991. Quaternary Ecology. A Paleoecological Perspective. Chapman & Hall, London. 243 pp.

DeLuca, T.H., Zackrisson, O., 2007. Enhanced soil fertility under *Juniperus communis* in arctic ecosystems. Plant Soil 294, 147-155.

DeLuca, T., Nilsson, M., Zackrisson, O., 2002. Nitrogen mineralization and phenol accumulation along a fire chronosequence in northern Sweden. Oecologia 133, 206-214.

Demesure, B., Comps, B., Petit, R.J., 1996. Chloroplast DNA Phylogeography of the Common Beech (*Fagus sylvatica* L.) in Europe. Evolution 50, 2515-2520.

Denèfle, M., Lézine, A.-M., Fouache, E., Dufaure, J.-J., 2000. A 12.000-year pollen record from Lake Maliq, Albania. Quat. Res. 54, 423-432.

Denk, T., Grimm, G., Stögerer, K., M, L., Hemleben, V., 2002. The evolutionary history of *Fagus* in western Eurasia: Evidence from genes, morphology and the fossil record. Plant Syst. Evol. 232, 213-236.

Denton, G.H., Broecker, W.S., Alley, R.B., 2006. The mystery interval 17.5 to 14.5 kyrs ago. PAGES News 14, 14-16.

Deslippe, J., Hartmann, M., Mohn, W., Simard, S., 2011. Long-term experimental manipulation of climate alters the ectomycorrhizal community of *Betula nana* in Arctic tundra. Global Change Biol. 17, 1625-1636.

Desprat, S., Díaz Fernández, P., Coulon, T., Ezzat, L., Pessarossi-Langlois, J., Gil, L., Morales-Molino, C., Sánchez Goñi, M., 2015. *Pinus nigra* (European black pine) as the dominant species of the last glacial pinewoods in south-western to central Iberia: a morphological study of modern and fossil pollen. J. Biogeogr. 42, 1998-2009.

Deza-Araujo, M., Morales-Molino, C., Conedera, M., Henne, P.D., Krebs, P., Hinz, M., Heitz, C., Hafner, A., Tinner, W., 2022. A new indicator approach to reconstruct agricultural land use in Europe from sedimentary pollen assemblages. Palaeogeogr. Palaeoclimatol. Palaeoecol. 599, 111051.

Dhillion, S.S., 1994. Ectomycorrhizae, Arbuscular Mycorrhizae, and Rhizoctonia sp. of Alpine and Boreal *Salix* spp. in Norway. Arc. Alp. Res. 26, 304-307.

Di Domenico, F., Lucchese, F., Magri, D., 2012. Buxus in Europe: Late Quaternary dynamics and modern vulnerability Perspect. Plant Ecol. Evol. Syst. 14, 354-362

Di Rita, F., Magri, D., 2009. Holocene drought, deforestation and evergreen vegetation development in the central Mediterranean: a 5500 year record from Lago Alimini Piccolo, Apulia, southeast Italy. Holocene 19, 295 - 306.

Di Rita, F., Molisso, F., Sacchi, M., 2018. Late Holocene environmental dynamics, vegetation history, human impact, and climate change in the ancient Literna Palus (Lago Patria; Campania, Italy). Rev. Palaeobot. Palynology 258, 48-61.

Dickie, I.A., Koide, R.T., Fayish, A.C., 2001. Vesicular-arbuscular mycorrhizal infection of *Quercus rubra* seedlings. New Phytol. 151, 257-264.

Dickie, I.A., Martínez-García, L., Koele, N., Grelet, G., Tylianakis, J., Peltzer, D., Richardson, S., 2013. Mycorrhizas and mycorrhizal fungal communities throughout ecosystem development. Plant Soil 367, 11-39.

Dickinson, W., 1975. Recurrence surfaces in Rusland Moss, Cumbria (formerly North Lancashire). J. Ecol. 63, 913-935.

Dickson, C., 1970. The study of plant macrofossils in British Quaternary deposits. In: Walker, D., West, R.G. (Eds.), Studies in the Vegetational History of the British Isles. Cambridge University Press, Cambridge, pp. 233-254.

Dickson, J., 1993. The Yew tree (*Taxus baccata* L.) in Scotland - native or early introduction or both? Diss. Bot. 196, 293-304.

Diehl, M., 2007. Palynologie und Sedimentologie der Interglazialprofile Döttingen, Bonstorf, Munster und Bilshausen. PhD thesis, University of Mainz.

Diekmann, M., 2003. Species indicator values as an important tool in applied plant ecology - a review. Basic Appl. Ecol. 4, 493-506.

Dierschke, H., 1989. Artenreiche Buchenwald-Gesellschaften Nordwest-Deutschlands. In: Pott, R. (Ed.), Buchenwald-Symposium, Göttingen, pp. 107-148.

Dierssen, K., Dierssen, B., 2001. Moore. Ulmer, Stuttgart.

Dietre, B., Walser, C., Kofler, W., Kothieringer, K., Hajdas, I., Lambers, K., Reitmaier, T., Haas, J.N., 2017. Neolithic to Bronze Age (4850–3450 cal. BP) fire management of the Alpine Lower Engadine landscape (Switzerland) to establish pastures and cereal fields. Holocene 27, 181-196.

Dietre, B., Reitmaier, T., Walser, C., Warnk, T., Unkel, I., Hajdas, I., Lambers, K., Reidl, D., Haas, J.N., 2020. Steady transformation of primeval forest into subalpine pasture during the Late Neolithic to Early Bronze Age (2300-1700 BC) in the Silvretta Alps, Switzerland. Holocene 30, 355-368.

Diez, C., Trujillo, I., Martinez-Urdiroz, N., Barranco, D., Rallo, L., Marfil, P., Gaut, B., 2015. Olive domestication and diversification in the Mediterranean Basin. New Phytol. 206, 436–447.

Digerfeldt, G., 1972. The post-glacial development of Lake Trummen. Regional vegetation history, water level changes and paleolomnology. Folia Limnol. Scand. 16, 1-104.

Digerfeldt, G., 1973. The post-glacial development of the bay Ranviken, Lake Immeln, I Regional vegetation history, II Water level changes. University of Lund Department of Quaternary Geology Report, Lund, pp. 1-59.

Digerfeldt, G., 1974. The post-glacial development of the Ranviken bay in Lake Immelen. I. The history of the regional vegetaion. II. The water-level changes. Geol. Fören. Stockholm Förh. 96, 3-32.

Digerfeldt, G., 1977. The Flandrian development of lake Flarken. Regional vegetation history and palaeolimnology. University of Lund Department of Quaternary Geology Report 13, 1-101.

Digerfeldt, G., 1986. Studies on past lake-level fluctuations. In: Berglund, B.E. (Ed.), Handbook of Holocene Palaeoecology and Palaeohydrology. J. Wiley & Sons, Chichester, pp. 127-143.

Digerfeldt, G., 1997. Reconstruction of Holocene lake-level changes in Lake Kalvsjön, southern Sweden, with a contribution to the local palaeohydrology at the Elm Decline. Veget. Hist. Archaeobot. 6, 9-14.

Digerfeldt, G., Olsson, S., Sandgren, P., 2000. Reconstruction of lake-level changes in lake Xinias, central Greece, during the last 40000 years. Palaeogeogr. Palaeoclimatol. Palaeoecol. 158, 65-82.

Diot, M.-F., Tastet, J.-P., 1995. Paléo-environnements holocènes et limites chronoclimatiques enregistrés dans un marais estuarien de la Gironde (France) [Holocene paleo-environments and chrono-climatic limits recorded in a reclaimed marsh of the Gironde estuary (France).]. Quaternaire 6, 63–75.

Djamali, M., Gambin, B., Marriner, N., Andrieu-Ponel, V., Gambin, T., Gandouin, E., Lanfranco, S., Medail, F., Pavon, D., Ponel, P., Morhange, C., 2013. Vegetation dynamics during the early to mid-Holocene transition in NW Malta, human impact versus climatic forcing. Veget. Hist. Archaeobot. 22, 367-380.

Dodson, J.R., 1990. Fine resolution pollen analysis of vegetation history in the Lough Adoon Valley, Co Kerry, Western Ireland. Rev. Palaeobot. Palynol. 64, 235-245.

Dokturovsky, W.S., 1925. Über die Stratigraphie der russischen Torfmooe (nebst Angaben zu interglazialen Flora). Geol. Fören. Förhandl. Stockholm, 47, 81-119.

Dokulil, M., Hamm, A., Kohl, J.-G. (Eds.), 2001. Ökologie und Schutz von Seen. UTB.

Doll, R., Pankow, H., 1989. Die Verbreitung der Sippen der Gattung *Najas* L. in den Nordbezirken der DDR. Feddes Repert., Berlin, 100, 431-438.

Dommain, R., Cobb, A.R., Joosten, H., Glaser, P.H., Chua, A.F.L., Gandois, L., Kay, F.-M., Noren, A., Salim, K.A., Su'ut, N.S., Harvey, C.F., 2015. Forest dynamics and tip-up pools drive pulses of high carbon accumulation rates in a tropical peat dome in Borneo (Southeast Asia). J. Geophys. Res. Biogeosci. 120, 617-640.

Donders, T., Hoeve, M. et al., 2007. Late Neogene continental stages in NW Europe revisited. Earth-Science Review 85/3-4, 61-86.

Donner, J., 1983. The identification of Eemian interglacial and Weichselian interstadial deposits in Finland. Annales Acad. Scientiarum Fennicae Ser. A III, 136.

Dörfler, W., 1989. Pollenanalytische Untersuchungen zur Vegetations- und Siedlungsgeschichte im Süden des Landkreises Cuxhaven, Niedersachsen. Probleme der Küstenforschung im südlichen Nordseegebiet 17, 1–75.

Dörfler, W., Feeser, I., van den Bogaard, C., Dreibrodt, S., Erlenkeuser, H., Kleinmann, A., Merkt, J., Wiethold, J., 2012. A high-quality annually laminated sequence from Lake Belau, Northern Germany: Revised chronology and its implications for palynological and tephrochronological studies. Holocene 22, 1413-1426.

Douda, J., Doudová, J., Drašnarová, A., Kuneš, P.,

Hadincová, V., Krak, K., Zákravský, P., Mandák, B., 2014. Migration patterns of subgenus *Alnus* in Europe since the Last Glacial Maximum: A systematic review. PLoS ONE 9, 99-112.

Dräger, N., Theuerkauf, M., Szeroczyńska, K., Wulf, S., Tjallingii, R., Plessen, B., Kienel, U., Brauer, A., 2017. Varve microfacies and varve preservation record of climate change and human impact for the last 6000 years at Lake Tiefer See (NE Germany). Holocene 27, 450-464.

Dransfield, J., Flenley, J.R., King, S.M., Harkness, D.D., Rapu, S., 1984. A recently extinct palm from Easter Island. Nature 312, 750-752.

Draxler, I., 1977. Pollanalytische Untersuchungen von Mooren zur spät- und postglazialen Vegetationsgeschichte im Einzugsgebiet der Traun. Jahrb. Geol. B.-A. 120, 131-163.

Draxler, I., 2000. Pollenanalytische Untersuchung der schieferkohlenführenden Sedimentfolge von Nieselach bei St. Stefan im unteren Gailtal, Kärnten. Mitt. Kommission f. Quartärforschung der Österr. Akad. Wiss. 12, 155 pp.

Drescher-Schneider, R., 1993. Funde nordafrikanischer Pollen in spät- und postglazialen Sedimenten am Südrand der Alpen. In: Brombacher, C., Jacomet, S., Haas, J.N. (eds.), 1993. Festschrift Zoller. Diss. Bot., Berlin/Stuttgart, 196, 415-425.

Drescher-Schneider, R., 2000a. Kiesgrube Thalgut: Pollen- und großrestanalytische Untersuchungen. In: Exkursionsführer DEUQUA 2000, Bern, pp. 128-136.

Drescher-Schneider, R., 2000b. Die Vegetations- und Klimaentwicklung im Riß/Würm-Interglazial und im Früh- und Mittelwürm in der Umgebung von Mondsee. Ergebnisse der pollenanalytischen Untersuchungen. Mitt. Kommission f. Quartärforschung der Österr. Akad. Wiss. 12, 39-92.

Drescher-Schneider, R., Jacquat, C., Schoch, W., 2007a. Palaeobotanical investigations at the mammoth site of Niederweningen (Kanton Zürich), Switzerland. Quat. Int. 164-165, 85-97.

Drescher-Schneider, R., de Beaulieu, J.-L., Magny, M., Walter-Simonnet, A., Bossuet, G., Millet, L., Brugiapaglia, E., Drescher, A., 2007b. Vegetation history, climate and human impact over the last 15,000 years at Lago dell'Accesa (Tuscany, central Italy). Veget. Hist. Archaeobot. 16, 279-299.

Drzymulska, D., 2018. On the history of *Brasenia* Schreb. in the European Pleistocene. Veget. Hist. Archaeobot. 27, 527-534.

Du Rietz, G.E., 1930. The fundamental units of biological taxonomy. Svensk Botaniska Foreningen, Uppsala.

Du Rietz, G.E., 1954. Die Mineralbodenwasserzeigergrenze als Grundlage einer natürlichen Zweigliederung der nord- und mitteleuropäischen Moore. Vegetatio 5-6, 571-585.

Dubois, A.D., Ferguson, D.K., 1985. The climatic history of pine in the Cairngorms based on radiocarbon dates and stable isotope analysis, with an account of the events leading up to its colonization. Rev. Palaeobot. Palynol. 46, 55-80.

Dubois, A.D., Ferguson, D.K., 1988. Additional evidence for the climatic history of pine in the Cairngorms, Scotland, based on radiocarbon dates and tree ring D/H ratios. Rev. Palaeobot. Palynol. 54, 181-185.

Duff, E., Laing, T., Smol, J., Lean, D., 1999. Limnological characteristics of lakes located across arctic treeline in northern Russia. Hydrobiologia 391, 205-222.

Duller, G.A.T., Whittington, G., Heijnis, H., Hall, A. M., Gordon, J.E, 2002. Sedimentology, palaeoecology and geochronology of Marine Isotope Stage 5 deposits on the Shetland Islands, Scotland. Journal Quat. Sciences 17, 51-67.

Dunwiddie, P.W., 1987. Macrofossil and pollen representation of coniferous trees in modern sediments from Washington. *Ecology* 68, 1-11.

Dupont, L., 1986. Temperature and rainfall variations in the Holocene based on comparative palaeoecology and isotope geology of a hummock and a hollow (Bourtangerveen, The Netherlands). Rev. Palaeobot. Palynol. 48, 71-159.

Duvigneaud, P. (ed.), 1971. Productivity of forest ecosystems. Paris.

Duvigneaud, P., 1974. La synthèse écologique: populations, communautés, écosystèmes, biosphère, noosphère. Doin, 296 pp.

Dynesius, M., Jansson, R., 2000. Evolutionary consequences of changes in species' geographical distributions driven by Milankovitch climate oscillations. Proc. Natl. Acad. Sci. USA 97, 9115-9120.

Eastwood, W.J., 2004. East Mediterranean vegetation and climate change. In: Griffiths, H.I., Kryštufek, B., Reed, J.M. (eds.), Balkan Biodiversity: Pattern and Process in the European Hotspot. Kluwer Academic Publishers, Dordrecht, pp. 25-48.

Eckmeier, E., Gerlach, R., Gehrt, E., Schmidt, M., 2007. Pedogenesis of Chernozems in Central Europe—A review. Geoderma 139, 288-299.

Eddudóttir, S.D., Erlendsson, E., Tinganelli, L., Gísladóttir, G., 2016. Climate change and human impact in a sensitive ecosystem: the Holocene environment of the Northwest Icelandic highland margin. Boreas 45, 715-728.

Eddudóttir, S.D., Erlendsson, E., Gísladóttir, G., 2020. Landscape change in the Icelandic highland: A long-term record of the impacts of land use, climate and volcanism. Quat. Sci. Rev. 240.

Edwards, K.J., 1981. The separation of *Corylus* and *Myrica* pollen in modern and fossil samples. Pollen et Spores 23, 205-218.

Edwards, M.E., 1986. Disturbance histories of four Snowdonian woodlands and their relation to Atlantic bryophyte distributions. Biol. Conserv. 37, 301-320.

Edwards, M.E., 2020. The maturing relationship between Quaternary paleoecology and ancient sedimentary DNA. Quat. Res. 96, 39-47.

Edwards, K.J., Fyfe, R.M., Jackson, S.T., 2017. The first 100 years of pollen analysis. Nature Plants 3, 17001.

Edwards, K.J., Bennett, K.D., Davies, A.L., 2019. Palaeoecological perspectives on Holocene environmental change in Scotland. Earth Environ. Sci.: Trans. R. Soc. Edinburgh 110, 199–217.

Ehlers, J., 1994. Allgemeine und historische Quartärgeologie. Enke, Stuttgart.

Ehlers, J., Eißmann, L., Lippstreu L., Stephan, H.-J., Wansa, S., 2004. Pleistocene glaciations in North Germany. In: Ehlers, J., Gibbard, P. L., (Eds.), Quaternary Glaciations – Extent and Chronology. Elsevier, Amsterdam, pp. 135-146.

Ehrenberg, C.G., 1838. Beobachtungen über neue Lager fossiler Infusorien und das Vorkommen von Fichtenblütenstaub neben deutlichem Fichtenholz, Haifischzähnen, Echiniten und Infusorien in wolhynischen Feuersteinen der Kreide. Verh. Preuß. Akad. Wiss. Berlin, 104 S.

Eicher, U., Siegenthaler, U., Wegmüller, S., 1981. Pollen and oxygen isotope analyses on late- and post-glacial sediments of the Tourbière de Chirens (Dauphiné, France). Quat. Res. 15, 160-170.

Eide, W., Birks, H.H., Bigelow, N.H., Peglar, S.M., Birks, H.J.B., 2006. Holocene forest development along the Setesdal valley, southern Norway, reconstructed from macrofossil and pollen evidence. Veget. Hist. Archaeobot. 15, 65-85.

Eißmann, L., Litt, T., Wansa, S., 1995. Elsterian and Saalian deposits in their type area in central Germany. In: Ehlers, J., Gibbard, P.L., Kosarski, S., Rose, J. (Eds.), Glacial Deposits in North-East Europe, Rotterdam, pp. 439-464.

Eldredge, N., 1989. Time Frames. Princeton University Press, Princeton. 340 pp.

Elena-Rossello, J., Lumaret, R., Cabrera, E., Michaud, H., 1992. Evidence for hybridization between sympatric holm-oak and cork-oak in Spain based on diagnostic enzyme markers. Vegetatio 99-100, 115-118.

Elias, S.A., Wilkinson, B., 1983. Lateglacial insect fossil assemblages from Lobsigensee (Swiss Plateau). Studies in the late Quaternary of Lobsigensee 3. Rev. Paléobiol. 2, 189-204.

Elina, G.A., 1981. Printsipy i metody rekonstruktsii i kartirovaniya rastitelonosti golotsena [Principles and methods for reconstruction and mapping of Holocene vegetation]. Nauka, Leningrad, Russia.

Elina, G.A., Filimonova, L.V., 1987. Late-Glacial vegetation on the territory of Karelia, in: Raukas, A., Saarse, L. (Eds.), Palaeohydrology of the Temperate Zone III: Mires and Lakes. Valgus, Tallinn, Estonia, pp. 53–69.

Elina, G.A., Filimonova, L.V., 1996. Russian Karelia - Dynamics of vegetation and natural conditions in type regions Kc and Kg during the Holocene. In: Berglund, B.E., Birks, H.J.B., Ralska-Jasiewiczowa, M., Wright, H.E. (Eds.), Palaeoecological Events During the Last 15 000 Years. Regional Syntheses of Palaeoecological Studies of Lakes and Mires in Europe. Wiley, Chichester, pp. 353-366.

Ellenberg, H., 1963. Vegetation Mitteleuropas mit den Alpen in ökologischer Sicht. Ulmer, Stuttgart, 943 pp.

Ellenberg, H., 1966. Leben und Kampf an den Baumgrenzen der Erde. Naturwiss. Rundschau 19, 133-139.

Ellenberg, H., 1988. Vegetation Ecology of Central Europe, 4th ed. Cambridge University Press, Cambridge, 731 pp.

Ellenberg, H., 2009. Vegetation Ecology of Central Europe, 4th ed. Cambridge University Press, Cambridge.

Ellenberg, H., Weber, H.E., Düll, R., Wirth, V., Werner, W., Paulissen, D., 1992. Zeigerwerte von Pflanzen in Mitteleuropa. Scripta Geobotanica 18, 1-248.

Ellis, E.C., Antill, E.C., Kreft, H., 2012. All is not loss: plant biodiversity in the Anthropocene. PLoS ONE 7, e30535.

Elovicheva, Y.K., Bogdel, I.I., 1985. Novye razrezy golosena Belarusi [New Holocene sections of Byelorussia], in: Geologicheskoe Stroenie Osadochnoi Tolshchi Belorussii [Geological Composition of Sedimentary Sequence of Byelorussia]. Nauka i Tekhnika, Minsk, Belarus, pp. 141–169.

Eneroth, O., 1951. Undersökning rörande möljigheter att i fossilt material urskilja de olika Betula-arternas pollen. Geol. Fören. Stockholm Förh. 73.

Engler, A., 1879. Versuch einer Entwicklungsgeschichte der Pflanzenwelt, insbesondere der Florengebiete, seit der Tertiärperiode. I. Theil: Die extratropischen Gebiete der nördlichen Hemisphäre, Leipzig.

Eppinga, M.B., de Ruiter, P.C., Wassen, M.J., Rietkerk, M., 2009. Nutrients and hydrology indicate the driving mechanisms of peatland surface patterning. Am. Nat. 173, 803-818.

Eppinga, M.B., Rietkerk, M., Belyea, L.R.,

Nilsson, M.B., de Ruiter, P.C., Wassen, M.J., 2010. Resource contrast in patterned peatlands increases along a climatic gradient. Ecology 91, 2344-2355.

Epstein, H.E., Yu, Q., Kaplan, J.O., Lischke, H., 2007. Simulating future changes in Arctic and subarctic vegetation. Comput. Sci. Eng. 9, 12-23.

Erd, K., 1973. Pollenanalytische Gliederung des Pleistozäns der Deutschen Demokratischen Republik. Zeitschr. Geol. Wissensch. 1, 1087-1103.

Erdogan, V., Mehlenbacher, S., 2000. Phylogenetic relationships of *Corylus* species (Betulaceae) based on nuclear ribosomal DNA ITS region and chloroplast matK gene sequences. Syst. Bot. 25, 727-737.

Erdtman, G., 1920: Einige geobotanische Resultate einer pollenanalytischen Untersuchung von südwestschwedischen Torfmooren. Svensk. Bot. Tidskr. 14.

Erdtman, G., 1921: Pollenanalytische Untersuchungen von Torfmooren und marinen Sedimenten in Südwest-Schweden. Arkiv for Botanik 1-173.

Erdtman, G., 1943. An Introduction to Pollen Analysis. Waltham, Massachusetts, USA.

Erdtman, G., Berglund, B.E., Praglowski, J., 1961. An Introduction to a Scandinavian Pollen Flora. I. Almqvist & Wiksell, Stockholm.

Erdtman, G., Praglowski, J., Nilsson, S., 1963. An Introduction to a Scandinavian Pollen Flora II. Almqvist & Wiksell, Stockholm.

Eriksson, B., 1966. Vesipähkinan (Trapa natans L.) postglasiaaliseta esiintymisestä Suomessa. Pro gradu (unpublished). Dept. Bot. Univ. Helsinki.

Erkkamo, V., 1956: Untersuchungen über die pflanzenbiologischen und einige andere Folgeerscheinungen der neuzeitlichen Klimaschwankung in Finnland. Ann. Bot. Soc. Vanamo 28, 3.

Eroglu, D., Marwan, N., Stebich, M., Kurths, J., 2018. Multiplex recurrence networks. Phys. Rev. E E97, 012312.

Eronen, M., 1979. The retreat of pine in Finnish Lapland since the Holocene climatic optimum: a general discussion with radiocarbon evidence from subfossil pines. Fennia 157, 93-114.

Eronen, M., Huttunen, P., 1987. Radiocarbon-dated subfossil pines from Finnish Lapland. Geograf. Annal. 69A, 297-304.

Eronen, M., Huttunen, P., 1993. Pine megafossils as indicators of Holocene climatic changes in Fennoscandia. Paläoklimaforschung 9, 29-40.

Eronen, M., Hyvärinen, H., 1982. Subfossil pine dates and pollen diagrams from northern Fennoscandia. Geol. Fören. Stockholm Förh. 103, 437-455.

Eronen, M., Zetterberg, P., 1996. Climatic changes in northern Europe since late glacial times, with reference to dendroclimatological studies in northern Finnish Lapland. Geophysica 2, 35-60.

Eronen, M., Hyvärinen, H., Zetterberg, P., 1999. Holocene humidity changes in northern Finnish Lapland inferred from lake sediments and submerged Scots pines dated by tree-rings. Holocene 9, 569-580.

Eurola, S., Hicks, S., Kaakinen, G., 1984. Key to Finnish mire types. In: Moore, P. (Ed.), European Mires. Academic Press, London, pp. 11-117.

Ewald, J., 2003. The calcareous riddle: why are there so many calciphilous species in the central European flora? Folia Geobot. 38, 357-366.

Fady, B., Forest, I., Hochu, I., Ribiollet, A., de Beaulieu, J.-L., Pastuzska, P., 1999. Genetic differentiation in *Abies alba* populations from south-eastern France. Forest Genetics 6, 129-138.

Fægri, K., 1940. Quartärgeologische Untersuchungen im westlichen Norwegen. II. Zur spätquartären Geschichte Jaerens. Bergen Museums Aarbok, Naturv. R. 7, 1-201.

Fægri, K., 1947. Heterodokse tanker om pollenanalysen. Geol. Fören. Stockholm Förhandl. 69, 55-66.

Fægri, K., 1960. The Coastal Plants. In: Fægri, K., Gjaerevoll, O., Lid, J., Nordhagen, R. (Eds.), Maps of Distribution of Norwegian Plants. Oslo University Press, Oslo.

Fægri, K., 1963. Problems of immigration and dispersal of the Scandinavian Flora. In: Löve, A., Löve, D. (Eds.), North Atlantic Biota and their History. Macmillan, New York, pp. 221-237.

Fægri, K., 1974. Quaternary pollen analysis - past, present and future. Adv. Pollen Spore Res. 1, 62-69.

Fægri, K., 1981. Some pages of the history of pollen analysis. Striae 14, 42-47.

Fægri, K., Iversen, J., 1950. Textbook of Modern Pollen Analysis. Munksgaard, Copenhagen. 168 pp.

Fægri, K., Iversen, J., 1964. Textbook of Pollen Analysis, 2nd ed. Munksgaard, Copenhagen.

Fægri, K., Iversen, J., 1989. Textbook of Pollen Analysis, 4th ed. (edited by Faegri, K., Kaland, P.E., Krzywinski, K.). Wiley & Sons, Chichester. 328 pp.

Fagerlind, F. 1952. The real signification of pollen diagrams. Botaniska Notiser 2, 185–224.

Fan, D., Chen, J., Meng, Y., Wen, J., Huang, J., Yang, Y., 2013. Molecular phylogeny of *Koenigia* L. (Polygonaceae: Persicarieae): Implications for classification, character evolution and biogeography. Mol. Phylogen. Evol. 69, 1093-1100.

Fang, K., Morris, J.L., Salonen, J.S., Miller, P.A., Renssen, H., Sykes, M.T., Seppä, H., 2013. How robust are Holocene treeline simulations? A model-data comparison in the European Arctic treeline region. J. Quat. Sci. 28, 595-604.

Fang, Y., Ma, C., Bunting, M.J., 2019. Novel methods of estimating relative pollen productivity: A key parameter for reconstruction of past land cover from pollen records. Prog. Phys. Geogr.: Earth Env. 43, 731-753.

Fankhauser, A., 1995. Pollenanalytische Untersuchungen zur jüngsten Vegetationsgeschichte der Franches Montagnes (PhD thesis). University of Bern, Bern, Switzerland.

Fărcaş, S., de Beaulieu, J.-L., Reille, M., Coldea, G., Diaconeasa, B., Goeury, C., Goslar, T., Jull, T., 1999. First ^{14}C datings of Late Glacial and Holocene pollen sequences from Romanian Carpathes. Comptes Rendus de l'Académie des Sciences. Sciences de la vie 322, 799-807.

Fărcaş, S., Lupsa, V., Tanţău, I., Bodnariuc, A., 2003. Reflectarea procesului de antropizare în diagramele sporo-polinice din Muntii Apuseni. Environ. Prog.

Fărcaş, S., Popescu, F., Tanţău, I., 2006. Dinamica spatiala si temporala a stejarului, frasinului si carpenului in timpul Tardi-si Postglaciarului pe teritoriul Romaniei.[Temporal and spatial dynamics of oak, ash and hornbeam in the Late-and Postglacial era in Romania.]. Presa Universitara Clujeana, Cluj-Napoca.

Farjon, A., 1990. Pinaceae: drawings and descriptions of the genera *Abies, Cedrus, Pseudolarix, Keteleeria, Nothotsuga, Tsuga, Cathaya, Pseudotsuga, Larix* and *Picea*. Koeltz Scientific Books, Königstein.

Farjon, A., 2005. A monograph of Cupressaceae and *Sciadopitys*. Kew, Royal Botanic Gardens.

Farrell, M., 2015. Later prehistoric vegetation dynamics and Bronze Age agriculture at Hobbister, Orkney, Scotland. Veget. Hist. Archaeobot. 24, 467-486.

Feeny, P.P., 1976. Plant apparency and chemical defense. Recent Advances in Phytochemistry 10, 1-40.

Feilberg, J., Fredskild, B., Holt, S., 1984. Grønlands Blomster. Flowers of Greenland. Regnbuen, Ringsted

FEIS, 2020. Fire Effects Information System. Species: *Pteridium aquilinum*, fire ecology or adaptations. https://www.fs.usda.gov/database/feis/plants/fern/pteaqu/all.html#FIRE%20ECOLOGY. Last visited 18.02.2023.

Fekete, G., Molnár, Z., Magyari, E., Somodi, I., Varga, Z., 2014. A new framework for understanding Pannonian vegetation patterns: regularities, deviations and uniqueness. Commun. Ecol. 15, 12-26.

Felde, V.A., Birks, H.H., 2019. Using species attributes to characterize Late-glacial and early-Holocene environments at Kråkenes, western Norway. J. Veg. Sci. 30, 1228-1238.

Felde, V.A., Bjune, A.E., Grytnes, J.-A., Birks, H.J.B., 2014a. A comparison of novel and traditional numerical methods for the analysis of modern pollen assemblages from major vegetation-landform types. Rev. Palaeobot. Palynol. 210, 22-36.

Felde, V.A., Peglar, S.M., Bjune, A.E., Grytnes, J.-A., Birks, H.J.B., 2014b. The relationship between vegetation composition, vegetation zones and modern pollen assemblages in Setesdal, southern Norway. Holocene 24, 985-1001.

Felde, V.A., Peglar, S.M., Bjune, A.E., Grytnes, J.-A., Birks, H.J.B., 2016. Modern pollen-plant richness and diversity relationships exist along a vegetational gradient in southern Norway. Holocene 26, 163-175.

Felde, V.A., Grytnes, J.-A., Bjune, A.E., Peglar, S.M., Birks, H.J.B., 2018. Are diversity trends in western Scandinavia influenced by post-glacial dispersal limitation? J. Veg. Sci. 29, 360-370.

Felde, V.A., Flantua, S.G.A., Jenks, C.R., Benito, B.M., de Beaulieu, J.-L., Kuneš, P., Magri, D., Nalepka, D., Risebrobakken, B., ter Braak, C.J.F., Allen, J.R.M., Granoszewski, W., Helmens, K.F., Huntley, B., Kondratienė, O., Kalniņa, L., Kupryjanowicz, M., Malkiewicz, M., Milner, A.M., Nita, M., Noryśkiewicz, B., Pidek, I.A., Reille, M., Salonen, J.S., Šeirienė, V., Winter, H., Tzedakis, P.C., Birks, H.J.B., 2020. Compositional turnover and variation in Eemian pollen sequences in Europe. Veget. Hist. Archaeobot. 29, 101-109.

Fernández, S., Fuentes, N., Carrión, J.S., González-Sampériz, P., Montoya, E., Gil, G., Vega-Toscano, G., Riquelme, J., 2007. The Holocene and Upper Pleistocene pollen sequence of Carihuela Cave, southern Spain. Geobios 49, 75-90.

Fernane, A., Penaud, A., Gandouin, E., Goslin, J., Van Vliet-Lanoë, B., Vidal, M., 2015. Climate variability and storm impacts as major drivers for human coastal marsh withdrawal over the Neolithic period (Southern Brittany, NW France). Palaeogeogr. Palaeoclimatol. Palaeoecol. 435, 136-144.

Feurdean, A., 2005. Holocene forest dynamics in northwestern Romnania. The Holocene 15, 435-446.

Feurdean, A., Bennike, O., 2004. Late Quaternary palaeoecological and palaeoclimatological reconstruction in the Gutaiului Mountains, northwest Romania. J. Quat. Sci. 19, 809-827.

Feurdean, A., Mosbrugger, V., Onac, B., Polyak, V., Veres, D., 2007a. Younger Dryas to mid-Holocene environmental history of the lowlands of NW Transylvania, Romania. Quat. Res. 68, 364-378.

Feurdean, A., Wohlfarth, B., Björkman, L., Tanţău, I., Bennike, O., Willis, K.J., Fărcaş, S., Robertson, A., 2007b. The influence of refugial population on Lateglacial and early Holocene vegetational changes in Romania. Rev. Palaeobot. Palynol. 145, 305-320.

Feurdean, A., Klotz, S., Mosbrugger, V., Wohlfarth, B., 2008. Pollen-based quantitative reconstructions of Holocene climate variability in NW Romania. Palaeogeogr. Palaeoclimatol. Palaeoecol. 260, 494-504.

Feurdean, A., Willis, K.J., Parr, C., Tanţău, I., Fărcaş, S., 2010. Post-glacial patterns in vegetation dynamics in Romania: homogenization or differentiation? J. Biogeogr. 37, 2197-2208.

Feurdean, A., Tanţău, I., Fărcaş, S., 2011. Holocene variability in the range distribution and abundance of *Pinus, Picea abies*, and *Quercus* in Romania; implications for their current status. Quat. Sci. Rev. 30, 3060-3075.

Feurdean, A., Tămaş, T., Tanţău, I., Fărcaş, S., 2012. Elevational variation in regional vegetation responses to late-glacial climate changes in the Carpathians. J. Biogeogr. 39, 258-271.

Feurdean, A., Parr, C.L., Tanţău, I., Fărcaş, S., Marinova, E., Perşoiu, I., 2013a. Biodiversity variability across elevations in the Carpathians: Parallel change with landscape openness and land use. Holocene 23, 869-881.

Feurdean, A., Bhagwat, S.A., Willis, K.J., Birks, H.J.B., Lischke, H., Hickler, T., 2013b. Tree migration-rates: narrowing the gap between inferred post-glacial rates and projected rates. PLOS One 8, e71797.

Feurdean, A., Perşoiu, A., Tanţău, I., Stevens, T., Magyari, E., Onac, B., Marković, S., Andrič, M., Connor, S., Fărcaş, S., Gałka, M., Gaudeny, T., Hoek, W., Kolaczek, P., Kuneš, P., Lamentowicz, M., Marinova, E., Michczyńska, D., Perşoiu, I., Płóciennik, M., Słowiński, M., Stancikaite, M., Sumegi, P., Svensson, A., Tămaş, T., Timar, A., Tonkov, S., Toth, M., Veski, S., Willis, K.J., Zernitskaya, V., 2014. Climate variability and associated vegetation response throughout Central and Eastern Europe (CEE) between 60 and 8 ka. Quat. Sci. Rev. 106, 206-224.

Feurdean, A., Marinova, E., Nielsen, A., Liakka, J., Veres, D., Hutchinson, S., Braun, M., Timar-Gabor, A., Astalos, C., Moosbrugger, V., Hickler, T., 2015. Origin of the forest steppe and exceptional grassland diversity in Transylvania (central-eastern Europe) J. Biogeogr. 42, 951-963.

Feurdean, A., Gałka, M., Tanţău, I., Geanta, A., Hutchinson, S.M., Hickler, T., 2016. Tree and timberline shifts in the northern Romanian Carpathians during the Holocene and the responses to environmental changes. Quat. Sci. Rev. 134, 100-113.

Feurdean, A., Florescu, G., Vannière, B., Tanţău, I., O'Hara, R.B., Pfeiffer, M., Hutchinson, S.M., Gałka, M., Moskal del Hoyo, M., Hickler, T., 2017. Fire has been an important driver of forest dynamics in the Carpathian Mountains during the Holocene. For. Ecol. Manage. 389, 15-26.

Figueiral, I., 1995. Charcoal analysis and the history of *Pinus pinaster* (cluster pine) in Portugal. Rev. Palaeobot. Palynol. 89, 441-454.

Figueiral, I., Terral, J., 2002. Late Quaternary refugia of Medietrranean taxa in the Portuguese Estremadura: charcoal based palaeovegetation and climatic reconstruction. Quat. Sci. Rev. 21, 549-558.

Filipova, M., 1985. Palaeoecological investigations of lake Shabla-Ezeretz in north-eastern Bulgaria. Ecologia Mediterranea 11, 147-158.

Filipova-Marinova, M., 2003. Paleoenvironmental changes along the Southern Black Sea Coast of Bulgaria during the last 29 000 years. Phytologia Balcanica 9, 275-292.

Filipova-Marinova, M., Bozilova, E., 1990. Palaeoecological investigation of Lake ShablaeEzeretz on the Bulgarian Black Sea coast, in: Krastev, T. (Ed.), Geological Evolution of the Western Part of the Black Sea in Neogen-Quaternary. Bulgarian Academy of Sciences, Sofia, pp. 41–87.

Filipova-Marinova, M., Kvavadze, E.V., Connor, S.E., Sjögren, P., 2010. Estimating absolute pollen productivity for some European Tertiary-relict taxa. Veget. Hist. Archaeobot. 19, 351-364.

Filipova-Marinova, M., Pavlov, D., Coolen, M., Giosan, L., 2013a. First high-resolution marinopalynological stratigraphy of Late Quaternary sediments from the central part of the Bulgarian Black Sea area. Quat. Int. 293, 170-183.

Filipova-Marinova, M., Pavlov, D., Vergiev, S., Slavchev, V., Giosan, L., 2013b. Palaeoecology and Geoarchaeology of Varna Lake, northeastern Bulgaria. Comptes rendus de l'Academie Bulgare des Sciences 66, 377-392.

Filipovitch, L., 1985. Palynological studies of peat bogs on the southern slopes of Vitosha mountain. For. Sci. 2, 3–16.

Filipovitch, L., 1988. Vegetation history of the high parts of Vitosha mountain during the Late Postglacial times. Fitologija 34, 3–27.

Fineschi, S., Taurchini, D., Villani, F., Vendramin, G.G., 2000. Chloroplast DNA polymorphism reveals little geographical structure in *Castanea sativa* Mill. (Fagaceae) throughout southern European countries. Mol. Ecol. 9, 1495-1503.

Fineschi, S., Taurchini, D., Grossoni, P., Petit, R.J., Vendramin, G.G., 2002. Chloroplast DNA variation of white oaks in Italy. For. Ecol. Managem. 156, 103-114.

Fineschi, S., Salvini, D., Taurchini, D., Carnevale, S., Vendramin, G.G., 2003. Chloroplast DNA variation of *Tilia cordata* Mill. (Tiliaceae). Can. J. Forest Res. 33, 2503-2508.

Finlayson, C., Giles Pacheco, F., Rodríguez-Vidal, J., Fa, D., Gutierrez López, J., Santiago Pérez, A., Finlayson, G., Allue, E., Baena Preysler, J., Cáceres, I., Carrión, J.S., Fernández Jalvo, Y., Gleed-Owen, C., Jimenez Espejo, F., López, P., López-Sáez, J., Riquelme

Cantal, J., Sánchez Marco, A., Giles Guzman, F., Brown, K., Fuentes, N., Valarino, C., Villalpando, A., Stringer, C., Martinez Ruiz, F., Sakamoto, T., 2006. Late survival of Neanderthals at the southernmost extreme of Europe. Nature 443, 850-853.

Finsinger, W., Tinner, W., 2005. Minimum count sums for charcoal-concentration estimates in pollen slides: accuracy and potential errors. Holocene 15, 293-297.

Finsinger, W., Tinner, W., 2006. Holocene vegetation and land-use changes in response to climatic changes in the forelands of the southwestern Alps, Italy. J. Quat. Sci. 21, 243–258.

Finsinger, W., Tinner, W., 2007. Pollen and plant macrofossils at Lac de Fully (2135 m asl): Holocene forest dynamics on a highland plateau in the Valais, Switzerland. Holocene 17, 1119-1127.

Finsinger, W., Tinner, W., van der Knaap, W.O., Ammann, B., 2006. The expansion of hazel (*Corylus avellana* L.) in the southern Alps: a key for understanding its early Holocene history in Europe? Quat. Sci. Rev. 25, 612-631.

Finsinger, W., Lane, C.S., van den Brand, G.J., Wagner-Cremer, F., Blockley, S.P.E., Lotter, A.F., 2011. The Lateglacial *Quercus* expansion in the southern European Alps – rapid vegetation response to a late Allerød climate warming? J. Quat. Sci. 26, 694-702.

Finsinger, W., Kelly, R., Fevre, J., Magyari, E.K., 2014. A guide to screening charcoal peaks in macrocharcoal-area records for fire-episode reconstruction. Holocene 24, 1002-1008.

Finsinger, W., Giesecke, T., Brewer, S., Leydet, M., 2017a. Emergence patterns of novelty in European vegetation assemblages over the past 15 000 years. Ecol. Lett. 20, 336-346.

Finsinger, W., Morales-Molino, C., Gałka, M., Valsecchi, V., Bojovic, S., Tinner, W., 2017b. Holocene vegetation and fire dynamics at Crveni Potok, a small mire in the Dinaric Alps (Tara National Park, Serbia). Quat. Sci. Rev. 167, 63-77.

Finsinger, W., Schwörer, C., Heiri, O., Morales-Molino, C., Ribolini, A., Giesecke, T., Haas, J.N., Kaltenrieder, P., Magyari, E.K., Ravazzi, C., Rubiales, J.M., Tinner, W., 2019. Fire on ice and frozen trees? Inappropriate radiocarbon dating leads to unrealistic reconstructions. New Phytol. 222, 657-662.

Firbas, F., 1923. Pollenanalytische Untersuchungen einiger Moore der Ostalpen. Lotos Prag 71, 187-242.

Firbas, F., 1927. Beiträge zur Kenntnis der Schieferkohlen des Inntals und der interglazialen Waldgeschichte der Ostalpen. Zeitschr Gletscherkunde 15, 261-277.

Firbas, F., 1934. Über die Bestimmung der Walddichte und der Vegetation waldloser Gebiete mit Hilfe der Pollenanalyse. Planta 22, 109-145.

Firbas, F., 1939. Vegetationsentwicklung und Klimawandel in der mitteleuropäischen Spät- und Nacheiszeit. Die Naturwissenschaften 27, 81-89, 104-108.

Firbas, F., 1949. Spät- und nacheiszeitliche Waldgeschichte Mitteleuropas nördlich der Alpen. Gustav Fischer, Jena.

Firbas, F., 1952. Waldgeschichte Mitteleuropas 2. Jena.

Firbas, F., 1954. Die Vegetationsentwicklung im Spätglazial von Wallensen im Hils. Nachr. Akad. Wiss., Göttingen.

Fischer, H., 1890. Beiträge zur vergleichenden Morphologie der Pollenkörner. Breslau.

Fischer, H., Meissner, K.J., Mix, A.C., Abram, N.J., Austermann, J., Brovkin, V., Capron, E., Colombaroli, D., Daniau, A.L., Dyez, K.A., Felis, T., Finkelstein, S.A., Jaccard, S.L., McClymont, E.L., Rovere, A., Sutter, J., Wolff, E.W., Affolter, S., Bakker, P., Ballesteros-Canovas, J.A., Barbante, C., Caley, T., Carlson, A.E., Churakova, O., Cortese, G., Cumming, B.F., Davis, B.A.S., de Vernal, A., Emile-Geay, J., Fritz, S.C., Gierz, P., Gottschalk, J., Holloway, M.D., Joos, F., Kucera, M., Loutre, M.F., Lunt, D.J., Marcisz, K., Marlon, J.R., Martinez, P., Masson-Delmotte, V., Nehrbass-Ahles, C., Otto-Bliesner, B.L., Raible, C.C., Risebrobakken, B., Goni, M.F.S., Arrigo, J.S., Sarnthein, M., Sjolte, J., Stocker, T.F., Alvarez, P.A.V., Tinner, W., Valdes, P.J., Vogel, H., Wanner, H., Yan, Q., Yu, Z.C., Ziegler, M., Zhou, L.P., 2018. Palaeoclimate constraints on the impact of 2 degrees C anthropogenic warming and beyond. Nature Geoscience 11, 474-485.

Flenley, J.R., 1973. The use of modern pollen rain samples in the study of the vegetational history of tropical regions. In: Birks, H.J.B., West, R.G. (Eds.), Quaternary Plant Ecology. Blackwell, Oxford, pp. 131-141.

Flenley, J.R., 2011. Why is pollen yellow? And why are there so many species in the tropical rain forest? J. Biogeogr. 38, 809-816.

Flenley, J.R., Maloney, B.K., Ford, D., Hallam, G., 1975. *Trapa natans* in the British Flandrian. Nature 257, 39 - 41.

Flessa, K.W., Jackson, S.T., 2005a. Forging a common agenda for ecology and paleoecology. BioScience 55, 1030-1031.

Flessa, K.W., Jackson, S.T., 2005b. The Geological Record of Ecological Dynamics. Understanding the biotic effects of future environmental change. National Research Council of the National Academies, Washington, D.C. 198 pp.

Fletcher, W.J., Boski, T., Moura, D., 2007. Palynological evidence for environmental and climatic change in the lower Guadiana valley, Portugal, during the last 13 000 years. Holocene 17, 481-494.

Fletcher, W.J., Sánchez Goñi, M.F., 2008. Orbital- and sub-orbital-scale climate impacts on vegetation of the western Mediterranean basin over the last 48,000 yr. Quat. Res. 70, 451-464.

Fletcher, W.J., Müller, U., Koutsodendris, A., Christanis, K., Pross, J., 2013. A centennial-scale record of vegetation and climate variability from 312 to 240 ka (Marine Isotopic Stages 9c-a, 8 and 7e) from Tenaghi Philippon, NE Greece. Quat. Sci. Rev. 78, 108-125.

Florin, M.B., 1957. Pollen-analytical evidence of prehistoric agriculture at Mogetorp Neolithic Settlement Sweden. The Institute of Quaternary Geology, University of Uppsala, 6, 1-25.

Florin, M.B., 1969. Late-glacial and Pre-boreal vegetation in Central Sweden, I. Records of pollen species. Svensk Bot. Tidskrift 63, 143.

Florschütz, F., 1958. Steppen- und Salzsumpfelemente aus den Floren der letzten und vorletzten Eiszeit in den Niederlanden. Flora 146, 489-492.

Florschütz, F., Menéndez Amor, J., Wijmstra, T.A., 1971. Palynology of a thick quaternary succession in southern Spain. Palaeogeogr. Palaeoclimatol. Palaeoecol. 10, 233-264.

Flynn, L.E., Mitchell, F.J.G., 2019. Comparison of a recent elm decline with the mid-Holocene Elm Decline. Veget. Hist. Archaeobot. 28, 391-398.

Follieri, M., Magri, D., Sadori, L., 1986. Late Pleistocene *Zelkova* extinction in central Italy. New Phytol. 103, 269-273.

Follieri, M., Magri, D., Sadori, L., 1988. 250,000-year pollen record from Valle di Castiglione (Roma). Pollen et Spores 30, 329-356.

Follieri, M., Magri, D., Narcisi, B., 1993. Palaeoenvironmental investigations on long sediment cores from volcanic lakes of Lazio (central Italy), an overview. In: Negendank, J.W.F., Zolitschka, B. (Eds.), Palaeolimnology of European Maar Lakes. Springer, Berlin, pp. 95-107.

Follieri, M., Giardini, M., Magri, D., Sadori, L., 1998. Palynostratigraphy of the last glacial period in the volcanic region of Central Italy. Quat. Int. 47/48, 3-20.

Forbes, E., 1846. On the Connexion between the Distribution of the existing Fauna and Flora of the British Isles etc. Mem. Geol. Survey of Great Britain I, London.

Forel, F.A., 1901. Handbuch der Seenkunde: Allgemeine Limnologie, Stuttgart.

Forrest, G., 1980. Genotypic variation among native Scots pine in Scotland based on monoterpene analysis. Forestry 53, 101 - 128.

Forrest, G., 1982. Relationship of some European Scots Pine populations to native Scottish woodlands based on monoterpene analysis. Forestry 55, 19-37.

Foster, D.R., Fritz, S.C., 1987. Mire development, pool formation and landscape processes on patterned fens in Dalarna, Central Sweden. J. Ecol. 75, 409-437.

Foster, D.R., Wright, H.E., 1990. Role of ecosystem development and climate change in bog formation in central Sweden. Ecology 71, 450-463.

Foster, D.R., King, G.A., Glaser, P.H., Wright, H.E., 1983. Origin of string patterns in boreal peatlands. Nature 306, 256-258.

Foster, D.R., Knight, D.H., Franklin, J.F., 1998. Landscape patterns and legacies resulting from large, infrequent forest disturbances. Ecosystems 1, 497-510.

Fouinat, L., Sabatier, P., David, F., Montet, X., Schoeneich, P., Chaumillon, E., Poulenard, J., Arnaud, F., 2018. Wet avalanches: long-term evolution in the Western Alps under climate and human forcing. Clim. Past 14, 1299-1313.

Fowler, P. J., 1983. The Farming of Prehistoric Britain, 1983. Cambridge University Press, Cambridge, 272 pp.

Fowler, P. J., 1984. Wildscape to Landscape: "Enclosure" in Prehistoric Britain. In: Mercer, R. (ed.), Farming Practice in British Prehistory. Edinburgh.

Fox, J.W., 2013a. The intermediate disturbance hypothesis is broadly defined, substantive issues are key: a reply to Sheil and Burslem. Trends Ecol. Evol. 28, 572-573.

Fox, J.W., 2013b. The intermediate disturbance hypothesis should be abandoned. Trends Ecol. Evol. 28, 86-92.

Franco-Múgica, F., García-Antón, M., Maldonado-Ruiz, J., Morla-Juaristi, C., Sainz Ollero, H., 2001. The Holocene history of *Pinus* forests in the Spanish Northern Meseta. Holocene 11, 343-358.

Franco-Múgica, F., García-Antón, M., Maldonado-Ruiz, J., Morla-Juaristi, C., Sainz-Ollero, H., 2005. Ancient pine forest on inland dunes in the Spanish northern meseta. Quat. Res. 63, 1-14.

Fredh, D., Mazier, F., Bgagee, P., Lagerås, P., Rundgren, M., Hammarlund, D., Broström, A., 2017. The effect of local land-use changes on floristic diversity during the past 1000 years in southern Sweden. Holocene 27, 694-711.

Fredh, D., Lagerås, P., Mazier, F., Bjorkman, L., Lindbladh, M., Broström, A., 2019. Farm establishment, abandonment and agricultural practices during the last 1,300 years: a case study from southern Sweden based on pollen records and the LOVE model. Veget. Hist. Archaeobot. 28, 529-544.

Fredskild, B., 1969. A postglacial standard pollen diagram from Peary Land, North Greenland (1). Pollen et Spores 11, 573-583.

Fredskild, B., 1973. Studies in the vegetational

history of Greenland. Palaeobotanical investigations of some Holocene lake and bog deposits. Meddelelser om Grönland 198, 245 pp.

Fredskild, B., 1975. A late-glacial and early post-glacial pollen-concentration diagram from Langeland, Denmark. Geol. Foren. Stockholm Forh. 97, 151-161.

Frelich, L.E., 2002. Forest Dynamics and Disturbance Regimes: Studies from temperate evergreen-deciduous forests. Cambridge University Press, Cambridge. 266 pp.

Frenzel, B., 1964. Über die offene Vegetation der letzten Eiszeit am Ostrande der Alpen. Verh. zool.-bot. Ges. Wien 103/104, 110-143.

Frenzel, B., 1966. Climatic change in the Atlantic/Sub-Boreal transition on the northern hemisphere: botanical evidence. In: Sawyer, J. (Ed.), World Climate from 8000 to 0 BC. Royal Meteorological Society, London.

Frenzel, B., 1968. Grundzüge der pleistozänen Vegetationsgeschichte Nord-Eurasiens. Steiner, Wiesbaden.

Frenzel, B., 1983. Mires - repositories of climatic information or self-perpetuating ecosystems. In: Gore, A.J.P. (Ed.), Mires: Swamp, Bog, Fen and Moor. General Studies. Elsevier, Amsterdam, pp. 35-65.

Frenzel, B., 1987. Grundprobleme der Vegetationsgeschichte Mitteleuropas während des Eiszeitalters. Mitt. Naturf. Ges. Luzern 29, 99-122.

Freund, H., Caspers, G. (Eds.), 1997. Vegetation und Paläoklima der Weichsel-Kaltzeit im nördlichen Mitteleuropa. Schr. Deutsche Geolog. Gesellschaft 4, Hannover.

Friedrich, W., 1978. Fossil plants from Weichselian Interstadials, Santorini (Greece). In: Doumas, C. (Ed.), Thera and the Aegean World. The Thera Foundation, London, pp. 741-744.

Friedrich, M., Kromer, B., Kaiser, K.F., Spurk, M., Hughen, K.A., Johnsen, S.J., 2001. High-resolution climate signals in the Bölling-Alleröd Interstadial (Greenland Interstadial 1) as reflected in European tree-ring chronologies compared to marine varves and ice-core records. Quat. Sci. Rev. 20, 1223-1232.

Friedrich, M., Remmele, S., Kromer, B., Hofmann, J., Spurk, M., Kaiser, K.F., Orcel, C., Kuppers, M., 2004. The 12,460-Year Hohenheim Oak and Pine tree-ring chronology from central Europe - a unique annual record for radiocarbon calibration and paleoenvironmental reconstruction. Radiocarbon 46, 1111-1122.

Fries, M., 1967. Lennart von Post's pollen diagram series of 1916. Rev. Palaeobot. Palynol. 4, 9-13.

Fritsche, J., 1834. Ueber denPollen der Pflanzen und das Pollenin. Annaler der Physik und Chemie 32, 479-480.

Fritz, A., 1976. Pollen im Sahara-Staub. Carinthia II, Klagenfurt, 166/86, 173-174.

Fritz, A., 1992. Fagus-reiche Waldbestände im Riss/Würm-Interglazial des südöstlichen Alpenraumes. Carinthia II, 182./102. Jg., 597-610.

Froyd, C.A., 2005. Fossil stomata reveal early pine presence in Scotland: implications for postglacial colonisation analyses. Ecology 86, 579-586.

Froyd, C.A., Willis, K.J., 2008. Emerging issues in biodiversity and conservation management: the need for a palaeoecological perspective. Quat. Sci. Rev. 27, 1723-1732.

Froyd, C.A., Coffey, E.E.D., van der Knaap, W.O., van Leeuwen, J.F.N., Tye, A., Willis, K.J., 2014. The ecological consequences of megafaunal loss: giant tortoises and wetland biodiversity. Ecol. Lett. 17, 144-154.

Fujiyoshi, M., Yoshitake, S., Watanabe, K., Murota, K., Tsuchiya, Y., Uchida, M., Nakatsubo, T., 2011. Successional changes in ectomycorrhizal fungi associated with the polar willow *Salix polaris* in a deglaciated area in the High Arctic, Svalbard. Polar Biol. 34, 667-673.

Fuller, D.Q., 2007. Contrasting patterns in crop domestication rates: recent archaeobotanical insights from the Old World. Annals Bot. 100, 903-924.

Fuller, D.Q., Allaby, R.G. and Stevens, C., 2010. Domestication as innovation: the entanglement of technique, technology and chance in the domestication of cereal crops. World Archeol. 41/1, 13-28.

Furrer, H., Graf, H.R., Mäder, A., 2007. The mammoth site of Niederweningen, Switzerland. Quat. Int. 164-165, 85-97.

Fusco, F., 2007. Vegetation response to Early Pleistocene climatic cycles in the Lamone valley (Northern Appennines, Italy). Rev. Palaeobot. Palynol. 145, 1-23.

Futuyma, D.J., 2005. Evolution. Sinauer Associates, Sunderland, MA. 656 pp.

Fyfe, R.M., 2018. 'Natural' vegetation in Britain: the pollen-eye view. British Wildlife 29, 339-349.

Fyfe, R.M., Woodbridge, J., 2012. Differences in time and space in vegetation patterning: analysis of pollen data from Dartmoor, UK. Landsc. Ecol. 27, 745-760.

Fyfe, R.M., Bruck, J., Johnston, R., Lewis, H., Roland, T.P., Wickstead, H., 2008. Historical context and chronology of Bronze Age land enclosure on Dartmoor, UK. J. Archaeol. Sci. 35, 2250-2261.

Fyfe, R.M., de Beaulieu, J.-L., Binney, H.A., Bradshaw, R.H.W., Brewer, S., Le Flao, A., Finsinger, W., Gaillard, M.-J., Giesecke, T., Gil-Romera, G., Grimm, E.C., Huntley, B., Kuneš, P., Kühl, N., Leydet, M., Lotter, A.F., Tarasov, P.E., Tonkov, S., 2009. The European

Pollen Database: past efforts and current activities. Veget. Hist. Archaeobot. 18, 417-424.

Fyfe, R.M., Twiddle, C.L., Sugita, S., Gaillard, M.-J., Barratt, P., Caseldine, C.J., Dodson, J., Edwards, K.J., Farrell, M., Froyd, C.A., Grant, M.J., Huckerby, E., Innes, J.B., Shaw, H., Waller, M., 2013. The Holocene vegetation cover of Britain and Ireland: overcoming problems of scale and discerning patterns of openness. Quat. Sci. Rev. 73, 132-148.

Fyfe, R.M., Ombashi, H., Davies, H., Head, K., 2018a. Quantified moorland vegetation and assessment of the role of burning over the past five millennia. J. Veg. Sci. 29, 1-17.

Fyfe, R.M., Woodbridge, J., Roberts, C.N., 2018b. Trajectories of change in Mediterranean Holocene vegetation through classification of pollen data. Veget. Hist. Archaeobot. 27, 351-364.

Gaillard, M.-J., 1983. On the occurrence of *Betula nana* L. pollen grains in the Late-Glacial deposits of Lobsigensee (Swiss Plateau). Studies in the late Quaternary of Lobsigensee 2. Rev. Paléobiol.e 2, 181-188.

Gaillard, M.-J., 1984a. Etude palynologique de l'Evolution Tardi- et Postglaciaire de la Végétation du Moyen-Pays Romand (Suisse). Diss. Bot. 77, 1-322.

Gaillard, M.-J., 1984b. A palaeohydrological study of Krageholmssjön (Scania, South Sweden). Regional vegetation history and water-level changes. Lundqua report 25, 40 pp.

Gaillard, M.-J., 1984c. Analyse pollinique et macrorestes végétaux des sédiments tardi- et postglaciaires du Grand Marais de Boussens, Moyen-Pays romand, Suisse. Diss. Bot. 72, 117-136.

Gaillard, M.-J., 1985. Postglacial palaeoclimatic changes in Scandinavia and central Europe. A tentative correlation based on studies of lake level fluctuations. Ecologia Mediterranea 11, 159-175.

Gaillard, M.-J., Birks, H.H., 2007. Plant Macrofossil Methods and Studies: Paleolimnological Applications, In: Elias, S.A. (Ed.), Encyclopedia of Quaternary Science. Elsevier, Amsterdam, pp. 2337-2356.

Gaillard, M.-J., Digerfeldt, G., 1991. Palaeohydrological studies and their contribution to palaeoecological and palaeoclimatic reconstructions. In: Berglund, B.E. (Ed.), The Cultural Landscape during 6000 years in Southern Sweden - the Ystad Project. Ecol. Bull. 41, 275-282.

Gaillard, M.-J., Lemdahl, G., 1994. Lateglacial insect assemblages from Grand-Marais, South-Western Switzerland – Climatic implications and comparison with pollen and plant macrofossil data. Diss. Bot. 234, 287-308.

Gaillard, M.-J., Birks, H.J.B., Emanuelsson, U., Berglund, B.E., 1992. Modern pollen/land-use relationships as an aid in the reconstruction of past land-uses and cultural landscapes: an example from south Sweden. Veget. Hist. Archaeobot. 1, 3-17.

Gaillard, M.-J., Birks, H.J.B., Emanuelsson, U., Karlsson, S., Lageras, P., Olausson, D., 1994. Application of modern pollen/land-use relationships to the interpretation of pollen diagrams - reconstructions of land-use history in south Sweden, 3000-0 BP. Rev. Palaeobot. Palynol. 82, 47-73.

Gaillard, M.-J., Birks, H.J.B., Karlsson, S., Lageras, P., 1996. Quantitative reconstruction of past land-use and soil conditions using the modern analogue approach - a case study in South Sweden. Pact 50, 431-442.

Gaillard, M.-J., Birks, H.J.B., Ihse, M., Runborg, S., 1998. Pollen/landscape calibrations based on modern pollen assemblages from surface-sediment samples and landscape mapping - a pilot study in South Sweden. Paläoklimaforschung 27.

Gaillard, M.-J., Sugita, S., Bunting, M.J., Middleton, R., Broström, A., Caseldine, C., Giesecke, T., Hellman, S.E.V., Hicks, S., Hjelle, K.L., Langdon, C., Nielsen, A.-B., Poska, A., von Stedingk, H., Veski, S., POLLANDCAL members, 2008. The use of modelling and simulation approach in reconstructing past landscapes from fossil pollen data: a review and results from the POLLANDCAL network. Veget. Hist. Archaeobot. 17, 419-443.

Gaillard, M.-J., Sugita, S., Mazier, F., Kaplan, J.O., Trondman, A.-K., Broström, A., Hickler, T., Kaplan, J.O., Kjellström, E., Kokfelt, U., Kuneš, P., Lemmen, C., Miller, P., Olofsson, J., Poska, A., Rundren, M., Smith, B., Standberg, G., Fyfe, R., Nielsen, A.B., Alenius, T., Balakauskas, L., Barnekow, L., Birks, H.J.B., Bjune, A., Björkman, L., Giesecke, T., Hjelle, K., Kalnina, L., Kangur, M., van der Knaap, W.O., Koff, T., Lagerås, P., Latałowa, M., Leydet, M., Lechterbeck, J., Lindbladh, M., Odgaard, B., Peglar, S., Segerström, U., von Stedingk, H., Seppä, H., 2010. Holocene land-cover reconstruction for studies on land cover-climate feedbacks. Clim. Past 6, 483-499.

Gál, E., Juhász, I., Sümegi, P. (eds.), 2005. Environmental Archaeology in North-Eastern Hungary. Varia Archaeologica Hungarica 19, Budapest.

Gale, R., Carruthers, W., 2001. Charcoal and charred seed remains from the Middle Palaeolithic Levels at Gorham's and Vanguards Caves. In: Stringer, C.B., Barton, R.N.E., Finlayson, C. (Eds.), Neanderthals on the Edge. Oxbow Books, Oxford, pp. 207-210.

Galenieks, P., 1925. Interglacials kudras slanis pie Deseles Lejniekiem, Kurzeme. Acta Univ. Latviensis, Riga 12, 265-280.

Gałka, M., Miotk-Szpiganowicz, G., Marczewska, M., Barabach, J., van der Knaap, W.O., Lamentowicz, M., 2015. Palaeoenvironmental changes in central Europe

(NE Poland) during the last 6200 years reconstructed from a high-resolution multi-proxy peat archive. Holocene 25, 421-434.

Galop, D., Carozza, L., Marembert, F., Bal, M.-C., 2004. Activités agropastorales et climat durant l'Âge du Bronze dans les Pyrénées: l'état de la question à la lumière des données environnementales et archéologiques. Presented at the Eclipse 2» emprises et déprises agricoles, expension et régression des sociétés entre 3500 et 2500 BP), CTHS, pp. 107–119.

Gams, H., 1926. Hydrocaryaceae. In: Hegi, 1926. Flora von Mitteleuropa. Band V/2. München. pp. 882-894.

Gams, H., 1927. Die Gattung *Trapa*. Pflanzenareale 1, Karte 25-27. Fischer, Jena.

Gams, H., 1943. Der Sanddorn (*Hippophaë rhamnoides* L.) im Alpengebiet. Beihefte zum Bot. Centralbl. Series B 62, 68-96.

Gams, H., 1952. Das Meerträubl (*Ephedra*) und seine Ausbreitung in Europa. Jahrb. Ver. Gesch. Alpenpfl. u. -tiere, München, 17, 34-40.

Gams, H., 1958. Beitrag zur Variabilität der Früchte von *Trapa natans*. Veröff. Geobot. Inst. Rübel 33, 108-115.

Garcés-Pastor, S., Canellas-Boltà, N., Pèlachs, A., Soriano, J.-M., Pérez-Obiol, R., Pérez-Haase, A., Calero, M.-A., Andreu, O., Escolà, N., Vegas-Villarrúbia, T., 2017. Environmental history and vegetation dynamics in response to climate variations and human pressure during the Holocene in Bassa Nera, central Pyrenees. Palaeogeogr. Palaeoclimatol. Palaeoecol. 479, 48-60.

García Esteban, L., de Palacios, P., Guinedo, A., García Fernández, F., 2007. Comparative anatomy of the wood of *Abies pinsapo* and its Moroccan varieties. IAWA Journal 28, 285-299.

García Esteban, L., de Palacios, P., García Fernández, F., Martín, J., 2009. Wood anatomical relationships within *Abies* spp. from the Mediterranean area: a phyletic approach. Sistemas y Recursos Forestales 18, 213-225.

García-Álvarez, S., García-Amorena, I., Rubiales, J.M., Morla, C., 2009a. The value of leaf cuticle characteristics in the identification and classification of Iberian Mediterranean members of the genus *Pinus*. Bot. J. Linn. Soc. 161, 436-448.

García-Álvarez, S., Morla Juaristi, C., Solana Gutiérrez, J., García-Amorena, I., 2009b. Taxonomic differences between *Pinus sylvestris* and *P. uncinata* revealed in the stomata and cuticle characters for use in the study of fossil material. Rev. Palaeobot. Palynol. 155, 61 – 68.

García-Amorena, I., Gómez Manzaneque, F., Rubiales, J.M., Granja, H., Soares de Carvalho, G., Morla, C., 2007. The Late Quaternary coastal forests of western Iberia: A study of their macroremains. Palaeogeogr. Palaeoclimatol. Palaeoecol. 254, 448-461.

García-Amorena, I., Morla, C., Rubiales, J.M., Manzaneque, F.G., 2008. Taxonomic composition of the Holocene forests of the northern coast of Spain, as determined from their macroremains. Holocene 18, 819-829.

García-Ruiz, J., Valero-Garcés, B., Martí-Bono, C., González-Sampériz, P., 2003. Asynchroneity of maximum glacier advances in the central Spanish Pyrenees. J. Quat. Sci. 18, 61-72.

Gardner, A.R., 2002. Neolithic to Copper Age woodland impacts in northeast Hungary? Evidence from the pollen and sediment chemistry records. Holocene 12, 541-553.

Gardner, A.R., 2005. Natural environment or human impact? A palaeoecological study of two contrasting sites in northeastern Hungary. Environmental Archaeology in North-Eastern Hungary 19, 87–106.

Gassner, S., Gobet, E., Schwörer, C., van Leeuwen, J.F.N., Vogel, H., Giagkoulis, T., Makri, S., Grosjean, M., Panajiotidis, S., Hafner, A., Tinner, W., 2020. 20,000 years of interactions between climate, vegetation and land use in Northern Greece. Veget. Hist. Archaeobot. 29, 75-90.

Gaudig, G., Couwenberg, J., Joosten, H. 2006. Peat accumulation in kettle holes: bottom up or top down? Mires and Peat 1: Art. 6.

Gaüzère, P., Iversen, L.L., Barnagaud, J.-Y., Svenning, J.-C., Blonder, B., 2018. Empirical predictability of community responses to climate change. Front. Ecol. Evol. 6, art186.

Gavin, D.G., Oswald, W.W., Wahl, E.R., Williams, J.W., 2003. A statistical approach to evaluating distance metrics and analog assignments for pollen records. Quat. Res. 60, 356-367.

Gavin, D.G., Fitzpatrick, M.C., Gugger, P.F., Heath, K.D., Rodriguez-Sanchez, F., Dobrowski, S.Z., Hampe, A., Hu, F.S., Ashcroft, M.B., Bartlein, P.J., Blois, J.L., Carstens, B.C., Davis, E.B., de Lafontaine, G., Edwards, M.E., Fernandez, M., Henne, P.D., Herring, E.M., Holden, Z.A., Kong, W.-S., Liu, J., Magri, D., Matzke, N.J., McGlone, M.S., Saltré, F., Stigall, A.L., Tsai, Y.-H.E., Williams, J.W., 2014. Climate refugia: joint inference from fossil records, species distribution models and phylogeography. New Phytol. 204, 37-54.

Geanta, A., Gałka, M., Tanţău, I., Hutchinson, S.M., Mindrescu, M., Feurdean, A., 2014. High mountain region of the Northern Romanian Carpathians responded sensitively to Holocene climate and land use changes: A multi-proxy analysis. Holocene 24, 944-956.

Gearey, B., Fyfe, R. 2016. Peatlands as knowledge archives. In: Bonn, A., Allott, T., Evans, M., Joosten, H., Stoneman, R. (Eds.), Peatland Restoration and Ecosystem Services: Science, policy and practice. Cambridge

University Press/ British Ecological Society, Cambridge, pp. 95-113.

Géhu, J., Julve, P., 1989. Die atlantischen Wälder mit Buche: Struktur, Pflanzengeographie, Ökologie, Dynamik und Syntaxonomie. In: Pott, R. (Ed.), Buchenwald-Symposium, Göttingen, pp. 93-105.

Geikie, J., 1867. On the buried forests and peat mosses of Scotland and the changes of climate they indicate. Trans. R. Soc. Edinburgh 24 363-384.

Genries, A., Mercier, L., Lavoie, M., Muller, S.D., Radakovitch, O., Carcaillet, C., 2009a. The effect of fire frequency on local cembra pine populations. Ecology 90, 476-486.

Genries, A., Muller, S.D., Mercier, L., Bircker, L., Carcaillet, C., 2009b. Fires control spatial variability of subalpine vegetation dynamics during the Holocene in the Maurienne valley (French Alps). Ecoscience 16, 13-22.

Gerasimenko, N.P., 2000. Late Pleistocene vegetational and soil evolution at the Kiev loess plain as recorded in the Stari Bezradychy section, Ukraine. Studia Quaternaria 17, 19-28.

Gerazimov, D.A., 1924. Several scientific and practical results of geobotanical investigations of bogs (in Russian). Peat Economy, Moscow I/12, 1-7.

Gervais, B.R., MacDonald, G.M., Snyder, J.A., Kremenetski, C.V., 2002. *Pinus sylvestris* treeline development and movement on the Kola Peninsula of Russia: pollen and stomate evidence. J. Ecol. 90, 627-638.

Gey, V., Saarnisto, M., Lunkka, J. P., Demidov, I., 2001. Mikulino and Valdai palaeoenvironment in the Vologda area, NW Russia. Glob. Planet. Change 31, 347-366.

Geyh, M.A., Schleicher, H., 1990. Absolute Age Determination. Springer, Berlin.

Geyh, M.A., Müller, H., 2005. Numerical ^{230}Th/U dating and a palynological review of the Holsteinian/Hoxnian Interglacial. Quat. Sci. Rev. 24, 1861-1872.

Geyh, M.A., Müller, H., 2007. Palynological and geochronological study of the Holsteinian/Hoxnian/Landos Interglacial. In: Sirocko, F., Claussen, M., Sanchez-Goñi, M.F., Litt, T. (Eds.), The Climate of Past Interglacials. Elsevier, Amsterdam, pp. 381-396.

Gibbard, P.L., Head, M.J., Walker, J.C., 2010. Formal ratification of the Quaternary System/Period and the Pleistocene Series/Epoch with a base at 2.58 Ma. J. Quat. Sci. 25, 96-102.

Giesecke, T., 2001. Pollenanalytische und sedimentchemische Untersuchungen zur natürlichen und anthropogenen Geschichte im Schlaubetal. Sitz Naturforsch Freunde Berl. 39, 89-112.

Giesecke, T., 2004. The Holocene Spread of Spruce in Scandinavia. PhD Thesis, Acta Universitatis Upsaliensis, Uppsala.

Giesecke, T., 2005a. Moving front or population expansion: How did *Picea abies* (L.) Karst. become frequent in central Sweden? Quat. Sci. Rev. 24, 2495-2509.

Giesecke, T., 2005b. Holocene dynamics of the southern boreal forest in Sweden. Holocene 15, 858-872.

Giesecke, T., 2005c. Holocene forest development in the central Scandes Mountains, Sweden. Veget. Hist. Archaeobot. 14, 133-147.

Giesecke, T., 2007. Changing plant distributions. In: Elias, S.A. (Ed.), Encyclopedia of Quaternary Science. Elsevier, Amsterdam, pp. 2544-2551.

Giesecke, T., 2013. Changing Plant Distributions and Abundances. In: Elias, S.A., Mock, C.A. (Ed.), Encyclopedia of Quaternary Science, 2nd ed. Elsevier, Amsterdam, pp. 854-860.

Giesecke, T., 2016. Did thermophilous trees spread into central Europe during the Late Glacial? New Phytol. 212, 15-18.

Giesecke, T., Bennett, K.D., 2004. The Holocene spread of *Picea abies* (L.) Karst. in Fennoscandia and adjacent areas. J. Biogeogr. 31, 1523-1548.

Giesecke, T., Brewer, S., 2018. Notes on the postglacial spread of abundant European tree taxa. Veget. Hist. Archaeobot. 27, 337-349.

Giesecke, T., Hickler, T., Kunkel, T., Sykes, M.T., Bradshaw, R.H.W., 2007. Towards an understanding of the Holocene distribution of *Fagus sylvatica* L. J. Biogeogr. 34, 118-131.

Giesecke, T., Bjune, A.E., Chiverrell, R., Seppä, H., Ojala, A., Birks, H.J.B., 2008. Exploring Holocene continentality changes in Fennoscandia using present and past tree distributions. Quat. Sci. Rev. 27, 1296-1308.

Giesecke, T., Fontana, S.L., van der Knaap, W.O., Pardoe, H.S, Pidek, I.A., 2010a. From early pollen trapping experiments to the Pollen Monitoring Programme. Veget. Hist. Archaeobot. 19, 247-258.

Giesecke, T., van der Knaap, Pim (W.O.), Bittmann, F., (Eds.) 2010b. Towards quantitative palynology: using pollen accumulation rates and models of pollen dispersal. Veget. Hist. Archaeobot.19 (4), pp. 387.

Giesecke, T., Bennett, K.D., Birks, H.J.B., Bjune, A.E., Bozilova, E., Feurdean, A., Finsinger, W., Froyd, C.A., Pokorný, P., Rösch, M., Seppä, H., Tonkov, S., Valsecchi, V., Wolters, S., 2011. The pace of Holocene vegetation change – testing for synchronous developments. Quat. Sci. Rev. 30, 2805-2814.

Giesecke, T., Wolters, S., Jahns, S., Brande, A., 2012. Exploring Holocene changes in palynological richness in northern Europe – Did postglacial immigration matter? PLoS ONE 7, e51624.

Giesecke, T., Ammann, B., Brande, A., 2014a. Palynological richness and evenness: insights from the taxa accumulation curve. Veget. Hist. Archaeobot. 23, 217-228.

Giesecke, T., Davis, B.A.S., Brewer, S., Finsinger, W., Wolters, S., Blaauw, M., de Beaulieu, J.-L., Binney, H.A., Fyfe, R.M., Gaillard, M.J., Gil-Romera, G., van der Knaap, W.O., Kuneš, P., Kühl, N., van Leeuwen, J.F.N., Leydet, M., Lotter, A.F., Ortu, E., Semmler, M.S.S., Bradshaw, R.H.W., 2014b. Towards mapping the late Quaternary vegetation change of Europe. Veget. Hist. Archaeobot. 23, 75-86.

Giesecke, T., de Beaulieu, J.-L., Leydet-Barbier, M., 2016. The European Pollen Database: Research tool and community. PAGES News 24, 48.

Giesecke, T., Brewer, S., Finsinger, W., Leydet, M., Bradshaw, R.H.W., 2017. Patterns and dynamics of European vegetation change over the last 15000 years. J. Biogeogr. 44, 1441-1456.

Giesecke, T., Wolters, S., van Leeuwen, J.F.N., van der Knaap, W.O., Leydet, M., Brewer, S., 2019. Postglacial change of the floristic diversity gradient in Europe. Nat. Commun. 10, 5422.

Gigov, A., 1956. Bisherige Ergebnisse über die postglaziale Geschichte der Wälder Serbiens. Zbor. Rad. Inst. Ekol. Biogeogr. SAN Beograd 6.

Giguet-Covex, C., Ficetola, G.F., Walsh, K., Poulenard, J., Bajard, M., Fouinat, L., Sabatier, P., Gielly, L., Messager, E., Develle, A.L., David, F., Taberlet, P., Brisset, E., Guiter, F., Sinet, R., Arnaud, F., 2019. New insights on lake sediment DNA from the catchment: importance of taphonomic and analytical issues on the record quality. Sci. Rep. 9, 14676.

Gil-Romera, G., Carrión, J.S., McClure, S., Schmich, S., Finlayson, C., 2009. Holocene vegetation dynamics in Mediterranean Iberia. Historical contingency and climate-human interactions. J. Anthropol. Res. 65, 271-285.

Gil-Romera, G., Carrión, J.S., Pausas, J.G., Sevilla-Callejo, M., Lamb, H.F., Fernández, S., Burjachs, F., 2010. Holocene fire activity and vegetation response in South-Eastern Iberia. Quat. Sci. Rev. 29, 1082–1092.

Gil-Romera, G., González-Sampériz, P., Lasheras-Álvarez, L., Sevilla-Callejo, M., Moreno, A., Valero-Garcés, B., López-Merino, L., Carrión, J.S., Pérez Sanz, A., Aranbarri, J., García-Prieto Fronce, E., 2014. Biomass-modulated fire dynamics during the Last Glacial–Interglacial transition at the central Pyrenees (Spain). Palaeogeogr. Palaeoclimatol. Palaeoecol. 402, 113-124.

Gil, E., Kotarba, A., Szczepanek, K., 1972. The site ll-3 the landslide at Szymbark-Kamionka. In Exe. Guide-Book, INQUA Holocene Symp (Vol. 1).

Gil, E., Gilot, E., Kotarba, A., Starkel, L., Szczepanek, K., 1974. An early Holocene landslide in the Niski Beskid and its significance for palaeogeographical reconstructions. Studia Geomorphologica Carpatho-Balcanica 8, 69–83.

Gill, J.L., 2013. Is pollen analysis dead? Paleoecology in the era of Big Data, The Contemplative Mammoth, https://contemplativemammoth.com/2013/07/10/is-pollen-analysis-dead-paleoecology-in-the-era-of-big-data/.

Gill, J.L., Williams, J.W., Jackson, S.T., Lininger, K.B., Robinson, G.S., 2009. Pleistocene megafaunal collapse, novel plant communities, and enhanced fire regimes in North America. Science 326, 1100-1103.

Gillson, L., Ekblom, A., Willis, K.J., Froyd, C.A., 2008. Holocene palaeo-invasions: the link between pattern, process and scale in invasion ecology. Landscape Ecol. 23, 757-769.

Gimingham, C.H., 1972. Ecology of Heathlands. Chapman and Hall, London, 266 pp.

Giraudi, C., Magny, M., Zanchetta, G., Drysdale, R.N., 2011. The Holocene climatic evolution of Mediterranean Italy: A review of the continental geological data. Holocene 21, 105-115.

Girling, M., 1988. The bark beetle *Scolytus scolytus* (Fabricius) and the possible role of elm disease in the early Neolithic. Oxford University Committee for Archaeology Monograph 14, 34-38.

Girling, M., Greig, J., 1985. A first fossil record for *Scolytus scolytus* (F.) (Elm bark beetle): its occurrence in elm decline deposits from London and the implications for Neolithic elm disease. J. Archaeol. Sci. 12, 347-351.

Gjaerevoll, O., 1956. The Plant Communities of the Scandinavian Alpine Snow-beds. Kong. Norske Vidensk.Selsk. Forh. 1, 1-405.

Glaser, P.H., 1998. The distribution and origin of mire pools. In: Standen, V., Tallis, J.H., Meade, R. (Eds.), Patterned Mires and Mire Pools. Origin and development; flora and fauna. British Ecological Society, Durham, UK, pp. 4-25.

Gleason, H.A., 1917. The structure and development of the plant association. Bull. Torrey Bot. Club 43, 463-481.

Gleason, H.A., 1926. The individualistic concept of the plant association. Bull. Torrey Bot. Club 53, 7-26.

Gleason, H.A., 1939. The individualistic concept of the plant association. Am. Midland Nat. 21, 92-110.

Glew, J.R., Smol, J.P., Last, W.M., 2001. Sediment core collection and extrusion. In: Last, W.M., Smol, J.P. (Eds.), Tracking Environmental Change Using Lake Sediments, Vol. 1: Basin Analysis, Coring, and Chronological Techniques. Kluwer, Dordrecht, pp. 73-105.

Gobet, E., Tinner, W., Hubschmid, P., Jansen, I., Wehrli, M., Ammann, B., Wick, L., 2000. Influence of human impact and bedrock differences on the vegetational history of the Insubrian Southern Alps. Veget. Hist. Archaeobot. 9, 175-178.

Gobet, E., Tinner, W., Hochuli, P.A., van Leeuwen, J.F.N., Ammann, B., 2003. Middle to Late

Holocene vegetation history of the Upper Engadine (Swiss Alps): the role of man and fire. Veget. Hist. Archaeobot. 12, 143-163.

Gobet, E., Tinner, W., Bigler, C., Hochuli, P.A., Ammann, B., 2005. Early-Holocene afforestation processes in the lower subalpine belt of the central Swiss Alps as inferred from macrofossil and pollen records. Holocene 15, 672-686.

Gobet, E., Schwörer, C., van Leeuwen, J.F.N., Abdel Wahab, S., Nielsen, E.H., Kotova, N., Makhortykh, S., Kiosak, D., Tinner, W., 2017. Vegetation shifts at the monumental Ukrainian site of Kamyana Mohyla during the Neolithisation period, Archaeology and Palaeoecology of the Ukrainian Steppe. IA NAS of Ukraine, Kyiv, pp. 51-60

Godwin, H., 1956. The History of the British Flora. A Factual Basis for Phytogeography (1st ed.). Cambridge University Press, Cambridge.

Godwin, H., 1960. Studies of the Post-glacial history of British vegetation. XIV. Late-glacial deposits at Moss Lake. Phil. Trans. Roy. Soc. London B 242, 127-149.

Godwin, H., 1975. The History of the British Flora. A Factual Basis for Phytogeography. 2nd ed. Cambridge University Press, Cambridge.

Godwin, H., 1981. The Archives of the Peat Bogs. Cambridge University Press, Cambridge, 229 pp.

Godwin, H., Mitchell, G., 1938. Stratigraphy and development of two raised bogs near Tregaron, Cardiganshire. New Phytol. 37, 425-450.

Godwin, H., Tallantire, P.A., 1951. Studies in the Post-Glacial History of British Vegetation: XII. Hockham Mere, Norfolk. J. Ecol. 39, 285-307.

Goetcheus, V.G., Birks, H.H., 2001. Full-glacial upland tundra vegetation preserved under tephra in the Beringia National Park, Seward Peninsula, Alaska. *Quat. Sci. Rev.* 20, 135-147.

Goetz, S.J., Epstein, H.E., Bhatt, U., Jia, G.J., Kaplan, J., Lischke, H., Yu, Q., Bunn, A., Lloyd, A., Alcaraz-Segura, D., Beck, P.S.A., Comiso, J.C., Raynolds, M.K., Walker, D.A., 2011. Recent changes in Arctic vegetation: satellite observations and simulation model predictions. In: Gutman, G., Reissell, A. (Eds.), Eurasian Arctic Land Cover and Land Use in a Changing Climate. Springer, Dordrecht, pp. 9-36.

Gómez-Orellana, L., Ramil-Rego, P., Muñoz Sobrino, C., 2007. The Würm in NW Iberia, a pollen record from Area Longa (Galicia). Quat. Res. 67, 438-452.

Gómez-Ortiz, A., Oliva, M., Palacios, D., Salvador-Franch, F., Vázquez-Selem, L., Salvá-Cataribeu, M., de Andrés, N., 2015. The deglaciation of Sierra Nevada (Spain), synthesis of knowledge and new contributions. Cuadernos de Investigación Geográfica 41, 409-426.

Gómez, A., Lunt, D., 2007. Refugia within refugia: patterns of phylogeographic concordance in the Iberian Peninsula. In: Weiss, S., Ferrand, N. (Eds.), Phylogeography in Southern European Refugia: evolutionary perspectives on the origins and conservation of European biodiversity Kluwer, Dordrecht, pp. 155-188.

Gömöry, D., Paule, L., Brus, R., Tomović, Z., Gračan, J., 1999. Genetic differentiation and phylogeny of beech on the Balkan Peninsula. J. Evol. Biol. 12, 746-754.

Gömöry, D., Longauer, R., Liepelt, S., Ballian, D., Brus, R., Kraigher, H., Parpan, V.I., Parpan, T.V., Paule, L., Stupar, V.I., Ziegenhagen, B., 2004. Variation patterns of mitochondrial DNA of *Abies alba* Mill. in suture zones of postglacial migration in Europe. Acta Soc. Bot. Polon. 73, 203-206.

Gömöry, D., Paule, L., Vyšný, J., 2007. Patterns of allozyme variation in western Eurasian Fagus. Bot. J. Linn. Soc. 154, 165-174.

Gömöry, D., Longauer, R., Paule, L., Krajmerová, D., Schmidtová, J., 2010. Across-species patterns of genetic variation in forest trees of central Europe. Biodivers. Conserv. 19, 2025-2038.

Gonzales, L.M., Grimm, E.C., Williams, J.W., Nordheim, E.V., 2009a. A modern plant-climate research dataset for modelling eastern North American plant taxa. Grana 48, 1-18.

Gonzales, L.M., Williams, J.W., Grimm, E.C., 2009b. Expanded response-surfaces: a new method to reconstruct paleoclimates from fossil pollen assemblages that lack modern analogues. Quat. Sci. Rev. 28, 3315-3332.

González-Martínez, S., Dubreuil, M., Riba, M., Vendramin, G.G., Sebastiani, F., Mayol, M., 2010. Spatial genetic structure of *Taxus baccata* L. in the western Mediterranean Basin: Past and present limits to gene movement over a broad geographic scale. Mol. Phylogen. Evol. 55, 805-815.

González-Sampériz, P., Montes, L., Utrilla, P., 2003. Pollen in hyena coprolites from Gabasa Cave (northern Spain). Rev. Palaeobot. Palynol 126, 7-15.

González-Sampériz, P., Valero-Garcés, B., Moreno, A., Jalut, G., García-Ruiz, J., Martí-Bono, C., Delgado-Huertas, A., Navas, A., Otto, T., Dedoubat, J., 2006. Climate variability in the Spanish Pyrenees during the last 30,000 yr revealed by the El Portalet sequence. Quat. Res. 66, 38-52.

González-Sampériz, P., Leroy, S., Carrión, J.S., Fernández, S., García Antón, M., Gil-García, M., Uzquiano, P., Valero-Garcés, B., Figueiral, I., 2010. Steppes, savannahs, forests and phytodiversity reservoirs during the Pleistocene in the Iberian Peninsula. Rev. Palaeobot. Palynol. 162, 427-457.

González-Sampériz, P., García-Prieto, E.,

Aranbarri, J., Valero-Garcés, B., Moreno, A., Gil-Romera, G., Sevilla-Callejo, M., Santos, L., Morellón, M., Mata Campo, M.P., Andrade, A., Carrión, J.S., 2013. Paleoenvironmental reconstruction of the last glacial cycle in the continental Iberia. Cuadernos de Investigacion Geografica 39, 49-76.

González-Trueba, J., Moreno, R., Martínez de Pisón, E., Serrano, E., 2008. 'Little Ice Age' glaciation and current glaciers in the Iberian Peninsula. Holocene 18, 551-568.

González-Vila, F.J., Polvillo, O., Boski, T., Moura, D., de Andrés, J.R., 2003. Biomarker patterns in a time-resolved Holocene/terminal Pleistocene sedimentary sequence from the Guadiana river estuarine area (SW Portugal/Spain border). Org. Geochem. 34, 1601–1613.

Göppert, H.R., 1836. De floribus in statu fossili commentatio.Nova Acta Akad. Leopold Carol. Nat. Cur 18/1, 547-572.

Göppert, H.R., 1841. Über das Vorkommen von Pollen im fossilen Zustande.N. Jahrb. Mineral. 1841, 338-340.

Göransson, H., 1977. The Flandrian vegetational history of southern Östergötland, Lundqua thesis. Department of quaternary geology [Kvartärgeologiska avd.], Univ. Lund.

Göransson, H., 1986. Man and the forests of nemoral broad-leaved trees during the Stone Age. STRIAE 24, 143 - 152.

Göransson, H., 1989. Dags mosse: Östergötlands förhistoriska kalender. Sven. Bot. Tidskr. 83, 371–407.

Gorham, E., 1953. Some early ideas concerning the nature, origin and development of peat lands. J. Ecol. 41, 257-274.

Gorham, E., Brush, G.S., Graumlich, L.J., Rosenzweig, M.L., Johnson, A.H., 2001. The value of paleoecology as an aid to monitoring ecosystems and landscapes, chiefly with reference to North America. Environ. Rev. 9, 99-126.

Goring, S.J., Lacourse, T., Pellatt, M.G., Mathewes, R.W., 2013. Pollen assemblage richness does not reflect regional plant species richness: a cautionary tale. J. Ecol. 101, 1137-1145.

Gortemaker, R., 1986. A method to identify pollen of some recent and fossil species of *Fagus* L. (Fagaceae). Rev. Palaeobot. Palynol. 47, 263-292.

Goslar, T., Pazdur, A., Pazdur, M.F., Arnold, M., Hajdas, I., 1998. Correlation of radiocarbon and varve chronologies of Lake Gościąż. In: Ralska-Jasiewiczowa, M., Goslar, T., Madeyska, T., Starkel, L. (Eds.), Lake Gościąż, Central Poland. A monographic study. W. Szafer Institute of Botany, Polish Academy of Sciences, Krakow, pp. 100-104.

Goslar, T., Arnold, M., Tisnerat-Laborde, N., Hatté, C., Paterne, M., Ralska-Jasiewiczowa, M., 2000. Radiocarbon calibration by means of varves versus 14C ages of terrestrial macrofossils from Lake Gościąż and Lake Perespilno, Poland. Radiocarbon 42, 335-348.

Göttlich, K., 1957. Über interglaziale, spät- und postglaziale Funde von *Isoëtes tenella*, *Ephedra* und *Armeria* in Oberschwaben. Ber. Deutsche Bot. Ges. 70, 139-144.

Göttlich, K., 1990. Moor- und Torfkunde. E. Schweizerbart'sche Verlagsbuchhandlung (Nägele und Obermiller), Stuttgart, p. 529.

Gould, S.J., 1985. The paradox of the first tier: An agenda for paleobiology. Paleobiology 11, 2-12.

Gould, S.J., 2002. The Structure of Evolutionary Theory. Harvard University Press, Cambridge, MA. 1464 pp.

Gradmann, R., 1933. Die Steppenheidetheorie. Geographische Zeitschrift 39, 265–278.

Graebner, P., 1912. Die Entwicklung der deutschen Flora. Leipzig.

Granlund, E., 1932. De svenska högmossarnas geologi. Sver. Geol. Unders. Ser. C. 373, 1-193.

Granoszewski, W., 2003. Late Pleistocene vegetation history and climatic changes at Horoszki Duze, Eastern Poland: a palaeobotanical study. Acta Palaeobot. Suppl. 4, 1-95.

Grant, M.J., Hughes, P.D.M., Barber, K.E., 2009. Early to mid-Holocene vegetation-fire interactions and responses to climatic change at Cranes Moor, New Forest, in: Briant, R.M., Bates, M.R., Hosfield, R.T., Wenban-Smith, F.F. (Eds.), The Quaternary of the Solent Basin and West Sussex Raised Beaches. Quaternary Research Association, pp. 198–214.

Grant, M.J., Hughes, P.D.M., Barber, K.E., 2014. Climatic influence upon early to mid-Holocene fire regimes within temperate woodlands: a multi-proxy reconstruction from the New Forest, southern England. J. Quat. Sci. 29, 175–188.

Gray, A., 1878. Forest Geography and Archaeology: a lecture delivered before the Harvard University Natural History Society. Am. J. Sci. Arts 16, 183-196.

Green, D.G., 1981. Time-series and post-glacial forest ecology. Quat. Res. 15, 265-277.

Green, D.G., 1982. Fire and stability in the postglacial forests of southwest Nova Scotia. J. Biogeogr. 9, 29-40.

Green, D.G., 1983. The ecological interpretation of fine resolution pollen records. New Phytol. 94, 459-477.

Gregory, P.H. 1973. The Microbiology of the Atmosphere. 2nd ed. Leonard Hill Books.

Greig, J., 1988. Some evidence of the development of grassland plant communities, Bot. Soc. British Isles, Conf. Report, pp. 39-52.

Greig, J., 1996. Great Britain - England. In: Berglund, B.E., Birks, H.J.B., Ralska-Jasiewiczowa, M., Wright, H.E. (Eds.), Palaeoecological Events During the Last 15 000 Years. Regional Syntheses of Palaeoecological Studies of Lakes and Mires in Europe. Wiley, Chichester, pp. 15-76.

Greig, J., Turner, J., 1974. Some pollen diagrams from Greece and their archaeological significance. J. Archaeol. Sci. 1, 177-194.

Greiser, C., Joosten, H., 2018. Archive value: measuring the palaeo-information content of peatlands in a conservation and compensation perspective. Int. J. Biodivers. Sci. Ecosyst. Serv. Managem. 14, 210-221.

Gremmen W., Hanss, C., Puisseguir, J.J., 1984. Die warmzeitlichen Ablagerungen am R[au] de l'Amourette (Trièves, französische Alpen). Eiszeitalter u. Gegenwart 34,87-103.

Grichuk, V.P., 1964. Comparative study on the interglacial and interstadial flora of the Russian plain. Report of the VIth International Congress on the Quaternary, Warsaw 1961, 2, 395-406.

Grichuk, V.P., 1984. Late Pleistocene Vegetation History. In: Velichko, A. A., (ed.): Late Quaternary Environments of the Soviet Union, 155-179, Minneapolis.

Grimm, E.C., 1983. Chronology and dynamics of vegetation change in the prairie woodland region of southern Minnesota, U.S.A. New Phytol. 93, 311-350.

Grimm, E.C., 1984. Fire and other factors controlling the Big Woods vegetation of Minnesota in the mid-19th century. Ecol. Monogr. 54, 291-311.

Grimm, E.C., 1993. TILIA, 2.0. Illinois State Museum, Springfield Illinois.

Grimm, E.C., 2001. Trends and palaeoecological problems in the vegetation and climate history of the northern Great Plains, USA. Biol. Environ.: Proc. R. Irish Acad. 101B, 47-64.

Grimm, E.C., Keltner, J., Cheddadi, R., Hicks, S., Lézine, A.M., Berrio, J.C., Williams, J.W., 2007. Pollen Methods and Studies: Databases and their Application. In: Elias, S.A. (Ed.), Encyclopedia of Quaternary Science. Elsevier, Amsterdam, pp. 2521-2528.

Grimm, E.C., Maher, L.J., Jr., Nelson, D.M., 2009. The magnitude of error in conventional bulk-sediment radiocarbon dates from central North America. Quat. Res. 72, 301-308.

Grimm, E.C., Donovan, J.J., Brown, K.J., 2011. A high-resolution record of climate variability and landscape response from Kettle Lake, northern Great Plains, North America. Quat. Sci. Rev. 30, 2626-2650.

Grimm, E.C., Bradshaw, R.H.W., Brewer, S., Flantua, S., Giesecke, T., Lézine, A.-M., Takahara, H., Williams, J.W., 2013. Databases and their application. In: Elias, S.A., Mock, C.J. (Eds.), Encyclopedia of Quaternary Science, 2nd ed. Elsevier, Amsterdam, pp. 831-838.

Grivet, D., Petit, R.J., 2002. Phylogeography of the common ivy (*Hedera* sp.) in Europe: genetic differentiation through space and time. Mol. Ecol. 11, 1351-1362.

Grivet, D., Petit, R.J., 2003. Chloroplast DNA phylogeography of the hornbeam in Europe: Evidence for a bottleneck at the outset of postglacial colonization. Conserv. Genet. 4, 47-56.

Groenman-van Waateringe, W., 1983. The early agricultural utilization of the Irish landscape: The last word on the elm decline? B.A.R., Oxford 116, 217-232.

Groenman-van Waateringe, W., 1992. Palynology and archaeology: the history of a plaggen soil from the Veluwe, The Netherlands. Rev. Palaeobot. Palynol. 73, 87-98.

Groenman-van Waateringe, W., 1998. Pollen in animal coats and bird feathers. Rev. Palaeobot. Palynol. 103, 11-16.

Gros-Louis, M., Bousquet, J., Pâques, E., Isabel, N., 2005. Species-diagnostic markers in *Larix* spp. based on RAPDs and nuclear, cpDNA, and mtDNA gene sequences, and their phylogenetic implications. Tree Genet. Genomes 1, 50-63.

Grosjean, M., Amann, B., Butz, C., Rein, B., Tylmann, W., 2014. Hyperspectral imaging: a novel, nondestructive method for investigating subannual sediment structures and composition. Past Glob. Chang. 24, 10-11.

Gross, H., 1935. Der Döhlauer Wald in Ostpreußen. Beih. Botan. Centralbl. 53B, 405-431.

Grosse-Brauckmann, G., 1972. Über pflanzliche Makrofossilien mitteleuropäischer Torfe. I: Gewebereste krautiger Pflanzen und ihre Merkmale. Telma 2, 19-55.

Grosse-Brauckmann, G., 1974. Über pflanzliche Makrofossilien mitteleuropäischer Torfe. II: Weitere Reste (Früchte und Samen, Moose u. a.) und ihre Bestimmungsmöglichkeiten. Telma 4, 51-117.

Grosvenor, M.J., Jones, R.T., Turney, C.S.M., Charman, D.J., Hogg, A., Coward, D., Wilson, R., 2017. Human activity was a major driver of the mid-Holocene vegetation change in southern Cumbria: implications for the elm decline in the British Isles. J. Quat. Sci. 32, 934-945.

Groves, J., Waller, M., Grant, M., Schofield, J., 2012. Long-term development of a cultural landscape: the origins and development of lowland heathland in southern England. Veget. Hist. Archaeobot. 21, 453-470.

Grubb, P.J., 1986. Problems posed by sparse and patchily distributed species in species-rich plant communities. In: Diamond, J., Case, T.J. (Eds.), Community Ecology. Harper & Row, New York, pp. 207-225.

Grubb, P.J., 1987. Global trends in species-richness in terrestrial vegetation: a view from the Northern

Hemisphere. In: Gee, J.H.R., Giller, P.S. (Eds.), Organization of Communities, Past and Present. Blackwell Scientific Publications, Oxford, pp. 99-118.

Grüger, E., 1967. Vegetationsgeschichtliche Untersuchungen an cromerzeitlichen Ablagerungen im nördlichen Randgebiet der deutschen Mittelgebirge. Eiszeitalter und Gegenwart 18, 204-235.

Grüger, J., 1968. Untersuchungen zur spätglazialen und frühpostglazialen Vegetationsentwicklung der Südalpen im Umkreis des Gardasees. Bot. Jahrb. 88, 163-199.

Grüger, E., 1977. Pollenanalytische Untersuchung zur würmzeitlichen Vegetationsgeschichte von Kalabrien (Süditalien). Flora 166, 475-489.

Grüger, E., 1979. Spätriss, Riss/Würm und Frühwürm am Samerberg in Oberbayern – ein vegetationsgeschichtlicher Beitrag zur Gliederung des Jungpleistozäns. Geol. Bavarica 80, 5-64.

Grüger, E., 1983. Untersuchungen zur Gliederung und Vegetationsgeschichte des Mittelpleistozäns am Samerberg in Oberbayern. Geologica Bavarica 84, 21-40.

Grüger, E., 1991. Late Quaternary biostratigraphy in northern Germany 150 000–15 000 years B.P. Striae 34, 7-14.

Grüger, E., 1996. Vegetation change, in: Chapman, J., Shiel, R., Batovic, S. (Eds.), The Changing Face of Dalmatia. Archaeological and Ecological Investigations in a Mediterranean Landscape. Leicester University Press, Leicester, UK.

Grüger, E., Schreiner, A., 1993. Riß/Würm- und würmzeitliche Ablagerungen im Wurzacher Becken (Rheingletschergebiet). Neues Jahrb. Geol. Paläontol. Abh. 189, 81-117.

Grüger, E., Thulin, B., 1998. First results of biostratigraphical investigations of Lago d'Averno near Naples relating to the period 800 BC-800 AD. Quat. Int. 47, 35-40.

Gubler, M., Henne, P.D., Schwörer, C., Boltshauser-Kaltenrieder, P., Lotter, A.F., Brönnimann, S., Tinner, W., 2018. Microclimatic gradients provide evidence for a glacial refugium for temperate trees in a sheltered hilly landscape of Northern Italy. J. Biogeogr. 45, 2564-2575.

Güemes, J., Sánchez-Gómez, P., 2015. *Pistacia* L. In Muñoz, F., Navarro, C., Quintanar, A., Buira, A. (eds.), Flora Iberica vol. 9, Real Jardín Botánico de Madrid-CSIC, Madrid, Spain, pp. 99-106.

Gugerli, F., Holderegger, R., 2001. Nunatak survival, tabula rasa and the influence of the Pleistocene ice-ages on plant evolution in mountain areas. Trends Plant Sci. 6, 397-398.

Gugerli, F., Sperisen, C., Büchler, U., Magni, F., Geburek, T., Jeandroz, S., Senn, J., 2001a. Haplotype variation in a mitochondrial tandem repeat of Norway spruce (*Picea abies*) populations suggests a serious founder effect during postglacial re-colonization of the western Alps. Mol. Ecol. 10, 1255-1263.

Gugerli, F., Senn, J., Anzidei, M., Madaghiele, A., Büchler, U., Sperisen, C., Vendramin, G.G., 2001b. Chloroplast microsatellites and mitochondrial nad1 intron2 sequences indicate congruent phylogenetic relationships among Swiss stone pine (*Pinus cembra*), Siberian stone pine (*Pinus sibirica*), and Siberian dwarf pine (*Pinus pumila*). Mol. Ecol. 10, 1489-1497.

Gugerli, F., Rüegg, M., Vendramin, G.G., 2009. Gradual decline in genetic diversity in Swiss stone pine populations (*Pinus cembra*) across Switzerland suggests postglacial re-colonization into the Alps from a common eastern glacial refugium. Bot. Helv. 119, 13-22.

Gugerli, F., Alvarez, N., Tinner, W., 2013. A deep dig – hindsight on Holocene vegetation composition from ancient environmental DNA. Mol. Ecol. 22 3433-3436.

Guggenheim, R., 1975. Rasterelektronenmikroskopische und morphometrische Untersuchungen an *Tilia* Pollen. Ein Beitrag zur Artunterscheidung von *Tilia platyphyllos* Scop. und *Tilia cordata* Mill. in der Palynologie. Flora 164, 287-338.

Guiot, J., Harrison, S.P., Prentice, I.C., 1993. Reconstruction of Holocene precipitation patterns in Europe using pollen and lake-level data. Quat. Res. 40, 139-149.

Guiter, F., Andrieu-Ponel, V., Digerfeldt, G., Reille, M., de Beaulieu, J.-L., Ponel, P., 2005. Vegetation history and lake-level changes from the Younger Dryas to the present in Eastern Pyrenees (France): pollen, plant macrofossils and lithostratigraphy from Lake Racou (2000 m asl). Veget. Hist. Archaeobot. 14, 99-118.

Gunova, V.S., 1972. Palynological characteristic of Holocene Nero lake sediments. Vestn Mosk Univ Ser 5 Geogr 3, 117–119.

Gunova, V.S., 1975. Istoriya ozera Nero po palinologicheskim dannym [The history of Lake Nero by the pollen record] (MSc thesis). Moscow State University, Moscow, Russia.

Gunova, V.S., Sirin, A.R., 1995. Paleogeographical condition of bog development in Zapadnodvinskaya lowland during Holocene, in: Palynology in Russia. Moscow, Russia, pp. 27–36.

Guyan, W., 1955. Das jungsteinzeitliche Moordorf von Thayingen-Weier. Das Phaflbauproblem. pp 223-272.

Haas, J.N., Richoz, I., Tinner, W., Wick, L., 1998. Synchronous Holocene climatic oscillations recorded on the Swiss Plateau and at timberline in the Alps. Holocene 8, 301-309.

Hadač, E., 1971. The vegetation of springs, lakes, and "flags" of Reykjanes Peninsula, S.W. Iceland. Plant Communities of Reykjanes Peninsula, Part 3. Folia Geobot. Phytotax. 6, 29 - 41.

Haesaerts, P., Damblon, F., Bachner, M., Trnka, G., 1996. Revised stratigraphy and chronology of the Willendorf II sequence, Lower Austria. Archaeol. Austriaca 80, 25-42.

Haeupler, H., Schönfelder, P. (eds.), 1988. Atlas der Farn- und Blütenpflanzen der Bundesrepublik Deutschland. Ulmer, Stuttgart. 768 pp.

Hafsten, U., 1953. Nyoppdagede pionérplanter i Norge etter istiden. Naturen 77, 501.

Hafsten, U., 1956. Pollen-analytic investigations on the late Quaternary development in the inner Oslofjord area. Univ. Bergen Aarbok 1956, Nat. R. 8, 1-161.

Hafsten, U., 1958. Finds of subfossil *Koenigia islandica* from Scandinavia. Bot. Notiser 111, 333.

Hafsten, U., 1961. Pleistocene development of vegetation and climate in Tristan da Cunha and Gough Island. Univ. Bergen Årbok, Mat. nat. 20, 1-48.

Hafsten, U., 1963. A Late-glacial pollen profile from Lista, South Norway. Grana Palynol. 4, 326.

Hafsten, U., 1966. Den senkvartaere forekomst av tindved (*Hippophae rhamnoides* L.) i Sör-Norge. The late glacial occurrence of *Hippophae rhamnoides* L. in south Norway. Blyttia 24, 196-215.

Hågvar, S., Ohlson, M., 2013. Ancient carbon from a melting glacier gives high 14C age in living pioneer invertebrates. Scientific Reports 3, 2820.

Hahne, J., 1992. Untersuchungen zur spät- und postglazialen Vegetationsgeschichte im nordöstlichen Bayern (Bayerisches Vogtland, Fichtelgebirge, Steinwald). Flora 187, 169-200.

Hahne, J., Kemle, S., Merkt, J., Meyer, K.-D., 1994. Eem-, weichsel- und saalezeitliche Ablagerungen der Bohrung „Quakenbrück GE 2". Geolog. Jahrbuch A 134, 9-69.

Hahne, J., Ellwanger, D., Stritzke, R., 2008. Evidence for a Waalian thermomer pollen record from the research borehole Heidelberg UniNord, Upper Rhine Graben, Baden-Württemberg. Eiszeitalter u. Gegenwart 57, 403-410.

Haines-Young, R., Potschin, M., 2013. Common international classification of ecosystem services. Report prepared following consultation on CICES Version 4. August-September 2012. EEA Framework Contract No EEA/IEA/09/003. European Environment Agency, London.

Hajkova, P., Jamrichova, E., Petr, L., Dudova, L., Rolecek, J., Galova, A., Dresler, P., Novak, J., Hajek, M., 2018. Persistence of a vegetation mosaic in a peripheral region: could turbulent medieval history disrupt Holocene continuity of extremely species-rich grasslands? Veget. Hist. Archaeobot. 27, 591-610.

Hall, A.R., 1980. Late Pleistocene deposits at Wing, Rutland. Phil. Trans. Royal Soc. London B, Biol. Sci. 289, 135-164.

Hamilton, H., 1997. Slash-and-burn in the history of the Swedish forests. Rural Develop. For. Network Paper 21, 19-24.

Hammarlund, D., Aravena, R., Barnekow, L., Buchardt, B., Possnert, G., 1997. Multi-component carbon isotope evidence of early Holocene environmental change and carbon-flow pathways from a hard-water lake in northern Sweden. J. Paleolimnol. 18, 219-233.

Hammer, Ø., Harper, D.A.T., Ryan, P.D., 2001. PAST: Paleontological statistics software package for education and data analysis. Palaeo. Elec. 4, 9 pp.

Hampe, A., Arroyo, J., Jordano, P., Petit, R., 2003. Rangewide phylogeography of a bird-dispersed Eurasian shrub: contrasting Mediterranean and temperate glacial refugia. Mol. Ecol. 12, 3415-3426.

Hannon, G.E., 1985. Late Quaternary vegetation of Sanabria Marsh, North West Spain (MSc thesis). Trinity College, Dublin, Ireland.

Hannon, G.E., Bradshaw, R.H.W., 2000. Impacts and timing of the first human settlement on vegetation of the Faroe Islands. Quat. Res. 54, 404-413.

Hannon, G.E., Gaillard, M.-J., 1997. The plant-macrofossil record of past lake-level changes. J. Paleolimnol. 18, 15-28.

Hannon, G.E., Bradshaw, R.H.W., Wastergaard, S., 2003. Rapid vegetation change during the early Holocene in the Faroe Islands detected in terrestrial and aquatic ecosystems. J. Quat. Sci. 18, 615-619.

Hannon, G.E., Bradshaw, R.H.W., Bradshaw, E., Snowball, I., Wastegård, S., 2005. Climate change and human settlement as drivers of late-Holocene vegetational change in the Faroe Islands. Holocene 15, 639-647.

Hänsel, B., 1998. Mensch und Umwelt in der Bronzezeit Europas. Oetker-Voges, Kiel, 576 pp.

Harff, J., Furmanczyk, K., von Storch, H., 2017. Coastline Changes of the Baltic Sea from South to East. Springer, Cham.

Harkness, D., Walker, M., 1991. The Devensian Lateglacial carbon isotope record from Llanilid, South Wales. Presented at the Quaternary Proceedings, pp. 35-43.

Harley, J.L., Harley, E., 1987a. A check-list of mycorrhiza in the British flora—addenda, errata and index. New Phytol. 107, 741-749.

Harley, J.L., Harley, E., 1987b. A check-list of mycorrhiza in the British flora. New Phytol. 105, Suppl, 1-102.

Harmata, K., 1987. Late-Glacial and Holocene history of vegetation at Roztoki and Tarnowiec near Jaslo (Jaslo-Sanok Depression). Acta Palaeobot. 27, 43-65.

Harmata, K., 1995. A Late-Glacial and Early Holocene profile from Jaslo and recapitulation of the studies on the vegetational history of the Jaslo-Sanok Depression in the last 13 000 years. Acta Palaeobot. 35, 15-46.

Harrington, G.J., 1995. Anthropogenic impact upon Eastern Hungarian landscape. MSc thesis. University of Cambridge, Cambridge.

Harrison, S.P., Digerfeldt, G., 1993. European lakes as palaeohydrological and palaeoclimatic indicators. Quat. Sci. Rev. 12, 233-248.

Harrison, S.P., Saarse, L., Digerfeldt, G., 1991. Holocene changes in lake levels as climate proxy data in Europe. In: Frenzel, B., Pons, A., Gläser, B. (Eds.), Evaluation of Climate Proxy Data in Relation to the European Holocene. ESF, Strasbourg, pp. 159-169.

Harrison, S.P., Prentice, I.C., Guiot, J., 1993. Climatic controls on Holocene lake-level changes in Europe. Clim. Dyn. 8, 189-200.

Hartz, N., Milthers, V., 1901. Det seniglaciale ler i Allerød Tæglværsgrav. Meddelelser fra Danks Geologisk Forening 8, 31-60.

Haskell, J., 2001. The latitudinal gradient of diversity through the Holocene as recorded by fossil pollen in Europe. Evol. Ecol. Res. 3, 345-360.

Hassler-Schwarz, J., 1999. Die Eibe (*Taxus baccata* L.). Eine Beschreibung unter besonderer Berücksichtigung des Kantons Graubünden., Haldenstein.

Hatté, C., Jull, A.J.T., 2007. Radiocarbon Dating: Plant Macrofossils. In: Elias, S.A. (Ed.), Encyclopedia of Quaternary Science. Elsevier, Amsterdam, pp. 2958-2965.

Hatziskakis, S., Papageorgiou, A., Gailing, O., Finkeldey, R., 2009. High chloroplast haplotype diversity in Greek populations of beech (*Fagus sylvatica* L.). Plant Biol. 11 425-433.

Hazler, K., Comps, B., Šugar, I., Melovski, L., Tashev, A., Gračan, J., 1997. Genetic structure of *Fagus sylvatica* L. populations in Southeastern Europe. Silvae Genetica 46, 229 - 236.

Hazzeldine Warren, F., 1922. The Mesvinian industry of Clacton-on-Sea, Essex. Proc. Prehist. Soc. East Anglia 3, 597-602.

He, F., Vavrus, S.J., Kutzbach, J.E., Ruddiman, W.F., Kaplan, J.O., Krumhardt, K.M., 2014. Simulating global and local surface temperature changes due to Holocene anthropogenic land cover change. Geophys. Res. Lett. 41, 623-631.

He, H.S., Gustafson, E.J., Lischke, H., 2017. Modeling forest landscapes in a changing climate: theory and application. Landsc. Ecol. 32, 1299-1305.

Hedberg, O., 1946. Pollen morphology in the genus *Polygonum* L. s.lat. and its taxonomical significance. Svensk Bot. Tidskrift 40, 371-404.

Hedberg, H.D., 1972. Summary of an international guide to stratigraphic cassification, terminology anduUsage. Boreas 1, 213-239.

Hedberg, H.D., 1976. International Stratigraphic Guide. John Wiley, New York.

Hedberg, O., 1997. The genus *Koenigia* L. emend. Hedberg (Polygonaceae). Bot. J. Linn. Soc. 124, 295-330.

Heeb, K., Welten, M., 1972. Moore und Vegetationsgeschichte der Schwarzenegg und des Molassevorlandes zwischen dem Aaretal unterhalb Thun und dem obern Emmental. Mitteilungen der Naturforschenden Gesellschaft in Bern NF 29, 1-54.

Heegaard, E., Birks, H.J.B., Telford, R.J., 2005. Relationships between calibrated ages and depth in stratigraphical sequences: an estimation procedure by mixed-effect regression. Holocene 15, 612-618.

Heer, O., 1858. Die Schieferkohlen von Utznach und Dürnten. Zürich.

Heer, O., 1865a. Die Urwelt der Schweiz. Zürich.

Heer, O., 1865b. Die Pflanzen der Pfahlbauten. Neujahrsblatt der Naturforschenden Gesellschaft in Zürich auf das Jahr 1866, 68, 1-54.

Heer, O., 1872. Über den Flachs und die Flachskultur im Altertum. Neujahrsblatt der Naturforschenden Gesellschaft in Zürich74, 26ff.

Heggen, M.P., Birks, H.H., Anderson, N.J., 2010. Long-term ecosystem dynamics of a small lake and its catchment in west Greenland. *Holocene* 20, 1207-1222.

Heijnis, H., Plicht, J. van der, 1992. Uranium/Thorium dating of Late Pleistocene peat deposits in MW Europe, uranium/thorium isotope systematics and open-system behaviour of peat layers. Chemical Geology, Isotope Geoscience Section 94, 161-171.

Heikkilä, M., Seppä, H., 2003. A 11,000 yr palaeotemperature reconstruction from the southern boreal zone in Finland. Quat. Sci. Rev. 22, 541-554.

Heikkilä, M., Seppä, H., 2010. Holocene climate dynamics in Latvia, eastern Baltic region: a pollen-based summer temperature reconstruction and regional comparison. Boreas 39, 705-719.

Heikkilä, M., Fontana, S.L., Seppä, H., 2009. Rapid Lateglacial tree population dynamics and ecosystem changes in the eastern Baltic region J. Quat. Sci. 24, 802-815.

Heim, J., 1970. Les relations entre les spectres polliniques récents et la végétation actuelle en Europe occidentale. Louvain, Louvain-la-Neuve.

Heiniger, U. (Ed.), 2001. *Castanea sativa*: pathology, genetic resources, ecology and silviculture. For. Snow Landsc. Res. 76, 1-190.

Heino, R., 1978. Climatic changes in Finland during the last hundred years. Fennia 150.

Heinsalu, A., Veski, S., 1991. History of Lake Ermistu (In Estonian), in: Inimene Ja Geografiline Kesk. Vaba, Tallinn, Estonia, pp. 14–18.

Heinz, T., Rein, B., Negendank, J.F.W., 1993. Sediments and basin analysis of Lake Schalkenmehrener Maar. Lecture Notes in Earth Sciences 49, 149 -161.

Heiri, O., Lotter, A.F., Hausmann, S., Kienast, F., 2003. A chironomid-based Holocene summer air temperature reconstruction from the Swiss Alps. Holocene 13, 477-484.

Heiri, O., Tinner, W., Lotter, A.F., 2004. Evidence for cooler European summers during periods of changing meltwater flux to the North Atlantic. Proc. Nat. Acad. Sci. USA, 101, 15285-15288.

Heiri, C., Bugmann, H., Tinner, W., Heiri, O., Lischke, H., 2006. A model-based reconstruction of Holocene treeline dynamics in the central Swiss Alps. J. Ecol. 94, 206-216.

Heiri, O., Koinig, K.A., Spotl, C., Barrett, S., Brauer, A., Drescher-Schneider, R., Gaar, D., Ivy-Ochs, S., Kerschner, H., Luetscher, M., Moran, A., Nicolussi, K., Preusser, F., Schmidt, R., Schoeneich, P., Schwörer, C., Sprafke, T., Terhorst, B., Tinner, W., 2014. Palaeoclimate records 60–8 ka in the Austrian and Swiss Alps and their forelands. Quat. Sci. Rev. 106, 186-205.

Heiri, O., Ilyashuk, B., Millet, L., Samartin, S., Lotter, A.F., 2015. Stacking of discontinuous regional palaeoclimate records: Chironomid-based summer temperatures from the Alpine region. Holocene 25, 137-149.

Hejcman, M., Hejcmanova, P., Pavlu, V., Benes, J., 2013. Origin and history of grasslands in Central Europe - a review. Grass Forage Sci. 68, 345-363.

Helbæk, H., 1950. Tollunds maandens sidste maltid. Aarböger for Nordisk Oldkyndighed og Historie, 311-341.

Helbæk, H., 1952. Early crops in southern England. Proc. Prehist. Soc. 18, 194-233.

Helbæk, H., 1959. Die Paläoethnobotanik des Nahen Ostens und Europas. Opuscula Memorial Ludovici Biro Sacra, 265-289.

Helle, M., Sønstegaard, E., Coope, G.R., Rye, N. 1981. Early Weichselian peat at Brumunddal, southeastern Norway. Boreas 10, 369-379.

Hellegers, M., Ozinga, W.A., Hinsberg van, A., Huijbregts, M.A.J., Hennekens, S.M., Schaminée, J.H.J., Dengler, J., Schipper, A.M., 2019. Evaluating the ecological realism of plant species distribution models with ecological indicator values. Ecography 43, 161-170.

Hellman, S., Gaillard, M.-J., Broström, A., Sugita, S., 2008a. The REVEALS model, a new tool to estimate past regional plant abundance from pollen data in large lakes - validation in southern Sweden. J. Quat. Sci. 23, 21-42.

Hellman, S., Gaillard, M.-J., Broström, A., Sugita, S., 2008b. Effects of the sampling design and selection of parameter values on pollen-based quantitative reconstructions of regional vegetation: a case study in southern Sweden using the REVEALS model. Veget. Hist. Archaeobot. 17, 445-459.

Hellman, S., Gaillard, M.-J., Bunting, J.M., Mazier, F., 2009. Estimating the relevant source area of pollen in the past cultural landscapes of southern Sweden – a forward modelling approach. Rev. Palaeobot. Palynol. 153: 259–271.

Helmens, K.F., 2014. The Last Interglacial-Glacial cycle (MIS 5-2) re-examined based on long proxy records from central and northern Europe. Quat. Sci. Rev. 86, 115-143.

Helmens, K.F., Räsänen, M. E., Johansson, P. W., Jungner, H., Korjonen, K., 2000. The Last Interglacial-Glacial cycle in NE Fennoscandia: a nearly continuous record from Sokli (Finnish Lapland). Quat. Sci. Rev. 19, 1605-1623.

Henne, P.D., Elkin, C.M., Reineking, B., Bugmann, H., Tinner, W., 2011. Did soil development limit spruce (*Picea abies*) expansion in the central Alps during the Holocene? Testing a palaeobotanical hypothesis with a dynamic landscape model. J. Biogeogr. 38, 933-949.

Henne, P.D., Elkin, C.M., Colombaroli, D., Samartin, S., Bugmann, H., Heiri, O., Tinner, W., 2013. Impacts of changing climate and land use on vegetation dynamics in a Mediterranean ecosystem: Insights from paleoecology and dynamic modeling. Landsc. Ecol. 28, 819-833.

Henne, P.D., Elkin, C.M., Franke, J., Colombaroli, D., Calò, C., La Mantia, T., Pasta, S., Conedera, M., Dermody, O., Tinner, W., 2015. Reviving extinct Mediterranean forests increases ecosystem potential in a warmer future. Front. Ecol. Environ. 13, 356-362.

Henriksen, M., Mangerud, J., Matiouchkov, A., Murray, A. S., Paus, A., Svendsen, J. I., 2008. Intriguing climatic shifts in a 90 kyr old lake record from northern Russia. Boreas 37, 20-37.

Henrion, I., 1982. Untersuchungen zu den Sattelmooren im Oberharz. Universiy of Göttingen, Göttingen, p. 167.

Herben, T., Chytrý, M., 2003. Calcium and plant species richness. Folia Geobot. 38, 355.

Herben, T., Chytrý, M., Klimešová, J., 2016. A quest for species-level indicator values for disturbance. J. Veg. Sci. 27, 628-636.

Hering, D., Carvalho, L., Argillier, C., Beklioglu, M., Borja, A., Cardoso, A., Duel, H., Ferreira, T., Globevnik, L., Hanganu, J., Hellsten, S., Jeppesen, E., Kodeš, V., Solheim, A., Nõges, T., Ormerod, S., Panagopoulos, Y., Schmutz, S., Venohr, M., Birk, S., 2015. Managing aquatic ecosystems and water resources under multiple stress — An introduction to the MARS project. Sci. Total Environ. 503-504, 10-21.

Hermsdorf, N., Strahl, J., 2008. Karte der Eem-Vorkommen des Landes Brandenburg. Brandenburger Geowiss. Beiträge 15, 23-55 + CD.

Herrera, C.M., 1984. A study of avian frugivores, bird-dispersed plants, and their interaction in Mediterranean scrublands. Ecol. Monogr. 54, 1-23.

Herrera, J., 1988. Pollination relationships in southern Spanish Mediterranean shrublands. J. Ecol. 76, 274-287.

Herrera, C.M., Jordano, P., López-Soria, L., Amat, J.A., 1994. Recruitment of a mast-fruiting, bird-dispersed tree: bridging frugivore activity and seedling establishment. Ecol. Monogr. 64, 315-344.

Herzschuh, U., 2019. Legacy of the Last Glacial on the present-day distribution of deciduous versus evergreen boreal forests. Global Ecol. Biogegor. 29, 198-206.

Herzschuh, U., Birks, H.J.B., Laepple, T., Andreev, A.A., Melles, M., Brigham-Grette, J., 2016. Glacial legacies on interglacial vegetation at the Pliocene-Pleistocene transition in Asia. Nat. Commun. 7, 11967.

Hess, H.E., Landolt, E., Hirzel, R., 1976. Flora der Schweiz und angrenzender Gebiete. Band I: Pteridophyta bis Caryophyllaceae. 2. Aufl. Birkhäuser, Basel/Stuttgart. 858 pp.

Hesse, M.B., 1966. Models and Analogies in Science. Paris

Hesse, M.B., Halbritter, H., Zetter, R., Weber, M., Buchner, R., Frosch-Radivo, A., Ulrich, S., 2009. Pollen Terminology. An illustrated handbook. Springer, Wien.

Heuertz, M., Carnevale, S., Fineschi, S., Sebastiani, F., Hausman, J.F., Paule, L., Vendramin, G.G., 2006a. Chloroplast DNA phylogeography of European ashes, *Fraxinus* sp. (Oleaceae): roles of hybridization and life history traits. Molecular Ecology 15, 2131-2140.

Heuertz, M., De Paoli, E., Källman, T., Larsson, G., Jurman, I., Morgante, M., Martin Lascoux, M., Gyllenstrand, N., 2006b. Multilocus patterns of nucleotide diversity, linkage disequilibrium and demographic history of Norway spruce [*Picea abies* (L.) Karst]. Genetics 174, 2095-2105.

Hicks, S., 1975. Variations in pollen frequency in a bog at Kangerjoki, northeastern Finland during Flandrian. Comment. Biol Soc Sci Fenn.

Hicks, S., 1996. The feasibility of using pollen deposition data as climatic indices. Paläoklimaforschung 20, 173-187.

Hicks, S., 2001. The use of annual arboreal pollen deposition values for delimiting tree-lines in the landscape and exploring models of pollen dispersal. Rev. Palaeobot. Palynology 117, 1-29.

Hicks, S., 2006. When no pollen does not mean no trees. Veget. Hist. Archaeobot. 15, 253-261.

Hicks, S., 2007. Pollen Methods and Studies: Surface Samples and Trapping. In: Elias, S.A. (Ed.), Encyclopedia of Quaternary Science. Elsevier, Amsterdam, pp. 2529-2535.

Hicks, S., Isaksson, E., 2006. Assessing source areas of pollutants from studies of fly ash, charcoal, and pollen from Svalbard snow and ice. J. Geophys. Res.: Atmospheres 111 (D2).

Hicks, S., Helander, M., Heino, S., 1994. Birch pollen production, transport and deposition for the period 1984-1993 at Kevo, northernmost Finland. Aerobiologia 10, 183-191.

Hicks, S., Tinsley, H., Huusko, A., Jensen, C., Hättestrand, M., Gerasimides, A., Kvavadze, E., 2001. Some comments on spatial variation in arboreal pollen deposition: first records from the Pollen Monitoring Programme (PMP). Rev. Palaeobot. Palynol. 117, 183-194.

Hill, M.O., Mountford, J., Roy, D., Bunce, R., 1999. Ellenberg's Indicator Values for British Plants. Natural Environment Research Council, Norwich.

Hill, M.O., Preston, C.D., Roy, D.B., 2004. PLANTATT - Attributes of British and Irish Plants: status, size, life history, geography and habitats. Centre for Ecology & Hydrology, Huntingdon. 73 pp.

Hill, M.O., Preston, C.D., Bosanquet, S.D.S., Roy, D.B., 2007. BRYOATT - Attributes of British and Irish mosses, liverworts and hornworts. Centre for Ecology and Hydrology, Huntingdon. 88 pp.

Hintikka, V., 1963. Ueber das Grossklima einiger Pflanzenareale, in zwei Klima-Koordinatensystemen dargestellt. Ann. Bot. Soc. Zool. Bot. Fenn. Vanamo 34, 1-64.

Hiscott, R., Aksu, A., Mudie, P., Marret, F., Abrajano, T., Kaminski, M., Evans, J., Çakiroğlu, A., Yaşar, D., 2007. A gradual drowning of the southwestern Black Sea shelf: Evidence for a progressive rather than abrupt Holocene reconnection with the eastern Mediterranean Sea through the Marmara Sea Gateway. Quat. Int. 167-168, 19-34.

Hiura, T., 1995. Gap formation and species diversity in Japanese beech forests: a test of the intermediate disturbance hypothesis on a geographic scale. Oecologia 104, 265-271.

Hjelle, K.L., Sugita, S., 2012. Estimating pollen productivity and relevant source area of pollen using lake sediments in Norway: How does lake size variation affect the estimates? Holocene 22, 312-323.

Hjelle, K.L., Mehl, I.K., Sugita, S., Andersen, G.L., 2015. From pollen percentage to vegetation cover – evaluation of the Landscape Reconstruction Algorithm in western Norway. J. Quat. Sci. 30, 312-324.

Hjelle, K.L., Halvorsen, L.S., Prøsch-Danielsen, L., Sugita, S., Paus, A., Kaland, P.E., Mehl, I.K., Overland, A., Danielsen, R., Høeg, H.I., Midtbø, I., 2018.

Long-term changes in regional vegetation cover along the west coast of southern Norway: The importance of human impact. J. Veg. Sci. 29, 404-415.

Hjelmqvist, H., 1955. Die älteste Geschichte der Kulturpflanzen in Schweden. Op. Bot. 1, 1-186.

Hobbs, R.J., Higgs, E.S., Hall, C.M. (Eds.), 2013. Novel Ecosystems: Intervening in the New Ecological World Order. Wiley-Blackwell, Chichester. 368 pp.

Hobbs, R.J., Higgs, E.S., Hall, C.M., Bridgewater, P., Chapin, F.S., Ellis, E.C., Ewel, J.J., Hallett, L.M., Harris, J., Hulvey, K.B., Jackson, S.T., Kennedy, P.L., Kueffer, C., Lach, L., Lantz, T.C., Lugo, A.E., Mascaro, J., Murphy, S.D., Nelson, C.R., Perring, M.P., Richardson, D.M., Seastedt, T.R., Standish, R.J., Starzomski, B.M., Suding, K.N., Tognetti, P.M., Yakob, L., Yung, L., 2014. Managing the whole landscape: historical, hybrid, and novel ecosystems. Front. Ecol. Environ. 12, 557-564.

Hobbs, R.J., Valentine, L.E., Standish, R.J., Jackson, S.T., 2018. Movers and stayers: novel assemblages in changing environments. Trends Ecol. Evol. 33, 116-128.

Hofstetter, S., Tinner, W., Valsecchi, V., Carraro, G., Conedera, M., 2006. Lateglacial and Holocene vegetation history in the Insubrian Southern Alps – New indications from a small-scale site. Veget. Hist. Archaeobot. 15, 87-98.

Höhn, M., Gugerli, F., Abran, P., Bisztray, G., Buonamici, A., Cseke, K., Hufnagl, L., Quintela-Sabaris, C., Sebastiani, F., Vendramin, G.G., 2009. Variation in the chloroplast DNA of Swiss stone pine (*Pinus cembra* L.) reflects contrasting post-glacial history of populations from the Carpathians and the Alps. J. Biogeogr. 36, 1798-1806.

Hollingworth, S., Allison, J., Godwin, H., 1950. Interglacial deposits from Histon Road, Cambridge. Quart. J. Geol. Soc. London 105, 495 pp.

Holmes, J.A., Tindall, J., Roberts, N., Marshall, W., Marshall, J.D., Bingham, A., Feeser, I., O'Connell, M., Atkinson, T., Jourdan, A.L., March, A., Fisher, E.H., 2016. Lake isotope records of the 8200-year cooling event in western Ireland: Comparison with model simulations. Quat. Sci. Rev. 131, 341-349.

Holtmeier, F.K., 1985. Die klimatische Waldgren ze- Linie oder Übergangssaum (Ökoton)? Ein Diskussionsbeitrag unter besonderer Berücksichtigung der Waldgrenzen in den mittleren und hohen Breiten der Nordhalbkugel. Erdkunde 39, 271-285.

Holtmeier, F.K., 1989: Ökologie und Geographie der oberen Waldgrenze. In: Pott, R. (Ed.), Rin- telner Symposium I.-Ber. Reinh.-Tüxen-Ges., Hannover, 1, 15-45.

Holzhauser, H., Magny, M., Zumbuühl, H.J., 2005. Glacier and lake-level variations in west-central Europe over the last 3500 years. Holocene 15, 789-801.

Holzner, W., Immonen, R., 1982. Europe: an overview. In: Holzner, W., Numata, M. (Eds.), Biology and Ecology of Weeds. Junk, The Hague, pp. 202-226.

Hood, E., Battin, T.J., Fellman, J., O'Neel, S., Spencer, R.G.M., 2015. Storage and release of organic carbon from glaciers and ice sheets. Nature Geosci. 8, 91-96.

Hoops, J., 1905. Waldbäume und Kulturpflanzen im germanischen Altertum. Strassburg.

Hopf, M., 1988. Zu den Anfängen des Ackerbaues im Vorderen Orient. In: Hrouda, B. (Ed.), Methoden der Archäologie. Munchen, pp. 280-297.

Hopkins, J., 1950. Differential flotation and deposition of coniferous and deciduous tree pollen. Ecology 31, 633-641.

Hörmann, H., 1929. Die pollenanalytische Unterscheidung von *Pinus montana, P. sylvetsris* und *P. cembra*. Oesterr. Bot.Z. 78, 215-228.

Hörnberg, G., Östlund, O., Zackrisson, O., Bergman, I., 1999. The genesis of two *Picea-Cladina* forests in northern Sweden. J. Ecol. 87, 800-814.

Horvat, I., Glavač, V., Ellenberg, H., 1974. Vegetation Südosteuropas, Stuttgart.

Hošek, J., Pokorný, P., Kubovčík , V., Horáček, I., Žáčková, P., Kadlec, J., Rojik, F., Lisá, L., Bučkuliaková, S., 2014. Late glacial climatic and environmental changes in eastern-central Europe: Correlation of multiple biotic and abiotic proxies from the Lake Švarcenberk, Czech Republic. Palaeogeogr. Palaeoclimatol. Palaeoecol. 396, 155-172.

Houmark-Nielsen, M., 2010. Extent, age and dynamics of Marine Isotopic Stage 3 glaciations in the southern Baltic Basin. Boreas 39, 343-359.

Houmark-Nielsen, M., Bennike, O., Lemdahl, G., Lüthgens, C., 2016. Evidence of ameliorated Weichselian climate and sub-arctic environment in the western Baltic region: coring lake sediments at Klintholm, Møn, Denmark. Boreas 45, 347-359.

Hrynowiecka-Czmielewska, A., 2010. History of vegetation and climate of the Mazovian (Holsteinian) interglacial and the Liviecian (Saalian) glaciation on the basis of pollen analysis of palaeolake sediments from Nowiny Zukowskie, SE Poland. Acta Palaeobot. 50, 17-51.

Hu, F.S, Finney, B.P, Brubaker, L.B, 2001. Effects of Holocene *Alnus* expansion on aquatic productivity, nitrogen cycling, and soil development in southwestern Alaska. Ecosystems 4, 358-368.

Hu, F.S., Hampe, A., Petit, R.J., 2009. Paleoecology meets genetics: deciphering past vegetational dynamics. Front. Ecol. Environ. 7, 371-379.

Hubbell, S.P., 2001. The Unified Neutral Theory of Biodiversity and Biogeography. Princeton University Press, Princeton. 448 pp.

Hughen, K.A., Overpeck, J.T., Peterson, L.C.,

Anderson, R.R., 1996. The nature of varved sedimentation in the Cariaco Basin, Venezuela, and its palaeoclimatic significance. Geological Society 116, 171-183.

Hughen, K.A., Baillie, M.G.L., Bard, E., Bayliss, A., Beck, J.W., Bertrand, C.J.H., Blackwell, P.G., Buck, C.E., Burr, G.S., Cutler, K.B., Damon, P., Edwards, R.L., Fairbanks, R.G., Friedrich, M., Guilderson, T.P., Kromer, B., McCormac, F., Manning, S., Bronk Ramsey, C., Reimer, P.J., Reimer, R.W., Remmele, S., Southon, J.R., Stuiver, M., Talamo, S., Taylor, F.W., van der Plicht, J., Weyhenmeyer, C.E., 2004a. Marine04 marine radiocarbon calibration, 26-0 kyr BP. Radiocarbon 46, 1059-1086.

Hughen, K.A., Southon, J.R., Bertrand, C.J.H., Frantz, B., Zermeno, P., 2004b. Cariaco Basin calibration update: revisions to calendar and ^{14}C chronologies for core PL07-58PC. Radiocarbon 46, 1161-1187.

Hughes, M.K., 2002. Dendrochronology in climatology - the state of the art. Dendrochronologia 20, 95-116.

Hughes, P.D.M., Barber, K.E., 2003. Mire development across the fen-bog transition on the Teifi floodplain at Tregaron Bog, Ceredigion, Wales, and a comparison with 13 other raised bogs. J. Ecol. 91, 253-264.

Hughes, P.D.M., Mauquoy, D., Barber, K.E., Langdon, P.G., 2000. Mire-development pathways and palaeoclimatic records from a full Holocene peat archive at Walton Moss, Cumbria, England. Holocene 10, 465-479.

Hughes, P.D., Gibbard, P.L., Ehlers, J., 2013. Timing of glaciation during the last glacial cycle: evaluating the concept of a global 'Last Glacial Maximum' (LGM). Earth Sci. Rev. 125, 171-198.

Huikari, O., 1956. Primaarisen soistumisen osuudesta Suomen soiden synnyssä (Untersuchungen über den Anteil der primären Versumpfung an der Entstehung der finnischen Moore). Communicationes Instituti Forestalis Fenniae 46, 1-79.

Hultén, E., 1964. The Circumpolar Plants. Almqvist & Wiksell, Stockholm.

Hultén, E., 1971. Atlas of the Distribution of Vascular Plants in Northern Europe. 2nd ed. Stockholm.

Huntley, B., 1988. Glacial and Holocene vegetation history - Europe. In: Huntley, B., Webb, T.I. (Eds.), Vegetation History. Kluwer, Dordrecht, pp. 341-383.

Huntley, B., 1990a. Dissimilarity mapping between fossil and contemporary pollen spectra in Europe for the past 13,000 years. Quat. Res. 33, 360-376.

Huntley, B., 1990b. European post-glacial forests: compositional changes in response to climatic change. J. Veg. Sci. 1, 507-518.

Huntley, B., 1993. Rapid early-Holocene migration and high abundance of hazel (*Corylus avellana* L.): alternative hypotheses. In: Chambers, F.M. (Ed.), Climate Change and Human Impact on the Landscape. Chapman & Hall, London, pp. 205-215.

Huntley, B., 1994. Late Devensian and Holocene palaeoecology and palaeoenvironments of the Morrone Birkwoods, Aberdeenshire, Scotland. J. Quat. Sci. 9, 311-336.

Huntley, B., 1996. Quaternary palaeoecology and ecology. Quat. Sci. Rev. 15, 591-606.

Huntley, B., Birks, H.J.B., 1979. The Past and Present Vegetation of the Morrone Birkwoods National Nature Reserve, Scotland: I. A Primary Phytosociological Survey. J. Ecol. 67, 417-446.

Huntley, B., Birks, H.J.B., 1983. An Atlas of Past and Present Pollen Maps for Europe: 0-13000 Years Ago. Cambridge University Press, Cambridge. 667 pp.

Huntley, B., Webb III, T. (Eds.), 1988. Vegetation History. Kluwer Academic Publishers, Dordrecht. 803 pp.

Huntley, B., Webb III, T., 1989. Migration: species' response to climatic variations caused by changes in the earth's orbit. J. Biogeogr. 16, 5-19.

Huntley, B., Bartlein, P.J., Prentice, I.C., 1989. Climatic control of the distribution and abundance of beech (*Fagus* L.) in Europe and North America. J. Biogeogr. 16, 551-560.

Huntley, B., Watts, W.A., Allen, J.R.M., Zolitschka, B., 1999. Palaeoclimate, chronology and vegetation history of the Weichselian Lateglacial: comparative analysis of data from three cores at Lago Grande di Monticchio, southern Italy. Quat. Sci. Rev. 18, 945-960.

Huntley, B., Long, A.J., Allen, J.R.M., 2013. Spatio-temporal patterns in Lateglacial and Holocene vegetation and climate of Finnmark, northernmost Europe. Quat. Sci. Rev. 70, 158-175.

Hüppe, J., Hofmeister, H., 1990. Syntaxonomische Fassung und Übersicht über die Ackerunkrautgesellschaften der Bundesrepublik Deutschland. Berichte R. Tüxen-Gesellschaft 2, 61-81.

Hustich, I., 1947. On variations in climate, in crop of cereals and in growth of pine in northern Finland 1890-1939. Fennia 70.

Hustich, I., 1948. The scotch pine in northernmost Finland and its dependance on the climate in the last decades. Acta Bot. Fenn. 42.

Hustich, I., 1958. On the recent expansion of the scotch pine in northern Europe. Fennia 82.

Huston, M.A., 2014. Disturbance, productivity, and species diversity: empiricism vs. logic in ecological theory. Ecology 95, 2382-2396.

Hutchinson, G.E., 1957. A Treatise of Limnology 1: Geography, Physics and Chemistry. Wiley, Chichester.

Hutchinson, G.E., 1975. A Treatise on Limnology 3: Limnological Botany. Wiley, Chichester.

Hutchinson, G.E., Löffler, H., 1956. The thermal

classification of lakes. Proc. Nat. Acad. Sci. USA 42, 84-86.

Huttunen, P., 1980. Early land use, especially the slash-and-burn cultivation in the commune of Lammi, southern Finland, interpreted mainly using pollen und charcoal analysis. Acta Bot. Fennica 113, 1-45.

Huttunen, A., 1990. Vegetation and palaeoecology of a bog complex in S Finland. Aquilo Ser. Bot. 28, 27-37.

Huttunen, A., Huttunen, R.-L., Vasari, Y., Panovska, H., Bozilova, E., 1992. Late-glacial and Holocene history of flora and vegetation in the Western Rhodopes Mountains, Bulgaria. Acta Bot. Fennica 144, 63-80.

Huusko, A., Hicks, S., 2009. Conifer pollen abundance provides a proxy for summer temperature: evidence from the latitudinal forest limit in Finland. J. Quat. Sci. 24, 522-528.

Hyvärinen, H., 1970. Flandrian pollen diagrams from Svalbard. Geografiska Annaler 52 A.

Hyvärinen, H., 1975a. Absolute and relative pollen diagrams from northernmost Fennoscandia. Fennia 142, 1-23.

Hyvärinen, H., 1975b. Lateglacial palaeogeography of Fennoscandia. Terra 52 A 155-166.

Hyvärinen, H., 1976. Flandrian pollen deposition rates and tree-line history in northern Fennoscandia. Boreas 5, 163-175.

Hyvärinen, H., 1985. Holocene pollen history of the Alta area, an isolated pine forest north of the general pine forest region in Fennoscandia. Ecol. Mediterr. 11, 69-71.

Hyvärinen, H., 1993. Holocene pine and birch limits near Kilpisjärvi, western Finnish Lapland: Pollen stratigraphical evidence. Paläoklimaforschung 9, 19-27.

Hyvärinen, H., Alhonen, P., 1994. Holocene lake-level changes in the Fennoscandian tree-line region, western Finnish Lapland: diatom and cladoceran evidence. Holocene 4, 251-258.

Hyvärinen, H., Mäkelä, E., 1996. Parallel trends of Holocene tree-line changes and lake and mire development in northwestern Finnish Lapland. In: Frenzel, B., Birks, H.H., Alm, T., Vorren, K-D, (Eds.), Holocene treeline oscillations, dendrochronology and palaeoclimate. Gustav Fischer, Stuttgart, pp. 293-300.

Ickert-Bond, S., Wojchiechowski, M.F., 2004. Phylogenetic Relationships in *Ephedra* (Gnetales): Evidence from Nuclear and Chloroplast DNA Sequence Data. Sys. Bot. 29, 834-849.

Iglesias, V., Yospin, G.I., Whitlock, C., 2015. Reconstruction of fire regimes through integrated palaeoecological proxy data and ecological modeling. Front. Plant Sci. 5, art785.

Ilon, G., Juhász, I., Sümegi, P., Jakab, G., Szegvári, G., Törőcsik, T., 2006. Mezőlak-Szélmező tőzegláp geoarcheológiai vizsgálatának eredményei (Effects of the geo-archeological survey on the Mezőlak-Szélmező peatbog). Savaria 29, 147-215.

Innes, J.B., Blackford, J.J., Rowley-Conwy, P.A., 2013. Late Mesolithic and early Neolithic forest disturbance: a high resolution palaeoecological test of human impact hypotheses. Quat. Sci. Rev. 77, 80-100.

IPCC, 2013. Climate Change 2013: The Physical Science Basis. Contribution of Working Group I to the Fifth Assessment Report of the Intergovernmental Panel on Climate Change. Cambridge University Press, Cambridge.

IPCC, 2018. Global warming of 1.5°C. An IPCC Special Report on the impacts of global warming of 1.5°C above pre-industrial levels and related global greenhouse gas emission pathways, in the context of strengthening the global response to the threat of climate change, sustainable development, and efforts to eradicate poverty. IPCC, Geneva.

Iriarte-Chiapusso, M.J., Muñoz Sobrino, C., Gómez-Orellana, L., Hernández-Beloqui, B., García-Moreiras, I., Fernández Rodriguez, C., Heiri, O., Lotter, A.F., Ramil-Rego, P., 2016. Reviewing the Lateglacial–Holocene transition in NW Iberia: A palaeoecological approach based on the comparison between dissimilar regions. Quat. Int. 403, 211-236.

Isarin, R., Bohncke, S., 1999. Mean July temperatures during the Younger Dryas in northwestern and central Europe as inferred from climate indicator plant species. Quat. Res. 51, 158-173.

Isarin, R., Renssen, H., 1999. Reconstructing and modelling Late Weichselian climates: the Younger Dryas in Europe as a case study. Earth-Sci. Rev. 48, 1-38.

Isenberg, E., 1986. Der pollenanalytische Nachweis von *Juglans*. Abhandlungen aus dem Landesmuseum für Naturkunde zu Münster in Westfalen 48, 457-469.

Isturiz, M., Sánchez-Goñi, M.F., 1990. Investigaciones palinológicas en la prehistoria vasca. Munibe Antropologia-Arkeologia 42, 277 - 285.

Iversen, J., 1934. Moorgeologische Untersuchungen auf Grönland. Dansk Geolog. Foren. 8, 341-358.

Iversen, J., 1936. Biologische Pflanzentypen als Hilfsmittel in der Vegetationsforschung. Ein Beitrag zur ökologischenCharakterisierung und Anordnung der Pflanzengesellschaften. thesis, University of Copenhagen.

Iversen, J., 1941. Land occupation in Denmark's Stone Age. A pollen-analytical study of the influence of farmer culture in the vegetational development. Danm. Geol. Unders. II Rke. 66, 1-68.

Iversen, J., 1942. En pollenanalytsik tidsfaestelse af ferskvandslagene ved Nørre Lyngby. Medd. Dan. Geol. Foren. 10, 130-151.

Iversen, J., 1944. *Viscum*, *Hedera* and *Ilex* as

climatic indicators. Geol. Fören. Stockholm Förh. 66, 463-483.

Iversen, J., 1947. *Centaurea cyanus* pollen i Danish Late-Glacial. Dansk. Geol. Fören. Copenhagen , 11, 197-200.

Iversen, J., 1949. The Influence of Prehistoric Man on Vegetation. Danm. Geol. Unders. IV R. 3, 25.

Iversen, J., 1951, Steppeelementer i den senglaciale flora og fauna. Medd. Dansk Geol. Foren. 12.

Iversen, J., 1954. The Late-glacial flora of Denmark and its relation to climate and soil. Danm. Geol. Unders. Rke 2, 80, 87-119.

Iversen, J., 1958. The bearing of glacial and interglacial epochs on the formation and extinction of plant taxa. Uppsala Universitets Årsskrift 6, 210-215.

Iversen, J., 1960. Problems of the early post-glacial forest development in Denmark. Danm. Geol. Unders. R. 4, 1-32.

Iversen, J., 1964. Retrogressive vegetational succession in the post-glacial. J. Ecol. 52, 59-70.

Iversen, J., 1973. The development of Denmark's nature since the Last Glacial. Danm. Geol. Unders. V-7 C, 1-126.

Iversen, J., Olsen, S., 1943. Die Verbreitung der Wasserpflanzen in Relation zur Chemie des Wassers. Botanisk Tidsskrift 46, 136-145.

Iversen, J., Troels-Smith, J., 1950. Pollenmorfologiske definitioner og typer. Danm. Geol. Unders. IV 3, 1-43.

Jabłońska, E., Michaelis, D., Tokarska, M., Goldstein, K., Grygoruk, M., Wilk, M., Wyszomirski, T., Kotowski, W., 2019. Alleviation of plant stress precedes termination of rich fen stages in peat profiles of lowland mires. Ecosystems.

Jackson, S.T., 1997. Documenting natural and human-caused plant invasions using paleoecological methods. In: Luken, J.O., Thieret, J.W. (Eds.), Assessment and Management of Plant Invasions. Springer-Verlag, New York, pp. 37-55.

Jackson, S.T., 2000. Out of the garden and into the cooler? A Quaternary perspective on deep-time paleoecology. The Paleontological Society Papers 6, 287-308.

Jackson, S.T., 2006. Vegetation, environment, and time: the origination and termination of ecosystems. J. Veg. Sci. 17, 549-557.

Jackson, S.T., 2013a. Natural, potential and actual vegetation in North America. J. Veg. Sci. 24, 772-776.

Jackson, S.T., 2013b. Perspective: ecological novelty is not new. In: Hobbs, R.J., Higgs, E.S., Hall, C.M. (Eds.), Novel Ecosystems: Intervening in the New Ecological World Order. Wiley-Blackwell, Chichester, pp. 63-65.

Jackson, S.T., 2016. Reinventing conservation – again. Front. Ecol. Environ. 14, 519-519.

Jackson, S.T., Blois, J.L., 2015. Community ecology in a changing environment: Perspectives from the Quaternary. Proc. Natl. Acad. Sci. USA 112, 4915-1921.

Jackson, S.T., Booth, R.K., 2013. Plant Macrofossils Methods and Studies: Validation of Pollen Studies. In: Elias, S.A. (Ed.), Encyclopedia of Quaternary Science, 2nd ed. Elsevier, Amsterdam, pp. 725-732.

Jackson, S.T., Kearsley, J.B., 1998. Quantitative representation of local forest composition in forest-floor pollen assemblages. J. Ecol. 86, 474-490.

Jackson, S.T., Lyford, M.E., 1999. Pollen dispersal models in Quaternary plant ecology: Assumptions, parameters, and prescriptions. Bot. Rev. 65, 39-75.

Jackson, S.T., Overpeck, J.T., 2000. Responses of plant populations and communities to environmental changes of the late Quaternary. Paleobiology 26 (Supplement), 194-220.

Jackson, S.T., Weng, C.Y., 1999. Late Quaternary extinction of a tree species in eastern North America. Proc. Natl. Acad. Sci. USA 96, 13847-13852.

Jackson, S.T., Williams, J.W., 2004. Modern analogs in Quaternary paleoecology: Here today, gone yesterday, gone tomorrow? Annu. Rev. Earth Planet. Sci. 32, 495-537.

Jackson, S.T., Webb, R.S., Anderson, K.H., Overpeck, J.T., Webb, T., Williams, J.W., Hansen, B.C.S., 2000. Vegetation and environment in eastern North America during the Last Glacial Maximum. Quat. Sci. Rev. 19, 489-508.

Jackson, S.T., Betancourt, J.L., Booth, R.K., Gray, S.T., 2009a. Ecology and the ratchet of events: Climate variability, niche dimensions, and species distributions. Proc. Natl. Acad. Sci. USA 106, 19685-19692.

Jackson, S.T., Gray, S.T., Shuman, B., 2009b. Paleoecology and resource management in a dynamic landscape: Case studies from the Rocky Mountain headwaters. In: Dietl, G.P., Flessa, K.W. (Eds.), Conservation Paleobiology: Using the Past to Manage for the Future. The Paleontological Society Papers, New York, pp. 61-80.

Jackson, S.T., Booth, R.K., Reeves, K., Andersen, J.J., Minckley, T.A., Jones, R.A., 2014. Inferring local to regional changes in forest composition from Holocene macrofossils and pollen of a small lake in central Upper Michigan. Quat. Sci. Rev. 98, 60-73.

Jacob-Friesen, K., 1949. Die Altsteinzeitfunde aus dem Leinetal bei Hannover. Veröffentlichungen der Urgeschichtlichen Sammlungen Landesmuseum Hannover 10, 1-133.

Jacobson, G.L., Bradshaw, R.H.W., 1981. The selection of sites for paleovegetational studies. Quat. Res. 16, 80-96.

Jacomet, S., 1985. Botanische Makroreste aus den

Sedimenten des neolithischen Siedlungsplatzes AKAD-Seehofstrasse am untersten Zürichsee. Die Reste der Uferpflanzen und ihre Aussagemöglichkeiten zu Vegetationsgeschichte, Schichtentstehung und Seespiegelschwankungen. Zürcher Studien zur Archäologie 3.

Jacomet, S., 1986. Zur Morphologie subfossiler Samen und Früchte aus postglazialen See- und Kulturschichtsedimenten der neolithischen Siedlungsplätze „AKAD-Seehofstrasse" und „Pressehaus" am unteren Zürichsee. Bot. Helvetica 96/2, 159-204.

Jacomet, S., 1998. Holozäne Seespiegelschwankungen. In: Burga, C., Perret, R. (Eds.), Vegetation und Klima der Schweiz seit dem jüngeren Eiszeitalter. Ott-verlag, pp. 725-730.

Jacomet, S., 2004. Archaeobotany. A vital tool in the investigation of lake-dwellings. In: Menotti, F. (Ed.), Living on the Lake in Prehistoric Europe. Routledge, Oxon, pp. 162-177.

Jacomet, S., Karg, S., 1996. Ackerbau und Umwelt der Seeufersiedlungen von Zug-Sumpf im Rahmen der mitteleuropäischen Spätbronzezeit. In: Seifert, M. (Ed.), Die spätbronzezeitlichen Ufersiedlungen von Zug-Sumpf. Zug, pp. 198-303.

Jacomet, S., Kreuz, A., 1999. Archäobotanik. UTB, Stuttgart, 368 pp.

Jacomet, S., Brombacher, C., Dick, M., 1989. Archäobotanik am Zürichsee. Ackerbau, Sammelwirtschaft und Umwelt von neolithischen und bronzezeitlichen. Ufersiedlungen im Raum Zürich.

Jacomet, S., Ebersbach, R., Akeret, Ö., Antolin, F., Baum, T., Bogaard, A., Brombacher, C., Bleicher, N., Heitz-Weniger, A., Hüster-Plogmann, A., Gross, E., Kühn, M., Rentzel, P., Steiner, B., Wick, L., Schibler, J., 2016. On-site data cast doubts on the hypothesis of shifting cultivation in the late Neolithic (c. 4300-2400 cal. BC): Landscape management as an alternative paradigm. Holocene 26, 1858-1874.

Jadwiszczak, K.A., Drzymulska, D., Banaszek, A., Jadwiszczak, P., 2012. Population history, genetic variation and conservation status of the endangered birch species *Betula nana* L. in Poland. Silva Fennica 46, 465–477.

Jäger, E.J., 1969. Die klimatischen Bedingungen des Areals der Dunklen Taiga und der sommergrünen Breitlaubwälder. Berichte Deutsche Botanische Gesellschaft 82, 397-408.

Jahn, C., 1977. Die Fichtenwaldgesellschaften in Europa. In: Schmidt-Vogt, H. (Ed.), Die Fichte. I. Taxonomie, Verbreitung, Morphologie, Oekologie, Waldgesellschaften. Parey, Hamburg/Berlin, pp. 468-560.

Jahns, S., 1993. On the Holocene vegetation history of the Argive Plain (Peloponnese, southern Greece). Veget. Hist. Archaeobot. 2, 187-203

Jahns, S., 2000. Late-glacial and Holocene woodland dynamics and land-use history of the Lower Oder valley, north-eastern Germany, based on two, AMS14C-dated, pollen profiles. Veget Hist Archaebot 9, 111–123.

Jahns, S., 2002. An improved time scale for the Holocene history of vegetation and environment on the South Dalmatian Island of Mljet. Veget Hist Archaeobot 11, 315 - 316.

Jahns, S., 2005. The Holocene history of vegetation and settlement at the coastal site of Lake Voulkaria in Acarnania, western Greece. Veget. Hist. Archaeobot. 14, 55-66.

Jahns, S., 2007. Palynological investigations into the Late Pleistocene and Holocene history of vegetation and settlement at the Löddigsee, Mecklenburg, Germany. Veg. Hist. Archaeobotany 16, 157–169.

Jahns, S., van den Bogaard, C., 1998. New palynological and tephra-stratigraphical investigations of two salt lagoons on the island of Mljet, south Dalmatia, Croatia. Veget. Hist. Archaeobot. 7, 219-234.

Jakab, G., Sümegi, P., Magyari, E., 2004. A new paleobotanical method for the description of Late Quaternary organic sediments (Mire-development pathways and paleoclimatic records from S Hungary). Acta Geol. Hungarica 47, 373-409.

Jalas, J., Suominen, J. (eds), 1973. Atlas Florae Europae. Distribution of vascular plants in Europe. 2. Gymnospermae. Helsinki, 40 pp.

Jalas, J., Suominen, J., 1976. Atlas Florae Europaeae. Distribution of vascular plants in Europe. Salicaceae to Balanophoraceae. Luomos, Helskinki.

Jalut, G., 1974. Evolution de la végétation et variations climatiques, durant les quinze derniers millénaires dans l'extrémité orientale des Pyrénées. (PhD thesis). University of Toulouse, Toulouse, France.

Jalut, G., Delibrias, G., Dagnac, J., Mardones, M., Bouhours, M., 1982. A palaeoecological approach to the last 21 000 years in the pyrenees: The peat bog of Freychinède (Alt. 1350 m, Ariège, South France). Palaeogeogr. Palaeoclimatol. Palaeoecol. 40(4), 321-359.

Jalut, G., Andrieu, V., Delibrias, G., Fontugne, M., Pages, P., 1988. Palaeoenvironment of the Valley of Ossau (western French Pyrénées) during the last 27000 years. Pollen spores 30, 357–393.

Jalut, G., Marti, J.M., Fontugne, M., Delibrias, G., Vilaplana, J.M., Julia, R., 1992. Glacial to interglacial vegetation changes in the northern and southern Pyrenees: Deglaciation, vegetation cover and chronology. Quaternary Science Reviews 11, 449–480.

Jalut, G., Aubert, S., Galop, D., Fontugne, M., Belet, J., 1996. Type Regions F-zg and F-r, northern slope of the Pyrenees. In: Berglund, B.E., Birks, H.J.B., Ralska-Jasiewiczowa, M., Wright, H.E. (Eds.), Palaeoecological Events During the Last 15 000 Years. Regional Syntheses of Palaeoecological Studies of Lakes and Mires in Europe. Wiley, Chichester, pp. 612-632.

Jalut, G., Esteban Amat, A., Riera i Mora, S., Fontugne, M., Mook, R., Bonnet, L., Gauquelin, T., 1997. Holocene climatic changes in the western Mediterranean: installation of the Mediterranean climate. Comptes Rendus del'Académie des Sciences - Earth and Planetary Sciences 325, 327-334.

Jalut, G., Esteban Amat, A., Bonnet, L., Gauquelin, T., Fontugne, M., 2000. Holocene climatic changes in the Western Mediterranean, from south-east France to south-east Spain. Palaeogeogr. Palaeoclimatol. Palaeoecol. 160, 255-290.

Jalut, G., Dedoubat, J., Fontugne, M., Otto, T., 2009. Holocene circum-Mediterranean vegetation changes: Climate forcing and human impact. Quat. Int. 200, 4–18.

Jamrichová, E., Szabó, P., Hédl, R., Kuneš, P., Bobek, P., Pelánková, B., 2013. Continuity and change in the vegetation of a central European oakwood. Holocene 23, 46-56.

Jamrichová, E., Potůčková, A., Horsák, M., Hajnalová, M., Barta, P., Tóth, P., Kuneš, P., 2014. Early occurrence of temperate oak-dominated forest in the northern part of the Little Hungarian Plain, SW Slovakia. Holocene 24, 1810-1824.

Jamrichová, E., Hédl, R., Kolar, J., Tóth, P., Bobek, P., Hajnalová, M., Prochazka, J., Kadlec, J., Szaboo, P., 2017. Human impact on open temperate woodlands during the middle Holocene in central Europe. Rev. Palaeobot. Palynol. 245, 55-68.

Jancyk-Kopikowa, Z. 1963. Analiza pylkowa osadów czwartorzędowych z Ferdinandowa. Kwartalnik Geologiczny 7, 365-371.

Jankovská, V., 1980. Paläogeobotanische Rekonstruktion der Vegetationsentwicklung im Becken Třeboňská pánev während des Spätglazials und Holozäns. Veg. ČSSR A11, 1–151.

Jankovská, V., 1984. Late Glacial finds of *Pinus cembra* in the Lubovnianská kotlina basin. Folia Geobot. Phytotax. 19, 310-321.

Jankovská, V., 1988. A reconstruction of the Late-Glacial and Early-Holocene evolution of forest vegetation in the Poprad Basin, Czechoslovakia. Folia Geobot. Phytotaxon. 23, 303–319.

Jankovská, V., Pokorný, P., 2008. Forest vegetation of the last full-glacial period in the Western Carpathians (Slovakia and Czech Republic). Preslia 80, 307-324.

Jankovská, V., Pokorný, P., 2015. Native occurrence of larch (*Larix*) in central Europe: Overview of currently available fossil record. In: Sázelová, S., Novák, J., Mizerová, A. (Eds.), Forgotten Times and Spaces: New Perspectives in paleoanthropological, paleoethnological and archaeological studies. Institute of Archaeology of the Czech Academy of Sciences, Masaryk University, Brno, pp. 80-90.

Jankovská, V., Chromy, P., Niznianska, M., 2002. Safarka – first palaeobotanical data of the character of the Last Glacial vegetation and landscape in the West Carpathians (Slovakia). Acta Palaeobotanica 42, 39-50.

Janssen, C.R., 1966. Recent pollen spectra from the deciduous and coniferous-deciduous forests of northeasten Minnesota: a study in pollen dispersal. Ecology 47, 805-825.

Janssen, C.R., 1967. A comparison between the recent regional pollen rain and the sub-recent vegetation in four major vegetation types in Minnesota (U.S.A.)1. Rev. Palaeobot. Palynology 2, 331-342.

Janssen, C.R., 1973. Local and regional pollen deposition. In: Birks, H.J.B., West, R.G. (Eds.), Quaternary Plant Ecology. Blackwell Scientific Publications, Oxford, pp. 31-42.

Janssen, C.R., 1974. Verkenningen in de Palynologie. Utrecht.

Janssen, C.R., 1994. Palynological indication for the extent of the impact of man during Roman times in the Western part of the Iberian peninsula. Palaeoclim. Res. 10, 15-22.

Jantz, N., Homeier, J., León-Yánez, S., Moscoso, A., Behling, H., 2013. Trapping pollen in the tropics — Comparing modern pollen rain spectra of different pollen traps and surface samples across Andean vegetation zones. Rev. Palaeobot. Palynol. 193, 57-69.

Járai-Komlódi, M., 1966a. Etudes palynologiques des couches de la dernière épisode glaciaire (Brörup, Pléniglaciaire) de la Grande Plaine Hongroise. Pollen et Spores 8, 479.

Járai-Komlódi, M., 1966b. Quaternary climatic changes and vegetational history of the Great Hungarian Plain I Botan. Közlem. 53, 191-201.

Járai-Komlódi, M., 1968. The Late glacial and Holocene flora of the Hungarian Great Plain. Ann. Univ. Sc. Budapest 9/10, 199-225.

Jastrzębska-Mamełka, M., 1985. Interglacjal Eemski I Wczesny Vistulian W Zgierz-Rudunkach na Wyznie Lodzkiej. Acta Geographica Lodziensia 53, 75 p.

Jeffers, E.S., Bonsall, M.B., Brooks, S.J., Willis, K.J., 2011a. Abrupt environmental changes drive shifts in tree–grass interaction outcomes. J. Ecol. 99, 1063-1070.

Jeffers, E.S., Bonsall, M.B., Willis, K.J., 2011b. Stability in ecosystem functioning across a climatic threshold and contrasting forest regimes. PLoS ONE 6, e16134.

Jeffers, E.S., Bonsall, M.B., Watson, J.E., Willis, K.J., 2012. Climate change impacts on ecosystem functioning: evidence from an *Empetrum* heathland. New Phytol. 193, 150-164.

Jeffers, E.S, Bonsall, M.B., Froyd, C.A., Brooks, S.J., Willis, K.J., 2015a. The relative importance of biotic and abiotic processes for structuring plant communities through time. J. Ecol. 103, 459-472.

Jeffers, E.S., Nogué, S., Willis, K.J., 2015b. The role of palaeoecological records in assessing ecosystem services. Quat. Sci. Rev. 112, 17-32.

Jeffers, E.S., Whitehouse, N.J., Lister, A., Plunkett, G., Barratt, P., Smyth, E., Lamb, P., Dee, M.W., Brooks, S.J., Willis, K.J., Froyd, C.A., Watson, J.E., Bonsall, M.B., 2018. Plant controls on Late Quaternary whole ecosystem structure and function. Ecol. Lett. 21, 814-825.

Jeffree, E.P., Jeffree, C.E., 1994. Temperature and the biogeographical distributions of species. Funct. Ecol. 8, 640-650.

Jensen, U., 1990. Die Moore des Hochharzes. Spezieller Teil. [The mires of the upper Harz mountains. Specific part]. Naturschutz und Landschaftspflege in Niedersachsen 23, 1-116. (in German)

Jensen, C., Kuiper, J., Vorren, K.D., 2002a. First post-glacial establishment of forest trees: early Holocene vegetation, mollusc settlement and climate dynamics in central Troms, North Norway. Boreas 31, 285-301.

Jensen, J., Gillies, A., Csaikl, U., Munro, R., Madsen, S., Roulund, H., Lowe, A., 2002b. Chloroplast DNA variation within the Nordic countries. For. Ecol. Managem. 156, 167-180.

Jerosch, M., 1903. Geschichte und Herkunft der Schweizerischen Alpenflora. Leipzig.

Jessen, K., 1920. Moseundersoegelser i det nordoestlige Sjaelland. Danm. Geol. Unders. II 34, 1-269.

Jessen, K., 1935. Archaeological dating in the history of North Jutlands vegetation. Acta Archaeol. 5, 185-214.

Jessen, K., 1949. Studies in the late quaternary deposits and flora history of Ireland. Proc. Roy. Irish Acad, 52, 85.

Jessen, K., 1955. Key to Subfossil *Potamogeton*. Bot. Tidskrift 52, 1-7.

Jessen, K., Farrington, A., 1938. The bogs at Balybetagh, near Dublin, with remarks on late-glacial conditions in Ireland. Proc. Roy. Irish Acad., Dublin, 44 B, 205-260.

Jessen, K., Jonassen, H., 1935. The composition of the forests in Northern Europe in Epipalæolithic time. Kgl. Danske Vidensk. Sels. Biol. Meddel. 12, 1-64.

Jessen, K., Milthers, V., 1928. Stratigraphical and palaeontological of interglacial fresh-water deposits in Jutland and north-west Germany. Danm. Geol. Unders. II 48, 1-379.

Jessen, K., Andersen, S.T., Farrington, A., 1959. The interglacial deposit near Gort, Co. Galway, Ireland. Proceed. R. Irish Acad. 60, Sect. B, No. 1, 3-77.

Jiménez-Sánchez, M., Rodríguez-Rodríguez, L., García-Ruiz, J., Domínguez-Cuesta, M., Farias, P., Valero-Garcés, B., Moreno, A., Rico, M., Valcárcel, M., 2013. A review of glacial geomorphology and chronology in northern Spain: Timing and regional variability during the last glacial cycle. Geomorphology 196, 50-64.

Joannin, S., Brugiapaglia, E., de Beaulieu, J.-L., Bernardo, L., Magny, M., Peyron, O., Goring, S., Vannière, B., 2012. Pollen-based reconstruction of Holocene vegetation and climate in southern Italy: the case of Lago Trifoglietti. Climate of the Past 8, 1973–1996.

Joannin, S., Magny, M., Peyron, O., Vannière, B., Galop, D., 2014. Climate and land-use change during the late Holocene at Lake Ledro (southern Alps, Italy). Holocene 24, 591-602.

Jochimsen, M., 1986. Zum Problem des Pollenfluges in den Hochalpen. Diss. Bot. 90, 1-249.

Jóhansen, J., 1968. Groðrarleivdir (sað)av dvørgabjørk (*Betula nana*) i Føroyum. A pollen diagram from the Faroe Islands showing former presence of Betula nana Froðskaparrit. Ann. Soc. Sci. Faer. 16, 119.

Jóhansen, J., 1975. Pollen diagrams from the Shetland and Faroe Islands. New Phytol. 75, 369-387.

Jóhansen, J., 1982. Vegetational development in the Faroes from 10 Kyr BP to the present. Dan. Geol. Unders. Årb. 111–136.

Jóhansen, J., 1985. Studies in the vegetational history of the Faroe and Shetland Islands. Ann. Soc. Sci. Faer., Suppl. 11, 1-117.

Jóhansen, J., 1996. Faroe Islands. In: Berglund, B.E., Birks, H.J.B., Ralska-Jasiewiczowa, M., Wright, H.E. (Eds.), Palaeoecological Events During the Last 15 000 Years. Regional Syntheses of Palaeoecological Studies of Lakes and Mires in Europe. Wiley, Chichester, pp. 145-152.

Joly, C., Visset, L., 2009. Evolution of vegetation landscapes since the Late Mesolithic on the French West Atlantic coast. Rev. Palaeobot. Palynol. 154, 124-179.

Jónasson, P., 2004. Benthic invertebrates. In: O'Sullivan, P., Reynolds, C.S. (Eds.), The Lakes Handbook. Limnology and Limnetic Ecology. Blackwell, Oxford, pp. 341-416.

Jones, M., 1988. The phytosociology of early arable weed communities with special reference to southern England. Forschungen u. Berichte zur Vor- und Frühgeschichte Baden-Württembergs 31, 43-51.

Jones, R., Jordan, J., 2007. Lake level studies. In: Elias, S.A. (Ed.), Encyclopedia of Quaternary Science. Elsevier, Amsterdam, pp. 1319-1336 (reprinted 2013).

Jones, G., Valamoti, S., Charles, M., 2000. Early crop diversity: a "new" glume wheat from northern Greece. Veget. Hist. Archaeobot. 9, 133-146.

Jonsgard, B., Birks, H.H., 1995. Late-glacial mosses and environmental reconstructions at Kråkenes, western Norway. *Lindbergia* 20, 64-82.

Joosten, H., 1993. Denken wie ein Hochmoor: Hydrologische Selbstregulation von Hochmooren und deren Bedeutung für Wiedervernässung und Restauration. Telma 23, 95-115.

Joosten, H., 1995. Time to regenerate: long-term perspectives of raised bog regeneration with special emphasis on palaeoecological studies. In: Wheeler, B., Shaw, S., Fojt, W., Robertson, R. (Eds.), Restoration of Temperate Wetlands. Wiley, Chichester, pp. 379-404.

Joosten, H., 2001. Identifying peatlands of international biodiversity importance. International Mire Conservation Group, Greifswald.

Joosten, H., 2019. Permafrost peatlands: Losing ground in a warming world. In: UNEP 2019. Frontiers 2018/19 Emerging Issues of Environmental Concern. United Nations Environment Programme, Nairobi, pp. 38-51.

Joosten, H., 2021. Ramsar Global Guidelines for Peatland Rewetting and Restoration. Ramsar Convention Secretariat, Gland.

Joosten, H., Clarke, D., 2002. Wise use of mires and peatlands - background and principles including a framework for decision-making. International Mire Conservation Group and International Peat Society, Saarijärvi.

Joosten, H., de Klerk, P., 2007. DAMOCLES: a DAshing MOnolith Cutter for fine sectioning of peats and sediments into LargE Slices. Boreas 36, 76-81.

Joosten, H., Moen, A., Couwenberg, J., Tanneberger, F., 2017a. Mire diversity in Europe: mire and peatland types. In: Joosten, H., Tanneberger, F., Moen, A. (Eds.), Mires and Peatlands of Europe – Status, distribution and conservation. Schweizerbart Science Publishers, Stuttgart, pp. 5-64.

Joosten, H., Couwenberg, J., Moen, A., Tanneberger, F., 2017b. Mire and peatland terms and definitions in Europe. In: Joosten, H., Tanneberger, F., Moen, A. (Eds.), Mires and Peatlands of Europe – Status, distribution and conservation. Schweizerbart Science Publishers, Stuttgart, pp. 65-96.

Joosten, H., Tanneberger, F., Moen, A., 2017c. Mires and peatlands of Europe. Status, distribution and conservation. Schweizerbart Science Publishers, Stuttgart, 780 pp.

Josefsson, T., Ramqvist, P.H., Hörnberg, G., 2014. The history of early cereal cultivation in northernmost Fennoscandia as indicated by palynological research. Veget. Hist. Archaeobot. 23, 821-840.

Jouffroy-Bapicot, I., Vannière, B., Gauthier, É., Richard, H., Monna, F., Petit, C., 2013. 7000 years of vegetation history and land-use changes in the Morvan Mountains (France): A regional synthesis. The Holocene 23, 1888–1902.

Jouffroy-Bapicot, I., Pedrotta, T., Debret, M., Field, S., Sulpizio, R., Zanchetta, G., Sabatier, P., Roberts, N., Tinner, W., Walsh, K., Vannière, B., 2021. Olive groves around the lake. A ten-thousand-year history of a Cretan landscape (Greece) reveals the dominant role of humans in making this Mediterranean ecosystem. Quat. Sci. Rev. 267.

Juggins, S., 2004. C2 User guide. Software for ecological and paleoecological data analysis and visualisation. University of Newcastle, Newcastle-upon-Tyne, UK.

Juhász, I., 2002. Reconstitution palynologique de la végétation depuis le Tardiglaciaire de la région de Zala, sud-ouest de la Hongrie. PhD thesis. University of Aix-Marseille and University of Pecs.

Juhász, I., 2005. Detecting antropogenic impacts in the palaeobotanical samples from Csaroda-Nyıres-tó, in: Gál, E., Juhász, I., Sümegi, P. (Eds.), Environmental Archaeology in North-Eastern Hungary, Varia Archaeologica Hungarica. Budapest, Hungary, pp. 55–66.

Juhász, I., Sümegi, P., Szántó, Z., Svingor, É., Molnár, M., Jakab, G., 2007. The Little Balaton region and the Balaton uplands. Environmental Archaeology in Transdanubia. Varia Archaeologica Hungarica 20, 27–175.

Jull, A.J.T., 2007. Radiocarbon Dating: AMS. In: Elias, S.A. (Ed.), Encyclopedia of Quaternary Science. Elsevier, Amsterdam, pp. 2911-2918.

Jumpponen, A., 2001. Dark septate endophytes - are they mycorrhizal? Mycorrhiza 11, 207-211.

Jumpponen, A., Trappe, J.M., 1998. Dark septate endophytes: a review of facultative biotrophic root-colonizing fungi. New Phytol. 140, 295-310.

Jumpponen, A., Brown, S.P., Trappe, J.M., Cázares, E., Strömmer, R., 2012. Twenty years of research on fungal-plant interactions on Lyman Glacier forefront - lessons learned and questions yet unanswered. Fungal Ecol. 5, 430-442.

Jung, W., Beug, H.-J., Dehm, R., 1972. Das Riss/Würm-Interglazial von Zeifen, Landkreis Laufen an der Salzach. Bayerische Akademie der Wissenschaften. Mathematisch-Naturwissenschaftliche Klasse 151, 1-131.

Juul, J.G., 1925. Furuens utbredelse i Finnmark og Troms. Tidsskr. for Skogsbruk 33, 359-440.

Kaakinen, A., Eronen, M., 2000. Holocene pollen stratigraphy indicating climatic and tree-line changes derived from a peat section at Ortino, in the Pechora Lowland, northern Russia. Holocene 10, 611-620.

Kaal, J., Criado-Boado, F., Costa-Casais, M., López-Sáez, J.A., López-Merino, L., Mighall, T., Carriòn, Y., Sanchez, N.S., Cortizas, A.M., 2013. Prehistoric land use at an archaeological hot-spot (the rock art park of

Campo Lameiro, NW Spain) inferred from charcoal, synanthropic pollen and non-pollen palynomorph proxies. J. Archaeol. Scie. 40, 1518-1527.

Kabailiene, M.K., 1965. On the stratigraphy and palaeogeography of the Holocene in south-eastern Lithuania [in Russian]. Tr. Instituta Geol. 2, 302–335.

Kabailiene, M.K., 1969. On formation of pollen spectra and restoration of vegetation. Transactions, Vol. 11. Ministry of Geology of the USSR, Institute of Geology (Vilnius), Vlinius. (in Russian with English summary).

Kabailiene, M.K., 1971. On the interpretation of pollen diagrams of the Holocene in Lithuania, in: Khotinsky (Ed.), Palinologiya Golotsena. Nauka, Moscow, Russia, pp. 63–70.

Kabailiene, M.K., 1986. Methods of pollen and spores analysis [in Russian], in: Istoriya Ozior v SSSR. Tallinn, Estonia, pp. 91–94.

Kabailiene, M.K., 1990. Lietuvos holocenas [Holocene of Lithuania]. Minkstas, minksti virseliai edition, Vilnius, Lithuania.

Kabailiene, M.K., 1996. Lithuania. In: Berglund, B.E., Birks, H.J.B., Ralska-Jasiewiczowa, M., Wright, H.E. (Eds.), Palaeoecological Events During the Last 15 000 Years. Regional Syntheses of Palaeoecological Studies of Lakes and Mires in Europe. Wiley, Chichester, pp. 395-401.

Kadereit, J., 2008. Systematik und Stammesgeschichte. In: Strasburger, E. (Ed.), Lehrbuch der Botanik, 36th ed., 609-945.

Kahlke, R.-D., Mania, D., 1994. Komplexe Interglazialfundstellen Thüringens. Altenburger Naturwiss. Forschungen 7, 357-377.

Kaland, P.E., 1986. The origin and management of Norwegian coastal heaths as reflected by pollen analysis. In: Behre, K.-E. (Ed.), Anthropogenic Indicators in Pollen Diagrams. Balkema, Rotterdam, pp. 19-36.

Kaland, P.E., 2014. Heathlands - Land-use, ecology and vegetation history as a source for archaeological interpretations. In: Gulløv, H.C. (Ed.), Northern Worlds, Landscapes, interpretations and dynamics. Publications from the National Museum of Denmark, Studies in Archaeology & History Copenhagen, pp. 19-47.

Kalis, A.J., 1984. Foret de la Bresse (Vogezen), Vegetatiekundige en pollenanalytische onderzoekingen naar de bosgeschiedenis van een Centraal-Europeese Middelgebergte. Thesis Utrecht.

Kalis, A.J., Schoch, W., 2019. Anthracological studies on the early Holocene sediments of the Grotta di Corbeddu (Nuoro, Sardinia). Archäologische Berichte 30, 193-210.

Kalm, V., 2006. Pleistocene chronostratigraphy in Estonia, southeastern sector of the Scandinavian glaciation. Quat. Sci. Rev. 25, 960-975.

Kalnina, L., 2001. Middle and Late Pleistocene environmental changes recorded in the Latvian part of the Baltic Sea basin. PhD Thesis, Stockholm University.

Kaltenrieder, P., Tinner, W., Ammann, B., 2005. Zur Langzeitökologie des Lärchen-Arvengürtels in den südlichen Walliser-Alpen. Bot. Helv. 115, 137-154.

Kaltenrieder, P., Tinner, W., Ammann, B., 2006. Lago della Costa, Arquà Petrarca, XXXth International Moor-Excursion 2006 Northern and Central Italy. Excursion Guide, 57-61.

Kaltenrieder, P., Belis, C.A., Hofstetter, S., Ammann, B., Ravazzi, C., Tinner, W., 2009. Environmental and climatic conditions at a potential Glacial refugial site of tree species near the Southern Alpine glaciers. New insights from multiproxy sedimentary studies at Lago della Costa (Euganean Hills, Northeastern Italy). Quat. Sci. Rev. 28, 2647-2662.

Kaltenrieder, P., Procacci, G., Vannière, B., Tinner, W., 2010. Vegetation and fire history of the Euganean Hills (Colli Euganei) as recorded by Lateglacial and Holocene sedimentary series from Lago della Costa (northeastern Italy). Holocene 20, 679-695.

Kaniewski, D., Van Campo, E., Boiy, T., Terral, J.-F., Khadari, B., Besnard, G., 2012. Primary domestication and early uses of the emblematic olive tree: palaeobotanical, historical and molecular evidence from the Middle East. Biol. Rev. 87, 885-899.

Kaplan, J.O., Krumhardt, K.M., Ellis, E.C., Ruddiman, W.F., Lemmen, C., Klein Goldewijk, K., 2011. Holocene carbon emissions as a result of anthropogenic land cover change. Holocene 21, 775-791.

Karlén, W., 1976. Lacustrine sediments and tree-limit variations as indicators of Holocene climatic fluctuations in Lappland, northern Sweden. Geogr. Ann. 58A, 1-34.

Karlén, W., 1982. Holocene glacier fluctuations in Scandinavia. Striae 18, 26-34.

Karpuz, N., K., Jansen, E., 1992. A high-resolution diatom record of the last deglaciation from the SE Norwegian Sea: Documentation of rapid climatic changes. Paleoceanography 7, 499-520.

Kasse, K., 1988. Early-Pleistocene tidal and fluvial environments in the southern Netherlands and northern Belgium. PhD Thesis, Free University Amsterdam.

Katz, N.J., Katz, S.V., 1933. Atlas der Pflanzenreste im Torf. Staatsverlag für Landwirtschaftliche Literatur, Moscow.

Katz, N.J., Katz, S.V., Kipiani, M., 1965. Atlas and Keys of Fruits and Seeds Occurring in the Quaternary Deposits of the USSR. Nauka, Moscow.

Katz, N.J., Katz, S.V., Skobeyeva, E.I., 1977. Atlas of Plant Remains in Peat. Nedra, Moscow.

Keeley, J., 1981. *Isoëtes howellii*: a submerged aquatic CAM plant? Am. J. Bot. 68, 420 - 424.

Keeley, J., 1998. CAM photosynthesis in submerged aquatic plants. The Botanical Review 64, 121-175.

Keeley, J., 2014. Aquatic CAM photosynthesis: A brief history of its discovery. Aquatic Botany 118, 38 - 44.

Keller, P., 1929. Analyse pollinique de la tourbière de Pinet. Archives de Botaniques 3, 57–63.

Keller, F., Lischke, H., Mathis, T., Mohl, A., Wick, L., Ammann, B., Kienast, F., 2002. Effects of climate, fire, and humans on forest dynamics: forest simulations compared to the palaeological record. Ecol. Model. 152, 109-127.

Kelly, M., Huntley, B., 1991. An 11 000-year record of vegetation and environment from Lago di Martignano, Latium, Italy. J. Quat. Sci. 6, 209-224.

Kelso, G.K., Solomon, A.M., 2006. Applying modern analogs to understand the pollen content pf coproliths. Paleogeogr. Paleoclimatol. Paleoecol. 237, 80-91.

Kéry, M., 2011. Towards the modelling of true species distributions. J. Biogeogr. 38, 617-618.

Khatab, I., Ishiyama, H., Inomata, N., Wang, X., Szmidt, A., 2008. Phylogeography of Eurasian *Larix* species inferred from nucleotide variation in two nuclear genes. Genes Genet. Syst. 83, 55-66.

Khomutova, V.I., 1976. Geokhronologiya ozernykh otlozhenii po palinologicheskim dannym (Geochronology of bottom sediments by palynological data), Palinologiya Onezhskogo ozera (Palynology of Onego Lake). Nauka, Leningrad, pp. 45-67.

Khomutova, V.I., 1989. Paleogeografiya i biostratigrafiya ozernykh otlozhenii lesnoi zony Evropeiskoi chasti SSSR (Paleogeography and biostratigraphy of lake sediments from the forest zone of the European part of USSR by spore-pollen data). PhD Thesis, Institute of Limnology, Leningrad, p. 491.

Khomutova, V.I., Elina, G.A., 1990. Stratigraphy of lake sediments by palynological data [in Russian], in: Istoria Ladozhskogo, Onezhskogo, Pskovsko- Chudskogo Ozer, Baikala i Khanki [The History of Ladoga, Onego, Pskovsko-Chudskoe, Baikal and Khanki Lakes]. Nauka, Leningrad, Russia, pp. 92–96.

Khotinsky, N.A., Bezusko, L.G., Cherkinsky, A.E., 1994. Changes in vegetation in central and western regions of the Russian Plain, in: Paleogeographical Basis of the Modern Landscapes. Nauka, Moscow, Russia, pp. 111–118.

Kilian, M.R., van Geel, B., van der Plicht, J., 2000. ^{14}C AMS wiggle matching of raised bog deposits and models of peat accumulation. Quat. Sci. Rev. 19, 1011-1033.

Kim, J.H., Jackson, R.B., 2012. A global analysis of groundwater recharge for vegetation, climate, and soils. Vadose Zone J. 11, 1–35.

King, W., 1685. On the bogs and loughs of Ireland. Philos. Trans. R. Soc. 15, 947-960.

King, R., Ferris, C., 1998. Chloroplast DNA phylogeography of *Alnus glutinosa* (L.) Gartn,. Mol. Ecol. 7, 1151-1161.

King, J., Peck, J., 2001. Use of paleomagnetism in studies of lake sediments. In: Last, W.M., Smol, J.P. (Eds.), Tracking Environmental Change Using Lake Sediments, Vol. 1: Basin Analysis, Coring, and Chronological Techniques. Kluwer, Dordrecht, pp. 371-389.

Kinloch, B., Westfall, R., Forrest, G., 1986. Caledonian Scots Pine: Origins and genetic structure. New Phytol. 104, 703-729.

Kirleis, W., Fischer, E., 2014. Neolithic cultivation of tetraploid free threshing wheat in Denmark and Northern Germany: implications for crop diversity and societal dynamics of the Funnel Beaker Culture. Veget. Hist. Archaeobot. 23, 81-96.

Kirleis, W., Klooß, S., Kroll, H., Müller, J., 2012. Crop growing and gathering in the northern German Neolithic: a review supplemented by new results. Veget. Hist. Archaeobot. 21, 221-242.

Kirpotin, S.N., Antoshkina, O.A., Berezin, A.E., Elshehawi, S., Feurdean, A., Lapshina, E.D., Pokrovsky, O.S., Peregon, A.M., Semenova, N.M., Tanneberger, F., Volkov, I.V., Volkova, I.V., Joosten, H., 2021. Great Vasyugan Mire: How the World's largest peatland helps addressing the World's largest problems. Ambio 50, 2038-2049.

Kitagawa, H., van der Plicht, J., 1998a. A 40,000-year varve chronology from Lake Suigetsu, Japan: Extension of the ^{14}C calibration curve. Radiocarbon 40, 505-515.

Kitagawa, H., van der Plicht, J., 1998b. Atmospheric radiocarbon calibration to 45,000 yr B.P.: Late glacial fluctuations and Cosmogenic isotope production. Science 279, 1187-1190.

Klaus, W., 1975a. Das Mondsee-Interglazial, ein neuer Florenfundpunkt der Ostalpen. Jahrb. oberösterr. Musealverein 120, 215-344.

Klaus, W., 1975b. Über bemerkenswerte morphologische Bestimmungsmerkmale an Pollenkörnern der Gattung *Pinus* L. Linzer Biol. Beitr. 7, 329-369.

Klaus, W., 1987. Das Mondsee-Profil: R/W-Interglazial und vier Würm-Interstadiale in einer geschlossenen Schichtfolge. Mitt. Komm. Quartärforschung Österr. Akad. Wiss. 7, 3-18.

Kleiber, H., 1974. Pollenanalytische Untersuchungen zum Eisrückzug und zur Vegetationsgeschichte im Oberengadin I. Bot. Jahrb. Syst. 94, 1-53.

Klimanov, V.A., Bezusko, L.G., 1981. Climate and vegetation of small Polessie in Holocene. Ukr. Bot. Zhurnal 38.

Klimaszewski, M., Szafer, W., Szafran, B., Urbanski, J., 1950. The Dryas flora of Kroscienko on the river Dunajec. Bull. Serv. Géol. Pologne, Bull. 24, 1-87.

Klinger, P.U., 1976. Feinstratigraphische Untersuchungen an Hochmooren. PhD Thesis, University of Kiel.

Kneblovà, V., 1954. Fytopaleontologicky rozbor uhliku z paleolitického sidliste v Dolnich Vestonicich. Anthropozoikum 3, 297-299.

Knipping, M., 2008. Early and Middle Pleistocene pollen assemblages of deep core drillings in the northern Upper Rhine Graben, Germany. Netherlands J. Geosci. 87, 51-65.

Knörzer, K., 1971a. Römerzeitliche Getreideunkräuter von kalkreichen Böden. Rheinische Ausgrabungen 10, 467-481.

Knörzer, K., 1971b. Urgeschichtliche Unkräuter im Rheinland. Ein Beitrag zur Entstehungsgeschichte der Segetalgesellschaften. Vegetatio 23, 89-111.

Knörzer, K., 1975. Entstehung und Entwicklung der Grünlandvegetation im Rheinland. Decheniana 127, 195-214.

Knörzer, K., 1996. Beitrag zur Geschichte der Grünlandvegetation am Niederrhein. Tuexenia 16, 627-636.

Knörzer, K., 2007. Geschichte der synanthropen Flora im Niederrheingebiet. Rheinische Ausgrabungen 61. von Zabern, Mainz am Rhein, 484 pp.

Koff, T., 1989. Reconstruction of Palaeogeographical Conditions in NE Estonia on the Basis of Bog and Lake Deposits. Presented at the Estonia-Finnish seminar on environmental questions, pp. 99-115.

Kofler, W., Oeggl, K., 2001. L'evoluzione del clima e dell'ambiente Alpino negli ultimi 15,000 anni. In: Cason, E. (Ed.), Uso dei valichi alpini orientali dalla preistoria ai pellegrinaggi medievali. Fondazione Giovanni Angelini, Belluno, pp. 19-28.

Koivula, L., 1987. Keski-Suomen viljelyhistoriaa: siitepölytutkimus viidestä Keski-Suomen kunnasta (MSc thesis). University of Jyväskylä, Jyväskylä, Finland.

Kolstrup, E., 1980. Climate and stratigraphy in northwestern Europe between 30,000 B.P. and 13,000 B.P. with special reference to the Netherlands. Med Rijks Geol Dienst 32, 181–253.

Kondratiene, O., 1996. The Quaternary stratigraphy and paleogeography of Lithuania based on paleobotanic studies. Academia Vilnius (in Russian with English summary). 213 pp.

Kondratiene, O., Damušytė, A., 2009. Pollen biostratigraphy and environmental pattern of Snaigupėlė Interglacial, Late Middle Pleistocene, western Lithuania. Quat. Int. 207, 4-13.

König, A., Ziegenhagen, B., van Dam, B., Csaikl, U., Coart, E., Degen, B., Burg, K., de Vries, S., Petit, R., 2002. Chloroplast DNA variation of oaks in western central Europe and genetic consequences of human influences. For. Ecol. Managem. 156, 147-166.

Königsson, L.-K., 1968. The Holocene history of the Great Alvar of Öland. Acta Phytogeogr. Suecica 55, 1-172.

Konnert, M., Bergmann, F., 1995. The geographical distribution of genetic variation of silver fir (*Abies alba*, Pinaceae) in relation to its migration history. Plant Syst. Evol. 196, 19-30.

Konradi, P., Larsen, B., Sørensen, A., 2005. Marine Eemian in the Danish eastern North Sea. Quat. Int. 133-134, 21-31.

Kooistra, M., Maas, G., 2008. The widespread occurrence of Celtic field systems in the central parts of the Netherlands. J. Archaeol. Sci. 35, 2318-2328.

Koperowa, W., 1962. Poznoglacjalna i holocenska historia roslinnosci Kotliny Nowotarskiej (The history of the Late-Glacial and Holocene vegetation in Nowy Targ basin). Acta Palaeobot. 2, 3-62.

Koperowa, W., 1970. Poznoglacjalna i holocenska historia roslinnosci wschodniej czesci Dolów Jasielsko-Sanockich (Late-Glacial and Holocene history of the vegetation of the eastern part of the Jaslo-Sanok Doly (Flysch Carpathians)). Acta Palaeobot. 11, 1-46.

Koperowa, W., Środoń, A., 1965. Pleniglacial deposits of the last glaciation at Zator (west of Krakow). Acta Paleobot. 6 1-31.

Körber-Grohne, U., 1964. Bestimmungsschlüssel für subfossile Juncus-Samen und Gramineen-Früchte. Probleme der Küstenforschung im südlichen Nordseegebiet 7. Lax, Hildesheim, 47 pp.

Körber-Grohne, U., 1987. Nutzpflanzen in Deutschland: Kulturgeschichte und Biologie. Theiss, Stuttgart, 490 pp.

Körber-Grohne, U., 1990. Gramineen und Grünlandvegetationen vom Neolithikum bis zum Mittelalter in Mitteleuropa. Biblio. Bot. 139, 1-108.

Körber-Grohne, U., 1991. Bestimmungsschlüssel für subfossile Gramineen-Früchte. Probleme der Küstenforschung im südlichen Nordseegebiet 18, 169-234.

Korhola, A., 1994. Radiocarbon evidence for rates of lateral expansion in raised mires in southern Finland. Quat. Res. 42, 299-307.

Korhola, A., 1995. Holocene climatic variations in southern Finland reconstructed from peat-initiation data. Holocene 5, 43-58.

Korhola, A., Weckström, J., Holmström, L., Erasto, P., 2000. A quantitative Holocene climatic record from diatoms in Northern Fennoscandia. Quat. Res. 54, 284-294.

Korhola, A., Vasko, K., Toivonen, H., Olander, H., 2002. Holocene temperature changes in northern

Fennoscandia reconstructed from chironomids using Bayesian modelling. Quat. Sci. Rev. 21, 1841-1860.

Körner, C., 1999. Alpine Plant Life. Springer, Berlin.

Körner, C., 2012. Alpine Treelines. Springer, Basel.

Korpela, K., 1969. Die Weichsel-Eiszeit und ihr Interstadial in Peräpohjola (nördliches Nordfinnland) im Licht von submoränen Sedimenten. Ann. Acad. Sci. Fenn. A III 99, 1-108.

Koska, I., Succow, M., Clausnitzer, U., Timmermann, T., Roth, S., 2001. Vegetationskundliche Kennzeichnung von Mooren (topische Betrachtung). In: Succow, M., Joosten, H. (Eds.), Landschaftsökologische Moorkunde. Schweizerbart, Stuttgart, pp. 112-184.

Kothieringer, K., Walser, C., Dietre, B., Reitmaier, T., Haas, J.N., Lambers, K., 2015. High impact: early pastoralism and environmental change during the Neolithic and Bronze Age in the Silvretta Alps (Switzerland/Austria) as evidenced by archaeological, palaeoecological and pedological proxies. Zeits. Geomorphol. 59, 177-198.

Kotowski, W., van Diggelen, R., Kleinke, J., 1998. Behaviour of wetland plant species along a moisture gradient in two geographically distant areas. Acta Bot. Neerland. 47, 337-349.

Kotthoff, U., Müller, U., Pross, J., Schmiedl, G., Lawson, I., van de Schootbrugge, B., Schulz, H., 2008. Lateglacial and Holocene vegetation dynamics in the Aegean region: an integrated view based on pollen data from marine and terrestrial archives. Holocene 18, 1019-1032.

Kouli, K., Dermitzakis, M.D., 2010. Lake Orestiás (Kastoria, northern Greece). Grana 49, 154–156.

Koutsodendris, A., Müller, U.C., Pross, J., Brauer, A., Kotthoff, U., Lotter, A.F., 2010. Vegetation dynamics and climate variability during the Holsteinian interglacial based on a pollen record from Dethlingen (northern Germany). Quat. Sci. Rev. 29, 3298-3307.

Kowarik, I., 2010. Biologische Invasionen. Neophyten und Neozoen in Mitteleuropa, 2nd ed. Eugen Ulmer, Stuttgart, 492 pp.

Kral, F., 1974. Grundzüge einer postglazialen Waldgeschichte des Ostalpenraums. In: Mayer, H. (Ed.), Wälder des Ostalpenraumes. Fischer, Stuttgart, pp. 236-273.

Kral, F., 1979. Spät- und postglaziale Waldgeschichte der Alpen auf Grund der bisherigen Pollenanalysen. Veröff. Inst. Waldbau Univ. Bodenkultur Wien. 175pp.

Krebs, P., Conedera, M., Pradella, M., Torriani, D., Felber, M., Tinner, W., 2004. Quaternary refugia of the sweet chestnut (*Castanea sativa* Mill.): an extended palynological approach. Veget. Hist. Archaeobot. 13, 145-160.

Krebs, P., Pezzatti, G.B., Beffa, G., Tinner, W., Conedera, M., 2019. Revising the sweet chestnut (*Castanea sativa* Mill.) refugia history of the last glacial period with extended pollen and macrofossil evidence. Quat. Sci. Rev. 206, 111-128.

Kreft, H., Jetz, W., 2007. Global patterns and determinants of vascular plant diversity. Proc. Natl. Acad. Sci. USA 104, 5925.

Kremenetski, C., 1995. Holocene vegetation and climate history of southwestern Ukraine. Rev. Palaeobot. Palynol. 85, 289-301.

Kremenetski, C., 1997a. Environment of the Razdorskoye multilayer settlement (Rostov region, southern Russia). ISKOS (Finska Fornminnesför) 11, 236-241.

Kremenetski, C., 1997b. The Late Holocene environmental and climate shift in Russia and surrounding lands. In: Dalfes, H., Kukla, G., Weiss, H. (Eds.), Third Millennium BC Climate Change and Old World Collapse. Springer, Berlin, pp. 351-370.

Kremenetski, C., Patyk-Kara, N., 1997. Holocene vegetation dynamics of the southeast Kola Peninsula, Russia. Holocene 7, 473-479.

Kremenetski, C., Vaschalova, T., Goriachkin, S., Cherkinsky, A., Sulerzhitsky, L., 1997. Holocene pollen stratigraphy and bog development in the western part of the Kola Peninsula, Russia. Boreas 26, 91–102.

Kremenetski, C., Sulerzhitsky, L., Hantemirov, R., 1998. Holocene history of the Northern Range limits of some trees and shrubs in Russia. Arc. Alp. Res. 30, 317-333.

Kremenetski, C., Chichagova, O., Shishlina, N., 1999a. Palaeoecological evidence for Holocene vegetation, climate and land-use change in the Don basin and Kalmuk area, southern Russia. Veget. Hist. Archaeobot. 8, 233-246.

Kremenetski, C., Vaschalova, T., Sulerzhitsky, L., 1999b. The Holocene vegetation history of the Khibiny Mountains: implications for the post-glacial expansion of spruce and alder on the Kola Peninsula, northwestern Russia J. Quat. Sci. 14, 29-43.

Kremenetski, C., Böttger, T., Junge, F., Tarasov, A., 1999c. Late- and postglacial environment of the Buzuluk area, middle Volga region, Russia. Quat. Sci. Rev. 18, 1185-1203.

Kremenetski, C., Borisova, O., Zelikson, E., 2000. The Late Glacial and Holocene history of vegetation in the Moscow region. Palaeontol. J. 34, 567–574.

Kreuz, A., 2012. Die Vertreibung aus dem Paradies? Archäobiologische Ergebnisse zum Frühneolithikum im westlichen Mitteleuropa. Berichte der Römisch-Germanischen Kommission 91, 23-196.

Kreuz, A., Marinova, E., Schäfer, E., Wiethold, J., 2005. A comparison of early Neolithic crop and weed assemblages from the Linearbandkeramik and the Bulgarian Neolithic cultures: differences and similarities. Veget. Hist. Archaeobot. 14, 237-258.

Krisai, R., 1975. Die Ufervegetation der Trumer Seen (Salzburg). Diss. Bot. 29, 1–202.

Krisai, R., Burgstaller, B., Ehmer-Künkele, U., Schiffer, R., Wurm, E., 1991. Die Moore des Ost-Lungaues - Heutige Vegetation, Enstehung, Waldgeschichte ihrer Umgebung. Sauteria 5, 1-240.

Kristiansen, K., 1998. The Construction of a Bronze Age Landscape. Cosmology, Economy and Social Organisation in Thy, Northwestern Jutland, In: Hänsel, B. (Ed.), Mensch und Umwelt in der Bronzezeit Europas. VML, Kiel, 281-291.

Krog, H., 1954. Pollen analytical investigation of a C14-dated Alleröd-section from Ruds-Vedby. In: Iversen, J. (ed.), 1954. Studies in Vegetational History in honour of Knud Jessen. Danm. Geol. Unders., København, II/80, 120-139.

Krupiński, K., Noryśkiewicz, A., Nalepka, D., 2004. *Taxus baccata* L. - Yew. In: Ralska-Jasiewiczowa, M., Latałowa, Wasylikowa, K., Tobolski, K., Madeysaka, E., Wright, H.E., Turner, C. (Eds.), Late Glacial and Holocene History of Vegetation in Poland based on Isopollen Maps. W. Szafer Institute of Botany, Polish Academy of Sciences, Kraków, pp. 209-215.

Kubiak-Martens, L., 2002. New evidence for the use of root foods in pre-agrarian subsistence recovered from the late Mesolithic site at Halsskov, Denmark. Veget. Hist. Archaeobot. 11, 23-31.

Kubitz, B., 2000. Die holozäne Vegetations- und Siedlungsgeschichte in der Westeifel am Beispiel eines hochauflösenden Pollendiagrammes aus dem Meerfelder Maar. Diss. Bot. 339, 1–106.

Kudrachov, W.W., 1924. Torfmoore der Beluchij Halbinsel (Nowaja Semlja). Ber. Wissensch. Meeresinst. 12, 112-119, Moscow.

Kühl, N., 2003. Die Bestimmung botanisch-klimatologischer Transferfunktionen und die Rekonstruktion des bodennahen Klimazustandes in Europa während der Eem-Warmzeit. Diss. Bot. 375, 1-149.

Kühl, N., Gobet, E., 2010. Climatic evolution during the Middle Pleistocene warm period of Bilshausen, Germany, compared to the Holocene. Quat. Sci. Rev. 29, 3736-3749.

Kühl, N., Litt, T., 2007. Quantitative time-series reconstructions of Holsteinian and Eemian temperatures using botanical data. In: Sirocko, F., Claussen, M., Sánchez Goñi, M.F., Litt, T. (Eds.), The Climate of Past Interglacials. Elsevier, pp. 239-254.

Kühl, N., Gebhardt, C., Litt, T., Hense, A., 2002. Probability Density Functions as botanical-climatological transfer functions for climate reconstruction. Quat. Res. 58, 381-392.

Kühl, N., Litt, T., Schölzel, C., Hense, A. 2007. Eemian and early Weichselian temperature and precipitation variability in central Europe. Quat. Sci. Rev. 26, 3311-3317.

Kulczyński, S., 1949. Torfowiska Polesia. Peat bogs of Polesie. Mémoires de l'Académie Polonaise des Sciences et des Lettres, Classe des Sciences Mathématiques et Naturelles, Séries B: Sciences Naturelles 15.,1-359.

Kullman, L., 1980. Radiocarbon dating of subfossil Scots pine (*Pinus sylvestris* L.) in the southern Swedish Scandes. Boreas 9, 101-106.

Kullman, L., 1981. Recent tree-limit dynamics of Scots pine (*Pinus sylvestris* L.) in the southern Swedish Scandes. Wahlenbergia 8, 1-67.

Kullman, L., 1987a. Sequences of Holocene forest history in the Scandes, inferred from megafossil *Pinus sylvestris*. Boreas 16, 21-26.

Kullman, L., 1987b. Little Ice Age decline of a cold marginal *Pinus sylvestris* forest in the Swedish Scandes. New Phytol. 106, 567-584.

Kullman, L., 1988. Holocene history of the forest-alpine tundra ecotone in the Scandes Mountains (central Sweden). New Phytol. 108, 101-110.

Kullman, L., 1989. Tree-limit history during the Holocene in the Scandes Mountains, Sweden, inferred from subfossil wood. Rev. Palaeobot. Palynology 58, 163–171.

Kullman, L., 1992a. Orbital forcing and tree-limit history: hypothesis and preliminary interpretation of evidence from Swedish Lappland. Holocene 2, 131-137.

Kullman, L., 1992b. The ecological status of grey alder (*Alnus incana* (L.) Moench) in the upper subalpine birch forest of the central Scandes. New Phytol. 120, 445-451.

Kullman, L., 1993. Pine (*Pinus sylvestris* L.) tree-limit surveillance during recent decades, central Sweden. Arc. Alp. Res. 25, 24-31.

Kullman, L., 1995. Holocene tree-limit and climate history from the Scandes Mountains, Sweden. Ecology 76, 2490–2502.

Kullman, L., 1999. Early Holocene tree growth at a high elevation site in the northernmost Scandes of Sweden (Lapland). A palaeobiogeographical case study based on megafossil evidence. Geografiska Annaler 81, 63-74.

Kullman, L., 2001. Immigration of *Picea abies* into North-Central Sweden. New evidence of regional expansion and tree-limit evolution Nord. J. Bot. 21, 39 54.

Kullman, L., 2002. Boreal tree taxa in the central Scandes during the Late-Glacial: implications for Late-Quaternary forest history. J. Biogeogr. 29, 1117–1124.

Kullman, L., 2004. Early Holocene appearance of Mountain Birch (*Betula pubescens* ssp. *tortuosa*) at unprecedented high elevations in the Swedish Scandes: megafossil evidence exposed by recent snow and ice recession. Arc. Antarc. Alp. Res. 36, 172-180.

Kullman, L., 2013. Ecological treeline history and palaeoclimate - review of megafossil evidence from the Swedish Scandes. Boreas 42, 555-567.

Kullman, L., 2017. Melting glaciers in the Swedish Scandes provide new insights into palaeotreeline performance. Int. J. Current Multidisc. Studies 3, 607-618.

Kullman, L., Kjällgren, L., 2000. A coherent postglacial tree-limit chronology (*Pinus sylvestris* L.) for the Swedish Scandes: aspects of paleoclimate and "recent warming," based on megafossil evidence. Arc. Antarc. Alp. Res. 32, 419-428.

Kullman, L., Öberg, L., 2015. New aspects of high-mountain palaeobiogeography: a synthesis of data from forefields of receding glaciers and ice patches in the Tärna and Kebnekaise Mountains, Swedish Lapland. Arctic 68, 141-152.

Kultti, S., Väliranta, M., Sarmaja-Korjonen, K., Solovieva, N., Virtanen, T., Kauppila, T., Eronen, M., 2003. Palaeoecological evidence of changes in vegetation and climate during the Holocene in the pre-Polar Urals, northeast European Russia. J. Quat. Sci. 18, 503-520.

Kultti, S., Oksanen, P., Väliranta, M., 2004. Holocene tree line, permafrost, and climate dynamics in the Nenets Region, East European Arctic. Can. J. Earth Sci. 41, 1141-1158.

Kuneš, P., Pelankova, B., Chytry, M., Jankovská, V., Pokorný, P., Petr, L., 2008a. Interpretation of the last-glacial vegetation of eastern-central Europe using modern analogues from southern Siberia. J. Biogeogr. 35, 2223-2236.

Kuneš, P., Pokorný, P., Sida, P., 2008b. Detection of the impact of early Holocene hunter-gatherers on vegetation in the Czech Republic, using multivariate analysis of pollen data. Veget. Hist. Archaeobot. 17, 269-287.

Kuneš, P., Abraham, V., Kovárík, O., Kopecký, M., PALYCZ contributors, 2009. Czech Quaternary Palynological Database - PALYCZ: review and basic statistics of the data. Preslia 81, 209-238.

Kuneš, P., Odgaard, B.V., Gaillard, M.-J., 2011. Soil phosphorus as a control of productivity and openness in temperate interglacial forest ecosystems. J. Biogeogr. 38, 2150-2164.

Kuneš, P., Abraham, V., Herben, T., 2019. Changing disturbance-diversity relationships in temperate ecosystems over the past 12000 years. J. Ecol. 107, 1678-1688.

Kunth, C., 1826. Examen botanique, in Passalacqua, J.: Catalogue raisonné et historique de antiquités découvertes en Egypte. Paris. (cited in Renfrew, 1973).

Kuoppamaa, M., Huusko, A., Hicks, S., 2009. *Pinus* and *Betula* pollen accumulation rates from the northern boreal forest as a record of interannual variation in July temperature. J. Quat. Sci. 24, 513-521.

Kuosmanen, N., Fang, K., Bradshaw, R.H.W., Clear, J.L., Seppä, H., 2014. Role of forest fires in Holocene stand-scale dynamics in the unmanaged taiga forest of northwestern Russia. Holocene 24, 1503-1514.

Kuparinen, A., Schurr, F.M., 2007. A flexible modelling framework linking the spatio-temporal dynamics of plant genotypes and populations: Application to gene flow from transgenic forests. Ecol. Model. 202, 476-486.

Kuparinen, A., Markkanen, T., Riikonen, H., Vesala, T., 2007. Modeling air-mediated dispersal of spores, pollen and seeds in forested areas. Ecol. Model. 208, 177-188.

Kupryjanowicz, M., 2008. Vegetation and climate of the Eemian and early Vistulian lakeland in northern Podlasie. Acta Palaeobot. 48, 3-130.

Kupryjanowicz, M., Nalepka, D., Madeyska, E., Turner, C. (eds.), 2018a. Eemian history of vegetation in Poland based on isopollen maps. W. Szafer Institute of Batany, Polish Academy of Science, Kraków.

Kupryjanowicz, M., Granoszewski, Filoc, M., 2018b. New finds of Eemian *Tilia tomentosa* Moench macroremains in NE Poland, and the reconstructed European range of this species during the last interglacial. Quat. Int. 467, 107-116.

Küster, H., 1993. Die Entwicklung der montanen und subalpinen Heiden Mitteleuropas in vegetationsgeschichtlicher Sicht. Ber. R.-Tüxen-Ges. 5, 77-90.

Küster, H., 1994. The economic use of *Abies* wood as timber in central Europe during Roman times. Veget. Hist. Archaeobot. 3, 25-32.

Küster, J., 1995. Postglaziale Vegetationsgeschichte Südbayerns. Akademie-Verlag, Berlin, 372 pp.

Küster, H., 1997. The role of farming in the postglacial expansion of beech and hornbeam in the oak woodlands of central Europe. Holocene 7, 239-242.

Küttel, M., 1984. Vuolep Allakasjaure - eine pollenanalytische Studie zur Vegetationsgeschichte der Tundra in Nordschweden. Diss. Bot. 72, 191-212.

Küttel, M., 1985. Biostratigraphische und paläoökologische Untersuchungen in SE-Småland, Schweden. Striae, Uppsala, 21, 1-34.

Lacourse, T., 2009. Environmental change controls postglacial forest dynamics through interspecific differences in life-history traits. Ecology 90, 2149-2160.

Lagercrantz, U., Ryman, N., 1990. Genetic

structure of Norway spruce (*Picea abies*): concordance of morphological and allozymic variation. Evolution 44, 38-53.

Lambdon, P.W., Pyšek, P., Basnou, C., Hejda, M., Arianoutsou, M., Essl, F., Jarošik, V., Pergl, J., Winter, M., Anastasiu, P., Andriopoulos, P., Bazos, I., Brundu, G., Celesti-Grapow, L., Chassot, P., Vilà, M, 2008. Alien flora of Europe: species diversity, temporal trends, geographical patterns and research needs. Preslia 80, 101-149.

Lambert, C., Vidal, M., Penaud, A., Combourieu-Nebout, N., Lebreton, V., Ragueneau, O., Gregoire, G., 2017. Modern palynological record in the Bay of Brest (NW France): Signal calibration for palaeo-reconstructions. Rev. Palaeobot. Palynol. 244, 13-25.

Lamentowicz, Ł., Lamentowicz, M., Gąbka, M., 2008. Testate amoebae ecology and local transfer function from a peatland in Western Poland. Wetlands 28, 164-175.

Lamoureux, S., 2001. Varve chronology techniques. In: Last, W.M., Smol J.P. (Eds.), Tracking Environmental Change Using Lake Sediments, Vol. 1: Basin Analysis, Coring, and Chronological Techniques. Kluwer, Dordrecht, pp. 247-260.

Landolt, E., 1992. Unsere Alpenflora, 6th ed. Schweizer Alpenclub, Brugg, Switzerland.

Landolt, E., Bäumler, B., Erhardt, A., Hegg, O., Klötzli, F., Lämmler, W., Nobis, M., Rudmann-Maurer, K., Schweingruber, F., Theurillat, J., Urmi, E., Vust, M., Wohlgemuth, T., 2010. Flora Indicativa. Ecological Indicator Values and Biological Attributes of the Flora of Switzerland and the Alps. Haupt, Bern.

Lang, G., 1951. Nachweis von *Ephedra* im südwestdeutschen Spätglazial. Naturwiss. 38, 334.

Lang, G., 1952a. Späteiszeitliche Pflanzenreste in Südwestdeutschland. Beitr. naturk. Forsch. Südwestdeutschl. 11, 89-110.

Lang, G., 1952b. Zur späteiszeitlichen Vegetations- und Florengeschichte Südwestdeutschlands. Flora 139, 243-294.

Lang, G., 1954. Neue Untersuchungen über die spät- und nacheiszeitliche Vegetationsgeschichte des Schwarzwaldes. I. Der Hotzenwald im Südschwarzwald. Beitr. naturk. Forsch. Südwestdeutschland 13, 3-42.

Lang, G., 1955. Über spätquartäre Funde von *Isoëtes* und *Najas flexilis* im Schwarzwald. Ber. Deutsche Bot. Ges. 68, 24-27.

Lang, G., 1958. Neue Untersuchungen über die spät- und nacheiszeitliche Vegetationsgeschichte des Schwarzwaldes. III. Der Schurmsee im Nordschwarzwald. Ein Beitrag zur Kieferntrage. Beitr. naturk. Forsch. Südwestdeutschl., Karlsruhe, 17, 20-34.

Lang, G., 1967. Über die Geschichte von Pflanzengesellschaften auf Grund quartärbotanischer Untersuchungen. In: Tüxen, R. (Ed.), Pflanzensoziologie und Palynologie. Junk, Den Haag, pp. 24-37.

Lang, G., 1970. Florengeschichte und mediterran-mitteleuropäische Florenbeziehungen. Feddes Repertorium 81, 315-335.

Lang, G., 1971. Die Vegetationsgeschichte der Wutachschlucht und ihrer Umgebung. Die Wutach, 323-349.

Lang, G., 1973. Die Vegetation des westlichen Bodenseegebietes. Pflanzensoziologie, Jena, 17, 1-451. 2nd edition 1990, Fischer, Stuttgart/New York, 462 pp.

Lang, G., 1975. Palynologische, grossrestanalytische und paläolimnologische Untersuchungen im Schwarzwald. Beitr. naturk. Forsch. Südwestdeutschl., Karlsruhe, 34, 201-208.

Lang, G., 1985. Palynologic and stratigraphic investigations of Swiss lake and mire deposits - A general view over a research programme. Diss. Bot. 87, 107-114.

Lang, G., 1990. Die Vegetation des westlichen Bodenseegebietes. Fischer, Stuttgart.

Lang, G., 1992. Some aspects of European late- and post-glacial flora history. Acta Bot. Fennica 144, 1-17.

Lang, G., 1994. Quartäre Vegetationsgeschichte Europas - Methoden und Ergebnisse. Gustav Fischer, Jena.

Lang, G., 2003. Immigration and expansion of *Tilia* in Europe since the last Glacial. In: Tonkov, S. (Ed.), Aspects of palynology and palaeoecology. Festschrift in honour of Elissaveta Bozilova. Pensoft, Sofia, pp. 21-41.

Lang, G., 2005. Seen und Moore des Schwarzwaldes als Zeugen spätglazialen und holozänen Vegetationswandels. Stratigraphische, pollenanalytische und grossrestanalytische Untersuchungen. Andrias 16, 1-160.

Lang, G., Tobolski, K., 1985. Hobschensee - Late-Glacial and Holocene environment of a lake near the timberline. Diss. Bot. 87, 209-228.

Lang, G., Trautmann, W., 1961. Zur spät- und nacheiszeitlichen Vegetationsgeschichte der Auvergne (Franz. Zentralmassiv). Flora 150, 11-42.

Langdon, P.G., Barber, K.E., 2005. The climate of Scotland over the last 5000 years inferred from multiproxy peatland records: inter-site correlations and regional variability. J. Quat. Sci. 20, 549-566.

Langdon, P.G., Barber, K.E., Hughes, P., 2003. A 7500-year peat-based palaeoclimatic reconstruction and evidence of an 1100-cyclicity in bog surface wetness from Temple Hill Moss, Pentland Hills, southeast Scotland. Quat. Sci. Rev. 22, 259-274.

Langdon, P.G., Barber, K.E., Lomas-Clarke, S.H., 2004. Reconstructing climate and environmental change in Northern England through chironomid and pollen analyses: evidence from Talkin Tarn, Cumbria. J. Paleolimnol. 32, 197-213.

Langgut, D., Cheddadi, R., Carrión, J.S., Cavanagh, M., Colombaroli, D., Eastwood, W., Greenberg, R., Litt, T., Mercuri, A., Miebach, A., Roberts, C., Woldring, H., Woodbridge, J., 2019. The origin and spread of olive cultivation in the Mediterranean Basin: The fossil pollen evidence. Holocene 29, 902-922.

Larcher, W., 2000. Temperature stress and survival ability of Mediterranean sclerophyllous plants. Plant Biosys. 134, 279-295.

Last, W., Smol, J.P. (Eds.), 2001. Tracking Environmental Change Using Lake Sediments, Vol. 2: Physical and Geochemical Methods. Kluwer, Dordrecht.

Latałowa, M., 1982. Postglacial vegetational changes in the eastern Baltic coastal zone of Poland. Acta Palaeobot. 22, 179–249.

Latałowa, M., 1992. Man and vegetation in the pollen diagrams from Wolin Island. Acta Palaeobot. 32, 123–249.

Latałowa, M., Tobolski, K., 1989. Type Region P-u: Baltic shore, W part Wolin Island. Acta Palaeobot. 29, 115–120.

Latałowa, M., van der Knaap, W.O., 2006. Late Quaternary expansion of Norway spruce *Picea abies* (L.) Karts. in Europe according to pollen data. Quat. Sci. Rev. 25, 2780-2805.

Latałowa, M., Swieta-Musznicka, J., Slowinski, M., Pedziszewska, A., Noryskiewicz, A.M., Zimny, M., Obremska, M., Ott, F., Stivrins, N., Pasanen, L., Ilvonen, L., Holmstrom, L., Seppä, H., 2019. Abrupt *Alnus* population decline at the end of the first millennium CE in Europe – The event ecology, possible causes and implications. Holocene 29, 1335-1349.

Latham, R., Ricklefs, R., 1993. Continental comparisons of temperate zone tree species diversity, In: Ricklefs, R., Schluter, D. (Eds.), Species Diversity in Ecological Communities. Historical and Geographical Perspectives. University of Chicago Press, Chicago, pp. 294-314.

Lauterbach, S., Brauer, A., Andersen, N., Danielopol, D.L., Dulski, P., Hüls, M., Milecka, K., Namiotko, T., Obremska, M., von Grafenstein, U., Declakes participants, 2011a. Environmental responses to Lateglacial climatic fluctuations recorded in the sediments of pre-Alpine Lake Mondsee (northeastern Alps). J. Quat. Sci. 26, 253-267.

Lauterbach, S., Brauer, A., Andersen, N., Danielopol, D.L., Dulski, P., Hüls, M., Milecka, K., Namiotko, T., Plessen, B., Grafenstein, U.V., Declakes participants, 2011b. Multi-proxy evidence for early to mid-Holocene environmental and climatic changes in northeastern Poland. Boreas 40, 57–72.

Lauterbach, S., Brauer, A., Litt, T., Schettler, G., 2012. Re-evaluation of the Bispingen Palaeolake record – a revised chronology for the Eemian in Northern Germany. Geophysical Research Abstracts 14, EGU 2012, 8613.

Lavrieux, M., Disnar, J.-R., Chapron, E., Bréheret, J.-G., Jacob, J., Miras, Y., Reyss, J.-L., Andrieu-Ponel, V., Arnaud, F., 2013. 6700 yr sedimentary record of climatic and anthropogenic signals in Lake Aydat (French Massif Central). The Holocene 23, 1317–1328.

Lawson, I., Frogley, M., Bryant, C., Preece, R., Tzedakis, P.C., 2004. The Lateglacial and Holocene environmental history of the Ioannina basin, north-west Greece. Quat. Sci. Rev. 23, 1599-1625.

Lawson, I., Al-Omari, S., Tzedakis, P.C., Bryant, C., Christanis, K., 2005. Lateglacial and Holocene vegetation history at Nisi Fen and the Boras mountains, northern Greece. Holocene 15, 873-887.

Lazarova, M., 1995. Human impact on the natural vegetation in the region of lake Srebarna and mire Garvan (northeastern Bulgaria). Palynological and Palaeoethnobotanical evidence. In: Bozilova, E., Tonkov, S. (Eds.), Advances in Holocene Palaeoecology in Bulgaria. Pensoft, Sofia, pp. 47-67.

Lazarova, M., Bozilova, E., 2001. Studies on the Holocene history of vegetation in the region of lake Srebarna (northeast Bulgaria). Veget Hist Archaeobot 10, 87–95.

Lazarova, M., Tonkov, S., Snowball, I., Marinova, E., 2009. Peat-bog Begbunar (Osogovo Mountains, south-west Bulgaria): Four millennia of vegetation history. Grana 48, 147–149.

Lazarova, M., Tonkov, S., Marinova, E., Ivanov, D., Bozilova, E., 2011. Western Rhodopes Mountains (Bulgaria): peat bog Beliya Kanton. Grana 50, 162–164.

Lazarova, M., Marinova, E., Tonkov, S., Snowball, I., 2015. A 5000-year pollen and plant macrofossil record from the Osogovo Mountain, Southwestern Bulgaria: Vegetation history and human impact. Rev. Palaeobot. Palynol. 223, 1–9.

Lehsten, V., Mischurov, M., Lindström, E., Lehsten, D., Lischke, H., 2019. LPJ-GM 1.0: simulating migration efficiently in a dynamic vegetation model. Geosci. Model Develop. 12, 893-908.

Leira, M., Santos, L., 2002. An early Holocene short climatic event in the northwest Iberian Peninsula inferred from pollen and diatoms. Quat. Int. 93–94, 3–12.

Leng, M.J. (Ed.) 2006. Isotopes in Palaeoenvironmental Research. Springer, Dordrecht. 307 pp.

Leng, M.J., Wagner, B., Boehm, A., Panagiotopoulos, K., Vane, C., Snelling, A., Haidon, C., Woodley, E., Vogel, H., Zanchetta, G., Baneschi, I., 2013. Understanding past climatic and hydrological variability in the Mediterranean from Lake Prespa sediment isotope and geochemical record over the Last Glacial cycle. Quat. Sci. Rev. 66, 123-136.

Leonelli, G., Pelfini, M., di Cella, U.M., Garavaglia, V., 2011. Climate warming and the recent treeline shift in the European Alps: the role of geomorphological factors in high-altitude sites. Ambio 40, 264-273.

Leroi-Gourhan, A., 1980. Interstades Würmiens: Laugerie et Lascaux. Bulletin de l'Association française pour l'Etude du Quaternaire 3, 95-100.

Leroi-Gourhan, A., 1986. The palynology of La Riera cave. Anthropol. Res. Papers 36, 59-64.

Leroy, S.A.G., Colman, S.M., 2001. Coring and drilling equipment and procedures for recovery of long lacustrine sequences. In: Last, W.M., Smol J.P. (Eds.), Tracking Environmental Change Using Lake Sediment, Vol. 1: Basin Analysis, Coring, and Chronological Techniques. Kluwer, Dordrecht, pp. 107-135.

Leunda, M., González-Sampériz, P., Gil-Romera, G., Bartolome, M., Belmonte-Ribas, A., Gomez-Garcia, D., Kaltenrieder, P., Rubiales, J.M., Schwörer, C., Tinner, W., Morales-Molino, C., Sancho, C., 2019. Ice cave reveals environmental forcing of long-term Pyrenean treeline dynamics. J. Ecol. 107, 814-828.

Leunda, M., Gil-Romera, G., Daniau, A.L., Benito, B.M., González-Sampériz, P., 2020. Holocene fire and vegetation dynamics in the central Pyrenees (Spain). Catena 188.

Levins, R., Lewontin, R., 1980. Dialectics and reductionsim in ecology. Synthese 43, 47-78.

Lewis, W.M., 1983. A Revised Classification of Lakes Based on Mixing. Can. J. Fish. Aqut. Sci. 40, 1779-1787.

Lézine, A.-M., von Grafenstein, U., Andersen, N., Belmecheri, S., Bordon, A., Caron, B., Cazet, J.-P., Erlenkeuser, H., Fouache, E., Grenier, C., Huntsman-Mapila, P., Hureau-Mazaudier, D., Manelli, D., Mazaud, A., Robert, C., Sulpizio, R., Tiercelin, J.-J., Zanchetta, G., Zeqollari, Z., 2010. Lake Ohrid, Albania, provides an exceptional multi-proxy record of environmental changes during the last glacial-interglacial cycle. Palaeogeogr. Palaeoclimatol. Palaeoecol. 287, 116-127.

Li, F., Gaillard, M.-J., Sugita, S., Mazier, F., Xu, Q., Zhou, Z., Zhang, Y., Li, Y., Faffly, D., 2017. Relative pollen productivity estimates for major plant taxa of cultural landscape in central eastern China. Veget. Hist. Archaeobot. 26, 587-605.

Li, F., Gao, X., Herzschuh, U., Jia, X., Sugita, S., Tarasov, P.E., Wagner, M., Xu, Q., Chen, F., Sun, A., Gaillard, M.-J., 2018. What do pollen-based quantitative reconstructions of plant cover tell us about past anthropogenic deforestation in eastern China? PAGES Magazine 26, 32-33.

Lian, O.B., 2007. Luminescence Dating: Optically-stimulated Luminescence. In: Elias, S.A. (Ed.), Encyclopedia of Quaternary Science. Elsevier, Amsterdam, pp. 1491-1505.

Lian, O.B., Huntley, D.J., 2001. Luminescence dating. In: Last, W.M., Smol, J.P. (Eds.), Basin Analysis, Coring, and Chronological Techniques. Kluwer Academic Publishers, Dordrecht, pp. 261-282.

Libby, W.F., 1952. Radiocarbon Dating. Chicago University Press, Chicago.

Liepelt, S., Bialozyt, R., Ziegenhagen, B., 2002. Wind-dispersed pollen mediates postglacial gene flow among refugia. Proc. National Acad. Sci. USA 99, 14590-14594.

Liepelt, S., Cheddadi, R., de Beaulieu, J.-L., Fady, B., Gömöry, D., Hussendörfer, E., Konnert, M., Litt, T., Longauer, R., Terhürne-Berson, R., Ziegenhagen, B., 2009. Postglacial range expansion and its genetic imprints in *Abies alba* (Mill.) - A synthesis from palaeobotanic and genetic data. Rev. Palaeobot. Palynol. 153, 139-149.

Liepelt, S., Mayland-Quellhorst, E., Lahme, M., Ziegenhagen, B., 2010. Contrasting geographical patterns of ancient and modern genetic lineages in Mediterranean *Abies* species. Plant Syst. Evol. 284, 141-151.

Linares, J., 2013. Shifting limiting factors for population dynamics and conservation status of the endangered English yew (*Taxus baccata* L., Taxaceae). For. Ecol. Managem. 291, 119-127.

Lindberg, H., 1910. Phytopaläontologische Beobachtungen als Belege für postglaziale Klimaschwankungen in Finnland. Stockholm.

Lindberg, H., 1916. Hvilka vittnesbörd lämnar fytopaeontologin om vårt lands och dess floras utvecklingshistoria sedan istiden samt rörande tiden för människans första uppträdande i landet? Öfvers. Finska Vet.-Soc. Förh., Helsingfors, 58/C.

Lindbladh, M., 1999. The influence of former land-use on vegetation and biodiversity in the boreonemoral zone of Sweden. Ecography 22, 485-498.

Linderholm, H.W., Folland, C.K., Walther, A., 2009. A multicentury perspective on the summer North Atlantic Oscillation (SNAO) and drought in the eastern Atlantic Region. J. Quat. Sci. 24, 415-425.

Lindner, L., Bogutsky, A., Gozhik, P., Marks, L., Łanczont, M., Wojtanowicz, J. 2006. Correlation of Pleistocene deposits in the area between the Baltic and the Black Sea, Central Europe. Geological Quaterly 50/1, 195-210.

Lindner, L., Morks, S., Nita, M., 2013. Climatostratigraphy of interglacials in Poland: Middle and Upper Pleistocene lower boundaries from a Polish perspective. Quat. Int. 292, 113-123.

Lindsay, R.A., 1995. Bogs: The Ecology, Classification and Conservation of Ombrotrophic Mires. Scottish Natural Heritage, Edinburgh.

Lindsay, R.A., Riggall, J., Burd, F., 1985. The use of small-scale surface patterns in the classification of

British peatlands. Aquilo Ser. Bot. 21, 69-79.

Lindsay, R.A., Charman, D., Everingham, F., O'Reilly, R., Palmer, M., Rowell, T., Stroud, D., 1988. The Flow Country. The Peatlands of Caithness and Sutherland. Nature Conservancy Council, Petersborough.

Linke, G., Hallik, R., 1993. Die pollenanalytischen Ergebnisse der Bohrungen Hamburg-Dockenhuden (qho 2) und Hamburg-Billbrook. Geol. Jahrb. A 138, 169-184.

Linkola, K., 1916. Studien über den Einfluss der Kultur auf die Flora in den Gegenden nördlich vom Ladogasee I. Allgemeiner Teil. Acta Soc. Fauna Flora Fennica 45, 1-432.

Lischke, H., 2004. Simulation of dynamic tree species patterns in the Alpine region of Valais (Switzerland) during the Holocene. In: Pahl-Wostl, C., Schmidt, S., Rizzoli, A.E., Jakeman, A.J. (Eds.). Complexity and Integrated Resources Management, Transactions of the 2nd Biennial Meeting of the International Environmental Modelling and Software Society. Osnabrück, pp. 908-913.

Lischke, H., 2005. Modeling tree species migration in the Alps during the Holocene: What creates complexity? Ecol. Complex. 2, 159-174.

Lischke, H., 2020. Simulation der Baumartenmigration im Klimawandel. Schweiz. Z. Forstwes. 171, 151-157.

Lischke, H., Lotter, A.F., Fischlin, A., 2002. Untangling a Holocene pollen record with forest model simulations and independent climate data. Ecol. Model. 150, 1-21.

Lischke, H., Zimmermann, N.E., Bolliger, J., Rickebusch, S., Löffler, T.J., 2006. TreeMig: A forest-landscape model for simulating spatio-temporal patterns from stand to landscape scale. Ecol. Modell. 199, 409-420.

Lischke, H., von Grafenstein, U., Ammann, B., 2013. Forest dynamics during the transition from the Oldest Dryas to the Bølling-Allerød at Gerzensee - a simulation study. Paleogeogr. Paleoclimatol. Paleoecol. 391, 60-73.

Lisitsyna, O.V., Hicks, S., 2012. Do moss samples, pollen traps and modern lake sediments all collect pollen in the same way? A comparison from the forest limit area of northernmost Europe. Veget. Hist. Archaeobot. 21, 187-199.

Lisitsyna, O.V., Giesecke, T., Hicks, S., 2011. Exploring pollen percentages threshold values as an indication for the regional presence of major European trees. Rev. Palaeobot. Palynol. 166, 311-324.

Litt, T., 1990. Stratigraphie und Ökologie des interglazialen Waldelefanten-Schlachtplatzes von Gröbern, Kreis Gräfenhainichen. Veröff. Landesmuseum f. Vorgeschichte Halle, 43, 193-208.

Litt, T., 1992. Fresh investigations into the natural and anthropogenically influenced vegetation of the earlier Holocene in the Elbe-Saale Region, Central Germany. Veget. Hist. Archaeobot. 1, 69-74.

Litt, T., 1994. Paläoökologie, Paläobotanik und Stratigraphie des Jungquartärs im nordmitteleuropäischen Tiefland, unter besonderer Berücksichtigung des Elbe-Saale-Gebietes. Diss. Bot. 227, 1-187.

Litt, T., 2007. The Quaternary as a chronostratigraphical unit. Eiszeitalter und Gegenwart 56, 5-6.

Litt, T., Stebich, M., 1999. Bio- and chronostratigraphy of the lateglacial in the Eifel region, Germany. Quat. Int. 61, 5-16.

Litt, T., Junge, F. W., Böttger, T., 1996. Climate during the Eemian in north-central Europe – a critical review of the palaeobotanical and stable isotope data from central Germany. Veget. Hist. Archaeobot. 5, 247-256.

Litt, T., Brauer, A., Goslar, T., Merkt, J., Balaga, K., Müller, H., Ralska-Jasiewiczowa, M., Stebich, M., Negendank, J., 2001. Correlation and synchronisation of Lateglacial continental sequences in northern central Europe based on annually laminated lacustrine sediments. Quat. Sci. Rev. 20, 1233-1249.

Litt, T., Schmincke, H.U., Kromer, B., 2003. Environmental response to climatic and volcanic events in central Europe during the Weichselian Lateglacial. Quat. Sci. Rev. 22, 7-32.

Litt, T., Behre, K.-E., Meyer, K.-D., Stephan, H.-J., Wansa, S., 2007. Stratigraphische Begriffe für das Quartär des norddeutschen Vereisungsgebietes. Eiszeitalter u. Gegenwart 56, 7-65.

Litt, T., Schmincke, H.-U., Frechen, M., Schlüchter, C., 2008. Quaternary. In: McCann, T. (Ed.), The Geology of Central Europe 2: Mesozoic and Cenozoic. The Geological Spociety, London, pp. 1287-1340.

Litt, T., Schölzel, C., Kühl, N., Brauer, A., 2009. Vegetation and climate history in the Westeifel Volcanic Field (Germany) during the past 11 000 years based on laminated lacustrine maar sediments. Boreas 38, 679 - 690.

Litt, T., Pickarski, N., Heumann, G., Stockhecke, M., Tzedakis, P. C., 2014. A 600 000 year long continental pollen record from Lake Van, eastern Anatolia (Turkey). Quat. Sci. Rev. 104, 30-41.

Livingstone, D.A. 1968. Some interstadial and postglacial pollen diagrams from eastern Canada. Ecological Monographs 38, 87-125.

Livingstone, D.M., Lotter, A.F., Walker, I.R., 1999. The decrease in summer surface water temperature with altitude in Swiss alpine lakes: A comparison with air temperature lapse rates. Arc. Antarc. Alp. Res. 31, 341-352.

Lóczy, D., 1988. Cultural landscape histories in Hungary - two case studies. In: Birks, H.H., Birks, H.J.B., Kaland, P.E., Moe, D. (Eds.), The Cultural Landscape - Past, Present and Future. Cambridge University Press, Cambridge, pp. 165-176.

Löffler, H., 2004. The origin of lake basins, In: O'Sullivan, P., Reynolds, C.S. (Eds.), The Lakes Handbook: Limnology and Limnetic Ecology. Blackwell, Oxford, pp. 8-60.

Lona, F., 1950. Contributi alla storia della vegetacione e del clima nella Val Padana.Analisi pollinica del giacimento villafranciano de Leffe (Bergamo). Atti Soc.It. Sci. Nat. 89, 123-178.

López de Heredia, U., Carrión, J.S., Jiménez, P., Collada, C., Gil, L., 2007. Molecular and palaeobotanical evidence for multiple glacial refugia for evergreen oaks in the Iberian Peninsula. J. Biogeogr. 34, 1505-1517.

López-Merino, L., López-Sáez, J.A., Ruiz Zapata, M.B., Gil García, M.J., 2008. Reconstructing the history of beech (*Fagus sylvatica* L.) in the north-western Iberian Range (Spain): From Late-Glacial refugia to the Holocene anthropic-induced forests. Rev. Palaeobot. Palynol. 152, 58-65.

López-Merino, L., Peña-Chocarro, L., Ruiz-Alonso, M., López-Sáez, J., Sánchez-Palencia, F., 2010. Beyond nature: the management of a productive cultural landscape in Las Médulas (León, Spain). Plant Biosyst. 144, 905-919.

López-Merino, L., Silva Sánchez, N., Kaal, J., López-Sáez, J.A., Martínez Cortizas, A., 2012. Post-disturbance vegetation dynamics during the Late Pleistocene and the Holocene: An example from NW Iberia. Glob. Planet. Change 92–93, 58–70.

López-Sáez, J., López-García, P., Cortés, M., 2007. Paleovgetatión del Cuaternario reciente: estudio arqueopalinológico. In: Cortés, M. (Ed.), Cueva Bajondillo (Torremolinos). Secuencuia cronocultural y paleoambiental del Cuaternario reciente en la Bahía de Málaga. Centro de Editiones de la Deputatión Provincial de Málaga, Málaga, pp. 139-156.

López-Sáez, J., Abel-Schaad, D., Pérez-Díaz, S., Blanco-González, A., Alba-Sánchez, F., Dorado, M., Ruiz-Zapata, M., Gil-García, M., Gómez-González, C., Franco-Múgica, F., 2014. Vegetation history, climate and human impact in the Spanish Central System over the last 9,000 years Quat. Int. 353, 98-122.

Lotter, A.F., 1988. Paläoökologische und paläolimnologische Studie des Rotsees bei Luzern. Pollen-, grossrest-, diatomeen- und sedimentanalytische Untersuchungen. Diss. Bot. 124, 1-187.

Lotter, A.F., 1999. Late-glacial and Holocene vegetation history and dynamics as evidenced by pollen and plant macrofossil analyses in annually laminated sediments from Soppensee (central Switzerland). Veget. Hist. Archaeobot. 8, 165-184.

Lotter, A.F., Birks, H.J.B., 1993. The impact of the Laacher See Tephra on terrestrial and aquatic ecosystems in the Black Forest, southern Germany. J. Quat. Sci. 8, 263-276.

Lotter, A.F., Birks, H.J.B., 2003a. The Holocene palaeolimnology of Sägistalsee and its environmental history - a synthesis. J. Paleolimnol. 30, 333-342.

Lotter, A.F., Birks, H.J.B., 2003b. Holocene sediments of Sägistalsee, a small lake at the present-day tree-line in the Swiss Alps. J. Paleolimnol. 30, 253-260.

Lotter, A.F., Hölzer, A., 1994. A high-resolution late-glacial and early Holocene environmental history of Rotmeer, southern Black Forest (Germany). Diss. Bot. 234, 365-388.

Lotter, A.F., Kienast, F., 1992. Validation of a forest succession model by means of annually laminated sediments. Geol. Survey Finland Special Paper 14, 25-31.

Lotter, A.F., Ammann, B., Sturm, M., 1992a. Rates of change and chronological problems during the late-glacial period. Climate Dynamics 6, 233-239.

Lotter, A.F., Eicher, U., Siegenthaler, U., Birks, H.J.B., 1992b. Late-glacial climatic oscillations as recorded in Swiss lake sediments. J. Quat. Sci. 7, 187-204.

Lotter, A.F., Walker, I.R., Brooks, S.J., Hofmann, W., 1999. An intercontinental comparison of chironomid palaeotemperature inference models: Europe vs North America. Quat. Sci. Rev. 18, 717-735.

Lotter, A.F., Birks, H.J.B., Eicher, U., Hofmann, W., Schwander, J., Wick, L., 2000. Younger Dryas and Allerod summer temperatures at Gerzensee (Switzerland) inferred from fossil pollen and cladoceran assemblages. Palaeogeogr. Palaeoclim. Palaeoecol. 159, 349-361.

Lotter, A.F., Heiri, O., Hofmann, W., van der Knaap, W.O., van der Leeuwen, J.F.N., Walker, I.R., Wick, L., 2006. Holocene timber-line dynamics at Bachalpsee, a lake at 2265 m a.s.l. in the northern Swiss Alps. Veget. Hist. Archaeobot. 15, 295-307.

Lotter, A.F., Heiri, O., Brooks, S.J., van Leeuwen, J.F.N., Eicher, U., Ammann, B., 2012. Rapid summer temperature changes during Termination 1a: high-resolution multi-proxy climate reconstructions from Gerzensee (Switzerland). Quat. Sci. Rev. 36, 103-113.

Lousley, J., 1951. Recent research and additions to the British flora. Nature 168.

Löve, Á., Löve, D., 1975. Cytotaxonomical Atlas of the Arctic Flora Cramer, Vaduz.

Lowe, S., 1981. Radiocarbon dating and stratigraphic resolution in Welsh lateglacial chronology. Nature 293, 210-212.

Lowe, J.J., Lowe, S., 1989. Interpretation of the pollen stratigraphy of Late Devensian lateglacial and early Flandrian sediments at Llyn Gwernan, near Cader

Idris, North Wales. New Phytol. 113, 391-408.

Lowe, J.J., Lowe, S., Fowler, A.J., Hedges, R.E.M., Austin, T.J.F., 1988. Comparison of accelerator and radiometric radiocarbon measurements obtained from Late Devesian Lateglacial lake sediments from Llyn Gwernan, North Wales, UK. Boreas 17, 355-369.

Lowe, J.J., Accorsi, C.A., Bandini Mazzanti, M., Bishop, A., van der Kaars, S., Forlani, L., Mercuri, A.M., Rivalenti, C., Torri, P., Watson, C., 1996. Pollen stratigraphy of sediment sequences from lakes Albano and Nemi (near Rome) and from the central Adriatic, spanning the interval from oxygen isotope Stage 2 to the present day. In: Guilizzoni, P., Oldfield, F. (Eds.), Palaeoenvironmental Analysis of Italian Crater Lake and Adriatic Sediments, pp. 71-98.

Lowe, J.J., Hoek, W.Z., and INTIMATE Members, 2001. Inter-regional correlation of palaeoclimatic records for the Last Glacial–Interglacial Transition: a protocol for improved precision recommended by the INTIMATE project group. Quat. Sci. Rev. 20, 1175-1187.

Lowe, J.J., Rasmussen, S.O., Björck, S., Hoek, W.Z., Steffensen, J.P., Walker, M.J.C., Yu, Z.C., and INTIMATE Members, 2008. Synchronisation of palaeoenvironmental events in the North Atlantic region during the Last Termination: a revised protocol recommended by the INTIMATE group. Quat. Sci. Rev. 27, 6-17.

Lucas, L., Colledge, S., Simmons, A., Fuller, D., 2012. Crop introduction and accelerated island evolution: archaeobotanical evidence from *'Ais Yiorkis* and Pre-Pottery Neolithic Cyprus. Veget. Hist. Archaeobot. 21, 117-129.

Lumaret, R., Ouazzani, N., 2001. Ancient wild olives in Mediterranean forests. Nature Commun. 413, 700.

Lumaret, R., Amane, M., Ouazzani, N., Baldoni, L., Debain, C., 2000. Chloroplast DNA variation in the cultivated and wild olive taxa of the genus *Olea* L. Theor. Appl. Genet. 101, 547-553.

Lumaret, R., Mir, C., Michaud, H., Raynal, V., 2002. Phylogeographical variation of chloroplast DNA in holm oak (*Quercus ilex* L.) Mol. Ecol.11, 2327-2336.

Lumaret, R., Ouazzani, N., Michaud, H., Vivier, G., Deguilloux, M., Di Giusto, F., 2004. Allozyme variation of oleaster populations (wild olive tree) (*Olea europaea* L.) in the Mediterranean Basin. Heredity 92, 343-351.

Lumaret, R., Tryphon-Dionnet, M., Michaud, H., Sanuy, A., Ipotesi, E., Born, C., Mir, C., 2005. Phylogeographical variation of chloroplast DNA in Cork Oak (*Quercus suber*). Annals Bot. 96, 853-861.

Lundqvist, G., 1927. Bodenablagerungen und Entwicklungstypen der Seen. Vol. 2. E. Schweizerbart.

Lundqvist, G., 1959. 14C-daterade tallstubbar fran fjällen. Sver. Geol. Unders. Ser. C 565, 1-21.

Lundqvist, J., 1967. Submoräna sediment i Jämtlands län. Sverig. Geol. Unders. C 618.

Lundqvist, J., 1969. Beskrivning till jordartskarta över Jämtlands län. Sver. Geol. Unders., Stockholm, Ser. Ca 45, 1-418.

Lundqvist, J., 1981. Weichselian (Vistulian) stratigraphy of Sweden. Quaternary Studies in Poland 3, 57-59.

Lüning, J., 2000. Steinzeitliche Bauern in Deutschland. Die Landwirtschaft im Neolithikum. Universitätsforschungen zur prähistorischen Archäologie, 58. Bonn, 285 pp.

Luther, H., 1945. Über die rezenten Funde von *Najas flexilis* (Willd.) Rostk. & Schmidt in Ostfennoskandien. Mem. Soc. Fauna Flora Fenn. 21.

Lyell, C., 1830-1833. Principles of Geology. 3 volumes. London.

Lynch, E.A., 1998. Origin of a park-forest vegetation mosaic in the Wind River Range, Wyoming. Ecology 79, 1320-1338.

Lyons, S.K., Amatangelo, K.L., Behrensmeyer, A.K., Bercovici, A., Blois, J.L., Davis, M., DiMichele, W.A., Du, A., Eronen, J.T., Faith, J.T., Graves, G.R., Jud, N., Labandeira, C., Looy, C.V., McGill, B.J., Miller, J.H., Patterson, D., Pineda-Munoz, S., Potts, R., Riddle, B., Terry, R., Tóth, A., Ulrich, W., Villaseñor, A., Wing, S.L., Anderson, H., Anderson, J., Waller, D., Gotelli, N.J., 2016. Holocene shifts in the assembly of plant and animal communities implicate human impacts. Nature 529, 80-83.

Lytle, D.E., Wahl, E.R., 2005. Palaeoenvironmental reconstructions using the modern analogue technique: the effects of sample size and decision rules. Holocene 15, 554-566.

Maberly, S.C., 1985. Photosynthesis by *Fontinalis antipyretica*. II. Assessment of environmental factors limiting photosynthesis and production. New Phytol. 100, 141-155.

Maberly, S.C., Madsen, T., 1998. Affinity for CO_2 in relation to the ability of freshwater macrophytes to use HCO_3^-. Funct. Ecol. 12, 99-106.

MacArthur, R.H., Wilson, E.O., 1967. The Theory of Island Biogeography. Princeton University Press, Princeton. 224 pp.

MacDonald, G.M., 1993. Fossil pollen analysis and the reconstruction of plant invasions. Adv. Ecol. Res. 24, 67-110.

MacDonald, G.M., 2010. Global warming and the Arctic: a new world beyond the reach of the Grinnellian niche? J. Exp. Biol. 213, 855-861.

Mackey, R.L., Currie, D.J., 2001. The diversity–disturbance relationship: is it generally strong and peaked? Ecology 82, 3479-3492.

Macklin, M.G., Bonsall, C., Davies, F.M., Robinson, M.R., 2000. Human-environment interactions during the Holocene: new data and interpretations from the Oban area, Argyll, Scotland. Holocene 10, 109-121.

Magny, M., 1992. Holocene lake-level fluctuations in Jura and the northern subalpine ranges, France: regional pattern and climatic implications. Boreas 21, 319-334.

Magny, M., 1993. Solar influences on Holocene climatic changes illustrated by correlations between past lake-level fluctuations and the atmospheric ^{14}C record. Quat. Res. 40, 1-9.

Magny, M., 1994. Holocene climatic oscillations in Jura and Alps: oceanic and solar forcing? C.R. Acad. Sci. Paris, sér. II, Géosciences de surface, Paléoclimatologie 319, 667-674.

Magny, M., 1995. Paleohydrological changes in Jura (France), and climatic oscillations around the North Atlantic from Alleröd to Preboreal. Géographie physiques et Quaternaire 49, 401-408.

Magny, M., 1998. Reconstruction of Holocene lake-level changes in the French Jura: methods and results. In: Frenzel, B. (Ed.), Palaeohydrology as Reflected in lake-level changes as climatic evidence for Holocene times. Gustav Fischer, Stuttgart, pp. 67-85.

Magny, M., 1999. Lake-level fluctuations in the Jura and French sub-alpine ranges associated with ice-rafting events in the North Atlantic and variations in the polar atmospheric circulation. Quaternaire 10, 61-64.

Magny, M., 2001. Palaeohydrological changes as reflected by lake-level fluctuations in the Swiss Plateau, the Jura Mountains and the northern French Pre-Alps during the Last Glacial-Holocene transition: a regional synthesis. Global Planet. Change 30, 85-101.

Magny, M., 2004. Holocene climate variability as reflected by mid-European lake-level fluctuations and its probable impact on prehistoric human settlements. Quat. Int. 133, 65-79.

Magny, M., 2007. Lake Level Studies/West-Central Europe, In: Elias, S.A. (Ed.), Encyclopedia of Quaternary Science. Elsevier, Amsterdam, pp. 1389-1399 (reprinted 2013).

Magny, M., Ruffaldi, P., 1995. Younger Dryas and early Holocene lake-level fluctuations in the Jura mountains, France. Boreas 24, 155-172.

Magny, M., Guiot, J., Schoellammer, P., 2001. Quantitative reconstruction of Younger Dryas to mid Holocene paleoclimates at Le Locle, Swiss Jura, using pollen and lake-level data. Quat. Res. 56, 170-180.

Magny, M., Bégeot, C., Guiot, J., Peyron, O., 2003. Contrasting patterns of hydrological changes in Europe in response to Holocene climate cooling phases. Quat. Sci. Rev. 22, 1589-1596.

Magny, M., Leuzinger, U., Bortenschlager, S., Haas, J., 2006. Tripartite climate reversal in Central Europe 5600-5300 years ago. Quat. Res. 65, 3-19.

Magny, M., Bossuet, G., Ruffaldi, P., Leroux, A., Mouthon, J., 2011. Orbital imprint on Holocene palaeohydrological variations in west-central Europe as reflected by lake-level changes at Cerin (Jura Mountains, eastern France). Quat. Sci. 26, 171-177.

Magny, M., Peyron, O., Sadori, L., Ortu, E., Zanchetta, G., Vannière, B., Tinner, W., 2012. Contrasting patterns of precipitation seasonality during the Holocene in the south- and north-central Mediterranean. J. Quat. Sci. 27, 290-296.

Magri, D., 1999. Late Quaternary vegetation history al Lagaccione near Lago di Bolsena (central Italy). Rev. Palaeobot. Palynol. 106, 171-208.

Magri, D., 2008. Patterns of post-glacial spread and the extent of glacial refugia of European beech (*Fagus sylvatica*). J. Biogeogr. 35, 450-463.

Magri, D., Sadori, L., 1999. Late Pleistocene and Holocene pollen stratigraphy at Lago Vico, central Italy. Veg. Hist. Archaeobot. 8, 247-260.

Magri, D., Vendramin, G.G., Comps, B., Dupanloup, I., Geburek, T., Gömöry, D., Latałowa, M., Litt, T., Paule, L., Roure, J., Tanţău, I., van der Knaap, W.O., Petit, R., de Beaulieu, J.-L., 2006. A new scenario of the Quaternary history of European beech populations: palaeobotanical evidence and genetic consequences. New Phytol. 171, 199-221.

Magri, D., Fineschi, S., Bellarosa, R., Buonamici, A., Sebastiani, F., Schirone, B., Simeone, M.C., Vendramin, G.G., 2007. The distribution of *Quercus suber* chloroplast haplotypes matches the palaeogeographical history of the western Mediterranean. Mol. Ecol. 16, 5259-5266.

Magri, D., Agrillo, E., Di Rita, F., Furlanetto, G., Pini, R., Ravazzi, C., Spada, F., 2015. Holocene dynamics of tree taxa populations in Italy Rev. Palaeobot. Palynol. 218, 267-284.

Magyari, E.K., Jakab, G., Rudner, E., Sümegi, P., 1999. Palynological and plant macrofossil data on late Pleistocene short-term climatic oscillations in north-eastern Hungary. Acta Palaeobot. Suppl. 2, 491-502.

Magyari, E.K., Jakab, G., Sümegi, P., Rudner, E., Molnár, 2000. Paleobotanikai vizsgálatok a keleméri Mohostavakon. [Palaeobotanical investigations on Mohos-lakes, Kelemér, N.-Hungary.] (in Hungarian). In: Szurdoki, E. (Ed.), Tözegmohás élöhelyek Magyarországon: kutatás, kezelés, védelem (Peat Bog Habitats in Hungary: Research, Conservation, Management). Miskolc, pp. 101-131.

Magyari, E.K., Sümegi, P., Braun, M., Jakab, G.,

Molnár, M., 2001. Retarded wetland succession: anthropogenic and climatic signals in a Holocene peat bog profile from the NE Carpathian Basin. J. Ecol. 89, 1019–1032.

Magyari, E.K., Buczkó, K., Jakab, G., Braun, M., Szántó, Z., Molnár, M., Pál, Z., Karátson, D., 2006. Holocene palaeohydrology and environmental history in the South Harghita Mountains, Romania. Földtani Közlöny 136, 249-284.

Magyari, E.K., Jakab, G., Sümegi, P., Szöör, G., 2008. Holocene vegetation dynamics in the Bereg Plain, NE Hungary - the Báb-tava pollen and plant macrofossil record. Acta GGM Debrecina Series 3, 33-50.

Magyari, E.K., Braun, M., Buczkó, K., Kern, Z., László, P., Hubay, K., Bálint, M., 2009a. Radiocarbon chronology of glacial lake sediments in the Retezat Mts (South Carpathians, Romania): a window to Late Glacial and Holocene climatic and paleoenvironmental changes. Cent. Eur. Geol. 52, 225–248.

Magyari, E.K., Buczkó, K., Jakab, G., Braun, M., Pál, Z., Karátson, D., Pap, I., 2009b. Palaeolimnology of the last crater lake in the Eastern Carpathian Mountains: a multiproxy study of Holocene hydrological changes. Hydrobiologia 631, 29-63.

Magyari, E.K., Chapman, J.C., Passmore, D.G., Allen, J.R.M., Huntley, J.P., Huntley, B., 2010. Holocene persistence of wooded steppe in the Great Hungarian Plain. J. Biogeogr. 37, 915-935.

Magyari, E.K., Major, A., Bálint, M., Nédli, J., Braun, M., Rácz, I., Parducci, L., 2011. Population dynamics and genetic changes of *Picea abies* in the South Carpathians revealed by pollen and ancient DNA analyses. BMC Evol. Biol. 11, 66.

Magyari, E.K., Chapman, J.C., Fairbairn, A., Francis, M., de Guzman, M., 2012a. Neolithic human impact on the landscapes of North-East Hungary inferred from pollen and settlement records. Veget. Hist. Archaeobot. 21, 278-302.

Magyari, E.K., Jakab, G., Bálint, M., Kern, Z., Buczkó, K., Braun, M., 2012b. Rapid vegetation response to Lateglacial and early Holocene climatic fluctuation in the South Carpathian Mountains (Romania). Quat. Sci. Rev. 35, 116-130.

Magyari, E.K., Veres, D., Wennrich, V., Wagner, B., Braun, M., Jakab, G., Karátson, D., Pál, Z., Ferenczy, G., St-Onge, G., Rethemeyer, J., Francois, J.-P., von Reumont, F., Schäbitz, F., 2014. Vegetation and environmental responses to climate forcing during the Last Glacial Maximum and deglaciation in the East Carpathians: attenuated response to maximum cooling and increased biomass burning. Quat. Sci. Rev. 106, 278-298.

Maher, L.J., 1964. *Ephedra* pollen in sediments of the Great Lakes Region. Ecology 45, 391-395.

Maher, L.J., 1981. Statistics for microfossil concentration measurements employing samples spiked with marker grains. Rev. Palaeobot. Palynol. 32, 153-191.

Mai, D.H., 1981. Entwicklung und klimatische Differenzierung der Laubwaldflora Mitteleuropas im Tertiär. Flora 171, 525-582.

Mai, D.H., 1988. Einige exotische Gehölze in den Interglazialfloren der Mitteleuropäischen Florenregion. Feddes Repertorium 99, 419-461.

Mai, D.H., 1989. Development and regional differentiation of the European vegetation during the Tertiary. Plant Syst. Evol. 162, 79-91.

Mai, D.H., 1995. Tertiäre Vegetationsgeschichte Europas: Methoden und Ergebnisse. Gustav Fischer, Jena.

Maier, U., 1996. Morphological studies of free-threshing wheat ears from a Neolithic site in southwest German, and the history of the naked wheats. Veget. Hist. Archaeobot. 5, 39-55.

Maier, U., 2001. Archäobotanische Untersuchungen in der neolithischen Ufersiedlung Hornstaad-Hönle IA am Bodensee. Forschungen und Berichte zur Vor- und Frühgeschichte in Baden-Württemberg, Stuttgart 74, 9-384.

Makhortykh, S., de Capitani, A., 2017. Archaeology and Palaeoecology of the Ukrainian Steppe. National Academy of Sciences of the Ukraine, Kyiv.

Mäkirinta, U., 1978.Ein neues ökomorphologisches Lebensformensystem der aquatischen Makrophyten. Phytocoenol. 4, 446-470.

Maliouchenko, O., Palme, A.E., Buonamici, A., Vendramin, G.G., Lascoux, M., 2007. Comparative phylogeography and population structure of European *Betula* species, with particular focus on *B. pendula* and *B. pubescens*. J. Biogeogr. 34, 1601-1610.

Malmström, C., 1920. *Trapa natans* L. i Sverige. Svensk Bot. Tidskr., Stockholm, 14.

Mamakowa, K., 1989. Late Middle Polish glaciation, Eemian and early Vistulian vegetation at Imbramowice near Wroclaw and the pollen stratigraphy of this part of the Pleistocene in Poland. Acta Palaeobot. 29, 11-176.

Mamakowa, K., Velichkevich, F.Y., 1993. *Aracites interglacialis* Wieliczk. - extinct plant found in the floras of the Mazovian (Alexandrian, Likhvinian) interglacial in Poland, Belarus, Russia, and the Ukraine. Acta Palaeobot. 33, 321-341.

Mangerud, J., 1970. Late Weichselian Vegetation and Ice-Front Oscillations in the Bergen District, Western Norway. Nor. Geogr. Tidsskr. - Nor. J. Geogr. 24, 121–148.

Mangerud, J., 1972. Radiocarbon dating of marine shells, including a discussion of apparent ages of Recent shells from Norway. Boreas 1, 143-172.

Mangerud, J., 1982. The chronostratigraphical subdivision of the Holocene in Norden: a review. Striae 16, 65-70.

Mangerud, J., Andersen, S.T., Berglund, B.E., Donner, J.J., 1974. Quaternary stratigraphy of Norden, a proposal for terminology and classification. Boreas 3, 109-128.

Mangerud, J., Gulliksen, S., Longvar, O., Miller, G.H., Sejrup, H.P., Sønstegaard, E., 1981. A middle-Weichselian ice-free period in Western Norway: the Ålesund Interstadial. Boreas 10, 447-462.

Mangerud, J., Astakhov, V., Svendsen, J.-I., 2002. The extent of the Barents-Kara ice sheet during the Last Glacial Maximum. Quat. Sci. Rev. 21, 111-119.

Mangerud, J., Lundqvist, J., Ehlers, J., Gibbard, P., 2007. Late Glacial Events in Northwest Europe. In: Elias, S.A. (Ed.), Encyclopedia of Quaternary Science. Elsevier, Amsterdam, pp. 1116-1122.

Manos, P., Doyle, J., Nixon, K., 1999. Phylogeny, biogeography, and processes of molecular differentiation in *Quercus* subgenus *Quercus* (Fagaceae). Mol. Phylogenet. Evol. 12, 333-349.

Manzano, S., Carrión, J.S., López-Merino, L., González-Sampériz, P., Munuera, M., Fernández, S., Martín-Lerma, I., del Carmen Gómez Ferreras, M., 2017. Mountain strongholds for woody angiosperms during the Late Pleistocene in SE Iberia. Catena 149, 701-712.

Mao, K., Gang, H., Liu, J., Adams, R., Milne, R.I., 2010. Diversification and biogeography of *Juniperus* (Cupressaceae): variable diversification rates and multiple intercontinental dispersals. New Phytol. 188, 254-272.

Marambat, L., 1992. Paleoenvironnements et empreinte anthropique dans l'ouest Aquitaine et la Saintonge a l'Holocene. L'apport de la palynologie. Université de Bordeaux I.

Marchant, R., Brewer, S., Webb, T., Turvey, S.T., 2009. Holocene deforestation: a history of human-environmental interactions, climate change, and extinction. In: Turvey, S.T. (Ed.), Holocene Extinctions. Oxford University Press, Oxford, pp. 213-233.

Marcisz, K., Tinner, W., Colombaroli, D., Kołaczek, P., Słowiński, M., Fiałkiewicz-Kozieł, B., Łokas, E., Lamentowicz, M., 2015. Long-term hydrological dynamics and fire history over the last 2000 years in CE Europe reconstructed from a high-resolution peat archive. Quat. Sci. Rev. 112, 138-152.

Marcisz, K., Jassey, V.E.J., Kosakyan, A., Krashevska, V., Lahr, D.J.G., Lara, E., Lamentowicz, Ł., Lamentowicz, M., Macumber, A., Mazei, Y., Mitchell, E.A.D., Nasser, N.A., Patterson, R.T., Roe, H.M., Singer, D., Tsyganov, A.N., Fournier, B., 2020. Testate amoeba functional traits and their use in paleoecology. Front. Ecol. Evol. 8, 340.

Mardones, M., Jalut, G., 1983. Paleoecological approach of the last 45,000 years to the peat bog of Biscaye (altitude 409 meters, Hautes Pyrenees). Pollen Spores 25, 163–211.

Margari, V., Gibbard, P., Bryant, C., Tzedakis, P., 2009. Character of vegetational and environmental changes in southern Europe during the last glacial period; evidence from Lesvos Island, Greece. Quat. Sci. Rev. 28, 1317–1339.

Margaritis, E., Jones, M., 2008a. Crop processing of *Olea europaea* L.: an experimental approach for the interpretation of archaeobotanical olive remains. Veget. Hist. Archaeobot. 17, 381–392.

Margaritis, E., Jones, M., 2008b. Olive oil production in Hellenistic Greece: the interpretation of charred olive remains from the site of Tria Platania, Macedonia, Greece (fourth–second century B.C.). Veget. Hist. Archaeobot. 17, 393-401.

Mariani, M., Connor, S.E., Theuerkauf, M., Kuneš, P., Fletcher, M.-S., 2016. Testing quantitative pollen dispersal models in animal-pollinated vegetation mosaic: An example from temperate Tasmania, Australia. Quat. Sci. Rev. 154, 214-225.

Mariani, M., Connor, S.E., Fletcher, M.-S., Theuerkauf, M., Kuneš, P., Jacobson, G., Saunders, K.M., Zawadzki, A., 2017. How old is the Tasmanian Cultural Landscape? A test of landscape openness using quantitative land-cover reconstructions. J. Biogeogr. 44, 2410-2420.

Maringer, J., Ascoli, D., Küffer, N., Schmidtlein, S., Conedera, M., 2016. What drives European beech (*Fagus sylvatica* L.) mortality after forest fires of varying severity? For. Ecol. Manage. 368, 81-93.

Maringer, J., Wohlgemuth, T., Hacket-Pain, A., Ascoli, D., Berretti, R., Conedera, M., 2020. Drivers of persistent post-fire recruitment in European beech forests. Sci. Total Environ. 699.

Marinova, E., 2003. The new pollen core Lake Durankulak-3: a contribution to the vegetation history and human impact in northeastern Bulgaria. In: Tonkov, S. (Ed.), Aspects of Palynology and Palaeoecology. Pensoft, Sofia, pp. 257-268.

Marinova, E., Tonkov, T., Bozilova, E., Vajsov, I., 2012. Holocene anthropogenic landscapes in the Balkans: the palaeobotanical evidence from southwestern Bulgaria. Veget. Hist. Archaeobot. 21, 413-427.

Maristo, L., 1941. Die Seentypen Finnlands auf floristischer und vegetationsphysiognomischer Grundlage. Ann. Bot. Soc. Zool. Bot. Fenn. Vanamo 15, 1-314.

Markgraf, V., 1969. Moorkundliche und vegetationsgeschichtliche Untersuchungen an einem Moorsee an der Waldgrenze im Wallis. Botanische Jahrbuecher 89, 1-63.

Markgraf, V., 1980. Pollen dispersal in a mountain area. Grana 19, 127-146.

Marks, L., Piotrowski, J.A., Stephan, H.-J., Fedorowicz, S., Butrym, J., 1995. Thermoluminescence indications of the middle Weichselian (Vistulian) Glaciation in Northern Germany. Meyniana 47, 69-82.

Marlon, J.R., Bartlein, P.J., Daniau, A.-L., Harrison, S.P., Maezumi, S.Y., Power, M.J., Tinner, W., Vannière, B., 2013. Global biomass burning: a synthesis and review of Holocene paleofire records and their controls. Quat. Sci. Rev. 65, 5-25.

Marquer, L., Gaillard, M.-J., Sugita, S., Trondman, A.-K., Mazier, F., Nielsen, A.B., Fyfe, R.M., Odgaard, B.V., Alenius, T., Birks, H.J.B., Bjune, A.E., Christiansen, J., Dodson, J., Edwards, K.J., Giesecke, T., Herzschuh, U., Kangur, M., Lorenz, S., Poska, A., Schult, M., Seppä, H., 2014. Holocene changes in vegetation composition in northern Europe: why quantitative pollen-based vegetation reconstructions matter. Quat. Sci. Rev. 90, 199-216.

Marquer, L., Gaillard, M.-J., Sugita, S., Poska, A., Trondman, A.-K., Mazier, F., Nielsen, A.B., Fyfe, R.M., Jönsson, A.M., Smith, B., Kaplan, J.O., Alenius, T., Birks, H.J.B., Bjune, A.E., Christiansen, J., Dodson, J., Edwards, K.J., Giesecke, T., Herzschuh, U., Kangur, M., Koff, T., Latałowa, M., Lechterbeck, J., Olofsson, J., Seppä, H., 2017. Quantifying the effects of land use and climate on Holocene vegetation in Europe. Quat. Sci. Rev. 171, 20-37.

Marquer, L., Mazier, F., Sugita, S., Galop, D., Houet, T., Faure, E., Gaillard, M.-J., Haunold, S., de Munnik, N., Simonneau, De Vleeschauwer, F., Le Roux, G., 2020. Pollen-based reconstruction of Holocene land-cover in mountain regions: evaluation of the Landscape Reconstruction Algorithm in the Vicdessos valley, northern Pyrenees, France. Quat. Sci. Rev. 228.

Marsicek, J.P., Shuman, B., Brewer, S., Foster, D.R., Oswald, W.W., 2013. Moisture and temperature changes associated with the mid-Holocene *Tsuga* decline in the northeastern United States. Quat. Sci. Rev. 80, 129-142.

Martin, C., Ménot, G., Thouveny, N., Peyron, O., Andrieu-Ponel, V., Montade, V., Davtian, N., Reille, M., Bard, E., 2020. Early Holocene Thermal Maximum recorded by branched tetraethers and pollen in Western Europe (Massif Central, France). Quat. Sci. Rev. 228.

Martinelli, E., Michetti, A.M., Colombaroli, D., Mazzola, E., De Carlo, S.M., Livio, F., Gilli, A., Ferrario, M.F., Hobig, N., Brunamonte, F., Castelletti, L., Tinner, W., 2017. Climatic and anthropogenic forcing of prehistorical vegetation succession and fire dynamics in the Lago di Como area (N-Italy, Insubria). Quat. Sci. Rev. 161, 45-67.

Martinetto, E., 2009. Palaeoenvironmental significance of plant macrofossils from the Pianico Formation, Middle Pleistocene of Lombardy, North Italy. Quat. Int. 204, 20-30.

Martínez-Cortizas, A., Pontevedra-Pombal, X., García-Rodeja, E., Nóvoa-Munoz, J.C., Shotyk, W., 1999. Mercury in a Spanish peat bog: Archive of climate change and atmospherical metal deposition. Science 284, 939-942.

Martínez-López, V., García, C., Zapata, V., Robledano, F., De la Rúa, P., 2020. Intercontinental long-distance seed dispersal across the Mediterranean Basin explains population genetic structure of a bird-dispersed shrub. Mol. Ecol. 29, 1408-1420.

Masing, V., 1975. Mire typology of the Estonian S.S.R. In: Laasimer, L. (Ed.), Some aspects of botanical research in the Estonian S.S.R. Tartu: Academy of Sciences of the Estonian SSR, Tartu, pp. 123-138.

Mateus, J.E., 1992. Holocene and present-day ecosystems of the Carvalhal Region, Southwest Portugal. PhD thesis, University of Utrecht, The Netherlands.

Mateus, J.E., Queiroz, P., 1993. Os estudos da vegetação quaternária em Portugal; contextos, balanço de resultados, perspectivas In: Carvalho, G., Ferreira, A., Senna-Martínez, J. (Eds.), Quaternário em Portugal. Balanço e perspectivas. Associação Portuguesa para o Estudo do Quaternário, Lisboa, pp. 105-131.

Matthias, I., Semmler, M.S.S., Giesecke, T., 2015. Pollen diversity captures landscape structure and diversity. J. Ecol. 103, 880-890.

Matuszkiewicz, W., 1989. Über die standörtliche und regionale Gliederung der Buchenwälder in ihrem ost-europäischen Rand-Areal. In: Pott, R. (Ed.), Buchenwald-Symposium, Göttingen, pp. 83-92.

Matveev, A.V., Krutos, E.A., Zernitskaya, V.P., 1993. Geochronology of the Holocene of the Byelorussian Polessie. Radiocarbon 35, 435-439.

Mauquoy, D., van Geel, B., Blaauw, M., Speranza, A.O.M., van der Plicht, J., 2004. Changes in solar activity and Holocene climate shifts derived from ^{14}C wiggle-matched dated peat deposits. Holocene 14, 45-52.

Mayer, H., 1966. Waldgeschichte des Berchtesgadener Landes (Salzburger Kalkalpen). Beih. Forstwiss. Cbl. 22.

Mayer, H., 1970. Waldgrenzen in den Berchtesgadener Kalkalpen. Mitt. Ostalp.-Dinar. Ges. Veg.-kde. 11, 109-120.

Mayer, H., 1984. Wälder Europas. Fischer, Stuttgart, New York.

Mayer, H., Aksoy, H., 1986. Wälder der Türkei. Fischer, Stuttgart.

Mazei, Y., Tsyganov, A.N., 2006. Freshwater Testate Amoebae. KMK, Moscow.

Mazier, F., Nielsen, A.B., Broström, A., Sugita, S.,

Hicks, S., 2012a. Signals of tree volume and temperature in a high-resolution record of pollen accumulation rates in northern Finland. J. Quat. Sci. 27, 564-574.

Mazier, F., Gaillard, M.-J., Kuneš, P., Sugita, S., Trondman, A.-K., Broström, A., 2012b. Testing the effect of site selection and parameter setting on REVEALS model estimates of plant abundance using the Czech Quaternary Palynological Database. Rev. Palaeobot. Palynol. 187, 38-49.

Mazier, F., Broström, A., Bragée, P., Fredh, D., Stenberg, L., Thiere, G., Sugita, S., Hammarlund, D., 2015. Two hundred years of land-use change in the South Swedish Uplands: comparison of historical map-based estimates with a pollen-based reconstruction using the Landscape Reconstruction Algorithm. Veget. Hist. Archaeobot. 24, 555-570

McAndrews, J.H., Berti, A.A., Norris, G., 1973. Key to the Quaternary Pollen and Spores of the Great Lakes Region. Toronto Royal Univ. Mus. Miscell. Publ.

McCarroll, D., Nesje, A., 1993. The vertical extent of ice sheets in Nordfjord, western Norway: measuring degree of rock surface weathering. Boreas 22, 255–265.

McCarroll, J., Chambers, F.M., Webb, J.C., Thom, T., 2016. Informing innovative peatland conservation in light of palaeoecological evidence for the demise of *Sphagnum imbricatum*: the case of Oxenhope Moor, Yorkshire, UK. Mires and Peat 18, Art. 08, 1-2.

McCarroll, J., Chambers, F.M., Webb, J.C., Thom, T., 2017. Application of palaeoecology for peatland conservation at Mossdale Moor, UK. Quat. Int. 432, 39-47.

McDermott, F., 2014. Palaeo-climate reconstruction from stable isotope variations in speleothems: a review. Quat. Sci. Rev. 23, 901-918.

McGarry, S.F., Caseldine, C., 2004. Speleothem Palynology: an undervalued tool in Quaternary studies. Quat. Sci. Rev. 23, 2389–2404.

McGill, B.J., Enquist, B.J., Weiher, E., Westoby, M., 2006. Rebuilding community ecology from functional traits. Trends Ecol. Evol. 21, 178-185.

McGill, B.J., Dornelas, M., Gotelli, N.J., Magurran, A.E., 2015. Fifteen forms of biodiversity trend in the Anthropocene. Trends Ecol. Evol. 30, 104-113.

McIntosh, R.P., 1998. The myth of community as organism. Perspect. Biol. Med. 41, 426-438.

McKeever, M.H., 1984. Comparative palynological studies of two lake sites in western Ireland and northwestern Spain. (MSc thesis). Trinity College, Dublin, Ireland.

McVean, D., 1953. *Alnus glutinosa* (L.) Gaertn. J. Ecol. 41, 447 - 466.

Meinke, H., 1927. Atlas und Bestimmungsschlüssel zur Pollenanalytik. Botanisches Archiv Königsberg, 19, 380-449.

Melchior, H., 1964. A. Engler's Syllabus der Pflanzenfamilien. Borntraeger, Berlin-Nikolassee, p. 666.

Mellars, P., Dark, O., 1998. Star Carr in context: new archaeological and palaeoecological investigations at the Early Mesolithic site of Star Carr, North Yorkshire. MacDonald Institute Monographs, Cambridge, 250 pp.

Meltsov, V., Poska, A., Reitalu, T., Sammul, M., Kull, T., 2013. The role of landscape structure in determining palynological and floristic richness. Veget. Hist. Archaeobot. 22, 39-49.

Menéndez Amor, J., Florschütz, F., 1963. Sur les éléments steppiques dans la végétation quaternaire de l'Espagne. Bol. R. Soc. Espanola Hist. Natural. 61, 121-133.

Menke, B., 1968. Beiträge zur Biostratigraphie des Mittelpleistozäns in Norddeutschland. Meyniana 18, 35-42.

Menke, B., 1969. Vegetationsgeschichtliche Untersuchungen an altpleistozänen Ablagerungen aus Lieth bei Elmshorn. Eiszeitalter u. Gegenwart 20, 76-83.

Menke, B., 1972. Wann begann die Eiszeit? Umschau in Wissenschaft und Technik, 1972, 214-218.

Menke, B., 1975. Vegetationsgeschichte und Florenstratigraphie Nordwestdeutschlands im Pliozän und Frühquartär. Mit einem Beitrag zur Biostratigraphie des Weichsel-Frühglazials. Geol. Jahrb. A 26, 3-151.

Menke, B., 1980. Lieth (Elmshorn), Ältest-Quartär. In: Stremme, H.-E., Menke, B. (Eds.), Quaternary-Excursions in Schleswig-Holstein. Kiel, pp. 9-20.

Menke, B., Ross, P.-H., 1967. Der erste Fund von Kieselgur in Schleswig-Holstein bei Brokenlande, südlich von Neumünster. Eiszeitalter und Gegenwart 18, 113-126.

Menke, B., Tynni, R., 1984. Das Eeminterglazial und das Weichselfrühglazial von Rederstall/Dithmarschen und ihre Bedeutung für die mitteleuropäische Jungpleistozän-Gliederung. Geol. Jahrb. A 76, 3-120.

Menotti, F., 2004. Living on the Lake in Prehistoric Europe: 150 years of lake-dwelling research. Routledge, London, 304 pp.

Mercuri, A.M., Mazzanti, M.B., Torri, P., Vigliotti, L., Bosi, G., Florenzano, A., Olmi, L., N'siala, M., 2012. A marine/terrestrial investigation for middle-late Holocene vegetation history and the development of the cultural landscape in the Po valley as the result of human impact and climatic change. Veget. Hist. Archaeobot. 21, 353-372.

Mercuri, A.M., Mazzanti, M.B., Florenzano, A., Montecchi, M., Rattighieri, E., 2013. *Olea, Juglans* and *Castanea*: The OJC group as pollen evidence of the development of human-induced environments in the Italian peninsula. Quat. Int. 303, 24-42.

Merkt, J., Müller, H., 1999. Varve chronology and palynology of the Lateglacial in Northwest Germany from lacustrine sediments of Hämelsee in Lower Saxony. Quat. Int. 61, 41-59.

Metcalf, C., 1958. Gorham's Cave: Report on the plant remains. Bull. Inst. Archaeol. 4, 219.

Meusel, H., Jäger, E., 1992. Vergleichende Chorologie der zentraleuropäischen Flora. Fischer, Jena.

Meusel, H., Niedermaier, K., 1985. Die Breitlaubwälder des Süd- und Ostkarpatenraumes in vergleichend- chorologischer Sicht. Siebenbürg. Archiv, 3. Folge 20, 1-42. Köln/Wien.

Meusel, H., Jäger, E., Weinert, E., 1965. Vergleichende Chorologie der zentraleuropäischen Flora. Gustav Fischer Jena.

Meusel, H., Jäger, E., Rauschert, S., Weinert, E., 1978. Vergleichende Chorologie der zentraleuropäischen Flora. Fischer, Jena.

Meyer, K.-J., 1974. Pollenanalytische Untersuchungen und Jahresschichtenzählungen an der holsteinzeitlichen Kieselgur von Hetendorf. Geolog. Jahrbuch A 21, 87-105.

Michaelis, D., 2002. Die spät- und nacheiszeitliche Entwicklung der natürlichen Vegetation von Durchströmungsmooren in Mecklenburg-Vorpommern am Beispiel der Recknitz. [The late- and postglacial development of the natural vegetation of the percolation mires of Mecklenburg-West Pomerania, using the example of the Recknitz valley] Diss. Bot. 365, 1-188.

Michaelis, D., 2019. The *Sphagnum* species of the world. Bibliotheca Botanica 162. Schweizerbart Science Publishers, Stuttgart.

Michalczyk, I., Opgenoorth, L., Luecke, Y., Huck, S., Ziegenhagen, B., 2010. Genetic support for periglacial survival of *Juniperus communis* L. in Central Europe. Holocene 20, 887–894.

Millar, C.I., Woolfenden, W., 1999. The role of climate change in interpreting historic variability. Ecol. Appl. 9, 1207-1216.

Millennium Ecosystem Assessment, 2005. Ecosystems and Human Well-being: Current State and Trends, Volume 1. Island Press, Washington DC.

Miller, P.A., Giesecke, T., Hickler, T., Bradshaw, R.H.W., Smith, B., Seppä, H., Valdes, P.J., Sykes, M.T., 2008. Exploring climatic and biotic controls on Holocene vegetation change in Fennoscandia. J. Ecol. 96, 247-259.

Miller, A.D., Roxburgh, S.H., Shea, K., 2011. How frequency and intensity shape diversity-disturbance relationships. Proc. Natl. Acad. Sci. USA 108, 5643-5648.

Milner, A.M., Müller, U.C., Roucoux, K.H., Collier, R.E.L., Pross, J., Kalaitzidis, S., Christianis, K., Tzedakis, P.C., 2013. Environmental variability during the Last Interglacial: a new high-resolution pollen record from Tenaghi Philippon, Greece. J. Quat. Sci. 28, 113-117.

Miola, A., 2012. Tools for Non-Pollen Palynomorphs (NPPs) analysis: A list of Quaternary NPP types and reference literature in English language (1972-2011). Rev. Palaeobot. Palynol. 186, 142-161.

Miotk-Szpiganowicz, G., 1992. The history of vegetation of Bory Tucholskie and the role of man in the light of palynological investigations. Acta Palaeobot. 32, 39–122.

Miras, Y., Ejarque, A., Riera, S., Palet, J., Orengo, H., Euba, I., 2007. Dynamique holocène de la végétation et occupation des Pyrénées andorranes depuis le Néolithique ancien d'après l'analyse pollinique de la tourbière de Bosc dels Estanyons (2180 m, Vall del Madriu, Andorre). Comptes Rendus de l'Académie des Science Paris, Palévol 6, 291-300.

Mitchell, G., 1956. Post-Boreal pollen-diagrams from Irish raised-bogs (Studies in Irish Quaternary Deposits: No. 11). Proc. R. Irish Acad. B: Biol. Geol. Chem. Sci. 57, 185-251.

Mitchell, F.J.G., 1986. The Shell Guide to Reading the Irish Landscape. Dublin.

Mitchell, F.J.G., 1990. The history and vegetation dynamics of a yew wood (*Taxus baccata* L.) in S.W. Ireland. New Phytol. 115, 573-577.

Mitchell, E.A.D., 1995. The postglacial developmental history of the Praz-Rodet Bog, Vallée de Joux, Swiss Jura (MSc thesis). Université de Neuchatel, Neuchatel, Switzerland.

Mitchell, F.J.G., 2005. How open were European primeval forests? Hypothesis testing using palaeoecological data. J. Ecol. 93, 168-177.

Mitchell, F.J.G., Bradshaw, R.H.W., Hannon, G., O'Connell, M., Pilcher, J., Watts, W., 1996. Ireland. In: Berglund, B.E., Birks, H.J.B., Ralska-Jasiewiczowa, M., Wright, H.E. (Eds.), Palaeoecological Events During the Last 15 000 Years. Regional Syntheses of Palaeoecological Studies of Lakes and Mires in Europe. Wiley, Chichester, pp. 1-13.

Mitchell, E.A.D., Charman, D.J., Warner, B.G., 2008a. Testate amoebae analysis in ecological and paleoecological studies of wetlands: past, present and future. Biodiv. Conserv. 17, 2115-2137.

Mitchell, E.A.D., Payne, R.J., Lamentowicz, M., 2008b. Potential implications of differential preservation of testate amoebae shells for paleoenvironmental reconstruction in peatlands. J. Paleolimnol. 40, 603-618.

Mitchell, E.G., Harris, S., Kenchington, C.G., Vixseboxse, P., Roberts, L., Clark, C., Dennis, A., Liu, A.G., Wilby, P.R., 2019. The importance of neutral over niche processes in structuring Ediacaran early animal communities. Ecol. Lett. 22, 2028-2038.

Mittelbach, G.G., McGill, B.J., 2019. Community

Ecology, 2nd ed. Oxford University Press, Oxford. 409 pp.

Moe, D., 1983. Studies in the Vegetation History of Vestvågøy, Lofoten, North-Norway. A preboreal recession and the vegetation optimum. Tromura, Tromsø, Naturvitenskap. Nr. 39, 1-28.

Moe, D., Vorren, K.D., Alm, T., Fimreite, S., Mørkved, B., Nilssen, E., Paus, A., Ramfjord, H., Selvik, S., Sørensen, R., 1996. Norway. In: Berglund, B.E., Birks, H.J.B., Ralska-Jasiewiczowa, M., Wright, H.E. (Eds.), Palaeoecological Events During the Last 15 000 Years. Regional Syntheses of Palaeoecological Studies of Lakes and Mires in Europe. Wiley, Chichester, pp. 153-213.

Moen, A., Joosten, H., Tanneberger, F., 2017. Mire diversity in Europe: mire regionality. In: Joosten, H., Tanneberger, F., Moen, A. (Eds.), Mires and Peatlands of Europe – Status, distribution and conservation. Schweizerbart Science Publishers, Stuttgart, pp. 97-149.

Mohl, H., 1834. Über den Bau und die Formen der Pollenkörner. Bern.

Möller, P., Östlund, O., Barnekow, L., Sandgren, P., Palmbo, F., Willerslev, E., 2012. Living at the margin of the retreating Fennoscandian Ice Sheet: The early Mesolithic sites at Aareavaara, northernmost Sweden. Holocene 23, 104-116.

Molloy, K., O'Connell, M., 1987. The nature of the vegetational changes at about 5000 B.P. with particular reference to the elm decline: fresh evidence from Connemara, western Ireland. New Phytol. 106, 203-220.

Molloy, K., O'Connell, M., 1991. Palaeoecological investigations towards the reconstruction of woodland and land-use history at Lough Sheeauns, Connemara, western Ireland. Rev. Palaeobot. Palynol. 671/2, 75-113.

Montserrat-Martí, J.M., 1992. Evolución glaciar y postglaciar del clima y la vegetación en la vertiente sur del Pirineo: estudio palinológico, Monografias del Instituto Pirenaico de Ecología. Consejo Superior de Investigaciones Científicas (España).

Moore, P.D., 1973. The influence of prehistoric cultures upon the initiation and spread of blanket bog in upland Wales. Nature 241, 350-353.

Moore, P.D., 1975. Origin of blanket mires. Nature 256, 267-269.

Moore, P.D., 1977. Stratigraphy and pollen analysis of Claish Moss, Scotland: significance for the origin of surface pools and forest history. J. Ecol. 65, 375-397.

Moore, P.D., 1982. Beneath the blanket bogs of Britain. Nat. Hist. 91, 49-54.

Moore, P.D., 1984. European Mires. Academic Press, London.

Moore, P.D., Bellamy, D.J., 1976. Peatlands. 2nd ed. Elek Science, London, 221 pp.

Moore, P.D., Webb, J.A., 1978. An Illustrated Guide to Pollen Analysis, 1st ed. Hodder & Stoughton, London.

Moore, P.D., Webb, J.A., Collinson, M.E., 1991. Pollen analysis, 2nd ed. Blackwell Scientific Publications., Oxford.

Moorman, B.J., 2001. Ground-penetrating radar applications in paleolimnology. In: Last, W.M., Smol J.P. (Eds.), Tracking Environmental Change Using Lake Sediments, Vol. 1: Basin Analysis, Coring, and Chronological Techniques. Kluwer, Dordrecht, pp. 23-47.

Morales-Molino, C., García-Antón, M., 2014. Vegetation and fire history since the last glacial maximum in an inland area of the western Mediterranean Basin (Northern Iberian Plateau, NW Spain). Quat. Res. 81, 63-77.

Morales-Molino, C., García Antón, M., Morla, C., 2011. Late Holocene vegetation dynamics on an Atlantic–Mediterranean mountain in NW Iberia. Palaeogeography, Palaeoclimatology, Palaeoecology 302, 323–337.

Morales-Molino, C., García-Antón, M., Postigo-Mijarra, J.M., Morla, C., 2013. Holocene vegetation, fire and climate interactions on the westernmost fringe of the Mediterranean Basin. Quat. Sci. Rev. 59, 5-17.

Morales-Molino, C., Vescovi, E., Krebs, P., Carlevaro, E., Kaltenrieder, P., Conedera, M., Tinner, W., Colombaroli, D., 2015. The role of human-induced fire and sweet chestnut (*Castanea sativa* Mill.) cultivation on the long-term landscape dynamics of the southern Swiss Alps. Holocene 25, 482-494.

Morales-Molino, C., Tinner, W., García-Antón, M., Colombaroli, D., 2017. The historical demise of *Pinus nigra* forests in the Northern Iberian Plateau (south-western Europe). J. Ecol. 105, 634-646.

Morales-Molino, C., Colombaroli, D., Tinner, W., Perea, R., Valbuena-Carabana, M., Carrión, J.S., Gil, L., 2018. Vegetation and fire dynamics during the last 4000 years in the Caballeros National Park (central Spain). Rev. Palaeobot. Palynol. 253, 110-122.

Morales-Molino, C., Tinner, W., Perea, R., Carrión, J.S., Colombaroli, D., Valbuena-Carabaña, M., Zafra, E., Gil, L., 2019. Unprecedented herbivory threatens rear-edge populations of *Betula* in southwestern Eurasia. Ecology 100: e02833.

Morales-Molino, C., Steffen, M., Samartin, S., van Leeuwen, J.F.N., Hürlimann, D., Vescovi, E., Tinner, W., 2021. Long-Term Responses of Mediterranean Mountain Forests to Climate Change, Fire and Human Activities in the Northern Apennines (Italy). Ecosystems 24, 1361–1377.

Morellón, M., Valero-Garcés, B., Moreno, A., González-Sampériz, P., Mata, P., Romero, O., Maestro, M., Navas, A., 2008. Holocene palaeohydrology and climate variability in northeastern Spain: The sedimentary record of Lake Estanya (Pre-Pyrenean range). Quaternary International, The Last 15ka of Environmental Change in Mediterranean Regions - Interpreting Different Archives 181, 15–31.

Morellón, M., Valero-Garcés, B., Vegas-Vilarrubia, T., González-Sampériz, P., Romero, O., Delgado-Huertas, A., Mata, P., Moreno, A., Rico, M., Corella, J.P., 2009. Lateglacial and Holocene palaeohydrology in the western Mediterranean region: the Lake Estanya record (NE Spain). Quat. Sci. Rev. 28, 2582-2599.

Morellón, M., Aranbarri, J., Moreno, A., González-Sampériz, P., Valero-Garcés, B.L., 2018. Early Holocene humidity patterns in the Iberian Peninsula reconstructed from lake, pollen and speleothem records. Quat. Sci. Rev. 181, 1-18.

Moreno, A., González-Sampériz, P., Morellón, M., Valero-Garcés, B., Fletcher, W., 2012. Northern Iberian abrupt climate change dynamics during the last glacial cycle: A view from lacustrine sediments. Quat. Sci. Rev. 36, 139e153.

Morris, P., Swindles, G., Valdes, P., Ivanovic, R., Gregoire, L., Smith, M., Tarasov, L., Haywood, A., Bacon, K., 2018. Global peatland initiation driven by regionally asynchronous warming. Proc. National Acad. Sci. USA 115, 4851-4856.

Mortensen, M.F., Birks, H.H., Christensen, C., Holm, J., Noe-Nygaard, N., Odgaard, B.V., Olsen, J., Rasmussen, K.L., 2011. Lateglacial vegetation development in Denmark - new evidence based on macrofossils and pollen from Slotseng, a small-scale site in southern Jutland. Quat. Sci. Rev. 30, 2534-2550.

Mortensen, M.F., Henriksen, P.S., Bennike, O., 2014. Living on the good soil: relationships between soils, vegetation and human settlement during the late Allerød period in Denmark. Veget. Hist. Archaeobot. 23, 195-205.

Moskal-del Hoyo, M., 2013. Mid-Holocene forests from Eastern Hungary: new anthracological data. Rev. Palaeobot. Palynol. 193, 70-81.

Moskal-del Hoyo, M., Lityńska-Zając, M., Raczky, P., Anders, A., Magyari, E.K., 2018. The character of the Atlantic oak woods of the Great Hungarian Plain. Quat. Int. 463, 337-351.

Mudie, P., Marret, F., Aksu, A., Hiscott, R., Gillespie, H., 2007. Palynological evidence for climatic change, anthropogenic activity and outflow of Black Sea water during the late Pleistocene and Holocene: Centennial- to decadal-scale records from the Black and Marmara Seas. Quat. Int. 167-168, 73-90.

Mühlmann, O., Peintner, P., 2008. Mycobionts of *Salix herbacea* on a glacier forefront in the Austrian Alps. Mycorrhiza 18, 171-180.

Müller, H., 1953. Zur spät- und nacheiszeitlichen Vegetationsgeschichte des mitteldeutschen Trockengebietes. Nova Acta Leopold. N.F. 16, 1-67.

Müller, H., 1962. Pollenanalytische Untersuchung eines Quartärprofils durch die spät- und nacheiszeitlichen Ablagerungen des Schleinsees (Südwestdeutschland). Geol. Jahrb. 79, 493-526.

Müller, H., 1965. Eine pollenanalytische Neubearbeitung des Interglazial-Profils von Bilshausen (Untereichsfeld). Geol. Jahrb. 83, 327-352.

Müller, H., 1974a. Pollenanalytische Untersuchungen und Jahresschichtenzählungen an der holstein-zeitlichen Kieselgur von Munster-Breloh. Geol. Jahrb. A21, 107-140.

Müller, H., 1974b. Pollenanalytische Untersuchungen und Jahresschichtenzählungen an der eemzeitlichen Kieselgur von Bispingen/Luhe. Geol. Jahrb. A 21, 149-169.

Müller, T., 1989. Die artenreichen Rotbuchenwälder Süddeutschlands. In: Pott, R. (Ed.), Buchenwald-Symposium, pp. 149-163.

Müller, H., 1992. Climate changes during and at the end of the interglacials of the Cromerian complex. In: Kukla, G.J., Went, E. (Eds.), Start of a Glacial. NATO ASI Series I 3, Berlin.

Müller, J., 2009. Die Jungsteinzeit. In: von Schnurbein, S. (Ed.), Atlas der Vorgeschichte. Theiss, Stuttgart, 58-105.

Müller, H., Höfle, H.-C., 1994. Die Holstein-Interglazialvorkommen bei Bossel westlich von Stade und Wanhöden nördlich Bremerhaven. Geolog. Jahrbuch A 134, 71-116.

Müller, U.C., Kukla, G.J., 2004. North Atlantic Current and European environments during the declining stage of the last glacial. Geology 32, 1009-1012.

Müller, G., Weber, C., 1904. Über eine frühglaziale und vorglaziale Flora bei Lüneburg. Königl. Preuss. Geol. Landesanst. Berlin 40.

Müller, U.C., Klotz, S., Geyh, M.A., Pross, J., Bond, G.C., 2005. Cyclic climate fluctuations during the last interglacial in Central Europe. Geology (USA) 33, 449-452.

Muller, S.D., Nakagawa, T., de Beaulieu, J.-L., Court-Picon, M., Carcaillet, C., Miramont, C., Roiron, P., Boutterin, C., Ali, A., Bruneton, H., 2007. Post-glacial migration of silver fir (*Abies alba* Mill.) in the south-western Alps. J. Biogeogr. 34, 876-899.

Muñoz Sobrino, C., Ramil-Rego, P., Gómez-Orellana, L., 2007. LateWürm and early Holocene in the mountains of northwest Iberia: biostratigraphy, chronology and tree colonization. Veget. Hist. Archaeobot. 16, 223-240.

Muñoz Sobrino, C., Ramil-Rego, P., Gómez-Orellana, L., Ferreiro da Costa, J., Díaz Varela, R., 2009. Climatic and human effects on the post-glacial dynamics of *Fagus sylvatica* L. in NW Iberia. Plant Ecol. 203, 317-340.

Muñoz Sobrino, C., Heiri, O., Hazekamp, M., van der Velden, D., Kirilova, E.P., Carcía-Moreiras I., Lotter, A.F., 2013. New data on the Lateglacial period of SW

Europe: a high resolution multiproxy record from Laguna de la Roya (NW Iberia). Quat. Sci. Rev. 80, 58-77.

Munsell, A.H., 1975. Munsell soil color charts. Baltimore, Maryland. U.S.A.

Muscheler, R., Kromer, B., Björck, S., Svensson, A., Friedrich, M., Kaiser, K.F., Southon, J., 2008. Tree rings and ice cores reveal ^{14}C calibration uncertainties during the Younger Dryas. Nature Geosci. 1, 263-267.

Musil, R., 2003. The Middle and Upper Palaeolithic game suite in central and south-eastern Europe. In: van Andel, T., Davies, S. (Eds.), Neanderthals and Modern Humans in the European Landscape during the Last Glaciation. McDonald Institute for Archaeological Research, Cambridge, pp. 167-190.

Nabel, J.E.M.S., Kirchner, J.W., Zurbriggen, N., Kienast, F., Lischke H., 2014. Extrapolation methods for climate time series revisited - Spatial correlations in climatic fluctuations influence simulated tree species abundance and migration. Ecol. Comp. 20, 315-324.

Nagy-Bodor, E., Cserny, T., Nagy, E., 1995. Lake Balaton (Southwest TO-25, Northeast TO-9, Centre TO-30). Unpublished report on the study performed for the European Pollen Database.

Nakagawa, T., 1998. Études palynologiques dans les Alpes françaises centrales et méridionales: histoire de la végétation tardiglaciaire et Holocène. PhD thesis, University of Aix-Marseille, 211 pp.

Nakagawa, T., Gotanda, K., Haraguchi, T., Danhara, T., Yonenobu, H., Brauer, A., Yokoyama, Y., Tada, R., Takemura, K., Staff, R.A., Payne, R., Bronk Ramsey, C., Bryant, C., Brock, F., Schlolaut, G., Marshall, M., Tarasov, P., Lamb, H., Members, S.P., 2012. SG06, a fully continuous and varved sediment core from Lake Suigetsu, Japan: stratigraphy and potential for improving the radiocarbon calibration model and understanding of late Quaternary climate changes. Quat. Sci. Rev. 36, 164-176.

Nardini, A., Ghirardelli, L., Salleo, S., 1998. Vulnerability to freeze stress of seedlings of *Quercus ilex* L.: an ecological interpretation. Annales Des Sciences Forestieres 55, 553-565.

Nash, D., De Cort, G., Chase, B., Verschuren, D., Nicholson, S., Shanahan, T., Asrat, A., Lézine, A., Grab, S., 2016. African hydroclimatic variability during the last 2000 years. Quat. Sci. Rev. 154, 1-22.

Nathorst, A., 1870. Om några arktiska växtlämningar i en sötvattenslera vid Alnarp i Skåne. Lunds Universitets Årsskrift 7, 17 pp.

Naydenov, K., Senneville, S., de Beaulieu, J.-L., Tremblay, F., Bousquet, J., 2007. Glacial vicariance in Eurasia: mitochondrial DNA evidence from Scots pine for a complex heritage involving genetically distinct refugia at mid-northern latitudes and in Asia Minor. BMC Evol. Biol. 7, 233.

Necesaný, V., 1951. Studie o diluvialni flore Dyjskos-vrateckého ulavu. PràceMoravskoslez. Akad. Ved. Prirod. 23, 291-306.

Neef, R., Cappers, R., Bekker, R., 2012. Digital Atlas of Economic Plants in Archaeology. Barkhuis, Groningen, 760 pp.

Nesbitt, M., 2002. When and where did domesticated cereals first occur in southwest Asia? In: Cappers, R., Bottema, S. (Eds.), The Dawn of Farming in the Near East. Berlin, pp. 113-132.

Neuhäusl, R., 1991. Vegetation Map of Europe: First results and current state. J. Veget. Sci. 2, 131-134.

Neustadt, M.I., 1957. Forest history and palaeogeography of the USSR in the Holocene (in Russian). Izd-vo Akademii nauk, Moscow.

Neuweiler, E., 1905. Prähistorische Pflanzenreste Mitteleuropas. Botanische Exkursionen und pflanzengeographische Studien in der Schweiz. 6, 1-111.

Newman, E.A., 2019. Disturbance ecology in the Anthropocene. Front. Ecol. Evol. 7, art147.

Nguetsop, V., Servant-Vildary, S., Servant, M., 2004. Late Holocene climatic changes in west Africa, a high-resolution diatom record from equatorial Cameroon. Quat. Sci. Rev. 23, 591-609.

Nichols, J.E., Walcott, M., Bradley, R., Pilcher, J., Huang, Y.S., 2009. Quantitative assessment of precipitation seasonality and summer surface wetness using ombrotrophic sediments from an Arctic Norwegian peatland. Quat. Res. 72, 443-451.

Nielsen, A.B., 2004. Modelling pollen sedimentation in Danish lakes at c. AD 1800: an attempt to validate the POLLSCAPE model. J. Biogeography 31, 1693-1709.

Nielsen, A.B., Odgaard, B.V., 2004. The use of historical analogues for interpreting fossil pollen records. Veget. Hist. Archaeobot. 13, 33-43.

Nielsen, A.B., Odgaard, B.V., 2010. Quantitative landscape dynamics in Denmark through the last three millennia based on the Landscape Reconstruction Algorithm approach. Veget. Hist. Archaeobot. 19, 375-387.

Nielsen, A.B., Sugita, S., 2005. Estimating relevant source area of pollen for small Danish lakes around AD 1800. Holocene 15, 1006-1020.

Nielsen, A.B., Möller, P., Giesecke, T., Stavngaard, B., Fontana, S., Bradshaw, R.H.W., 2010. The effect of climate conditions on inter-annual flowering variability monitored by pollen traps below the canopy in Draved Forest, Denmark. Veget. Hist. Archaeobot. 19, 309-323.

Nielsen, A.B., Giesecke, T., Theuerkauf, M., Feeser, I., Behre, K.-E., Beug, H.-J., Chen, S.-H., Christiansen, J., Dörfler, W., Endtmann, E., Jahns, S., de Klerk, P., Kühl, N., Latałowa, M., Odgaard, B.V., Rasmussen, P.,

Stockholm, J., Voigt, R., Wiethold, J., Wolters, S., 2012. Quantitative reconstructions of changes in regional openness in north-central Europe reveal new insights into old questions. Quat. Sci. Rev. 47, 131-149.

Niinemets, W., Saarse, L., 2006. Holocene forest dynamics and human impact in southeastern Estonia. Veget. Hist. Archaeobot. 16, 1-13.

Nilssen, E.J., 1983. Klima - og vegetasjonshistoriske undersøkelser i Lofoten (MSc thesis). University of Tromsø, Tromsø, Norway.

Nilssen, E.J., 1988. Development of the cultural landscape in the Lofoten area, north Norway. In: Birks, H.H., Birks, H.J.B., Kaland, P.E., Moe, D. (Eds.), The Cultural Landscape - Past, Present and Future. Cambridge University Press, Cambridge, pp. 369-380.

Nilsson, T., 1935. Die pollenanalytische Zonengliederung der spät- und postglazialen Bildungen Schonens. Geol. Fören. iStockholm Förh. 57, 385-562.

Nilsson, T., 1964. Standardpollendiagramme und C14 Datierungen aus dem Ageröds Mosse im mittleren Schonen. Publications from the Institutes of Mineralogy, Paleontology and Quaternary Geology, University of Lund, Sweden 124, 1-52.

Nilsson, T., 1983. The Pleistocene. Reidel Publishing, Dordrecht.

Nilsson, S., Praglowski, J., Nilsson, L., 1977. Atlas of Airborne Pollen Grains and Spores in Northern Europe. Natur och Kultur, Stockholm.

Nitychoruk, J., Binka, K., Hoefs, J., Ruppert, H., Schneider, J., 2005. Climate reconstruction for the Holsteinian Interglacial in eastern Poland and its comparison with isotopic data from Marine Isotope Stage 11. Quat. Sci. Rev. 24, 631-644.

Nobis, M.P., Normand, S., 2014. KISSMig - a simple model for R to account for limited migration in analyses of species distributions. Ecography 37, 1282-1287.

Nogué, S., de Nascimento, L., Froyd, C.A., Wilmshurst, J.M., de Boer, E.J., Coffey, E.E.D., Whittaker, R.J., Fernández-Palacios, J.M., Willis, K.J., 2017. Island biodiversity conservation needs palaeoecology. Nat. Ecol. Evol. 1, 0181.

Nogués-Bravo, D., Veloz, S., Holt, B.G., Singarayer, J., Valdes, P., Davis, B., Brewer, S.C., Williams, J.W., Rahbek, C., 2016. Amplified plant turnover in response to climate change forecast by Late Quaternary records. Nat. Clim. Change 6, 1115.

Nogués-Bravo, D., Rodríguez-Sánchez, F., Orsini, L., de Boer, E., Jansson, R., Morlon, H., Fordham, D.A., Jackson, S.T., 2018. Cracking the code of biodiversity responses to past climate change. Trends Ecol. Evol. 33, 765-776.

Noryskiewicz, B., 1978. The Eemian interglacial at Naklo on the river Notec, N. Poland (Polish, English summary). Acta Palaeobotanica 19, 67-112.

Noryskiewicz, B., 1987. History of vegetation during the Late-Glacial Holocene in the Brodnica Lake District in light of pollen analysis of Lake Strazym deposits. Acta Palaeobot. 27, 283-304.

Noryskiewicz, B., 1995. Zmiany szaty roslinnej okolic Jeziora Biskupinskiego pod wplywem czynnikow naturalnych i antropogenicznych w poznym glacjale i holocenie [Changes in vegetation of the Biskupin (Biskupinskie) lake area during the Late-Glacial and the Holocene, caused by natural and anthropogenic factors.]. Turpress, Torun, pp. 147-180.

Noti, R., van Leeuwen, J.F.N., Colombaroli, D., Vescovi, E., Pasta, S., La Mantia, T., Tinner, W., 2009. Mid- and late-Holocene vegetation and fire history at Biviere di Gela, a coastal lake in southern Sicily, Italy. Veget. Hist. Archaeobot., 18, 371-387.

Novenko, E.Y., Volkova, E.M., Glasko, M.P., 2012. Palaeoecological evidence for the middle and late Holocene vegetation, climate and land use in the upper Don River basin (Russia). Veget. Hist. Archaeobot. 21, 337-352.

Novenko, E.Y., Tsyganov, A.N., Volkova, E.M., Babeshko, K.V., Lavrentiev, N.V., Payne, R.J., Mazei, Y.A., 2015. The Holocene paleoenvironmental history of central European Russia reconstructed from pollen, plant macrofossil, and testate amoeba analyses of the Klukva Peatland, Tula Region. Quat. Res. 83, 459-468.

Novenko, E.Y., Tsyganov, A.N., Payne, R.J., Mazei, N.G., Volkova, E.M., Chernyshov, V.A., Kupriyanov, D.A., Mazei, Y.A., 2018. Vegetation dynamics and fire history at the southern boundary of the forest vegetation zone in European Russia during the middle and late Holocene. The Holocene 28, 308-322.

Novenko, E.Y., Tsyganov, A.N., Mazei, N.G., Kupriyanov, D.A., Rudenko, O.V., Bobrovsky, M.V., Erman, N.M., Nizovtsev, V.A., 2019. Palaeoecological evidence for climatic and human impacts on vegetation in the temperate deciduous forest zone of European Russia during the last 4200 years: A case study from the Kaluzhskiye Zaseki Nature Reserve. Quat. Int. 516, 58-69.

Ntinou, M., 2010. Wood charcoal analysis at Klissoura Cave 1 (Prosymna, Peloüonese): The Upper Palaeolithic vegetation. Eurasian Prehistory 7, 47-69.

O'Connell, M., 1986. Reconstruction of local landscape development in the post-Atlantic based on palaeoecological investigations at Carrownaglogh prehistoric field system, county Mayo, Ireland. Rev. Palaeobot. Palynol. 49, 117-176.

O'Connell, M., 2007. European cultural landscapes: a view from Ireland informed by PAN (EU 5fp network). Initiativen zum Umweltschutz 67, 113-130.

O'Connell, M., Molloy, K., 2001. Farming and

woodland dynamics in Ireland during the Neolithic. Biology and Environment: Proc. Roy. Irish Acad. 101**B**, 99-128.

O'Connell, M., Molloy, K., Bowler, M., 1988. Post-glacial landscape evolution in Connemara, Western Ireland with particular reference to woodland history. In: Birks, H.H., Birks, H.J.B., Kaland, P., Moe, D. (Eds.), The Cultural Landscape - Past, Present and Future. Cambridge University Press, Cambridge, pp. 267-287.

O'Connell, M., Molloy, M., McMahon, H., 2001. Reconstructing prehistoric farming and human impact at a fine spatial resolution: palaeoecological investigations at Mooghaun, Co. Clare, western Ireland. GFG-Graduiertenkolleg University of Regensburg 462, 161-185.

O'Sullivan, P., Reynolds, C.S., 2004. The Lakes Handbook: Limnology and Limnetic Ecology. Blackwell, Oxford.

Oberdorfer, E., 1931. Die postglaziale Klima- und Vegetationsgeschichte des Schluchsees (Schwarzwald). Ber. Naturf. Ges. Freiburg i. Br. 31, 1 85.

Oberdorfer, E., 1983. Süddeutsche Pflanzengesellschaften, Teil III. Gustav Fischer Verlag, Jena.

Oberdorfer, E., 1990. Pflanzensoziologische Exkursionsflora, 6th ed., E. Ulmer, Stuttgart, 1050 pp.

Oberdorfer, E., 1992. Süddeutsche Pflanzengesellschaften. Fischer, Jena/Stuttgart.

Obidowicz, A., 1988. The Puścizna Rękowiańska raised bog. In: Late glacial and Holocene environmental changes Vistula Basin. Excursion Guide Book-Symposium, Cracow, 15-21.

Obidowicz, A., 1989. Type region Pa: Inner West Carpathians-Nowy Targ Basin. Acta Palaeobot. 29, 11-15.

Obidowicz, A., 1990. Eine pollenanalytische und moorkundliche Studie zur Vegetationsgeschichte des Podhale-Gebietes (West-Karpaten). Acta Palaeobot. 30, 147-219.

Obidowicz, A., 1993. Wahania górnej granicy lasu w póznym plejstocenie i holocenie w Tatrach [Fluctuations of the forest timberline in the Tatra Mountains during the last 12 000 years]. Dok. Geogr. 4, 31-43.

Obidowicz, A., Madeysaka, E., Turner, C., 2013. Postglacial History of Vegetation in the Polish Part of the Western Carpathians Based on Isopollen Maps. W. Szafer Institute of Botany, Polish Academy of Sciences, Kraków.

Ochando, J., Carrión, J.S., Rodríguez-Vidal, J., Jiménez-Arenas, J., Fernández, S., Amorós, G., Munuera, M., Scott, L., Stewart, J., Knul, M., Toro Moyano, I., Ponce de León, M., Zollikofer, C., 2020. Palynology and chronology of hyaena coprolites from the Piñar karstic Caves Las Ventanas and Carihuela, southern Spain. Palaeogeogr. Palaeoclim. Palaeoecol. 552.

Odgaard, B.V., 1985. Kulturlandskabets historie i Vestjylland. Fortidsminder 48-59.

Odgaard, B.V., 1988. Heathland history in western Jutland, Denmark, in: Birks, H.H., Birks, H.J.B., Kaland, P.E., Moe, D. (Eds.), The Cultural Landscape-Past, Present and Future. Cambridge, UK, pp. 311-319.

Odgaard, B.V., 1992. The fire history of Danish heathland areas as reflected by pollen and charred particles in lake sediments. Holocene 2, 218-226.

Odgaard, B.V., 1993. Wind-determined sediment distribution and Holocene sediment yield in a small, Danish, kettle lake. J. Paleolimn. 8, 3-13.

Odgaard, B.V., 1994. The Holocene vegetation history of northern West Jutland, Denmark. Nordic Journal of Botany 14, 546-546.

Odgaard, B.V., 1999. Fossil pollen as a record of past biodiversity. J. Biogeogr. 26, 7-17.

Odgaard, B.V., 2001. Palaeoecological perspectives on pattern and process in plant diversity and distribution adjustments: a comment on recent developments. Divers. Distrib. 7, 197-201.

Odgaard, B.V., 2007. Pollen Methods and Studies: Reconstructing Past Biodiversity Development. In: Elias, S.A. (Ed.), Encyclopedia of Quaternary Science. Elseveir, Amsterdam, pp. 2508-2514.

Odgaard, B.V., Rasmussen, P., 1998. The use of historical data and sub-recent (A.D. 1800) pollen assemblages to quantify vegetation/pollen relationships. Akademie der Wissenschaften und der Literatur. Paläoklimforschung 27, 67-75.

Odgaard, B.V., Rasmussen, P., 2000. Origin and temporal development of macro-scale vegetation patterns in the cultural landscape of Denmark. J. Ecol. 88, 733-748.

Odum, E.P., Odum, H.T., 1953. Fundamentals of Ecology. Saunders, Philadelphia.

Oeggl, K., 1988. Beiträge zur Vegetationsgeschichte Tirols VII: Das Hochmoor Schwemm bei Walchsee. Berichte Naturwissenschaftlich-Med. Ver. Innsbr. 75, 37-60.

Oeggl, K., 2009. The significance of the Tyrolean Iceman for the archaeobotany of Central Europe. Veget. Hist. Archaeobot. 18, 1-11.

Ogle, K., Barber, J.J., Barron-Gafford, G.A., Patrick Bentley, L., Young, J.M., Huxman, T.E., Loik, M.E., Tissue, D.T., 2015. Quantifying ecological memory in plant and ecosystem processes. Ecol. Lett. 18, 221-235.

Ohlson, M., Korbol, A., Okland, R.H., 2006. The macroscopic charcoal record in forested boreal peatlands in southeast Norway. Holocene 16, 731-741.

Ohlson, M., Brown, K.J., Birks, H.J.B., Grytnes, J.-A., Hornberg, G., Niklasson, M., Seppä, H., Bradshaw, R.H.W., 2011. Invasion of Norway spruce diversifies the fire regime in boreal European forests. J. Ecol. 99, 395-403.

Ohlson, M., Ellingsen, V.M., del Olmo, M.V., Lie, M.H., Nybakken, L., Asplund, J., 2017. Late-Holocene fire history as revealed by size, age and composition of the soil charcoal pool in neighbouring beech and spruce forest landscapes in SE Norway. Holocene 27, 397-403.

Oksanen, P.O., Kuhry, P., Alekseeva, R.N., 2001. Holocene development of the Rogovaya River peat plateau, European Russian Arctic. Holocene 11, 25-40.

Okuda, M., Yasuda, Y., Setoguchi, T., 2001. Middle to Late Pleistocene vegetation history and climatic changes at Lake Kopais, Southeast Greece. Boreas 30, 73-82.

Olalde, M., Herrán, A., Espinel, S., Goicoechea, P., 2002. White oaks phylogeography in the Iberian Peninsula. For. Ecol. Managem. 156, 89-102.

Oldfield, F., 1964. Late Quaternary vegetational history in South West France. Pollen et Spores 6, 157-168.

Oldfield, F., 1970. The ecological history of Blelham Bog National Nature Reserve. In: Walker, D., West, R.G. (Eds.), Studies in the Vegetational History of the British Isles. Cambridge University Press, Cambridge, pp. 141-157.

Oldfield, F., 2005. Environmental Change: Key Issues and Alternative Approaches. Cambridge University Press, Cambridge. 386 pp.

Oldfield, F., Stratham, D., 1963. Pollen analytical data from Urswick Tarn and Ellerside Moss, northern Lancashire. New Phytol. 62, 53.

Olsson, I.U., 1986. Radiometric dating. In: Berglund, B.E. (Ed.), Handbook of Holocene Palaeoecology and Palaeohydrology. John Wiley & Sons, Chichester, pp. 273-312.

Opravil, E., 1994. Vegetation. In: Svoboda, J. (Ed.), Pavlov-I Excavation 1952-1983. The Dolní Vestonice Studies 2, Liège, pp. 177-180.

Orbán, I., Birks, H.H., Vincze, I., Finsinger, W., Pal, I., Marinova, E., Jakab, G., Braun, M., Hubay, K., Biro, T., Magyari, E.K., 2018. Treeline and timberline dynamics on the northern and southern slopes of the Retezat Mountains (Romania) during the late glacial and the Holocene. Quat. Int. 477, 59-78.

Ordonez, A., Williams, J.W., Svenning, J.-C., 2016. Mapping climatic mechanisms likely to favour the emergence of novel communities. Nat. Clim. Change 6, 1104-1109.

Orion, L., Nadal, L., Heiri, L., Ercolino G., Tanner, W., 2020. Effects of grazing by new world camelids and old world felids on grassland diversity and human well-being in the Andes and the Alps. Int. J Grassland Sci. 34, 249-267.

Osvald, H., 1923. Die Vegetation des Hochmoores Komosse. Svensk Växtosociologiska Sällsk. Handl. 1. Almqvist & Wiksell, Uppsala, 436 pp.

Osvald, H., 1925. Zur Vegetation der ozeanischen Hochmoore in Norwegen. [The vegetation of oceanic raised bogs in Norway]. Svenska Växtsociol. Sällsk. Handl. 7, 1-106. (in German)

Osvald, H., 1949. Notes on the vegetation of British and Irish mosses. Acta Phytogeogr. Suecica 26, 1-62.

Osvald, H., 1970. Vegetation and stratigraphy of peatlands in North America. Acta Universitatis Upsaliensis, 1-96.

Overballe-Petersen, M.V., Nielsen, A.B., Bradshaw, R.H.W., 2012. Quantitative vegetation reconstruction from pollen analysis and historical inventory data around a Danish small forest hollow. J. Veg. Sci. 24, 755-771.

Overbeck, F., 1975. Botanisch-geologische Moorkunde unter besonderer Berücksichtigung der Moore Nordwestdeutschlands als Quellen zur Vegetations- Klima- und Siedlungsgeschichte. Wachholtz, Neumünster.

Overbeck, F., Schneider, S., 1938. Mooruntersuchungen bei Lüneburg und Bremen und die Reliktnatur von *Betula nana* L. in Nordwestdeutschland. Zeitschr. Bot. 33, 154.

Overbeck, F., Münnich, K., Aletsee, L., Averdieck, F., 1957. Das Alter des „Grenzhorizonts" norddeutscher Hochmoore nach Radiocarbon-Datierungen. Flora 145, 37-71.

Overpeck, J.T., Webb, T., Prentice, I.C, 1985. Quantitative interpretation of fossil pollen spectra: dissimilarity coefficients and the method of modern analogs. Quat. Res. 23, 87-108.

Overpeck, J.T., Whitlock, C., Huntley, B., 2003. Terrestrial biosphere dynamics in the climate system: past and future. In: Alverson, K.D., Bradley, R.S., Pedersen, T.F. (Eds.), Paleoclimate, Global Change and the Future. Springer, Berlin, pp. 81-103.

Ozenda, P., 1979. Vegetation map of the Council of Europe member states: scale: 1/3 000 000 Council of Europe, Strasbourg

Ozenda, P., 1983. La végétation de l'arc alpin. Conseil de l'Europe, Strasbourg.

Ozenda, P., 1985. La végétation de la chaine alpine dans l'espace montagnard européen. Masson, Paris. 344 pp.

Paciorek, C.J., McLachlan, J.S., 2009. Mapping ancient forests: Bayesian inference for spatio-temporal trends in forest composition using the fossil pollen proxy record. J. Am. Stat. Assoc. 104(486), 608-622.

Paganelli, A., Miola, A., 1991. Chestnut (*Castanea sativa* Mill.) as an indigenous species in Northern Italy. Il Quaternario, Italian J. Quat. Sci. 4, 99-106.

Palacios, D., de Marcos, J., Vázquez-Selem, L., 2011. Last Glacial Maximum and deglaciation of Sierra de Gredos, central Iberian Peninsula. Quat. Int. 233, 16-26.

Palmé, A., Vendramin, G.G., 2002. Chloroplast

DNA variation, postglacial recolonization and hybridization in hazel, *Corylus avellana*. Mol. Ecol. 11, 1769-1779.

Palmisano, A., Woodbridge, J., Roberts, C., Bevan, A., Fyfe, R.M., Shennan, S., Cheddadi, R., Greenberg, R., Kaniewski, D., Langgut, D., Leroy, S., Litt, T., Miebach, A., 2019. Holocene landscape dynamics and long-term population trends in the Levant. Holocene 29, 708-727.

Panagiotopoulos, K., Aufgebauer, A., Schäbitz, F., Wagner, B., 2013. Vegetation and climate history of the Lake Prespa region since the Lateglacial. Quat. Int 293, 157-169.

Panagiotopoulos, K., Böhm, A., Leng, M.J., Wagner, B., Schäbitz, F., 2014. Climate variability over the last 92 ka in SW Balkans from analysis of sediments from Lake Prespa. Climate of the Past 10, 643–660.

Pantaléon-Cano, J., Yll, E.-I., Pérez-Obiol, R., Maria Roure, J., 2003. Palynological evidence for vegetational history in semi-arid areas of the western Mediterranean (Almería, Spain). Holocene 13, 109-119.

Papageorgiou, A., Tsiripidis, I., Mouratidis, T., Hatziskakis, S., Gailing, O., Eliades, N.-G.H., Vidalis, A., Drouzas, A., Finkeldey, R., 2014. Complex fine-scale phylogeographical patterns in a putative refugial region for *Fagus sylvatica* (Fagaceae). Bot. J. Linn. Soc. 174, 516-528.

Pape, J., 1970. Plaggen soils in the Netherlands. Geoderma 4, 229-255.

Pardoe, H.S., 2001. The representation of taxa in surface pollen spectra on alpine and sub-alpine glacier forelands in southern Norway. Rev. Palaeobot. Palynol. 117, 63-78.

Pardoe, H.S., Giesecke, T., van der Knaap, W.O., Svitavská-Svobodová, H., Kvavadze, E.V., Panajiotidis, S., Gerasimidis, A., Pidek, I.A., Zimny, M., Swieta-Musznicka, J., Latałowa, M., Noryskiewicz, A.M., Bozilova, E., Tonkov, S., Filipova-Marinova, M., van Leeuwen, J.F.N., Kalnina, L., 2010. Comparing pollen spectra from modified Tauber traps and moss samples: examples from a selection of woodlands across Europe. Veget. Hist. Archaeobot. 19, 271-283.

Parducci, L., Jørgensen, T., Tollefsrud, M.M., Elverland, E., Alm, T., Fontana, S.L., Bennett, K.D., Haile, J., Matetovici, I., Suyama, Y., Edwards, M.E., Andersen, K., Rasmussen, M., Boessenkool, S., Coissac, E., Brochmann, C., Taberlet, P., Houmark-Nielsen, M., Larsen, N.K., Orlando, L., Gilbert, M.T.P., Kjær, K.H., Alsos, I.G., Willerslev, E., 2012. Glacial survival of boreal trees in Northern Scandinavia. Science 335, 1083–1086.

Parsons, R.W., Prentice, I.C., 1981. Statistical approaches to R-values and the pollen—vegetation relationship. Rev. Palaeobot. Palynol. 32, 127-152.

Parsons, R.W., Prentice, I.C., 1983. Statistical uncertainty in forest composition estimates obtained from fossil pollen spectra via the R-value model. Rev. Palaeobot. Palynol. 40, 177-189.

Pärtel, M., 2002. Local plant diversity patterns and evolutionary history at the regional scale. Ecology 83, 2361-2366.

Pasquill, F., Smith, F.B., 1983. Atmospheric Diffusion, 3rd edition. Ellis Horwood Limited, Chichester.

Pasta, S., De Rigo, D., Caudullo, G., 2016a. *Acer pseudoplatanus* in Europe: distribution, habitat, usage and threats. In: San-Miguel-Ayanz, J., De Rigo, D., Caudullo, G., Houston Durrant, T., Mauri, A. (Eds.), European Atlas of Forest Tree Species. Publication Office of the European Union, Luxembourg, pp. 56-57.

Pasta, S., De Rigo, D., Caudullo, G., 2016b. *Ostrya carpinifolia* in Europe: distribution, habitat, usage and threats. In: San-Miguel-Ayanz, J., de Rigo, D., Caudullo, G., Houston Durrant, T., Mauri, A. (Eds.), European Atlas of Forest Tree Species. Publication Office of the European Union, Luxembourg, pp. 112-113.

Patin, C., 1663. Traité des tourbes combustibles. Paris.

Patzelt, G., 1977. Der zeitliche Ablauf und das Ausmass postglazialer Klimaschwankungen in den Alpen. In: Frenzel, B. (Ed.), Dendrochronologie und postglaziale Klimaschwankungen in Europa. F. Steiner Verlag, Wiesbaden, pp. 248-259.

Patzelt, G., Bortenschlager, S., 1973. Die postglazialen Gletscher- und Klimaschwankungen in der Venedigergruppe (Hohe Tauern, Ostalpen). Zeit. Geomorph. N.F. Suppl. 16, 25-72.

Paus, A., 1982. Paleo-økologiske undersøkelser på Farøya, Sør-Trøndelag. Den vegetasjonshistoriske utviklingen fra senistiden og fram til idag. PhD Thesis, University of Trondheim, 234 pp.

Paus, A., 1988. Late Weichselian vegetation, climate, and floral migration at Sandvikvatn, North Rogaland, southwestern Norway. Boreas 17, 113–139.

Paus, A., 1996. Type region N-mo, southern coastal area in Møre and Trøndelag (Central Norway). In: Berglund, B.E., Birks, H.J.B., Ralska-Jasiewiczowa, M., Wright, H.E. (Eds.), Palaeoecological Events During the Last 15 000 Years. Regional Syntheses of Palaeoecological Studies of Lakes and Mires in Europe. Wiley, Chichester, pp. 177-183.

Paus, A., 2010. Vegetation and environment of the Rødalen alpine area, Central Norway, with emphasis on the early Holocene. Veget. Hist. Archaeobot. 19, 29-51.

Paus, A., 2013. Human impact, soil erosion, and vegetation response lags to climate change: challenges for the mid-Scandinavian pollen-based transfer-function temperature reconstructions. Veg. Hist. Archaeobotany 22, 269–284.

Paus, A., Haugland, V., 2017. Early- to mid-Holocene forest-line and climate dynamics in southern Scandes mountains inferred from contrasting megafossil and pollen data. Holocene 27, 361-383.

Paus, A., Svendsen, J.I., Matiouchkov, A., 2003. Late Weichselian (Valdaian) and Holocene vegetation and environmental history of the northern Timan Ridge, European Arctic Russia. Quat. Sci. Rev. 22, 2285-2302.

Pears, N.V., 1968. Post-glacial tree-lines of the Cairngorm Mountains, Scotland. Trans. Bot. Soc. Edinburgh 40, 361-394.

Pears, N.V., 1972. Interpretation problems in the study of tree-line fluctuations. In: Taylor, J.A. (Ed.), Forest Meteorology. Cambrian News, Aberystwyth. pp. 31-45.

Pears, N.V., 1988. Pine stumps, radiocarbon dates and stable isotope analysis in the Cairngorm Mountains: some observations. Rev. Palaeobot. Palynol. 54, 175-180.

Pearsall, D., 1989. Palaeoethnobotany. A handbook of procedures. Routledge, San Diego.

Peck, R.M., 1973. Pollen budget studies in a small Yorkshire catchment. In: Birks, H.J.B., West, R.G. (Eds.), Quaternary Plant Ecology. John Wiley & Sons, pp. 43-60.

Pedersen, M.W., Ruter, A., Schweger, C., Friebe, H., Staff, R.A., Kjeldsen, K.K., Mendoza, M.L.Z., Beaudoin, A.B., Zutter, C., Larsen, N.K., Potter, B.A., Nielsen, R., Rainville, R.A., Orlando, L., Meltzer, D.J., Kjær, K.H., Willerslev, E., 2016. Postglacial viability and colonization in North America's ice-free corridor. Nature 537, 45-49.

Pedrotta, T., Gobet, E., Schwörer, C., Beffa, G., Butz, C., Henne, P.D., Morales-Molino, C., Pasta, S., van Leeuwen, J.F.N., Vogel, H., Zwimpfer, E., Anselmetti, F.S., Grosjean, M., Tinner, W., 2021. 8,000 years of climate, vegetation, fire and land-use dynamics in the thermo-mediterranean vegetation belt of northern Sardinia (Italy). Veg. Hist. Archaeobotany 30, 789-813.

Peet, R.K., Fridley, J.D., Gramling, J.M., 2003. Variation in species richness and species pool size across a pH gradient in forests of the southern Blue Ridge Mountains. Folia Geobot. 38, 391-401.

Peev, P., 2009. The Neolithisation of the Eastern Balkan Peninsula and fluctuations of the Black Sea level. Quat. Int. 197, 87-92.

Peglar, S.M., 1993a. The development of the cultural landscape around Diss Mere, Norfolk, UK, during the past 7000 years. Rev. Palaeobot. Palynol. 76, 1-47.

Peglar, S.M., 1993b. The mid-Holocene *Ulmus* decline at Diss Mere, Norfolk, UK: a year-by year pollen stratigraphy from annual laminations. Holocene 3, 1-13.

Peglar, S.M., Birks, H.J.B., 1993. The mid-Holocene *Ulmus* fall at Diss Mere, south-east England - disease and human impact? Veget. Hist. Archaeobot. 2, 61-68.

Peglar, S.M., Fritz, S., Birks, H.J.B., 1989. Vegetation and land-use history at Diss, Norfolk, U.K. J. Ecol. 77, 203-222.

Pèlachs, A., Pérez-Obiol, R., Ninyerola, M., Nadal, J., 2009. Landscape dynamics of *Abies* and *Fagus* in the southern Pyrenees during the last 2200 years as a result of anthropogenic impacts. Rev. Palaeobot. Palynol. 156, 337-349.

Peñalba, M.C., 1989. Dynamique de la végétation Tardiglaciaire et Holocène du centre-nord de l'Espagne d'après l'analyse pollinique. PhD Thesis, University of Aix-Marseille, 151 pp.

Peñalba, M.C., 1994. The history of the Holocene vegetation in northern Spain from pollen analysis. J. Ecol. 82, 815-832.

Peñalba, M.C., Arnold, M., Guiot, J., Duplessy, J., de Beaulieu, J.-L., 1997. Termination of the last glaciation in the Iberian Peninsula inferred from the pollen sequence of Quintanar de la Sierra. Quat. Res. 48, 205-214.

Penck, A., 1879. Die Geschiebeformation Norddeutschlands. Zeitschr. Deutsche Geolog. Gesellschaft 31, 117-203.

Penck, A., Brückner, E., 1901-1909. Die Alpen im Eiszeitalter, Leipzig.

Penkman, K.E.H., Preece, R.C., Bridgland, D.R., Keen, D.H., Meijer, T., Parfitt, S.A., White, T. S., Collins, M.J., 2013. An aminostratigraphy for the British Quaternary based on *Bithynia* opercula. Quat. Sci. Rev. 61, 111-134.

Pennington, W., 1973. Absolute pollen frequencies in the sediments of lakes of different morphometry. In: Birks, H.J.B., West, R.G. (Eds.), Quaternary Plant Ecology. John Wiley & Sons, pp. 79-104.

Pennington, W., 1974. The History of British Vegetation. The English Universities Press, London.

Pennington, W., 1979. The origin of pollen in lake sediments: an enclosed lake compared with one receiving inflow streams. New Phytol. 83, 189-213.

Pennington, W., 1986. Lags in adjustment of vegetation to climate caused by the pace of soil development: Evidence from Britain. Vegetatio 67, 105-118.

Pennington, W., Haworth, E.Y., Bonny, A.P., Lishman, J.P., 1972. Lake sediments in northern Scotland. Phil.Trans. R. Soc. London, B 264, 191-294.

Peñuelas, J., 1985. HCO_3^- as an exogenous carbon source for aquatic bryophytes *Fontinalis antipyretica* and *Fissidens grandifrons*. J. Exper. Bot. 36, 441-448.

Perea, R., Delibes, M., Polko, M., Suárez-Esteban, A., Fedriani, J.M., 2013. Context-dependent fruit-frugivore interactions: partner identities and spatio-temporal variations. Oikos 122, 943-951.

Pérez-Díaz, S., López-Sáez, J.A., Galop, D., 2015. Vegetation dynamics and human activity in the Western Pyrenean Region during the Holocene. Quat. Int. 364, 65-77.

Pérez-Obiol, R., Julià, R., 1994. Climatic change on the Iberian Peninsula recorded in a 30,000-yr pollen record from Lake Banyoles. Quat. Res. 41, 91-98.

Pérez-Obiol, R., Jalut, G., Julià, R., Pèlachs, A., Iriarte, J., Otto, T., Hernández-Beloqui, B., 2011. Mid-Holocene vegetation and climatic history of the Iberian Peninsula. Holocene 21, 75-93.

Pérez-Sanz, A., González-Sampériz, P., Moreno, A., Valero-Garcés, B., Gil-Romera, G., Rieradevall, M., Tarrats, P., Lasheras-Álvarez, L., Morellón, M., Belmonte, A., Sancho, C., Sevilla-Callejo, M., Navase, A., 2013. Holocene climate variability, vegetation dynamics and fire regime in the central Pyrenees: the Basa de la Mora sequence (NE Spain). Quat. Sci. Rev. 73, 149-169.

Pernaud, J.-M., 2001. Postglacial vegetation history in Luxembourg: new charcoal data from the cave of la Karelslé (Waldbillig, Eastern Gutland). Veget. Hist. Archaeobot. 10, 219-225.

Perring, F., Walters, S., 1976. Atlas of the British Flora, 2nd ed. BSBI, Wakefield.

Persson, H., Wide, B., Andersson, S., Svensson, L., 2004. Allozyme diversity and genetic structure of marginal and central populations of *Corylus avellana* L. (Betulaceae) in Europe. Plant Syst. Evol. 244, 157–179.

Petit, R.J., Brewer, S., Bordács, S., Burg, K., Cheddadi, R., Coart, E., Cottrell, J., Csaikl, U., van Dam, B., Deans, J., Espinel, S., Fineschi, S., Finkeldey, R., Glaz, I., Goicoechea, P., Jensen, J., König, A., Lowe, A., Madsen, S., Mátyás, G., Munro, R., Popescu, F., Slade, D., Tabbener, H., de Vries, S., Ziegenhagen, B., de Beaulieu, J.-L., Kremer, A., 2002a. Identification of refugia and post-glacial colonisation routes of European white oaks based on chloroplast DNA and fossil pollen evidence. For. Ecol. Managem. 156, 49-74.

Petit, R.J., Csaikl, U., Bordács, S., Burg, K., Coart, E., Cottrell, J., van Dam, B., Deans, J., Dumolin-Lapègue, S., Fineschi, S., Finkeldey, R., Gillies, A., Glaz, I., Goicoechea, P., Jensen, J., König, A., Lowe, A., Madsenn, S., Mátyás, G., Munro, R., Olalde, M., Pemonge, M.H., Popescu, F., Slade, D., Tabbener, H., Taurchini, D., de Vries, S., Ziegenhagen, B., Kremer, A., 2002b. Chloroplast DNA variation in European white oaks: Phylogeography and patterns of diversity based on data from over 2600 populations. For. Ecol. Managem. 156, 5-26.

Petit, R.J., Latouche-Hallé, C., Pemonge, M., Kremer, A., 2002c. Chloroplast DNA variation of oaks in France and the influence of forest fragmentation on genetic diversity. For. Ecol. Managem. 156, 115-129.

Petit, R.J., Aguinagalde, I., de Beaulieu, J.L., Bittkau, C., Brewer, S., Cheddadi, R., Ennos, R., Fineschi, S., Grivet, D., Lascoux, M., Mohanty, A., Muller-Starck, G.M., Demesure-Musch, B., Palme, A., Martin, J.P., Rendell, S., Vendramin, G.G., 2003. Glacial refugia: Hotspots but not melting pots of genetic diversity. Science 300, 1563-1565.

Petit, R.J., Hampe, A., Cheddadi, R., 2005. Climate changes and tree phylogeography in the Mediterranean. Taxon 54, 877–885.

Petr, L., 2005. Vývoj vegetace pozdního glaciálu a raného holocénu v centrální části České kotliny [Late Glacial and Early Holocene vegetation development in the central part of Czech basin] (MSc thesis). Charles University, Prague.

Petr, L., Novák, J., 2014. High vegetation and environmental diversity during the Late Glacial and Early Holocene on the example of lowlands in the Czech Republic. Biologia (Bratisl.) 69, 847–862.

Petr, L., Petrik, J., Chattova, B., Jamrichova, E., Rohovec, J., Matouskova, S., Hajnalova, M., 2020. The history of a Pannonian oak woodland – palaeoecological evidence from south-eastern Slovakia. Folia Geobot. 55, 29-40.

Peyron, O., Bégeot, C., Brewer, S., Heiri, O., Magny, M., Millet, L., Ruffaldi, P., van Campo, E., Yu, G., 2005. Late-Glacial climatic changes in Eastern France (Lake Lautrey) from pollen, lake-levels, and chironomids. Quat. Res. 64, 197-211.

Peyron, O., Combourieu-Nebout, N., Brayshaw, D., Goring, S., Andrieu-Ponel, V., Desprat, S., Fletcher, W., Gambin, B., Ioakim, C., Joannin, S., Kotthoff, U., Kouli, K., Montade, V., Pross, J., Sadori, L., Magny, M., 2017. Precipitation changes in the Mediterranean basin during the Holocene from terrestrial and marine pollen records:a model–data comparison. Clim. Past 13, 249-265.

Pfadenhauer, J.S., Klötzli, F.A., 2014. Vegetation der Erde. Springer, Berlin.

Phillips, G., Pietiläinen, O., Carvalho, L., Solimini, A., Lyche Solheim, A., Cardoso, A., 2008. Chlorophyll–nutrient relationships of different lake types using a large European dataset. Aquat. Ecol. 42, 213–226.

Pickett, S.T.A., White, P.S. (Eds.), 1987. The Ecology of Natural Disturbance and Patch Dynamics. Academic Press, Orlando. 488 pp.

Pidek, I.A., 2003. Mesopleistocene vegetation history in the northern foreland of the Lublin upland based on palaeobotanical studies of the profiles from Zdany and Brus sites. Lublin.

Pidek, I.A., 2015. Palynostratigraphy and vegetation changes during the early Middle Pleistocene, based on new studies of deposits from Ferdynandów (central eastern Poland). Acta Palaeobot. 55, 53-66.

Pidek, I.A., Malek, M., 2010. A bi-partite Ferdinandowian succession from Lukow, Eastern Poland: a new palynostratigraphic approach. Geol. Quart. 54, 69-84.

Pidek, I.A., Svitavská-Svobodová, H., van der Knaap, W.O., Noryskiewicz, A.M., Filbrandt-Czaja, A.,

Norsykiewicz, B., Latałowa, M., Zimny, M., Swieta-Musznicka, J., Bozilova, E., Tonkov, S., Filipova-Marinova, M., Poska, A., Giesecke, T., Gikov, A., 2010. Variation in annual pollen accumulation rates of *Fagus* along a N-S transect in Europe based on pollen traps. Veget. Hist. Archaeobot. 19, 259-270.

Pignatti, S. 1997. Ecologia del paesaggio. Unione Tipografico-Editrice Torinese, Torino, 228 pp.

Pignatti, S., 1989. Chorological and ecological information as basis for the syntaxonomy of beech forests in Italy. In: Pott, R. (Ed.), Buchenwald-Symposium, Göttingen, pp. 73-82.

Pignatti, S., 2005. Valori di bioindicazione delle piante vascolari della flora d'Italia. Braun-Blanquetia 39, 3-97.

Pigott, C.D., Huntley, J.P., 1978. Factors controlling the distribution of *Tilia cordata* at the northern limits of its geographical range. I. Distribution in north-west England. New Phytol. 81, 429-441.

Pigott, C.D., Huntley, J.P., 1980. Factors controlling the distribution of *Tilia cordata* at the northern limits of its geographical range II. History in north-west England. New Phytol. 84, 145-164.

Pigott, C.D., Huntley, J.P., 1981. Factors controlling the distribution of *Tilia cordata* at the northern limits of its geographical range III. Nature and causes of seed sterility. New Phytol. 87, 817-839.

Pilcher, J.R., Hall, V.A., 1992. Towards a tephrochronology for the Holocene of the north of Ireland. Holocene 2, 255-259.

Pilcher, J.R., Hall, V.A., 1996. Tephrochronological studies in northern England. Holocene 6, 100-105.

Pilcher, J.R., Baillie, M., Schmidt, B., Becker, B., 1984. A 7,272-year tree-ring chronology for Western Europe. Nature 312, 150-152.

Pillans, B., Naish, T., 2004. Defining the Quaternary. Quat. Sci. Rev. 23, 2271-2282.

Pini, R., 2002. A high-resolution Late-Glacial - Holocene pollen diagram from Pian di Gembro (central Alps, northern Italy). Veget. Hist. Archaeobot. 11, 251-262.

Pini, R., Ravazzi, C., Donegana, M., 2009. Pollen stratigraphy, vegetation and climate history of the last 215 ka in the Azzano Decimo core (plain of Friuli, north-eastern Italy). Quaternary Science Reviews 28, 1268-1290.

Pini, R., Ravazzi, C., Reimer, P.J., 2010. The vegetation and climate history of the last glacial cycle in a new pollen record from Lake Fimon (southern Alpine foreland, N-Italy). Quat. Sci. Rev. 29, 3115-3137.

Pini, R., Ravazzi, C., Raiteri, L., Guerreschi, A., Castellano, L., Comolli, R., 2017. From pristine forests to high-altitude pastures: an ecological approach to prehistoric human impact on vegetation and landscapes in the western Italian Alps. J. Ecol. 105, 1580-1597.

Piovesan, G., Presutti Saba, E., Franco Biondi, F., Alessandrini, A., Di Filippo, A., Schirone, B., 2009. Population ecology of yew (*Taxus baccata* L.) in the central Apennines: spatial patterns and their relevance for conservation strategies. Plant Ecol. 205, 23-46.

Pirazzoli, P., 1991. World Atlas of Holocene Sea-level Changes. Elsevier, Amsterdam.

Pirrus, R., Rouk, A.M., Liiva, A., 1987. Geology and stratigraphy of the reference site of Lake Raigastvere in Saadjärv drumlin field. Palaeohydrology Temp. Zone II Lakes 101-122.

Pirzamanbin, B., Lindström, J., Poska, A., Sugita, S., Trondman, A.-K., Fyfe, R., Mazier, F., Nielsen, A.B., Kaplan, J.O., Bjune, A.E., Birks, H.J.B., Giesecke, T., Kangur, M., Latałowa, M., Marquer, L., Smith, B., Gaillard, M.-J., 2014. Creating spatially continuous maps of past land cover from point estimates: a new statistical approach applied to pollen data. Ecol. Complex. 20, 127-141.

Planchais, N., 1962. Le pollen de quelques chênes de domaine méditerranéen occidental. Pollen et Spores 4, 87-93.

Planchais, N., 1970. Tardiglaciaire et Postglaciaire à Mur-de -Sologne (Loir-et-Cher). Pollen Spores 12, 381-428.

Plantifa, S., Pflänzibänzi, N., 2022. Amore e anarchia: the secret life of plants between ice-age hibernation and global-change challenges. Reports of Long-Term Enthalpy 7, 30-41.

Poher, Y., Ponel, P., Medail, F., Andrieu-Ponel, V., Guiter, F., 2017. Holocene environmental history of a small Mediterranean island in response to sea-level changes, climate and human impact. Paleogeogr. Paleoclim. Paleoecol. 465, 247-263.

Pokorný, P., 2002. A high-resolution record of Late-Glacial and Early-Holocene climatic and environmental change in the Czech Republic. Quat. Int. 91, 101-122.

Pokorný, J., Květ, J., 2004. Aquatic plants and lake ecosystems. In: O'Sullivan, P., Reynolds, C.S. (Eds.), The Lakes Handbook: Limnology and Limnetic Ecology. Blackwell, Oxford, pp. 309-340.

Pokorný, P., Klimesova, J., Klimeš, L., 2000. Late Holocene history and vegetation dynamics of a floodplain alder carr: a case study from eastern Bohemia, Czech Republic. Folia Geobot. 35, 43-58.

Pokorný, P., Chytrý, M., Juřičková, L., Sádlo, J., Novák, J., Ložek, V., 2015. Mid-Holocene bottleneck for central European dry grasslands: Did steppe survive the forest optimum in northern Bohemia, Czech Republic? Holocene 25, 716-726.

Pokorný, P., Novák, J., Šída, P., Divišová, M., Kozáková, R., Abraham, V., 2017. Vývoj vegetace severočeských pískovcových území od pozdního glaciálu po střední holocén. (Vegetation development of

Northern Bohemian sandstone areas from the Late Glacial to the Middle Holocene In: Svoboda, J. (Ed.), Mezolit severních Čech II. Komplexní výzkum skalních převisů na Českolipsku a Děčínsku, 2003-2015. (Mesolithic of northern Bohemia II. Complex excavation of rockshelters in the Česká Lípa and Děčín areas 2003-2015. Dolnověstonické studie, sv. 22, Archeologický ústav AVČR, Brno, pp. 11-37.

Pollegioni, P., Woeste, K., Olimpieri, I., Marandola, D., Cannata, F., Malvolti, M., 2011. Long-term human impacts on genetic structure of Italian walnut inferred by SSR markers. Tree Genet. Genomes 7, 707-723.

Pollegioni, P., Woeste, K., Chiocchini, F., Olimpieri, I., Tortolano, V., Clark, J., Hemery, G., Mapelli, S., Malvolti, M., 2014. Landscape genetics of Persian walnut (*Juglans regia* L.) across its Asian range. Tree Genet. Genomes 10, 1027-1043.

Ponel, P., Court-Picon, M., Badura, M., Guiter, F., de Beaulieu, J.-L., Andrieu-Ponel, V., Djamali, M., Leydet, M., Gandouin, E., Buttler, A., 2011. Holocene history of Lac des Lauzons (2180 m asl), reconstructed from multiproxy analyses of Coleoptera, plant macroremains and pollen (Hautes-Alpes, France). Holocene 21, 565-582.

Ponel, P., Guiter, F., Gandouin, E., Pailles, C., Rioual, P., Djamali, M., Andrieu-Ponel, V., Leydet, M., Van der Putten, N., de Beaulieu, J.-L., 2016. Novel insights from coleopteran and pollen evidence into the Lateglacial/Holocene transition in Aubrac, French Massif Central. Palaeogeogr. Palaeoclimatol. Palaeoecol. 463, 83-102.

Pons, A., 1980. The History of Mediterranean Shrublands. In: Goodall, D. (Ed.), Ecosystems of the World. Elsevier, Amsterdam, pp. 131-138.

Pons, A., Reille, M., 1988. The Holocene- and upper Pleistocene pollen record from Padul (Granada, Spain) - a new study. Palaeogoegr. Palaeoclimatol. Palaeoecol. 66, 243-263.

Pons, A., Guiot, J., De Beaulieu, J.-L., Reille, M., 1992. Recent contributions to the climatology of the last glacial-interglacial cycle based on French pollen sequences. Quat. Sci. Rev. 11, 439-448.

Popov, P., 2003. Structure and differentiation of spruce populations in eastern Europe and western Siberia. Russian J. Ecol. 34, 27-33.

Popov, P., 2012. Reference populations for discriminant analysis in the continuous range of Norway and Siberian spruces. Russian J. Ecol. 43, 13-18.

Popov, P., 2015. Cluster groups, population structure, and relative localization of spruce populations in eastern Europe. Russian J. Ecol. 46, 128-135.

Poschlod, P., 2015. Geschichte der Kulturlandschaft. Ulmer, Stuttgart, 320 pp.

Poska, A., 1994. Three pollen diagrams from coastal Estonia. Sediments, vegetational development, cilmatic events and the environmental impact of man (MSc thesis). University of Uppsala, Uppsala, Sweden.

Poska, A., Saarse, L., 2002. Biostratigraphy and ^{14}C dating of a lake sediment sequence on the north-west Estonian carbonaceous plateau, interpreted in terms of human impact in the surroundings. Veget. Hist. Archaeobot. 11, 191-200.

Poska, A., Saarse, L., 2006. New evidence of possible crop introduction to north-eastern Europe during the Stone Age. Cerealia pollen finds in connection with the Akali Neolithic settlement, East Estonia. Veget. Hist. Archaeobot. 15 169-179.

Poska, A., Saarse, L., Koppel, K., Nielsen, A.B., Avel, E., Vassiljev, J., Väli, V., 2014. The Verijärv area, South Estonia over the last millennium: A high resolution quantitative land-cover reconstruction based on pollen and historical data. Rev. Palaeobot. Palynol. 207, 5-17.

Postigo-Mijarra, J.M, Gómez Manzaneque, F., Morla, C., 2008. Survival and long-term maintenance of Tertiary trees in the Iberian Peninsula during the Pleistocene: first record of *Aesculus* L. (Hippocastanaceae) in Spain. Veget. Hist. Archaeobot. 17, 351-364.

Postigo-Mijarra, J.M, Morla, C., Barrón, E., Morales-Molino, C., García, S., 2010. Patterns of extinction and persistence of Arctotertiary flora in Iberia during the Quaternary. Rev. Palaeobot. Palynol. 62, 416-426.

Pott, R., 1986. Der pollenanalytische Nachweis extensiver Waldbewirtschaftungen in den Haubergen des Siegerlandes. In: Behre, K.-E. (Ed.), Anthropogenic Indicators in Pollen Diagrams. Balkema, Rotterdam, pp. 125-134.

Pott, R., 1995. Die Pflanzengesellschaften Deutschlands, 2nd ed.. Ulmer, Stuttgart, 427 pp.

Pott, R., 1997. Invasion of beech and establishment of beech forests in Europe. Annal. Bot. 55, 27-58.

Pott, R., 2000. Palaeoclimate and vegetation - long-term vegetation dynamics in central Europe with particular reference to beech. Phytocoenologia 30, 285-333.

Pott, R., Hüppe, J., 1991. Die Hudelandschaften Nordwestdeutschlands. Abhandl. Westfäl. Museum f. Naturkunde 53, 1-313.

Power, M.J., Marlon, J., Ortiz, N., Bartlein, P.J., Harrison, S.P., Mayle, F.E., Ballouche, A., Bradshaw, R.H.W., Carcaillet, C., Cordova, C., Mooney, S., Moreno, P.I., Prentice, I.C., Thonicke, K., Tinner, W., Whitlock, C., Zhang, Y., Zhao, Y., Ali, A.A., Anderson, R.S., Beer, R., Behling, H., Briles, C., Brown, K.J., Brunelle, A., Bush, M., Camill, P., Chu, G.Q., Clark, J., Colombaroli, D., Connor, S.E., Daniau, A.L., Daniels, M., Dodson, J.,

Doughty, E., Edwards, M.E., Finsinger, W., Foster, D., Frechette, J., Gaillard, M.-J., Gavin, D.G., Gobet, E., Haberle, S., Hallett, D.J., Higuera, P., Hope, G., Horn, S., Inoue, J., Kaltenrieder, P., Kennedy, L., Kong, Z.C., Larsen, C., Long, C.J., Lynch, J., Lynch, E.A., McGlone, M., Meeks, S., Mensing, S., Meyer, G., Minckley, T., Mohr, J., Nelson, D.M., New, J., Newnham, R., Noti, R., Oswald, W., Pierce, J., Richard, P.J.H., Rowe, C., Sanchez Goñi, M.F., Shuman, B.N., Takahara, H., Toney, J., Turney, C., Urrego-Sanchez, D.H., Umbanhowar, C., Vandergoes, M., Vannière, B., Vescovi, E., Walsh, M., Wang, X., Williams, N., Wilmshurst, J., Zhang, J.H., 2008. Changes in fire regimes since the Last Glacial Maximum: an assessment based on a global synthesis and analysis of charcoal data. Clim. Dynam. 30, 887-907.

Praglowski, J., 1982. Fagaceae L., Fagoideae. In: Nilsson, S. (Ed.), World Pollen and Spore Flora, Stockholm, pp. 1-29.

Prebble, M., Dowe, J.L., 2008. The late Quaternary decline and extinction of palms on oceanic Pacific islands. Quat. Sci. Rev. 27, 2546-2567.

Prentice, H.C., 1981. A Late Weichselian and early Flandrian pollen diagram from Østervatnet, Varanger peninsula, NE Norway. Boreas 10, 53-70.

Prentice, I.C. 1985. Pollen representation, source area, and basin size: Toward a unified theory of pollen analysis. Quat. Res. 23, 76-86.

Prentice, I.C., 1986a. Forest-composition calibration of pollen data. In: Berglund, B.E. (Ed.), Handbook of Holocene Palaeoecology and Palaeohydrology. J Wiley & Sons, Chichester, pp. 799-816.

Prentice, I.C., 1986b. Vegetation responses to past climatic variation. Vegetatio 67, 131-141.

Prentice, I.C., 1988a. Paleoecology and plant-population dynamics. Trends Ecol. Evol. 3, 343-345.

Prentice, I.C., 1988b. Records of vegetation in time and space: the principles of pollen analysis. In: Huntley, B., Webb, T. (Eds.), Vegetation History. Kluwer Academic Publishers, Dordrecht, pp. 17-42.

Prentice, I.C., Parsons, R.W., 1983. Maximum likelihood linear calibration of pollen spectra in terms of forest composition. Biometrics 39, 1051-1057.

Prentice, I.C., Webb III, T., 1986. Pollen percentages, tree abundances and the Fagerlind effect. J. Quat. Sci. 1, 35-43.

Prentice, I.C., Webb III, T., 1998. BIOME 6000: reconstrucing global mid-Holocene vegetation patterns from palaeoecological records. J. Biogeogr. 25, 997-1005.

Prentice, I.C., Berglund, B.E., Olsson, T., 1987. Quantitative forest-composition sensing characteristics of pollen samples from Swedish lakes. Boreas 16, 43-54.

Prentice, I.C., Cramer, W., Harrison, S.P., Leemans, R., Monserud, R., Solomon, A., 1992a. A global biome model based on plant physiology and dominance, soil properties and climate. J. Biogeogr. 19, 117-134.

Prentice, I.C., Guiot, J., Harrison, S.P., 1992b. Mediterranean vegetation, lake levels and palaeoclimate at the Last Glacial Maximum. Nature 360, 658-660.

Prentice, I.C., Guiot, J., Huntley, B., Jolly, D., Cheddadi, R., 1996. Reconstructing biomes from palaeoecological data: a general method and its application to European pollen data at 0 and 6 ka. Clim. Dyn. 12, 185-194.

Prentice, I.C., Harrison, S.P., Jolly, D., Guiot, J., 1998. The climate and biomes of Europe at 6000 yr BP: comparison of model simulations and pollen-based reconstructions. Quat. Sci. Rev. 17, 659-668.

Preusser, F., Drescher-Schneider, R., Fiebig, M., Schlüchter, C., 2005. Re-interpretation of the Meikirch pollen record, Swiss Alpine Foreland, and implications for Middle Pleistocene chronostratigraphy. J. Quat. Sci. 20, 607-620.

Prøsch-Danielsen, L., Simonsen, A., 2000. Palaeoecological investigation towards the reconstruction of the history of forest clearances and coastal heathlands in south-western Norway. Veget. Hist. Archaeobot. 9, 189-204.

Punt, W., Bos, J.A.A., Hoen, P.P., 1991. The Northwest European Pollen Flora, 45. Oleaceae. Rev. Palaeobot. Palynol. 69, 23-47.

Punt, W., et al., 1976-2009. The Northwest European Pollen Flora. 9 Volumes. Elsevier, Amsterdam.

Pyhäjärvi, T., Salmela, M., Savolaine, O., 2008. Colonization routes of *Pinus sylvestris* inferred from distribution of mitochondrial DNA variation. Tree Genet. Genomes 4, 247-254.

Qin, B., Harrison, S.P., Kutzbach, J., 1998. Evaluation of modelled regional water balance using lake status data: a comparison of 6 ka simulations with the NCAR CCM. Quat. Sci. Rev. 17, 535-548.

Quézel, P., Médail, F., 2003. Écologie et biogéographie des forêts du bassin méditerranéen. Elsevier, Paris.

Rackham, O., 1976. Trees and Woodland in the British Landscape. Dent, London, 204 pp.

Rackham, O., 1980. Ancient Woodland: its history, vegetation and uses in England. Edward Arnold, London, 402 pp.

Rackham, O., Moody, J., 1996. The Making of the Cretan Landscape. Manchester University Press, Manchester.

Radeloff, V.C., Williams, J.W., Bateman, B.L., Burke, K.D., Carter, S.K., Childress, E.S., Cromwell, K.J., Gratton, C., Hasley, A.O., Kraemer, B.M., Latzka, A.W., Marin-Spiotta, E., Meine, C.D., Munoz, S.E., Neeson, T.M., Pidgeon, A.M., Rissman, A.R., Rivera, R.J.,

Szymanski, L.M., Usinowicz, J., 2015. The rise of novelty in ecosystems. Ecol. Appl. 25, 2051-2068.

Raghoebarsing, A.A., Smolders, A.J.P., Schmid, M.C., Rijpstra, W.I.C., Wolters-Arts, M., Derksen, J., Jetten, M.S.M., Schouten, S., Sinninghe Damsté, J.S., Lamers, L.P.M., Roelofs, J.G.M., Op den Camp, H.J.M., Strous, M., 2005. Methanotrophic symbionts provide carbon for photosynthesis in peat bogs. Nature 436, 1153-1156.

Ralska-Jasiewiczowa, M., 1966. Bottom sediments of the Mikolajki Lake (Masurian Lake District) in the light of palaeobotanical investigations. Acta Palaeobot. 7, 3-118.

Ralska-Jasiewiczowa, M., 1972a. Remarks on the Late-glacial and Holocene History of Vegetation in the Eastern Part of Polish Carpathians. Berichte Dtsch. Bot. Ges. 85, 101-112.

Ralska-Jasiewiczowa, M., 1972b. The Forest of the Polish Carpathians in the Late Glacial and Holocene. Studia Geomorphologica Carpatho-Balcanica 6.

Ralska-Jasiewiczowa, M., 1980. Late-glacial and Holocene vegetation of the Bieszczady Mts (Polish Eastern Carpathians). PWN, Warszawa-Krakow.

Ralska-Jasiewiczowa, M., 1983. Isopollen maps for Poland: 0–11000 years B.P. New Phytol. 94, 133-175.

Ralska-Jasiewiczowa, M. (ed.), 1987. Environmental Changes recorded in Lakes and Mires of Poland during the last 13'000 years. Part two, Acta Palaeobot. 29.

Ralska-Jasiewiczowa, M., 1989. Type region P-e: The Bieszczady Mts. Acta Palaeobot. 29, 31-35.

Ralska-Jasiewiczowa, M., Latałowa, M., 1996. Poland, in: Berglund, B.E., Birks, H.J.B., Ralska-Jasiewiczowa, M., Wright, H.E. (Eds.), Palaeohydrological Events during the Last 15,000 Years. Regional Syntheses of Palaeoecological Studies of Lakes and Mires in Europe. Wiley, Chichester, UK, pp. 403–472.

Ralska-Jasiewiczowa, M., van Geel, B., Demske, D., 1998. Holocene regional vegetation history recorded in the Lake Gościąż sediments. In: Ralska-Jasiewiczowa, M., Goslar, T., Madeyska, T., Starkel, L. (Eds.). Lake Gościąż, Central Poland. W Szafer Institute of Botany, Polish Academy of Sciences, Kraków, pp. 202-219.

Ralska-Jasiewiczowa, M., Nalepka, D., Goslar, T., 2003. Some problems of forest transformation at the transition to the oligocratic/Homo sapiens phase of the Holocene interglacial in northern lowlands of central Europe. Veget. Hist. Archaeobot. 12, 233-247.

Ralska-Jasiewiczowa, M., Latałowa, M., Wasylikowa, K., Tobolski, K., Madeysaka, E., Wright, H.E., Turner, C., 2004. Late Glacial and Holocene History of Vegetation in Poland based on Isopollen Maps. W. Szafer Institute of Botany Polish Academy of Sciences, Kraków, 444 pp.

Rameau, J.-C., Mansion, D., Dumé, G., Gauberville, C., 2008. Flore forestière française: Guide écologique illustré. 3. Région méditerranéene. Institut pour le développement forestier – CNPPF, Paris, France.

Ramfjord, H., 1982. On the late Weichselian and Flandrian shoreline displacement in Nærøy, Nord-Trøndelag, Norway. Nor. Geol. Tidsskr. 62, 191–205.

Ramil-Rego, P., Muñoz-Sobrino, C., Rodríguez-Guitián, M.A., Gómez-Orellana, L., 1998. Differences in the vegetation of the North Iberian Peninsula during the last 16,000 years. Plant Ecol. 138, 41–62.

Ramil-Rego, P., Rodríguez Guitián, M.A., Muñoz Sobrino, C., Gómez-Orellana, L., 2000. Some considerations about the postglacial contribution of *Fagus sylvatica* in the Iberian peninsula Folia Geobot. 35, 241-271.

Ramos-Román, M.J., Jiménez-Moreno, G., Camuera, J., García-Alix, A., Anderson, R.S., Jiménez-Espejo, F.J., Sachse, D., Toney, J.L., Carrión, J.S., Webster, C., Yanes, Y., 2018. Millennial-scale cyclical environment and climate variability during the Holocene in the western Mediterranean region deduced from a new multi-proxy analysis from the Padul record (Sierra Nevada, Spain). Global Planet. Change 168, 35-53.

Ramsar Convention, 2004. Peatlands. Guidelines for Global Action on Peatlands. Ramsar handbooks for the wise use of wetlands 14. 2nd edition.

Ran, E.T.H., 1990. Dynamics of vegetation and environment during the middle pleniglacial in the Dinkel valley (The Netherlands). Mededel. Rijks Geol. Dienst. 33, 141-208.

Randall Hughes, A., Byrnes, J.E., Kimbro, D.L., Stachowicz, J.J., 2007. Reciprocal relationships and potential feedbacks between biodiversity and disturbance. Ecol. Lett. 10, 849-864.

Rankama, T., Vuorela, I., 1988. Between inland and coast in Metal Age Finland - human impact on the primeval forests of Southern Häme during the Iron Age. Inland Coast Met. Age Finl. - Hum. Impact Primeval For. South. Häme Iron Age 64, 25–34.

Rapacciuolo, G., Rominger, A.J., Morueta-Holme, N., Blois, J.L., 2019. Ecological non-equilibrium in the Anthropocene. Front. Ecol. Evol. 7, art428.

Räsänen, S., Hicks, S., Odgaard, B.V., 2004. Pollen deposition in mosses and in a modified 'Tauber trap' from Hailuoto, Finland: what exactly do the mosses record? Rev. Palaeobot. Palynology 129, 103-116.

Rasmussen, P., 1989. Leaf-foddering of livestock in the Neolithic: Archaeobotanical evidence from Weier, Switzerland. J. Danish Archaeol. 8, 51-71.

Rasmussen, P., 1990. Leaf foddering in the earliest Neolithic agriculture. Evidence from Switzerland and Denmark. Acta Archaeol. 60, 71-86.

Rasmussen, P., 1993. Analysis of goat/sheep

faeces from Egolzwil 3, Switzerland: evidence for branch and twig foddering of livestock in the Neolithic. J. Archaeol. Sci. 20, 479-502.

Rasmussen, P., 2005. Mid-to late-Holocene land-use change and lake development at Dallund Sø, Denmark: vegetation and land-use history inferred from pollen data. The Holocene 15, 1116–1129.

Rasmussen, S.O., Birks, H.H., Blockley, S.P.E., Brauer, A., Hajdas, I., Hoek, W.Z., Lowe, J.J., Moreno, A., Renssen, H., Roche, D.M., Svensson, A.M., Valdes, P., Walker, M.J.C., 2014a. Dating, synthesis, and interpretation of palaeoclimatic records of the Last Glacial cycle and model-data integration: advances by the INTIMATE (INTegration of Ice-core, MArine and TErrestrial records) COST Action ES0907. Quat. Sci. Rev. 106, 1–13.

Rasmussen, S.O., Bigler, M., Blockley, S.P., Blunier, T., Buchardt, S.L., Clausen, H.B., Cvijanovic, I., Dahl-Jensen, D., Johnsen, S.J., Fischer, H., Gkinis, V., Guillevic, M., Hoek, W.Z., Lowe, J.J., Pedro, J.B., Popp, T., Seierstad, I.K., Steffensen, J.P., Svensson, A.M., Vallelonga, P., Vinther, B.M., Walker, M.J.C., Wheatley, J.J., Winstrup, M., 2014b. A stratigraphic framework for abrupt climatic changes during the Last Glacial period based on three synchronized Greenland ice-core records: refining and extending the INTIMATE event stratigraphy. Quat. Sci. Rev. 106, 14-28.

Ratcliffe, D., 1959. The habitat of *Koenigia islandica* in Scotland. Trans. Bot. Soc. Edin. 37.

Raunkiær, C., 1905. Types biologiques pour la géographie botanique. Kong. Danske Vidensk. Selsk. Forh. 1905, 347-438.

Raunkiær, C., 1907. Planterigels Livsformer og dered Betydning for Geografien. Gylendalske Boghandel-Nordisk Forlag, København & Kristiania. 132 pp.

Raunkiær, C., 1934. The Life Forms of Plants and Statistical Plant Geography, being the collected papers of C. Raunkiær. Clarenden Press, Oxford.

Ravazzi, C., 2002. Late Quaternary history of spruce in southern Europe. Rev. Palaeobot. Palynol. 120, 131-177.

Ravazzi, C. (Ed.), 2003. Gli antichi bacini lacustri e i fossili di Leffe, Ranica e Pianico-Sèllere (Prealpi Lombarde). Bergamo.

Ravazzi, C., Moscariello, A., 1998. Sedimentation, palaeoenvironmental evolution and time duration of earliest Pleistocene climatic cycles in the 24-56 m FM core interval (Leffe Basin, Northern Italy). Mededelingen Nederlandse Instituut voor toegepaste Geowetenschappen, TNO 60, 467-490.

Ravazzi, C., Pini, R., Breda, M., Martinotto, E., Muttoni, G., Chiesa, S., Confortini, S., Egli, R., 2005. The lacustrine deposits of Fornaci di Ranica (late Early Pleistocene, Italian Pre-Alps): stratigraphy, palaeoenvironment and geological evolution. Quat. Int. 131, 35-58.

Ravazzi, C., Donegana, M., Vescovi, E., Arpenti, E., Caccianiga, M., Kaltenrieder, P., Londeix, L., Marabini, S., Mariani, S., Pini, R., Vai, G.B., Wick, L., 2006. A new Late-glacial site with *Picea abies* in the northern Apennine foothills: an exception to the model of glacial refugia of trees. Veget. Hist. Archaeobot. 15, 357-371.

Ravazzi, C., Pini, R., Breda, M., 2009. Reconstructing the palaeoenvironments of the Early Pleistocene mammal faunas from the pollen preserved on fossil bones. Quat. Sci. Rev. 28, 2940-2954.

Read, D.J., 1993a. Mycorrhiza in plant communities. Adv. Plant Pathol. 9, 1-31.

Read, D.J., 1993b. Plant-microbe mutualisms and community structure. In: Schulze, E.-D., Moony, H. (Eds.), Biodiversity and Ecosystem Function. Springer, Berlin, pp. 181-209.

Read, D.J., Kianmehr, H., Malibari, A., 1977. The Biology of Mycorrhiza in *Helianthemum* Mill. New Phytol. 78, 305-312.

Reed, J.M., Stevenson, A.C., Juggins, S., 2001. A multi-proxy record of Holocene climatic change in southwestern Spain: the Laguna de Medina, Cádiz. The Holocene 11, 707–719.

Regattieri, E., Zanchetta, G., Drysdale, R.N., Isola, I., Hellstrom, J.C., Dallai, L., 2014. Lateglacial to Holocene trace element record (Ba, Mg, Sr) from Corchia Cave (Apuan Alps, central Italy): paleoenvironmental implications. J. Quat. Sci. 29, 381-392.

Regnell, M., 2012. Plant subsistence and environment at the Mesolithic site Tagerup, southern Sweden: new insights on the "Nut Age". Veget. Hist. Archaeobot. 21, 1-16.

Reich R., 1987. The archaeologist's dilemma: On some aspects of archaeology as a 'destructive' scientific discipline. In: 8th General Assembly and International Symposium 'Old Cultures in New Worlds'. International Council on Monuments and Sites (ICOMOS), Washington DC, p. 1009–1015. Available from: http://international.icomos.org/publications/wash133.pdf

Reid, C., 1882. The geology of the country around Cromer. Mem. Geol. Survey England and Wales, Sheet mem. 68 E.

Reid, C., 1890. The Pliocene deposits of Britain. Mem. Geol. Survey England and Wales.

Reid, C., 1899. The Origin of the British Flora. Dulau & Co., London. 191 pp.

Reid, P.C., Allitt, U., 1981. Comments on the pseudo-microfossil Linotolypen. *Rev. Palaeobot. Palynol.* 34(2), 263-267.

Reid, C., Reid, C.M., 1915.The Pliocene floras of the Dutch-Prussian border. Mededel. Rijksopspor. Delftstoffen 6, 1-178.

Reille, M., 1975. Contribution pollenanalytique a

l'histoire tardiglaciaire et holocène de la végétation de la montagne Corse. Aix-Marseille, p. 206.

Reille, M., 1988. Nouvelles analyses polliniques de la tourbière de la Moulinasse: versant nord-oriental des Pyrenées (France). Lab. Bot. Hist. Pal., 21 pp.

Reille, M., 1989. L'origine du Pin à crochets dans le Massif Central français. Bull. Soc. Bot. France 136, 61-70.

Reille, M., 1990a. The significance of *Erica arborea* and *Quercus ilex* in Corsica. International workshop on *Quercus ilex* L. Ecosystems: Function, Dynamics and Management, Barcelona/Montpellier, pp. 41-42.

Reille, M., 1990b. La tourbière de La Borde (Pyrénées orientales, France): un site clé pour l'étude du Tardiglaciaire sud-européen. C.R. Acad. Sci. Paris. 310, 823-829.

Reille, M., 1990c. Recherches pollenanalytiques dans l'extrémité orientale des Pyrénées: données nouvelles, de la fin du Glaciaire à l'Actuel. Ecologia Mediterranea 16, 317-357.

Reille, M., 1991a. Exemples de quelques séquences polliniques polluées par de la matière organique intruse. Conséquences pour l'histoire de la végétation des Pyrénées (France). Palynosciences 1, 113–138.

Reille, M., 1991b. L'origine de la station de pin à crochets de la tourbière de Pinet (Aude) et de quelques stations isolées de cet arbre dans les Vosges. Bull. Soc. Bot. France - Lettres Botaniques 138, 123-148.

Reille, M., 1992a. New pollen-analytical researches in Corsica: the problem of *Quercus ilex* L., *Erica arborea* L., the origin of *Pinus halepensis* Miller forests. New Phytol. 122, 359-378.

Reille, M., 1992b. Pollen et Spores d'Europe et d'Afrique du Nord. Laboratoire de Botanique Historique et Palynologie, Marseille.

Reille, M., 1993a. L'interface Tardiglaciaire-Holocène dans un site du littoral atlantique sud-européen: Le Moura (Pyrénées Atlantiques, France). C.R. Acad. Sci. Paris 316, 463-468.

Reille, M., 1993b. Nouvelles recherches pollenanalytiques dans la region de Freychinede, Pyrénées ariègeoises, France. Palynosciences 2, 109-131.

Reille, M., 1995. Pollen et spores d'Europe et d'Afrique du Nord, Supplément 1. Laboratoire de Botanique Historique et Palynologie, Marseille.

Reille, M., 1998. Pollen et spores d'Europe et d'Afrique du Nord, Supplément 2. Laboratoire de Botanique Historique et Palynologie, Marseille.

Reille, M., Andrieu, V., 1991. Données nouvelles sur l'histoire postglaciaire de la végétation des Pyrénées occidentales (France). C.R. Acad. Sci. Paris 312, 97-103.

Reille, M., Andrieu, V., 1995. The late Pleistocene and Holocene in Lourdes Basin, western Pyrénées, France: new pollen analytical and chronological data. Veget. Hist. Archaeobot. 4, 1-21.

Reille, M., de Beaulieu, J.-L., 1988. History of the Würm and Holocene vegetation in Western Velay (Massif Central, France): A comparison of pollen analysis from three corings at Lac du Bouchet. Rev. Palaeobot. Palynol. 54, 233-248.

Reille, M., de Beaulieu, J.-L., 1990. Pollen analysis of a long upper Pleistocene continental sequence in a Velay maar (Massif Central, France). Palaeogeogr. Palaeoclimatol. Palaeoecol. 80, 35-48.

Reille, M., de Beaulieu, J.-L., 1995. Long Pleistocene pollen records from the Praclaux crater, south-central France. Quat. Res. 44, 205-215.

Reille, M., Lowe, J.J., 1993. A re-evaluation of the vegetation history of the eastern Pyrenees (France) from the end of the last glacial to the present. Quat. Sci. Rev. 12, 47-77.

Reille, M., Pons, A., 1990. The real significance of sclerophyllous Oak Forests in the Western Mediterranean Basin. International workshop on *Quercus ilex* L. Ecosystems: Function, Dynamics and Management, Barcelona, Montpellier, pp. 36-37.

Reille, M., Pons, A., 1992. The ecological significance of sclerophyllous oak forests in the western part of the Mediterranean basin: a note on pollen analytical data. Vegetatio 99-100, 13-17.

Reille, M., Guiot, J., de Beaulieu, J.-L., 1992. The Montaigu event: an abrupt climatic change during the early Würm in Europe. In: Kukla, D.J., Went, E. (Eds.), Start of a Glacial. NATO ASI Ser. I, 3, 85-95.

Reille, M., Gamisans, J., de Beaulieu, J.-L., Andrieu, V., 1997. The late-glacial at Lac de Creno (Corsica, France): a key site in the western Mediterranean basin. New Phytol. 135, 547-559.

Reille, M., Andrieu, V., de Beaulieu, J.-L., Guenet, P., Goeury, C., 1998. A long pollen record from Lac du Bouchet, Massif Central, France: For the period ca 325 to 100 ka BP (OIS 9c to OIS 5e). Quat. Sci. Rev. 17, 1107-1123.

Reille, M., Gamisans, J., Andrieu-Ponel, V., de Beaulieu, J.-L., 1999. The Holocene at Lac de Creno, Corsica, France: a key site for the whole island. New Phytol. 141, 291-307.

Reille, M., de Beaulieu, J.-L., Svobodova, H., Andrieu-Ponel, V., Goeury, C., 2000. Pollen analytical biostratigraphy of the last five climatic cycles from a long continental sequence from the Velay region (Massif Central, France). J. Quat. Sci. 15, 665-685.

Reimer P.J., Reimer, R.W., 2001. A Marine Reservoir Correction Database and Online Interface. Radiocarbon 43, 461-463.

Reimer, P.J., Reimer, R.W., 2007. Radiocarbon Dating: Calibration. In: Elias, S.A. (Ed.), Encyclopedia of Quaternary Science. Elsevier, Amsterdam, pp. 2941-2950.

Reisigl, H., Danesch, E., Danesch, O., 1992. Mittelmeerflora. Parkland, Stuttgart.

Reitalu, T., Seppä, H., Sugita, S., Kangur, M., Koff, T., Avel, E., Kihno, K., Vassiljev, J., Renssen, H., Hammarlund, D., Heikkilä, M., Saarse, L., Poska, A., Veski, S., 2013. Long-term drivers of forest composition in a boreonemoral region: the relative importance of climate and human impact. J. Biogeogr. 40, 1524-1534.

Reitalu, T., Gerhold, P., Poska, A., Pärtel, M., Väli, V., Veski, S., 2015. Novel insights into post-glacial vegetation change: functional and phylogenetic diversity in pollen records. J. Veg. Sci. 26, 911-922.

Reitalu, T., Bjune, A.E., Blaus, A., Giesecke, T., Helm, A., Matthias, I., Peglar, S.M., Salonen, J.S., Seppä, H., Väli, V., Birks, H.J.B., 2019. Patterns of modern pollen and plant richness across northern Europe. J. Ecol. 107, 1662-1677.

Rejmánek, M., Rejmáneková, E., Holzner, W., 2004. Species diversity of plant communities on calcareous screes: the role of intermediate disturbance. Preslia 76, 207-222.

Renault-Miskovsky, J., Bui-Thi-Mai, M., Girard, M., 1984. A propos de l'indigénat ou de l'introduction de *Juglans* et *Platanus* dans l'ouest de l'Europe au Quaternaire. Rev. Paléobiol. spécial, 155-178.

Renberg, I., Segerstrom, U., 1981. Applications of varved lake sediments in palaeoenvironmental studies. Wahlenbergia 7, 125-133.

Rendell, S., Ennos, R.A., 2003. Chloroplast DNA diversity of the dioecious European tree *Ilex aquifolim* L. (English holly). Mol. Ecol. 12, 2681-2688.

Renfrew, J.N., 1973. Palaeoethnobotany: The Prehistoric Food Plants of the Near East and Europe. Methuen, London. 248 pp.

Renner, S., Grimm, G., Kapli, P., Denk, T., 2016. Species relationships and divergence times in beeches: new insights from the inclusion of 53 young and old fossils in a birth-death clock model. Philos. Trans. R. Soc. London B 371, 20150135.

Rennie, R., 1807. Essays on the natural history and origin of peat moss: the peculiar qualities of that substance; the means of improving it as a soil; the methods of converting it into a manure; and the other economical purposes to which it may be made subservient. Essays I & II. Archibald Constable, Edinburgh, 233 pp.

Renssen, H., Moberg, A., Holmgren, K., Sundqvist, H., Zhang, G. (eds.), 2010. Holocene climate variability over Scandinavia. Clim. Past, 6.

Reu, B., Zaehle, S., Bohn, K., Pavlick, R., Schmidtlein, S., Williams, J.W., Kleidon, A., 2014. Future no-analogue vegetation produced by no-analogue combinations of temperature and insolation. Global Ecol. Biogegor. 23, 156-167.

Revelles, J., Ghilardi, M., Rossi, V., Curras, A., Lopez-Bulto, O., Brkojewitsch, G., Vacchi, M., 2019. Coastal landscape evolution of Corsica island (W. Mediterranean): palaeoenvironments, vegetation history and human impacts since the early Neolithic period. Quat. Sci. Rev. 225.

Rey, F., Schwörer, C., Gobet, E., Colombaroli, D., van Leeuwen, J.F.N., Schleiss, S., Tinner, W., 2013. Climatic and human impacts on mountain vegetation at Lauenensee (Bernese Alps, Switzerland) during the last 14,000 years. Holocene 23, 1415-1427.

Rey, F., Gobet, E., van Leeuwen, J.F.N., Gilli, A., van Raden, U.J., Hafner, A., Wey, O., Rhiner, J., Schmocker, D., Zund, J., Tinner, W., 2017. Vegetational and agricultural dynamics at Burgäschisee (Swiss Plateau) recorded for 18,700 years by multi-proxy evidence from partly varved sediments. Veget. Hist. Archaeobot. 26, 571-586.

Rey, F., Gobet, E., Schwörer, C., Wey, O., Hafner, A., Tinner, W., 2019a. Causes and mechanisms of synchronous succession trajectories in primeval Central European mixed *Fagus sylvatica* forests. J. Ecol. 107, 1392-1408.

Rey, F., Gobet, E., Szidat, S., Lotter, A.F., Gilli, A., Hafner, A., Tinner, W., 2019b. Radiocarbon wiggle matching on laminated sediments delivers high-precision chronologies. Radiocarbon 61, 265-285.

Rey, F., Gobet, E., Schwörer, C., Hafner, A., Szidat, S., Tinner, W., 2020. Climate impacts on vegetation and fire dynamics since the last deglaciation at Moossee (Switzerland). Clim. Past 16, 1347-1367.

Reynaud, C., 1974. The pioneer vegetational phase in Central Lapland (Finland) based on new palynological data. Ann. Bot. Fenn. 11, 197-201.

Reynaud, C., 1976. Palaeoecological significance of *Hippophaë rhamnoides*, with the example of the protocratic vegetation stage in NE-Fennoscandia. Boreas 5, 9-24.

Richard, J.L., 1987. Dynamique de la végétation sur les marges glaciares récentes de la réserve d'Aletsch (Alpes valaisiannes). Botanica Helvetica 97, 265-275.

Riera, S., Wansard, G., Julià, R., 2004. 2000-year environmental history of a karstic lake in the Mediterranean Pre-Pyrenees: the Estanya lakes (Spain). Catena 55, 293-324.

Riera, S., López-Sáez, J., Julià, R., 2006. Lake responses to historical land use changes in northern Spain: The contribution of non-pollen palynomorphs in a multiproxy study. Rev. Palaeobot. Palynol. 141, 127-137.

Rietkerk, M., Dekker, S.C., Wassen, M.J., Verkroost, A.W.M., Bierkens, M.F.P., 2004. A putative mechanism for bog patterning. Am. Nat 163, 699-708.

Rigling, D., Prospero, S., 2018. *Cryphonectria parasitica*, the causal agent of chestnut blight: invasion history, population biology and disease control. Molec. Plant Pathol. 19, 7-20.

Rintanen, T., 1996. Changes in the flora and vegetation of 113 Finnish lakes during 40 years. Ann. Bot. Fenn. 33, 101-122.

Rinterknecht, V., Clark, P., Raisbeck, G., Yiou, F., Bitinas, A., Brook, E., Marks, L., Zelčs, V., Lunkka, J.-P., Pavlovskaya, I., Piotrowski, J., Raukas, A., 2006. The last deglaciation of the southeastern sector of the Scandinavian ice sheet. Science 311, 1449-1452.

Ritchie, J.C., 1985. Late-Quaternary climatic and vegetational change in the lower MacKenzie Basin, northwest Canada. Ecology 66, 612-621.

Ritchie, J.C., Lichti-Federovich, S., 1967. Pollen dispersal phenomena in arctic-subarctic Canada. Rev. Palaeobotan. Palynol. 3, 255-266.

Rivas-Martínez, S., 1987. Mapa de series de vegetación de España. ICONA Serie Técnica, Madrid.

Rivas-Martínez, S., Fernández-González, F., Loidi, J., Lousã, M., Penas, A., 2001. Worldwide Bioclimatic Classification System. Itinera Geobot. 14, 5-341.

Roberts, N., Eastwood, W., Kuzucuoğlu, C., Fiorentino, G., Caracuta, V., 2011. Climatic, vegetation and cultural change in the eastern Mediterranean during the mid-Holocene environmental transition. Holocene 21, 147-162.

Roberts, N., Fyfe, R.M., Woodbridge, J., Gaillard, M.-J., Davis, B.A.S., Kaplan, J.O., Marquer, L., Mazier, F., Nielsen, A.B., Sugita, S., Trondman, A.-K., Leydet, M., 2018. Europe's lost forests: a pollen-based synthesis for the last 11,000 years. Sci. Rep. 8, 716.

Robin, V., Nadeau, M.J., Grootes, P.M., Bork, H.R., Nelle, O., 2016. Too early and too northerly: evidence of temperate trees in northern central Europe during the Younger Dryas. New Phytol. 212, 259-268.

Robles-Lopez, S., Luelmo-Lautenschlaeger, R., Perez-Diaz, S., Abel-Schaad, D., Alba-Sanchez, F., Ruiz-Alonso, M., López-Sáez, J.A., 2017. Vulnerability and resilience of high-mountain pine forests of the Gredos range (Avila, Spanish Central System): two thousand years of socio-ecological dynamics. Cuaternario Y Geomorfologia 31, 51-71.

Roces-Díaza, J., Jiménez-Alfaro, B., Chytrý, M., Díaz-Varela, E., Álvarez-Álvarez, P., 2018. Glacial refugia and mid-Holocene expansion delineate the current distribution of *Castanea sativa* in Europe. Palaeogeogr. Palaeoclimatol. Palaeoecol. 491, 152-160.

Rodríguez-Sánchez, F., Hampe, A., Jordano, P., Arroyo, J., 2010. Past tree range dynamics in the Iberian Peninsula inferred through phylogeography and palaeodistribution modelling: A review. Rev. Palaeobot. Palynol. 162 507-521.

Roe, H.M., Coope, R., Devoy, R.N.J., Harrison, J. C.O., Penkman, K.E.H., Preece, R.C., Schreve, D., 2009. Differentiation of MIS 9 and MIS 11 in the continental record: vegetational, faunal, aminostratigraphic and sea-level evidence from coastal sites in Essex, UK. Quat. Sci. Rev. 28, 2342-2373.

Romane, F., Terradas, J. (Eds.), 1992. Quercus ilex L. ecosystems: function, dynamics and management. Advances in Vegetation Science. Kluwer, Dordrecht, 376 pp..

Ronikier, M., 2011. Biogeography of high-mountain plants in the Carpathians: An emerging phylogeographical perspective. Taxon 60, 373-389.

Roos-Barraclough, F., Martinez-Cortizas, A., Garcia-Rodeja, E., Shotyk, W., 2002. A 14 500 year record of the accumulation of atmospheric mercury in peat: volcanic signals, anthropogenic influences and a correlation to bromine accumulation. Earth Planet. Sci. Lett. 202, 435-451.

Röpke, A., Stobbe, A., Oeggl, K., Kalis, A.J., Tinner, W., 2011. Late-Holocene land-use history and environmental changes at the high altitudes of St Antonien (Switzerland, Northern Alps): Combined evidence from pollen, soil and tree-ring analyses. Holocene 21, 485-498.

Rösch, M., 1983. Geschichte der Nussbaumer Seen/Kt. Thurgau und ihrer Umgebung seit dem Ausgang der letzten Eiszeit aufgrund quartärbotanischer, stratigraphischer und sedimentologischer Untersuchungen. Mitt. Thurg. Naturf. Ges. 45, 3-110.

Rösch, M., 1986. Zwei Moore im westlichen Bodenseegebiet als Zeugen prähistorischer Landschaftsveränderung. TELMA 16, 83-111.

Rösch, M., 1992. Human impact as registered in the pollen record: some results from the western Lake Constance region, Southern Germany. Veget. Hist. Archaeobot. 1, 101-109.

Rösch, M., 1998. The history of crops and crop weeds in south-western Germany from the Neolithic period to modern times, as shown by archaeobotanical evidence. Veget. Hist. Archaeobot. 7, 109-125.

Rösch, M., Fischer, E., 2000. A radiocarbon dated Holocene pollen profile from the Banat mountains (Southwestern Carpathians, Romania). Flora 195, 277-286.

Rösch, M., Tserendorj, G., 2011a. Der Nordschwarzwald-früher besiedelt als gedacht? Pollenprofile belegen ausgedehnte vor-geschichtliche Besiedlung und Landnutzung. Denkmalpfl. Baden-Württ. Landesdenkmalpflege 40, 66-73.

Rösch, M., Tserendorj, G., 2011b. Florengeschichtliche Beobachtungen im Nordschwarzwald (Südwestdeutschland). Hercynia 44, 53-71.

Rösch, M., Kleinmann, A., Lechterbeck, J., Wick, L., 2014. Botanical off-site and on-site data as indicators of different land use systems: a discussion with examples from southwest Germany. Veget. Hist. Archaeobot. 23, S121-S133.

Rose, N.L., 1994. A note on further refinements to a procedure for the extraction of carbonaceous fly-ash particles from sediments. J. Paleolimnol. 11, 201-204.

Rose, N.L., 2001. Fly-ash particles. In: Last, W.M. and Smol, J.P. (eds.), Tracking Environmental Change Using Lake Sediments, Vol. 2: Physical and Geochemical Methods. Kluwer Academic Publishers, Dordrecht, pp. 319-349.

Rose, N.L., 2008. Quality control in the analysis of lake sediments for spheroidal carbonaceous particles. Limnol. Oceanogr.: Methods 6, 172-179.

Rose, N.L., 2015. Spheroidal carbonaceous fly-ash particles provide a globally synchronous stratigraphic marker for the Anthropocene. Environ. Sci. Tech. 49, 4155-4162.

Rose, N.L., Appleby, P.G., 2005. Regional applications of lake sediment dating by spheroidal carbonaceous particle analysis I. United Kingdom. J. Paleolimnol. 34, 349-361.

Rosén, P., Segerström, U., Eriksson, L., Renberg, I., Birks, H.J.B., 2001. Holocene climatic change reconstructed from diatoms, chironomids, pollen and near-infrared spectroscopy at an alpine lake (Sjuodjijaure) in northern Sweden. Holocene 11, 551-562.

Rosén, P., Segerström, U., Eriksson, L., Renberg, I., 2003. Do diatom, chironomid, and pollen records consistently infer Holocene July air temperature? A comparison using sediment cores from four alpine lakes in northern Sweden. Arctic AntArc. Alp. Res. 35, 279-290.

Rosselló, J., Lázaro, A., Cosín, R., Molins, A., 2007. A phylogeographic split in *Buxus balearica* (Buxaceae) as evidenced by nuclear ribosomal markers: when ITS paralogues are welcome. J. Mol. Evol. 64, 143-157.

Rossi S., 2003. Le piante raccontano la storia di un intero interglaciale. In: Ravazzi, C. (Ed.), Gli antichi bacini lacustri e i fossili di Leffe, Ranica e Pianico-Sèllere. Quaderni di Geodinamica Alpina e Quaternaria, numero speciale. C.N.R. Bergamo.

Rossignol-Strick, M., 1995. Sea-land correlation of pollen records in the eastern Mediterranean for the glacial-interglacial transition: Biostratigraphy versus radiometric time-scale. Quat. Sci. Rev. 14, 893-915.

Rossignol-Strick, M., Planchais, N., Paterne, M., Duzer, D., 1992. Vegetation dynamics and climate during the deglaciation in the south adriatic basin from a marine record. Quat. Sci. Rev. 114, 415-423.

Roucoux, K., de Abreu, L., Shackleton, N., Tzedakis, P.C., 2005. The response of NW Iberian vegetation to North Atlantic climate oscillations during the last 65 kyr. Quat. Sci. Rev. 24, 1637-1653.

Roulleau, E., Pinti, D.L., Rouchon, V., Quidelleur, X., Gillot, P.-Y., 2009. Tephro-chronostratigraphy of the lacustrine interglacial record at Pianico, Italian Southern Alps: Identifying the volcanic sources using radiogenic isotopes and trace elements. Quat. Int. 204, 31-43.

Rousi, A., 1971. The genus *Hippophaë* L. A taxonomic study. Annal. Bot. Fenn. 8, 177-227.

Rowley-Conwy, P., 1982. Forest grazing and clearance in temperate Europe with special reference to Denmark: an archaeological view. In: Bell, M., Limbrey, S. (Eds.), Archaeological Aspects of Woodland Ecology. Association for Environmental Archaeology, pp. 199-215.

Roxburgh, S.H., Shea, K., Wilson, J.B., 2004. The intermediate disturbance hypothesis: Patch dynamics and mechanisms of species coexistence. Ecology 85, 359-371.

Rozanski, K., Weislo, D., Harmata, K., Norys-kiewicz, B., Ralska-Jasiewiczowa, M., 1985. Palynological and isotopes studies on carbonate sediments from some Polish lakes: Preliminary results. Presented at the INQUA/IGCP 158 meeting on the palaeohydrological changes during the last 15 000 years, pp. 41-49.

Rozenzweig, M.L., 1995. Species Diversity in Space and Time. Cambridge University Press, Cambridge. 460 pp.

Ruddiman, W.F., Fuller, D.Q., Kutzbach, J.E., Tzedakis, P.C., Kaplan, J.O., Ellis, E.C., Vavrus, S.J., Roberts, C.N., Fyfe, R., He, F., Lemmen, C., Woodbridge, J., 2016. Late Holocene climate: Natural or anthropogenic? Rev. Geophys. 54, 93 - 118.

Rudner, Z., Sümegi, P., 2001. Recurring taiga forest-steppe habitats in the Carpathian Basin in the Upper Weichselian. Quat. Int. 76/77, 177-189.

Rudolph, K., 1928. Die bisherigen Ergebnisse der botanischen Mooruntersuchungen in Böhmen. Beihefte zum Botanischen Centralblatt 45, 1-180.

Rudolph, K., 1930. Grundzüge der nacheiszeitlichen Waldgeschichte Mitteleuropas. Beihefte zum Botanischen Centralblatt 47, 111-176.

Rudolph, K., Firbas, F., 1922. Pollenanalytische Untersuchungen böhmischer Moore (Vorl. Mitt.). Ber. Deutsche Botan. Ges. 40, 393 - 405.

Rudolph, K., Firbas, F., 1924. Die Hochmoore des Erzgebirges. Beihefte zum Botanischen Centralblatt 41, 1-162.

Rudow, A., Conedera, M., 2001. Blüte und Sortenerkennung bei der Edelkastanie (*Castanea sativa* Mill.) auf der Alpensüdseite der Schweiz. Bot. Helv. 111, 1-23.

Ruiz, G., Carlton, J.-T. (Eds.), 2003. Invasive Species. Vectors And Management Strategies. Island

Press, Washington, 518 pp.

Ruiz Zapata, M., Gil García, M., Dorado Valiño, M., Valdeolmillos Rodríguez, A., Vegas, J., Pérez-González, A., 2002. Clima y vegetación durante el Tardiglaciar y el Holoceno en la Sierra de Neila (Sistema Ibérico Noroccidental). Revista Cuaternario y Geomorfologia 16, 9-20.

Ruiz-Alonso, M., Pérez-Díaz, S., López-Sáez, J., 2019. From glacial refugia to the current landscape configuration: permanence, expansion and forest management of *Fagus sylvatica* L. in the Western Pyrenean Region (Northern Iberian Peninsula). Veget. Hist. Archaeobot. 28, 481-496.

Rundgren, M., Beerling, D.J., 1999. A Holocene CO_2 record from the stomatal index of subfossil *Salix herbacea* L. leaves from northern Sweden. Holocene 9, 509-513.

Rundgren, M., Björck, S., 2003. Late-glacial and early Holocene variations in atmospheric CO_2 concentrations indicated by high-resolution stomatal index data. Earth Planet. Sci. Lett. 213, 191-204.

Ruosch, M., Spahni, R., Joos, F., Henne, P.D., van der Knaap, W.O., Tinner, W., 2016. Past and future evolution of *Abies alba* forests in Europe - comparison of a dynamic vegetation model with palaeo data and observations. Global Change Biol. 22, 727-740.

Ruprecht, H., Dhar, A., Aigner, B., Oitzinger, G., Klumpp, R., Vacik, H., 2010. Structural diversity of English yew (*Taxus baccata* L.) populations. Eur. J. For. Res. 129, 189-198.

Ruskál, A., Diaconu, A., Grindean, R., Tanţău, I., 2020. Early to mid-Holocene hydroclimate trends in the western Carpathians of Romania. Palaeogeogr. Palaeoclim. Palaeoecol. 543, 109608.

Ruttner, F., 1962. Grundriss der Limnologie. De Gruyter, Berlin.

Ruuhijärvi, R., 1983. The Finnish mire types and their regional distribution. In: Gore, A.J.P. (Ed.), Ecosystems of the World 4A: Mires, Swamp, Bog, Fen, Moor. General studies. Elsevier, Amsterdam, pp. 47-67.

Ryan, W., 2007. Status of the Black Sea flood hypothesis. In: Yanko-Hombach, V., Gilbert, A., Panin, N., Dolukhanov, P. (Eds.), The Black Sea Flood Question: Changes in Coastline, Climate and Human Settlement. Springer, Dordrecht, The Netherlands, pp. 63-88.

Ryan, W., Pitman, W., 1998. Noah's Flood: The New Scientific Discoveries about the Event that Changed History. Simon Schuster, New York.

Ryan, W., Pitman, W., Major, C., Shimkus, K., Moskalenko, V., Jones, G., Dimitrov, P., Gorür, N., Saking, M., Yüce, H., 1997. An abrupt drowning of the Black Sea shelf. Mar. Geol. 138, 119-126.

Rybníček, K., 1973. A comparison of the present and past mire communities of Central Europe, In: Birks, H.J.B. (Ed.), Quaternary Plant Ecology. Blackwell, Cambridge, pp. 237-261.

Rybníček, K., 1984. The Vegetation and Development of Central European Mires. In: Moore, P.D. (Ed.), European Mires. Academic Press, London, pp. 177-201.

Rybníček, K., Rybníčková, E., 1968. The history of the flora and vegetation on the Blato mire in south-eastern Bohemia, Czechoslovakia. Folia Geobot. Phytotaxon. 3, 117-142.

Rybníček, K., Rybníčková, E., 1985. A palaeoecological reconstruction of precultural vegetation in the intermontane basins of the Western Carpathians. Ecol. Mediterr. 11, 27-31.

Rybníček, K., Rybníčková, E., 2002. Vegetation of the upper Oravo region (NW Slovakia) in the last 11000 years. Acta Palaeobot. 42, 153-169.

Rybníčková, E., 1974. Die Entwicklung der Vegetation und Flora im südlichen Teil der böhmisch-mährischen Höhe während des Spätglazials und Holozäns. Vegetace CSSR A7, 1-163.

Rybníčková, E., Rybníček, K., 1972. Erste Ergebnisse paläobotanischer Untersuchungen des Moores Vracov, Südmähren. Folia Geobot. Phytotax. 7, 285-308.

Rybníčková, E., Rybníček, K., 1988. Isopollen maps of *Picea abies, Fagus sylvatica* and *Abies alba* in Czechoslovaria - Their application and limitations. In: Lang, G., Schlüchter, C. (Eds.), Lake, Mire and River Environments - During the last 15 000 years. Balkema, Rotterdam, pp. 51-66.

Rybníčková, E., Rybníček, K., 1992. The Environment of the Pavlovian - Palaeoecological Results from Bulhary, South Moravia. Acta naturw. Mus. Wien, 73-79.

Rybníčková, E., Rybníček, K., 1993. Late Quaternary forest line oscillations in the West Carpathians. In: Frenzel, B. (Ed.), Oscillation of Alpine and Polar Tree Limits in the Holocene. Fischer, Stuttgart.

Rybníčková, E., Rybníček, K., 1996. Czech and Slovak Republics. In: Berglund, B.E., Birks, H.J.B., Ralska-Jasiewiczowa, M., Wright, H.E. (Eds.), Palaeoecological Events During the Last 15 000 Years. Regional Syntheses of Palaeoecological Studies of Lakes and Mires in Europe. Wiley, Chichester, pp. 473-505.

Rybníčková, E., Rybníček, K., 2014. Palaeovegetation in the Pavlovské vrchy hills region (South Moravia, Czech Republic) around 25,000 BP: the Bulhary core. Veget. Hist. Archaeobot. 23, 719-728.

Rydin, H., Barber, K.E., 2001. Long-term and fine-scale coexistence of closely related species. Folia Geobot. 36, 53-61.

Rydin, H., Jeglum, J., 2008. The Biology of Peatlands. Oxford University Press, Oxford.

Rzechowski, J., 1996. The Ferdinandowian Interglacial and its stratigraphical position in the Middle Pleistocene of Europe. In: Turner, C. (Ed.), The Early Middle Pleistocene in Europe. CRC Press, Boca Raton, pp. 279-293.

Saarse, L., Liiva, A., 1995. Geology of the Äntu group of lake. Proc. Acad Sci Est. Geol. 44, 119–132.

Saarse, L., Rajamäe, R., 1997. Holocene vegetation and climatic change on the Haanja Heights, SE Estonia. Proc Est Acad Sci Geol 46, 75–92.

Saarse, L., Veski, S., 2001. Spread of broad-leaved trees in Estonia. Estonian Acad. Sci. Geol. 50 1, 51-65.

Saarse, L., Veski, S., Rajamäe, R., Sarv, A., Heinsalu, A., 1990a. Geology of Lake Maardu. Eesti Teaduste Akademia Geoloogia Instituut edition, Tallinn, Estonia.

Saarse, L., Vishnevskaya, E., Sarv, A., 1990b. Geological development of lakes, Island Saaremaa. Proc. Est. Acad. Sci. Biol. 39.

Saarse, L., Veski, S., Heinsalu, A., Rajamäe, R., Martma, T., 1995. Litho-and biostratigraphy of lake Päidre, south Estonia. Proc Est Acad Sci Geol 44, 45–59.

Saarse, L., Mäemets, H., Pirrus, R., Rôuk, A., Sarv, A., Ilves, E., 1996. Estonia. In: Berglund, B.E., Birks, H.J.B., Ralska-Jasiewiczowa, M., Wright, H.E. (Eds.), Palaeoecological Events During the Last 15 000 Years. Regional Syntheses of Palaeoecological Studies of Lakes and Mires in Europe. Wiley, Chichester, pp. 367-393.

Sadori, L., 2013. Pollen Records, Postglacial, Southern Europe. In: Elias, S.A., Mock, C.A. (Eds.), Encyclopedia of Quateernary Science, 2nd ed. Elsevier, Amsterdam, pp. 179–188.

Sadori, L., Narcisi, B., 2001. The Postglacial record of environmental history from Lago di Pergusa, Sicily. Holocene 11, 655-670.

Sadori, L., Zanchetta, G., Giardini, M., 2008. Last Glacial to Holocene palaeoenvironmental evolution at Lago di Pergusa (Sicily, Southern Italy) as inferred by pollen, microcharcoal, and stable isotopes. Quat. Int. 181, 4-14.

Sadori, L., Giardini, M., Chiarini, E., Mattei, M., Papasodaro, F., Porreca, M., 2010. Pollen and macrofossil analyses of Pliocene lacustrine sediments (Salto river valley, Central Italy). Quat. Int. 225, 44-57.

Sainz Ollero, H., Sánchez de Dios, R., García-Cervigón Morales, A., 2010. La cartografía sintética de los paisajes vegetales Españoles: Una asignatura pendiente en geobotánica. Ecología 23, 249-272.

Salonen, J.S., Seppä, H., Väliranta, M., Jones, V., Self, A., Heikkilä, M., Kultti, S., Yang, H., 2011. The Holocene thermal maximum and late-Holocene cooling in the tundra of NE European Russia. Quat. Res. 75, 501-511.

Salvador, L., Alía, R., Agúndez, D., Gil, L., 2000. Genetic variation and migration pathways of maritime pine (Pinus pinaster Ait) in the Iberian Peninsula. Theor. Appl. Genet. 100, 89-95.

Samartin, S., Heiri, O., Lotter, A.F., Tinner, W., 2012. Climate warming and vegetation response after Heinrich event 1 (16 700–16 000 cal yr BP) in Europe south of the Alps. Clim. Past 8, 1913-1927.

Samartin, S., Heiri, O., Vescovi, E., Brooks, S.J., Tinner, W., 2012. Lateglacial and early Holocene summer temperatures in the southern Swiss Alps reconstructed using fossil chironomids. Journal of Quaternary Science 27, 279-289.

Samartin, S., Heiri, O., Kaltenrieder, P., Kühl, N., Tinner, W., 2016. Reconstruction of full glacial environments and summer temperatures from Lago della Costa, a refugial site in northern Italy. Quat. Sci. Rev. 143, 107-119.

Samartin, S., Heiri, O., Joos, F., Renssen, H., Franke, J., Brönnimann, S., Tinner, W., 2017. Warm Mediterranean mid-Holocene summers inferred from fossil midge assemblages. Nature Geoscience 10, 207-214.

Samuelsson, G., 1934. Die Verbreitung der höheren Wasserpflanzen in Nordeuropa (Fennoskandien und Dänemark). Acta Phytogeogr. Suecica, Uppsala 6, 1-211.

Sánchez-Goñi, M.F., 1988. A propos de la présence du pollen de *Castanea* et de *Juglans* dans les sédiments archéologiques würmiens anciens du Pays Basque espagnol, Symposium APLF. Institut fr. Pondichéry traveaux section scientifique et technique, Bordeaux, pp. 73-82.

Sánchez-Goñi, M.F., 1993. De la taxonomie pollinique à la reconstitution de l'environnement. L'example de la région cantabrique. British Archaeological Reports, International Series 44-45.

Sánchez-Goñi, M.F., Hannon, G.E., 1999. High-altitude vegetational pattern on the Iberian Mountain Chain (north-central Spain) during the Holocene. The Holocene 9, 39–57.

Sánchez-Goñi, M.F., Cacho, I., Turon, J.L., Guiot, J., Sierro, F., Peypouquet, J.P., Grimalt, J., Shackleton, N.J., 2002. Synchroneity between marine and terrestrial responses to millennial scale climatic variability during the last glacial period in the Mediterranean region. Clim. Dynam. 19, 95–105.

Sánchez-Goñi, M.F., Landais, A., Fletcher, W., Naughton, F., Desprat, S., Duprat, J., 2008. Contrasting impacts of Dansgaard–Oeschger events over a western European latitudinal transect modulated by orbital parameters. Quat. Sci. Rev. 27, 1136-1151.

Sand-Jensen, K., Krause-Jensen, D., 1997.

Broad-scale comparison of photosynthesis in terrestrial and aquatic plant communities. Forum 80, 203-208.

Sandegren, R., 1920. *Najas flexilis* i Fennoskandia under postglacialtiden. Sensk Bot. Tidskr., Stockholm, 14.

Sandegren, R., 1943. *Hippophaë rhamnoides* L. i Sverige under senkvartär tid. Svensk Bot. Tidskrift 37, 1-26.

Sandel, B., 2019. Disequilibrium in trait-climate relationships of trees and birds. Front. Ecol. Evol. 7, art138.

Sandom, C.J., Ejrnaes, R., Hansen, M.D.D., Svenning, J.C., 2014. High herbivore density associated with vegetation diversity in interglacial ecosystems. Proc. Nat. Acad. Sci. USA 111, 4162-4167.

Sanz, R., Pulido, F., Nogués-Bravo, D., 2009. Predicting mechanisms across scales: amplified effects of abiotic constraints on the recruitment of yew *Taxus baccata*. Ecography 32, 993-1000.

Sarmaja-Korjonen, K., Vasari, Y., Haeggström, C., 1991. *Taxus baccata* and influence of Iron Age man on the vegetation in Aland, SW Finland. Annal. Bot. Fenn. 28, 143-159.

Sarmaja-Korjonen, K., Krystyna S., Gąsiorowski, M., 2003. Subfossil chydorid taxa and assemblages from lake sediments in Poland and Finland with special reference to climate. Studia Quaternaria.

Sarv, A., Ilves, E., 1975. On the age of the Holocene deposits at the mouth of Emajögi River (based on material from Saviku) [in Russian]. Eesti NSV Tead Akad Toim 24, 64-69.

Sass-Klaasen, U., 2002. Dendroarchaeology: successes in the past and challenges for the future. Dendrochronologia 20, 87-93.

Saulnier-Talbot, É., 2016. Paleolimnology as a tool to achieve environmental sustainability in the Anthropocene: an overview. Geosciences 6, 26.

Scarnati, L., Attorre, F., De Sanctis, M., Farcomeni, A., Francesconi, F., Mancini, M., Bruno, F., 2009. A multiple approach for the evaluation of the spatial distribution and dynamics of a forest habitat: the case of Apennine beech forests with *Taxus baccata* and *Ilex aquifolium*. Biodivers. Conserv. 18, 3099-3113.

Schaffers, A.P., Sýkora, K.V., 2000. Reliability of Ellenberg indicator values for moisture, nitrogen and soil reaction: a comparison with field measurements. J. Veg. Sci. 11, 225-244.

Scharfetter, R., 1938. Das Pflanzenleben der Ostalpen. Deuticke, Wien.

Scheeder, T., 1994. Die Eibe (*Taxus baccata* L.). Hoffnung für ein fast verschwundenes Waldvolk. IHW-Verlag, Eching.

Schenk, F., Väliranta, M., 2020. Definition of climate indicator plant species and their common minimum July temperature limits. Dataset version 1.0. Bolin Centre Database.

Schenk, F., Bennike, O., Väliranta, M., Avery, R., Björck, S., Wohlfarth, B., 2020. Floral evidence for high summer temperatures in southern Scandinavia during 15–11 cal ka BP. Quat. Sci. Rev. 233, 106243.

Schimel, D.S., Asner, G.P., Moorcroft, P., 2013. Observing changing ecological diversity in the Anthropocene. Front. Ecol. Environ. 11, 129-137.

Schippers, P., Vermaat, J.E., de Klein, J., Mooij, W.M., 2004. The effect of atmospheric carbon dioxide elevation on plant growth in freshwater ecosystems. Ecosystems 7, 63-74.

Schloss, S., 1979. Pollenanalytische und stratigraphische Untersuchungen im Sewensee. Ein Beitrag zur spät- und postglazialen Vegetationsgeschichte der Südvogesen. Diss. Bot., Vaduz, 52, 1-138.

Schlüchter, C., Maisch, M., Suter, J., Fitze, P., Keller, W. A., Burga, C. A., Wynistorf, E., 1987. Das Schieferkohlen-Profil von Gossau (Kanton Zürich) und seine stratigraphische Stellung innerhalb der letzten Eiszeit. Vjschr. Naturforsch. Ges. Zürich 132, 135-174.

Schmid, S., Genevest, R., Gobet, E., Suchan, T., Sperisen, C., Tinner, W., Alvarez, N., 2017. HyRAD-X, a versatile method combining exome capture and RAD sequencing to extract genomic information from ancient DNA. Methods Ecol. Evol. 8, 1374-1388.

Schmidt, E., 1974. Ökosystem See. Das Beziehungsgefüge der Lebensgemeinschaft im eutrophen See und die Gefährdung durch zivilisatorische Eingriffe. Quelle & Meyer, Heidelberg.

Schmidt, R., 1981. Grundzüge der spät- und postglazialen Vegetations- und Klimageschichte des Salzkammergutes (Österreich) aufgrund palynologischer Untersuchungen von See- und Moorprofilen. Mitteilungen der Kommission für Quartärforschung der Österreichischen Akademie der Wissenschaften 3, 1-96.

Schmidt, R., Müller, J., Drescher-Schneider, R., Krisai, R., Szeroczynska, Baric, A., 2000. Changes in lake level and trophy at Lake Vrana, a large karstic lake on the Island of Cres (Croatia), with respect to palaeoclimate and anthropogenic impacts during the last approx. 16,000 years. J. Limnol. 59, 113-130.

Schmidt-Vogt, H., 1977. Die Fichte - Ein Handbuch in zwei Bänden. Band I: Taxonomie - Verbreitung - Morphologie - Waldgesellschaften. Paul Parey, Hambug.

Schneider, R., 1978. Pollenanalytische Untersuchungen zur Kenntnis der Spät- und postglazialen Vegetationsgeschichte am Südrand der Alpen zwischen Turin und Varese (Italien). Bot. Jahrb. Syst. 100, 26-109.

Schneider, R., 1985a. Analyse palynologique dans l'Aspromante en Calabre (Italie Meridional). Extrait des: Cahiers Ligures de Préhistoire et de Protohistoire. 2, 279-288.

Schneider, R., 1985b. Palynology research in the southern and southeastern Alps between Torino and Trieste. In: Lang, G. (Ed.), Swiss Lake and Mire Environments during the last 15000 Years. Diss. Bot. 87, 83-103.

Schneider, R., Tobolski, K., 1985. Lago di Ganna – Late-glacial and Holocene environments of a lake in the southern Alps. Diss. Bot. 87, 229-271.

Schoch, W., Heller, I., Schweingruber, F.H., Kienast, F., 2004. Wood anatomy of central European Species. Online version: www.woodanatomy.ch

Scholz, C.A., 2001. Application of seismic sequence stratigraphy in lacustrine basins. In: Last, W.M., Smol J.P. (Eds.), Tracking Environmental Change Using Lake Sediments, Vol. 1: Basin Analysis, Coring, and Chronological Techniques. Kluwer, Dordrecht, pp. 7-22.

Schookius, M., 1658. Tractatus de turfis, ceu cespitibus bituminosus. Groningen.

Schroedter, T.M., Nelle, O., 2015. New insights into Mid-Holocene vegetation in the southern Marmara region: Charcoal from the Late Neolithic to Early Chalcolithic settlement site Aktopraklik, northwestern Turkey. Quat. Int. 366, 81-95.

Schumacher, S., Bugmann, H., Mladenoff, D.J., 2004. Improving the formulation of tree growth and succession in a spatially explicit landscape model. Ecol. Model. 180, 175-194.

Schuster, T., Reveal, J., Bayly, M., Kathleen A. Kron, K., 2015. An updated molecular phylogeny of Polygonoideae (Polygonaceae): Relationships of *Oxygonum, Pteroxygonum*, and *Rumex*, and a new circumscription of *Koenigia*. Taxon 64, 1188 - 1208.

Schütrumpf, R., 1936. Paläobotanisch-pollenanalytische Untersuchungen der paläolithischen Rentierjägerfundstätte von Meiendorf bei Hamburg. Veröff. Archäol. Reichsinst. 1, 1 - 54.

Schütt, P., 1991. Tannenarten Europas und Kleinasiens. Birkhäuser, Basel.

Schütt, P., Schuck, H.J., Stimm, B., 2002. Lexikon der Baum- und Straucharten. Nikol, Hamburg.

Schütt, P., Weisgerber, H., Schuck, H.J., Lang, U.M., Stimm, B., Roloff, A., 2006. Enzyklopädie der Laubbäume. Nikol, Hamburg.

Schwartz, D., Thinon, M., Goepp, S., Schmitt, C., Casner, J., Rosique, T., Wuscher, P., Alexandre, A., Dambrine, E., Martin, C., Guillet, B., 2005. Premieres datations directes de défrichements protohistoriques sur les chaumes secondaires des Vosges (Rossberg, Haut-Rhin). Approche pédoanthracologique. Comp. Rend. Geosci. 337, 1250-1256.

Schweingruber, F.H., 1978. Microscopic wood anatomy. Structural variability of stems and twigs in recent and subfossil woods from central Europe. Zürcher, Zug.

Schweingruber, F.H., 1990. Anatomy of European woods. An atlas for the identification of European trees, shrubs and dwarf shrubs. Verlag Paul Haupt, Bern.

Schweingruber, F.H., Börner, A., Schulze, E., 2006. Atlas of Woody Plant Stems. Evolution, Structure, and Environmental Modifications. Springer, Berlin.

Schwoerbel, J., Brendelberger, H., 2005. Einführung in die Limnologie. Spektrum Akademischer Verlag.

Schwörer, C., Kaltenrieder, P., Glur, L., Berlinger, M., Elbert, J., Frei, S., Gilli, A., Hafner, A., Anselmetti, F., Grosjean, M., Tinner, W., 2014a. Holocene climate, fire and vegetation dynamics at the treeline in the Northwestern Swiss Alps. Veget. Hist. Archaeobot. 23, 479–496.

Schwörer, C., Henne, P.D., Tinner, W., 2014b. A model-data comparison of Holocene timberline changes in the Swiss Alps reveals past and future drivers of mountain forest dynamics. Global Change Biol. 20, 1512-1526.

Schwörer, C., Colombaroli, D., Kaltenrieder, P., Rey, F., Tinner, W., 2015. Early human impact (5000–3000 BC) affects mountain forest dynamics in the Alps. J. Ecol. 103, 281-295.

Schwörer, C., Gobet, E., van Leeuwen, J.F.N., Bögli, S., Imboden, R., van der Knaap, W.O., Kotova, N., Makhortykh, S., Tinner, W., 2022. Holocene vegetation, fire and land use dynamics at Lake Svityaz, an agriculturally marginal site in northwestern Ukraine. Veg. Hist. Archaeobotany 31, 155–170.

Scott, E.M., 2007. Radiocarbon Dating: Sources of Error. In: Elias, S.A. (Ed.), Encyclopedia of Quaternary Science. Elsevier, Amsterdam, pp. 2918-2923.

Seddon, B., 1962. Late-glacial deposits at Llyn Dwythwch and Nant Ffrancon, Caernarvonshire. Phil. Trans. Roy. Soc. London B244, 459-481.

Seddon, A.W.R., Mackay, A.W., Baker, A.G., Birks, H.J.B., Breman, E., Buck, C.E., Ellis, E.C., Froyd, C.A., Gill, J.L., Gillson, L., Johnson, E.A., Jones, V.J., Juggins, S., Macias-Fauria, M., Mills, K., Morris, J.L., Nogués-Bravo, D., Punyasena, S.W., Roland, T.P., Tanentzap, A.J., Willis, K.J., Palaeo50 working group, 2014. Looking forward through the past. Identification of fifty priority research questions in palaeoecology. J. Ecol. 102, 256-267.

Seddon, A.W.R., Macias Fauria, M., Willis, K.J., 2015. Climate and abrupt vegetation change in Northern Europe since the last deglaciation. Holocene 25, 25-36.

Sefström, F. G., 1836: Undersökning af de räfflor, hvaraf Skandinaviens berg äro med bestämd riktning färade, samt om deras sannolika uppkomst. K. Svenska Vetenskapsakad. Handl., 141-255.

Segerström, U., 1990. The post-glacial history of

vegetation and agriculture in the Lukälv River Valley. Archaeol. Environ. 7, 1-76.

Segerström, U., 1997. Long-term dynamics of vegetation and disturbance of a southern boreal spruce swamp forest. J. Veg. Sci. 8, 295-306.

Seidl, R., Rammer, W., Scheller, R.M., Spies, T.A., 2012. An individual-based process model to simulate landscape-scale forest ecosystem dynamics. Ecol. Model. 231, 87-100.

Semerikov, V., Lascoux, M., 2003. Nuclear and cytoplasmatic variation within and between Eurasian *Larix* (Pinaceae) species. Am. J. Bot. 90, 1113-1123.

Seppä, H., 1996. Post-glacial dynamics of vegetation and tree-lines in the far north of Fennoscandia. Fennia 174, 1-96.

Seppä, H., 1998. Postglacial trends in palynological richness in the northern Fennoscandian tree-line area and their ecological interpretation. Holocene 8, 43-53.

Seppä, H., 2007. Pollen Analysis, Principles. In: Elias, S.A. (Ed.), Encyclopedia of Quaternary Science. Elsevier, Amsterdam, pp. 2486-2497.

Seppä, H., 2013. Pollen Analysis, Principles. In: Elias, S.A., Mock, C.J. (Eds.), Encyclopedia of Quaternary Science, 2nd ed. Elsevier, Amsterdam, pp. 794-804.

Seppä, H., Birks, H.J.B., 2001. July mean temperature and annual precipitation trends during the Holocene in the Fennoscandian tree-line area: pollen-based climate reconstructions. Holocene 11, 527-539.

Seppä, H., Birks, H.J.B., 2002. Holocene climate reconstruction from Fennoscandian tree-line area based on pollen data from Toskaljavri. Quat. Res. 57, 191-199.

Seppä, H., Hicks, S., 2006. Integration of modern and past pollen accumulation rate (PAR) records across the arctic tree-line: a method of more precise vegetation reconstruction. Quat. Sci. Rev. 25, 1501-1516.

Seppä, H., Poska, A., 2004. Holocene annual mean temperature changes in Estonia and their relationship to solar insolation and atmospheric circulation patterns. Quat. Res. 61, 22-31.

Seppä, H., Nyman, M., Korhola, A., Weckström, J., 2002. Changes of treelines and alpine vegetation in relation to post-glacial climate dynamics in northern Fennoscandia based on pollen and chironomid records. J. Quat. Sci. 17, 287-301.

Seppä, H., Alenius, T., Bradshaw, R.H.W., Giesecke, T., Heikkilä, M., Muukkonen, P., 2009a. Invasion of Norway spruce (*Picea abies*) and the rise of the boreal ecosystem in Fennoscandia. J. Ecol. 97, 629-640.

Seppä, H., Alenius, T., Muukkonen, P., Giesecke, T., Miller, P.A., Ojala, A.E.K., 2009b. Calibrated pollen accumulation rates as a basis for quantitative tree biomass reconstructions. Holocene 19, 209-220.

Seppä, H., Bjune, A.E., Telford, R.J., Birks, H.J.B., Veski, S., 2009c. Last nine-thousand years of temperature variability in Northern Europe. Clim. Past 5, 523-535.

Seppä, H., Schurgers, G., Miller, P., Bjune, A., Giesecke, T., Kühl, N., Renssen, H., Salonen, J.S., 2015. Trees tracking a warmer climate: The Holocene range shift of hazel (*Corylus avellana*) in northern Europe. Holocene 25, 53-63.

Seppälä, M., Koutaniemi, L., 1985. Formation of a string and pool topography as expressed by morphology, stratigraphy and current processes on a mire in Kuusamo, Finland. Boreas 14, 287-309.

Šercelj, A., 1971. Die Postglaziale Entwicklung der Gebirgswälder im Nordwestlichen Jugoslawien. Razprave IV. Razreda sazu 9, 267-293.

Šercelj, A., 1984. The development of the Late Glacial vegetation on the borders of the Pannonian Plain. Radovi Akademije Navka Bih 23, 165-170.

Serebryanny, L., Malyasova, E., 1998. The Quaternary vegetation and landscape evolution of Novaya Zemlya in the light of palynological records. Quat. Int. 45/46, 59-70.

Serebryanny, L., Andreev, A., Malyasova, E., Tarasov, P., Romanenko, F., 1998. Lateglacial and early Holocene environments of Novaya Zemlya and the Kara Sea Region of the Russian Arctic. Holocene 8, 323-330.

Sernander, R., 1909. De scandinaviska Torfmossarnas stratigrafi. Geol. Fören. Stockholm Förh. 31, 423-448.

Sernander, R., 1910. Die schwedischen Torfmoore als Zeugen postglazialer Klimaschwankungen. 11. Intern. Geologenkongress Stockholm.

Serrano, E., González-Trueba, J., Pellitero, R., González-García, M., Gómez-Lende, M., 2013. Quaternary glacial evolution in the Central Cantabrian Mountains (Northern Spain). Geomorphology 196, 65–82.

Servera-Vives, G., Riera, S., Picornell-Gelabert, L., Moffa-Sánchez, P., Llergo, Y., Garcia, A., Mus-Amezquita, M., García Álvarez, S., Calvo Trías, M., 2018. The onset of islandscapes in the Balearic Islands: A study-case of Addaia (northern Minorca, Spain). Palaeogeogr. Palaeoclimatol. Palaeoecol. 498, 9-23.

Shackleton, N.J., Chapman, M., Sánchez-Goñi, M.F., Pailler, D., Lancelot, Y., 2002. The classic Marine Isotope Substage 5e. Quat. Res. 58, 14-16.

Shala, S., Helmens, K.F., Luoto, T.P., Salonen, J.S., Valiranta, M., Weckström, J., 2017. Comparison of quantitative Holocene temperature reconstructions using multiple proxies from a northern boreal lake. Holocene 27, 1745-1755.

Shalaboda, V. L., 2001. Characteristic features of Muravian (Eemian) pollen succession from various

regions of Belarus. Acta Palaeobotanica 41, 27-41.
Sheil, D., Burslem, D.F.R.P., 2013. Defining and defending Connell's intermediate disturbance hypothesis: a response to Fox. Trends Ecol. Evol. 28, 571-572.
Shennan, I., Andrews, J., 2000. Holocene Land-Ocean Interaction and Environmental Change around the North Sea. Geological Society, Special Publication 166, London, 326 pp.
Shifley, S.R., He, H.S., Lischke, H., Wang, W., Jin, W., Gustafson, E.J., Thompson, J.R., Thompson, F.R., Dijak, W.D., Yang, J., 2017. The past and future of modeling forest dynamics: From growth and yield curves to forest landscape models. *Landsc. Ecol.* **32**, 1307-1325.
Shopov, V., Bozilova, E., Atanassova, J., 1992. Biostratigraphy and radiocarbon data of upper Quaternary sediments from the western part of the Black Sea. Geologica Balcanica 22, 59-70.
Shotyk, W., Weiss, D., Appleby, P.G., Cheburkin, A.K., Frei, R., Gloor, M., Kramers, D., Reese, S., van der Knaap, W.O., 1998. History of atmospheric lead deposition since 12,370 ^{14}C yr BP from a peat bog, Jura Mountains, Switzerland. Science 281, 1635-1640.
Shotyk, W., Blaser, P., Grünig, A., Cheburkin, A.K., 2000. A new approach for quantifying cumulative, anthropogenic, atmospheric lead deposition using peal cores from bogs: Pb in eight Swiss peat bog profiles. Sci. Tot. Env. 249, 281-295.
Shulija, K.S., Lujanas, V.J., Kibilda, Z.A., Banys, J.J., Genutiene, I.K., 1967. Stratigraphy and chronology of lacustrine and bog deposits of the Bebrukas lake hollow [in Russian]. Tr. Instituta Geol. 5, 231–239.
Shuman, B.N., 2013. Paleoclimate Reconstruction | Approaches. In: Elias, S. (Ed.), Encyclopedia of Quaternary Science. Elsevier, Amsterdam, pp. 1942-1948.
Shumilovskikh, L.S., Arz, H.W., Wegwerth, A., Fleitmann, D., Marret, F., Nowaczyk, N., Tarasov, P., Behling, H., 2013. Vegetation and environmental changes in Northern Anatolia between 134 and 119ka recorded in Black Sea sediments. Quat. Res. 80, 349-360.
Shumilovskikh, L.S., Rodinkova, V., Rodionova, A., Troshina, A., Ershova, E., Novenko, E., Zazovskaya, E., Sycheva, S., Kiselev, D., Schlütz, F., Schneeweiss, J., 2019. Insights into the late Holocene vegetation history of the East European forest-steppe: case study Sudzha (Kursk region, Russia). Veget. Hist. Archaeobot. 28, 513-528.
Siegenthaler, U., Eicher, U., 1979. Stable oxygen and carbon isotope analysis on lacustrine carbonate sediments. International Geological Correlation Programme, Project 158 II, 99-110.
Sierralta, M.E., Frechen, M.A., Urban, B.R., 2012. 230Th/U dating results from opencast mine Schöningen. Die chronologische Einordnung der paläolithischen Fundstellen von Schöningen, Römisch-Germanisches Zentralmuseum, 143-54.
Sikkerna, R., Caudullo, G., 2016. *Carpinus orientalis* in Europe: distribution, habitat, usage and threats. In: San-Miguel-Ayanz, J., de Rigo, D., Caudullo, G., Houston Durrant, T., Mauri, A. (Eds.), European Atlas of Forest Tree Species. Publication Office of the European Union, Luxembourg, pp. 76-77.
Sillasoo, Ü., Poska, A., Seppä, H., Blaauw, M., Chambers, F., 2009. Linking past cultural developments to palaeoenvironmental changes in Estonia. Veget. Hist. Archaeobot. 18, 315-327.
Silvertown, J., 1985. History of a latitudinal diversity gradient: woody plants in Europe 13,000-1000 years B.P. J. Biogeogr. 12, 519-525.
Simberloff, D., 2000. Global climate change and introduced species in United States forests. Sci. Total Environ. 262, 253-261.
Simmons, I.G., Innes, J.L., 1985. Late Mesolithic land-use and its impact in the English uplands. Biogeogr. Monogr. 2, 7-17.
Simova, A., Panek, T., Gałka, M., Zernitskaya, V., Hajkova, P., Brodska, H., Jarmrichova, E., Hajek, M., 2019. Landslides increased Holocene habitat diversity on a flysch bedrock in the Western Carpathians. Quat. Sci. Rev. 219, 68-83.
Simpson, I.M., West, R., 1958. On the stratigraphy and palaeobotany of a late-Pleistocene organic deposit at Chelford, Chesire. New Phytologist 57, 239-250.
Sinclair, W., Morman, J., Ennos, R., 1999. The postglacial history of Scots pine (*Pinus sylvestris* L.) in western Europe: evidence from mitochondrial DNA variation. Mol. Ecol. 8, 83–88.
Sjögersten, S., Wookey, P.A., 2004. Decomposition of mountain birch leaf litter at the forest-tundra ecotone in the Fennoscandian mountains in relation to climate and soil conditions. Plant Soil 262, 215-227.
Sjögren, P., 2009. Climate, cod and crops: coastal land use in the SW Barents Sea region during the past 2.5 ka. Holocene 19, 703-716.
Sjögren, P., van Leeuwen, J.F.N., van der Knaap, W.O., van der Borg, K., 2006. The effect of climate variability on pollen productivity, AD 1975-2000, recorded in a *Sphagnum* peat hummock. Holocene 16, 277-286.
Sjögren, P., Connor, S.E., van der Knaap, W.O., 2010. The development of composite dispersal functions for estimating absolute pollen productivity in the Swiss Alps. Veget. Hist. Archaeobot. 19, 341-349.
Sjörs, H., 1948. Myrvegetation i Bergslagen. [Mire vegetation in Bergs lagen, Sweden] Acta Phytogeogr. Suecica 21, 1-299. (in Swedish with English summary)

Skene, M., 1915. The acidity of *Sphagnum* and its relation to chalk and mineral salts. Ann. Bot. Lond. 29, 65-87.

Smettan, H., 1986. Pollenanalytische Untersuchungen zur Vegetations- und Siedlungsgeschichte der Umgebung von Sersheim, Kreis Ludwigsburg. Fundberichte aus Baden-Württemberg 10, 367-421.

Smit, A., 1973. A scanning electron microscopical study of the pollen morphology in the genus Quercus. Acta Bot. Neerland. 12, 525-532.

Smith, A.G., 1961. The Atlantic – Sub-Boreal Transition. Proc. Linn. Soc. London 172, 38-49.

Smith, A.G., 1965. Problems of inertia and threshold related to post-glacial habitat changes. Proc. R. Soc. B: Biol. Sci. 161, 331-342.

Smith, A.B., 2010. Caution with curves: Caveats for using the species–area relationship in conservation. Biol. Conserv. 143, 555-564.

Smith, A.G., 1970. The influence of Mesolithic and Neolithic man on British vegetation: A discussion. In: Walker, D., West, R.G. (Eds.), Studies in the Vegetation History of the British Isles. Cambridge University Press, Cambridge, pp. 81-96.

Smith, A.G., Goddard, I.C., 1991. A 12 500-year record of vegetational history at Sluggan Bog, Co. Antrim, N. Ireland (incorporating a pollen zone scheme for the non-specialist). New Phytol. 118, 167-187

Smith, A.G., Pilcher, J., 1973. Radiocarbon dates and vegetational history of the British Isles. New Phytol. 72, 903-914.

Smol, J.P., 2008. Pollution of Lakes and Rivers: A Paleoenvironmental Perspective, 2nd ed. Blackwell, Oxford. 396 pp.

Smolander, A., Sarsa, M.L., 1990. Frankia strains of soil under *Betula pendula*: Behaviour in soil and in pure culture. Plant Soil 122, 129-136.

Snyder, J.A., Macdonald, G.M., Forman, S.L., Tarasov, G.A., Mode, W.N., 2000. Postglacial climate and vegetation history, north-central Kola Peninsula, Russia: pollen and diatom records from Lake Yarnyshnoe-3. Boreas 29, 261–271.

Soepboer, W., Sugita, S., Lotter, A.F., van Leeuwen, J.F.N., van der Knaap, W.O., 2007. Pollen productivity estimates for quantitative reconstruction of vegetation cover on the Swiss Plateau. Holocene 17, 65-77.

Soepboer, W., Sugita, S., Lotter, A.F., 2010. Regional vegetation-cover changes on the Swiss Plateau during the past two millennia: a pollen-based reconstruction using the REVEALS model. Quat. Sci. Rev. 29, 472-483.

Sollesnes, A., Fægri, K., 1951. Pollenanalytisk undersökelse av to avleiringer fra Norges sydvestlige kyst. Blyttia 9, 41.

Solovieva, N., Tarasov, P.E., MacDonald, G., 2005. Quantitative reconstruction of Holocene climate from the Chuna Lake pollen record, Kola Peninsula, northwest Russia. The Holocene 15, 141–148.

Soranzo, N., Alia, R., Provan, J., Powell, W., 2000. Patterns of variation at a mitochondrial sequence-tagged-site locus provides new insights into the postglacial history of European *Pinus sylvestris* populations. Mol. Ecol. 9, 1205-1211.

Sorsa, P., 1965. Pollenanalytische Untersuchungen zur spätquartären Vegetations- und Klimaentwicklung im östlichen Nordfinnland. Ann. Bot. Fenn. 2, 301-413.

Sorsa, P., 1971. Pollen morphological study of the genus *Hippophaë* L., including the new taxa recognized by A. Rousi. Ann. Bot. Fenn. 50, 95 - 100.

Spek, T., 2004. Het Drentse esdorpen-landschap. Een historisch-geografische studie. Matrijs, Utrecht, 1104 pp.

Speranza, A., van Geel, B., van der Plicht, J., 2002. Evidence for solar forcing of climate change at ca. 850 cal BC from a Czech peat sequence. Global Planet. Change 35, 51-65.

Sperisen, C., Büchler, U., Gugerli, F., 2001. Tandem repeats in plant mitochondrial genomes: application to the analysis of population differentiation in the conifer Norway spruce. Mol. Ecol. 10, 257-263.

Sperl, G., 1992. Das Beil vom Hauslabjoch, Der Mann im Eis 1. Veröff. Universität Innsbruck, Innsbruck, 454-461.

Spinner, H., 1925. Analyse pollinique de la tourbe de deux marais de la Vallée de la Brévine. Bull. Soc. Neuchat. Sci. Nat. 50, 95-100.

Środoń, A., 1960 Pollen spectra from Spitsbergen. Folia Quaternaria 3, 1-17.

Stachowicz-Rybka, R., 2011. Flora and vegetation changes on the basis of plant macroremains analysis from an early Pleistocene lake of the Augustów Plain, NE Poland. Acta Palaeobot. 51, 39-103.

Stachowicz-Rybka, R., 2015a. Environmental and climate changes reflected in the Domuraty 2 section (NE Poland) based on analysis of plant macroremains. Acta Palaeobot. 55, 213.

Stachowicz-Rybka, R., 2015b. Record of environmental and climatic changes in middle Pleistocene sediments from Łuków (eastern Poland) on the basis of plant macroremains analysis. Acta Palaeobot. 55, 68.

Stachowicz-Rybka, R., 2015c. Vegetation of the Ferdynandovian interglacial (MIS 13–15) based on plant macrofossils from a new profile of the stratotype site. Acta Palaeobot. 55, 233.

Stähli, M., Finsinger, W., Tinner, W., Allgöwer, B., 2006. Wildfire history and fire ecology of the Swiss

National Park (Central Alps): new evidence from charcoal, pollen and plant macrofossils. Holocene 16, 805-817.

Stalling, H., 1987. Untersuchungen zur spät- und postglazialen Vegetationsgeschichte im Bayerischen Wald. Diss. Bot. 105, 1-202.

Stankevica, K., Kalnina, L., Klavins, M., Cerina, A., Ustupe, L., 2015. Reconstruction of the Holocene palaeoenvironmental conditions accordingly to the multiproxy sedimentary records from Lake Pilvelis, Latvia. Quat. Int. 386, 102-115.

Stapf, O., 1889. Die Arten der Gattung *Ephedra*. Denkschr. Math.-Natuw. Cl. Kaiserl. Akad. Wiss., Wien, 56, 1-112.

Stark, P., 1923. Zur Entwicklungsgeschichte der badischen Bodenseemoore. Berichte Deutsche Botan. Ges. 31, 361-373.

Stark, P., 1925. Der gegenwärtige Stand der pollenanalytischen Forschung. Zeitschrift für Botanik 17, 89-125.

Stebich, M., 1999. Palynologische Untersuchungen zur Vegetationsgeschichte des Weichsel-Spätglazial und Frühholozän an jährlich geschichteten Sedimenten des Meerfelder Maares (Eifel). Diss. Bot. 320, 1-127.

Steenstrup, J.J.S., 1841. Geognostik-Geologisk Undersögelse af Skovmoserne Vidnesdam- og Lillemose i det nordlige Sjelland, ledsaget af sammenlignende Bemaerkinger, hentede fra Danmarks Skov-, Kjaer- og Lyngmoser Ialmindelighed. Kong. Danske Videnskap. Selkabs Naturvid. og Mathem. Afhandlinger 9, 17-120.

Stefanova, I., 1997. New data on the late Holocene vegetative succession in the northern Pirin Mts.: pollen and macrofossil analysis of depositions from peat bogs Goce Delchev and Visokata Ela. Pyt. Balcanica 3, 53-63.

Stefanova, I., Ammann, B., 2003. Lateglacial and Holocene vegetation belts in the Pirin Mountains (southwestern Bulgaria). Holocene 13, 97-107.

Stefanova, I., Bozilova, E., 1995. Studies on the Holocene history of vegetation in the Northern Pirin Mts. southwestern Bulgaria. In: Bozilova, E., Tonkov, S. (Eds.), Advances in Holocene Palaeoecology in Bulgaria. Pensoft, Sofia-Moscow, pp. 9-31.

Stefanova, I., Oeggl, K., 1993. Zur holozänen Vegetationsgeschichte SW-Bulgariens: Das Moor Praso im Pirin-Gebirge. Ber. Nat.med. Ver. Inssbruck 80, 69-80.

Stefanova, I., Atanassova, J., Delcheva, M., Wright, H.E., 2006a. Chronological framework for the Late Glacial pollen and macrofossil sequence in the Pirin Mountains, Bulgaria: Lake Besbog and Lake Kremensko-5. Holocene 16, 877-892.

Stefanova, V., Lazarova, M., Wright, H.E., 2006b. Elevational gradients during the Late-Glacial/Holocene vegetational transition in southern Bulgaria. Veget. Hist. Archaeobot. 15, 333-343.

Steffen, W., Crutzen, P., McNeill, J.R., 2007. The Anthropocene: are humans now overwhelming the great forces of nature? Ambio 36, 614-621.

Stehlik, I., 2003. Resistance or emigration? Response of alpine plants to the ice ages. Taxon 52, 499-510.

Stehlik, I., Schneller, J., Bachmann, K., 2002. Immigration and in situ glacial survival of the low-alpine *Erinus alpinus* (Scrophulariaceae). Biol. J. Linnean Soc. 77, 67-103.

Steinbauer, M.J., Grytnes, J.-A., Jurasinski, G., Kulonen, A., Lenoir, J., Pauli, H., Rixen, C., Winkler, M., Bardy-Durchhalter, M., Barni, E., Bjorkman, A.D., Breiner, F.T., Burg, S., Czortek, P., Dawes, M.A., Delimat, A., Dullinger, S., Erschbamer, B., Felde, V.A., Fernández-Arberas, O., Fossheim, K.F., Gómez-García, D., Georges, D., Grindrud, E.T., Haider, S., Haugum, S.V., Henriksen, H., Herreros, M.J., Jaroszewicz, B., Jaroszynska, F., Kanka, R., Kapfer, J., Klanderud, K., Kühn, I., Lamprecht, A., Matteodo, M., di Cella, U.M., Normand, S., Odland, A., Olsen, S.L., Palacio, S., Petey, M., Piscová, V., Sedlakova, B., Steinbauer, K., Stöckli, V., Svenning, J.-C., Teppa, G., Theurillat, J.-P., Vittoz, P., Woodin, S.J., Zimmermann, N.E., Wipf, S., 2018. Accelerated increase in plant species richness on mountain summits is linked to warming. Nature 556, 231-234.

Steiner, G., 2005. Moore von Sibirien bis Feuerland. Biologiezentrum Oberösterreichisches Landesmuseum, Linz, 626 pp.

Stevenson, A.C., 2000. The Holocene forest history of the Montes Universales, Teruel, Spain. The Holocene 10, 603–610.

Stewart, D.A., Walker, A., Dickson, J.H., 1984. Pollen diagrams from Dubh Lochan, near Loch Lomond. New Phytol. 98, 531-549.

Stewart, J., Lister, A., Barnes, I., Dale, L., 2010. Refugia revisited: individualistic responses of species in space and time. Proc. R. Soc. B 277, 661–671.

Stivrins, N., Buchan, M.S., Disbrey, H.R., Kuosmanen, N., Latałowa, M., Lempinen, J., Muukkonen, P., Slowinski, M., Veski, S., Seppä, H., 2017. Widespread, episodic decline of alder (*Alnus*) during the medieval period in the boreal forest of Europe. J. Quat. Sci. 32, 903-907.

Stobbe, A., 1996. Die holozäne Vegetationsgeschichte der nördlichen Wetterau - Paläoökologische Untersuchungen unter besonderer Berücksichtigung anthropogener Einflüsse. Diss. Bot. 260, 1-216.

Stockmarr, J., 1970. Species identification of *Ulmus* pollen. Danm. Geol. Unders. 4, 19.

Stockmarr, J., 1971. Tablets with spores used in

absolute pollen analysis. Pollen et Spores 13, 615-621.

Stockmarr, J., 1974. SEM studies on pollen grains of Northern European *Ulmus* species. Grana 14, 103-107.

Stoel, B., Borman, T., 2008. A comparison of wood density between classical Cremonese and modern violins. PLoS ONE 3, e2554.

Strahl, J., 2000. Detailergebnisse pollenanalytischer Untersuchungen an saalespätglazialen bis weichselfrühglazialen Sedimenten aus dem Kiestagebau Hinterste Mühle bei Neubrandenburg (Mecklenburg-Vorpommern). Brandenburgische Geowiss. Beiträge 7, 29-40. Kleinmachnow.

Straka, H., 1952. Zur Feinmorphologie des Pollens von Salix und von Artemisia. Grana Palynologica 14, 204-227.

Straka, H., 1975. Die spätquartäre Vegetationsgeschichte der Vulkaneifel. Pollenanalytische Untersuchungen an vermoorten Maaren. Beitr. Landesk. Rheinl.-Pfalz, Beih. 3, 1-163.

Streif, H., 1990. Quaternary sea-level changes in the North Sea, an analysis of amplitudes and velocities. In: Brosche, P., Sündermann, P. (Eds.), Earth's Rotation from Eons to Days. Springer, Berlin, pp. 201-214.

Streif, H., 2004. Sedimentary record of Pleistocene and Holocene marine inundations along the North Sea coast of Lower Saxony, Germany. Quaternary International 112, 3-28.

Stroud, J.T., Bush, M.R., Ladd, M.C., Nowicki, R.J., Shantz, A.A., Sweatman, J., 2015. Is a community still a community? Reviewing definitions of key terms in community ecology. Ecol. Evol. 5, 4757-4765.

Stuart, A.J., 1976. The history of the mammal fauna during the Ipswichian/Last interglacial in England. Phil. Trans. Royal Soc. London B 276, 221-250.

Stuiver, M., Braziunas, T.F., 1993a. Modeling atmospheric ^{14}C influences and ^{14}C ages of marine samples to 10000 BC. Radiocarbon 35, 137-191.

Stuiver, M., Braziunas, T.F., 1993b. Sun, ocean, climate and atmospheric $^{14}CO_2$: an evaluation of causal and spectral relationships. Holocene 3, 289-305.

Stuiver, M., Kromer, B., Becker, B., Ferguson, C.W., 1986. Radiocarbon age calibration back to 13,300 years BP and the ^{14}C age matching of the German oak and US bristlecone pine chronologies. Radiocarbon 28/2B, 969-979.

Stuiver, M., Grootes, P.M., Braziunas, T.F., 1995. The GISP2 $\delta^{18}O$ climate record of the past 16,500 years and the role of the sun, ocean, and volcanoes. Quat. Res. 44, 341-354.

Stuiver, M., Reimer, P.J., Bard, E., Beck, J.W., Burr, G.S., Hughen, K.A., Kromer, B., McCormac, G., van der Plicht, J., Spurk, M., 1998. INTCAL98 radiocarbon age calibration, 24,000-0 cal BP. Radiocarbon 40, 1041-1084.

Sturlodottir, S., Turner, J., 1985. The elm decline at Pawlaw mire: an anthropogenic interpretation. New Phytol. 99, 323-329.

Sturm, M., Lotter, A.F., 1995. Lake sediments as environmental archives. A means of distinguishing natural events from human activity. EAWAG News 38, 6-9.

Succow, M., 1988. Landschaftsökologische Moorkunde. Jena, Fischer.

Succow, M., Joosten, H., 2001. Landschaftsökologische Moorkunde, 2nd ed. Schweizerbart, Stuttgart, 622 pp.

Succow, M., Lange, E., 1984. The mire types of the German Democratic Republic. In: Moore, P.D. (Ed.), European Mires. Academic Press, London, pp. 149-175.

Sudakova, N.G., Rychagov, G.I., Antonov, S.I., 2007. Topical problems of the middle Neopleistocene statigraphy and palaeogeography of the central Russian Plain. In: Gladenkov, J.B. (Ed.), Geologic events of the Neogene and Quaternary of Russia: modern state of stratigraphic schemes and paleogeographic reconstructions. GEOS, Moscow, pp. 86-90 (in Russian).

Sugita, S., 1993. A model of pollen source area for an entire lake surface. Quat. Res. 39, 239-244.

Sugita, S., 1994. Pollen representation of vegetation in Quaternary sediments: theory and method in patchy vegetation. J. Ecol. 82, 881-897.

Sugita, S., 2007a. Theory of quantitative reconstruction of vegetation I: pollen from large sites REVEALS regional vegetation composition. Holocene 17, 229-241.

Sugita, S., 2007b. Theory of quantitative reconstruction of vegetation II: all you need is LOVE. Holocene 17, 243-257.

Sugita, S., 2013. POLLSCAPE Model: Simulation Approach for Pollen Representation of Vegetation and Land Cover. In: Elias S.A., Mock, C.J. (Eds.) Encyclopedia of Quaternary Science, Vol. 3, 2nd ed. Elsevier, Amsterdam, pp. 871-879.

Sugita, S., Gaillard, M.-J., Broström, A., 1999. Landscape openness and pollen records: a simulation approach. Holocene 9, 409-421.

Sugita, S., Parshall, T., Calcote, R., 2006. Evaluating differences in plant abundance between pollen sites. Holocene 16, 1123-1135.

Sugita, S., Hicks, S., Sormunen, H., 2010a. Absolute pollen productivity and pollen-vegetation relationships in northern Finland. J. Quat. Sci. 25, 724-736.

Sugita, S., Parshall, T., Calcote, R., Walker, K., 2010b. Testing the Landscape Reconstruction Algorithm for spatially explicit reconstruction of vegetation in northern Michigan and Wisconsin. Quat. Res. 74, 289-300.

Sümegi, P., 1999. Reconstruction of flora, soil and landscape evolution, and human impact on the Bereg

Plain from late-glacial up to the present, based on palaeoecological analysis., in: Hamar, J., Sárkány-Kiss, A. (Eds.), In the Upper Tisza Valley, Tiscia Monograph Series, Szeged. pp. 171–203.

Sümegi, P., 2007. Palaeoenvironmental studies at Mezőlak. Varia Archaeologica Hungarica 20, 311-332.

Sümegi, P., Gulyás, S., 2004. The Geohistory of Bátorliget Marshland – An example for the reconstruction of Late Quaternary environmental changes and past human impact from the NE part of the Carpathian Basin. Archaeolingua 16, Budapest.

Sümegi, P., Hertelendi, E., 1998. Reconstruction of microenvironmental changes in the Kopasz Hill loess area at Tokaj (Hungary) between 15 and 70 ka BP. Radiocarbon 40, 855-863.

Sümegi, P., Magyari, E., Daniel, P., Hertelendi, E., Rudner, E., 1999. A kardoskúti Fehr-tó negyedidőszaki fejlődéstörténetének rekonstrukciója. (A reconstruction of the Quaternary geohistory of Fehér lake at Kardoskút) (in Hungarian). Földtani Közlöny 129, 479-519.

Sümegi, P., Juhász, I., Magyari, E., Jakab, G., Rudner, E., Szántó, Z., Molnár, M., 2008. A keleméri Mohos-tavak fejlődéstörténetének rekonstrukciója paleobotanikai vizsgálatok alapján, in: Doldogh, S., Farkas, T.G. (Eds.), A Keleméri Mohos-Tavak Kutatás, Kezelés, Védelem. (The Mohos Peat Bogs in Kelemér Research, Conservation, Management). ANP Füzetek IV. Aggteleki Nemzeti Park Igazgatóság. Jósvafo. pp. 35–57.

Sümegi, P., Magyari, E., Dániel, P., Molnár, M., Törőcsik, T., 2013. Responses of terrestrial ecosystems to Dansgaard-Oeshger cycles and Heinrich events: A 28,000-year record of environmental changes from SE Hungary. Quat. Int. 293, 34-50.

Surmely, F., Miras, Y., Guenet, P., Nicolas, V., Savignat, A., Vannière, B., Walter-Simonnet, A.-V., Servera, G., Tzortzis, S., 2009. Occupation and land-use history of a medium mountain from the Mid-Holocene: A multidisciplinary study performed in the South Cantal (French Massif Central). Comptes Rendus Palevol 8, 737–748.

Surova, T.G., Chernavskaya, M.M., 1997. Paleobotanical description of peat bog Chistic in connection with climatic change in the Golotsen [in Russian]. Data Meteorol. Stud. 16.

Suter, P., Schlichtherle, H., 2009. Pfahlbauten, Palafittes, Palafitte, Pile dwellings, Kolišča. Gassmann, Biel, 107 pp.

Sutton, O.G. 1953. Micrometeorology. Mc-Graw-Hill, London.

Svendsen, J., Astakhov, V., Bolshiyanov, D., Demidov, I., Dowdeswell, J., Gataullin, V., Hjort, C., Hubberten, H., Larsen, E., Mangerud, J., Melles, M., Möller, P., Saarnisto, M., Siegert, M., 1999. Maximum extent of the Eurasian ice sheets in the Barents and Kara Sea region during the Weichselian. Boreas 28, 234-242.

Svendsen, J., Alexanderson, H., Astakhov, V., Demidovd, I., Dowdeswelle, J., Funder, S., Gataullin, V., Henriksen, M., Hjort, C., Houmark-Nielsen, M., Hubberten, W., Ingólfsson, Ó., Jakobsson, M., Kjær, K., Larsen, E., Lokrantz, H., Lunkka, J., Lyså, V., Mangerud, J., Matiouchkov, A., Murray, A., Möller, P., Niessen, F., Nikolskaya, O., Polyak, L., Saarnisto, M., Siegert, C., Siegert, M., Robert F. Spielhagen, R., Stein, R., 2004. Late Quaternary ice sheet history of northern Eurasia. Quat. Sci. Rev. 23, 1229-1271.

Svenning, J.-C., 2002. A review of natural vegetation openness in north-western Europe. Biol. Cons. 104, 133-148.

Svenning, J.-C., 2003. Deterministic Plio-Pleistocene extinctions in the European cool-temperate tree flora. Ecol. Lett. 6, 646-653.

Svenning, J.-C., Magård, E., 1999. Population ecology and conservation status of the last natural population of English yew *Taxus baccata* in Denmark. Biol. Conserv. 88, 173-182.

Svenning, J.-C., Sandel, B., 2013. Disequilibrium vegetation dynamics under future climate change. Am. J. Bot. 100, 1266-1286.

Svenning, J.-C., Skov, F., 2004. Limited filling of the potential range in European tree species. Ecol. Lett. 7, 565-573.

Svenning, J.-C., Skov, F., 2007. Ice age legacies in the geographical distribution of tree species richness in Europe. Global Ecol. Biogegor. 16, 234-245.

Svenning, J.-C., Eiserhardt, W.L., Normand, S., Ordonez, A., Sandel, B., 2015. The influence of paleoclimate on present-day patterns in biodiversity and ecosystems. Annu. Rev. Ecol., Evol. Syst. 46, 551-572.

Svensson, G., 1988. Fossil plant communities and regeneration patterns on a raised bog in south Sweden. J. Ecol. 76, 41-59.

Svensson, N.O., 1989. Late Weichselian and early Holocene shore displacement in the central Baltic, based on stratigraphical and morphological records from eastern Smaland and Gotland, Sweden (PhD thesis). Lund University, Lund, Sweden.

Svoboda, J., Svobodová, H., 1985. Les industries de type Bohunice dans leur cadre stratigraphique et écologique. L'Anthropologie 89, 505-514.

Svoboda, J., Pokorný, P., Horáček, I., Sázelová, S., Abraham, V., Divišová, M., Ivanov, M., Kozáková, R., Novák, J., Novák, M., Šída, P., Perri, A., 2018. Late Glacial and Holocene sequences in rockshelters and adjacent wetlands of Northern Bohemia, Czech Republic: Correlation of environmental and archaeological records. Quat. Int. 465, 234-250.

Swain, A.M., 1973. A history of fire and vegetation in northeastern Minnesota as recorded in lake sediments. Quat. Res. 3, 383-396.

Swain, A.M., 1978. Environmental changes during the past 2000 years in north-central Wisconsin: analysis of pollen, charcoal, and seeds from varved lake sediments. Quat. Res. 10, 55-68.

Sweeney, C.A., 2004. A key for the identification of stomata of the native conifers of Scandinavia. Rev. Palaeobot. Palynol. 128, 281-290.

Swenson, U., Bartish, I., 2003. Taxonomic synopsis of *Hippophae* (Elaeagnaceae). Nordic J. Bot. 22, 369-374.

Szafer, W., 1935. The significance of isopollen lines for the investigation of the geographical distribution of trees in the post-glacial period. Bull. Int. Acad. Pol. Sci et Lettr. Cl- sci. math. nat. Sér. B 1 Bot. 1935, 235-239.

Szafer, W. (ed), 1966. The Vegetation of Poland. (English translation of the Polish book). Pergamon, Oxford etc. 738 pp.

Szafer, W., Jaron, B., 1935. Pleistocenskie jezioro pod Jaslem. Pleistocene lake near Jaslo. Starunia 8.

Szafer, W., Trela, J., 1928. Interglaziale Flora von Schilling bei Posen mit bes. Berücksichtigung der Pollenanalyse. LXIII Spraw. Kom. Fiz. Polsk. Ak. Um. 71-82 (in Polish and German).

Szafran, B., 1926. Der Bau und das Alter des Moores von Pakoslaw bei Ilza in Mittelpolen. Bull. Acad. Polon. Sc. Et Lettres, Cl. Sc. Math. et Nat., Ser. B, Crakow, 751-768.

Szczepanek, K., 1971. Kras staszowski w swietle badan paleobotanicznych [The Staszow karst in the light of palaeobotanical studies - south Poland]. Acta Palaeobot. 12, 60–140.

Szczepanek, K., 1987. Late-Glacial and Holocene pollen diagrams from Jasiel in the Low Beskid Mts.(The Carpathians). Acta Palaeobot. 27, 9–26.

Szczepanek, K., 1989. Type region P-c: Low Beskidy Mts. Acta Palaeobot. 29, 17-23.

Takahashi, M., 1995. Development of structure-less pollen wall in *Ceratophyllum demersum* L. (Ceratophyllaceae). J. Plant Res. 108, 205-208.

Takhtajan, A., 1986. Floristic Regions of the World. University of California Press, Berkeley.

Tallantire, P.A., 2002. The early-Holocene spread of hazel (*Corylus avellana* L.) in Europe north and west of the Alps: an ecological hypothesis. Holocene 12, 81-96.

Tallis, J.H., 1975. Tree remains in southern Pennine peats. Nature 256, 482-484.

Tallis, J.H., 1983. Changes in wetland communities. In: Gore, A.J.P. (Ed.), Ecosystems of the World 4a: Mires, Swamp, Bog, Fen, Moor. General studies. Elsevier, Amsterdam, pp. 311-347.

Tallis, J.H., 1991a. Forest and moorland in the South Pennine uplands in the mid-Flandrian period. III. The spread of moorland - local, regional and national. J. Ecol. 79, 401-415.

Tallis, J.H., 1991b. Plant Community History. Long-term changes in plant distribution and diversity. Chapman and Hall, London, 398 pp.

Tanneberger, F., Tegetmeyer, C., Busse, S., Barthelmes, A., Shumka, S., Moles Mariné, A., Jenderedjian, K., Steiner, G.M., Essl, F., Etzold, J., Mendes, C., Kozulin, A., Frankard, P., Milanović, Đ., Ganeva, A., Apostolova, I., Alegro, A., Delipetrou, P., Navrátilová, J., Risager, M., Leivits, A., Fosaa, A.M., Tuominen, S., Muller, F., Bakuradze, T., Sommer, M., Christanis, K., Szurdoki, E., Oskarsson, H., Brink, S.H., Connolly, J., Bragazza, L., Martinelli, G., Aleksāns, O., Priede, A., Sungaila, D., Melovski, L., Belous, T., Saveljić, D., de Vries, F., Moen, A., Dembek, W,. Mateus, J., Hanganu, J., Sirin, A., Markina, A., Napreenko, M., Lazarević, P., Šefferová Stanová, V., Skoberne, P., Heras Pérez, P., Pontevedra-Pombal, X., Lonnstad, J., Küchler, M., Wüst-Galley, C., Kirca. S., Mykytiuk, O., Lindsay, R., Joosten, H., 2017. The peatland map of Europe. Mires and Peat 19, Art. 22, 1-17.

Tanneberger, F., Moen, A., Barthelmes, A., Lewis, E., Miles, L., Tegetmeyer, C., Joosten, H., 2021. Regional mire diversity in Europe – nature, condition and protection. Diversity 13, 381.

Tansley, A.G., 1953. The British Islands and their vegetation. Cambridge University Press, Cambridge.

Tanţău, I., 2003. Pollen analytic researches in the Eastern Romanian Carpathians. History of vegetation and human impact (PhD thesis). Aix-Marseille III University and Babes-Bolyai of Cluj-Napoca University, Marseille, France and Cluj-Napoca, Romania.

Tanţău, I., Fărcaş, S., 2005. Cronologie de l'histoire de la végetation holocène de Monts Rodnei (Carpates orientales). Contributii Botanice 40, 241–250.

Tanţău, I., Fărcaş, S., Reille, M., de Beaulieu, J.-L., 2003a. L'analyse palynologique de la séquence de Luci: nouvelle données concernant l'histoire de la végétation tardiglaiaire et holocène de Monts Harghitei. Contributii Botanice 38, 155-161.

Tanţău, I., Reille, M., de Beaulieu, J.-L., Fărcaş, S., Goslar, T., Paterne, M., 2003b. Vegetation history in the Eastern Romanian Carpathians: pollen analysis of two sequences from the Mohos crater. Veget. Hist. Archaeobot. 12, 113-125.

Tanţău, I., Reille, M., Fărcaş, S., de Beaulieu, J.-L., 2003c. Aspects de l'histoire de la végétation tardiglaciaire et Holocène dans la région des Subcarpates de la Courbure (Subcarpates de Buzau). Studia UBB Geologia 48, 15–26.

Tanţău, I., Reille, M., de Beaulieu, J.-L., Fărcaş, S., 2006. Late Glacial and Holocene vegetation history in the southern part of Transylvania (Romania): pollen analysis of two sequences from Avrig. J. Quat. Sci. 21, 49-61.

Tanţău, I., Reille, M., de Beaulieu, J.-L., Fărcaş, S., Brewer, S., 2009. Holocene vegetation history in Romanian Subcarpathians. Quat. Res. 72, 164–173.

Tanţău, I., Feurdean, A., de Beaulieu, J.-L., Reille, M., Fărcaş, S., 2011. Holocene vegetation history in the upper forest belt of the Eastern Romanian Carpathians. Palaeogeogr. Palaeoclimatol. Palaeoecol. 309, 281–290.

Tanţău, I., Feurdean, A., de Beaulieu, J.-L., Reille, M., Fărcaş, S., 2014. Vegetation sensitivity to climate changes and human impact in the Harghita Mountains (Eastern Romanian Carpathians) over the past 15 000 years. J. Quat. Sci. 29, 141-152.

Tarasov, P.E., Webb, T., Andreev, A.A., Afanas'eva, N.B., Berezina, N.A., Bezusko, L.G., Blyakharchuk, T.A., Bolikhovskaya, N.S., Cheddadi, R., Chernavskaya, M.M., Chernova, G.M., Dorofeyuk, N.I., Dirksen, V.G., Elina, G.A., Filimonova, L.V., Glebov, F.Z., Guiot, J., Gunova, V.S., Harrison, S.P., Jolly, D., Khomutova, V.I., Kvavadze, E.V., Osipova, I.M., Panova, N.K., Prentice, I.C., Saarse, L., Sevatyanov, D.V., Volkova, V.S., Zernitskaya, V.P., 1998. Present-day and mid-Holocene biomes reconstructed from pollen and plant macrofossil data from the former Soviet Union and Mongolia. J. Biogeogr. 25, 1029-1053.

Tauber, H., 1965. Differential pollen dispersal and the interpretation of pollen diagrams. Danm. Geol. Unders. R. 2. 89, 1-69.

Tauber, H., 1967. Differential pollen dispersion and filtration. In: Cushing, E., Wright, H.E. (Eds.), Quaternary Paleoecology. Yale University Press, New Haven, pp. 131-141.

Tauber, H., 1977. Investigations of aerial pollen transport in a forested area. Dansk Botanisk Arkiv 32, 1-121.

Taylor, R.E., Aitken, M.J. (Eds.), 1997. Chronometric Dating in Archaeology. Springer, Berlin.

Telford, R.J., Heegaard, E., Birks, H.J.B., 2004. All age-depth models are wrong: but how badly? Quat. Sci. Rev. 23, 1-5.

ten Cate, C., 1972. Wan god mast gift... Bilder aus der Geschichte der Schweinezucht im Walde. Centre for Agricultural Publishing and Documentation, Wageningen.

ten Hove, H., 1968. The *Ulmus* fall at the transition Atlanticum-Subboreal in the pollen diagrams. Palaeogeogr. Palaeoclimatol. Palaeoecol. 5, 359-369.

Terasmae, J., 1951. On the pollen morphology of *Betula nana*. Svensk Bot. Tidskrift 45, 358-361.

Terberger, T., 2009. Die Alt- und Mittelsteinzeit. In: von Schnurbein, S. (Ed.), Atlas der Vorgeschichte, Stuttgart. pp. 10-57.

Terberger, T., 2019. Der eiszeitliche Mensch. In: Bánffy, E., Hofmann, K. P., v. Rummel, P., (eds.), Spuren des Menschen. Darmstadt.

Terhürne-Berson, R., 2005. Changing distribution patterns of selected conifers in the Quaternary of Europe caused by climate variations. PhD Thesis, University of Bonn, Bonn, 141 pp.

Terhürne-Berson, R., Litt, T., Cheddadi, R., 2004. The spread of *Abies* throughout Europe since the last glacial period: combined macrofossil and pollen data. Veget. Hist. Archaeobot. 13, 257-268.

Terrab, A., Schönswetter, P., Talavera, S., Vela, E., Stuessy, T., 2008. Range-wide phylogeography of *Juniperus thurifera* L., a presumptive keystone species of western Mediterranean vegetation during cold stages of the Pleistocene. Mol. Phylogen. Evol. 48, 94-102.

Terral, J., 1996. Wild and cultivated olive (*Olea europaea* L.): a new approach to an old problem using inorganic analyses of modern wood and archaeological charcoal Rev. Palaeobot. Palynol. 91, 383-397.

Terral, J., Arnold-Simard, G., 1996. Beginnings of olive cultivation in eastern Spain in relation to Holocene bioclimatic changes. Quat. Res. 46, 176-185.

Terral, J., Alonso, N., Buxó i Capdevila, R., Chatti, N., Fabre, L., Fiorentino, G., Marinval, P., Pérez Jordá, G., Pradat, B., Rovira, N., Alibert, P., 2004. Historical biogeography of olive domestication (*Olea europaea* L.) as revealed by geometrical morphometry applied to biological and archaeological material. J. Biogeogr. 31, 63-77.

Terral, J., Badal, E., Heinz, C., Roiron, P., Thiébault, S., Vernet, J., Figueiral, I., 2005. Paléoécologie de l'olivier et paléoclimats au Quaternaire récent en Méditerranée nord-occidentale: la mémoire du bois. Archéo-Plantes 2, 5-28.

Thammina, C., Olsen, R., Kramer, M., Pooler, M., 2016. Genetic relationships of boxwood (*Buxus* L.) accessions based on genic simple sequence repeat markers. Genet. Resources Crop Evol. 64, 1281-1293.

Thatcher, D.L., Wanamaker, A.D., Denniston, R.F., Asmerom, Y., Polyak, V.J., Fullick, D., Ummenhofer, C.C., Gilliken, D.P., Haws, J.A. 2020. Hydroclimate variability from western Iberia (Portugal) during the Holocene: Insights from a composite stalagmite isotope record. Holocene 30, 966-981.

Theuerkauf, M., Couwenberg, J., 2018. ROPES reveals past land cover and PPEs from single pollen records. Front. Earth Sci. 6.

Theuerkauf, M., Joosten, H., 2009. Substrate dependency of Lateglacial forests in north-east Germany: untangling vegetation patterns, ecological amplitudes and pollen dispersal in the past by downscaling regional pollen. J. Biogeogr. 36, 942-953.

Theuerkauf, M., Kuparinen, A., Joosten, H., 2012. Pollen productivity estimates strongly depend on assumed pollen dispersal. Holocene 23, 14-24.

Theuerkauf, M., Dräger, N., Kienel, U., Kuparinen, A., Brauer, A., 2015. Effects of changes in land management practices on pollen productivity of open vegetation during the last century derived from varved lake sediments. Holocene 25, 733-744.

Theuerkauf, M., Couwenberg, J., Kuparinen, A., Liebscher, V., 2016. A matter of dispersal: REVEALSinR introduces state-of-the-art dispersal models to quantitative vegetation reconstruction. Veget. Hist. Archaeobot. 25, 541-553.

Thiel-Egenter, C., Alvarez, N., Holderegger, R., Tribsch, A., Englisch, T., Wohlgemuth, T., Colli, L., Gaudeul, M., Gielly, L., Jogan, N., Linder, H., Negrini, R., Niklfeld, H., Pellecchia, M., Rioux, D., Schönswetter, P., Taberlet, P., van Loo, M., Winkler, M., Gugerli, F., IntraBioDiv Consortium, 2011. Break zones in the distributions of alleles and species in alpine plants. J. Biogeogr. 38, 772-782.

Thieme, H. (ed.), 2007. Die Schöninger Speere. Theiss, Stuttgart.

Thöle, L., Schwörer, C., Colombaroli, D., Gobet, E., Kaltenrieder, P., van Leeuwen, J.F.N., Tinner, W., 2016. Reconstruction of Holocene vegetation dynamics at Lac de Bretaye, a high-mountain lake in the Swiss Alps. Holocene 26, 380-396.

Thom, T., Hanlon, A., Lindsay, R.A., Richards, J., Stoneman, R., Brooks, S.J., 2019. Conserving bogs: The management handbook, 2nd ed. IUCN UK Peatland programme

Thomas, P.A., Polwart, A., 2003. *Taxus baccata*. J. Ecol. 91, 489-524.

Thomas, P.A., El-Barghathi, M., Polwart, A., 2007. Biological flora of the British Isles: *Juniperus communis* L. J. Ecol. 95, 1404-1440.

Thomson, P., 1925. Die Pollenflora der Torflager in Estland (mit Nachtrag). Bot. Archiv Königsberg 3+1 pp.

Thornalley, D.J.R., Elderfield, H., McCave, I., N., 2010. Intermediate and deep water paleoceanography of the northern North Atlantic over the past 21,000 years. Paleoceanography 25.

Thornalley, D.J.R., Oppo, D.W., Ortega, P., Robson, J.I., Brierley, C.M., Davis, R., Hall, I.R., Moffa-Sanchez, P., Rose, N.L., Spooner, P.T., Yashayaev, I., 2018. Anomalously weak Labrador Sea convection and Atlantic overturning during the past 150 years. Nature *556* (7700), 227-230.

Thorpe, I.G.N., 2004. The Megalithic World. In: Bogucki, P., Crabtree, T.J. (eds.). Ancient Europe, 398-419.

Thuiller, W., Lavorel, S., Araújo, M., Sykes, M.T., Prentice, I.C., 2005. Climate change threats to plant diversity in Europe. Proc. Natl. Acad. Sci. USA 102, 8245-8250.

Thórsson, Æ.T., Pálsson, S., Lascoux, M., Anamthawat-Jónsson, K., 2010. Introgression and phylogeography of *Betula nana* (diploid), *B. pubescens* (tetraploid) and their triploid hybrids in Iceland inferred from cpDNA haplotype variation. J. Biogeogr. 37, 2098-2110.

Tian, F., Cao, X., Dallmeyer, A., Ni, J., Zhao, Y., Wang, Y., Herzschuh, U., 2016. Quantitative woody cover reconstructions from eastern continental Asia of the last 22 kyr reveal strong regional peculiarities. Quat. Sci. Rev. 137, 33-44.

Tibbett, M., Gil-Martinez, M., Fraser, T., Green, I.D., Duddigan, S., De Oliveira, V.H., Raulund-Rasmussen, K., Sizmur, T., Diaz, A., 2019. Long-term acidification of pH neutral grasslands affects soil biodiversity, fertility and function in a heathland restoration. Catena 180, 401-415.

Tinner, W., 2007. Treeline studies. In: Elias, S.A. (Ed.), Encyclopedia of Quaternary Science. Elsevier, Amsterdam, pp. 2374-2384.

Tinner, W., Ammann, B., 2005. Long-term responses of mountain ecosystems to environmental changes: Resilience, adjustment, and vulnerability. In: Huber, U.M., Bugmann, H., Reasoner, M. (Eds.), Global Change and Mountain Research – state of knowledge overview. Springer, Dordrecht, pp. 133-144.

Tinner, W., Hu, F.S., 2003. Size parameters, size-class distribution and area-number relationship of microscopic charcoal: Relevance for fire reconstruction. Holocene 13, 499-505.

Tinner, W., Kaltenrieder, P., 2005. Rapid responses of high-mountain vegetation to early Holocene environmental changes in the Swiss Alps. J. Ecol. 93, 936-947.

Tinner, W., Lotter, A.F., 2001. Central European vegetation response to abrupt climate change at 8.2 ka. Geology 29, 551-554.

Tinner, W., Lotter, A.F., 2006. Holocene expansions of *Fagus silvatica* and *Abies alba* in Central Europe: where are we after eight decades of debate? Quat. Sci. Rev. 25, 526-549.

Tinner, W., Theurillat, J.P., 2003. Uppermost limit, extent, and fluctuations of the timberline and treeline ecocline in the Swiss Central Alps during the past 11,500 years. Arc. Antarc. Alp. Res. 35, 158 169.

Tinner, W., Vescovi, E., 2007. Ecologia e oscillazioni del limite degli alberi nelle Alpi dal Pleniglaciale al presente. In: Frisia, S., Filippi, M.L., Borsato, A. (Eds.), Cambiamenti climatici e ambientali in Trentino: dal passato prospettive per il futuro. Museo Tridentino di Scienze Naturali., Trento, pp. 5-13.

Tinner, W., Ammann, B., Germann, P., 1996. Treeline fluctuations recorded for 12,500 years by soil profiles, pollen, and plant macrofossils in the Central Swiss Alps. Arc. Alp. Res. 28, 131-147.

Tinner, W., Hubschmid, P., Wehrli, M., Ammann, B., Conedera, M., 1999. Long-term forest fire ecology and dynamics in southern Switzerland. J. Ecol. 87, 273-289.

Tinner, W., Conedera, M., Gobet, E., Hubschmid, P., Wehrli, M., Ammann, B., 2000. A palaeoecological attempt to classify fire sensitivity of trees in the southern Alps. Holocene 10, 565-574.

Tinner, W., Conedera, M., Ammann, B., Lotter, A.F., 2005a. Fire ecology north and south of the Alps since the last ice age. Holocene 15, 1214-1226.

Tinner, W., Kaltenrieder, P., Soom, M., Zwahlen, P., Schmidhalter, M., Boschetti, A., Schluchter, C., 2005b. The postglacial rockfall in the Kander valley (Switzerland): age and effects on palaeo-environments. Eclogae Geologicae Helvetiae 98, 83-95.

Tinner, W., Nielsen, E., Lotter, A.F., 2007. Mesolithic agriculture in Switzerland? A critical review of the evidence. Quat. Sci. Rev. 26, 1416-1431.

Tinner, W., Bigler, C., Gedye, S., Gregory-Eaves, I., Jones, R.T., Kaltenrieder, P., Krahenbuhl, U., Hu, F.S., 2008. A 700-year paleoecological record of boreal ecosystem responses to climatic variation from Alaska. Ecology 89, 729-743.

Tinner, W., Nielsen, E.H., Lotter, A.F., 2008. Evidence for Late-Mesolithic agriculture? A reply to Karl-Ernst Behre. Quat. Sci. Rev. 27, 1468-1470.

Tinner, W., van Leeuwen, J.F.N., Colombaroli, D., Vescovi, E., van der Knaap, W.O., Henne, P.D., Pasta, S., D'Angelo, S., La Mantia, T., 2009. Holocene environmental and climatic changes at Gorgo Basso, a coastal lake in southern Sicily, Italy. Quat. Sci. Rev. 28, 1498-1510.

Tinner, W., Colombaroli, D., Heiri, O., Henne, P.D., Steinacher, M., Untenecker, J., Vescovi, E., Allen, J.R.M., Carraro, G., Conedera, M., Joos, F., Lotter, A.F., Luterbacher, J., Samartin, S., Valsecchi, V., 2013. The past ecology of *Abies alba* provides new perspectives on future responses of silver fir forests to global warming. Ecol. Monogr. 83, 419-439.

Tinner, W., Beer, R., Bigler, C., Clegg, B.F., Jones, R.T., Kaltenrieder, P., van Raden, U.J., Gilli, A., Hu, F.S., 2015. Late-Holocene climate variability and ecosystem responses in Alaska inferred from high-resolution multiproxy sediment analyses at Grizzly Lake. Quat. Sci. Rev. 126 41–56.

Tinner, W., Vescovi, E., van Leeuwen, J.F.N., Colombaroli, D., Henne, P.D., Kaltenrieder, P., Morales-Molino, C., Beffa, G., Gnaegi, B., van der Knaap, W.O., La Mantia, T., Pasta, S., 2016. Holocene vegetation and fire history of the mountains of Northern Sicily (Italy). Veget. Hist. Archaeobot. 25, 499-519.

Tittensor, R., 1980. Ecological history of yew *Taxus baccata* L. in Southern England. Biol. Conserv. 17, 243-265.

Tobolski, K., 1990. Paleoekologia i paleolimnologia postglacjału niżu polskiego (na przykładzie środkowej Wielkopolski i Borów Tucholskich) (Palaeoecology and palaeolimnology of the postglacial in the Polish lowland). Wydawnictwo Szkoły Głównej Gospodarstwa Wiejskiego-Akademia Rolnicza, Warszawa.

Tobolski, K., 1991. Biostratigraphy and palaeoecology of the Eemian interglacial and the Vistulian glaciation of the Konin region. (Polish with English summary). In: Stankowski, W. (ed.) Przemiany srodowiska geograficznego obszaru Konin-Turek. Wyd. UAM, Poznan, 45-87.

Tobolski, K., Ammann, B., 2000. Macrofossils as records of plant responses to rapid Late Glacial climatic changes at three sites in the Swiss Alps. Palaeogeogr. Palaeoclimatol. Palaeoecol. 159, 251-259.

Tollefsrud, M., Kissling, R., Gugerli, F., Johnsen, O., Skröppa, T., Chedaddi, R., van der Knaap, W.O., Latałowa, M., Terhürne-Berson, R., Litt, T., Geburek, T., Brochmann, C., Sperisen, C., 2008. Genetic consequences of glacial survival and postglacial colonization in Norway spruce: combined analysis of mitochondrial DNA and fossil pollen. Mol. Ecol. 17, 4134-4150.

Tollefsrud, M., Sønstebø, J., Brochmann, C., Johnsen, Ø., Skrøppa, T., Vendramin, G.G., 2009. Combined analysis of nuclear and mitochondrial markers provides new insight into the genetic structure of North European *Picea abies*. Heredity 102, 549-562.

Tollefsrud, M., Latałowa, M., van der Knaap, W.O., Brochmann, C., Sperisen, C., 2015. Late Quaternary history of North Eurasian Norway spruce (*Picea abies*) and Siberian spruce (*Picea obovata*) inferred from macrofossils, pollen and cytoplasmic DNA variation. J. Biogeogr. 42, 1431-1442.

Tolonen, K., 1966. Stratigraphic and rhizopod analyses on an old raised bog, Varrassuo, in Hollola, South Finland. Ann. Bot. Fenn. 3, 147-166.

Tolonen, K., 1967. Über die Entwicklung der Moore im finnischen Nordkarelien. Annal. Bot. Fennici 4, 219-416.

Tolonen, K., 1968. Zur Entwicklung der Binnenfinnland-Hochmoore. Annal. Bot. Fennici 5, 17-33.

Tolonen, K., 1987. Natural history of raised bogs and forest vegetation in the Lammi area, southern Finland, studied by stratigraphical methods. Annal. Acad. Sci. Fennicae A III 144, 1-46.

Tolonen, M., 1980. Identification of fossil *Ulmus* pollen in sediments of Lake Lamminjärvi, S Finland. Ann. Bot. Fenn. 17, 7-10.

Tonkov, S., Bozilova, E., Possnert, G., Velev, A., 2008. A contribution to the postglacial vegetation history of the Rila mountains, Bulgaria: The pollen record of Lake Trilistnika. Quat. Int. 190, 58-70.

Tonkov, S., Possnert, G., Bozilova, E., 2011. The Lateglacial in the Rila Mountains (Bulgaria) re-visited: The pollen record of Lake Ribno (2184 m). Rev. Palaeobot. Palynology 166, 1-11.

Tonkov, S., Lazarova, M., Bozilova, E., Ivanov, D., Snowball, I., 2014. A 30 000-year pollen record from Mire Kupena, Western Rhodopes Mountains (south Bulgaria). Rev. Palaeobot. Palynol. 209, 41-51.

Toulemonde, F., Durand, F., Berrio, L., Bonnaire, E., Daoulas, G., Wiethold, J., 2015. Records of "new" glume wheat in France: a review. Veget. Hist. Archaeobot. 24, 197-206.

Toumi, L., Lumaret, R., 2001. Allozyme characterisation of four Mediterranean evergreen oak species. Biochem. System. Ecol. 29, 799-817.

Tralau, H., 1963. The recent and fossil distribution of some boreal and arctic montane plants in Europe. Arkiv för Botanik 2/5, 533-582.

Tranquillini, W., 1979. Physiological ecology of the alpine timberline. Springer, Berlin.

Trautmann, W., 1953. Zur Unterscheidung fossiler Spaltöffnungen der mitteleuropäischen Coniferen. Flora 140, 523-533.

Traveset, A., 1994. Reproductive biology of *Phillyrea angustifolia* L. (Oleaceae) and effect of galling-insects on its reproductive output. Bot. J. Linn. Soc. 114, 153-166.

Tribsch, A., 2004. Areas of endemism of vascular plants in the Eastern Alps in relation to Pleistocene glaciation. J. Biogeogr. 31, 747-760.

Tribsch, A., Schönswetter, P., 2003. Patterns of endemism and comparative phylogeography confirm palaeoenvironmental evidence for Pleistocene refugia in Eastern Alps. Taxon 52, 477-497.

Troels-Smith, J., 1953. Ertebøllekultur-Bondekultur. Arbøger Nordisk Oldkyndiget Historie 1953, 5-62.

Troels-Smith, J., 1955. Karakterisering af lose jordarter [Characterization of Unconsolidated Sediments]. Danmarks Geol. Undersog. IV. R., 3, No. 10, 39-71.

Troels-Smith, J., 1960. Ivy, Mistletoe and Elm climate indicators – fodder plants. A contribution to the interpretation of the pollen zone border VII-VIII. Danm. Geol. Unders. 4, 32.

Troels-Smith, J., 1981. Naturwissenschaftliche Beiträge zur Pfahlbauforschung. Archäol. Schweiz 4, 98-111.

Troels-Smith, J., 1984. Stall-feeding and field-manuring in Switzerland about 6000 years ago. Tools and Tillage 5, 13-25.

Troll, K., 1925. Die Landbauzonen Europas in ihrer Beziehung zur natürlichen Vegetation. Geogr. Z. 31, 265-280.

Troll, C., 2015. Ozeanische Züge im Pflanzenkleid Mitteleuropas. Festschr. für E. v. Drygalski, München, 307-335.

Trondman, A.-K., Gaillard, M.-J., Mazier, F., Sugita, S., Fyfe, R., Nielsen, A.B., Twiddle, C., Barratt, P., Birks, H.J.B., Bjune, A.E., Björkman, L., Broström, A., Caseldine, C., David, R., Dodson, J., Dörfler, W., Fischer, E., van Geel, B., Giesecke, T., Hultberg, T., Kalnina, L., Kangur, M., van der Knaap, P., Koff, T., Kuneš, P., Lagerås, P., Latałowa, M., Lechterbeck, J., Leroyer, C., Leydet, M., Lindbaldh, M., Marquer, L., Mitchell, F., Odgaard, B.V., Peglar, S.M., Persson, T., Poska, A., Rösch, M., Seppä, H., Smith, A., Wick, L., 2015. Pollen-based quantitative reconstructions of Holocene regional vegetation cover (plant functional types and land-cover types) in Europe suitable for climate modelling. Glob. Change Biol. 21, 676-697.

Trondman, A.-K., Gaillard, M.-J., Sugita, S., Björkman, L., Greisman, A., Hultberg, T., Lagerås, P., Lindbladh, M., Mazier, F., 2016. Are pollen records from small sites appropriate for REVEALS model-based quantitative reconstruction of past regional vegetation? An empirical test in southern Sweden. Veget. Hist. Archaeobot. 25, 131-151.

Tsuda, Y., Chen, J., Stocks, M., Källman, T., Sønstebø, J., Parducci, L., Semerijiov, V., Sperisen, C., Politov, D., Ronkainen, T., Väliranta, M., Vendramin, G., Tollefsrud, M., Lascoux, M., 2016. The extent and meaning of hybridization and introgression between Siberian spruce (*Picea obovata*) and Norway spruce (*Picea abies*): cryptic refugia as stepping-stones to the west? Mol. Ecol. 25, 2773-2789.

Tuittila, E.-S., Väliranta, M., Laine, J., Korhola, A., 2007. Quantifying patterns and controls of mire vegetation succession in a southern boreal bog in Finland using partial ordinations. J. Veg. Sci. 18, 891-902.

Turner, C., 1970. The Middle Pleistocene deposits at Marks Tey, Essex. Philos. Trans. R. Soc. London, B 257, 373-440.

Turner, C., 1975. The correlation and duration of Middle Pleistocene interglacial periods in northwest Europe. In: Butzer, K., Isaak, D. (Eds.), After the Australopithecines. De Gruyter, The Hague, pp. 259-308.

Turner, C. (Ed.), 1996. The Early Middle Pleistocene in Europe. CRC Press, Boca Raton.

Turner, C., 2000. The Eemian interglacial in the North European plain and adjacent areas. Geologie en Mijnbouw 79, 217-231.

Turner, C., 2005. Abrupt changes within the Hoxnian/Holsteinian Interglacial. Abstracts DEKLIM/Pages conference Mainz 2005, 31.

Turner, J., Greig, J.R.A., 1975. Some Holocene pollen diagrams from Greece. Rev. Palaeobot. Palynol 20, 171-204.

Turner, C., Hannon, G.E., 1988. Vegetational evidence for late Quaternary climatic changes in southwest Europe in relation to the influence of the North Atlantic Ocean. Philosophical Transactions of the

Royal Society of London. B, Biological Sciences 318, 451–485.

Turney, C.S.M., Lowe, J.J., 2001. Tephrochronology. In: Last, W.M., Smol, J.P. (Eds.), Tracking Environmental Change Using Lake Sediments, Vol. 1: Basin Analysis, Coring, and Chronological Techniques. Kluwer, Dordrecht, pp. 451-471.

Tutin, T., Burges, N., Chater, A., Edmondson, J., Heywood, V., Moore, D., Valentine, D., Walters, S., Webb, D., 2010. Flora Europaea. Cambridge University Press, Cambridge.

Tüxen, R., 1937. Die Pflanzengesellschaften Nordwestdeutschlands. Mitteilungen Florist.-Soziolog. Arbeitsgemeinschaft 3, 1-170.

Tüxen, R., 1956. Die heutige potentielle natürlich Vegetation als Gegenstand der Vegetationskartierung. Angewandte Pflanzensoziologie 13, 5-42.

Tweiten, M.A., Hotchkiss, S.C., Booth, R.K., Calcote, R.R., Lynch, E.A., 2009. The response of a jack pine forest to late-Holocene climate variability in northwestern Wisconsin. Holocene 19, 1049-1061.

Tweiten, M.A., Calcote, R.R., Lynch, E.A., Hotchkiss, S.C., Schuurman, G.W., 2015. Geophysical features influence the climate change sensitivity of northern Wisconsin pine and oak forests. Ecol. Appl. 25, 1984-1996.

Tynni, R., 1959. Über das Vorkommen von *Ephedra distachya* L. während des Spät- und Postglazials in Finnland. Arch. Soc. 'Vanamo' 13, 123-132.

Tzedakis, P.C., 1993. Long-term tree populations in in northwest Greece through multiple Quaternary climatic cycles. Nature 364, 437-440.

Tzedakis, P.C., 1994. Vegetation change through glacial-interglacial cycles: a long pollen sequence perspective. Phil. Trans. Royal Soc. London B 345, 403-432.

Tzedakis, P.C., 1999. The last climatic cycle at Kopais, central Greece. J. Geol. Soc. London 156, 425-434.

Tzedakis, P.C., 2000. Vegetation variability in Greece during the last interglacial. Geologie en Mijnbouw 79, 355-367.

Tzedakis, P.C., 2005. Towards an understanding of the response of southern European vegetation to orbital and suborbital climate variability. Quat. Sci. Rev. 24, 1585–1599.

Tzedakis, P.C., 2007. Seven ambiguities in the Mediterranean palaeoenvironmental narrative. Quat. Sci. Rev. 26, 2042-2066.

Tzedakis, P.C., Bennett, K.D., 1995. Interglacial vegetation succession: A view from southern Europe. Quat. Sci. Rev. 14, 967-982.

Tzedakis, P.C., Andrieu, V., de Beaulieu, J.-L., Crowhurst, S., Follieri, M., Hooghiemstra, H., Magri, D., Reille, M., Sadori, L., Shackleton, N.J., Wijmstra, T.A., 1997. Comparison of terrestrial and marine records of changing climate of the last 500,000 years. Earth Planet. Sci. Lett. 150, 171-176.

Tzedakis, P.C., Andrieu, V., de Beaulieu, J.-L., Birks, H.J.B., Crowhurst, S., Follieri, M., Hooghiemstra, H., Magri, D., Reille, M., Sadori, L., Shackleton, N., Wijmstra, T., 2001. Establishing a terrestrial chronological framework as a basis for biostratigraphical comparisons. Quat. Sci. Rev. 20, 1583-1592.

Tzedakis, P.C., Lawson, I., Frogley, M.R., Hewitt, G., Preece, R.C., 2002. Buffered tree population changes in a Quaternary refugium: evolutionary implications. Science 297, 2044-2047.

Tzedakis, P.C., McManus, P., Hooghiemstra, H., Oppo, D.W., Wijmstra, T., 2003. Comparison of changes in vegetation in northeast Greece with records of climate variability on orbital and suborbital frequencies over the past 450,00 years. Earth Planet. Sci. Lett. 212, 197-212.

Tzedakis, P.C., Hooghiemstra, H., Pälike, H., 2006. The last 1.35 million years at Tenaghi Philippon: revised chronostratigraphy and long-term vegetation trends. Quat. Sci. Rev. 25, 3416-3430.

Tzedakis, P.C., Channell, J.E.T., Hodell, D.A., Kleiven, H.F., Skinner, L.C., 2012. Determining the natural length of the current interglacial. Nat. Geosci. 5, 138-141.

Tzedakis, P.C., Emerson, B.C., Hewitt, G.M., 2013. Cryptic or mystic? Glacial tree refugia in northern Europe. Trends Ecol. Evol. 28, 696-704.

Tzedakis, P.C., Crucifix, M., Mitsui, T., Wolff, E.W., 2017. A simple rule to determine which insolation cycles lead to interglacials. Nature 542, 427-432.

Urban, B., 1978. Vegetationsgeschichtliche Untersuchungen zur Gliederung des Altquartärs der Niederrheinischen Bucht. Sonderveröffentl. Geologisches Institut der Universität Köln 34.

Urban, B., 1995. Palynological evidence of younger Middle Pleistocene Interglacials (Holsteinian, Reinsdorf and Schöningen) in the Schöningen open cast lignite mine (eastern Lower Saxony, Germany). Mededelingen Rijks Geologische Dienst 52, 175-195.

Urban, B., Sierralta, M., 2012. New palynological evidence and correlation of Early Palaeolithic sites Schöningen 12 B and 13 II, Schöningen open lignite mine. In: Behre, K.-E. (Ed.), The Chronological Setting of the Palaeolithic Sites of Schöningen. Romisch-Germanisches Zentralmuseum, Mainz, pp. 77-96.

Urban, B., Sierralta, M., Frechen, M., 2011. New evidence for vegetation development and timing of Upper Middle Pleistocene interglacials in Northern Germany and tentative correlations. Quat. Int. 241, 125-142.

Usinger, H., 1978a. Bölling-Interstadial und Laacher Bimstuff in einem Spätglazial-Profil aus dem Vallensgard Mose/Bornholm. Mit grössenstatistischer Trennung der Birken. Danm. Geol. Unders. Aarbog 1977, 5-29.

Usinger, H., 1978b. Pollen- und grossrestanalytische Untersuchungen zur Frage des Bölling-Interstadials und der spätglazialen Baumbirken-Einwanderung in Schleswig-Holstein. Schr. naturw. Ver. Schlesw.-Holst. 48, 41-61.

Usinger, H., 1985. Pollenstratigraphische, vegetations- und klimageschichtliche Gliederung des Bölling-Alleröd-Komplexes in Schleswig-Holstein und ihre Bedeutung für die Spätglazial-Stratigraphie in benachbarten Gebieten. Flora 177, 1-43.

Uzquiano, P., 1992. L'Homme et le bois au Paléolithique en région Cantabrique, Espagne. Exemples d'Altamira et d'Buxu. Bull. Soc. Bot. France 139, 361-372.

Uzquiano, P., 2008. Domestic fires and vegetation cover among Neanderthalians and anatomically modern human groups (>53-30 kyr BP) in the Cantabrian Region (Cantabria, Northern Spain). In: Fiorentino, G., Magri, D. (Eds.), Charcoals from the Past. BAR Int. Ser., pp. 273-285.

Uzquiano, P., Allué, E., Antolín, F., Burjachs, F., Picornel, L., Piqué, R., Zapata, L., 2015. All about yew: on the trail of *Taxus baccata* in southwest Europe by means of integrated palaeobotanical and archaeobotanical studies. Veget. Hist. Archaeobot. 24, 229-247.

Valamoti, S., Gkatzogia, E., Ntinou, M., 2018. Did Greek colonisation bring olive growing to the north? An integrated archaeobotanical investigation of the spread of *Olea europaea* in Greece from the 7th to the 1st millennium BC. Veget. Hist. Archaeobot. 27, 177-195.

Valcàrcel, V., Fiz, O., Vargas, P., 2003. Chloroplast and nuclear evidence for multiple origins of polyploids and diploids of *Hedera* (Araliaceae) in the Mediterranean basin. Mol. Phylogen. Evol. 27, 1-20.

Väliranta, M., Kaakinen, A., Kuhry, P., 2003. Holocene climate and landscape evolution East of the Pechora Delta, East-European Russian Arctic. Quat. Res. 59, 335-344.

Väliranta, M., Kultti, S., Nyman, M., Sarmaja-Korjonen, K., 2005. Holocene development of aquatic vegetation in shallow Lake Njargajavri, Finnish Lapland, with evidence of water-level fluctuations and drying. J. Paleolimnol. 34, 203-215.

Väliranta, M., Kultti, S., Seppä, H., 2006. Vegetation dynamics during the Younger Dryas-Holocene transition in the extreme northern taiga zone, northeastern European Russia. Boreas 35, 202-212.

Väliranta, M., Korhola, A., Seppä, H., Tuittila, E., Sarmaja-Korjonen, K., Laine, J., Alm, J., 2007. High-resolution reconstruction of wetness dynamics in a southern boreal raised bog, Finland, during the late Holocene: a quantitative approach. Holocene 17, 1093-1107.

Väliranta, M., Kaakinen, A., Kuhry, P., Kultti, S., Salonen, J.S., Seppä, H., 2011. Scattered late-glacial and early Holocene tree populations as dispersal nuclei for forest development in north-eastern European Russia. J. Biogeogr. 38, 922-932.

Väliranta, M., Salonen, J.S., Heikkilä, M., Amon, L., Helmens, K., Klimaschewski, A., Kuhry, P., Kultti, S., Poska, A., Shala, S., Veski, S., Birks, H.H., 2015. Plant macrofossil evidence for an early onset of the Holocene summer thermal maximum in northernmost Europe. Nature Commun. 6, 6809.

Valle-Hernandez, M., Rivas-Carballo, M., Lucini, M., Ortiz, J., Torres, T., 2003. Interpretación paleoecológica y paleoclimática del tramo superior de la turbera de Padul (Granada, España). Pollen 13, 85-95.

Valsecchi, V., Finsinger, W., Tinner, W., Ammann, B., 2008. Testing the influence of climate, human impact and fire on the Holocene population expansion of *Fagus sylvatica* in the southern Prealps (Italy). Holocene 18, 603-614.

Valsecchi, V., Carraro, G., Conedera, M., Tinner, W., 2010. Late-Holocene vegetation and land-use dynamics in the Southern Alps (Switzerland) as a basis for nature protection and forest management. Holocene 20, 483-495.

van Breemen, N., 1995. How *Sphagnum* bogs down other plants. Trends Ecol. Evol. 10, 270-275.

van Campo, M., Jalut, G., 1969. Analyse pollinique de sédiments des Pyrénées orientales, lac de Balcère (1764 m). Pollen et spores 11, 117-126.

van den Bogaard, C., Dörfler, W., Sandgren, P., Schmicke, H.-U., 1994. Correlating the Holocene records: Icelandic tephra found in Schleswig-Holstein (Northern Germany). Naturwissenschaften 81, 554-556.

van den Bogaard, C., Schmincke, H.U., 2002. Linking the North Atlantic to central Europe: a high-resolution Holocene tephrochronological record from northern Germany. J. Quat. Sci. 17, 3-20.

van den Brink, L., Janssen, C., 1985. The effect of human activities during cultural phases on the development of montane vegetation in the Serra de Estrela, Portugal. Rev. Palaeobot. Palynol. 44, 193-215.

van der Hammen, T., 1995. The Dinkel valley revisited: Pleniglacial stratigraphy of the eastern Netherlands and global climatic changes. Mededel. Rijks Geolog. Dienst 52, 343-355.

van der Hammen, T., Iversen, J., 1951. Pollen of *Sanguisorba officinalis* and *Sanguisorba minor* in Danish and Dutch Late-Glacial deposits. Danm. Geol. Unders. 3.

van der Hammen, T., Maarleveld, G.C., Vogel, J.

C., Zagwijn, W.H., 1967. Stratigraphy, climatic succession and radiocarbon dating of the last Glacial in the Netherlands. Geologie en Mijnbouw 46, 79-95.

van der Hammen, T., Wijmstra, T.A., Zagwijn, W.H., 1971. The floral record of the Late Cenozoic of Europe. In: Turekian, K.K. (Ed.), The Late Cenozoic Glacial Ages. Yale University Press, New Haven, pp. 391-424.

van der Knaap, W.O., 1987. Long-distance transported pollen and spores on Spitsbergen and Jan Mayen. Pollen et Spores 24, 449-453.

van der Knaap, W.O., 1988. A pollen diagram from Brøggerhalvøya, Spitsbergen: changes in vegetation and environment from c. 4400 to c. 800 BP. Arc. Alp. Res. 20, 106-116.

van der Knaap, W.O., 1989. Past vegetation and reindeer on Edgeoya (Spitsbergen) between c. 7900 and c. 3800 BP, studied by means of peat layers and reindeer faecal pellets. J. Biogeography 16, 379-394.

van der Knaap, W.O., 2009. Estimating pollen diversity from pollen accumulation rates: a method to assess taxonomic richness in the landscape. Holocene 19, 159-163.

van der Knaap, W.O., van Leeuwen, J.F.N., 1995. Holocene vegetation succession and degradation as responses to climatic change and human activity in the Serra da Estrela, Portugal. Rev. Palaeobot. Palynol. 89, 153-211.

van der Knaap, W.O., van Leeuwen, J.F.N., 1997. Late Glacial and early Holocene vegetation succession, altitudinal vegetation zonation, and climatic change in the Serra da Estrela, Portugal. Rev. Palaeobot. Palynol. 97, 239-285.

van der Knaap, W.O., van Leeuwen, J.F.N., Fankhauser, A., Ammann, B., 2000. Palynostratigraphy of the last centuries in Switzerland based on 23 lake and mire deposits: chronostratigraphic pollen markers, regional patterns, and local histories. Rev. Palaeobot. Palynol. 108, 85–142.

van der Knaap, W.O., van Leeuwen, J.F.N., Finsinger, W., Gobet, E., Pini, R., Schweizer, A., Valsecchi, V., Ammann, B., 2005. Migration and population expansion of *Abies, Fagus, Picea*, and *Quercus* since 15000 years in and across the Alps, based on pollen-percentage threshold values. Quat. Sci. Rev. 24, 645-680.

van der Knaap, W.O., van Leeuwen, J.F.N., Svitavská-Svobodová, H., Pidek, I.A., Kvavadze, E., Chichinadze, M., Giesecke, T., Kaszewski, B.M., Oberli, F., Kalnina, L., Pardoe, H.S., Tinner, W., Ammann, B., 2010. Annual pollen traps reveal the complexity of climatic control on pollen productivity in Europe and the Caucasus. Veget. Hist. Archaeobot. 19, 285-307.

van der Knaap, W.O., Lamentowicz, M., van Leeuwen, J.F.N., Hangartner, S., Leuenberger, M., Mauquoy, D., Goslar, T., Mitchell, E.A.D., Lamentowicz, L., Kamenik, C., 2011. A multi-proxy, high-resolution record of peatland development and its drivers during the last millennium from the subalpine Swiss Alps. Quat. Sci. Rev. 30, 3467-3480.

van der Knaap, W.O., van Leeuwen, J.F.N., Fahse, L., Szidat, S., Studer, T., Baumann, J., Heurich, M., Tinner, W., 2020. Vegetation and disturbance history of the Bavarian Forest National Park, Germany. Veg. Hist. Archaeobotany 29, 277–295.

van der Maarel, E., Franklin, J., 2013. Vegetation Ecology, 2nd ed. Wiley-Blackwell, Chichester.

van der Meer, J.J.M., Slotboom, R.T., de Vires-Bruynsteen, I.M.E., 1984. Lithology and palynology of Weichselian alluvial fan deposits near Eerbeek, The Netherlands. Boreas 13, 393-402.

van der Merwe, M., Winfield, M.O., Arnold, G.M., Parker, J.S., 2000. Spatial and temporal aspects of the genetic structure of *Juniperus communis* populations. Mol. Ecol. 9, 379-386.

van der Plicht, J., 2007. Radiocarbon dating and the Dead Sea scrolls: A comment on "Redating". Dead Sea Disc. 14(1), 77-89.

van der Wiel, A.M., Wijmstra, T.A., 1987a. Palynology of the 112.8-197.8 m interval of the core Tenaghi Philippon III, Middle Pleistocene of Macedonia. Rev. Palaeobot. Palynol. 52, 89-117.

van der Wiel, A.M., Wijmstra, T.A., 1987b. Palynology of the lower part (78-120 m) of the core Tenaghi Philippon II, Middle Pleistocene of Macedonia, Greece. Rev. Palaeobot. Palynol. 52, 73-88.

van Dinter, M., Birks, H.H., 1996. Distinguishing fossil *Betula nana* and *B. pubescens* using their wingless fruits: implications for the late-glacial vegetational history of western Norway. Veget. Hist. Archaeobot. 5, 229-240.

van Geel, B., 1976. Fossil spores of Zygnemataceae in ditches of a prehistoric settlement in Hoogkarspel (The Netherlands). Rev. Palaeobot. Palynol. 22, 337-344.

van Geel, B., 1978. A palaeoecological study of Holocene peat bog sections in Germany and the Netherlands. Rev. Palaeobot. Palynol. 25, 1-120.

van Geel, B., 2001. Non-pollen palynomorphs. In: Smol, J.P., Birks, H.J.B., Last, W.M. (Eds.), Tracking Environmental Change Using Lake Sediments, Vol. 3: Terrestrial, Algal and Silicaceous Indicators. Kluwer, Dordrecht, pp. 99-119.

van Geel, B., Aptroot, A., 2006. Fossil ascomycetes in Quaternary deposits. Nova Hedwigia 82, 313-329.

van Geel, B., Grenfell, H.R., 1996. Spores of Zygnemataceae. In: Jansonius, J., McGregor, D.C. (Eds.), Palynology: Principles and Applications, Vol. 1. American Association of Stratigraphic Palynologists Foundation, pp. 173-179.

van Geel, B., Renssen, H., 1998. Abrupt climate change around 2,650 BP in North-West Europe: evidence for climatic teleconnections and a tentative explanation. In: Issar, A.S., Brown, N. (Eds.), Water, Environment and Society in Times of Climatic Change. Kluwer, Dordrecht, pp. 21-41.

van Geel, B., Mur, L.R., Ralska-Jasiewiczowa, M., Goslar, T., 1994. Fossil akinetes of *Aphanizomenon* and *Anabaena* as indicators for medieval phosphate-eutrophication of Lake Gosciaz (Central Poland). Rev. Palaeobot. Palynol. 83, 97-105.

van Geel, B., Buurman, J., Waterbolk, H.T., 1996. Archaeological and palaeoecological indications of an abrupt climate change in The Netherlands, and evidence for climatological teleconnections around 2650 BP. J. Quat. Sci. 11, 451-460.

van Geel, B., Buurman, J., Brinkkemper, O., Schelvis, J., Aptroot, A., van Reenen, G., Hakbijl, T., 2003. Environmental reconstruction of a Roman Period settlement site in Uitgeest (The Netherlands), with special reference to coprophilous fungi. J. Archaeol. Sci. 30, 873-883.

van Geel, B., Guthrie, R.D., Altmann, J.G., Broekens, P., Bull, I.D., Gill, F.L., Jansen, B., Nieman, A.M., Gravendeel, B., 2011. Mycological evidence for coprophagy from the feces of an Alaskan Late Glacial mammoth. Quat. Sci. Rev. 30, 2289-2303.

van Geel, B., Protopopov, A., Protopopova, V., Pavlov, I., van der Plicht, J., van Reenen, G.B.A., 2017. *Larix* during the mid-pleniglacial (Greenland Interstadial 8) on Kotelny Island, northern Siberia. Boreas 46, 338-345.

van Kolfschoten, T., 1992. Aspects of the migration of mammals to northwestern Europe during the Pleistocene, in particular the reimmigration of *Arvicola terrestris*. Courier Forschungsinstitut Senckenberg 153, 213-220.

van Laere, K., Herman, D., Leus, L., van Huylenbroeck, J., 2011. Genetic relationships in European and Asiatic *Buxus* species based on AFLP markers, genome sizes and chromosome numbers. Plant Syst. Evol. 293, 1-11.

van Leeuwen, J.F.N., Schafer, H., van der Knaap, W.O., Rittenour, T.M., Björck, S., Ammann, B., 2005. Native or introduced? Fossil pollen and spores may say. An example from the Azores Islands. Neobiota 6, 27-34.

van Leeuwen, J.F.N., Froyd, C.A., van der Knaap, W.O., Coffey, E.E., Tye, A., Willis, K.J., 2008. Fossil pollen as a guide to conservation in the Galápagos. Science 322, 1206-1206.

van Raden, U.J., Colombaroli, D., Gilli, A., Schwander, J., Bernasconi, S.M., van Leeuwen, J.F.N., Leuenberger, M., Eicher, U., 2013. High-resolution late-glacial chronology for the Gerzensee lake record (Switzerland): d-^{18}O correlation between a Gerzensee-stack and NGRIP. Paleogeogr. Paleoclimatol. Paleoecol. 391, 13-24.

van Zeist W., 1964. A palaeobotanical study of some bogs in Western Brittany (Finistère), France. Palaeohistoria 10, 157-180.

van Zeist, W., 1967. Archaeology and palynology in the Netherlands. Rev. Palaeobot. Palynology 4, 45-65.

van Zeist, W., 1980. Aperçu de la diffusion des végétaux cultivés dans la région méditerranéenne. La mise en place, l'évaluation et la caractérisation de la flore et des végétations circum méditerranéennes. Naturalia Monspeliena, Coll. Fondation L. Emberger, 129-145.

van Zeist, W., 1992. The origin and development of plant cultivation in the Near East. Japan Review 3, 149-165.

van Zeist, W., Bottema, S., 1982. Vegetational history of the Eastern Mediterranaen and the near East during the last 20,000 years., In: Bintliff, J., van Zeist, W. (Eds.), Paleoclimates, Paleoenvironments and Human Communities in the Eastern Mediterranean Region in Later Prehisory. British Archaeol. Rep., International Series, pp. 277-321.

van Zeist, W., Bottema, S., 1991. Late Quaternary Vegetation of the Near East. L. Reichert, Wiesbaden, 156 pp.

van Zeist, W., van der Spoel-Walvius, M.R., 1980. A palynological study of the Late-glacial and the Postglacial in the Paris basin. Palaeohistoria 22, 67-109.

van Zeist, W., Wasylikowa, C., Behre, K.-E. (Eds.), 1991. Progress in Old World Palaeoethnobotany. Balkema, Rotterdam.

Vandenberghe, J., 1985. Paleoenvironment and stratigraphy during the last glacial in the Belgian-Dutch border region. Quat. Res. 24, 23-38.

Vannière, B., Magny, M., Joannin, S., Simonneau, A., Wirth, S.B., Hamann, Y., Chapron, E., Gilli, A., Desmet, M., Anselmetti, F.S., 2013. Orbital changes, variation in solar activity and increased anthropogenic activities: controls on the Holocene flood frequency in the Lake Ledro area, Northern Italy. Clim. Past 9, 1193-1209.

Vannière, B., Blarquez, O., Rius, D., Doyen, E., Brucher, T., Colombaroli, D., Connor, S., Feurdean, A., Hickler, T., Kaltenrieder, P., Lemmen, C., Leys, B., Massa, C., Olofsson, J., 2016. 7000-year human legacy of elevation-dependent European fire regimes. Quat. Sci. Rev. 132, 206-212.

Vargas, P., McAllister, H.A., Morton, C., Jury. S.L., Wilkinson, M.J., 1999. Polyploid speciation in *Hedera* (Araliaceae): phylogenetic and biogeographic insights based on chromosome counts and ITS sequences. Plant Syst. Evol. 219, 165-179.

Vasari, Y., 1962. A study of the vegetational history of the Kuusamo district (North East Finland) during the Late-Quaternary period. Ann. Bot. Soc. Zool. Bot. Fenn. 'Vanamo', Helsinki, 33, 1-140 + Appendix.

Vasari, Y., 1965. Studies on the vegetational history of the Kuusamo district (North East Finland) during the Late-quaternary period. IV. The age and origin of some present-day vegetation types. Ann. Bot. Fenn. 2, 248-273.

Vasari, Y., Väänänen, K., 1986. Stratigraphical indications of the former use of wetlands, In: Behre, K.-E. (Ed.), Anthropogenic Indicators in Pollen Diagrams. Balkema, Rotterdam, pp. 65-71.

Vasari, Y., Vasari, A., 1968. Late- and Post-glacial macrophytic vegetation in the Lochs of Northern Scotland. Acta Bot. Fenn., Helsinki, 80, 1-120.

Vasari, Y., Vasari, A., 1985. Relic occurrence of *Betula nana* L. in Massif Central, France. Ecol. Mediterranea 11, 65-68.

Vasari, Y., Vasari, A., 1990. L'histoire holocène des lacs islandais. In: Devers, S. (Ed.), Pour Jean Malaurie. 102 témoignages en hommage à quarante ans d'études arctiques. Éditions Plon, Paris, pp. 277-293.

Vasari, Y., Glückert, G., Hicks, S., Hyvärinen, H., Simola, H., Vuorela, I., 1996. Finland. In: Berglund, B.E., Birks, H.J.B., Ralska-Jasiewiczowa, M., Wright, H.E. (Eds.), Palaeoecological Events During the Last 15 000 Years. Regional Syntheses of Palaeoecological Studies of Lakes and Mires in Europe. Wiley, Chichester, pp. 281-351.

Vassio, E., Martinetto, E., 2012. Biases in the frequency of fruits and seeds in modern fluvial sediments in northwestern Italy: the key to interpreting analogous fossil assemblages. Palaios 27, 779-797.

Vegas-Vilarrúbia, T., González-Sampériz, P., Morellón, M., Gil-Romera, G., Pérez-Sanz, A., Valero-Garcés, B., 2013. Diatom and vegetation responses to Late Glacial and Early Holocene climate changes at Lake Estanya (Southern Pyrenees, NE Spain). Palaeogeogr. Palaeoclimatol. Palaeoecol. 392, 335-349.

Velichkevich, F.Y., Granoszewski, W., 1996. *Potamogeton sukaczevii* Wieliczk. in the NeoPleistocene floras of Poland, Belarus and Lithuania. Acta Palaeobot. 36, 97-105.

Velichkevich, F.Y., Lesiak, M.A., 1996. Fossil *Potamogeton* species of Mizerna. Acta Palaeobot. 36, 79-95.

Velichkevich, F.Y., Mamakowa, K., 1999. Taxonomic revision of the collections of plant macrofossils from some localities of Poland now referred to the Vistulian glaciation. Acta Palaeobot. 39, 29-37.

Velichkevich, F.Y., Mamakowa, K., 2003. Revision of plant macrofossils from the Mazovian interglacial locality Nowiny Żukowskie (SE Poland). Acta Palaeobot. 43, 61-76.

Velichkevich, F.Y., Zastawniak, E., 2003. The Pliocene flora of Kholmech, south-eastern Belarus and its correlation with other Pliocene floras of Europe. Acta Palaeobot. 43, 137-259.

Velichkevich, F.Y., Zastawniak, E., 2006. Pleistocene Vascular Plant Macrofossils of Central and Eastern Europe, I. Pteridophytes and Monocotyledons. W Szafer Institute of Botany, Polish Academy of Sciences, Kraków. 525 pp.

Velichkevich, F.Y., Mamakowa, K., Stuchlik, L., 2004. Revision of some Mazovian interglacial macrofossil floras of Poland. Acta Palaeobot. 44, 93-104.

Velichko, A.A., Andrev, A.A., Klimanov, V.A., 1997. Climate and vegetation dynamics in the tundra and forest zone during the Late Glacial and Holocene. Quat. Int. 41/42, 71-96.

Velle, G., Brooks, S.J., Birks, H.J.B., Willassen, E., 2005a. Chironomids as a tool for inferring Holocene climate: an assessment based on six sites in southern Scandinavia. Quat. Sci. Rev. 24, 1429-1462.

Velle, G., Larsen, J., Eide, W., Peglar, S.M., Birks, H.J.B., 2005b. Holocene environmental history and climate of Råtåsjøen, a low-alpine lake in south-central Norway. J. Paleolimnol. 33, 129-153.

Vellend, M., 2010. Conceptual syntheses in community ecology. Q. Rev. Biol. 85, 183-206.

Vellend, M., 2016. The Theory of Ecological Communities. Princeton University Press, Princeton. 224 pp.

Veloz, S.D., Williams, J.W., Blois, J.L., He, F., Otto-Bliesner, B., Liu, Z., 2012. No-analog climates and shifting realized niches during the late Quaternary: implications for 21st-century predictions by species distribution models. Global Change Biol. 18, 1698-1713.

Vendramin, G.G., Ziegenhagen, B., 1997. Characterisation and inheritance of polymorphic plastid microsatellites in *Abies*. Genome 40, 857-864.

Vendramin, G.G., Anzidei, M., Madaghiele, A., Bucci, G., 1998. Distribution of genetic diversity in *Pinus pinaster* Ait. as revealed by chloroplast microsatellites. Theor. Appl. Genet. 97, 456-463.

Vendramin, G.G., Anzidel, M., Madaghiele, A., Sperisen, C., Bucci, G., 2000. Chloroplast microstellite analysis reveals the presence of population subdivision in Norway spruce (*Picea abies* K.). Genome 43, 68-78.

Vendramin, G.G., Fady, B., González-Martínez, S., Hu, F., Scotti, I., Sebastiani, F., Soto, Á., Petit, R., 2008. Genetically depauperate but widespread: the case of an emblematic Mediterranean pine. Evolution 62, 680-688.

Vera, F., 1997. Metaforen voor de wildernis: eik, hazelaar, rund en paard. Thesis, Landbouwuniversiteit Wageningen, 440 pp.

Vera, F., 2000. Grazing Ecology and Forest History. CABI, Wallingford. 528 pp.

Verbruggen, C., 1979. Vegetational and palaeoecological history of the Late-Glacial period in Sandy Flanders (Belgium). Acta Univ. Ouluensis 82, 133–142.

Verbruggen, C., Denys, L., Kiden, P., 1996. Palaeoecological events in Belgium during the last 13,000 years with special reference to sandy Flanders. In: Berglund, B.E., Birks, H.J.B., Ralska-Jasiewiczowa, M., Wright, H.E. (Eds.), Palaeoecological Events During the Last 15 000 Years. Regional Syntheses of Palaeoecological Studies of Lakes and Mires in Europe. Wiley, Chichester, pp. 553-574.

Vernet, J., 1980. La végétation du bassin de l'Aude, entre Pyrenées et Massif Central, au tardiglaciaire et au postglaciaire d'après l'analyse anthracologique. Rev. Palaeobot. Palynol. 30, 33-55.

Vescovi, E., Ravazzi, C., Arpenti, E., Finsinger, W., Pini, R., Valsecchi, V., Wick, L., Ammann, B., Tinner, W., 2007. Interactions between climate and vegetation during the Lateglacial period as recorded by lake and mire sediment archives in Northern Italy and Southern Switzerland. Quat. Sci. Rev. 26, 1650-1669.

Vescovi, E., Ammann, B., Ravazzi, C., Tinner, W., 2010a. A new Late-glacial and Holocene record of vegetation and fire history from Lago del Greppo, northern Apennines, Italy. Veget. Hist. Archaeobot. 19, 219-233.

Vescovi, E., Kaltenrieder, P., Tinner, W., 2010b. Late-Glacial and Holocene vegetation history of Pavullo nel Frignano (Northern Apennines, Italy). Rev. Palaeobot. Palynol. 160, 32-45.

Vescovi, E., Tinner, W., Beer, R., van Leeuwen, J.F.N., Steinhauser, U., Ziegler, S., Gilli, A., Wirth, S.B., Samartin, S., Jacquat, C., 2018. Vegetation history of the upper Leventina Valley from Neolithic to Roman times, recorded by pollen, spores, plant macrofossils, and charcoal. In: Della Casa, P. (Ed.), The Leventina Prehistoric Landscape (Alpine Ticino Valley, Switzerland), Zurich Studies in Archaeology. Chronos Verlag, Zürich, pp. 207–225.

Veski, S., Amon, L., Heinsalu, A., Reitalu, T., Saarse, L., Stivrins, N.V., J, 2012. Lateglacial vegetation dynamics in the eastern Baltic region between 14,500 and 11,400 cal yr BP: A complete record since the Bolling (GI-1e) to the Holocene Quat. Sci. Rev. 40, 39-53.

Veski, S., Seppä, H., Stančikaitė, M., Zernitskaya, V., Reitalu, T., Gryguc, G., Heinsalu, A., Stivrins, N., Amon, L., Vassiljev, J., Heiri, O., 2015. Quantitative summer and winter temperature reconstructions from pollen and chironomid data between 15 and 8 ka BP in the Baltic–Belarus area. Quat. Int. 388, 4-11.

Vidal-Matutano, P., Hernández, C.M., Galván, B., Mallol, C., 2015. Neanderthal firewood management: evidence from Stratigraphic Unit IV of Abric del Pastor (Eastern Iberia). Quat. Sci. Rev. 111, 81-93.

Viinikka, Y., 1976. *Najas marina* L. (Najadaceae). Karyotypes, cultivation and morphological variation. Ann. Bot. Fenn., Helsinki, 13, 119-131.

Villemur, P., Musho, S., Delmas, J., Maamar, M., Ouksili, A., 1984. Contribution à létude de la biologie florale de l'olivier: stérilité mâle, flux pollinique et période effective de pollinisation. Fruits 39, 467-473.

Vincze, I., Orbán, I., Birks, H.H., Pal, I., Finsinger, W., Hubay, K., Marinova, E., Jakab, G., Braun, M., Biro, T., Toth, M., Danau, C., Ferencz, I.V., Magyari, E.K., 2017. Holocene treeline and timberline changes in the South Carpathians (Romania): Climatic and anthropogenic drivers on the southern slopes of the Retezat Mountains. Holocene 27, 1613-1630.

Virah-Sawmy, M., Willis, K.J., Gillson, L., 2009. Threshold response of Madagascar's littoral forest to sea-level rise. Global Ecol. Biogegor. 18, 98-110.

Virah-Sawmy, M., Willis, K.J., Gillson, L., 2010. Evidence for drought and forest declines during the recent megafaunal extinctions in Madagascar. J. Biogeogr. 37, 506-519.

Visset, L., Aubert, S., Belet, J., David, F., Fontugne, M., Galop, D., Jalut, G., Janssen, C., Voeltzel, D., Huault, M., 1996. France. In: Berglund, B.E., Birks, H.J.B., Ralska-Jasiewiczowa, M., Wright, H.E. (Eds.), Palaeoecological Events During the Last 15 000 Years. Regional Syntheses of Palaeoecological Studies of Lakes and Mires in Europe. Wiley, Chichester, pp. 575-645.

von Chamisso A., 1824. Untersuchung eines Torfmoors bei Greifswald und ein Blick auf die Insel Rügen. Karsten Arch für Bergbau und Hüttenwes 8, 129-139.

von der Goltz, H., Müller, G., Raisch, W., Wipf, J., 1976. Die Bodengesellschaften im oberen Haslachtal bei Neuglashütten (Hochschwarzwald). University of Freiburg im Breisgau, 249 pp.

von Engelbrechten, S., 1998. Late-glacial and Holocene vegetation and environmental history of the Sierra de Urbion, North-Central Spain. University of Dublin, Trinity College, Dublin.

von Grafenstein, U., Erlenkeusen, H., Müller, J., Kleinmann-Eisenmann, A., 1992. Oxygen isotope records of benthic Ostracods in Bavarian lake sediments. Naturwissenschaften 79, 145-152.

von Grafenstein, U., Erlenkeuser, H., Müller, J., Jouzel, J., Johnsen, S., 1998. The cold event 8200 years ago documented in oxygen isotope records of precipitation in Europe and Greenland. Clim. Dynam. 14, 73-81.

von Grafenstein, U., Belmecheri, S., Eicher, U., van Raden, U.J., Erlenkeuser, H., Andersen, N., Ammann, B., 2013. The oxygen and carbon isotope signature of biogenic carbonates in Gerzensee, Switzerland, during the rapid warming around 14.685 ka BP and the following interstadial. Palaeogeogr. Palaeoclimatol. Palaeoecol. 391, 25-32.

von Holle, B., Delcourt, H.R., Simberloff, D., 2003. The importance of biological inertia in plant community resistance to invasion. J. Veg. Sci. 14, 425-432.

von Koenigswald, W., 2007. Mammalian faunas from the interglacial periods in Central Europe and their stratigraphical correlation. In: Sirocko, F. Claussen, M., Sánchez-Goñi, M.F, Litt, T. (Eds.), The Climate of Past Interglacials. Elsevier, Amsterdam, 445-454.

von Koenigswald, W., 2010. Lebendige Eiszeit. Klima und Tierwelt im Wandel. Darmstadt.

von Koenigswald, W., van Kolfschoten, T., 1996. The *Mimomys-Arvicola* boundary and the enamel thickness quotient (SDQ) of *Arvicola* as stratigraphic markers in the Middle Pleistocene. In: Turner, C. (Ed.), The Early Middle Pleistocene in Europe. Balkema, Rotterdam,CRC Press, Boca Raton, pp. 211-226.

von Morlot, A., 1858. Sur le terrain quaternaire du bassin du Léman. Bull. Soc. Vaud. Sci. Nat. 4, 101..

von Post, L., 1916a. Mötet den 2 november 1916. Geol. Fören. Stockholm Förhandl. 38, 383-394.

von Post, L., 1916b. Om skogsträdpollen I sydsvenska torvmosslagerföljder. Geol. Fören. Stockholm Förhandl. 38, 435-465.

von Post, L., 1924. Ur de sydsvenska skogarnas regionala historia under post-arktisk tid. Geol. Fören. Stockholm Förhandl. 46, 83-128.

von Post, L., 1926. Einige Aufgaben der regionalen Moorforschung. [Some challenges of regional mire research] Sveriges Geol. Unders. Ser. C 337, 1-41. (in German)

von Post, L., 1929. Die Zeichenschrift der Pollenstatistik. Geol. Fören. Stockholm Förh. 51, 543-565.

von Post, L., 1946. The prospect for pollen analysis in the study of the Earth's climatic history. New Phytol. 45, 193-217.

von Post, L., 1967. Forest tree pollen in south Swedish peat bog deposits. Translated by M.B. Davis and K. Fægri with introduction by K. Fægri and J. Iversen. Pollen et Spores 9, 375-401.

von Post, L., Sernander, R., 1910. Pflanzenphysiognomische Studien auf Torfmooren in Närke, Livret-guide des excursions en Suède du 11e Congrès géol. internat., Stockholm, p. 14.

von Schnurbein, S., 2009. Atlas der Vorgeschichte. Theiss, Stuttgart, 242 pp.

von Stedingk, H., Fyfe, R.M., 2009. The use of pollen analysis to reveal Holocene treeline dynamics: a modelling approach. Holocene 19, 273-283.

Vorren, K.-D., 1978. Late and Middle Weichselian stratigraphy of Andøya, north Norway. Boreas 7, 19-38.

Vorren, K.-D., 1986. The impact of early agriculture on the vegetation of Northern Norway. A discussion of anthropogenic indicators in biostratigraphical data. In: Behre, K.-E. (Ed.), Anthropogenic Indicators in Pollen Diagrams. Balkema, Rotterdam, pp. 1-18.

Vorren, K.-D., Alm, T., 1985. An attempt at synthesizing the Holocene biostratigraphy of a «type area» in northern Norway by means of recommended methods for zonation and comparison of biostratigraphical data. Ecol. Mediterr. 11, 53–64.

Vorren, K.-D., Plassen, L., 2002. Deglaciation and palaeoclimate of the Andfjord-Vågsfjord area, North Norway. Boreas 31, 97-125.

Vorren, K.-D., Elverland, E., Blaauw, M., Ravna, E., Jensen, C., 2009. Vegetation and climate c. 12 300-9000 cal. yr BP at Andøya, NW Norway. Boreas 38, 401-420.

Vorren, T.O., Vorren, K.-D., Aasheim, O., Dahlgren, K.I.T., Forwick, M., Hassel, K., 2013. Palaeoenvironment in northern Norway between 22.2 and 14.5 cal. ka BP. Boreas 42, 876-895.

Voznyachuk, L.M. et al. (Eds.), 1978. Study of the Anthropogene of Belarus. Academy of Sciences Belarussian SSR, Nauka I Tekhnica, Minsk (in Russian).

Vuorela, I., 1975. Pollen analysis as a means of tracing settlement history in SW-Finland. Acta Bot. Fenn. 104, 1-48.

Vuorela, I., 1986. Palynological and historical evidence of slash-and-burn cultivation in South Finland. In: Behre, K.-E. (Ed.), Anthropogenic Indicators in Pollen Diagrams. Balkema, Rotterdam, pp. 53-64.

Vuorela, I., 1991. Turvetutkimus Suomen asutushistoian selvittämisessä. Suo 42, 101–108.

Vuorela, I., Aalto, M., 1982. Paleobotanical investigations at a Neolithic dwelling site in southern Finland, with special reference to *Trapa natans*. Ann. Bot. Fenn. 19, 81-92.

Vuorela, I., Kankainen, T., 1991. Siitepölyanalyyttinen tutkimus asutuksen vaikutuksesta kasvillisuuteen Puolangan kunnan Kotilan kylässä. Tutkimusraportti P 34, 21.

Wacnik, A., 1995. The vegetational history of local flora and evidence of human activities recorded in the pollen diagram from site Regetovka, NE Slovakia. Acta Palaeobot. 35, 253-274.

Wada, S., Sasaki, N., Takahara, H., Sugita, S., 2018. Pollen production of six species in Poaceae - baseline information toward quantitative reconstruction of vegetation in Japan. Japanese J. Palynol. 63, 37-51.

Wagenitz, G., 1981. Fagaceae, Illustrierte Flora von Mitteleuropa, 3rd ed, Berlin.

Wagner, S., Liepelt, S., Gerber, S., Petit, R.J., 2015a. Within-range translocations and their consequences in European Larch. PLoS One 10, e0127516.

Wagner, S., Litt, T., Sánchez-Goñi, M., Petit, R.J.,

2015b. History of *Larix decidua* Mill. (European larch) since 130 ka. Quat. Sci. Rev. 124, 224-247.

Wahlmüller, N., 1985. Beiträge zur Vegetationsgeschichte Tirols V: Nordtiroler Kalkalpen. Ber. nat.-med. Verein Innsbruck 72, 101-144.

Wahlmüller, N., 1990. Spät- und postglaziale Vegetationsgeschichte des Tschögglberges (Südtirol). Berichte Naturwissenschaftlichen-Med. Ver. Innsbr. 77, 7-16.

Wahlmüller, N., 1992. Beitrag der Pollenanalyse zur Besiedlungsgeschichte des Haidbergs bei Bischofshofen/Salzburg, in: Lippert, A. (Ed.), Der Götschenberg Bei Bischofshofen, Mittl. Prähist. Komm. Österr. Akad. Wiss. pp. 129-142.

Wahlmüller, N., 1993. Palynologische Forschung in den Ostalpen und ihren vorgelagerten Gebieten. Berichte der natur-medizinischen Verein Innsbruck 80, 81-95.

Walanus, A., Nalepka, D., 1997. Palynological diagram drawing in Polish POLPAL for windows. INQUA Working Group on Data-Handling Methods Newsletter 16, 8-10.

Walker, D., 1961. Peat stratigraphy and bog regeneration. Proc. Linn. Soc. London 172, 29-33.

Walker, D., 1966. The Late Quaternary History of the Cumberland Lowland. Phil. Trans. R. Soc. London B251, 1-210.

Walker, D., 1970. Direction and rate in some British Post-glacial hydroseres, In: Walker, D., West R.G. (Eds.), Studies in the Vegetational History of the British Isles. Cambridge University Press, Cambridge, pp. 117-139.

Walker, D., 1982. Vegetation's fourth dimension. New Phytol. 90, 419-429.

Walker, M.J.C., 2005. Quaternary Dating Methods. Wiley.

Walker, M.J.C., Harkness, D.D., 1990. Radiocarbon dating the Devensian Lateglacial in Britain: New evidence from Llanilid, South Wales. J. Quat. Sci. 5, 135-144.

Walker, J., Peet, R.K., 1983. Composition and species diversity of pine-wiregrass savannas of the Green Swamp, North Carolina. Vegetatio 55, 163-179.

Walker, D., Walker, P.M., 1961. Stratigraphic evidence of regeneration in some Irish bogs. J. Ecol. 49, 169-185.

Walker, M.J.C., Merrit, J.W., Auton, C.A., Coope, G.R., Field, M.H., Heijnis, H., Taylor, B.J., 1992. Allt Odhar and Dalcharn: two pre-Late Devensian/Late Weichselian sites in northern Scotland. J. Quat. Sci. 7, 69-86.

Walker, M.J.C., Berkelhammer, M., Björck, S., Cwynar, L.C., Fisher, D.A., Long, A.J., Lowe, J.J., Newnham, R.M., Rasmussen, S.O., Weiss, H., 2012. Formal subdivision of the Holocene Series/Epoch: a Discussion Paper by a Working Group of INTIMATE (Integration of ice-core, marine and terrestrial records) and the Subcommission on Quaternary Stratigraphy (International Commission on Stratigraphy). J. Quat. Sci. 27, 649 - 659.

Walker, M.J.C., Head, M.J., Berkelhammer, M., Björck, S., Cheng, H., Cwynar, L., Fisher, D., Gkinis, V., Long, A., Lowe, J., Newnham, R., Rasmussen, S.O., Weiss, H., 2018. Formal ratification of the subdivision of the Holocene Series/Epoch (Quaternary System/Period): two new Global Boundary Stratotype Sections and Points (GSSPs) and three new stages/subseries. Episodes J. Int. Geosci. 41, 213-223.

Waller, M.P., 1987. The Flandrian vegetational history and environmental development of the Brede and Panel Valleys East Sussex. (PhD thesis). Polytechnic of North London, London, United Kingdom.

Waller, M.P., 1993. Flandrian vegetational history of southeastern England. Pollen data from Pannel Bridge, East Sussex. New Phytol. 124, 345-369.

Waller, M.P., 2013. Drought, disease, defoliation and death: forest pathogens as agents of past vegetation change. J. Quat. Sci. 28, 336-342.

Waller, M.P., Hamilton, S., 2000. Vegetation history of the English chalklands: a mid-Holocene pollen sequence from the Caburn, East Sussex. J. Quat. Sci. 15, 253-272.

Walter, H., 1974. Die Vegetation Osteuropas, Nord- und Zentralasiens. Fischer, Stuttgart.

Walter, H., Breckle, S.-W., 1986. Spezielle Ökologie der Gemäßigten und Arktischen Zonen Euro-Nordasiens. Band 3 der Ökologie der Erde. Fischer, Stuttgart. 587 pp.

Walter, H., Straka, H., 1970. Arealkunde. Floristisch-historische Geobotanik. Ulmer, Stuttgart.

Walther, G.R., Post, E., Convey, P., Menzel, A., Parmesan, C., Beebee, T.J.C., Fromentin, J.M., Hoegh-Guldberg, O., Bairlein, F., 2002. Ecological responses to recent climate change. Nature 416, 389-395.

Walther, G.R., Gritti, E.S., Berger, S., Hickler, T., Tang, Z.Y., Sykes, M.T., 2007. Palms tracking climate change. Global Ecol. Biogeogr. 16, 801-809.

Walzak, I.W., Baldini, J.U.L., Baldini, L.M., McDermott, F., Marsden, S., Standish, C.D., Richards, D.A., Andreo, B., Slater, J., 2015. Reconstructing high-resolution climate using CT scanning of unsectioned stalagmites: A case study identifying the mid-Holocene onset of the Mediterranean climate in southern Iberia. Quat. Sci. Rev. 127, 117-128.

Wang, Y., Heintzman, P.D., Newsom, L., Bigelow, N., Wooller, M., Shapiro, B., Williams, J.W., 2017. The southern coastal Beringian land bridge: cryptic refugium or pseudorefugium for woody plants during the Last Glacial Maximum? J. Biogeogr. 44, 1559-1571.

Wardle, D., Hörnberg, G., Zackrisson, O., Kalela-Brundin, M., Coomes, D., 2003. Long-term effects of wildfire on ecosystem properties across an island area gradient. Science 300, 972-975.

Warming, E., 1904. Den Danske Planteverdens Historie efter Istiden. Copenhagen.

Wassen, M.J., Joosten, J.H.J., 1996. In search of a hydrological explanation for vegetation changes along a fen gradient in the Biebrza Upper Basin (Poland). Vegetatio 124, 191-209.

Wasylikowa, K., 1978. Plant remains from early and late medieval time found on the Wawel Hill in Cracow. Acta Palaeobot. 19, 115-198.

Wasylikowa, K., 1981. The role of fossil weeds for the study of former agriculture. Z. Arch. Berlin 15, 11-23.

Wasylikowa, K., Starkel, L., Niedzialkowska, E., Skiba, S., Stworzewicz, E., 1985. Environmental changes in the Vistula valley at Pleszow caused by neolithic man. Przeglad Archeol. 33, 19-55.

Watson, C.S., 1996. The vegetational history of the northern Apennines, Italy: information from three new sequences and a review of Regional vegetational change. Journal of Biogeography 23, 805–841.

Watts, W.A., 1959. Interglacial Deposits at Kilbeg and Newton, Co. Waterford. Proc. Roy. Irish Acad., Dublin, 60 B, 79-134.

Watts, W.A., 1961. Post-Atlantic forests in Ireland. Proc. Linn. Soc. London. 172, 33-38.

Watts, W.A., 1973. Rates of change and stability in vegetation in the perspective of long periods of time. In: Birks, H.J.B., West, R.G. (Eds.), Quaternary Plant Ecology. Blackwell Scientific Publications, Oxford, pp. 195-206.

Watts, W.A., 1985. Quaternary vegetation cycles. In: Edwards, K.E., Warren, W.P. (Eds.), The Quaternary History of Ireland. Academic Press, London, pp. 155-185.

Watts, W.A., 1988. Europe. In: Huntley, B., Webb, T. (Eds.), Vegetation History. Handbook of Vegetation Science 7, pp. 155-192.

Watts, W.A., Allen, J.R.M., Huntley, B., 1996a. Vegetation history and palaeoclimate of the last glacial period at Lago Grande di Monticchio, Southern Italy. Quat. Sci. Rev. 15, 133-153.

Watts, W.A., Allen, J.R.M., Huntley, B., Fritz, S., 1996b. Vegetation history and climate of the last 15,000 years at Laghi di Monticchio, Southern Italy. Quat. Sci. Rev. 15, 113-132.

Watts, W.A., Allen, J.R.M., Huntley, B., 2000. Palaeoecology of three interstadial events during oxygen-isotope Stages 3 and 4: a lacustrine record from Lago Grande di Monticchio, southern Italy. Palaeogeogr. Palaeoclimatol. Palaeoecol. 155, 83-93.

Webb III, T., 1986. Is vegetation in equilibrium with climate? How to interpret late-Quaternary pollen data. Vegetatio 67, 75-91.

Webb III, T., Bryson, R.A., 1972. Late- and postglacial climatic change in the northern Midwest, U.S.A., quantitative estimates derived from fossil pollen spectra by multivariate statistical analysis. Quat. Res. 2, 70-115.

Webb, D.A., 1954. Is the classification of plant communities either possible or desirable? Botanisk Tidsskrift 51, 362-370.

Weber, B., 1977. Contribution à l'étude du tardiglaciaire de la région lémanique (le profil de St-Laurent è Lausanne) III. Étude des macrorestes végétaux. Bulletin de la Société botanique suisse 87, No.3/4, 207-226.

Weber, B., 1979. Morphologie des macrorestes végétaux de Vidy. I. Les fruits et graines. Ber. Schweiz. Bot. Ges. 89, 159-186.

Weber, B., 1980. Morphologie des macrorestes végétaux de Vidy. II. Les feuilles. Ber. Schweiz. Bot. Ges. 90, 78-96.

Weber, C.A., 1893. Über die diluviale Vegetation von Klinge in Brandenburg und ihre Herkunft. Englers Botanische Jahrbücher 17, Beiblatt 1.

Weber, C.A., 1896. Die fossile Flora von Honerdingen und das nordwestdeutsche Diluvium. Abhandl. Naturwiss. Ver. Bremen 13, 413-486.

Weber, C.A., 1898. Über eine omorikaartige Fichte aus einer dem ältern Quartäre Sachsens angehörenden Moorbildung. Englers Bot. Jahrb. 24, 510-543.

Weber, C.A., 1900. Über die Moore, mit besonderer Berücksichtigung der zwischen Unterweser und Unterelbe liegenden. Jahresber. Männer v. Morgenstern 3, 3-23.

Weber, C.A., 1904. Über eine frühdiluviale und vorglaziale Flora bei Lüneburg. Abh. Königl. Preuss. Geolog. Landesanstalt, N.F. 40.

Weber, C.A., 1906. Die Geschichte der Pflanzenwelt des norddeutschen Tieflandes seit der Tertiärzeit. Wiss. Ergebn. Int. Bot. Kongr. Wien 1905, 98-116.

Weber, C.A., 1910. Was lehrt der Aufbau der Moore Norddeutschlands über den Wechsel des Klimas in postglazialer Zeit? Z. Deutsche Geol. Ges. 62, 2.

Weber, C.A., 1914. Die Mammutflora von Borna. Abhandlungen des Naturwissenschaftlichen Vereins zu Bremen 23, 1-69.

Weber, H.A., 1918. Über spät- und postglaziale lakustrine und fluviatile Ablagerungen in der Wyhraniederung bei Lobstädt und Borna und die Chronologie der Postglazialzeit Mitteleuropas. Abhandl. Naturwiss. Verein Bremen 24, 189-267.

Weber, H., Moravec, J., Theurillat, J., 2000. International Code of Phytosociological Nomenclature. 3rd ed. J. Veget. Sci. 11, 739 - 768.

Wegener, A., 1915. Die Entstehung der Kontinente

und Ozeane, Braunschweig.

Wegmüller, S., 1977. Pollenanalytische Untersuchungen zur spät- und postglazialen Vegetationsgeschichte der französischen Alpen (Dauphine). Haupt, Bern. 185 pp.

Wegmüller, S., 1984. Zur Ausbreitungsgschichte von *Buxus sempervirens* L. im Spät- und Postglazial in Süd- und Mitteleuropa. In: Lang, G. (Ed.), Festschrift Max Welten, pp. 333-344.

Wegmüller, S., 1992. Vegetationsgeschichtliche und stratigraphische Untersuchungen an Schieferkohlen des nördlichen Alpenvorlandes. Denkschr. Schweiz. Akad. Naturwiss. 102.

Wegmüller, S., Lotter, A.F., 1990. Palynostratigraphische Untersuchungen zur spät- und postglazialen Vegetationsgeschichte der nordwestlichen Kalkvoralpen. Botanica Helvetica 100, 37-73.

Weiner, J., 1998. Drei Brunnenkästen, aber nur zwei Brunnen: Eine neue Hypothese zur Baugeschichte des Brunnens von Erkelenz-Kükhoven. Materialien zur Bodendenkmalpflege im Rheinland 11, 95-112.

Weiss, D., Shotyk, W., Kempf, O., 1999. Archives of atmospheric lead pollution. Naturwissenschaften 86, 262-275.

Welc, M., Bünemann, E.K., Fließbach, A., Frossard, E., Jansa, J., 2012. Soil bacterial and fungal communities along a soil chronosequence assessed by fatty acid profiling. Soil Biol. Biogeochem. 49, 184-192.

Welc, M., Frossard, E., Egli, S., Bünemann, E.K., Jansa, J., 2014. Rhizosphere fungal assemblages and soil enzymatic activities in a 110-years alpine chronosequence. Soil Biol. Biogeochem. 74, 21-30.

Welten, M., 1944. Pollenanalytische, stratigraphische und geochronologische Untersuchungen aus dem Faulenseemoos bei Spiez. Veröff. Geobot. Inst. Rübel Zürich 21, 1-201.

Welten, M., 1950. Beobachtungen über den rezenten Pollenniederschlag in alpiner Vegetation. Bericht Geobot. Forschungsinstitut Rübel Zürich, 47-57.

Welten, M., 1952. Über die spät- und postglaziale Vegetationsgeschichte des Simmentals sowie die frühgeschichtliche und historische Wald- und Weiderodung auf Grund pollenanalytischer Untersuchungen. Veröff. Geobot. Inst. Rübel Zürich 26, 1-135.

Welten, M., 1955. Pollenanalytische Untersuchung von Bodenprofilen: Historische Entwicklung des Bodens und säkulare Sukzession der örtlichen Pflanzengesellschaften. Verh. Schweiz. Naturf. Ges. 1955, 148-150.

Welten, M., 1957. Über das glaziale und spätglaziale Vorkommen von *Ephedra* am nordwestlichen Alpenrand. Ber. Schweiz. Bot. Ges. 67, 33-54.

Welten, M., 1967. Ein Brachsenkraut, *Isoëtes setacea* Lam., fossil im schweizerischen Molasseland. Diskussionsbeitrag zu den Begriffen des Areals und des historischen Elements. Bot. Jahrb. Syst. 86, 527-536.

Welten, M., 1979. Late Glacial and Late Weichselian (Spätglazial und Spät-Würm): a comment. Boreas 8, 396.

Welten, M., 1982a. Pollenanalytische Untersuchungen im Jüngeren Quartär des nördlichen Alpenvorlandes der Schweiz. Beitr. Geol. Karte Schweiz, NF 156.

Welten, M., 1982b. Vegetationsgeschichtliche Untersuchungen in den westlichen Schweizer Alpen: Bern-Wallis. Denkschr. Schweizer Naturforsch. Ges. 95, 1-104.

Weninger, B., 1990. Theoretical radiocarbon discrepancies. In: Hardy, D., Renfrew, A. (Eds.), Thera and the Aegean World. The Thera Foundation, London.

Wenner, C., 1953. Investigations into the possibilities of distinguishing the pollen of various species of *Betula* in fossil material. Geol. Fören. i Stockholm Förh. 75, 103.

West, R.G., 1956. The Quaternary deposits at Hoxne, Suffolk.- Phil. Trans. Roy. Soc. London B 239, 265-356.

West, R.G., 1957. Interglacial deposits at Bobbitshole, Ipswich. Phil. Trans. Royal Soc. London B 241, 1-31.

West, R.G., 1980. The pre-glacial Pleistocene of the Norfolk and Suffolk coasts. Cambridge Unviersity Press, Cambridge.

Westerhoff, W.E., Cleveringa, P., Meijer, T., van Kolfschoten, T., Zagwijn, W.H., 1998: The Lower Pleistocene fluvial (clay) deposits in the Maalbeek pit near Tegelen, The Netherlands. Mededelingen TNO 60, 35-69.

Wetzel, R.G., 1983. Limnology, 2d ed. Saunders College Publishing, Fort Worth.

Wetzel, R.G., 2001. Limnology: Lake and River Ecosystems, 3rd ed. Saunders College Publishing, Fort Worth.

White, P.S., 1979. Pattern, process, and natural disturbance in vegetation. Bot. Rev. 45, 229-299.

Whitlock, C., Larsen, C., 2001. Charcoal as fire proxy. In: Smol, J.P., Birks, H.J.B., Last, W.M. (Eds.), Tracking Environmental Change Using Lake Sediments, Vol. 3: Terrestrial, Algal, and Siliceous Indicators. Kluwer, Dordrecht, pp. 75-97.

Whitlock, C., Millspaugh, S.H., 1996. Testing the assumptions of fire history studies: An examination of modern charcoal accumulation in Yellowstone National Park, USA. Holocene 6, 7-15.

Whitlock, C., Shafer, S.L., Marlon, J.R., 2003. The role of climate and vegetation change in shaping past and present fire regimes in the northwestern US and implications for ecosystem management. For. Ecol. Manage. 178, 5-21.

Whitmore, J., Gajewski, K., Sawada, M., Williams, J.W., Shuman, B., Bartlein, P.J., Minckley, T., Viau, A.E., Webb, T., Shafer, S., Anderson, P., Brubaker, L., 2005. Modern pollen data from North America and Greenland for multi-scale paleoenvironmental applications. Quat. Sci. Rev. 24, 1828-1848.

Whittaker, R.H., 1967. Gradient analysis of vegetation. Biol. Rev. 42, 207-264.

Whittington, G., 1994. Bruckenthalia spiculifolia (Salisb.) Reichenb. (Ericaceae) in the Late Quaternary of Western Europe. Quaternary Science Reviews 13, 761-768.

Wick, L., Möhl, A., 2006. The mid-Holocene extinction of silver fir (*Abies alba*) in the Southern Alps: a consequence of forest fires? Palaeobotanical records and forest simulations. Veget. Hist. Archaeobot. 15, 435-444.

Wick, L., Tinner, W., 1997. Vegetation changes and timberline fluctuations in the Central Alps as indicator of Holocene climatic oscillations. Arc. Alp. Res. 29, 445-458.

Wick, L., Lemcke, G., Sturm, M., 2003a. Evidence of Lateglacial and Holocene climatic change and human impact in eastern Anatolia: high-resolution pollen, charcoal, isotopic and geochemical records from the laminated sediments of Lake Van, Turkey. Holocene 13, 665-675.

Wick, L., van Leeuwen, J.F.N., van der Knaap, W.O., Lotter, A.F., 2003b. Holocene vegetation development in the catchment of Sagistalsee (1935 m asl), a small lake in the Swiss Alps. J. Paleolimnol. 30, 261-272.

Widmer, A., Lexer, C., 2001. Glacial refugia: sanctuaries for allelic richness, but not for gene diversity. Trends Ecol. Evol. 16, 267-269.

Wieckowska-Lüth, M., Kirleis, W., Dörfler, W., 2017. Holocene history of landscape development in the catchment of Lake Skogstjern, southeastern Norway, based on a high-resolution multi-proxy record. The Holocene 27, 1928–1947.

Wieder, R.K., Turetsky, M.R., Vile, M.A., 2009. Peat as an archive of atmospheric, climatic and environmental conditions. In: Maltby, E., Barker, T. (Eds.), The Wetlands Handbook. Blackwell Publishing, Oxford, pp. 96-112.

Wijmstra, T.A., 1969. Palynology of the first 30 meters of a 120 m deep section in northern Greece. Acta Bot. Neerl. 18, 511-527.

Wilkinson, D.M., 1999. The disturbing history of intermediate disturbance. Oikos 84, 145-147.

Willcox, G., 1996. Evidence of plant exploitation and vegetation history from three Early Neolithic pre-pottery sites on the Euphrates (Syria). Veget. Hist. Archaeobot. 5, 143-152.

Willcox, G., Nesbitt, M., Bittmann, F., 2012. From collecting to cultivation: transition to a production economy in the Near East. Veget. Hist. Archaeobot. 21, 81-83.

Willerding, U., 1968. Beiträge zur Geschichte der Eibe (Taxus baccata L.) - Untersuchungen über das Eibenvorkommen im Plesswald bei Göttingen. Plesse-Archiv 3, 96-155.

Willerding, U., 1978. Die Paläo-Ethnobotanik und ihre Stellung im System der Wissenschaften. Berichte der Deutschen Botanischen Gesellschaft 91, 3-30.

Willerding, U., 1986. Zur Geschichte der Unkräuter Mitteleuropas. Göttinger Schriften zur Vor- und Frühgeschichte 22.

Willerding, U., 1988. Zur Entwicklung von Ackerunkrautgesellschaften im Zeitraum vom Neolithikum bis in die Neuzeit. Forschungen und Berichte zur Vor- und Frühgeschichte in Baden-Württemberg, 31-41.

Willerding, U., 2007. Wiese. In: Hoops' Reallexikon der Germanischen Altertumskunde, 2nd ed. 34, pp. 38-50.

Williams, W., 1977. The Flandrian vegetational history of the Isle of Skye and the Morar Peninsula. PhD thesis, University of Cambridge.

Williams, J.W., Post, D.M., Cwynar, L.C., Lotter, A.F., Levesque, A.J., 2002. Rapid and widespread vegetation responses to past climate change in the North Atlantic region. Geology 30, 971-974.

Williams, J.W., Shuman, B.N., Webb III, T., Bartlein, P.J., Leduc, P.L., 2004. Late-Quaternary vegetation dynamics in North America: Scaling from taxa to biomes. Ecol. Monogr. 74, 309-334.

Williams, J.W., Shuman, B.N., Bartlein, P.J., 2009. Rapid responses of the prairie-forest ecotone to early Holocene aridity in mid-continental North America. Global Planet. Change 66, 195-207.

Williams, J.W., Blois, J.L., Shuman, B.N., 2011a. Extrinsic and intrinsic forcing of abrupt ecological change: case studies from the late Quaternary. J. Ecol. 99, 664-677.

Williams, J.W., Tarasov, P.E., Brewer, S., Notaro, M., 2011b. Late Quaternary variations in tree cover at the northern forest-tundra ecotone. J. Geophys. Res.: Biogeosci. 116, G01017.

Williams, J.W., Blois, J.L., Gill, J.L., Gonzales, L.M., Grimm, E.C., Ordonez, A., Shuman, B.N., Veloz, S.D., 2013a. Model systems for a no-analog future: species associations and climates during the last deglaciation. Ann. N. Y. Acad. Sci. 1297, 29-43.

Williams, J.W., Kharouba, H.M., Veloz, S., Vellend, M., McLachlan, J.S., Liu, Z., Otto-Bliesner, B., He, F., 2013b. The ice age ecologist: testing methods for reserve prioritization during the last global warming. Global Ecol. Biogegor. 22, 289-301.

Williams, J.W., Grimm, E.C., Blois, J.L., Charles, D.F., Davis, E.B., Goring, S.J., Graham, R.W., Smith, A.J.,

Anderson, M., Arroyo-Cabrales, J., Ashworth, A.C., Betancourt, J.L., Bills, B.W., Booth, R.K., Buckland, P.I., Curry, B.B., Giesecke, T., Jackson, S.T., Latorre, C., Nichols, J., Purdum, T., Roth, R.E., Stryker, M., Takahara, H., 2018. The Neotoma Paleoecology Database, a multiproxy, international, community-curated data resource. Quat. Res. 89, 156-177.

Willis, K.J., 1992. The late Quaternary vegetational history of Northwest Greece. 1. Lake Gramousti. New Phytol. 121, 101-117.

Willis, K.J., 1994. The vegetational history of the Balkans. Quat. Sci. Rev. 13, 769-788.

Willis, K.J., Bhagwat, S.A., 2010. Questions of importance to the conservation of global biological diversity: Answers from the past. Clim. Past 6, 759-769.

Willis, K.J., Birks, H.J.B., 2006. What is natural? The need for a long-term perspective in biodiversity conservation. Science 314, 1261-1265.

Willis, K.J., McElwain, J.C., 2002. The Evolution of Plants. Oxford University Press, Oxford.

Willis, K.J., van Andel, T., 2004. Trees or no trees? The environments of central and eastern Europe during the Last Glaciation. Quat. Sci. Rev. 23, 2369-2387.

Willis, K.J., Sümegi, P., Braun, M., Tóth, A., 1995. The late Quaternary environmental history of Bátorliget, N.E. Hungary. Palaeogeogr. Palaeoclimatol. Palaeoecol. 118, 25-47.

Willis, K.J., Braun, M., Sümegi, P., Tóth, A., 1997. Does soil change cause vegetation change or vice versa? A temporal perspective from Hungary. Ecology 78, 740-750.

Willis, K.J., Sümegi, P., Braun, M., Bennett, K.D., Tóth, A., 1998a. Prehistoric land degradation in Hungary: who, how and why? Antiquity 72, 101–113.

Willis, K.J., Bennett, K.D., Birks, H.J.B., 1998b. The late Quaternary dynamics of pines in Europe. In: Richardson, D. (Ed.), Ecology and Biogeography of Pinus. Cambridge University Press, Cambridge, pp. 107-121.

Willis, K.J., Rudner, E., Sümegi, P., 2000. The full-glacial forests of central and southeastern Europe. Quat. Res. 53, 203-213.

Willis, K.J., Gillson, L., Brncic, T.M., 2004. How "virgin" is virgin rainforest? Science 304, 402-403.

Willis, K.J., Bailey, R.M., Bhagwat, S.A., Birks, H.J.B., 2010. Biodiversity baselines, thresholds, and resilience: testing predictions and assumptions using palaeoecological data. Trends Ecol. Evol. 25, 583-591.

Willis, K.J., Jeffers, E.S., Tovar, C., 2018. What makes a terrestrial ecosystem resilient? Science 359, 988-989.

Wilmshurst, J.M., McGlone, M.S., 2005. Origin of pollen and spores in surface lake sediments: Comparison of modern palynomorph assemblages in moss cushions, surface soils and surface lake sediments. Rev. Palaeobotany Palynology 136, 1-15.

Wilson, J.B., Agnew, A.D., Roxburgh, S.H., 2019. The Nature of Plant Communities. Cambridge University Press, Cambridge. 370 pp.

Witte, H., 1905. *Stratiotes aloides* L. funnen I Sveriges postglacialer aflagringar. Geol. Fören. Förh. Stockholm 27, 432ff.

Wohlfarth, B., Björck, S., Possnert, G., Lemdahl, G., Brunnberg, L., Ising, J., Olsson, S., Svensson, N.-O., 1993. AMS dating Swedish varved clays of the last glacial/interglacial transition and the potential/difficulties of calibrating Late Weichselian "absolute" chronologies. Boreas 22, 113-128.

Wohlfarth, B., Hannon, G., Feurdean, A., Ghergari, L., Onac, B.P., Possnert, G., 2001. Reconstruction of climatic and environmental changes in NW Romania during the early part of the last deglaciation (ca 15,000-13,6000 cal yr BP). Quat. Sci. Rev. 20, 1897-1914.

Wohlfarth, B., Björck, S., Funder, S., Houmark-Nielsen, M., Ingólfsson, Ó, Lunkka, J.-P., Mangerud, J., Saarnisto, M., Vorren, T., 2008. Quaternary of Norden. Episodes 31, 73-81.

Wohlgemuth, T., Gigon, A., 2003. Calcicole plant diversity in Switzerland may reflect a variety of habitat templets. Folia Geobot. 38, 443-452.

Woillard, G., 1975. Recherches palynologiques sur le Pleistocène dans l'est de la Belgique et dans les Vosges Lorraines.Acta Geogr. Lovaniensia 14.

Woillard, G., 1978. Grande Pile peat bog: A continuous pollen record for the last 140 000 years. Quat. Res. 9, 1-21.

Woldstedt, P., 1961. Das Eiszeitalter, 3rd ed., Stuttgart.

Wołejko, L., 2000. Plant-sociological and ecological dynamics of spring ecosystems in NW Poland under conditions of decreasing intensity of agriculture. Rozprawy AR Szczecin, 195, 112.

Wolters, S., 2002. Vegetationsgeschichtliche Untersuchungen zur spätglazialen und holozänen Landschaftsenwicklung in der Döberitzer Heide (Brandenburg). Diss. Bot. 366, 1–157.

Woodward, F., 1987. Climate and Plant Distribution. Cambridge University Press, Cambridge.

Woodward, F., Williams, B., 1987. Climate and plant distribution at global and local scales. Vegetatio 69, 189-197.

Wright, J., 1955. Species crossability in spruce in relation to distribution and taxonomy. Forest Sci. 1, 319-349.

Wright, H.E., 1967. The use of surface samples in Quaternary pollen analysis. Rev. Palaeobot. Palynology 2, 321-330.

Wright, H.E., 1974. Landscape development, forest fires, and wilderness management. Science 186, 487-495.

Wright, H.E., 1991. Coring tips. J. Paleolimn. 6, 37-49.

Wright, H.E., Coffin, B.A. (Eds.) 1992. Patterned Peatlands of Northern Minnesota. University of Minnesota Press, Minneapolis.

Wright, H.E., McAndrews, J.H., van Zeist, W., 1967. Modern pollen rain in western Iran, and its relation to plant geography and Quaternary vegetational history. J. Ecol. 55, 415-443.

Wright, H.E., Ammann, B., Stefanova, I., Atanassova, J., Margalitadze, N., Wick, L., Blyakharchuk, T., 2003. Late-glacial and early-Holocene dry climates from the Balkan peninula to Southern Sibiria. In: Tonkov, S. (Ed.), Aspects of Palynology and Palaeoecology. Festschrift in Honour of Elissaveta Bozilova. Pensoft, Sofia, pp. 127-136.

Wulf, S., Kraml, M., Brauer, A., Keller, J., Negendank, J.F.W., 2004. Tephrochronology of the 100 ka lacustrine sediment record of Lago Grande di Monticchio (southern Italy). Quat. Int. 122, 7-30.

Yang, H., Rose, N.L., Boyle, J.F., Battarbee, R.W., 2001. Storage and distribution of trace metals and spheroidal carbonaceous particles (SCPs) from atmospheric deposition in the catchment peats of Lochnagar, Scotland. *Environ. Pollut.* 115, 231-238.

Yanko-Hombach, V., Yanina, T., 2019. Towards an understanding of human responses to environmental change in the Caspian-Black Sea-Mediterranean Corridors (IGCP 610 final report). Episodes 41, 213-223.

Yanko-Hombach, V., Gilbert, A., Panin, N., Dolukhanov, P., 2007. The Black Sea Flood Question: Changes in Coastline, Climate and Human Settlement. Springer, The Netherlands.

Yasuda, Y., Kitagawa, H., Nakagawa, T., 2000. The earliest record of major anthropogenic deforestation in the Ghab Valley, northwest Syria: a palynological study. Quat. Int. 73/74, 127-136.

Yates, K.L., Bouchet, P.J., Caley, M.J., Mengersen, K., Randin, C.F., Parnell, S., Fielding, A.H., Bamford, A.J., Ban, S., Barbosa, A.M., Dormann, C.F., Elith, J., Embling, C.B., Ervin, G.N., Fisher, R., Gould, S., Graf, R.F., Gregr, E.J., Halpin, P.N., Heikkinen, R.K., Heinänen, S., Jones, A.R., Krishnakumar, P.K., Lauria, V., Lozano-Montes, H., Mannocci, L., Mellin, C., Mesgaran, M.B., Moreno-Amat, E., Mormede, S., Novaczek, E., Oppel, S., Ortuño Crespo, G., Peterson, A.T., Rapacciuolo, G., Roberts, J.J., Ross, R.E., Scales, K.L., Schoeman, D., Snelgrove, P., Sundblad, G., Thuiller, W., Torres, L.G., Verbruggen, H., Wang, L., Wenger, S., Whittingham, M.J., Zharikov, Y., Zurell, D., Sequeira, A.M.M., 2018. Outstanding challenges in the transferability of ecological models. Trends Ecol. Evol. 33, 790-802.

Yi, T., Wen, J., Golan-Goldhirsh, A., Parfitt, D.E., 2008. Phylogenetics and reticulate evolution in *Pistacia* (Anacardiaceae). Am. J. Bot. 95, 241-251.

Yll, E.-I., Roure, J.M., Pantaleón-Cano, J., Pérez-Obiol, R., 1994. Análisis polínico de una secuencia holocénica en Roquetas de Mar (Almería). Trabajos de palinología básica y aplicada, 189-198.

Yll, E.-I., Pérez-Obiol, R., Pantaleón-Cano, J., Roure, J., 1995. Dinámica del paisaje vegetal en la vertiente mediterránea de la Península Ibérica e Islas Baleares desde el Tardiglaciar hasta el presente. In: Aleixandre, T., Pérez-González, A. (Eds.), Reconstrucción de paleoambientes y cambios climáticos durante el Cuaternario. Monografías del Centro de Ciencias Medioambientales-CSIC, Madrid, pp. 319-328.

Yll, E.-I., Pérez-Obiol, R., Pantaleón-Cano, J., Roure, J., 1997. Palynological evidence for climatic change and human activity during the Holocene on Minorca (Balearic Islands). Quat. Res. 48, 339-347.

Yll, E.-I., Zazo, C., Goy, J., Pérez-Obiol, R., Pantaleón-Cano, J., Civis, J., Dabrio, C., González, A., Borja, F., Soler, V., Lario, J., Luque, L., Sierro, F., González-Hernández, F., Lézine, A., Denèfle, M., Roure, J., 2004. Quaternary palaeoenvironmental changes in south Spain. In: Ruiz-Zapata, M., Dorado, A., Valdeolmillos, A., Gil, M., Bardaji, T., de Bustamante, I., Martínez, I. (Eds.), Quaternary Climatic Changes and Environmental Crises in the Mediterranean Region, Universidad de Alcalá de Henares.

Yll, R., Carrión, J.S., Marra, A.C., Bonfiglio, L., 2006. Vegetation reconstruction on the basis of pollen in Late Pleistocene hyena coproliths from San Teodoro Cave (Sicily, Italy). Paleogeogr. Paleoclimatol. Paleoecol. 237, 32-39.

Yu, Z., 2007. Rapid response of forested vegetation to multiple climatic oscillations during the last deglaciation in the northeastern United States. Quaternary Research 67, 297-303.

Yu, Z., 2017. Rapid response of forested vegetation to multiple climatic oscillations during the last deglaciation in the northeastern United States. Quat. Res. 67(2), 297-303.

Yu, Z., Eicher, U., 1998. Abrupt climate oscillations during the last deglaciation in central North America. Science 282, 2235-2238.

Yu, G., Harrison, S.P., 1995a. Holocene changes in atmospheric circulation patterns as shown by lake status changes in northern Europe. Boreas 24, 260-268.

Yu, G., Harrison, S.P., 1995b. Lake status records from Europe: Data base documentation. NOAA Paleoclimatology Publication Series 3. Boulder, Colorado, 451 pp.

Zaccone, C., Cocozza C., Cheburkin, A.K, Shotyk, W., Miano, T.M., 2007. Highly organic soils as "witnesses" of anthropogenic Pb, Cu, Zn, and ^{137}Cs

inputs during centuries. Water Air Soil Pollut. 186, 263-271.

Zagwijn, W.H., 1952. Pollenanalytische Untersuchung einer spätglazialen Seeablagerung aus Tirol. Geol. en Mijnbouw 7, 235-239.

Zagwijn, W.H., 1957. Vegetation, climate and time-correlations in the Early Pleistocene of Europe. Geol. en Mijnbouw 19, 233-244.

Zagwijn, W.H., 1960. Aspects of the Pliocene and Early Pleistocene vegetation in the Netherlands. Mededel. Geol. Stichting Ser. C III-I, 1-78.

Zagwijn, W.H., 1961. Vegetation, climate and radiocarbon datings in the Late Pleistocene of The Netherlands, Part I. Eemian and early Weichselian. Mededelingen Rijks Geologische Dienst, N. S. 14, 15-45.

Zagwijn, W.H., 1963. Pollen-analytic investigations in the Tiglian of the Netherlands. Mededel. Geol. Stichting N. S. 16, 49-71.

Zagwijn, W.H., 1973. Pollenanalytic studies of the Holsteinian and Saalian Beds in the Northern Netherlands. Mededel. Rijks Geolog. Dienst NS 24, 139-156.

Zagwijn, W.H., 1974. Vegetation, climate and radiocarbon datings in the Late Pleistocene of the Netherlands, Part II. Middle Weichselian. Mededelingen Rijks Geologische Dienst NS 25,101-111.

Zagwijn, W.H., 1975. Variations in climate as shown by pollen analysis, especially in the Lower Pleistocene of Europe. In: Wright, A.E., Moseley, F., (Eds.), Ice Ages: Ancient and Modern. Seel House Press, Liverpool, pp. 137-152.

Zagwijn, W.H., 1992. The beginning of the Ice Age in Europe and its major subdivisions. Quat. Sci. Rev. 11, 583-591.

Zagwijn, W.H., 1996a. An analysis of Eemian climate in Western and Central Europe. Quat. Sci. Rev. 15, 451-469.

Zagwijn, W.H., 1996b. The Cromerian Complex Stage of the Netherlands and correlation with other areas in Europe. In: Turner, C. (Ed.), The Early Middle Pleistocene in Europe. CRC Press, Boca Raton, pp. 145-172.

Zagwijn, W.H., de Jong, J., 1984. Die Interglaziale von Bavel und Leerdam und ihre stratigraphische Stellung im niederländischen Früh-Pleistozän. Mededelingen Rijks Geologische Dienst 37, 155-169.

Zagwijn, W.H., van Montfrans, H.H., Zandstra, J.G., 1971. Subdivision of the "Cromerian" in the Netherlands; Pollen analysis, palaeomagnetism and sedimentary petrology. Geologic en Mijnbouw 50, 41-58.

Zakynthinos, G., Rouskas, D., 1996. Wild and cultivated *Pistacia* species in Greece. In: Padulosi, S., Caruso, T., Barone, E. (eds.), Taxonomy, distribution, conservation and uses of *Pistacia* genetic resources. Report of a workshop, 29-30 June 1995, Palermo, Italy. International Plant Genetic Resources Institute, Rome, Italy, pp. 88-92.

Zanchetta, G., van Welden, A., Baneschi, I., Drysdale, R., Sadori, L., Roberts, N., Giardini, M., Beck, C., Pascucci, V., Sulpizio, R., 2012. Multiproxy record for the last 4500 years from Lake Shkodra (Albania/ Montenegro). J. Quat. Sci. 27, 780-789.

Zapata, L., Peña-Chocarro, L., Pérez, G., Stika, H., 2004. Early Neolithic agriculture in the Iberian Peninsula. J. World Archaeol. 18, 285-326.

Zarrina, E.P., Krasnov, I.I., Spiridinova, E.A., 1981. Detailed chronostratigraphic scale of the late Pleistocene of the North-West and the centre of the Russian Plain. In: Gorbunov, G. I. (ed.), Glacial Deposits and Glacial History in Eastern Fennoscandia. Kola Branch, Acad. Sci. USSR, Apatity, 90-102.

Zatykó, C., Juhász, I., Sümegi, P. (eds.), 2007. Environmental archaeology in Transdanubia. Varia archaeologica Hungarica 20, Budapest.

Zecchin, B., Caudullo, G., De Rigo, D., 2016. *Acer campestre* in Europe: distribution, habitat, usage and threats. In: San-Miguel-Ayanz, J., De Rigo, D., Caudullo, G., Houston Durrant, T., Mauri, A. (Eds.), European Atlas of Forest Tree Species. Publication Office of the European Union, Luxembourg, pp. 52-53.

Zelikson, E., 1971. Palynological investigation of a Holocene peat deposit from Spitsbergen, in: Neustadt, M.I. (Ed.), . Presented at the Palynology of pliocene, pleistocene and holocene. 3rd International Palynological Conference (Novosibirsk, USSR, 1971), Academy of Sciences of the USSR, Institute of Geography, Moscow, Russia, pp. 199–212.

Zernitskaya, V., 1985. Razvitie rastitel'nosti zapadnoi chasti Kopyl'skoi gryady v pozdnelednikovye i golotsene (Development of vegetation in the western part of the Kopyl upland during the Late Glacial time and Holocene), Geograficheskie aspekty ratsional'nogo prirodopol'zovaniya (Geographical aspects of rational nature use). Nauka i Tekhnika, Minsk, pp. 110-116.

Zernitskaya, V., 1991. Palaeogeography of the Byelorussian Polessie in the Late Glacial and Holocene (PhD thesis). Minsk University, Minsk, Belarus.

Zernitskaya, V., Krutous, E.A., Klimanov, V.A., 1988. Studies of mire for the purpose of reconstruction of the climatic peculiarities of the Byelorussian Poles'e, in: Problem of the Practical Geomorphology. Nauka i Tekhnika, Minsk, Belarus, pp. 68–73.

Zernitskaya, V., Stančikaitė, M., Vlasov, B., Šeirienė, V., Kisielienė, D., Gryguc, G., Skipitytė, R., 2015. Vegetation pattern and sedimentation changes in the context of the Lateglacial climatic events: Case study of Staroje Lake (Eastern Belarus). Quat. Int. 386, 70-82.

Zhao, Y., Sayer, C.D., Birks, H.H., Hughes, M., Peglar, S.M., 2006. Spatial representation of aquatic vegetation by macrofossils and pollen in a small and shallow lake. J. Paleolimnol. 35, 335-350.

Ziegenhagen, B., Fady, B., Kuhlenlamp, V., Liepelt, S., 2005. Differentiating groups of *Abies* species with a simple molecular marker. Silvae Genetica 54, 123-126.

Zimmermann, W.H., 1976. Die eisenzeitliche Ackerfluren - Typ "Celtic Field' - von Flögeln-Haselhörn, Kr. Wesermünde. Probleme der Küstenforschung im südlichen Nordseegebiet 11, 79-90.

Zobel, M., 1988. Autogenic succession in boreal mires - a review. Folia Geobot. Phytotax. 23, 417-445.

Zobel, K., 2001. On the species-pool hypothesis and on the quasi-neutral concept of plant community diversity. Folia Geobot. 36, 3-8.

Zohary, D., Hopf, M., 2000. Domestication of Plants in the Old World, 3rd ed. Oxford University Press, Oxford.

Zohary, D., Hopf, M., Weiss, E., 2012. Domestication of Plants in the Old World. The origin and spread of domesticated plants in Southwest Asia, Europe and the Mediterranean Basin, 4th ed. Oxford University Press, Oxford.

Zolitschka, B., 2007. Varved Lake Sediments. In: Elias, S.A. (Ed.), Encyclopedia of Quaternary Science. Elsevier, Amsterdam, pp. 3105-3114.

Zolitschka, B., 2013. Varved Lake Sediments. In: Elias, S.A., Mock, C.J. (Eds.), Encyclopedia of Quaternary Science, 2nd ed. Elsevier, Amsterdam, pp. 573-581.

Zolitschka, B., Negendank, J.F.W., 1996. Sedimentology, dating and palaeoclimatic interpretation of a 76.3 ka record from Lago Grande di Monticchio, Southern Italy. Quat. Sci. Rev. 15, 101-112.

Zolitschka, B., Brauer, A., Negendank, J.F.W., Stockhausen, H., Lang, A., 2000. Annually dated late Weichselian cntinental paleoclimate record from the Eifel, Germany. Geology 28, 783-786.

Zoller, H., 1960. Pollenanalytische Untersuchungen zur Vegetationsgeschichte der insubrischen Schweiz. Denkschri. Schweiz. Naturfors. Gesells. 83, 45-156.

Zoller, H., 1967. Holocene fluctuations of cold climate in the Swiss Alps. Rev. Palaeobot. Palynol. 2, 267-269.

Zoller, H., 1981. Gymnospermae. In: Hegi, G. (Ed.), Illustrierte Flora Mitteleuropas, pp. 104-120.

Zoller, H., 1987. Zur Geschichte der Vegetation im Spätglazial und Holozän der Schweiz. Mitt. Naturf. Ges. Luzern 29, 123-149.

Zoller, H., Brombacher, C., 1984. Das Pollenprofil "Chalavus" bei St. Moritz. ein Beitrag zur Wald- und Landschaftsgeschichte im Oberengadin. Diss. Bot. 72, 377-398.

Zoller, H., Kleiber, H., 1967. Über die postglaziale Einwanderung und Ausbreitung der Rotbuche (*Fagus silvatica* L.) am südlichen Alpenrand. Bauhinia 3, 255-264.

Zólyomi, B., 1953. Die Entwicklungsgeschichte der Vegetation Ungarns seit dem letzten Interglazial. Acta Biol. Acad. Sci. Hunger. 4.

Zólyomi, B., 1995. Opportunities for pollen stratigraphic analysis of shallow lake sediments: the example of Lake Balaton. GeoJournal 36, 237-241.

Zólyomi, B., Járainé Komlódi, M., 2008. Fosszilis szubarktikus forrásláp Magyarországon. (A fossil subarctic peat bog in Hungary), in: A Keleméri Mohos-Tavak Kutatás, Kezelés, Védelem. (The Mohos Peat Bogs in Kelemér Research, Conservation, Management). ANP Füzetek IV. Aggteleki Nemzeti Park Igazgatóság. Jósvafo. pp. 25-34.

Zuber, D., 2004. Biological flora of central Europe: *Viscum album* L. Flora 199, 181-203.

Zuber, D., Widmer, A., 2000. Genetic evidence for host specificity in the hemi-parasitic *Viscum album* L. (Viscaceae). Mol. Ecol. 9, 1069-1073.

Zuber, D., Widmer, A., 2009. Phylogeography and host race differentiation in the European mistletoe (*Viscum album* L.). Mol. Ecol. 18, 1946-1962.

Editors

Gerhard Lang was Professor of Geobotany at the University of Bern. He studied biology, chemistry and physics at the University of Freiburg in Breisgau. After his PhD in biology with Franz Firbas at the University of Göttingen, co-supervised by Harry Godwin at the University of Cambridge, he was curator at the Museum of Natural History in Karlsruhe and Adjunct Professor at the University of Karlsruhe. He applied plant sociological relevés and sedimentary pollen and macrofossil analyses to investigate biotic and non-biotic controls of vegetation change in space and time. He was particularly interested in the long-term responses of vegetation to climate change. He died on June 19th, 2016.

Brigitta Ammann was Professor of Palaeoecology at the University of Bern. She studied botany and zoology at the Universities of Bern and Bergen. After her PhD at the University of Bern with Max Welten and a postdoctoral position at the Duke University in Durham, North Carolina she became associated with the research group of Gerhard Lang. Her main interests are interdisciplinary approaches, for example studying biotic responses to climate change based on records of stable isotopes and biological remains.

Karl-Ernst Behre was head of the Lower Saxony Institute for Historical Coastal Research in Wilhelmshaven/Germany. He studied biology, geography and chemistry at the Universities of Marburg, Innsbruck and Göttingen. He joined the Lower Saxony Institute after his PhD in botany with Franz Firbas and Heinz Ellenberg and also held part-time professorships in Göttingen and, later, Amsterdam. His main field of research is vegetation history (pollen, macroremains and charcoal) during the Pleistocene and the Holocene with special attention to human impact, and also coastal research.

Willy Tinner is Professor of Palaeoecology at the University of Bern. He studied geography, botany, geology and archaeology at the same university. After his PhD in biology with Brigitta Ammann and Marco Conedera, also at the University of Bern, he had a postdoctoral position at the University of Illinois in Urbana-Champaign, was lecturer at the University of Bern and later SNF-professor at ETH Zurich. He uses Quaternary sedimentary sequences (pollen, spores, macrofossils and charcoal) and modelling approaches to study the long-term interactions between climate, the biosphere and society.

The editors and authors would like to thank the sponsors of the printing of this book:

UNIVERSITÄT BERN

OESCHGER CENTRE
CLIMATE CHANGE RESEARCH

fondation johanna dürmüller-bol

SEBASTIANA STIFTUNG

Burgergemeinde Bern